FOOD CHEMISTRY

FOOD SCIENCE AND TECHNOLOGY

A Series of Monographs and Textbooks

Editors

STEVEN R. TANNENBAUM
Department of Nutrition and Food Science
Massachusetts Institute of Technology
Cambridge, Massachusetts

PIETER WALSTRA
Department of Food Science
Wageningen Agricultural University
Wageningen, The Netherlands

1. Flavor Research: Principles and Techniques, *R. Teranishi, I. Hornstein, P. Issenberg, and E. L. Wick (out of print)*

2. Principles of Enzymology for the Food Sciences, *John R. Whitaker*

3. Low-Temperature Preservation of Foods and Living Matter, *Owen R. Fennema, William D. Powrie, and Elmer H. Marth*

4. Principles of Food Science
    Part I: Food Chemistry, *edited by Owen R. Fennema*
    Part II: Physical Methods of Food Preservation, *Marcus Karel, Owen R. Fennema, and Daryl B. Lund*

5. Food Emulsions, *edited by Stig Friberg*

6. Nutritional and Safety Aspects of Food Processing, *edited by Steven R. Tannenbaum*

7. Flavor Research: Recent Advances, *edited by R. Teranishi, Robert A. Flath, and Hiroshi Sugisawa*

8. Computer-Aided Techniques in Food Technology, *edited by Israel Saguy*

9. Handbook of Tropical Foods, *edited by Harvey T. Chan*

10. Antimicrobials in Foods, *edited by Alfred Larry Branen and P. Michael Davidson*

11. Food Constituents and Food Residues: Their Chromatographic Determination, *edited by James F. Lawrence*

12. Aspartame: Physiology and Biochemistry, *edited by Lewis D. Stegink and L. J. Filer, Jr.*

13. Handbook of Vitamins: Nutritional, Biochemical, and Clinical Aspects, *edited by Lawrence J. Machlin*

14. Starch Conversion Technology, *edited by G. M. A. van Beynum and J. A. Roels*

15. Food Chemistry: Second Edition, Revised and Expanded, *edited by Owen R. Fennema*

Other Volumes in Preparation

FOOD CHEMISTRY

SECOND EDITION, REVISED AND EXPANDED

Edited by Owen R. Fennema

Department of Food Science
University of Wisconsin–Madison
Madison, Wisconsin

MARCEL DEKKER, INC.   New York and Basel

Library of Congress Cataloging in Publication Data
Main entry under title:

Food Chemistry.

   (Food science and technology ; 15)
   Includes index.
   1. Food--Analysis. 2. Food--Composition. I. Fennema,
Owen R. II. Series: Food science and technology
(Marcel Dekker, Inc.) ; 15.
TX541.F65 1985   664'.001'54   85-4511
ISBN 0-8247-7271-7 hardcover
ISBN 0-8247-7449-3 softcover

Marcel Dekker, Inc.
270 Madison Avenue, New York, New York 10016

Current printing (last digit):
10 9 8 7 6 5 4 3 2 1

Printed in the United States of America

PREFACE TO THE SECOND EDITION

Considerable time has passed since publication of the favorably received first edition so a new edition seems appropriate. The purpose of the book remains unchanged—it is intended to serve as a textbook for upper division undergraduates or beginning graduate students who have sound backgrounds in organic chemistry and biochemistry, and to provide insight to researchers interested in food chemistry. Although the book is most suitable for a two-semester course on food chemistry, it can be adapted to a one-semester course by specifying selective reading assignments. It should also be noted that several chapters are of sufficient length and depth to be useful as primary source materials for graduate-level speciality courses.

This edition has the same organization as the first, but differs substantially in other ways. The chapters on carbohydrates, lipids, proteins, flavors, and milk and the concluding chapter have new authors and are, therefore, entirely new. The chapter on food dispersions has been deleted and the material distributed at appropriate locations in other chapters. The remaining chapters, without exception, have been substantially modified, and the index has been greatly expanded, including the addition of a chemical index. Furthermore, this edition, in contrast to the first, is more heavily weighted in the direction of subject matter that is unique to food chemistry, i.e., there is less overlap with materials covered in standard biochemistry courses. Thus the book has undergone major remodeling and refinement, and I am indebted to the various authors for their fine contributions and for their tolerance of my sometimes severe editorial guidance.

This book, in my opinion, provides comprehensive coverage of the subject of food chemistry with the same depth and thoroughness that is characteristic of the better quality introductory textbooks on organic chemistry and biochemistry. This, I believe, is a significant achievement that reflects a desirable maturation of the field of food chemistry.

Owen R. Fennema

PREFACE TO THE FIRST EDITION

For many years, an acute need has existed for a food chemistry textbook that is suitable for food science students with backgrounds in organic chemistry and biochemistry. This book is designed primarily to fill the aforementioned need, and secondarily, to serve as a reference source for persons involved in food research, food product development, quality assurance, food processing, and in other activities related to the food industry.

Careful thought was given to the number of contributors selected for this work, and a decision was made to use different authors for almost every chapter. Although involvement of many authors results in potential hazards with respect to uneven coverage, differing philosophies, unwarranted duplication, and inadvertent omission of important materials, this approach was deemed necessary to enable the many facets of food chemistry to be covered at a depth adequate for the primary audience. Since I am acutely aware of the above pitfalls, care has been taken to minimize them, and I believe the end product, considering it is a first edition, is really quite satisfying—except perhaps for the somewhat generous length. If the readers concur with my judgment, I will be pleased but unsurprised, since a book prepared by such outstanding personnel can hardly fail, unless of course the editor mismanages the talent.

Organization of the book is quite simple and I hope appropriate. Covered in sequence are major constituents of food, minor constituents of food, food dispersions, edible animal tissues, edible fluids of animal origin, edible plant tissues and interactions among food constituents—the intent being to progress from simple to more complex systems. Complete coverage of all aspects of food chemistry, of course, has not been attempted. It is hoped, however, that the topics of greatest importance have been treated adequately. In order to help achieve this objective, emphasis has been given to broadly based principles that apply to many foods.

Figures and tables have been used liberally in the belief that this approach facilitates understanding of the subject matter presented. The number of references cited should be adequate to permit easy access to additional information.

To all readers I extend an invitation to report errors that no doubt have escaped my attention, and to offer suggestions for improvements that can be incorporated in future (hopefully) editions.

Since enjoyment is an unlikely reader response to this book, the best I can hope for is that readers will find it enlightening and well suited for its intended purpose.

Owen R. Fennema

CONTRIBUTORS

MICHAEL C. ARCHER, Department of Medical Biophysics, University of Toronto, Ontario Cancer Institute, Toronto, Ontario, Canada

JEAN CLAUDE CHEFTEL, Laboratoire de Biochimie et Technologie Alimentaires, Université des Sciences et Techniques du Languedoc, Montpellier, France

JEAN-LOUIS CUQ, Laboratoire de Biochimie et Technologie Alimentaires, Université des Sciences et Techniques du Languedoc, Montpellier, France

JAMES R. DANIEL, Department of Foods and Nutrition, Purdue University, West Lafayette, Indiana

OWEN R. FENNEMA, Department of Food Science, University of Wisconsin-Madison, Madison, Wisconsin

FREDERICK J. FRANCIS, Department of Food Science and Nutrition, University of Massachusetts, Amherst, Massachusetts

NORMAN F. HAARD, Department of Biochemistry, Memorial University of Newfoundland, St. John's, Newfoundland, Canada

HERBERT O. HULTIN, Department of Food Science and Nutrition, Marine Foods Laboratory, University of Massachusetts Marine Station, Gloucester, Massachusetts

DOUGLAS B. HYSLOP, Department of Food Science, University of Wisconsin-Madison, Madison, Wisconsin

THEODORE P. LABUZA, Department of Food Science and Nutrition, University of Minnesota, St. Paul, Minnesota

ROBERT C. LINDSAY, Department of Food Science, University of Wisconsin-Madison, Madison, Wisconsin

DENIS LORIENT, Départment de Biochimie et Toxicologie Alimentaires, Ecole Nationale Supérieure de Biologie Appliquée à la Nutrition et à l'Alimentation, Dijon, France

MICHAEL A. MARLETTA, Department of Nutrition and Food Science, Massachusetts Institute of Technology, Cambridge, Massachusetts

SHURYO NAKAI, Department of Food Science, University of British Columbia, Vancouver, British Columbia, Canada

WASSEF W. NAWAR, Department of Food Science and Nutrition, University of Massachusetts, Amherst, Massachusetts

WILLIAM D. POWRIE, Department of Food Science, University of British Columbia, Vancouver, British Columbia, Canada

THOMAS RICHARDSON,* Department of Food Science, University of Wisconsin-Madison, Madison, Wisconsin

HAROLD E. SWAISGOOD, Department of Food Science, North Carolina State University, Raleigh, North Carolina

STEVEN R. TANNENBAUM, Department of Nutrition and Food Science, Massachusetts Institute of Technology, Cambridge, Massachusetts

ROY L. WHISTLER, Department of Biochemistry, Purdue University, West Lafayette, Indiana

GERALD N. WOGAN, Department of Nutrition and Food Science, Massachusetts Institute of Technology, Cambridge, Massachusetts

VERNON R. YOUNG, Department of Nutrition and Food Science, Massachusetts Institute of Technology, Cambridge, Massachusetts

---

*Current affiliation: Department of Food Science and Technology, University of California-Davis, Davis, California

CONTENTS

Preface to the Second Edition   iii
Preface to the First Edition   v
Contributors   vii

1. Introduction to Food Chemistry   1
   Owen R. Fennema and Steven R. Tannenbaum

2. Water and Ice   23
   Owen R. Fennema

3. Carbohydrates   69
   Roy L. Whistler and James R. Daniel

4. Lipids   139
   Wassef W. Nawar

5. Amino Acids, Peptides, and Proteins   245
   Jean Claude Cheftel, Jean-Louis Cuq, and Denis Lorient

6. Enzymes   371
   Thomas Richardson and Douglas B. Hyslop

7. Vitamins and Minerals   477
   Steven R. Tannenbaum, Vernon R. Young, and Michael C. Archer

8. Pigments and Other Colorants   545
   Frederick J. Francis

9. Flavors   585
   Robert C. Lindsay

10. Food Additives   629
    Robert C. Lindsay

11. Undesirable or Potentially Undesirable Constituents of Foods   689
    Gerald N. Wogan and Michael A. Marletta

12. Characteristics of Muscle Tissue   725
     Herbert O. Hultin

13. Characteristics of Edible Fluids of Animal Origin: Milk   791
     Harold E. Swaisgood

14. Characteristics of Edible Fluids of Animal Origin: Eggs   829
     William D. Powrie and Shuryo Nakai

15. Characteristics of Edible Plant Tissues   857
     Norman F. Haard

16. An Integrated Approach to Food Chemistry: Illustrative Cases   913
     Theodore P. Labuza

Subject and Chemical Index   939

FOOD CHEMISTRY

1

INTRODUCTION TO FOOD CHEMISTRY

Owen R. Fennema   University of Wisconsin-Madison, Madison, Wisconsin

Steven R. Tannenbaum   Massachusetts Institute of Technology, Cambridge, Massachusetts

I.	What Is Food Chemistry?	1
II.	History of Food Chemistry	2
III.	Approach to the Study of Food Chemistry	7
	A. Quality and Safety Attributes	7
	B. Chemical and Biochemical Reactions	8
	C. Effect of Reactions on the Quality and Safety of Food	9
	D. Analysis of Situations Encountered During the Storage and Processing of Food	10
IV.	Societal Role of Food Chemists	13
	A. Why Should Food Chemists Become Involved in Societal Issues?	13
	B. Type of Involvement—Committees and Professional Societies	14
	C. Type of Involvement—Individual Initiatives	14
	D. Perspectives	17
	References	20

I. WHAT IS FOOD CHEMISTRY?

Concern about food exists throughout the world, but the aspects of concern differ with location. In underdeveloped regions of the world, the bulk of the population is involved in food production, yet attainment of adequate amounts and kinds of basic nutrients remains an ever-present problem. In developed regions of the world, food production is highly mechanized and efficient, a small portion of the population is involved in food production, food is available in abundance, and much of it is processed or has been altered by the addition of chemicals. In these localities, concern is directed mainly to the cost of food, its quality, its variety, and its ease of preparation, and to the effects of processing and added chemicals on its wholesomeness and nutritive value. All these concerns are important, and they fall within the realm of food science—the science that deals with the nature of food and the principles underlying its spoilage, preservation, and modification.

Food science is an interdisciplinary subject involving primarily bacteriology, chemistry, biology, and engineering. Food chemistry, a major aspect of food science, is the science that deals with the composition and properties of food and the chemical changes it undergoes. Food chemistry is intimately related to chemistry, biochemistry, physiological chemistry, botany, zoology, and molecular biology. The food chemist relies heavily on knowledge of the aforementioned sciences to effectively study and control biological substances as sources of human food. Knowledge of the innate properties of biological substances and mastery of the means of studying them are common interests of both food chemists and other biological scientists. It is important to note, however, that food chemists have specific interests distinct from those of other biological scientists. The primary interests of biological scientists include reproduction, growth, and changes biological substances undergo under environmental conditions that are compatible or almost compatible with life. On the other hand, food chemists are concerned primarily with biological substances that are dead or dying (postharvest physiology of plants and postmortem physiology of muscle) and are exposed to a wide range of environmental conditions. For example, conditions suitable for sustaining residual life processes are of concern to food chemists during the marketing of fresh fruits and vegetables, whereas conditions incompatible with life processes are of major interest when long-term preservation of food is attempted, that is, during thermal processing, freezing, concentration, dehydration, and irradiation and during the addition of chemical preservatives. In addition, food chemists are concerned with the chemical properties of disrupted food tissues (flour, fruit and vegetable juices, isolated and modified constituents, and manufactured foods), single-cell sources of food (eggs and microorganisms), and one major biological fluid (milk). In summary, food chemists have much in common with other biological scientists, yet they also have interests that are distinctive and of the utmost importance to humankind.

II. HISTORY OF FOOD CHEMISTRY

The origins of food chemistry are obscure, and its general history has not yet been properly analyzed and recorded. This is not surprising, since food chemistry did not acquire an identity until the twentieth century and its history is deeply entangled with that of agricultural chemistry for which historical documentation is not considered exhaustive (5,17). Thus, the following brief excursion into the history of food chemistry is incomplete and somewhat selective. Nonetheless, the available information is sufficient to impart needed perspectives with regard to when, where, and why certain notable food-related events transpired, and with regard to changes that have occurred in the quality of the food supply since the early 1800s.

Although the origin of food chemistry, in a sense, extends to antiquity, the most significant discoveries, as we judge them today, began in the late 1700s. The best accounts of developments during this period are those of Filby (15) and Browne (5), and the authors are indebted to these sources for much of the information presented here.

During the period of 1780-1850 a number of famous chemists made important discoveries, many of which related directly or indirectly to the chemistry of food. In the writings of Scheele, Lavoisier, de Saussure, Gay-Lussac, Thenard, Davy, Berzelius, Thomson, Beaumont, and Liebig lie the origins of modern food chemistry. Some may question whether these scientists, whose most famous discoveries bear little relationship to food chemistry, were in fact involved to a significant degree with the origins of modern food chemistry. Whereas it is admittedly difficult to categorize early scientists as chemists, bacteriologists, or food chemists, for example, it is relatively easy to determine whether a given scientist made substantial contributions to a given field of science. From the following brief examples it is clearly evident that many of these scientists did in fact study foods intensively and did make discoveries of such fundamental importance to food chemistry that their inclusion in any historical account of food chemistry cannot be questioned.

Carl Wilhelm Scheele (1742-1786), a Swedish pharmacist, was one of the greatest chemists of all time. In addition to his more famous discoveries of chlorine, glycerol, and oxygen (3 years before Priestly, but unpublished) he isolated and studied the properties of lactic acid (1780), prepared mucic acid by oxidation of lactic acid (1780), devised a means of preserving vinegar by means of heat (1782, well in advance of Appert's "discovery"), isolated citric acid from lemon juice (1784) and gooseberries (1785), isolated malic acid from apples (1785), and tested twenty common fruits for citric, malic, and tartaric acids (1785). His isolation of various new chemical compounds from plant and animal substances is considered the beginning of accurate analytical research in agricultural and food chemistry.

The French chemist *Antoine Laurent Lavoisier* (1743-1794) was instrumental in the final rejection of the phlogiston principle and in formulating the principles of modern chemistry. With respect to food chemistry, he established the fundamental principles of combustion organic analysis, he was the first to show that the process of fermentation could be expressed as a balanced equation, he made the first attempt to determine the elemental composition of alcohol (1784), and he presented one of the first papers (1786) on organic acids of various fruits.

(Nicolas) Théodore de Saussure (1767-1845), a French chemist, did much to formalize and clarify the principles of agricultural and food chemistry provided by Lavoisier. He also studied CO_2 and O_2 changes during plant respiration (1804), studied the mineral contents of plants by ashing, and made the first accurate elemental analysis of alcohol (1807, by the combusion technique).

Joseph Louis Gay-Lussac (1778-1850) and *Louis-Jacques Thenard* (1777-1857) devised in 1811 the first method for quantitatively determining the percentages of carbon, hydrogen, and nitrogen in dry vegetable substances. Their oxidative combusion technique did not, however, provide a procedure for estimating the quantity of water formed.

The English chemist *Sir Humphry Davy* (1778-1829) in the years 1807 and 1808 isolated the elements K, Na, Ba, Sr, Ca, and Mg. His contributions to agricultural and food chemistry came largely through his books on agricultural chemistry, of which the first edition (1813) was *Elements of Agricultural Chemistry, in a Course of Lectures for the Board of Agriculture* (11). His books served to organize and clarify knowledge existing at that time. In the first edition he stated, "all the different parts of plants are capable of being decomposed into a few elements. Their uses as food, or for the purpose of the arts, depend upon compound arrangements of these elements, which are capable of being produced either from their organized parts, or from the juices they contain; and the examination of the nature of these substances is an essential part of agricultural chemistry" (p. 64). In the fifth edition (Ref. 12, p. 121) he stated that plants are usually composed of only seven or eight elements, and that "the most essential vegetable substances consist of hydrogen, carbon, and oxygen in different proportion, generally alone, but in some few cases combined with azote [nitrogen] ."

The works of the Swedish chemist *Jons Jacob Berzelius* (1779-1848) and the Scottish chemist *Thomas Thomson* (1773-1852) resulted in the beginnings of organic formulas, "without which organic analysis would be a trackless desert and food analysis an endless task" (Ref. 15, p. 189). Berzelius determined by analysis the elemental components of about 2000 compounds, thereby verifying the law of definite proportions. He also devised a means of accurately determining the water content of organic substances, a deficiency in the method of Gay-Lussac and Thenard. Moreover, Thomson showed that the laws governing the composition of inorganic substances apply equally well to organic substances, a point of immense importance.

In a book entitled *Considérations générales sur l'analyse organique et sur ses applications* (6), *Michel Eugene Chevreul* (1786-1889), a French chemist, listed the elements known at that time to exist in organic substances (O, Cl, I, N, S, P, C, Si, H, Al, Mg, Ca, Na, K, Mn, Fe) and cited the processes then available for organic analysis: (a) extraction with a neutral solvent, such as water, alcohol, or aqueous ether, (b) slow distillation, or fractional distillation, (c) steam distillation, (d) passing the substance through a tube heated to incandescence, and (e) analysis with oxygen.

Chevreul was a pioneer in the analysis of organic substances, and his classic research on the composition of animal fat led to the discovery and naming of stearic and oleic acids.

Dr. William Beaumont (1785-1853), an American Army surgeon stationed at Fort Mackinac, Michigan, performed classic experiments on gastric digestion that destroyed the concept existing from the time of Hippocrates that food contained a single nutritive component. His experiments were performed during the period 1825-1833 (4) on a Canadian, Alexis St. Martin, whose musket wound afforded direct access to the stomach interior, thereby enabling food to be introduced and subsequently examined for digestive changes.

Among his many notable accomplishments, *Justus von Liebig* (1803-1873) studied vinegar fermentation (1837) and showed that acetaldehyde was an intermediate between alcohol and acetic acid. In 1842 he classified foods as either nitrogenous (vegetable fibrin, albumin, casein, and animal flesh and blood) and nonnitrogenous (fats, carbohydrates, and alcoholic beverages). Although this classification was not correct in detail, it served to distinguish differences among various foods. He also perfected methods for the quantitative analysis of organic substances, especially by combustion, and he published in 1847 what is apparently the first book on food chemistry, entitled *Researches on the Chemistry of Food* (22). Included in this book are accounts of his research on the water soluble constituents of muscle (creatine, creatinine, sarcosine, inosinic acid, lactic acid, and so on).

It is interesting that the developments just reviewed paralleled the beginning of serious and widespread adulteration of food, and it is no exaggeration to state that the need to detect impurities in food was a major stimulus for the development of analytical chemistry in general and analytical food chemistry in particular. Unfortunately, it was also true that advances in chemistry contributed somewhat to the adulteration of food since unscrupulous purveyors of food were able to profit from the availability of chemical literature, including formulas for adulterated food, and could replace older and less effective empirical approaches to food adulteration by more efficient approaches based on scientific principles. Thus, the history of food chemistry and the history of food adulteration are closely interwoven by the threads of several causative relationships (Ref. 15, pp. 18-19), and it is therefore appropriate to consider the matter of food adulteration.

The history of food adulteration in the more developed countries of the world falls into three distinct phases. From ancient times to about 1820 food adulteration was not a serious problem and there was little need for methods of detection. The most obvious explanation for this situation was that food was procured from small businesses or individuals, and the transactions involved a large measure of personal accountability. The second phase began in the early 1800s, when intentional food adulteration increased greatly in both frequency and seriousness. This development can be attributed primarily to increased centralization of food processing and distribution, with a corresponding decline in personal accountability, and partly to the rise of modern chemistry, as already mentioned. Intentional adulteration of food remained a serious problem until about 1920, which marks the end of phase two and the beginning of phase three. At this point regulatory pressures and effective methods of detection reduced the frequency and seriousness of intentional food adulteration to respectable levels, and the situation has gradually improved up to the present time.

Some would argue that a fourth phase of food adulteration began about 1950, when foods containing legal chemical additives became increasingly prevalent, when the use of highly processed foods increased to a point where they represented a major part of the diet of persons in most industrialized countries, and when contamination of some foods with uncontrolled by-products of industrialization, such as mercury and pesticides, became a threat to food safety. The validity of this contention is hotly debated and a consensus is not possible at this time. Nevertheless the course of action in the next few years seems clear. Public concern over the safety of the food supply has already led to some remedial actions, both voluntary and enforced, and more such actions can be anticipated.

The early 1800s also were a period of public concern over the quality of the food supply. Public concern, or more properly public indignation, was aroused in England by Frederick Accum's publication *A Treatise on Adulterations of Food* (1) and by an anonymous publication entitled *Death in the Pot* (3). Accum claimed that, "Indeed, it would be difficult to mention a single article of food which is not to be met with in an adulterated state; and there are some substances which are scarcely ever to be procured genuine" (p. 14). He further remarked, "It is not less lamentable that the extensive application of chemistry to the useful purposes of life, should have been perverted into an auxiliary to this nefarious traffic" (p. 20). Although Filby (15) asserted that Accum's accusations were somewhat overstated, the seriousness of intentional adulteration of food that prevailed in the early 1800s is clearly exemplified by the following not uncommon adulterants cited by Accum and Filby:

Annatto. Adulterants included turmeric, rye, barley, wheat flour, calcium sulfate and carbonate, salt, and Venetian red (ferric oxide, which in turn was sometimes adulterated with red lead and copper).

Pepper, black. This important product was commonly adulterated with gravel, leaves, twigs, stalks, pepper dust, linseed meal, and ground parts of plants other than pepper (e.g., wheat flour, mustard husks, pea flour, sago, and rice flour).

Pepper, cayenne. Substances such as vermilion (α-mercury sulfide), ocher (native earthy mixtures of metallic oxides and clay), and turmeric were commonly added to overcome bleaching that resulted from exposure to light.

Essential oils. Oil of turpentine, other oils, and alcohol.

Vinegar. Sulfuric acid.

Lemon juice. Sulfuric and other acids.

Coffee. Roasted grains, occasionally roasted carrots or scorched beans and peas; also, baked horse liver.

Tea. Spent, redried tea leaves, and leaves of many other plants.

Milk. Watering was the main form of adulteration; also common were chalk, starch, turmeric (color), gums, and soda. Occasionally encountered were gelatin, dextrin, glucose, preservatives [borax, boric acid, salicylic acid, sodium salicylate, potassium nitrate, and sodium fluoride or benzoate (formalin)], and such colors as annatto, saffron, caramel, and some sulfonated dyes.

Beer. "Black extract," obtained by boiling the poisonous berries of *Cocculus indicus* in water and concentrating the fluid, was apparently a common additive. This extract imparted flavor, narcotic properties, and additional intoxicating qualities to the beverage.

Wine. Colorants: alum, husks of elderberries, Brazil wood, and burnt sugar, among others. Flavors: bitter almonds, tincture of raisin seeds, sweet-brier, oris root, and others. Aging agents: bitartrate of potash, "oenathic" ether (heptyl ether), and lead salts. Preservatives: salicylic acid, benzoic acid, fluoborates, and lead salts. Antacids: lime, chalk, gypsum, and lead salts.

Sugar. Sand, dust, lime, pulp, and coloring matters.

Butter. Excessive salt and water, potato flour, and curds.

Chocolate. Starch, ground sea biscuits, tallow, brick dust, ocher, Venetian red (ferric oxide), and potato flour.

Bread. Alum and flour made from products other than wheat.

Confectionary products. Colorings containing lead and arsenic.

Once the seriousness of food adulteration in the early 1800s was made evident to the public, remedial forces gradually increased. These took the form of new legislation to make adulteration unlawful and greatly expanded efforts by chemists to learn the native properties of foods, the chemicals commonly used as adulterants, and the means of detecting them. Thus, during the period 1820-1850, chemistry and food chemistry began to assume importance in Europe. This was possible because of the work of the scientists already cited, and was stimulated largely by the establishment of chemical research laboratories for young students in various universities, and by the founding of new journals for chemical research (5). Since then, advances in food chemistry have continued at an accelerated pace, and some of these advances and some of the causative factors are mentioned below.

Microscopic analysis of food was raised to a position of importance through the efforts of *Arthur Hill Hassall* in England during the middle 1800s. He and his associates produced an extensive set of diagrams illustrating the microscopic appearance of pure and adulterated foodstuffs.

In 1860, the first publicly supported agricultural experiment station was established in Weede, Germany, and *W. Hanneberg* and *F. Stohmann* were appointed director and chemist, respectively. Based largely on the work of earlier chemists, they developed an important procedure for the routine determination of major constituents in food. By dividing a given sample into several portions they were able to determine moisture content, "crude fat," ash, and nitrogen. Then, by multiplying the nitrogen value by 6.25, they arrived at its protein content. Sequential digestion with dilute acid and dilute alkali yielded a residue termed "crude fiber." The portion remaining after removal of protein, fat, ash, and crude fiber was termed "nitrogen-free extract," and this was believed to represent utilizable carbohydrate. Unfortunately, for many years chemists and physiologists wrongfully assumed that like values obtained by this procedure represented like nutritive value regardless of the kind of food (24).

In 1871, *Jean Baptiste Dumas* (1800-1884) suggested that a diet consisting of only protein, carbohydrate, and fat was inadequate for support of life.

In 1862, the Congress of the United States passed the Land-Grant College Act, authored by Justin Smith Morrill. This act helped establish colleges of agriculture in the United States and provided considerable impetus for the training of agricultural and food chemists. Also in 1862, the United States Department of Agriculture was established and Isaac Newton was appointed the first commissioner.

In 1863, Harvey Washington Wiley became chief chemist of the U.S. Department of Agriculture from which office he led the campaign against misbranded and adulterated food, culminating in the passage of the first Pure Food and Drug Act in the United States (1906).

In 1887, agriculture experiment stations were established in the United States as a result of enactment of the Hatch Act. Representative William H. Hatch of Missouri, Chairman of the House Committee on Agriculture, was author of the measure. As a result, the world's largest national system of agricultural experiment stations came into existence and this had a great impact on food research in the United States.

During the first half of the twentieth century, most of the essential dietary substances were discovered and characterized, namely, vitamins, minerals, fatty acids, and some amino acids.

The development and extensive use of chemicals to aid in the growth, manufacture, and marketing of foods was an especially noteworthy and contentious event in the middle 1900s, and more will be said about this subject in a later section.

This historical review, although brief, makes the current food supply seem almost perfect in comparison to that which existed in the 1800s.

III. APPROACH TO THE STUDY OF FOOD CHEMISTRY

It is desirable to establish an analytical approach to the chemistry of food formulation, processing, and storage, in which facts derived from the study of one food or model system can enhance our understanding of the spoilage of other products. There are four components to this approach: (a) determining those properties that are important characteristics of safe, high-quality foods, (b) determining those chemical and biochemical reactions that have important influences on loss of quality and/or wholesomeness of foods, (c) integrating the first two points so that one understands how the key chemical and biochemical reactions influence quality and safety, and (d) applying this understanding to the various situations encountered during the formulation, storage, and processing of food.

A. Quality and Safety Attributes

It is essential to reiterate that safety is the first requisite of any food. In the broad sense, this is taken to mean that a food is free of any harmful chemical or microbial contaminant at the time of its consumption. Often this concept is used in its operational sense for practical reasons. In the canning industry, commercial sterility as applied to low-acid foods is taken to mean the absence of viable spores of *Clostridium botulinum*. This in turn can be translated into a specific set of heating conditions for a specific product in a specific package. Given this information one can then approach optimization of retention of other quality attributes through the techniques discussed in Chap. 7. Similarly, in such a product as peanut butter, operational safety may be taken as the absence of aflatoxins, carcinogenic substances produced by certain species of molds (Chap. 11). Steps taken to prevent growth of the mold in question may or may not interfere with the retention of some other quality attribute; nevertheless, the conditions for safety must be satisfied.

A simple list of some quality attributes of food and some undesirable changes they can undergo is given in Table 1. The major attributes are texture, flavor, color, and nutritive value. The changes that can occur, with the exception of those involving nutritive value, are readily evident to the consumer. That is, they are macroscopic changes arising from the entire gestalt of microscopic or chemical changes that take place in the product during processing or storage. The major subheadings of Table 1 are linked later in this chapter to specific chemical reactions.

B. Chemical and Biochemical Reactions

Many reactions can lead to the deterioration of food quality or impairment of food safety. Some of the more important classes of these reactions are listed in Table 2. Each reaction class can involve different reactants or substrates depending on the specific food and the particular conditions for processing or storage. They are treated as reaction classes because the general nature of the substrates or reactants are similar for all foods. Thus, nonenzymic browning involves reactions of carbonyl compounds, which can arise from diverse prior reactions, such as oxidation of ascorbic acid or hydrolysis of starch. Lipid oxidation may involve primarily triacylglycerols in one food or phospholipids in another, but in both, autoxidation of unsaturated fatty acids is the primary event. It is unnecessary to discuss each reaction in Table 2 since each, as indicated, has been treated in some depth elsewhere in this volume.

Table 1 Classification of Undesirable Changes That Can Occur in Food

Attribute	Undesirable change
Texture	a. Loss of solubility b. Loss of water-holding capacity c. Toughening d. Softening
Flavor	Development of e. Rancidity (hydrolytic or oxidative) f. Cooked or caramel flavors g. Other off-flavors
Color	h. Darkening i. Bleaching j. Development of other off-colors
Nutritive value	Loss or degradation of k. Vitamins l. Minerals m. Proteins n. Lipids

Table 2 Chemical and Biochemical Reactions That Can Lead to
Deterioration of Food Quality or Impairment of Safety

Class of reaction	Chapters
Nonenzymic browning	3, 5
Enzymic browning	6
Lipid hydrolysis	4, 6
Lipid oxidation	4, 5, 6, 9
Protein denaturation	5, 6, 12-14
Protein cross-linking	5, 12
Oligo- and polysaccharide hydrolysis	3, 6, 12
Protein hydrolysis	5, 6, 12
Polysaccharide synthesis	3, 6, 12, 13
Degradation of specific natural pigments	8, 12, 15
Glycolytic changes	3, 6, 12, 13, 15

C. Effect of Reactions on the Quality and Safety of Food

The reactions listed in Table 2 can lead to deterioration of the quality attributes described in Table 1, and sometimes to impairment of food safety. Integration of the information contained in both tables can lead to an understanding of the causes of food deterioration. Deterioration of food usually consists of a series of primary events, each with a set of consequences, and these in turn ultimately manifest themselves as one or more of the macroscopic changes listed in Table 1. Examples of sequences of this type are shown in Table 3. Note particularly that a given quality attribute can be altered as a result of several different primary events. If Table 3 could be enlarged to include most of the information in this volume, and if this information were assembled with the primary events arranged horizontally at the bottom of a page, the intermediate consequences at the center of the page, and the quality manifestations at the top, the assembled data would approximate the shape of a triangle.

Note that the sequences in Table 3 can be applied in two directions. Operating from left to right one can consider a particular primary event and its consequences, and then predict the possible macroscopic changes. Alternatively, one can determine the probable cause(s) of an observed quality change (column 3, Table 3) by considering all primary events that are possibly involved and then isolating, by appropriate chemical tests, the key primary event. The interested reader might attempt to conceive, from his or her own experience, additions to Table 3. The utility of constructing such sequences is that they encourage one to approach problems of food deterioration in an analytical manner.

Figure 1 is a summary of the reactions and interactions of the major constituents of food. The major cellular pools of carbohydrates, lipids, proteins, and their intermediary metabolites are shown on the left-hand side of the diagram. The exact nature of these pools is dependent on the physiological state of the tissue at the time of processing or storage, or the constituents present in or added to nontissue foods. Each class of compound can undergo its own characteristic type of deterioration. Of great interest is the common role that carbonyl compounds play in the deterioration process. They arise from lipid oxidation and carbohydrate degradation, and lead to the destruction of nutritional value, to off-colors, and to off-flavors. Of course these same reactions also lead to desirable flavors and colors during the cooking of foods.

Table 3 Cause and Effect in the Deterioration of Food

Some primary events	Consequence	Quality change (see Table 1)
Hydrolysis of lipids	Free fatty acids react with protein	Texture: a, b, c Flavor: e, g Nutritive value: m
Hydrolysis of polysaccharides	Sugars react with proteins	Texture: a, b, c Flavor: f Color: h Nutritive value: k, m
Oxidation of lipids	Oxidation products react with many other constituents	Texture: a, b, c Flavor: 3 Color: h and/or i Nutritive value: k, m, n
Bruising of fruit	Cells break, enzymes are released, oxygen accessible	Texture: d Flavor: g Color: h Nutritive value: k
Heating of green vegetables	Cell walls and membranes lose integrity, acids and enzymes are released	Texture: d Flavor: g Color: j Nutritive value: k, l
Heating of muscle tissue	Proteins denature and aggregate, enzymes become inactive	Texture: b and c or d Flavor: f Color: j Nutritive value: k

D. Analysis of Situations Encountered During the Storage and Processing of Food

Having before us a description of the attributes of high-quality, safe foods, the significant chemical reactions involved in the deterioration of food, and the relation of the two, we can now begin to consider how to apply this information to situations encountered during the storage and processing of food.

The variables that are important during the storage and processing of food are listed in Table 4. Temperature is perhaps the most important of these variables because of its broad influence on all types of chemical reactions. The effect of temperature on an individual reaction can be expressed by the Arrhenius equation, $k = Ae^{-\Delta E/RT}$. Data conforming to the Arrhenius equation yield a straight line when log k is plotted versus $1/T$. The Arrhenius plots in Fig. 2 represent reactions important in food deterioration. It is evident that food reactions generally conform to the Arrhenius relationship over a certain intermediate temperature range but that deviations from this relationship can occur at high or low temperatures (25). Thus, it is important to remember that the Arrhenius relationship for food systems can be used only over a range of temperatures that has been experimentally tested. Deviations from the Arrhenius relation-

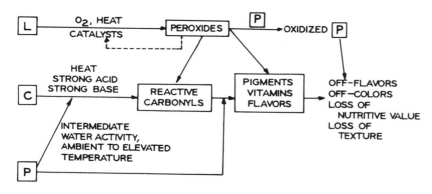

Figure 1 Summary of chemical interactions among major food constituents. L, Lipid pool (triacylglycerols, fatty acids, and phospholipids); C, carbohydrate pool (polysaccharides, sugars, organic acids, and so on); P, protein pool (proteins, peptides, amino acids, and other N-containing substances).

ship can occur because of the following events, most of which are induced by either high or low temperatures: (a) enzyme activity may be lost, (b) the reaction pathway may change or may be influenced by a competing reaction(s), (c) the physical state of the system may change, or (d) one or more of the reactants may become depleted.

The second variable in Table 4 is time, and this should be considered together with the rate at which temperature changes with time. During storage of a food product, one frequently wants to know how long the food can be expected to remain above a certain level of quality. Therefore, one is interested in time with respect to the integral of chemical and/or microbiological changes that occur during a specified storage period, and in the way these changes combine to determine a specified storage life for the product.

During processing, attention should be given primarily to the time variable as it appears in the rate expression, usually the rate of temperature change in a process (dT/dt), since this determines the relative rates of competing chemical reactions and the rate at which microorganisms are destroyed.

Time also can have an important influence on the relative importance of concurrent reactions. For example, if a given food can deteriorate by both lipid oxidation and non-enzymic browning, and if the products of the browning reaction are antioxidants, it is

Table 4 Important Variables in the Processing and Storage of Foods

Temperature (T)
Time (t)
Rate (dT/dt)
pH
Composition of product
Composition of gaseous phase
Water activity (a_w)

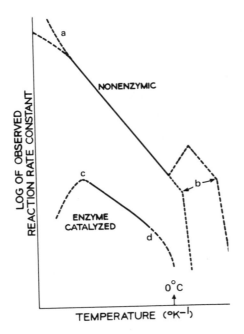

Figure 2 Conformity of important deteriorative reactions in food to the Arrhenius relationship. (a) Above a certain value of T there may be deviations from linearity due to a change in the path of the reaction. (b) As the temperature is lowered below the freezing point of the system, the ice phase (essentially pure) enlarges and the fluid phase, which contains all the solutes, diminishes. This concentration of solutes in the unfrozen phase can decrease reaction rates (supplement the effect of decreasing temperature) or increase reaction rates (oppose the effect of declining temperature), depending on the nature of the system (see Chap. 2). (c) For an enzymic reaction there is a temperature range in which denaturation of the enzyme competes with formation of reaction product. (d) In the vicinity of the freezing point of water, subtle changes, such as the dissociation of an enzyme complex, can lead to a sharp decline in reaction rate.

important to know whether the time scales for the two reactions overlap sufficiently to cause a significant interaction.

Another variable, pH, influences the rates of many chemical and enzymic reactions. Extreme pH values are usually required for severe inhibition of microbial growth or enzymic processes, and these conditions can result in acceleration of acid- or base-catalyzed reactions. In contrast, relatively small pH changes can cause profound changes in the quality of some foods, for example, muscle (Chap. 12).

The composition of the product is important since this determines the reactants available for chemical transformation. Particularly important from a quality standpoint in this regard is the relationship that exists between the composition of the raw material and the composition of the finished product. For example, (a) the manner in which fruits and vegetables are handled postharvest can influence sugar content, and this, in turn, influences the degree of browning obtained during dehydration or deep-fat frying (Chap. 15). (b) The manner in which animal tissues are handled postmortem influences the extents and rates of proteolysis, glycolysis, and ATP degradation, and these in turn can influence storage life, water-holding capacity, toughness, flavor, and color. (c) The blending of raw materials may cause unexpected interactions; for example, the rate of oxidation can be accelerated or inhibited depending on the amount of salt present.

Furthermore, in fabricated foods, the composition can be controlled by adding approved chemicals, such as acidulants, chelating agents, or antioxidants, or by removing undesirable reactants, for example, the removal of glucose from dehydrated egg albumen.

The composition of the gaseous phase is important mainly with respect to the availability of oxygen as a reactant. In situations in which it is desirable to limit oxygen it is unfortunately almost impossible to achieve complete exclusion. The detrimental consequences of a small amount of residual oxygen sometimes become apparent during product storage. For example, the early formation of a small amount of dehydroascorbic acid can lead to browning during storage.

One of the more important variables controlling reaction rates in foods is water activity (a_w). Numerous investigators have shown a_w to be an important factor in enzyme reactions (2), lipid oxidation (20,26), nonenzymic browning (13,20), sucrose hydrolysis (27), chorophyll degradation (21), anthocyanin degradation (14), and in many other reactions. As is discussed in Chap. 2, most reactions tend to decrease in rate below an a_w corresponding to the range for intermediate moisture foods (0.75-0.85), and this is primarily due to the reduced solvent capacity of the decreased water phase. Oxidation of lipids and associated secondary effects, such as carotenoid decoloration, are exceptions to this rule; that is, these reactions accelerate at the lower end of the a_w scale.

Food chemists must integrate information about quality attributes of foods, deteriorative reactions to which foods are susceptible, and the factors governing the kinds and rates of these deteriorative reactions to solve problems related to food formulation, processing, and storage. The details of how this integration is accomplished is dealt with in the final chapter of this book, after the reader had had the opportunity to assimilate needed information in the intervening chapters.

IV. SOCIETAL ROLE OF FOOD CHEMISTS

A. Why Should Food Chemists Become Involved in Societal Issues?

Food chemists, for the following reasons, should feel obligated to become involved in societal issues that encompass pertinent technological aspects (technosocietal issues).

Food chemists have had the privilege of receiving a high level of education and of acquiring special scientific skills, and these privileges and skills carry with them a corresponding high level of responsibility.

Many activities of food chemists have a bearing on the general welfare of the population—development and use of food additives and packaging materials, disposal practices for food manufacturing wastes, development of food regulations, and so on—and food chemists should, therefore, feel a responsibility to have these activities diected to the benefit of society.

If food chemists do not become involved in technosocietal issues the opinions of others—professional lobbyists, persons in the news media, consumer activists, charlatans, antitechnology zealots, and others—will prevail. Many of these individuals are less well qualified than food chemists to speak on food-related issues, and some are obviously *far* less well qualified.

B. Type of Involvement—Committees and Professional Societies

The societal obligations of food chemists include good job performance, good citizenship, and guarding the ethics of the scientific community, but the fulfillment of these very necessary roles is not enough. An additional role of great importance, and one that often goes unfulfilled by food chemists, is that of helping determine how scientific knowledge is interpreted and used by society. Although food chemists and other food scientists cannot, and should not, be the sole arbitrators of these decisions, they must, in the interest of wise decision making, have their views heard and considered. Acceptance of this position, which is surely indisputable, leads to the obvious question, what exactly should food chemists do to properly discharge their responsibilities in this regard?

One course of action involves participation in pertinent professional societies and serving on governmental advisory committees, when invited. Unfortunately, most professional societies are content to publish journals, sponsor an annual meeting featuring technical papers, and grant a few awards to worthy members. A few engage actively in the dissemination of educational materials to the public, and still fewer actively provide a forum for the discussion of contentious societal issues. Conscientious scientists should attempt to overcome this shortcoming.

Service on governmental advisory committees can be fraught with problems. A question that should be answered before accepting such an appointment is, why has the committee been formed? In some instances, the committee's function will be to provide sound scientific advice, and if so, this is commendable. Sometimes, however, the committee's role, although never proclaimed as such, is to legitimize governmental actions that have already occurred or are contemplated. One should carefully evaluate, of course, whether service on a committee of the latter type is in the best interests of society.

Additional problems connected with governmental advisory committees and, for that matter, with committees in general, is that they tend to move rapidly from the realm of hard facts to the realm of educated speculation, they tend to give more weight to easily quantifiable views than to qualitative views that may have equal merit (ethics, quality, aesthetics, and others) and they almost inevitably arrive at a consensus that does not deviate far from conventional viewpoints that are neither profound nor creative. Committee members need to be aware of these undesirable tendencies and attempt to surmount them.

C. Type of Involvement—Individual Initiatives

A second course of action, and no doubt the most important and most demanding for the average food chemist, is to undertake personal initiatives of a public service nature. This can involve letters to newspapers, journals, legislators, government regulators, company executives, university administrators, and others, and speeches before various civic groups. The major objectives are education and enlightenment of the public. Since education and enlightenment are eternal needs that require continual attention, the remainder of this section will bedevoted to the problems in this area and the approaches that food chemists can use to aid in their solution.

With food, the matters of education and enlightenment are especially acute, since to many persons, food is more than a source of nutrients and a means of satisfying hunger. It is an integral part of religious practice, cultural heritage, ritual, and social symbolism and a route to physiological well-being, not only because of its nutrients, but because of its impact on state of mind.

An especially important need in the area of consumer enlightenment is the development of skills that will enable consumers to intelligently evaluate information about foods. This aspect of consumer enlightenment is not easy because a significant portion of the populace seems to have ingrained false notions or ignorance about foods and wise dietary practices. Some still believe, for instance, that red wine is a "tonic" wine, that red beets are "good for the blood," that "natural" vitamins are superior to synthetic vitamins, and that "natural" foods are more healthful than those foods not carrying the natural label. In addition, it is a combination of ignorance, tradition, and religious belief that leads most persons of the Western world to regard dogs, cats, insects, snakes, and horses as unacceptable foods.

Perhaps the most contentious food issue involves modification of foods through the addition of chemicals. "Chemophobia," the fear of chemicals, has afflicted a significant portion of the populace, causing food additives, in the minds of many, to represent hazards inconsistent with fact. This point is of sufficient importance to deserve elaboration.

One can, with ease, find in the popular literature numerous articles by authors who stridently proclaim how the American food supply is sufficiently laden with poisons to render it unwholesome at best, and life threatening at worst. Truly shocking, they say, is the manner in which greedy industrialists poison our foods for profit while an ineffectual Food and Drug Administration watches with placid unconcern. Should authors holding this viewpoint be believed? It is advisable to apply the following criteria when evaluating the validity of any journalistic account dealing with issues of this kind.

Credibility of the author. Is the author, by virtue of formal education, experience, and acceptance by reputable scientists, qualified to write on the subject? The writer should, to be considered authoritative, have been a frequent publisher of articles in respected scientific journals, especially those requiring peer review. If the writer has contributed only to the popular literature, particularly in the form of articles with sensational titles and "catch" phrases, this is cause to exercise special care in assessing whether the presentation is scholarly and accurate.

Appropriateness of literature citations. A lack of literature citations does not constitute proof of irresponsible or inaccurate writing, but it should provoke a feeling of interpretive caution in the reader. In trustworthy publications, literature citations will almost invariably be present and will direct the reader to highly regarded scientific publications. When "popular" articles constitute the bulk of the literature citations, the author's views should be regarded with caution.

Credibility of the publisher. Is the publisher of the article, book, or magazine regarded, by reputable scientists, as a consistent publisher of quality scientific materials? If not, an extra measure of caution is appropriate when considering the data.

Based on these criteria, information in the poison-pen types of books and articles, which are abundant in the popular press, should be evaluated critically and largely dismissed as unreliable. However, even if this approach is followed, knowledgeable persons disagree concerning the safety of foods. A few examples are in order.

Some support the view that our food supply is acceptably safe and nutritious and that food additives pose no special risks.

Our food supply in the United States is in its golden age of safety, adequacy, variety, and abundance. It has reached this stage first as a result of a free and open marketplace in which producers and processors compete for the attention of consumers and second, because of the flowering of science and technology during the past century. These facts are evident to anyone who compares the contents of a modern supermarket with the flyblown grocery shop of past years with its cracker-barrel, salt pork, rancid lard, and wormy apples, or with the meager, contaminated, and nutritionally inadequate diet that has to suffice for the majority of the world's population. The only really important problem with the diet of the United States is that people eat too much. (Jukes, Ref. 19, p. 772)

What with frequent headlines touching on the alleged dangers of certain additives, I keep running into people who profess to believe that there's a massive plot afoot to poison the population of the United States. Are there good reasons for these fears?

There are not. Our food supply today is much safer than it was in the past. Spoilage and microbe infestation once exposed the public to the constant threat of gastroenteritis, not to mention typhoid, cholera, tuberculosis and a variety of other food-borne diseases.

Still, we can expect a great deal of continued controversy over the safety of many specific additives. This will occur partly because some countries have imposed restrictions on certain chemicals that are permitted in the United States. Why? Different countries have different laws and different procedures and, of course, different officials have different opinions. (Mayer, Ref. 23, pp. 199-200).

It can be argued that, not only has there not been any noticeably untoward effect from the consumption of food additives in food, but rather that, during a period of expanding use, there has been a consistent improvement in the general health and life expectation of people. This is not to claim a cause and effect situation, but merely to reinforce the argument that additives have enabled food to make maximum nutritional impact. (Taylor, Ref. 29, p. 10).

For all of us recognize that the U.S. has the most abundant and safest food supply in the world. We also know that it is impossible to require that any food be absolutely safe. Everything, even water, poses some risk to some consumers. So it follows that the safety of food must be determined on the basis of an acceptable level of risk. (Clausi, Ref. 7, p. 66).

We do have good foods in this country—as good as, if not better than any country in the world. The Department of Agriculture, Food and Drug Administration, Public Health Service, National Academy of Sciences, to say nothing of the excellent laboratories of many of the food and chemical companies—and let's not forget our universities—have done and are doing an outstanding job to give us good, safe food—and plenty of it. (Stare and Whelan, Ref. 28, pp. 11-12).

The benefits of additives are often ignored, and the risks are often overstated. At this point in time, quite a few additives have been banned because of questions about their safety. Yet, there is no known case of any illness other than the kind of allergenic reactions that many foods can cause. Even mishaps from improperly used additives—huge overdoses taken by mistake—are so rare as to be medical curiosities (Clydesdale and Francis, Ref. 8, p. 111).

The chief fact is that additives, being more tested than the ingredients nature puts in our food supply, are much safer and, in any remotely normal diet, have never hurt anybody. Of course, there is always the possibility that new tests or epidemiologic research will show that an additive causes some danger or disease, but additives are, for lack of a better expression, safe by every scientific measurement (Hall, Ref. 16, p. 148).

Others, with equal sincerity, believe that our food supply is unnecessarily hazardous, particularly with regard to food additives.

It is no longer possible for any knowledgeable person to argue that the American food supply is without potential hazard. Since we have not been able to eliminate serious potential health hazards from the food supply, we now face the problem of how to manage the food supply to minize these health hazards. (Turner, Ref. 31, p. 63).

Not all additives have been systematically tested. Many chemicals have not been studied for their ability to cause cancer. Even fewer have been tested on pregnant animals to see if they cause birth defects. Virtually none have been screened for genetic effects.

Food additives can save consumers time, raise corporate profits, create new products and improve old ones and make foods more nutritious. On the other hand, many additives have not been adequately tested, some persons may be sensitive to certain additives, many factory made foods made with additives are devoid of nutrients and some chemicals may be used solely for cosmetic, deceptive, or irrational purposes. (Jacobson, Ref. 18, pp. 15, 20).

If we had to take a dangerous drug, we would want to know its potential benefit compared with the risk. But with food additives we often take risks for little or nothing. The benefit-risk ratio for food chemicals often boils down to a great big health risk to you and a great big economic benefit to industry.

Why doesn't the FDA clean up the food supply? Why must we be subjected to so many untested or unsafe chemicals in our diet, many of which if taken in pill form could be obtained only by prescription? The simple answer is that the FDA is not performing its public function. Legally, it has the authority to end food pollution, or at least reduce it and the risks drastically, but it doesn't. Often the FDA seems more preoccupied with perpetuating the use of food additives than with getting rid of them—even though they are found unsafe. Some observers of the Washington scene laughingly say "FDA" stands for "foot-dragging artists" (Verrett and Carper, Ref. 32, pp. 21, 52).

This serious dichotomy of opinion cannot be resolved here, but a few additional comments, developed for their relevance and uncontentious nature, should help undecided individuals arrive at a soundly based personal perspective on food additives and food safety.

D. Perspectives

1. Laws and Regulations Governing the Food Supply

It is almost universally agreed that laws and regulations are needed to ensure that foods are processed, handled, and stored in a manner that ensures virtual freedom from known risks for "normal" individuals on reasonable diets.

A century ago, food processing and handling practices were often poorly regulated, and serious food adulteration, many times intentional, was commonplace. Now, in developed countries, detailed laws and regulations exist to govern food processing and handling practices and the most serious problems of food adulteration and unsafe handling practices have been largely eradicated. Almost everyone agrees that proper laws and regulations have resulted in a food supply that has become progressively safer.

2. Testing for Food Safety

The potential hazard of food chemicals has been assessed historically by means of animal feeding studies. This has been, and will continue to be, necessary, especially for assessing moderate to long-term effects, since obvious constraints exist on using humans as test subjects. The animal approach involves at least two formidable shortcomings.

1. The assessment of sublethal, adverse effects in animals is an inexact science.
2. Animal data can be extrapolated to humans only with considerable uncertainty, even under the best of circumstances. This uncertainty increases when a test substance is fed to animals at levels far in excess of those normally encountered in human diets. It is believed possible that massive doses in animals may produce "metabolic overload" with the formation of metabolites that would not occur at lower doses (10).

Interactions among dietary components may also increase the uncertainty of extrapolation. Interactions may lessen or intensify the effect of a given test substance, and the response direction is not always known; even when it is, it is difficult to quantify.

3. Cost and Time of Testing

Thorough testing of food chemicals for potential hazard is expensive and time consuming. Many chemicals, especially those having GRAS (generally recognized as safe) status, as well as many naturally occurring components of foods, have not been rigorously tested. These chemicals have received a low priority for testing primarily because they have long histories of use without evidence of harm.

4. What Remains to Be Learned About the Safety of Chemical Constituents of Food?

Screening of chemicals for acute and subchronic (determined, in part, by animal studies spanning several months and involving multiple, very large dose levels) hazards is relatively easy and straightforward, and it is generally believed that the American food supply has been effectively freed of significant hazards of these kinds as they relate to normal individuals on reasonable diets. Testing of food chemicals for chronic hazards (determined by animal studies spanning a major portion of the test animal's normal life span and involving low-level doses) remains in its infancy with respect to test procedures and numbers of chemicals tested. As test procedures become perfected, and more extensive testing is done, additional chemicals, probably constituting marginal hazards, will no doubt be banned or restricted in use. To some, this is alarming; to others, it represents simply the expected and acceptable result of scientific endeavor. Statistics concerning the general state of healthfulness and life expectancy of the U.S. population, for the most part, show encouraging trends, which would not be the case if food constitutents represented important chronic risks.

5. Absolute Safety

Absolute safety of a food constituent, either added or naturally occurring, cannot be proven, nor is a goal of absolute safety attainable. Consider a food additive fed at a controlled concentration to 1000 animals over several generations, and no evidence of harm is found using standard test procedures. A critic can properly argue that this does not prove safety; it merely proves the absence of harm in animals tested under specific circumstances. The critic can also argue that harm was not demonstrated because the chosen concentration of food additive was too low, the exposure time was too short, the number of animals tested was too small, the generations spanned were too few, the test procedures chosen were insufficiently sensitive, and/or the species of animal was inappropriate, and this view cannot be rationally challenged. Virtually every chemical component of food, be it natural or synthetic, will cause harmful effects in some individual provided the intake is sufficiently high and/or the consumption period is sufficiently long. This concept is far from new, having first been stated by the Swiss alchemist and physician, Paracelsus, some 500 years ago. Irrefutable proof of this concept was not available, however, until recent years.

The unattainability of absolute safety becomes evident when one considers such matters as dietary habits, state of health, age, and biological diversity. What may be safe for one individual may not be safe for another, leading to the unfortunate, but inescapable conclusion, that absolute safety, or even acceptable safety, is simply not attainable for all persons under all circumstances.

6. Relative Importance of Potential Food-Derived Hazards

It is useful to consider just where food additives rank as a potential hazard in the food supply. The briefest and most pertinent statement on this subject was drafted at the Marabou Symposium on Food and Cancer (30).

> Of the potential sources of harm in foods the largest by far are, first, microbiological contamination and, next, nutritional imbalance. Risks from environmental contamination are 1,000 times less and risks from pesticide residues and food additives can be estimated as about a further 100 times smaller again. Naturally occurring compounds in food are far more likely to cause toxicity than intentional food additives.

With regard to food additives, this statement surely must be reassuring to all but the most skeptical individuals. Moreover, the potential hazard from toxicants occurring naturally in foods, a safety aspect given little attention by regulatory agencies and almost totally ignored by the most vocal critics of food additives, is placed in an interesting and realistic perspective (9).

7. Judgmental Issues

Some troublesome questions ultimately arise when considering the safety of food constituents.

1. How many and what kinds of sublethal, food-related incidences should be tolerated annually in our population before a given chemical constitutent is declared safe?
2. Should the potential risks of allowing a given chemical in the food supply be weighed against the benefits (health, sensory quality, and cost) of using the chemical and the risk of not using the chemical? For example, the banning of nitrites from certain meat products would presumably result in increased hazard from *C. botulinum*, yet allowing nitrites to remain may lead to the development of carcinogenic nitrosamines. Which risk is greater? Further, the banning of cyclamates resulted in greater consumption of saccharin, a compound also suspected of representing some risk. Are we sure

risk was reduced by the decision to ban cyclamates? Moreover, significant increases in the cost of bread would presumably occur if certain additives were disallowed. If some of these bread additives were one day deemed to represent a small risk for a tiny percentage of the population, should these additives be banned despite the economic consequences and the indirect result that a higher bread price might cause a portion of the population to consume a diet of lower nutritional quality (less bread)? If these risk/benefit and risk/risk considerations are to be involved in the formulation of public policy dealing with safety of the food supply, a troublesome issue arises as to how these risks and benefits are to be quantified, and how final decisions will be determined.

3. Should the imposed safety standards be less severe for important nutritional additives and preservatives than for additives used for cosmetic or cost-reduction purposes?

These are policy issues, and the views of food chemists should be carefully considered when relevant policies are formulated. However, these are also societal issues of a judgmental nature, about which food chemists possess no special expertise. Thus, all members of society, not just food chemists, should have a voice in formulating policies of this kind—policies that transcend simplistic right or wrong answers, and have full impact on the agonizingly difficult issue of what is acceptable and reasonable. Consensus on these issues may never develop, but if it does, the process will surely be fraught with strident disagreement unless food chemists and other scientists have done an effective job of consumer education and enlightenment.

In summary, scientists and technologists have greater obligations than do individuals without formal scientific education. The generation of scientific knowledge in a productive, ethical manner is one additional responsibility of considerable significance, but even more important is the obligation to ensure that scientific knowledge is used in a manner that will yield the greatest benefits for society. Fulfillment of this obligation requires that scientists not be content with excellence in their day-to-day occupational activities, but that they develop a deep-seated concern for the well-being and scientific enlightenment of all humans.

REFERENCES

1. Accum, F. (1966). *A Treatise on Adulterations of Food, and Culinary Poisons*, 1920, Ab'm Small, Philadelphia. Fascimile reprint by Mallinckrodt Chemical Works, St. Louis, Missouri.
2. Acker, L. W. (1969). Water activity and enzyme activity. *Food Technol. 23*(10): 1257-1270.
3. Anonymous (1831). *Death in the Pot*. Cited by Filby, 1934 (Ref. 15).
4. Beaumont, W. (1833). *Experiments and Observations of the Gastric Juice and the Physiology of Digestion*, F. P. Allen, Plattsburgh, N.Y.
5. Browne, C. A. (1944). *A Source Book of Agricultural Chemistry*, Chronica Botanica Co., Waltham, Mass.
6. Chevreul, M. E. (1824). Considérations générales sur l'analyse organique et sur ses applications. Cited by Filby, 1934 (Ref. 15).
7. Clausi, A. S. (1979). Revising the U.S. food safety policy: An industry viewpoint. *Food Technol. 33*(11):65-67.
8. Clydesdale, F. M., and F. J. Francis (1977). *Food, Nutrition and You*, Prentice-Hall, Englewood Cliffs, N.J.
9. Committee on Food Protection (1973). *Toxicants Occurring Naturally in Foods*, 2nd ed., National Academy Press, Washington, D.C.

10. Committee on Food Protection (1980). *Risk Assessment/Safety Evaluation of Food Chemicals*, National Academy Press, Washington, D. C.
11. Davy, H. (1813). *Elements of Agricultural Chemistry, in a Course of Lectures for the Board of Agriculture*, Longman, Hurst, Rees, Orme and Brown, London. Cited by Browne, 1944 (Ref. 5).
12. Davy, H. (1936). *Elements of Agricultural Chemistry*, 5th ed., Longman, Rees, Orme, Brown, Green and Longman, London.
13. Eichner, K., and M. Karel (1972). The influence of water content and water activity on the sugar-amino browning reaction in model systems under various conditions. *J. Agr. Food Chem. 20*(2):218-223.
14. Erlandson, J. A., and R. E. Wrolstad (1972). Degradation of anthocyanins at limited water concentration. *J. Food Sci. 37*(4):592-595.
15. Filby, F. A. (1934). *A History of Food Adulteration and Analysis*, George Allen and Unwin, Ltd., London.
16. Hall, R. L. (1982). Food additives, in *Food and People* (D. Kirk and I. K. Eliason, eds.), Boyd and Fraser, San Francisco, pp. 148-156.
17. Ihde, A. J. (1964). *The Development of Modern Chemistry*, Harper and Row, New York.
18. Jacobson, M. (1976). *Eater's Digest*. Doubleday and Co., Garden City, N.Y.
19. Jukes, T. H. (1978). How safe is our food supply? *Arch Intern. Med. 138*:772-774.
20. Labuza, T. P., S. R. Tannenbaum, and M. Karel (1970). Water content and stability of low-moisture and intermediate-moisture foods. *Food Technol. 24*(5):543-550.
21. LaJollo, F., S. R. Tannenbaum, and T. P. Labuza (1971). Reaction at limited water concentration. 2. Chlorophyll degradation. *J. Food Sci. 36*(6):850-853.
22. Liebig, J. von (1847). *Researches on the Chemistry of Food*, edited from the author's manuscript by William Gregory; Londson, Taylor and Walton, London. Cited by Browne, 1944 (Ref. 5).
23. Mayer, J. (1975). *A Diet for Living*, David McKay, Inc., New York.
24. McCollum, E. V. (1959). The history of nutrition. *World Rev. Nutr. Diet 1*:1-27.
25. McWeeny, D. J. (1968). Reactions in food systems: Negative temperature coefficients and other abnormal temperature effects. *J. Food Technol. 3*:15-30.
26. Quast, D. G., and M. Karel (1972). Effects of environmental factors on the oxidation of potato chips. *J. Food Sci. 37*(4):584-588.
27. Schoebel, T., S. R. Tannenbaum, and T. P. Labuza (1969). Reaction at limited water concentration. 1. Sucrose hydrolysis. *J. Food Sci. 34*(4):324-329.
28. Stare, F. J., and E. M. Whelan (1978). *Eat OK—Feel OK*, Christopher Publ. House, North Quincy, Mass.
29. Taylor, R. J. (1980). *Food Additives*, John Wiley & Sons, New York.
30. Truswell, A. S., N. G. Asp, W. P. T. James, and B. MacMahon (1978). Conclusions, in *Marabou Symposium on Food and Cancer, Proceedings*, Carlson Press, Stockholm, pp. 112-113.
31. Turner, J. M. (1979). Putting food safety policy into proper focus. *Food Technol. 33*(11):63-64.
32. Verrett, J., and J. Carper (1975). *Eating May Be Hazardous to Your Health*, Simon and Schuster, New York.

2

WATER AND ICE

Owen R. Fennema   University of Wisconsin-Madison, Madison, Wisconsin

	Prologue—Water: the Deceptive Matter of Life and Death	24
I.	Introduction	24
II.	Physical Constants of Water and Ice	25
III.	The Water Molecule	25
IV.	Association of Water Molecules	28
V.	Structure of Ice	30
	A. Pure Ice	30
	B. Ice in the Presence of Solutes	34
VI.	Structure of Water	35
VII.	Water-Solute Interactions	36
	A. General Concepts	36
	B. Interaction of Water with Ions and Ionic Groups	39
	C. Interaction of Water with Neutral Groups Possessing Hydrogen-Bonding Capabilities	43
	D. Interaction of Water with Nonpolar Substances	43
VIII.	Water Activity	46
	A. Definition and Measurement	46
	B. Temperature Dependence	47
IX.	Isotherms	50
X.	Water Activity and Food Stability	55
XI.	Solute Mobility and Food Stability	60
XII.	Role of Ice in the Stability of Foods at Subfreezing Temperatures	61
	References	64
	General Bibliography	67

PROLOGUE—WATER: THE DECEPTIVE MATTER OF LIFE AND DEATH

Unnoticed in the darkness of a subterranean cavern, a water droplet trickles slowly down a stalactite, following a path left by countless predecessors, imparting as did they, a small but almost magical touch of mineral beauty. Pausing at the tip, the droplet grows slowly to full size, then plunges quickly to the cavern floor, as if anxious to perform other tasks or to assume different forms. For water, the possibilities are countless. Some droplets assume roles of quiet beauty—on a child's coat sleeve, where a snowflake of unique design and exquisite perfection lies unnoticed, on a spider's web, where dew drops burst into sudden brilliance at the first touch of the morning sun, in the countryside, where a summer shower brings refreshment, or in the city, where fog gently permeates the night air, subduing harsh sounds with a glaze of tranquility. Others lend themselves to the noise and vigor of a waterfall, to the overwhelming immensity of a glacier, to the ominous nature of an impending storm, or to the persuasiveness of a tear on a woman's cheek. For others the role is less obvious but far more critical. There is life—initiated and sustained by water in a myriad subtle and poorly understood way—or death inevitable, catalyzed under special circumstances by a few hostile crystals of ice, or decay at the forest's floor, where water works relentlessly to dissassemble the past so life can begin anew. But the form of water most familiar is none of these; rather it is simple, ordinary, and uninspiring, unworthy of special notice as it flows forth in cool abundance from a household tap. "Humdrum," galunks a frog in concurrence, or so it seems, as it views with stony indifference the water milieu on which its very life depends. Surely, then, water's most remarkable feature is deception, for it is in reality a substance of infinite complexity, of great and unassessable importance, and one endowed with a strangeness and beauty sufficient to excite and challenge anyone making its acquaintance.

I. INTRODUCTION

On this planet, water is the only substance that occurs abundantly in all three physical states. It is our only common liquid and is our most widely distributed pure solid, ever present somewhere in the atmosphere as suspended ice particles, or on the earth's surface as various types of snow and ice. It is essential to life: as a stabilizer of body temperature, as a carrier of nutrients and waste products, as a reactant and reaction medium, as a stabilizer of biopolymer conformation, as a likely facilitator of the dynamic behavior of macromolecules, including their catalytic (enzymic) properties, and in other ways as yet unknown.

As is evident from Table 1, water is the major component of most foods, and each has its own characteristic water content. Water, in the proper amount, location, and orientation, is crucial for life processes, and it also profoundly influences the structure, appearance and taste of food, and its susceptibility to spoilage. Because most kinds of fresh foods contain large amounts of water, effective forms of preservation are needed if long-term storage is desired. It is important to note that removal of water, either by conventional dehydration or by separation locally in the form of pure ice crystals (freezing), greatly alters the native properties of foods and biological matter. Furthermore, all attempts to return water to its original status (rehydration, or thawing) are never more than partially successful. Ample justification exists, therefore, to study water and ice with considerable care.

Table 1   Water Contents of Various Foods

Food	Water content (%)
Meat	
Pork, raw, composite of lean cuts	53-60
Beef, raw, retail cuts	50-70
Chicken, all classes, raw meat without skin	74
Fish, muscle proteins	65-81
Fruit	
Berries, cherries, pears	80-85
Apples, peaches, oranges, grapefruit	85-90
Rhubarb, strawberries, tomatoes	90-95
Vegetables	
Avocado, bananas, peas (green)	74-80
Beets, broccoli, carrots, potatoes	80-90
Asparagus, beans (green), cabbage,	
cauliflower, lettuce	90-95

II. PHYSICAL CONSTANTS OF WATER AND ICE

As a first step in becoming familiar with water, it is appropriate to consider its physical constants, as shown in Table 2. By comparing water's properties with those of molecules of similar molecular weight and atomic composition (CH_4, NH_3, HF, H_2S, H_2Se, and H_2Te), it is possible to determine if water behaves in a normal fashion. When this is done, water is found to melt and boil at unusually high temperatures; to exhibit unusually large values for surface tension, dielectric constant, heat capacity, and heats of phase transition (heats of fusion, vaporization, and sublimation); to have a moderately low value for density; to exhibit an unusual attribute of expanding upon solidification; and to possess a viscosity, which, in the light of the above oddities, is strangely normal. In addition, the thermal conductivity of water is large compared to that of other liquids, and the thermal conductivity of ice is moderately large compared to that of other non-metallic solids. Of greater interest is that the thermal conductivity of ice at 0°C is approximately four times that of water at the same temperature, indicating that ice will conduct heat energy at a much faster rate than immobilized water (e.g., in tissue). The thermal diffusivities of water and ice are of yet greater interest since these values indicate the rate at which the solid and liquid forms of HOH undergo changes in temperature. Ice has a thermal diffusivity approximately nine times greater than that of water, indicating that ice, in a given environment, will undergo a temperature change at a much greater rate than water. These sizable differences in thermal conductivity and thermal diffusivity values of water and ice provide a sound basis for explaining why tissues freeze more rapidly than they thaw, when equal, but reversed, temperature differentials are employed.

III. THE WATER MOLECULE

Water's unusual properties suggest the existence of strong attractive forces among water molecules and uncommon structures for water and ice. These features are best explained by first considering the nature of a single water molecule and then small groups of molecules. To form a molecule of water, two hydrogen atoms approach the two sp³ bonding

Table 2 Physical Constants of Water and Ice

Molecular weight	18.01534
Phase transition properties	
Melting point at 101.3 kPa (1 atm)	0.000°C
Boiling point at 101.3 kPa (1 atm)	100.000°C
Critical temperature	374.15°C
Critical pressure	22.14 MPa (218.6 atm)
Triple point	0.0099°C and 610.4 kPa (4.579 mm Hg)
Heat of fusion at 0°C	6.012 kJ (1.436 kcal)/mol
Heat of vaporization at 100°C	40.63 kJ (9.705 kcal)/mol
Heat of sublimation at 0°C	50.91 kJ (12.16 kcal)/mol

Other properties at	20°C	0°C	0°C (ice)	−20°C (ice)
Density (kg/liter)	0.998203	0.999841	0.9168	0.9193
Viscosity (Pa · s)	1.002×10^{-3}	1.787×10^{-3}	–	–
Surface tension against air (N/m)	72.75×10^{-3}	75.6×10^{-3}	–	–
Vapor pressure (Pa)	2.337×10^3	6.104×10^2	6.104×10^2	1.034×10^2
Heat capacity (J/kg · K)	4.1819	4.2177	2.1009	1.9544
Thermal conductivity (J/m · s · k)	5.983×10^2	5.644×10^2	22.40×10^2	24.33×10^2
Thermal diffusivity (m^2/s)	1.4×10^{-5}	1.3×10^{-5}	$\sim 1.1 \times 10^{-4}$	$\sim 1.1 \times 10^{-4}$
Dielectric constant, static[a]	80.36	80.00	91[b]	98[b]
at 3×10^9 Hz	76.7 (25°C)	80.5 (1.5°C)	–	3.2 (−12°C)

[a]Limiting value at low frequencies.
[b]Parallel to c-axis of ice; values about 15% larger if perpendicular to c-axis.
Source: From Refs. 23 and 61.

orbitals of oxygen (ϕ_3^1, ϕ_4^1) and form two covalent sigma (σ) bonds (40% partial ionic character), each of which has a dissociation energy of 4.614×10^2 kJ/mol (110.2 kcal/mol). The localized molecular orbitals remain symmetrically oriented about the original orbital axes, thus retaining an approximate tetrahedral structure. A schematic orbital model of a water molecule is shown in Fig. 1a and the appropriate van der Waals radii are shown in Fig. 1b.

The bond angle of the isolated water molecule (vapor state) is 104.5°, and this value is near the perfect tetrahedral angle of 109°28'. The O–H internuclear distance is 0.96 Å, and the van der Waals radii for oxygen and hydrogen are, respectively, 1.40 and 1.2 Å.

At this point, it is important to emphasize that the picture so far presented is oversimplified. Pure water contains not only ordinary HOH molecules, but many other constituents in trace amounts. In addition to the common isotopes ^{16}O and ^1H, also present are ^{17}O, ^{18}O, ^2H (deuterium), and ^3H (tritium), giving rise to 18 isotopic variants of

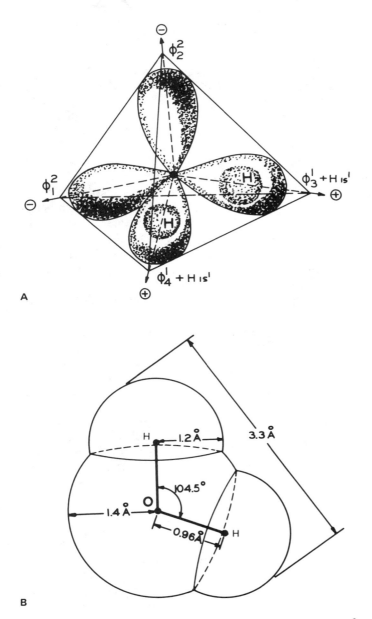

Figure 1 Schematic model of a single HOH molecule. (A) sp^3 configuration. (B) van der Waals radii for an HOH molecule in the vapor state.

molecular HOH. Water also contains ionic particles, such as hydrogen ions (existing as H_3O^+), hydroxyl ions, and their isotopic variants. Water therefore consits of more than 33 chemical variants of HOH. Fortunately, the isotopic variants occur only in small amounts and they can, in most instances, be ignored.

IV. ASSOCIATION OF WATER MOLECULES

The V-like form of a HOH molecule and the polarized nature of the O—H bond result in an unsymmetrical charge distribution and a vapor-state dipole moment of 1.84D for pure water. Polarity of this magnitude produces intermolecular attractive forces, and water molecules therefore associate with considerable tenacity. Water's unusually large intermolecular attractive force cannot, however, be fully accounted for on the basis of its large dipole moment. This is not surprising, since dipole moments give no indication of the degree to which charges are exposed or of the geometry of the molecule, and these aspects, of course, have an important bearing on the intensity of molecular association.

Water's large intermolecular attractive forces can be explained quite adequately on its ability to engage in multiple hydrogen bonding on a three-dimensional basis. Compared to covalent bonds (average bond energy of about 335 kJ/mol), hydrogen bonds are weak (typically 2-40 kJ/mol) and have greater and more variable lengths. The oxygen-hydrogen hydrogen bond has a dissociation energy of about 13-25 kJ/mol.

Since electrostatic forces make a major contribution to the energy of the hydrogen bond (perhaps the largest contribution), and since an electrostatic model of water is simple and leads to an essentially correct geometric picture of HOH molecules as they are known to exist in ice, further discussion of the geometric patterns formed by association of water molecules will emphasize electrostatic effects. This simplified approach, although entirely satisfactory for the present purpose, must be modified if other behavioral characteristics of water are to be explained satisfactorily.

The highly electronegative oxygen of the water molecule can be visualized as partially drawing away the single electrons from the two covalently bonded hydrogen atoms, thereby leaving each hydrogen atom with a partial positive charge and a minimal electron shield; that is, each hydrogen atom assumes some characteristics of a bare proton. Since the hydrogen-oxygen bonding orbitals are located on two of the axes of an imaginary tetrahedron (Fig. 1a), these two axes can be thought of as representing lines of positive force (hydrogen-bond donor sites). Oxygen's two lone-pair orbitals can be pictured as residing along the remaining two axes of the imaginary tetrahedron, and these then represent negative lines of force (hydrogen-bond acceptor sites). By virtue of these four lines of force, each water molecule is able to hydrogen bond with a maximum of four others. The resulting tetrahedral arrangement is depicted in Fig. 2. Because each water molecule has an equal number of hydrogen-bond donor and receptor sites, arranged to permit three-dimensional hydrogen bonding, it is found that the attractive forces among water molecules are unusually large, even when compared with those existing among other small molecules that also engage in hydrogen bonding (e.g., NH_3 or HF). Ammonia, with its tetrahedral arrangement of three hydrogens and one receptor site, and hydrogen fluoride, with its tetrahedral arrangement of one hydrogen and three receptor sites, do not have equal numbers of donor and receptor sites and therefore can form only two-dimensional hydrogen-bonded networks involving fewer hydrogen bonds per molecule than water.

Conceptualizing the association of a few water molecules becomes considerably more complicated when one considers isotopic variants and hydronium and hydroxyl ions. The hydronium ion (Fig. 3), because of its positive charge, would be expected to exhibit a greater hydrogen-bond donating potential than nonionized water (38). The hydroxyl ion (Fig. 4), because of its negative charge, would be expected to exhibit a greater hydrogen-bond acceptor potential than un-ionized water (38).

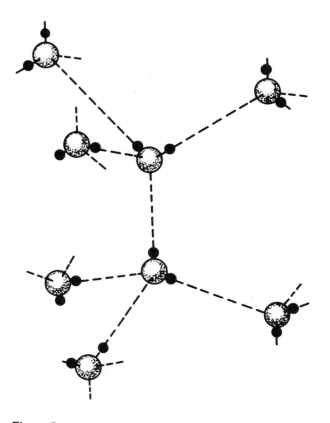

Figure 2 Hydrogen bonding of water molecules in a tetrahedral configuration. Hydrogen bonds are represented by dashed lines.

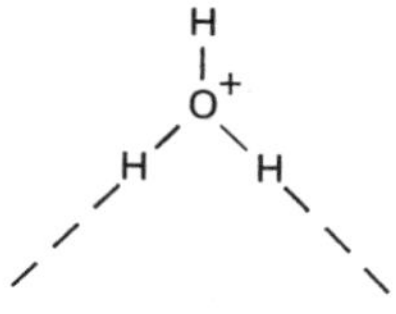

Figure 3 Structure and hydrogen-bond possibilities for a hydronium ion. Dashed lines are hydrogen bonds.

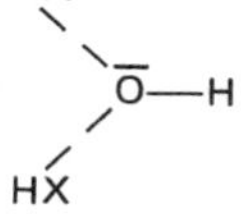

Figure 4 Structure and hydrogen-bond possibilities for a hydroxyl ion. Dashed lines are hydrogen bonds, and XH represents a solute or a water molecule.

Water's ability to engage in three-dimensional hydrogen bonding provides a logical explanation for many of its unusual properties. For example, its large values for heat capacity, melting point, boiling point, surface tension, and heats of various phase transitions are all related to the extra energy needed to break intermolecular hydrogen bonds.

The dielectric constant of water is also influenced by hydrogen bonding. Although water is a dipole, this alone does not explain the magnitude of its dielectric constant. Hydrogen-bonded groups of molecules apparently give rise to multimolecular dipoles that effectively increase the dielectric constant of water. Water's viscosity is discussed in a later section.

V. STRUCTURE OF ICE

The structure of ice will be considered before the structure of water because the former is far better understood than the latter, and because ice's structure represents a logical extension of the information presented in the previous section.

A. Pure Ice

Water, with its tetrahedrally directed forces, crystallizes in an open (low-density) structure that has been accurately elucidated by studies involving x-ray, neutron, and electron diffraction, and infrared and Raman spectroscopy. As shown in Fig. 5, the O—O internuclear nearest-neighbor distance in ice is 2.76 Å, and the O—O—O bond angle is about 109°, or very close to the perfect tetrahedral angle of 109°28'. The manner in which each

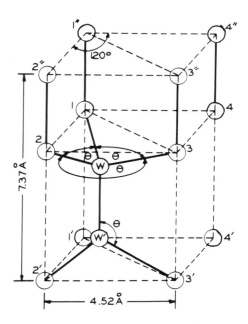

Figure 5 Unit cell of ordinary ice at 0°C. Circles represent oxygen atoms of water molecules. Nearest-neighbor internuclear O—O distance is 2.76 Å; $\theta = 109°$.

HOH molecule can associate with four others (coordination number of four) is easily visualized by considering molecule W and its four nearest neighbors 1, 2, 3, and W'.

When several unit cells are combined and viewed from the top (along the c-axis) the hexagonal symmetry of ice becomes apparent. This is shown in Fig. 6a. The tetrahedral substructure is evident from molecule W and its four nearest neighbors, with 1, 2, and 3 visible, and the fourth lying below the plane of the paper directly under molecule W. When Fig. 6a is viewed in three dimensions, as in Fig. 6b, it is evident that two planes of molecules are involved. These two planes are parallel and very close together and they

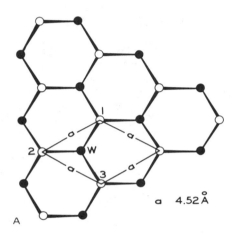

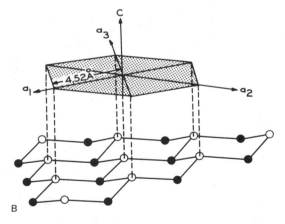

Figure 6 The "basal plane" of ice (combination of two planes of slightly different elevation). Each circle represents the oxygen atom of a water molecule. Open and shaded circles, respectively, represent oxygen atoms in the upper and lower layers. (A) Hexagonal structure viewed parallel to c axis. Numbered molecules relate to the unit cell in Fig. 5. (B) Three-dimensional view of the basal plane. The front edge of view (B) corresponds to the bottom edge of view (A). The crystallographic axes have been positioned in accordance with external (point) symmetry.

move as a unit during the "slip" or flow of ice under pressure (glacier). Pairs of planes of this type comprise the "basal planes" of ice.

By stacking several basal planes, an extended structure of ice is obtained. Three basal planes have been combined to form the structure shown in Fig. 7. Viewed parallel to the c-axis, the appearance is exactly the same as that shown in Fig. 6a, indicating that the basal planes are perfectly aligned. Ice is monorefringent in this direction, whereas it is birefringent in all other directions. The c-axis is therefore the optical axis of ice.

The location of hydrogen atoms in ice was established during the late 1950s by means of diffraction studies of ice containing deuterium (47). It is generally agreed that

1. Each line connecting two nearest-neighbor oxygen atoms is occupied by one hydrogen atom located 1 ± 0.01 Å from the oxygen to which it is covalently bonded and 1.76 ± 0.01 Å from the oxygen to which it is hydrogen bonded. This configuration is shown in Fig. 8a.
2. If the locations of the hydrogen atoms are viewed over a period of time, a somewhat different picture is obtained. A hydrogen atom on a line connecting two nearest-neighbor oxygen atoms, X and Y, can situate itself in one of two possible positions: either 1 Å from X or Å from Y. As predicted by Pauling (45) and later confirmed by Peterson and Levy (47), the two positions have an equal probability of being occupied. Expressed in another way, each position will, on the average, be occupied one-half the time. This is possible because HOH molecules, except at extremely low

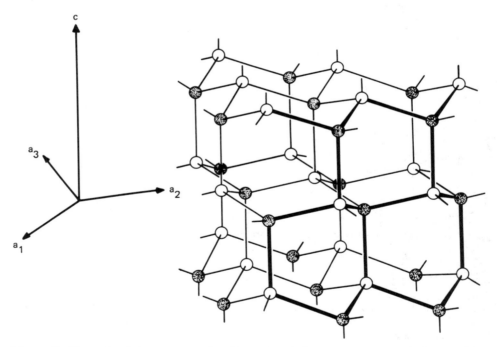

Figure 7 The extended structure of ordinary ice. Only oxygen atoms are shown. Open and shaded circles, respectively, represent oxygen atoms in upper and lower layers of a basal plane.

temperatures, can cooperatively rotate, and additionally, hydrogen atoms can "jump" between adjacent oxygen atoms. The resulting mean structure, known also as the half-hydrogen, Pauling, or statistical structure, is shown in Fig. 8b.

With respect to crystal symmetry, ordinary ice belongs to the dihexagonal bipyramidal class of the hexagonal system. In addition, ice can exist in nine other crystalline polymorphic structures, and also in an amorphous or vitreous state of rather uncertain structure. However, of the 11 total structures, only ordinary hexagonal ice is stable under normal pressure at 0°C.

It is appropriate at this point to deal with the fact that ice is not a static system consisting solely of HOH molecules precisely arranged so that one H atom resides on a line between each pair of oxygen atoms. First, pure ice contains not only ordinary HOH molecules, but ionic and isotopic variants of HOH. Fortunately, the isotopic variants occur in such small amounts that they can, in most instances, be ignored, leaving for major consideration only HOH, H^+ (H_3O^+), and OH^-.

Second, ice crystals are never perfect, and the defects encountered are usually of the orientational (caused by proton dislocation accompanied by neutralizing orientations) or ionic types (caused by proton dislocation with formation of H_3O^+ and OH^-) (see Fig. 9). The presence of these defects provides a means for explaining the much greater mobility of protons in ice than in water, and the small decrease in direct current electrical conductivity that occurs when water is frozen.

In addition to the atomic mobilities involved in crystal defects, there are other types of activity in ice. Each HOH molecule in ice is believed to vibrate, with a root mean amplitude (assuming each molecule vibrates as a unit) of about 0.4 Å at −10°C. Furthermore, HOH molecules, which presumably exist in some of the interstitial spaces in ice, can apparently diffuse slowly through the lattice.

Ice, therefore, is far from static or homogenous, and its characteristics are dependent on temperature. Although the HOH molecules in ice are four-coordinated at all temperatures, it is necessary to lower the temperature greatly, probably to −183°C or lower, to "fix" the hydrogen atoms in one of the many possible configurations. Therefore, only at temperatures near −183°C or lower will all hydrogen bonds be intact, and

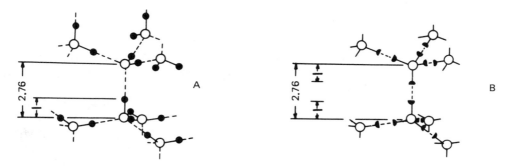

Figure 8 Location of hydrogen atoms (●) in the structure of ice. (A) Instantaneous structure. (B) Mean structure (known also as the half-hydrogen (◉), Pauling, or statistical structure). (○) Oxygen.

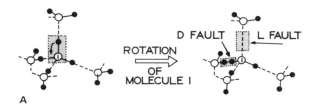

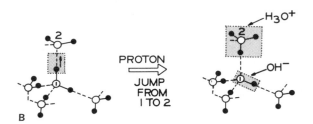

Figure 9 Schematic representation of proton defects in ice. (A) Formation of orientational defects. (B) Formation of ionic defects. Open and shaded circles represent oxygen and hydrogen atoms, respectively. Solid and dashed lines, respectively, represent chemical bonds and hydrogen bonds.

as the temperature is raised, the mean number of intact (fixed) hydrogen bonds will decrease gradually. It is likely that the amount of "activity" in ice has some relationship to the rate of deterioration of foods and biological matter stored at low temperatures.

B. Ice in the Presence of Solutes

The amount and kind of solutes present can influence the quantity, size, structure, location, and orientation of ice crystals. Consideration here is given only to the effects of solutes on ice structure. Luyet and coworkers (39,40) studied the nature of ice crystals formed in the presence of various solutes, including sucrose, glycerol, gelatin, albumin, myosin, and polyvinylpyrrolidone. They devised a classification based on morphology, elements of symmetry, and the cooling velocity required for development of various types of ice structure. Four major types of ice structure were observed: hexagonal forms, irregular dendrites, coarse spherulites, and evanescent spherulites. A variety of intermediate types was also observed.

The hexagonal form, the only form of importance in most foods, is the normal and most highly ordered type, and it was found to occur provided the specimen was frozen in a coolant of only moderately low temperature (to avoid extremely rapid freezing) and the solute was of a nature and concentration that it did not interfere unduly with the mobility of water molecules. Hexagonal crystals were producible in all specimens except those containing large amounts of gelatin. The ice structures observed in gelatin possessed greater disorder than the hexagonal form.

Frozen gelatin solutions were also studied by Dowell et al. (10), and they found that cubic and vitreous ice became more prevalent with increasing rates of freezing or increasing concentrations of gelatin. Apparently, gelatin, a large, complex, hydrophilic molecule, is able to greatly restrict the movement of water molecules and their ability to form highly ordered hexagonal crystals. Although ice crystals other than hexagonal types can be formed in foods and biological matter, they are uncommon.

VI. STRUCTURE OF WATER

Elucidation of the structure of pure water is an extremely complex problem that has, in recent years, attracted increasing attention from capable investigators. Many theories have been set forth, but all are incomplete, overly simple, and subject to weaknesses quickly cited by supporters of rival theories. This, of course, is a healthy situation that will eventually result in an accurate structural picture (or pictures) of water. In the meantime, few statements can be made with any assurance they will stand essentially unmodified in years to come.

To some, it may seem strange to speak of structure in a liquid when fluidity is the major feature of the liquid state. Yet it is an old and well-accepted idea (51) that liquid water has structure, obviously not sufficiently established to produce long-range rigidity, but certainly far more organized than molecules in the vapor state, so that the orientation and mobility of a given water molecule is influenced significantly by its neighbors. Considerable evidence has accumulated in support of the belief that substantial order exists in water. For example, water is an "open" liquid, being only 60% as dense as would be expected on the basis of the close packing that can prevail in nonstructured liquids. Partial retention of the open, hydrogen-bonded, tetrahedral arrangement of ice can easily account for water's density. Furthermore, the heat of fusion of ice, although unusually high, is sufficient to break only about 15% of the hydrogen bonds believed to exist in ice. Although this does not necessarily require that 85% of the total possible hydrogen bonds remain in water (for example, more could be broken, but the change in energy could be masked by a simultaneous increase in van der Waals interactions), it appears likely that sizable numbers of hydrogen bonds and extensive association of water molecules do indeed remain. Further evidence of association can be gained from the many additional unusual properties of water, some of which were discussed earlier, from direct evidence gained from experiments involving such methods as x-ray, infrared, and Raman spectroscopy, and from computer simulation studies of water.

A detailed discussion of the many models of water structure would be of little benefit to readers of this chapter, since all models are highly speculative and of little direct application of food chemistry. For a relatively recent update of this subject, the interested reader should refer to a review by Stillinger (54). According to Stillinger, liquid water at and below room temperature contains uninterrupted three-dimensional paths of hydrogen bonds that prevade the entire sample. This network of hydrogen bonds has a local preference for tetrahedral geometry, but contains a large proportion of strained and broken bonds, thus, in principle, allowing for the presence of unconnected clusters of a few water molecules, but not allowing for the dominance of disconnected "icebergs" as proposed by some authors. This arrangement is dynamic, permitting molecules to easily and rapidly terminate one hydrogen bond in exchange for a new one, thus rapidly altering the bonding arrangement of individual molecules, while maintaining, at constant temperature, an intact network with a constant degree of hydrogen bonding for the entire system.

The degree of hydrogen bonding is, of course, temperature dependent. Ice at 0°C has a coordination number (number of nearest neighbors) of four, with the nearest neighbors at a distance of 2.76 Å. Upon melting, some hydrogen bonds are broken (distance between the nearest neighbors increases). Simultaneously, the rigid structure is destroyed and the water molecules rearrange themselves in more compact network patterns involving substantial hydrogen-bond strain and the input of latent heat of fusion. As the temperature is raised, the coordination number of water increases from 4.0 in ice at $0°C$ to 4.4 in water at $1.50°C$ and to 4.9 at $83°C$, and the distance between nearest neighbors increases from 2.76 Å at $0°C$ (ice), to 2.9 Å in water at $1.5°C$, and to 3.05 Å at $83°C$ (7, 43).

It is evident, therefore, that the ice-to-water transformation is accompanied by an increase in the distance between nearest neighbors (decreases density) and by an increase in the average number of nearest neighbors (increases density), with the latter factor predominating to yield the familiar net increase in density. Further warming above the melting point causes the density value to pass through a maximum at $3.98°C$, then gradually decline. It is apparent, then, that the effect of an increase in coordination number must predominate at temperatures between 0 and $3.98°C$, and that the effect of increasing distance between nearest neighbors (thermal expansion) must predominate above $3.98°C$.

The low viscosity of water is readily reconcilable with the type of structure that has been described, since the hydrogen-bond network is highly dynamic, allowing individual molecules, within the time frame of nanoseconds to picoseconds, to alter their hydrogen-bonding relationships with neighboring molecules, thereby enhancing molecular mobility and fluidity.

Discussion of the effects of solutes on water structure is deferred to a later section.

VII. WATER-SOLUTE INTERACTIONS

A. General Concepts

The addition of various substances to water results in altered properties for the added substance and for water itself. Hydrophilic substances interact strongly with water by ion-dipole or dipole-dipole mechanisms causing changes in water structure and mobility and changes in the structure and reactivity of the hydrophilic substances. Hydrophobic groups of added substances interact only weakly with adjacent water, preferring a non-aqueous environment. Water adjacent to hydrophobic groups assumes a greater degree of structure than in pure water, a change thermodynamically unfavorable because of the decrease in entropy. To minimize this unfavorable thermodynamic occurrence, hydrophobic groups, when possible, aggregate to minimize their contact with water, a process known as "hydrophobic interaction."

Before dealing with the specifics of water-solute interactions it is first appropriate to discuss general phenomena referred to by various terms such as water binding, hydration, bound water, and water-holding capacity.

The terms "water binding" and "hydration" are all-encompassing terms referring to the tendency of water to associate, with various degrees of tenacity, to hydrophilic substances. The extent and tenacity of water binding or hydration depends on a number of factors, including the nature of the nonaqueous constituent, salt composition, pH, and temperature.

The term "bound water" cannot be dealt with so easily. Bound water is not a homogenous, easily identifiable entity, and because of this, descriptive terminology is difficult, and a generally accepted definition does not exist—nor is one likely to emerge in the future. This term is controversial, frequently misused, and, in general, poorly understood, causing some to suggest that its use be terminated. The latter course of action seems inadvisable since abandonment of the term will not eradicate the phenomenon.

If all water is regarded as either "free" or "bound," one can reasonably argue that all water in tissue is bound since it does not flow freely from tissue when moderate force is applied, and all of it is to some degree under the influence of biological structures or solutes and would therefore behave differently than pure water. Furthermore, "boundness" can be viewed from the standpoints of water structure (average position of water molecules in relation to each other and in relation to solutes and biological structures), mobility of water molecules (translational and rotational motion), bond-dissociation energies (water-water, water-solute, or water-ion), or water activity (defined as p/p_0 where p is the partial pressure of water in a sample and p_0 is the vapor pressure of pure water at the same temperature). Although water activity and bond-dissociation energy may, at first, appear to follow a strict inverse relationship, this is not necessarily so since Raoult's law requires that solutes decrease the vapor pressure of water and water activity even when the solution behaves ideally (i.e., when the water-water bond dissociation energy equals the water-solute bond dissociation energy).

A few authors stand guilty of the inexcusable act of using the term "bound water" without benefit of definition, and others have created such a broad array of definitions that confusion is rampant. The following examples illustrate amply how confusion can arise from the existence of multiple definitions (3,29).

1. Bound water is the equilibrium water content of a sample at some appropriate temperature and relative humidity (low).
2. Bound water does not contribute significantly to the dielectric constant at high frequencies and therefore has its rotational mobility restricted by the substance with which it is associated.
3. Bound water does not freeze at some arbitrary low temperature (usually $-40°C$ or lower).
4. Bound water is unavailable as a solvent for additional solutes.
5. Bound water produces line broadening in experiments involving proton nuclear magnetic resonance (NMR).
6. Bound water moves with a macromolecule in experiments involving sedimentation rates, viscosity, or diffusion.
7. Bound water exists in the vicinity of solutes and other nonaqueous substances and has properties differing significantly from those of "bulk" water in the same system.

All these definitions are valid, but few produce the same value when a given sample is analyzed. It is the author's preference to combine definitions 3 and 7 so that bound water is defined as that which exists in the vicinity of solutes and other non-aqueous constituents, exhibits reduced molecular mobility and other significantly altered properties as compared with "bulk water" in the same system, and does not freeze at $-40°C$. This definition has two desirable attributes. First, it produces a conceptual picture of bound water, and second, it provides a realistic approach to quantifying bound water. Water unfreezable at $-40°C$ can be measured with equally satisfying results by either proton NMR or calorimetric procedures.

Finally, it is important that all the following points be kept in mind when bound water is considered.

1. The *apparent* amount of bound water will often vary depending on the method of measurement.
2. The *actual* amount of bound water will vary depending on the product.
3. Various degrees of water "boundness" exist in complex systems. The most tenaciously bound water is that which is an integral part of a nonaqueous substance. This water, which may be thought of as "constitutional" water, represents a small fraction of the water in high-moisture foods and is situated, for example, in interstitial regions of proteins or as a part of chemical hydrates. The next most tenaciously bound water, walled "vicinal water," occupies first-layer sites at the most hydrophilic groups of nonaqueous constituents. Water associated in this manner with ions and ionic groups is the most tightly bound kind of vicinal water. "Multilayer" water occupies the remaining first-layer sites and forms several layers beyond the vicinal water. Although multilayer water is less tenaciously bound than vicinal water, it is still close enough to nonaqueous constituents that its properties are significantly altered from those of pure water. Thus, bound water, as defined here, consists of constitutional water, vicinal water, and almost all the multilayer water, with the degree of boundness varying within and among these categories.
4. The manner in which a given water molecule is bound to other molecules can change as the total water content (especially in low-moisture foods) of the system is altered.
5. Water that is bound should not be thought of as totally immobilized. As boundness increases, the rate at which a water molecule changes place with a neighboring water molecule decreases, but usually does not decline to zero.
6. Water bound to hydrophilic substances is more structured than ordinary water, but its structure differs from that of ordinary ice.
7. In addition to chemically bound water, to which the previous discussion related, small amounts of water in some cellular systems can exhibit reduced mobility and vapor pressure because of physical confinement in small capillaries. The reduction in vapor pressure (and a_w) becomes substantial when capillaries with radii of less than 0.1 μm are abundant. According to Bluestein and Labuza (6), most foods have capillaries ranging in size from 10 to 100 μm, thus eliminating this mechanism as an important means of a_w reduction in foods.
8. Although bound water is important, it is present in relatively small amounts in most high-moisture foods. For example, reliable estimates of bound water attached to proteins range from 0.3 to 0.5 g H_2O per gram dry protein.
9. In low moisture foods, water activity is a more meaningful concept than bound water.

"Water-holding capacity" is a term frequently employed to describe the ability of a matrix of molecules, usually macromolecules, to entrap large amounts of water in a manner such that exudation is prevented. Food examples include gels of pectin and starch, and tissues, both plant and animal. In these situations, a small amount of organic material physically entraps large amounts of water.

Although the structure of water entrapped in cells or macromolecular matrices is still a matter of some controversy, the behavior of this kind of water in food systems and its importance to food quality is quite clear. Entrapped water does not flow from foods even when considerable mechanical damage (e.g., cutting) occurs. On the other hand, this water behaves almost like pure water during food processing; that is, it is easily removed during drying and is easily converted to ice during freezing. Thus, its bulk-flow properties are severely restricted, but the movement of individual molecules is essentially the same as that of water molecules in a dilute salt solution.

Entrapped water constitutes the major fraction of water in cells and gels, and alterations in its amount, that is, the water-holding capacity of foods, have profound effects on food quality. For example, storage of gels frequently results in quality loss because of decreased water-holding capacity—an occurrence known as *syneresis*. The freeze preservation of tissue often results in an undesirable decrease in water-holding capacity, evident as thaw exudate. Also, the decline in pH that accompanies postmortem physiological changes in muscle causes a decrease in water-holding capacity that can adversely affect the quality of sausage. Gel structures and water-holding capacity are discussed more fully in Chaps. 3, 5, and 12.

It is appropriate to conclude this section with a classification system for various kinds of water in food and a listing of the properties associated with these various classes of water (Tables 3, 4, and 5).

B. Interaction of Water with Ions and Ionic Groups

As noted earlier, water interacting with ions and ionic groups represents some of the most tightly bound water in food. The normal structure of pure water (based on a hydrogen-bonded, tetrahedral arrangement) is disrupted by the addition of dissociable solutes. For simple inorganic ions that do not possess either hydrogen-bond donor or receptor sites, the bonding is merely polar. An example is shown in Fig. 10 for the hydrated ion pair of NaCl. Only those water molecules existing in the innermost (vicinal) layer and in the plane of the paper are illustrated.

Water in the multilayer environment of ions (Table 4) is believed to exist in a structurally disrupted state because of conflicting structural influences of the innermost vicinal water and the outermost bulk-phase water. Bulk-phase water has properties similar to those of water in a dilute salt solution.

Table 3 Constitutional Water in Foods and Its Properties

General description	Water that is an integral part of a nonaqueous constituent, e.g., in interstitial sites of proteins
Freezing point compared with pure water	Unfreezable at $-40°C$ (bound)
Solvent capability	None
Translational mobility (molecular level) compared with pure water	None
Enthalpy of vaporization compared with pure water	Increased greatly
% Total water in a high-moisture food (90% H_2O or 9 g H_2O per g dry matter)	<0.03%
Relationship to sorption isotherm (Fig. 17) Isotherm zone	Constitutional water exhibits essentially zero water activity and thus exists at the extreme left end of zone I
Common deteriorative consequences	Autoxidation

Source: Compiled from many sources, including Refs. 2, 18a, 23, and 29.

Table 4 Categories of Vicinal and Multilayer Water in Foods and Their Properties

Properties	Vicinal water	Multilayer water
General description	Water that strongly interacts with specific hydrophilic sites of nonaqueous constituents by water-ion and water-dipole associations; when this type of water is at the maximum level, it is sufficient to provide single-layer coverage of accessible, strongly hydrophilic groups of nonaqueous constituents; also includes water in microcapillaries (<0.1 μm diameter)	Water that occupies remaining first-layer sites and forms several additional layers around hydrophilic groups of nonaqueous constituents; water-water and water-solute hydrogen bonds predominate
Freezing point compared with pure water	Unfreezable at $-40°C$ (bound)	Mostly unfreezable at $-40°C$ (bound); remainder freezable with greatly reduced freezing point
Solvent capability	None	Slight to moderate
Translational mobility (molecular level) compared with pure water	Reduced greatly	Reduced slightly to greatly
Enthalpy of vaporization compared with pure water	Increased greatly	Increased slightly to moderately
% Total water in high-moisture food (90% H_2O or 9 g H_2O per g dry matter)	$0.5 \pm 0.4\%$	$3 \pm 2\%$
Relationship to sorption isotherm (Fig. 17) Isotherm zone	Water in zone I of the isotherm consists of a minute amount of constitutional water with the remainder vicinal water; upper boundary of zone I is not distinct and varies somewhat with product and temperature	Water in zone II of the isotherm consists of water present in zone I plus water added or removed within the confines of zone II; the latter water is entirely multilayer water; boundaries of zone II are not distinct and vary somewhat with product and temperature
Common deteriorative consequences	Optimum overall stability at the "monolayer" value (0.2–0.3 a_w)	As water content is increased over the lower portion of this zone, rates of almost all reaction increase

Source: Compiled from many sources, including Refs. 2, 18a, 23, and 29.

Table 5 Categories of Bulk-Phase Water in Foods and Its Properties

Properties	Bulk-phase water	
	Free	Entrapped
General description	Water that occupies positions furthest removed from nonaqueous constituents; water-water hydrogen bonds predominate; has properties similar to water in a dilute salt solution; macroscopic flow is unimpeded	Water that occupies positions furthest removed from nonaqueous constituents; water-water hydrogen bonds predominate; has properties similar to water in a dilute salt solution, except macroscopic flow is impeded by matrix of gel or tissue
Freezing point compared with pure water	Freezable with slight to moderate reduction of freezing point	Freezable with slight to moderate reduction of freezing point
Solvent capability	Large	Large
Translational mobility (molecular level) compared with pure water	Reduced very slightly	Reduced very slightly
Enthalpy of vaporization compared with pure water	Essentially no change	Essentially no change
% Total water in high-moisture food (90% H_2O or 9 g H_2O per g dry matter)	~96%	~96%
Relationship to sorption isotherm (Fig. 17): Isotherm zones	Water in zone III consists of water present in zones I and II plus water added or removed within the confines of zone III; in the absence of gels and cellular structures, the latter water is entirely "free water"; lower boundary of zone III is not distinct and varies somewhat with product and temperature	Water in zone III consists of water present in zones I and II plus water added or removed within the confines of zone III; in the presence of gels or cellular structures, the latter water is entirely "entrapped" water; lower boundary of zone III is not distinct and varies somewhat with product and temperature
Common deteriorative consequences	Rapid rates of most chemical reactions; microbial growth	Rapid rates of most chemical reactions; microbial growth

Source: Compiled from many sources including Refs. 2, 18a, 23, and 29.

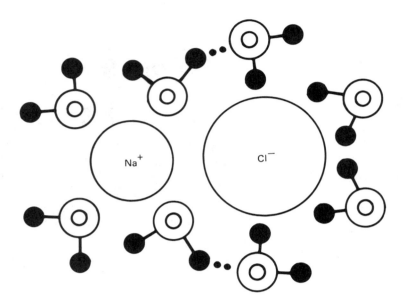

Figure 10 Likely arrangement of water molecules adjacent to sodium chloride. Only water molecules in plane of paper are shown.

In concentrated salt solutions the bulk-phase water would no doubt be eliminated, and water structures common in the vicinity of the ions would predominate.

There is abundant evidence indicating that some ions in dilute aqueous solution have a net structure-breaking effect (solution is more fluid than pure water), whereas others have a net structure-forming effect (solution is less fluid than pure water). It should be understood that the term "net structure" refers to *all* kinds of structures, either the normal or new types of water structure. From the standpoint of "normal" water structure, all ions are disruptive and all will, for example, pose a hindrance to the formation of ice.

The ability of a given ion to alter net structure is related closely to its polarizing power (charge divided by radius) or simply the strength of its electric field. Ions that are small and/or multivalent (mostly positive ions), such as Li^+, Na^+, H_3O^+, Ca^{2+}, Ba^{2+}, Mg^{2+}, Al^{3+}, F^-, and OH^-, have strong electrical fields and are net structure formers. The structure imposed by these ions more than compensates for any loss in normal water structure, and these ions strongly bind (the bound water is less mobile and more dense than HOH molecules in pure water) the four to six water molecules adjacent to each of them.

Ions that are large and monovalent (most of the negatively charged ions and large positive ions), such as K^+, Rb^+, Cs^+, NH_4^+, Cl^-, Br^-, I^-, NO_3^-, BrO_3^-, IO_3^- and ClO_4^- have rather weak electrical fields and are found to be net structure breakers, although the effect is very slight with K^+. These ions disrupt the normal structure of water and fail to impose a compensating amount of new structure.

Ions, of course, have effects that extend well beyond their influence on water structure. Through their varying abilities to hydrate (compete for water), alter water structure, influence the dielectric constant of the aqueous medium, and govern the thickness of the electrical double-layer around colloids, ions profoundly influence the "degree of hospitality" extended to other nonaqueous solutes and to substances suspended in the medium. Thus, the conformation of proteins and stability of colloids (salting-in, salting-out in accord with the Hofmeister or lyotropic series) are greatly influenced by the kinds and amounts of ions present (12,38).

C. Interaction of Water with Neutral Groups Possessing Hydrogen-Bonding Capabilities

Water-solute hydrogen bonds are weaker than water-ion interactions. Nevertheless, water directly engaged in hydrogen bonding to solutes would be classed as either constitutional or vicinal water, based on its specific location, and would exhibit reduced mobility compared with bulk-phase water.

Solutes capable of hydrogen bonding might be expected to enhance or at least not disrupt the normal structure of pure water. However, in some instances it is found that the distribution and orientation of the solute's hydrogen-bonding sites are geometrically incompatible with those believed to exist in normal water. Thus, these kinds of solutes frequently have a disruptive influence on the normal structure of water. Urea is a good example of a small hydrogen-bonding solute that for geometric reasons has a marked disruptive effect on the normal structure of water. For similar reasons, most hydrogen-bonding solutes would be expected to hinder freezing.

It should be noted that the total number of hydrogen bonds per mole of solution may not be significantly altered by the addition of a hydrogen-bonding solute that disrupts the normal structure of water. This is possible since disrupted water-water hydrogen bonds may be replaced by water-solute hydrogen bonds. Solutes that behave in this manner have little influence on net structure as defined in the previous section.

Hydrogen bonding of water can occur with various potentially eligible groups (e.g., hydroxyl, amino, carbonyl, amide, or imino), and two examples are given in Fig. 11. "Water bridges" also have been suggested as possibilities, whereby one or more water molecules form a bridge between two sites on a single macromolecule or between two macromolecules. The example in Fig. 12 involves a three-molecule water bridge between backbone peptide units in papain.

It has been observed that hydrophilic groups in many crystalline macromolecules are separated by distances identical to the nearest-neighbor oxygen spacing in pure water (60). If this spacing prevails in hydrated macromolecules, this could encourage cooperative hydrogen bonding in vicinal and multilayer water.

D. Interaction of Water with Nonpolar Substances

The introduction into water of hydrophobic substances, such as hydrocarbons, rare gases, and the apolar groups of fatty acids, amino acids, and proteins, is thermodynamic-

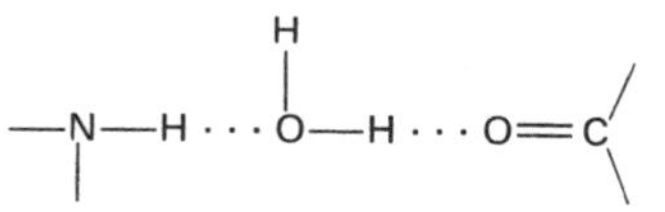

Figure 11 Hydrogen bonding (dotted lines) of water to two kinds of functional groups occurring in proteins. [From Lewin (38); courtesy of Academic Press.]

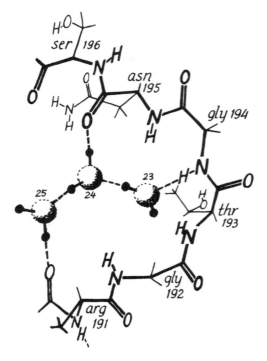

Figure 12 Example of a three-molecule water bridge in papain. [From Berendsen (4); courtesy of Plenum Press.]

ally unfavorable because of the decrease in entropy. The decrease in entropy arises from the increase in water-to-water hydrogen bonding (increase in water structure) adjacent to the apolar entities. Two aspects of the structure-forming response of water to apolar solutes are especially worthy of mention: the formation of clathrate hydrates and hydrophobic interactions in proteins.

Clathrate hydrates are icelike inclusion compounds wherein water, the "host" substance, forms hydrogen-bonded, cagelike structures that physically entrap molecules of a second molecular species, known as the "guest."

Clathrate crystals can be easily grown to a visible size, and some are stable at temperatures above 0°C provided the pressure is sufficient. Mention is made of them because they represent the most extreme structure-forming response of water to apolar substances and because microstructures of a similar type may occur naturally in biological matter.

The guest molecules of clathrate hydrates are low-molecular-weight compounds with sizes and shapes compatible with the dimensions of host cages comprised of 20-74 water molecules. Typical guests include low-molecular-weight hydrocarbons and halogenated hydrocarbons; rare gases; carbon dioxide; sulfur dioxide; ethylene oxide; ethyl alcohol; short-chain primary, secondary, and tertiary amines; and alkyl ammonium, sulfonium, and phosphonium salts. Interactions between water and guest often involve weak van der Waals forces, but electrostatic interaction occurs in some instances.

There is evidence that structures similar to crystalline clathrate hydrates may exist naturally in biological matter, and if so, these structures would be of far greater importance than crystalline hydrates, since they would likely influence the conformation, reactivity, and stability of such molecules as proteins. It is also possible that clathratelike structures of water have a role in the anesthetic action of inert gases, such as xenon (42, 46). For further information on clathrates and their applications, the interested reader should refer to reviews by Fennema (16) and Davidson (9).

Hydrophobic interactions, the association of apolar groups in an aqueous environment, do occur and are important because about 40% of the total amino acids in most proteins have nonpolar side chains. These include the methyl group of alanine, the benzyl group of phenylalanine, the isopropyl group of valine, the mercaptomethyl group of cysteine, and the secondary butyl and isobutyl groups of the leucines. The nonpolar groups of other compounds such as alcohols, fatty acids, and free amino acids, also can participate in hydrophobic interactions, but the consequences of these interactions are undoubtedly less important than those involving proteins. Since, as previously mentioned, exposure of these nonpolar groups to water is thermodynamically unfavorable, the association of hydrophobic groups or "hydrophobic interaction" is encouraged.

It is of interest that hydrophobic interactions exist because of the aqueous environment, and that something of the order of one-third the total nonpolar groups in proteins remain exposed to water despite hydrophobic interactions (62). This immediately poses a question concerning the nature of water adjacent to the exposed hydrophobic groups. The arrangement depicted in Fig. 13 has been suggested by Lewin (38) and is compatible with the requirement that water and nonpolar groups are not mutually attracted (other than by weak van der Waals forces). That this scheme would tend to repel cations and attract anions is in accord with the fact that many proteins above their isoelectric points (net negative charge) do in fact bind some anions.

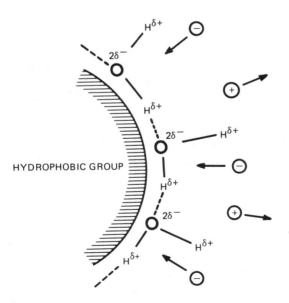

Figure 13 Proposed water orientation at a hydrophobic surface. [Adapted from Lewin (38); courtesy of Academic Press.]

Hydrophobic interactions are regarded as of primary importance in maintaining the tertiary structure of most proteins (13,44,55). Thus, it is clear that water and water structure play an important role in protein conformation.

VIII. WATER ACTIVITY

A. Definition and Measurement

It has long been recognized that a relationship, although imperfect, exists between the water content of food and its perishability. Concentration and dehydration processes are conducted primarily for the purpose of decreasing the water content of a food, simultaneously increasing the concentration of solutes, and thereby decreasing perishability.

However, it has also been observed that various foods with the same water content differ significantly in perishability. Thus, water content alone is not a reliable indicator of perishability. This inadequacy can be attributed, in part, to differences in the intensity with which water associates with nonaqueous constituents—water engaging in strong associations is less able to support degradative activities, such as growth of microorganisms and hydrolytic chemical reactions. The term "water activity" (a_w) was developed to take this factor into account. This term, although a far better indicator of food perishability than is water content, is still not perfect, since other factors, such as oxygen concentration, pH, water mobility, and the type of solute present, can, in some instances, have strong influences on the rate of degradation. Nonetheless, water activity correlates sufficiently well with the rates of many degradative reactions to make it worthwhile, in many situations, to measure and use. As an indication of its usefulness, the U.S. Federal Government has included water activity values in regulations dealing with good manufacturing practices for food (32).

Water activity is defined in the following manner.

$$a_w = \frac{p}{p_0}$$

where a_w is water activity, p is the partial pressure of water above the sample, and p_0 is the vapor pressure of pure water at the same temperature (must be specified).

This expression is an approximation of the original activity expression of Lewis, $a_w = f/f_0$, where f is the fugacity of the solvent (fugacity is the escaping tendency of a solvent from solution) and f_0 is the fugacity of the pure solvent.

At low pressures (e.g., ambient) the difference between f/f_0 and p/p_0 is so small (less than 1%) that defining a_w in terms of p and p_0 is clearly justifiable.

Water activity is also related to several other terms, and these interrelations are useful.

$$a_w = \frac{p}{p_0}$$
$$= \frac{ERH}{100}$$
$$= N$$
$$= \frac{n}{n_1 + n_2} \tag{1}$$

where ERH is the equilibrium relative humidity (%) surrounding the product, N is the mole fraction of solvent (water), n_1 is the moles of solvent, and n_2 is the moles of solute. The term n_2 can be determined by measuring the freezing point of the sample and then employing the relation

$$n_2 = \frac{G \Delta T_f}{1000 K_f} \tag{2}$$

where G is the grams of solvent in the sample, T_f is the freezing point depression (C°), and K_f is the molal freezing point depression constant for water (1.86).

It is worth emphasizing that water activity is an intrinsic property of the sample, whereas equilibrium relative humidity is a property of the atmosphere in equilibrium with the sample. Furthermore, attainment of equilibrium between the sample and its environment is a time-consuming process in very small samples (less than 1 g) and almost impossible in large samples.

The relation between the water activity and water content of a sample is often of interest, and the general approaches used to gather this information are as follows (24, 30,49,57).

1. Freezing point. Measure the freezing point depression and moisture content of the sample and calculate a_w according to the relationship in Eqs. (1) and (2). The error involved in measuring freezing point at a low temperature and calculating an a_w value for a higher temperature is small (<0.001 a_w/C°) (21).
2. Relative humidity sensors. Place the sample of known water content in a small, closed chamber at constant temperature, allow equilibration to occur, then measure the ERH of the sample atmosphere by any one of several electronic or psychrometric techniques (a_w = ERH/100).
3. Equilibrium chamber at constant relative humidity. Place the sample in a small, closed chamber at constant temperature, maintain the sample atmosphere at constant relative humidity by means of an appropriate saturated salt solution, allow equilibration to occur, and determine the water content of the sample.

B. Temperature Dependence

In the definition of water activity just presented it was indicated that the measurement temperature must be specified. This is so because a_w values are temperature dependent. The Clausius-Clapeyron equation in modified form describes the temperature dependence of a_w most accurately (59).

$$\frac{d \ln a_w}{d(1/T)} = \frac{-\Delta H}{R} \tag{3}$$

where T is absolute temperature, R is the gas constant, and ΔH is the isosteric net heat of sorption at the water content of the sample. By rearrangement, this equation can be made to conform to the generalized equation for a straight line, in which form it is evident that a plot of $\ln a_w$ versus $1/T$ (at constant water content) should be linear. Plots of this kind for native potato starch at various moisture contents are shown in Fig. 14. For these examples, good linearity is obtained and the degree of temperature dependence is a function of moisture content. At a starting water activity of 0.5, the temperature coefficient

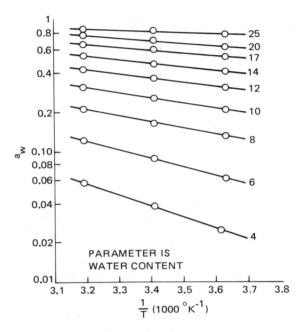

Figure 14 Clausius-Clapeyron relationship between water activity and temperature for native potato starch. Water content in g HOH per g dry starch. [From van den Berg and Leniger (59); courtesy of the Agricultural University of Wageningen, The Netherlands.]

is $0.0034°C^{-1}$ over the temperature range 2-40°C. Based on the work of other investigators, temperature coefficients for a_w (temperature range 5-50°C; starting a_w, 0.5) range from 0.003 to $0.02°C^{-1}$ for high-carbohydrate or high-protein foods (59). Thus, depending on the product, a 10°C change in temperature can cause a change in a_w ranging from 0.03 to 0.2. Thus, a temperature change, through its effect on a_w, can influence the stability of foods sealed in pouches or cans.

Plots of ln a_w versus $1/T$ are not always linear over broad temperature ranges and they generally exhibit sharp breaks with the onset of ice formation. Before showing data at subfreezing temperatures, it is appropriate to reexamine the definition of water activity as applied to subfreezing temperatures, since a question arises as to whether the denominator term (p_0) should be equated to the vapor pressure of supercooled water or to the vapor pressure of ice. The vapor pressure of supercooled water turns out to be the proper choice since (a) values of a_w at subfreezing temperatures can then, and only then, be accurately compared to a_w values at above-freezing temperatures, and (b) choice of the vapor pressure of ice as p_0 would result, for samples that contain ice, in a meaningless situation whereby a_w would be unity for all subfreezing temperatures. The second point results because the partial pressure of water in a frozen food is equal to the vapor pressure of ice at the same temperature (19a,54a). Since the vapor pressure of supercooled water has been measured down to −15°C, and the vapor pressure of ice has been measured to much lower temperatures, it is possible to accurately calculate a_w values for frozen foods. This is clearly apparent when one considers the following relationship.

$$a_w = \frac{P_{ff}}{P_{0(SCW)}}$$

$$= \frac{P_{ice}}{P_{0(SCW)}} \tag{4}$$

where p_{ff} is the partial pressure of water in partially frozen food, $p_{0(SCW)}$ is the vapor pressure of pure supercooled water, and p_{ice} is the vapor pressure of pure ice.

Presented in Table 6 are a_w values for frozen food, calculated from the vapor pressure of ice and supercooled water. Figure 15 is a plot of log a_w versus $1/T$ illustrating that (a) that the relationship is linear at subfreezing temperatures, (b) the influence of temperature on a_w is typically far greater at subfreezing temperatures than at above-freezing temperatures, and (c) a sharp break occurs in the plot at the freezing point of the sample.

Three important distinctions should be noted when comparing a_w values at above-freezing and below-freezing temperatures. First, at above-freezing temperatures, a_w is a function of sample composition and temperature, with the former factor predominating. At below-freezing temperatures, a_w becomes independent of sample composition and depends solely on temperature; that is, in the presence of an ice phase, a_w values are not influenced by the kind or ratio of solutes present. As a consequence, any subfreezing event influenced by the kind of solute present (e.g., diffusion-controlled processes, catalyzed reactions, and reactions affected by the absence or presence of cryoprotective agents, by antimicrobial agents, and/or by chemicals that alter pH and oxidation-reduction potential) cannot be accurately forecast based on the a_w value (19a). Thus, a_w values at subfreezing temperatures are far less valuable indicators of physical, chemical, and physiological events than are a_w values at above-freezing temperatures.

Table 6   Vapor Pressures and Water Activities of Water, Ice, and Food at Various Subfreezing Temperatures

Temperature (°C)	Vapor pressure, liquid water[a]		Vapor pressure, ice[b] or food containing ice		a_w (P_{ice}/P_{water})
	kPa	mm Hg	kPa	mm Hg	
0	0.6104[b]	4.579[b]	0.6104	4.579	1.00[d]
−5	0.4216[b]	3.163[b]	0.4016	3.013	0.953
−10	0.2865[b]	2.149[b]	0.2599	1.950	0.907
−15	0.1914[b]	1.436[b]	0.1654	1.241	0.864
−20	0.1254[c]	0.941[c]	0.1034	0.776	0.82
−25	0.0806[c]	0.605[c]	0.0635	0.476	0.79
−30	0.0509[c]	0.382[c]	0.0381	0.2859	0.75
−40	0.0189[c]	0.142[c]	0.0129	0.0966	0.68
−50	0.0064[c]	0.048[c]	0.0039	0.02955	0.62

[a]Supercooled at all temperatures except 0°C.
[b]Observed data (61).
[c]Calculated data (41).
[d]Applies only to pure water.

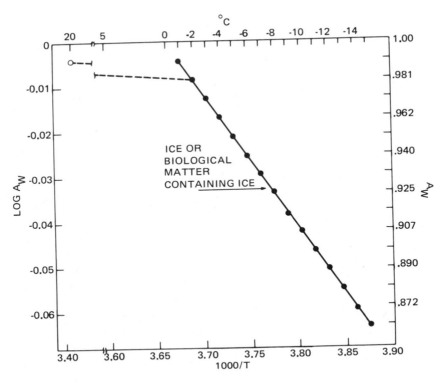

Figure 15 Relationship between water activity and temperature for samples above and below freezing. [From Fennema (19); courtesy of Academic Press.]

Second, as the temperature is changed sufficiently to form or melt ice, the meaning of a_w, in terms of food stability, also changes. For example, in a product at $-15°C$ (a_w = 0.86), microorganisms will not grow and chemical reactions will occur slowly. However, at $20°C$ and a_w 0.86, some chemical reactions will occur rapidly and some microorganisms will grow at moderate rates.

Third, knowledge of a_w at a below-freezing temperature cannot be used to predict the a_w of the same food at an above-freezing temperature. This is true since a_w values at subfreezing temperatures are independent of sample composition and dependent only on temperature.

IX.   ISOTHERMS

Plots interrelating water content (expressed as mass of water per unit mass of dry material) of a food with its water activity at constant temperature are known as sorption isotherms. The information that can be derived from such a plot is useful (a) in concentration and dehydration processes, because the ease or difficulty of water removal is related to water activity, and (b) for assessing stability of the food (see next section).

Shown in Fig. 16 is a schematic sorption isotherm for a high-moisture food. The entire range of moisture contents encountered during dehydration is shown on this plot. This kind of plot is not very useful because the data of greatest interest—those in the low-moisture region—are not shown in sufficient detail. Expansion of the low-moisture region yields a more useful plot, as illustrated in Fig. 17.

Real sorption isotherms for substances yielding various isotherm shapes are shown in Fig. 18. These are resorption (or adsorption) isotherms prepared by adding water to previously dried samples. Desorption isotherms are also common. The isotherms (Fig. 18) with a sigmoidal shape are characteristic of most foods. Such foods as fruits, confections, and coffee extract that contain large amounts of sugar and other small, soluble molecules and are not rich in polymeric materials exhibit the J-type isotherm in Fig. 18, curve 1.

As an aid to understanding the meaning and usefulness of sorption isotherms, it is sometimes appropriate to divide them into three zones as depicted in Fig. 17. The properties of water, newly incorporated (resorption) as one moves from zone I (dry) to zone III (high moisture), differ greatly, and the major attributes of the water added in each zone are described below and summarized in Tables 3-5.

Water present in zone I of the isotherm is the most strongly absorbed and most immobile water in food. This water is absorbed to accessible polar sites by water-ion or water-dipole interactions. The enthalpy of vaporization of this water is much greater than that of pure water, and it is unfreezable at −40°C. This water is not able to serve as a solvent, and it is not present in sufficient amount to have a plasticizing effect on the solid. It behaves simply as part of the solid.

The high-moisture end of zone I (boundary of zones I and II) corresponds to the "monolayer" moisture content of the food. Contrary to what the name infers, monolayer does not mean coverage of all dry matter with a closely packed single layer of water molecules, and its exact meaning at the molecular level is not fully understood. Probably the best interpretation is to regard the monolayer value as approximating the amount of water needed to form a monolayer over the accessible, highly polar groups of the dry matter, perhaps on the approximate basis of 1 mol water to 1 mol of highly polar groups. In another sense, the monolayer value corresponds to the maximum amount of water that can be very strongly bound to the dry matter, and thus represents the constitutional and

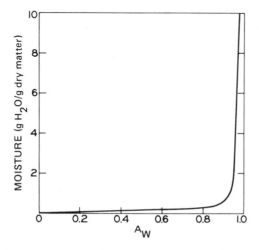

Figure 16 Schematic moisture sorption isotherm encompassing a broad range of moisture contents.

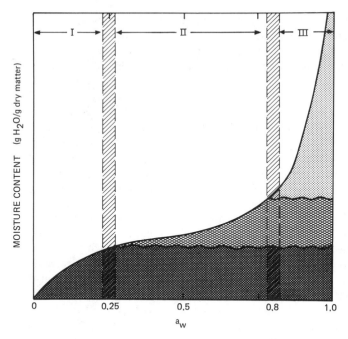

Figure 17 Generalized moisture sorption isotherm for the low-moisture range of a food (20°C).

vicinal categories of water in Tables 3 and 4. Zone I water constitutes a very small fraction of the total water in a high-moisture food material.

Water in zone II of the isotherm consists of zone I water plus water added (resorption procedure) within the confines of zone II. Water added in zone II occupies the remaining first-layer sites and several additional layers around hydrophilic groups of the solid and is designated multilayer water (Table 4). Multilayer water associates with neighboring molecules primarily by water-water and water-solute hydrogen bonding. The enthalpy of vaporization for multilayer water is slightly to moderately greater than that of pure water, depending on the closeness of the water to nonaqueous constituents, and most of this water is unfreezable at −40°C.

As water is added to a food with a water content corresponding to the boundary of zone I and II, this water will initiate solution processes, act as a plasticizing agent, and promote incipient swelling of the solid matrix. The onset of solution processes will mobilize reactants, thereby leading to an acceleration in the rate of most reactions. Water in zones II and I usually constitutes less than 5% of the water in a high-moisture food material.

Water in zone III of the isotherm consists of zone I and II water plus water added (resorption procedure) within the confines of zone III. Water added in zone III is the least strongly bound and most mobile (molecularly) water in foods and is designated bulk-phase water (Table 5). In gels or cellular systems, bulk-phase water is physically entrapped so that macroscopic flow is impeded. In all other respects this water has properties similar to that of water in a dilute salt solution. This is reasonable, since a typical water molecule added in zone III is "insulated" from the effects of nonaqueous

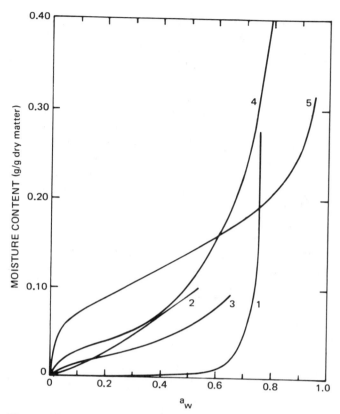

Figure 18 Various types of resorption isotherms obtained with foods and biological substances. Temperature 20°C, except for 1, which was 40°C. (1) confection (main component powdered sucrose), (2) spray-dried chicory extract, (3) roasted Columbian coffee, (4) pig pancreas extract powder, and (5) native rice starch. [From van den Berg and Bruin (58); courtesy of Academic Press.]

molecules by several layers of zone I and II water molecules. Water added or removed within the confines of zone III exhibits an enthalpy of vaporization essentially the same as that of pure water, is freezable, is available as a solvent, and is sufficiently abundant and normal to allow chemical reactions and microbial growth to occur readily. The bulk-phase water of zone III, either entrapped or free, usually constitutes more than 95% of the total water in a high-moisture food material.

It should be emphasized that the boundaries separating the zones of the isotherm cannot be accurately established, and that water, with the possible exception of constitutional water (Table 3), can interchange within and between zones. Furthermore, although the addition of water to a dry material can alter somewhat the properties of water that is already present (matrix swelling and solution processes) it is conceptually helpful to view zone I water as remaining almost constant as zone II water is added, and zone II water remaining almost constant as zone III water is added (Fig. 17).

The important consequences of these differences in water properties on stability of foods will be discussed in a later section. At this point, it will suffice to say it is the least bound fraction of water existing in any food sample that governs stability.

As mentioned earlier, water activity is temperature dependent; thus, moisture sorption isotherms must also exhibit temperature dependence. An example involving potato slices is shown in Fig. 19. At any given moisture content, a_w increases with increasing temperature, in agreement with the Clausius-Clapeyron equation, and in accord with the behavior of foods in general.

An additional complication is that a moisture sorption isotherm prepared by the addition of water (resorption) to a dry sample will not necessarily be superimposable with an isotherm prepared by desorption. This lack of superimposibility is referred to as *hysteresis*, and a schematic example is shown in Fig. 20. Moisture sorption isotherms of many foods exhibit hysteresis. The magnitude of hysteresis, the shape of the curves, and the inception and termination points of the hysteresis loop can vary considerably depending on such factors as the nature of the food, the physical changes it undergoes when water is removed or added, temperature, the rate of desorption, and the degree of water removed during desorption (26). Typically, at any given a_w, the water content of the food will be greater during desorption then during resorption.

A number of largely qualitative theories have been advanced to explain sorption hysteresis (57). These theories include such factors as swelling phenomena, metastable

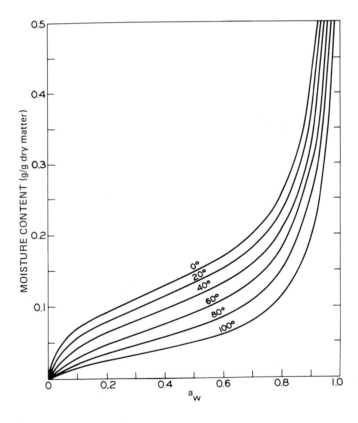

Figure 19 Moisture desorption isotherms of potatoes at various temperatures. [Redrawn from Görling (25).]

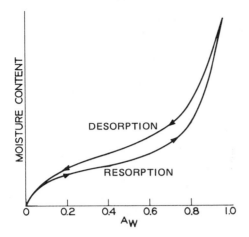

Figure 20 Hysteresis of moisture sorption isotherm.

local domains, presumed barriers of diffusion, capillary phenomena, and time dependency of equilibria. A definitive explanation of sorption hysteresis has yet to be formulated.

Sorption hysteresis is more than a laboratory curiosity. Labuza et al. (35) have conclusively established that lipid oxidation in strained meat from chicken and pork, at a_w values in the range 0.75-0.84, proceeds much more rapidly if the samples are adjusted to the desired a_w by desorption rather than by resorption. The desorption samples, as already noted, would contain more water at a given a_w than the resorption samples. According to Labuza et al. (35), this would cause the high-moisture sample to have a lower viscosity, which in turn would cause greater catalyst mobility, greater exposure of catalytic sites because of the swollen matrix, and somewhat improved diffusion of oxygen as compared to the lower moisture (resorption) sample.

In another study, Labuza et al. (36) found that, to stop the growth of several microorganisms, a product's a_w must be significantly lower if prepared by desorption then if prepared by resorption.

X. WATER ACTIVITY AND FOOD STABILITY

From what has already been said, it should be clear that food stability and water activity are closely related in many (but not all) situations. The data in Fig. 21 and Table 7 provide examples of these relationships. Shown in Table 7, are various common microorganisms and the range of water activity permitting their growth. Also shown in this table are common foods categorized according to their water activity.

The data in Fig. 21 are intended to represent typical reaction rate-a_w relationships for several important kinds of reactions in the temperature range 25-45°C. For comparative purposes a typical isotherm, Fig. 21f, is also shown. It is important to remember that the exact reaction rates and the positions and shapes of the curves (Fig. 21a-e) can be altered by the composition, physical state, and structure (capillarity) of the sample, by the composition of the atmosphere (especially oxygen), by temperature, and by hysteresis effects.

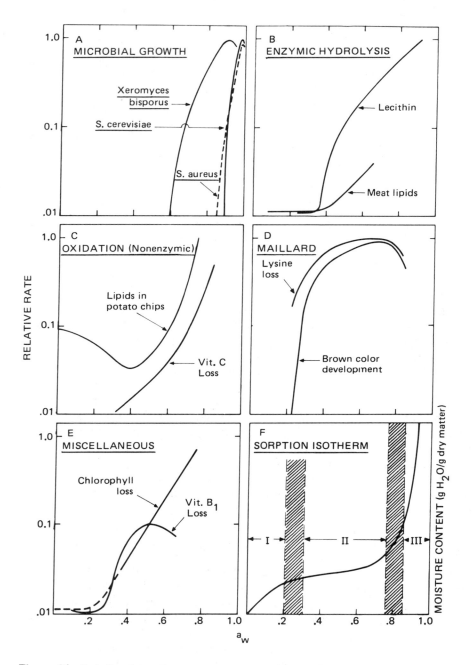

Figure 21 Relationships among water activity, food stability, and sorption isotherms. (A) Microbial growth versus a_w (52,56). (B) Enzymic hydrolysis versus a_w (1,48). (C) Oxidation (nonenzymic) versus a_w (37,50). (D) Maillard browning versus a_w (14). (E) Miscellaneous reaction rates versus a_w (28,31). (F) Water content versus a_w. All ordinates are "relative rate," except for F.

Table 7 Water Activity and Growth of Microorganisms in Food

Range a_w	Microorganisms generally inhibited by lowest a_w in this range	Foods generally within this range
1.00-0.95	*Pseudomonas, Escherichia, Proteus, Shigella, Klebsiella, Bacillus, Clostridium perfringens*, some yeasts	Highly perishable (fresh) foods and canned fruits, vegetables, meat, fish, and milk; cooked sausages and breads; foods containing up to approximately 40% (w/w) sucrose or 7% sodium chloride
0.95-0.91	*Salmonella, Vibrio parahaemolyticus, C. botulinum, Serratia, Lactobacillus, Pediococcus*, some molds, yeasts (*Rhodotorula, Pichia*)	Some cheeses (cheddar, Swiss, Muenster, provolone), cured meat (ham), some fruit juice concentrates; foods containing 55% (w/w) sucrose or 12% sodium chloride
0.91-0.87	Many yeasts (*Candida, Torulopsis, Hansenula*), *Micrococcus*	Fermented sausage (salami), sponge cakes, dry cheeses, margarine; foods containing 65% (w/w) sucrose (saturated) or 15% sodium chloride
0.87-0.80	Most molds (mycotoxigenic penicillia), *Staphylococcus aureus*, most *Saccharomyces* (*bailii*) spp., *Debaryomyces*	Most fruit juice concentrates, sweetened condensed milk, chocolate syrup, maple and fruit syrups; flour, rice, pulses containing 15-17% moisture; fruit cake, country-style ham, fondants, high-ratio cakes
0.80-0.75	Most halophilic bacteria, mycotoxigenic aspergilli	Jam, marmalade, marzipan, glacéed fruits, some marshmallows
0.75-0.65	Xerophilic molds (*Aspergillus chevalieri, A. candidus, Wallemia sebi*), *Saccharomyces bisporus*	Rolled oats containing approximately 10% moisture, grained nougats, fudge, marshmallows, jelly, molasses, raw cane sugar, some dried fruits, nuts
0.65-0.60	Osmophilic yeasts (*Saccharomyces rouxii*), few molds (*Aspergillus echinulatus, Monascus bisporus*)	Dried fruits containing 15-20% moisture, some toffees and caramels; honey
0.50	No microbial proliferation	Pasta containing approximately 12% moisture; spices containing approximately 10% moisture
0.40	No microbial proliferation	Whole-egg powder containing approximately 5% moisture
0.30	No microbial proliferation	Cookies, crackers, bread crusts, and so on containing 3-5% moisture
0.20	No microbial proliferation	Whole milk powder containing 2-3% moisture, dried vegetables containing approximately 5% moisture, corn flakes containing approximately 5% moisture, country-style cookies, crackers

Source: From Beauchat (5); courtesy of The American Association of Cereal Chemists.

The lipid oxidation-a_w relation in Fig. 21c deserves comment. Starting at very low a_w values it is apparent that the rate of oxidation decreases as water is added up to an a_w value approximating the boundary of zones I and II of the isotherm (Fig. 21f). Further addition of water then results in increased rates of oxidation until an a_w value approximating the boundary of zones II and III is attained. Further increases in water can cause a reduction in rate of oxidation (not shown). Karel and Yong (27) have offered the following interpretative suggestions regarding this behavior. The first water added to a very dry sample obviously interferes with oxidation. This water (zone I) is believed to bind hydroperoxides, interfere with their decomposition, and thereby hinder progress of oxidation. In addition, this water hydrates metal ions that catalyze oxidation, apparently reducing their effectiveness.

Additions of water beyond the boundary of zones I and II (Fig. 21c and f) results in increased rates of oxidation. Karel and Yong (27) have suggested that water added in this region of the isotherm may accelerate oxidation by increasing the solubility of oxygen and by allowing macromolecules to swell, thereby exposing more catalytic sites. At still greater a_w values (more than about 0.80) the added water may retard rates of oxidation, and the suggested explanation is that dilution of catalysts reduces their effectiveness.

In Fig. 21a, d, and e it should be noted that curves for the Maillard reaction, vitamin B_1 degradation, and microbial growth all exhibit rate maxima at intermediate to high a_w values. Two possibilities have been advanced to account for the decline in reaction rate that sometimes accompanies increases in water activity in foods with moderate to high moisture contents (14,34).

1. For those reactions in which water is a product, an increase in water content can result in product inhibition.
2. When the water content of the sample is such that solubility, accessibility (surfaces of macromolecules), and mobility of rate-enhancing constituents are no longer rate limiting, then further addition of water will dilute rate-enhancing constituents and decrease the reaction rate.

For the chemical reactions shown in Fig. 21, large or maximum rates typically occur in the range of intermediate moisture foods (0.7-0.9 a_w), which is clearly undesirable.

Finally, for the chemical reactions in Fig. 21, minimum reaction rates are typically first encountered (desorption) at the boundary of zones I and II of the isotherm (a_w, 0.20-0.30), and all but oxidative reactions remain at this minimum as a_w is further reduced. The water content corresponding to this first-encountered minimum (desorption) is the so-called monolayer water content.

Since the monolayer value of a food provides a good first estimate of the water content providing maximum stability of a dry product, knowledge of this value is of considerable practical importance. Determining the monolayer value for a specific food can be done with moderate ease if data from the low-moisture end of the sorption isotherm are available. One can then use an equation developed by Brunauer et al. (8) to compute the monolayer value.

$$\frac{a_w}{m(1-a_w)} = \frac{1}{m_1 c} + \frac{C-1}{m_1 c} a_w \tag{5}$$

where a_w is water activity, m is water content (g H_2O per g dry matter), m_1 is the monolayer value, and C is a constant.

From this equation, it is apparent that a plot of $a_w/[m(1-a_w)]$ versus a_w, known as a BET plot, should yield a straight line. An example for native potato starch is shown in Fig. 22. The linear relationship, as is generally acknowledged, begins to deteriorate at a_w values greater than about 0.35.

The monoloayer value can be calculated as follows.

Monolayer value = m_1

$$= \frac{1}{(y - \text{intercept}) + \text{slope}}$$

From Fig. 22, the y intercept is 0.6. Calculation of the slope from Fig. 22 yields a value of 10.7. Thus,

$$m_1 = \frac{1}{0.6 + 10.7}$$

$$= 0.088 \text{ g } H_2O \text{ per g dry matter}$$

In this particular instance the monolayer value corresponds to an a_w of 0.2.

In addition to chemical reactions and microbial growth, a_w also influences the texture of dry and semidry foods. For example, suitably low water activities are necessary if crispness of crackers, popped corn, and potato chips is to be retained; if caking of

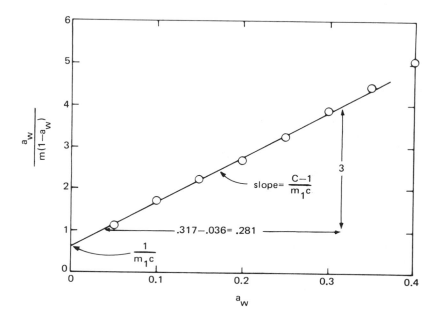

Figure 22 BET plot for native potato starch (resorption data, 20°C). [Data from van den Berg (57).]

granulated sugar, dry milk and instant coffee is to be avoided; and if stickiness of hard candy is to be prevented (33). The maximum a_w that can be tolerated in dry materials, without incurring loss of desirable properties, ranges from 0.35 to 0.5, depending on the product. Furthermore, suitably high water activities of soft-textured foods are needed to avoid undesirable hardness.

XI. SOLUTE MOBILITY AND FOOD STABILITY

It is generally agreed that solute mobility, as determined by wide-line NMR, has an important bearing on rates of chemical reactions and microbial growth, and some have argued it is a better indicator of these changes than is water activity. Solute mobility, as used here, includes both positional (diffusional) and nonpositional (e.g., a change in molecular conformation) aspects, since both can influence rates of chemical reactions.

Conceptually it is believed that reactions do not proceed in very dry food because reactants are immobile. Exceptions include free radicals (autoxidation) and reactants fluid in the absence of water (oils can be enzymically hydrolyzed in samples of very low water content). In general, however, reactant mobility and the ability of water to serve as a solvent does not become significant until the water content becomes slightly greater than the monolayer value. From Table 4, it is apparent that water contributing to the incipient mobility of reactants is unfreezable and is the least mobile fraction of the multilayer category.

Examples of the relationship between reaction rate and reactant mobilization are shown for nonenzymic browning of a simple amino acid-glucose-starch mixture in Fig. 23, and for ascorbic acid degradation in Fig. 24. In all these instances, rather abrupt increases in reaction rate are found to accompany increases in moisture content beyond that needed to mobilize the reactants. Noteworthy in Fig. 24 is that addition of glycerol

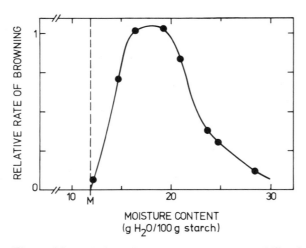

Figure 23 Relationship between reactant mobilization and rate of nonenzymic browning. System consisted of γ-aminobutyric acid, glucose, starch, and water. M is mobilization point as determined by wide-line NMR. [Redrawn from Duckworth et al. (11); courtesy of Applied Science Publishers, London.]

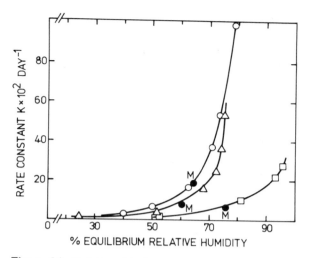

Figure 24 Relationship between mobilization point of ascorbic acid and its rate of loss. M is the mobilization point of ascorbic acid, 0 represents data for savoy cabbage at 25°C, □ represents data for potato starch at 45°C, and △ represents data for glycerated potato starch gel at 45°C. [Data from Seow as reported by Duckworth et al. (11); courtesy of Applied Science Publishers, London.]

to potato starch gel lowers the equilibrium relative humidity (and a_w) at which mobilization and rapid onset of ascorbic acid degradation occur. Glycerol apparently excerts its effect simply by increasing the solvent pool. This result provides some support for the view that food stability can be more accurately predicted based on reactant mobility than on water activity (53). However, even if solute mobility proves to be superior to a_w as an indicator of food stability, the former may not achieve widespread use until methodology is available for quality assurance situations.

XII. ROLE OF ICE IN THE STABILITY OF FOODS AT SUBFREEZING TEMPERATURES

Although freezing is regarded as the best method of long-term preservation for most kinds of foods, the benefits of this preservation technique derive primarily from low temperature, as such, not from ice formation. The formation of ice in cellular foods and food gels has two important adverse consequences: (a) nonaqueous constituents become concentrated in the unfrozen phase (an unfrozen phase exists in foods at all storage temperatures used commercially), and (b) all water converted to ice increases 9% in volume. Both these occurrences deserve further comment.

During freezing of aqueous solutions, cellular suspensions, or tissues, water from solution is transferred into ice crystals of a variable but rather high degree of purity. Nearly all the nonaqueous constituents are therefore concentrated in a diminished quantity of unfrozen water. The net effect is similar to conventional dehydration, except in this instance the temperature is lower and the separated water is deposited locally in the form of ice. The extent of concentration is influenced mainly by the final temperature, and to a lesser degree by the eutectic temperatures of the solutes present, agitation, and the rate of cooling.

Because of the freeze concentration effect, the unfrozen phase changes significantly in such properties as pH, titratable acidity, ionic strength, viscosity, freezing point (and all other colligative properties), surface and interfacial tension, and oxidation-reduction potential. In addition, eutectic mixtures may form, oxygen and carbon dioxide may be expelled from solution, water structure and water-solute interactions may be drastically altered, and macromolecules will be forced closer together, making interactions more probable. These changes in concentration-related properties often favor increases in reaction rates. Thus, freezing has two opposing effects on reaction rate: lowering temperature, as such, will decrease reaction rates, and freeze concentration, as such, will sometimes increase reaction rates. It should not be surprising, therefore, that reaction rates at subfreezing temperatures deviate considerably from the Arrhenius relationship.

A few examples are in order. Listed in Table 8 are a number of nonenzymic reactions that have been observed to increase in rate during freezing. Oxidative reactions and protein insolubilization frequently behave in this manner, and have an especially important bearing on food quality. Data in Fig. 25 illustrate the amount of protein insolubilization occurring during a 30 day period at various subfreezing temperatures. These data are shown to illustrate two important generalities: (a) when a freeze-induced rate acceleration occurs, it is generally most evident just a few degrees below the initial freezing point of the sample, and (b) reaction rates at normal temperatures of frozen storage ($-18°C$) are almost always substantially less than those at $0°C$. The last point reconciles how freezing can be an effective method of preservation even though the acceleration of reaction rates (at high subfreezing temperatures) sometimes occurs.

The information in Table 9 are presented to help the reader better visualize how the effects of temperature lowering and freeze concentration combine to govern rates of reaction in noncellular systems.

Some enzyme-catalyzed reactions in cellular system also accelerate in rate during freezing (Table 10), but this is believed to be caused by freeze-induced dislocation of

Table 8 Examples of Nonenzymatic Reactions That Can Accelerate in Foods During Freezing

Type of reaction	Substrate
Acid-catalyzed hydrolysis	Sucrose
Oxidation	Ascorbic acid Butterfat Lipids in cooked beef Tocopherol in fried potato products β-Carotene and vitamin A in fat Tuna and beef oxymyoglobin Milk
Protein insolubilization	Beef, rabbit, and fish protein
Formation of nitric oxide myoglobin or hemoglobin (cured meat color)	Myoglobin or hemoglobin

Source: Adapted from Fennema (18).

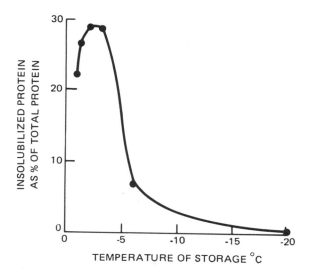

Figure 25 Effect of storage temperature on the insolubilization of proteins in liquid expressed from ox muscle. Storage time, 30 days. [Redrawn from Finn (22); courtesy of the Royal Society, London.]

enzymes, substrates, and/or enzyme activators, rather than to solute concentration. The increase in volume that accompanies ice formation, as well as freeze-concentration effects, are responsible for these dislocations. Further information on the consequences of freezing can be obtained in publications by Farrant (15) and Fennema et al. (20).

In conclusion, it can be said that water is not only the most abundant constituent in foods but that it also contributes greatly to food's desirable native qualities, is the cause of food's perishable nature, governs the rates of many chemical reactions, is instrumental in the undesirable side effects of freezing, is associated with nonaqueous food constituents in such a complex manner that once these relationships are disturbed by

Table 9   Rates of Reactions as Influenced by Temperature and Concentration of Solutes During Freezing

Situation	Change in rate of reaction caused by		Relative influence of the two effects[a]	Total effect of freezing on reaction rate
	Lowering temperature	Concentration of solutes and other effects of ice		
1	Decrease	Decrease	Cooperative	Decrease
2	Decrease	Slight increase	$T > S$	Slight decrease
3	Decrease	Moderate increase	$T = S$	None
4	Decrease	Great increase	$T < S$	Increase

[a]T = effect of temperature;
S = effect of solute concentration.

Table 10 Some Instances in Which Cellular Systems Exhibit Increased Rates of Enzyme-Catalyzed Reactions During Freezing

Type of reactions	Sample	Temperature (°C) at which increased reaction rate was observed
Glycogen loss and/or accumulation of lactic acid	Frog, fish, beef, or poultry muscle	−2.5 to −6
Degradation of high-energy phosphates	Fish, beef, and poultry muscle	−2 to −8
Hydrolysis of phospholipids	Cod	−4 (T_{max})
Decomposition of peroxides	Catalase in rapidly frozen potatoes and slowly frozen peas	−0.8 to −5
Oxidation of L-ascorbic acid	Rose hips, strawberries, Brussels sprouts	−10 −6 −2.5 to −5

Source: Adapted from Fennema (17).

some means, such as drying or freezing, they can never again be completely reinstated, and is, above all else, frustratingly complex, inadequately studied, and poorly understood.

ACKNOWLEDGMENT

The helpful comments of Theodore Labuza are gratefully acknowledged.

REFERENCES

1. Acker, L. (1962). Enzymic reactions in foods at low moisture content. *Advan. Food Res. 11*:263-330.
2. Almási, E. (1979). Dependence of the amount of bound water of foods on temperature. *Acta Aliment. 8*(1):41-56.
3. Berendsen, H. J. C. (1971). The molecular dynamics of water in biological systems. *Proc. First Eur. Biophys. Cong.*, Baden, Austria, *1*:483-488.
4. Berendsen, H. J. C. (1975). Specific interactions of water with biopolymers, in *Water—A Comprehensive Treatise*, vol. 5 (F. Franks, ed.), Plenum Press, New York, pp. 293-349.
5. Beauchat, L. R. (1981). Microbial stability as affected by water activity. *Cereal Foods World 26*(7):345-349.
6. Bluestein, P., and T. P. Labuza (1972). Kinetics of water sorption in a model freeze-dried food. *A. I. Ch. E. (Amer. Inst. Chem. Eng.) Journal 18*:706-712.
7. Brady, G. W., and W. J. Romanov (1960). Structure of water. *J. Chem. Phys. 32*:106.
8. Brunauer, S., H. P. Emmett, and E. Teller (1938). Adsorption of gases in multimolecular layers. *J. Amer. Chem. Soc. 60*:309-319.
9. Davidson, D. W. (1973). Clathrate hydrates, in *Water—A Comprehensive Treatise*, vol. 2 (F. Franks, ed.), Plenum Press, New York, pp. 115-234.
10. Dowell, L. G., S. W. Moline, and A. P. Rinfret (1962). A low-temperature X-ray

diffraction study of ice structures formed in aqueous gelatin gels. *Biochim. Biophys. Acta 59*:158-167.

11. Duckworth, R. B., J. Y. Allison, and H. A. A. Clapperton (1976). The aqueous environment for chemical change in intermediate moisture foods, in *Intermediate Moisture Foods* (R. Davies, G. G. Birch, and K. J. Parker, eds.), Applied Science Publ., London, pp. 89-99.

12. Eagland, D. (1975). Nucleic acids, peptides and proteins, in *Water—A Comprehensive Treatise*, vol. 5 (F. Franks, ed.), Plenum Press, New York, pp. 305-518.

13. Edelhoch, H., and J. C. Osborne, Jr. (1976). The thermodynamic basis of the stability of proteins, nucleic acids, and membranes. *Advan. Protein Chem. 30*: 183-250.

14. Eichner, K. (1975). The influence of water content on non-enzymic browning reactions in dehydrated foods and model systems and the inhibition of fat oxidation by browning intermediates, in *Water Relations of Foods* (R. B. Duckworth, ed.), Academic Press, London, pp. 417-434.

15. Farrant, J. (1977). Water transport and cell survival in cryobiological procedures. *Phil. Trans. Royal Soc. London B278*:191-205.

16. Fennema, O. (1973). Water and ice, in *Low-Temperature Preservation of Foods and Living Matter* (O. Fennema, W. D. Powrie, and E. H. Marth, eds.), Marcel Dekker, Inc., New York, pp. 1-77.

17. Fennema, O. (1975). Activity of enzymes in partially frozen aqueous systems, in *Water Relations of Foods* (R. B. Duckworth, ed.), Academic Press, London, pp. 397-413.

18. Fennema, O. (1975). Reaction kinetics in partially frozen aqueous systems, in *Water Relations of Foods* (R. B. Duckworth, ed.), Academic Press, London, pp. 539-556.

18a. Fennema, O. (1977). Water and protein hydration. in *Food Proteins* (J. R. Whitaker and S. R. Tannenbaum, eds.), AVI Publishing Co., Westport, Conn., pp. 50-90.

19. Fennema, O. (1978). Enzyme kinetics at low temperature and reduced water activity, in *Dry Biological Systems* (J. H. Crowe and J. S. Clegg, eds.), Academic Press, New York, pp. 297-322.

19a. Fennema, O., and L. A. Berny (1974). Equilibrium vapor pressure and water activity of food at subfreezing temperatures. in *Proceedings of the Fourth International Congress of Food Science and Technology*, vol. 2, Madrid, Spain. Instituto Nacional de Ciencia y Technologia de Aliments, Valencia, Spain, pp. 27-35.

20. Fennema, O., W. D. Powrie, and E. H. Marth (1973). *Low-Temperature Preservation of Foods and Living Matter*, Marcel Dekker, Inc., New York.

21. Ferro Fontan, C., and J. Chirife (1981). The evaluation of water activity in aqueous solutions from freezing point depression. *J. Food Technol. 16*:21-30.

22. Finn, D. B. (1932). Denaturation of proteins in muscle juice by freezing. *Proc. Royal Soc. B111*:396-411.

23. Franks, F. (ed.) (1972-1979). *Water—A Comprehensive Treatise*, 6 volumes, Plenum Press, New York.

24. Gal, S. (1981). Recent developments in techniques for obtaining complete sorption isotherms, in *Water Activity: Influences on Food Quality* (L. B. Rockland and G. F. Stewart, eds.), Academic Press, New York, pp. 89-110.

25. Gorling, P. (1958). Physical phenomena during the drying of foodstuffs. in *Fundamental Aspects of the Dehydration of Foodstuffs*, Society of Chemical Industry, London, pp. 42-53.

26. Kapsalis, J. G. (1981). Moisture sorption hysteresis, in *Water Activity: Influences on Food Quality* (L. B. Rockland and G. F. Stewart, eds.), Academic Press, New York, pp. 143-177.

27. Karel, M., and S. Yong (1981). Autoxidation-initiated reactions in food, in *Water Activity: Influences on Food Quality* (L. B. Rockland and G. F. Stewart, eds.), Academic Press, New York, pp. 511-529.

28. Kirk, J. R. (1981). Influence of water activity on stability of vitamins in dehydrated foods, in *Water Activity: Influences on Food Quality* (L. B. Rockland and G. F. Stewart, eds.), Academic Press, New York, pp. 531-566.

29. Kuntz, Jr., I. D., and W. Kauzmann (1974). Hydration of proteins and polypeptides. *Advan. Protein Chem. 28*:239-345.

30. Labuza, T. P. (1968). Sorption phenomena in foods. *Food Technol. 22*(3):15-24.

31. Labuza, T. P. (1971). Properties of water as related to the keeping quality of foods, in *Proceedings SOS/70, Third International Congress of Food Science and Technology*, Washington, D.C., Institute of Food Technologists, Chicago, pp. 618-635.

32. Labuza, T. P. (1980). The effect of water activity on reaction kinetics of food deterioration. *Food Technol 34*(4):36-41.

33. Labuza, T. P., and R. Contreras-Medellin (1981). Prediction of moisture protection requirements for foods. *Cereal Foods World 26*(7):335-344.

34. Labuza, T. P., and M. Saltmarch (1981). The nonenzymatic browning reaction as affected by water in foods, in *Water Activity: Influences on Food Quality* (L. B. Rockland and G. F. Stewart, eds.), Academic Press, New York, pp. 605-650.

35. Labuza, T. P., L. McNally, D. Gallagher, J. Hawkes, and F. Hurtado (1972). Stability of intermediate moisture foods. 1. Lipid oxidation. *J. Food Sci. 37*:154-159.

36. Labuza, T. P., S. Cassil, and A. J. Sinskey (1972). Stability of intermediate moisture foods. 2. Microbiology. *J. Food Sci. 37*:160-162.

37. Lee, S. H., and T. P. Labuza (1975). Destruction of ascorbic acid as a function of water activity. *J. Food Sci. 40*:370-373.

38. Lewin, S. (1974). *Displacement of Water and Its Control of Biochemical Reactions*, Academic Press, London.

39. Luyet, B. J. (1966). Anatomy of the freezing process in physical systems, in *Cryobiology* (H. T. Meryman, ed.), Academic Press, New York, pp. 115-138.

40. Luyet, B. J., and G. Rapatz (1958). Patterns of ice formation in some aqueous solutions. *Biodynamica 8*:1-68.

41. Mason, B. J. (1957). *The Physics of Clouds*, Clarendon, Oxford, p. 445.

42. Miller, S. L. (1961). A theory of gaseous anesthetics. *Proc. Nat. Acad. Sci. U.S. 47*:1515-1524.

43. Morgan, J., and B. E. Warren (1938). X-ray analysis of the structure of water. *J. Chem. Phys. 6*:666-673.

44. Oakenfull, D., and D. W. Fenwick (1977). Thermodynamics and mechanism of hydrophobic interaction. *Aust. J. Chem. 30*:741-752.

45. Pauling, L. (1935). The structure and entropy of ice and other crystals with some randomness of atomic arrangement. *J. Amer. Chem. Soc. 57*:2680-2684.

46. Pauling, L. (1961). A molecular theory of general anesthesia. *Science 134*:15-21.

47. Peterson, W. W., and H. A. Levy (1957). A single-crystal neutron diffraction study of heavy ice. *Acta Crystallogr. 10*:70-76.

48. Potthast, K., R. Hamm, and L. Acker (1975). Enzymic reactions in low moisture foods, in *Water Relations of Foods* (R. B. Duckworth, ed.), Academic Press, London, pp. 365-377.

49. Prior, B. A. (1979). Measurement of water activity in foods: A review. *J. Food Protection 42*:668-674.

50. Quast, D. G., and M. Karel (1972). Effects of environmental factors on the oxidation of potato chips. *J. Food Sci 37*:584-588.

51. Röntgen, W. C. (1882). VIII. Ueber die Constituion des flüssigen Wassers. *Ann. Phys. Chem. 281*:91-97.

52. Scott, W. J. (1957). Water relations of food spoilage microorganisms. *Advan. Food Res. 7*:83-127.

53. Simatos, D., M. Le Meste, D. Petroff, and B. Halphen (1981). Use of electron spin resonance for the study of solute mobility in relation to moisture content in model

food systems, in *Water Activity: Influences on Food Quality* (L. B. Rockland and G. F. Stewart, eds.), Academic Press, New York, pp. 319-346.

54. Stillinger, F. H. (1980). Water revisited. *Science 209*:451–457.
54a. Storey, R. M., and G. Stainsby (1970). The equilibrium water vapor pressure of frozen cod. *J. Food Technol. 5*:157-163.
55. Taborsky, G. (1979). Protein alterations at low temperatures: An overview. in *Proteins at Low Temperatures* (O. Fennema, ed.), Advances in Chemistry Series 180, American Chemical Society, Washington, D.C., pp. 1-26.
56. Troller, J. A., and J. H. B. Christian (1978). *Water Activity and Food*, Academic Press, New York.
57. van den Berg, C. (1981). *Vapour Sorption Equilibria and Other Water-Starch Interactions; A Physico-chemical Approach*, Agricultural University Wageningen, The Netherlands.
58. van den Berg, C., and S. Bruin (1981). Water activity and its estimation in food systems: Theoretical aspects, in *Water Activity: Influences on Food Quality* (L. B. Rockland and G. F. Stewart, eds.), Academic Press, New York, pp. 1-61.
59. van den Berg, C., and H. A. Leniger (1978). The water activity of foods. in *Miscellaneous Papers* 15, Landbouwhogeschool Wageningen, The Netherlands, pp. 231-242.
60. Warner, D. T. (1981). Theoretical studies of water in carbohydrates and proteins. in *Water Activity: Influences on Food Quality* (L. B. Rockland and G. F. Stewart, eds.), Academic Press, New York, pp. 435–465.
61. Weast, R. C., and M. J. Astle (eds.) (1981). *CRC Handbook of Chemistry and Physics*, 63rd ed., CRC Press, Boca Raton, Florida.
62. Wetlaufer, D. B. (1973). Protein structure and stability: Conventional wisdom and new perspectives. *J. Food Sci. 38*:740-743.

GENERAL BIBLIOGRAPHY

Dorsey, N. E. (1968). *Properties of Ordinary Water-Substance*, Hafner Publishing Co., New York (facsimile of the 1940 edition).
Duckworth, R. B. (1975). *Water Relations of Foods*, Academic Press, London.
Eisenberg, D., and W. Kauzmann (1969). *The Structure and Properties of Water*, Oxford University Press, London.
Fennema, O. (1973). Water and ice, in *Low-Temperature Preservation of Foods and Living Matter* (O. Fennema, W. D. Powrie and E. H. Marth), Marcel Dekker, Inc., New York, pp. 1-77.
Fennema, O. (1977). Water and protein hydration, in *Food Proteins* (J. R. Whitaker and S. R. Tannebaum, eds.), AVI Publishing Co., Westport, Conn., pp. 50-90.
Fletcher, N. H. (1970). *The Chemical Physics of Ice*, Cambridge University Press, London.
Franks, F. (ed.) (1972-1979). *Water—A Comprehensive Treatise*, 6 volumes, Plenum Press, New York.
Labuza, T. P. (1977). The properties of water in relationship to water binding in foods: A review. *J. Food Proc. Preserv. 1*:167-190.
Lewin, S. (1974). *Displacement of Water and Its Control of Biochemical Reactions*, Academic Press, London.
Richards, R. E., and F. Franks (eds.) (1977). A discussion on water structure and transport in biology. *Phil. Trans. Royal Soc. London. B278*:1-205.
Rockland, L. G., and G. F. Stewart (eds.) (1981). *Water Activity: Influences on Food Quality*. Academic Press, New York.
Troller, J. A., and J. H. B. Christian (1978). *Water Activity and Food*, Academic Press, New York.

3

CARBOHYDRATES

Roy L. Whistler and James R. Daniel Purdue University, West Lafayette, Indiana

I.	Introduction	70
	A. Classes of Carbohydrates	70
	B. Dietary Utilization of Carbohydrates	72
	C. Carbohydrates and Dental Caries	73
	D. Carbohydrates in Foods	74
II.	Carbohydrate Structure	76
	A. Monosaccharides	76
	B. Glycosides	80
	C. Oligosaccharides	85
	D. Polysaccharides	87
III.	Reactions of Carbohydrates	90
	A. Hydrolysis	90
	B. Acyclic Carbohydrate Reactions	92
	C. Dehydration and Thermal Degradation of Carbohydrates	94
	D. Browning Reactions	96
IV.	Functions of Monosaccharides and Oligosaccharides in Foods	105
	A. Hydrophilicity	105
	B. Binding of Flavor Ligands	106
	C. Carbohydrate Browning Products and Food Flavors	106
	D. Sweetness	108
V.	Functions of Polysaccharides in Foods	108
	A. Structure-Function Relations of Polysaccharides	108
	B. Starch	112
	C. Glycogen	120
	D. Cellulose	121
	E. Hemicelluloses	123
	F. Pectic Substances	123
	G. Plant Gums	125

VI. Conclusion   133
     References   135
     General Bibliography   137

I. INTRODUCTION

A. Classes of Carbohydrates

Carbohydrates constitute three-quarters of the biological world and about 80% of the caloric intake of humankind. The most abundant carbohydrate is cellulose, the principal structural component of trees and other plants. The major food ingredient consumed by the human is starch, providing 75-80% of the total caloric intake. In the United States carbohydrates supply 46% of the calories, fats supply 42%, and proteins supply 12%.

Carbohydrate, a term derived from the German *kohlenhydrate* and similar to the French *hydrate de carbone*, expresses the early determined elemental composition of $C_x(H_2O)y$, which signified a composition containing carbon along with hydrogen and oxygen in the same ratio as in water. As chemical methods evolved, this empirical formula was replaced by a more descriptive graphic formulation that gave a clear visualization of spatial or steric structure. Thus, the formulation of D-glucose, the central metabolic sugar of most living organisms, was originally depicted as an open chain, then later as a more stable hemiacetal in ring form and is presently shown in its most thermodynamically stable ring conformation.

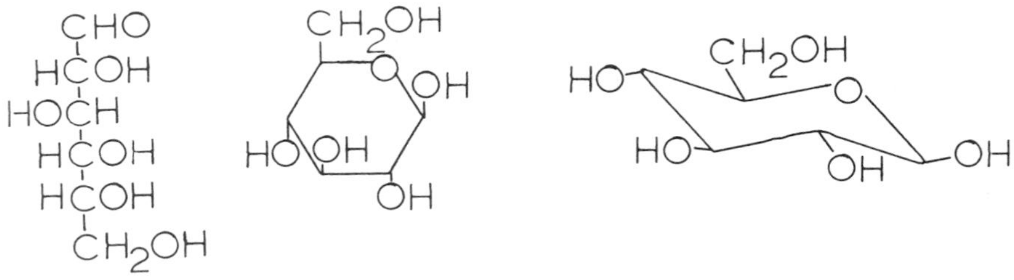

D-glucose   β-D-glucopyranose

D-Glucose has the empirical formula $C_6H_{12}O_6$, but the structural formula shows that it has aldehyde and polyhydroxyl functionality. D-Glucose is a monosaccharide with one saccharose group

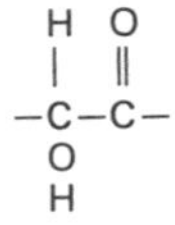

One monosaccharide may be combined with another to produce a disaccharide (as maltose), or it may be combined in a chain of monosaccharide units to produce trisaccharides, tetrasaccharides, or higher polymers with chains of a thousand to several thousand sugar units. When 2-10 monosaccharide units are joined they form the class of oligosaccharides (*oligo*: few). When more than 10 monosaccharide units are joined together they produce a polysaccharide (*poly*: many).

Monosaccharides occur naturally in only small amounts. Rather than as free sugars they are present almost entirely as units in polysaccharides. A small percentage occur as units in disaccharides and a still smaller amount as units in higher oligosaccharides. D-Glucose, the most abundant sugar in the world, is present as a monosaccharide in a very small quantity in plants, but represents 0.8-1.0% of the blood, where it is the main energy source for cells in humans and animals. D-Glucose units, in chains with $(1{\to}4)$-β-D-linkages, constitute cellulose, the most abundant carbohydrate in the plant world, present in the cell walls as the main structural component. D-Glucose units, joined by $(1{\to}4)$-α-D-linkages, constitute starches, and D-glucose units combined with other sugar types are present in numerous other polysaccharides. Other sugar units commonly found in polysaccharides are those containing six carbon atoms (hexoses), such as D-galactose and L-galactose and D-mannose, and those containing five carbon atoms (pentoses), such as D-xylose and L-arabinose. Two other pentoses, D-ribose and 2-deoxy-D-ribose, are found in small amounts as components of nucleic acids.

Common disaccharides in foods are sucrose (cane or beet sugar), maltose (from corn syrup), and lactose (milk sugar). Disaccharides, like the polysaccharides, are not taken into the bloodstream from the small intestines and, hence, are not available as energy sources to the human body unless they are first hydrolyzed to monosaccharides, which can be rapidly transported from the intestinal lumen to the blood. Nonhydrolyzed oligosaccharides and polysaccharides pass through the small intestine to the large intestine where they provide beneficial bulk (fiber). The oligosaccharides may be attacked by the microflora of the large intestine to give such metabolic products as acetic and lactic acids. When produced in sufficient quantities these acids have a laxative effect or even cause diarrhea. Less commonly the metabolic products may be gases that produce flatulence. The latter effect can result from ingestion of certain di-, tri-, and tetrasaccharides, such as the tetrasaccharide stachyose present in beans and certain soybean fractions. Even larger oligosaccharides are encountered in nature, such as verbascose, which is isolated from the roots of the mullein, *Verbascum thapsus*. This nonreducing pentasaccharide has the structure O-α-D-galactopyranosyl-$(1{\to}6)$-O-α-D-galactopyranosyl-$(1{\to}6)$-O-α-D-galactopyranosyl-$(1{\to}6)$-O-α-D-glucopyranosyl-$(1{\to}2)$-β-D-fructofuranoside (18).

Although polysaccharides are classified as polymers containing more than 10 sugar units, there are not many with less than 100 units, and most have from 100 to several thousand sugar units. Some polysaccharides contain only one type of monosaccharide, and these are termed "homopolysaccharides," or more technically, "homoglycans." Glycose is a general name for any monosaccharide, and the *an* ending denotes a polysaccharide. Examples of homoglycans are cellulose and starch. Cellulose and starch do not have a proper *-an* ending because they were named before systematic nomenclature was instituted. Both are polymers of D-glucose units (D-glucopyranosyl units) and may more specifically be designated homoglucans. Xylan is a common plant cell wall polysaccharide composed of D-xylose (D-xylopyranosyl) units. Sometimes two different types of sugars are combined to produce a diheteroglycan, or three sugar units may be joined to produce a triheteroglycan, and up to six sugar types to produce a hexaheteroglycan. No polysaccharide is known to contain more than six or seven types of sugar units.

Some polysaccharides are linear chains of sugar units, and others are branched; some have single sugar units as branches on a main chain; and some possess a branch on a branch, or bushlike, structure. All structures can be homoglycans composed of only one type of sugar unit, or they can be heteroglycans. When more than two types of sugar units are present in a heteroglycan the structure is usually of the branch on branch (bush-shaped) type.

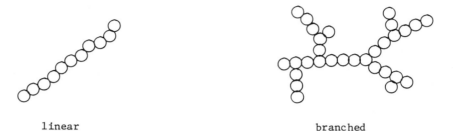

linear   branched

B. Dietary Utilization of Carbohydrates

Carbohydrates provide the majority of dietary calories for the human population. In addition, they can provide desirable texture, pleasant mouth sensation, and universally enjoyed sweetness.

Of the numerous variety and enormous amounts of polysaccharides in the biological world, humans can digest only starch, glycogen, and certain dextrans. Glycogen is similar to starch amylopectin but is more highly branched and is the carbohydrate energy reserve of the body, stored mainly in the liver with small concentrations in muscle. Because of its small quantity, it contributes little as a food source of metabolic energy. Dextrans are rare and are not normally present in natural foods. In the United States, 47% of the carbohydrate-supplied calories are from starch and 52% from sucrose.

Since the human body can not assimilate intact polysaccharides, or even disaccharides, these carbohydrates must be hydrolyzed to monosaccharides before they can be transported into the blood from the small intestine. Starch and glycogen begin to hydrolyze in the mouth where they are subjected to salivary α-amylase. This enzyme, a calcium metalloenzyme with optimum catalytic activity at pH 6, attaches to starch molecules to induce cleavage of a six-unit fragment before it dissociates to encounter another chain. These fragments, malto-oligosaccharides, are also hydrolyzed to produce mainly maltose, maltotriose, and maltotetraose. No D-glucose is produced unless the enzyme has nearly completed its activity. During normal eating of starchy foods there is insufficient time for D-glucose to be produced, but if bread is chewed for an extended time a slight sweetness will develop in the mouth.

On passage of starchy food to the stomach, slight enzyme action may continue, as well as slight acid hydrolysis due to the pH of 0.8-1.0. But no significant hydrolysis takes place until the food reaches the duodenum, the beginning of the small intestine, where pancreatic α-amylase and β-amylase are added. β-Amylase attacks starch and the fragments from α-amylase hydrolysis, hydrolyzing off maltose units from the nonreducing ends. Maltose is hydrolyzed to D-glucose by maltase liberated into the intestinal lumen by brush-border cells. D-Glucose is actively transported from the luminal brush-border cells to the portal blood supply. The transport rate is phenomenal, the calculated potential maximum rate being 22 lb/day, equivalent to 40,000 kcal, or 13 times that normally needed. D-Galactose is also actively transported, but other monosaccharides are more slowly transported by facilitated diffusion.

Sucrose, on entering the small intestine, is hydrolyzed by invertase at the brush border. D-Glucose is absorbed rapidly, but free D-fructose is absorbed more slowly. Some D-fructose is converted to D-glucose in the intestine, but 85% is converted to D-glucose or directly metabolized only after it reaches the liver by way of the portal blood. If sucrose is intravenously injected, it is quantitatively excreted by the kidneys since the cardiovascular system has no invertase.

Lactose is hydrolyzed by intestinal lactase to D-glucose and D-galactose, both actively transported to the portal bloodstream. Lactase synthesis tends to be deficient in blacks, some children and some persons of Asiatic origin, and synthesis after childhood tends to decrease with age. Deficiencies in lactase synthesis allow lactose to pass into the large intestine where it is attacked by indigenous microflora and is converted to acetic and lactic acid and other products that bind water. If these products accumulate in sufficient quantity, they cause diarrhea.

Polysaccharides other than starch and glycogen are not hydrolyzed by gastrointestinal enzymes and pass to the large intestines more or less intact, only insignificantly hydrolyzed by stomach acid. Normally, ingested polysaccharides, such as cellulose, hemicelluloses, and pectin of plant cell walls, have a highly beneficial effect in that they provide bulk for peristaltic action and facilitate the passage of material through the digestive system. This improved bowel movement causes more rapid removal of unabsorbed breakdown products that might otherwise produce irritation and conditions perhaps conducive to the development of cancer. In addition, these polysaccharides bind bile acids and thereby decrease their reabsorption. Thus, they tend to lower the cholesterol level in the blood and presumably retard atherosclerosis.

The group of nondigestible polysaccharides in the diet is termed "dietary fiber." Some, such as pectin and other vegetable gums, if eaten in excessive amounts, cause diarrhea due to their absorption of large amounts of water to produce a viscous solution or gel. Some of the polysaccharides are attacked by the intestinal microflora, causing degradation to smaller molecular fragments, but in general, these polysaccharides are not metabolized in the gastrointestinal tract.

Digestible carbohydrates provide about 4 kcal energy per gram, equivalent to that provided by 1 g protein and less than the 9 kcal provided by 1 g fat. However, utilization of carbohydrates as an energy source is of significant advantage in that they may, under normal conditions, promote utilization of fat. This will tend to reduce adipose stores and obesity. Dietary carbohydrate also permits protein to be used for other important purposes, such as the replenishment of tissue protein. Of course, carbohydrates are a much more economical and abundant source of dietary calories than either fats or proteins.

In addition to the consumption of carbohydrate for caloric needs, humans and animals alike appreciate the sweetness of low-molecular-weight sugars, such as sucrose and D-fructose. The tendency toward greater consumption of sweet sugars in place of starchy foods may contribute to physiological changes, such as an increase in dental caries (5). This has led to a search for the basis of sweetness and a search for safe, noncariogenic, or even nonmetabolizable sweeteners.

C. Carbohydrates and Dental Caries

Dental caries result from acid production on tooth surfaces leading to dissolution of the enamel. That carbohydrates are the causative factor has been known since the time of

Aristotle. Easily metabolized carbohydrates are required, and sucrose is the most cariogenic sugar—the more finely powdered the more active—and sticky, sugary foods are the most cariogenic of all.

Dental caries, a surface periodontal disease, results from the growth and acid production of plaque-forming microorganisms that are common inhabitants of the mouth (5). Prime among these organisms is *Streptococcus mutans*, but *Steptococcus sanguis* and members of *Actinomyces* are commonly cariogenic. They metabolize sucrose, consuming the D-fructose component and transforming the D-glucose, by transferase action on sucrose, into a chain or branched-chain polysaccharide (dextran). This substance adheres to the enamel, protecting the microorganism and providing an oxygen-depressed, hypoxic, or anaerobic condition in which much of the sugar is metabolized to lactic, pyruvic, and acetic acids along with lesser amounts of other acids. The low pH results in localized severe dissolution of the enamel. The older and harder the plaque, the greater its pathogenicity. Fluoride ion inhibits enolase, and thus formation of D-glucose-6-phosphate is also inhibited. This reduces sugar transport in the microorganism. Carbohydrates that are slowly metabolized or are nonmetabolized result in less plaque development, and foods without sucrose greatly lessen plaque formation. The sugar alcohol D-xylitol, developed in Finland, is very sweet and does not contribute to plaque formation (40). Oral tests spanning 2 years have shown that caries development is almost nonexistent when D-xylitol is the only carbohydrate sweetener ingested.

D. Carbohydrates in Foods

Carbohydrates constitute three-quarters of the dry weight of all land plants and seaweeds and are present in all grains, vegetables, fruits, and other plant parts eaten by humans. Furthermore, it is natural that plants for human consumption be selected from those richest in metabolizable carbohydrate. Since most of the metabolizable carbohydrate used by humans comes from sucrose or starch, plants rich in these carbohydrates are most frequently and abundantly consumed as foods. It should be noted, however, that nondigestible polysaccharides (fiber), mainly plant cell walls, are unavoidably consumed when natural plants are consumed, and that these polysaccharides are beneficial to healthy intestinal activity.

Since sucrose is present in relatively minor quantities in most plant foods, it is consumed in only small quantities when natural plant products are eaten. Most of the dietary sucrose is not, therefore, derived from the consumption of unmodified plant foods, but is derived from the consumption of fabricated foods containing sucrose isolated from sugar beets or sugarcane. The amount of sucrose added to commercial foods is greater than generally realized (Table 1). Fruits and vegetables contain small amounts of sucrose, D-glucose, and D-fructose, as seen in Tables 2, 3, and 4.

Cereals contain only small amounts of sugars, since most of the sugar normally transported to the seed is converted to starch. Thus, corn grain contains 0.2-0.5% D-glucose, 0.1-0.4% D-fructose, and 1-2% sucrose; wheat grain contains <0.1%, 0.1%, and about 1% of these sugars, respectively.

Sweet corn is sweet because it is picked before all the sucrose has been converted to starch. During growth, the corn plant, as well as other cereals, converts much of its photogenerated energy in the leaves to making sucrose, the normal sugar of transport in plants. A large part of the generated sucrose is transported to the seed, where it is transformed into starch. Starch is the food reserve the plant embryo uses as energy until it

Table 1 Sugar in Common Foods

Food	Sugar present (%)
Coke	9
Cracker	12
Ice cream	18
Ready-to-eat cereals (dry)	1-50
Orange juice	10
Catsup	29
Cake (dry mix)	36
Non-dairy creamer (dry)	65
Jello (dry)	83

Table 2 Free Sugars in Fruit (% Fresh Basis)

Fruit	D-Glucose	D-Fructose	Sucrose
Apple	1.17	6.04	3.78
Grape	6.86	7.84	2.25
Peach	0.91	1.18	6.92
Pear	0.95	6.77	1.61
Cherry	6.49	7.38	0.22
Strawberry	2.09	2.40	1.03

Table 3 Free Sugars in Vegetables (% Fresh Basis)

Vegetable[a]	D-Glucose	D-Fructose	Sucrose
Beet	0.18	0.16	6.11
Broccoli	0.73	0.67	0.42
Carrot	0.85	0.85	4.24
Cucumber	0.86	0.86	0.06
Endive	0.07	0.16	0.07
Onion	2.07	1.09	0.89
Spinach	0.09	0.04	0.06
Sweet corn	0.34	0.31	3.03
Sweet potato	0.33	0.30	3.37
Tomato	1.12	1.34	0.01

[a]Layman's classification.

Table 4 Free Sugars in Legumes (% Fresh Basis)

Legume	D-Glucose	D-Fructose	Sucrose
Lima bean	0.04	0.08	2.59
Snap bean	1.08	1.20	0.25
Pea	0.32	0.23	5.27

can germinate and begin photogenerating its own carbohydrate source. Sweet corn has an abundance of sucrose present for conversion to starch. If the immature corn ear is picked and quickly boiled or frozen to inactivate the enzyme system for converting the sucrose, a large amount of sugar remains to provide a delicious meal for humans. If, however, the sweet corn is picked when mature or if the enzyme system of sweet corn is not inactivated, and a delay occurs between harvest and consumption, most of the sucrose will be used and the corn on the cob will have lost sweetness and will become firm or tough.

Ripe fruits are sweet to attract bird and animals, thereby leading to dissemination of the seeds. Fruits marketed commercially are picked before full ripeness so that they may be shipped while firm and before microbial spoilage becomes significant. Enzymic processes associated with ripening lead to increasing softness and an increase in sucrose content. During storage and distribution, commercial fruits generally ripen and become sweet due to the generation of sucrose or other sweet carbohydrates from their reserve starch. This ripening phenomenon is a reversal of the sugar-to-starch conversion found in grains, tubers, and roots.

Starch is the most common carbohydrate energy reserve of plants, present even in the wood of trees but stored most copiously in seeds, roots, and tubers. Starch evolved naturally, perhaps because it is compact, easily dries when exposed to low relative humidities, quickly regains softness on exposure to water, and can be rapidly depolymerized to yield energy from D-glucose. It is because of the omnipresence of starch in the plant world that humans and some animals developed enzyme systems that can use this rich and widely occurring energy source.

A seed storage carbohydrate that humans cannot metabolize is the galactomannan of leguminous plants, such as guar, alfalfa, and locust trees. This diheteroglycan will be discussed later, but one should recognize that galactomannan gums are soluble additives used in ice cream, salad dressings, and many other foods.

Animal products contain less metabolizable carbohydrate than other foods. Muscle and liver glycogen, glucans with a structure similar to starch amylopectin, are metabolized in the same way as starch.

Lactose, present in milk—4.8% in cow's milk and 6.7% in human—constitutes 5% of commercial liquid whey. Lactose is commercially crystallized from whey, with some 900 million pounds produced annually.

II. CARBOHYDRATE STRUCTURE

A. Monosaccharides

Important food monosaccharides are relatively low molecular weight molecules containing five or six carbon atoms with the general empirical formula $C_n(H_2O)_n$. The known monosaccharides can be easily generated from D-glyceraldehyde, as shown in Fig. 1.

Sugars readily form acetals and ketals. The carbonyl group can react with one of its own alcohol groups to form half an acetal or ketal. These intramolecular hemiacetals or hemiketals usually involve an alcohol group

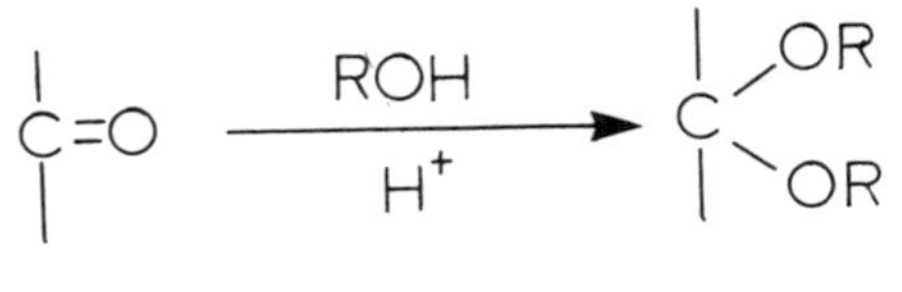

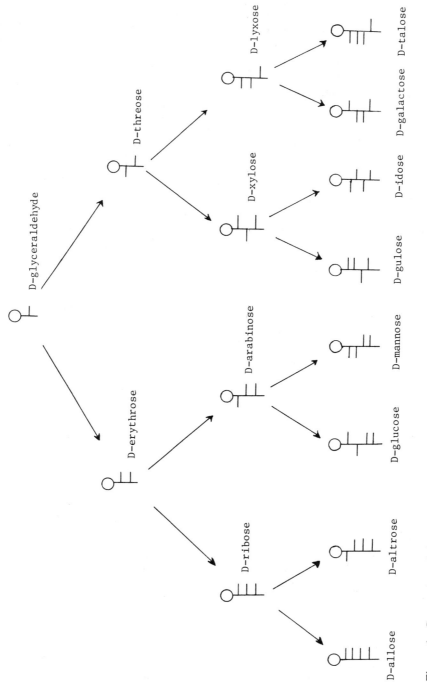

Figure 1 Generation of the eight D-hexoses from D-glyceraldehyde.

located to produce a five-membered furanose (furanlike) ring or a more stable six-membered pyranose (pyranlike) ring. For example, in D-glucose, the hydroxyl oxygen at carbon C5 can react with the aldehyde group at carbon C1 (Fig. 2). This causes carbon C5 to rotate 180° to bring the oxygen atom into the major plane of the ring and forces C6 above the plane. When carbon C1 of D-glucose becomes involved in hemiacetal formation it then has four different groups attached to it and is thus chiral and exists in two stereochemical forms called anomers. In the D-isomeric series to which natural glucose belongs, the α-D-form is that in which the oxygen at carbon C1, the *anomeric* carbon, is on the same side of the molecule as the oxygen on the highest numbered chiral carbon atom, C5 for D-glucose. This places the oxygen of the anomeric carbon on the opposite side from C6 when shown in the Haworth ring structure. Thus, in β-D-glucopyranose the anomeric oxygen is on the same side of the Haworth ring as carbon C6, the hydroxymethyl group. The structure of β-D-glucopyranose is easy to remember since all groups attached to the ring are trans (E in organic designation).

Sugars differing in configuration at any single chiral center other than C1 are called epimers. Thus, D-mannose is the C2 epimer of D-glucose and D-galactose is the C4 epimer of D-glucose.

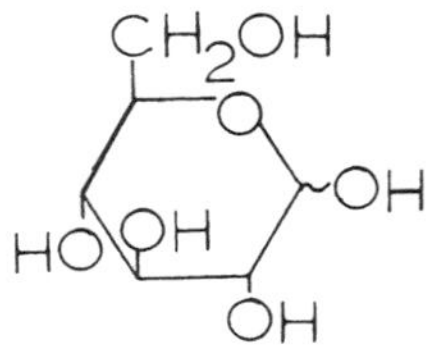

D-glucose 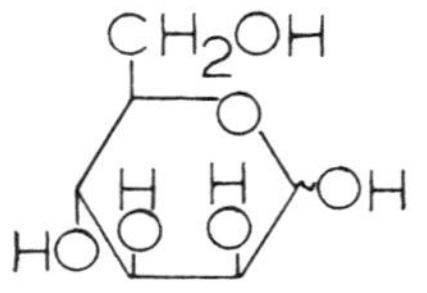   D-mannose 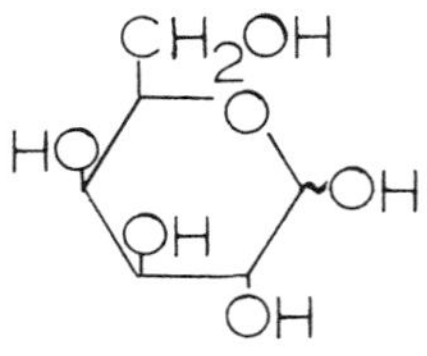   D-galactose

Actual sugar rings, as they occur in nature, are not flat as represented in the Haworth projections. Pyranoses may adopt different conformations that include the chair and the boat.

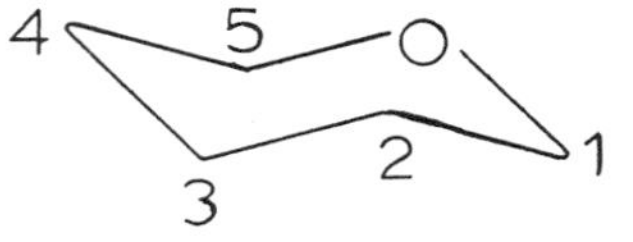

$^4C_1$ chair 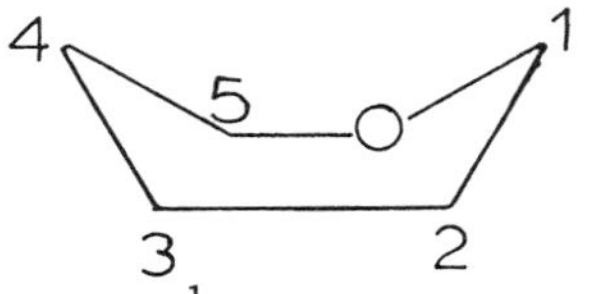 ^1B boat

Many hexoses occur mainly in the fairly rigid chair form; fewer exist in the more flexible boat forms. There are other forms, such as half-chairs and skew arrangements, but these are of higher energy and are less frequently encountered.

As a general rule the most stable ring conformation is that in which all or most of the bulky groups are equatorial to the axis of the ring and the smallest ring substituents (hydrogens) are in axial positions.

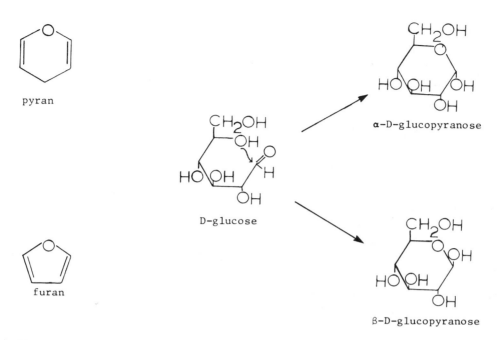

pyran

D-glucose

α-D-glucopyranose

furan

β-D-glucopyranose

Figure 2   Ring forms and anomeric structures for D-glucose.

The furanoses are less stable ring systems than the pyranoses and occur as rapidly equilibrating mixtures of so-called envelope and twist forms (44). The conformation of furanose and pyranose rings

envelope   twist

has been examined by nuclear magnetic resonance spectroscopy, which has provided much of the knowledge about the conformation of monosaccharides in solution. Model building is an aid to visualization and understanding of the structures in three-dimensions.

Although most of the discussion has revolved around the D-series of sugars because of the central importance of D-glucose, the mirror image L-series also has been well examined. All L-sugars, in a Fischer projection in which the aldehyde or ketone group is at the top of the vertically drawn carbon chain, have the hydroxyl group of the highest numbered chiral atom written on the left. Thus, in hexoses this chiral center is carbon atom C5 and in pentoses it is C4.

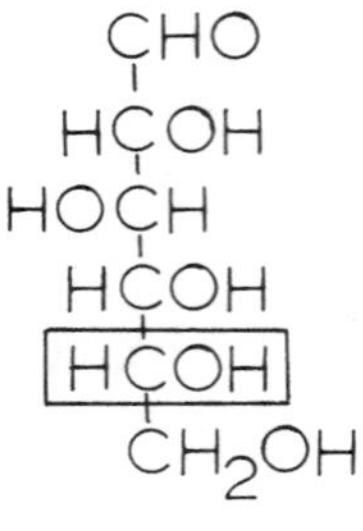

CHO
HCOH
HOCH
HCOH
| HCOH |
CH₂OH

D-glucose

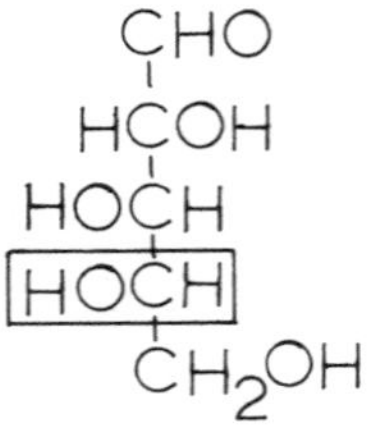

CHO
HCOH
HOCH
| HOCH |
CH₂OH

L-arabinose

Whenever applicable, D and L descriptors should be used in naming sugars and an anomeric designation (α or β) should *never* be used in a monosaccharide name without an associated configurational descriptor (D or L).

B. Glycosides

As indicated, sugars react intramolecularly, wherein the carbonyl group combines with an alcohol group of the same sugar to form a hemiacetal or hemiketal. This yields a ring structure and a new chiral center at the original location of the carbonyl group. If the sugar is dissolved in an alcohol and made slightly acidic (16), a reaction between the two components takes place to form a complete acetal or ketal. This produces a mixed acetal or ketal in which solvent alcohol is one part and one of the sugar's own alcohol groups is the second part. The product formed with the elimination of water is called a glycoside.

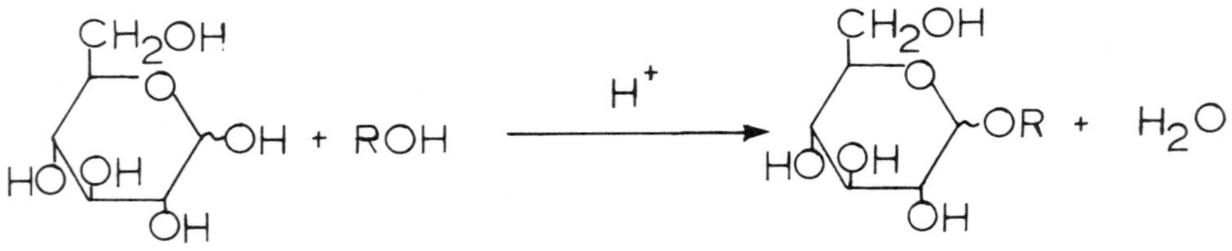

D-glucose   +   ROH $\xrightarrow{H^+}$ alkyl D-glucopyranoside   + H_2O

It usually contains a furanose or a pyranose ring. Although larger or smaller rings are possible, they are very unstable because of bond strains. The newly formed chiral center can be α or β. The specific illustrations given here show D-glucopyranose, as a mixture of α-D and β-D anomers, converted to a mixture of α-D- and β-D-glucopyranosides. In nomenclature, the sugar ending *e* is replaced by *ide*. The parent alcohol group, which is combined with the sugar to form the complete acetal, is called the *aglycon*.

Such glycosides are easy to form with acid catalyst, but the reaction is reversible, and for a good yield of glycoside, water must be removed as it is produced. Since the pyranoside ring is most stable it will predominate, almost to the exclusion of furanosides, in reactions that have run for some time. During the early stages of the glycosylation reaction, however, furanosides will generally predominate. The *glycosyl* group is the sugar group remaining after removal of the hydroxyl at the anomeric carbon.

Glycosides may contain atoms other than oxygen at the anomeric carbon. Thus, reaction of sugar with a thiol, RSH, gives a thioglycoside, and with an amine, RNH_2, an aminoglycoside.

Natural glycosides are produced by transfer of glycosyl groups from nucleotides, such as adenosine and uridine diphosphates, to an appropriate aglycon. The resulting glycoside can be either α or β depending on the specific enzyme effecting the exchange (see p. 82).

With the exception of the oligo- and polysaccharides, only minor amounts of glycosides occur in the human diet. However, their importance frequently lies not in the amount in the diet, but in their physiological potency. Some examples of naturally occurring glycosides include the cardiac glycosides digitalin and digitoxin, the saponins (triterpene or steroid glycosides), which are strong foam formers and stabilizers, and flavonoid glycosides, which may confer bitterness and/or other flavors and color to foods. It has been postulated that the function of glycoside formation in natural systems is to provide a measure of water solubility to otherwise quite insoluble aglycons. This may be especially important for the flavonoid and steroidal glycosides, in which glycoside formation allows their transport in aqueous media.

Although a few complex glycosides are exceptionally sweet, such as steveoside, osladin, and glycyrrhizic acid (4), most glycosides, especially when the aglycon moiety is larger than methyl, have astringent, bitter tastes that may range from slight to extreme. Glycosides may be formed with either aldoses or ketoses. For example, D-mannose forms an acetal, and D-fructose, a ketose, forms a ketal.

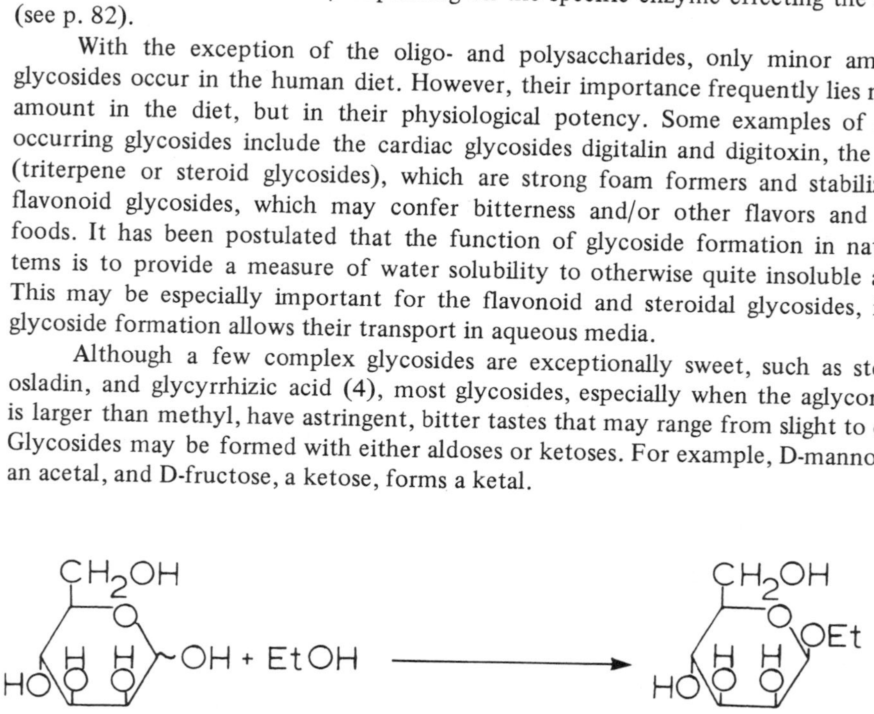

D-mannose   ethyl β-D-mannopyranoside

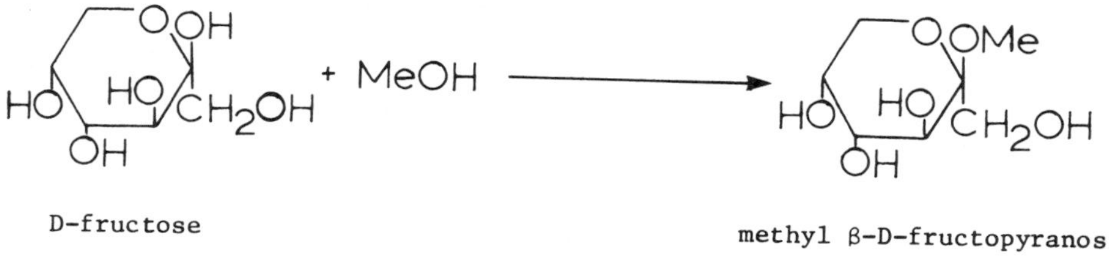

D-fructose   methyl β-D-fructopyranoside

alkyl
D-glucopyranoside

O-Glycosides are generally stable at neutral and basic pH but are hydrolyzed under acidic conditions. Most food glycosides are stable in all but the most acidic foods. Glycosides may also be hydrolyzed by action of various food enzymes (glycosidases). The action of pectinases and amylases are good examples of this type of degradation.

New glycosides may be characterized by chemical reaction or, more conveniently, by various spectroscopic methods. The most useful of these methods is nuclear magnetic resonance, which allows the determination of anomeric configuration, ring conformation, and ring size. Such information can be obtained nondestructively on 1-50 mg of material.

N-Glycosides are less stable than *O*-glycosides and are readily hydrolyzed in water. However, some *N*-glycosides are fairly stable, notably *N*-glycosylamides and some *N*-glycosylpurines and pyrimidines. Specific examples of importance are the 5'-monophosphate flavor potentiators of inosine, xanthosine, and guanosine (24).

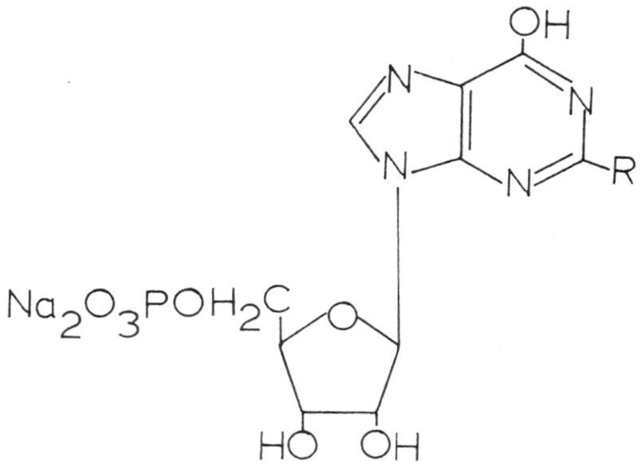

Inosine 5'-monophosphate, R = H
Xanthosine 5'-monophosphate, R = OH
Guanosine 5'-monophosphate, R = NH_2

Normal, unstabilized *N*-glycosides (glycosylamines) decompose in water via a complex series of reactions with concurrent development of highly colored solutions, initially yellow, then dark brown. These reactions are responsible for Maillard browning and are discussed later.

S-Glycosides have a sulfur atom between the glycosyl group and the aglycon. Such compounds occur naturally as components of mustard seed and horseradish. These *S*-glycosides are termed "glucosinolates." Enzymic action of natural thioglucosidases causes cleavage of the aglycon and molecular rearrangement as shown in Fig. 3. The well-known mustard oils are isothiocyanates in which R = allyl, 3-butenyl, 4-pentenyl, benzyl, or other organic radicals. The allylglucosinolate is probably the best known of this class of *S*-glycosides and is named *sinigrin*. It is obvious that selected foods owe their distinctive flavors to these compounds, but some workers have suggested that the *S*-glycosides and/or their breakdown products may reasonably be regarded as food toxicants and this point of view has been discussed in the literature (47).

When, in the formation of *O*-glycosides, the *O*-donor group is a hydroxyl group in the same sugar molecule, an intramolecular glycoside is formed and the sugar is said

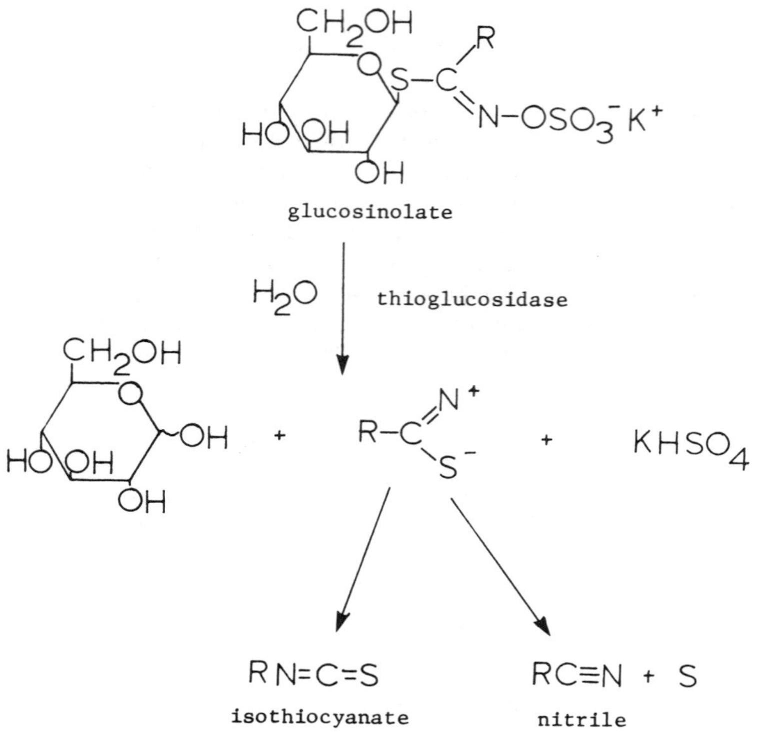

Figure 3 Enzymic decomposition of glucosinolate.

to have an anhydro ring. This is illustrated by the formation of 1,6-anhydro-β-D-gluco-pyranose from D-glucose during pyrolysis.

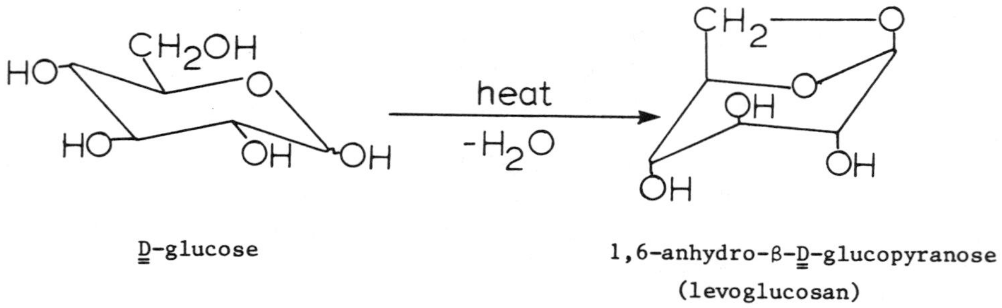

Note that, to effect this reaction, the sugar ring flips from the stable $^4C_1$ conformation of D-glucose to the highly unstable $^1C_4$ conformation of the product. In this notation $^4C_1$ indicates that carbon C4 is above the plane of four atoms and carbon Cl is below the plane. Levoglucosan may be prepared by pyrolysis of D-glucose, cellulose, or starch or by heating phenyl β-D-glucopyranoside in the presence of a strong base. Some small amount of levoglucosan is formed under the pyrolytic conditions of toasting and baking or of heating sugars or syrups to high temperatures. Large amounts are undesirable in foods because of the bitter taste.

Another class of glycosides of importance in some foods is the cyanogenetic glycosides (32). These compounds, which produce hydrogen cyanide on degradation in vivo, occur widely in nature, especially in almonds, cassava, sorghum, bamboo, and lima beans.

Amygdalin, a glucoside of mandelonitrile, is perhaps the best known cyanogenetic glycoside and yields, on complete hydrolysis, D-glucose, benzaldehyde, and hydrogen cyanide. Other cyanogenetic glycosides include dhurrin, a glycoside of *p*-hydroxybenzaldehyde cyanohydrin, and linamarin, a glycoside of acetone cyanohydrin. The cyanide formed on degradation of these compounds is generally detoxified by conversion to thiocyanate. This reaction involves the cyanide ion, the sulfite ion, and the enzyme rhodanase, a sulfur transferase. However, if this detoxification pathway is overwhelmed by the ingestion of large amounts of cyanogenetic glycosides, cyanide toxicity can occur. Human poisonings have been reported as a result of ingestion of lima beans (*Phaseolus lumatus*), cassava, bamboo shoots, and bitter almonds. Poisoning of cattle has been observed after consumption of unripe millet or sorghum. Symptoms of fatal doses of cyanide-generating foods include mental confusion and stupor, cyanosis, involuntary twitching and convulsions, and terminal coma. Lower, nonfatal doses of cyanogenetic foods may produce headaches, throat and chest tightness, muscular weakness, and palpitations.

Prevention of cyanide poisoning ideally involves nonconsumption or consumption of only small amounts of cyanogenetic foods. If this is not possible or practical, these foods should be stored for only short periods of time, care should be taken not to bruise them after harvest, and they should be thoroughly cooked and then thoroughly washed after cooking to remove as much cyanide as possible.

C. Oligosaccharides

This class of carbohydrate compounds, containing 2-10 sugar units, are water soluble and ubiquitous in nature. Natural synthesis occurs by transfer of glycosyl units from nucleotides or from enzymic cleavage and degradation (fragmentation) of polysaccharides.

Disaccharides consist of two monosaccharide units condensed with the concomitant loss of one molecule of water.

$$2 \text{ monosaccharides} \longrightarrow \text{disaccharide} + H_2O$$

Disaccharides may be homogeneous or heterogenous with respect to their monomer composition. Homogeneous disaccharides from D-glucose include cellobiose, maltose, isomaltose, gentiobiose, and trehalose (Fig. 4).

The first four of these D-glucose disaccharides contain a free hemiacetal group that can react to produce a carbonyl function. Such carbohydrates are called reducing sugars because of their ability to reduce metal ions, such as silver or copper, while the sugar is oxidized to a carboxylic acid. α,α-Trehalose does not contain a free hemiacetal function; therefore, it is not easily oxidized and is termed a "nonreducing sugar."

Examples of heterogeneous oligosaccharides are sucrose, lactose, lactulose, and melibiose (Fig. 5). Sucrose is a nonreducing disaccharide; the others are reducing disaccharides. As is discussed later, the reducing or nonreducing properties of a sugar may have an important bearing on in its use in foods, especially when the food also contains

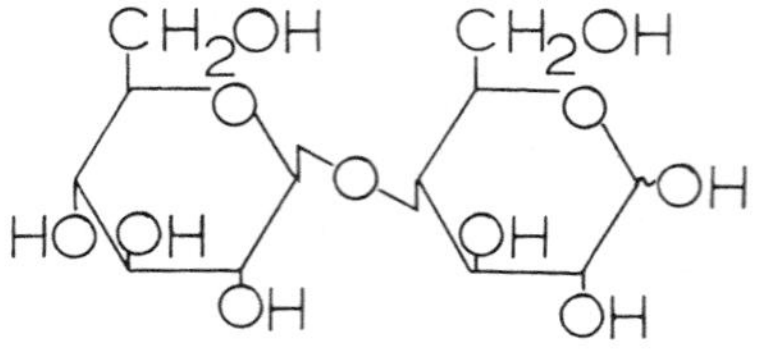

cellobiose

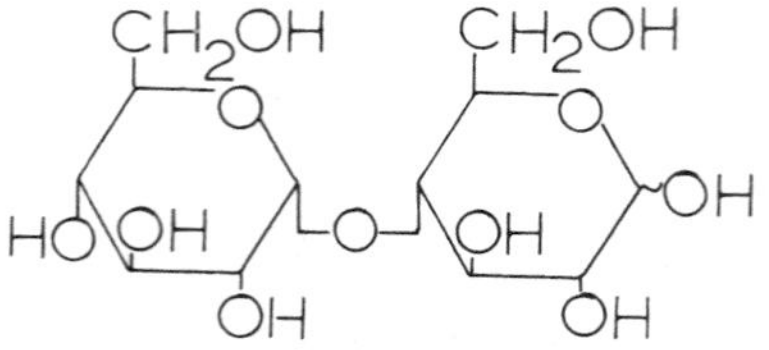

maltose

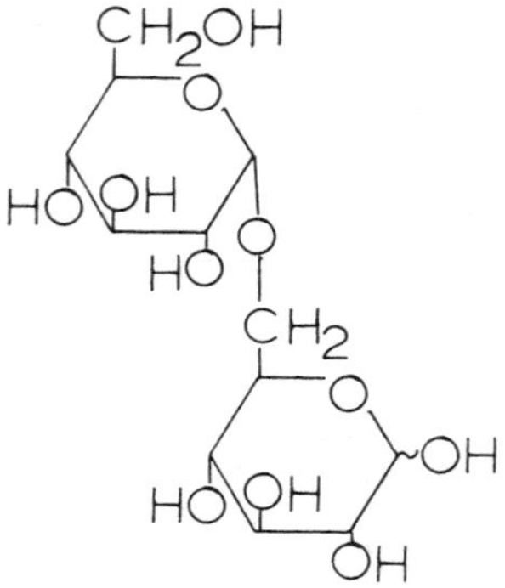

isomaltose

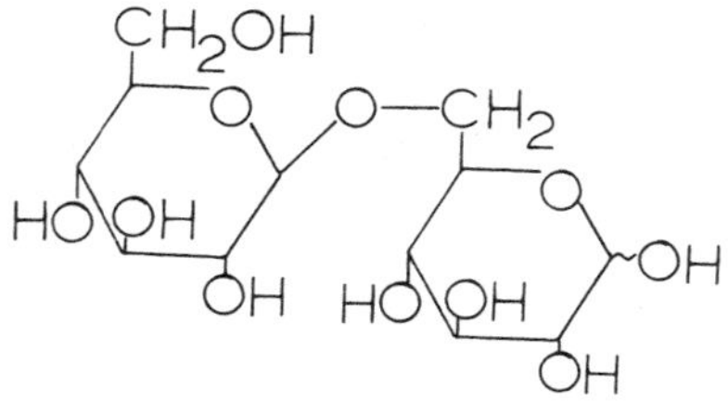

gentiobiose

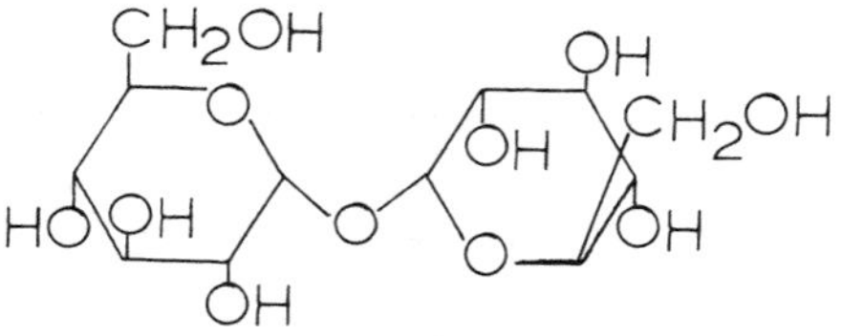

α,α-trehalose

Figure 4 Examples of homogeneous disaccharides.

protein or other amino-containing compounds and is subjected to heat during preparation, processing, or preservation.

Trisaccharides also occur in foods and may be homogeneous, heterogeneous, reducing, or nonreducing. Some examples of trisaccharides are maltotriose (a homogeneous, reducing D-glucose oligomer), manninotriose (a heterogeneous, reducing oligomer of D-galactose and D-glucose), and raffinose (a heterogeneous, nonreducing trisaccharide containing D-galactosyl, D-glucosyl, and D-fructosyl units) (Fig. 6).

Larger oligosaccharides also occur, notably the maltose oligomers (degree of polymerization, DP, or number of monosaccharide residues = 4-10) in corn syrups and the cyclic α-D-glucopyranosyl oligomers of 6-10 units called Schardinger dextrins (16). The

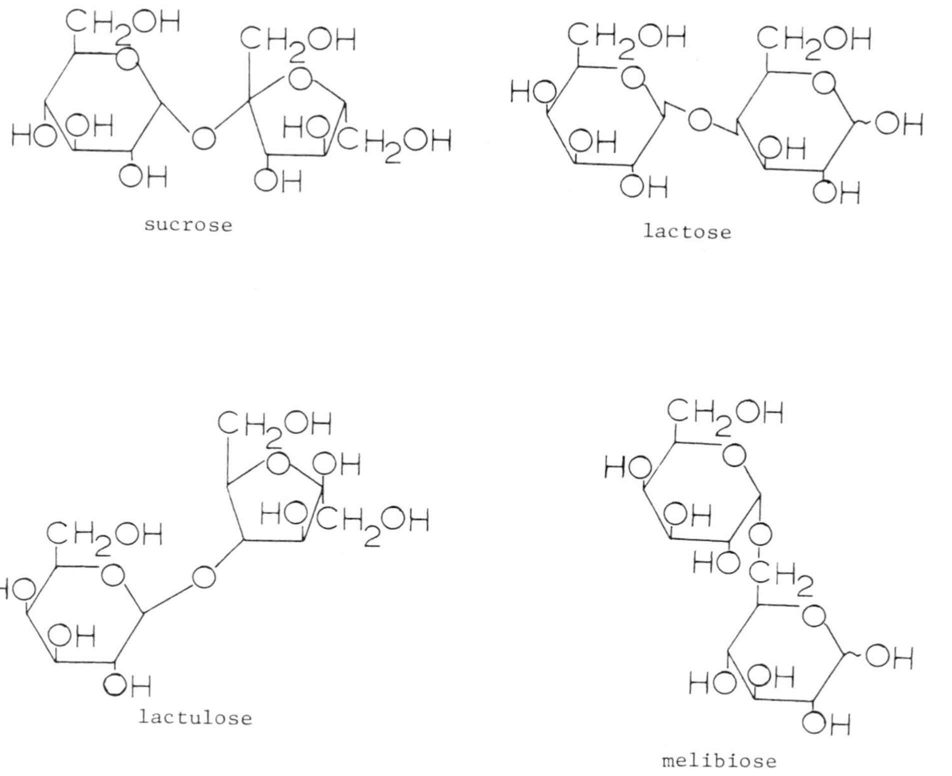

Figure 5   Examples of heterogeneous disaccharides.

latter nonreducing cyclic oligosaccharides are obtained from starch by the action of *Bacillus macerans* amylase. They have the ability to complex guest compounds in a nonstoichiometric inclusion structure, with the guest compound entrapped in a carbohydrate ring (see Fig. 21). This type of complex is useful for fragrance stabilization.

Methods for synthesizing many oligosaccharides are well known, but no useful synthesis of sucrose has yet been developed. Oligosaccharides are interesting principally because of their presence in foods, but they also have been extremely useful in elucidating the structure of high-molecular-weight polysaccharides, of which they constitute structural portions. For example, the isolation of isomaltose from the hydrolysate of amylopectin, under conditions in which acid-catalyzed disaccharide formation is insignificant, was of crucial importance in identifying the nature of the branch point in starch amylopectin.

D. Polysaccharides

Later sections deal in greater detail with specific polysaccharides. They are high-molecular-weight polymers (DP = 10 to several thousand) and they contain various glycosyl units. The polymers are generically referred to as glycans. Homopolymers of a particular monosaccharide are named using the monosaccharide name as a prefix with the suffix *an*. For example, polymers of D-glucose are called D-glucans. Some polysaccharides

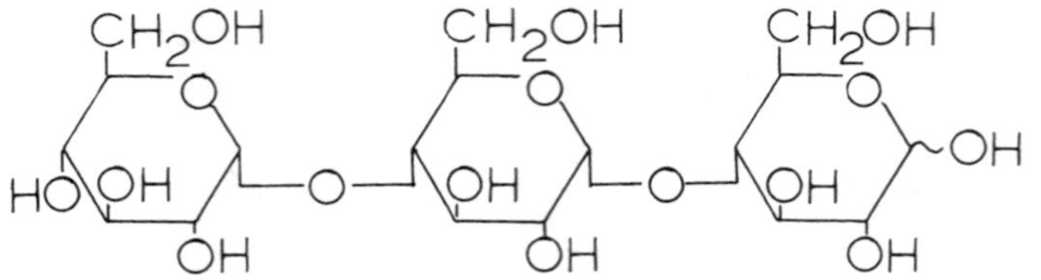

maltotriose

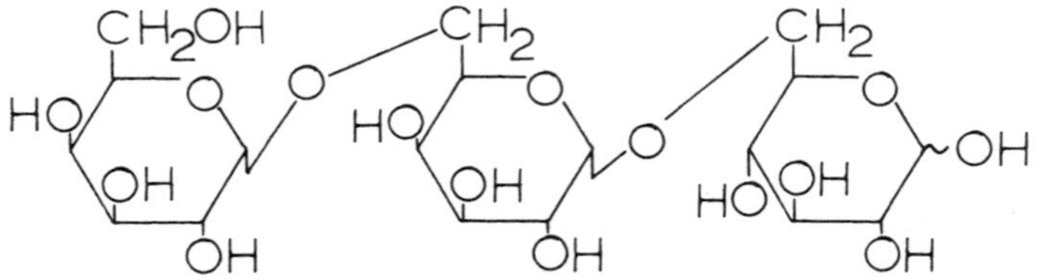

manninotriose

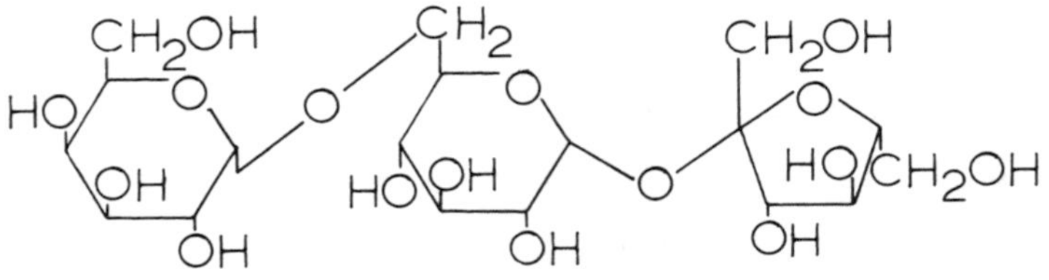

raffinose

Figure 6 Homogeneous and heterogeneous trisaccharides.

named long ago have the improper *ose* ending, such as cellulose and amylose. The names of others have been changed by modern chemists, thus, agarose to agaran. Some older names also end in the suffix *in*, such as pectin and inulin.

Polysaccharides do not have a uniform degree of polymerization but rather a Gaussian distribution of molecular weight. A few have a narrow range of molecular weights. Some polysaccharides occur as mixtures. For example, almost all starches are a mixture of linear and branched glucans termed amylose and amylopectin, respectively. Commercial pectin is a mixture of mostly galacturonan with some araban and galactan. The repeating units or representative structures of cellulose, amylose, amylopectin, pectin, and guaran are shown in Fig. 7.

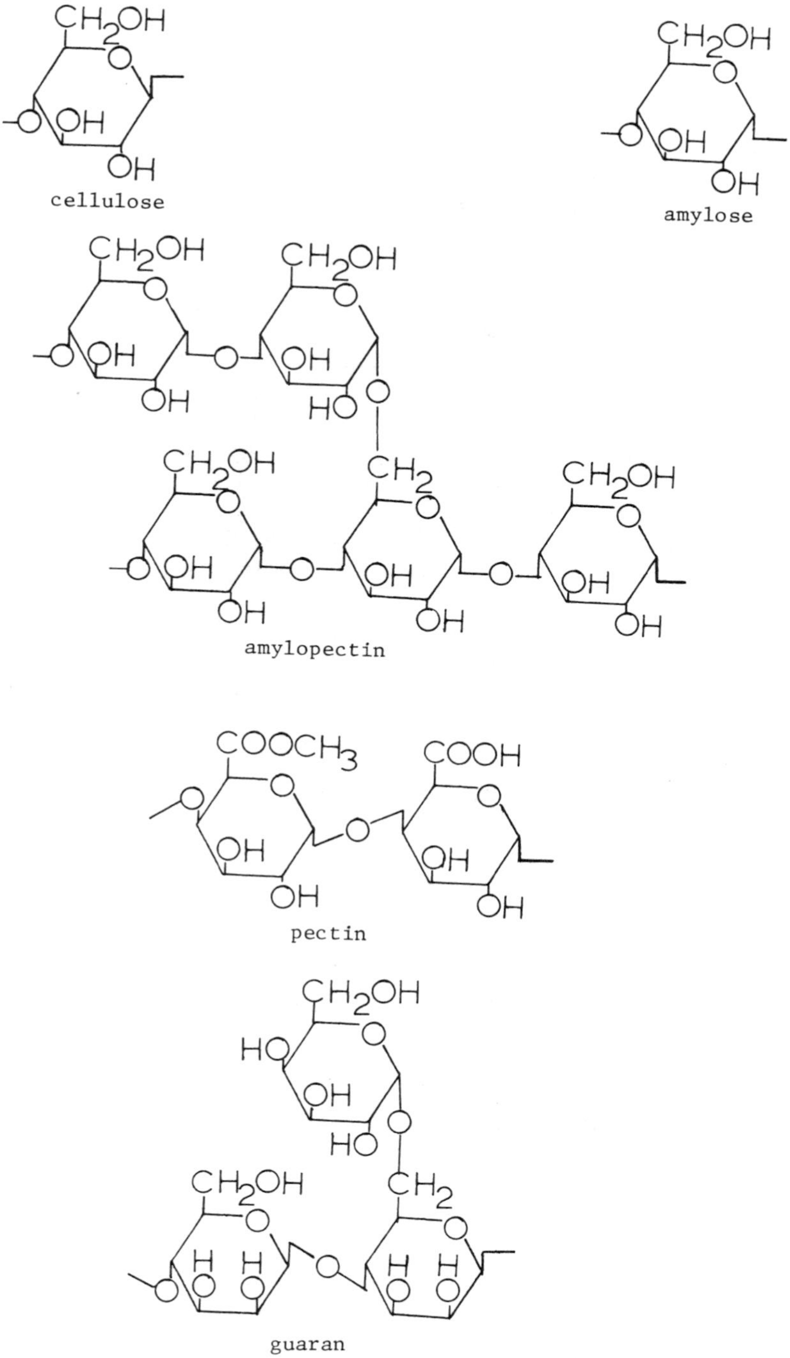

Figure 7 Representative repeating units of some food polysaccharides.

III. REACTIONS OF CARBOHYDRATES

A. Hydrolysis

The hydrolysis of food glycosides, oligosaccharides, and polysaccharides is influenced by numerous factors, including pH, temperature, anomeric configuration, and the size of the glycosyl ring. The hydrolysis of carbohydrates may be important in some food processing or preservation techniques and may lead to undesirable changes in color or, with polysaccharides, to their inability to form a gel.

Glycosidic linkages are more readily cleaved in acidic media than in alkaline media, where they are fairly stable. The hydrolysis of glycosides is believed to follow the mechanism shown in Fig. 8 (7). The rate-determining step is loss of ROH and generation of the resonance-stabilized carbonium ion. As a result of acid sensitivity, certain carbohydrates may be unstable in acidic foods, especially if they are subjected to high temperatures.

Increasing temperature greatly increases the rate of glycoside hydrolysis as reflected by changes in the first-order rate constant (Table 5). As seen in Table 6, anomers differ in hydrolysis rates, with β-D-glycosides hydrolyzed less rapidly than the α-D-anomers. Rate differences due to structural variations and differences in degree of association are noted also among oligo- and polysaccharides. In polysaccharides the rate of hydrolysis is markedly reduced in proportion to the degree of association between the polysaccharide molecules.

Beginning in the 1970s it became commercially feasible to hydrolyze low-cost corn starch to nearly pure D-glucose by using a sequence of α-amylase and glucoamylase and then to isomerize the D-glucose, with isomerase, to an equilibrium mixture of 54% D-glucose and 42% D-fructose. This low-cost sweetener quickly replaced nearly 25% of the 23 billion pound per year sucrose market in the United States, and this percentage is increasing in the United States and in other countries. Primary use of the new sweet-

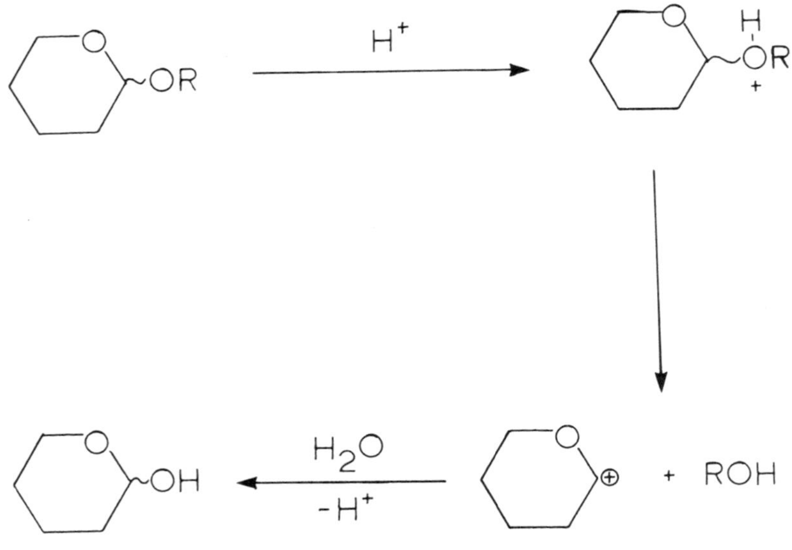

Figure 8 Mechanism of acid-catalyzed hydrolysis of alkyl pyranosides.

Table 5 Effect of Temperature on Rate of Glycoside Hydrolysis

Glycoside in 0.5 M sulfuric acid solution	Temperature[a]		
	70°C	80°C	93°C
Methyl α-D-glucopyranoside	2.82	13.8	76.1
Methyl β-D-glucofuranoside	6.01	15.4	141.0

[a] First-order rate constant, $K \times 10^6$ sec^{-1}.

ener is in confections and soft drinks. The composition and relative sweetness of several typical industrial high-fructose corn syrups (HFCS) available commercially are shown in Table 7.

Production of corn syrups from starch is accomplished by three distinct methods (35). The degree of conversion of starch to D-glucose (dextrose) is measured in terms of dextrose equivalent (DE). This is defined as the percentage of reducing sugars in a corn syrup, calculated as dextrose, on a dry weight basis.

The first method is the acid conversion method. Starch (30-40% in aqueous slurry) is treated with hydrochloric acid (approximately 0.12%) and the mixture is cooked at 140-160°C for 15-20 min or until the desired DE is reached. At the end of the hydrolysis, heating is discontinued and the mixture is neutralized with soda ash and adjusted to pH 4-5.5. After centrifugation, filtration, and concentration, purified acid-converted corn syrup is obtained.

Acid-enzyme-converted corn syrups employ the same acid treatment as just described, followed by enzyme treatment. Enzymes employed in this step include α-amylase, β-amylase, and glucoamylase, depending on the final product desired. Production of 62 DE corn syrup involves acid conversion until the DE reaches 45-50. Neutralization and clarification of the mixture are followed by enzyme addition, usually α-amylase. This conversion is allowed to proceed until the DE is about 62. The enzyme is deactivated by heating.

High-maltose corn syrups are also acid-enzyme-converted syrups. Acid treatment proceeds until a DE of about 20 is achieved. Then, after neutralization and clarification, β-amylase is added and the enzyme is allowed to act until the desired DE is reached. Again, enzyme deactivation is achieved by heat treatment.

Enzyme-enzyme-converted corn syrups are produced by first gelatinizing the corn starch so that the starch polymers are more accessible to the enzymes. Initial treatment

Table 6 Effect of Anomeric Form on the Rate of Hydrolysis of Various Glycosides[a]

α-D-Anomer		K*	β-D-Anomer		K*
Kojibiose	1→2	1.46	Sophorose	1→2	1.17
Nigerose	1→3	1.78	Laminaribiose	1→3	0.99
Maltose	1→4	1.55	Cellobiose	1→4	0.66
Isomaltose	1→6	0.40	Gentiobiose	1→6	0.58

[a] First order rate constant, $K^* = K \times 10^5$. At 80°C, 0.1 M HCl.

Table 7 Composition and Sweetness of Typical High-Fructose Corn Syrups Available Commercially

Components	Type		
	Normal	55% Fructose	90% Fructose
Glucose	52	40	7
Fructose	42	55	90
Oligosaccharides	6	5	3
Relative sweetness	100	105	140

of the gelatinized starch with α-amylase (and/or glucoamylase) is followed by treatment with a second enzyme, the nature of which depends on the type of corn syrup desired. The most important syrup is high-fructose corn syrup. The enzyme used for the D-glucose to D-fructose conversion is D-glucose isomerase, normally immobilized on a support.

Starch amylose, which is $(1{\rightarrow}4)$-α-D-linked, is more susceptible to acid hydrolysis than cellulose which is, $(1{\rightarrow}4)$-β-D-linked, but the slow reaction of insoluble celluose is due at least partly to close intermolecular packing. Furanoside rings are far more easily hydrolyzed than pyranosides, as exemplified by the easy removal of L-arabinofuranosyl units from hemicelluloses and the ease with which the D-fructofuranoside linkage in sucrose is cleaved by mild acid. Although pyranoside hydrolysis is unimolecular, the hydrolysis of furanosides is thought to be bimolecular, because of the negative entropy of activation (7). Hydrolysis is kinetically first order due to the large excess of water.

The exceptional lability of sucrose to hydrolysis must be taken into consideration whenever the sugar is used. On heating, such as in caramelization or candy making, small quantities of food acid or even a high temperature may cause hydrolysis, liberating D-glucose and D-fructose. These reducing sugars may then undergo dehydration reactions to produce, eventually, characteristic odors and colors that may or may not be desirable. Proteins, if present, will also partially lose their nutritive value because of the Maillard reaction.

B. Acyclic Carbohydrate Reactions

Although sugars, especially reducing sugars, are usually represented in the ring form, the open-chain form is present to a small extent and is required in some reactions, such as transition in ring size, mutarotation, and enolization.

For aldoses, such as D-glucose, five structures may be in equilibrium when the sugar is dissolved in water. The isomers present are the α- and β-D-glucopyranoses, the α- and β-D-glucofuranoses, and the open-chain *aldehydo* form, which for D-glucose is present to the extent of 0.006%. The amount of the various isomers present in aqueous equilibrium of a specific sugar depends on the relative stabilities of the various isomeric forms. Pure isomers, such as crystalline α-D-glucopyranose, on dissolution in water, adopt an equilibrium composed of the five species previously mentioned (Fig. 9).

An example of maturotation and the conversion of one anomer to another is represented by the equilibration of α- or β-D-glucose to a mixture of definite proportions. The change is easily followed polarimetrically. β-D-Glucose, on dissolution in pure water, has a polarimetric value of $[\alpha]_D$ + 18.7° initially and after several hours reaches a

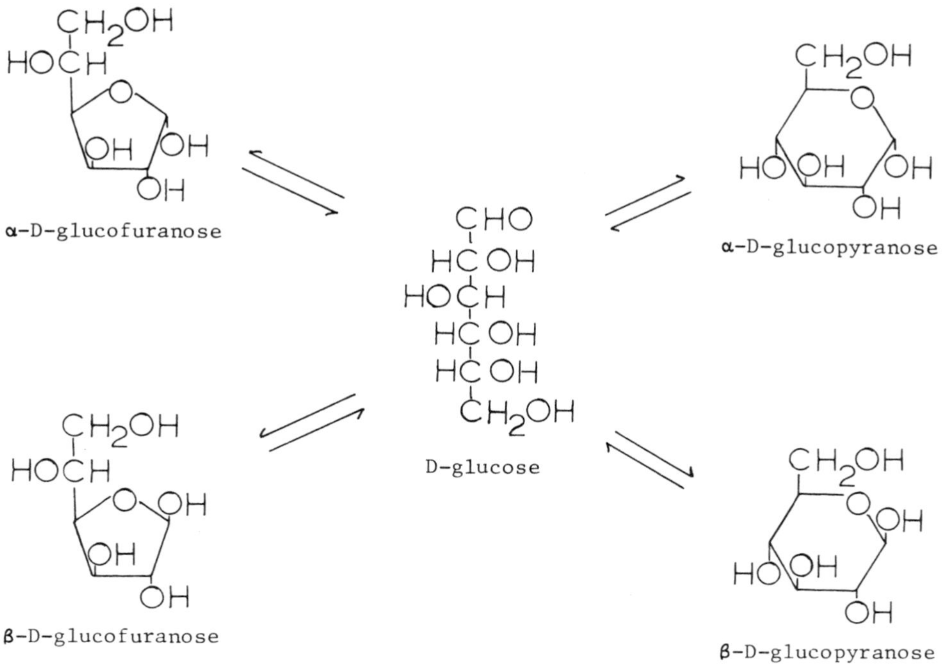

Figure 9   Possible isomers of D-glucose in aqueous solution.

constant value of +53°. α-D-Glucose initially exhibits a value of $[\alpha]_D$ +112°, which on standing decreases to +53°. This equilibrium rotation corresponds to the presence of 36.2% α-D-glucose and 63.8% β-D-glucose. Acid or alkali serves as catalyst and thereby greatly increases the rate of mutarotation.

Alkali-labile glycosides comprise phenolic or flavonoid glycosides, glycosides of enols conjugated with carbonyl groups, and those glycosides in which the aglycon is able to undergo β-elimination. Alkaline reactions of this type may have significant degradative effects, but most glycosides are alkali stable, and fairly drastic conditions, such as 10% sodium hydroxide at 75°C, are needed to cause significant cleavage. Furanosides are more labile than pyranosides.

In the presence of more acid or alkali than is necessary to produce mutarotation of reducing sugars, other phenomena take place. One, *enolization*, is effectively catalyzed by alkali and occurs through the open-chain form to yield an enediol as illustrated for D-glucose in Fig. 10. Loss of asymmetry at C2 in the enediol means that D-glucose can undergo the forward and reverse reactions to yield a mixture of D-glucose and its C2 epimer, D-mannose. If the pair of double-bond electrons in the enediol move down the chain to give the C2 carbonyl, D-fructose is produced from either D-glucose or D-mannose. Similarly, D-fructose may participate in the enolization reaction to yield its C3 epimer, D-mannose. If the pair of double-bond electrons in the enediol moves down the chain to give the C2 carbonyl, D-fructose is produced from either D-glucose or D-manstein transformation. The enediol intermediate has not been isolated. Bases employed synthetically in this transformation include calcium hydroxide, pyridine, sodium aluminate, and, recently, a mixture of boric acid and triethylamine (19). In the pH range 3-4 most reducing sugars are stable.

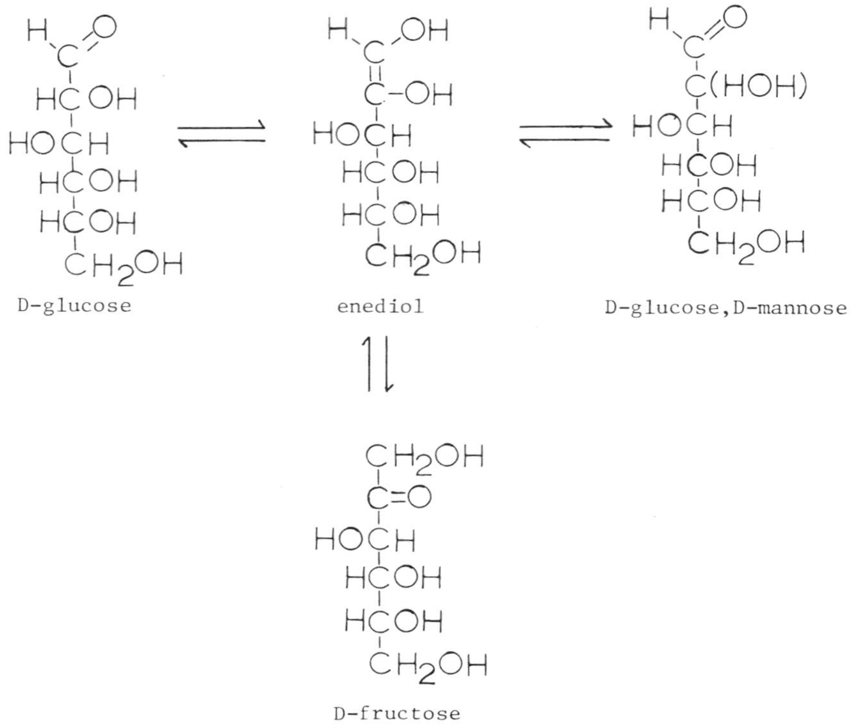

Figure 10   Lobry de Bruyn-Alberda van Ekenstein reaction of D-glucose.

C. Dehydration and Thermal Degradation of Carbohydrates

Dehydration and thermal degradation of sugars are important reactions in foods and are catalyzed by acid or base and many are of the β-elimination type. Pentoses yield 2-furaldehyde as the main dehydration product; hexoses yield 5-hydroxymethyl-2-furaldehyde (HMF) and other products, such as 2-hydroxyacetylfuran and isomaltol. Fragmentation of the carbon chain of these primary dehydration products leads to other chemical species, such as levulinic acid, formic acid, acetol, acetoin, diacetyl, lactic acid, pyruvic acid, and acetic acid. Some of these degradation products are highly odorous and may have desirable or undesirable flavors. Higher temperatures promote such reactions. 2-Furaldehyde and HMF develop in thermally processed fruit juice, and the toxicities of these compounds have been studied in rats. Interestingly, furfural is much more toxic than HMF. A feeding experiment in rats found no adverse effects of HMF in the diet, even at levels as high as 450 mg/kg body weight (25). A key intermediate in intramolecular dehydration reactions is the 3-deoxyosone, the formation of which from D-glucose is shown in Fig. 11.

The β-elimination reaction may continue with the enol form of the 3-deoxyglucosone (Fig. 12). The *cis*-3,4-ene sugar may then undergo ring closure and dehydration to yield HMF (Fig. 13).

The principal of β-elimination may be used to predict the primary dehydration products of most aldoses and ketoses. In the case of ketoses, the 2,3-enediol formed by

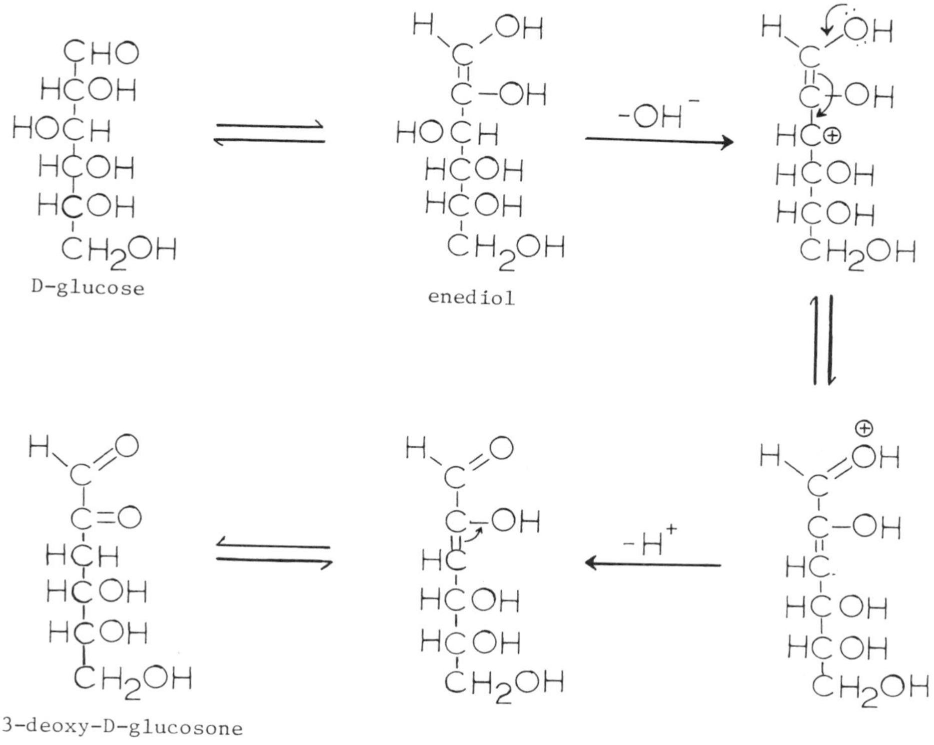

Figure 11 Decomposition of D-glucose leading to 3-deoxy-D-glucosone.

tautomerization of 2-ketoses has two possibilities for β-elimination, one route leading to 2-hydroxylacetylfuran and the other to isomaltol.

Reactions that occur during heating of sugars may be divided into those that occur without carbon-carbon bond cleavage and those that occur with cleavage. In the former category are anomerization on melting, aldose-ketose isomerization, and inter- and intramolecular dehydrations.

Anomerization:

$$\alpha\text{- or }\beta\text{-D-glucose} \xrightarrow{\text{melt}} \alpha/\beta \text{ equilibrium}$$

Aldose-ketose interconversion:

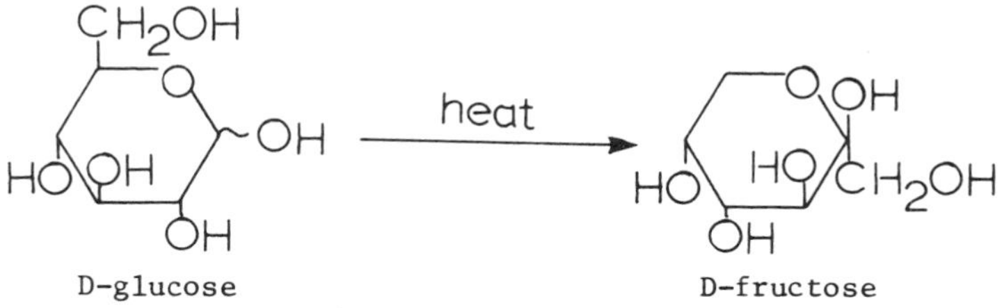

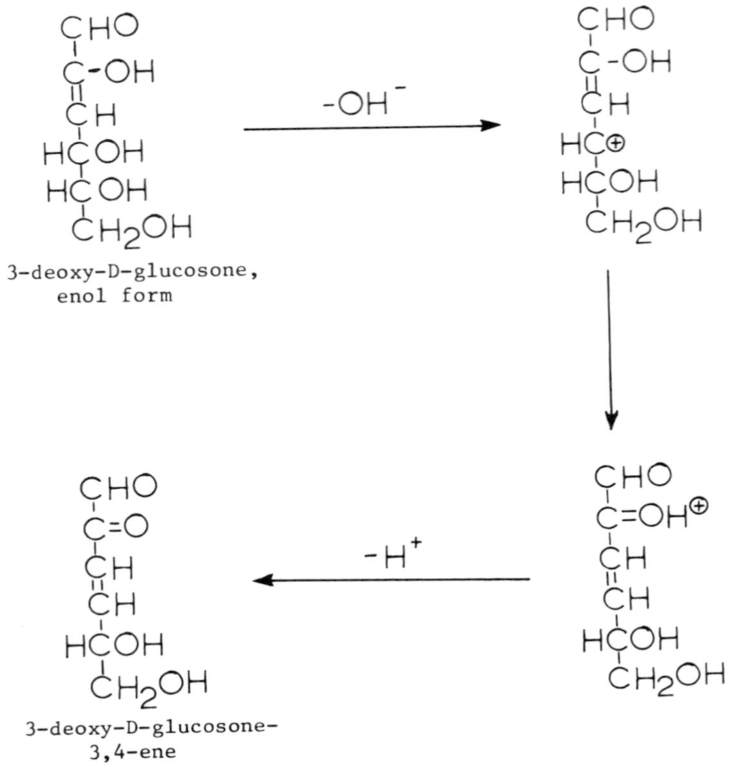

Figure 12 Production of 3-deoxy-D-glucosone-3,4-ene from the enol form of 3-deoxy-D-glucosone.

In more complex carbohydrates, such as starch, transglycosylation may be important as when starch is pyrolyzed at 200°C. At this temperature the number of (1→4)-α-D-linkages decreases with time while concomitantly (1→6)-α-D- and (1→6)-β-D-, and even (1→2)-β-D-linkages, are produced (48).

Anhydro sugar formation may occur to a significant extent in the treatment of some foods and is particularly favored on dry heating, especially of D-glucose or polymers containing D-glucose. Some common products are shown in Fig. 14.

Thermal reactions occurring with carbon-carbon bond cleavage yield as primary products volatile acids, aldehydes, ketones, diketones, furans, alcohols, aromatics, carbon monoxide, and carbon dioxide. Products produced by these reactions can be identified by gas chromatography (GC) or combined gas chromatography-mass spectroscopy (GC-MS).

D. Browning Reactions

Browning of foods is due to oxidative or nonoxidative reactions. Oxidative, or enzymic browning, is a reaction between oxygen and a phenolic substrate catalyzed by polyphenol oxidase. This is the common browning that occurs in cut apples, bananas, pears, and even lettuce, and it does not involve carbohydrates. Nonoxidative or nonenzymic browning is

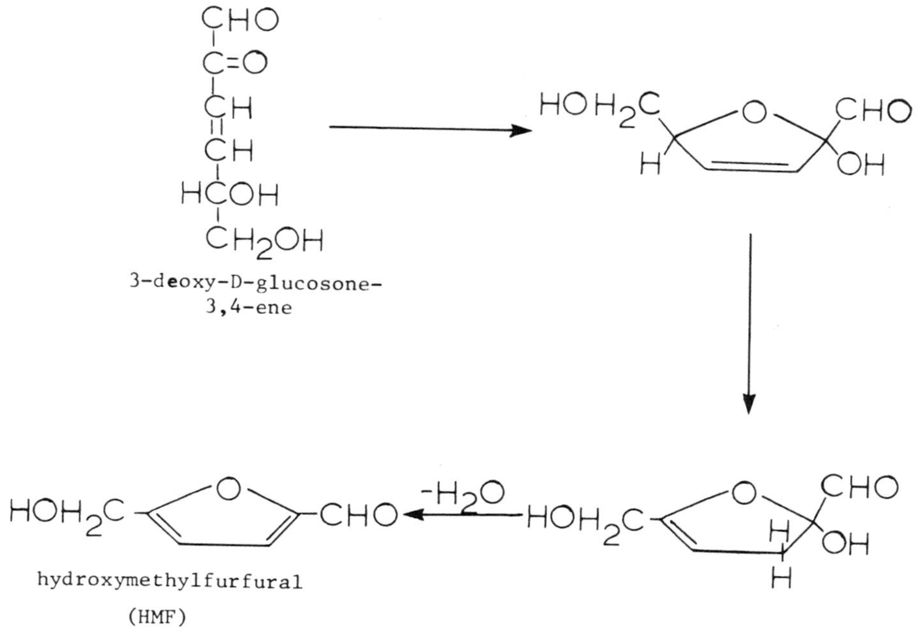

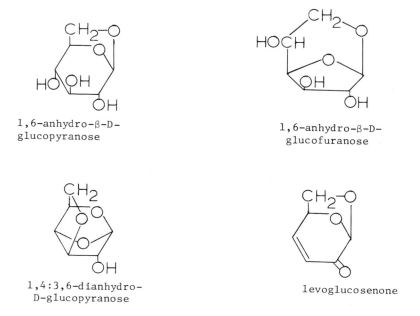

Figure 13 Cyclization and dehydration of 3-deoxy-D-glucosone-3,4-ene to produce hydroxymethylfurfural.

Figure 14 Thermal degradation products of D-glucose or polymers containing D-glucose.

of widespread importance in foods. It involves the phenomena of carmelization and/or the interaction of proteins or amines with carbohydrates. The latter reaction is the *Maillard reaction.*

Direct heating of carbohydrates, particularly sugars and sugar syrups, produces a complex group of reactions termed "caramelization." Reaction is facilitated by small amounts of acids and certain salts. Mild or initial thermolysis causes anomeric shifts, ring size alterations, and breakage of glycosidic bonds, when they exist, and the formation of new glycosidic bonds. Mostly, however, thermolysis causes dehydration with formation of anhydro rings, as in levoglucosan, or introduction of double bonds into sugar rings. The latter produces intermediates to unsaturated rings, such as furans. Conjugated double bonds absorb light and produce color. Often in unsaturated ring systems, condensation will occur to polymerize ring systems, yielding useful colors or flavors. Catalysts speed up the reactions and often are used to direct the reaction to specific types of caramel colors, solubilities, and acidities.

Sucrose is commonly used for making caramel colors and flavors. It is heated in solution with acid or acidic ammonium salts to produce a variety of products used in food, candies, and beverages. Commercially, three types of caramel colors are produced. The most abundant is acid-fast caramel made with ammonium bisulfite catalyst to produce the color for cola drinks. Another is a brewer's color for beer, made by heating a sucrose solution with ammonium ion, and the third is a baker's color produced by direct pyrolysis of sucrose to give a burnt sugar color. Heating D-glucose at about pH 4 produces polymeric or condensed-ring particles of 0.46-4.33 nm diameter.

Caramel pigments contain hydroxyl groups of varying acidity, carbonyl, carboxyl, enolic, and phenolic hydroxyl groups. Increasing temperatures and increasing pH increase the reaction rate, with the rate at pH 8.0 ten times that at pH 5.9. In the absence of buffering salts, large quantities of humic substances are formed. The humin is of high molecular weight (average molecular formula $= C_{125}H_{188}O_{80}$) and has a bitter taste. Thus, its production in flavoring caramelizations is to be avoided.

Certain pyrolytic reactions produce unsaturated ring systems that have unique tastes and fragrances. Thus, maltol, 3-hydroxy-2-methylpyran-4-one, and isomaltol, 3-hydroxy-2-acetylfuran, contribute to the flavor of baked bread. 2-H-4-Hydroxy-5-methyl-furan-3-one has a burnt flavor, as in cooked meat, and can be used to enhance various flavors and sweeteners.

maltol   isomaltol   2-H-4-hydroxy-5-methyl-
                                    furan-3-one

Nonoxidative browning, the Maillard reaction, is not well defined. The minimum reactant requirements for Maillard browning are the presence of an amino-bearing compound, usually a protein, a reducing sugar, and some water. Methods of detection include observation of the formation of a yellow or brown color monitored quantitatively at 420 or 490 nm, observation of products separated by chromatography, observation of carbon dioxide evolution, and analysis of the ultraviolet (UV) and infrared spectra.

The initial Maillard reaction is characterized by colorless solutions exhibiting no UV absorption but possessing increased reducing power. As the reaction progresses, the solution turns yellow and manifests increasing near-UV absorption. There is also some dehydration of sugar to HMF, chain cleavage, formation of α-dicarbonyl compounds, and incipient pigment formation. In many food processes, amino sugars are detected early as reducing power increases. Some decolorization can be effected by addition of a reducing agent, such as sulfite. In the final stages of Maillard browning the product or solution is red-brown to very dark brown in color and added sulfite does not remove the color. At this point a caramel-like aroma is evident and colloidal and insoluble melanoidins are present as a result of complex aldol condensations and polymerization. Additionally, some evolution of carbon dioxide occurs. The reactions involved in the early stages of the Maillard reaction are shown in Fig. 15. The carbonyl carbon of the reducing sugar, in open-chain form, first undergoes nucleophilic attack by the amino nitrogen lone-pair electrons. This is followed by loss of water and ring closure to form a glycosylamine. In the presence of excess reducing sugar, a diglycosylamine may be formed. The glycosylamine undergoes the Amadori rearrangement to produce a 1-amino-2-keto sugar, which has been identified in browned, freeze-dried apricots (Fig. 16).

If the initial sugar reactant is a ketose, a glycosylamine is formed by the same mechanism as for aldoses, but it can then undergo a reverse-Amadori (Heyns) rearrangement to yield a 2-amino aldose (Fig. 17).

The Amadori compound formed may be degraded in at least two distinct ways, one proceeding through a 3-deoxyosone intermediate and the other through a methyl-α-dicarbonyl compound, as shown in Fig. 18. Both paths produce melanoidin pigments, which have pyrazine and imidazole rings in addition to HMF and reductones.

When an amino acid or part of a protein chain reacts in the Maillard reaction it is obvious that the amino acid is lost from a nutritional standpoint. This destructiveness is especially important for essential amino acids, of which lysine with its free ε amino group is most susceptible.

Although loss of lysine is important because of its essentiality, other amino acids are also susceptible to degradation in Maillard browning. These include the other basic amino acids, L-arginine and L-histidine. The basic amino acids are more susceptible to degradation than others because of the presence of a relatively basic nitrogen atom in the side chain. It is useful to note that, if a food has undergone Maillard browning, some loss of amino acids and nutritive value must have occurred. However, the absence of Maillard browning does not ensure that no loss of nutritive value has occurred. This is because amino acid degradation and, hence, loss of nutritive value, occurs well before the development of colored pigments. Moreover, the Strecker degradation contributes to the loss of amino acids without browning.

That relatively mild conditions are required for the loss of nutrients, especially L-lysine, is shown in Table 8 (1). It should be obvious from the preceding that whenever foods containing protein and reducing sugars are heated, even at relatively low temperatures and for short periods of time, we may expect losses of amino acids, especially the basic amino acids. This loss will be most important for L-lysine, which is essential for humans, is generally decreased the most during browning, and may be limiting in some foods before browning, for example, cereals.

The Maillard reaction is not the only pathway for the destruction of essential amino acids in food preparation, processing, or preservation. Another pathway, known as the Strecker degradation, involves the interaction of α-dicarbonyl compounds and

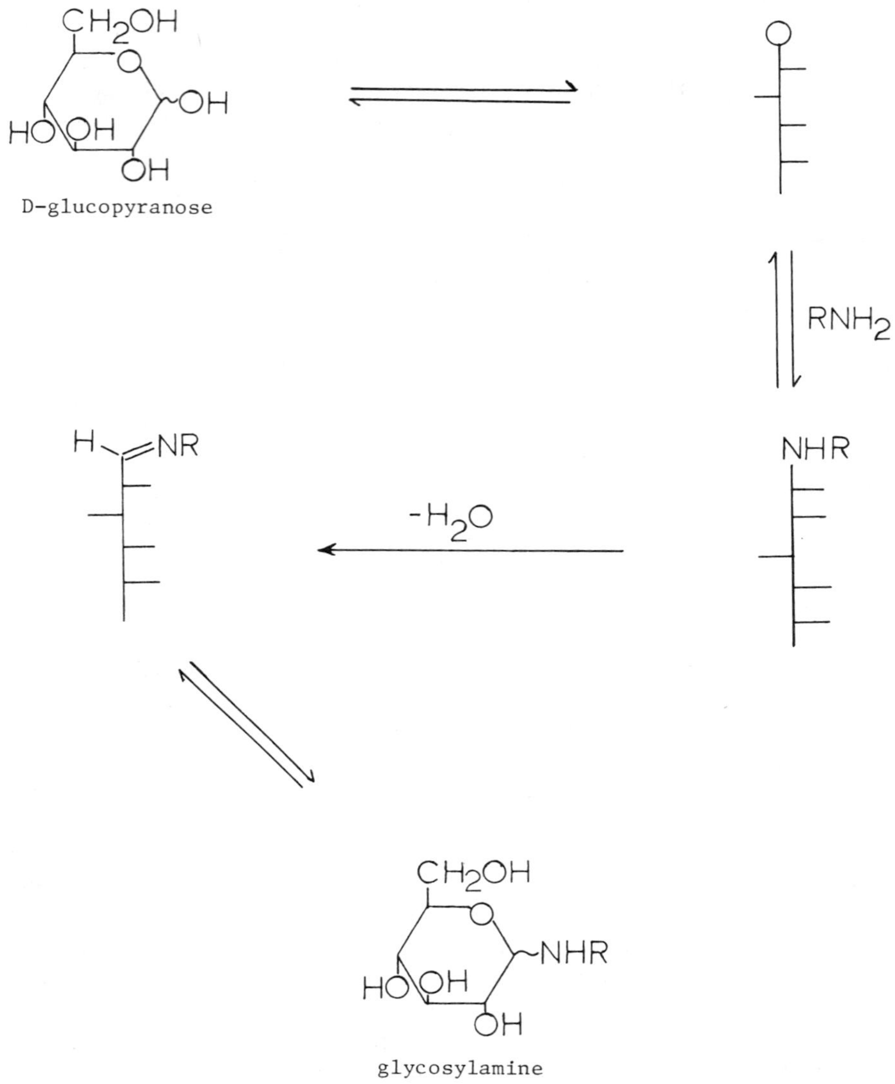

D-glucopyranose

glycosylamine

Figure 15 Formation of a glycosylamine, the initial step in Maillard browning.

α-amino acids (Fig. 19). Volatile products, such as aldehydes, pyrazines, and sugar fragmentation products from the Strecker reaction, may contribute to aromas and flavor. Commercially, the Strecker degradation is used to produce the distinctive flavors of chocolate, honey, maple syrup, and bread. Thus, at times, the Maillard and Strecker reactions are desirable, but in other situations they may be undesirable. In either case, it is desirable for a food scientist to understand the effect of reaction variables on the nature and extent of the reaction. Such variables include temperature, pH, moisture content, presence or absence of metal ions, and effect of sugar structure.

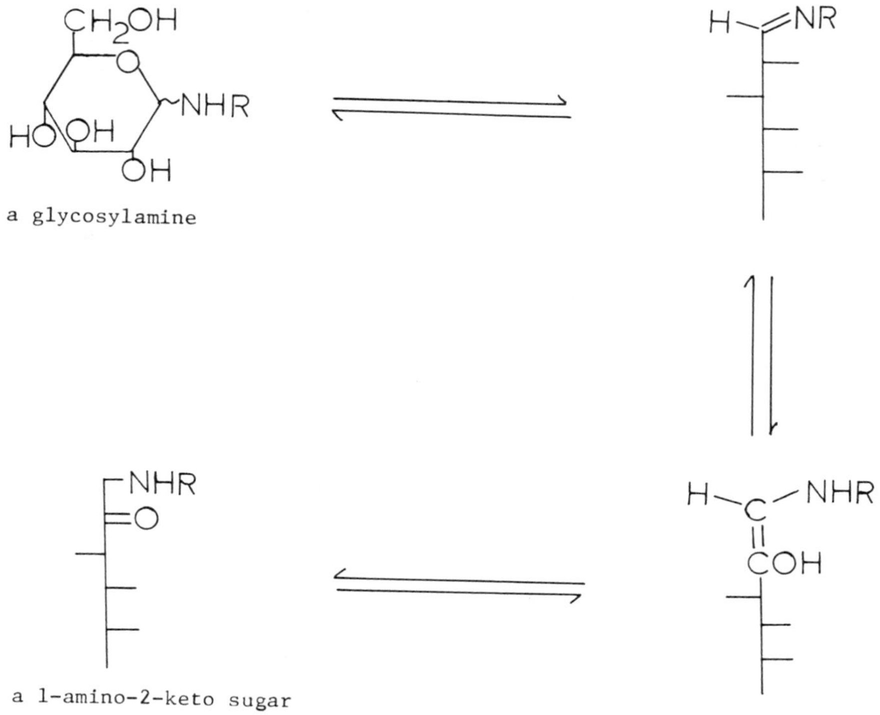

Figure 16   Amadori rearrangement of a glycosylamine.

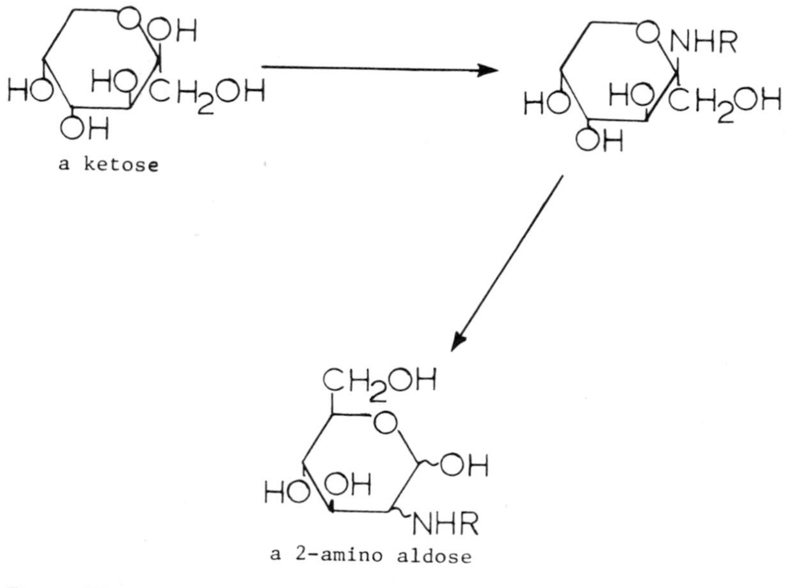

Figure 17   Glycosylamine formation with a ketose and a Heyns rearrangement.

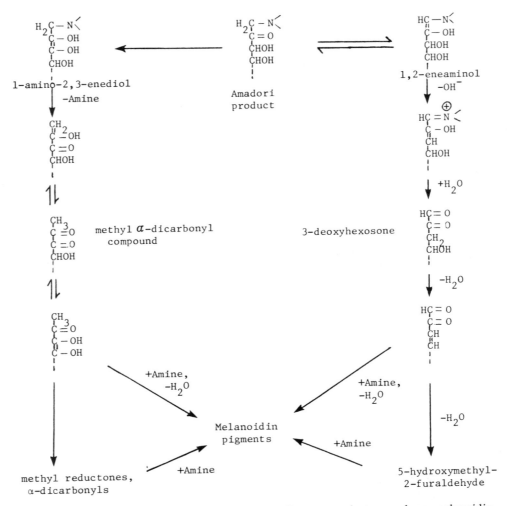

Figure 18 Decomposition pathways of Amadori compounds to produce melanoidin pigments.

Table 8 L-Lysine Degradation in Dairy Products

Product	Temperature (°C)	Time	L-Lysine degradation (%)
Fresh milk	100	Several minutes	5
Condensed milk	—	—	20
Nonfat dry milk	150	Several minutes	40
Nonfat dry milk	150	3 h	80

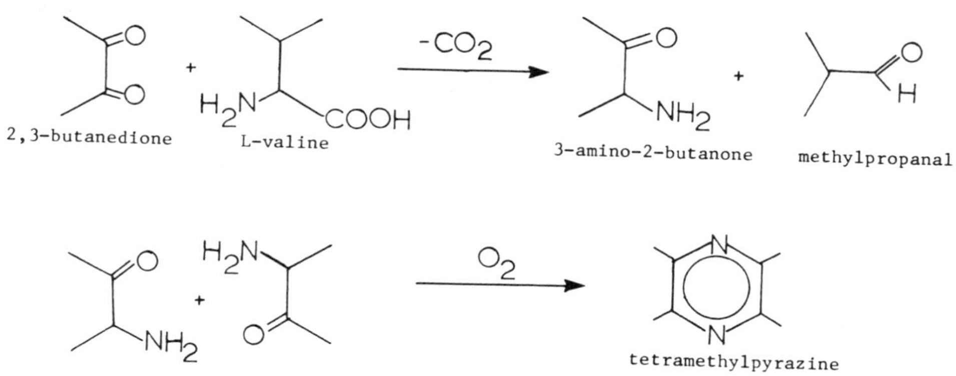

Figure 19 Strecker degradation of L-valine by reaction with 2,3-butanedione.

It is important to control the extent of both the Maillard and Strecker reactions, not only because of adverse contribution to flavor and odor if these reactions go too far, but also because of possible toxicity of the degradation products. There is some concern that the "premelanoidin" products of these reactions may contribute to nitrosamine formation or that they are mutagenic. However, more work needs to be conducted on the toxicity of these products.

The effect of pH on Maillard browning is significant, with little, if any, browning occurring in solutions of pH 6 or less (10). The greatest decrease in amino nitrogen occurs with increasing pH in the range 7.8-9.2. It is obvious from the proposed mechanism that Maillard browning is insignificant in strongly acidic solutions, since under these conditions, the amino group is protonated and consequently glycosylamine formation is prevented.

Early investigation of the effect of moisture content on browning was stimulated by the color changes occurring in dried eggs on storage. Investigators studied pigment production from preformed D-glucosylamines in dry methanol with various amounts of added hydrochloric acid and water. In all experiments, the rate of browning was greatest under anhydrous conditions, for a given concentration of acid. Increased water concentration led to a decreased rate of browning. Browning in systems involving D-xylose and glycine powders showed that no browning occurred in the solid state when the relative humidity (RH) was 0 or 100% but maximum browning occurred at 30% RH. (See also Chap. 2.) Considering the many steps in the browning reaction one might expect the maximum reaction rate to occur, as it does, at intermediate water levels.

Copper and iron enhance browning, with Fe(III) more effective than Fe(II). The sodium ion is without effect (21). Metal ion catalysis of browning suggests that later steps in the Maillard reaction, perhaps those contributing to pigment formation, may be oxidation-reduction reactions.

The effect of sugar structure on extent of browning was noted by Maillard (29). He found that the decrease in the extent to which common sugars brown was in the order D-xylose > L-arabinose > hexoses (D-galactose, D-mannose, D-glucose, and D-fructose) > disaccharides (maltose, lactose, and sucrose). D-Fructose is much less reactive than the aldoses because of the different mechanism followed by ketose sugars.

The degree of pigment formation from a particular sugar is directly proportional to the amount of open-chain (free carbonyl) sugar in the equilibrium solution (10). This strongly suggests that the amine reacts with the open-chain form, as shown in the proposed mechanism.

When Maillard browning is undesirable in food systems, it can be inhibited by decreasing moisture to very low levels or, if the food is liquid, increasing dilution or lowering pH or temperature. Browning can be reduced also by removing one of the substrates, usually the sugar. For dried eggs, this can be accomplished by adding D-glucose oxidase (prior to drying) to degrade D-glucose. Browning in fish can be minimized by adding *Lactobacillus pentoaceticum*, a bacterium with D-ribose oxidase activity. The most widely used chemical to inhibit browning is sulfur dioxide or sulfites, and the reaction sequence is shown in Fig. 20.

Although sulfur dioxide or sulfites inhibit browning, it is not likely that they prevent loss of nutritive value of the amino acid involved in the Maillard reaction. This is because the utilization and subsequent degradation of amino acids, such as L-lysine, occur before the point of action of sulfur dioxide in browning inhibition. Moreover, sulfur dioxide and sulfites have little effect on the Strecker degradation, which is a significant pathway for the loss of essential amino acids and, hence, nutritive value.

The importance of controlling Maillard browning in foods is threefold. First, browning in many foods may be aesthetically desirable or undesirable because of the dark colors and strong odors and flavors that can develop. Second, avoidance of browning is desirable to prevent antinutritional effects, especially the loss of essential amino acids, such as lysine (37). This loss is critical in lysine-limited foods, such as cereal grains. Significant loss of lysine from soy proteins occurs when soy flour or soy isolate is heated with D-glucose. Such losses may also occur during toasting of cereals, in the crust of baked bread, and in baked beans. Third, it has been reported that some of the products of the Maillard reaction are mutagenic. Powrie and coworkers (36) recently demon-

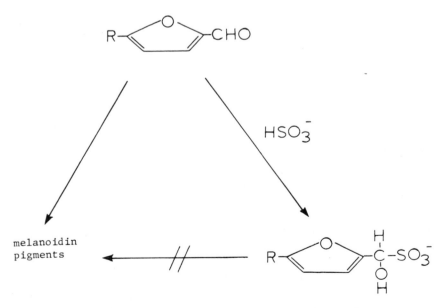

Figure 20 Possible mechanism of bisulfite prevention of browning.

strated the mutagenicity of some products of the reaction of D-glucose or D-fructose with L-lysine or L-glutamic acid. This effect was demonstrated in the *Salmonella* TA 100 strain. However, other workers, using different model systems, have been unable to demonstrate mutagenicity (46).

IV. FUNCTIONS OF MONOSACCHARIDES AND OLIGOSACCHARIDES IN FOODS

A. Hydrophilicity

Attraction of water to carbohydrates is one of their basic and most useful physical properties. Hydrophilicity is expected because of their numerous hydroxyl groups. Hydroxyl groups interact with water molecules by hydrogen bonding, and this leads to solvation and/or solubilization of sugars and many of their polymers. The structure of the carbohydrate can greatly affect the rate of water binding and the amount of water bound, as shown for several sugars in Table 9.

D-Fructose is much more hygroscopic than D-glucose even though they both have the same number of free hydroxyl groups. At 100% relative humidity, sucrose and maltose bind the same amount of water but isomeric lactose binds much less. Interestingly, the two sugar hydrates form stable crystal structures that do not readily absorb water from their surroundings. In fact, well-crystallized sugars are not very deliquescent because most of their hydrogen-bonding sites are already involved in sugar-sugar-hydrogen bonds.

Impure sugars or syrups generally absorb more water and at a faster rate than pure sugars. This is evident even when the "impurity" is the anomeric form of the sugar, and is even more evident when small amounts of oligosaccharides are present, for example, when malto-oligosaccharides are present in commercial corn syrups. The impurities function by interferring with the orderly intermolecular forces, mainly hydrogen bonds, that form between sugar molecules. The sugar hydroxyl groups then become more available to participate in hydrogen bonding to ambient water.

The ability to bind water and control water activity in foods is one of the most important properties of carbohydrates. This water-binding capacity is frequently referred to as *humectancy*. Depending on the particular product, it may be desirable to limit entry of water into the food or to control exit of water. Baked icings provide an example of

Table 9 Water (%) Absorbed by Sugars from Moist Air

Sugar	Water (%) absorption at various RH and times (20°C)		
	60%, 1 hr.	60%, 9 days	100%, 25 days
D-Glucose	0.07	0.07	14.5
D-Fructose	0.28	0.63	73.4
Sucrose	0.04	0.03	18.4
Maltose, anhydrous	0.80	7.0	18.4
Maltose, hydrate	5.05	5.1	—
Lactose, anhydrous	0.54	1.2	1.4
Lactose, hydrate	5.05	5.1	—

Source: From Ref. 6.

the former situation, in that the icing should not become sticky after packaging. Use of sugars with limited water-absorption capacity, such as lactose or maltose, is therefore desirable in situations of this kind. In other situations, control of water activity, especially avoidance of water loss, is important. Examples are confectionaries and baked goods. In such instances it may be necessary to add more hygroscopic sugars, such as corn syrup, high-fructose corn syrup, or invert sugar.

B. Binding of Flavor Ligands

In many foods, especially those subjected to water removal by spray or freeze drying, carbohydrates may be important for retaining colors and volatile flavor components. In these instances there is a shift from sugar-water interaction to sugar-flavorant interaction.

$$\text{SUGAR–WATER} \ + \ \text{FLAVORANT} \ \rightleftharpoons \ \text{SUGAR–FLAVORANT} \ + \ \text{WATER}$$

Volatiles include numerous carbonyl (aldehydes and ketones) and carboxylic acid derivatives (principally esters) and they are more effectively retained in foods by disaccharides than by monosaccharides. Disaccharides and larger oligosaccharides are also effective binders for flavors. Schardinger dextrins, by means of their ability to form inclusion structures, are very effective for trapping flavorants and other small molecules (Fig. 21) (44).

Larger carbohydrate molecules are effective flavor fixatives; one of the most common and most widely used is gum arabic (13). Gum arabic forms a thick film around flavor particles, protecting them from moisture absorption, loss by evaporation, and chemical oxidation. Gum arabic-gelatin mixtures are employed in the technique of microencapsulation, an important advance in food flavor fixation. Gum arabic is also used as a flavor emulsifier in emulsions of lemon, lime, orange, and cola.

C. Carbohydrate Browning Products and Food Flavors

As mentioned, nonoxidative browning reactions, in addition to producing the highly colored melanoidin pigments, yield a variety of volatile flavorants. These volatiles may or may not be desirable but they are usually responsible for the distinctive flavors of heat-processed foods. Browning flavors occur, for example, during roasting of coffee beans and peanuts.

Browning products that contribute to flavors may have specific flavors themselves and/or they may enhance other flavors. Dual-function examples include the carmelization products maltol and ethyl maltol:

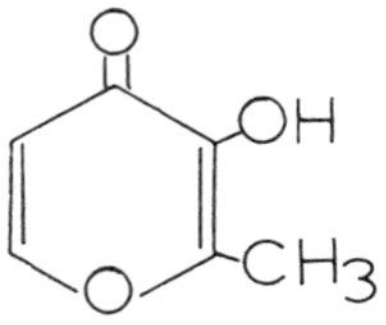

maltol 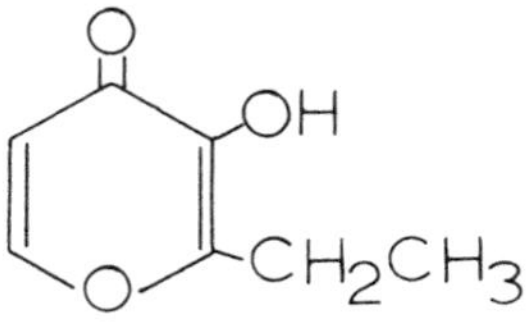   ethyl maltol

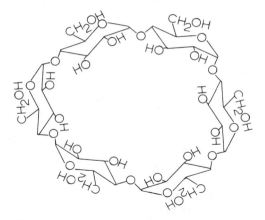

Figure 21   Schardinger dextrin.

These compounds have a distinct, strong caramel odor and are also sweetness enhancers. Maltol lowers the detectable threshold concentration of sucrose sweetness to one-half its normal value (44). Additionally, maltol can affect texture and produce a more "velvety" sensation. Isomaltol

isomaltol

is reportedly six times as effective as maltol as a sweetness enhancer. Thermal decomposition products of sugar are not limited to pyrones and furans but include furanones, lactones, carbonyls, acids, and esters. The sum of the flavor and odor characteristics of these compounds produces the distinctive aromas of some foods.

When amine-sugar reactions are involved in browning, volatile flavorants also may be produced (20). These are principally pyridines, pyrazines, imidazoles, and pyrroles. The flavor character of 1:1 amino acid:D-glucose mixtures heated to 100°C range from a caramel aroma for glycine, to a rye bread aroma for valine, to a chocolate aroma for glutamine. Furthermore, the characteristic aroma of an amine-carbonyl browning reaction changes with temperature. Thus, valine gives a rye bread flavor at 100°C but a penetrating chocolate flavor when heated to 180°C. Proline gives a burnt protein aroma at 100°C but a pleasant, bakery aroma at 180°C. Histidine produces no aroma at 100°C, but aromas developed at 180°C have been variously described as cornbread-like, buttery, or similar to burnt sugar.

The sulfur-containing amino acids, in combination with D-glucose, provide different aromas than do the other amino acids. Methionine, at both 100 and 180°C, reacts with D-glucose to produce a potato aroma, cysteine hydrochloride yields a meaty, sulfurous aroma, and cystine gives an aroma best described as burnt turkey skin.

In browning-derived flavorings, the general volatility and pungency of the products requires that their development in food be restricted to the level necessary for consumer acceptance. Over development of aroma may be highly objectionable.

D. Sweetness

The sweetness of low-molecular-weight carbohydrates is one of their most recognizable and pleasant properties. Sweetness, as a result of sugar concentration, has shaped our diet over the course of history. The pleasantness of honey and most fruits is strongly dependent on their content of sucrose, D-fructose, and/or D-glucose. The perceived sweetness of sugars varies with constitution, configuration, and physical form, as shown in Table 10.

Sugar alcohols have also been used as sweetening agents. Sugar alcohols sometimes possess advantages over the parent sugar in sweetness, caloric reduction, and/or noncariogenicity. The relative sweetness of some sugar alcohols are given in Table 11. For further information on sweetness of carbohydrates, refer to Chap. 9.

V. FUNCTIONS OF POLYSACCHARIDES IN FOODS

A. Structure-Function Relations of Polysaccharides

Of the great number of polysaccharides present in foods, all perform some useful service based on their molecular architecture, their size, and their secondary molecular forces, mainly those of hydrogen bonding. The bulk of dietary polysaccharides are insoluble and undigestible, mainly cellulose and hemicellulose components of cell walls from vegetables, fruits, and seeds. These more or less inert structures give physical compactness, crispness, and good mouth feel to many foods. In addition, as fiber, they are beneficial for health and are an important aid to intestinal motility. The remaining polysaccharides in foods are water soluble or water dispersable and serve numerous diverse

Table 10 Relative Sweetness (RS) of Sugars (w/w%)

Sugar	Solution RS	Crystalline RS
β-D-Fructose	100-175	180
Sucrose[a]	100	100
α-D-Glucose	40-79	74
β-D-Glucose	<α-Anomer	82
α-D-Galactose	27	32
β-D-Galactose	—	21
α-D-Mannose	59	32
β-D-Mannose	Bitter	Bitter
α-D-Lactose	16-38	16
β-D-Lactose	48	32
β-D-Maltose	46-52	—
Raffinose	23	1
Stachyose	—	10

[a]Reference sugar, arbitrarily given a value of 100. [From Shallenberger and Acree (43).]

Table 11   Relative Sweetness of Sugar
Alcohols[a]

Sugar alcohol	Relative sweetness[b]
Xylitol	90
Sorbitol	63
Galactitol	58
Maltitol	68
Lactitol	35

[a]Sugar alcohols presented in tap water at 25°C.
[b]Sucrose = 100.

roles, such as providing hardness, crispness, compactness, thickening quality, viscosity, adhesiveness, gel-forming ability and mouth feel. They permit food to be designed into shapes and structures that may be brittle or soft, swellable or gellable, or completely soluble.

Basically, polysaccharides should be soluble, since after all they are chains of glycosyl units made from hexoses and pentoses. Each glycosyl unit in the chain has several points for hydrogen bonding. Thus, in a glucan there are five oxygens per chain unit that can form hydrogen bonds.

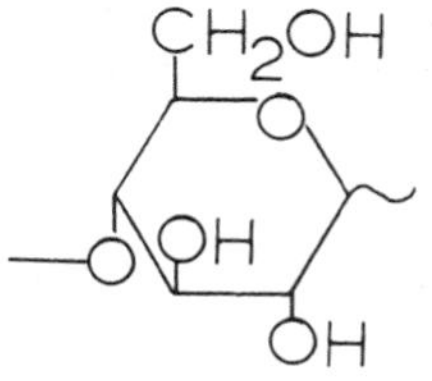

Each hydroxyl hydrogen or oxygen can potentially bond to a water molecule, and thus each chain unit conceivably can be fully solvated and consequently can contribute to the water solubility of the entire molecule. The only time the polysaccharide is not soluble in water is when access to water does not readily occur. Such a decrease in water accessibility occurs maximally in perfectly uniform, linear molecules where chains are fully extended and can fit closely together over much of their length. Such is the case in cellulose, where the β-D-glucopyranosyl units are arranged in optimal order and the β-D-linkages allow for excellent linear extension. Thus, a long section of one molecule can bond to a like section of another molecule, producing a parallel array of cellulose molecules in a crystalline arrangement (Fig. 22). Cellulose-cellulose binding occurs mainly by hydrogen bonds; consequently, these sites cannot participate in water-cellulose hydrogen bonds, and these crystalline regions are therefore insoluble in water and very stable. This property has enabled giant trees to grow and exist for centuries. Nevertheless, only part of each cellulose molecule is involved in crystalline arrangements. Some sections intertwine with sections of other cellulose molecules in a spaghettilike fashion, causing interchain fit to be poor or nonexistent, thus precluding a crystalline arrangement in these areas and allowing these areas to hydrogen bond strongly to water. Consequently, the nonstructured, amorphous regions are highly hydrated.

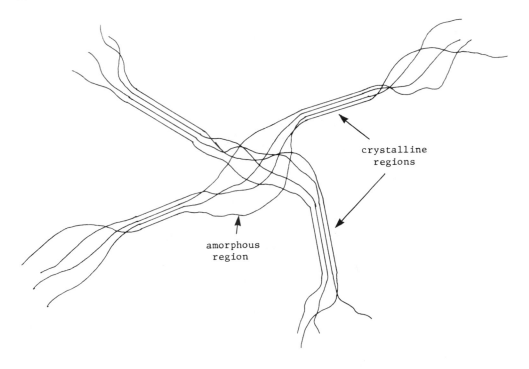

crystalline
regions

amorphous
region

Figure 22 Examples of amorphous and crystalline regions in cellulose.

One can thus state the general principle that a uniform linear molecule, if brought into solution at neutral pH, will, at one or more locations, interact in a crystalline manner with neighboring molecules of like kind, and that water will be excluded from these crystalline junction zones. If the ambient temperature is not sufficiently high to energetically pull apart the combined chain segments the junction zone will remain. It may even grow by extending in length as the thermal motion of the chain causes neighboring chain units to move together in a zipperlike manner. If more molecules contribute chain segments to the junction zone, a particle forms that in time can reach a size at which gravitational effects cause it to precipitate. This insolubilizing effect, initiated when the long and somewhat unwieldly molecules begin to crystallize, is *retrogradation* when it occurs in starch. The process of water exclusion that accompanies retrogration is *syneresis*.

In some instances, the junction zone does not grow in size as just described, but instead it remains limited to segments of only two molecules. When this situation arises, it is likely that another junction zone, involving one of the original two molecules and a new molecule, will form elsewhere. Thus, each polysaccharide molecule generally participates in two or more junction zones, these molecular associations culminating in a three-dimensional network with solvent water molecules dispersed throughout. The effect is to give a unique structure to what was originally a solution—the result being a gel.

The strength of the gel will depend entirely on the strength of the junction zones that hold the entire structure together. If the length of the junction zones are short and the chains are not held together very strongly, the molecules will separate under physical pressure or by a slight increase in temperature that increases the movement of the polymer chains. Such gels are weak and thermally unstable. If, however, long segments of chains are present in the junction zones, the interchain forces will be sufficiently strong to withstand the application of pressure or thermal excitation. Such gels are firm and thermally stable. Thus, many degrees of firmness and stability can be developed by proper control of junction zones, and this is utilized by food scientists.

Branched molecules or heteroglycans do not fit together well and hence do not form junction zones of sufficient size and strength to produce gels. Such molecules simply form viscous, stable solutions. The same holds true for molecules with charged groups, such as polysaccharides that contain carboxyl groups. The negative charges cause coulombic repulsion between approaching chain segments and prevent junction zone formation.

All soluble polysaccharides produce viscous solutions because of their large molecular size. Among the natural food-grade polysaccharides, solutions (equal weight percent per volume) of gum arabic are the least viscous, and solutions of guaran or guar gum are the most viscous. Viscosity depends on molecular size, shape, and charge. If the molecule carries a formal charge resulting from ionization of such groups as carboxyl, the charge effect can be very large at all but very acidic pH values. For carboxyl-containing polysaccharides, charge effects are minimal at pH 2.8 where ionization of the carboxyl is repressed, and the polymer behaves like an uncharged molecule. Viscosity is influenced by the presence of polyelectrolytes since they affect conformation, size of the macromolecule, and the nature of the counterions, which act as a brake on polymer flow. Further, because of coulombic repulsion, a charged polymer approaching or overtaking another similar molecule in a flow pattern alters its path to bypass the overtaken molecule, thus increasing resistance to flow. Probably of greatest importance with respect to viscosity is that charged polysaccharides tend to exist in a fully expanded state if branched and in a fully extended state if linear, thus occupying maximum space in the solution and offering maximum resistance to flow.

For steric reasons, all linear molecules, whether they bear a charge or not, require more space for gyration than highly branched or bush-shaped molecules of the same molecular weight. Thus, as a general rule, solutions of linear polysaccharides will be more viscous than solutions of branched polysaccharides. Since a primary reason for using polysaccharides in foods is to provide viscosity, structure, or gelling, linear polysaccharides are generally most useful.

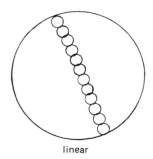

linear

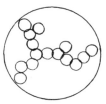

branched

Polysaccharide gels can also be produced by cross-links that do not involve normal junction zones. One such method involves synergistic interaction of different polysaccharides to produce a viscosity increase and eventually gel formation. This can occur, for example, between locust bean gum and carrageenan, and is discussed later. A second cross-linking procedure is to chemically connect polysaccharide molecules by a difunctional bridging reagent, such as a difunctional acid or epoxide. Cross-links of this kind are commonly produced from phosphorylchloride or maleic or succinic anhydrides. Less commonly, epichlorohydrin can be used to form an ether linkage between two polysaccharides. Chemical cross-links are most often used in starches, and are discussed later.

Anything that causes a dissolved linear molecule to become more extended causes a viscosity increase, and conversely any action that causes a linear molecule to become less linear, that is, more compact or coiled, lowers solution viscosity. In foods, non-gel-forming components can influence viscosity by their influence on polysaccharides. Thus, sugars compete for available water molecules, leaving less water for polysaccharides. This causes polysaccharide molecules to coil, that is, to form more polysaccharide-polysaccharide hydrogen bonds in place of water-polysaccharide hydrogen bonds. This may lead to gel formation or gel strengthening, as with pectin gels. Salts compete for water in the same way but they also act as counterions for polysaccharides, thereby lowering repulsive effects, producing molecular coiling, desolvation, and perhaps even precipitation of polysaccharide molecules.

B. Starch

1. Characteristics of Starch Granules

The major reserve of most plants is starch, most abundant in seeds, roots, and tubers. Of all polysaccharides, starch is the only one universally produced in small individual packets called *granules*. Since these are biosynthesized in plant cells, they assume a size and shape prescribed by the biosynthetic system of the host plant and by the physical constraints imposed by the tissue environment. For example, the most abundant commercial starch, that from corn grain, has granules that are somewhat rounded when present in the central floury endosperm and are very angular in shape when present in the outer, highly proteinacious, horny endosperm. Thus, the size and shape of starch granules vary from plant to plant, each different and recognizable as to plant source when examined under the microscope. In addition to corn, other commercial starches are wheat, rice, tapioca (cassava), potato, and sago, with corn (the least expensive), wheat, tapioca and potato starch used in the United States. Potato has the largest granules and rice the smallest of the commercial starches, although canna starch granules are much larger than those of potato, and dasheen starch granules, 3-7 μm in diameter, are the smallest.

All granules show a cleft, called the hilum, which is the nucleation point around which the granule developed. In polarized light between crossed Nicol lenses, granules are birefringent, showing a black Maltese cross with its center at the hilum. This is indicative of a spherocrystalline arrangement, whereby most of the starch molecules radiate out from the hilum toward the periphery. Even the main chains and many of the branches of amylopectin are radially arranged. In addition, most starch granules show more or less distinct lamina or growth rings around the central hilum. It is not known how the amylose and amylopectin molecules are arranged relative to each other, but the two types of molecules appear to be distributed rather uniformily throughout the granules.

Normal starches contain roughly 25% amylose (Fig. 23). Mutant varieties of corn, called high-amylose corn, produce starch with amylose contents ranging up to 85%, although commercial varieties usually have a maximum of 65% amylose. As expected, these starches are difficult to gelatinize, some requiring temperatures in excess of 100°C. At the other extreme, some starches consist only of amylopectin. These are waxy corn, waxy barley, and waxy rice, or glutinous rice. They usually produce clear pastes, somewhat resembling pastes of root or tuber starches. With no amylose to retrograde, these starch pastes are fairly stable.

Amylose molecules tend to form helix structures that entrap other molecules, such as fatty acids or properly sized hydrocarbons. These nonstoichiometric complexes are referred to as inclusion compounds. In fact, amylose can be cleanly separated from a well-disintegrated starch paste by addition of *n*-butanol and allowing the mixture to cool slowly from near boiling to ambient temperature. During cooling, inclusion crystals of butanol-amylose separate and can be removed by filtration or centrifugation (12).

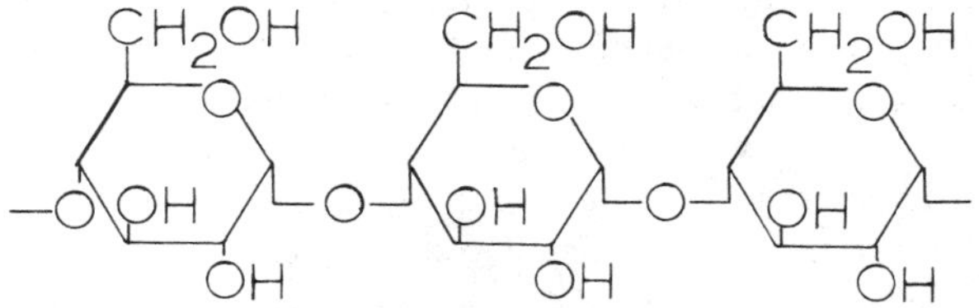

amylose

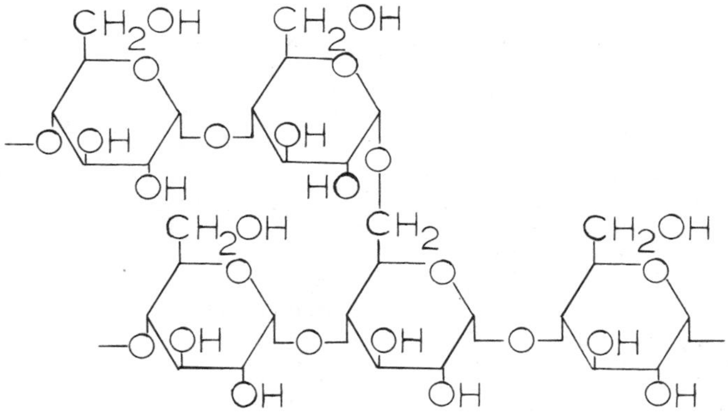

amylopectin

Figure 23 Representative partial structures of amylose and amylopectin.

There is some evidence that amylose exists as a double helix in solution and may even occur in this state in starch granules. Separated amylose can be formed into strong, transparent films and fibers, with appearance and properties similar to items from cellulose, except for the greater swellability of amylose and its solubility under alkaline conditions. Amylose molecules have degrees of polymerization of 350-1000 compared with amylopectin with a DP of several thousand or more.

2. Gelatinization and Other Properties of Starch

Undamaged starch granules are not soluble in cold water, but can reversibly imbibe water and swell slightly. The percentage increase in granule diameter ranges from 9.1% for normal corn starch to 22.7% for waxy corn. However, as the temperature is increased, the starch molecules vibrate more vigorously, breaking intermolecular bonds and allowing their hydrogen-bonding sites to engage more water molecules. This penetration of water, and the increased separation of more and longer segments of starch chains, increases randomness in the general structure and decreases the number and size of crystalline regions. Continued heating in the presence of abundant water results in a complete loss of crystallinity as judged by loss of birefringence and the nature of the x-ray diffraction pattern. The point at which birefringence first disappears is regarded as the *gelatinization point* or gelatinization temperature. It usually occurs over a narrow temperature range, with larger granules gelatinizing first and smaller granules later, although this is not a universal pattern (Table 12). During gelatinization, granules swell extensively. Thus, a 1% slurry of starch granules in cold water has a low viscosity, but on heating, a thick paste is produced in which almost all the water has entered the granules. This causes them to swell and push tightly against each other in a honeycomblike manner. The viscosity of the paste results from the flow resistance of the enlarged granules that now occupy the entire sample volume. Highly swollen granules can be easily broken and disintegrated by mild stirring, and this results in a large decrease in paste viscosity.

In the native state, starch granules do not have membranes; their surfaces consist simply of tightly packed chain ends resembling the bottom of a broom with the straws pressed tightly together. During the early stages of gelatinization, pressure is created within the granules as water enters among the starch molecules. This pressure causes some of the starch molecules near the granule periphery to be drawn tangentially along the surface. Some of these molecules align themselves at the surface to create a membrane.

Table 12 Starch Granule Characteristics

Source	Diameter		Gelatinization temperature (°C)
	Range (μm)	Mean (μm)	
Corn	21-96	15	61-72
White potato	15-100	33	62-68
Sweet potato	15-55	25-50	82-83
Tapioca	6-36	20	59-70
Wheat	2-38	20-22	53-64
Rice	3-9	5	65-73

Amylose molecules, which are linear and less bulky than branched amylopectin molecules, can, during the early stage of the gelatinization process, diffuse to and through the surface membrane and thus emerge in the extragranular solution. If a long diffusion time is allowed, left behind will be wrinkled, collapsed granule membranes appearing like empty or deflated balloons. In fact, one way to separate amylose from starch granules is to allow a dilute suspension of swollen starch granules to stand for an extended period at 80-90°C and then centrifuge the suspension, leaving most of the amylose in the supernatant fluid. Unfortunately, the separation is not complete and some amylopectin molecules also appear in the supernatant fluid.

This partial fractionation of starch types can also occur during food preparation and affects the character of the starch paste and the resulting food. Because of the segregation of amylose molecules, retrogradation is more severe when the gelatinized mixture is stored at a temperature less than about 65°C. Nevertheless, it is the ability of starch to form thick pastes that makes it a valuable ingredient in many foods. The properties of amylose and amylopectin are summarized in Table 13.

Starch gelatinization, the viscosity of starch solutions, and the characteristics of starch gels depend not only on temperature, but also on the kinds and amounts of other constitutents present. In many situations, starch exists in the presence of such substances as sugars, proteins, fats, food acids, and water.

In foods, water is not just a medium for reaction, but is also an active ingredient used to control reactions, texture, and general physical and biological behavior. It is not the total amount of water that is important, but rather the availability of water or water activity. Water activity is influenced by salts, sugars, and other strong water-binding agents (see Chap. 2). Thus, if these types of constituents are present in large amounts, water activity will be low and gelatinization will not occur or will occur to only a limited extent. In essence, strong water-binding constituents retard starch gelatinization by binding water in competition with starch.

High sugar concentrations decrease the rate of starch gelatinization, the peak viscosity, and gel strength (Fig. 24). Disaccharides are more effective in delaying gelatinization and in reducing peak viscosity than are monosaccharides (34). Sugars decrease gel strength by exerting a plasticizing action and interfering with the formation of junction zones.

Lipids, such as triacylglycerols (fats and oils), and lipid-associated materials, such as mono- and diacylglycerol emulsifiers, also occur in foods and affect starch gelatinization.

Table 13   Properties of Amylose and Amylopectin

Property	Amylose	Amylopectin
Molecular weight	50,000-200,000	1 to several million
Glycosidic linkages	Mainly α-D-(1→4)	α-D-(1→4), α-D-(1→6)
Susceptibility to retrogradation	High	Low
Products of action of β-amylose	Maltose	Maltose, β-limit dextrin
Products of action of glucoamylase	D-Glucose	D-Glucose
Molecular shape	Essentially linear	Bush-shaped

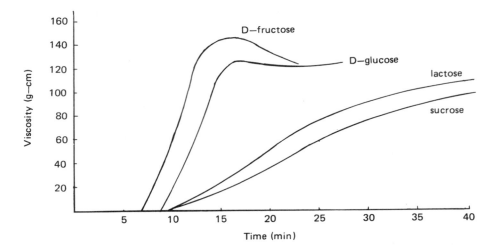

Figure 24 Effect of various sugars on viscosity development during starch gelatinization (5% cornstarch; 100°C).

Fats that can complex with amylose retard swelling of granules. Thus, in white bread, which is low in fat, 96% of the starch is usually fully gelatinized as evidenced by microscopic examination and by determination of the amount of starch quickly attacked by glucoamylase. Unswollen starch granules are only slowly attacked by amylases. Pie crust and cookies, both high in fat and low in water, have large portions of ungelatinized starch. In systems where starch gelatinization does occur, added fat, in the absence of emulsifiers, will not influence the maximum viscosity obtained but will lower the temperature at which the maximum viscosity occurs. For example, during gelatinization of a corn starch-water suspension, maximum viscosity is attained at 92°C, whereas in the presence of 9-12% fat, maximum viscosity is attained at 82°C.

The addition of monoacylglycerols, with a fatty acid component of 16-18 carbon atoms, causes an increase in gelatinization temperature, an increase in the temperature at which maximum viscosity is obtained, a decrease in the temperature of gel formation, and a decrease in gel strength. Fatty acids or the fatty acid component of monoacylglycerols can form inclusion complexes with helical amylose, and possibly with the longer outer chains of amylopectin as shown below.

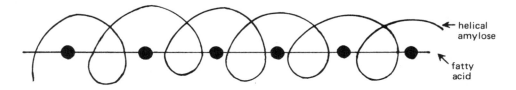

Such complexes are less easily leached from the granule and they resist entry of water into the granule. The lipid-amylose complex also interferes with the formation of junction zones.

Due to the neutral character of starch, low concentrations of salts have little effect on gelatinization or gel formation. Exceptions to this occur with potato starch amylopectin, which contains some phosphate groups, and with manufactured ionic starches. With these salt-sensitive starches, salts may either increase or decrease swelling, depending on the conditions (34), and it is important to consider such charge effects when establishing processing times, temperatures, and procedures for starch-thickened foods.

Acids are prevalent in many starch-thickened foods. Fortunately, most foods have pH values in the range 4-7, and this acid concentration has little effect on starch swelling or gelatinization. The rate of starch swelling is greatly increased at pH 10.0, but this is outside the range of foods. At low pH, such as in salad dressings and fruit pie fillings, there occurs a marked reduction in peak viscosity of starch pastes, along with a rapid loss of viscosity on cooking (Fig. 25). At low pH, extensive hydrolysis of the starch occurs, yielding nonthickening dextrins. To avoid acid thinning in acidic, starch-thickened foods, a cross-linked starch is normally used. Since these molecules are so enormous, extensive hydrolysis is required before the viscosity decreases significantly.

Starch-protein interactions are important in many foods, notably batters and doughs, in which wheat starch and gluten interact to give structure. It is the combination of gluten formation due to mixing, and starch gelatinization and protein denaturation due to heating in the presence of water, that give baked goods their structure. However, the exact nature of the interaction between starch and protein in food systems remains unclear due to the inherent difficulties of studying the interaction of two unlike macromolecules. The development of experimental tools that will allow such investigations in both model systems and real foods would be very desirable.

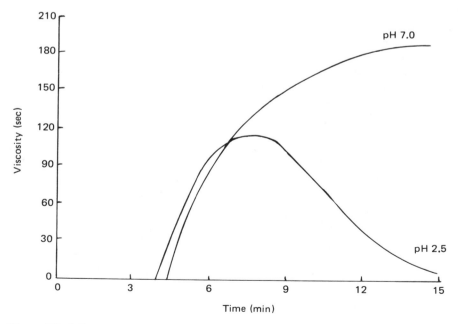

Figure 25 Effect of pH on the viscosity of a heated starch slurry (5% cornstarch; 90°C).

Starch-thickened foods and starch gravies and pastes have poor freeze-thaw stability due mainly to retrogradation of amylose. Cherry pie filling, thickened with normal starch and exposed to a freeze-thaw cycle, acquires a fibrous or grainy texture. Waxy starches have much better performance characteristics in frozen foods than do starches containing large amounts of amylose. Chemical modification to produce phosphate cross-links also improves the ability of starch to function well in frozen products.

Staling in starchy foods, at least in the early stages, results from the association of amylose molecules. This association can be hindered by complex formation between amylose and fatty materials, including natural lipids. Staling that occurs over a longer period may involve the association of the longer branches of amylopectin molecules. Staling effects can be somewhat reversed in bread by heating. The heat energy, plus lubrication by moisture, allows thermal movement of starch molecules, partially restoring a more amorphous structure that is less tightly bound and hence softer in texture.

Pregelatinized starch is a common ingredient in food formulation. It is normally prepared commercially by placing a starch slurry in a trough between two counterrotating horizontal rolls heated internally by steam. The starch slurry is heated above the gelatinization temperature and is then carried downward between the nip of the hot rolls to form a thin surface film. This film dries as the rolls turn and is scraped off and ground to a fine powder. Pregelatinized starch also can be prepared by spray drying a gelatinized paste. Pregelatinized starch quickly rehydrates in water. In food products it can be uniformly incorporated without heating to provide thickening, binding, and other useful properties. Pregelatinized starch is a useful component in foods in which cooking is not normally employed, such as in instant puddings, pie fillings, and cake frostings.

3. Chemically Modified Starches

Acid-modified starches have been reviewed by Shildneck and Smith (45). The technique of acid treating starch without gelatinization of the granule was explored late in the nineteenth century by Lintner (28), and patents on acid modifications were obtained later by Duryea (9). Acid hydrolysis below the gelatinization temperature takes place in amorphous regions of the granules leaving the crystalline regions relatively intact (33). In corn starch, acid hydrolyzes amylopectin more extensively than amylose. Some properties of acid-treated starches, relative to the parent starches, include decreased hot-paste viscosity (11), decreased intrinsic viscosity (26), decreased gel strength (3), and increased gelatinization temperature (27).

Commercial acid-modified starch is produced by reacting a 40% slurry of corn or waxy maize starch with hydrochloric or sulfuric acid at 25-55°C. Time of treatment is determined by the viscosity decrease and may vary from 6 to 24 hr. The mixture is neutralized with soda ash or dilute sodium hydroxide, and the modified starch is then filtered and dried. These starches are used in the manufacture of gum candies and confections because of their ability to form hot, concentrated pastes that gel firmly on cooling.

Starches undergo many of the reactions of alcohols, such as esterification and etherification. Since the D-glucopyranosyl monomers contain three free hydroxyl groups, the degree of substitution (DS) may vary from 0 to a maximum of 3. Commercially, the more important starch derivatives are very lightly derivatized (DS less than 0.1). Such modifications may produce distinct changes in colloidal properties and generally produce polymers with properties useful in a variety of applications.

Hydroxyethylstarch ethers of DS 0.05-0.10 have been produced by reacting starch with ethylene oxide at 50°C. Wet-milled but undried starch is used for hydroxyethylation. The derivative is easily filterable and can be produced economically in a fairly pure form.

The introduction of hydroxyethyl groups at low DS results in extensive modification of physical properties. Among them are reduced gelatinization temperature (41), increased rate of granule swelling (17), and lowered tendency of starch pastes to gel and retrograde. Other hydroxylalkylstarches, such as hydroxypropylstarches, find use as food additives in salad dressings, pie fillings, and other food-thickening applications.

Starch phosphate monoesters may be prepared by reacting a dry mixture of starch and acid salts of ortho-, pyro-, or tripolyphosphate at elevated temperature. Typical reaction conditions involve heating for 1 h at 50-60°C. The degree of substitution obtained is generally less than 0.25, but higher DS derivatives can be prepared by elevating the temperature and concentration of phosphate salt and by extending the reaction time.

Compared with their parent starches, the monophosphate esters have a lower gelatinization temperature and swell in cold water at DS 0.07 and higher. In common with other derivatives, starch phosphates exhibit increased paste viscosity and clarity and decreased setback or retrogradation. Their characteristics are very similar to those of native potato starch, which also contains phosphate groups.

Starch monophosphates are useful in frozen foods because of their excellent freeze-thaw stability. They are often employed as thickeners in frozen gravy and frozen cream pie preparations, in which they are superior to other unmodified starches. A pre-gelatinized starch phosphate has been developed (23) that is dispersable in cold water and is useful in instant dessert powders and icings.

In contrast to the starch monophosphate esters, starch phosphate diesters have the phosphate esterified with two hydroxyl groups, very often with one hydroxyl on each of two neighboring starch chains. Thus, a chemical bridge is formed between adjacent chains and these starches are referred to as "cross-linked" starch. This covalent linkage of one starch chain with another prevents the starch granule from swelling normally and gives it greater stability to heat, agitation, damage from hydrolysis, and a reduced tendency to rupture.

Cross-linking may be produced by reacting starch in aqueous suspension with phosphorylchloride, by reacting dry starch with trimetaphosphate (22), or by reacting a starch slurry with 2% trimetaphosphate (pH 10-11) for 1 h at 50°C.

The most notable change effected by phosphate cross-linking is an increase in stability of the swollen granule. Depending on the degree of derivatization, the hot-paste viscosity may be greater or less than that of the parent starch. In contrast to starch phosphate monoesters, pastes of the diesters are not clear. More highly cross-linked starches are very stable at high temperatures, low pH, and conditions of mechanical agitation. If the starch is sufficiently cross-linked, swelling can be inhibited, even in boiling water.

Cross-linked starch is used mainly in baby foods, salad dressings, fruit pie filling, and cream-style corn, in which it functions principally as a thickener and stabilizer. Starch phosphate diesters are superior to unmodified starches for these applications because of their ability to keep food in suspension after cooking, because they provide resistance to gelling and retrogradation, show good freeze-thaw stability, and do not synerese on standing.

Starch acetates of a low degree of substitution produce stable solutions since the presence of only a few acetyl groups inhibits the association of both amylose molecules and the long outer chains of amylopectin. Low-DS starch acetates are produced by treatment of granular starch with acetic acid or, preferably, acetic anhydride, either alone or in the presence of a catalyst, such as acetic acid or an aqueous alkaline solution. A common commercial product is made by exposing starch to acetic anhydride at pH 7-11 and 25°C to give DS 0.5.

Low-DS starch acetates have low gelatinization temperatures and good resistance to retrogradation after pasting and cooling. They are used in foods because of the clarity and stability of their pastes. Applications include frozen fruit pies and gravies, baked goods, instant puddings, pie fillings, and gravies. Higher DS products exhibit a decreased ability to form gels.

In Table 14, the properties of various starches are summarized.

C. Glycogen

Glycogen occurs in foods in minor amounts because of its low concentration in muscle and liver. Glycogen is a homoglucan, similar in structure to starch amylopectin, containing α-D-(1→4) and α-D-(1→6) glycosidic linkages. However, glycogen differs from amylo-

Table 14 Properties of Various Corn Starches

Type	Amylose/amylopectin	Gelatinization temperature range (°C)	Distinguishing properties
Normal	1:3	62-72	Poor freeze-thaw stability
Waxy	0:1	63-72	Minimal retrogradation
High amylose	3:2-4:1	66-92	Granules less birefringent than normal starches
Acid-modified	Variable	69-79	Decreased hot-paste viscosity compared with unmodified
Hydroxyethylated	Variable	58-68 (DS 0.04)	Increased paste clarity, reduced retrogradation
Phosphate, monoesters	Variable	56-66	Reduced gelatinization temperature, reduced retrogradation
Cross-linked	Variable	Higher than unmodified depends on degree of cross-linking	Reduced peak viscosity, increased paste stability
Acetylated	Variable	55-65	Good paste clarity and stability

pectin in having a much greater molecular weight and a greater degree of branching. A small amount of "phytoglycogen," a low-molecular-weight, highly branched material can be isolated from corn starch and perhaps other starches.

Glycogen is the principal storage carbohydrate in muscle and liver tissue, but after death of the animal much of it is degraded to D-glucose and then to lactate (see Chap. 12).

D. Cellulose

Cellulose occurs in all plants as the principal structural component of cell walls. It is usually associated with various hemicelluloses and lignin, and the type and extent of these associations contributes greatly to the characteristic texture of plant foods. Most of the textural changes during the maturation and ripening of plants are, however, due to changes in pectic substances.

Cellulose is a homoglucan composed of linear chains of $(1\rightarrow)$-β-D-glucopyranosyl units (16). The essential linearity of cellulose makes it easy for molecules to strongly associate in a side-by-side manner, as occurs extensively in native plant cellulose, especially in trees and the woody parts of other plants. Cellulose has both amorphous and crystalline regions, and it is the amorphous regions that are attacked first by solvents and chemical reagents. This differential reaction is used in making microcrystalline cellulose in which the amorphous regions are hydrolyzed by acid, leaving only the tiny, acid-resistant crystalline regions. The product, known commercially as Avicel, is used as a non-metabolizable, bulking and rheological control agent in low-calorie foods.

More drastic chemical modifications of cellulose are used in the preparation of cellulose-based food gums. The most widely used cellulose derivative is the sodium salt of carboxymethylcellulose (CMC). It is made by treating cellulose with sodium hydroxide-chloroacetic acid (2). The anionic cellulose reacts nucleophilicly, displacing the chlorine atom to produce an ether linkage (Fig. 26).

Carboxymethylcellulose of DS 0.7-1.0 is mainly used to increase the viscosity of foods. It dissolves in water to form a non-Newtonian solution, the viscosity of which decreases with increasing temperature. Solutions are stable at pH 5-10, with maximum stability at pH 7-9. Carboxymethylcellulose forms soluble salts with monovalent cations, but solubility decreases in the presence of divalent cations and hazy dispersions result. Trivalent cations can induce gelation or precipitation.

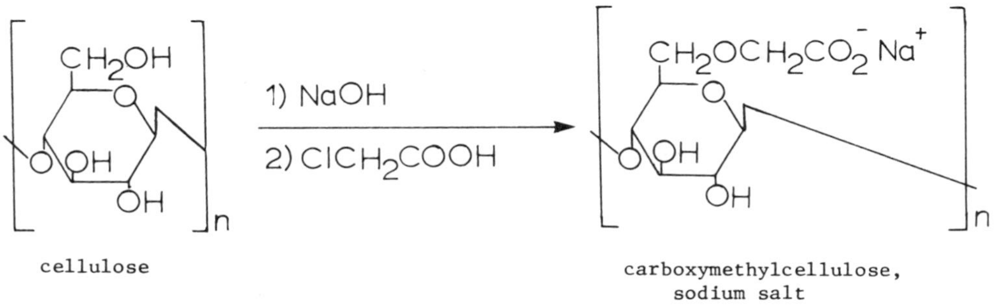

Figure 26 Preparation of carboxymethylcellulose, sodium salt.

Carboxymethylcellulose helps solubilize common food proteins, such as gelatin, casein, and soy proteins. It solubilizes by forming a CMC-protein complex observable by an increase in viscosity.

Due to its desirable rheological properties and lack of toxicity and digestibility, CMC has found broad use in foods. It acts as a binder and thickener in pie fillings, puddings, custards, and cheese spreads, and its water-binding capacity makes it useful in ice cream and other frozen desserts in which it retards ice crystal growth. It retards the growth of sugar crystals in confectionaries, glazes, and syrups and has beneficial effects on volume and shelf life of cakes and other baked goods. It helps stabilize emulsions in salad dressings and is used in dietetic foods to provide the bulk, body, and mouth feel that would normally be contributed by sucrose. In low-calorie carbonated beverages, CMC helps retain carbon dioxide.

Another cellulose ether useful in foods is methylcellulose (Fig. 27) (15). It is prepared by a method similar to that for CMC, in which cellulose and sodium hydroxide are reacted with methylchloride. Maximum water solubility occurs at DS 1.64-1.92, with viscosity principally dependent upon molecular chain length.

Methylcellulose, unlike other gums, exhibits thermogelation. That is, a solution, when heated, forms a gel that reverts to a normal solution on cooling. When an aqueous solution of methylcellulose is heated, the viscosity initially decreases, but soon thereafter the viscosity increases rapidly and a gel develops. This phenomenon is due to an increase in interpolymer hydrophobic bonding caused by thermal disruption of the hydration shell around the individual molecules. Electrolytes, such as sodium chloride, and nonelectrolytes, such as sucrose or sorbitol, depress the gelation point, perhaps because of their competition for the water. Methylcellulose is not digestible and thus makes no caloric contribution to the diet.

In baked goods methylcellulose increases water absorption and retention and confers a degree of resistance to oil absorption in deep-fried foods, such as doughnuts. In some dietetic foods, methylcellulose acts as a syneresis inhibitor and bulking agent, and in gluten-free products, it provides texture and structure. It is employed to inhibit syneresis in frozen foods, especially in sauces, meats, fruits, and vegetables, and as a thickener and stabilizer of emulsions in salad dressings. Methylcellulose is used in edible coatings for a variety of food products.

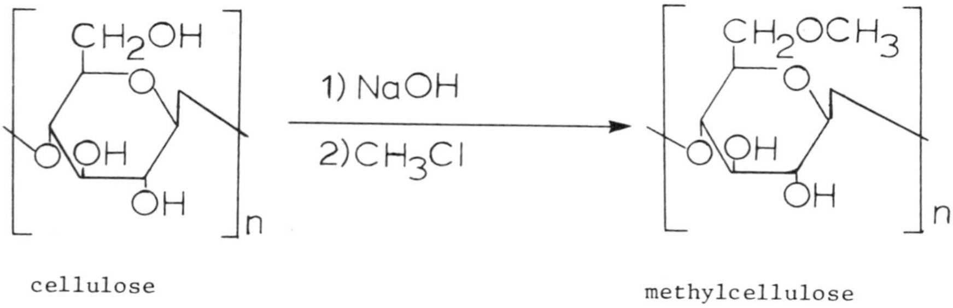

cellulose   methylcellulose

Figure 27 Preparation of methylcellulose.

E. Hemicelluloses

As indicated, the cell wall material of plants is a complex matrix of cellulose, lignin, and hemicelluloses. Hemicelluloses are a class of polymers generally distinguised by yielding, on hydrolysis, an abundance of pentoses, glucuronic acids, and some deoxy sugars. Details of the binding interrelationships among cell wall components is not clear, but both physical and covalent, chemical bonding occurs. The most prevalent hemicellulose present in foodstuffs has a xylan backbone composed of $(1{\rightarrow}4)$-β-D-xylopyranosyl units. The polymer frequently contains β-L-arabinofuranosyl side chains attached to the 3 position of some D-xylosyl units. Other typical constituents are the 4-*O*-methylether of D-glucuronic acid, D- or L-galactose, and acetyl ester groups. A typical, representative structure is shown below.

$$
\begin{array}{c}
O-Acetyl \\
3 \\
\rightarrow 4)-\beta-D-Xylp-(1{\rightarrow}4)-\beta-D-Xylp-(1{\rightarrow}4)-\beta-D-Xylp-(1{\rightarrow}4)-\beta-D-Xylp\ (1\rightarrow \\
\quad 2 \qquad\qquad\qquad\qquad\qquad\qquad\qquad\qquad 3 \qquad\qquad 3 \\
4-O-Me-\alpha-D-GlupA \qquad\qquad\qquad L-Araf \qquad\qquad R
\end{array}
$$

$$
\begin{array}{l}
R = \beta-D-Xylp-(1{\rightarrow}2)-L-Araf-(1 \quad , \\
\quad \alpha-D-Xylp-(1{\rightarrow}3)-L-Araf-(1 \quad , \quad or \\
\quad D-Galp-(1{\rightarrow}4)-D-Xylp-(1{\rightarrow}2)-L-Araf-(1 \quad)
\end{array}
$$

Hemicelluloses probably have their greatest effect in baked goods where they improve the water binding of flour. In bread doughs, they improve mixing quality, reduce mixing energy, aid in incorporation of protein, and improve loaf volume. Most interesting and useful is the observation that plant hemicellulose greatly retards staling of bread as compared with bread with no hemicellulose (8).

The importance of dietary hemicellulose is not accurately known, but since it is a good source of dietary fiber, being part of the indigestible complex comprising plant cell walls, it is likely it has beneficial nutritional and physiological effects on intestinal motility, stool weight, and stool transit time. The effects of these polysaccharides on bile acid and steroid metabolism have not been fully elucidated, but they may have beneficial effects in this regard by facilitating the elimination of bile acids and lowering the level of cholesterol in the blood. There are indications that dietary fiber, including hemicelluloses, lessens risk of cardiovascular diseases and colonic disorders, especially colonic cancer. Also, diabetic patients on high-fiber diets may experience reduced insulin requirements. On the negative side, polysaccharide gums and fiber may decrease the absorption of some vitamins and essential trace minerals in the small intestine.

F. Pectic Substances

Pectic substances are polymers composed mainly of $(1{\rightarrow}4)$-α-D-galacturonopyranosyl units, and they are found in the middle lamella of plant cells. Some galactans and arabans are also included in this definition. One aspect of difference among the pectic substances is their content of methyl esters, or degree of esterification, which decreases somewhat as plant ripening takes place. The degree of esterification (DE) is defined as the (number of esterified D-galacturonic acid residues per total number of D-galacturonic acid residues) $\times$ 100. The pectic substance in Fig. 28, for example, has a degree of esterification of 50%.

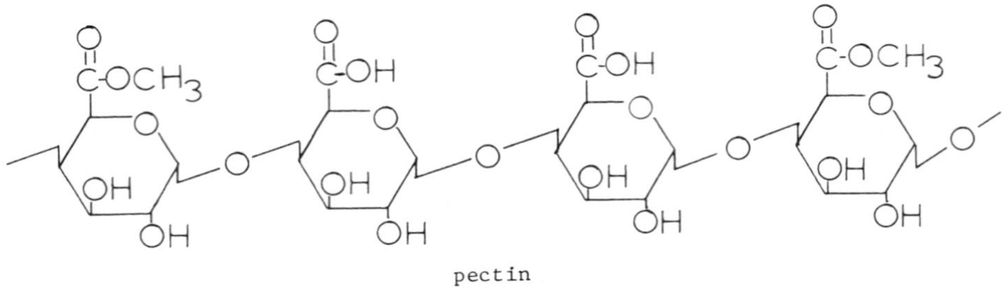

pectin

Figure 28 Representative structure of pectin.

Protopectin is the pectic substance in the flesh of immature fruits and vegetables. It is highly esterified with methanol, is insoluble in water, and produces the hard texture of unripe fruits and vegetables. *Pectinic acids* are less highly methylated pectic substances and are derived from protopectin by the action of protopectinase and pectin methylesterase. Depending on the DP and degree of methylation, pectinic acids may be colloidal or water soluble. Water-soluble pectinic acids are termed "low-methoxyl pectins." The continued action of pectin methylesterase on the pectinic acids leads to complete removal of methyl ester groups and formation of *pectic acids*.

Pectic enzymes contribute to the development of the desirable texture produced during ripening of plants. During this period, *protopectinase* converts protopectin to colloidal pectin or water-soluble pectinic acids. *Pectin methylesterase* (pectase) cleaves methyl esters from pectin to produce poly-D-galacturonic acid, or pectic acid, and this substance is then partially degraded to monomeric D-galacturonic acid by *polygalacturonase*. These enzymes act in concert during maturation and are crucial in determining fruit and vegetable texture.

A special property of pectin is its ability to produce strong gels. If a hot water dispersion of 0.3-0.4% pectin is cooled to room temperature, no gel forms. However, if the pH is adjusted to 2.0-3.5 and sucrose is present at a concentration of 60-65%, a gel forms on cooling, and this gel retains its character even when reheated to near 100°C. Adjustment of the pH to 2.0-3.5 prevents ionization of the carboxylate groups that would be ionized at pH 7.0. Sucrose at 60-65% concentration dehydrates the neutralized pectin molecules to allow formation of intermolecular hydrogen bonds, and a gel. The hydrogen bonds existing between the pectin molecules may be hydroxyl-hydroxyl, carboxyl-carboxyl, or hydroxyl-carboxyl as depicted below.

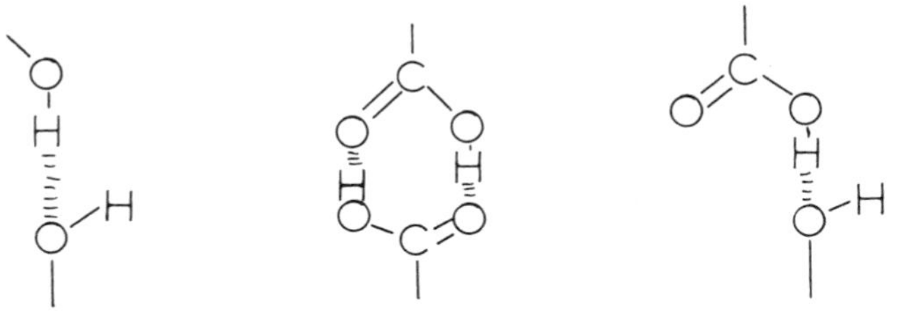

Normal pectins produce optimum gels when the concentration of pectin is about 1%, although the concentration varies with the type of pectin.

High gel strength correlates positively with high-molecular-weight pectin molecules and extensive intermolecular association. The effect of DE on gel setting time has been the subject of some study and much confusion. Generally, setting times increase for pectins with DE values increasing from 30 to 50. This is probably explained as due to increased steric interference by the methyl ester groups with intermolecular hydrogen-bonding interactions. From DE 50-70 the setting time decreases, and this may stem from increased hydrophobic interaction between pectin molecules due to the relatively high degree of methylation. Gelation characteristics as a function of pectin DE are shown in Table 15.

Low-DE pectic substances (low-methoxy) can form stable gels in the absence of sugars but require the presence of divalent ions, such as calcium, which results in molecular cross-linking. Gels of this type are produced for sugarless or low-sugar dietetic jams and jellies. Low-methoxy pectin (LMP) is less sensitive to pH changes than are standard pectin gels. Low-methoxy pectin gels may be formed in the pH range 2.5-6.5; normal pectin gels are limited to the pH range 2.7-3.5, with 3.2 optimal. Although LMP gels do not require sugar, the addition of 10-20% sucrose provides a gel with better textural properties. Without the addition of sugar or some plasticizer, the LMP gels tend to be brittle and less elastic than those of normal pectin. The gel-firming action of calcium ions is used in firming canned tomatoes and pickled cucumbers and in the preparation of dietetic jams and jellies with low-methoxyl pectin.

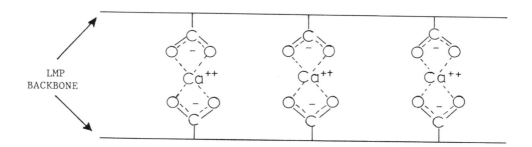

When pectin gels are subjected to low mechanical pressure, they undergo plastic flow. Under more stress the gel fractures. This suggests that the gel structure may contain two types of bonds. One is weak and easily ruptured, but capable of reformation. A second is stronger and randomly distributed throughout the gel. This behavior could be due to a mixture of short and long junction zones.

G. Plant Gums

A gum may be generally defined as any water-soluble polysaccharide that is extractable from land or marine plants or from microorganisms and that possesses the ability to contribute viscosity or gelling ability to their dispersions. Such a definition, however, excludes starches and pectins, which have already been discussed. Common gums from plants include the seed galactomannans of guar and locust bean; the plant exudates, gum arabic and gum tragacanth; and the seaweed-derived gums, agar, carrageenan, and algin. These polysaccharide gums are used widely for thickening foods.

Table 15 Effect of Degree of Esterification of Pectin on Gel Formation

Degree of esterification (%)[a]	Requirments for gel formation			Rapidity of gel formation
	pH	Sugar (%)	Divalent ion	
>70	2.8-3.4	65	No	Rapid
50-70	2.8-3.4	65	No	Slow
<50 (Low-methoxy)	2.5-6.5	None	Yes	Rapid

[a]Degree of esterification = (number of esterified D-galacturonic acid residues per total number of D-galacturonic acid residues) × 100.

Guar gum, or guaran, is the endosperm polysaccharide of the seed of *Cyamopsis tetragonolobus*, family Leguminosae, which grows naturally in India and Pakistan and has more recently been introduced as a cash crop in the United States. It is a galactomannan, containing a backbone of (1→4)-β-D-mannopyranosyl units with every second unit bearing a (1→6)-α-D-galactopyranosyl unit (14). The polymer is relatively large, with a molecular weight of about 220,000 daltons (Fig. 29).

Guaran hydrates rapidly in cold water to give a highly viscous, thixotropic solution. A 1% dispersion has a viscosity of about 6000 cps, depending on temperature, ionic strength, and the presence of other food components. Dissolution of the gum is hastened by heating a dispersion, but the gum is degraded at very high temperatures. Because guar can produce solutions of high viscosity it is usually used at a concentration of 1% or less in foods.

Since the gum is neutral its solution viscosity is little affected by changes in pH. Guaran is compatible with most other food components. Salts have little effect on its

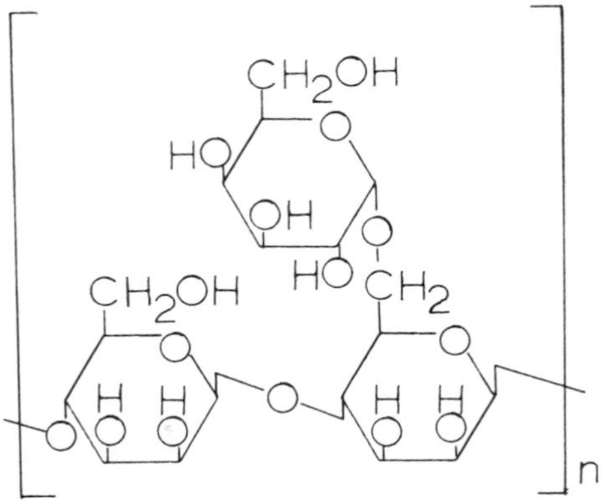

Figure 29 Repeating unit of guaran.

solution viscosity, but large amounts of sucrose may reduce viscosity and delay attainment of maximum viscosity.

Guaran exhibits viscosity synergism with wheat starch and some other gums. It is used in cheeses, in which it eliminates syneresis, and in ice cream, in which it contributes body, chewiness, and resistance to heat shock. In baked goods it promotes longer shelf life, and in pastry icings it lessens absorption of water by sucrose. Guaran is also used in meat products, such as sausage, to improve casing stuffing. When used in dressings and sauces at a level of 0.2-0.8% it increases viscosity and contributes to a pleasant mouth feel.

Locust bean gum, another plant seed galactomannan, is derived from the carob seed, *Ceratonia siliqua*. The plant grows predominately in the Near East and Mediterranean areas. It consists of a D-mannopyranosyl backbone with attached D-galactopyranosyl units, the two components existing in a ratio of 4:1, respectively (Fig. 30) (38). However, its D-galactopyranosyl units are not uniformly distributed, leaving long stretches of the mannan chain devoid of D-galactopyranosyl units. This leads to unique synergistic properties, especially with the seaweed polymer carrageenan, where the two cross-link to form a gel.

Locust bean gum is employed in frozen desserts to bind water, and provide body, smoothness, and chewiness. In soft cheese manufacture it speeds curd formation and reduces the loss of solids. In composite meat products, such as salami, bologna, and sausages, it acts as a binder.

Among the plant exudate polysaccharides, *gum arabic* is the oldest and best known. It is produced as teardrop-shaped globules exuding from bark wounds of *Acacia* trees (13). Production of the gum is stimulated by intentional removal of bark, analogous to the harvest of turpentine from pine trees. Unfortunately, production of gum arabic is likely to decline because of high labor costs, and comparable replacement substances will be difficult to find.

Gum arabic is a complex heteroglycan with a molecular weight of 250,000-1,000,000 daltons. A likely partial structure is shown in Fig. 31. The polymer usually contains some minor amounts of L-arabinose and L-rhamnose. Molecules exist as short, stiff spirals of 1050-2400 Å in length, depending on molecular charge.

Gum arabic dissolves readily in water to produce solutions of low viscosity. Arabic can be dissolved to an extent of 50% w/w to form a high-solids gel similar to that from starch. At concentrations of less than 40% its solutions exhibit Newtonian rheology; above 40% concentration the dispersions are pseudoplastic. High-quality types of this gum form colorless, tasteless solutions.

Due to the presence of ionic charges, the viscosity of solutions of gum arabic change with changes in pH. Viscosity is low at low and high pHs and reaches a maximum at pH 6-8. The addition of electrolytes causes the viscosity to decrease in proportion to the valence and concentrations of the electrolyte cation. Gum arabic is incompatible with

Figure 30 Idealized structure of locust bean gum. ○ D = mannopyranosyl; ● D = galactopyranosyl.

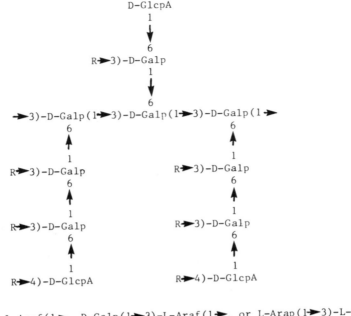

R = L-Rhap(1➤, L-Araf(1➤, D-Galp(1➤3)-L-Araf(1➤, or L-Arap(1➤3)-L-Araf(1➤.

```
D-GlcpA = D-glucopyranosiduronic acid
D-Galp  = D-galactopyranose
L-Rhap  = L-rhamnopyranose
L-Arap  = L-arabinopyranose
L-Araf  = L-arabinofuranose
```

Figure 31 Proposed partial structure of gum arabic.

some food polymers, such as gelatin and sodium alginate, but is compatible with most other gums.

Gum arabic retards or prevents sugar crystallization in confectionaries, stabilizes emulsions, and contributes to viscosity. In bakery toppings it prevents excess moisture absorption by the icings or glazings. In frozen dairy products, such as ice cream, ices, and sherbets, its aids in the formation and retention of small ice crystals. In beverages the gum functions as an emulsion and foam stabilizer. Gum arabic is a useful flavor fixative in powdered or dried beverage mixes. It is especially important in spray-dried citrus drink mixes, where it functions to retain volatile flavor components.

Gum tragacanth, another plant exudate, is derived from several species of *Astragalus*. It is an ancient gum, like gum arabic, the use of which extends back more than 2000 years. It is collected from a low, bushy shrub growing principally in Iran, Syria, and Turkey. It occurs in bark breaks or wounds and is hand collected, like gum arabic.

Gum tragacanth has a complex structure, and on hydrolysis yields D-galacturonic acid, L-fucose, D-galactose, D-xylose, and L-arabinose (31). On mixing in water, a soluble portion called tragacanthin, comprising 60-70% by weight of the gum and with a molecular weight of about 800,000, is solubilized. The insoluble portion, the polymer bassorin, has a molecular weight of about 840,000 daltons. Aqueous dispersions of tragacanth have a high viscosity at concentrations as low as 0.5%.

Gum tragacanth is used in salad dressings and sauces because of its stability to heat and acidic conditions. In frozen desserts the gum provides desirable body, texture, and mouth feel. It is also used in frozen fruit pie fillings, in which it provides clarity and brilliance to the thickened pie filling.

Seaweed gums of importance in foods include agar, carrageenan, and algin. *Agar*, probably best known as a bacteriological culture medium, is obtained from a variety of red marine algae of the class Rhodophyceae, found mainly on the coast of Japan. Agar, like normal starch, can be separated into two fractions called agaran and agaropectin (42). The basic repeating disaccharide unit of agarin is believed to be a β-D-galactosyl unit linked (1→4) to a 3,6-anhydro-α-L-galactopyranosyl unit.

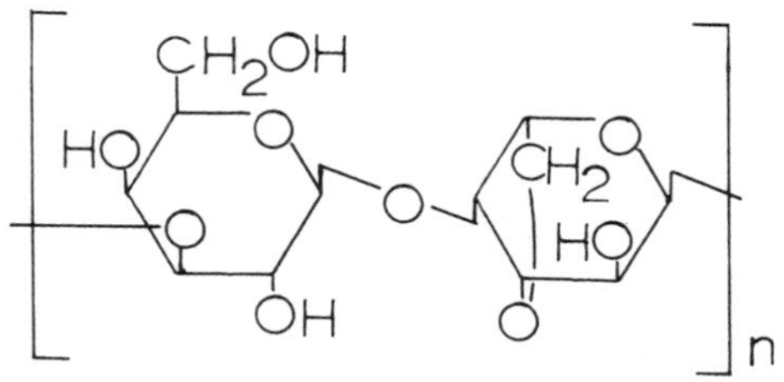

Repeating unit of agaran

Agaropectin has a similar repeating unit and contains 5-10% sulfate esters, some D-glucuronic acid residues, and some pyruvate esters.

Probably the most unique property of agar gels is their ability to remain stable at temperatures far above the initial gelation temperature. As an example, a 1.5% aqueous dispersion of agar gels at 30°C but has a melting point of 85°C. Agar forms one of the strongest gels known.

Food applications of agar include frozen desserts, in which it inhibits syneresis and provides desirable texture; process cheeses and cream cheeses, in which it provides stability and desirable texture; and baked goods and icings, in which it controls water activity and retards staling. It is also used with meat products as a canning jelly. Frequently, agar is used with other polymers, such as gum tragacanth, locust bean gum, and/or gelatin.

The polysaccharide gum *carrageenan* is extracted from Irish moss (*Chondrus crispus*) that grows along the shores of Ireland, England, France, Spain, and, in greatest profusion, on the coast of Halifax-Prince Edward Island. Carrageenan is a complex mixture of at least five distinct polymers designated kappa (κ), lambda (λ), mu (μ), iota (ι), and nu (ν). Of these, κ and λ-carrageenan are of major importance in foods.

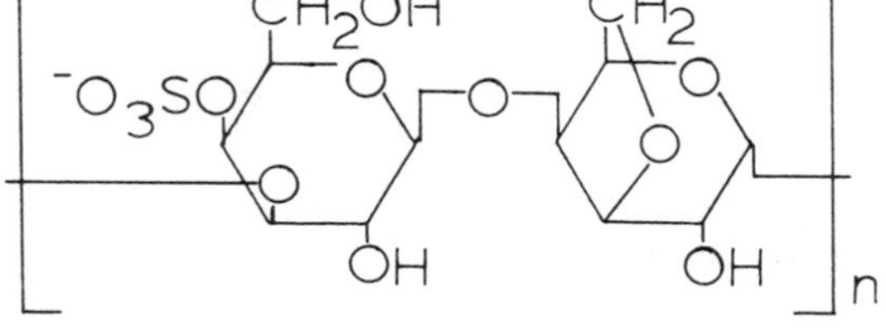

Repeating unit of κ-carrageenan

R = H, SO₃⁻

Repeating unit of λ-carrageenan

The properties of these sulfated polymers depend markedly on the associated cation. For instance, when the associated cation is potassium, a firm gel is produced, but when the cation is sodium, the polymer is soluble in cold water and is nongelling. Carrageenan interacts synergistically with many other food gums, notably locust bean gum, in which, depending on concentration, it increases viscosity, gel strength, and gel elasticity. At high concentrations, carrageenan increases the gel strength of guaran, but at low concentrations it produces only an increase in viscosity. Decreases in viscosity occur when carrageenan is added to solutions of gum ghatti, gum tragacanth, alginate, and pectin. Carrageenan is used in water-based and milk-based food systems to stabilize suspensions.

Although commercial carrageenan is a mixture of various types, it is approximately 60% κ (gelling) and 40% λ (nongelling). The polymer is stable above pH 7, degrades slightly in the pH region 5-7, and rapidly degrades below pH 5. The potassium salt of κ-carrageenan is the best gel former, but the gels are brittle and prone to syneresis. The tendency to brittleness can be reduced by the addition of small amounts of locust bean gum. Interestingly, this synergism does not occur with guaran because of structural differences.

The ability of carrageenan to stabilize milk is dependent on the number and position of sulfate groups in the carrageenan. The carrageenan anion reacts with protein to form a protein-carrageenanate complex that can exist as a stable colloidal dispersion. Carrageenan is added to chocolate milk to prevent chocolate precipitation. Its stabilizing power is also used in milk puddings and "eggless" custards. In cheese products the gum acts as an emulsion stabilizer, and in frozen desserts, it is an inhibitor of ice crystallization, and in these it is commonly used in combination with carboxymethylcellulose, locust bean gum, or guaran. In bakery products, it acts as a dough conditioner, contributing to increased loaf volume. In cakes it produces improved appearance and crumb texture, and in deep-fried products it reduces fat absorption.

Alginates are extracted from brown algae, of the class Phaeophyceae (39). The principal source of commercial alginate is the giant kelp, *Macrocystis pyrifera*. Alginates contain D-mannopyranosyluronic acid units (M) and L-gulopyranosyluronic acid units (G). The M/G ratio differs from source to source and influences solution properties of the alginates. The molecule is a copolymer of homogeneous sections of poly-M and poly-G units connected by sections of alternating M–G. (See p. 131).

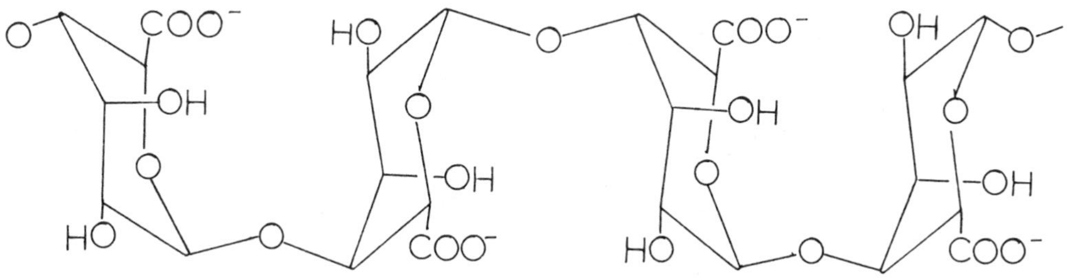

poly(L-guluronic acid) segment of alginate

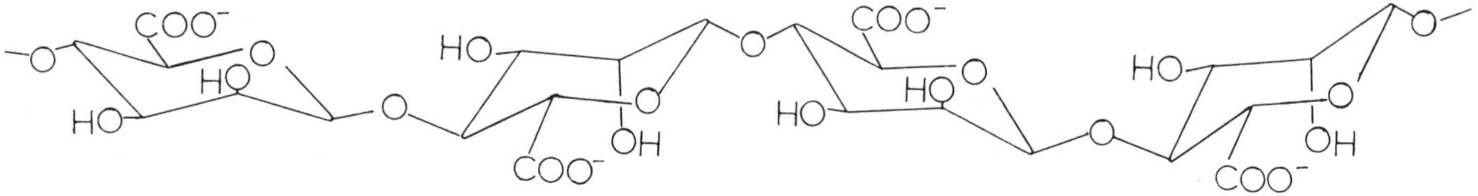

poly(D-mannuronic acid) segment of alginate

Alginate salts of alkali metals, ammonia, and low-molecular-weight amines are readily soluble in hot or cold water, but salts of di- or trivalent cations are insoluble. Solutions of alginates are viscous, with properties depending on the M/G ratio, molecular weight, and the electrolytes present in solution. The viscosity of alginate solutions decreases as the temperature is raised, and is influenced only slightly by pH changes in the range 4-10. Alginate solutions in the pH range 5-10 are stable for long periods at room temperature. Alginates gel at room temperature in the presence of small amounts of calcium ion or other di- or trivalent metal ions, or in the absence of ions at pH 3 or less. Gel strength increases with alginate concentration and DP and can be controlled to produce soft, elastic or hard, rigid gels.

Alginates are used in ice cream, to which they contribute body, texture, and resistance to the formation of large ice crystals. In bakery products alginate is used in icings, cake fillings, meringues, glazes, and pie fillings, mainly for texture and gel-forming properties. Alginates are also used in French dressing for thickening and emulsion stabilization, in dessert puddings for thickening, in beer for foam stabilization, and in many other foods for similar reasons.

Several food gums are products of microbial biosynthesis, for example, *dextran* and *xanthan gum*. Several organisms produce dextrans, but those organisms of commercial importance are *Leuconostoc mesenteroides* and *L. dextranicum* (39). Dextrans are composed solely of α-D-glucopyranosyl units, but the type of glycosidic linkages and the amount of each varies from dextran to dextran. *L. mesenteroides* NRRL B-512 dextran is reported to have about 95% (1→6) linkages; the rest are divided between (1→3) and (1→4). Because of these molecular differences, some dextrans are water soluble and others are insoluble.

Uses recommended for dextran include confections, in which it improves moisture retention, viscosity, and inhibits sugar crystallization. In gum and jelly candies it acts as a gelling agent, and in icings it is a crystallization inhibitor. In ice cream it acts as a crystallization inhibitor, and in pudding mixes it provides desirable body and mouth feel.

Xanthan gum is an extracellular polysaccharide elaborated by several *Xanthomonas* species, but *X. campestris* is the organism used commercially (30). Examination of its structure shows that it is a cellulose chain with attached oligosaccharide groups. (See page 133.)

Xanthan gum is readily soluble in hot or cold water and produces highly viscous solutions at a low gum concentration. The viscous suspension are pseudoplastic and exhibit marked shear thinning, a property of value to food technologists. Temperature changes in the vicinity of 60-70°C have little effect on the viscosity of xanthan gum, and the gum itself is stable. Acidity in the pH range 6-9 also has little effect on viscosity, and viscosity changes are small even beyond this pH range. Xanthan gum is compatible with most food salts and acids. The gum functions with guaran to increase viscosity and with locust bean gum to produce a thermoreversible gel.

Xanthan gum is employed in beverages to enhance mouth feel and flavor release, and in orange juice to achieve cloud stabilization. Because of its thermal stability, it is used as a suspending and stabilizing agent in various canned foods. In frozen, starch-thickened foods, such as fruit pie fillings, the addition of xanthan markedly improves freeze-thaw stability and decreases syneresis. It is also used to advantage in relishes containing high levels of salts and/or acids because of its stability. Xanthan gum-locust bean

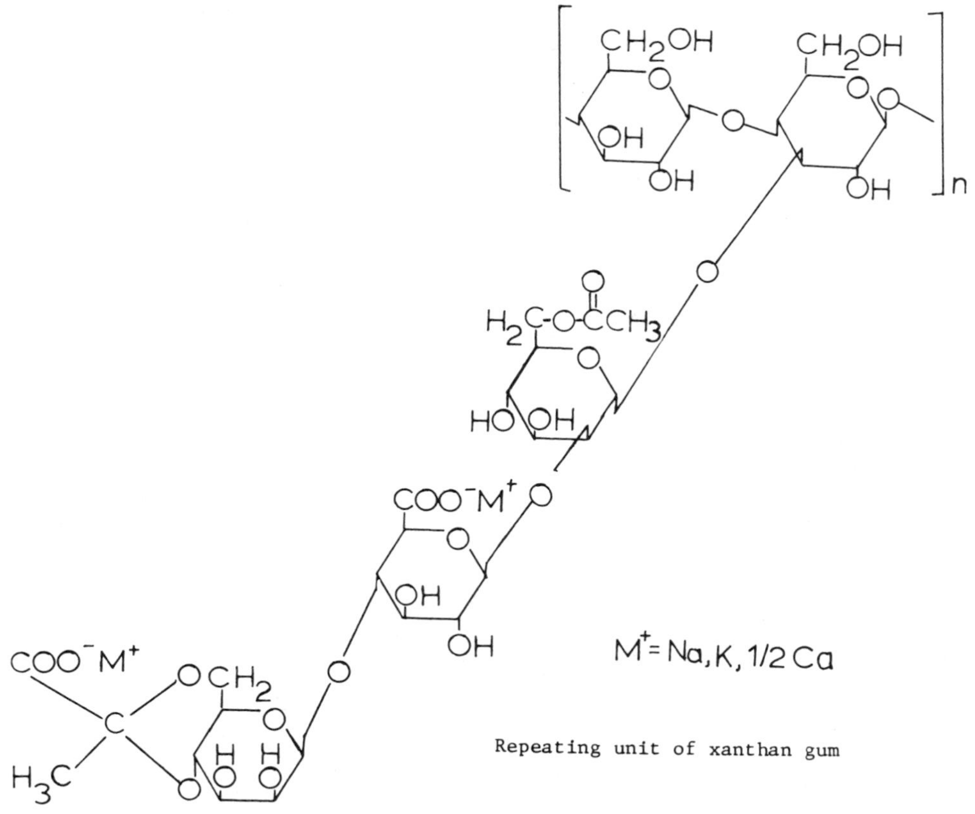

Repeating unit of xanthan gum

$M^+ = Na, K, 1/2 Ca$

gum gels can be used to produce milk-based instant puddings. The puddings are not gummy, have excellent mouth feel, and due to pseudoplastic shear thinning in the mouth, have excellent release of pudding flavor.

The properties of polysaccharide gums are summarized in Table 16.

VI. CONCLUSION

Carbohydrates, both simple and complex, are of great importance, not only as digestible or nondigestible components of the diet, but also with regard to their role in food formulation and processing. Carbohydrates, especially sucrose and starch, provide the bulk of caloric intake for most of the world's population, and this fact is unlikely to change, although consumption of more complex carbohydrates at the expense of sucrose has been suggested in recent dietary guidelines.

Starch and other carbohydrates are important contributors to the sensory attributes of foods. They have profound effects on consistency, texture, and mouth feel, through their ability to influence viscosity, colligative properties, ice crystallization, gelation, and the stability of dispersions. They also influence color and flavor through their ability to undergo browning reactions with the concomitant production of flavors, through their ability to impart sweetness and through their ability to influence flavor retention and release.

Table 16 Selected Properties of Some Polysaccharide Gums

Name	Principal monosaccharide constituents	Source	Distinguishing properties
Guar	D-Mannose, D-galactose	*Cyamopsis tetragonolobus*	High solution viscosity at low concentration
Locust bean gum	D-Mannose, D-galactose	*Ceratonia siliqua*	Synergistic with carrageenan
Gum arabic	D-Galactose, D-glucuronic acid	*Acacia*	High water solubility
Gum tragacanth	D-Galacturonic acid and D-galactose, L-fucose, D-xylose, L-arabinose	*Astragalus*	Stable over wide pH range
Agar	D-Galactose, 3,6-anhydro-L-galactose	Red seaweed	Forms extremely strong gels
Carrageenan	Sulfated D-galactose, sulfated 3,6-anhydro-D-galactose	Irish moss (*Chondrus crispus*)	Forms chemically set gels with K^+
Alginate	D-Mannuronic acid, L-guluronic acid	Brown seaweed (*Macrocystis pyrifera*)	Forms chemically set gels with Ca^{2+}
Dextran	D-Glucose	*Leuconostoc mesenteroides*	Crystallization inhibitor in candies and frozen desserts
Xanthan	D-Glucose, D-mannose, D-Glucuronic acid	*Xanthomonas campestris*	Dispersions are high pseudoplastic

REFERENCES

1. Baltes, W. (1982). Chemical changes in food by the Maillard reaction. *Food Chem.* 9:59-73.

2. Batdorf, J. B., and J. M. Rossman (1973). Sodium carboxymethylcelluose, in *Industrial Gums, Polysaccharides and Their Derivatives* (R. L. Whistler and J. N. BeMiller, eds.), Academic Press, New York, pp 695-729.

3. Bechtel, W. G. (1950). Measurements of properties of corn starch gels. *J. Colloid Sci. 5*:260-270.

4. Beck, K. M. (1978). Essential properties and potential uses of synthetic sweeteners, in *Proceedings Sweeteners and Dental Caries*, Sp. Suppl, *Feeding, Weight and Obesity* (abstracts) (J. H. Shaw and G. G. Roussos, eds.), Information Retrieval, Inc., Washington, D.C., pp. 241-252.

5. Bowen, W. H. (1978). Role of carbohydrates in dental caries, in *Proceedings Sweeteners and Dental Caries*, Sp. Supp., *Feeding, Weight and Obesity* (abstracts). (J. H. Shaw and G. G. Roussos, eds.), Information Retrieval, Inc., Washington, D.C., pp. 147-156.

6. Browne, C. A. (1922). Moisture absorptive power of different sugars and carbohydrates under varying conditions of atmospheric humidity. *J. Ind. Eng. Chem. 14*: 712-714.

7. Capon, B. (1967). Mechanism in carbohydrate chemistry. *Chem. Rev. 69*:407-498.

8. Casier, J. (1971). Pentosans, additives for accelerating bread-making. Ger. Offen. 2, 047, 504; *Chem. Abstr. 75*:117359p.

9. Duryea, C. B. (1902). Method of making thin boiling starch. U.S. Pat. 696, 949 April 8 Abst.; *Offic. Gaz. U.S. Patent Office 99*:252.

10. Ellis, G. P. (1959). The Maillard reaction, in *Advances in Carbohydrate Chemistry* (M. L. Wolfrom, ed.), Academic Press, New York, pp. 63-134.

11. Gallay, W., and A. C. Bell (1936). The effect of various starches on the stability of baking powders. *Can. J. Res. 14B*:204-215.

12. Gilbert, L. M., G. A. Gilbert, and S. P. Spragg (1964). Amylose and amylopectin from potato starch, in *Methods in Carbohydrate Chemistry*, vol. IV (R. L. Whistler, ed.), Academic Press, New York, pp. 25-27.

13. Glicksman, M., and R. E. Sand (1973). Gum arabic, in *Industrial Gums, Polysaccharides and Their Derivatives* (R. L. Whistler and J. N. BeMiller, eds.), Academic Press, New York, pp. 197-263.

14. Goldstein, A. M., E. N. Alter, and J. K. Seaman (1973). Guar gum, in *Industrial Gums, Polysaccharides and Their Derivatives* (R. L. Whistler and J. N. BeMiller, eds.), Academic Press, New York, pp. 303-321.

15. Greminges, G. K., and A. B. Savage (1973). Methylcellulose and its derivatives, in *Industrial Gums, Polysaccharides and Their Derivatives* (R. J. Whistler and J. N. BeMiller, eds.), Academic Press, New York, pp. 619-647.

16. Guthrie, R. D. (1974). *Introduction to Carbohydrate Chemistry*, Clarendon Press, Oxford.

17. Harsveldt, A. (1962). Starch requirements for paper coating. *TAPPI 45*:85-89.

18. Hassid, W. Z., and C. E. Ballou (1957). Oligosaccharides, in *The Carbohydrates: Chemistry, Biochemistry, Physiology* (W. Pigman, ed.), Academic Press, New York, pp. 478-535.

19. Hicks, K. B., and F. W. Parrish (1980). A new method for the preparation of lactulose from lactose. *Carbohyd. Res. 82*:393-397.

20. Hodge, J. E., and E. Osman (1976). Carbohydrates, in *Principles of Food Science*, Part 1, *Food Chemistry* (O. Fennema, ed.), Marcel Dekker, Inc., New York, pp. 41-138.

21. Kato, Y., K. Watanabe, and Y. Sato (1981). Effect of some metals on the Maillard reaction of ovalbumin. *J. Agr. Food Chem. 29*:540-543.

22. Keir, R. A., and F. C. Cleveland, Jr. (to Corn Products Co.) (1960). Distarch phosphate. U.S. Pat. 2, 938, 901 Division of U.S. 2, 801, 242 August 25; *Chem. Abst. 51*:18666h.

23. Korth, J. A. (1959). Pregelatinized starch. U.S. Pat. 2, 884, 346 July 10; *Chem. Abstr. 53*:12719.

24. Kuninaka, A. (1966). Recent studies of 5'-Nucleotides as new flavor enhancers, in *Flavor Chemistry* (I. Hornstein, ed.), Advances in Chemistry Series 56, American Chemical Society Publications, Washington, D. C., pp. 261-274.

25. Lang, K. (1970). Influence of cooking on foodstuffs, in *World Review of Nutrition and Dietetics*, vol. 12, (G. F. Bourne, ed.), S. Karger, New York, pp. 266-317.

26. Lansky, S., M. Kooi, and T. S. Schoch (1949). Properties of the fractions and linear subfractions from various starches. *J. Amer. Chem. Soc. 71*:4066-4075.

27. Leach, H. W., and T. J. Schoch (1962). Structure of the starch granule. III. Solubilities of granular starches in dimethyl sulfoxide. *Cereal Chem. 39*:318-327.

28. Lintner, C. J. (1886). Studien über diastase. *J. Prakt. Chem. 34*:378-394.

29. Maillard, L. C. (1912). Action of amino acids on sugars. Formation of melanoidins in a methodical way. *Compt. Rend. 154*:66-68.

30. McNeely, W. H., and K. S. Kang (1973). Xanthan and some other biosynthetic gums, in *Industrial Gums, Polysaccharides and Their Derivatives* (R. L. Whistler and J. N. BeMiller, eds.), Academic Press, New York, pp. 473-497.

31. Meir, G., W. A. Meer, and T. Guard (1973). Gum tragacanth, in *Industrial Gums, Polysaccharides and Their Derivatives* (R. L. Whistler and J. N. BeMiller, eds.), Academic Press, New York, pp. 289-299.

32. Montgomery, R. D. (1980). Cyanogens, in *Toxic Constituents of Plant Foodstuffs* (I. E. Liener, ed.), Food Science and Technology, A series of monographs. Academic Press, New York, pp. 143-160.

33. Musselman, W. C., and J. A. Wagoner (1968). Electron microscopy of unmodified and acid-modified corn starches. *Cereal Chem. 45*:162-171.

34. Osman, E. M. (1967). Starch in the food industry, in *Starch: Chemistry and Technology*, vol. II, *Industrial Aspects* (R. L. Whistler and E. F. Paschall, eds.), Academic Press, New York, pp. 163-215.

35. Pancoast, H. M., and W. R. Junk (1980). Manufacture of corn syrups and sugars, in *Handbook of Sugars* (H. M. Pancoast and W. R. Junk, eds.), AVI Publishing Co., Inc., Westport, Conn., pp. 157-170.

36. Powrie, W. D., C. H. Wu, M. P. Rosin, and H. F. Stich (1981). Clastogenic and mutagenic activities of Maillard reaction model systems. *J. Food Sci. 46*:1433-1438, 1445.

37. Rhee, K. S., and K. C. Rhee (1981). Nutritional evaluation of the protein in oilseed products heated with sugars. *J. Food Sci. 46*:164-168.

38. Rol, F. (1973). Locust bean gum, in *Industrial Gums, Polysaccharides and Their Derivatives* (R. L. Whistler and J. N. BeMiller, eds.), Academic Press, New York, pp. 323-337.

39. Sanderson, G. R. (1981). Polysaccharides in food. *Food Technol. 7*:50-57, 83.

40. Scheinin, A., K. K. Makinen, E. Tammisalo, and M. Rekola (1975). Turku sugar studies. XVIII. Incidence of dental caries in relation to 1-year consumption of xylitol chewing gum. *Acta Odontol. Scand. (Suppl. 33) 70*:307-316.

41. Schoch, T. J., and E. C. Maywald (1956). Microscopic examination of modified starches. *Anal. Chem. 28*:382-387.

42. Selby, H. H., and T. A. Selby (1973). Agar, in *Industrial Gums, Polysaccharides*

and Their Derivatives (R. L. Whistler and J. N. BeMiller, eds.), Academic Press, New York, pp. 29-48.

43. Shallenberger, R. S., and T. E. Acree (1967). Molecular theory of sweet taste. *Nature 216*:480-481.

44. Shallenberger, R. S., and G. G. Birch (1975). *Sugar Chemistry*, AVI Publishing Co., Inc., Westport, Conn.

45. Shildneck, P., and C. E. Smith (1967). Production and uses of acid-modified starch, in *Starch: Chemistry and Technology*, vol. II, *Industrial Aspects* (R. L. Whistler and E. F. Paschall, eds.), Academic Press, New York, pp. 217-235.

46. Toda, H., J. Sekiyawa, and T. Shibamoto (1981). Mutagencity of the L-rhamnose-ammonia-hydrogen sulfide browning reaction mixture. *J. Agr. Food Chem. 29*: 381-384.

47. Van Etten, C. H., and I. A. Wolff (1973). Natural sulfur compounds, in *Toxicants Occurring Naturally in Foods* (National Research Council, ed), National Academy of Sciences, Washington, D.C., pp. 210-234.

48. Wolfrom, M. L., A. Thompson, and R. B. Ward (1959). The composition of pyro-dextrins. II. Thermal polymerization of levoglucosan. *J. Amer. Chem. Soc. 81*: 4623-4625.

GENERAL BIBLIOGRAPHY

Banks, W., and C. T. Greenwood (1975). *Starch and Its Components*, Edinburgh University Press, Edinburgh.

Birch, G. G., and L. F. Green (eds.) (1973). *Molecular Structure and Function of Food Carbohydrate*, John Wiley and Sons, New York.

Blanshard, J. M. V., and J. R. Mitchell (eds.) (1979). *Polysaccharides in Food*, Butterworths, London.

Glicksman, M. (1969). *Gum Technology in the Food Industry*, Academic Press, New York.

Guthrie, R. D. (1974). *Introduction to Carbohydrate Chemistry*, 4th ed., Clarendon Press, Oxford.

Pigman, W., and D. Horton (eds.) (1970). *The Carbohydrates: Chemistry and Biochemistry*, Vols. IIA and IIB, 2nd ed., Acadmic Press, New York.

Pigman, W., and D. Horton (eds.) (1972). *The Carbohydrates: Chemistry and Biochemistry*, vol. IA, 2nd ed., Academic Press, New York.

Shallenberger, R. S., and G. G. Birch (eds.) (1975). *Sugar Chemistry*, AVI Publishing Co., Inc., Westport, Conn.

Sipple, H. L., and K. W. McNutt (eds.) (1974). *Sugars in Nutrition*, Academic Press, New York.

Whistler, R. L., and J. N. BeMiller (eds.) (1973). *Industrial Gums, Polysaccharides and Their Derivatives*, 2nd ed., Academic Press, New York.

Whistler, R. L., and E. F. Paschall (eds) (1965). *Starch: Chemistry and Technology*, vol. I, *Fundamental Aspects*, Academic Press, New York.

Whistler, R. L., and E. F. Paschall (eds.) (1967). *Starch: Chemistry and Technology*, vol. II, *Industrial Aspects*, Academic Press, New York.

4

LIPIDS

Wassef W. Nawar University of Massachusetts, Amherst, Massachusetts

I. Introduction	140
II. Nomenclature	140
A. Fatty Acids	140
B. Acylglycerols	143
C. Phospholipids	145
III. Classification	145
A. Milk Fats	146
B. Lauric Acids	146
C. Vegetable Butters	146
D. Oleic-Linoleic Acids	147
E. Linolenic Acids	147
F. Animal Fats	147
G. Marine Oils	147
IV. Physical Aspects	147
A. Theories of Triacylglycerol Distribution Patterns	147
B. Positional Distribution of Fatty Acids in Natural Fats	149
C. Consistency	151
D. Emulsions and Emulsifiers	166
V. Chemical Aspects	175
A. Lipolysis	175
B. Autoxidation	176
C. Thermal Decomposition	205
D. Chemistry of Frying	210
E. Effects of Ionizing Radiation on Fats	213
VI. Chemistry of Fat and Oil Processing	217
A. Refining	217
B. Hydrogenation	218
C. Interesterification	223

VII. Role of Food Lipids in Flavor   226
      A. Physical Effects   226
      B. Lipids as Flavor Precursors   226
VIII. Biological Aspects   229
      A. Consumption and Trends   229
      B. Nutritional Functions   229
      C. Safety of Heated and Oxidized Fats   229
      D. Safety of Hydrogenated Fats   232
      E. Dietary Fat and Coronary Heart Disease   233
      References   238
      Bibliography   244

I. INTRODUCTION

Lipids consist of a broad group of compounds that are generally soluble in organic solvents but only sparingly soluble in water. They are major components of adipose tissue, and together with proteins and carbohydrates, they constitute the principal structural components of all living cells. Glycerol esters of fatty acids, which make up to 99% of the lipids of plant and animal origin, have been traditionally called fats and oils. Based solely on whether the material is solid or liquid at room temperature, distinction between a fat and an oil is of little practical importance, and the two terms are often used interchangeably.

Food lipids are either consumed in the form of "visible" fats, which have been separated from the original plant or animal source, for example, butter, lard, shortening, or salad oils, or as constituents of basic foods, such as milk, cheese, or meat. The largest supply of vegetable oil comes from the seeds of soybean, cottonseed, and peanut, and the oil-bearing trees of palm, coconut, and olive.

Lipids in food exhibit unique physical and chemical properties. The composition, crystalline structure, melting and solidifying behavior, and association with water and other nonlipid molecules are especially important with regard to the various textural properties they impart, and to their functionality in bakery and confectionary products and in many other products that are cooked. They undergo complex chemical changes and react with other food constituents, producing numerous compounds both desirable and deleterious to food quality.

Dietary lipids play an important role in nutrition. They supply calories and essential fatty acids, act as vitamin carriers, and increase the palatability of food, but for decades they have been at the center of controversy with respect to toxicity and disease.

II. NOMENCLATURE

Lipid nomenclature can be understood more readily if simple nomenclature of the various classes of organic compounds, as covered in most texts of organic chemistry, is reviewed first. Updated recommendations for the nomenclature of lipids have been recently reviewed (50).

A. Fatty Acids

This term refers to any aliphatic monocarboxylic acid that can be liberated by hydrolysis from naturally occurring fats.

1. Saturated Fatty Acids

Fatty acids may be named in five different ways.

1. The acids are named in accord with hydrocarbons of the same number of carbon atoms (CH_3 replaced by COOH). The terminal letter *e* in the name of the parent hydrocarbon is replaced with *oic*. Thus,

$$CH_3CH_2CH_2CH_2CH_2CH_3$$

alkan<u>e</u>

hexan<u>e</u>

$$\overset{6}{C}H_3\overset{5}{C}H_2\overset{4}{C}H_2\overset{3}{C}H_2\overset{2}{C}H_2\overset{1}{\underline{C}OOH}$$

alkan<u>oic</u>

hexan<u>oic</u>

If the acid contains two carboxyl groups the suffix becomes *dioic* (e.g., hexanedioic). The terminal carboxyl carbon is regarded as carbon number 1, as shown above.

2. The acids can be numbered on the basis of the carboxyl group as a substitute of the corresponding hydrocarbon (H replaced by COOH), and the suffix *carboxylic acid* is used:

$$CH_3CH_2CH_2CH_2\overset{H}{\underset{H}{C}}-H$$

Pentane

$$\overset{5}{C}H_3\overset{4}{C}H_2\overset{3}{C}H_2\overset{2}{C}H_2\overset{1}{C}H_2COOH$$

1-Pentanecarboxylic acid

In this system carbon number 1 is the carbon atom adjacent to the terminal carboxyl group. This convention corresponds to the long-practiced use of Greek letters α, β, γ, δ, and so on, in which the α-carbon atom is that adjacent to the carboxyl carbon.

3. Acids of long standing can be designated by common or "trivial" names, such as butyric, stearic, and oleic.

4. Acids can be represented by a numerical designation giving the number of carbon atoms and the number of double bonds, with a colon between, for example, 4:0, 18:1, 18:3.

5. For use in triacylglycerol abbreviations, each acid can be given a standard letter abbreviation, such as P for palmitic, and L for linoleic.

Thus, the acid $CH_3CH_2CH_2COOH$ can be referred to as 4:0, *n*-butanoic, 1-propanecarboxylic, or butyric acid. Similarly,

$$CH_3-\underset{\underset{CH_3}{|}}{CH}-CH_2-COOH$$

is named 3-methylbutanoic, 2-methyl-1-propanecarboxylic, or β-methylbutyric. Table 1 gives a list of some of the fatty acids commonly found in natural fats, with their systematic and common names.

2. Unsaturated Fatty Acids

As in the case of the saturated fatty acids, the unsaturated acids are named after the parent unsaturated hydrocarbons. Replacement of the terminal *anoic* by *enoic* indicates

Table 1   Nomenclature of Some Common Fatty Acids

Abbreviation	Systematic name	Common name	Symbol
4:0	Butanoic	Butyric	B
6:0	Hexanoic	Caproic	H
8:0	Octanoic	Caprylic	Oc
10:0	Decanoic	Capric	D
12:0	Dodecanoic	Lauric	La
14:0	Tetradecanoic	Myristic	M
16:0	Hexadecanoic	Palmitic	P
18:0	Octadecanoic	Stearic	St[a]
20:0	Arachidic	Eicosanoic	Ad
16:1	9-Hexadecenoic	Palmitoleic	Po
18:1	9-Octadecenoic	Oleic	O
18:2	9,12-Octadecadienoic	Linoleic	L
18:3	9,12,15-Octadecatrienoic	Linolenic	Ln
20:4	5,8,11,14-Eicosatetraenoic	Arachidonic	An
22:1	13-Docosenoic	Erucic	E

[a]Some authors use S for stearic, but this can be confusing, since S is also used for saturated whenever triacylglycerol composition is expressed in terms of saturated (S) and unsaturated (U) fatty acids. For example, S_3 or SSS = all three fatty acids saturated; SU_2 or SUU = diunsaturated-monounsaturated, and so on.

unsaturation, and the *di, tri,* and so on represent the number of double bonds present. Hence, hexadecenoic for 16:1, octadecatrienoic for 18:3, and so on.

The simplest way to specify the location of double bonds is to put, before the name of the acid, one number for each unsaturated linkage. Oleic acid, for example, with one double bond between carbons 9 and 10, is named 9-octadecenoic acid. In certain cases it is convenient to distinguish unsaturated fatty acids by the location of the first double bond from the methyl end of the molecule, that is, the omega carbon. Linoleic acid (9,12-octadecadienoic acid) is therefore an $18:2\omega6$ acid.

The geometric configuration of double bonds is usually designated by the use of cis (Latin, *on this side*) and trans (Latin, *across*), indicating whether the alkyl groups are on the same side or on opposite sides of the molecule.

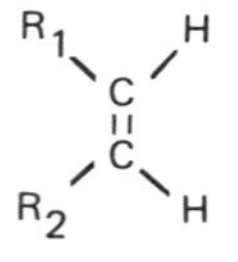

cis 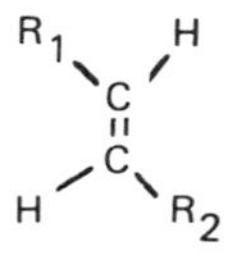 -   trans-

The cis configuration is the naturally occurring form, but the trans configuration is thermodynamically favored. Linoleic acid, with both double bonds in the cis configuration, is named *cis*-9, *cis*-12-octadecadienoic acid. Difficulty arises, however, if the four attached groups are all different, as in

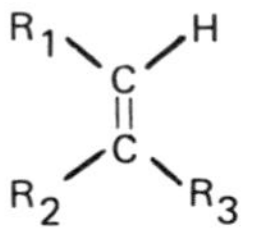

In this case the two atoms or groups attached to each carbon are assigned "priorities" in accord with the Cahn-Ingold-Prelog procedure (described under "R/S system"). If the high-priority groups (greater atomic number) lie on the same side of both carbons, the letter Z (German, *zusammen*) is used to designate the configuration. If the two high-priority groups are on opposite sides, the letter E (German, *entgegen*) is used.

B. Acylglycerols

Neutral fats are mono-, di-, and triesters of glycerol with fatty acids, and are termed monoacylglycerol, diacylglycerol, and triacylglycerol, respectively. Use of the old terms mono-, di-, and triglyceride is discouraged. The compound

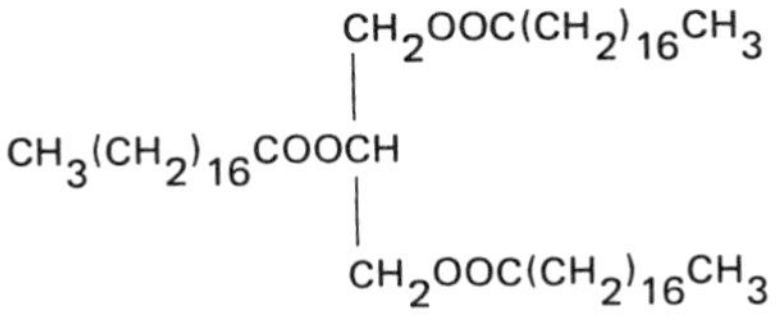

can be named any of the following: tristearoylglycerol, glycerol tristearate, tristearin, or StStSt.

Although glycerol by itself is a completely symmetrical molecule, the central carbon atom acquires chirality (asymmetry) if one of the primary hydroxyl groups (on carbons 1 and 3) is esterified, or if the two primary hydroxyls are esterified to different acids. Several methods have been used to specify the absolute configuration of glycerol derivatives.

1. R/S System

Use of the prefixes R and S was proposed by Cahn et al. (15). A sequence of priority is assigned to the four atoms or groups of atoms attached to a chiral carbon, with the atom of greatest atomic number assigned the highest priority. The molecule is oriented so that the group of lowest priority is directed straight away from the viewer, and the remaining groups are directed toward the viewer in a tripodal fashion. If the direction of decrease in order of priority is clockwise, the configuration is R (Latin, *rectus*); if counterclockwise, it is S (Latin, *sinister*). For an application of this system to acylglycerols, consider the structures in Fig. 1. In both instances, the H atom on asymmetric carbon 2 is the substituent of lowest priority and thus is depicted as beneath the plane of the paper. Among the substituents of carbon 2, oxygen has the highest rank. The remaining two substituents next to the center of asymmetry are —CO— (hydrogens on carbons 1 and 3 are disregarded since 0 has a higher priority). Thus, a comparison must be made of the atoms attached to these —CO— groups. Long saturated acyls have higher priority than short. Unsaturated chains outrank saturated chains, a double bond outranks single branching, two double bonds outrank one, cis outranks trans, and a branched chain out-

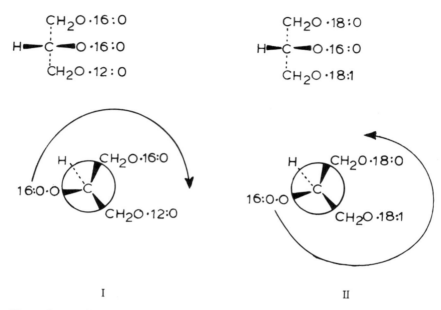

Figure 1 Application of the R/S system of nomenclature for triacylglycerols.

ranks an unbranched chain. Accordingly, the groupings in configuration I have a clockwise arrangement in order of decreasing priority, and configuration I is regarded as R. Similarly, configuration II is S.

Although the R/S system does convey the stereochemical configuration of acylglycerols, it is obvious that the designated structure depends on the nature of the acyl groups in positions 1 and 3 of the triacylglycerol molecule. Consequently, this system cannot be applied in situations in which these positions contain mixtures of fatty acids, as in the case of natural fats, unless individual triacylglycerols are separated.

2. Stereospecific Numbering

The sn-system as proposed by Hirschmann (47) is simple and applicable to both synthetic and natural fats and has now been universally adopted. The usual Fischer planar projection of glycerol is utilized with the middle hydroxyl group positioned on the left side of the central carbon. The carbon atoms are numbered 1 to 3 in the conventional top-to-bottom sequence.

$$
\begin{array}{ll}
CH_2OH & 1 \\
HO\!\!-\!\!C\!\!-\!\!H & 2 \\
CH_2OH & 3 \\
\end{array}
$$

If, for example, stearic acid is esterified at the sn-1 position, oleic at sn-2, and myristic at sn-3, the acylglycerol would appear as

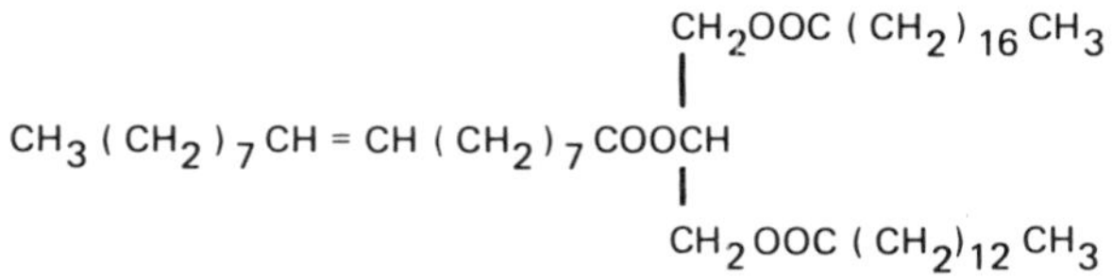

$$CH_2OOC(CH_2)_{16}CH_3$$
$$|$$
$$CH_3(CH_2)_7CH=CH(CH_2)_7COOCH$$
$$|$$
$$CH_2OOC(CH_2)_{12}CH_3$$

and would be designated 1-stearoyl-2-oleoyl-3-myristoyl-sn-glycerol, sn-glycerol-1-stearate-2-oleate-3-myristate, sn-StOM, or sn-18:0-18:1-16:0.

The following prefixes are now widely used with abbreviations to designate the positional distribution of fatty acids within triacylglycerol molecules.

sn: used immediately preceding the term "glycerol," indicates that the sn-1, sn-2, and sn-3 position are listed in that order.

rac: racemic mixture of two enantiomers. The middle acid in the abbreviation is attached at the sn-2 position, but the remaining two acids are equally divided between sn-1 and sn-3 (e.g., rac-StOM indicates equal amounts of sn-StOM and sn-MOSt).

β: middle acid in the abbreviation is at the sn-2 position but positioning of the remaining two is unknown (e.g., β-StOM indicates a mixture of sn-StOM and sn-MOSt in any proportion.

No prefix is given in case of monoacid acylglycerols (e.g., MMM) or if the positional distribution of the acids is unknown, and hence any mixture of isomers is possible (e.g., StOM indicates a possible mixture of sn-StOM, sn-MOSt, sn-OStM, sm-MStO, sn-StMO, and sn-OMSt in any proportion).

C. Phospholipids

The term "phospholipid" may be used for any lipid containing phosphoric acid as a mono- or diester. "Glycerophospholipid" signifies any derivative of glycerophosphoric acid that contains an *O-acyl, O*-alkyl, or *O*-alkenyl group attached to the glycerol residue. The common glycerophospholipids are named as derivatives of phosphatidic acid, such as 3-sn-phosphatidylcholine (trivial name, lecithin), or by their systematic name, similar to the system for triacylglycerols. The term "phospho" is used to indicate the phosphodiester bridge; for example, 1-palmitoyl-2-linoleyl-sn-glycero-3-phosphocholine is the designation for the compound

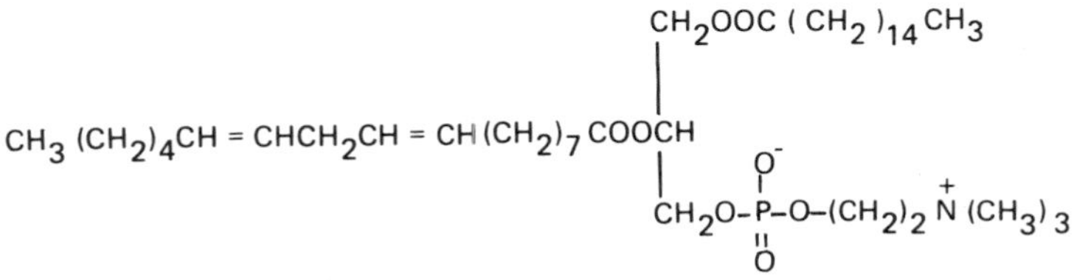

$$CH_2OOC(CH_2)_{14}CH_3$$
$$|$$
$$CH_3(CH_2)_4CH=CHCH_2CH=CH(CH_2)_7COOCH$$
$$|\quad\quad O^-$$
$$CH_2O-P-O-(CH_2)_2\overset{+}{N}(CH_3)_3$$
$$\underset{O}{\overset{||}{}}$$

III. CLASSIFICATION

A general classification of lipids based on their structural components is presented in Table 2. Such a classification, however, may be too rigid for a group of compounds so di-

Table 2 Classification of Lipids

Major class	Subclass	Description
Simple lipids	Acylglycerols	Glycerol + fatty acids
	Waxes	Long-chain alcohol + long-chain fatty acid
Compound lipids	Phosphoacylglycerols (or glycerophospholipids)	Glycerol + fatty acids + phosphate + another group usually containing nitrogen
	Sphingomyelins	Sphingosine + fatty acid + phosphate + choline
	Cerebrosides	Sphingosine + fatty acid + simple sugar
	Gangliosides	Sphingosine + fatty acid + complex carbohydrate moiety that includes sialic acid
Derived lipids	Materials that meet the definition of a lipid but are not simple or compound lipids	Examples: carotenoids, steroids, fat-soluble vitamins

verse as lipids, and should be used only as a guide. It also should be recognized that other classifications sometimes may be more useful. For example, the phosphoacylglycerols and the sphingomyelins can be classed as phospholipids because of the presence of phosphate. Similarly, the cerebrosides and the gangliosides also can be classed as glycolipids because of the presence of carbohydrate, and the sphingomyelins and the glycolipids also can be classed as sphingolipids because of the presence of sphingosine.

The most abundant class of food lipids is the acylglycerols, which dominate the composition of animal or vegetable fats and oils. Acylglycerols are traditionally classified into the following subgroups.

A. Milk Fats

Fats of this group are derived from the milk of ruminants, particularly dairy cows. Although the major fatty acids of milk fat are palmitic, oleic, and stearic, this fat is unique among animal fats in that it contains appreciable amounts of the shorter chain acids C4 to C12, and small amounts of branched and odd-numbered acids.

B. Lauric Acids

Fats of this group are derived from certain species of palm, such as coconut and babasu. The fats are characterized by their high content of lauric acid (40-50%), moderate amounts of C6, C8, and C10 fatty acids, low content of unsaturated acids, and their relatively low melting points.

C. Vegetable Butters

Fats of this group are derived from the seeds of various tropical trees and are distinguished by their narrow melting range, which is mainly due to the arrangement of fatty acids in their triacylglycerol molecules. In spite of their large ratio of saturated to unsaturated fatty acids, trisaturated glycerides are not present. The vegetable butters are extensively used in the manufacture of confections, with cocoa butter the most important member of the group.

D. Oleic-Linoleic Acids

Fats in this group are the most abundant. The oils are all of vegetable origin and contain large amounts of oleic and linoleic acids, and less than 20% saturated fatty acids. The most important members of this group are cottonseed, corn, peanut, sunflower, safflower, olive, palm, and sesame oils.

E. Linolenic Acids

Fats in this group contain substantial amounts of linolenic acid. Examples are soybean oil, wheat germ, hempseed, and perilla oils, with soybean the most important. The abundance of linolenic acid in these oils is responsible for the development of an off-flavor problem known as flavor reversion.

F. Animal Fats

This group consists of depot fats from domestic land animals (e.g., lard and tallow), all containing large amounts of C16 and C18 fatty acids, medium amounts of unsaturated acids, mostly oleic and linoleic, and small amounts of odd-numbered acids. These fats also contain appreciable amounts of fully saturated triacylglycerols and exhibit relatively high melting points.

G. Marine Oils

These oils typically contain large amounts of long-chain polyunsaturated fatty acids, with up to six double bonds, and they are usually rich in vitamins A and D. Because of their high degree of unsaturation, they are less resistant to oxidation than other animal or vegetable oils.

IV. PHYSICAL ASPECTS

A. Theories of Triacylglycerol Distribution Patterns

1. Even or Widest Distribution

This theory resulted from a series of systematic studies initiated by Hilditch and Williams on the quantitative triacylglycerol composition of natural fats (46). Component fatty acids in natural fat tend to be distributed as broadly as possible among all the triacylglycerol molecules. Accordingly, if an acid S forms less than one-third of the total fatty acids present, it should not appear more than once in any triacylglycerol. If X refers to the other acids present, only XXX and SXX species will be found. If an acid forms between one-third and two-thirds of the total fatty acids, it should occur at least once, but never three times in any one molecule; that is, only SXX and SSX will be present. If an acid forms more than two-thirds of the total acids, it should occur at least twice in every molecule; that is, only SSX and SSS will be present.

Shortcomings of the even distribution method were soon realized. Analysis of many natural fats, especially those of animal origin, revealed marked deviation from the theory. Trisaturated acylglycerols were found in fats containing less than 67% saturated fatty acids. The theory can be applied only to two component systems and does not take into account positional isomers. Thus, this theory is no longer considered valid.

2. Random (1,2,3-Random) Distribution

According to this theory, fatty acids are distributed randomly both within each triacylglycerol molecule and among all the triacylglycerols. Thus, the fatty acid composition of all three positions should be the same and also equivalent to the fatty acid composition of the total fat. The proportion of any given fatty acid species expected on the basis of this theory can be calculated according to the equation

$$\%sn\text{-}XYZ = [mol\% \ X \ \text{in total fat}] \times [mol\% \ Y \ \text{in total fat}] \times [mol\% \ Z \ \text{in total fat}] \times 10^{-4}$$

where X, Y, and Z are the component fatty acids at positions 1, 2, and 3 of the acylglycerol. For example, if a fat contains 8% palmitic acid, 2% stearic, 30% oleic, and 60% linoleic, 64 triacylglycerol species ($n = 4$, $n^3 = 64$) would be predicted. The following examples illustrate calculations for such species.

$$\%sn\text{-}OOO = 30 \times 30 \times 30 \times 10^{-4} = 2.7$$

$$\%sn\text{-}PLSt = 8 \times 60 \times 2 \times 10^{-4} = 0.096$$

$$\%ns\text{-}LOL = 60 \times 30 \times 60 \times 10^{-4} = 10.8$$

Most fats do not conform to a completely random distribution pattern. For example, the proportion of fully saturated triacylglycerols expected on the basis of random distribution exceeds, in many cases, that found experimentally. Modern techniques of analysis have revealed that in natural fats the fatty acid composition of the sn-2 position is always different from that of the combined 1,3 positions. Calculations of random distribution are, however, useful for understanding other hypotheses and for predicting distribution patterns of fatty acids in fats randomized by interesterification.

3. Restricted Random

According to this hypothesis, first proposed by Kartha (56), saturated and unsaturated fatty acids in animal fats are distributed randomly. However, fully saturated triacylglycerols (SSS) may be present only to the extent that the fat can remain fluid in vivo. Excess SSS, according to this theory, can be exchanged with UUS and UUU to form SSU and SUU. Kartha's calculations do not account for positional isomers or positioning of individual acids.

4. 1,3-Random-2-Random

The fatty acid composition at the 2 position is known to be different from that of the 1 or 3 positions. This theory assumes that two different pools of fatty acids are separately and randomly esterified to the 2 and 1,3 positions. Thus, composition at positions 1 and 3 will, presumably, be identical.

On the basis of this hypothesis, the amount of a given triacylglycerol can be computed as

%sn-XYZ = [mol% X at 1,3] × [mol% Y at 2] × [mol% Z at 1,3] × 10^{-4}

Composition of the sn-2 and/or the combined 1,3 positions can be obtained by analysis of the mono- or diacylglycerols derived from partial deacylation by chemical or enzymic methods (23,64).

5. 1-Random-2-Random-3-Random

According to this theory, three different pools of fatty acids are separately but randomly distributed at each of the three positions of the triacylglycerol molecules of a natural fat. Thus, the relative fatty acid amounts in the sn-1, sn-2, and sn-3 positions should all be different. The content of a given triacylglycerol species can be calculated as

%sn-XYZ = [mol% X at sn-1] × [mol% Y at sn-2] × [mol% Z at sn-3] × 10^{-4}

To calculate the range of molecular species expected to be present in a natural fat on the basis of this theory, the fatty acid compositions of the sn-1 and the sn-3 position must be distinguished. This can be determined by the techniques of Brockerhoff (14) or Lands et al. (60).

B. Positional Distribution of Fatty Acids in Natural Fats

In earlier studies, major classes of triacylglycerols were separated on the basis of unsaturation (i.e., trisaturated, disaturated, diunsaturated, and triunsaturated) via fractional crystallization and oxidation-isolation methods. More recently, the techniques of stereospecific analysis made possible the detailed determinations of individual fatty acid distribution in each of the three positions of the triacylglycerols of many fats. The data listed in Table 3 clearly indicate the differences in distribution patterns among plant and animal fats.

1. Plant Triacylglycerols

In general, seed oils containing common fatty acids show preferential placement of unsaturated fatty acids at the sn-2 position. Linoleic acid is especially concentrated at this position. The saturated acids occur almost exclusively at the 1,3 positions. In most cases, the individual saturated or unsaturated acids are distributed in approximately equal quantities between the sn-1 and the sn-3 positions.

The more saturated plant fats show a different distribution pattern. Approximately 75% of the triacylglycerols in cocoa butter are disaturated, with 18:1 concentrated in the 2 position and the saturated acids almost exclusively located in the primary positions (β-POS constitutes the major species). There is approximately 1½ times as much oleic in the sn-1 as in the sn-2 position.

Approximately 80% of the triacylglycerols in coconut oil are trisaturated, with lauric acid concentrated at the sn-2 position, octanoic at the sn-3 position, and myristic and palmitic at the sn-1 position.

Plants containing erucic acid, such as rapeseed oil, show considerable positional selectivity in placement of their fatty acids. Erucic acid is preferentially located at the 1,3 position but more of it is present at the sn-3 position than at the sn-1 position.

Table 3 Positional Distribution of Individual Fatty Acids in Triacylglycerols of Some Natural Fats

Source of	Position	4:0	6:0	8:0	10:0	12:0	14:0	16:0	18:0	18:1	18:2	18:3	20:0	20:1	22:0	24:0
								Fatty acid (mol %)								
Cow's milk	1	5	3	1	3	3	11	36	15	21	1					
	2	3	5	2	6	6	20	33	6	14	3					
	3	43	11	2	4	3	7	10	4	15	0.5					
Coconut	1		1	4	4	39	29	16	3	4						
	2		0.3	2	5	78	8	1	0.5	3	2					
	3		3	32	13	38	8	1	0.5	3	2					
Cocoa butter	1							34	50	12	1					
	2							2	2	87	9					
	3							37	53	9						
Corn	1							18	3	28	50					
	2							2	–	27	70					
	3							14	31	52	1					
Soybean	1							14	6	23	48	9				
	2							1	–	22	70	7				
	3							13	6	28	45	8				
Olive	1							13	3	72	10	0.6				
	2							1	–	83	14	0.8				
	3							17	4	74	5	1				
Peanut	1							14	5	59	19	–	1	1	–	1
	2							2	–	59	39	–	–	–	–	0.5
	3							11	5	57	10	–	4	3	6	3
Beef (depot)	1						4	41	17	20	4	1				
	2						9	17	9	41	5	1				
	3						1	22	24	37	5	1				
Pig (outer back)	1						1	10	30	51	6					
	2						4	72	2	13	3					
	3						–	–	7	73	18					

2. Animal Triacylglycerols

Triacylglycerol distribution patterns differ among animals and vary among parts of the same animal. Depot fat can be altered by changing dietary fat. In general, however, the saturated acid content of the sn-2 position is greater than in plant fats, and the difference in composition between the sn-1 and sn-2 positions is also greater. In most animal fats the 16:0 acid is preferentially esterified at the sn-1 position and the 14:0 at the sn-2 position. The short-chain acids in milk fat are selectively associated with the sn-3 position. The major triacyglycerols of beef fat are of the SUS type.

Pig fat is unique among animal fats. The 16:0 acid is significantly concentrated at the central position, the 18:0 acid is primarily located at the sn-1 position, 18:2 at the sn-3 position, and large amounts of oleic acid occur at positions 3 and 1. Major triacylglycerol species in lard are sn-SPS, OPO, and POS.

The long polyunsaturated fatty acids, typical of marine oils, are preferentially located at the sn-2 position.

C. Consistency

1. Crystallographic Structure

Most of our present knowledge regarding the crystalline structure and behavior of fats has resulted from x-ray diffraction studies. However, significant insights also have been gained from the application of other techniques, such as nuclear magnetic resonance, infrared spectroscopy, calorimetry, dilatometry, microscopy, and differential thermal analysis.

In the solid or crystalline state, atoms or molecules assume rigid positions forming a repeatable, highly ordered, three-dimensional pattern. If reference points representing this regularity of structure are chosen (e.g., the center of a certain atom or a convenient point in a molecule), the resulting three-dimensional arrangement in space is known as a *space lattice*. This network of points embodies all the symmetry properties of the crystal. If the points of a space lattice are joined, a series of parallel-sided unit cells is produced, each of which contains all elements of the lattice. A complete crystal can thus be regarded as unit cells packed side by side in space. In the example of the simple space lattice given in Fig. 2, each unit cell has an atom or molecule at each of its corners. However, since each corner is shared by eight other adjacent cells, there is only one atom (or molecule) per unit cell. It can be seen that each point in the space lattice is similar in its environment to all other points. The ratios a:b:c (called axial ratios), as well as the angles between the crystallographic axes OX, OY, and OZ, are constant values usually used to distinguish different lattice arrangements.

Organic compounds having long chains pack side by side in the crystal to obtain maximum van der Waals interaction. In the unit cell three spacings can be identified, two short and one long. Thus, the long spacing of normal alkanes increases steadily with an increase in carbon number, but the short spacings remain constant. The molecular end groups (e.g., methyl or carboxyl) associate with each other to form planes. If the chains are tilted with respect to the base of the unit cell, the long spacing will be smaller to an extent depending on the angle of tilt. Fatty acids tend to form double molecules oriented head to head by sharing hydrogen bonds between carboxyl groups. Consequently, the long spacings of fatty acids are almost twice as great as those of hydrocarbons of equal carbon number (Fig. 3).

When lipid mixtures of different but similar compounds are present, crystals containing more than one kind of molecule can be formed. In case of medium or low-molecu-

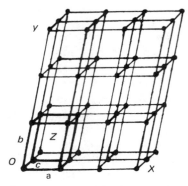

Figure 2   Crystal lattice.

lar-weight fatty acids differing in chain length by one carbon atom, compound crystals are formed in which dissimilar pairs are bound carboxyl to carboxyl but otherwise arranged as in crystals containing only one acid. Also common are solid solutions in which component molecules of one type are distributed at random into the crystal lattice of another. Slow cooling can result under certain conditions in the formation of layer crystals in which layers of one type of crystal deposit on the crystal surfaces of another.

2. Polymorphism

Polymorphic forms are solid phases of the same chemical composition that differ among themselves in crystalline structure but yield identical liquid phases upon melting. Dia-

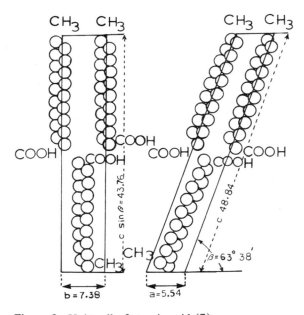

Figure 3   Unit cell of stearic acid (7).

monds and black carbon are polymorphs. Each polymorphic form, sometimes termed "polymorphic modification," is characterized by specific properties, such as x-ray spacing, specific volume, and melting point, distinguishing it from all other forms of the same compound. Several factors determine the polymorphic form assumed upon crystallization of a given compound. These include purity, temperature, rate of cooling, presence of crystalline nuclei, and type of solvent.

Depending of their particular stabilities, transformation of one polymorphic form into another can take place in the solid state without melting. Two crystalline forms are said to be "monotropic" if one is stable and the other metastable throughout their existence and regardless of temperature change. Transformation will take place only in the direction of the more stable form. Two crystalline forms are "enantiotropic" when each has a definite range of stability. Either modification may be the stable one, and transformation in the solid state can go in either direction, depending upon the temperature. The temperature at which their relative stability changes is known as the *transition point*. Natural fats are invariably monotropic, although enantiotropism is known to occur in some fatty acid derivatives.

With long-chain compounds, polymorphism is associated with different packing arrangements of the hydrocarbon chains or different angles of tilt. The mode of packing can be described by using the subcell concept.

Subcells. A subcell is the smallest spatial unit of repetition along the chain axes within the main unit cell. The schematic in Fig. 4 represents a "subcell lattice" in a fatty acid crystal. In this case, each subcell contains one ethylene group, and the height of the subcell is equivalent to the distance between alternate carbon atoms in the hydrocarbon chain, that is, 2.54 Å. The methyl and acid groups are not part of the subcell lattice.

Seven packing types for hydrocarbon subcells have been observed. The most common types are the three shown in Fig. 5. In the triclinic (T//) packing, also called β, the two methylene units together make up the ethylene repeat unit of which there is one per subcell, and all zigzag planes are parallel. This subcell packing occurs in *n*-hydrocarbons, fatty acids, and triacylglycerols.

The common orthorhombic (O⊥) packing, also called β', has two ethylene units in each subcell. Alternate-chain planes are perpendicular to their adjacent planes. This packing occurs with *n*-paraffins and with fatty acids and their esters.

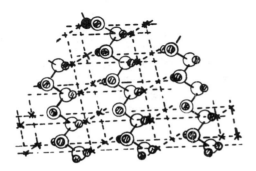

Figure 4 Subcell lattice in a fatty acid crystal (76).

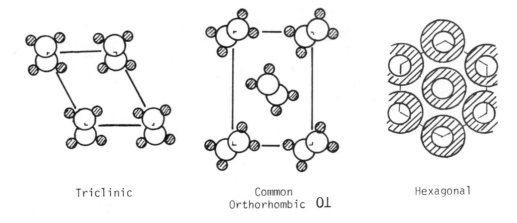

Triclinic   Common   Hexagonal
           Orthorhombic OⱢL

Figure 5   Common types of hydrocarbon subcell packing (76).

The hexagonal packing (H), generally called α, occurs with hydrocarbons just below their melting points. The chains are randomly oriented and exhibit rotation about their long vertical axes. This type of packing is observed with hydrocarbons, alcohols, and ethyl esters.

Fatty Acids. Depending on the method by which they are obtained, the even-numbered saturated fatty acids can crystallize in any of the polymorphic forms. In decreasing length of long spacings (or increasing tilt of the chains), these are designated A, B, and C. Similarly, the acids with odd numbers of carbon atoms are designated A′, B′, and C′. The A and A′ forms have triclinic subcell chain packing (T∥); the remaining forms are packed in the common orthorhombic (OⱢL) manner.

The β form of stearic acid has been studied in detail. The unit cell is monoclinic and contains four molecules. Its axial dimensions are: a = 5.54 Å, b = 7.38 Å, and c = 48.84 Å. However, the c-axis is inclined at an angle of 63°38′ from the a-axis, which results in a long spacing of 43.76 Å (Fig. 3).

In the case of oleic acid, the low-melting form has double molecules per unit cell length, but the hydrocarbon portions about the cis double bond are tilted in opposite directions (Fig. 6).

Triacylglycerols. The nomenclature used in earlier literature to designate the different polymorphic forms of triacylglycerols is extremely confusing. Different authors used different criteria, such as melting points or x-ray spacings, as the basis for their nomenclature. Often the same symbols were used by different investigators to designate different polymorphic forms, and much debate took place regarding the number of forms exhibited by certain triacylglycerol species. Much of the disagreement was finally resolved based on results from infrared spectroscopy. In general, triacyglycerols, due to their relatively long chains, take on many of the features of hydrocarbons. They exhibit, with some exceptions, three principal polymorphic forms: α, β', and β. Characteristics typical of each form are summarized in Table 4.

If a monoacid triacylglycerol, such as StStSt, is cooled from the melt, it crystallizes in the least dense, lowest melting form, α. On further cooling of the α form, the

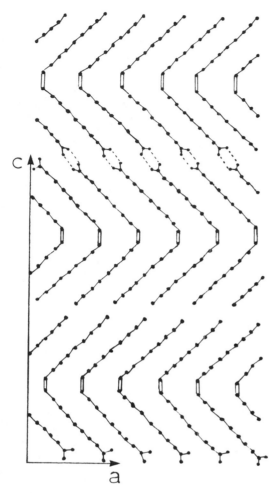

Figure 6 Crystal structure of oleic acid (1).

Table 4 Characteristics of the Polymorphic Forms of Monoacid Triacylglycerols

Characteristic	α Form	β' Form	β Form
Chain packing	Hexagonal	Orthorhombic	Triclinic
Short spacing	4.2 Å	3.8 and 4.2 Å	4.6 Å
Characteristic infrafred spectrum	Single band at 720 cm^{-1}	Doublet at 727 and 719 cm^{-1}	Single band at 717 cm^{-1}
Density	Least dense	Intermediate	Most dense
Melting point	Lowest	Medium	Highest

chains pack more tightly and gradual transition into the β form takes place. If the α form is heated to its melting point, a rapid transformation to the most stable β form occurs. The β' form can also be obtained directly by cooling the melt and maintaining the temperature a few degrees above the melting point of the α form. On heating the β' form to its melting point, some melting takes place and transition to the stable β form occurs.

The general molecular arrangement in the lattice of monoacid triacylglycerols is a double chain-length modified tuning fork, or chair structure, as shown in Fig. 7 for trilaurin. The chains in the 1 and 3 positions of the glycerol point opposite the chain in the 2 position.

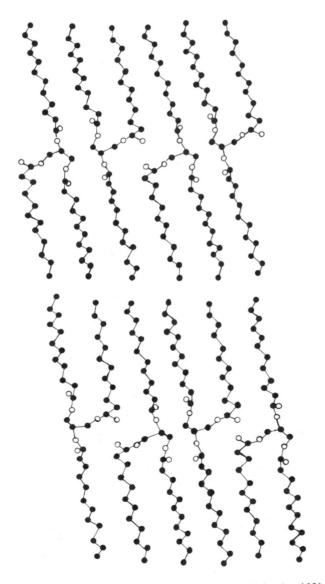

Figure 7   Molecular arrangement in trilaurin lattice (62).

Because triacylglycerols contain a variety of fatty acids, departure from the simple polymorphic classification outlined above is bound to occur. In the case of triacylglycerols containing different fatty acids, some polymorphic forms are more difficult to obtain than others. In some cases, β' forms have been observed that have higher melting points than the β form. PStP glycerides, as in cottonseed hardstock, tend to crystallize in a β' form of relatively high density. This form results in greater stiffening power when added to oils than the expanded (snowlike) β form commonly obtained from the StStSt glycerides of soybean hardstock.

The polymorphic structure of mixed triacylglycerols is further complicated by a tendency for carbon chains to segregate according to length or degree of unsaturation to form structures in which the long spacing is made up of triple chain lengths. Various tuning fork structures have been proposed for mixed triacyglycerols containing fatty acids of different chain lengths. If the middle chain is shorter or longer than the other two by four or more carbons, there may be a segregation of chains, as in Fig. 8a. In the case of unsymmetrical triacylglycerols, a chain-type arrangement similar to Fig. 8b may result. Sorting of chains may also arise on the basis of unsaturation, as in Fig. 8c. Such structures are designated by a number following the Greek letter. For example, β-3 would indicate the β modification with a triple chain length.

Polymorphic Behavior in Commercial Fats. It is evident from the above that the polymorphic behavior of a fat is largely influenced by the composition of its fatty acids and their positional distribution in the glycerides. In general, fats that consist of relatively few closely related triacyglycerol species tend to transform rapidly to stable β forms. Conversely, heterogenous fats tend to transform more slowly to stable forms. For example, highly randomized fats exhibit β' forms that transform slowly.

Fats that tend to crystallize in β forms include soybean, peanut, corn, olive, coconut, and safflower oils, as well as cocoa butter and lard. On the other hand, cottonseed, palm, and rapeseed oils, milk fat, tallow, and modified lard tend to produce β' crystals that tend to persist for long periods. The β' crystals are desirable in the preparation of

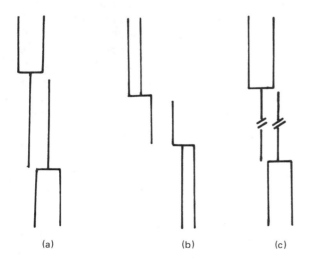

(a)　　　　　　(b)　　　　　(c)

Figure 8 Arrangement of molecules in triacylglycerol crystals.

shortenings, margarine, and baked products since they aid in the incorporation of a large amount of air in the form of small air bubbles, giving rise to products of better plastic and creaming properties.

In the case of cocoa butter, in which the two main glycerides are StOSt and POSt, four polymorphic forms (typical of POSt) have been identified (22,83). These are α-2, β'-2, β-3 V, and β-3 VI, in increasing order of melting point. The β-3 V form is the desired structure, since it produces the bright glossy appearance of chocolate coatings. It is obtained by proper tempering, that is, warming or holding of partially crystallized coating at about 32°C before molding, rapid chilling, and storage at about 16°C. Improper tempering or storage at high temperatures results in the development of "chocolate bloom," a condition involving the deposition of small fat crystals on the chocolate surface giving it a white or gray appearance. The development of bloom has been associated with a change from the β-3 V form to the different but very closely related, β-3 VI form.

3. Formation of Crystals and Solidification

The formation of a solid from a solution or a melt is a complicated process in which molecules must first come in contact, orient, and then interact to form highly ordered structures. It can be assumed, as with chemical reactions, that an energy barrier exists to oppose the aggregation of molecules to form crystals. The more complex and more stable the polymorphic form (i.e., highly ordered, compact, and with a high melting point) the more difficult it is to form the crystal. Thus, stable crystals generally do not form at temperatures just below their melting points, and the lipid persists for some time in a metastable, supercooled state. On the other hand, the least stable, least ordered form (α) crystallizes readily at a temperature very slightly below its melting point.

Although the magnitude of the energy barrier decreases as the temperature is lowered, the rate of formation of crystal nuclei does not increase indefinitely as the temperature is lowered. At some point, decreasing the temperature increase the viscosity of the lipid sufficiently to seriously interfere with the crystallization process.

Crystallization begins in a supercooled liquid with the formation of submicroscopic crystal nuclei. Once this difficult step has occurred, growth of these nuclei (crystal growth) progresses at a rate dependent on the temperature. Nucleus formation can be encouraged by stirring, or the nucleation process can be circumvented by seeding the supercooled liquid with tiny crystals of a form similar to one of the desired natural forms.

4. Melting

Shown in Fig. 9 are schematic heat content curves for the stable β form and the metastable α form of a simple triacylglycerol. Heat is absorbed upon changing from solid to liquid. The curve ABC represents the increase in heat content of the β form with increase in temperature. At the melting point, heat is absorbed with no rise in temperature (heat of fusion) until all solid is transformed into liquid (final melting at point B). On the other hand, transformation from an unstable to a stable polymorphic form (beginning at point E in Fig. 9 and extending to curve ABC) involves an evolution of heat.

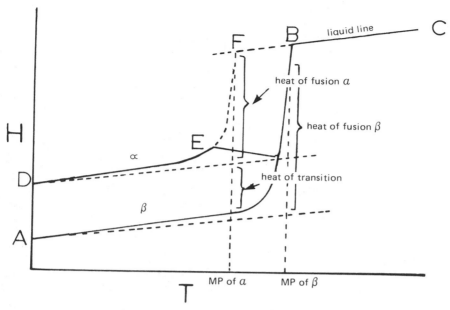

Figure 9 Heat content (H) melting curves of stable (β) and unstable (α) polymorphic forms.

Fats similarly expand upon melting and contract upon polymorphic transformation. Consequently, if the change in specific volume (dilation) is plotted against temperature, dilatometric curves very similar to calorimetric curves are obtained, with the melting dilation corresponding to heat of fusion and the coefficient of expansion corresponding to specific heat. Since dilatometric measurements involve very simple instruments, they are much more practical than calorimetric methods. Dilatometry is widely used to determine the melting behavior of fats. If several components of different melting points are present, melting takes place over a wide range of temperature, and dilatometric or calorimetric curves similar to the schematic in Fig. 10 are obtained.

Point X represents the beginning of melting; below this point, the system is completely solid. Point Y represents the end of melting; above this point the fat is completely liquid. The curve XY represents the gradual melting of the solid components of the system. If the fat melts over a narrow range of temperature, the slope of the melting curve is steep. Conversely, a fat is said to have a wide plastic range if the difference in temperature between the beginning and the end of melting is large. Thus, the plastic range of a fat can be extended on either side of the melting curve by adding a relatively high melting or low melting component.

Solid Fat Index. The proportion of solids and liquids in a plastic fat at different temperatures can be estimated by constructing calorimetric or, preferably, melting dilation curves similar to that of Fig. 10. Using a dilatometer, for example, measurements of specific volume can be made starting at temperatures sufficiently low to establish a solid line, at other temperatures high enough to establish a liquid line, and at intervals between these to determine the melting curve. The solid and liquid lines can then be extrapolated, and the solid or liquid fraction at any temperature can be calculated as

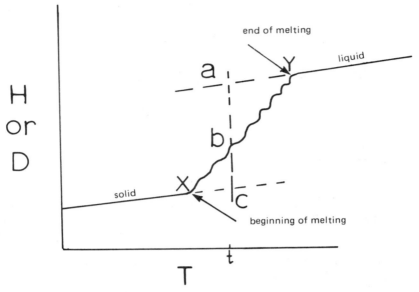

Figure 10 Heat content (H) or dilatometric (D) melting curve of a glyceride mixture.

shown in Fig. 10, where ab/ac represents the fraction of solids, and bc/ac is the fraction of liquid at temperature t. The ratio of solid to liquid is known as the solid fat index (SFI), and this has relevance to the functional properties of fats in foods.

Phase Diagrams. A phase diagram is a graphic representation of conditions enabling equilibrium between phases in a system of one or more components. A diagram of a simple binary system consisting of two components, A and B, completely miscible in the liquid state but with no intersolubility in their solid states, is given in Fig. 11. The diagram is obtained by determining the solidification temperature of various mixtures of the two liquids. Points A and B represent the melting points of the two pure compounds. Above ABC only liquid can exist; below DCE only solid can exist. The regions ACD and BCE represent conditions under which liquid mixtures of varying compositions are in equilibrium with solid A or solid B, respectively. Point C is the eutectic point. The relevance of this figure will become evident in the next section.

The reader is referred to any text on physical chemistry for a detailed understanding of the phase behavior of other multicomponent systems exhibiting compound formation, polymorphic transformation, and partial or complete miscibility in the solid state.

5. Consistency of Commercial Fats

Although natural fats and products derived from them contain exceedingly complex mixtures of large numbers of individual acylglycerols of different composition and structure, they show a remarkable tendency to behave as simple mixtures of only a few components. Each group of similar compounds appears to act as a single component, so only the distinctly different groups are apparent in melting behavior. This tendency for simplification in the melting behavior of complex mixtures is indeed fortunate since it permits the application of rules governing simple mixtures to the more complex natural or

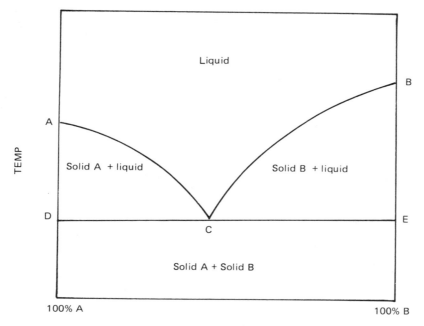

Figure 11 Phase diagram of a simple binary system.

processed fats. An initial reduction in the solidification point of certain mixtures of commercial fats has been observed upon hydrogenation or the addition of a high-melting component. Such systems give rise to eutectic-type curves of striking resemblance to the liquid curve of Fig. 11.

The dilatometric curves of plastic fats do not show a smooth melting line but rather they consist of a series of somewhat linear segments with visible changes in slope (Fig. 12). The inflection points between the segments are sometimes designated K points with K the point of final melting. The K points obviously correspond to specific boundaries in the phase diagrams of the complex fat; that commercial fats give rise to only a few K points emphasizes the tendency for different individual components in a narrow melting range to behave like a single component.

If a high-melting fraction is added to a natural fat, melting of the "hard" component is clearly reflected in the dilatometric curve. The vertical distance (d in Fig. 13) provides a relative measure of the amount of the hard component.

It can be seen from Figs. 12, 13, and 14 that much useful information regarding the melting characteristics of plastic fats can be obtained from dilatometric curves.

Hard butters melt over a relatively narrow range of temperature because their triacylglycerols are mainly β-POSt, β-StOSt, and β-POP. This abrupt melting, which occurs at the temperature of the mouth, makes such fats particularly suitable for confectionary coatings.

A typical dilatometric curve for milk fat, showing almost complete melting at the temperature of the mouth, is quite different from that of lard, which exhibits a more gradual course of melting (Fig. 14). The difference in solid content of the two fats at any given temperature is clearly evident.

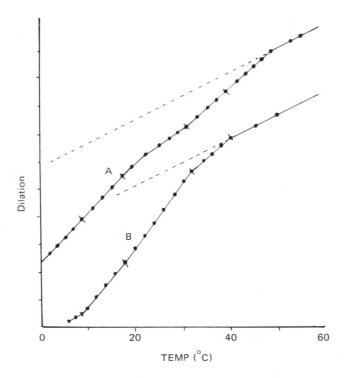

Figure 12 Dilatometric curves of (a) a typical all-hydrogenated shortening and (b) a typical American margarine (7).

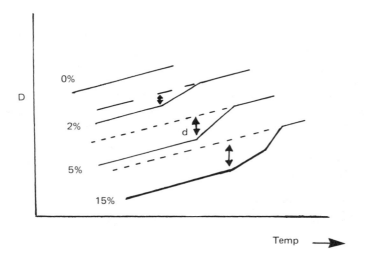

Figure 13 Upper portions of dilatometric curves of cottonseed oil mixed with 0-15% highly hardened cottonseed oil (7).

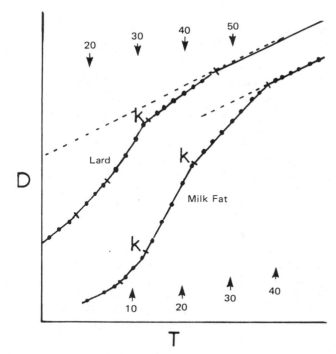

Figure 14 Dilatometric curves of lard and milk fat (7).

A new approach to the measurement of solids in fats involves the use of broadline nuclear magnetic resonance (NMR). This technique is now being used in commercial SFI control practices.

6. Mesomorphic Phases (Liquid Crystals)

As discussed earlier, lipids in the solid state exist as highly ordered structures (crystalline) with regular three-dimensional arrangements of their molecules. In the liquid state, the intermolecular forces are weakened and the molecules acquire freedom of movement and assume a state of almost complete disorder. In addition, phases with properties intermediate between those of the liquid and the crystalline states are known to occur, and these mesomorphic phases consist of so-called liquid crystals. Typically amphiphilic compounds, that is, compounds with polar and nonpolar portions of their molecules, give rise to mesomorphic phases. For example, when an amphiphilic crystalline compound is heated, the hydrocarbon region may melt (i.e., it is transformed into a state of disorder similar to that in the liquid state), before the true melting point is reached. This occurs because relatively weak van der Waals forces exist among the hydrocarbon chains as compared with the somewhat stronger hydrogen bonding that exists among polar groups. Pure crystals that form liquid crystals upon heating are said to be *thermotropic*. In the presence of water, at temperatures above the melting point of the hydrocarbon region (the so-called Krafft temperature), hydrocarbon chains of triacylglycerols transform into a disordered state and water penetrates among the ordered polar groups. Liquid crystals formed in this manner, that is, with the aid of a solvent, are said to be *lyotropic*.

Mesomorphic structure depends on such factors as the concentration and chemical structure of the amphiphilic compound, water content, temperature, and the presence of other components in the mixture. The principle kinds of mesomorphic structures are lamellar, hexagonal, and cubic.

Lamellar or Neat. This structure corresponds to that existing in biological bilayer membranes. It is made up of double layers of lipid molecules separated by water (Fig. 15a). Structures of this kind are usually less viscous and less transparent than the other mesomorphic structures.

The capacity of a lamellar phase to retain water depends on the nature of the lipid constituents. Lamellar liquid crystals of monoacylglycerols, for example, can accommodate up to approximately 30% water, corresponding to a water layer thickness of about 16 Å between the lipid bilayers. However, almost infinite swelling of the lamellar phase can occur if small amounts of an ionic surface-active substance is added to the water. If the water content is increased above the swelling limit of the lamellar phase, a dispersion of spherical aggregates consisting of concentric, alternating layers of lipid and water gradually forms.

In general, a lamellar liquid crystalline phase tends to transform upon heating into hexagonal II or cubic mesophases. On the other hand, if the lamellar liquid-crystal phase is cooled below the Krafft temperature, a metastable "gel" will form in which water remains between the lipid bilayers and the hydrocarbon chains recrystallize. During extended holding the water is expelled and the gel phase transforms into a microcrystalline suspension in water, called a coagel.

Hexagonal. In this structure the lipids form cylinders arranged in a hexagonal array. The liquid hydrocarbon chains fill the interior of the cylinders, and the space between the cylinders is taken up by water (Fig. 15b). This type of liquid crystal is termed "hexagonal I" or "middle." A reversed hexagonal structure, "hexagonal II," is also possible in which the water fills the interior of the cylinders and is surrounded by the polar groups of the amphiphile. The hydrocarbon chains extend outward making up the continuous phase between the cylinders (Fig. 15c). If hexagonal I liquid crystals are diluted with water, spherical micelles form. However, dilution of hexagonal II liquid crystals with water is not possible.

Cubic or Viscous Isotropic. Although this state is encountered with many long-chain compounds, it is not as well characterized as are the lamellar and hexagonal liquid crystals. Cubic-phase structures were studied in monoglyceride-water systems by Larsson (61), who observed the existence of closed water regions in these systems and proposed a cubic-phase model based on space-filling polyhedra packed in a body-centered cubic lattice (Fig. 15d). The cubic phases are usually very viscous and completely transparent.

In biological systems mesomorphic states are of great importance in many physiological processes; for example, they influence the permeability of cell membranes. Liquid crystals also play a significant role with regard to the stability of emulsions, as is discussed later.

7. Factors Influencing Consistency

The following is a summary of the main factors affecting the consistency of commercial fats.

Proportion of Solids in the Fat. The greater the solids content, the firmer the fat.

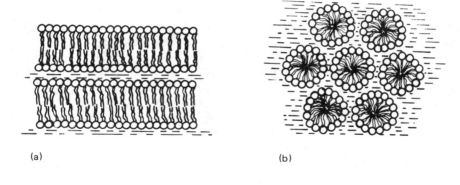

(a)   (b)

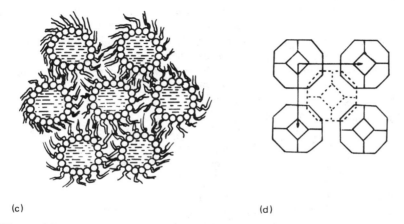

(c)   (d)

Figure 15 Mesomorphic structures of lipids: (a) lamellar, (b) hexagonal I, (c) hexagonal II, (d) cubic (62).

Number, Size, and Kind of Crystals. At a given solids content, an abundance of small crystals produces a harder fat than do a few coarse crystals. Large, soft crystals are typically produced by slow cooling. Crystals composed of high-melting glycerides provide greater stiffening power than those of lower melting glycerides.

Viscosity of the Liquid. Part of the change in consistency in response to temperature is associated with changes in viscosity of the melt.

Temperature Treatment. If a fat tends to supercool excessively, this can be overcome by melting it at the lowest possible temperature, holding it for an extended period of time at a temperature just above its melting point, and then cooling it. This facilitates formation of a large number of crystal nuclei, numerous small crystals, and a firm consistency.

Mechanical Working. Crystallized fats are generally thixotropic; that is, they become reversibly softer after vigorous agitation.

D. Emulsions and Emulsifiers

An emulsion has generally been described as a system containing two immisible liquid phases, one of which is dispersed in the other as droplets varying between 0.1 and 50 μm in diameter. The phase present in the form of droplets is called the "internal" or "dispersed" phase; the matrix in which the droplets are dispersed is termed the "external" or "continuous" phase. The importance of mesomorphic or liquid crystalline phases to the properties of emulsions has been realized only recently and is reflected in the 1972 definition of an emulsion by the IUPAC: "In an emulsion, liquid droplets and/or liquid crystals are dispersed in a liquid" (48). The abbreviations O/W and W/O are frequently used to indicate the type of emulsion, that is, oil-in-water and water-in-oil, respectively.

The formation of small dispersed droplets is associated with an increase in interfacial area between the two liquids, the increase occurring exponentially with a decrease in droplet diameter. In practical situations, the interfacial area can become almost unbelievably large. For example, if 1 ml oil is dispersed as 1 μm diameter particles in water, 1.9×10^{12} globules are created and the total interfacial area is 6 m^2.

The volume percentage of the dispersed phase can vary from a very small value, such as 2-3% in milk, to a large value, such as 65-80% in a mayonnaise, or even to 99% in experimental emulsions. It is interesting to note that nondeformable, perfect spheres of a uniform size, when packed to maximum density, occupy 75% of the sample volume. Thus, emulsions with dispersed-phase volumes of greater than 74% can exist only because of the diversity in globule size and/or the ability of the globules to deform.

The work W required to increase the interfacial area by an amount A can be represented by the relationship

$$W = \alpha \, \Delta A$$

where α is the interfacial tension. Due to the large positive free energy at the interface of the two liquids, emulsions are thermodynamically unstable. Thus, many emulsions tend to destabilize by one or more of the following three mechanisms.

Creaming or *sedimentation* can result from the action of gravitational force on phases that differ in density. The rate of this occurrence obeys Stokes law.

$$V = \frac{2r^2 g \, \Delta p}{9\mu}$$

where V is the velocity of the globule, r is its radius, g is the force of gravity, Δp is the difference in density between the two phases, and μ is the viscosity of the continuous phase. A major deviation from this equation can occur when globules form clusters. When this happens, the radius of the cluster, not the radius of individual globules, must be used to make the equation function accurately.

Flocculation or *clustering* is a second destabilizing mechanism for emulsions. Following flocculation, fat globules move as groups rather than individually. Fat globules in unhomogenized milk are prone to flocculate, and as just mentioned, this enhances the rate of creaming. Flocculation does not involve rupture of the interfacial film that normally surrounds each globule and it therefore does not involve a change in the size of the original globules. An inadequate electrostatic charge at the globule surface is the main cause of flocculation.

Coalescence is the third and most serious form of emulsion destabilization. This involves rupture of the interfacial film, a joining of globules, and a reduction in interfacial area. When carried to the extreme, a planar interface exists between the homogeneous lipid phase and the homogeneous aqueous phase. Contact of globules must precede coalescence, and this can occur through flocculation, creaming or sedimentation, and/or Brownian movement.

To produce stable emulsions, the tendency to minimize interfacial area through coalescence must be counteracted, and this is usually accomplished by adding substances known as *emulsifiers*. These are usually surface-active compounds that adsorb at the interface to lower interfacial tension to provide a physical resistance to coalescence, and sometimes, to increase surface charge.

Emulsions and emulsifiers are of immense importance to the food industry. Milk, cream, mayonnaise, salad dressings, ice cream mix, and cake batters are all O/W emulsions. Butter and margarine are W/O emulsions. The term "meat emulsion" involves a more complex system in which the dispersed phase is fat (solid) in the form of fine particles and the continuous phase is an aqueous matrix containing salts, soluble and insoluble proteins, and particles of muscle fibers and connective tissue.

The expanding development of new types of food products and the continuing mechanization of food processes have increased the use of food emulsions and the need for a better understanding of their properties. Today, numerous emulsifying agents, tailored to satisfy a variety of specific applications, are commercially available.

1. Emulsion Stability

Only a brief discussion of the factors contributing to emulsion stability is given below. For a more comprehensive treatment of the subject the reader is referred to books by Becher (9), Sherman (74), and Friberg (35).

Interfacial Tension. As indicated above, most emulsifying agents are amphiphilic compounds. They concentrate at the oil-water interface, causing a significant lowering of interfacial tension and a reduction in the energy needed to form emulsions. Although lowering of interfacial tension has, historically, been regarded as an important means of achieving emulsion stability, it is only one of several contributing factors. Despite a lowering of interfacial tension when surface-active agents are added, the free energy of the interface remains positive; thus, a state of thermodynamic instability persists.

Repulsion by Electrical Charge. Emulsion stability is often attributable, in large measure, to the presence of repulsive electrical charges on the surfaces of emulsion droplets. Thus, the classic DLVO theory of colloidal stability often has been applied to emulsions. According to this theory, so named from the initials of its principal authors, dispersed particles are subject to two independent forces: the van der Waals force of attraction, and the electrostatic force of repulsion arising from the presence of electrical double-layers at the particle surfaces. The net interaction between the particles is obtained by summing these two terms. If the repulsion potential exceeds the attraction potential, an energy barrier opposing collision results. If the magnitude of this energy barrier exceeds the kinetic energy of the particles, the suspension is stable. The van der Waals potential (negative) becomes significant only when the distance between the particles is quite small. At intermediate distances, the repulsive potential is larger than the attractive potential.

Caution must be exercised in applying the DLVO theory, which was originally developed for inorganic sols (in which the dispersed phase consists of submicroscopic spherical solid particles), to emulsions (in which the dispersed phase consists of oil droplets stabilized by adsorbed emulsifying agents). For example, in emulsions, coalescence involves disruption of an adsorbed film around the droplets, and calculations of the potential energy barrier opposing the collision of oil globules must take into account such factors as the distortion or flattening of the oil droplets upon close approach. Nonetheless, the DLVO theory does provide a good approximation of the electrostatic contribution to emulsion stability.

Ionic surfactants contribute significantly to the stability of O/W emulsions by contributing to the establishment of electric double layers in the aqueous phase adjacent to each oil droplet. On the other hand, this mechanism is of little importance in the stabilization of W/O emulsions, since the oil phase does not generally supply counterions in sufficient amounts to establish a strong potential gradient (2).

Stabilization by Finely Divided Solids. Solid particles of very small size, as compared with the size of the dispersed oil droplet, can stabilize an emulsion by adsorbing at the interface to form a physical barrier around the droplets. Furthermore, energy is required to dislodge solid particles from the interface, since the oil/water interface must be increased to do so. Powdered silica, various clays, basic salts of metals, and plant cell fragments are examples of such agents.

The emulsion type produced and its stability depend largely on the relative abilities of the two phases to wet the solid particles. The phase that preferentially wets the solid particle tends to become the continuous phase. If the interfacial tension between solid and oil (γ_{SO}) is greater than that between solid and water (γ_{SW}), the contact angle (θ) of the solid with the aqueous phase is less than 90°, and the major portion of the solid particle resides in the water phase, thus favoring an O/W emulsion (Fig. 16). The converse takes place if $\gamma_{SW} > \gamma_{SO}$. Obviously, if solid particles remain exclusively in either of the two phases, they have no stabilizing effect. On the other hand, the most stable emulsion is formed when the angle of contact between the two liquids and the solid surface is in the vicinity of 90°. The surface of the solid, and in turn its contact angle, can be modified by adjusting pH and by adsorbing various amphiphilic compounds to its surface. Concentration and chain length of the amphiphile's hydrophobic group are important in this regard.

Based on these consideration, it has been recommended that, for the preparation of emulsions stabilized by solid particles, a surface-active substance be added that is soluble in the least-wetting (discontinuous) phase, and that the concentration of the surface-active agent be adjusted to give a contact angle close to 90° between the powder and the two liquids (36). It is also important that the surface-active agent adsorb strongly

OIL

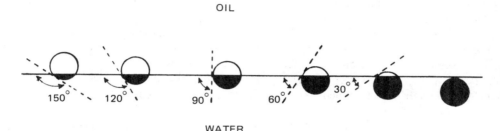

WATER

Figure 16 Contact angles at solid particles.

at the solid/liquid interface. Adsorption at the liquid/liquid interface decreases interfacial tension, thereby reducing the energy needed to displace the solid particles and decreasing emulsion stability.

Stabilization by Macromolecules. Various substances of relatively high molecular weight, including some gums and proteins, are capable of forming thick films around droplets of an emulsion, thus providing a physical barrier to coalescence. When proteins are adsorbed, they unfold and orient at the interface so that their nonpolar groups align toward the oil and their polar groups align toward the aqueous phase (see Chap. 5). The stabilizing effect on the emulsion depends mainly on the rheological (viscoelastic) properties and thickness of the protein film.

Most macromolecular emulsifiers are water-soluble proteins, and these proteins usually promote and stabilize O/W emulsions. Dispersions of protein solutions in oil are generally unstable unless other stabilizing mechanisms are used. It appears, therefore, that to be most effective in promoting emulsion stability, the main portion of the adsorbed film of macromolecules must be in the continuous phase and this film must have elastic, gellike properties.

Stabilization by Liquid Crystals. The role of liquid crystals in relation to emulsion stability has been demonstrated only in recent years, largely through the studies of Friberg and coworkers (37,38). In an emulsion (O/W or W/O), weak interactions between the emulsifier, oil, and water result in the formation of a liquid-crystalline multilayer around the droplets. This barrier at the interface causes a reduction in van der Waals forces and an improvement in emulsion stability. The importance of such structures to the stability of the emulsion is further appreciated when one considers the extremely high viscosity of liquid crystals as compared with that of water.

The type of liquid-crystal multilayer formed depends to a great extent on the nature of the emulsifier. For example, polysorbates, which form hexagonal I liquid crystals with water, transform into lamellar liquid crystals when triacylglycerols are introduced. Similarly, both sodium stearoyl lactate (which forms lamellar liquid crystals with water) and unsaturated monoacylglycerols (which form viscous isotropic structures with water) give hexagonal II liquid crystals when both water and oil are present. If excess oil is introduced, an O/W emulsion forms with liquid crystals at the interface.

Stabilization by Increasing the Viscosity of the Continuous Phase. Any factor that contributes to an increase in the viscosity of the continuous phase of an emulsion causes a significant delay in flocculation and coalescence. Gelatin and many gums, some of which are not surface active, are extremely useful in stabilizing O/W emulsions because of their effect on the viscosity of the aqueous phase.

In addition to these factors, a minimum density difference between the phases and the existence of dispersed globules that are small favors emulsion stability. This is evident from examination of Stoke's law.

2. Emulsifiers

Numerous emulsifying agents, varying widely in their structure and properties, are at present available to the food industry. Emulsifiers have been classified according to whether they are anionic, cationic, or nonionic; whether they are naturally occurring or synthetic; whether they function as surface-active agents, viscosity enhancers, or solid absorbents; or whether their hydrophobic or hydrophilic properties are most prominent.

These classifications, however, are arbitrary and overlapping, and they often obscure the function and applicability of the emulsifier. Several methods have been introduced to aid in the selection of an appropriate emulsifier, or blend of emulsifiers, for a given purpose. Two methods are discussed, with the most prominent of these being the one based on the relative importance of the hydrophilic and hydrophobic properties of the molecules (HLB system).

The HLB System for Selecting Emulsifiers. Since the principal emulsifying agents are compounds containing both hydrophobic and hydrophilic groups, and since the phase in which the emulsifier is more soluble is generally the continuous phase, the type of emulsion produced (i.e., O/W or W/O) can be predicted on the basis of the relative hydrophilic-lipophilic properties of the emulsifier. According to the hydrophilic-lipophilic balance (HLB) concept, each surface-active agent can be assigned a numerical value representing its hydrophilic-lipophilic balance (43).

Experimental determination of the HLB number for a given emulsifier is a tedious process. However, this value can be calculated with satisfactory accuracy based on easily determined characteristics of the emulsifier. The following equation was suggested by Griffin (44) for polyhydric alcohol, fatty acid esters.

$$HLB = 20 \left(1 - \frac{S}{A} \right)$$

where S is the saponification number of the ester and A is the acid number of the acid. In certain cases, where accurate determination of the saponification number is difficult, the relationship

$$HLB = \frac{E + P}{5}$$

is used, where E is the weight percent of oxyethylene and P is the weight percent of polyhydric alcohol. When ethylene oxide is the only hydrophilic group present the equation reduces to

$$HLB = \frac{E}{5}$$

HLB numbers for some common emulsifiers are presented in Table 5. The solubility of emulsifiers in water generally follows their HLB rank. As a rule, emulsifiers with HLB values in the range 3-6 promote W/O emulsions; values between 8 and 18 promote O/W emulsions.

It has also been suggested that HLB values are algebraically additive so that the HLB of a blend of two or more emulsifiers can be obtained by simple calculation and that the blend of emulsifiers needed to produce maximum emulsion stability can be easily obtained. This, however, is not always the case. Although the HLB concept is useful as a guide for comparing emulsion-forming or stabilizing properties, it suffers from a number of limitations. First, commercial emulsifiers usually consist of a group of compounds rather than a single component. This makes direct calculation based on chemical properties very difficult. Furthermore, the HLB method does not take into consideration such factors as emulsifier concentration, mesomorphic behavior, temperature, ionization of the emulsifier, interaction with other compounds present, or properties and relative concentrations of the oil and aqueous phases. Pure monoglycerides, for example, have an HLB value of approximately 3.8. Accordingly they would be expected to form only W/O

Table 5 HLB[a] and ADI[b] Values of Some Common Food Emulsifiers

Emulsifier	HLB value	ADI (mg/kg body weight)
Glycerol monostearate	3.8	Not limited
Diglycerol monostearate	5.5	0–25
Tetraglycerol monostearate	9.1	0–25
Succinic acid ester of monoglycerides	5.3	
Diacetyl tartaric acid ester of monoglycerides	9.2	0–50
Sodium stearoyl-2-lactylate	21.0	0–20
Sorbitan tristerate (SPAN 15[c])	2.1	0–25
Sorbitan monostearate (SPAN 60[c])	4.7	0–25
Sorbitan monooleate (SPAN 80[c])	4.3	
Polyoxyethylene sorbitan mono-stearate (TWEEN 60[c])	14.9	0–25
Propylene glycol monostearate	3.4	0–25
Polyoxyethylene sorbitan mono-oleate (TWEEN 80[c])	15.0	0–25

[a]Hydrophilic-lipophilic balance (9,58).
[b]Acceptable daily intake (53).
[c]Brand name by Atlas.

emulsions. However, at emulsifier concentrations that permit the formation of protective mesomorphic layers around the fat globules, pure monoacylglycerols promote O/W emulsions. Moreover, it is well known that O/W emulsions prepared from a blend of emulsifiers are usually more stable than those prepared from a single agent having the same HLB.

PIT as a Basis for Selecting Emulsifiers. It is obvious that temperature is an important factor in relation to the emulsion-forming characteristics of a surface-active agent. An emulsifier that tends to be preferentially soluble in water at relatively low temperatures may become preferentially soluble in oil at higher temperatures at which hydrophobic interactions become stronger. Determination of the temperature at which this inversion occurs provides a useful basis for emulsifier selection. A strong positive correlation has been observed between the phase-inversion temperature (PIT) of emulsifiers and emulsion stability.

Toxicology. Based on extensive toxicological studies, including metabolic tests and long- and short-term feeding experiments with animals, Acceptable Daily Intake (ADI) values have been assigned to most food emulsifiers by the FAO-WHO Codex Alimentarius Committee, and some of these values are given in the last column of Table 5.

Specific Food Emulsifiers. A compilation of commercially available emulsifiers, together with their composition, manufacturing procedures, and possible applications, is available elsewhere (9). Only a brief discussion of the more common types of food emulsifiers is given here.

1. *Glycerol esters* are a class of nonionic emulsifiers extensively used in the food industry. Monoglycerides (more properly monoacylglycerols) are prepared by direct reaction of glycerol with fatty acids or refined fats in the presence of an alkaline catalyst. Commercial monoglycerides usually contain a mixture of mono-, di-, and triesters of fatty acids with a monoglyceride content of about 45%. However, concentrated products containing more than 90% monoester can be prepared by molecular distillation. Distilled monoacylglycerols are commonly used in the manufacture of margarine, snack foods, low-caloric spreads, whipped frozen dessert, and pasta products.

The hydrophilic nature of a monoester can be increased by increasing the number of free hydroxyl groups in the alcoholic moiety of the molecule. Polyglycerol esters with a wide range of HLB values are thus produced by esterification of fatty acids with polyglycerols. Polyglycerol chains containing up to 30 glycerol units can be prepared by polymerization of glycerol.

The phase behavior of monoacylglycerol-water systems is critical for optimum functionality of monoacylglycerols in aqueous systems. With pure monoacylglycerols, the lamellar-type liquid crystal dominates for esters of the 12:0 and 16:0 fatty acids; hexagonal II or cubic type liquid crystals are usually produced from fatty acid esters with longer chains. When the water content is low, unsaturated monoacylglycerols yield lamellar-type liquid crystals at room temperature. By increasing the water content to approximately 20%, a viscous isotropic phase forms that transforms into a hexagonal II phase at temperatures above 70°C. If the water content is increased above 40% the viscous isotropic phase will separate as gelatinous lumps, making uniform distribution very difficult (58).

Commercially produced distilled monoacylglycerols are frequently used in the form of aqueous mixtures to facilitate their distribution in food products. As pointed out earlier, the swelling capacity of distilled monoacylglycerols can be increased very significantly by neutralization of the free fatty acids commonly present, or by the addition of trace amounts of ionic substances. Dilute dispersions of the commercial products, when buffered to pH 7, gave clear homogenous dispersions that form a stable gel upon cooling (57).

Industrial products known as crystalline hydrates are prepared by heating a mixture of about 25% saturated distilled monoacylglycerols in water to about 65°C, acidifying the resulting mesophase with acetic or propionic acid to pH 3, and cooling with a scraped-surface heat exchanger. The product is a stable dispersion of tiny monoacylglycerol β crystals in water. The so-called hydrates, which possess unusually smooth texture, are commonly used in the baking industry.

2. The hydrophobic character of monoacylglycerols can be enhanced by the addition of various organic acid radicals yielding *esters of monoacylglycerols with hydroxycarboxylic acids*. Lactylated monoacylglycerols, for example, are prepared from glycerol, fatty acids, and lactic acid.

$$
\begin{array}{ccccccc}
& & CH_2OH & & CH_3 & & CH_2{-}OCO{-}R \\
& & | & & | & & | \\
RCOOH & + & CHOH & + & CHOH & \longrightarrow & CHOH \\
& & | & & | & & | \\
& & CH_2OH & & COOH & & CH_2{-}OCO{-}CHOH{-}CH_3
\end{array}
$$

Fatty acid Glycerol Lactic acid   Lactylated monoacylglycerol

Succinic and malic esters can be obtained in a similar fashion. Acetylated tartaric acid monoacylglycerols are produced by reacting the monoacylglycerol with diacetyl tartaric acid anhydride.

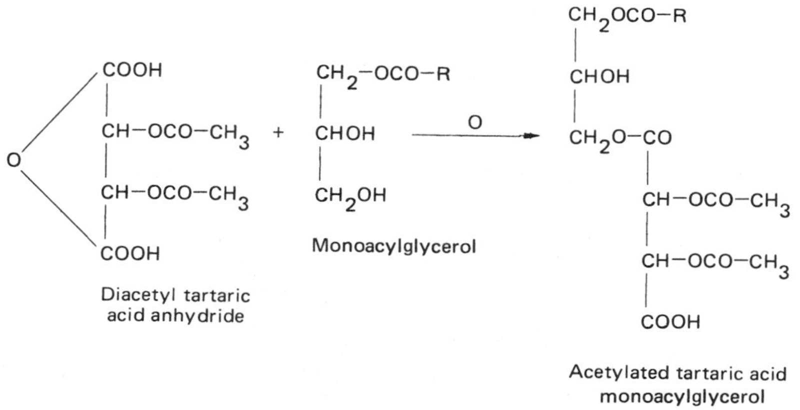

Acetylated tartaric acid
monoacylglycerol

The diacetyl tartaric acid esters, as well as the succinic acid esters, form lamellar liquid crystals that have limited swelling capacity in water. However, as is true of distilled monoacylglycerols, their capacity to imbibe water can be increased drastically by the addition of NaOH. Malic acid esters form cubic mesomorphic phases with water contents of up to 20%, and hexagonal II phases at higher temperatures and water concentrations. Succinic acid esters do not form mesomorphic phases with water, but they do exhibit mesomorphism.

3. Sodium stearoyl-2-lactylate (SSL), an ionic emulsifier, is a strongly hydrophilic surface-active agent capable of forming stable liquid crystalline phases between oil droplets and water, and thus can be used to promote very stable O/W emulsions. It is obtained from the interaction of stearic acid, 2 molecules of lactic acid, and NaOH.

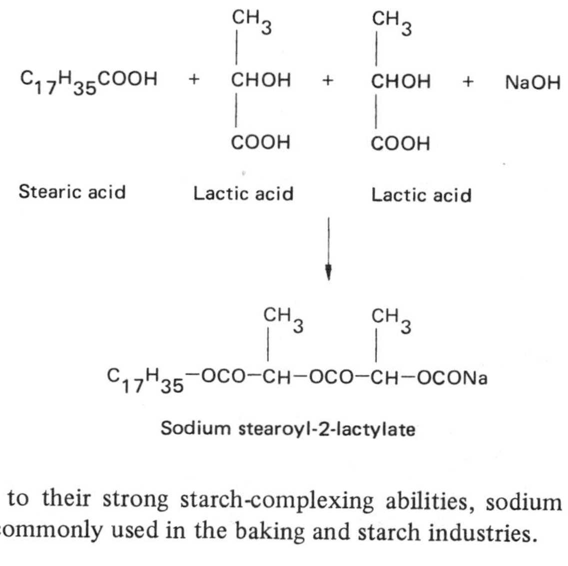

Sodium stearoyl-2-lactylate

Due to their strong starch-complexing abilities, sodium (and calcium) stearoyl lactates are commonly used in the baking and starch industries.

4. Fatty acid *monoesters of ethylene or propylene glycol* are also widely used in baking. A more hydrophilic ester can be prepared from a fatty acid and an alcohol, such as nonaethylene glycol.

.Proplylene glycol monoester   Nonaethylene glycol monoester

5. *Sorbitan fatty acid esters* are usually mixed esters of fatty acids with sorbitol anhydride or sorbitan. Sorbitol is dehydrated first to form hexitans and hexides, which are then esterified with fatty acids. The resulting products are known commercially as Spans (9).

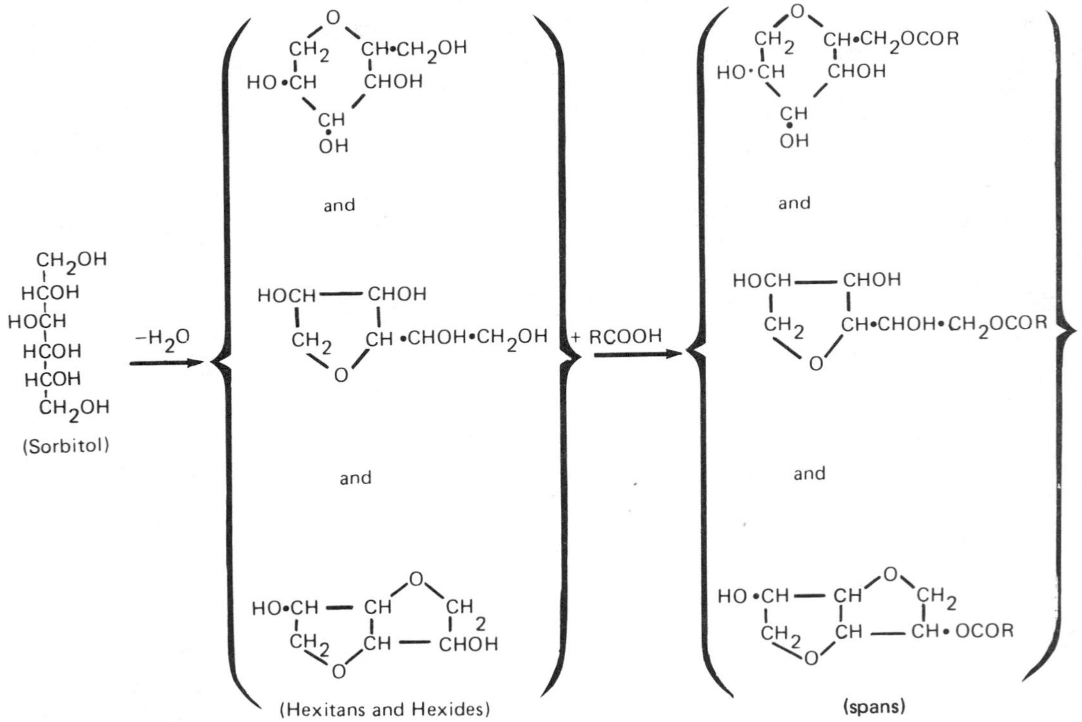

These agents tend to promote W/O emulsions. Compounds that are more hydrophilic can be produced by reacting sorbitan esters with ethylene oxide. Polyoxyethylene chains add to the hydroxyl groups through ether linkages. The resulting polyoxyethylene sorbitan fatty acid esters are commercially known as Tweens. In general, these compounds form hexagonal I liquid crystals in water, and they can solubilize small quantities of triacylglycerols. With larger amounts of triacylglycerols, transformation to a lamellar-type liquid crystal takes place. The ability of an emulsifier to solubilize nonpolar lipids is important to the formation of phase equilibria at the emulsion interface.

6. *Phospholipids* such as soybean lecithin and those in egg yolk are natural emulsifiers that promote mainly O/W emulsions. Egg yolk contains 10% phospholipid and is used to help form and stabilize emulsions in mayonnaise, salad dressing, and cake. Commercial soybean lecithin contains approximately equal amounts of phosphatidylcholine, phosphatidylethanolamine, and inositol. It is used to help form and stabilize emulsions in ice cream, cakes, candies, and margarine. Lecithin emulsifiers of different phospholipid composition and HLB characteristics can be obtained from commercial lecithin by fractionation based on solubility in alcohol.

The mesomorphic behavior of phospholipid-water systems and its relevance to biological membranes have been reviewed in detail (65,85) and consequently will not be described here.

7. Water-soluble *gums*, derived from a variety of plants, are effective in stabilizing O/W emulsions. They inhibit coalescence by increasing the viscosity of the continuous phase, and/or by forming strong films around the oil droplets. Materials in this class include gum arabic, tragacanth, xanthan, agar, pectin, methyl- and carboxymethylcellulose, and carrageenan (see Chap. 3).

8. Proteins (see Chap. 5).

V. CHEMICAL ASPECTS

A. Lipolysis

Hydrolysis of ester bonds in lipids (lipolysis) may occur by enzyme action or by heat and moisture, resulting in the liberation of free fatty acids. Free fatty acids are virtually absent in the fat of living animal tissue. They can form, however, by enzyme action after the animal is killed. Since edible animal fats are not usually refined to reduce their free fatty acid content, prompt rendering is of particular importance. The temperatures commonly used in the rendering process are capable of inactivating the enzymes responsible for hydrolysis.

The release of short-chain fatty acids by hydrolysis is responsible for the development of a rancid flavor (hydrolytic rancidity) in raw milk. On the other hand, certain typical cheese flavors are produced by deliberate addition of microbial and milk lipases. Controlled and selective lipolysis is also used in the manufacture of other food items, such as yoghurt and bread.

In contrast to animal fats, oils in mature oil seeds may have undergone substantial hydrolysis by the time they are harvested, giving rise to significant amounts of free fatty acids. Neutralization with alkali is thus required for most vegetable oils after they are extracted.

Lipolysis is a major reaction occurring during deep-fat frying due to the large amounts of water introduced from the food, and the relatively high temperatures at which the oil is maintained. The development of high levels of free fatty acids in the course of frying is usually associated with decreases in the smoke point and surface tension of the oil and a reduction in the quality of the fried food. Furthermore, free fatty acids are more susceptible to oxidation than fatty acids esterified to glycerol.

Enzymic lipolysis is used extensively as an analytical tool in lipid research. As discussed earlier in this chapter, pancreatic lipase and snake venom phospholipase are used to determine the positional distribution of fatty acids in acylglycerol molecules. The specificity of these enzymes make them particularly useful in the preparation of intermediates in the chemical synthesis of certain lipids.

B. Autoxidation

Lipid oxidation is one of the major causes of food spoilage. It is of great economic concern to the food industry because it leads to the development, in edible oils and fat-containing foods, of various off-flavors and off-odors generally called rancid, which renders these foods unacceptable or reduces their shelf life. In addition, oxidative reactions can decrease the nutritional quality of food, and certain oxidation products are potentially toxic. On the other hand, a limited degree of lipid oxidation under certain conditions is sometimes desirable, as in the production of typical cheeses or fried-food aromas.

For these reasons, extensive research has been done not only to identify the products of lipid oxidation and the conditions that influence their production, but also to study the mechanisms involved. Since oxidative reactions in food lipids are exceedingly complex, simpler model systems, such as oleate, linoleate, and linolenate, have been used to ascertain mechanistic pathways. Although useful information, applicable to the more complex system of food lipids has thus been obtained, it must be clear that such generalizations are not always justified.

It is generally agreed that "autoxidation," that is, the reaction with molecular oxygen, is the main reaction involved in oxidative deterioration of lipids. Although photochemical reactions have been known for a long time, only recently has the role of photosensitized oxidation and its interaction with autoxidation begun to be realized. In foods, the lipids can be oxidized by both enzymic and nonenzymic mechanisms.

1. General Characteristics of the Autoxidation Reaction

Our present knowledge regarding the fundamental mechanisms of lipid oxidation resulted largely from the pioneering work of Farmer and his coworkers (30), Bolland and Gee (12), and Bateman et al. (8).

Much evidence has been introduced to show that autoxidation of fats proceeds via typical free radical mechanisms as characterized by (a) marked inhibition in rate by chemical species known to interfere with other well-established free radical reactions; (b) catalysis by light and by free radical-producing substances; (c) high yields of the hydroperoxide, ROOH; (d) quantum yields exceeding unity when the oxidation reactions are initiated by light; and (e) a relatively long induction period observed when starting with the pure substrate.

Based on experimental results, mostly with ethyl linoleate, the rate of oxygen absorption can be expressed as

$$\text{Rate} = -\frac{d[O_2]}{dt} = \frac{K_a[RH][ROOH]}{1 + \lambda[RH]/p}$$

where RH is the substrate fatty acid (H is an α-methylenic hydrogen atom easily detachable due to the activating influence of the neighboring double bond or bonds), ROOH is the hydroperoxide formed, p is the pressure of oxygen, and λ and K_a are empirical constants.

To explain the experimental results, a three-step simplified, free radical scheme has been postulated as follows:

$$\text{Initiator} \xrightarrow{\ k_1\ } \text{free radicals (R}^{\cdot}\text{,ROO}^{\cdot}) \qquad\qquad\qquad \text{INITIATION} \quad (1)$$

$$\text{R}^{\cdot} + \text{O}_2 \xrightarrow{\ k_2\ } \text{ROO}^{\cdot} \qquad\qquad (2)$$

PROPAGATION

$$\text{ROO}^{\cdot} + \text{RH} \xrightarrow{\ k_3\ } \text{ROOH} + \text{R}^{\cdot} \qquad\qquad (3)$$

$$\text{R}^{\cdot} + \text{R}^{\cdot} \xrightarrow{\ k_4\ } \qquad\qquad (4)$$

$$\text{R}^{\cdot} + \text{ROO}^{\cdot} \xrightarrow{\ k_5\ } \left.\begin{array}{c}\\ \end{array}\right\} \text{Nonradical products} \qquad \text{TERMINATION} \quad (5)$$

$$\text{ROO}^{\cdot} + \text{ROO}^{\cdot} \xrightarrow{\ k_6\ } \qquad\qquad (6)$$

At high oxygen pressure ($\lambda[\text{RH}]/p$ much smaller than 1) reactions (4) and (5) can be neglected to give

$$\text{Rate} = k_3 \left(\frac{k_1}{k_6}\right)^{\!1/2} [\text{ROOH}]\,[\text{RH}]$$

Thus, the rate of oxygen absorption is independent of oxygen pressure.

At low oxygen pressure ($\lambda[\text{RH}]/p$ greater than 1), steps (5) and (6) can be neglected to give

$$\text{Rate} = k_2 \left(\frac{k_1}{k_4}\right)^{\!1/2} [\text{ROOH}]\,[\text{O}_2]$$

Since the reaction RH + $\text{O}_2 \longrightarrow$ free radicals, is thermodynamically difficult (activation energy of about 35 kcal/mol), the production of the first few radicals necessary to start the propagation reaction normally must occur by some catalytic means. It has been proposed that the initiation step may take place by hydroperoxide decompostion, by metal catalysis, or by exposure to light. More recently, it has been postulated that singlet oxygen is the active species involved, with plant and tissue pigments, such as chlorophyll or myoglobin, acting as sensitizers.

Upon the formation of sufficient free radicals, the chain reaction is propagated by the abstraction of hydrogen atoms at positions α to double bonds. Oxygen addition then occurs at these locations (P$^{\cdot}$), resulting in the production of peroxy radicals ROO$^{\cdot}$, and these in turn abstract hydrogen from α-methylenic groups RH of other molecules to yield hydroperoxides ROOH and R· groups. The new R· groups react with oxygen, and the sequence of reactions just described is repeated.

Due to resonance stabilization of the R· species, the reaction sequence is usually accompanied by a shift in the position of double bonds, resulting in the formation of isomeric hydroperoxides often containing conjugated diene groups (atypical of unoxidized, natural acylglycerols).

Hydroperoxides, the primary initial products of lipid autoxidation, are relatively unstable. They enter into numerous and complex breakdown and interaction mechanisms responsible for the production of myriad compounds of various molecular weights, flavor thresholds, and biological significance.

A general scheme summarizing the overall picture of lipid autoxidation is given in Fig. 17, and some aspects of the reaction sequence are discussed in more detail in the next section.

2. Formation of Hydroperoxides

Qualitative and quantitative analyses of the isomeric hydroperoxides from oleate, linoleate, and linolenate were recently made possible through the use of modern analytical tools (18,19,33).

Oleate. Hydrogen abstraction at carbons 8 and 11 of oleate results in the formation of two allylic radical intermediates. Oxygen attack at the end carbons of each radical produces an isomeric mixture of 8-, 9-, 10-, and 11-allylic hydroperoxides.

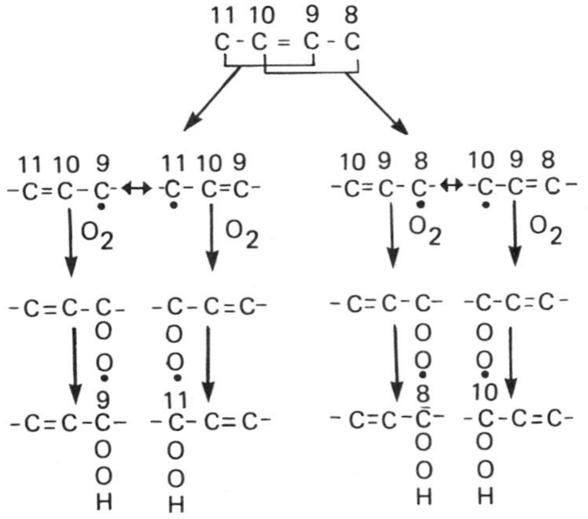

The amounts of the 8- and 11-hydroperoxides formed are slightly greater than the 9- and 10-isomers. At 25°C the amounts of cis and trans 8- and 11-hydroperoxides are similar, but the 9- and 10-isomers are mainly trans.

Linoleate. The 1,4-pentadiene structure in linoleates makes them much more susceptible (by a factor of about 20) to oxidation than the propene system of oleate. The methylene group at position 11 is doubly activated by the two adjacent double bonds.

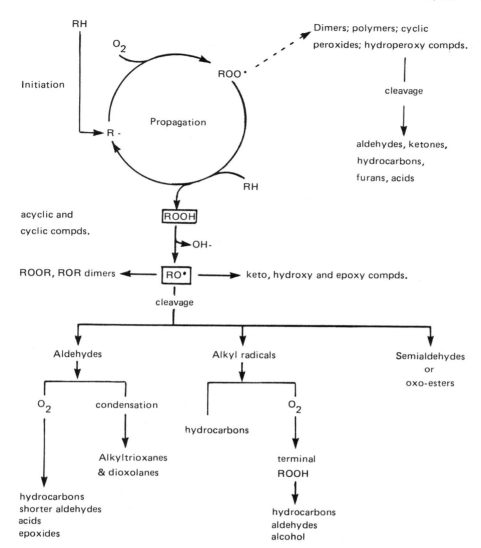

Figure 17   Generalized scheme for autoxidation of lipids.

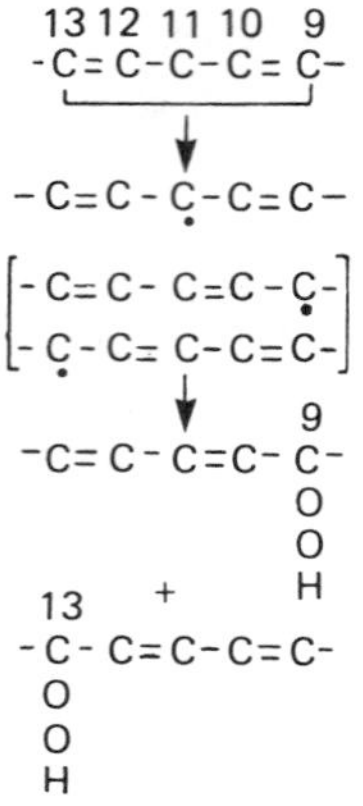

```
      13 12 11 10  9
      -C = C - C - C = C -
         └──────┬──────┘
                ↓
      - C = C - C - C = C -
                ·
      ⎡- C = C - C = C - C· -⎤
      ⎣- C· - C = C - C = C -⎦
                ↓          9
      - C = C - C = C - C -
                         O
                         O
      13        +        H
      - C - C = C - C = C -
        O
        O
        H
```

Hydrogen abstraction at this position produces a pentadienyl radical intermediate, which upon reaction with molecular oxygen produces an equal mixture of conjugated 9- and 13-diene hydroperoxides. Evidence in the literature indicates that the 9- and 13-*cis, trans*-hydroperoxides undergo interconversion, along with some geometric isomerization, forming *trans,trans*-isomers. Thus, each of the two hydroperoxides (9- and 13-) is found in both the cis, trans and the trans, trans forms.

Linolenates. In linolenates, two 1,4-pentadiene structures are present. Hydrogen abstraction at the two active methylene groups of carbons 11 and 14 produces two pentadienyl radicals.

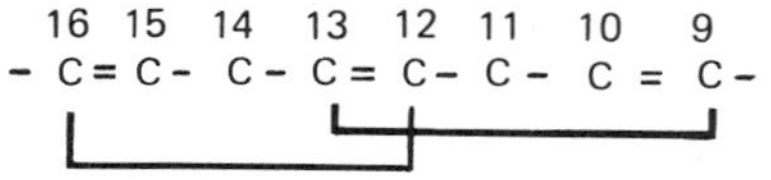

```
      16  15  14  13  12  11  10   9
    - C = C - C - C = C - C - C = C -
      └──────────┘   └──────────────┘
```

Oxygen attack at the carbon end of each radical results in the formation of a mixture of isomeric 9-, 12-, 13-, and 16-hydroperoxides. For each of the four hydroperoxides, geometric isomers exist, each having a conjugated diene system in either the cis, trans or the trans, trans configuration, and an isolated double bond that is always cis.

The 9- and 16-hydroperoxides are formed in significantly greater amounts than the 12- and 13-isomers. This has been attributed to (a) a preference of oxygen to react with carbons 9 and 16; (b) faster decomposition of the 12- and 13-hydroperoxides, or (c) a tendency of the 12- and 13-hydroperoxides to form six-membered peroxide hydroperoxides via 1,4-cyclization, as shown below, or prostaglandinlike endoperoxides via 1,3-cyclization. The cyclization mechanisms are thought to be the most likely explanations.

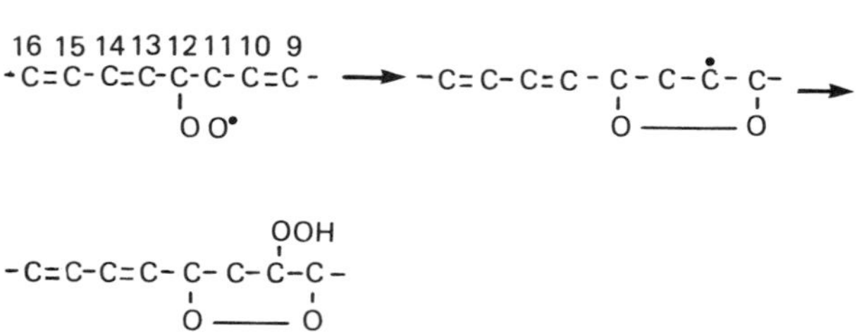

```
  16 15 14 13 12 11 10 9
  ·C=C- C=C- C- C- C=C -   ⟶   - C=C- C=C- C - C- C· - C-   ⟶
                  │                                 ·
                  O O·                          O ——— O
```

```
                             OOH
                             │
  - C=C- C=C- C - C - C· - C -
               │        │
               O ——— O
```

3. Oxidation with Singlet Oxygen

As just discussed, the major pathway for the oxidation of unsaturated fatty acids involves a self-catalytic free radical mechanism (autoxidation) that accounts for the chain reaction of hydroperoxide (ROOH) formation and decomposition. However, the origin of the initial free radicals necessary to begin the process has been difficult to explain. It is unlikely that initiation occurs by direct attack of oxygen, in its most stable form (triplet state), on double bonds of fatty acids (RH) since the C=C bonds in RH and ROOH are in singlet states and thus such a reaction does not obey the rule of spin conservation. A more satisfactory explanation is that singlet oxygen ($^1O_2$), believed to be the active species in photooxidative deterioration, is responsible for initiation.

Since electrons are charged, they act like magnets that can exist in two different orientations, but with equal magnitude of spin, +1 and −1. The total angular momentum of the electrons in an atom is described by the expression 2S + 1 where S is total spin. If an atom such as oxygen has two unpaired electrons in its outer orbitals, they can align their spins parallel or antiparallel with respect to each other, giving rise to two different multiplicities of state; that is, $2(\frac{1}{2} + \frac{1}{2}) + 1 = 3$ and $2(\frac{1}{2} - \frac{1}{2}) + 1 = 1$. These are called the triplet ($^3O_2$) and singlet state ($^1O_2$), respectively. In the triplet state, the two electrons in the antibonding $2p\pi$ orbitals have the same spin but are in different orbitals. These electrons are kept apart via "Pauli exclusion" and, therefore, have only a small repulsive electrostatic energy.

In the singlet state, the two electrons have opposite spins, and therefore, electrostatic repulsion will be great, resulting in an excited state. Singlet oxygen can be found in two energy states: $^1\Delta$ with an energy of 22 kcal above ground state, and $^1\Sigma$, with an energy of 37 kcal above ground state.

$^1\Delta$ $^1\Sigma$

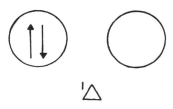

Singlet state oxygen is more electrophilic than triplet state oxygen. It can thus react rapidly (∼1500 times faster than $^3O_2$) with moieties of high electron density, such as C=C bonds. The resulting hydroperoxides can then cleave to initiate a conventional free radical chain reaction.

Singlet oxygen can be generated in a variety of ways; probably the most important is via photosensitization by the natural pigments in foods. Two pathways have been proposed for photosensitized oxidation (17). In the type 1 pathway, the sensitizer presumably reacts, after light absorption, with substrate (A) to form intermediates, which then react with ground state (triplet) oxygen to yield the oxidation products.

$$\text{Sens} + \text{A} + h\nu \longrightarrow \text{intermediates} - \text{I}$$

$$\text{Intermediates} - \text{I} + O_2 \longrightarrow \text{products} + \text{sens}$$

In the type 2 pathway, molecular oxygen rather than the substrate is presumably the species that reacts with the sensitizer upon light absorption.

$$\text{Sens} + O_2 + h\nu \longrightarrow \text{intermediates} - \text{II}$$

$$\text{Intermediates} - \text{II} + \text{A} \longrightarrow \text{products} + \text{sens}$$

Several substances are commonly found in fat-containing foods that can act as photosensitizers to produce $^1O_2$. These include natural pigments, such as chlorophyll-a, pheophytin-a, and hematoporphyrin, the pigment portion of hemoglobin, and myoglobin. The synthetic colorant, erythrosine, also acts as an active photosensitizer.

β-Carotene is the most effective $^1O_2$ quencher, and tocopherols are somewhat effective. Synthetic quenchers such as BHA and BHT are also effective and permissible in foods.

The formation of hydroperoxides by singlet oxygen proceeds via mechanisms different than that for free radical autoxidation. The most important of these is the "ene" reaction, which involves the formation of a six-membered ring transition state. Oxygen is thus inserted at the ends of the double bond, which then shifts to yield an allylic hydroperoxide in the trans-configuration. Accordingly, oleate produces the 9- and 10-hydroperoxides (instead of the 8-, 9-, 10-, and 11-hydroperoxides occurring by free radical autoxidation), linoleate produces 9-, 10-, 12-, and 13-hydroperoxides (instead of 9- and 13-), and linolenate produces a mixture of 9-, 10-, 12-, 13-, 15-, and 16-hydroperoxides (instead of 9-, 12-, 13-, and 16-).

In addition to the role of singlet oxygen as an active initiator of free radicals, it has been suggested that certain products of lipid oxidation can be explained only on the basis of the decomposition of hydroperoxides typical of those produced by singlet oxygen. However, there is general agreement that, once the initial hydroperoxides are formed, the free radical chain reaction prevails as the main mechanism. Further research is needed to clarify the extent to which singlet oxygen oxidation may be involved in the production of oxidative decomposition products.

4. Decomposition of Hydroperoxides

Hydroperoxides break down in several steps yielding a wide variety of decomposition products. Each hydroperoxide produces a set of initial breakdown products that are typical of the specific hydroperoxide and depend on its position in the parent molecule. Such products can themselves undergo further oxidation and decomposition, thus contributing to a large and varied free radical pool. The multiplicity of the many possible reaction pathways results in a pattern of autoxidation products so complex that in most cases their hydroperoxide origin is completely obliterated.

It should also be pointed out here that hydroperoxides begin to decompose as soon as they are formed. In the first stages of autoxidation, their rate of formation exceeds their rate of decomposition. The reverse takes place at later stages.

The first step in hydroperoxide decomposition is scission at the oxygen-oxygen bond of the hydroperoxide group, giving rise to an alkoxy radical and a hydroxy radical.

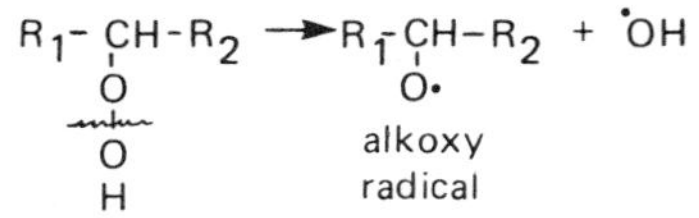

$$R_1- \underset{\underset{\underset{H}{\overset{|}{O}}}{\overset{|}{O}}}{CH}-R_2 \longrightarrow R_1- \underset{\underset{\text{radical}}{\overset{|}{O\cdot}}}{\underset{\text{alkoxy}}{CH}}-R_2 \; + \; \overset{\cdot}{O}H$$

Carbon-carbon bond cleavage on either side of the alkoxy group is the second step in decomposition of the hydroperoxides. In general, cleavage on the acid side (i.e., the carboxyl or ester side) results in the formation of an aldehyde and an acid (or ester); scission on the hydrocarbon (or methyl) side produces a hydrocarbon and an oxoacid (or oxoester). If, however, a vinylic radical results from such cleavage, an aldehydic functional group is formed.

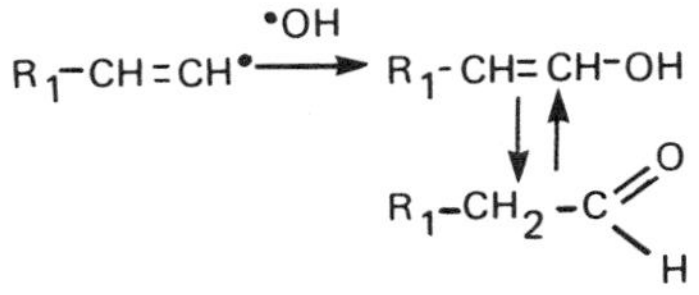

$$R_1-CH=CH\overset{\cdot}{\underset{}{}}\overset{\cdot OH}{\longrightarrow} R_1-CH=CH-OH$$
$$\downarrow\uparrow$$
$$R_1-CH_2-C\overset{\overset{\displaystyle O}{\diagup\!\!\diagup}}{\diagdown_H}$$

For example, with the 8-hydroperoxide isomer from methyl oleate, cleavage on the hydrocarbon side (at a) would produce decanal and methyl-8-oxooctanoate, and scission on the ester side (at b) forms 2-undecenal and methyl heptanoate.

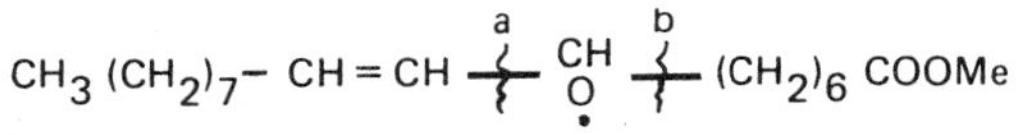

$$CH_3(CH_2)_7-CH=CH \overset{a}{\dashv} \underset{\underset{\cdot}{\overset{|}{O}}}{CH} \overset{b}{\dashv} (CH_2)_6 \, COOMe$$

In the same manner, each of the remaining three oleate hydroperoxides would be expected to produce four typical products; that is, the 9-hydroperoxide

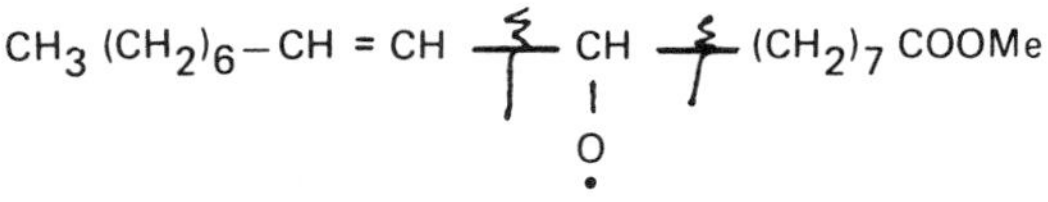

$$CH_3(CH_2)_6-CH=CH \dashv CH \dashv (CH_2)_7 \, COOMe$$
$$\underset{\underset{\cdot}{\overset{|}{O}}}{}$$

would produce nonanal, methyl-9-oxononanoate, 2-decenal and methyl octanoate; the 10-hydroperoxide

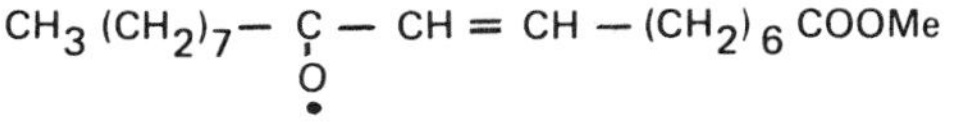

$$CH_3(CH_2)_7- \underset{\underset{\cdot}{\overset{\|}{O}}}{C}-CH=CH-(CH_2)_6 \, COOMe$$

would produce octane, methyl-10-oxo-8-decenoate, nonanal, and methyl-9-oxononanoate; and the 11-hydroperoxide

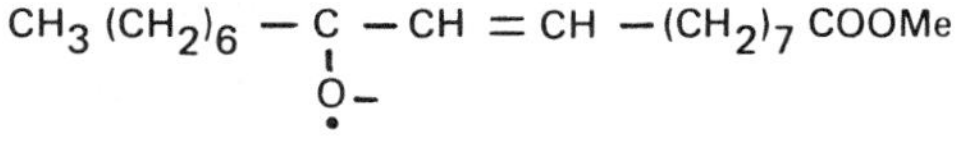

$$CH_3(CH_2)_6 - \underset{\underset{\cdot}{\overset{\|}{O}-}}{C}-CH=CH-(CH_2)_7 \, COOMe$$

would produce heptane, methyl-11-oxo-9-undecenoate, octanal, and methyl-10-oxo-decanoate.

As mentioned above, autoxidation of linoleate produces two conjugated hydroperoxides, the 9- and the 13-hydroperoxides. Fig. 18 shows typical cleavage of the 9-alkoxy radical. It was also pointed out that oxidation by singlet oxygen leads to the formation of four isomeric hydroperoxides, two of which are not typical of autoxidation. The 10-hydroperoxide would be expected to yield 2-octene, methyl-10-oxo-8-decenoate, 3-nonenal, and methyl-9-oxononanoate; the 12-hydroperoxide would give hexanal, methyl-12-oxo-9-dodecenoate, 2-heptenal and methyl-9-undecenoate.

In the case of linolenates, although some of the cleavage products expected from classic cleavage of the 9-, 12-, 13-, and 16-hydroperoxides have been reported, several others, particularly the longer chain and the polyunsaturated oxoesters, have not been detected. It is possible such compounds rapidly undergo further decomposition.

The cyclic peroxides or the hydroperoxy cyclic peroxides, which are commonly formed during the oxidation of polyunsaturated fatty acids, also decompose, yielding a variety of compounds. Formation of 3,5-octadiene-2-one is shown below.

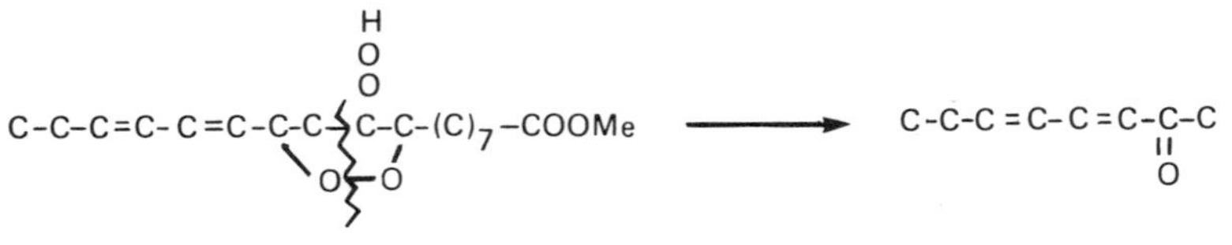

5. Further Decomposition of the Aldehydes

As demonstrated earlier, aldehydes are a major class of compounds typical of fat oxidation. Many of the aldehydes found in oxidized fats, however, cannot be explained solely on the basis of the classic hydroperoxide cleavage just discussed. This is not surprising in view of the ease and variety of reactions that aldehydes can undergo.

Saturated aldehydes can easily oxidize to form the corresponding acids and they can participate in dimerization and condensation reactions. Three molecules of hexanal, for example, can combine to form tripentyltrioxane:

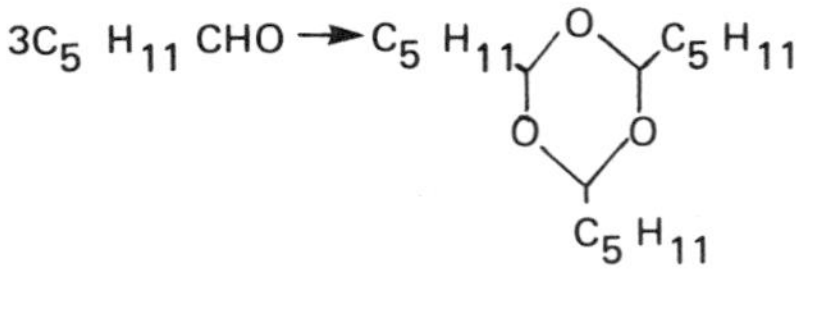

Trialkyltrioxanes possessing relatively strong odors have been reported as secondary oxidation products of linoleate.

In a study of oleic acid and nonanal, a speculative but interesting mechanism was proposed to account for the aldehydes, alcohols, alkyl formates, and hydrocarbons produced from the autoxidation of oleic acid (63). Abstraction of hydrogen from nonanal, the aldehyde formed from the 10-hydroperoxide of oleate, results in the formation of a resonance equilibrium between two forms of the carbonyl free radical (Fig. 19). This leads to the formation of the peracid and an α-hydroperoxyaldehyde. Carbon-carbon

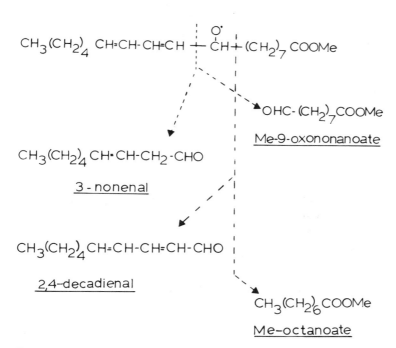

The 9-OOH

Figure 18 Decomposition of methyl-9-hydroperoxy-10,12-octadecadienoate.

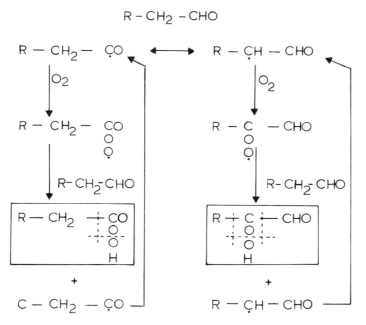

$R - CH_2 - CHO$

Figure 19 Speculative mechanism for the autoxidation of nonanal (63).

and oxygen-oxygen cleavages can produce a variety of free radicals that can initiate chain reactions or combine to form oxidation products.

Unsaturated aldehydes can undergo classic autoxidation with oxygen attack at α-methylenic positions giving rise to short-chain hydrocarbons, aldehydes, and dialdehydes:

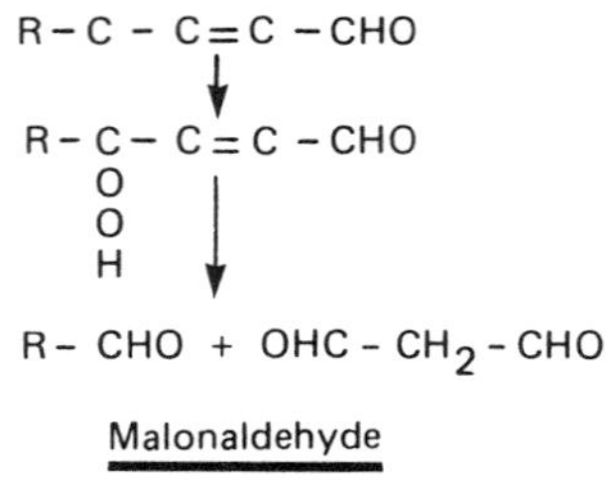

R – C – C = C – CHO

R – C – C = C – CHO
O
O
H

R – CHO + OHC – CH$_2$ – CHO

Malonaldehyde

Formation of malonaldehyde is the basis for the well-known TBA method used for measuring fat oxidation, as is discussed later.

For oxidation of aldehydes with conjugated double bonds, a mechanism was proposed in which epoxides are formed by oxygen attack at the olefinic centers (66). In the case of 2,4-decadienal, the major aldehyde from the 9-hydroperoxide of linolenate, either the 2,3-epoxy or the 4,5-epoxy derivative, can be produced as an intermediate. Fig. 20 shows the formation and decomposition of the 2,3-epoxide.

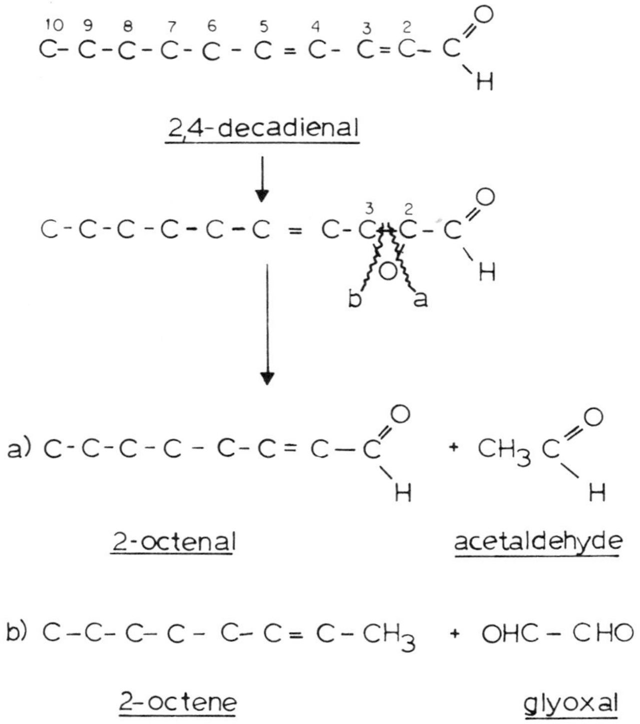

Figure 20 Formation and decomposition of the 2,3-epoxide from 2,4-decadienal (66).

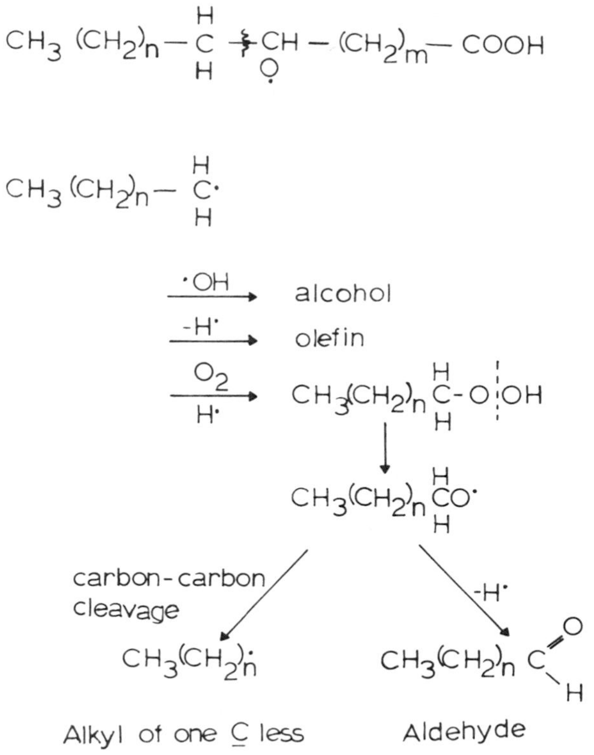

$$CH_3 \; (CH_2)_n - \overset{\overset{\displaystyle H}{|}}{\underset{\underset{\displaystyle H}{|}}{C}} \; \nmid \; \overset{}{\underset{\underset{\displaystyle \cdot}{O}}{CH}} - (CH_2)_m - COOH$$

$$CH_3 \; (CH_2)_n - \overset{\overset{\displaystyle H}{|}}{\underset{\underset{\displaystyle H}{|}}{C}} \cdot$$

$\xrightarrow{\cdot OH}$ alcohol

$\xrightarrow{-H^\cdot}$ olefin

$\xrightarrow[H^\cdot]{O_2}$ $CH_3(CH_2)_n \; \overset{\overset{\displaystyle H}{|}}{\underset{\underset{\displaystyle H}{|}}{C}} - O \vdots OH$

$\downarrow$

$CH_3(CH_2)_n \; \overset{\overset{\displaystyle H}{|}}{\underset{\underset{\displaystyle H}{|}}{C}} O \cdot$

carbon-carbon cleavage $-H^\cdot$

$CH_3(CH_2)_n \cdot$ $CH_3(CH_2)_n \; C\overset{\displaystyle O}{\underset{\displaystyle H}{\diagup\!\!\!\!\backslash}}$

Alkyl of one C̲ less   Aldehyde

Figure 21 Some reactions of the alkyl radical.

Similarly, hexanal, 2-butenal, hexane, and 2-butene-1,4-dial can be formed from decomposition of the 4,5-epoxide.

6. Other Reactions of Alkyl and Alkoxy Radicals

The alkyl radical resulting from cleavage on the methyl side of the alkoxy group can enter into a variety of reactions (Fig. 21). It can combine with a hydroxyl radical to give an alcohol, abstract a hydrogen to form 1-alkene, or peroxidize to form a terminal hydroperoxide. The hydroperoxide can decompose to the corresponding alkoxyradical, which in turn may form an aldehyde, or it can be further cleaved to yield another alkyl radical.

The alkoxy radical can also abstract a hydrogen atom from the α-methylene group of another molecule, producing a hydroxy acid, or can lose a hydrogen to give a keto acid.

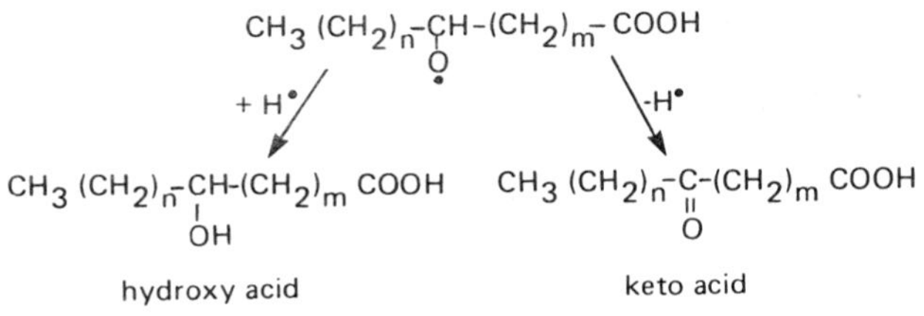

$$CH_3 \; (CH_2)_n - \underset{\underset{\displaystyle \cdot}{O}}{CH} - (CH_2)_m - COOH$$

$+ H^\cdot \diagup$ $\diagdown -H^\cdot$

$CH_3 \; (CH_2)_n - \underset{\underset{\displaystyle OH}{|}}{CH} - (CH_2)_m \; COOH$ $CH_3 \; (CH_2)_n - \underset{\underset{\displaystyle O}{\|}}{C} - (CH_2)_m \; COOH$

hydroxy acid   keto acid

Reaction of alkoxy and peroxy radicals with double bonds can produce epoxides as follows.

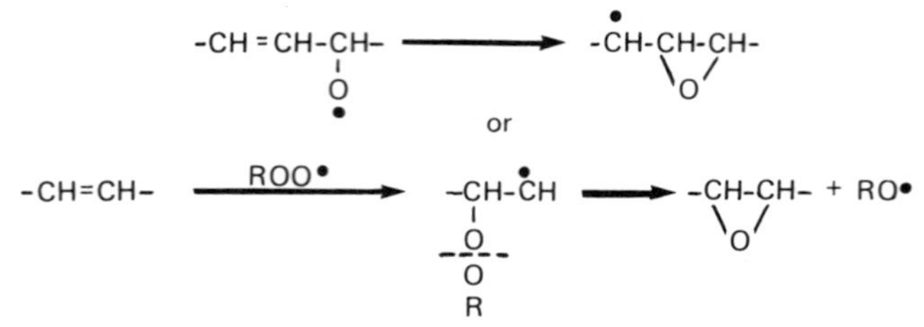
-CH = CH-CH-
O
or
-CH-CH-CH-
O
-CH=CH-
ROO•
-CH-CH
O
O
R
-CH-CH- + RO•
O

7. Formation of Dimeric and Polymeric Compounds

Dimerization and polymerization are major reactions that occur in lipids by thermal and oxidative mechanisms. Such changes are usually accompanied by a decrease in iodine value and increase in molecular weight, viscosity, and refractive index.

Under conditions of low oxygen tension, carbon-carbon bond formation between two acyl groups can occur in several ways:

A Diels-Alder Reaction Between a Double Bond and a Conjugated Diene to Produce a Tetrasubstituted Cyclohexene

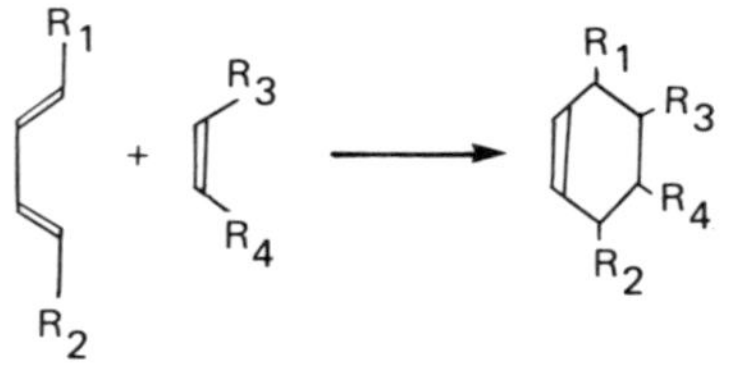
R1
R2
R3
R4
R1
R3
R4
R2

1, 4 Diels - Alder Reaction

Linoleate, for example, can develop a conjugated double-bond system during thermal oxidation and then react with another molecule of linoleate (or with oleate) to produce a cyclic dimer:

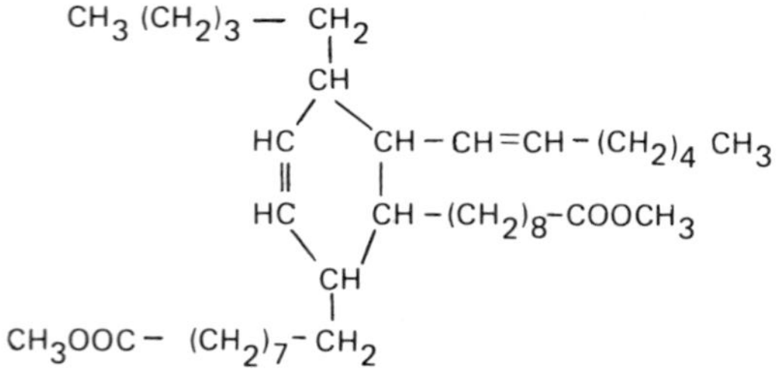
CH3 (CH2)3 — CH2
CH
HC
CH − CH=CH − (CH2)4 CH3
HC
CH − (CH2)8−COOCH3
CH
CH3OOC − (CH2)7 − CH2

In the case of acylglycerols, dimerization can take place between acyl groups in two triacylglycerol molecules or between two acyl groups in the same molecule.

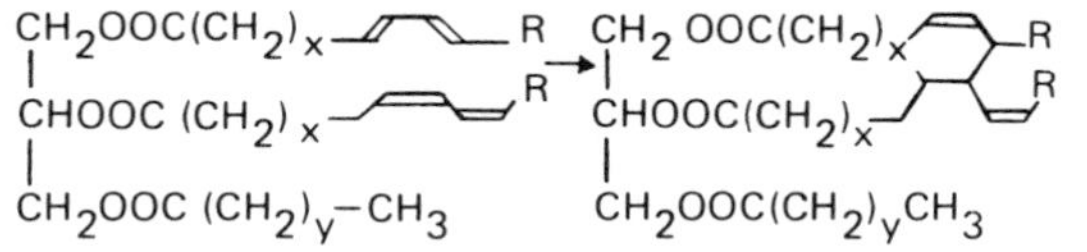

$$
\begin{array}{l}
\text{CH}_2\text{OOC(CH}_2)_x \diagup\diagdown\text{R} \\
| \\
\text{CHOOC (CH}_2)_x \diagup\diagup\diagdown\text{R} \\
| \\
\text{CH}_2\text{OOC (CH}_2)_y\text{--CH}_3
\end{array}
\quad \xrightarrow[\text{R}]{}\quad
\begin{array}{l}
\text{CH}_2\text{ OOC(CH}_2)_x\diagup\diagdown\text{R} \\
| \\
\text{CHOOC(CH}_2)_x\diagdown\diagdown\text{R} \\
| \\
\text{CH}_2\text{OOC(CH}_2)_y\text{CH}_3
\end{array}
$$

Combination of Free Radicals to Give Noncyclic Dimers. Oleate, for example, can give rise to a mixture of dimers (dehydrodimers) coupled at 8, 9, 10, or 11 positions.

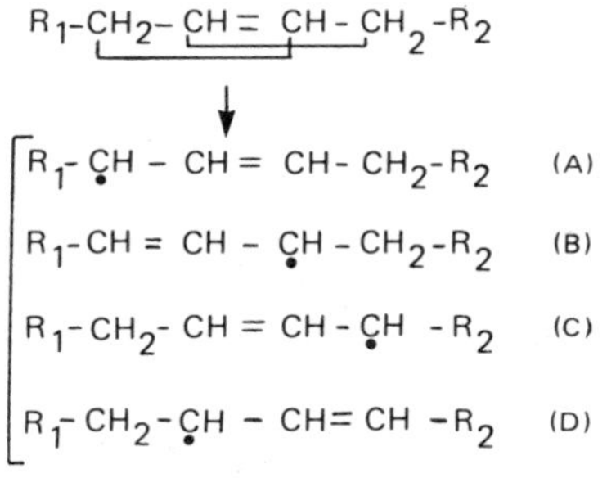

$$
\text{R}_1\text{-CH}_2\text{- CH} = \text{CH - CH}_2\text{-R}_2
$$

$$\downarrow$$

$$
\left[
\begin{array}{ll}
\text{R}_1\text{- }\overset{\bullet}{\text{C}}\text{H - CH} = \text{CH- CH}_2\text{-R}_2 & \text{(A)} \\
\\
\text{R}_1\text{-CH} = \text{CH - }\overset{\bullet}{\text{C}}\text{H - CH}_2\text{-R}_2 & \text{(B)} \\
\\
\text{R}_1\text{- CH}_2\text{- CH} = \text{CH - }\overset{\bullet}{\text{C}}\text{H - R}_2 & \text{(C)} \\
\\
\text{R}_1\text{- CH}_2\text{-}\overset{\bullet}{\text{C}}\text{H - CH= CH -R}_2 & \text{(D)}
\end{array}
\right.
$$

AA, AB, AC, AD, BB, B**C**, BD, CC, CD, or DD

Addition of a Free Radical to a Double Bond. This yields a dimeric radical, which may abstract hydrogen from another molecule or attack other double bonds to produce acyclic or cyclic compounds. Figures 22 and 23 show dimerization of oleate and linoleate, respectively. Similar reactions can occur between acyl groups in different acylglycerols to produce dimeric and trimeric triacylglycerols.

In the presence of a plentiful supply of oxygen, combinations between alkyl, alkoxy, and peroxy free radicals may result in a variety of dimeric and polymeric acids and acylglycerols with carbon-oxygen-carbon or carbon-oxygen-oxygen-carbon crosslinks (Fig. 24).

The addition of a free radical to a double bond may take place in the same molecule, giving rise to cyclic monomers. Cyclization occurs more readily in case of the longer chain polyunsaturated acids, as shown for arachidonic acid (Fig. 25).

8. Oxidation in Biological Systems

The preceding discussion has dealt mainly with the oxidation of pure lipid substrates in a liquid bulk phase. In biological systems, including food, the lipid molecules often exist in a highly ordered state, are relatively restricted in terms of distance between molecules and mobility, and are closely associated with neighboring nonlipid material, such as proteins, carbohydrates, water, enzymes, salts, vitamins, and pro- and antioxidants. The compostion of the lipids, the degree of their molecular order, and their association with nonlipid components vary considerably depending on the plant or animal species and the location of the lipid within the individual organism. Obviously, the consequences and

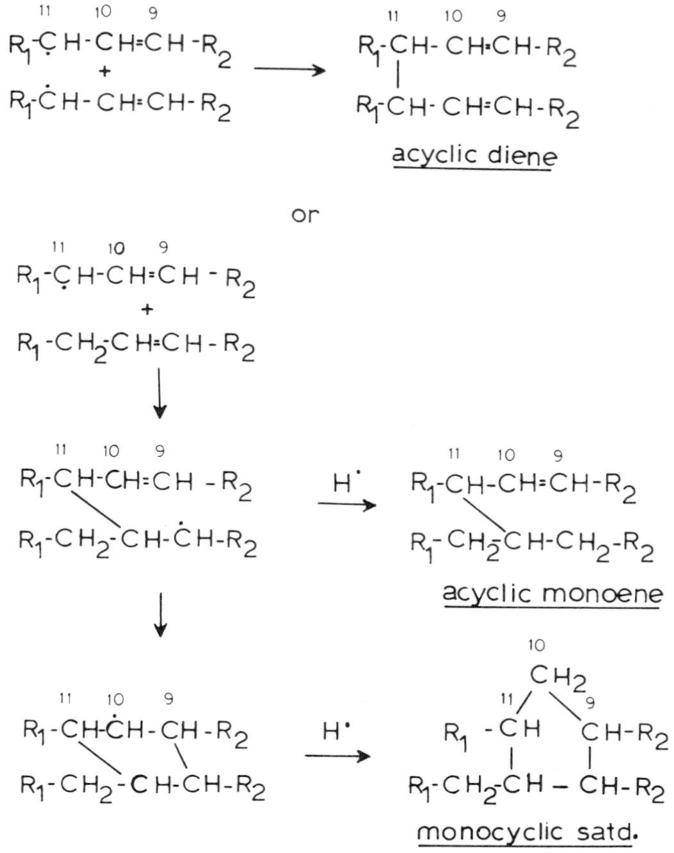

$$R_1\overset{\cdot}{\underset{+}{C}}H\text{-}CH\text{=}CH\text{-}R_2$$

$$\overset{11\quad\ 10\quad\ 9}{R_1\text{-}CH\text{-}CH\text{=}CH\text{-}R_2}$$

$$R_1\overset{\cdot}{C}H\text{-}CH\text{=}CH\text{-}R_2 \longrightarrow \underset{R_1\text{-}CH\text{-}CH\text{=}CH\text{-}R_2}{R_1\text{-}CH\text{-}CH\text{=}CH\text{-}R_2}$$

acyclic diene

or

$$R_1\text{-}\overset{\cdot}{\underset{+}{C}}H\text{-}CH\text{=}CH\text{-}R_2$$

$$R_1\text{-}CH_2CH\text{=}CH\text{-}R_2$$

$$\downarrow$$

$$\underset{R_1\text{-}CH_2\text{-}CH\text{-}\overset{\cdot}{C}H\text{-}R_2}{R_1\text{-}CH\text{-}CH\text{=}CH\text{-}R_2} \quad \overset{H^\cdot}{\longrightarrow} \quad \underset{R_1\text{-}CH_2CH\text{-}CH_2\text{-}R_2}{R_1\text{-}CH\text{-}CH\text{=}CH\text{-}R_2}$$

acyclic monoene

$$\downarrow \qquad\qquad\qquad\qquad\qquad\qquad \downarrow$$

$$\underset{R_1\text{-}CH_2\text{-}CH\text{-}CH\text{-}R_2}{R_1\text{-}CH\text{-}\overset{\cdot}{C}H\text{-}CH\text{-}R_2} \quad \overset{H^\cdot}{\longrightarrow} \quad R_1\text{-}CH \overset{CH_2}{\underset{CH_2}{\diagup\diagdown}} CH\text{-}R_2$$

$$R_1\text{-}CH_2CH-CH\text{-}R_2$$

monocyclic satd.

Figure 22 Dimerization of oleate.

mechanisms of oxidative reactions in such natural systems can be quite different from those carried out in pure bulk-lipid models.

Since the mechanistic assessment of oxidative reactions of lipids in intact biological systems is exceedingly difficult, recent interest has focused on the use of simpler nonbulk lipid systems with the hope that the results obtained can be interpretable in the more complex biological systems. Such nonbulk lipid models include monolayers (on the surface of an aqueous phase or adsorbed on silica), artificial lipid vesicles, and phospholipid bilayers. In experiments with such systems, Mead and coworkers (67) suggested that, unlike oxidation in bulk phase (in which hydroperoxides are the main intermediates), oxidation of linoleic acid esters in monolayers results primarily in cis or trans epoxy compounds. Kinetically, the reaction is apparently first order, in contrast to bulk-phase oxidation, in which the reaction is more complex. When the resistant saturated chains are interposed between the reactive unsaturated acids, the rate of peroxidation is decreased provided the chains are long enough to interfere with oxygen transfer. Much more work remains to be done to ascertain the influence of molecular orientation on the mechanisms and products of lipid oxidation.

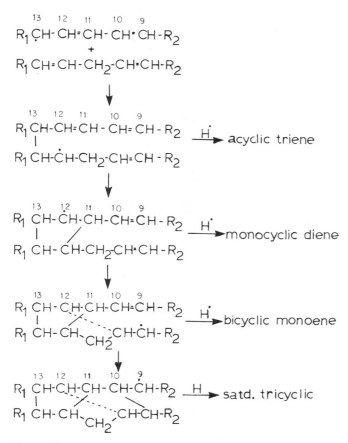

Figure 23 Dimerization of linoleate.

Gardner (41,42) reviewed the enzymic pathways by which lipids are oxidized. Sequential enzyme action starts with lypolysis. Released polyunsaturated fatty acids are then oxidized by either lipoxygenase or cyclo-oxygenase to form hydroperoxides or endoperoxides, respectively. Until recently, lipoxygenases were thought to be restricted to plants. However, the existence of animal lipoxygenase systems has now been demonstrated. Plant and animal lipoxygenases are both regiospecific (catalyze oxygenation of specific carbons) and stereospecific (produce enantiomeric hydroperoxides). The next sequence of events involves enzymic cleavage of the hydroperoxides and endoperoxides to yield a varieity of breakdown products, which are often responsible for the characteristic flavors of many natural products.

In the presence of other components (e.g., proteins and antioxidants), oxidative reactions (both enzymic and nonenzymic) may terminate by reactions with compounds other than those originating from the oxidation of the lipid substrate, and this can influence reaction rates. For example, some products of nonenzymic browning are known to act as antioxidants. Also, the basic groups in proteins may catalyze aldol condensation of carbonyls produced from lipid oxidation, resulting in the formation of brown pigments. Furthermore, lipid hydroperoxides may induce oxidative changes in sulfur-con-

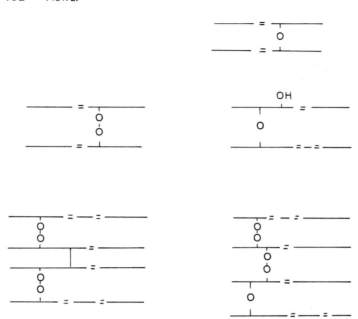

Figure 24   Polymerization of acylglycerols.

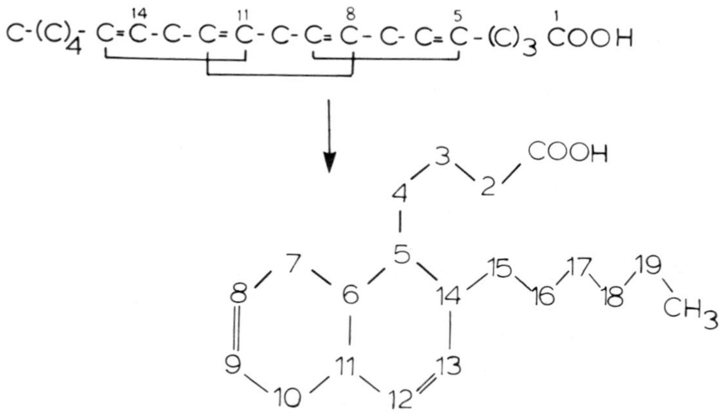

C-(C)₄⁻ C=C-C- C=C- C- C= C- C- C= C-(C)₃ COOH

Figure 25   Cyclization of arachidonic acid.

taining protiens (55), causing significant nutritional losses, and secondary oxidation products from lipids can initiate free radical reactions in proteins, or form Schiff-base addition products with the ε amino groups of lysine. It has also been reported that free carbohydrates may increase the rates of oxidation of lipids in emulsions.

9. Techniques for the Measurement of Lipid Oxidation

It is obvious from this discussion that lipid oxidation is an exceedingly complex process involving numerous reactions that give rise to a variety of chemical and physical changes. Although these reactions appear to follow recognized stepwise pathways, they often occur simultaneously and competitively. The nature and the extent of each of these changes in influenced to a great degree by a very large number of variables, many of which have already been mentioned. Since oxidative decomposition is of major significance in regard to both the acceptability and nutritional quality of food products, many methods have been devised for assessing the extent of oxidation. No single test, however, can possibly measure all oxidative events at once; no single test can be equally useful at all stages of the oxidative process; and no single test can be applicable to all fats, all foods, or all conditions of processing. At best, a test can monitor one or a few physical changes that may provide important information for specific systems and under specific conditions. Of course, more reliability can be obtained when a combination of tests is employed. Keeping these considerations in mind, we list some of the methods commonly used.

Peroxide Value. Peroxides are the main initial products of autoxidation. They can be measured by techniques based on their ability to liberate iodine from potassium iodide, or to oxidize ferrous to ferric ions. Their content is usually expressed in terms of milliequivalents of oxygen per kilogram of fat (69). Various other colorimetric techniques are available. Although the peroxide value is applicable for following peroxide formation at the early stages of oxidation, it is, nevertheless, highly empirical. The accuracy is questionable, the results vary with details of the procedure used, and the test is extremely sensitive to temperature changes. During the course of oxidation, peroxide values reach a peak and then decline.

Various attempts have been made to correlate peroxide values with the development of rancid flavors. Good correlations are sometimes obtained, but very often the results are inconsistent. It should be pointed out that the amount of oxygen that must be absorbed, or peroxides that must be formed, to produce rancidity vary with the composition of the oil (the more saturated fats require less oxygen absorption to become rancid), the presence of antioxidants and trace metals, and conditions of oxidation.

Thiobarbituric Acid (TBA) Test. This is one of the most widely used tests for evaluating the extent of lipid oxidation. Oxidation products of unsaturated systems produce a color reaction with TBA. It is believed that the chromagen results from condensation of two molecules of TBA with one molecule of malonaldehyde. However, malonaldehyde is not always found in all oxidized systems. Many alkanals, alkenals, and 2,4-dienals produce a yellow pigment (450 nm) in conjuction with TBA, but only dienals produce a red pigment (530 nm). It has been suggested that measurement at both absorption maxima is desirable.

In general, TBA-reactive material is produced in substantial amounts only from fatty acids containing three or more double bonds. The reaction mechanism shown in

Fig. 26 was offered by Dahle and coworkers (26). They suggested that radicals with a double bond β-γ to the carbon bearing the peroxy groups (which can only arise from acids containing more than two double bonds) cyclize to form peroxides with five-membered rings, which decompose to give malonaldehyde (Fig. 26). More recently, however, Pryor et al. (73) concluded that malonaldehyde arises, at least in part, from decomposition of prostaglandinlike endoperoxides produced during the autoxidation of polyunsaturated fatty acids.

Various compounds, other than those found in oxidized systems, have been found to interfere with the TBA test by producing the characteristic red pigment upon reaction with the reagents. For example, sucrose and some compounds in wood smoke have been reported to give a red color upon reaction with TBA, thus requiring corrections in the case of cured and smoked meats. On the other hand, abnormally low values may result if some of the malonaldehyde reacts with proteins in an oxidizing system. In addition, flavor scores for different systems cannot be consistently estimated from TBA values since the relative level of TBA produced from a given amount of oxidation varies from product to product. In many cases, however, the TBA test is applicable for comparison of samples of a single material at different states of oxidation.

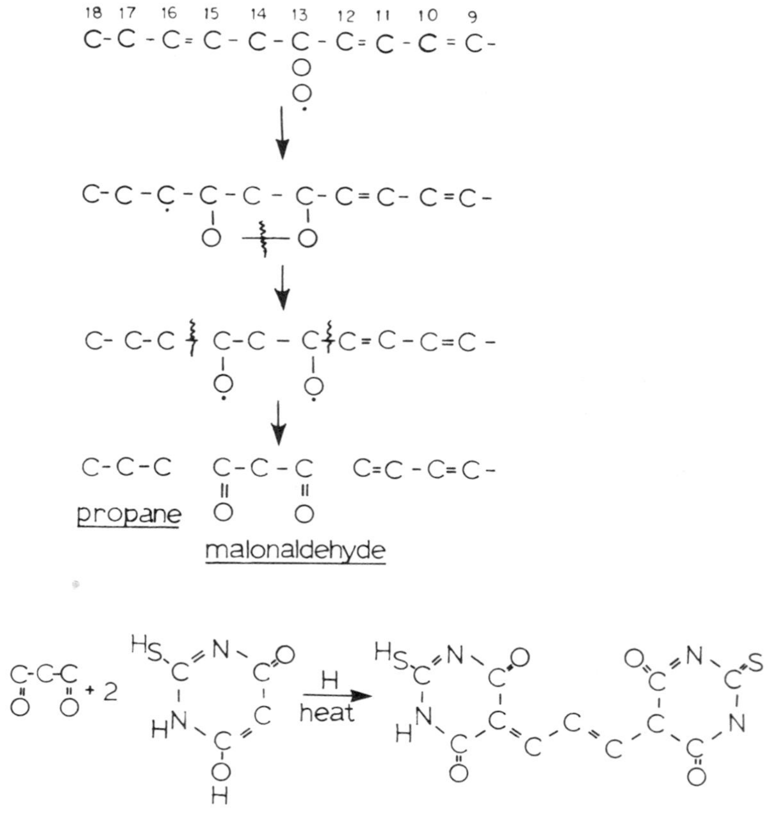

Figure 26   Speculative mechanism for malonaldehyde formation (73).

Total and Volatile Carbonyl Compounds. Methods for the determination of total carbonyl compounds are usually based on measurement of the hydrazones arising from reaction of aldehydes and ketones (oxidation products) with 2,4-dinitrophenylhydrazine (45). However, under the experimental conditions used for these tests, carbonyl compounds may be generated by decomposition of unstable substrates, for example, hydroperoxides, thus interfereing with quantitative results. Attempts to minimize such interference have involved reduction of the hydroperoxides to noncarbonyl compounds prior to determination of carbonyls or carrying out the reaction at a low temperature.

Since the major portion of the carbonyl compounds in oxidized fat are of high molecular weight and thus do not contribute directly to flavor, a variety of techniques has been developed to separate and measure the volatile carbonyl compounds. Most commonly the volatile decomposition products are recovered by distillation under atmospheric or reduced pressure and the carbonyls are then determined by reaction of the distillate with appropriate reagents.

Anisidine Value (79). In the presence of acetic acid, *p*-anisidine reacts with aldehydes producing a yellowish color. The molar absorbance at 350 nm increases if the aldehyde contains a double bond conjugated to the carbonyl double bond. Thus, the anisidine value is mainly a measure of 2-alkenals. An expression termed the Totox or oxidation value (OV), which is equivalent to 2 X peroxide value + ansidine value, has been suggested for the assessment of oxidation in oils.

Kreis Test. This is one of the first tests used commercially to evluate oxidation of fats. The procedure involves measurement of a red color believed to result from reaction of epihydrin aldehyde (an isomer of malonaldehyde) or other oxidation products with phloroglucinol. However, fresh samples free of oxidized flavor have been frequently found to develop some color upon reaction with the Kreis reagent, and consistent results among different laboratories have been difficult to obtain.

Ultraviolet Spectrophotometry. Measurement of absorbance at 234 nm (conjugated dienes) and 268 nm (conjugated trienes) is sometimes used to monitor oxidation. However, the magnitude of the change is not easily related to the degree of oxidation except in the early stages.

Oxirane Test. This method is based on the addition of hydrogen halides to the oxirane group. Epoxide content is determined by titrating the sample with HBr in acetic acid, in the presence of crystal violet, to a bluish green end point (69). The test, however, is not sensitive and lacks specificity. Hydrogen halides also attack α,β-unsaturated carbonyls and conjugated dieneols, and this reaction is not quantitative with some trans epoxides. A colorimetric method, based on the reaction of the oxirane group with picric acid, was reported to be more sensitive and not subject to these interferences (31).

Iodine Value. This test is a measure of the unsaturated linkages in a fat and is expressed in terms of percentage of iodine absorbed (69). The decline in iodine value is sometimes used to monitor the reduction of dienoic acids during the course of the autoxidation.

Fluorescence. Fluorescent compounds may develop from the interaction of carbonyl compounds (produced by lipid oxidation) and certain cellular constituents possessing free amino groups (27). The detection of oxidation products in biological tissues by fluorescent methods is a relatively sensitive technique.

Chromatographic Methods. Various techniques of liquid, thin-layer, high-performance liquid, size-exclusion, and gas chromatography have been used to determine oxidation of oils or lipid-containing foods. This approach is based on the separation and quantitative measurement of certain fractions—volatile (28), polar (11), or polymeric (82), for example—or individual components, such as pentane or hexanal (39), which are known to be typically produced in the course of autoxidation.

Organoleptic Evaluation. The ultimate judgment regarding the development of oxidized flavor in foods always requires sensory testing. The value of any objective chemical or physical method depends largely on how well it correlates with organoleptic evaluation. The testing of flavor is usually conducted by trained taste panels using special flavor score forms (52).

Several "accelerated" tests are also available to measure the resistance of a lipid to oxidation under adverse conditions.

Schaal Oven Test. The sample is stored at about 65°C and periodically tested until oxidative rancidity is detected. This can be done organoleptically or by measuring the peroxide value.

Active Oxygen Method (AOM). This test is widely used. It involves maintenance of the sample at 97.8°C while air is continously bubbled through it at a constant rate. The time required to obtain a specific peroxide value is determined (69).

Oxygen Bomb Method (24). The sample is dispersed in filter pulp in the glass liner of a sealed bomb at 50 psig oxygen pressure. The bomb is placed in boiling water and the time required to reduce the pressure to 2 psig is noted. This test is faster than the original AOM and is claimed to be more precise and accurate.

Oxygen Absorption. The amount of oxygen absorbed by the sample, as determined by the time to produce a specific pressure decline in a closed chamber, or the time to absorb a preestablished quantity of oxygen under specific oxidizing conditions, is taken as a measure of stability. This test has been particularly useful in studies of antioxidant activity.

10. Factors Influencing the Rate of Lipid Oxidation in Foods

Food fats contain mixtures of fatty acids that differ significantly in their susceptibility to oxidation. In addition, foods contain numerous nonlipid components that affect the rate of lipid oxidation. The complicated interactions and their influence on the various autoxidation steps make any precise analysis of oxidation kinetics in food nearly impossible. An excellent review of this subject by Labuza (59) is highly recommended. Only a brief discussion of the factors involved is given here.

Fatty Acid Composition. The number, position, and geometry of double bonds affect the rate of oxidation. Relative rates of oxidation for arachidonic, linolenic, linoleic, and oleic acids are approximately 40:20:10:1, respectively. Cis acids oxidize more readily than their trans isomers, and conjugated double bonds are more reactive than nonconjugated. Autoxidation of saturated fatty acids is extremely slow; at room temperature, they remain practically unchanged when the oxidative rancidity of unsaturates is detectable. At high temperatures, however, saturated acids can undergo significant oxidative changes (see Temperature).

Free Fatty Acids Versus the Corresponding Acylglycerols. Fatty acids oxidize at a slightly greater rate when free than when esterified to glycerol. Randomizing the fatty acid distribution of a natural fat reduces the rate of oxidation.

The existence in a fat or oil of a small amount of free fatty acids does not have a marked effect on oxidative stability. In some commercial oils, however, the presence of relatively large amounts of free acids can increase incorporation of catalytic trace metals from equipment or storage tanks and thereby increase the rate of lipid oxidation.

Oxygen Concentration. As noted earlier, if the supply of oxygen is unlimited, the rate of oxidation is independent of oxygen pressure, but at very low oxygen pressure the rate is approximately proportional to oxygen pressure. However, the effect of oxygen pressure on rate is influenced by other factors, such as temperature and surface area.

Temperature. In general, the rate of oxidation increases as the temperature is increased. Temperature is also important in terms of the effect of oxygen partial pressure on the rate of oxidation. As the temperature increases, the increase in rate with increasing oxygen concentration become less evident, since oxygen becomes less soluble as the temperature is raised.

Surface Area. The rate of oxidation increases in proportion to the surface area of the lipid exposed to air. However, as the surface-volume ratio increases, reducing the oxygen partial pressure becomes less effective in decreasing the rate of oxidation. In oil-in-water emulsions the rate of oxidation is governed by the rate at which oxygen diffuses into the oil phase.

Moisture. In studies of model lipid systems and various fat-containing foods, it has been shown that oxidation rate depends strongly on water activity (see also Chap. 2) (54). In dried foods with very low moisture contents (a_w values of less than about 0.1) oxidation proceeds very rapidly. Increasing the water content to an a_w of about 0.3 retards lipid oxidation and often produces a minimum rate. This protective effect of water is believed to occur by reducing the catalytic activity of metal catalysts, by quenching free radicals, by promoting nonenzymic browning (which produces compounds with antioxidant activities), and/or by impeding the access of oxygen to the food.

At somewhat higher water activities (a_w = 0.55-0.85), the rate of oxidation increases again, probably as a result of increased mobilization of the catalysts present.

Pro-oxidants. The transition metals, particularly those possessing two or more valency states with a suitable oxidation-reduction potential between them (e.g., cobalt, copper, iron, manganese, and nickel) are major pro-oxidants. If present, even at concentrations as low as 0.1 ppm, they can decrease the length of the induction period and increase the rate of oxidation. Trace amounts of heavy metals are encountered in most edible oils, and they originate from the soil in which the oil-bearing plant was grown, from the animal, or from metallic equipment used in processing or storage. Trace metals are also naturally occurring components of all food tissues and of all fluid foods of biological origin (eggs, milk, and fruit juices), present in both free and bound forms.

Several mechanisms for metal catalysis have been postulated. These include

1. *Acceleration of hydroperoxide decomposition*

$$M^{n+} + ROOH \longrightarrow M^{(n+1)+} + OH^- + RO^\cdot$$

$$M^{n+} + ROOH \longrightarrow M^{(n-1)+} + H^+ + ROO^\cdot$$

2. *Direct reaction with the unoxidized substrate*

$$M^{n+} + RH \longrightarrow M^{(n-1)^+} + H^+ + R^.$$

3. *Activation of molecular oxygen to give singlet oxygen and peroxy radicals*

$$M^{n+} + O_2 \longrightarrow M^{(n+1)^+} + O_2^- \begin{cases} \xrightarrow{-e^-} {}^1O_2 \\ \xrightarrow{+H^+} H\dot{O}_2 \end{cases}$$

Hematin compounds, present in many food tissues, are also important pro-oxidants.

Radiant Energy. Visible, u.v., and γ radiation are effective promoters of oxidation.

Antioxidants. Due to their importance in lipid chemistry, this subject is treated separately in the following section.

11. Antioxidants

Antioxidants are substances that can delay the onset or slow the rate of oxidation of autoxidizable materials. Literally hundreds of compounds, both natural and synthesized, have been reported to possess antioxidant properties. Their use in food, however, is limited by certain obvious requirements not the least of which is adequate proof of safety. The main antioxidants currently used in food are monohydric or polyhydric phenols with various ring substitutions (Fig. 27). For maximum efficiency, primary antioxidants are often used in combination with other phenolic antioxidants or with various metal sequestering agents (Table 6).

Although much is known regarding the mechanisms by which many antioxidants impart stability to pure oils, further investigation is needed to clarify their function in complex foods.

Effectiveness and Mechanisms of Action. Several reviews of antioxidant kinetics and mechanism of action have been published (72,75,80). As should be clear from the autoxidative mechanisms outlined above, a substance delays the autoxidation reaction if it inhibits the formation of free radicals in the initiation step, or if it interrupts the propagation of the free radical chain. The initiation of free radicals can be retarded by the use of peroxide decomposers, or metal chelating agents, or singlet oxygen inhibitors. Since traces of peroxides and metal initiators, however, cannot be entirely eliminated, most studies have concentrated on the action by free radical acceptors.

The first detailed kinetic study of antioxidant action was conducted in 1947 by Bolland and ten Have using a model system of autoxidizing ethyl linoleate containing hydroquinone as an inhibitor (13). They postulated that antioxidants inhibit the chain reaction by acting as hydrogen donors or free radical acceptors and concluded that the free radical acceptor (AH) reacts primarily with $RO_2^.$ and not with $R^.$ radicals.

$$RO_2^. + AH \longrightarrow ROOH + A^.$$

These authors also concluded that the most probable number of oxidation chains terminated by one inhibitor molecule is two, and that the reaction can, therefore, be represented as occurring in two stages.

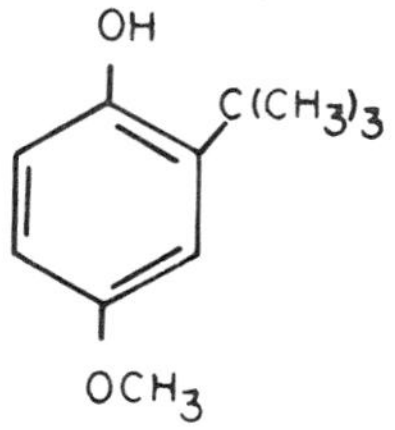

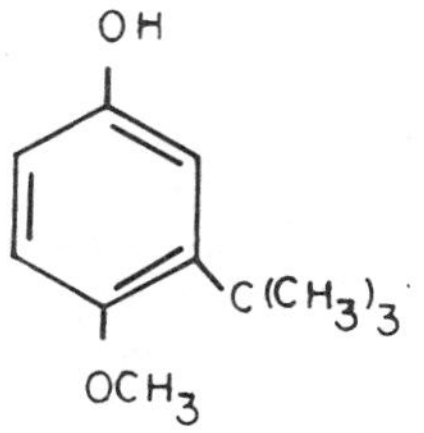

Figure 27 Major antioxidants used in food.

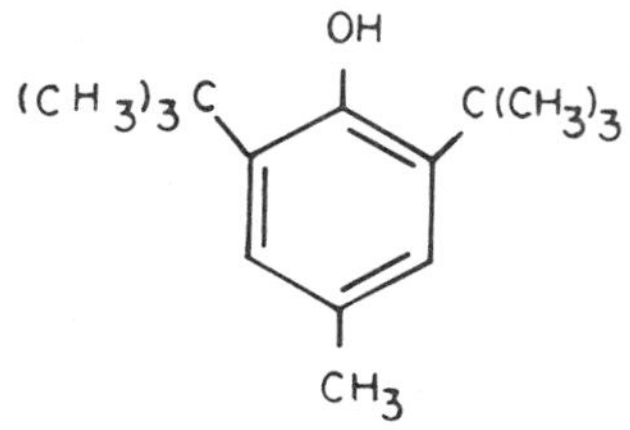

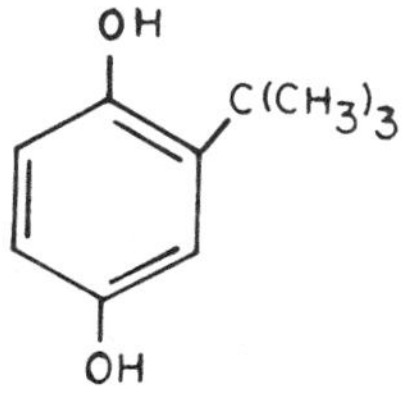

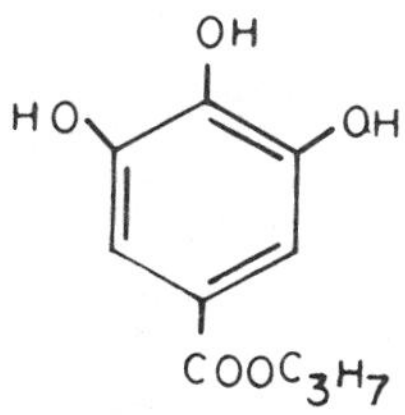

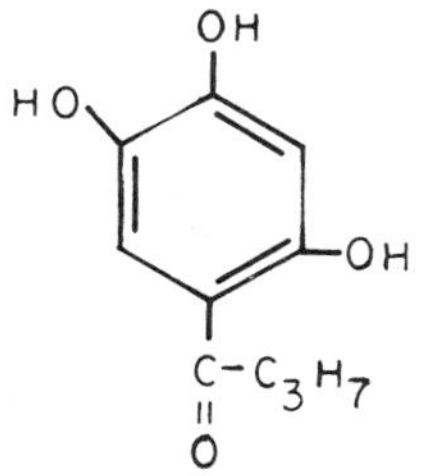

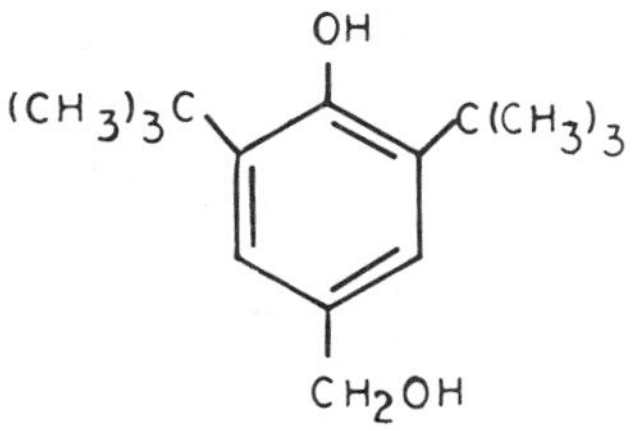

Table 6 Antioxidants Permitted in Foods in the United States

Primary antioxidants	Synergists
Tocopherols	Citric acid and isopropyl
Gum guaiac	citrate
Propyl gallate	Phosphoric acid
Butylated hydroxyanisole (BHA)	Thiodipropionic acid and
Butylated hydroxytoluene (BHT)	its didodecyl, dilauryl,
2,4,5-Trihydroxybutyrophenone (THBP)	and dioctadecyl esters
4-Hydroxymethyl-2,6-di-*tert*-butylphenol	Ascorbic acid and ascorbyl palmitate
tert-Butylhydroquinone (TBHO)	Tartaric acid
	Lecithin

$$RO_2{}^\bullet \;+\; AH_2 \longrightarrow ROOH \;+\; AH\bullet$$

$$AH\bullet \;+\; AH\bullet \longrightarrow A \;+\; AH_2$$

Different versions of antioxidation mechanisms were suggested by various other workers. It was proposed, for example, that the free radical intermediate $AH^\bullet$ forms stable products by reacting with an $RO_2^\bullet$ radical, or that a complex between the $RO_2^\bullet$ radical and the inhibitor is formed followed by reaction of the complex with another $RO_2^\bullet$ radical to yield stable products

$$RO_2{}^\bullet \;+\; \text{inh.} \longrightarrow [RO_2 - \text{inh.}]$$

$$[RO_2 - \text{inh.}] \;+\; RO_2^\bullet \longrightarrow \text{stable products}$$

Although all these reactions may be taking place, the original concept of Bolland and ten Have is considered to be the most important, and the basic mechanism can thus be visualized as a competition between the "inhibitor reaction"

$$RO_2{}^\bullet \;+\; AH \longrightarrow ROOH \;+\; A\bullet$$

and the chain propagation reaction

$$RO_2{}^\bullet \;+\; RH \longrightarrow ROOH \;+\; R\bullet$$

The effectiveness of an antioxidant is related to many factors, including activation energy, rate constants, oxidation reduction potential, ease of antioxidant loss or destruction, and solubility properties. For the two competing reactions discussed above (i.e., the chain propagation and the inhibitor reactions), both of which are exothermic, the activation energy increases with increasing A–H and R–H bond dissociation energies and, therefore, the efficiency of the antioxidant (AH) increases with decreasing A–H bond strength. Ideally, however, the resulting antioxidant free radical must not itself initiate new free radicals or be subject to rapid oxidation by a chain reaction. In this regard phenolic antioxidants occupy a favored status. They are excellent hydrogen or electron donors and, in addition, their radical intermediates are relatively stable due to resonance delocalization and to the lack of positions suitable for attack by molecular oxygen. Hydroquinones, for example, react with hydroperoxy radicals forming stable semiquinone resonance hybrids.

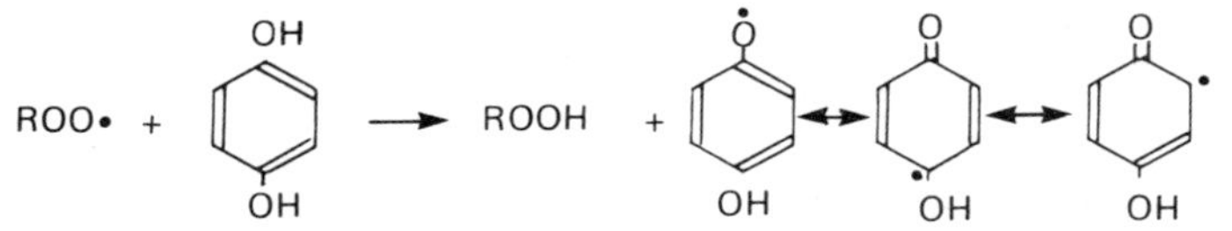

The semiquinone radical intermediates may undergo a variety of reactions forming more stable products. They can react with one another to form dimers, dismutate, producing quinones with renewed formation of the original inhibitor molecule,

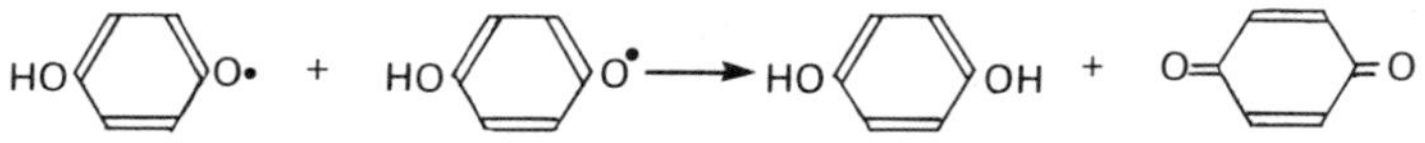

or react with another $RO_2^{\cdot}$ radical.

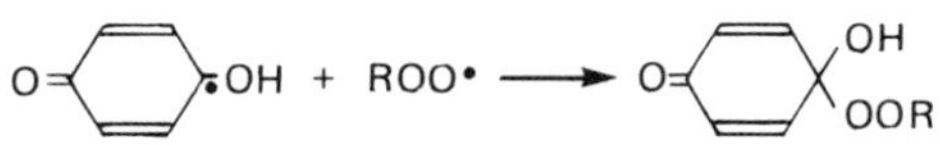

The monohydric phenols cannot form semiquinones or quinones, as such, but they give rise to radical intermediates with moderate resonance delocalization, and they can be sterically hindered, as for example, by the two *t*-butyl groups in BHT. The *t*-butyl groups reduce further chain initiation by the alkoxy radical after the initial hydrogen donation. The net reaction between linoleate and BHT can be represented as

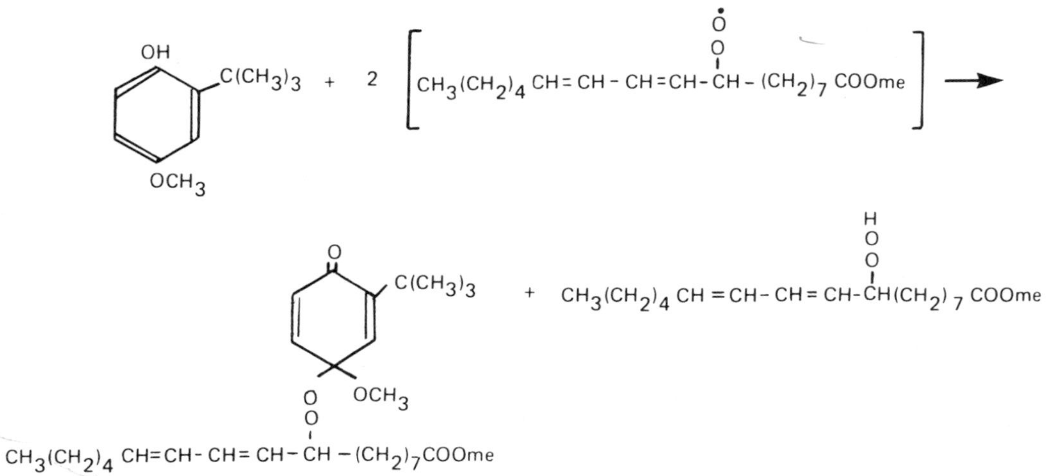

In addition to the chemical potency of the antioxidant, its solubility in oil and its volatility influence its effectiveness. Its solubility affects accessibility to the peroxy

radical sites, and its volatility affects its permanence during storage or heating. In the last few years, emphasis has shifted to the study of antioxidant action at surfaces, especially to such systems as membranes, micelles, and emulsions. The importance of the amphiphilic character of antioxidant molecules to their effectiveness in biphasic and multiphasic systems is demonstrated in a provocative discussion by Porter (72).

Synergism. Synergism occurs when a mixture of antioxidants produces a more pronounced activity than the sum of the activities of the individual antioxidants when used separately. Two categories of synergism are recognized. One involves the action of mixed free radical acceptors; the other involves the combined action of a free radical acceptor and a metal chelating agent. It is often the case, however, that a so-called synergist may play more than one role. Ascorbic acid, for example, may function as an electron donor, a metal chelator, an oxygen scavenger, and a contributor to the formation of browning products with antioxidant activity.

1. Uri (80) introduced a hypothesis to explain the action of two *mixed free radical acceptors*, say, AH and BH, where it is assumed that the bond dissociation energy of B–H is smaller than that of A–H. It is further assumed that BH reacts slowly with $RO_2^{\cdot}$ because of steric hindrance. The following reaction would then take place.

$$RO_2^{\cdot} + AH \longrightarrow ROOH + A^{\cdot}$$

$$A^{\cdot} + BH \longrightarrow B^{\cdot} + AH$$

Thus, the presence of BH will have a sparing effect since it results in regeneration of the primary antioxidant. In addition, the tendency of $A^{\cdot}$ to disappear via a chain reaction is substantially diminished. An example of such a system is the combination of a phenolic antioxidant and ascorbic acid. Since the phenolic compound, if used by itself, is the more effective of the two, it is customarily termed the primary antioxidant and ascorbic acid is called the synergist. In a similar way, it is also possible for two phenolic antioxidants to exhibit synergism.

2. In general, *metal complexing agents* partly deactivate trace metals, often present as salts of fatty acids. The antioxidant properties of a free radical acceptor can be enhanced so substantially by the action of a metal complexing agent that a synergistic effect is produced. Citric acid, phosphoric acid, polyphosphates, and ascorbic acid are typical examples of metal chelating agents.

Choice of Antioxidant. Due to differences in their molecular structure, the various antioxidants exhibit substantial differences in effectiveness when used with different types of oils or fat-containing foods and when used under different processing and handling conditions. In addition to the potency of the antioxidants or mixture of antioxidants in any particular application, due consideration must be given to other factors, such as ease of incorporation into the food, carry-through characteristics, sensitivity to pH, tendency to discolor or produce off-flavor, availability, and cost. The problem of selecting the optimum antioxidant or combination of antioxidants is further complicated by the difficulty of predicting how the added antioxidant(s) will function in the presence of pro-oxidants and anti-oxidants already present in the food (or produced in the course of processing).

As indicated above, the implications of the hydrophilic-lipophilic properties of the various antioxidants, with regard to their effectiveness in different applications, have recently received considerable attention. Porter (72) proposed that two basic situations call for two different types of antioxidants. The first involves a small surface-volume ratio, such as occurs with bulk oils (lipid-gas interface). In this case antioxidants with relative large values for hydrophilic-lipophilic balance (e.g., PG or TBHQ) are most effective since they concentrate at the surface of the oil where reaction of fat with molecular oxygen is most prevalent.

The second situation involves a large surface-volume ratio, such as occurs with intact tissue foods with polar lipid membranes, with intracellular micelles of neutral lipids, and with micelles of emulsified oils (e.g., salad dressing). These are multiphasic systems in which the concentration of water is large and the lipid is often in a mesophasic state. For these situations, the more lipophilic antioxidants (e.g., BHA, BHT, higher alkyl gallates, and tocopherols) are most effective.

Characteristics of Some Commonly Used Primary Antioxidants. *Tocopherols* are the most widely distributed antioxidants in nature and they constitute the principal antioxidant in vegetable oils. The small amounts of tocopherols present in animal fats originate from vegetable components in the animal diet. Seven tocopherol structures, all methyl-substituted forms of tocol, are known. Of these α, γ, and δ predominate in vegetable oils.

Tocopherols : α = 5, 7, 8 - trimethyl - ; γ = 7, 8 - dimethyl - ; ζ = 8 - methyltocol

In general, tocopherols with high vitamin E activity are less effective as antioxidants than those with low vitamin E activity. The order of antioxidant activity is thus $\delta > \gamma > \beta > \alpha$. However, the relative activity of these compounds is significantly influenced by temperature and light.

A relatively high proportion of the tocopherols present in crude vegetable oils survive the oil processing steps and remain in sufficient quantities to provide oxidative stability in the finished product. As antioxidants, tocopherols exert their maximum effectiveness at relatively low levels, approximately equal to their concentration in vegetable oils. If used at a very high concentration, they may actually act as pro-oxidants.

Gum quaiac is a resinous exudate from a tropical tree. Its effectiveness, due mainly to its appreciable content of phenolic acids, is more pronounced in animal fats than in vegetable oils. Gum guaiac has a reddish brown color, is very slightly soluble in oil, and gives rise to some off-flavor.

Both *butylated hydroxyanisole* (BHA) which is commercially available as a mixture of two isomers (Fig. 27), and *butylated hydroxytoluene* (BHT) have found wide commercial use in the food industry. Both are highly soluble in oil and exhibit weak antioxidant activity in vegetable oils, particularly those rich in natural antioxidants. BHT and BHA are relatively effective when used in combination with other primary antioxidants. BHA has a typical phenolic odor that may become noticeable if the oil is subjected to high heat.

Nordihydroquaiaretic acid (NDGA) is extracted from a desert plant, *Larrea divaricata*. Its solubility in oils is limited (0.5-1%), but greater amounts can be dissolved if the oil is heated. NDGA has few carry-through properties and tends to darken slightly on storage, in the presence of iron, or when subjected to high temperatures. The antioxidant activity of NDGA is markedly influenced by pH and it is readily destroyed under highly alkaline conditions. NDGA is reported to be very effective in the prevention of hematin-catalyzed oxidation in fat-aqueous systems and in certain meats. At present, NDGA is not readily available in the United States due to its high cost.

As would be expected from their phenolic structure and three hydroxyl groups, *gallic acid* and *alkyl gallates* exhibit considerable antioxidant activity. Gallic acid is soluble in water but nearly insoluble in oil. Esterification of the carboxyl group with alcohols of varying chain length produces alkyl gallates with increased oil solubility. Of these, propyl gallate is widely used in the United States. It has also been reported to be effective in retarding lipoxygenase oxidation of linoleate. In the presence of traces of iron, the gallates give rise to a blue-black discoloration at alkaline conditions; they are lost rapidly during baking or frying.

Tertiary butylhydroquinone (TBHQ) is one of the more recently introduced antioxidants. It has undergone extensive testing and was cleared by the U. S. Government in 1972. TBHQ is moderately soluble in oil and slightly soluble in water. In many cases, TBHQ is more effective than other common antioxidants in providing oxidative stability to crude and refined polyunsaturated oils, without encountering problems of color or flavor stability. TBHQ is also reported to exhibit good carry-through characteristics in the frying of potato chips. Based on extensive feeding and biological studies, TBHQ has been judged to be safe for use as an antioxidant with estimated safety margins of 1000-10,000 times the normal use level.

As would be expected from their similarity in structure, *2,4,5-trihydroxybutyrophenone* (THBP) and the gallates exhibit similar antioxidant properties. However, THBP is not widely used in the United States.

4-Hydroxymethyl-2,6-ditertiarybutylphenol is produced by substituting a hydroxyl for one hydrogen in the methyl group of BHT. It is therefore less volatile than BHT but otherwise behaves similarly as an antioxidant.

Safety and Control. Antioxidants are food additives. In the United States their use is subject to regulation under the Federal Food, Drug and Cosmetic Act. Antioxidants for food products are also regulated under the Meat Inspection Act, the Poultry Inspection Act, and various state laws. Antioxidants permitted for use in foods are listed in Table 6. In general, the total concentration of authorized antioxidants, added singly or in combination, must not exceed 0.02% by weight based on the fat content of the food. Certain exceptions exist in the case of standardized foods and products covered by special regulations. Under the Meat Inspection Act concentrations up to 0.01% are permitted for single antioxidants, based on fat content, with a combined total of no more than 0.02%.

Although some antioxidants reportedly cause certain pathological effects if given to animals at excessive doses, rigorous testing has shown the authorized antioxidants to be safe when used at levels of at least 100 times the permitted concentration. The tocopherols and the major acid synergists are unregulated.

The general public concern with chemical additives and food safety has triggered a continuing search for new antioxidants that may occur naturally in food or may form inadvertently during processing. Compounds with antioxidant properties have been found in many spices, oil seeds, citrus pulp and peel, cocoa shells, oats, soybean, hydrolyzed plants, animal and microbial proteins, and in products from heating and nonenzymic browning.

Another result of the effort to achieve "antioxidant safety" is the recent development of nonabsorbable polymeric antioxidants. These are generally hydroxyaromatic polymers with various alkyl and alkoxyl substitutions. Such compounds are usually very large molecules and their absorption from the intestinal tract is practically nil. In addition to their reportedly high antioxidant activity they are nonvolatile under deep-fat frying conditions, which results in nearly quantitative carry-through to the fried items. They have not yet received FDA approval.

Modes of Application. Antioxidants can be added directly to vegetable oils or to melted animal fats after they are rendered. In some cases, however, better results are achieved when the antioxidant is administered in a diluent. Examples are mixtures of monoacylglycerols and glycerol in propylene glycol, monoacylglycerol-water emulsions, and mixtures of antioxidants in volatile solvents. Food products can also be sprayed with, or dipped in solutions or suspensions of antioxidants, or they can be packaged in films containing antioxidants.

C. Thermal Decomposition

Heating of food produces various chemical changes, some of which can be important from the standpoint of flavor, appearance, nutritive value, and toxicity. Not only do the different nutrients in food undergo decomposition reactions but these nutrients also interact among themselves in extremely complex ways to form a very large number of new compounds.

The chemistry of lipid oxidation at high temperature is complex since both thermolytic and oxidative reactions are simultaneously involved. Both saturated and unsaturated fatty acids undergo chemical decomposition when exposed to heat in the presence of oxygen. A schematic summary of these mechanisms is shown in Fig. 28.

1. Thermal Nonoxidative Reactions of Saturated Fats

In general, very high temperatures of heating are required to produce substantial nonoxidative decomposition of saturated fatty acids. Thus, heating of saturated triacylglycerols and methyl esters of fatty acids at temperatures of 200-700°C yields detectable amounts of decomposition products consisting mostly of hydrocarbons, acids, and ketones. In a recent study, however, using very sensitive measurement techniques, thermolytic products were detected in triacylglycerols after heating in vacuum for only 1 hr at 180°C.

Shown in Fig. 29 are the products arising from anaerobic heating of tributyrin (Tri-4), tricaproin (Tri-6), and tricaprylin (Tri-8). Each triacylglycerol produces the following products (n is the number of carbons in the fatty acid molecule): a series of normal alkanes and 1-alkenes with the C_{n-1} alkane predominating; a C_n fatty acid; a C_{2n-1} symmetrical ketone; a C_n oxopropyl ester; C_n propene- and propanediol diesters; and C_n diacylglycerols. Acrolein, CO, and CO_2 are also formed. Quantitatively, the component fatty acids are the major compounds produced from thermolytic decomposition of triacylglycerols. In the absence of moisture the free fatty acid can be formed via a "six-atom-ring-closure," as follows.

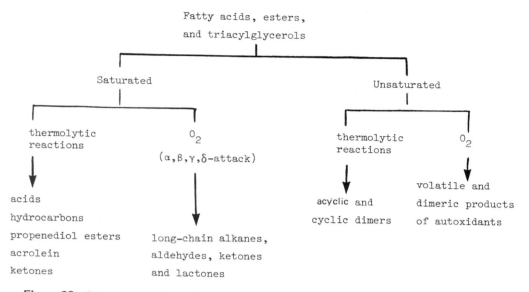

Figure 28   Generalized scheme for thermal decomposition of lipids.

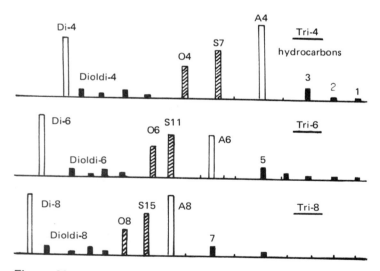

Figure 29   Nonoxidative decomposition products of triacylglycerols: A, free acid; S, symmetrical ketone; O, oxopropyl ester; Dioldi, propane- and propenediol diesters; Di, diglycerides. Numbers refer to hydrocarbon or fatty acid carbon chain.

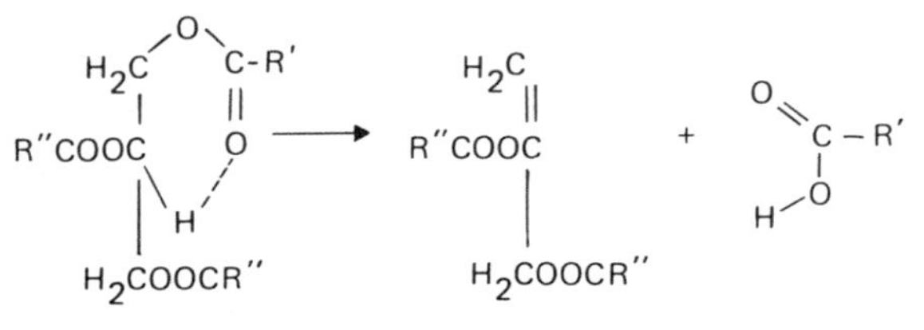

This mechanism also explains the formation of propenedioldiesters. Expulsion of the acid anhydride from the triacylglycerol molecule produces 1- or 2-oxopropyl esters and the acid anhydride:

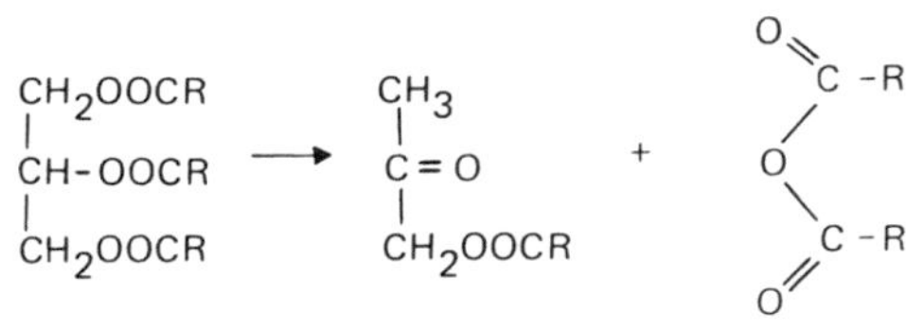

Decomposition of the 1-oxopropyl ester gives rise to acrolein and the C_n fatty acid; decarboxylation of the acid anhydride intermediate produces the symmetrical ketone. Free radical mechanisms similar to those proposed for radiolysis of triacylglycerols may also play a significant role in the formation of thermolytic products, particularly when relatively high temperatures are used.

2. Thermal Oxidative Reactions of Saturated Fats

Saturated fatty acids and their esters are considerably more stable than their unsaturated analogs. However, when heated in air at temperatures higher than 150°C, even saturated lipids undergo oxidation, giving rise to a complex decomposition pattern. The major oxidative products consist of a homologous series of carboxylic acids, 2-alkanones, *n*-alkanals, lactones, *n*-alkanes, and 1-alkenes.

In a recent study, a series of triacylglycerols containing the even-numbered fatty acid chains C_6 to C_{16} were heated in air for 1 hr at either 180 or 250°C and their oxidation products examined (25). The hydrocarbon series was the same as that produced in the absence of oxygen, but the amounts formed were much greater under the oxidative conditions. In all cases the C_{n-1} alkane was the major hydrocarbon produced.

In general, the 2-alkanones are produced in larger quantities during oxidative heating than the alkanals, with the C_{n-1} methyl ketone the most abundant carbonyl compound. All the lactones found were γ-lactones, except for those with a carbon number equal to that of the parent fatty acid, in which case both C_n, γ-, and δ-lactones were produced in relatively large amounts.

It is generally accepted that thermal oxidation of a saturated fatty acid involves the formation of monohydroperoxides as a principal mechanism, and that oxygen attack can occur at all the methylene groups of the fatty acid. There is, however, controversy regarding whether hydroperoxide formation is favored at certain locations along the alkyl chain. Since the dominant oxidative products of saturated fatty acids are those with chain lengths near or equal to the parent fatty acids, it is likely that oxidation occurs preferentially at the α, β, and γ positions.

Oxidative attack at the β-carbon of the fatty acid, for example, results in the formation of β-keto acids, which in turn yield C_{n-1} methyl ketones upon decarboxylation. Cleavage between the α- and β-carbons of the alkoxy radical intermediate gives rise to C_{n-2} alkanal; scission between the β- and γ-carbons produces C_{n-3} hydrocarbons:

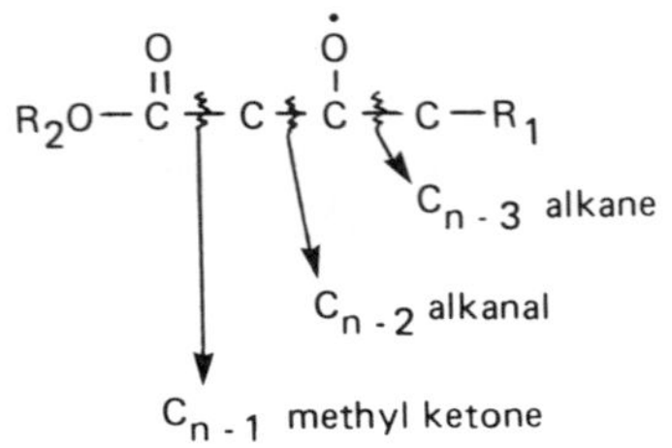

$$R_2O - \overset{\overset{O}{\|}}{C} \, \ddagger \, C \, \ddagger \, \overset{\overset{\cdot}{O}}{C} \, \ddagger \, C - R_1$$

$$C_{n-3} \text{ alkane}$$
$$C_{n-2} \text{ alkanal}$$
$$C_{n-1} \text{ methyl ketone}$$

Oxygen attack at the γ position yields a C_{n-4} hydrocarbon, a C_{n-3} alkanal, and a C_{n-2} methyl ketone.

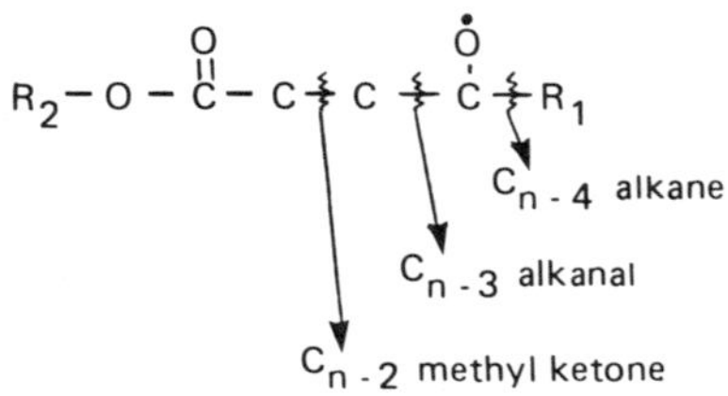

$$R_2 - O - \overset{\overset{O}{\|}}{C} - C \, \ddagger \, C \, \ddagger \, \overset{\overset{\cdot}{O}}{C} \, \ddagger \, R_1$$

$$C_{n-4} \text{ alkane}$$
$$C_{n-3} \text{ alkanal}$$
$$C_{n-2} \text{ methyl ketone}$$

In addition, hydroperoxide formation at the γ position is believed to be responsible for the production of the C_n γ-lactones via cyclization of the resulting hydroxy acids.

Oxygen attack at the α-carbon accounts for the production of a C_{n-1} fatty acid (via formation of the α-keto acid), a C_{n-1} alkanal, and a C_{n-2} hydrocarbon.

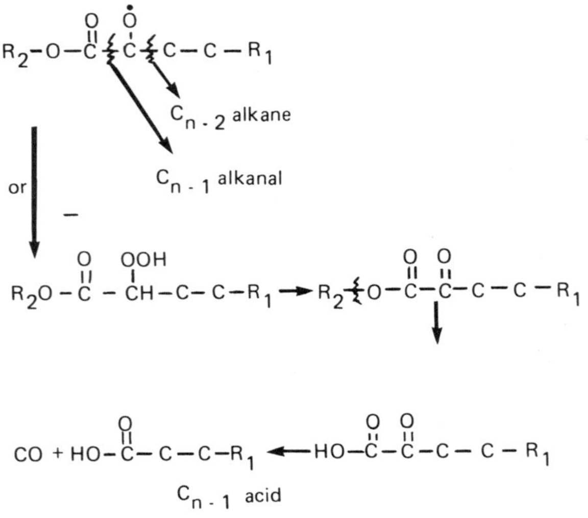

$$R_2 - O - \overset{\overset{O}{\|}}{C} \, \ddagger \, \overset{\overset{\cdot}{O}}{C} \, \ddagger \, C - C - R_1$$

$$C_{n-2} \text{ alkane}$$
$$C_{n-1} \text{ alkanal}$$

or

$$R_2O - \overset{\overset{O}{\|}}{C} - \overset{\overset{OOH}{|}}{CH} - C - C - R_1 \longrightarrow R_2 \ddagger O - \overset{\overset{O}{\|}}{C} - \overset{\overset{O}{\|}}{C} - C - C - R_1$$

$$CO + HO - \overset{\overset{O}{\|}}{C} - C - C - R_1 \longleftarrow HO - \overset{\overset{O}{\|}}{C} - \overset{\overset{O}{\|}}{C} - C - C - R_1$$

$$C_{n-1} \text{ acid}$$

Further stepwise oxidations give rise to a series of smaller acids, which may themselves undergo oxidation, producing typical decomposition products of their own. This may explain the formation, in small amounts, of the shorter chain hydrocarbons, lactones, and carbonyl compounds.

3. Thermal Nonoxidative Reactions of Unsaturated Fatty Acid Esters

Formation of dimeric compounds appears to be the predominant reaction of unsaturated fatty acids when heated in the absence of oxygen. In addition, certain other substances of lower molecular weight are formed. Relatively severe heat treatment, however, is required for these reactions to take place. Thus, no significant decomposition of methyl oleate can be detected at temperatures below 220°C, but when samples are heated at 280°C for 65 hr under argon, hydrocarbons, short- and long-chain fatty acid esters, and straight-chain dicarboxylic dimethyl esters, as well as dimers, are produced. The occurrence of many of these compounds has been explained on the basis of the formation and/or combination of free radicals resulting from homolytic cleavage of C–C linkages near the double bond. The dimeric compounds, which include alicyclic monoene and diene dimers, as well as saturated dimers which cyclopentane structure, are believed to arise via allyl radicals resulting from hydrogen abstraction at methylene groups α to the double bond. Such radicals may undergo disproportionation into monoenoic and dienoic acids, or inter- and intramolecular addition to C–C double bonds (Fig. 22).

Methyl linoleate, heated under the same conditions, produces a more complex mixture of dimers consisting of saturated tricyclic, monounsaturated bicyclic, diunsaturated monocyclic, and triunsaturated acyclic dimers, as well as dehydrodimers containing one or two double bonds (Fig. 23). At a less severe heat treatment of 180°C for 1 hr under vacuum, ethyl linoleate yields only traces of volatile decomposition products. These compounds appear to be hydrocarbons, but the amounts produced are so small that positive identification has not been accomplished.

4. Thermal Oxidative Reactions of Unsaturated Fatty Acid Esters

Unsaturated fatty acids are much more susceptible to oxidation than their saturated analogs. At elevated temperatures their oxidative decomposition proceeds very rapidly. Although certain specific differences between high- and low-temperature oxidations have been observed by some investigators, the evidence accumulated to date indicates that in both cases the principal reaction pathways are the same. The formation and decomposition of hydroperoxide intermediates, predictable according to location of double bonds, appear to occur over a wide temperature range. Many decomposition products have been isolated from heated fats. The major compounds produced at high temperature are qualitatively typical of those ordinarily produced at room temperature autoxidation.

At elevated temperatures, hydroperoxide decomposition and secondary oxidations occur at extremely rapid rates. The amount of a given decomposition product, at a given time during the autoxidation process, is determined by the net balance between the complex effects of many factors. Hydroperoxide structure, temperature, the degree of autoxidation, and the stability of the decomposition products themselves undoubtedly exert major influences on the final quantitative pattern. The picture is further complicated, since these factors influence not only C–C bond scission, but also a large number of other possible decomposition reactions that occur simultaneously and in competition with C–C cleavage. The latter reactions include C–O scission, which may lead to positional isomerization of hydroperoxides (20), epoxidation, formation of dihydroperoxides, intramolecular cyclization, and dimerization.

We have already discussed the basic mechanisms of thermal and oxidative polymerization of unsaturated fatty acids. Heating of unsaturated fatty acids in air at elevated temperatures leads to the formation of oxydimers or polymers with hydroperoxide, hydroxide, epoxide, and carbonyl groups, as well as ether and peroxide bridges. The precise structures of many of these compounds, as well as the effects of the various oxidation parameters on reactions leading to their formation, are still largely unknown. However, with the development of highly efficient chromatographic techniques for the separation of high-molecular-weight compounds, renewed interest has been generated in the study of oxidative polymerization.

D. Chemistry of Frying

Deep-fat fried foods are becoming more and more significant as contributors to the total caloric intake in an average U. S. diet because of current trends in the eating habits of the American consumer. In the course of frying, food contacts hot oil (about 180°C) in the presence of air for various periods of time. The finished product usually contains 5-40% absorbed oil. Thus frying, more than any other standard process or handling method normally administered to lipid-containing foods, has the greatest potential for causing chemical changes in fat, and it therefore deserves discussion here.

1. Behavior of the Frying Oil

The following classes of compounds are produced from the oil during frying.

Volatiles. As expected from oxidative reactions involving the formation and decomposition of hydroperoxides, such compounds as saturated and unsaturated aldehydes, ketones, hydrocarbons, lactones, alcohols, acids and esters, are produced during frying (21). After the first 30 min of heating in air at 180°C, the primary volatile oxidation products can be detected by gas chromatography. Although the amounts of the volatiles produced in the oil vary widely, depending on the type of oil, the food, and the heat treatment they generally reach a plateau value, probably as a result of a balance between formation of the volatiles and their evaporation or decomposition. In Table 7 the amounts of some volatiles produced during typical frying operations (commercial chicken frying and frying of a variety of food items in a university kitchen) are compared with the volatiles produced in corn oil heated under two different conditions. These data show clearly that far greater quantities of most of these volatiles are produced (or are at least present) during these conditions of continuous heating than during intermittent frying. It is also evident that the amounts of volatiles produced cannot be used as an indicator of the extent to which a commercial frying oil had been used.

Nonpolymeric polar compounds of moderate volatility (e.g., hydroxy - and epoxy acids). These compounds are produced according to the various oxidative pathways involving the alkoxy radical as discussed earlier.

Dimeric and polymeric acids, and dimeric and polymeric glycerides. These compounds occur as expected from thermal and oxidative free radical combinations. Polymerization results in a substantial increase in the viscosity of the frying oil.

Free fatty acids. These compounds arise from hydrolysis of triacylglycerols in the presence of heat and water.

Table 7 Concentration of Some Volatiles in Used Frying Shortenings as Compared to Those in Corn Oil Heated Under Controlled Conditions

	Concentration (mg/kg)			
Volatile	Corn oil (1 hr, 180°C in air)	Corn/H_2O (70 hr, 180°C in air)	University kitchens (after 12 weeks)[a]	Commercial frying (10 days)[b]
Hexanal	13	11	1.7	2.2
Heptenal	10	4.2	2.1	2.2
Octenal	4.3	4.9	2.2	2.7
Decadienal (t,c)	6.5	7.4	5.0	5.8
Decadienal (t,t)	32	20	20	12
Octane	1.8	1.2		4.4
Undecane	0.24	0.12	0.48	0.6
Pentylfuran	1.9	3.6	0.4	0.9
Pentadecane	0.3	0.2	1.4	1.3

[a]Intermittent frying of various food items, composition of shortening unknown.
[b]Intermittent frying of chicken, composition of shortening unknown.

These reactions are responsible for a variety of physical and chemical changes that can be observed in the oil during the course of the frying process. Such changes include increases in viscosity and free fatty acid content, development of a dark color, a decrease in iodine value, changes in refractive index, a decrease in surface tension, and an increased tendency of the oil to foam.

2. Behavior of the Food During Frying

Food is involved in the following occurrences during frying:

Water is continuously released from the food into the hot oil. This produces a steam-distillation effect, sweeping volatile oxidative products from the oil. The released moisture also agitates the oil and hastens hydrolysis, resulting in increased amounts of free fatty acids. The blanket of steam formed above the surface of the oil may reduce the amount of oxygen available for oxidation.

During frying, volatiles (e.g., sulfur compounds and pyrazine derivatives from potato) may develop in the food itself or from the interaction between the food and the oil.

Food absorbs varying amounts of oil from deep-fat fryers (potato chips contain approximately 40% fat), resulting in the need for frequent or continuous addition of fresh oil. In continuous fryers, this results in the rapid attainment of a steady-state condition for oil properties.

The food itself releases some of its endogenous lipids into the fryer (e.g., chicken fat). The oxidative stability of the new mixture may be different from that of the original frying fat.

The presence of food causes the oil to darken at an accelerated rate.

3. Chemical and Physical Changes

As described above, a variety of chemical and physical changes take place in both the oil and the food during the frying process. It should be pointed out that these changes are not to be automatically construed as signs of undesirable or harmful deteriorations. In fact, some of these changes are sought by the frying process to provide the sensory qualities typical of fried food. On the other hand, extensive decomposition, resulting from lack of adequate control of the frying operation, can be a potential source of damage not only to the senory quality of the fried product but also to its nutritional value.

The chemical and physical changes in the frying fat are influenced by a number of frying parameters. Obviously, the compounds formed depend on the composition of both the oil and the food being fried. High temperature, long frying times, and metal contaminants favor extensive decomposition of the oil. The design and type of the fryer (pan, deep-fat batch, or deep-fat continuous) is also important. Oxidation of the oil will occur faster at large surface-volume ratios. Other factors of concern are turnover rate of the oil, the heating pattern (continuous or intermittent), and whether antioxidants are present.

4. Tests for Assessing the Quality of Frying Oils

Several of the methods just discussed for the measurement of fat oxidation are commonly used to monitor thermal and oxidative decomposition of oils during the frying process. Thus, techniques to measure viscosity, free fatty acids, organopeptic quality, smoke point, foaming, polymer formation, and specific degradation products have been applied with various degrees of success. The changes occurring during frying are numerous and variable. One test may be adequate for one set of conditions but totally unsatisfactory for other conditions. Some newer tests, generally based on the relative polarity of oxidative decomposition products, are described below.

Petroleum Ether Insolubles. This method was developed in Germany and later recommended by the German Society for Fat Research. It was suggested that a used frying fat be regarded as "deteriorated" if the petroleum ether insolubles are $\geqslant 0.7\%$ and the smoke point is less than $170°C$, or if the petroleum ether insolubles are $\geqslant 1.0\%$ regardless of smoke point. The method is tedious and somewhat inaccurate, since oxidation products are partially soluble in petroleum ether.

Column Chromatography of Polar Compounds (11). The heated fat is fractionated on a silica gel column and the nonpolar fraction is eluted with a petroleum ether-diethylether mixture. The percentage weight of the polar fraction is calculated by difference. A value of 27% polar components, as obtained by this method, is suggested as the maximum tolerable for usable oil.

Dielectric Constant (40). Changes in the dielectric constant of the oil are measured using an instrument known as the Foodoil Sensor. The dielectric constant increases with an increase in polarity. This technique has several advantages. The unit is compact and portable and can be operated in the field by unskilled personnel, and analysis time is less than 5 min. However, caution must be exercised in interpreting data obtained by this method (70). Dielectric constant readings represent the net balance between polar and nonpolar components, both of which develop upon frying. The increase in the polar fraction usually predominates, but the net difference between the two fractions depends on a variety of complex factors, some of which may be unrelated to oil quality (e.g., moisture).

Gas Chromatography of Dimer Esters (71). This technique involves complete conversion of the oil to its methyl esters followed by analysis on a short column. Parameters are adjusted to provide a pattern in which the dimeric esters emerge as a doublet peak with a retention time of about 3 min, whereas all other monomeric esters elute with the solvent peak. As indicated above, an increase in dimeric compounds is associated with thermal decomposition, particularly after heating has been prolonged.

5. Control Measures

While it is economically important to maximize the useful life of the frying oil, every effort must be made from the standpoints of food acceptability and wholesomeness to avoid extensive oxidative decomposition, the development of objectionable flavors, or the excessive formation of cyclic or polymeric material. Good manufacturing practices include

1. Choice of frying oil of good quality and consistent stability.
2. Use of properly designed equipment.
3. Selection of the lowest frying temperature consistent with producing a fried product of good quality.
4. Frequent filtering of the oil to remove food particles.
5. Frequent shutdown and cleaning of equipment.
6. Replacement of oil as needed to maintain high quality.
7. Consideration of antioxidant use. In some cases the level of antioxidant decreases significantly after a short processing time.
8. Adequate training of personnel.
9. Frequent testing of the oil throughout the frying process.

E. Effects of Ionizing Radiation on Fats

The objective of food irradiation is comparable to that of other methods of preservation. Its primary purpose is to destroy microorganisms and to prolong shelf life of the food. The process is capable of sterilizing meats or meat products [at high doses, e.g., 10-50 kGy (1 kilogray = 100 krads)]; of extending the shelf life of refrigerated fresh fish, chicken, fruits, and vegetables (at medium doses, e.g., 1-10 kGy); and of preventing sprouting of potatoes and onions, delaying fruit ripening, and killing insects in cereals, peas, and beans (at low doses, e.g., up to 1 kGy). Radiation preservation of food is becoming increasingly attractive to industry both from food quality and economic points of view.

In November 1980, the joint FAO/WHO/IAEA Expert Committee on Wholesomeness of Irradiated Food concluded that "the irradiation of any food commodity up to an overall average dose of 10 kGy causes no toxocological hazard and hence toxicological testing of foods so treated is no longer required."

As with heating, chemical changes are induced in food as a result of the radiation treatment. Indeed, in both cases, the primary objective of the treatment is achieved as a result of these chemical changes. On the other hand, treatment conditions must be controlled such that the nature and extent of such changes do not compromise the wholesomeness or quality of the food.

Considerable effort has been devoted to determining the effects of irradiation on various food components. Most of the early work on irradiation of lipids was concerned with the study of radiation-induced oxidation in natural fats or in certain synthetic lipid systems. Only recently have strictly radiolytic (i.e., nonoxidative) changes in fats been investigated.

1. Radiolytic Products

Listed in Table 8 are the classes of compounds that have been found to result from irradiation of natural fats or model systems of fatty acids and their derivatives. This listing of compounds is derived from studies of beef and pork fats, mackerel oil, oils from corn, soybean, olive, safflower, and cottonseed, and pure triacylglycerols, all treated under vacuum with radiation doses ranging from 5 to 60 kGy (68). Hydrocarbons, aldehydes, methyl and ethyl esters, and free fatty acids are the major volatile compounds produced by irradiation. Found in all irradiated fats studied are hydrocarbons consisting of complete homologous series of C_1-C_{17} n-alkanes, C_2-C_{17} 1-alkenes, certain internally unsaturated monoenes, a series of alkadienes, and, in some cases, several polyenes. The aldehydes and esters found include normal alkanals containing 16 and 18 carbon atoms, and the methyl and ethyl esters of C_{16} and C_{18} fatty acids. In these studies, unsaturated hydrocarbons larger than C_{17} arose only in irradiated fish oil, and aldehydes of chain length shorter than C_{16} were found only in irradiated coconut oil.

The quantitative pattern of the radiolytic products is strictly dependent on the fatty acid composition of the original fat. Although most members of the hydrocarbon series are produced in only small amounts, those containing one or two carbon atoms less than the major component fatty acids of the original fat are formed in the greatest quantities. For example, in the case of pork fat, in which 18:1 is the major fatty acid, C_{17} and C_{16} hydrocarbons are the most abundant hydrocarbons; for coconut oil, in which lauric acid is the main fatty acid, undecane and decene are the hydrocarbons produced in the greatest quantities (Table 9). The major aldehydes in the radiolytic mixture are those with the same chain length as those of the major component fatty acids of the original fat, and the methyl and ethyl esters produced in the largest quantity are those corresponding to the most abundant saturated acids in the original fat. If the fats are partially hydrogenated before irradiation, the change in their original fatty acid composition results in a corresponding change in the composition of the radiolytic compounds. The radiolytic compounds of high molecular weight have been identified in a recent study with beef fat (81).

Table 8 Classes of Compounds[a] Identified in Irradiated Fats

n-Alkanes	Methyl and ethyl esters
1-Alkenes	Propane and propenediol diesters
Alkadienes	Ethanediol diesters
Alkynes	Oxopropanediol diesters
Aldehydes	Glyceryl ether diesters
Ketones	Long-chain hydrocarbons
Fatty acids	Long-chain alkyl esters
Lactones	Long-chain alkyldioldiesters
Monoacylglycerols	Long-chain ketones
Diacylglycerols	Dimers
Triacylglycerols	Trimers

[a]Approximately 150 compounds were identified in various studies in the author's laboratory. Natural fats and model systems were irradiated at various doses in the absence of oxygen.

Table 9 Relationship Between the Major Radiolytic Compounds Formed at 60 kGy and the Major Fatty Acids in the Original Fats

| | Most abundant fatty acids | | Radiolytic products | | | |
| | | | Most abundant hydrocarbons | | Most abundant aldehyde | |
Fat	Carbon No.	% in fat	Carbon No.	Concentration (mg per 100 g fat)	Carbon No.	Concentrations (mg per 100 g fat)
Pork	18:1	60.6	16:2 17:1	6.84 3.96	18:1	
Safflower	18:2	78.0	17:2 16:3	19.3 18.1	18:2	11.9
Hydrogenated safflower	18:0	90.5	17:0 16:1	59.0 10.0	18:0	2.2
Coconut	12:0	55.8	11:0 10:1	9.0 5.8	12:0	4.5

2. Mechanisms of Radiolysis

General Principles. Ions and excited molecules are the first species formed when ionizing radiation is absorbed by matter. Chemical breakdown is brought about by the decomposition of the excited molecules and ions, or by their reaction with neighboring molecules. The fate of the excited molecules includes dissociation into free radicals. Among the reactions of ions is neutralization, producing excited molecules that may then dissociate into smaller molecules or free radicals. The free radicals formed by dissociation of excited molecules and by ion reactions may combine with each other in regions of high radical concentrations, or may diffuse into the bulk of the medium and react with other molecules.

Radiolysis of Fats. The reactions induced by irradiation are not the result of a statistical distribution of random cleavages of chemical bonds, but rather they follow preferred pathways largely influenced by molecular structure. In the case of saturated fats, as with all oxygen-containing compounds, the high localization of the electron deficiency on the oxygen atom directs the location of preferential cleavage in the vicinity of the carbonyl group.

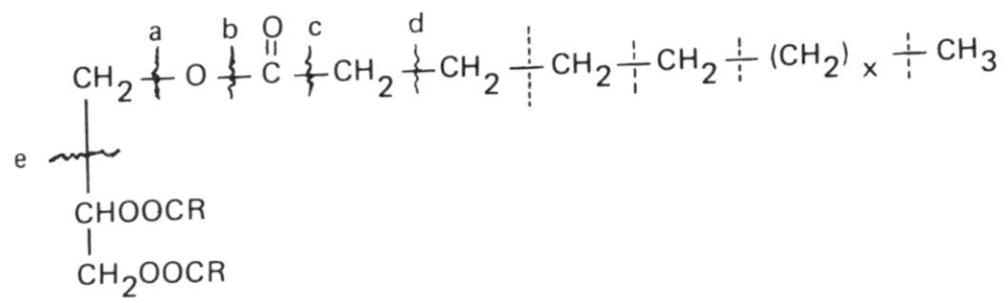

Based on the results available to date, the mechanisms of radiolysis in triacylglycerols appear to proceed in accord with the concepts of Williams, regarding the location of primary ionization events in oxygen-containing compounds (84). In a triacylglycerol molecule, radiolytic cleavage occurs preferentially at five locations in the vicinity of the carbonyl group (wavy solid lines), and randomly at all the remaining carbon-carbon bonds in the fatty acid moiety (dotted lines).

These cleavages result in the formation of alkyl, acyl, acyloxy, and acyloxymethylene free radicals and, in addition, free radicals representing the corresponding glyceryl residues. Free radical termination may occur by hydrogen abstraction, and to a lesser degree by loss of hydrogen with the formation of an unsaturated linkage. Thus, scission of an acyloxy-methylene bond (location a) produces a free fatty acid and a propane- or propenediol diester; cleavage at the acyl-oxygen bond (at b) produces an aldehyde of equal chain length to the parent fatty acid and a diacylglycerol; cleavage at c or d produces a hydrocarbon with one or two carbon atoms less than the parent fatty acid and a triacylglycerol; and cleavage between carbons of the glycerol skeleton (e.g., at e) produces a methyl ester of the parent fatty acid and an ethanediol diester.

Alternatively, the free radicals may recombine, giving rise to a variety of radiolytic products. Thus, combination of alkyl radicals with other alkyl radicals results in the production of longer-chain or dimeric hydrocarbons

$$R\cdot + R'\cdot \longrightarrow R-R'$$

combination of an acyl radical with an alkyl radical produces a ketone

$$\underset{RC\cdot}{\overset{O}{\overset{\|}{}}} + R'\cdot \longrightarrow \underset{RCR'}{\overset{O}{\overset{\|}{}}}$$

combination of an acyloxy radical with an alkyl radical produces an ester

$$\underset{RCO\cdot}{\overset{O}{\overset{\|}{}}} + R'\cdot \longrightarrow \underset{RCOR'}{\overset{O}{\overset{\|}{}}}$$

and combination of alkyl radicals with various glyceryl residues produces alkyl diglycerides

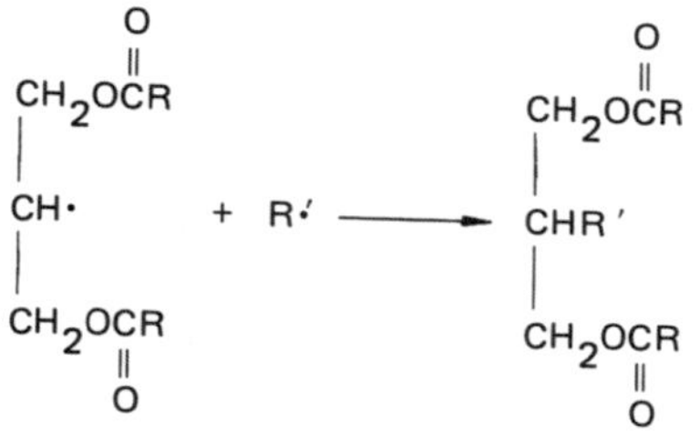

$$
\begin{array}{ccc}
\overset{O}{\overset{\|}{CH_2OCR}} & & \overset{O}{\overset{\|}{CH_2OCR}} \\
| & & | \\
CH\cdot & + R'\cdot \longrightarrow & CHR' \\
| & & | \\
\underset{O}{\underset{\|}{CH_2OCR}} & & \underset{O}{\underset{\|}{CH_2OCR}}
\end{array}
$$

and glyceryl ether diesters

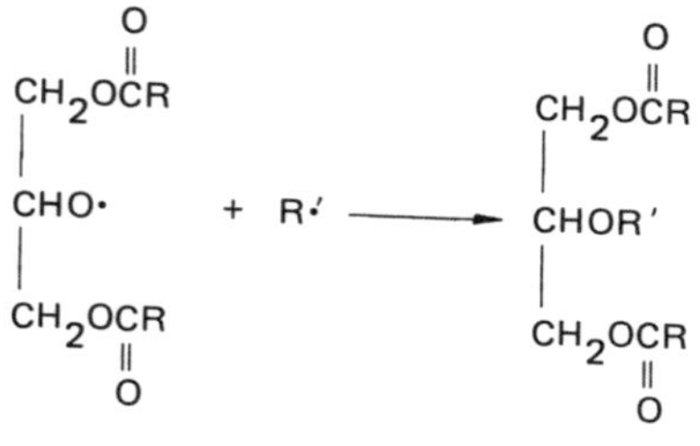

$$
\begin{array}{ccc}
\overset{O}{\overset{\|}{CH_2OCR}} & & \overset{O}{\overset{\|}{CH_2OCR}} \\
| & & | \\
CHO\cdot & + R'\cdot \longrightarrow & CHOR' \\
| & & | \\
\underset{O}{\underset{\|}{CH_2OCR}} & & \underset{O}{\underset{\|}{CH_2OCR}}
\end{array}
$$

It is indeed likely that other free radical combinations take place to produce dimeric or high-molecular-weight compounds that may have escaped detection by current analytical techniques. It should also be understood that mechanisms other than those presented here are probably involved in the radiolysis of fats. In particular, radiolytic reactions resulting from unsaturation of the glyceride molecule (e.g., cross-linking, dimerization, and cyclization) have not been thoroughly investigated.

In the presence of oxygen, irradiation will also accelerate the autoxidation of fats. This may occur because radiation accelerates one or more of the following reactions: (a) formation of free radicals, which may combine with oxygen to form hydroperoxides, (b) breakdown of hydroperoxides, giving rise to a variety of decomposition products, particularly carbonyl compounds, and (c) destruction of antioxidants.

Complex Foods Containing Fats. The irradiation of isolated fats does not have a practical value in terms of industrial application. Food lipids are normally exposed to irradiation as constituents coexisting with other major and minor components of complex food systems. The radiolytic products obtained from an isolated fat are usually also observed when the complex food containing such a fat is irradiated. However, the concentration of these substances in the irradiated food will be considerably reduced by the diluting effect of the other substances present. Of course, additional changes can be anticipated from radiolysis of the nonlipid constituents and from any interaction between these and the lipids.

Some researchers have concluded that irradiation reduces the stability of fatty foods by destroying antioxygenic factors, and they have suggested that irradiation be performed in the absence of air and that antioxidants be added after irradiation. On the other hand, others have found that, in certain cases, irradiation resulted in the formation of new protective factors that can improve product stability.

3. Comparison with Heat Effects

Although the mechanisms involved are different, many of the compounds produced from fats by irradiation are similar to those formed by heating. Far more decomposition products, however, have been identified from heated or thermally oxidized fats than from irradiated fats. Recent research in our laboratory with fatty acid esters and triacylglycerols appears to indicate that, even when irradiation is conducted at doses as high as 250 kGy, both the volatile and the nonvolatile patterns of products formed are much simpler than those obtained by heating at frying temperatures (180°C) for 1 hr.

4. Biological Effects

Irradiation results in partial destruction of fat-soluble vitamins, and tocopherol is particularly sensitive. Extensive testing of fat-containing foods irradiated at pasteurization doses has failed to demonstrate any toxicological hazard.

VI. CHEMISTRY OF FAT AND OIL PROCESSING

A. Refining

Crude oils and fats contain varying amounts of substances that may impart undesirable flavor, color, or keeping quality. These substances include free fatty acids, phospholipids, carbohydrates, proteins and their degradation products, water, pigments (mainly carotenoids and chlorophyll), and fat oxidation products. Crude oils are subjected to a number of commercial refining processes designed to remove these materials.

1. Settling and Degumming

Settling involves heating the fat and allowing it to stand until the aqueous phase separates and can be withdrawn. This rids the fat of water, proteinaceous material, phospholipids, and carbohydrates. In some cases, particularly with oils containing substantial amounts of phospholipids (e.g., soybean oil), a preliminary treatment known as *degumming* is applied by adding 2-3% water, agitating the mixture at about 50°C, and separating the hydrated phospholipids by settling or centrifugation.

2. Neutralization

To remove free fatty acids, caustic soda in the appropriate amounts and strength is mixed with the heated fat and the mixture is allowed to stand until the aqueous phase settles. The resulting aqueous solution, called foots or soapstock, is separated and used for making soap. Residual soapstock is removed from the neutral oil by washing it with hot water, followed by settling or centrifugation.

Although free fatty acid removal is the main purpose of the alkali treatment, this process also results in a significant reduction of phospholipids and coloring matter.

3. Bleaching

An almost complete removal of coloring materials can be accomplished by heating the oil to about 85°C and treating it with adsorbants, such as Fuller's earth or activated carbons. Precautions should be taken to avoid oxidation during bleaching. Other materials, such as phospholipids, soaps, and some oxidation products, are also adsorbed along with the pigments. The bleaching earth is then removed by filtration.

4. Deodorization

Volatile compounds with undesirable flavors, mostly arising from oxidation of the oil, are removed by steam distillation under reduced pressure. Citric acid is often added to sequester traces of pro-oxidant metals. It is believed that this treatment also results in thermal destruction of nonvolatile off-flavor substances, and that the resulting volatiles are distilled away.

Although the oxidative stability of oils is generally improved by refining, this is not always the case. Crude cottonseed oil, for example, has a greater resistance to oxidation than its refined counterpart, due to the greater amounts of gossypol and tocopherols in the crude oil. On the other hand, there can be little doubt as to the remarkable quality benefits that accrue from refining edible oils. An impressive example is the upgrading of palm oil quality that has occurred in the last decade. Furthermore, in addition to the obvious improvements in color, flavor, and stability, powerful toxicants (e.g., aflatoxins in peanut oil and gossypol in cottonseed oil) are effectively eliminated during the refining process.

B. Hydrogenation

Hydrogenation of fats involves the addition of hydrogen to double bonds in the chains of fatty acids in triacylglycerols. The process is of major importance in the fats and oils industry since it accomplishes two main objectives. First, it allows the conversion of liquid oils into semisolid or plastic fats more suitable for specific applications, such as in shortenings and margarine, and second, it improves the oxidative stability of the oil.

In practice, the oil is first mixed with a suitable catalyst (usually nickel), heated to the desired temperature (140-225°C), then exposed to hydrogen at pressures up to 60 psig and agitated. Agitation is necessary to aid in dissolving the hydrogen, to achieve uniform mixing of the catalyst with oil, and to help dissipate the heat of the reaction. The starting oil must be refined, bleached, low in soap, and dry; the hydrogen gas must be dry and free of sulfur, CO_2, or ammonia; and the catalyst must exhibit long-term activity, function in the desired manner with respect to selectivity of hydrogenation and isomer formation, and be easily removable by filtration. The course of the hydrogenation reaction is usually monitored by determining the change in refractive index, which is related to the degree of saturation of the oil. When the desired end point is reached, the hydrogenated oil is cooled and the catalyst removed by filtration.

1. Selectivity

During hydrogenation, not only are some of the double bonds saturated, but some may also be relocated and/or transformed from the usual cis to a trans configuration. The isomers produced are commonly called iso acids. Partial hydrogenation thus may result in the formation of a relatively complex mixture of reaction products, depending on which of the double bonds are hydrogenated, the type and degree of isomerization, and the relative rates of these various reactions. A simplistic scheme showing the possible reactions that linolenate can undergo during hydrogenation is shown below.

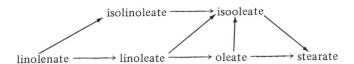

In the case of natural fats the situation is further complicated by the fact that they already contain an extremely complex mixture of starting materials.

The term "selectivity" refers to the relative rate of hydrogenation of the more unsaturated fatty acids as compared with that of the less saturated acids. When expressed as a ratio (selectivity ratio), a quantitative measure of selectivity can be obtained in more absolute terms. The term "selectivity ratio," as defined by Albright (3), is simply the ratio: rate of hydrogenation of linoleic to produce oleic acid ÷ rate of hydrogenation of oleic to produce stearic acid. Reaction rate constants can be calculated from the starting and ending fatty acid compositions and the hydrogenation time (Fig. 30). For the reactions just mentioned, the selectivity ratio (SR) is $K_2/K_3 = 0.159/0.013 = 12.2$, which means that linoleic acid is being hydrogenated 12.2 times faster than oleic acid.

Since calculations of SR for every oil hydrogenated would be quite tedious, Albright (3) prepared a series of graphs for various oils by calculation of the fatty acid compostion at constant SR. In these graphs the decrease in iodine value (ΔIV) is plotted against the fraction of linoleic that remains unhydrogenated (L/L_0). Although the curves are calculated with the assumption that the reaction rates are first order, and that isooleic acid is hydrogenated at the same rate as oleic, they are nonetheless very useful for determining selectivity. Selectivity ratio curves for soybean oil (K_2/K_3) are shown in Fig. 31). Of course, linolenic acid selectivity can be similarly expressed; that is, LnSR =

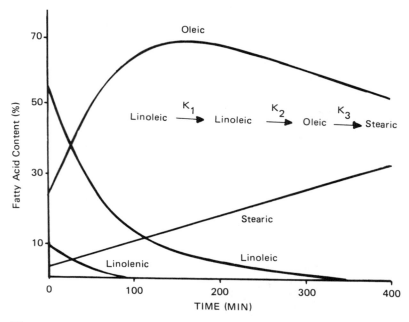

Figure 30 Reaction rate constants for hydrogenation of soybean oil (4).

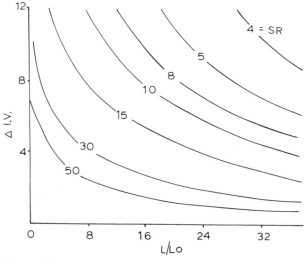

Figure 31 SR curves for soybean oil (3), ΔIV = decrease in iodine value; L/L_0 = fraction of linoleic acid unhydrogenated.

K_1/K_2, where K_1 and K_2 are as defined in Fig. 30). This is relevant to the hydrogenation of soybean oil, since flavor reversion in this oil is believed to arise from its linolenate content.

Different catalysts result in different selectivities, and operating parameters also have a profound effect on selectivity. As shown in Table 10, larger SR values result from high temperatures, low pressures, high catalyst concentration, and low intensity of agitation. The effects of processing conditions on the rate of hydrogenation and on the formation of trans acids are also given. A number of mechanistic speculations have been advanced to explain the observed influence of process conditions on selectivity and rate of hydrogenation, and these are discussed next.

2. Mechanism

The mechanism involved in fat hydrogenation is believed to be the reaction between unsaturated liquid oil and atomic hydrogen adsorbed on a metal catalyst. First, a carbon-metal complex is formed at either end of the olefinic bond (complex a in Fig. 32). This intermediate complex is then able to react with an atom of catalyst-adsorbed hydrogen to form an unstable half-hydrogenated state (b or c in Fig. 32) in which the olefin is attached to the catalyst by only one link and is thus free to rotate. The half-hydrogenated compound now may either: (a) react with another hydrogen atom and dissociate from the catalyst to yield the saturated product (d in Fig. 32), or (b) lose a hydrogen atom to the nickel catalyst to restore the double bond. The regenerated double bond can be either in the same position as in the unhydrogenated compound or a positional and/or geometric isomer of the original double bond (e and f in Fig. 32).

In general, evidence seems to indicate that the concentration of hydrogen adsorbed on the catalyst is the factor that determines selectivity and isomer formation (4). If the catalyst is saturated with hydrogen, most of the active sites hold hydrogen atoms and the chance is greater that two atoms are in the appropriate position to react with any double bond upon approach. This results in low selectivity, since the tendency will be toward saturation of any double bond approaching the two hydrogens. On the other hand, if the hydrogen atoms on the catalyst are scarce, it is more likely that only one hydrogen atom reacts with the double bond, leading to the half-hydrogenation-dehydrogenation sequence and a greater likelihood of isomerization. Thus, operating conditions (hydrogen pressure, intensity of agitation, temperature, and kind and concentration of catalyst) influence selectivity through their effect on the ratio of hydrogen to catalyst sites. An increase in temperature, for example, increases the speed of the reaction and causes a faster removal of hydrogen from the catalyst, giving rise to increased selectivity.

Table 10 Effects of Processing Parameters on Selectivity and Rate of Hydrogenation

Processing parameter	SR	Trans acids	Rate
High temperature	High	High	High
High pressure	Low	Low	High
High catalyst concentration	High	High	High
High-intensity agitation	Low	Low	High

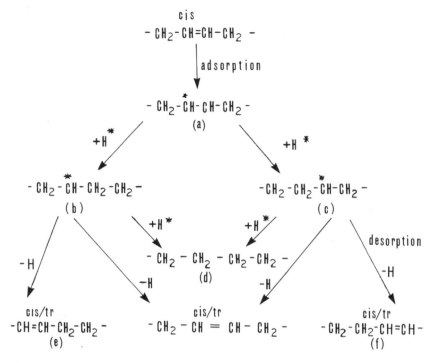

Figure 32 Half-hydrogenation–hydrogenation reaction scheme. Asterisk indicates metal link.

The ability to change the SR by changing the processing conditions enables processors to exert considerable control over the properties of the final oil. A more selective hydrogenation, for example, allows linoleic acid to be decreased and stability to be improved while minimizing the formation of fully saturated compounds and avoiding excessive hardness. On the other hand, the more selective the reaction, the greater will be the formation of trans isomers, which are of concern from a nutritional standpoint. For many years manufacturers of food fats have been trying to devise hydrogenation processes that minimize isomerization while avoiding the formation of excessive amounts of fully saturated material.

3. Catalysts

As indicated earlier, catalysts vary with regard to the degree of selectivity they provide. Nickel on various supports is almost invariably used commercially to hydrogenate fats. Other catalysts, however, are available. These include copper, copper/chromium combinations, and platinum. Palladium has been found to be considerably more efficient (in terms of the amount of catalyst required) than nickel, although it produces a high proportion of trans isomers. The so-called homogeneous catalysts, which are soluble in oil, provide greater contact between oil and catalyst and better control of selectivity.

A host of different compounds are capable of poisoning the catalyst used, and these compounds are often the major source of problems encountered during commercial hydrogenation. Poisons include phospholipids, water, sulfur compounds, soaps, partial glycerol esters, CO_2, and mineral acids.

C. Interesterification

It has been mentioned that natural fats do not contain a random distribution of fatty acids among the glyceride molecules. The tendency of certain acids to be more concentrated at specific sn positions varies from one species to another and is influenced by such factors as environment and location in the plant or animal. The physical characteristics of a fat are greatly affected not only by the nature of constituent fatty acids (i.e., chain length and unsaturation) but also by their distribution in the triacylglycerol molecules. Indeed, unique fatty acid distribution patterns of some natural fats limit their industrial applications. Interesterification is one of the processes that can be applied to improve the consistency of such fats and to improve their usefulness. This process involves rearranging the fatty acids so they become distributed randomly among the triacylglycerol molecules of the fat.

1. Principle

The term "interesterification" refers to the exchange of acyl radicals between an ester and an acid (acidolysis), an ester and an alcohol (alcoholysis), or an ester and an ester (transesterification). It is the latter reaction that is relevant to industrial interestification of fat (also known as randomization), since it involves ester interchange within a single triacylglycerol molecule (intraesterification) as well as ester exchange among different molecules.

If a fat contains only two fatty acids (A and B), eight possible triacylglycerol species (n^3) are possible according to the rule of chance.

$$
\begin{bmatrix} A \\ A \\ A \end{bmatrix} \begin{bmatrix} A \\ A \\ B \end{bmatrix} \begin{bmatrix} A \\ B \\ A \end{bmatrix} \begin{bmatrix} A \\ B \\ B \end{bmatrix} \begin{bmatrix} B \\ A \\ A \end{bmatrix} \begin{bmatrix} B \\ A \\ B \end{bmatrix} \begin{bmatrix} B \\ B \\ A \end{bmatrix} \begin{bmatrix} B \\ B \\ B \end{bmatrix}
$$

Regardless of the distribution of the two acids in the original fat (e.g., AAA and BBB or ABB, ABA, BBA), interesterification results in the "shuffling" of fatty acids within a single molecule and among triacylglycerol molecules until an equilibrium is achieved in which all possible combinations are formed. The quantitative proportions of the different species depend on the amount of each acid in the original fat and can be predicted by simple calculation as already discussed under the 1,2,3-random distribution hypothesis.

2. Industrial Process

Interesterification can be accomplished by heating the fat at relatively high temperatures ($<200°C$) for a long period. However, catalysts are commonly used that allow the reaction to be completed in a short time (e.g., 30 min) at temperatures as low as $50°C$. Alkali metals and alkali metal alkylates are effective low-temperature catalysts, with sodium methoxide the most popular. Approximately 0.1% catalyst is required. Higher concentrations may cause excessive losses of oil resulting from the formation of soap and methyl esters.

The oil to be esterified must be extremely dry and low in free fatty acids, peroxides, and any other material that may react with sodium methoxide. Minutes after the catalyst is added, the oil acquires a reddish brown color due to the formation of a complex between the sodium and the glycerides. This complex is believed to be the "true catalyst." After esterification, the catalyst is inactivated by addition of water or acid and removed.

3. Mechanisms

Two mechanisms for interesterification have been proposed (78).

Enolate Ion Formation. According to this mechanism, an enolate ion (II), typical of the action of a base on an ester, is formed. The enolate ion reacts with another ester group in the triacylglycerol molecule to produce a β-keto ester (III) which in turn reacts further to give another β-keto ester (IV). Intermediate IV yields the intramolecularly esterified product V. The

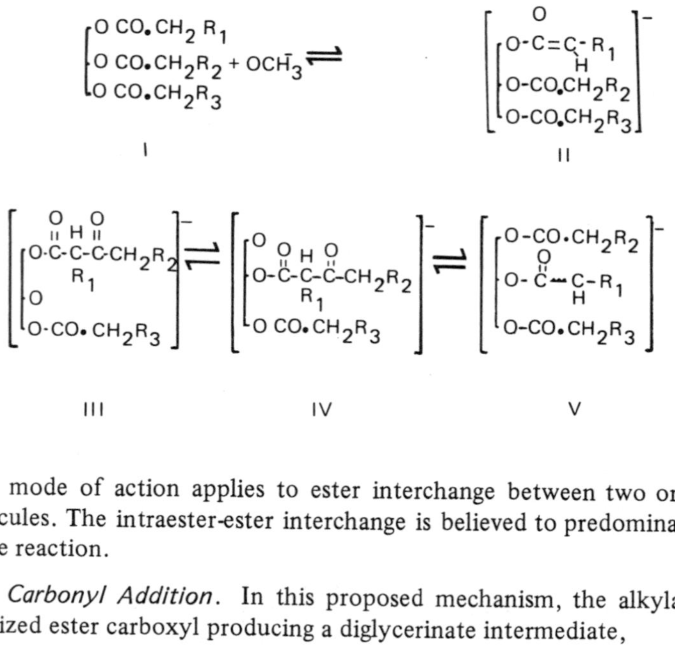

$$\begin{bmatrix} O\ CO.CH_2\ R_1 \\ O\ CO.CH_2R_2 \\ O\ CO.CH_2R_3 \end{bmatrix} + OCH_3 \rightleftharpoons \begin{bmatrix} O-C=\underset{H}{\overset{O}{C}}-R_1 \\ O-CO.CH_2R_2 \\ O-CO.CH_2R_3 \end{bmatrix}^-$$

I   II

$$\begin{bmatrix} O\ H\ O \\ O\text{-}\overset{\parallel}{C}\text{-}\overset{\mid}{C}\text{-}\overset{\parallel}{C}\text{-}CH_2R_2 \\ \qquad R_1 \\ O \\ O\text{-}CO.\ CH_2R_3 \end{bmatrix}^- \rightleftharpoons \begin{bmatrix} O\ \ O\ H\ O \\ O\text{-}\overset{\parallel}{C}\text{-}C\text{-}\overset{\mid}{C}\text{-}CH_2R_2 \\ \qquad\ R_1 \\ O\ CO.\ CH_2R_3 \end{bmatrix}^- \rightleftharpoons \begin{bmatrix} O\text{-}CO.CH_2R_2 \\ \qquad O \\ O\text{-}\ \overset{\parallel}{C}\text{--}\overset{\mid}{C}\text{-}R_1 \\ \qquad\ H \\ O\text{-}CO.CH_2R_3 \end{bmatrix}^-$$

III   IV   V

same mode of action applies to ester interchange between two or more triacylglycerol molecules. The intraester-ester interchange is believed to predominate in the initial stages of the reaction.

Carbonyl Addition. In this proposed mechanism, the alkylate ion adds on to a polarized ester carboxyl producing a diglycerinate intermediate,

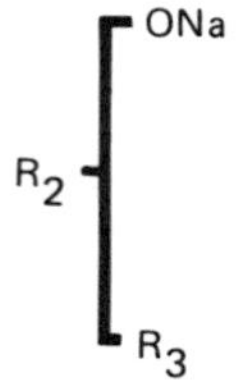

$$R_2 \begin{bmatrix} ONa \\ \\ \\ \\ R_3 \end{bmatrix}^-$$

This intermediate reacts with another glyceride by abstracting a fatty acid, thus forming a new triacylglycerol and regenerating a diglycerinate for further reaction. Ester interchange between fully saturated S_3 and unsaturated U_3 molecules is shown below.

$$S_3 + U_2ONa \xrightleftharpoons[k]{3k} SU_2 + S_2ONa$$

$$SU_2 + U_2ONa \xrightleftharpoons[3k]{2k} U_3 + SUONa$$

$$U_3 + S_2ONa \xrightleftharpoons[k]{3k} S_2U + U_2ONa$$

$$S_2U + S_2ONa \xrightleftharpoons[3k]{2k} S_3 + SUONa$$

$$S_2U + U_2ONA \xrightleftharpoons[k]{2k} SU_2 + SUONa$$

$$SU_2 + S_2ONa \xrightleftharpoons[k]{2k} S_2U + SUONa$$

4. Directed Interesterification

A random distribution, such as that produced by interesterification, is not always the most desirable. Interesterification can be directed away from randomness if the fat is maintained at a temperature below its melting point. This results in selective crystallization of the trisaturated glycerides, which has the effect of removing them from the reaction mixture and changing the fatty acid equilibrium in the liquid phase. Interesterification proceeds with the formation of more trisaturated glycerides than would have otherwise occurred. The newly formed trisaturated glycerols crystallize and precipitate, thus allowing the formation of still more trisaturated glycerides, and the process continues until most of the saturated fatty acids in the fat have precipitated. If the original fat is a liquid oil containing a substantial amount of saturated acids, it is possible, by this method, to convert the oil into a product with the consistency of shortening without resorting to hydrogenation or blending with a hard fat. The procedure is relatively slow due to the low temperature used, the time required for crystallization, and the tendency of the catalyst to become coated. A dispersion of liquid sodium-potassium alloy is commonly used to slough off the coating as it forms.

Rearrangement can also be selectively controlled during interesterification by adding excess fatty acids and continuously distilling out the liberated acids that are highly volatile. This impoverishes the fat of its acids of lower molecular weight. The content of certain acids in a fat also can be reduced by using suitable solvents to extract appropriate acids during the interesterification process.

5. Applications

Interesterification finds its greatest application in the manufacture of shortenings. Lard, due to its high proportion of disaturated triacylglycerols with palmitic acid in the 2 position, forms relatively large and coarse crystals, even when rapidly solidified in commercial chilling machines. Shortenings made from natural lard possess a grainy consistency and exhibit poor performance in baking. Randomization of lard improves its plastic range and makes it a better shortening. Directed interesterification, however, produces a product with a higher solids content at high temperatures (Fig. 33) and thus extends its plastic range.

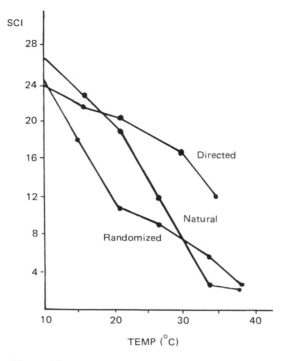

Figure 33   Effect of interesterification on solid content index (78).

Salad oil of a relatively low cloud point can be made from palm oil by fractiona-tion after directed interesterification. The use of interesterification has also been applied to the production of high-stability margarine blends and hard butters that have highly desirable melting qualities.

Using a column with countercurrent flow of dimethylformamide, a process has been developed to selectively reduce the content of linolenic acid in soybean oil by direct interesterification.

VII.   ROLE OF FOOD LIPIDS IN FLAVOR

A.   Physical Effects

Pure food lipids are nearly odorless. However, apart from their major contributions as precursors of flavor compounds, they modify the overall flavor of many foods through their effect on mouth feel (e.g., the richness of whole milk and the smooth or creamy nature of ice cream) and on the volatility and threshold value of the flavor components present.

B.   Lipids as Flavor Precursors

In the preceding discussion we have seen that lipids can undergo a variety of reactions and give rise to a multitude of intermediates and decomposition end products. These compounds vary widely in their chemical and physical properties. They also vary in their chemical and physical properties. They also vary in their impact on flavor. Some are responsible for pleasant aromas, as is typical in fresh fruits and vegetables; others produce offensive odors and flavors, often causing major problems in the storage and processing of foods. A review by Forss (32) provides detailed information regarding the characteristics of volatile flavor compounds derived from lipids. In the following, some typical off-flavors of lipid origin are discussed (see also Chap. 9).

1. Rancidity

Hydrolytic rancidity results from the release of free fatty acids by lipolysis. Since only the short-chain fatty acids have unpleasant odors, the problem is typically encountered in milk and dairy products.

  The word *rancid*, however, is a general term that also commonly refers to off-flavors resulting from lipid oxidation. The qualitative nature of oxidation flavors, however, varies significantly from product to product and even in the same food item. For example the rancid off-flavors developed from the oxidation of fat in meat, walnuts, or butter are not quite the same. Fresh milk develops various off-flavors described as rancid, cardboard, fishy, metallic, stale, or chalky, and all are believed to be oxidative. Although considerable advances have been made in understanding the basic mechanisms of lipid oxidation and the flavor significance of many individual compounds, progress in correlating specific descriptions of rancid flavors with individual compounds or combinations of selected compounds has been slow, and little progress has been made in preventing undesirable off-flavors in foods that contain unsaturated fatty acids. This is not surprising if one considers: (a) the huge number of oxidative decomposition products identified so far (literally in the thousands), (b) their vast range of concentrations, volatilities, and flavor potency, (c) the many possible interactions among the lipid decomposition products or between these products and other nonlipid food components, (d) the different food environments in which these flavor compounds reside, (e) the subjective nature of flavor description, (f) the complex influence of oxidative conditions on the multitude of possible reaction pathways and reaction products, and (g) the always persisting suspicion that some trace but important flavor components may still escape detection by the most elegant analysis. Unfortunately, much more extensive research has been done on identification of volatiles in oxidized fats than correlating these volatiles with flavor. With the continuous improvements of sensitivity and resolution in analytical instrumentation, the detection of whole series of new compounds is made possible, and interpretations regarding flavor become more complex. Perhaps new approaches are needed to establish more precise relationships between volatile (and pertinent nonvolatile) lipid-derived components and specific oxidized flavors.

2. Flavor Reversion

This problem is unique to soybean oil and other linolenate-containing oils. The off-flavor has been described as beany and grassy and usually develops at low peroxide values (about 5 mEq/kg). Several compounds have been suggested as responsible for or contributing to reversion flavors.

  Smouse and Chang (77) postulated that 2-*n*-pentylfuran, one of the compounds identified in reverted soybean oil, and found to produce reversionlike flavors in other oils if added at 2 ppm levels, is formed from the autoxidation of linoleate by the following mechanism.

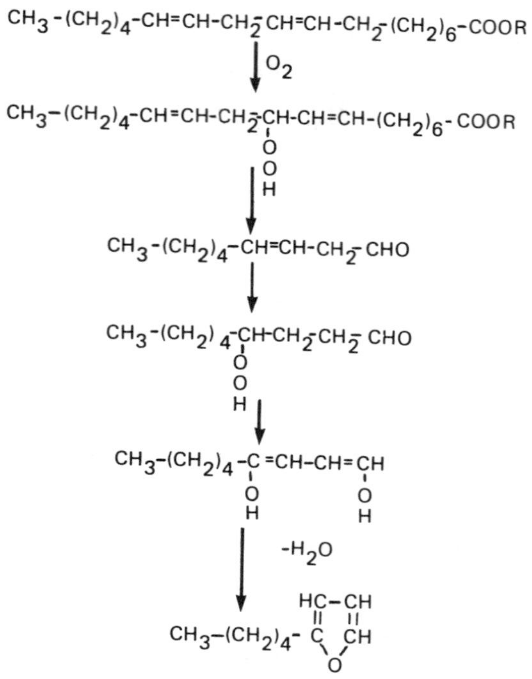

$$CH_3-(CH_2)_4-CH=CH-CH_2-CH=CH-CH_2-(CH_2)_6-COOR$$

$$\downarrow O_2$$

$$CH_3-(CH_2)_4-CH=CH-CH_2-\underset{\underset{\underset{H}{O}}{\overset{O}{|}}}{CH}-CH=CH-(CH_2)_6-COOR$$

$$\downarrow$$

$$CH_3-(CH_2)_4-CH=CH-CH_2-CHO$$

$$\downarrow$$

$$CH_3-(CH_2)_4-\underset{\underset{\underset{H}{O}}{\overset{O}{|}}}{CH}-CH_2-CH_2-CHO$$

$$\downarrow$$

$$CH_3-(CH_2)_4-\underset{\underset{\underset{H}{O}}{\overset{O}{|}}}{C}=CH-CH=\underset{\underset{\underset{H}{O}}{\overset{O}{|}}}{CH}$$

$$\downarrow \quad -H_2O$$

$$CH_3-(CH_2)_4-\underset{O}{\overset{HC-CH}{\underset{\diagdown \diagup}{C \quad CH}}}$$

Presence of the 18:3 acid catalyzes this reaction. It should be pointed out that the hydroperoxide intermediate involved here is the 10-hydroperoxide, which is not typical of linoleate autoxidation. It can arise, however, from singlet oxygen. An alternative mechanism involving singlet oxygen and the formation of a hydroperoxy cyclic peroxide (34) is given below.

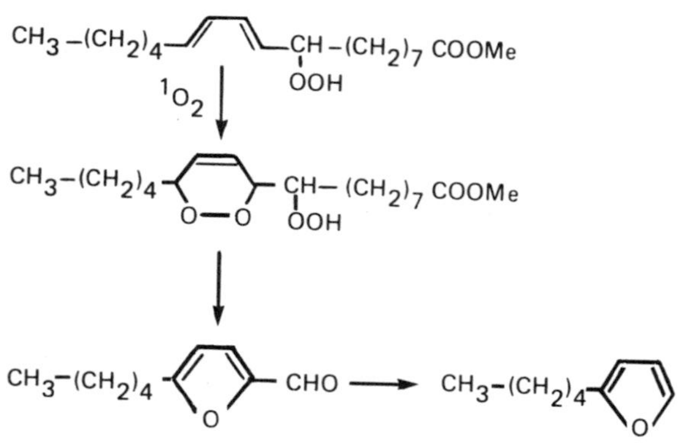

$$CH_3-(CH_2)_4 \diagup=\diagdown \underset{\underset{OOH}{|}}{CH}-(CH_2)_7 \, COOMe$$

$$^1O_2 \downarrow$$

$$CH_3-(CH_2)_4 \underset{O-O}{\diagup \diagdown} \underset{\underset{OOH}{|}}{CH}-(CH_2)_7 \, COOMe$$

$$\downarrow$$

$$CH_3-(CH_2)_4 \underset{O}{\diagup \diagdown}-CHO \longrightarrow CH_3-(CH_2)_4 \underset{O}{\diagup \diagdown}$$

More recently, Chang's group suggested that *cis* and *trans*-2-(1-pentenyl) furans are possible contributors to the reversion flavor. Other compounds reported by various workers to be significant in soybean reversion are 3-*cis*- and 3-*trans*-hexenals, phosphatides, and nonglyceride components.

3. Hardening Flavor

This off-flavor develops in hydrogenated soybean and marine oils during storage. Compounds reported to contribute to this defect include 6-*cis*- and 6-*trans*-nonenal, 2-*trans*-6-*trans*-octadecadienal, ketones, alcohols, and lactones. These compounds are presumed to arise from autoxidation of the isomeric dienes, known as isolinoleates, formed during the course of hydrogenation.

VIII. BIOLOGICAL ASPECTS

A. Consumption and Trends

In the United States, the average daily intake of dietary fat is approximately 160 g per person (50 kg/year), which accounts for 1400 kcal/day or about 45% of the total energy intake. Of this amount, nearly 43% is consumed as isolated or visible fats, such as butter, margarine, lard, shortening, and salad oils. This is typical of industrialized countries and is in sharp contrast to the situation in less developed nations, where in some instances the annual consumption of fat varies from 2 to 14 kg/year per person, of which only a very small percentage is in the form of isolated fats.

Trends in U. S. consumption over the last 70 years include increased consumption of total fat; greater consumption of vegetable fats, especially salad and cooking oils; decreased consumption of animal fats, except for edible beef fat; a shift from lard to shortening; and a shift from butter to margarine.

B. Nutritional Functions

Fats serve as a concentrated source of energy. As compared with proteins and carbohydrates, they supply twice as many calories per gram (9 versus 4 kcal/g). They confer a feeling of satiety and contribute to the palatability of foods. The essential fatty acids (EFA) linoleic and arachidonic, as well as the fat-soluble vitamins A, D, E, and K, are obtained from the lipid fraction of the diet. Linoleic acid is converted in the body to arachidonic, which is the precursor of a group of hormones known as prostaglandins.

C. Safety of Heated and Oxidized Fats

The various changes induced in fats by both heating and oxidation, and the factors that influence such changes, were discussed earlier in this chapter. Since, in practice, frying or cooking is done in the presence of air, resulting in an overlap of oxidative and thermal reactions, the safety of heated and oxidized fats is considered together.

The possibility that consumption of heated and/or oxidized fats may produce adverse effects has been a major concern and has stimulated extensive research. Several reviews on this subject are available (3a,6,16,71a). Nutritional and toxicity studies have been conducted with animals involving (a) fats heated or oxidized under controlled laboratory conditions, (b) used frying oils, or (c) certain fractions or pure compounds typical of those identified in such oils.

It is frustrating to attempt to draw decisive conclusions from the available literature on this subject, since experimental conditions have varied widely, different oils and different foods have been used, and the interplay of the many pertinent variables usually has not been considered. At times, the oils were heated continuously at excessive temperatures and without contact with food. Often, synthetic compounds or fractions of heated oils were fed to animals at unreasonably high levels in their diets. This has resulted in confusing and contradictory results inapplicable to "real-life" frying situations.

In the discussion below we attempt to give a general idea of some of the specific findings in this field of research, including some of the more realistic experiments which, unfortunately, are relatively scarce.

1.   Unheated but Highly Oxidized Fats (5a,53a,67a)

In general, the feeding of highly oxidized fats (at levels of 10-20% in the diet of rats) results in loss of appetite, growth retardation, enlargement of livers and kidneys, and accumulation of peroxides in adipose tissue. However, in some instances, highly oxidized fats produce no harmful effects if the required vitamins are furnished according to standard practice and the peroxide level of the entire diet does not exceed 40 mEq/kg body weight.

2.   Fats Heated Under Anaerobic Conditions (3b,65a)

Highly unsaturated fats heated for 10-40 hr at 225-250°C and fed at a 20% level cause weight loss and high mortality rates in rats. Toxicity is apparently attributable to cyclic esters.

3.   Thermally Oxidized Fats (52b,56b,71d,71e,73a)

In most cases, feeding of heated fats that have become severely oxidized produces various detrimental effects in animals. However, when the experimental conditions approach those of normal frying much milder symptoms are observed. In one comparative experiment, highly unsaturated oil heated in the absence of air was found to be more toxic than oil heated in air. In other studies the reverse was noted.

4.   Specific Compounds (1c,5a,51,71b)

Hydrocarbons. Addition of hydrocarbons with chains of more than nine carbon atoms (e.g., decene) to the diet of rats at a level of 1 ml per 5 g of basal ration caused the death of all animals.

Peroxides. In certain studies lipid peroxides were found to be toxic in rats in doses of 16 mEq O_2/kg body weight. In other experiments, rats tolerated daily doses of 75 mg of concentrated peroxide for 6 weeks.

Hydroxy Fatty Acids. Methyl ricinoleate fed to rats at 5% of the diet for 4 weeks has been reported to cause growth depression, reduced protein efficiency, an elevated respiratory quotient, increased plasma triacylglycerols, and accumulation of ricinoleate in depot fat. In contrast, other studies have shown that dihydroxystearic acid is poorly absorbed and not deposited in the tissues.

Cyclic Monomers. In general, aromatic compounds cause acute toxicity when administered to rats in large doses. When material containing 60% aromatics, prepared by cylization of fish oil and subsequent fractionation, was fed to rats at a level of 15%, the rats survived and adapted during an 18 week study. The aromatic compounds were excreted in urine. When the same material was fed at a 9-30% level, these animals died within 5 weeks.

Iwaoka and Perkins (51) reported that incorporation of cyclic monomers into the diet of rats caused low weight gains and low feed consumption in animals fed low levels of protein. A level of only 0.15% of cyclic fatty acids in diets containing 8% protein produced fatty livers. In adddition, decreased rates of lipogenesis have been observed in the liver of rats fed 8% protein and higher levels of cyclic fatty acids. An increased rate of lipogenesis occurred in the adipose tissue of these animals.

Dimers and Polymers. Many workers have fractionated heated and oxidized fats, and reported that the non-urea-adductable fractions are toxic. These, however, were crude fractions that surely contained mixtures of many other compounds. In some studies, polar dimers were reported to be toxic, but the nonpolar dimers were not. A prevalent view is that polymeric material of relatively high molecular weight is not absorbed in the body and thus is not toxic.

In short-term experiments, the presence of noncyclic dimeric fatty acids, when fed to rats at moderate levels, did not exhibit any profound effects on growth or metabolism.

5. Frying Fats (3a,6,10,68a,73e)

Decreased feed intake, lower fat absorbability, decreased growth, and liver enlargement are among the various adverse effects reported in several animal feeding studies involving highly abused frying oils. In contrast, other investigators have reported that used frying fats produce no ill effects. In one experiment for example, rapeseed oil was used for frying 10 batches of potatoes at 180°C, then incorporated at a 15% level in the diet of rats, and no biochemical or histological effects were observed. In another study, fat at 182°C was used for frying potatoes, onions, and fish until the point of severe foaming occurred and then this was fed to rats as 15% of the diet for 2 years. These fats resulted in slightly poorer growth than that achieved with unheated control fats, but did not cause any increase in mortality, tumor incidence, or any biochemical or histological irregularities.

Perhaps one of the most relevant reports on the biological effects of frying oils is that published recently by Billek (10). Sunflower oil used for industrial production of fish-fingers was taken at the end of a production period when the oil was usually discarded and fractionated by column chromatography into a polar fraction (which presumably contained most of the oxidation and decomposition products) and a nonpolar fraction (which consisted mostly of unaltered acylglycerols). The animal group fed the polar fraction exhibited growth retardation. The greatest increase in weight was observed with the unheated sunflower oil control; the group fed the nonpolar fraction, or the completely used oil, showed somewhat less gain in weight. A rather extensive analysis using various techniques of clinical chemistry and hematology was also conducted. In most cases, no significant deviation from the normal values or difference among the four types of diets could be observed. On the other hand, values for serum glutamic pyruvic transaminase were significantly higher for the group fed the polar fraction, indicating a certain amount of liver damage. The polar fraction also produced an increase in serum glutamic oxaloacetic transaminase. In addition, the average weights of the livers and kidneys were significantly greater in animals receiving the polar fraction than they were in animals receiving other diets. Histological investigation of various organs showed no major irregularities.

Billek points out that, although the used oil at the time of discarding contained 30% decomposition products, the polar fraction contained 90%. Furthermore, the average daily intake of this fraction was 10 g/kg body weight, but humans usually consume only about 0.1 g frying oil per kilogram of body weight. This author concluded that, in commercial practice, frying oils reach the point of deterioration (accumulation of decomposition products, development of odor and color defects, foaming, and so on) and are usually discarded, long before they start to be toxic.

6. Carcinogenicity Studies (73g,79a)

A number of investigators have reported tumors in animals fed severely heated fats; however, the temperatures used in most of these studies were 150-220°C above the temperature of normal frying. Various other studies have failed to show evidence of carcinogenic effects from normally used frying fats. Furthermore, in a study of the mutagenicity of used frying fat, negative results were obtained when a polar fraction, isolated by column chromatography, and a so-called oxidized fatty acid fraction, obtained by petroleum ether fractionation, were subjected to the *Salmonella*/microsome mutagenicity test (Ames test).

In a more recent study, it was found that severely abusive frying conditions (abnormally long frying times, high frying temperatures, or repeated reuse of frying oil) were necessary to produce appreciable levels of mutagenic activity in french-fired potatoes or fish fillets (79a).

In spite of the disparity of results and the contradictory conclusions existing in the literature, it is evident that toxic compounds can be generated in fat by severe heating and/or oxidation; however, it also appears reasonable to conclude that no significant hazard to health is to be expected from moderate ingestion of fried foods, provided high-quality oils are used and recommended practices are followed.

D. Safety of Hydrogenated Fats (24a,28a,72a,83a)

As pointed out earlier, a certain amount of double-bond isomerization takes place during industrial hydrogenation of oils, resulting in the formation of positional and geometric isomers. Trans fatty acids constitute 20-40% of the total acids of some margarines and shortenings. Small amounts of trans fatty acids are found in some animal fats, for example, in butter, which contains less than 2.5% (produced by partial biohydrogenation of fatty acids in the rumen of the cow).

Although it is recognized that trans fatty acids are not biologically equivalent to their cis isomers, precise knowledge regarding their physiology and metabolism and their long-term effects on health is lacking.

We have already mentioned that linoleic and arachidonic acids are "essential fatty acids." However, the trans isomers of these compounds have no essential fatty acid activity; that is, they cannot cure EFA deficiency symptoms. It has also been observed that the trans dienes do serve as a source of calories, and they are in fact metabolized more rapidly than the cis isomers. It has been suggested that the trans dienes are used exclusively as a source of energy, whereas the cis acids of the appropriate kind are retained in the body as EFA. However, in cardiac muscle this pattern is not followed; that is, oleic acid is used more effectively as a source of energy than elaidic acid (the trans isomer of oleic acid). This may be of special relevance during heart attacks, since under stress there is greater need for fatty acids as a source of energy.

It has also been observed that the distribution of trans acids in phospholipid molecules and the manner of incorporation of trans acids into adipose tissue is somewhat different than that for cis isomers.

Feeding studies with both rats and dogs have failed to show any effect on reproductive processes when hydrogenated fats are fed at 15% in the diet for long periods of time. However, much more research is needed in this area.

E. Dietary Fat and Coronary Heart Disease (13a,56a,56d,56h,66a,73c,73d)

In modern times, cardiovascular disease has been the major public heath problem in the United States as well as in other prosperous societies. There is general agreement that atherosclerosis, a disease in which plaques of fatty material build up in the arteries, is the underlying cause in most cases of degenerative cardiovascular disease and that changes in lipid patterns in the blood serum are associated with atherosclerosis. It is also recognized, that instead of a single cause of atherogenesis, it is necessary to consider the contribution of a number of influences, many of which may not exert their effect alone but must be combined with other coexistent factors. Synergistic relationships among these multiple factors make interpretation of results from many single studies on atherosclerosis, and its pathogenesis, a difficult and frustrating endeavor. Despite the enormous flow of information accumulated in recent years, many gaps in knowledge about atherosclerosis still remain; a clear cause-and-effect relationship has not been established, and substantial controversy among leading scientists continues to exist. Factors that have been correlated directly or indirectly to atherosclerosis include amounts and types of blood lipids; blood pressure; smoking; diet (e.g., lipids, proteins, sugar, salt, vitamins, minerals, fiber, alcohol, and coffee); age; heredity; obesity; physical activity; stress; personality traits; sex; hormonal balance; and the occurrences of certain other diseases.

During the last three decades researchers have been preoccupied with the theory that dietary fats are an important cause, if not the principal cause, of atherosclerosis. Cholesterol and saturated fats are presumed to raise the level of blood serum cholesterol, which in turn leads to the deposition of atherosclerotic plaques on the arterial walls. Such theories are based on large numbers of epidemiological studies, as well as on short- and long-term experiments with animals and humans. Based on these theories, specific dietary recommendations have been made by various concerned organizations. For example, the Inter-Society Commission for Heart Disease Resources (49) suggested that total fat intake be reduced to less than 35% of total calories, that saturated fatty acids be reduced to less than 10% of total calories, that unsaturated fatty acids be maintained at a minimum level of 10% of total calories, and that dietary cholesterol be reduced to no more than 300 mg/day. The American Medical Association has been more conservative (5). It recommended routine measurement of the plasma lipid profile and maintenance of a desirable body weight by an appropriate combination of physical activity and caloric intake. For "risk categories" they advised a substantial decrease in the intake of both cholesterol and saturated fats (this entails substituting polyunsaturated vegetable oils for part of the saturated fat in the diet). The AMA also cautioned that such advice must not compromise the intake of essential nutrients.

It should be emphasized that a notable number of investigators in this country and abroad are strongly opposed to these recommendations since they are skeptical of the evidence upon which they are based. Obviously, a critical evaluation of the contradictory literature is beyond the scope of the present discussion. The following is a brief analysis of the dietary fat-heart disease controversy.

1. Endothelial Cell Injury (73f)

According to this theory, physical injury to the endothelium (lining of the arteries), by mechanical, chemical, or toxic factors, initiates atherogenesis by promoting release and passage of platelet and plasma constituents into the artery wall. It is believed that, if the injury were a single event, the lesions would be reversible. On the other hand, repeated or chronic endothelial injury, as may occur in chronic hyperlipidemia (high levels of lipids in blood), would presumably results in the formation of more complicated lesions that calcify and eventually lead to thrombosis and infarction.

2. Cholesterol (1b,53b,53c,56c,56g,66a,78a)

Cholesterol is an essential nutrient necessary for the maintenance of life and normal body functions. It is a part of the structure of all cell membranes and is the starting material from which the body makes its supply of sex and adrenal hormones. Approximately two-thirds of the cholesterol found in the body pool of the human adult is synthesized in the liver; about one-third comes from the diet. The cholesterol level in the body is controlled by the body's own regulatory mechanism, a mechanism that varies not only from species to species but also among individuals of the same species.

High levels of blood serum cholesterol have been, and still are, regarded as a risk factor for cardiovascular disease. A number of studies have shown that individuals with blood cholesterol level higher than average stand a greater than average chance of developing coronary heart disease. Furthermore, positive correlations have been observed between the consumption of foods rich in cholesterol and mortality from this disease. In addition to epidemiological observations, many of the early experiments leading to such a hypothesis were made with rabbits and other animal species that received high levels of cholesterol in their diets. There is general agreement concerning the positive association between high cholesterol concentration in blood serum and cardiovascular disease. However, the notion that cholesterol causes cardiovascular disease in humans, or that reduction in the consumption of cholesterol-rich foods reduces the incidence of this disease, is still highly questionable.

Experimental evidence for the cholesterol theory has been criticized on the grounds of errors in analytical methods, inapplicability of observations obtained with the rabbit (a herbivorous animal whose diet contains no cholesterol), variability in the "normal" serum cholesterol concentrations of humans, and lack of consensus as to what "normal," "optimal," and "desirable" serum cholesterol levels actually are. In addition, the tenuous association between cholesterol in the diet and heart disease loses much of its credibility when the exceedingly complex interplay among all the other influential factors is considered.

3. Serum Lipoproteins (15a,15b,24b,34a,42a,47a,56i)

The lipids circulating in the bloodstream (triacylglycerols, phospholipids, and free and esterified cholesterol) are closely associated with proteins in the form of macromolecular complexes that serve as the major vehicle for the transport of lipids from sites of their absorption or degradation to sites of their utilization or elimination. These lipoproteins are commonly classified, according to density, into four main categories: chylomicrons, very low density lipoproteins (VLDL), low-density lipoproteins (LDL), and high-density lipoproteins (HDL). Some characteristics of these groups are given in Table 11. Chylomicrons are the largest and the least dense of the lipoprotein particles, whereas HDL are the

Table 11 Classification and Composition of Blood Serum Lipoproteins

Class	$S_f{}^2$	Total lipids (%)	Total cholesterol (% of lipids)	Apoprotein
Chylomicrons	>400	98	6	A, B, C
VLDL[b]	20–40	91	20	B, C
LDL[c]	0–20	80	43	B
HDL[d]	–	52	20	A

[a]Flotation rate in a dense salt solution ("Svedberg flotation").
[b]Very low density lipoproteins (previously termed pre-β).
[c]Low-density lipoproteins (β).
[d]High-density lipoproteins (α).

smallest and most dense. The increase in density of the lipoproteins is associated with a corresponding decrease in lipid content and an increase in protein content. The VLDL contain the largest amount of cholesterol and the HDL the largest amount of phospholipids. The protein fractions (apolipoproteins) in each category are identified by the letters A, B, C, and D, based on differences in their chemical and metabolic features.

Interest in lipoproteins arose in the 1950s when a relationship was suspected between heart disease in certain subjects and the levels of various types of lipoproteins in blood plasma. Various studies indicated a correlation between the amount of VLDL or triacylglycerols in blood and the risk of coronary heart disease.

In 1967, Fredrickson and coworkers (34a) described five different types of hyperlipoproteinemia, designated I-V, based on elevation of specific lipoprotein classes or specific lipids. The most common of these are type II, characterized by increased levels of LDL, and type IV, characterized by abnormal accumulation of VLDL.

More recently, various studies (24b,47a) have shown that the association of heart disease and VLDL can be explained when other risk factors (e.g., blood cholesterol, HDL, body mass, cigarette smoking, and blood pressure) are taken into account, and concluded that "the widespread practice of identifying and treating hypertriglyceridemia in apparently healthy persons for the purpose of preventing coronary heart disease is inappropriate unless more persuasive evidence becomes available."

Since LDL are the major carriers of cholesterol in the blood, factors that influence LDL levels in blood tend to also affect total cholesterol levels. Thus, the discussion above concerning serum cholesterol and heart disease, or diet and serum cholesterol, is applicable to LDL cholesterol as well.

Based on tissue culture studies it has been suggested that, in contrast to LDL, which is presumed to promote the entry of cholesterol into the cell, HDL play a role in transporting cholesterol from the body cells to the liver, where they are transformed into bile acids and excreted into the intestine. This opinion, and the observation that HDL levels in blood tend to be higher in women, lean people, nonsmokers, moderate drinkers, and people who exercise, led to the belief that HDL may have a protective effect against atherosclerosis. Unfortunately, the effect of diet on serum HDL remains unclear.

4. Saturated and Polyunsaturated Fats (44a,52a,56c,56e,56f,58a,71c,73c)

Saturation or unsaturation of a fat is a matter of degree. One fat is more saturated than another when it contains a greater amount of saturated fatty acids, and fat that contains relatively large amounts of fatty acids with more than one double bond is referred to as a polyunsaturated fat. A common inaccuracy has been to refer to animal fats as "saturated fats," and to fats of plant origin as "polyunsaturated fats."

The inconsistency in results and controversy regarding the effects of saturated or unsaturated fats on patterns of blood lipids and incidence of artherosclerosis is even more disturbing than that involving cholesterol. The conclusion drawn from numerous experiments, and cited repeatedly in review papers, textbooks, lectures, newspapers, and other forms of media communication, has been that diets containing animal fats or hydrogenated vegetable fats tend to raise serum cholesterol levels. Unfortunately, and in spite of a serious lack of reliable evidence, such conclusions have been accepted as fact by many authoritative bodies.

Typical examples of existing inconsistencies follow. Although many vegetable oils have been shown to lower serum cholesterol levels, cocoa butter, a relatively saturated fat, has been found by most investigators to be neutral with respect to its effect on serum cholesterol level. Since cocoa butter has a large content of the fully saturated C18 fatty acid, some scientists have suggested that stearic acid may be an atypical saturated fatty acid with respect to its effect on serum cholesterol in humans, perhaps because of its position in the acylglycerol molecules of this natural fat. Another group has concluded that stearic acid may even counteract the hypercholesterolemic effect of palmitic acid. If stearic acid does not raise serum cholesterol, and this appears to be the consensus, it would be difficult to accept the argument that the observed hypercholesterolemic effect of hydrogenated fats is due to their relatively high content of saturated fatty acids, since most of the unsaturated fatty acids in natural fats are those containing 18 carbons, and would obviously produce stearic acid if hydrogenated.

Results pertaining to other individual saturated fatty acids are even more puzzling. Researchers at Minnesota, for example, have maintained that the hypercholesterolemic effect of saturated fatty acids is twice as potent as the hypocholesterolemic effect of comparable unsaturated fatty acids. Accordingly, they devised equations to predict the change in serum cholesterol concentration in response to changes in dietary fat. They later concluded that the C18 acid and acids of fewer than 12 carbon atoms were inconsequential, acknowledged the effect of dietary cholesterol, and revised their equation (56e,56f). According to these authors, the C12, C14, and C16 fatty acids are indistinguishable with respect to their effects on serum cholesterol levels. A group at Harvard, on the other hand, concluded that the C18 fatty acid, and all fatty acids with chains of 12 carbons or fewer, are ineffective in raising levels of serum cholesterol. In 1965, they claimed that the C14 and C16 fatty acids are the major hypercholesterolemic agents and that the C14 acid is four times more effective than the C16 (44a). However, when they incorporated myristic (C14) and palmitic (C16) in dietary safflower oil, no change in serum cholesterol was observed. The same authors revised their stand in 1970 and concluded that palmitic acid is as effective as myristic acid and that the C12 saturated fatty acid is also hypercholesterolemic, but that its effect is only one-third that of the C14 or C16 acids. They also observed that stearic acid can be hypercholesterolemic in a semisynthetic diet but is neutral when administered as a component of cocoa butter.

Similarly, the evidence that dietary polyunsaturated fats are hypocholesterolemic has been persuasive. A number of scientists, however, have criticized both the scientific community and commercial advertisers for overzealously promoting the benefits of polyunsaturated fats. They warned that polyunsaturates are being improperly promoted as both a curative and preventative agent for heart disease. Furthermore, evidence exists that millions of people are deliberately using polyunsaturated oils strictly for purposes of cardiovascular healthfulness without the advice of their physician and without any scientific evidence that such a course will, in fact, reduce heart disease mortality. These critics not only question the validity of the hypothesis linking polyunsaturated acids with cardiovascular well-being, but also emphasize the potential hazard of consuming excessive amounts of polyunsaturated fatty acids. According to these scientists, the possible hazards are increased incidence of cancer, free radical damage to cells, skin lesions, an increase in serum uric acid, and enhancement of ceroid production.

5. Phytosterols (1a,9a,73b)

Plant sterols are common components of human and animal diets, and they possess structures very similar to that of cholesterol. A number of reports have demonstrated a hypocholesterolemic property of plant sterols, both in animals and humans, and indeed certain preparations of plant sterols have been proposed as therapeutic agents for hypercholesterolemia. Some scientists suggested that the often observed hypocholesterolemic effect of corn oil and the hypercholesterolemic effect of coconut oil is due to a high content of phytosterols in the former and a low content of phytosterols in the latter, rather than to the obvious difference in fatty acid composition between these two oils. Similarly, the increase in serum cholesterol concentration after the ingestion of hydrogenated versus natural vegetable oils may be due to the destructive effect of hydrogenation on phytosterols rather than to the formation of saturated fatty acids. On the other hand, it has been argued that the amount of plant sterols needed to produce a discernible reduction in serum cholesterol is very large (several grams daily) and that their use is thus impractical.

It is evident that the relationship between dietary fats and atherosclerosis is still one of the most controversial areas of nutrition. Exceedingly complex metabolic interrelationships are involved, and it is thus quite understandable that experimental work in this field is complicated and difficult. We do not, however, wish to leave the impression that all published material on this topic should be dismissed, or that the recommendations of the National Research Council and the Council of Nutrition be ignored (5). One cannot quarrel with dietary modifications directed at correcting excesses and nutritional imbalances, and one cannot argue against adjustments in total caloric intake to avoid obesity. A reasonable reduction in cholesterol intake and a balance between saturated and unsaturated fatty acids may be justified. But it must be understood that the role of diet in heart disease is far from clear. The general public may well be advised not to go overboard and not to indulge in self-abuse through self-prescribed dietary extremes. Probably the best advice for the majority of human subjects at the present time is to consume a well-balanced diet that includes ample amounts of all essential nutrients, to exercise regularly, and to avoid tobacco products. Furthermore, a more vigorous and well-coordinated research effort is needed in the area of human nutrition in conjunction with an intensive program of nutrition education at all levels.

ACKNOWLEDGMENTS

The author wishes to thank Regina Whiteman and Muriel Scarborough for their kind assistance in the preparation of this manuscript.

REFERENCES

1. Abrahamsson, S., and I. Ryderstedt-Nahringbauer (1962). The crystal structure of the low-melting form of oleic acid. *Acta Crystallogr. 15*:1261-1268.

1a. Abrams, W. B., and M. A. Schwartz (1967). The pursuit of new antilipemic agents, in *Atherosclerotic Vascular Disease* (A. N. Brest and J. H. Moyer, eds.), Appleton-Century-Crofts, New York, p. 260.

1b. Ahrens, Jr., E. H. (1976). The management of hyperlipidemias: Whether, rather than how. *Ann. Intern. Med. 85*:87-93.

1c. Akiya, T., and T. Shimizu (1965). Toxicity of hydrocarbons found in the distillable products. *Yukagaku 14*:520-522.

2. Alberts, W., and J. T. Overbeek (1960). Stability of emulsions of water in oil. III. Flocculation and redispersion of water droplets covered by amphipolar monolayers. *J. Colloid Sci. 15*:489-502.

3. Albright, L. F. (1965). Quantitative measure of selectivity of hydrogenation of triglycerides. *J. Amer. Oil Chem. Soc. 42*:250-253.

3a. Alexander, J. C. (1978). Biological effects due to changes in fats during heating. *J. Amer. Oil Chem. Soc. 55*:711-717.

3b. Alfin-Slater, R. B., S. Auerback, and L. Aftergood (1959). Nutritional evaluation of some heated oils. *J. Amer. Oil Chem. Soc. 36*:638-641.

4. Allen, R. R. (1978). Principles and catalysts for hydrogenation of fats and oils. *J. Amer. Oil Chem. Soc. 55*:792-795.

5. AMA Council on Foods and Nutrition and Foods and Nutrition Board of the NAS/ NRC (1972). Joint statement. Diet and coronary heart disease. *J. Amer. Med. Assoc. 222*:1647.

5a. Andrews, J. S., W. H. Griffith, J. F. Mead, and R. A. Stein (1960). Toxicity of air-oxidized soybean oil. *J. Nutr. 70*:199-210.

6. Artman, N. R. (1969). The chemical and biological properties of heated and oxidized fats. *Adv. Lipid Res. 7*:245-330.

7. Bailey, A. E. (1950). *Melting and Solidification of Fats*, Interscience, New York.

8. Bateman, L., H. Hughes, and A. L. Morris (1953). Hydroperoxide decomposition in relation to the initiation of radical chain reactions. *Discussion Faraday Soc. 14*: 190-199.

9. Becher, B. (1959). *Emulsions: Theory and Practice*. Reinhold Publishing, New York.

9a. Beveridge, J. M. R., H. L. Haust, and W. Ford Connell (1964). Magnitude of the hypocholesterolemic effect of dietary sitosterol in man. *J. Nutr. 83*:119-122.

10. Billek, G. (1979). Heated oils—chemistry and nutritional aspects. *Nutr. Metab. 24* (Suppl. 1):200-210.

11. Billek, G., G. Guhr, and J. Waibel (1978). Quality assessment of used frying fats: A comparison of four methods. *J. Amer. Oil Chem. Soc. 55*:728-733.

12. Bolland, J. L., and G. Gee (1946). Kinetic studies in the chemistry of rubber and related materials. *Trans. Faraday Soc. 42*:236-252.

13. Bolland, J. L., and P. ten Have (1947). Kinetic studies in the chemistry of rubber and related materials. IV. The inhibitory effect of hydroquinone on the thermal oxidation of ethyl linoleate. *Trans. Faraday Soc. 43*:201-210.

13a. Brisson, G. J. (1981). *Lipids in Human Nutrition*. J. K. Burgess, Englewood Cliffs, N. J.

14. Brockerhoff, H. (1965). A stereospecific analysis of triglycerides. *J. Lipid Res. 6*: 10-15.

15. Cahn, R. S., C. K. Ingold, and V. Prelog (1956). The specification of asymmetric configuration in organic chemistry. *Experientia 12*:81-94.

15a. Carlson, L. A., and A. G. Olsson (1979). Serum-lipo-protein-cholesterol distribution in healthy men with high serum-cholesterol concentrations: Extrapolation to clofibrate trial. *Lancet 1*:869-870.

15b. Castelli, W. P., and R. F. Morgan (1971). Lipid studies for assessing the risk of cardiovascular disease and hyperlipidemia. *Hum. Pathol. 2*:153-164.

16. Causeret, J. (1982). Chauffage des corps gras et risques de toxicité. (1) *Cahiers Nutr. Diet 17*:19-33.

17. Chan, H. W. S. (1977). Photo-sensitized oxidation of unsaturated fatty acid methyl esters. The identification of different pathways. *J. Amer. Oil Chem. Soc. 54*:100-104.

18. Chan, H. W. S., and G. Levett (1977). Autoxidation of methyl linoleate. Separation and analysis of isomeric mixtures of methyl linoleate hydroperoxides and methyl hydroxylinoleates. *Lipids 12*:99-104.

19. Chan, H. W. S., and G. Levett (1977). Autoxidation of methyl linolenate: Analysis of methyl hydroxylinolenate isomers by high performance liquid chromatography. *Lipids 12*:837-840.

20. Chan, H. W. S., C. T. Costaras, F. A. A. Prescott, and P. A. T. Swoboda (1975). Specificity of lipoxygenases. Thermal isomerisation of linoleate hydroperoxides, a phenomenon affecting the determination of isomeric ratios. *Biochim. Biophys. Acta 398*:347-350.

21. Chang, S. S., R. J. Peterson, and C-T. Ho (1978). Chemistry of deep fat fried flavor, in *Lipids as a Source of Flavor* (M. K. Supton, ed.), ACS Symposium Series 75, ACS, Washington, D. C., pp. 18-41.

22. Chapman, G. M., E. E. Akehurst, and W. B. Wright (1971). Cocoa butter and confectionary fats. Studies using programmed temperature X-ray diffraction and differential scanning calorimetry. *J. Amer. Oil Chem. Soc. 48*:824-830.

23. Christie, W. W., and J. H. Moore (1969). A semimicro method for the stereospecific analysis of triglycerides. *Biochim. Biophys. Acta 176*:445-452.

24. Cooper, A. R., D. P. Matzinger, and T. E. Furia (1979). An improved apparatus for oxygen bomb testing. *J. Amer. Oil Chem. Soc. 56*:1-5.

24a. Coots, R. H. (1964). A comparison of the metabolism of elaidic, oleic, palmitic, and stearic acids in the rat. *J. Lipid Res. 5*:468-472.

24b. Coronary Drug Project Research Group (1978). Natural history of myocardial infarction in the coronary drug project: Long-term prognostic importance of serum lipid levels. *Am. J. Cardiol. 42*:489-498.

25. Crnjar, E. D., A. Witchwoot, and W. W. Nawar (1981). Thermal oxidation of a series of saturated triacylglycerols. *J. Agr. Food Chem. 29*:39-42.

26. Dahle, L. K., E. G. Hill, and R. T. Holman (1962). The thiobarbituric acid reaction and the autoxidations of polyunsaturated fatty acid methyl esters. *Arch. Biochem. Biophys. 98*:253-267.

27. Dillard, C. J., and A. L. Tappel (1971). Fluorescent products of lipid peroxidation and mitochondria and microsomes. *Lipids 6*:715-721.

28. Dupuy, H. P., E. T. Rayner, J. I. Wadsworth, and M. L.Legendre (1977). Analysis of vegetable oils for flavor quality by direct gas chromatography. *J. Amer. Oil Chem. Soc. 54*:445-449.

28a. Enig, M. G., L. A. Pallansch, H. E. Walker, J. Sampugna, and M. Keeney (1979). Trans fatty acids: Concerns regarding increasing levels in the American diet and possible health implications, in Proceedings, Maryland Nutritional Conference for Feed Manufacturers. pp. 9-17.

29. FAO/WHO Expert Committee on Food Additives (1974). Toxicological evaluation of certain food additives with a review of general principles and specifications. 17th and 18th reports, WHO Technical Report Series, No. 539 and 557, Geneva.

30. Farmer, E. H., G. F. Bloomfield, A. Sundralingam, and D. A. Sutton (1942). The

course and mechanism of autoxidation reactions in olefinic and polyolefinic substances, including rubber. *Trans. Faraday Soc. 38*:348-356.

31. Fioriti, J. A., A. P. Bentz, and R. J. Sims (1966). The reaction of picric acid with epoxides. II. The detection of epoxides in heated oils. *J. Amer. Oil Chem. Soc. 43*: 487-490.

32. Forss, D. A. (1972). Odor and flavor compounds from lipids, in *Progress in the Chemistry of Fats and Other Lipids*, Vol. 13 (R. T. Holman, ed.), Pergamon Press, London, pp. 177-258.

33. Frankel, E. N. (1979). Autoxidation, in *Fatty Acids* (E. H. Pryde, ed.), Amer. Oil Chem. Soc., Champaign, Ill., pp. 353-390.

34. Frankel, E. N. (1982). Volatile lipid oxidation products. *Prog. Lipid Res. 22*:1-33.

34a. Fredrickson, D. S., R. I. Levy, and R. S. Lees (1967). Fat transport in lipoproteins—an integrated approach to mechanisms and disorders. *N. Engl. J. Med. 276*: 34-44, 94-103, 148-156, 215-225, 273-281.

35. Friberg, S. (ed.) (1976). *Food Emulsions*, Marcel Dekker, Inc., New York.

36. Friberg, S. (1976). Emulsion stability, in *Food Emulsions* (S. Friberg, ed.), Marcel Dekker, Inc., New York, pp. 1-37.

37. Friberg, S., and L. Mandell (1970). Influence of phase equilibria on properties of emulsions. *J. Pharm. Sci. 59*:1001-1004.

38. Friberg, S., L. Mandell, and M. Larsson (1969). Mesomorphous phases, a factor of importance for the properties of emulsions. *J. Colloid Interfac Sci. 29*:155-156.

39. Fritsch, C. W., and J. A. Gale (1977). Hexanal as a measure of rancidity in low fat foods. *J. Amer. Oil Chem. Soc. 54*:225-228.

40. Fritsch, C. W., D. C. Egberg, and J. S. Magnuson (1979). Changes in dielectric constant as a measure of frying oil deterioration. *J. Amer. Oil Chem. Soc. 56*: 746-750.

41. Gardner, H. W. (1975). Decomposition of linoleic acid hydroperoxides. Enzymatic reactions compared with nonenzymic. *J. Agr. Food Chem. 23*:129-136.

42. Gardner, H. W. (1980). Lipid enzymes: Lipases, lipoxygenases and hydroperoxidases, in *Autoxidation in Food and Biological Systems* (M. G. Simic and M. Karel, eds.), Plenum Press, New York, pp. 447-504.

42a. Gordon, T., W. P. Castelli, M. C. Hjortland, W. B. Kannel, and T. R. Dawber (1977). High density lipoprotein as a protective factor against coronary heart disease. The Framingham study. *Am. J. Med. 62*:707-714.

43. Griffin, W. C. (1949). Classification of surface-active agents by "HLB". *J. Soc. Cosmet. Chem. 1*:311-326.

44. Griffin, W. C. (1954). Calculation of HLB values of non-ionic surfactants. *J. Soc. Cosmet. Chem. 5*:249-256.

44a. Hegsted, D. M., R. B. McGandy, M. L. Myers, and F. J. Stare (1965). Quantitative effects of dietary fat on serum cholesterol in man. *Am. J. Clin. Nutr. 17*:281-295.

45. Henick, A. S., M. F. Benca, and J. H. Mitchell (1954). Estimating carbonyl compounds in rancid fats and foods. *J. Amer. Oil Chem. Soc. 31*:88-91.

46. Hilditch, T. P., and P. N. Williams (1964). *The Chemical Constitution of Natural Fats*, 4th ed., Chapman & Hall, London, p. 19.

47. Hirschmann, H. (1960). The nature of substrate asymmetry in stereoselective reactions. *J. Biol. Chem. 235*:2762-2767.

47a. Hulley, S. B., R. H. Rosenman, R. D. Bawol, and R. J. Brand (1980). Epidemiology as a guide to clinical decisions. The association between triglyceride and coronary heart disease. *N. Engl. J. Med. 302*:1383-1389.

48. International Union of Pure and Applied Chemistry (1972). *Manual on Colloid and Surface Science*, Butterworths, London.

49. Inter-Society Commission for Heart Disease Resources (1970). Primary prevention of the artherosclerotic diseases. *Circulation 42*:A55-A95.

50. IUPAC-IUB Commission on Biochemical Nomenclature (1977). The nomenclature of lipids. Recommendations (1976). *Lipids 12*:455-468.

51. Iwaoka, W. T., and E. G. Perkins (1978). Metabolism and lipogenic effects of the cyclic monomers of methyl linolenate in the rat. *J. Amer. Oil Chem. Soc. 55*:734-738.

52. Jackson, H. W. (1981). Techniques for flavor and odor evaluation. *J. Amer. Oil Chem. Soc. 58*:227-231.

52a. Jackson, R. L., O. D. Taunton, J. D. Morrisett, and A. M. Gotts (1978). The role of dietary polyunsaturated fat in lowering blood cholesterol in man. *Circ. Res. 42*: 447-453.

52b. Johnson, O. C., E. Perkins, M. Sugia, and F. A. Kummerow (1957). Studies on the nutritional and physiological effects of thermally oxidized oils. *J. Amer. Oil Chem. Soc. 34*:594-597.

53. Joint FAO/WHO Expert Committee on Food Additives (1974). Toxicological evaluation of certain food additives with a review of general principles and specifications. WHO Technical Report Series No. 539 and 557, Geneva.

53a. Keneda, T., H. Sakai, and S. Ishil (1955). Nutritive value or toxicity of highly unsaturated fatty acids. *J. Biochem. (Japan) 42*:561-573.

53b. Kannel, W. B. (1976). Some lessons in cardiovascular epidemiology from Framingham. *Am. J. Cardiol. 37*:269-282.

53c. Kannel, W. B., D. McGee, and T. Gordon (1976). A general cardiovascular risk profile: The Framingham study. *Am. J. Cardiol. 38*:46-51.

54. Karel, M. (1980). Lipid oxidation, secondary reactions, and water activity of foods, in *Autoxidation in Food and Biological Systems* (M. S. Simic and M. Karel, eds.), Plenum Press, New York, pp. 191-206.

55. Karel, M., K. Schaich, and B. R. Roy (1975). Interaction of peroxidizing methyl linoleate with some proteins and amino acids. *J. Agr. Food Chem. 23*:159-163.

56. Kartha, A. R. S. (1953). The glyceride structure of natural fats. II. The rule of glyceride type distribution of natural fats. *J. Amer. Oil Chem. Soc. 30*:326-329.

56a. Kaunitz, H. (1975). Dietary lipids and arteriosclerosis. *J. Amer. Oil Chem. Soc. 52*:293-297.

56b. Kaunitz, H., C. A. Slanetz, and R. E. Johnson (1959). Influence of feeding fractioned esters of autoxidized lard and cotton seed oil on growth, thirst, organ weights, and liver lipids of rats. *J. Amer. Oil Chem. Soc. 36*:611-615.

56c. Keys, A. (1970). Coronary heart disease in seven countries. American Heart Association Monograph 29. *Circulation 41*:Suppl. 1.

56d. Keys, A. (1974). Bias and misrepresentation revisited: Perspective on saturated fat. *Am. J. Clin. Nutr. 27*:188-212.

56e. Keys, A., J. T. Anderson, and F. Grande (1965). Serum cholesterol response to changes in the diet. I. Iodine value of dietary fat versus 2S-P. *Metab., Clin. Exp. 14*:747-758.

56f. Keys, A., J. T. Anderson, and F. Grande (1965). Serum cholesterol response to changes in the diet. IV. Particular saturated fatty acids in the diet. *Metab., Clin. Exp. 14*:776-787.

56g. Keys, A., J. T. Anderson, O. Mickelsen, S. F. Adelson, and F. Fidanza (1956). Diet and serum cholesterol in man: Lack of effect of dietary cholesterol. *J. Nutr. 59*:39-56.

56h. Kritchersky, D. (1976). Diet and atherosclerosis. *Amer. J. Pathol. 84*:615-632.

56i. Kritchevsky, D., and S. Czarnechi (1980). Lipoproteins. *Contemp. Nutr.* (General Mills, Inc., Minneapolis) *5*(5):1-2.

57. Krog, N., and K. Larsson (1968). Phase behavior and rheological properties of aqueous systems of industrial distilled monoglycerides. *Chem. Phys. Lipids 2*: 129-143.

58.   Krog, N., and J. Lauridson (1976). Food emulsifiers and their association with water, in *Food Emulsions* (S. Friberg, ed.), Marcel Dekker, Inc., New York, pp. 67-139.

58a.   Kummerow, F. A. (1981). Saturated fat and cholesterol: Dietary "risk factors" or essential to human life? *Food Nutr. News* (National Live Stock and Meat Board) *53* (1):1-4.

59.   Labuza, T.P. (1971). Kinetics of lipid oxidation. in CRC *Crit. Rev. Food Technol.* 355-404.

60.   Lands, W. E. M., R. A. Pieringer, P. M. Slakey, and A. Zschocke (1966). A micro-method for the stereospecific determination of triglyceride structure. *Lipids 1*: 444-448.

61.   Larsson, K. (1972). On the structure of isotropic phases in lipid-water systems. *Chem. Phys. Lipids 9*:181-195.

62.   Larsson, K. (1976). Crystal and liquid crystal structures of lipids, in *Food Emulsions* (S. Friberg, ed.), Marcel Dekker, Inc., New York, pp. 39-66.

63.   Loury, M. (1972). Possible mechanisms of autoxidative rancidity. *Lipids 7*:671-675.

64.   Luddy, F. E., R. A. Barford, S. F. Herb, P. Magidman, and R. Riemenschneider (1964). Pancreatic lipase hydrolysis of triglycerides by a semimicro technique. *J. Amer. Oil Chem. Soc. 41*:693-696.

65.   Luzatti, V. (1968). In *X-ray Diffraction Studies of Lipid-Water Systems in Biological Membranes* (D. Chapman, ed.), Academic Press, London, pp. 71-123.

65a.   Matsuo, N. (1957). Toxicity of fish oil polymerized by heating in carbon dioxide. *Seikagaku 29*:885-891.

66.   Matthews, R. F., R. A. Scanlon, and L. M. Libbey (1971). Autoxidation products of 2,4-decadienal. *J. Amer. Oil Chem. Soc. 48*:745-747.

66a.   McGill, Jr., H. C. (1979). The relationship of dietary cholesterol to serum cholesterol concentration and to atherosclerosis in man. *Am. J. Clin. Nutr. 32*:2664-2702.

67.   Mead, J. F., R. A. Stein, G. S. Wu, A. Sevanian, and M. Gan-Elepnano (1980). Peroxidation of lipids in model systems and in biomembranes, in *Autoxidation in Food and Biological Systems* (M. G. Simic and M. Karel, eds.), Plenum Press, New York, pp. 413-428.

67a.   Nakamura, M., H. Tanaka, Y. Hattori, and M. Watanabe (1973). Biological effects of autoxidized safflower oils. *Lipids 8*:566-572.

68.   Nawar, W. W. (1977). Radiation chemistry of lipids, in *Radiation Chemistry of Major Food Components* (P. S. Elias and A. J. Cohen, eds.), Elsevier, Amsterdam, pp. 21-61.

68a.   Nolen, G. A., J. C. Alexander, and N. R. Artman (1967). Long-term rat feeding study with used frying fats. *J. Nutr. 93*:337-348.

69.   Official and Tentative Methods of the Amer. Oil Chem. Soc. (1980). Peroxide value, Cd 8-53; oxirane test, Cd 9-57; iodine value Cd 1-25; AOM, CD 12-57. *J. Amer. Oil Chem. Soc.*

70.   Paradis, A. J., and W. W. Nawar (1981). Evaluation of new methods for the assessment of used frying oils. *J. Food Sci. 46*:449-451.

71.   Paradis, A. J., and W. W. Nawar (1981). A gas chromatographic method for the assessment of used frying oils: Comparison with other methods. *J. Amer. Oil Chem. Soc. 58*:635-638.

71a.   Perkins, E. G. (1976). Chemical, nutritional and metabolic studies of heated fats. II. Nutritional aspects. *Rev. Fr. Corps Gras 23*:313-322.

71b.   Perkins, E. G., and R. Taubold (1978). Nutritional and metabolic studies of non-cyclic dimeric fatty acid methyl esters in the rat. *J. Amer. Oil Chem. Soc. 55*: 632-634.

71c. Pickney, E. R. (1973). The potential toxicity of excessive polyunsaturates. *Am. Heart J. 85*:723-726.

71d. Poling, C. E., E. Eagle, and E. E. Rice (1969). Long-term responses of rats to heat-treated dietary fats. IV. Weight gains, food and energy efficiencies, longevity, and histopathology. *Lipids 5*:128-136.

71e. Poling, C. E., W. D. Warner, P. E. Mone, and E. E. Rice (1962). The influence of temperature, heating time, and aeration upon the nutritive value of fats. *J. Amer. Oil Chem. Soc. 39*:315-320.

72. Porter, W. L. (1980). Recent trends in food applications of antioxidants, in *Autoxidation in Food and Biological Systems* (M. G. Simic and M. Karel, eds.), Plenum Press, New York, pp. 295-365.

72a. Privett, O. S., F. Phillips, H. Shimasaki, T. Nazawa, and E. C. Nickell (1977). Studies of effects of trans fatty acids in the diet on lipid metabolism in essential fatty acid deficient rats. *Am. J. Clin. Nutr. 30*:1009-1017.

73. Pryor, W. A., J. P. Stanley, and E. Blair (1976). Autoxidation of polyunsaturated fatty acids. II. A suggested mechanism for the formation of TBA-reactive materials from prostaglandin-like endoperoxides. *Lipids 11*:370-379.

73a. Raju, N. V., M. Narayana Rao, and R. Rajagopalan (1965). Nutritive value of heated vegetable oils. *J. Amer. Oil Chem. Soc. 42*:774-776.

73b. Ravi Subbiah, M. T. (1973). Dietary plant sterols: Current status in human and animal sterol metabolism. *Am. J. Clin. Nutr. 26*:219-225.

73c. Reiser, R. (1973). Saturated fat in the diet and serum cholesterol concentration: A critical examination of the literature. *Am. J. Clin. Nutr. 26*:524-555.

73d. Reiser, R. (1978). Oversimplification of diet: Coronary heart disease relationship and exaggerated diet recommendations. *Am. J. Clin. Nutr. 31*:865-875.

73e. Rice, E. E., C. E. Poling, P. E. Mone, and W. D. Warner (1960). A nutritive evaluation of over-heated fats. *J. Amer. Oil Chem. Soc. 37*:607-613.

73f. Ross, R., and L. Harker (1976). Hyperlipidemia and atherosclerosis. *Science 193*: 1094-1100.

73g. Scheutwinkel-Reich, M., G. Ingerowski, and H. J. Stan (1980). Microbiological studies investigating mutagenicity of deep frying fat fractions and some of their components. *Lipids 15*:849-852.

74. Sherman, P. (ed.) (1968). *Emulsion Science*, Academic Press, New York.

75. Sherwin, E. R. (1976). Antioxidants for vegetable oils. *J. Amer. Oil Chem. Soc. 53*: 430-436.

76. Simpson, T. D. (1979). Crystallography, in *Fatty Acids* (E. H. Pryde, ed.), Amer. Oil Chem. Soc., Champaign, Ill., pp. 157-172.

77. Smouse, T. H., and S. S. Chang (1967). A systematic characterization of the reversion flavor of soybean oil. *J. Amer. Oil Chem. Soc. 44*:509-514.

78. Sreenivasan, B. (1978). Interesterification of fats. *J. Amer. Oil Chem. Soc. 55*: 796-805.

78a. Stamler, J. (1979). Population studies, in *Nutrition, Lipids, and Coronoary Heart Disease. A Global View* (R. I. Levy, B. M. Rifkind, B. H. Dennis, and N. D. Ernst, eds.), Raven Press, New York, p. 57.

79. *Standard Methods for the Analysis of Oils, Fats and Derivatives* (1979). Determination of the p-anisidine value. Method 2.504, 6th ed. Pergamon Press, London, pp. 143-144.

79a. Taylor, S. L., C. M. Berg, N. H. Shoptaugh, and E. Traisman (1983). Mutagen formation in deep-fat fried foods as a function of frying conditions. *J. Amer. Oil Chem. Soc. 60*:576-580.

80. Uri, N. (1961). Mechanism of antioxidation, in *Autoxidation and Antioxidants* (W. O. Lundberg, ed.), Interscience, New York, pp. 133-169.

81. Vajdi, M., and W. W. Nawar (1979). Identification of radiolytic compounds from beef. *J. Amer. Oil Chem. Soc. 56*:611-615.
82. Waltking, A. E., W. E. Seery, and G. W. Bleffert (1975). Chemical analysis of polymerization products in abused fats and oils. *J. Amer. Oil Chem. Soc. 52*:96-100.
83. Wille, R. L., and E. S. Lutton (1966). Polymorphism of cocoa butter. *J. Amer. Oil Chem. Soc. 43*:491-496.
83a. Willebrands, A. F., and K. J. Van der Veen (1966). The metabolism of elaidic acid in the perfused rat heart. *Biochim. Biophys. Acta 116*:583-585.
84. Williams, T. F. (1962). Specific elementary processes in the radiation chemistry of organic oxygen compounds. *Nature (London) 194*:348-351.
85. Williams, R. M., and D. Chapman (1970). In *Progress in the Chemistry of Fats and Other Lipids* (P. T. Holman, ed.), vol. 11, part 1, *Phospholipids, Liquid Crystals and Cell Membranes*, Pergamon Press, London.

BIBLIOGRAPHY

Allen, R. R. (1978). Principles and catalysts for hydrogenation of fats and oils. *J. Amer. Oil Chem. Soc. 55*:792-795.

Bailey, A. E. (1950). *Melting and Solidification of Fats*, Interscience, New York.

Brisson, G. J. (1981). *Lipids in Human Nutrition*, J. K. Burgess, Englewood Cliffs, N. J.

Frankel, E. N. (1979), Autoxidation, in *Fatty Acids* (E. H. Pryde, ed.), J. Amer. Oil Chem. Soc., Champaign, Ill., pp. 353-390.

Friberg, S. (ed.) (1976). *Food Emulsions*, Marcel Dekker, Inc., New York.

Labuza, T. P. (1971). Kinetics of lipid oxidation. *CRC Crit. Rev. Food Technol.* pp. 355-404.

Litchfield, C. (1972). *Analysis of Triglycerides*, Academic Press, New York.

Lundberg, W. O. (ed.) (1961, 1962). *Autoxidation and Antioxidants*, vols. I and II, John Wiley & Sons, New York.

Pryde, E. H. (ed.) (1979). *Fatty Acids*, American Oil Chem. Soc., Champaign, Ill.

Schultz, H. W., E. A. Day, and R. O. Sinnhuber (eds.) (1962). *Symposium on Foods: Lipids and Their Oxidation*, AVI Publishing Co., Westport, Conn.

Simic, M. G., and M. Karel (1980). *Autoxidation in Food and Biological Systems*, Plenum Press, New York, pp. 659.

Swern, D. (ed.) (1979). *Bailey's Industrial Oil and Fat Products*, John Wiley & Sons, New York.

Weiss, T. J. (1983). *Food Oils and Their Uses*, 2nd ed., AVI Publishing Co., Westport, Conn.

5

AMINO ACIDS, PEPTIDES, AND PROTEINS

Jean Claude Cheftel and Jean-Louis Cuq Université des Sciences et Techniques du Languedoc, Montpellier, France

Denis Lorient Ecole Nationale Supérieure de Biologie Appliquée à la Nutrition et à l'Alimentation, Dijon, France

I.	Introduction	246
II.	Physicochemical Properties of Amino Acids and Proteins	247
	A. General Properties of Amino Acids	247
	B. General Properties of Proteins	254
	C. Chemical Reactions and Interactions of Amino Acids and Proteins	266
III.	Denaturation of Proteins	274
	A. Physical Agents	275
	B. Chemical Agents	278
	C. Energetics of Denaturation	279
IV.	Functional Properties of Proteins	282
	A. Hydration Properties	283
	B. Solubility	288
	C. Viscosity	288
	D. Gelation	290
	E. Texturization	294
	F. Dough Formation	296
	G. Emulsifying Properties	298
	H. Foaming Properties	303
	I. Flavor Binding	310
	J. Binding of Other Compounds	313
V.	Nutritional Attributes of Proteins	313
	A. Protein Metabolism	313
	B. Human Requirements for Proteins and Amino Acids	316
	C. Protein Nutritive Value of Foods	319
VI.	Unconventional Protein Sources	325
	A. Need for Increased Protein Production	325
	B. Approaches to Utilization of Plant Proteins	327

C. Separation and Purification of Proteins from Plants   327
D. Single-Cell Proteins   330
E. Chemical Synthesis and Genetic Engineering of Amino Acids
       and Proteins   331
VII. Modifications of Food Proteins Through Processing and Storage   332
A. Changes in Nutritive Value and Toxic Effects   332
B. Changes in the Functional Properties of Proteins   349
References   363
Bibliography   369

I. INTRODUCTION

Proteins are complex macromolecules that may constitute 50% or more of the dry weight of living cells. They play a fundamental role in the structure and function of cells. Numerous proteins have been isolated and purified, and their molar mass varies from about 5000 to many million daltons. These biopolymers are made up of carbon, hydrogen, oxygen, nitrogen, and, usually, sulfur. Some of them also contain iron, copper, phosphorus, or zinc. The complete hydrolysis (acid, alkaline, or enzymic) of proteins leads to α-amino acids of L configuration, and they differ from one another by the nature of their side chains. For most proteins, the component amino acids belong to a restricted group of 20 different amino acids. The component amino acids are linked together by substituted amide bonds called peptide bonds, forming polypeptide chains containing up to several hundred units (amino acid residues). The molar mass of these polymers, as well as the sequence of the amino acids in the chains, is known for many proteins.

Proteins can be classified into one of two groups: *homoproteins*, which contain only amino acids, and *heteroproteins*, which consist of amino acids and various other nonproteinaceous compounds, collectively called *prosthetic groups*. According to the chemical nature of the prosthetic group, one can distinguish *nucleoproteins* (ribosomes and viruses), *lipoproteins* (plasmatic and β-lipoprotein), *glycoproteins* (γ-globulin and orosomucoid), *phosphoproteins* (caseins), *hemoproteins* (hemoglobin, cytochrome C, catalase, and myoglobin), and *metalloproteins* (alcohol dehydrogenase and carbonic anhydrase).

Each protein is characterized by its conformation, that is, by its three-dimensional organization. Thus, *fibrous proteins* are composed of polypeptide chains assembled along a common straight axis, which leads to the formation of fibers (collagen, keratin, elastin, and fibroin). *Globular proteins*, on the other hand, are composed of one or several polypeptide chains folded upon themselves to form in space a three-dimensional structure (spheric and globular forms). Some molecules possess the properties of both fibrous and globular proteins (actin and fibrinogen).

Primary structure refers to the sequential order of amino acids in a protein. *Secondary* and *tertiary structures* relate to the three-dimensional organization of the polypeptide chain. *Quaternary structure* refers to the geometric arrangement among various polypeptide chains, these chains linked together by bonds that in most cases are not covalent.

Proteins possess an extraordinary diversity of functions, which arbitrarily may be classified in three main categories: structural proteins, proteins with biological activity, and food proteins.

Structural proteins (keratin, collagen, elastin, and so on) are present in all tissues, such as muscle, bone, skin, internal organs, cellular membranes, and intracellular organelles. Their functionality is largely related to their fibrous structure.

Proteins endowed with *biological activity* fulfill an active role in all biological processes. Enzymes are the most important proteins in this group. Over 2000 enzymes have been identified, and they are highly specific catalyzers. Other biologically active proteins include *hormones*, which regulate metabolic reactions (insulin and somatotrophin), *contractile proteins*, (myosin, actin, and tubulin), *transfer proteins* (hemoglobin, myoglobin, and transferrin), proteins that *protect the blood* of vertebrates (immunoglobulins, fibrinogen, and thrombin), and *storage proteins* (ovalbumin, gliadin, and zein). Proteins also exist that are *toxic* for higher animals (botulinum toxin, staphylococcal toxin, venom of certain snakes, and ricin) or for microorganisms (some antibiotics), and some possess antinutritional properties (e.g., trypsin inhibitors). One should also note that many allergic reactions to foods result from modification of the defense mechanisms of the consumer due to the presence in foods of proteinaceous antigens that promote the synthesis of protein antibodies.

It is worthwhile to stress that all the proteins just mentioned are composed of the same amino acids. It is the ratio and sequence of the amino acids that are specific for each protein and that provide it with the particular conformational and chemical properties necessary for expressing a unique structural, biological, or toxic activity.

Food proteins do not represent a unique group, because many of the structural or biologically active proteins described above are food proteins. Food proteins are simply those that are palatable, digestible, nontoxic, and available economically for humans. The nutritional needs of essential amino acids for humans are known, and they vary with the age and physiological condition of the individual (pregnancy, lactation, and so on). The deficit in nutritional proteins is unfortunately very large for some segments of the world population, and this situation may become still more serious in the future, as the world population increases. Indeed, the production of food proteins in sufficient amounts poses many problems, especially since they are more expensive to produce than carbohydrates or lipids. In order to satisfy this steadily growing demand for proteins, new protein sources must be found and methods developed for their technological utilization. Also, conventional proteins must be used more efficiently. It is for these reasons that it is so important to have available as much data as possible on the physical, chemical, and biological properties of food proteins. Knowing the effects of technological treatments on these proteins is also important, since such knowledge allows their properties, particularly nutritional quality and functional characteristics, to be improved.

II. PHYSICOCHEMICAL PROPERTIES OF AMINO ACIDS AND PROTEINS (12,65,70,81,104)

A. General Properties of Amino Acids

1. General Structure and Classification

Amino acids are monomers of protein molecules. A total of 20 kinds of amino acids are usually present in protein hydrolyzates, and other less common amino acids also exist naturally and have biological functions. To understand the properties of proteins, it is necessary to first understand the properties of amino acids.

Amino acids contain in their molecular structure at least one primary amino group ($-NH_2$) and one carboxyl group ($-COOH$). In amino acids derived from proteins, the primary amino group occupies an α position with respect to the carboxyl group. Natural α-amino acids have the following structure.

$$R \underset{\underset{NH_2}{|}}{\overset{\overset{H}{|}}{C}} COOH$$

R is a side chain of variable composition. Proline and hydroxyproline, which are derived from pyrrolidine, do not conform to this general structure (Table 1).

Every amino acid has a characteristic side chain R, which influences its physicochemical properties and, as a result, the properties of the protein to which it belongs. According to the polarity of this side chain, it is possible to group amino acids into four classes.

Amino acids with nonpolar or hydrophobic side chains (Ala, Ile, Leu, Met, Phe, Pro, Trp, and Val) are less soluble in water than polar amino acids. Their hydrophobicity increases with the length of the aliphatic side chain.

Amino acids with polar, uncharged (hydrophilic) side chains possess neutral, polar, functional groups able to establish hydrogen bonds with appropriate molecules, such as water. The polarity of serine, threonine, and tyrosine is related to their hydroxyl groups $(-OH)$, the polarity of asparagine and glutamine to their amide group $(-CO-NH_2)$, and the polarity of cysteine to its thiol group $(-SH)$. Glycine is sometimes included in this class. Cysteine and tyrosine possess the most polar functional groups in this class, since both thiol and phenol groups may undergo partial ionization at pH values close to neutrality. In proteins, cysteine is often present in an oxidized state, that is, as cystine. This occurs when the thiol groups of two cysteine molecules oxidize and form a disulfide cross-link. Asparagine and glutamine easily hydrolyze in the presence of an acid or an alkali to give aspartic and glutamic acids, respectively.

Amino acids with positively charged side chains (at pH close to 7) consist of lysine, arginine, and histidine. The $\epsilon-NH_2$ group is responsible for the charge on lysine and the guanidino group for the charge on arginine. The imidazole group of histidine is 10% protonated at pH 7 and 50% protonated at pH 6.

Amino acids with negatively charged side chains (at a pH close to 7) consist of aspartic and glutamic acid.

Besides these 20 frequently encountered amino acids, other amino acids have been isolated from protein hydrolyzates. Thus, hydroxyproline and 5-hydroxylysine are present in collagen; desmosine and isodesmosine in elastin; and methylhistidine, $\epsilon-N-$ methyllysine, and $\epsilon-N-$trimethyllysine in muscle proteins.

In total, more than 150 amino acids have been shown to exist in various animal, vegetable, or microbial cells either in a free or bound form. In the majority of cases, these amino acids are either important metabolic intermediates (or precursors) or chemical mediators participating in the transmisssion of nerve impulses.

Amino acids with the D-configuration are found in some antibiotics.

2. Acidobasic Properties of Amino Acids: Ionization

The ability of amino acids to ionize is very important biologically and also facilitates their quantitative analysis. Furthermore, several properties of amino acids (melting point, solubility in water, the large dipole moments, and large dielectric constants in aqueous solution) are attributable to the uneven distribution of electric charges on amino acids in aqueous solution. Thus, all amino acids in aqueous solution, at pH values approaching neutrality, exist as Zwitterions:

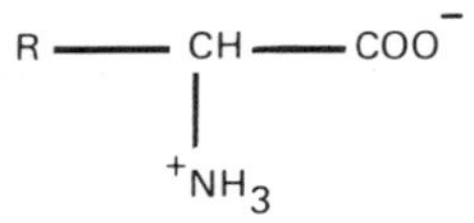

$$R \text{———} CH \text{———} COO^-$$
$$\overset{|}{{}^+NH_3}$$

When an amino acid is dissolved in water, it may behave as an acid:

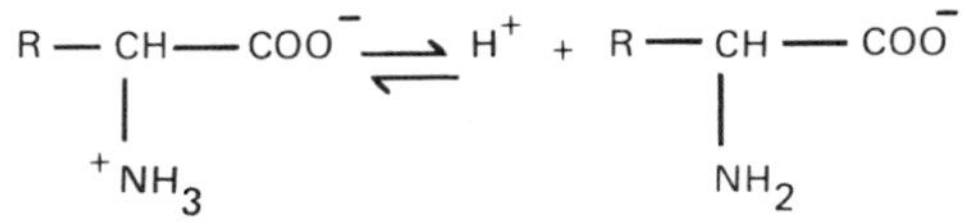

$$R \text{—} CH \text{—} COO^- \rightleftharpoons H^+ + R \text{—} CH \text{—} COO^- \qquad (1)$$

with $\overset{|}{{}^+NH_3}$ on the left and $\overset{|}{NH_2}$ on the right.

or as a base:

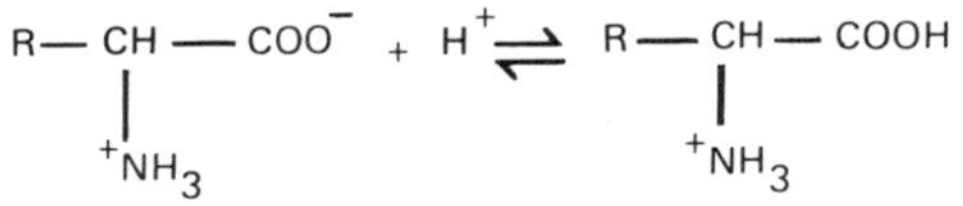

$$R \text{—} CH \text{—} COO^- + H^+ \rightleftharpoons R \text{—} CH \text{—} COOH \qquad (2)$$

with $\overset{|}{{}^+NH_3}$ on both sides.

depending on the pH. In other words, these molecules are amphoteric. In its totally protonated form, an α-amino acid (monoamino monocarboxylic) may give up two protons during its titration with a base. The isoelectric point (pI) is the pH at which the global charge of the amino acid in solution is zero.

The pK_a and pI values* of the most important amino acids are given in Table 2. The carboxyl groups of the amino acids have pK_{a_1} values lower than those of the aliphatic carboxylic acids (pK_a of acetic acid is 4.74). This is due to the presence on amino acids of a positively charged amino group on the carbon atom carrying the carboxyl group.

3. Hydrophobicity of Amino Acids

The solubility of proteins in water is related essentially to the distribution of the polar (charged or not) and apolar (hydrophobic) groups of the side chains of the amino acid constituents. The hydrophobicity of amino acids, but also of peptides and of proteins, may be determined from the relative solubilities of the amino acids in water and in a less polar solvent (e.g., ethanol), respectively. The free energy for transferring 1 mol of the amino acid from the aqueous solution into the ethanol solution is, neglecting the activity coefficients, given by

---

*The pK_a values are the negative logarithms of the dissociation constants of reactions 2 and 1. Thus

$$pK_{a_2} = -\log \frac{[H^+][\text{amino acid}^-]}{[\text{amino acid}^\pm]} \qquad (4)$$

and

$$pK_{a_1} = -\log \frac{[H^+][\text{amino acid}^\pm]}{[\text{amino acid}^+]} \qquad (5)$$

Table 1 α-Amino Acids Frequently Encountered in Proteins

Amino acid			Molecular Weight	Chemical name	Structure
name	Symbol 3 letters or 1 letter				
Alanine	Ala	A	89.1	α-amino-propionic acid	$CH_3-CH-COO^-$ $\underset{+NH_3}{\mid}$
Arginine	Arg	R	174.2	α-amino-δ- ureinovalerianic acid	$H_2N-\underset{\underset{+NH_2}{\parallel}}{C}-NH-(CH_2)_3-\underset{\underset{+NH_3}{\mid}}{CH}-COO^-$
Asparagine	Asn	N	132.1	amide of Asp	$H_2N-\underset{\underset{O}{\parallel}}{C}-CH_2-\underset{\underset{+NH_3}{\mid}}{CH}-COO^-$
Aspartic acid	Asp	D	133.1	α-amino-succinic acid	$^-O-\underset{\underset{O}{\parallel}}{C}-CH_2-\underset{\underset{+NH_3}{\mid}}{CH}-COO^-$
Cysteine	Cys	C	121.1	α-amino-β- mercaptopropionic acid	$HS-CH_2-\underset{\underset{+NH_3}{\mid}}{CH}-COO^-$
Glutamine	Glu	Q	146.1	amide of Glu	$H_2N-\underset{\underset{O}{\parallel}}{C}-(CH_2)_2-\underset{\underset{+NH_3}{\mid}}{CH}-COO^-$
Glutamic acid	Glu	E	147.1	α-amino-glutaric acid	$^-O-\underset{\underset{O}{\parallel}}{C}-(CH_2)_2-\underset{\underset{+NH_3}{\mid}}{CH}-COO^-$
Glycine	Gly	G	75.1	α-amino-acetic acid	$H-\underset{\underset{+NH_3}{\mid}}{CH}-COO^-$
Histidine	His	H	155.2	α-amino-β-imidazol propionic acid	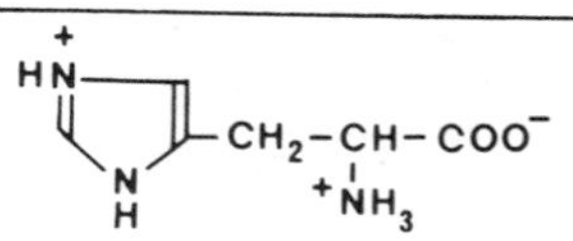 $CH_2-\underset{\underset{+NH_3}{\mid}}{CH}-COO^-$
Isoleucine	Ile	I	131.2	α-amino-β-methyl valerianic acid	$CH_3-CH_2-\underset{\underset{CH_3}{\mid}}{CH}-\underset{\underset{+NH_3}{\mid}}{CH}-COO^-$

Table 1 (Continued)

Amino acid			Molecular Weight	Chemical name	Structure
name	Symbol 3 letters or 1 letter				
Leucine	Leu	L	131.2	α-amino-isocaproic acid	$CH_3-CH-CH_2-CH-COO^-$ $\underset{CH_3}{\vert} \quad \underset{^+NH_3}{\vert}$
Lysine	Lys	K	146.2	α-ε-diamino-caproic acid	$NH_2-\left(CH_2\right)_4-CH-COO^-$ $\underset{^+NH_3}{\vert}$
Methionine	Met	M	149.2	α-amino-ȣ-methyl thiol–n–butyric acid	$CH_3-S-\left(CH_2\right)_2-CH-COO^-$ $\underset{^+NH_3}{\vert}$
Phenylalanine	Phe	F	165.2	α-amino-β-phenyl propionic acid	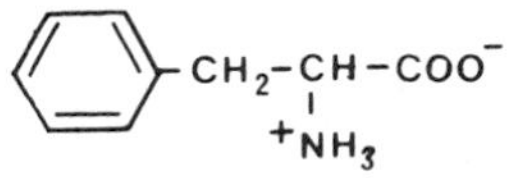 $\bigcirc-CH_2-CH-COO^-$ $\underset{^+NH_3}{\vert}$
Proline	Pro	P	115.1	pyrrolidine–2–carboxylic acid	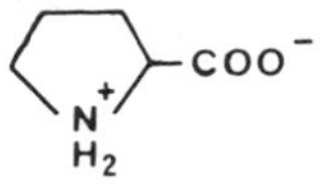 $\bigcirc\!\!\!-COO^-$ $\underset{H_2}{\overset{+}{N}}$
Serine	Ser	S	105.1	α-amino-β-hydroxy propionic acid	$HO-CH_2-CH-COO^-$ $\underset{^+NH_3}{\vert}$
Threonine	Thr	T	119.1	α-amino-β-hydroxy-n-butyric acid	$CH_3-CH-CH-COO^-$ $\underset{OH}{\vert}\ \underset{^+NH_3}{\vert}$
Tryptophan	Trp	W	204.2	α-amino-β-3-indolyl-propionic acid	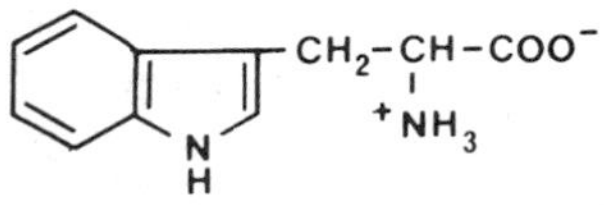 $\overset{indole}{\bigcirc}-CH_2-CH-COO^-$ $\underset{^+NH_3}{\vert}$
Tyrosine	Tyr	Y	181.2	α-amino-β-(p-hydroxy-phenyl propionic) acid	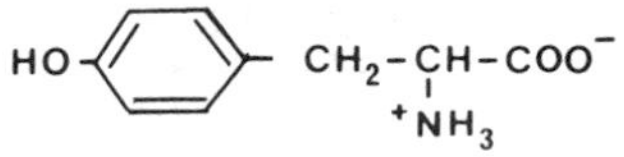 $HO-\bigcirc-CH_2-CH-COO^-$ $\underset{^+NH_3}{\vert}$
Valine	Val	V	117.1	α-amino-isovalerianic acid	$CH_3-CH-CH-COO^-$ $\underset{CH_3}{\vert}\underset{^+NH_3}{\vert}$

Table 2 pK$_a$ and pI Values of Amino Acids (at 25°C)[a]

Amino acid	pK$_{a1}$ (α – CÖO$^-$)	pK$_{a2}$ (α – $^+$NH$_3$)	pK$_{aR}$ (R = side chain)	pI
Alanine	2.35	9.69		6.02
Arginine	2.17	9.04	12.48	10.76
Asparagine	2.02	8.80		5.41
Aspartic acid	2.09	9.82	3.86	2.97
Cysteine	1.96	10.28	8.18	5.07
Glutamine	2.17	9.13		5.65
Glutamic acid	2.19	9.67	4.25	3.22
Glycine	2.34	9.78		6.06
Histidine	1.82	9.17	6.00	7.58
Isoleucine	2.36	9.68		6.02
Leucine	2.36	9.64		6.00
Lysine	2.18	8.95	10.53	9.74
Methionine	2.28	9.21		5.75
Phenylalanine	1.83	9.24		5.53
Proline	1.99	10.6		6.30
Serine	2.21	9.15		5.68
Threonine	2.71	9.62		6.16
Tryptophan	2.38	9.39		5.89
Tyrosine	2.20	9.11	10.07	5.65
Valine	2.32	9.62		5.97

[a]pK$_a$ is the colog of the apparent dissociation constant of the equilibria. pI (isoelectric point) is the pH at which the global charge of the amino acid in solution is zero.

$$\Delta G° = - RT \ln \frac{S_{eth}}{S_{H_2O}} \qquad (3)$$

where S_{eth} and S_{H_2O} are the solubilities in ethanol and in water (moles per liter).

If the amino acid has many functional groups, $\Delta G°$ is an additive function of the various groups in the amino acid:

$$\Delta G° = \Sigma \Delta G°'$$

For example, phenylalanine may be arbitrarily divided into two parts with respect to the free energy of the water to ethanol transfer: one part with a toluyl group and another part with an aminocarboxyl group.

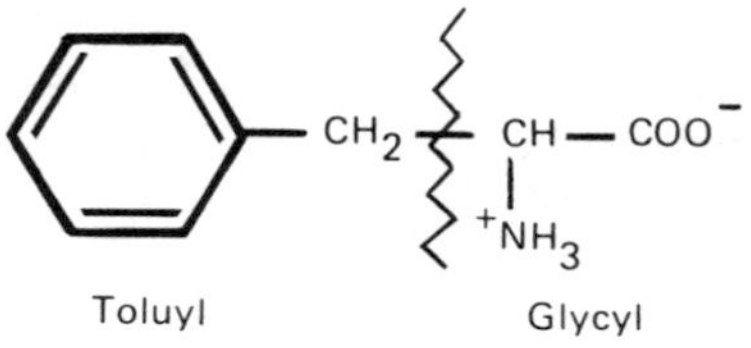

This second part is analogous to the transfer free energy of glycine given by Eq. (3). The hydrophobicity of the side chain is then given by the difference between the transfer free energies of the amino acid and of glycine:

$$\Delta G° \text{ (side chain)} = \Delta G° \text{ (amino acid)} - \Delta G° \text{ (glycine)} \qquad (6)$$

Presented in Table 3 are values for the hydrophobicity of the side chains of some amino acids. These data allow one to predict the fractionation pattern of amino acids (as a function of their hydrophobicity) on hydrophobic carriers, such as polystyrene, or silica grafted with aliphatic C8 or C18 chains. The adsorption coefficient is proportional to hydrophobicity.

4. Stereochemistry of Amino Acids

Except for glycine, all amino acids resulting from mild protein hydrolysis (acid or enzymic) possess rotatory optical activity. This property (chirality) is due to the presence of an asymmetric carbon atom, the orbitals of which are in an sp_3 hybridization state. According to the position of the tetrahedron of the four different substituents, one obtains two stereoisomers (or enantiomorphs). Thus, using Fisher's representation, and by analogy with D- and L-glyceraldehydes, the stereoisomers of alanine are

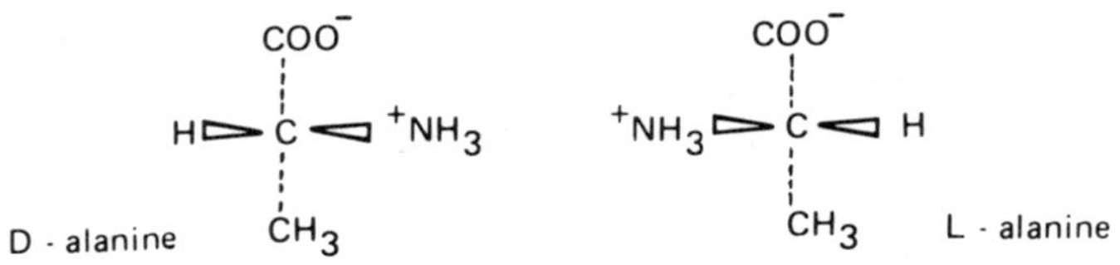

$$(7)$$

In protein hydrolyzates from plant or animal tissues, only isomers of the L form can be found. This structural uniformity of amino acids is a major governing factor in the construction of proteins. Four amino acids—isoleucine, threonine, hydroxylysine, and hydroxyproline—have a second asymmetry center, and therefore, each one has four stereoisomers.

Table 3 Side-Chain Hydrophobicity of Amino Acids (Ethanol $\longrightarrow$ Water)

Amino acid	$\Delta G°$ Side Chain (J/mol)	Amino acid	$\Delta G°$ Side chain (J/mol)
Alanine	3,100	Leucine	10,100
Arginine	3,100	Lysine	6,250
Asparagine	− 40	Methionine	5,450
Aspartic acid	2,250	Phenylalanine	11,100
Half-Cystine	4,200	Proline	10,850
Glutamine	− 400	Serine	170
Glutamic acid	2,300	Threonine	1,850
Glycine	0	Tryptophan	12,550
Histidine	2,100	Tryosine	12,000
Isoleucine	12,400	Valine	7,050

Source: From Refs. 7 and 83.

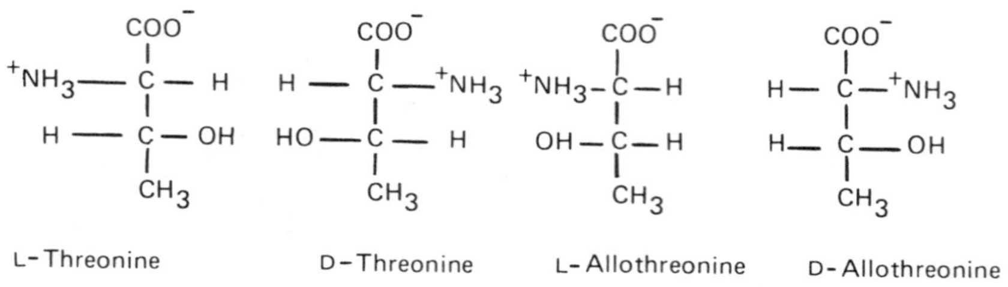

$$
\begin{array}{cccc}
\text{COO}^- & \text{COO}^- & \text{COO}^- & \text{COO}^- \\
{}^+\text{NH}_3\!-\!\text{C}\!-\!\text{H} & \text{H}\!-\!\text{C}\!-\!{}^+\text{NH}_3 & {}^+\text{NH}_3\!-\!\text{C}\!-\!\text{H} & \text{H}\!-\!\text{C}\!-\!{}^+\text{NH}_3 \\
\text{H}\!-\!\text{C}\!-\!\text{OH} & \text{HO}\!-\!\text{C}\!-\!\text{H} & \text{OH}\!-\!\text{C}\!-\!\text{H} & \text{H}\!-\!\text{C}\!-\!\text{OH} \\
\text{CH}_3 & \text{CH}_3 & \text{CH}_3 & \text{CH}_3 \\
\text{L-Threonine} & \text{D-Threonine} & \text{L-Allothreonine} & \text{D-Allothreonine}
\end{array}
\qquad (8)
$$

The D isomers of certain amino acids are present, for instance, in the cell walls of certain microorganisms and in polypeptides endowed with antibiotic action (actinomycin D, gramicidin, and tyrocidin A).

5. Absorption Spectra: Fluorescence

Among natural amino acids from proteins, tryptophan, tyrosine, and phenylalanine are the only ones that absorb ultraviolet light and have a maximum absorbance at 278, 274.5, and 260 nm, respectively. Cystine shows a slight absorption at 238 nm, and all amino acids absorb at a wavelength near 210 nm.

Tryptophan, tyrosine, and phenylalanine are, of all the amino acids, the only ones that show measurable natural fluorescence. The fluorescence of tryptophan remains, even when the amino acid is protein bound (excitation at 287 nm; maximum of fluorescence at 348 nm).

B. General Properties of Proteins

1. Primary Structure

The primary structure of a protein relates to the sequential order of amino acid residues bound together by covalent bonds (called peptide bonds). The primary structures of many proteins have been established. The shortest known protein chains (secretin and glucagon) contain 20-100 amino acids, and most proteins contain 100-500. Some rare chains contain thousands of amino acids. The primary structures of various milk proteins are presented in Chap. 13.

The peptide bond linking two amino acids is a substituted amide bond (Fig. 1). It is stabilized by resonance of two mesomeric forms: The C$-$N linkage of the peptide bond possesses about 40% of the character of a double bond, and the C$=$O linkage some 40% of a simple bond. This has two consequences. On the one hand, the $-$NH$-$ group does not protonate between pH 0 and 14. On the other hand, the C$-$N link is not able to rotate freely. The four atoms involved in the peptide link and the two α carbons are in the same plane. The oxygen and the hydrogen of the CO$-$NH group are in the trans position as a consequence of the stabilization by resonance. The polypeptidic chain may accordingly be represented by a series of rigid planes separated by $-$HCR$-$ groups. The simple covalent bonds of α-carbon atoms are the only ones that may rotate freely. The C$-$N linkage of the peptide bond is very stable (over 400 J/mol).

2. Secondary Structure

The secondary structure of a protein is the spatial structure the polypeptide chain assumes exclusively along the axis

$$-HCR-C-NH-CHR'-$$
$$\alpha \quad \| \quad \alpha$$
$$O$$

Because the substituents of the α carbons can rotate around the axes constituted by simple covalent bonds, there are numerous possibilities for the conformation of a polypeptide chain (Fig. 1).

However, at normal conditions, especially pH and temperature, each polypeptide chain assumes one specific conformation, called *native*. Thermodynamically this corresponds to a stable and organized system with a minimal free energy ΔG. Such a conformation is closely related to the polarity, the hydrophobicity, and the steric hindrance of the side chains R. The main secondary structures that have been found in proteins are the helices, α, α_{II}, γ, and 3_{10}; the β structures, such as the β-pleated sheets; the β bends; and other structures found, for instance, in polyproline or in collagen. There is also one ill-defined structure without any plane or axis of symmetry—the random coil.

The α *helix* (Fig. 2) is an ordered and particularly stable structure that contains 3.6 amino acid residues per turn (with a "step" of 0.54 nm). The side chains are located on the outside of the helix. The apparent diameter of the helix, without taking into account the side chains, is about 0.6 nm. There are many hydrogen bonds, especially between the hydrogen of a ($-NH-CO-$) group and the oxygen of a peptide bond located on the lower turn. Since each peptide bond is engaged in the formation of hydrogen bonds, and since the electric dipoles thus formed are oriented in the same direction, the helical structure has a very high stability. Moreover, this structure has a high density that restricts the interactions with other molecules (absence of hydrogen bonding with molecules of water).

As a consequence of its pyrrolidine structure, proline is not compatible with a α helix; its presence interrupts the helix and gives rise to a particular bending of the chain. A good example to mention is casein, in which proline is probably instrumental in determining the random coil structure and the thermal stability. The same occurs with some other amino acids when the electrostatic or steric properties of their side chains make impossible the establishment of the helix. When this occurs, the polypeptide chain

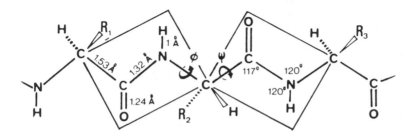

Figure 1 Structure of a fragment of an α-L-polypeptide chain (in the trans configuration). Interatomic distances (Å) and angles between links (°). The six atoms in the rectangle are in the same plane. The terms ϕ and ψ represent the possible torsion angles around an α-carbon. The two peptide bonds adjacent to the α-carbon are located each in a plane (rectangle). R_1, R_2, and R_3 are in a trans position ($\phi = \psi = 180°C$).

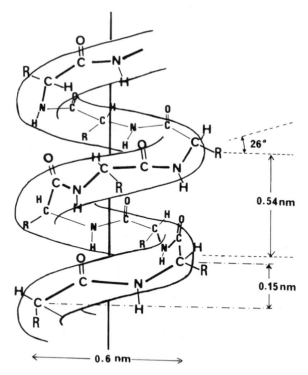

Figure 2 Schematic three-dimensional structure of an α helix (right-handed) (R are amino acid side chains).

assumes a structure in which the distance between groups carrying the same charge is maximal and the repulsive electrostatic free energy is minimal. This structure is the *random coil*. Steric hindrance from bulky side chains may also prevent the setting up of the α helix, as occurs with polyisoleucine.

The β structure (Fig. 3) is a zigzag structure, more stretched than the α helix. Transitions from α helix to β structure can result from suppression of intrachain hydrogen bonds (e.g., simply by heat). The stretched chains combine to form structures called *pleated sheets*. They are linked together by interchain hydrogen bonds, and all peptide bonds take part in their formation. Peptide chains are called parallel (Fig. 3a and b) or antiparallel (Fig. 3b and c). The side chains (R) of the amino acid residues are located above and beneath the plane of the sheets, and their charges and steric hindrances generally have little effect on the existence of such structures. However, a few amino acids cannot be integrated (Asn, Glu, His, Lys, Pro, and Ser).

The 3_{10} helix is a secondary structure sometimes found in certain regions of globular proteins. It is a variety of the α helix with three amino acid residues per turn. There are also certain infrequent helices, such as π and γ, which possess 4.4 and 5.2 residues per turn, respectively. β bends, on the contrary, are frequent (Fig. 4). They allow the polypeptide chains to bend over themselves with a stable angular conformation provided by a hydrogen bond. The β bends may be considered the starting point of a helix with a step equal to zero.

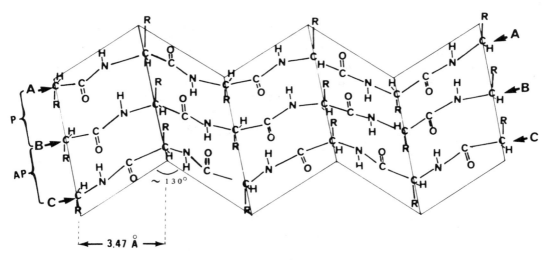

Figure 3 Schematic three-dimensional structure of β-pleated sheets. P: two parallel β-sheet strands (between A and B chains); AP: two antiparallel β-sheet strands (between B and C chains).

Finally, one should mention the helical structures *polyproline I* (left-handed; 3.3 residues per turn; cis peptide links) and *polyproline II* (left handed; 3 residues per turn; trans peptide links; distance between two residues, evaluated from their projection upon the axis is 0.31 nm). The interconversion of these two structures is possible, and form II is the more stable in an aqueous medium. Such a helical structure is found in collagen. This protein of connective tissues is the most abundant protein in vertebrates (skin, tendons, bone, cornea, and so on).

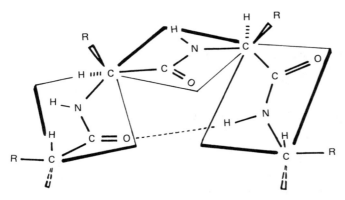

Figure 4 Schematic representation of one β bend. Parallelograms indicate the place of each peptide bond.

3. Tertiary Structure

This sort of structure relates to the three-dimensional organization of a polypeptide chain containing regions of well-defined (α helix, β bends, and β-pleated sheets) or ill-defined (random coil) secondary structures. The tertiary structure of many proteins is well known. It is, however, very difficult to represent such structures in a simple way. Most proteins contain over 100 amino acid residues, and a planar representation of their three-dimensional structure, however impressive it may be, does not provide much information. Many representations are merely schematic (Fig. 5), and the best are prepared with the aid of a computer.

In most globular, water-soluble proteins of known tertiary structure, it is found that hydrophobic amino acids tend to locate toward the inside of the molecule, and polar amino acids are mainly at the surface with a rather uniform distribution. Exceptions to this pattern can be found; for example, anisotropic distribution of electric charges may occur and cause the protein to have precise biological functions (e.g., with proteases). In the case of some proteins that are insoluble in water and soluble in certain organic solvents (e.g., lipoproteins acting as carrier for lipids), one finds a larger distribution of hydrophobic amino acids at the surface of the molecule.

Within the protein structure, hydrogen bonds are established with most atoms able to bond in this manner. Some hydrogen bonding sites are at the surface, which allows for interactions with other molecules, especially water, whereas others are intramolecular and help stabilize various structures (helices, pleated sheets, and random coil).

4. Quaternary Structure

The quaternary structure of proteins is the result of noncovalent associations of protein units. These "subunits" may be identical or not, and their arrangement may not be symmetrical. The forces or linkages that stabilize quaternary structures are the same, with the exception of disulfide cross-links, as those that stabilize tertiary structures.

The "actomyosin" system of muscle can be taken as an example of quaternary structure. Myosin and actin are proteins that constitute the main components of myofibrils. Myosin represents 55% (w/w) of the myofibril, has a molar mass of 475,000 daltons, and has six subunits that contribute to its quaternary structure. Two of the subunits are heavy polypeptide chains with a molar mass of 200,000 daltons; four are light chains, including one of 20,700 daltons, two of 19,050 daltons, and one of 16,500 daltons (Fig. 6a, b). The heavy chains possess regions with α-helix structure and have a globular and more voluminous end to which the four light chains are associated (Fig. 6c, d). A filament of myosin is composed of myosin molecules arranged in parallel with the "head" of one molecule longitudinally shifted about 60 Å with respect to the head of a neighboring molecule. The total length of the myosin filament is about 1.5 μm (Fig. 6e).

Actin constitutes 20% (w/w) of the myofibril. There are two sorts of actin: a globular form (G-actin) with a molar mass of 42,000 and a fibrous form (F-actin) resulting from the polymerization of G-actin into a double-helix filament (with 13 monomers per turn), each filament containing 300–400 monomers, all oriented the same way (Fig. 6f). The F-actin filament has a length of about 1.2 μm. Other proteins, such as tropomyosin, the various troponins, and α-actinin, are located along the helix of F-actin. Tropomyosin is composed of two subunits and has a molar mass of 70,000 daltons. There are three troponins: troponin T of 30,500 daltons, troponin I of 20,885, and troponin C of 17,845

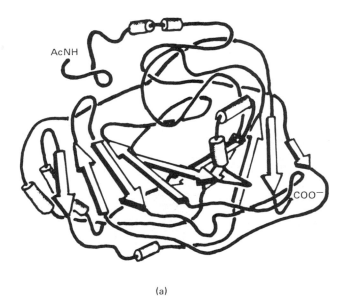

(a)

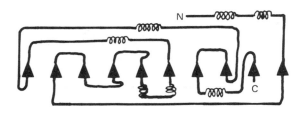

(b)

Figure 5 Schematic tertiary structure of carbonic anhydrase. (a) Helices are drawn as cylinders, β-sheet strands as arrows with C at the tail and N at the point of each arrow. [After K. K. Kannan et al., *Cold Spring Harbor Symp. Quant. Biol. 36*:221 (1971).] (b) Coils indicate α helices; triangles indicate β sheets.

daltons (Fig. 6f). Actomyosin results from the association of many filaments of myosin (Fig. 6e) and actin (Fig. 6f). This complicated arrangement represents the quaternary structure of the muscle protein system. It has biological activity, and it is the basis of muscular contraction (see also Chap. 12).

Some of the major food proteins are listed in Table 4, along with their distinguishing structural properties. Further information on specific food proteins can be found in Chaps. 6, 12, 13, 14, and 15.

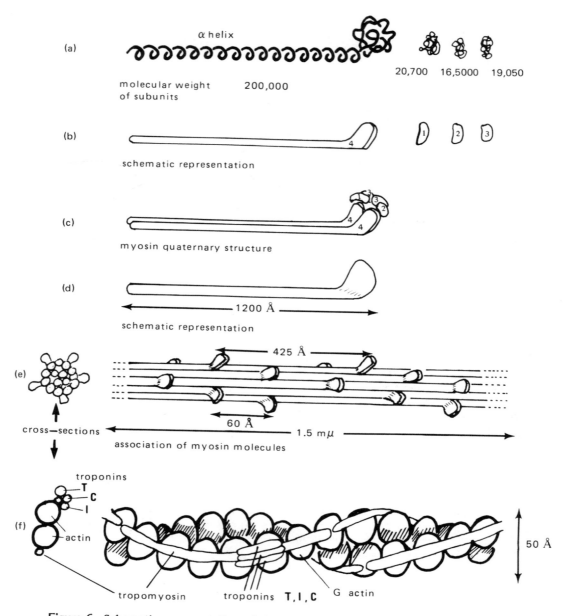

α helix

molecular weight   200,000
of subunits

20,700 16,5000 19,050

(a)

(b)   schematic representation

(c)   myosin quaternary structure

(d)   1200 Å
schematic representation

(e)   425 Å
60 Å
cross—sections   1.5 mμ
association of myosin molecules

(f)   troponins
T
C
I
actin
50 Å
tropomyosin   troponins T,I,C   G actin

Figure 6 Schematic representation of the quaternary structure of myofibrillar proteins. (a-e) Myosin subunits and spatial organization of molecules. (f) Quaternary structure of F-actin.

Table 4 Structural Properties of Some Major Food Proteins

Protein	Molecular weight (daltons)	Type: globular (G) fibrous (F) random coil (RC)	Secondary structure α-Helix (%)	Secondary structure β Sheets (%)	Number of residues	Number of S-S	Number of SH	pI	Number of subunits	Sequence known (K) unknown (U)	Prosthetic group (%, w/w)	Average hydrophobicity[a] (kJ/mol)
Myosin[b]	475,000	F	High		4500	0	40	4-5	6	K Partially	Phosphorus	4.25 (rabbit)
Actin[b]	42,000	G →F				0	5-6	4-5	1-300			4.4 (rabbit)
Collagen (tropocollagen)[b]	300,000	F	Collagen helix					~9	3	K Partially		4.5 (chicken)
αS1-Casein B[b]	23,500	R C	10		199	0	0	5.1	1	K	Phosphorus 1.1	5.0
β-Casein A[b]	24,000	R C			209	0	0	5.3	1	K	Phosphorus 0.56	
K-Casein B[b]	19,000	R C			169	0	2	4.1-4.5	1	K	Carbohydrate 5 Phosphorus 0.22	
β-Lactoglobulin A[b]	18,400	G	10	30	162	2	1	5.2	1	K		5.15
α-Lactalbumin B[b]	14,200	G	26	14	123	4	0	5.1	1	K		4.8
Ovalbumin[c]	45,000	G				1 or 2	4	4.6			Carbohydrate 3.3 Phosphorus	4.65
Serum albumin[b]	69,000	G				17	1	4.8	1	K		4.7
Gliadins[d] (α,β,γ)	30,000–45,000	G →F	30			2-4			1	K Partially		4.5
Glutenins[d]	≥ 1,000,000	F	15	35		50			15			
Glycinin[e]	350,000	G	5	35		23	2	4.6	12	U		
Conglycinin[e]	200,000	G	5	35		2		4.6	9	U	Carbohydrate 4	

[a] According to C. C. Bigelow, *J. Theoret. Biol 16*:187 (1967).
[b] Bovine.
[c] Hen's egg.
[d] Wheat.
[e] Soybean.

261

5. Interactions and Linkages Involved in Protein Structure

Several interactions and linkages are known to contribute to the formation of secondary, tertiary, and quaternary structures, as follows.

Steric Strains. As a consequence of their structure (Fig. 1) peptide bonds have a freedom of rotation which, provided there is no steric hindrance, should give the angles ϕ and ψ the possibility of assuming all values between $180°$ and $-180°$. However, with most amino acid residues certain torsion angles are not possible because of the presence of more or less bulky side chains.

van der Waals Interactions. van der Waals interactions exist between atoms in proteins, but these forces are small, typically 1-9 kJ/mol, as compared with other kinds of interactions (Table 5). Furthermore, the nature of the interaction (attractive or repulsive) is related to the distance between the atoms, and this in turn, in the case of proteins, is related to the torsion angles (ϕ and ψ, Fig. 4) around the α-carbon atoms. At large distances no interaction exists. As the distance is lessened, an attractive force develops, and as the distance is lessened even more, a repulsive force will develop.

The attractive van der Waals forces existing between atoms include dipole-dipole interactions (e.g., the peptide bond and serine are dipoles), dipole induced-dipole interactions, and London dispersion forces, with the latter the most important.

Electrostatic Interactions. Proteins may be considered polyelectrolytes since ionizable groups from amino acid side chains (Asp, Glu, Tyr, Lys, His, Arg, and Cys) and from C and N terminal amino acids participate in the acid-base equilibrium. α-Amino and α-carboxyl groups involved in peptide bonds have but a very minor part in the ionic properties of proteins.

The titration curves of proteins are complex since ionizable groups are numerous (30-50% of the total number of residues). Moreover, ionization of a given group is affected by the proximity of ionized groups, by hydrophobic residues, or by hydrogen bonds. The pK_a values of the ionizable groups from amino acid side chains may vary by more than one unit depending on the environment. In aqueous solution, most of the ionizable groups are located at the protein surface.

During titration of a protein from the anionic (protein⁻) to the cationic (protein⁺) form, both polycharged forms, there is a pH at which the average charge is zero—the "isoelectric point" (pI). At this point the molecule does not migrate when placed in an electrical field. The isoelectric point can be calculated, or rather estimated, from the component amino acids, and it varies with the medium in which the protein is placed (for example, certain ions bind to proteins and modify the pK_a of their ionizable groups).

The "isoionic point" is the pH of the protein solution when no other electrolyte is present.

The ionic properties of proteins aid in their purification by such methods as electrophoresis, ion-exchange chromatography, and electrofocusing, and are important determinants of structure and interactions among proteins in foods.

The ionizable groups are responsible for the attractive or repulsive forces that contribute to stabilize a secondary or tertiary structure. For example, the β and γ-carboxyl groups of aspartic and glutamic acids and the carboxyl group of the C-terminal amino acid often carry negative charges. The ϵ-amino group of lysine, the α-amino group of the N-terminal amino acid, the guanidino group of arginine, and the imidazole group of histidine may carry positive charges. Electrostatic interactions involve energies in the range of 42-84 kJ/mol (Table 5).

Some ion-protein interactions are known to contribute to the stabilization of a protein's quaternary structure. Thus, electrostatic interactions of the type protein$^-$-Ca^{2+}-protein$^-$ contribute to the stability of casein micelles (see Chap. 13). In some cases, the ion-protein complex results in biological activity, such as iron transport or enzymic activity. Usually, the ion binds at a well-defined site and several amino acid residues are involved in the binding. For instance, ions from transition metals (Cr, Mn, Fe, Cu, Zn, Hg, and so on) may bind simultaneously to imidazole and sulfhydryl groups through partially ionic linkages.

Hydrogen Bonding. A hydrogen bond links an electronegative atom, possessing at least one electronic doublet, to an hydrogen atom which is itself covalently bonded to a different electronegative atom. In proteins, hydrogen bonds may appear between the oxygen of the carbonyl group of a peptide bond and the hydrogen of the NH of another peptide bond. The hydrogen-bond distance O. . . H is about 1.75 Å and the energy of these bonds is of the order of 8-40 kJ/mol (Table 5).

$$\begin{array}{c}\diagdown \\ \diagup \end{array} C = O \ldots H - N \begin{array}{c}\diagup \\ \diagdown \end{array}$$

This type of bond plays a fundamental role in the stabilization of secondary structures, such as the α helix and the β-pleated sheets, and also in the stabilization of tertiary structures. Polar groups of amino acids, located at the protein's surface, may form a large number of hydrogen bonds with water molecules, thereby contributing to the specific structure and solubility of some proteins.

Hydrophobic Interactions. The native structure of proteins depends upon the solvent in which they reside. For example, the addition of 1 mol urea per 10 mol water completely disrupts the structure a protein would normally have in an aqueous environment. On the other hand, it is possible to encourage the formation of more ordered protein structures by using protein solvents (e.g., 2-chloroethanol) that are less able than water to form hydrogen bonds.

The hydrophobicity of the side chains of amino acid residues is due to their apolar chemical structure; thus, side chains of this kind do not interact (except for van der Waals interactions) with polar molecules like water. These side chains prefer to avoid contact with water and therefore tend to associate (hydrophobic interaction) in internal hydrophobic regions of the protein. Some characteristics of these interactions are indicated in Table 5.

Disulfide Cross-Links. The establishment of covalent cross-links between cysteine residues limits the number of possible protein structures and contributes to the stabilization of those that form. Thus, protein molecules that possess five to seven disulfide cross-links for 100 amino acids are particularly stable, especially in conditions that lead, for most proteins, to irreversible denaturation (extreme pH or high temperatures).

Some proteins contain both cysteine and cystine residues and are able to undergo "thiol-disulfide interchange" reactions of the type

Table 5 Protein-Protein Linkages and Interactions

Type	Energy (kJ/mol)	Interaction distance (Å)	Functional groups involved	Disrupting solvents	Enhancing conditions
Covalent bonding	330–380	1–2	Cystine S–S	Reducing agents: mercaptoethanol cysteine, dithiothreitol, sulfites	
Hydrogen bonding	8–40	2–3	Amide, $-NH\cdots O=C$; hydroxyl; phenol, $-OH\cdots O=C$	Solutions of urea, guanidine hydrochloride, detergents, heating	Cooling
Hydrophobic interactions	4–12	3–5	Amino acid residues with long aliphatic or aromatic side chains	Detergents, organic solvents	Heating
Electrostatic interactions	42–84	2–3	Carboxyl (COO^-), amino ($NH_3{}^+$), etc.	Salt solutions, high or low pH	
van der Waals	1–9		Permanent, induced, and instantaneous dipoles		

$$\overset{}{\underset{/}{>}}Cys_1-S-S-Cys_2^{/} + Cys_3S^- \rightleftharpoons \overset{}{\underset{/}{>}}Cys_1-S-S-Cys_3^{/} + Cys_2S^- \qquad (9)$$

These reactions may be intra- or intermolecular.

By way of summary, the various interactions and linkages that contribute to the secondary and tertiary structures of polypeptide chains are depicted schematically in Fig. 7.

6. Hydrophobicity of Proteins

The calculation of the average hydrophobicity of a protein from its amino acid composition can enable prediction, for example, of the bitterness of its hydrolyzate, or, for a membrane protein, of its location (external or internal) in phospholipidic membranes.

An estimate of hydrolyzate bitterness can be obtained from the expression

Average hydrophobicity $= \overline{G}^\circ$

$$\overline{G}^\circ = \frac{\Sigma_n^1 \, \Delta G^\circ \text{ side chains}}{n} \qquad (10)$$

where n is the number of amino acid residues. The ΔG° values are listed in Table 3. If $\overline{G}^\circ$ is larger than 5.85 kJ per residue, the hydrolyzate is likely to be bitter, and if $\overline{G}^\circ$ is less than 5.43 kJ per residues, the hydrolyzate is unlikely to be bitter. The term $\overline{G}^\circ$ is 6.44 kJ per residue for a soybean protein isolate, 6.71 for casein, 6.19 for zein, and 5.38 for collagen (83).

The average hydrophobicity of various food proteins as calculated by Bigelow (7) is given in Table 4. No precise correlation exists between the hydrophobicity of a protein and its structure.

Knowing the amino acid sequence in the polypeptide chain, one can establish the distribution patterns of electric charges and side-group hydrophobicities along this chain. Such patterns are shown in Fig. 8 for β-A_2-bovine casein, a single-chain protein with 209 amino acid residues and an average hydrophobicity of 5.56 kJ per residue (91). It can be seen that regions with a high density of electrical charges are located between amino

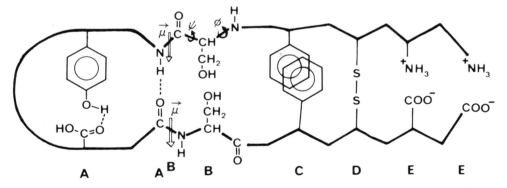

Figure 7 Bonds or interactions that determine secondary and tertiary structure of proteins: (A), hydrogen bond; (B), dipolar interaction; (C), hydrophobic interaction; (D), disulfide linkage, and (E), ionic interaction.

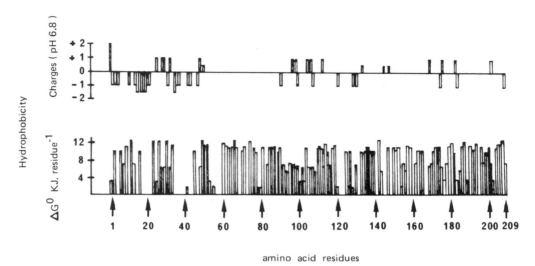

Figure 8 Distribution of hydrophobic residues and electric charges in β-A$_2$-casein at pH 6.8. [According to B. Ribadeau-Dumas, G. Brignon, F. Grosclaude, and J. C. Mercier, *Eur. J. Biochem. 25*:505 (1972).]

acid residues 1 and 50, whereas the distribution of hydrophobic amino acids is quite even. The uniform distribution of the numerous (35) residues of proline (not shown in Fig. 8) prevents the formation of extensive α helices or poly-L-proline structures. These data suggest that βA$_2$-bovine casein exists, at ambient temperatures, as a loose random coil structure stabilized mainly by hydrophobic interactions. These data also help explain the functional properties of caseinates.

7. Volume and Volumic Mass of Proteins

The volume of a protein can be estimated by summing the volumes of constituent atoms (e.g., 18,100 Å^3 for lysozyme). The resulting value is slightly greater than the value estimated by measuring density. This results because the protein has many charges on the surface that produce, by binding water molecules, a phenomenon of electrostriction. The volumic mass of proteins varies between 0.8 and 1.5 g/cm^3 and, for a given protein, is a function of the distance between the outermost regions and its center of mass.

The volume ratio of hydrophilic versus hydrophobic amino acids is often used to characterize the hydrophilicity or hydrophobicity of proteins.

C. Chemical Reactions and Interactions of Amino Acids and Proteins

1. Reactions of Various Functional Groups on Amino Acids and Proteins

The main reactions of analytical or food interest are summarized in Table 6. Additional information concerning the reactions of the amino group for amino acid determination is given as follows.

Reaction with Ninhydrin (2,2-Dihydroxy-1,3-indanedione). Ninhydrin reacts with amino acids, thus allowing their colorimetric (and even fluorimetric) determination. Reaction takes place upon heating and leads to the formation of complexes, most frequently of a blue or violet color. Proline gives yellow derivatives. Numerous reactions are known to occur, for example

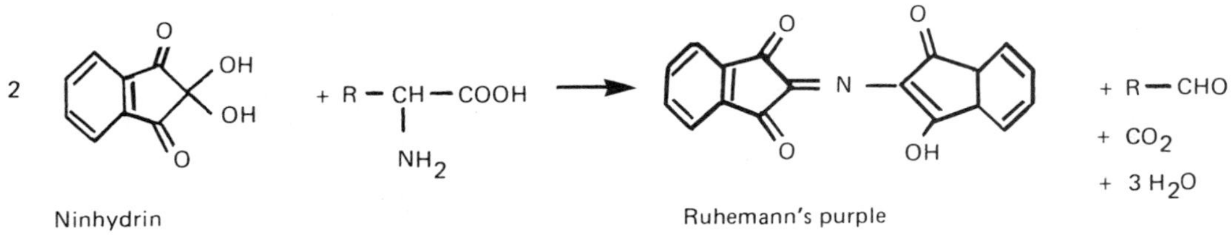

Ninhydrin   Ruhemann's purple

$$(11)$$

The formation of purple, blue, or violet derivatives depends upon the nature of the amino acid. Identification and determination of amino acids is carried out by reacting the chromatographic eluate of a protein hydrolyzate with ninhydrin and measuring the absorbance at two wavelengths (e.g., 440 for proline and hydroxyproline and 570 nm for all other major amino acids).

Reaction with Fluorescamine (4-Phenylspiro[furan-2(3H), 1'-phthalan]3,3'-dione). This compound reacts with primary amines, with the formation of highly fluorescent derivatives. This allows the rapid and sensitive quantitative determination of amino acids, peptides, and proteins (λ_{exc} = 390 nm; λ_{emis} = 475 nm):

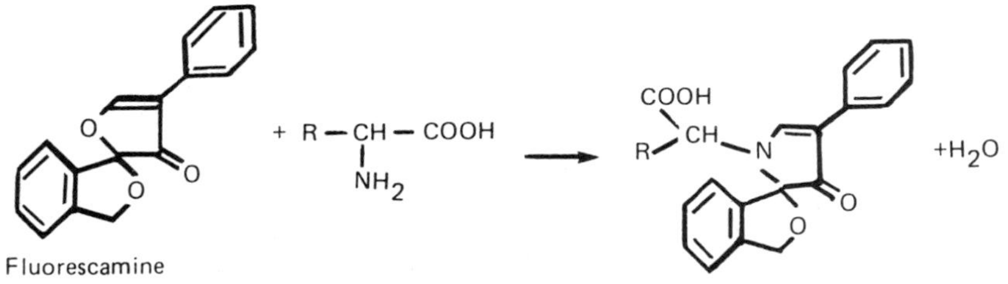

Fluorescamine

$$+H_2O \quad (12)$$

*Reaction with 1,2-Benzene dicarbonal.** 1,2-Benzene dicarbonal reacts with amino acids to give highly fluorescent isoindole derivatives (λ_{exc} = 380 nm; λ_{emis} = 450 nm).

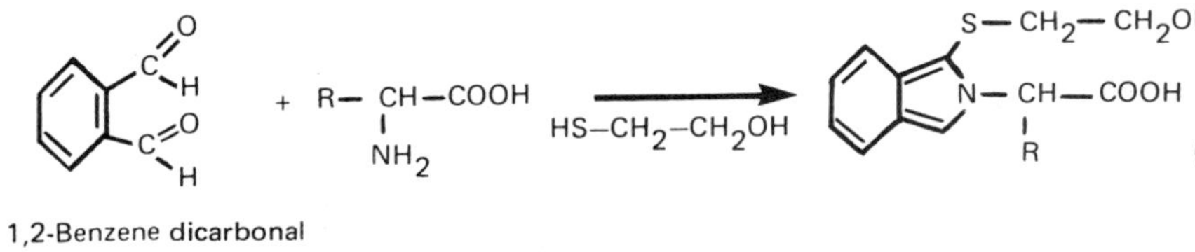

1,2-Benzene dicarbonal

$$(13)$$

---

*Also called o-phtaldialdehyde.

Table 6 Chemical Reactions of Functional Groups on Amino Acids and Proteins

Type of reaction	Details	Comments
A. General reactions of amino acids		
Oxidation with hypochlorite	$R{-}CH{-}COOH + NaOCl + H_2O \longrightarrow R{-}CHO + NH_3 + NaCl + CO_2$ $\quad\;\;\vert$ $\quad NH_2$	
Reaction with phosgene	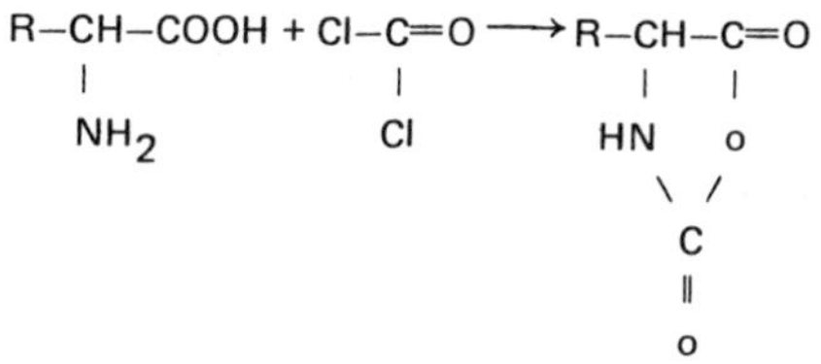 $R{-}CH{-}COOH + Cl{-}C{=}O \longrightarrow R{-}CH{-}C{=}O$ $\quad\;\;\vert \qquad\qquad\vert \qquad\qquad\vert\quad\;\;\vert$ $\quad NH_2 \qquad\quad Cl \qquad\quad HN \quad\;\; O$ $\qquad\qquad\qquad\qquad\qquad\quad\; \backslash\,C\diagup$ $\qquad\qquad\qquad\qquad\qquad\qquad || $ $\qquad\qquad\qquad\qquad\qquad\qquad O$	The N-carboxyanhydrides formed are able to react with the ϵ-NH_2 of lysine residues (see Sec. VII.B)
B. α–COOH groups		
Esterification	$R{-}COOH + R'OH \xrightarrow{\;HCl\;} R{-}COOR' + H_2O$ at boiling temperature	Protection of carboxyl groups during peptide synthesis
Reduction	$R{-}COOH \xrightarrow{\;NaBH_4\;} R{-}CH_2OH$	Identification of the C-terminal amino acid of a protein
Decarboxylation	$R{-}CH{-}COOH \longrightarrow R{-}CH_2{-}NH_2 + CO_2$ $\quad\;\;\vert$ $\quad NH_2$	By enzyme, heat, acid, or alkaline treatment
Amidation	$R{-}COOH \xrightarrow{\;NH_3\;} R{-}CO{-}NH_2 + H_2O$	
C. α-Amino groups		
Acylation	$R{-}NH_2 + R'{-}\underset{\|\|\,O}{C}{-}Cl \longrightarrow R{-}NH{-}CO{-}R' + HCl$	Protection of amino groups during peptide synthesis

Reaction with aldehydes	$R-NH_2 + R'-\underset{\underset{O}{\|}}{C}-H \longrightarrow R-N=CH-R' + H_2O$	Schiff's base formed is labile and represents the first step of the Maillard reaction
Deamination	$R-NH_2 + HNO_2 \longrightarrow R-OH + N_2 + H_2O$	N_2 released permits the determination of amino acids
Reactions with ninhydrine, fluorescamine, 1,2-benzene dicarbonal, phenylisothiocyanate, dansyl chloride		These reactions are used for the separation (HPLC) and the determination of amino acids (for more details see text)

D. Reactions of side chains
Reactions of the thiol group with:

Iodoacetic acid	$R-SH + ICH_2COOH \longrightarrow R-S-CH_2COOH$	Determination of cysteine as S-carboxymethyl derivative; protection against oxidation
Performic acid	$R-SH + HCOOOH \longrightarrow R-SO_3H$	Determination of cysteine as cysteic acid
p-Chloromercuribenzoate	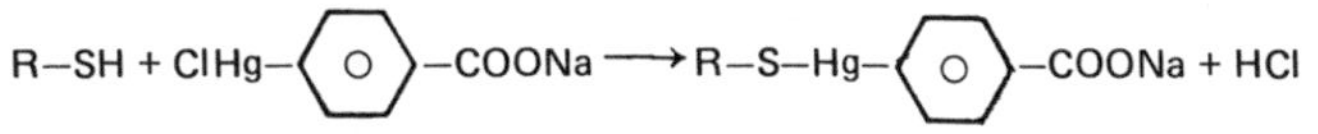 $R-SH + ClHg-$⟨⟩$-COONa \longrightarrow R-S-Hg-$⟨⟩$-COONa + HCl$	Determination of cysteine
5,5'-Dithio-bis-(2-nitrobenzoic acid) (Ellman's reagent)	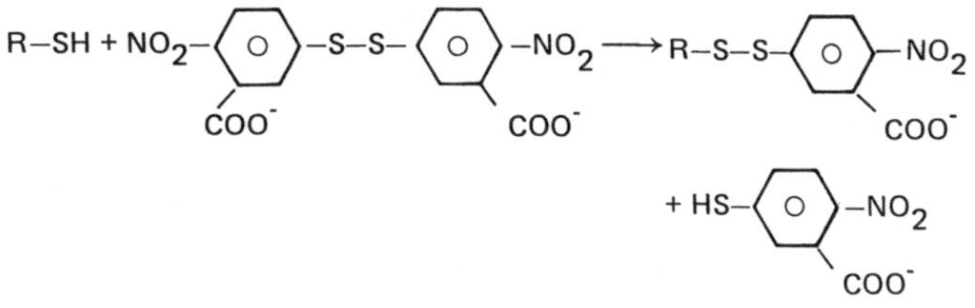 $R-SH + NO_2-$⟨⟩$-S-S-$⟨⟩$-NO_2 \longrightarrow R-S-S-$⟨⟩$-NO_2 + HS-$⟨⟩$-NO_2$	Determination of cysteine

Table 6 (Continued)

Type of reaction	Details	Comments
Oxidation	$2R\text{–}SH \rightleftharpoons R\text{–}S\text{–}S\text{–}R$	Oxidation of cysteine gives cystine (as a disulfide cross-link); reaction is reversible upon reduction with β-mercaptoethanol or dithiothreitol
Reactions of the amino group of lysine with		
1-Fluoro-2,4-dinitro-benzene	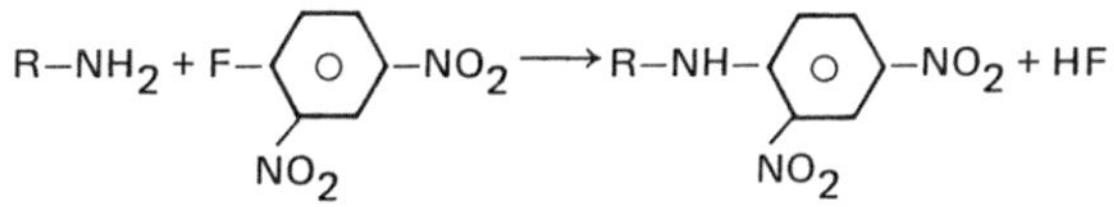 $R\text{–}NH_2 + F\text{–}\langle\text{C}_6\text{H}_3\rangle(NO_2)\text{–}NO_2 \longrightarrow R\text{–}NH\text{–}\langle\text{C}_6\text{H}_3\rangle(NO_2)\text{–}NO_2 + HF$	Determination of the ϵ-dinitrophenyl derivative of lysine may be correlated with the biological availability of this amino acid
Orange G, 2,4,6-trinitrobenzene sulfonic acid, or o-methyl isourea		Determination of ϵ-NH$_2$ (correlation with the biological availability
Reactions of the thioether group with		
Iodoacetic acid	$R\text{–}S\text{–}CH_3 + ICH_2COOH \longrightarrow R\overset{+}{\underset{CH_2COOH}{\text{–}S\text{–}}}CH_3$	Sulfonium derivative formed resists oxidation of the sulfur atom
Performic acid	$R\text{–}S\text{–}CH_3 + HCOOOH \longrightarrow R\text{–}SO_2\text{–}CH_3$	Determination of methionine as methionine sulfone

Reaction with Phenylisothiocyanate (Edman's Reaction)

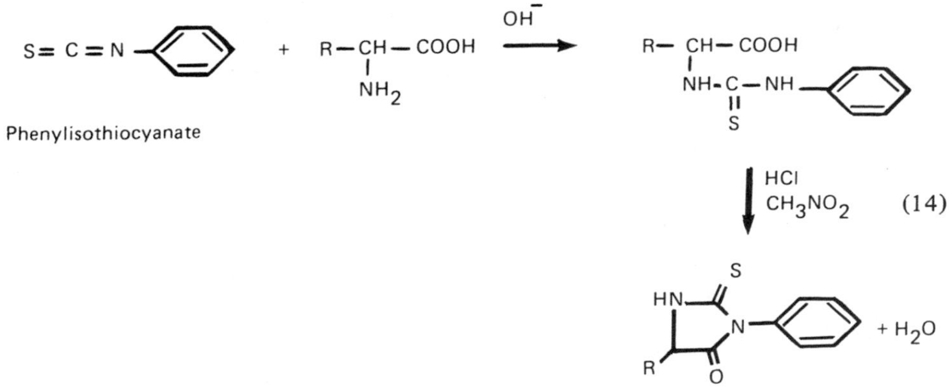

Phenylisothiocyanate

(14)

Reaction with Dansyl Chloride (1-Dimethylaminonaphthalene-5-sulfonyl Chloride).

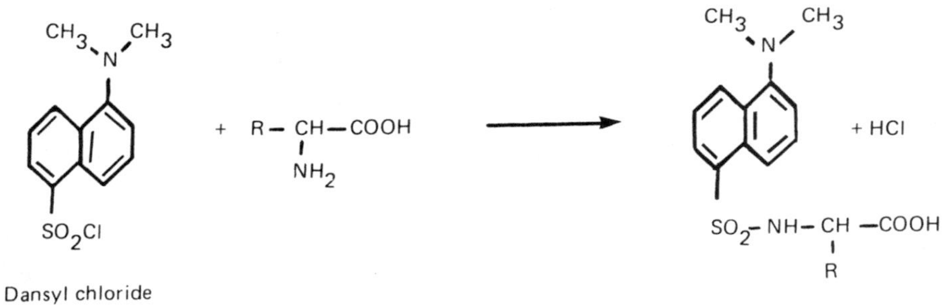

Dansyl chloride

(15)

This reaction and the previous one enable one to identify the N-terminal amino acid of a peptide or a protein. Dansyl derivatives of amino acids are easily separated by high-performance liquid chromatography on apolar columns.

2. Protein-Water Interactions

General Nature of the Interaction. Schematically, water associated with proteins can assume different states (Chap. 2). Constitutional and vicinal water, located inside the protein or strongly adsorbed to specific surface sites, represents up to 0.3 g/g dry protein. Water that occupies the remaining first-layer sites on the protein surface and another one to two layers adjacent to the protein can represent an additional 0.3 g/g dry protein.

Proteins interact with water through their peptide bonds (dipole-dipole or hydrogen bonds) or through their amino acid side chains (interactions with ionized, polar, and even nonpolar groups). Some of these interactions are represented schematically in Fig. 9 and are discussed more fully in Chap. 2.

Water Solubility of Proteins. The water solubility of proteins is a function of numerous parameters. From a thermodynamic standpoint, solubilization corresponds to separating the molecules of solvent, separating the molecules of proteins, and dispersing

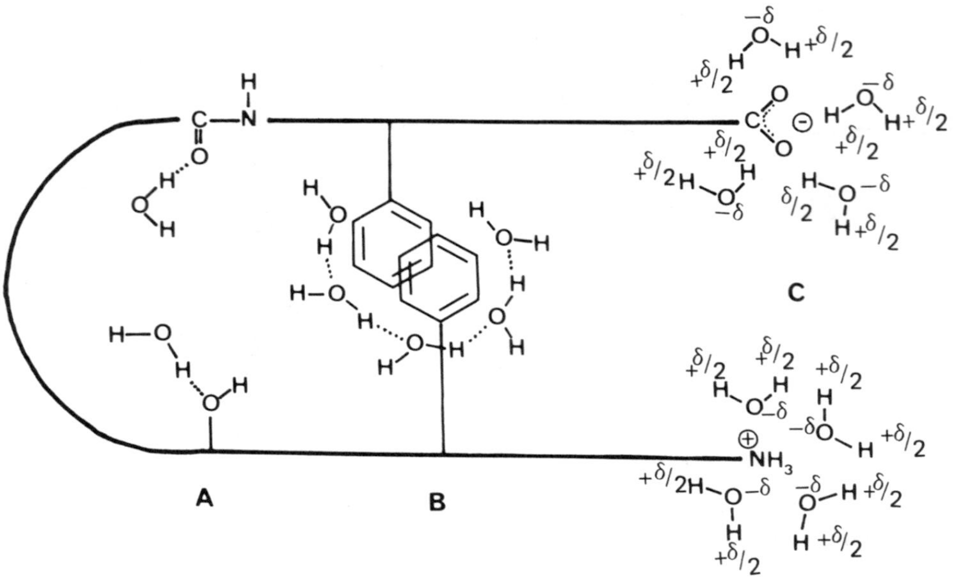

Figure 9 Schematic representation of some interactions of water with proteins. (A), hydrogen bonds; (B), hydrophobic interactions; (C), ionic interactions.

the latter in the solvent with maximum interaction between the protein and solvent. To be soluble a protein should therefore be able to interact as much as possible with the solvent (hydrogen-bond, dipole-dipole, and ionic interactions). The solubility depends mainly on pH, ionic strength, the type of solvent, and temperature.

a. Influence of pH. At pH values higher or lower than the isoelectric point, the protein carries a negative or positive electric charge and water molecules may interact with these charges, thus contributing to solubilization. Moreover, protein chains carrying electrical charges of the same sign have a tendency to repel each other and to dissociate or unfold. If the solubility of a given protein is plotted as a function of the pH, one usually obtains a V- or U-shaped curve, the minimum of which corresponds closely to the pI (see Fig. 15). This behavior is put to use for dissolving a number of proteins, especially seed proteins (soybean, sunflower, and so on). Solubility, and therefore the yield of extraction, is greater at alkaline than at acid pH. Indeed, the number of negatively charged residues at pH > pI (aspartic and glutamic acid) is larger than the number of positively charged residues at pH < pI (e.g., lysine). The solubility and extractability at alkaline (or neutral) pH can be enhanced by increasing the net electrical charge of the proteins. This can be done, for example, by succinylation or maleylation of lysine residues (see Sec. VII.B), which then become carriers of ionizable carboxyl groups. It is also possible to react proteins with amphipolar molecules that have hydrophobic as well as ionized zones (e.g., sodium dodecylsulfate). In this case, the hydrophobic residues of amino acids become, through these compounds, carriers of negative charges.

For pH values not far from the pI, protein molecules show minimal interactions with water and their net charges are sufficiently small to allow polypeptide chains to approach each other. Sometimes aggregates are formed, and this may lead to protein precipitation. The rate of precipitation is enhanced when the bulk densities of the aggregates differ greatly from that of the solvent, and when the diameters of the newly formed aggregates are large.

    b. Influence of the ionic strength μ.

$$\mu = \frac{1}{2} \sum c_i z_i^2$$

where C is concentration and Z is valence. The ions of neutral salts, at molarities of the order of 0.5-1 M, may increase the solubility of proteins. This effect is called "salting in." The ions react with the charges of proteins and decrease the electrostatic attraction between opposite charges of neighboring molecules. Moreover, the solvation connected with these ions serves to increase the solvation of the proteins and thereby increase their solubility. In this range of molarity the logarithm of the solubility is a function of $\sqrt{\mu}$. Ions differ in their ability to effect salting in, and this is discussed at the end of this section.

    If the concentration of neutral salts is greater than 1 M, the protein displays a decreased solubility, which may result in precipitation. This "salting out" effect results from the competition between the protein and the salt ions for the water molecules necessary for their respective solvations. At high salt concentrations, there are not enough water molecules available for protein solvation, since the majority of the water molecules are strongly bound to the salts. Thus, protein-protein interactions become more powerful than protein-water interactions and this may lead to aggregation followed by precipitation of the protein molecules. During salting out, the solubility of the effected protein is given by

$$\log S = -k'\mu + \log S_0 \tag{16}$$

where S_0 is the solubility at zero ionic strength, μ is ionic strength, and k' is the salting out constant, which depends on the protein and still more on the salt. The salting out action of different salts increases with their hydration energy and with their steric hindrance.

    According to the Hofmeister series, ions may be ordered as follows: $SO_4^{2-} < F^- < CH_3COO^- < Cl^- < Br^- < NO_3^- < I^- < ClO_4^- < SCN^-, NH_4^+ < K^+ < Na^+ < Li^+ < Mg^+ < Ca^{2+}$. Ions at the left promote salting out, aggregation, and stabilization of the native conformation. Ions to the right promote unfolding, dissociation, and salting in.

    c. Influence of nonaqueous solvents. Certain solvents, such as ethanol or acetone, lower the dielectric constant of the aqueous medium in which a protein is dissolved. As a consequence, the electrostatic forces of repulsion among protein molecules decrease, which contributes to their aggregation and precipitation. These solvents also compete for water molecules, thus further reducing the solubility of proteins.

    d. Influence of temperature (at constant pH and ionic strength). As a rule, the solubility of proteins increases with temperature between 0 and 40-50°C. Above 40-50°C, molecular motion becomes sufficient to disrupt bonds involved in the stabilization of secondary and tertiary structures. This denaturation is often followed by aggregation (see sec. III), in which case the solubility of the protein becomes less than that of the native protein. However, the ability of the aggregated protein to bind water is modified only slightly, sometimes even increased (mainly by water absorption in the capillaries of the resulting coagulum or gel).

    e. Classification of proteins based on solubility. Osborne, in 1907, classified proteins according to their solubility. Accordingly, *albumins* (e.g., serum albumin,

ovalbumin, and α-lactalbumin) are soluble in water at pH 6.6, *globulins* (e.g., β-lactoglobulin) are soluble in dilute salt solutions at pH 7, *prolamins* (zein and gliadins) are soluble
in 70% ethanol, and *glutelins* (e.g., wheat glutenins) are insoluble in the above-mentioned
solvents but are soluble in acids (pH 2) or alkalis (pH 12).

3. Protein-Lipid Interactions

Interactions between proteins and lipids, notably phospholipids, take place in many food
and biological systems, especially at the level of cellular and intracellular membranes.
The relative proportions of proteins and lipids may vary greatly, and many types of
arrangements are possible. In membranes, proteins may be located at the surface or more
or less integrated within the membrane structure.

The protein-lipid interactions that exist in biological systems (plasmatic lipoproteins, and membranes) or food systems (e.g., emulsions) do not involve covalent bonds,
but rather they involve hydrophobic interactions between apolar aliphatic chains of the
lipid and the apolar regions of the protein. In model systems, the energy of the protein-
lipid interaction reaches a maximum in the neighborhood of the pI of the protein, which
is confirmation of the major role played by hydrophobic interactions in this phenomenon. Triacylglycerols and methyl or ethyl esters of fatty acids react like free fatty acids,
which tends to show that the carboxyl or carboxylate group plays only a minor role in
the interaction. Moreover, the interaction energy is both a function of the lipid, particularly of the length of its aliphatic chain(s), and of the protein. Proteins that have a high
average hydrophobicity per residue (see Sec. II.B.6) usually interact strongly with lipids.
This interaction may modify the protein structure.

High-pressure homogenization increases the degree of protein-lipid interaction by
increasing interfacial area. Also, the propensity to foam decreases when the protein-lipid
interaction is strong. Finally, association of a protein with lipids protects the protein
against thermal denaturation, mainly because of the presence of groups with a high heat
capacity and the relative absence of water.

III. DENATURATION OF PROTEINS (12,65,70,89)

The conformation a protein derives from its secondary and tertiary structures is fragile.
As a result, treatment of proteins with acids, alkalis, concentrated saline solutions, solvents, heat, and radiations, for example, may modify this conformation to varying extents. Protein *denaturation* is any modification in conformation (secondary, tertiary, or
quaternary) not accompanied by the rupture of peptide bonds involved in primary structure. Denaturation is an elaborate phenomenon during which new conformations appear,
although often intermediary and short-lived. The ultimate step in denaturation might
correspond to a totally unfolded polypeptide structure, in which intraprotein and sol-
vent-protein interactions are transient (random coil). However, increases in structure
over those in the native structure must also be regarded as a form of denaturation. Certain proteins are already unfolded in their native state (casein monomers), which explains their stability toward certain denaturing agents, including heat.

The effects of protein denaturation are numerous and the following deserve special
attention.

1. Decreased solubility, resulting from the unmasking of hydrophobic groups
2. Altered water-binding capacity
3. Loss of biological activity (e.g., enzymic or immunological)

4. Increased susceptibility to attack by proteases, due to the unmasking of peptide bonds specifically vulnerable to these proteases
5. Increased intrinsic viscosity
6. Inability to crystallize

Protein denaturation can be monitored by measuring sedimentation behavior in an ultracentrifuge, viscosity, protein migration in an electrical field (by electrophoresis), optical rotatory dispersion, circular dichroism, thermodynamic properties (by differential scanning calorimetry), biological or immunological properties, and/or the reactivity of certain functional groups.

Denaturation may be reversible or irreversible. When disulfide cross-links contribute to the conformation of the protein, and if these are broken, denaturation is often irreversible.

The sensitivity of a protein to denaturation is related to the readiness with which a denaturing agent breaks the interactions or linkages that stabilize the protein's secondary, tertiary, or quaternary structures. Since these structures vary from one protein to another, the effects of denaturing agents will vary depending on the protein. An interesting case is the extraordinary thermal resistance of enzymes present in some thermophilic microorganisms.

The causative agents of denaturation are discussed next.

A. Physical Agents

1. Heat

Heat is the most common physical agent capable of denaturing proteins. The extent of protein unfolding that accompanies heat denaturation can be considerable. For example, a molecule of native serum albumin is elliptical in shape, with a ratio of length to width of 3.1. After thermal denaturation this ratio becomes 5.5.

The rate of denaturation depends on the temperature. It should be noted that, for many reactions, the rate increases about two fold with a 10°C increase in temperature. However, in the case of protein denaturation, the rate increases about 600 times when the temperature rises by 10°C in the range typical of denaturation. This results from the low energy involved in each of the interactions stabilizing the secondary, tertiary, and quaternary structures.

The susceptibility of proteins to denaturation by heat depends on many factors, such as the nature of the protein, protein concentration, water activity, pH, ionic strength, and the kind of ions present. This denaturation is very often followed by a decrease in solubility, due to the exposure of hydrophobic groups and to the aggregation of the unfolded protein molecules, and by an increase in the water absorption capacity of the protein. Many proteins, both native and denatured, have tendencies to migrate to interfaces, with the hydrophilic groups remaining in the aqueous phase and the hydrophobic groups situating themselves in the apolar nonaqueous phase (Fig. 10). Since native proteins become denatured in this process, it is important if one desires to maintain the native conformation of proteins, to avoid creating interfacial structures, such as those present in foams or emulsions.

Although extensive dehydration of proteins, even by mild methods such as freeze drying, may cause some protein denaturation, it is well known that proteins (or enzymes or microorganisms) are more resistant to heat denaturation when dry than when wet. Thus, the presence of water facilitates denaturation.

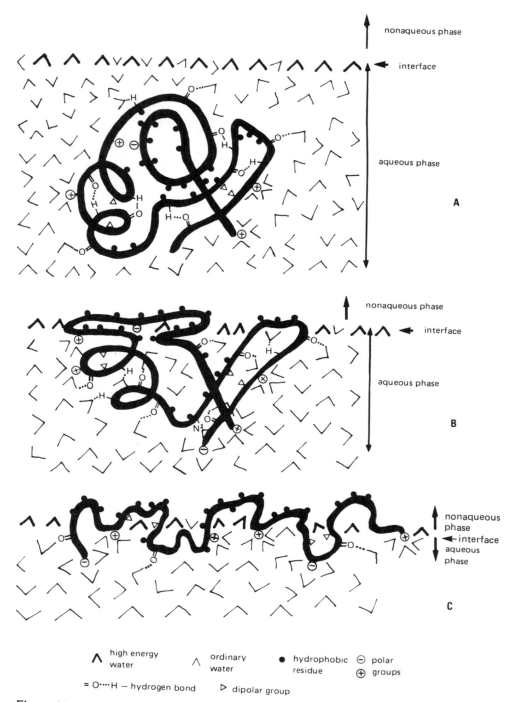

Figure 10 Schematic conformation of a protein at an interface: (A), globular native protein immersed in an aqueous solution; (B), globular protein close to the interface; (C), adsorbed, unfolded, and hydrated protein molecule.

Other consequences of thermal denaturation should also be noted. For example, the splitting of the disulfide cross-links sometimes results in the release of hydrogen sulfide. Also, heat can cause the chemical alteration of amino acid residues (dehydration of serine or deamidation of glutamine and asparagine) and new intra- or intermolecular covalent cross-links may form (e.g., γ-glutamyl-ε-N-lysine) (see Sec. IV.A). These changes can alter the nutritive and functional properties of proteins.

2. Cold (33)

Low temperature can result in the denaturation of some proteins. Some enzymes stable at room temperature may become less stable at 0°C (L-threonine deaminase), and certain proteins aggregate and precipitate when brought to cold or freezing temperature (11S soy proteins, gliadins, and some egg and milk proteins). In contrast, low temperatures can result in the dissociation of oligomers and in the rearrangement of the subunits of oligomers. Also, some lipases and oxidases are not only resistant to freezing temperatures, but they remain active at such temperatures. In the case of cellular systems, some oxidases are activated by freezing, due to their release from membrane structures.

The resistance of certain plants and marine animals to freezing temperatures should also be mentioned. Since proteins with large hydrophobic-polar amino acid ratios and with structures dependent on hydrophobic interactions are subject to denaturation at low temperatures, these types of proteins are usually not present in cold-tolerant plants and animals.

3. Mechanical Treatments

Some mechanical treatments, such as the kneading or rolling applied to bread or other doughs, may denature proteins as a result of shearing forces. The repeated stretchings modify the protein network principally by disruption of α helices.

4. Hydrostatic Pressure

Hydrostatic pressure can have a denaturing effect, but this is not evident below 50 kPa. Ovalbumin and trypsin are denatured at pressures of 50 and 60 kPa, respectively.

5. Irradiation

The effects of electromagnetic radiation on proteins vary depending on the wavelength and energy involved. Ultraviolet radiation is absorbed by aromatic amino acid residues (Trp, Tyr, and Phe). This can result in modification of conformation and, if the energy level is sufficiently great, in rupture of disulfide cross-links. γ Radiation and other ionizing radiations can also produce changes in conformation, together with oxidation of amino acid residues, rupture of covalent bonds, ionization, formation of protein free radicals, and recombination and polymerization reactions. Many of these reactions are mediated by the radiolysis of water.

6. Interfaces

Protein molecules that adsorb at interfaces of water and air, or water and a nonaqueous liquid or solid phase, usually become irreversibly denatured. The rate of adsorption, which stops when the interface is saturated (about 2 mg/m^2) with denatured proteins

(P_D), is controlled by the rate at which the native protein (P_N) diffuses toward the interface. Figure 10 represents the transition of a globular protein in aqueous solution from a native form (Fig. 10 A) to a denatured form adsorbed at the interface of water and a nonaqueous phase (Fig. 10 C).

To interpret this phenomenon, some investigators have suggested that water molecules exist at different energy levels: (a) those far from the interface which are in a low energy state and which interact with the ionic and polar sites of the protein as well as with other water molecules, and (b) those present near the interface, which are in a high energy state and react mainly with other water molecules.

Protein denaturation begins with the diffusion of the macromolecule toward the interface. At this stage the protein would interact with high-energy, interfacial water molecules, many protein-protein hydrogen bonds would be simultaneously disrupted, and "microunfoldings" of the structure would take place (Fig. 10 B). At this next stage, the partially unfolded protein is hydrated and activated (P^*), and also unstable, since many hydrophobic groups are exposed to the aqueous phase. It is by further unfolding and spreading of the protein at the interface that hydrophilic and hydrophobic residues can attempt to orient themselves, respectively, in the aqueous and nonaqueous phases. The protein thus adsorbed at the interface is denatured.

Some proteins, which do not possess regions that are mainly hydrophobic or mainly hydrophilic, or have a structure stabilized by disulfide cross-links, do not readily adsorb at the interface.

When adsorption occurs, the two steps that correspond to adsorption and to denaturation at the interface may be written as

$$P_N m(H_2O) + n(H_2O)^* \underset{k_{-1}}{\overset{k_1}{\rightleftharpoons}} P^* m + n(H_2O) \xrightarrow{k_2} P_D \qquad (17)$$

where P_N is native protein, P^* is hydrated activated protein, P_D is denatured protein, (H_2O) is ordinary water and (H_2O)* is high-energy water.

The properties of proteins at interfaces are important in various food systems; for instance, the formation and stabilization of emulsions and foams may be assisted by proteins that adsorb at interfaces (see Secs. IV.G, H).

B. Chemical Agents

1. Acids and Alkalis

The pH of the medium in which the protein is placed has a considerable influence on the denaturation process. Most proteins are stable over characteristic pH ranges, and if exposed to extremely high or low pH they usually denature. At extreme pH values, strong electrostatic repulsions of ionized groups inside the molecule occur, and this encourages unfolding (denaturation) of the molecule. However, in some cases, the protein may recover its native structure when the pH is brought back to the initial stability range.

2. Metals

Alkaline metals (such as sodium and potassium ions) react only to a limited extent with proteins, whereas alkaline earths, such as calcium and magnesium, are somewhat more reactive. Transition metals, for example, ions of Cu, Fe, Hg, and Ag, react readily with proteins, many forming stable complexes with thiol groups. Ca^{2+} ions (also Fe^{2+}, Cu^{2+}, and Mg^{2+}) may be an integral part of certain protein molecules or molecular associations. Their removal by dialysis or by sequestrants appreciably lowers the stability of the protein structure toward heat and proteases.

3. Organic Solvents

Most organic solvents may be regarded as denaturating agents. They alter the dielectric constant of the medium and therefore the electrostatic forces that contribute to the stability of proteins. Apolar organic solvents are able to penetrate into hydrophobic regions, disrupting hydrophobic interactions and promoting denaturation of the protein. Their denaturing action may also result from their interaction with water.

Certain solvents, such as chloro-2-ethanol, increase the prevalence of α helices, and this is regarded as a form of denaturation (alteration of secondary, tertiary, or quaternary structure) since ovalbumin has 31% α helices in an aqueous medium and 85% in chloro-2-ethanol.

4. Aqueous Solutions of Organic Compounds

Some organic compounds, such as urea and guanidine salts, when in concentrated aqueous solution (4-8 M), contribute to disruption of hydrogen bonds and cause varying degrees of protein denaturation. These compounds also decrease hydrophobic interactions by increasing the solubility of hydrophobic amino acid residues in the aqueous phase.

Surface-active agents, such as sodium dodecylsulfate, are very powerful denaturing agents. These compounds act as intermediates between the hydrophobic regions of proteins and the hydrophilic environment, thus disrupting hydrophobic interactions and contributing to the unfolding of native proteins. Moreover, anionic detergents, because of the pK_a of their ionizable groups, give proteins a large net negative charge at pH values close to neutrality, increased internal repulsive forces, and an increased tendency to unfold.

Reducing agents (cysteine, ascorbic acid, β-mercaptoethanol, and dithiothreitol) reduce disulfide cross-links and therefore modify the conformation of proteins.

C. Energetics of Denaturation

Protein denaturation is often irreversible, especially when a strong denaturating treatment is administered and the molar mass of the protein is large (see also Chap. 6). In some cases, however, "renaturation" of the protein takes place, more or less rapidly.

The thermodynamics of the denaturation process are complicated. It is nevertheless possible to consider it with a simplified formulation:

Native protein (P_N) $\rightleftharpoons$ denatured protein (P_D)

The equilibrium constant can be written

$$K_D = \frac{[P_D]}{[P_N]} \tag{18}$$

The thermodynamic parameters $\Delta G°$, $\Delta H°$, $\Delta S°$, and $\Delta C°_p$ are obtained from the usual equations.

$$\Delta G° = - RT \ln K_D \tag{19}$$

$$\Delta H° = - R \left[\frac{\delta(\ln K_D)}{\delta(1/T)} \right]_p \tag{20}$$

Also, from van't Hoff's law,

$$\frac{\delta(\ln K_D)}{\delta T} = \frac{\Delta H°}{RT^2}$$

$$\Delta C°_p = \left(\frac{\delta \Delta H°}{\delta T} \right)_p$$

$$= T \left(\frac{\delta \Delta S°}{\delta T} \right)_p \tag{21}$$

and

$$\Delta S° = \frac{\Delta H° - \Delta G°}{T} \tag{22}$$

where $\Delta G°$ is the change in standard free energy (difference between the free energy of the native system containing all reactants at equimolecular concentration and the free energy of the same system at equilibrium after denaturation), R is the gas constant, T is absolute temperature, $\Delta H°$ is the change in enthalpy at constant pressure, $\Delta S°$ is the change in entropy, and $\Delta C°_p$ is the change in heat capacity at constant pressure. Large values of $\Delta C°_p$ indicate that $\Delta H°$ and $\Delta S°$ are highly dependent on temperature. The values of these parameters for several proteins are presented in Table 7. The values of $\Delta H°$ and $\Delta S°$ for ribonuclease, chymotrypsinogen, and myoglobin are positive, and the values of $\Delta G°$ are relatively small. For ribonuclease, the increase in enthalpy upon unfolding indicates that the native form possesses a lower free energy than the denatured form. The increase in entropy corresponds to the disorder accompanying unfolding, and the increase of $\Delta C°_p$ is related to the transfer, during denaturation, of aliphatic or aromatic apolar groups to an aqueous environment, with an associated disruption of water structure.

To be more accurate, denaturation should be considered a stepwise phenomenon, with many more or less unfolded intermediates between the native (P_N) and fully denatured (P_D) state. These intermediate structural "microstates" correspond to various stages in the alteration of the protein conformation and, in an aqueous medium, to various distributions of water molecules in the protein. Thus, the denaturation reaction should be more properly written as

$$P_N \rightleftharpoons X_1 \rightleftharpoons X_2 \cdots \rightleftharpoons P_D \tag{23}$$

where the X_i states represent molecules with structures intermediate between those of the native and fully denatured states. It is then possible to calculate an apparent equilibrium constant for the system.

Table 7 Thermodynamic Parameters of Protein Denaturation

Protein and conditions of denaturation	Temperature of maximum stability ($^\circ$C)	ΔG° (kJ/mol)	ΔH° (kJ/mol)	ΔS° (J/degree per mol)	ΔCp° (kJ/degree per mol)
Ribonuclease unfolding (30°C)					
pH 1.13		−4.6	251	836	8.7
pH 2.5	−9	3.8	238	773	8.3
pH 3.15		12.9	222	690	8.3
Chymotrypsinogen (25°C), pH 3	10	30.5	163	439	10.9
Myoglobin (25°C) pH 9	<0	56.8	176	397	5.9
β-Lactoglobulin (25°C), pH 3, 5 M urea	35	2.5	−88	−301	9

Source: From C. Tanford, *Adv. Protein Chem.* 24:1-95 (1970).

The denaturation rate may be described by starting from a kinetic equation of the first order. For the simple model

$$P_N \underset{k_{-1}}{\overset{k_1}{\rightleftarrows}} P_D \tag{24}$$

and the rate equation is then

$$-\frac{d[P_N]}{dt} = k_1 [P_N] - k_{-1} [P_D] \tag{25}$$

The activation energy may be calculated from the Arrhenius law,

$$\frac{\delta(\ln k)}{\delta T} = \frac{E_a}{RT^2} \tag{26}$$

where k is the rate constant, T is absolute temperature, R is the gas constant, and E_a is the activation energy.

The E_a values related to denaturation of proteins are large when compared with those of other chemical reactions. For example, activation energies for the thermal denaturation of trypsin, ovalbumin, and peroxidase are, respectively, 167, 552, and 773 kJ/mol. Although covalent bonds (except disulfide cross-links) are not broken during denaturation, numerous low-energy noncovalent interactions and linkages are broken. Since the bond energies involved are similar and small and since the reaction takes place over a small temperature range or with a small change in concentration of the denaturant, the transition from the native to the denatured state appears to be a simple reaction.

In reality it is an extremely complex cooperative process (given bonds can be broken only after others have been ruptured first).

IV. FUNCTIONAL PROPERTIES OF PROTEINS

The term "functionality" as applied to food ingredients, is defined as any property, aside from nutritional attributes, that influences an ingredient's usefulness in food. Most functional properties affect the sensory character of food (especially their textural attributes) but also can play a major role in the physical behavior of foods or food ingredients during their preparation, processing, or storage (1,36,44,59,63,82).

The functional properties of proteins are those physicochemical properties that enable proteins to contribute to the desirable characteristics of food. Several functional properties of a protein are usually evident in any single food (Table 8). Although the ultimate aim of many food scientists is to provide a mechanistic explanation of functional behavior, the present state of knowledge, together with the complexity of the various food systems, does not permit a clear understanding of how a given protein structure will, for example, determine a final food texture. One of the difficulties is that the initial protein structure is usually modified as the protein ingredient or the food protein is processed into a complex final food.

Functional properties of food proteins can be classified into three main groups: (a) hydration properties (dependent on protein-water interactions), (b) properties related to protein-protein interactions, and (c) surface properties. The first group encompasses such properties as water absorption and retention, wettability, swelling, adhesion, dispersibility, solubility, and viscosity (the latter often referred to as a hydrodynamic property). The second group of properties is operative during such occurrences as precipitation, gelation, and the formation of various other structures (e.g., protein doughs and fibers). The third group of properties relates primarily to surface tension, emulsification, and foaming characteristics of the protein. These groups are not totally independent; for example, gelation involves not only protein-protein interactions but also protein-water interactions; and viscosity and solubility both depend on protein-water and protein-protein interactions.

Since the prediction of the functional properties of a protein from its structural characteristics is often unsuccessful, experimental evaluation is necessary. Evaluation tests range from accurate measurements of well-defined physicochemical properties (viscosity, surface tension, and solubility) to simple end-use (utility) tests, such as measurement of loaf volume following baking of bread, or measurement of water loss during cooking of hamburgers. Between lies a whole series of tests that can be applied to model systems, for example, measurement of emulsion capacity and/or stability of a simple oil-water-protein system, or water absorption of a protein powder under controlled conditions.

The best understanding of protein functionality can be gained when the protein constituent, in a model system being tested, is a single purified protein of known native structure. However, most protein ingredients available for industrial use are mixtures of proteins and contain appreciable amounts of carbohydrates, lipids, mineral salts, polyphenols, and so on. Although protein isolates contain fewer nonprotein constituents than do most other proteins, they generally have been subjected to extensive processing and this may affect both their initial structure and their functionality.

Since utility tests are costly and time consuming, much emphasis has recently been put on simple tests of model systems. Problems attached to this approach are

Table 8 Functional Properties of Proteins Required in Various Foods

Food	Functionality
Beverages	Solubility at different pH, heat stability, viscosity
Soups, sauces	Viscosity, emulsification, water retention
Dough formation, baked products (e.g., bread, cakes)	Formation of a matrix and film with viscoelastic properties, cohesion, heat denaturation, gelation, water absorption, emulsification, foaming, browning
Dairy products (e.g., processed cheese, ice cream, desserts)	Emulsification, fat retention, viscosity, foaming, gelation, coagulation
Egg substitutes	Foaming, gelation
Meat products (e.g., sausage)	Emulsification, gelation, cohesion, water and fat absorption and retention
Meat extenders (e.g., texturized vegetable protein)	Water and fat absorption and retention, insolubility, hardness, chewiness, cohesion, heat denaturation
Food coatings	Cohesion, adhesion
Confectionary products (e.g., milk chocolate)	Dispersibility, emulsification

twofold: (a) these tests currently are lacking in standardization, and (b) results from tests on model systems often correlate poorly with those from real systems (utility tests).

On the following pages, the main functional properties of food proteins are discussed. The changes these functional properties undergo during physical processing and/or chemical modifications of proteins are reported later.

A. Hydration Properties

1. General Considerations

It has already been indicated that the conformation of individual proteins in solution is largely dependent on interactions with water. Most foods are hydrated solid systems, and the physicochemical, rheological behavior of proteins and other constituents of the food is strongly influenced not only by the presence of water but also by the activity of the water. Furthermore, dry protein concentrates or isolates must be hydrated when used. For these reasons, the hydration or rehydration properties of food proteins are of great practical interest. When one considers the progressive hydration of proteins starting from the dry state, the sequential steps shown in Fig. 11 can be postulated.

Many functional properties of a protein preparation are related to this progressive hydration. *Water absorption* (also called water uptake, affinity, or binding), *swelling, wettability, water-holding capacity* (or water retention), and also *cohesion* and *adhesion* are related to the first four steps, whereas *dispersibility* and *viscosity* (or thickening power) involve also the fifth step. The final state of the protein, either soluble or insoluble (partially or totally), is also related to important functional properties, such as *solubility* or *instant solubility* (in which the first five steps take place rapidly). *Gela-*

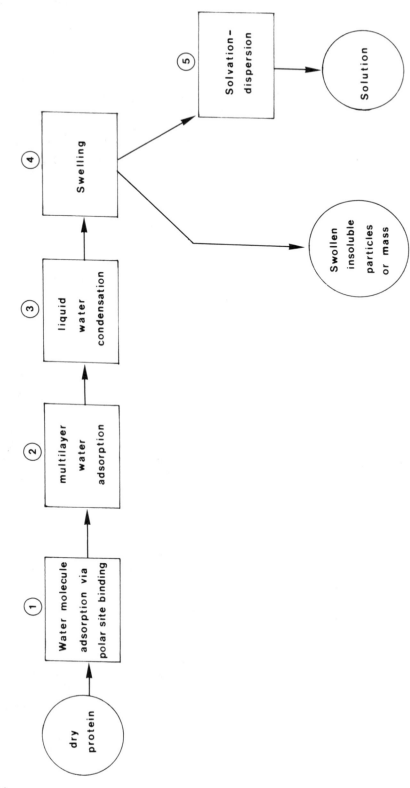

Figure 11 Sequence of protein-water interactions for dry protein. [After D. H. Chou and C. V. Morr, *J. Amer. Oil Chem. Soc. 56*:53A (1979).]

tion implies the formation of a well-hydrated insoluble mass, but specific protein-protein interactions are also required. Finally, *surface properties, such as emulsification and foaming*, necessitate a high degree of protein hydration and dispersion in addition to other characteristics.

Details concerning the nature of water-protein interactions are discussed in Chap. 2.

2. Methods for the Practical Determination of Hydration Properties

Four groups of methods are commonly used for the practical determination of the water absorption and the water-holding capacity of protein ingredients (54). (a) The *relative humidity method* (or equilibrium moisture content method) measures the amount of water adsorbed at a given a_w (or vice versa). This method is useful for assessing the hygroscopicity of a protein powder and the risk of caking phenomena, with the resulting deterioration of flow properties and solubility. (b) The *swelling method* uses a device (Baumann apparatus) consisting of a graduated capillary tube attached to a sintered glass filter. The protein powder is placed on the filter and allowed to spontaneously absorb water located in the capillary tube beneath the filter. Both the rate and the extent of hydration can be determined, as seen in Fig. 12. (c) *Excess water methods* involve exposing the protein sample to water in excess of that the protein can bind, followed by filtration or application of mild centrifugal or compression force to separate the excess water from the retained water. This method is only applicable to poorly soluble proteins, and a correction is necessary to compensate for the loss of soluble protein. (d) In the *water saturation method*, the amount of water needed to establish a state of protein saturation (maximum water retention as determined by centrifugation) is determined.

Methods b, c, and d measure bound water, unfreezable water, and also the capillary water held physically between protein molecules.

3. Environmental Factors Influencing Hydration Properties

Such factors as protein concentration, pH, temperature, time, ionic strength, and presence of other components affect the forces underlying protein-protein and protein-water

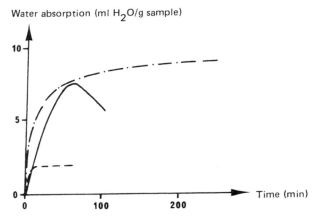

Figure 12 Water absorption (ml H_2O/g sample) in various proteins as a function of time (Baumann apparatus). —— Sodium caseinate; ·— ·— soy isolate; ——— whey protein concentrate (gel filtration). [A. M. Hermansson, *Lebensm. Wiss. Technol.* 5:24 (1972).]

interactions (18,59). Most functional properties are determined by the interrelations among these forces.

Total water absorption increases with increasing protein concentration.

Changes in pH, by affecting ionization and the magnitude of the net charge on the protein molecule, alter the attractive and repulsive interactive forces of proteins and their ability to associate with water. At the isoelectric point, protein-protein interactions are maximal and the associated, shrunken proteins exhibit minimal hydration and swelling. For example, the water-holding capacity of raw beef muscle (or beef muscle homogenate) markedly decreases when the pH declines from 6.5 to near 5.0 (isoelectric point) during the postmortem prerigor period (Fig. 13). This results in reduced juiciness and tenderness.

Water binding by proteins generally decreases as the temperature is raised, because of decreased hydrogen bonding. Denaturation and aggregation occur during heating, and the latter may reduce protein surface area and the availability of polar amino groups for water binding. On the other hand, when proteins with a very compact structure are heated, ensuing dissociation and unfolding may bring to the surface previously buried peptide bonds and polar side chains with a resulting improvement in water binding. Furthermore, several proteins, such as whey proteins, can undergo irreversible gelation when heated. If the gel is then dried, the protein displays markedly enhanced water absorption as a result of increased capillary forces within the insoluble protein network. The size, surface porosity, and internal porosity of the dried protein particles also influence the rate and the extent of water absorption.

Figure 14 shows how the water absorption of a soy isolate is influenced simultaneously by pH and by heat.

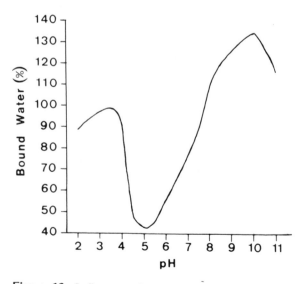

Figure 13 Influence of pH on the water-holding capacity (% "bound water") of beef muscle. [R. Hamm, in *Recent Advances in Food Science*, vol. III (Leitch and Rhodes, eds.), Butterworths, London, 1963, p. 218.]

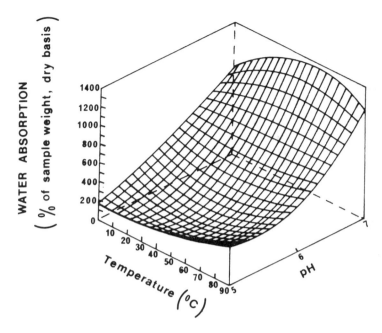

Figure 14 Water absorption response surface for Promine D soy isolate with variations in pH and temperature. [C. W. Hutton and A. M. Campbell, *J. Food Sci. 42*:454 (1977).] Copyright © by Institute of Food Technologists.

The kind and concentration of ions have a significant effect on the water absorption, swelling, and solubility of proteins. Competitive binding among water, salt, and amino acid side groups generally occurs. At low salt concentrations, protein hydration may increase. At higher salt concentrations, water-salt interactions predominate over water-protein interactions and the protein may be "dehydrated."

4. Relationships Between Hydration and Other Functional Properties

If meaningful correlations existed between water absorption and other functional properties, these would be useful for explaining and predicting the behavior of proteins. It is, however, obvious from the preceding pages that no consistent relationship exists between protein solubility and water absorption. Also, the effect of variations in pH and temperature on the water absorption and viscosity of dry proteins or proteins in solution differs depending on the protein being tested. This suggests that water absorption and viscosity are not always positively correlated, although they often are.

The ability of protein ingredients to absorb and retain water plays a major role in the texture performance of various foods, especially comminuted meats and baked doughs (Table 8). Water absorption without protein dissolution results in swelling (expansion) and will impart such characteristics as body, viscosity, and adhesion. Other functional properties of proteins (e.g., emulsification or gelation) also contribute to desirable properties of the same foods.

B. Solubility

The mechanisms responsible for the solubility or insolubility of proteins in either their native or altered states have already been discussed (Sec. II.C.2.b.). Solubility equilibria of proteins are attained slowly. Also, solubility values may vary with the pathway followed to establish the final conditions of pH, ionic strength, temperature, and protein concentration.

From a practical standpoint, data on solubility characteristics are very useful for determining optimum conditions for the extraction and purification of proteins from natural sources, and for the separation of protein fractions. Solubility behavior under various conditions also provides a good index of the potential applications of proteins. This is so because the degree of insolubility is probably the most practical measure of protein denaturation + aggregation, and because proteins that initially exist in a denatured, partially aggregated state often exhibit impaired ability to participate effectively in gelation, emulsification, and foaming. Finally, solubility is also an important attribute of proteins selected for use in food beverages (59).

Thus, protein solubility at neutral or isoelectric pH is often the first functional property measured at each stage of preparation and processing of a protein ingredient. Such tests as the nitrogen solubility index (NSI) and solubility profile as a function of pH (Fig. 15), of ionic strength, or of heat treatment (see Fig. 20) are the most frequently used.

The solubility of most proteins is markedly and irreversibly reduced when heating is involved. Nevertheless, heat treatments may be unavoidable to achieve other objectives (microbial inactivation, removal of off-flavors, removal of water, and others). Milder processing, such as that used for protein extraction and purification, may also induce a certain degree of denaturation and insolubilization. Thus, commercial soy flours, concentrates, and isolates are found to have NSI values varying from 10 to 90%.

Various techniques are available for improving the solubility of proteins. They are reviewed in Sec. VII.B.

The assumption that proteins must have a high *initial* solubility as a prerequisite for other functional properties is not always correct. It has already been noted that water absorption of a protein ingredient sometimes can be improved by prior denaturation and insolubilization. Also, gelation ability is sometimes retained after denaturation and partial insolubilization. This is in accord with the fact that the formation of emulsions, foams, and gels can involve various degrees of protein unfolding, aggregation, and insolubilization.

On the other hand, whey proteins and some other proteins must have reasonably high initial solubility if they are to function well in emulsions, foams, and gels. Also, soluble caseinates have better thickening and emulsifying properties than isoelectric (less soluble) casein. Perhaps the main advantage of initial solubility is that it permits rapid and extensive dispersion of protein molecules or particles. This leads to a finely dispersed colloidal system, with homogeneous macroscopic structure and a smooth texture. Also, initial solubility facilitates protein diffusion to air/water and oil/water interfaces, thus improving their surface activity.

C. Viscosity

The viscosity of a fluid reflects its resistance to flow. It is expressed as the viscosity coefficient μ, which is the ratio of the shear stress τ to the relative site of shear $\dot{\gamma}$ (or rate of flow):

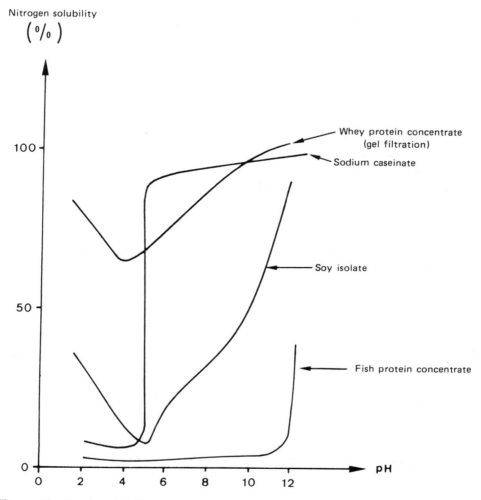

Nitrogen solubility $(^o/_o)$

Whey protein concentrate (gel filtration)

Sodium caseinate

Soy isolate

Fish protein concentrate

pH

Figure 15 Protein solubility as a function of pH in 0.2 M NaCl. [With permission from A. M. Hermansson, in *Protein in Human Nutrition* (Porter and Rolls, eds.), Academic Press, London, 1973, p. 409. [Copyright: Academic Press Inc. (London) Ltd.]

$$\tau = \mu\dot{\gamma} \tag{27}$$

Newtonian fluids have a constant viscosity coefficient independent of shear stress or shear rate. Solutions, dispersions (slurries or suspensions), emulsions, pastes, or gels of most hydrophilic macromolecules, including proteins, do not behave like Newtonian fluids, as their viscosity coefficient decreases when the flow rate increases. This behavior is called pseudoplastic or shear thinning, expressed as

$$\tau = m\dot{\gamma}^n \tag{28}$$

where m is the consistency coefficient and n is the flow behavior index.

The main single factor influencing the viscosity behavior of protein fluids is the apparent diameter of the dispersed molecules or particles. This diameter depends on the following parameters: (a) Intrinsic characteristics of the protein molecule, such as molar mass, size, volume, structure and asymmetry, electric charges and ease of deformation (environmental factors, such as pH, ionic strength, and temperature, can modify these characteristics through unfolding); (b) protein-solvent interactions, which influence swelling, solubility, and the hydrodynamic hydration sphere surrounding the molecule, and (c) protein-protein interactions, which determine the size of aggregates. Protein ingredients are generally used at high concentrations, at which protein-protein interactions predominate.

Shear thinning can be explained by the following phenomena: (a) progressive orientation of molecules in the direction of flow, so that frictional resistance is reduced, (b) deformation of the protein hydration sphere in the direction of flow (if the protein is highly hydrated and dispersed); and (c) rupture of hydrogen and other weak bonds resulting in dissociation of protein aggregates or networks. In all cases, the apparent diameter of molecules or particles in the direction of flow is reduced (110).

The rupture of weak bonds may take place slowly so that sometimes protein fluids experience a decrease, with time, in shear stress and apparent viscosity (constant shear rate and temperature) before equilibrium is reached. When the shearing treatment is stopped, the original aggregates or network may or may not be re-formed. If they are, the decrease in viscosity coefficient is reversible, and the system is *thixotropic*. For instance, dispersions of soy protein isolate and whey protein concentrates are thixotropic.

The viscosity coefficient of most protein fluids increases exponentially with protein concentration, because of protein-protein interactions. Such interactions also explain why shear thinning is more marked at high protein concentrations. When protein-protein interactions are numerous enough, as in protein pastes or gels, a plastic viscoelastic behavior is displayed, and fluid flow occurs only above a "yield stress" value necessary for the rupture of some of the interactions.

The viscosity and consistency of protein systems are important functional properties in fluid foods, such as beverages, soups, sauces, and creams. Knowledge of flow properties of protein dispersions is also of practical significance in optimizing operations, such as pumping, mixing, heating, cooling, and spray drying, that involve mass and/or heat transfers.

Correlations between viscosity and solubility are not simple, as can be understood from the preceding discussion. Insoluble heat-denatured protein powders do not develop high viscosity when placed in an aqueous medium. Highly soluble protein powders with low water absorption and swelling capacities (whey proteins) also exhibit low viscosity at neutral or isoelectric pH. Soluble protein powders with high initial water absorption (sodium caseinate, and some soy protein preparations) develop a high viscosity. Thus, for many proteins, a positive correlation is observed between water absorption and viscosity.

D. Gelation

1. General Aspects of Protein Gelation

Gelation must first be differentiated from other related phenomena in which the degree of dispersion of a protein solution decreases, that is association, aggregation, polymerization, precipitation, flocculation, and coagulation. Protein *association* reactions generally refer to changes occurring at the subunit or molecular level; *polymerization* or *aggregation* reactions generally involve the formation of large complexes. *Precipitation* includes all aggregation reactions leading to a total or partial loss of solubility. *Flocculation* refers to random aggregation reactions in the absence of denaturation. This often occurs because of suppression of electrostatic repulsions between chains. Random aggregation reactions with denaturation—and aggregation reactions where protein-protein interactions predominate over protein-solvent interactions—are defined as *coagulation*, and this leads to the formation of a coarse coagulum. When denatured molecules aggregate to form an ordered protein network, the process is referred to as *gelation* (48,49,95). Tofu gel, examined by means of scanning electron microscopy, is shown in Fig. 16. The size, shape, and arrangement of the protein aggregates and/or strands (filaments) and the size of pores are evident.

Gelation is a very important functional property of several proteins. It plays a major role in the preparation of many foods, including various dairy products, coagulated egg white, gelatin gels, various heated, comminuted meat or fish products, soybean protein gels, vegetable proteins texturized by extrusion or spinning, and bread doughs. Protein gelation is utilized not only for the formation of solid viscoelastic gels, but also for improved water absorption, thickening, particle binding (adhesion), and emulsion or foam-stabilizing effects.

Practical conditions for the gelation of various proteins are fairly well established, although optimization with respect to environmental factors, pretreatments of the protein, use of protein mixtures, and so on, is not easily attained.

In most cases, a thermal treatment is a requisite for gelation. Subsequent cooling may be necessary, and slight acidification may be helpful. Also, the addition of salts, especially calcium ions, may be necessary or may enhance the rate of gelation and/or gel

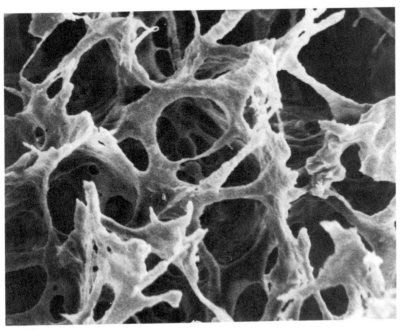

Figure 16 Scanning electron micrograph of tofu, a soybean gel. 1.5 cm = 2 μm. (Courtesy of K. Saio, Tsukuba.)

strength (with soy proteins, whey proteins, and serum albumin). Some proteins, however, may gel without heating, that is, with only mild enzymic hydrolysis (casein micelles, egg white, and fibrin), or with the simple addition of calcium ions (casein micelles), or with mere alkalinization followed by return to a neutral or isoelectric pH (soy proteins). Although many gels are formed from proteins in solution (ovalbumin and other egg white proteins, β-lactoglobulin and other whey proteins, casein micelles, serum albumin, and soy proteins), aqueous or saline dispersions of insoluble or sparingly soluble proteins may also form gels (collagen, myofibrillar proteins, such as actomyosin, partly or totally denatured soy protein isolates, and others). Thus protein solubility is not always necessary for gelation.

2. Characteristics of Gel Formation and Gel Structure

The mechanisms and interactions underlying the formation of the three-dimensional protein networks characteristic of gels are not fully understood. However, practically all studies point to the necessity of protein denaturation and unfolding prior to the step of ordered protein-protein interaction and aggregation. This explains, for instance, why soy protein isolates previously denatured by heat, solvent, or alkali may gel without further heating.

The formation of the protein network is considered to result from a balance between protein-protein and protein-solvent (water) interactions, and between attractive and repulsive forces between adjacent polypeptide chains. Hydrophobic interactions (enhanced at high temperatures), electrostatic interactions (such as bridges with Ca^{2+} and other divalent ions), hydrogen bonding (enhanced by cooling), and/or disulfide cross-links are known to represent the attractive forces. Their relative contribution may vary with the nature of the protein, the environmental conditions, and the various steps in the gelation process. Electrostatic repulsions (especially at pH values far from the isoelectric point) and protein-water interactions tend to keep polypeptide chains apart. Intermolecular protein attraction (and gelations) takes place more readily at high protein concentrations because of the greater probability of intermolecular contacts. At high protein concentrations, gelation may take place even in environmental conditions that are not especially favorable for aggregation (without heating, at pH values away from the isoelectric point, and so on).

The establishment of covalent disulfide cross-links usually leads to the formation of heat-irreversible gels, as is the case with ovalbumin and β-lactoglobulin gels. Gelatin gels, which are mainly stabilized by hydrogen bonding, will, on the other hand, melt upon heating (at about 30°C), and the setting-melting cycle can be repeated many times.

Some kinds of different proteins can form gels when heated together (cogelation). Proteins can also form gels through interactions with polysaccharide gelling agents (68). Nonspecific ionic interactions between positively charged gelatin and negatively charged alginates or pectates give gels with high melting points (80°C). Also, specific ionic interactions are known to take place between the positively charged region of κ-casein and the polysulfated κ-carrageenan at the pH of milk. Thus, casein micelles can be entrapped in carrageenan gels.

Many gels exist as highly expanded (open) and hydrated structures, with more than 10 g water per gram protein and various other food constituents entrapped in the protein network. Some protein gels can even contain 98% water. Although most of this water has properties similar to that of water in a dilute salt solution, it is physically entrapped and cannot be easily squeezed out. Many hypotheses exist concerning the forces responsible for the large water-retention capacity of gels. It is possible that, after heat denaturation of the secondary structures, unmasked CO and NH groups of peptide bonds become, respectively, negative and positive polarized centers along the polypeptide chain, and these may create an extensive system of water multilayers. Upon cooling, protein molecules of this kind may interact, via hydrogen bond reformation, to provide the structure necessary to immobilize (entrap) the free water. It is also likely that the pores in the protein network hold water through capillary mechanisms.

When starting from an aqueous protein solution, the first steps in heat-set gelation are usually [Eq. (29)]

a. Reversible dissociation of the quaternary structure into subunits or monomers (irreversible dissociation of native polymers may also take place as the first step of denaturation)
b. Irreversible denaturation of secondary and tertiary structures (unfolding generally remains partial)

$$(P_N)_n \rightleftharpoons nP_N \longrightarrow nP_D \tag{29}$$

where P_N is native protein, P_D is denatured protein, and n is a small known number.

Although the final gel state corresponds to aggregates of partly denatured proteins $(P_D)_x$, it is not always clear which of the following pathways is involved:

$$xP_N \xrightleftharpoons{heat} (P_N)_x \xrightarrow{heat} (P_D)_x \tag{30}$$

or

$$xP_N \xrightarrow{heat} xP_D \xrightarrow{heat \ and/or \ cooling} (P_D)_x \tag{31}$$

The first part of Eq. (30) is often regarded as applicable to flocculation reactions, the second part is generally considered applicable to coarse coagulation. In conditions favoring denaturation over aggregation reactions (high net protein charge at low or high pH, very low ionic strength, presence of certain salts, presence of bond-breaking agents, such as urea, guanidine, and detergents), heating will cause pathway (31) to take place.

The slower the aggregation step, as compared with denaturation, the better the partially unfolded polypeptides can orient themselves before aggregation. This, in turn, favors a gel that is ordered, homogeneous, of smooth consistency, highly expanded, highly elastic, transparent, and stable toward syneresis and exudation. Gels consisting of coarsely aggregated protein particles are opaque, lacking in elasticity, and especially susceptible to destabilization (syneresis and weeping).

Dissociation and/or unfolding of protein molecules generally increase the exposure of reactive groups, especially the hydrophobic groups of globular proteins. Protein-protein hydrophobic interactions are therefore favored and are usually the main cause of subsequent aggregation. Proteins with a high molar mass and a high percentage of hydrophobic amino acids therefore tend to establish strong networks. Hydrophobic interactions are enhanced at high temperatures; formation of hydrogen bonds is favored during cooling. Heating may also unmask internal SH groups and promote formation or interchange of disulfide bonds. The presence of a large number of SH and SS groups strengthens the intermolecular network and tends to make the gel thermally irreversible. Calcium bridges improve the firmness and stability of many gels.

The pH range over which gelation occurs generally increases with increasing protein concentration. This indicates that the numerous hydrophobic and disulfide bonds formed at high protein concentrations can compensate for the repulsive electrostatic forces induced by the large net charge of the proteins at pH values well removed from their isoelectric points. At the isoelectric pH, the lack of repulsive forces leads to a less expanded, less hydrated, and less firm gel. Proteins with large molar percentages of hydrophobic amino acids (>31.5%, such as hemoglobin, catalase, egg albumin, and urease) generally exhibit gelation pH ranges dependent on protein concentration, whereas those with small molar percentages of hydrophobic amino acids (22-31.5%, such as γ-globulins, α-chymotrypsin, prothrombin, serum albumin, conalbumin, ovomucoid, gelatin, and soy proteins) exhibit no change in the pH range of gelation as protein concentration is altered.

This difference in behavior has been suggested as a criterion for the classification of heat-set gels: (a) proteins like egg albumin heat precipitate at a low protein concentration and give an opaque gel at a high concentration; and (b) proteins like gelatin remain soluble when heated at a low protein concentration and give a clear and thermoreversible gel at a high concentration. Such proteins as caseins, β-lactoglobulin, and pepsin behave like proteins in group b, although their molar percentages of hydrophobic amino acids are 38, 34.6, and 34%, respectively. This behavior is perhaps attributable to their small molecular weights. Pure ovalbumin behaves like the caseins, but when this protein is mixed with conalbumin, they behave as group a proteins, due to the formation of protein-protein bonds and an increased apparent molecular weight (97-99).

E. Texturization

Proteins constitute the basis of structure and texture in several foods, whether these come from living tissues (myofibrils in meat and fish) or from fabricated substances (bread dough and crumb, soy or gelatin gels, cheese curds, sausage meat emulsions, and so on). There are also a number of texturization processes that begin with soluble vegetable or milk proteins and that lead to film or fiberlike products with chewiness and good water-holding characteristics, and that have the ability to retain these properties during subsequent hydration and heat treatment. These "texturized proteins" are often used as meat substitutes and/or extenders. Also, some texturization processes are done for the purpose of retexturizing or re-forming animal proteins, such as beef or poultry meat. The known physicochemical bases of some of these texturization processes are presented as follows (51,57,61,90,96).

1. Thermal Coagulation and Film Formation

Concentrated soy protein solutions can be thermally coagulated on a flat metallic surface, such as that of a drum-dryer. The resulting thin, hydrated, protein films can be folded, pressed together, or cut.

Thin protein-lipid films form on the surface of soy milk held for a few hours at 95°C. This results from surface evaporation of water and thermal coagulation of the proteins. Films can be prepared repeatedly by simply removing the previously formed film from the surface. This is a traditional approach to the making of yuba in Japan.

2. Fiber Formation

Fiber spinning of vegetable (especially soy) and milk proteins bears many similarities to the spinning of synthetic textile fibers. It is usually necessary to start from isolates containing 90% protein or more. Four to five successive operations are necessary, but these can be done continuously.

a. A solution of high protein concentration (10-40%), the "dope," is prepared, usually by raising the pH above 10 (such a high pH can alter the nutritive value; see Sec. VII.A). The electrostatic repulsions promote complete dissociation into subunits and extensive unfolding of the individual polypeptide chains. A high viscosity is therefore obtained. The dope must be degassed and clarified to avoid fiber interruption during subsequent spinning.

b. The dope is forced under pressure through a die plate containing over 1000 holes, each with a diameter of 50-150 μm. Streaming orientation of the unfolded protein molecules takes place as they flow through these holes. Thus, the molecules tend to extend and align themselves in a parallel manner.

c. The liquid "filaments" coming out of the die enter an acid-sodium chloride bath where the proteins are coagulated by the isoelectric pH and by the salting out effect. Because of their elongated, parallel orientation, the protein molecules of each filament interact strongly with each other through hydrogen, ionic, and disulfide bonds to form a hydrated protein fiber.

d. The coagulated protein fibers are removed from the bath on rollers. The speed of roller rotation is such that the fibers are stretched and the individual polypeptide chains achieve still better alignment, associate more closely, and form more intermolecular bonds. This partial "crystallization" increases the mechanical resistance and chewiness of the fibers, but may decrease their water-holding capacity.

e. The fibers are then compressed and/or heated between rollers to remove some water, promote adhesion, and increase toughness. Binding agents, such as gelatin, egg white, gluten, or gelling polysaccharides, may be added before heating. Many other food additives, such as flavoring agents or lipids, can also be incorporated. Bundles of fibers are cut, assembled, compressed, and so on, yielding products similar to ham, poultry meat, or fish muscle.

It appears that most globular proteins (with molecular weights above 10,000) can be spun into fibers, provided they are largely free of nonprotein constituents. The necessary initial protein unfolding can be obtained without alkali treatment, for instance by heating, or by dissolving proteins in organic solvents. In the latter case, the solvent is evaporated as the filaments come out of the spinning die.

It is of interest to note that many cheeses can be "melted" by heating in the presence or even in the absence of sodium phosphates. Such heating causes destruction of the casein micellar structure and leads to the specific gelled and emulsified texture of processed cheese. Some cheeses with a low degree of proteolysis tend to give threads when melted. Whether this approach to fiber formation is applicable to other proteins is apparently not known.

3. Thermoplastic Extrusion

The major technique used at present for texturization of vegetable proteins is that of thermoplastic extrusion. It leads to dry, fibrous, porous granules or chunks (rather than fibers), which, upon rehydration, possess a chewy texture. The starting material need not be protein isolates; thus, less costly protein concentrates or flours (containing 45-70% protein) can be used. Caseins or gluten can be added, or texturized as such. The addition of small amounts of starch or of amylose improves the final texture, but a lipid content above 5-10% is clearly detrimental. The addition of up to 3% sodium or calcium chloride also firms the texture.

The technology can be briefly described as follows. The hydrated (10-30% water) protein-polysaccharide mix is moved through a cylinder by action of a turning screw, while being exposed to high pressure (10,000-20,000 kPa), intensive shear forces, and high temperatures. Over a period of 20-150 sec, the mix is elevated to a temperature of 150-200°C, converted to a viscous consistency, and then extruded rapidly through a die into a normal-pressure environment. Flash vaporization of internal water takes place, with the formation of expanding steam bubbles. After cooling, the protein-polysaccharide matrix possesses a highly expanded, dry structure. Microscopic examination shows a foam structure with fibers oriented in the direction of flow. When rehydrated in water at about 60°C, the porous material absorbs twice to four times its weight of water, giving a fibrous, spongy, and partly elastic structure with a chewiness not unlike that of meat. These structures are stable under sterilization conditions. A product of this kind is used in hamburgers, meat balls, ravioli, and cured meat products.

Good texturization by thermoplastic extrusion requires proteins with appropriate initial solubility, high molecular weight, and the development of proper plasticizing and viscosity properties of the protein-polysaccharide mix within the die. The physical chemistry of extrusion texturization is not precisely known. However, protein bodies and cell walls must be disrupted and proteins are likely to unfold at the high temperatures and high shear forces developed within the extruder. Molecules probably orient in the direction of flow while passing through the extrusion die. Thermal aggregation and coagulation occur, since proteins that are initially water soluble become insoluble. The ground, extruded material can be totally solubilized in buffers containing sodium dodecylsulfate and dithiothreitol. This indicates that hydrogen bonds, hydrophobic interactions, and disulfide cross-links are involved in texturization. Disulfide interchange appears to be an important reaction for adequate texturization to take place, since oxidizing agents are clearly detrimental. Other covalent bonds [lysinoalanyl, lanthionyl, and ε-N-(γ-glutamyl)-lysyl] do not form in detectable amounts during extrusion. Small losses of cysteine and arginine are observed.

Proteins at higher moisture contents than previously discussed can also be texturized by thermal coagulation in an extruder. This results in the formation of hydrated, nonexpanded films or gels. The addition of cross-linking agents, such as glutaraldehyde, increases the firmness of the resulting product. This technique could be of use for the texturization of blood, mechanically deboned meat or fish, and other animal by-products.

It can be concluded that several processes, traditional or new, are available for the texturization of most proteins. A better understanding of the physical and chemical modifications underlying these processes will lead to further improvements in products generated by this means.

F. Dough Formation (58,112)

A unique property of the "gluten" proteins of wheat grain endosperm (and to a lesser extent of rye and barley grains) is their ability to form a strongly cohesive and viscoelastic paste or "dough" when mixed and kneaded in the presence of water at ambient temperatures. This is the basis for the transformation of wheat flour into bread dough and its further transformation into bread through fermentation and baking. In addition to the gluten proteins (gliadins and glutenins; for further details, see Chap. 15), wheat flour contains starch granules, pentosans, polar and nonpolar lipids, and soluble proteins, all of which contribute to the formation of the dough network and/or to the final texture of bread.

The composition and the large molecular size of gliadins and of glutenins explain much of the behavior of gluten. Due to their low content of ionizable amino acids, the gluten proteins are poorly soluble in neutral aqueous solutions. Rich in glutamine (over 33% by weight) and in hydroxy amino acids, they are prone to hydrogen bonding. This accounts largely for the water absorption capacity and for the cohesion-adhesion properties of gluten. The latter properties also derive in part from the presence of many apolar amino acids and the resulting hydrophobic interactions that contribute to protein aggregation and to the binding of lipids and glycolipids. Finally, the ability to form numerous disulfide cross-links accounts for the ease with which these proteins interlink tenaciously in dough.

Why are the gluten proteins of some wheat varieties especially suited to give a good bread dough? Why are the proteins of most other cereals and plants unsuited for bread making, and often detrimental when added to wheat flour? Can this situation be improved? Although full answers are not yet possible, progress has been made in explaining how gluten proteins of wheat behave during the formation of dough and during bread making.

When the hydrated bread flour is mixed and kneaded, the gluten proteins orient, align, and partially unfold. This enhances hydrophobic interactions and the formation of disulfide cross-links through disulfide interchange reactions. A three-dimensional viscoelastic protein network is established as the initial gluten particles transform into thin membranes (or films), thus serving to entrap the starch granules and other flour components. The cleavage of disulfide cross-links by reducing agents, such as cysteine, destroys the cohesive structures of hydrated gluten and bread dough; the addition of oxidizing agents, such as bromate, increases toughness and elasticity. "Strong" flours from certain wheat varieties require long mixing times and give very cohesive doughs. "Weak" flours are less effective, and the gluten network breaks down when the energy or duration of mixing exceeds a certain level, probably because disulfide bonds are ruptured (especially in the absence of air). Dough strength appears to be related to a large content of high-molecular-weight glutenins, including the totally insoluble "residue proteins." From experiments with "reconstituted" wheat flours of varying gliadin-glutenin ratios, it can be concluded that glutenins are responsible for the elasticity, cohesiveness, and mixing tolerance of the dough, and gliadins facilitate fluidity, extensibility, and expansion of the dough, thus contributing to a large bread-loaf volume. A proper balance of both types of protein is essential for bread making. Excessive cohesion (glutenins) inhibits the expansion of trapped CO_2 bubbles during fermentation, the rise of the dough, and the subsequent presence of open air "cells" in the bread crumb. Excessive extensibility (gliadins) results in gluten films that are weak and permeable; thus, retention of CO_2 is poor and dough collapse may occur.

Neutral and polar lipids from the flour, or added to the dough, interact with gliadins and glutenins and can either impair or improve the gluten network.

Baking apparently does not induce extensive additional denaturation of the gluten proteins. This is probably so because gliadins and glutenins are already partly unfolded in the flour and become even more unfolded during kneading of the dough, and/or because they are resistant to further unfolding at the temperatures of normal bread baking (<100°C). Above 70-80°C, gluten proteins release some moisture, and this water is absorbed by the partly gelatinized starch granules. Gluten proteins nevertheless play a major role in retaining moisture during baking, leading to a soft crumb (40-50% water).

Soluble wheat proteins (albumins and globulins) are denatured and aggregated by baking, and this partial gelation contributes to the setting of the bread crumb.

It is often desirable to add foreign proteins to bakery products, for example, for nutritional enrichment. Not all these foreign proteins are compatible with the formation of an adequate gluten network. Water-soluble globulins are highly detrimental to loaf volume, and this undesirable effect can be avoided by using heat-denatured soy, whey, or milk proteins rather than their native counterparts. The addition of polar lipids, either wheat flour glycolipids (mono- and digalactosyl diglycerides), or synthetic surfactants (nonionic sucrose esters or ionic sodium stearoyl-2-lactylate) permits the incorporation of more foreign protein without deterioration of the bread structure. This reflects the importance of glycolipids for the establishment of hydrophobic and hydrogen bonds in the gluten-dough network.

Isolated gluten can also be added to wheat flour to reinforce the dough network. Gluten, because of its adhesive properties, is also used as a binding agent in various meat products.

G. Emulsifying Properties

1. Protein-Stabilized Food Emulsions

Many food products are emulsions (milk, cream, ice cream, butter, processed cheese, mayonnaise, and comminuted meats), and protein constituents often play a major role in stabilizing these colloidal systems. The natural milk emulsion is stabilized by the fat globule "membrane" (see Chap. 13). This membrane is made of successive adsorbed layers of triacylglycerols, phospholipids, insoluble lipoproteins, and soluble proteins. In fresh milk, the latter consist of immunoglobulins. Homogenization of milk increases emulsion stability because it reduces the size of fat globules, and because newly formed casein submicelles displace the immunoglobulins and adsorb to the fat globules.

The fundamentals of emulsion formation, the mechanisms of emulsion breakdown (creaming, flocculation, and coalescence), and the factors that contribute to stabilize emulsions have been reviewed in Chap. 4 (see also Ref. 39).

Proteins adsorb at the interface between dispersed oil droplets and the continuous aqueous phase and contribute physical and rheological properties (thickness, viscosity, and elasticity-rigidity) (11) that determine resistance to droplet coalescence. Ionization of the amino acid side chain may also take place, depending on the pH, and this provides electrostatic repulsive forces that favor emulsion stability.

Proteins are generally poor stabilizers of water/oil (W/O) emulsions. This may be attributable to the predominantly hydrophilic nature of most proteins, causing the bulk of an adsorbed protein molecule to reside on the water side of the interface (45).

2. Methods for Determining the Emulsifying Properties of Proteins

The determination of droplet size and size distribution (and hence of the total interfacial area) is necessary to characterize food emulsions, and this can be done by microscopy, light scattering, centrifugal sedimentation, or the use of such devices as the Coulter counter (passage of droplets through orifices of known sizes).

Measurement of the amount of protein adsorbed at the interace (= surface concentration) requires prior separation of the oil droplets (in the case of O/W emulsions), for instance, by repeated centrifugation and washing to remove loosely bound protein. Protein concentrations of a few milligrams per milliliter emulsion generally lead to a few milligrams protein adsorbed per square meter. When the dispersed phase volume fraction ϕ is large and the droplet size is small, more protein is needed to obtain an adsorbed protein film that is an effective emulsion stabilizer.

Two tests are widely used for comparing the emulsification properties of proteins.

Emulsion capacity (EC) is the volume of oil (milliliters) that can be emulsified per gram of protein before phase inversion occurs. For this test, an aqueous (or saline) solution or dispersion of the protein is stirred while oil or melted fat is added continuously at constant speed. Phase inversion is detected by the sudden drop in viscosity, the change in color (especially if an oil-soluble dye is present), or the increase in electrical resistance. The driving force for inversion increases particularly when ϕ exceeds 0.74. The ϕ values at inversion (ϕ_i) are close to 0.5 in the absence of protein and usually range from 0.65 to 0.85 in the presence of proteins. Inversion may not lead immediately to a continuous oil phase, but first to a W/O/W double emulsion. Because it is expressed per gram protein, the EC value decreases with increasing protein concentration, whereas the ϕ_i value first increases steeply then reaches a plateau.

Emulsion stability (ES) is most often expressed as

$$ES = \frac{\text{volume final emulsion} \times 100}{\text{volume initial emulsion}} \tag{32}$$

after the emulsion has been centrifuged at low speed or let stand for several hours, with or without prior heating. Emulsion breakdown may or may not result in separation of an aqueous and/or an oil layer. Creaming (without coalescence) often occurs as the oil droplets move up to form a densely packed layer. Simultaneously, the aqueous phase migrates downward, resulting in a water layer at the bottom. The test is therefore a measure of the rate of drainage, and may not be related to the emulsifying properties of the protein tested (45). However, many food emulsions destabilize by means of creaming. Maximum stability for O/W emulsions is usually found with ϕ values between 0.77 and 0.88. Water drainage occurs with lower ϕ values; coalescence and separation of an oil layer tend to take place at greater ϕ values.

The two tests, emulsion capacity and emulsion stability, reflect the fact that proteins perform two functions; (a) they aid in the formation of emulsions, mainly by decreasing interfacial tension, and (b) they help stabilize the emulsion by forming a physical barrier at the interface. There is no strict correlation between these two functions.

3. Factors Influencing Emulsification

Many factors influence the characteristics of emulsions and the results of emulsion tests: equipment type and geometry, intensity of energy input, rate of oil addition, oil phase volume, temperature, pH, ionic strength, presence of sugars, presence of low-molecular-weight surfactants, exposure to oxygen, kind of oil (melting point), concentration of soluble protein, and emulsifying properties of the protein. These factors explain, in the absence of standardization, why a given emulsion does not yield the same test results when examined by different investigators, and why the reported emulsifying performance of various proteins differs depending on the investigator.

A positive correlation often exists between protein solubility and either emulsion capacity or emulsion stability. Undissolved protein contributes very little to emulsification, probably because, in the tests used, proteins must dissolve and migrate to the interface before their surface properties come into play. In comminuted meat emulsions (pH 4-8) the presence of sodium chloride (0.5-1 M) increases the emulsion capacity of the proteins, probably because myofibrillar proteins are salted in, with an increase both in solubility and in unfolding. Heat-aggregated insoluble soy proteins are less efficient emulsifying agents than their soluble counterparts. However insoluble protein particles can often play a role in the stabilization of emulsions, once formed.

The pH influences emulsification properties of proteins in various ways. Some proteins, at their pI values, are sparingly soluble, and this alone detracts from the ability to help form emulsions. Also, proteins in this state cannot contribute to the stabilizing (repulsive) surface charge of oil droplets. Moreover, at the pI, or at certain ionic strengths, proteins adopt compact structures with high viscoelasticity. This may either prevent unfolding and adsorption at the interface (undesirable with regard to emulsion formation) or stabilize an already adsorbed protein film against surface deformation or desorption. The latter effect favors emulsion stability, since deformation or desorption of the interfacial protein film precedes emulsion destabilization. In addition, hydrophobic interactions between lipids and proteins are enhanced at the pI of the protein. Thus, experimental data are conflicting in that some proteins have optimal emulsifying properties at the pI (gelatin and egg white proteins), and others perform better at pH values away from the pI (soy proteins, peanut proteins, caseins, whey proteins, bovine serum albumin, and myofibrillar proteins). However, the apparent protein behavior depends to some extent on the testing conditions.

The emulsion capacity of peanut proteins as a function of pH and of sodium chloride concentration is shown in Fig. 17. In this instance, there is a strong positive correlation between emulsion capacity and protein solubility (not shown).

Heating usually decreases the viscosity and rigidity of the protein film adsorbed to the interface, and therefore decreases emulsion stability. However, gelation of the highly hydrated interfacial protein film increases its surface viscosity and rigidity and stabilizes the emulsion. Thus, gelation of myofibrillar proteins contributes to the heat stability of meat emulsions, such as sausages, and this results in improved water and fat retention and greater cohesion.

The addition of surfactants of small molecular size is usually detrimental to the stability of protein-stabilized emulsions because they decrease the firmness of protein films and lessen the forces causing proteins to remain at the interface.

Since the rate of diffusion of some proteins from the bulk of the aqueous phase to the interface may be slow and since the concentration of protein in the aqueous phase is lessened by adsorption of the protein to the oil droplets, high initial protein concentration may be required for the formation of a protein film of appropriate thickness and proper rheological properties. In practice, protein concentrations of 0.5-5% (w/w emulsion) are used, leading to interfacial protein concentrations of 0.5-20 mg/m^2.

4. Surface Properties of Proteins

The most important attribute responsible for the emulsifying property of a soluble protein is its ability to diffuse toward, and adsorb at the oil/water interface (see Protein-Lipid Interactions, Sec. II.C.3). It is generally assumed that once a portion of a protein has contacted the interface, the nonpolar amino acid residues orient toward the nonaqueous phase, the free energy of the system decreases, and the remainder of the protein adsorbs spontaneously. During adsorption, most proteins unfold extensively and, if a large surface is available, spread to a monomolecular layer (≈ 1 mg/m^2, 10-20 Å thick). According to some investigators, the more hydrophobic the protein, the greater the concentration of protein at the interface, the lower the interfacial tension, and the more

Emulsification Capacity

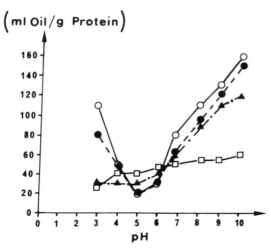

Figure 17 Effects of pH and sodium chloride concentrations on the emulsion capacity of peanut protein isolate. ○ —, 0.1 M NaCl; ● ———, 0.2 M NaCl; ▲ —·—, 0.5 M NaCl; □ —, 1.0 M NaCl. [G. Ramanatham, L. H. Ran, and L. N. Urs, *J. Food Sci. 43*:1270 (1978). Copyright © by Institute of Food Technology.]

stable the emulsion. However, overall protein hydrophobicity [assessed from the volume ratio of hydrophilic versus hydrophobic amino acid residues (= p), or from the average hydrophobicity (= $\overline{G}°$); see Sec. II.B.6] is not strongly correlated with emulsifying properties. Kato and Nakai (56) suggest that increasing "surface hydrophobicity" of proteins, as determined by hydrophobic affinity chromatography, hydrophobic partition, or with an hydrophobic reagent, significantly correlates with decreasing interfacial tension and an increasing index of emulsifying activity (Fig. 18). However, in a study of emulsions stabilized with a whey protein concentrate, it was found that β-lactoglobulin was the main protein adsorbed at the oil interface at an alkaline pH, α-lactalbumin was adsorbed at an acid pH, and neither of these proteins, at the pH of adsorption, exhibited a meaningful correlation between surface hydrophobicity and adsorption. It is likely, therefore, that flexible proteins, able to unfold and spread when in contact with the lipid surface, readily establish hydrophobic interactions with lipid droplets, produce adsorbed films with desirable viscoelasticity properties, and successfully stabilize emulsions (whether they possess a large $\overline{G}°$ or a large initial value for surface hydrophobicity).

Globular proteins with a stable structure and great surface hydrophilicity, such as whey proteins, lysozyme, and ovalbumin, are poor emulsifying agents unless they can be unfolded by prior treatment without loss of solubility. Moderate heat treatments appear to function in this manner (see Sec. VII.B). Caseinates are better emulsifiers than the proteins just mentioned because they possess, in addition to their high solubility, a dissociated and naturally unfolded structure (random coil), together with a relatively high overall hydrophobicity and a separation of highly hydrophobic and highly hydrophilic regions of the polypeptide chain. Casein micelles (and nonfat dry milk), actomyosin (meat and fish proteins), soy proteins (especially soy isolates), and both the plasma and the globin protein components of blood have good emulsifying properties (Table 9).

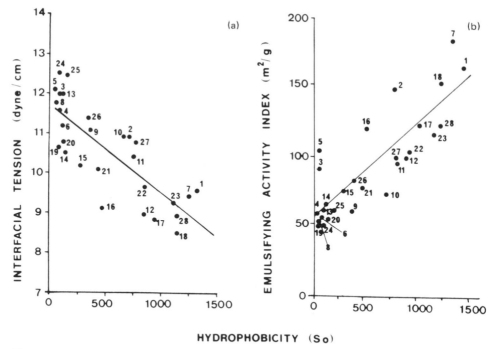

HYDROPHOBICITY (So)

Figure 18 Correlations of surface hydrophobicity of various proteins with (a) oil/water interfacial tension, and (b) emulsifying activity index. [From A. Kato and S. Nakai, *Biochim. Biophys. Acta 624*:13 (1980)]. Surface hydrophobicity was determined from the amount of a hydrophobic fluorescent probe bound per unit weight of protein. Emulsifying activity index is the interfacial surface area created per gram protein: (1) bovine serum albumin; (2) β-lactoglobulin; (3) trypsin; (4) ovalbumin; (5) conalbumin; (6) lysozyme; (7) κ-casein; (8-12) ovalbumin denatured by heating at 85°C for 1, 2, 3, 4, or 5 min, respectively; (13-18) lysozyme denatured by heating at 85°C for 1, 2, 3, 4, 5, or 6 min, respectively; (19-23) ovalbumin bound to 0.2, 0.3, 1.7, 5.7, or 7.9 mol dodecyl sulfate per mol protein, respectively; (24-28) ovalbumin bound to 0.3, 0.9, 3.1, 4.8, or 8.2 mol linoleate per mol protein, respectively.

Many theoretical and experimental studies have been carried out on the behavior of proteins at water/oil or water/air interfaces (69,86). However, much uncertainty remains as to the conformation that different proteins adopt at these interfaces and the relationship between initial conformation and conformation at the interface, versus emulsifying or foaming properties. Adsorbed protein films that are thick, highly hydrated, and electrically charged probably result in optimal emulsion or foam stability.

5. Other Consequences of Protein-Lipid Interactions

Protein-lipid interactions may have detrimental effects, especially when one attempts to purify proteins from lipid-rich materials, such as oilseeds or fish. For example, the direct extraction of proteins from oilseeds by aqueous or alkaline solutions is made impossible by the formation of protein-stabilized emulsions that are resistant to centrifugation. Neutral triacylglycerols bind to proteins via hydrophobic interactions and can be removed

Table 9 Emulsifying Activity Index Values for Various Proteins[a]

Protein	Emulsifying activity index (m^2 interface stabilized per g protein used in test)	
	pH 6.5	pH 8.0
Succinylated (88%) yeast protein	322	341
Bovine serum albumin	–	197
Sodium caseinate	149	166
β-Lactoglobulin	–	153
Whey protein powder	119	142
Soy protein isolate	41	92
Hemoglobin	–	75
Yeast protein	8	59
Lysozyme	–	50
Egg albumin	–	49

[a]Proteins were dispersed at a 0.5% concentration in phosphate buffer, pH 6.5, ionic strength 0.1. Succinylation (%) denotes the number of lysine groups succinylated in the yeast proteins.
Source: Reprinted with permission from K. N. Pearce and J. E. Kinsella, *J. Agric. Food Chem. 26*: 716 (1978). Copyright 1982 American Chemical Society.

by apolar solvents, such as hexane. Phospholipids, however, are even more tightly bound through polar linkages and their separation requires polar solvents, such as ethanol or isopropanol.

It is sometimes desirable that dry protein ingredients adsorb a certain amount of oil and retain it when submitted to centrifugal or other forces. Soy concentrates and isolates can bind 70 and 170 ml oil per gram, respectively, sunflower protein isolates bind up to 400 ml/g, and textured rapeseed proteins bind up to 150 ml/g. It appears that insoluble and the more hydrophobic proteins bind the greatest amount of oil. Low-density protein powders with a small particle size adsorb and/or entrap more oil than do high-density protein powders. Oil binding appears to decrease as the temperature increases, possibly because of lower oil viscosity. The carbohydrate constituents of vegetable protein flours and concentrates do not contribute markedly to oil binding. Some similarity exists between the binding of oil by proteins and the binding of apolar volatile compounds by proteins (see Sec. IV.I).

Oxidizing lipids may interact with, and damage, food proteins. This is discussed in Sec. VII.A.

H. Foaming Properties

1. Basic Aspects of Food Foams (Formation and Breakdown)

Food foams are usually dispersions of gas bubbles in a continuous liquid or semisolid phase that contains a soluble surfactant. A large variety of food foams exist with widely differing textures, such as meringue, cakes, marshmallow and some other confectionery products, whipped cream, whipped toppings, ice cream, soufflés, beer froth, mousses, and bread. In many cases, the gas is air (occasionally carbon dioxide) and the continuous phase is an aqueous solution or suspension containing proteins. Some food foams are very complex colloidal systems. Ice cream, for example, contains an emulsion (or suspension) of dispersed and clustered fat globules (mostly solid), a suspension of dispersed ice crystals, a polysaccharide gel, a concentrated solution of sugars and proteins, and air bubbles.

In foams, a continuous phase of thin liquid layers, called lamellae, separates the gas bubbles. The gas/liquid interface may measure 1 m^2/ml liquid. As with emulsions, mechanical energy is required for the creation of this interface. Maintaining the interface against coalescence of gas bubbles usually necessitates the presence of surface-active agents. These agents lower the interfacial tension* and form an elastic protective barrier between entrapped gas bubbles. Some proteins are able to form a protective film by adsorbing at the gas/liquid interface. In this case, the lamella between two adjacent bubbles consists of two adsorbed protein films separated by a thin liquid layer.

Gas bubbles in a foam can vary greatly in size, varying in diameter from 1 μm to several centimeters, depending on numerous factors, such as surface tension and viscosity of the liquid phase and energy input. Uniform distribution of fine bubbles usually imparts body, smoothness, and lightness to the food, and increased dispersion and perceptibility of flavors.

One method of forming a foam is to bubble gas through a porous sparger (such as sintered glass) into an aqueous solution of low protein concentration (0.01-2% w/v). An initial "gas emulsion" breaks down by bubble rise and drainage, and an upper layer of "true foam" separates. The latter has a large dispersed phase volume (ϕ), with bubbles distorted by compression into polyhedral shapes (Fig. 19). If very large amounts of gas are introduced, the liquid may be completely converted to foam. Very large foam volumes can be obtained, even from dilute protein solutions. For example, a 10-fold expansion (1000% when expressed as 100 $\times$ volume of foam over initial volume of liquid) is common; a 100-fold expansion can be obtained in some cases. The corresponding ϕ values are 0.9 and 0.99, respectively (assuming all the liquid has been converted to foam), and the foam density varies accordingly.

Foams can also be produced by whipping (beating) or shaking an aqueous protein solution in the presence of a bulk gas phase. Whipping is the preferred means of gas introduction in most aerated food products. Compared with sparging, whipping results in a more severe mechanical stress and shear action, and a more uniform dispersion of the gas. The more severe mechanical stress affects both coalescence and formation of bubbles and typically hinders adsorption of protein at the interface, resulting in a greater protein requirement (1-40% w/v). During whipping, the volume of air incorporated usually passes through a maximum (reflecting a dynamic equilibrium), and the increase in sample volume usually ranges from 300 to 2000%.

A third procedure for forming foams is to suddenly release the pressure from a previously pressurized solution. This technique is used, for instance, when whipped cream is formed during dispensing from an aerosol container.

A major difference between emulsions and foams is that, in foams, the volume fraction occupied by the dispersed phase (gas) varies over a much broader range than in emulsions.

Because many foams have very large interfacial areas, they are often unstable. There are three main *destabilizing mechanisms*.

---

*When the gas consists of air, the interfacial tension corresponds to the surface tension of the liquid phase.

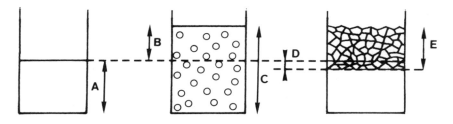

liquid   **gas emulsion**   **foam**

Figure 19 Schematic representation of foam formation. A, volume of liquid; B, volume of gas incorporated; C, total volume of dispersion; D, volume of liquid in foam (= E − B); E, volume of foam. Foam volume defined as 100 × E/A. Overrun defined as 100 × B/A = 100 (C − A)/A. Foaming power defined as 100 × B/D. Foam phase volume defined as 100 × B/E. [P. J. Halling. Reprinted with permission from *Critical Reviews in Food Science and Nutrition*, 1981, Vol. 15, p. 155 (1981). Copyright CRC Press, Inc., Boca Raton, Fla.]

a. Drainage (or leakage) of lamella liquid due to gravity, pressure differences, and/or evaporation. The internal pressure P within bubbles is given by Laplace's capillary pressure equation

$$P = P_{atm} + \frac{2\gamma}{R} \tag{33}$$

where P_{atm} is the atmospheric pressure (Pa), γ is the interfacial tension (N/m), and R is the bubble radius of curvature (m).
In low-density foams, bubbles tend to press tightly together, and this enhances liquid drainage from the lamellae. Low interfacial tensions and large bubble diameters decrease internal pressure and drainage. Drainage occurs during foam formation, and a large degree of expansion (large ϕ value) favors continuance of drainage. Drainage after foam formation further increases the ϕ value and reduces the thickness and strength of liquid lamellae.
Drainage is lessened when the bulk liquid phase is viscous (as can be obtained by adding sugar) and the surface viscosity of the adsorbed protein film is great. The surface viscosity depends on the strength of protein-protein and protein-water interactions.
b. Gas diffusion from small to large bubbles. Such disproportionation results from solubility of the gas in the aqueous phase.
c. Rupture of the liquid lamellae separating gas bubbles. Such ruptures result in an increase in bubble size through coalescence, and ultimately lead to a collapse of the foam. There is an interdependency between drainage and rupture, since rupture increases drainage, and drainage reduces the thickness and strength of lamellae. If the two adsorbed protein films of the lamellae approach to within 50-150 Å of each other because of drainage or stress thinning due to collisions, rupture takes place. It is not known whether electrostatic repulsions and/or molecular attractions between the two adsorbed protein films are of importance at these distances. Adsorbed protein films that are thick and elastic resist rupture.

The three most important attributes that serve to stabilize foams are low interfacial tension, high viscosity of the bulk liquid phase, and strong, elastic films of adsorbed proteins.

2. Assessment of Foaming Properties

Various methods are available to assess the foaming properties of proteins, the choice depending on whether bubbling, whipping, or shaking is used for foam formation.

Average bubble size can be determined, and this allows a rough estimation of total interfacial area.

Various expressions can be used to describe *foaming capacity*: (a) "*steady-state*" *foam volume* (100 × volume of foam/volume of initial liquid phase); (b) "*overrun*" [generally defined as 100 × (the total volume of the dispersion − volume of the original fluid)/volume of the original fluid] ; (c) "*foaming power*" (100 × volume of gas in foam/ volume of liquid in foam); (d) the gas in foam-gas-sparged ratio (bubbling method); or (e) *density of the foam* (see Fig. 19). The time necessary to reach a given foam volume or overrun may also be of interest.

The foaming power (F_p) generally increases with protein concentration (P_c) in the liquid phase until some maximum value is attained. The ability of various proteins to promote foaming can be compared by measuring the maximum F_p and the P_c corresponding to half the maximum F_p, as is done in Table 10. These values illustrate the high foaming power of gelatin at low protein concentration, which cannot be deduced from the classic determination of foaming power at a single relatively high protein concentration (1% w/v, for example).

Foam stability can be assessed by measuring: (a) the degree of liquid drainage or foam collapse (volume reduction) reached after a given time, (b) the time for total or half-drainage (or half-reduction in volume), or (c) the time before drainage starts.

The *strength* or *stiffness* of the foam can be assessed by measuring the ability of a column of foam to support a specific weight or by determining foam viscosity.

The behavior of the foam during heating is also of interest in some situations, for example, in meringues or angel food cakes. In these instances, sucrose is added to the protein solution before or during foaming, and the foam is subsequently heated.

Performances of protein samples in these various tests are very dependent on the equipment used and the test conditions. For comparisons among proteins, it is therefore necessary to use standard procedures and to include a protein standard. Egg white is the most frequently used standard since it possesses excellent foaming properties, and there are many occasions when one wishes to compare the foaming properties of inexpensive proteins with those of egg white.

Table 10 Comparative Foamability of Protein Solutions (pH 7, Whipping)

Protein	Max foaming power[a] (at 2-3% prot., w/v)	Prot. conc. (w/v) at 1/2 max. F_p	Foaming power[a] at 1% prot. (w/v)
Gelatin	228	0.04%	221
Sodium caseinate	213	0.10%	198
Soy protein isolate	203	0.29%	154

[a]Foaming power is defined as 100 × gas volume in foam/liquid volume in foam.
Source: From N. Kitabatake and E. Doi, *Agr. Biol. Chem. 46*:2177 (1982).

3. Environmental Factors Influencing Foam Formation and Stability (17,45,62,64)

Although many studies stress the importance of high protein solubility as a prerequisite to good foaming capacity and stability, it also appears that insoluble protein particles (such as myofibrillar proteins, micellar, and other proteins at pI) can play a beneficial role in stabilizing foams, probably by increasing surface viscosity. Although foam over-run is generally not great at the protein pI, foam stability is often quite good. This is the case with globin (pH 5-6), gluten (pH 6.5-7.5), and whey (pH 4-5) proteins and can be explained by the fact that electrostatic intermolecular attractions at the pI increase the thickness and rigidity of the protein films adsorbed at the air/water interface. With some proteins, however, an increase in foam stability is also observed at extremes of pH, possibly because of increases in viscosity. Egg white proteins display maximum foaming performances, both at their "natural" pH (pH 8-9), and near their pI (4-5). Most food foams are prepared at pHs different from the pIs of their protein constituents.

Salts also can affect the solubility, viscosity, unfolding, and aggregation of proteins, and this can alter foaming properties. Sodium chloride often increases overrun and reduces foam stability. This is probably mediated through a decrease in the viscosity of the protein solution. Calcium ions may improve foam stability by forming bridges between carboxyl groups of the protein.

Sucrose and other sugars often depress foam expansion but improve foam stability, because they increase the bulk viscosity. Thus, in making meringues and other foams, it is preferable to add sugar at a late stage, when foam expansion has already taken place. The glycoproteins of egg white (ovomucoid and ovalbumin) help stabilize foams because they absorb and retain water in the lamellae.

It is well known that low concentrations of contaminating lipids (down to 0.1%) seriously impair the foaming performances of proteins. Thus, soy protein preparations free from phosphatides, egg white proteins free from yolk lipids, and "clarified" whey proteins, or whey protein isolates with low lipid contents, display improved foaming properties as compared with their lipid-contaminated counterparts. It appears that surface-active, polar lipids interfere with the most desirable conformation of adsorbed protein films by situating themselves at the air/water interface.

As protein concentration is increased over a wide range (up to 10%), foam stability increases more than foam volume. Maximum overrun is generally achievable when protein concentration in the initial liquid phase is in the range 2-8% w/v (whipping method). This apparently results in favorable viscosity of the liquid phase and appropriate thickness of the adsorbed film. For a typical foam with 150 μm bubble diameters and $\phi = 0.95$, a protein concentration of 0.1% in the liquid before foaming gives an interfacial protein concentration of 1 mg/m^2, if completely adsorbed. This results in a film of surface rigidity sufficient to provide foam stability (45). An increase in protein concentration leads to smaller bubbles and stiffer foams. The aging of protein solutions prior to foaming may be beneficial to foam stability, probably because protein-protein interactions are enhanced, leading to thicker adsorbed films.

To form an adequate foam, the duration and intensity of stirring must be such that adequate protein unfolding and adsorption can occur. However, overly intense agitation can lead to a decrease in overrun and foam stability. Egg white is particularly sensitive to "overbeating." Whipping egg white or ovalbumin for more than 6-8 min causes partial aggregation-coagulation of the proteins at the air/water interface. These insolubilized proteins do not adsorb properly at the interface, causing the viscosity of the liquid lamellae to be insufficient for good foam stability.

Moderate heat treatments prior to foam formation are found to improve certain foaming properties of soy (70-80°C), whey (40-60°C), egg white (ovalbumin and lysozyme), and blood (serum albumin) proteins. These heat treatments cause overrun to increase, but foam stability may decrease. More severe thermal treatments than those described impair foaming capacity (Fig. 20). Heating of foams causes air expansion, decreased viscosity, bubble rupture, and foam collapse, unless gelation of the protein contributes sufficient rigidity to the adsorbed film to stabilize the foam. Egg white foams typically retain their structure when heated, whereas foams from whey proteins are less heat resistant.

4. Specific Foaming Properties of Proteins

Experimental evidence indicates that foam formation and foam stabilization require somewhat different protein properties. Foam formation involves the diffusion of soluble proteins toward the air/water interface, where they must unfold, concentrate, and spread quickly to lower interfacial tension. It has been well established that flexible molecules with few secondary and tertiary structures (e.g., β-casein) perform efficiently as surfactants. Prior unfolding of globular proteins through mild heating, exposure to denaturing

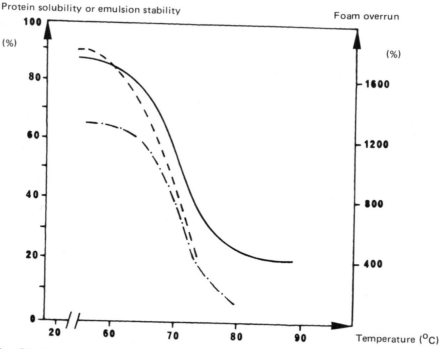

Fig. 20 Protein solubility, emulsion stability, and foam overrun of dried-rehydrated whey protein as influenced by heat treatment prior to drying. Whey proteins were concentrated by ultrafiltration, heated for 30 min at various temperatures (abscissa), dried, and then rehydrated. —, Protein solubility at pH 4.6 of heated WPC (%); —·—·—·, stability of emulsions prepared at pH 6.6 with heated WPC (%); – – – –, overrun of foams prepared at pH 6.6 with heated WPC (%). [J. N. De Wit and R. De Boer, *Neth. Milk Dairy J. 29*:198 (1975).]

agents (such as disulfide reducing agents), or partial proteolysis results in improved orientation at the interface and greater foam capacity, provided unfolding is not accompanied by aggregation and loss of solubility. As previously indicated, a direct correlation is observed between the surface hydrophobicity of a protein and its ability to lower surface and interfacial tensions (Fig. 18; Table 11). Hydrophobic derivatives of caseins and other proteins display improved foaming capacity, due to their better orientation and spreading at the air/water interface.

As mentioned, adequate foam stability requires protein characteristics somewhat different from those needed for foam formation. To stabilize a foam, a thick, cohesive, elastic, continuous, air-impermeable protein film must be present around each gas bubble. It appears that high-molecular-weight globular proteins that partly resist surface unfolding produce thick adsorbed films with good surface rheological properties and foams with good stability. It is also likely that, for such stable films to form, several layers of partly unfolded proteins must first associate at the interface through hydrophobic, and possibly hydrogen and electrostatic interactions. A critical balance apparently must exist between the protein's ability to undergo intermolecular cohesive interactions that lead to stable films, and the protein's tendency to self-associate excessively, resulting in protein aggregation and rupture of the lamellae. On the other hand, the protein must adsorb strongly at the air/water interface via hydrophobic interactions. This is necessary to prevent protein desorption and subsequent loss of drained liquid. Furthermore, sufficient flexibility and mobility of the protein molecules is required to counteract stress deformation, interface extension, and lamella thinning. Any extension of the interface results in a decrease in the concentration of molecules adsorbed at the interface and a rise in interfacial tension. Proteins, to provide effective foam stabilization, must be able to move from a region of low interfacial tension to a region of high interfacial tension, dragging with them underlying water molecules, and thereby restoring the initial thickness of the lamella ("Marangoni effect"). Finally, it is also desirable that polar side chains (or polypeptide loops) of the protein film should interact with water within the lamella and reduce its drainage.

Proteins with good foaming properties include egg white proteins, the globin part of hemoglobin, bovine serum albumin, gelatin, whey proteins, casein micelles, β-casein,

Table 11 Surface Activity and Hydrophobicity of Various Proteins

Protein	Water/air surface tension (dyne/cm)	Water/oil interfacial tension (dyne/cm)	Relative surface hydrophobicity
Bovine serum albumin	58	10.3	1400
κ_1-casein	54	9.5	1300
β-Lactoglobulin	60	11.0	750
Lysozyme	64	11.2	100
Trypsin	64	12.0	90
Conalbumin	64	12.1	70
Ovalbumin	61	11.6	60

Source: From A. Kato and S. Nakai (56).

wheat proteins (especially glutenins), soy proteins, and some protein hydrolyzates (with a low degree of hydrolysis). Structurally disordered β-casein (flexible random coil) rapidly lowers surface and interfacial tension and promotes quick foam formation. The adsorbed protein films, however, are thin and foam stability is poor. κ-Casein unfolds slowly during foam formation, probably because it is stabilized by an intramolecular disulfide bridge and spreads less at the interface than does β-casein. Consequently, foam formation is not rapid, but the resulting protein films are thick and strong and the final foam exhibits good stability. The highly ordered globular structure of serum albumin is flexible enough to allow partial unfolding and adsorption at the foam interface, and the residual structure of the adsorbed molecule is sufficient to result in good foam stability. In the case of egg white, the different physicochemical properties of the various protein components appear to complement each other, resulting in the rapid formation of foams that are low in density, stable and heat resistant.

Many similarities exist between emulsion and foam formation, but there is no strict correlation between the emulsifying and foaming abilities of proteins. Perhaps this can be explained by the fact that foam stability has a greater requirement for residual protein structure than does emulsion stability.

I. Flavor Binding

Some protein preparations, although acceptable from a functional and nutritional standpoint, necessitate a deodorization step to remove bound off-flavors. Various substances, such as aldehydes, ketones, alcohols, phenols, and oxidized fatty acids, may cause beany or rancid odors and bitter or astringent tastes. When bound to proteins or to other constituents, these substances are released and become perceptible after cooking and/or mastication. Some are so strongly bound that even steam or solvent extraction do not remove them.

Quite different from the problem of off-flavor removal, it may be useful to use proteins as carriers for desirable flavors. It is of interest, for example, to impart a meat flavor to textured vegetable proteins. Ideally, all the volatile constituents of the desirable flavor must remain bound during storage, possibly also during processing, and then be released quickly and totally in the mouth, without distortion. The problems mentioned above can only be solved through investigations of the mechanisms by which volatile compounds bind to proteins.

1. Interactions Between Volatile Substances and Proteins

The flavor of a food results from very low concentrations of volatile compounds near its surface, and these concentrations are dependent on a partition equilibrium between the bulk of the food and its "headspace." Franzen and Kinsella (37) have shown that the addition of proteins to a water-flavor model system lowered the concentration of volatile compounds in the headspace.

Flavor binding may involve adsorption at the surface of the food or penetration to the food interior by diffusion ("absorption"). Two types of adsorption onto solids can be differentiated: (a) reversible physical adsorption via van der Waals interactions, and (b) chemical adsorption via covalent or electrostatic linkages. In the first case the heat released by the reaction is less than 20 kJ/mol; in the second case it is at least 40 kJ/mol. Flavor binding by absorption implies the mechanisms mentioned, plus hydrogen and hydrophobic interactions. Polar compounds, such as alcohols, are bound via hydrogen linkages, but hydrophobic interactions with nonpolar amino acid residues predominate in the binding of low-molecular-weight volatile compounds.

In some cases, volatiles bind to proteins by covalent linkages and the process is usually irreversible. For example, binding of aldehydes or ketones to amino groups, and binding of amines to carboxyl groups, are usually irreversible, although reversible Schiff bases may form between carbonyl volatiles and ϵ- or α-amino groups of proteins and amino acids. Irreversible fixation is more likely to occur with volatiles of higher molecular weights (10% of octanal as compared to 50% of 2-dodecanal bind irreversibly to soy proteins, when tested at the same concentration), and this occurrence eliminates the flavor of the original volatile compound (21,22).

Binding of volatiles by proteins is possible only when binding sites are available; that is, the sites are not engaged in protein-protein or other interactions. The reversible, noncovalent, binding of a volatile compound to a protein obeys the Scatchard equation:

$$\frac{V_{bound}}{V} = K(n - V_{bound}) \quad \text{at equilibrium} \tag{34}$$

where V_{bound} is the moles of volatile compound bound per mole protein, V is the molar concentration of the free volatile compound, K is the association constant, and n is the number of binding sites per mole of protein.

Using the Scatchard equation, K and n can be calculated from experimental determinations of V_{bound} at equilibrium for different values of V. The V_{bound} increases when the concentration V of the free volatile increases. The Scatchard equation shows no dependence between V_{bound} and the protein concentration. This is true for proteins with a single polypeptide chain, such as bovine serum albumin, but is not true for oligomeric proteins, such as soy glycinins, in which the number of volatile molecules bound per mole of protein decreases as the protein concentration is raised (at a constant concentration of volatile compound). This results because protein-protein interactions increase with protein concentration. Sorption measurements (headspace analysis and dialysis equilibrium) have indicated that the number of binding sites increases as the binding of the volatile compound progresses, the curve from the Scatchard plot showing a decreasing slope with increasing V_{bound}. During binding, unfolding of the protein takes place (as can be followed by differential spectrophotometry), and more hydrophobic amino acid residues become available for further binding (21,22).

According to Solms et al. (102), apolar volatile compounds penetrate and interact with the hydrophobic core of the protein, thus displacing intra- or intermolecular protein-protein hydrophobic interactions. This occurrence favors protein destabilization and can result in a change in the protein solubility.

2. Methods for Assessing the Binding of Volatile Compounds

Sorption isotherms of a volatile compound on a protein carrier can be determined by equilibrating the protein in a closed space, with varying known initial concentrations of the volatile compound in the atmosphere. The equilibrium concentration of the free volatile compound, at a given temperature, can then be measured by gas-liquid chromatography. The capacity of the carrier to retain the bound volatile can be determined by extraction or distillation followed by measurement of the desorbed volatile.

3. Factors Influencing Binding

Any factor modifying protein conformation influences the binding of volatile compounds.

Water enhances the binding of polar volatiles but barely affects that of apolar compounds. In a dry protein ingredient, the diffusion of volatile compounds remains limited. Even a slight increase in water activity increases the mobility of the polar volatiles and enhances their ability to find binding sites. In more hydrated media, or in solutions, the availability of polar or apolar amino acid residues for the binding of volatiles is influenced by a number of factors. Casein binds more carbonyl, alcohol, or ester volatiles at neutral or alkaline pH than at acid pH. Ions, such as chloride and sulfate, that normally stabilize the native structure of globular proteins, may, at high concentrations, change the structure of water, weaken hydrophobic interactions, and lead to protein unfolding. This results in an improved binding of carbonyl compounds. Reagents that tend to dissociate proteins or to open up disulfide bonds usually improve the binding of volatiles. Dissociation of oligomers into their subunits, however, often decreases the binding of apolar volatiles, since the former intermolecular hydrophobic zones tend to become buried as the monomer changes its conformation. The effects of protein concentration have already been mentioned.

Extensive proteolysis decreases the binding of volatiles. For example, 6.7 mg 1-hexanal can be bound per kilogram soybean protein, whereas only 1 mg remains bound after proteolysis with an acid bacterial protease. Thus, proteolysis can be used to lessen

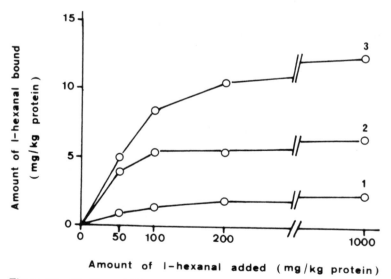

Figure 21 Binding of 1-hexanal by soy protein. Ten grams of acid-precipitated native soy proteins were dissolved in 100 ml water and various amounts of 1-hexanal were added. (1) Sample stirred for 5 hr at 20°C under N_2. (2) Sample stirred for 1 hr at 90°C under reflux. (3) Sample refluxed for 24 hr at 90°C. All samples were then freeze-dried, the 1-hexanal bound to the protein was released by dissolving the protein in NaOH, pH 13, and the 1-hexanal measured. [S. Arai, M. Noguchi, M. Yamashita, H. Kato, and M. Fujimaki. *Agr. Biol. Chem. 34*:1569 (1970).]

the beany off-flavor of soy proteins. Conversion of bound 1-hexanal to caproic acid by aldehyde dehydrogenase also reduces the off-flavor.

Protein denaturation by heat, on the contrary, generally results in increased binding of volatiles. For example, when a 10% aqueous solution of soy protein isolate is heated at 90°C for 1 or 24 hr in the presence of 1-hexanal and then freeze-dried, these heat treatments result, respectively, in three and six times greater binding of hexanal than that exhibited by the unheated control (Fig. 21) (5).

Dehydration processes, such as freeze-drying, often release over 50% of the volatiles initially bound to a protein, such as casein. Retention of volatiles is better when their vapor pressures are low and when they are present at low concentrations.

It should also be mentioned that the presence of lipids improves the binding and retention of various carbonyl volatiles, including those that may result from lipid oxidation.

J. Binding of Other Compounds

In addition to water, ions, metals, lipids, and volatile flavors, food proteins can bind a number of other substances through weak interactions or through covalent bonds, depending on their chemical structure. Examples include pigments, synthetic dyes (which may be used for the analytical determination of proteins), and substances with mutagenic, sensitizing, or other biological activities. Such binding may result either in enhanced toxicity or detoxification, and in some instances the nutritional value of the protein can be adversely affected. Some of these interactions are discussed in Sec. VII.A.

V. NUTRITIONAL ATTRIBUTES OF PROTEINS

A. Protein Metabolism

The primary function of dietary protein is to supply nitrogen and amino acids for the synthesis of body proteins and other nitrogen-containing substances.

1. Protein Digestion and Absorption

Food proteins are digested by proteolytic enzymes in the gastrointestinal tract; first by pepsin, present in the gastric juice, then by proteases secreted by the pancreas (trypsin, chymotrypsin, carboxypeptidases A and B, and elastase) and by the cells from the intestinal mucosa (aminopeptidases and dipeptidases). Most of these enzymes catalyze the hydrolysis of specific peptide bonds.

Free amino acids and small peptides are absorbed through the brush-border cells of the intestinal mucosa. Specific absorption mechanisms are operative for neutral, acid, or basic amino acids and for peptides. Most absorbed peptides are hydrolyzed within the intestinal cells. The absorbed amino acids then pass into the portal vein for transport to the liver.

For 100 g of food proteins ingested per day, only about 10 g is eliminated in the feces (Fig. 22), although an additional 50-70 g of "endogenous" proteins (digestive enzymes and proteins from the epithelial cells of the intestinal mucosa) are secreted daily into the gastrointestinal tract. These endogenous proteins also are digested and absorbed for the most part. This mechanism provides optimal amounts and proportions of amino acids for the synthesis of body proteins.

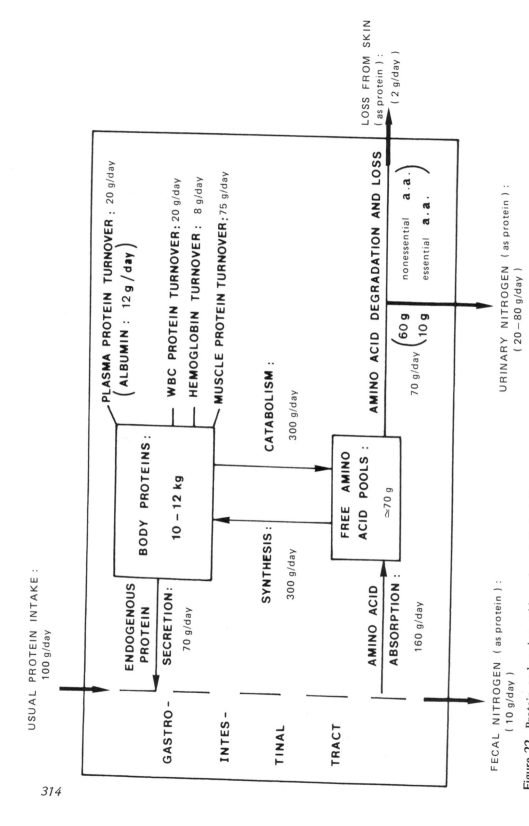

Figure 22 Protein and amino acid pools and their daily movements in a 70 kg man. [Adapted from H. N. Munro, in *Clinical Nutrition Update: Amino Acids* (Green, Holliday, and Munro, eds.), Am. Medical Assoc., Chicago, 1977, p. 141.]

314

2. Protein Anabolism and Catabolism (19,77-79,115)

Body proteins (10-12 kg in a 70 kg man) are continuously broken down into amino acids in situ (catabolism) and must therefore be resynthesized in corresponding amounts (anabolism). This "turnover" (dynamic equilibrium) is rapid (3-4 days) for some tissues, such as the liver and the intestinal mucosa, and less rapid (several months) for muscular proteins and for collagen from bone and connective tissues. Free amino acid "pools" present in the plasma and the tissues, and available for protein synthesis, are therefore provided both from ingested food proteins and from catabolized body proteins. Protein catabolism takes place primarily at the intracellular level, where various proteases, known as cathepsins, are located within the lysosomal organelles. It is believed that protein turnover is necessary to enable the level of biologically active proteins to adjust in accord with changing body needs. Because there is no perfect time and/or space synchronization between protein catabolism and anabolism, excess amino acids present at given times are degraded through the carbohydrate pathways (gluconeogenesis), with the production of energy. The nitrogen released is excreted into the urine, mainly as urea (Fig. 22). Some of the mechanisms responsible for these events are summarized as follows.

Plasmatic Free Amino Acids. The liver controls the flux of amino acids into blood and hence to the whole body. After a protein-rich meal, if the amino acids arriving at the liver are in excess of body requirements, amino acid-degrading enzymes are quickly induced and synthesized in the liver cells. Degradation occurs by transamination and oxidative deamination. The resulting glutamic acid and ammonia are further converted into urea in the liver.

The level of free amino acids in the peripheral blood (plasmatic amino acids) therefore only partly reflects the amino acid content of the meal. Dietary protein deficiency (as observed in some children) reduces the plasma level of most essential amino acids, especially "branched-chain" isoleucine, leucine, and valine. The plasma level of tryptophan influences its uptake by the brain, where it is converted into serotonin, a neurotransmitter.

Free Amino Acids in Tissue. The body pools of free amino acids amount to about 0.5% of the total protein-bound amino acids. The concentration of free amino acids in tissue depends not only on protein catabolism, protein anabolism, and ingested protein, but also on the fact that some nonessential amino acids (glycine, glutamic acid, glutamine, alanine, aspartic acid, and serine, which participate in numerous metabolic pathways) are synthesized in large amounts in tissue.

Muscular Protein Metabolism. Skeletal muscles represent about 45% of the body weight of adult men, and they contain a major proportion of the body proteins. In a 70 kg man, some 75 g of muscular proteins turn over each day. Muscular proteins constitute carbon reserves that can be used during fasting or other periods of caloric deprivation.

When the nitrogen intake is low, and the caloric supply is adequate, the breakdown of body proteins is minimal. This adaptation mechanism complicates the evaluation of daily protein requirements. Whereas a deficiency in dietary protein reduces the rate of protein catabolism, a protein-calorie deficiency initially increases this rate. Insulin and some other hormones enhance the synthesis of muscular proteins. The regulation of protein turnover is also affected by the level of essential amino acids—leucine and other branched-chain amino acids promote protein synthesis and inhibit protein breakdown in muscle. Muscular protein turnover decreases with age.

Metabolism of Plasma Proteins. Protein deficiency also decreases the hepatic synthesis, the blood level, and the catabolism of plasma proteins. The blood level of thyroxin-binding prealbumin, retinol-binding protein, total albumins, and/or prealbumins can apparently be used as an index of protein malnutrition.

Urinary Nitrogen. The final products of protein and amino acid catabolism are excreted in the urine in the form of urea (normally representing 80% of the urinary nitrogen), uric acid, creatinine, and ammonia. Fasting, or protein deficiency, markedly reduces nitrogen excretion in the urine, although excretion of ammonia increases. Creatinine production reflects muscular catabolism. Protein intakes in excess of body needs increase urinary nitrogen and do not lead to the accumulation of protein.

Protein and Amino Acid Fluxes in Humans. Summarized in Fig. 22 are the protein and amino acid fluxes in a young adult man. About 3% of the body proteins, that is, 4-5 g/kg per day turn over daily. This turnover decreases with age ($\sim$ 20 g/kg per day in the newborn infant; 7 in the 1-year-old child, and 3-5 in the adult).

B. Human Requirements for Proteins and Amino Acids (29,35,80,120)

1. Protein Requirements

The body metabolism of proteins can be expressed by the difference between nitrogen intake and nitrogen elimination. This is called nitrogen balance. If this difference is positive, as occurs during growth or repletion, then nitrogen retention occurs through tissue deposition and protein synthesis. If it is negative, as occurs during protein malnutrition, injury, or infection, for example, then nitrogen is lost. In the normal adult, the nitrogen balance is zero. Nitrogen equilibrium is maintained even when the protein intake largely exceeds body requirements, because excess protein is converted to energy and urea.

Protein requirements in adults can be assessed by measuring the minimal protein intake that will maintain nitrogen equilibrium, and in infants or children by measuring the minimal protein intake that will provide an optimal rate of growth. More frequently, however, protein requirements for adults are determined from the "obligatory" nitrogen loss during exposure to a protein-free diet ("factorial method"). After adult men have been on such a diet for a few days, the urinary loss of nitrogen falls to 37 mg/day per kilogram body weight, the fecal loss of nitrogen to about 12 mg, the skin, sweat, and other minor losses of nitrogen to about 5 mg (sometimes greater), and the total loss of nitrogen to about 54 mg/day per kg (29). These nitrogen losses are taken to represent the obligatory nitrogen requirement for maintenance of an adult. However, losses may vary between individuals, and a safety factor of 1.3 is incorporated in the final values. Moreover, it is known that about 1.3 g of an easily absorbed, efficiently utilized protein, such as egg or milk protein, is necessary to compensate for 1 g protein lost through catabolism. The minimum requirement of egg proteins (or equivalent) can therefore be estimated as $0.054 \times 1.3 \times 1.3 \times 6.25 = 0.57$ g protein per day per kg body weight. Finally, since the average food proteins of Western diets, unlike egg proteins, are absorbed and utilized only at an extent of 75%, the daily protein requirement is raised to 0.8 g protein per day per kg (roughly 56 g protein per day for a 70 kg man).

The daily protein requirements for adult men and women, for infants aged 0-6 months, and for children aged 10-12 years, are indicated in Table 12. These values, which are in agreement with the results of N-balance experiments, are recommended by the Food and Agriculture Organization and the World Health Organization of the United Nations (29), and by the National Academy of Sciences of the United States (35) as "safe protein intakes" for healthy people. Adults require less protein per kilogram than children, since protein turnover decreases with age and growth stops.

Adults in most Western countries exceed the safe protein intake by a factor of about 2, typically deriving 11-14% of their calories from protein, as compared with the recommended value of 8%.* There is evidence to indicate that very large protein intakes, in excess of 150-200 g/day for prolonged periods, may be detrimental to health, especially if the protein consumption involves meat rich in saturated fats.

Protein requirements are influenced by caloric intake. If the caloric intake is not sufficient, a part of the dietary protein is used for energy production. In adult men consuming the minimum recommended level of protein, nitrogen equilibrium can be maintained at somewhat lower nitrogen intakes if the nonprotein caloric content of the diet is increased. The protein-sparing effect of nonprotein calories varies from 0 to 1 mg N per kJ per kg body weight, becoming less effective as the diet becomes increasingly deficient in protein.

Protein requirements are increased during growth (infants and pregnant women) and product secretion (lactating women, lactating cows, and laying hens). Injury, burns, infections, parasitic attacks, previous malnutrition, and other factors may also increase protein requirements.

The value recommended for safe protein intake by adults is somewhat controversial. In developing countries, some adults have been found to remain in nitrogen equilibrium when given a rice protein intake as low as 0.44 g/day per kg (111). This may be possible because of a metabolic adaptation to habitually low protein intakes.

The long-term effects of minimal protein intake on health (organ function and resistance to diseases) and on work productivity are not known.It is known however, that young men fed amounts of egg protein just sufficient to maintain nitrogen equilibrium over extended periods of time undergo some biochemical abnormalities. For these reasons, many countries recommend protein intakes higher than those advocated by FAO and the NAS and more in line with usual Western diets. These recommendations are usually about 1 g protein per kg body weight per day, causing proteins to contribute about 12% of the calories in the diet. The value finally agreed upon is of great importance since it will influence the level of concern directed to the "protein gap" of developing countries, the planning goals for national food supplies, the dietary standards for assistance programs, the information contained in nutrition education programs, how data from food consumption surveys are interpreted, the requirements for nutrition labeling, and so on (114).

2. Requirements for Essential Amino Acids

Although nonessential amino acids are efficiently synthesized in the body, either from intermediary metabolites or from essential amino acids (cysteine from methionine and tyrosine from phenylalanine), eight "essential" amino acids cannot be synthesized by adult humans, or can be synthesized only at a negligible rate. Thus, these eight essential amino acids must be supplied in the diet (Table 12). The human infant also requires a dietary supply of histidine.

---

*This value results from the above-mentioned 56 g protein and from the recommended 3000 kcal (12556 kJ) per day for the "reference" man.

Table 12 Human Requirements for Essential Amino Acids and Provisional Amino Acid Patterns for Ideal Proteins

Amino acid	Infant requirement (0-6 months; mg/day per kg)	Provisional, ideal pattern (infant; mg/g protein)	Child requirement (10-12 years; mg/day per kg)	Provisional, ideal pattern (child; mg/g protein)	Adult requirement (mg/day per kg)	Provisional ideal pattern (adult; mg/g protein)
Histidine	28	14	0	0	0	0
Isoleucine	70	35	30	37	10	18
Leucine	161	80	45	56	14	25
Lysine	103	52	60	75	12	22
Methionine (+ cysteine)[a]	58	29	27	34	13	24
Phenylalanine (+ tyrosine)[b]	125	63	27	34	14	25
Threonine	87	44	35	44	7	13
Tryptophan	17	8.5	4	4.6	3.5	6.5
Valine	93	47	33	41	10	18
Total essential amino acids	742	372.5	261	325.6	83.5	151.5
Total protein requirement (egg or milk proteins)	2000		800		550 570: man 520: woman[c]	

[a]Cysteine may supply up to one-third the need for total sulfur amino acids.
[b]Tyrosine may supply up to one-third the need for total aromatic amino acids.
[c]Pregnant women need an extra 1-10 g protein per day. A lactating woman secretes about 850 ml milk per day. This contains about 10 g protein, thus necessitating an extra 17 g protein per day (egg protein or equivalent) in the diet. This calculation is based on the product of the two previously mentioned correction factors (1.3 × 1.3 = 1.7).

Source: From Ref. 29. Copyright Food and Agriculture Organization of the United Nations.

It is well established that diets deficient in one or several essential amino acids do not support proper growth and may lead to increased morbidity or mortality and early brain damage with impaired learning ability. The exact daily requirements in essential amino acids, however, are not known. Their determination has been attempted by use of the nitrogen balance method together with diets in which the nitrogen was supplied by free amino acids. The provisional values advocated by FAO (29) are shown in Table 12. Using these values together with information about daily protein requirements, FAO has calculated provisional, essential amino acid patterns of ideal protein diets for infants, children, and adults (Table 12). These protein diets are believed to correspond to the body's need for each essential amino acid, and the amino acids should, therefore, be fully utilized.

Protein synthesis at the ribosomal level ceases or is greatly slowed when one or more of the amino acids are deficient (mean concentration is too low or rate of intake is too variable). As a consequence, the other amino acids that are adequately provided in the diet will be partly utilized as a source of energy rather than for tissue growth or maintenance.

When the ratio of total essential amino acids required to total protein required (E/T) is considered, it is found to be 37% for infants and 15% for adults. Essential amino acid requirements for adults are low because adults are able to efficiently recycle these amino acids. This explains why proteins from cereals are better utilized by adults than by infants. During growth, however, the essential amino acids must be provided by the diet for tissue deposition. The E/T ratio of the overall body proteins is close to 45%. Age and disease influence the nutritional adequacy of a given dietary protein source. The influences that level of protein intake, and possible interactions among amino acids, have on requirements for essential amino acids are not precisely known (see Sec. C.1.b).

Recent research indicates that the rate of whole-body protein synthesis (and turn-over) increases as diet levels of lysine or leucine are increased well above their requirement values (Table 12). Since the optimal rate of protein turnover is not known, it is difficult to determine *optimal* requirements for essential amino acids.

C. Protein Nutritive Value of Foods (10,40,41)

1. Factors Influencing Protein Value

The protein nutritive value of food corresponds to its ability to meet nitrogen and amino acid requirements of the consumer, and to ensure proper growth and maintenance. This ability is a function of several factors.

Protein Content. Staple foods with protein contents below 3% (cassava and potato) do not meet the protein requirements of humans even when ingested in amounts supplying more than the caloric requirements. On the contrary, a diet of cereals (with an 8-10% protein content) meet the protein requirements of adults, provided enough is eaten to supply the caloric requirements. This probably explains why, in developing countries, protein-calorie malnutrition (marasmus) is much more common than malnutrition attributable solely to protein deficiency (kwashiorkor).

Protein Quality. The quality, value, or balance of a food protein depends on the kinds and amounts of amino acids it contains, and represents a measure of the efficiency with which the body can utilize the protein. A balanced or high-quality protein contains essential amino acids in ratios commensurate with human needs. Proteins of animal origin generally are of higher quality than those of plant origin. This can be determined by comparing the amino acid contents of various proteins with the FAO reference pattern (Table 13). The FAO reference pattern was chosen to satisfy the requirements of the young child.

Cereal proteins are often low in lysine, and in some instances they lack tryptophan and threonine. Oilseeds and nuts are often deficient in methionine and lysine, whereas legumes often lack methionine. Those essential amino acids in greatest deficit with respect to requirements are designated "limiting" amino acids.

When the dietary pattern of amino acids differs greatly from the ideal pattern, this situation is referred to as "amino acid imbalance." It can lead to reduced efficiency of amino acid utilization, depressed growth, increased susceptibility to diseases, and/or permanent impairment of mental capabilities in children.

It is possible to overcome, or partially overcome, amino acid deficiencies by providing a diet containing several proteins with complementary amino acid patterns, and this can be accomplished even if each protein alone is of low quality. Thus, cereal proteins can be efficiently complemented with small amounts of soy or milk proteins, but soy proteins are less efficiently complemented by small amounts of animal proteins. Improvement of essential amino acid balance, and of overall protein quality, also can be achieved by supplementing the diet with free amino acids (usually L-lysine or DL-methionine), or possibly with proteins to which amino acids have been covalently attached (see Sec. VII. A).

Supplementation and complementation are widely used in animal feeding. Complementation widely occurs in the mixed diets of the Western world and in some other countries. It is generally true that a diet in which 30-40% of the proteins comes from animal sources contains a proper balance of amino acids. FAO surveys of protein supplies in 1977 indicated that the mean daily availability of animal and vegetable proteins was, respectively, 73 and 34 g per person in the United States, compared with 12 and 44 g per person in developing countries. Attempts have been made to add lysine to cereal diets consumed by some human populations of Africa and the Middle East. However, these attempts have not resulted in physiological improvements, possibly because cereals alone met the protein requirements.

Excessive supplementation may lead to "amino acid antagonism" or even toxicity. Antagonism results in increased requirements for some amino acids because the dietary level of another amino acid has been increased. Thus, a leucine intake well in excess of requirement (as can result from the ingestion of corn and sorghum proteins by chickens) depresses tryptophan and isoleucine utilization, and therefore increases the requirements of these two amino acids. Large excesses of methionine, cysteine, tyrosine, tryptophan, and histidine are known to cause growth inhibition, reduced food intake, and certain pathological conditions in animals.

Amino Acid Availability. Amino acids present in dietary proteins are not necessarily fully "available," since digestion of the protein or absorption of the amino acids may be incomplete. Amino acids from animal proteins are generally digested and absorbed to an extent of 90%, whereas those from certain plant proteins may be digested and absorbed to an extent of only 60-70%. The lower utilization of certain proteins may be due to several factors: (a) *Protein conformation*: fibrous insoluble proteins are less readily attacked by proteases than are soluble globulins. However, protein denaturation by mild heating often enhances digestion. (b) *Binding* of metals, lipids, nucleic acids, cellulose, or other polysaccharides to proteins may partially impair digestion. (c) The presence of *antinutritional factors*, such as trypsin and chymotrypsin inhibitors, can impair digestion of proteins. Other inhibitors are able to impair amino acid absorption. (d) The *size* and the *surface area* of the ingested protein particles can influence digestibility. For instance, fine milling of flours tends to improve the digestibility of cereal proteins. (e) *Processing* at high temperatures, at alkaline pH, or in the presence of reducing carbohydrates often decreases protein digestibility and the biological availability of several amino acids, especially lysine (see Sec. VII.A). (f) *Biological differences* exist among individuals, and this can influence their ability to digest proteins and absorb amino acids.

The timing of amino acid release may be as important as the extent of digestion, since this can influence the efficiency with which amino acids are utilized. For instance, fast release and absorption of amino acids can lead to their increased degradation in the liver. Competition may also exist among amino acids at absorption sites, particularly if one amino acid is present in large excess. These factors may explain why proteins possess a greater nutritive value than equal weights of corresponding mixtures of free amino acids. However, excessively slow release of amino acids in the digestive system can also be undesirable, since absorption mechanisms do not exist in the lower bowel (where the intestinal flora metabolizes the remaining nutrients).

2. Determination of Protein Nutritive Value

Since proteins differ in nutritive value, evaluation of this aspect is useful (a) to allow prediction of the amount of food protein, or mixture of food proteins, necessary to meet the amino acid requirements for growth or maintenance, (b) to allow the ranking of proteins as a function of their potential nutritive value, and (c) to allow the detection of nutritive changes food proteins may undergo during processing and storage.

Biological or chemical means are used to evaluate protein nutritive value.

Bioassays. Biological assays are based on measurements of growth or nitrogen retention in experimental animals, such as rats, or in humans, as a function of protein intake. For reliable accuracy and meaningfulness of the data, several animals must be used per test, and the results must be analyzed statistically. Test conditions also must be standardized. The protein level of the diet is generally kept low ($\sim$ 10% by weight) so that protein intake remains below requirements, and the supply of energy and other nutrients must be adequate. Under these conditions, growth is slow, the protein is efficiently utilized (little protein is degraded to energy), and the experimental results emphasize differences in nutritive values among proteins and reflect the maximum nutritive value of each protein tested. However, the test value obtained overstates how the protein will perform under the practical conditions of human consumption.

The *protein efficiency ratio* (PER) is the weight (grams) gained by rats per gram protein consumed. Because this value is easy to determine, it is the most commonly used method. Inaccuracies arise because the result obtained depends on the amount of protein actually consumed by the rats, and because rats grow faster than children. Thus protein quality is understated. When the weight loss of a group of rats fed a protein-free diet is included in the calculation, proteins can be evaluated for their ability to support maintenance as well as growth, and a value known as net protein ratio (NPR) is obtained.

$$\text{NPR} = \frac{(\text{weight gain}) + (\text{weight loss of protein-free group})}{\text{protein ingested}} \qquad (35)$$

These methods can be improved if (a) values are determined at several different levels of protein in the diet, a plot of change in weight versus protein intake is prepared, and the slope of this plot is measured on the straight line portion of the response curve (multipoint slope ratio), and (b) the values obtained with the test protein are expressed as a percentage of those obtained with egg or milk protein (relative NPR). When both modifications are adopted, the protein values obtained (relative protein values, RPV) are more accurate and more useful in practical situations.

Another approach involves measuring the uptake of nitrogen and loss of nitrogen in the feces and urine. By this means, it is then possible to calculate the percent of ingested nitrogen absorbed (*coefficient of protein digestibility*), and the percentage of absorbed nitrogen retained in the body (*biological value*). The biological value reflects the balance of essential amino acids in the absorbed protein digest. The product of the coefficient of digestibility and of the biological value is called the *net protein utilization* (NPU) or the percentage of dietary nitrogen (or protein) retained. For "true" (versus apparent) values, the endogenous fecal and urinary losses of nitrogen (measured on a protein-free diet) should be taken into account as follows.

$$NPU = \frac{N_i - [(N_f - N_{f,e}) + (N_u - N_{u,e})]}{N_i} \times 100 \tag{36}$$

where i is the ingested, f is the fecal, and u is the urinary nitrogen, and e stands for endogenous. This method is easily applicable to nongrowing humans. For small growing animals, the nitrogen retained can be determined directly by total carcass analysis:

$$NPU = \frac{\text{body N} - \text{body N of animal on protein-free diet}}{N_i} \times 100 \tag{37}$$

When a bioassay is carried out by making one amino acid nutritionally limiting while all other free essential amino acids are added to the diet in appropriate amounts, the result indicates the bioavailability of the single limiting amino acid in the test protein.

Protein quality and amino acid availability also can be determined in humans or animals by a number of other methods: by repletion tests (regeneration of plasma proteins, or of body weight); by measuring changes in enzyme activity levels; by monitoring changes in the plasma level of free essential amino acids after ingestion of the test protein under standard conditions; and by measuring levels of urea in the plasma and the urine.

The various bioassays tend to provide essentially the same ranking with respect to the nutritive value of different proteins (Table 13). Most of these assays can also detect deterioration of protein quality during processing, provided the process affects the limiting amino acid. All bioassays with the exception of the one that involves measurement of plasma amino acids, give a result based on a limiting amino acid and do not provide information on the balance of other amino acids. Furthermore, none of the bioassays can be used to predict the complementary value of proteins in mixed diets (47). Also, bioassays do not identify the limiting amino acid of the test protein or diet, unless several additional experiments are carried out in the presence of added free amino acids. Finally, bioassays are expensive and time consuming, and extrapolating animal results to humans involves considerable uncertainty.

Table 13 Essential Amino Acid Contents and Nutritional Values of Some Protein Foods

Amino acid (mg/g protein)	Human milk	Cow's milk	Hen's egg	Meat (beef)	Fish (all types)	Wheat grain	Rice, brown	Soy-beans	Reference patterns FAO, 1973	Reference patterns NAS, 1980
Histidine	26	27	22	34	35	25	26	28	0	17
Isoleucine	46	47	54	48	48	35	40	50	40	42
Leucine	93	95	86	81	77	72	86	85	70	70
Lysine	66	78	70	89	91	31[a]	40[a]	70	55	51
Methionine + cysteine	42	33[a]	57	40	40	43	36	28[a]	35	26
Phenylalanine + tyrosine	72	102	93	80	76	80	91	88	60	73
Threonine	43	44	47	46	46	31	41	42	40	35
Tryptophan	17	14	17	11	11	12	13	14	10	11
Valine	55	64	66	50	61	47	58	53	50	48
Total essential amino acids without histidine	434	477	490	445	450	351	405	430	360	356
Protein content (%)	1.2	3.5	12	18	19	12	7.5	40		
Chemical score[b] (%) (based on FAO ref. pattern)	100	94	100	100	100	56	73	80		
Protein efficiency[b] ratio (PER) (on rats)	(4)	3.1	3.9	(3)	3.5	1.5	2	2.3[c]		
Biological value[b,d] (%) (BV) (on rats)	(95)	84	94	74	76	65	73	73[c]		
Net protein utilization[b] (%) (NPU) (on rats)	(87)	82	94	67	79	40	70	61[c]		

[a]Limiting amino acids in diet.
[b]Chemical score, PER, biological value, and NPU are devined in Sec. V.C.2.
[c]Heat-treated soybeans.
[d]The best Western diets have a biological value of 0.8; those of most developing countries vary between 0.7 and 0.6.
Source: From Refs. 28 and 29. (Copyright Food and Agriculture Organization of the United Nations.) Values in parentheses are from sources other than those cited.

Chemical Methods. In most chemical methods, a protein's nutritional value is assessed on the basis of its content of essential amino acids as compared with the human requirements for these amino acids. This is a logical approach but has shortcomings, as is mentioned later. A protein "chemical score" is defined as

$$\frac{\text{mg primary limiting amino acid per g test protein}}{\text{mg same amino acid per g reference protein}} \times 100 \qquad (38)$$

The 1973 FAO reference amino acid pattern (Table 13), which is based on the essential amino acid requirements of young children, is now considered the preferred reference protein (replaces the formerly used whole-egg protein). The 1980 NAS-NRC reference essential amino acid pattern is equally suitable. Chemical scores based on lysine, sulfur amino acids, tryptophan, or threonine are probably the only ones of practical importance since these amino acids appear to be the only limiting ones in most human diets. The chemical scores of various food proteins are indicated in Table 13.

The chemical score allows one to calculate the requirement of a single food protein or of a protein mix. For example, the requirement of a given protein mix would be calculated as

$$\begin{array}{c}\text{Required intake} \\ \text{of protein mix}\end{array} = \frac{\text{recommended intake of egg protein}}{\text{chemical score of protein mix}} \times 100 \qquad (39)$$

Knowledge of the chemical scores (for all essential amino acids) of various proteins permits calculation of the complementary value of different proteins in a mixture. Widely and successfully used in animal feeding, the chemical score also has been used to develop protein-rich food mixes for developing countries. However, this approach may underestimate the quality of a protein for adults, since it is based on the amino acid requirements of the young child (Table 12).

Another drawback of the chemical score method lies in the fact that accurate analyses of tryptophan and sulfur amino acids (and eventually of the proportions of L- and D-amino acids) require special techniques (18). It also does not account for the negative effects of excess amino acids or of antinutritional factors present in the protein food. Furthermore, it does not compensate for differences in the digestibility of proteins or in the biological availability of specific amino acids.

The correlation between the results of bioassays and the results of chemical scores is improved when the chemical scores are corrected based on overall protein digestibility (85). The latter can be determined by rapid in vitro enzymic tests. Such corrected amino acid scores can be correlated with PER values to give a computed PER (C-PER). Data on available lysine (see Sec. VII.A.) obtained from the chemical method of Carpenter, or from dye-binding methods, also can be used to correct chemical scores.

Several biological experiments recently performed on growing children in developing countries have shown that the chemical score (based on the FAO reference pattern) does permit a correct prediction of the amount of protein (or of a protein mix) needed to meet the essential amino acid requirements of growth (47).

Enzymic and Microbial Methods. Enzymic methods for evaluating protein quality are based on the measurement of free essential amino acids released after the test protein has been exposed to the action of one or more proteases under standard conditions. Such methods can provide estimates of protein digestibility, protein value, and/or bioavailability of specific amino acids (72,106). Their use is of great interest for rapidly evaluating the damage incurred by protein foods or feeds (milk powder, blood meal, and others) during industrial processing and storage.

Some microorganisms, especially the protozoan *Tetrahymena pyriformis*, possess essential amino acid requirements similar to those of humans and the rat. These microorganisms also possess their own proteolytic enzymes. Their growth on test proteins can therefore be used as an index of protein quality (T-PER) or amino acid availability.

Although of wide interest, concerns about protein quality and amino acid availability are of practical importance only when the intakes of total protein are marginal or inadequate. They are of greater importance in animal feeding, especially when fast growth is required.

VI. UNCONVENTIONAL PROTEIN SOURCES

A. Need for Increased Protein Production (3,4,71,75,103)

It is generally believed that about 1 billion humans currently suffer from chronic malnutrition. This drastic situation may become even worse, since the world population is expected to increase from 4.4 to 6.4 billion people between the years 1980 and 2000. Thus the United Nations foresees for the year 2000 a deficit of 0.4 billion tons of cereals based on the expected production. A large proportion of this deficit will be ascribable to the needs of animal feeding, since the projected increase in individual incomes is expected to increase the consumption of animal proteins. Indeed, the expected needs indicate, for the year 2000, a deficit of some 100 millions tons of protein-rich meals (expressed in terms of soybean meal containing 44% protein).

It is obvious that the world production of protein must be increased, and it is from this viewpoint that the possibilities of new protein resources made possible by modern chemistry and technology are briefly reviewed in the following pages. To provide a beginning perspective for this discussion, data in Table 14 are useful. They include the approximate world production, the yield per hectare per year, and the price per kilogram of the main plant and animal proteins. The yield of animal proteins per hectare is low, since 3-20 kg vegetable proteins are necessary to produce 1 kg animal protein, depending on the animal species. It is accordingly true that a reduction in the production of animals would lead to an increase in the direct consumption of vegetable proteins.

In many rich countries, there is an overconsumption of animal proteins, with total protein consumption in excess of human needs. When it is suggested that we "vegetalize" our diets, one should not forget, however, that considerable pleasure is derived from the consumption of animal foods, and that other advantages accrue from the raising of animals. For example, animal production encourages the setting up of stocks of cereals, is not very dependent on soil fertility and rainfall, and allows types of feeds to be used (animal by-products, plant by-products, proteins from microorganisms, urea, and ammonia) that otherwise probably would not be used in food production.

Cereals represent the main protein and energy supply for both human and animal nutrition, and attempts to increase the production of cereals has met with considerable success. Genetic selection and improved agricultural practices have led to greatly increased cereal yields and to higher protein or lysine contents of wheat, corn, and rice grains. A new cereal species resulting from the crossing of wheat and rye, triticale, has been created. It yields more protein per hectare than the best varieties of wheat. This "green revolution" has greatly enhanced the cereal production of many developing countries. Recent investigations aim at the selection of varieties that can give reasonable yields without irrigation or the addition of fertilizers.

The production of animal proteins (milk, meat, and eggs) has benefited from the selection of new breeds, artificial insemination, improved control of disease, and better

Table 14 World Production, Yield, and Cost of Important Food Proteins

	Cereals (grains)	Oilseeds	Leguminosae (except oil seeds)	Legumes (roots and tubers)	Alfalfa (leaves)	Meats	Fish	Milk and dairy products	Eggs
Production (10^6 tons protein/year) (1978-1979)	140	40	8.6	8.3	—	18	13	15	3
Yield (kg protein/ha per year) (1978-1979)	200-700	500-1200	200-1000	—	1800-3000	50-200	—	50-400	—
Cost ($U.S./kg protein) (1978-1979)	1	0.8	1	7	0.7	17	11	12	10

feeding practices. New biological techniques (production of growth hormones, transfer of embryos, synchronizing of births, production of twins, and increase in animal size through genetic manipulations) also hold great promise. The development of aquaculture, the fishing of unexploited oceanic species (e.g., krill), new fishing techniques (electronic detection, light or chemical attraction, and others) are also likely to increase the availability of less costly animal proteins (105). Furthermore, modern food technology aids in the more efficient use of proteins through such practices as mechanical deboning to better recover meat and fish flesh, ultrafiltration to recover proteins from whey, and modified procedures for manufacturing cheese so that more of the whey proteins are retained.

B. Approaches to Utilization of Plant Proteins (9,23,55,84,101,117-119)

The most significant approaches to making proteins more widely available, both for humans and for animals, relate to improved procedures to increase utilization of leguminous and other protein-rich plants and to the production of single-cell proteins. Shown in Table 15 are the compositions of some vegetable sources of food proteins.

Very large amounts of defatted meals from soy and other oilseeds, containing 40-45% protein, are used to enhance the growth of animals, especially pigs and poultry.

Seeds from leguminous plants have been traditionally consumed by humans, and they constitute an important protein complement to cereals and starchy foods. The use of oilseeds as a source of proteins for humans is, however, less well developed, except in the Far East, where soybeans are used in traditional foods, such as tofu, yuba, miso, and soy sauces. These foods are prepared directly from the seeds, either by grinding into "soy milk" and precipitating the proteins, or by fermentation. In Japan, 1 million tons of soy seeds are processed in this way each year, and they contribute significantly to the protein intake.

C. Separation and Purification of Proteins from Plants

In the United States and Europe, soybeans intended for human consumption are first ground and then treated with an organic solvent (*n*-hexane) to remove the oil. This is done under mild thermal conditions to retain the functional properties of the proteins. The resulting soy flour, with a protein content of about 45%, is further treated as follows, to obtain a product that is still richer in protein.

1. Soluble carbohydrates (oligosaccharides) and minerals can be extracted with acidified water, with a water-ethanol mixture, or with hot water. Most soy proteins remain desirably insoluble under these conditions, but the use of water acidified at the protein isoelectric pH is the best approach to minimizing unfolding, aggregation, and loss of functional properties. The result is a *protein concentrate* that contains, after drying, about 65-75% protein, 15-25% insoluble polysaccharides, 4-6% minerals, and 0.3-1.2% lipids.
2. Alternatively, the proteins in the defatted soy flour can be solubilized in alkaline water followed by filtration or centrifugation to eliminate the insoluble polysaccharides (including fibers). Reprecipitation at the pI (4.5), followed by centrifugation and washing of the protein curd, removes soluble carbohydrates and salts. After drying (usually spray drying), a *protein isolate* is obtained, which contains 90% protein, or more. Protein isolates are more expensive than protein concentrates, even on on a protein basis, because of the extra processing and chemicals, and because recovery of the initial flour proteins rarely exceeds 75%.

Table 15   Composition and Nutritional Value of Various Protein-Rich Vegetables

Plant	Dry solids (g per 100 g)			Moisture (g per 100 g)	Protein efficiency ratio[a]	Limiting amino acid	Toxic or antinutritive factors
	Protein	Lipids	Carbohydrates				
Soybean	45	25	15	7	2.3 (cooked)	met	Trypsin inhibitors, hemagglutinins, saponin, phytate, α-galactosides
Cotton seed	53	30	10	6	2.3	lys, met	Gossypol
Peanut	25	50	20	9	1.7	met	Aflatoxins
Sunflower seed	30	40	26	6	2.1	lys	Polyphenols
Rapeseed	26	33	35	3	2.6	Balanced	Glucosinolates, sinapine
Lupine	40	10	45	10	0.7	met	Alkaloids
Pea	23	2	70	12	2	met	Trypsin inhibitors, phytate
Fieldbean	30	2	60	12	1.8	met	Vicine, convicine, tannins, trypsin inhibitors
Alfalfa (leaf)	25	5	>45	80	1.6	met	Coumarins, estrogens, tannins, glycosides, alkaloids
Wheat (grain)	14	2	78	10	1.8	lys	
Oats	13	10	72	9	2.2	lys	
Corn	10	5	80	10	1.2	lys, trp	
Rice	8	1	87	12	1.7	lys	
Millet	11	3	81	11	0.5	lys, thr	
Sorghum	11	4	80	12	0.7	lys, thr	
Potato (tuber)	9	~0	73	78	2	lys, thr	Solanine
Cassava	3	0.8	90	63	–	met	Cyanogenetic factors
Yam	9	0.7	84	73	–	met, lys	

[a]Calculated on the basis of a PER = 2.5 for casein.

A similar "wet" extraction and purification of protein constituents can be carried out with defatted meals from other oilseeds (peanut, cottonseed, sunflower seed, and rapeseed, for example), or with flours from oil-poor seeds (fieldbean, pea, chickpea, and some other leguminous plants). Alternatively, a dry process, called air classification, can be applied to finely ground flours from oil-poor seeds, leading to the separation of a protein-rich concentrate and a starch-rich fraction. Such separation is based on the differences in size and density between "protein bodies" and starch granules.

Removal of undesirable low-molecular-weight constituents and knowledge of the protein solubility curve (see Fig. 15) are the main factors to be considered when devising a wet extraction process for the efficient recovery and purification of plant proteins. Thus, for the preparation of sunflower protein concentrates, it is necessary to prevent oxidation of polyphenols into pigments that bind covalently to the protein. This is accomplished by removing polyphenol-rich hulls, extracting oil from the kernel under mild conditions, and treating the defatted flour with a 50:50 v/v ethanol-water mixture to efficiently remove remaining polyphenols (Table 16).

With most meals or flours, wet concentration processes remove a large proportion of the low-molecular-weight toxic or antinutritional factors (gossypol, α-galactosides, aflatoxin, and others), but concentrate antinutritional factors of a protein nature (trypsin inhibitors and hemagglutinins; see Table 15). It is therefore necessary to inactivate these compounds by heat at a high water activity (see Sec. VII.A). Since this treatment affects the functional properties of the major proteins, it is usually carried out after the protein concentrate has been incorporated into the food (i.e., during baking, cooking, or processing).

It is technically feasible to prepare protein flours, concentrates, and isolates from the various seeds just mentioned. So far, however, for economic reasons, soybeans are the primary commercial source of such products. These products are used for protein enrichment of various foods, for the partial replacement of meat (at low cost) in meat products, and, more generally, for the functional properties they impart to foods in which they are incorporated (see Sec. IV).

It should be noted that plant leaves, where photosynthesis and protein synthesis take place, are a potentially unexhaustible source of protein (87). The vegetative green parts from many graminaceous and leguminous crop plants (such as cereals, sugar cane, soy, and alfalfa) contain 80% water and 2-4% protein. When fresh leaves or grasses are

Table 16 Composition of a Dehulled, Defatted Sunflower Flour and of the Protein Concentrate Resulting from Extraction of This Flour with an Ethanol-Water Mixture

Constituent	Defatted sunflower flour (% w/w)	Sunflower protein concentrate (% w/w)
Protein	58	67
Ethanol-soluble carbohydrates	10	0.3
Ash	8.2	0.3
Polyphenols	4.6	0.2
Lipids	2	2

Source: From Gueguen (1982), personal communication.

cut, ground, and pressed, they yield a green juice that contains 10% dry solids, 40-60% of the total initial proteins, and almost no fiber. The fibrous residue, partially dehydrated by pressing, is an excellent fodder for ruminants. The juice contains insoluble proteins linked to chloroplasts and soluble proteins. To remove low-molecular-weight growth-depressing factors, a protein curd is formed by heating the juice to 90°C. The washed and dried curd contains about 60% protein, 10% lipids, 10% minerals, and various pigments and vitamins (chlorophylls, xanthophylls, and carotenes). It is used commercially as a poultry feed since it imparts a yellow coloration to the skin and flesh, and to the egg yolks. Similar protein concentrates have been tested with success on children suffering from kwashiorkor.

D. Single-Cell Proteins (20,38,109)

It is possible to produce proteins from microorganisms on a commercial scale. Starches, sucrose from molasses, or lactose from whey can be used as fermentation substrates.

Certain yeasts of the genus *Candida* are able to grow on paraffinic substrates (*n*-alkanes) in the form of purified paraffin or even as crude gas-oil. Continuous fermentors, able to produce yearly some 100,000 tons of yeast biomass, have been built. The productivity can reach 2 g dry yeast per hour per liter of fermentor working volume. The alkane is first oxidized to the corresponding fatty acid, then degraded by β oxidation. The formation of biomass corresponds approximately to the global equation

$$C_{20}H_{42} + 1.84NH_3 + 32O_2 \longrightarrow 13.6C_1H_{1.66}N_{0.135}O_{0.5} + 12.46H_2O$$
$$+ 6.4CO_2 + 10.5 \text{ kJ} \tag{40}$$

where $C_1H_{1.66}N_{0.135}O_{0.5}$ represents the average cellular composition.

The yeast cells are recovered by centrifugation, washed with several solvents, and then dried. It is possible to obtain 1.2 ton dry yeast (containing 63% protein) from 1 ton alkanes and 0.11 ton ammonia.

At the present time, it is found more economical to grow *Pseudomonas* bacteria on 5% solutions of methanol in water than to grow yeast. The yield per ton of methanol is close to 0.4 ton dry bacteria containing 80% protein. The productivity for continuous culture is 3.5 g dry bacteria per hour per liter of fermentor working volume.

With yeasts and bacteria, continuous fermentation at a medium growth rate in the presence of excess nitrogen (such as ammonia) results in optimal cell composition, with high levels of proteins and lysine and low levels of nucleic acids. Sulfur amino acids are, however, somewhat deficient in amount. Thermal and/or alkaline treatments of the cells may be necessary to increase protein digestibility, nucleic acid removal, and amino acid availability. The nutritive value of yeasts and bacteria so treated has been checked on several animal species, and no toxic effects have been reported, even after long-term tests. These preparations are used to replace or supplement soy or fish meal for pigs and poultry, and to replace or supplement milk for preruminant calves. Considerable work has been carried out on isolating and purifying proteins from microbial sources, with the aim of producing materials suitable for human consumption.

Some molds can grow on inexpensive cellulosic substrates (from agricultural or other by-products), but the most promising substrates for filamentous molds, such as amylolytic strains of *Aspergillus, Rhizopus,* or *Fusarium,* are crushed wet cereals or heat-gelatinized starch doughs (from cassava, potatoes, and bananas, for example). After a few days of incubation at 30°C, the mycelium on the surface contains substantial amounts of protein.

Photosynthetic unicellular algae (*Chlorella, Scenedesmus*, or *Spirulina*) also have been proposed as sources of protein.

Large-scale production of microorganisms is independent of climate and land quality. If microorganisms were extensively used as animal feed, large amounts of the cereals, soybeans, and fish now fed to animals would be available as human food. However, this approach is discouraged because protein from defatted soy meal is currently less expensive than protein from microorganisms.

E. Chemical Synthesis and Genetic Engineering of Amino Acids and Proteins

Amino acids, synthesized either chemically (DL-methionine) or by fermentation (L-lysine, L-threonine, and L-tryptophan), are useful since they can be added to vegetable proteins to achieve an improvement in nutritive quality of the latter substances. Methionine, especially in combination with soy protein, is commonly used in the production of animal feeds.

Feeding of chemically synthesized urea allows the protein content of feeds for ruminants to be reduced. Microorganisms present in the rumen first transform urea into ammonia, then into microbial proteins, which these animals are able to hydrolyze and assimilate at a later point in the gastrointestinal tract.

When feeding proteins of high nutritional value to ruminants, it is useful to protect these proteins against microbial attack (decarboxylation and deamination) in the rumen by treating them with small amounts of aldehydes or tannins prior to feeding. This results in partial cross-linking of the proteins, allowing them to reach the intestine undegraded and thereby enabling them to contribute more fully to the yields of milk, meat, or wool. For methionine or lysine enrichment of ruminant feeds, it is also preferable to covalently attach these amino acids to protein constituents prior to feeding.

The possibility of growing differentiated vegetable or animal tissues or cells in vitro has engendered great interest with respect to the production of metabolites, antibodies, and so on. With regard to the cultivation of isolated vegetable cells, it is already possible to build a whole plant from a single cell or a meristem. This vegetative multiplication in vitro has brought about the rapid development of new hybrid varieties. It is currently possible to obtain haploid plants and homozygote diploid plants and to effect in vitro mutagenesis of isolated cells. The information acquired on the genetic behavior of the cytoplasm (fusion of the protoplasts and male cytoplasmic sterility) has allowed interspecific hybrids to be created. A very powerful method is thus available for giving birth to new varieties with better quantitative and qualitative performance.

The techniques of genetic recombination also provide the possibility of transferring, to isolated cells of a host plant, the genes responsible for rate of growth, resistance to pathogenic agents, the ability to fix nitrogen from the atmosphere, and so on. The genes governing the fixation of nitrogen are present in certain microorganisms: bluegreen algae, photosynthetic bacteria, *Spirillum* sp. associated with the roots of some tropical graminaceous plants used as fodder and with rice and corn (2-50 kg nitrogen fixed per hectare per year), and *Rhizobium* sp. associated with the root nodules of leguminosae (alfalfa, soy, peanut, beans, and pea; 50-500 kg nitrogen fixed per hectare per year). A bacterial enzyme, nitrogenase, mediates the energy-intensive reduction of N_2 to NH_3. The required energy is provided through plant photosynthesis. The selection of bacterial variants that can hyperproduce in situ is a possibility in the near future. One may further contemplate the transformation of certain *Rhizobium* sp. so as to enhance the efficiency of their symbiosis with leguminosae, or to give them the aptitude to recognize surface receptors present on rootlets of other plants, such as cereals, potatoes, and sugar beet. The yields, and eventually the levels of proteins, might be much increased if the soil were inoculated with this kind of bacterium. The successful introduction into a plant of the genes responsible for nitrogen fixation would negate the need for symbiosis with bacteria and the need for expensive nitrogen fertilizers. This would revolutionize agricultural production. So far, *nif* (nitrogen fixing) genes have been cloned, and it is possible to transfer them from one bacterium to another, causing the recipient organism to become a nitrogen fixer (26).

Another approach involving recombinant DNA is to modify bacteria so they can synthesize animal or plant proteins. Ovalbumin has already been synthesized in this way. The practical use of this approach is, however, still distant, because the amounts of protein produced per bacterial cell are small (current yields of recombinant proteins may only represent 0.001% of the total cell proteins). It also has been suggested that the sweet peptide aspartame, and the sweet proteins thaumatin and monellin, might be synthesized in this manner. Also, the bacterial synthesis of animal growth hormones could greatly enhance meat production.

Various fermentation processes, including the production of single-cell protein, could be improved by genetically modifying the producing microorganisms. By circumventing unnecessary metabolic pathways that use energy (ATP), better yields of desirable products might be obtained. Also, the introduction of genes coding for cellulases would facilitate utilization of cheap cellulosic substrates.

Achievement of controlled artificial photosynthesis, and in vitro enzymic syntheses of amino acids by the action of solar energy, are potentially important, but even more remote possibilities.

VII. MODIFICATIONS OF FOOD PROTEINS THROUGH PROCESSING AND STORAGE

A. Changes in Nutritive Value and Toxic Effects

The effects of domestic cooking and industrial processing on food proteins have been studied extensively, and a number of general reviews are available (6,14,16,31,34,52,53, 73,88,107,108).

Food processing has overall beneficial effects since it decreases food spoilage and enables a wide variety of seasonal foods to be available worldwide in shelf-stable forms that are appealing, nutritious, wholesome, and safe. In most cases food processing has little or no adverse effect on the nutritional value of proteins, and in some instances improvements may occur. However, some unfavorable reactions also can take place, usually at the level of the primary structure, resulting in a decrease in the content of essential amino acids or in the formation of antinutritional and possibly toxic derivatives. The decrease in nutritional value is of no significance when the damaged amino acid does not constitute the limiting nutritional factor in the diet, or when the damaged protein represents only a small part of the protein intake of the diet. Such damage can, however, be very detrimental when the diet consists of a restricted number of foods, such as milk, some cereals, or pulses, and/or has a minimal content of protein. These situations are encountered most frequently with infants, elderly persons, and poor populations.

1. Denaturation by Moderate Heat Treatments

Most food proteins display biological or functional properties only within a narrow temperature range. Heat causes structure modifications of proteins, which in the case of aqueous solutions of globular proteins, reduces solubility. When thermal treatments remain moderate, there is no disruption or formation of covalent bonds, and the primary structure remains unaffected. Changes resulting from these mild heat treatments are usually beneficial from a nutritional standpoint.

Blanching or cooking leads to the inactivation of enzymes, (lipases, lipoxygenases, proteases, polyphenoloxidases, and glycolytic enzymes, among others) that otherwise could cause the formation of undesirable colors or flavors, undesirable changes in texture, and/or a decrease in vitamin content. Also, the thermal destruction of myrosinase in rapeseed prevents the formation of a goitrogenic compound (5-vinyl-2-thiooxazoli-done) from the endogenous glucosinolates.

Most protein toxins or antinutritional factors naturally present in foods are denatured and inactivated by heat. This is the case with most protein toxins produced by contaminating microorganisms (botulinum toxin is readily inactivated at 100°C, but not the enterotoxin from *Staphylococcus aureus*). Seeds or leaves from leguminosae (soy, peanut, beans, broadbeans, peas, and alfalfa, for example) contain proteins that bind to and inhibit proteolytic enzymes in vivo and consequently reduce the digestion and nutritional value of ingested proteins (67). For example, trypsin inhibitors (Kunitz and Bowman inhibitors) and chymotrypsin inhibitors are present in soybean seeds (66). These heat-labile, proteinaceous inhibitors cause pancreatic hypersecretion and hyperplasia, together with growth reduction in several animal species (Table 17). It appears that excretion of the inhibitor-trypsin complex in the feces leads to severe losses of sulfur amino acids (trypsin is rich in these amino acids). Such losses increase the deficiency of sulfur amino acids in diets of soy proteins (small content of cysteine and methionine) (66).

Phytohemagglutinins (or lectins) from leguminosae are heat-labile proteins that can bind to polyosidic molecules. Their ingestion decreases the nutritional value of native vegetable proteins, presumably by forming a complex with membrane polyosides from

Table 17 Biological Effects of Raw Soybean Meal on Various Animals

Species	Growth inhibitors	Pancreas	
		Size	Enzyme secretion
Rat[a]	Yes	Increase	Increase
Chicken[a]	Yes	Increase	Increase
Pig[a]	Yes	No change	Decrease
Calf	Yes	No change	Decrease
Dog	No	No change	Temporary increase
Human[b]	Unknown	Unknown	Unknown

[a]Adult animals maintain body weight, but pancreas effects still occur.
[b]Two adults in a 9-day feeding trial had positive nitrogen balance for both raw and autoclaved soy flours.
Source: From J. J. Rackis, *J. Amer. Oil. Chem. Soc. 51*:161A-174A (1974).

the enterocytes (intestinal border cells), thus impairing amino acid transfer and digestibility. Toxic effects also have been observed.

All these antinutritional factors are denatured and inactivated when the seeds, the flours, or the protein concentrates are heated in moist conditions by such methods as autoclaving, extrusion cooking, sterilization, baking, or domestic cooking (Fig. 23). Since a moderate heat treatment increases markedly the nutritional value of vegetable proteins for some animal species, heating of the plant protein component of animal feeds is routinely practiced.

Finally, a number of proteins (such as soy glycinins, collagen, and ovalbumin) become more readily digestible after moderate heat processing. This is attributable to protein unfolding and the exposure of previously buried amino acid residues, thus enabling proteases specific for these amino acids to act more quickly and extensively.

2. Amino Acid Losses Through Protein Fractionation

Techniques used for purification, concentration, or separation of individual proteins may lead to protein ingredients with an overall amino acid content different from that of the raw material. This can occur during the preparation of protein isolates from soy and other vegetable sources. For example, the soluble protein constituents lost during isoelectric precipitation contain more sulfur amino acids than the other protein fractions. For similar reasons, protein inhibitors and toxic factors may be either eliminated or concentrated in the final "purified" protein preparation.

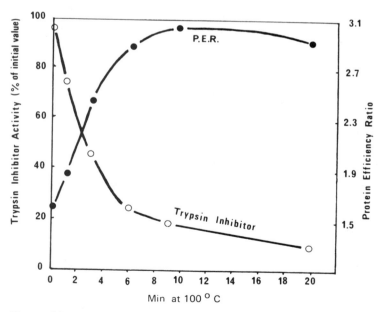

Figure 23 Effect of atmospheric steaming on trypsin inhibitor activity and protein efficiency ratio of soybean meal fed to rats. [From J. J. Rackis, *J. Amer. Oil Chem. Soc.* *51*:161A-174A (1974).]

3. Destruction of Amino Acids

Thermal treatments of proteins or proteinaceous foods in the absence of any added substance can lead, depending upon their intensity, to desulfuration, deamidation, isomerization and other chemical modifications of amino acids, and this is sometimes followed by the formation of toxic compounds.

Thus, thermal treatments, like sterilization at temperatures above 115°C, bring about the partial destruction (irreversible chemical modification) of cysteine and cystine residues and the formation of hydrogen sulfide, dimethylsulfide, and cysteic acid. These reactions have been observed with meats, fish muscle, milk, and a number of protein model systems. Hydrogen sulfide and other volatile compounds produced contribute to the flavor of these heated foods.

Deamidation reactions take place during the heating of proteins at temperatures above 100°C. The ammonia released comes mainly from the amide groups of glutamine and asparagine, and these reactions do not impair the nutritional value of the proteins. However, as a consequence of the unmasking of carboxyl groups, the isolectric pH and therefore the functional properties of the proteins are modified. Deamidation may be followed by the establishment of new covalent bonds between amino acid residues.

Thermal treatments carried out in the presence of oxygen lead to the partial destruction of the tryptophan residues in proteins. The kinetics of this reaction appear to be of the first order with respect to tryptophan concentration, and the activation energy is close to 110 kJ/mol, in a model system consisting of a 5% casein solution at pH 7 (see Sec. V.C).

Severe heat treatments at temperatures above 200°C, as well as heat treatments at alkaline pH, cause isomerization of amino acid residues. This involves the β-elimination reaction and formation of a carbanion, which after protonation leads randomly to L or D forms. When the reaction is complete, a racemic mixture of the amino acid residue is obtained.

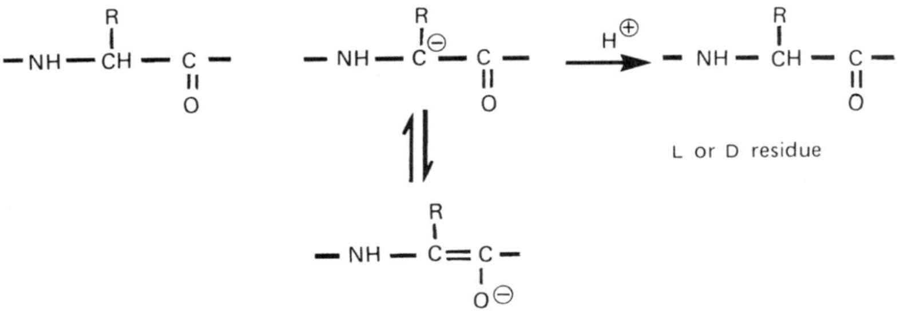

$$-NH-\underset{\underset{O}{\overset{\|}{C}}}{\overset{\overset{R}{|}}{CH}}-C- \qquad -NH-\underset{\underset{O}{\overset{\|}{C}}}{\overset{\overset{R}{|}}{\overset{\ominus}{C}}}-C- \qquad \xrightarrow{H^{\oplus}} \qquad -NH-\underset{\underset{O}{\overset{\|}{C}}}{\overset{\overset{R}{|}}{CH}}-C-$$

L or D residue

$$-NH-\underset{\underset{\ominus}{\overset{|}{O}}}{\overset{\overset{R}{|}}{C}}=C-$$

(41)

Since most D-amino acids have no nutritional value, racemization of the residue of an essential amino acid thus reduces its nutritional value to about 50%. The presence of D-isomers also reduces the digestibility of the protein, because peptide bonds involving D residues are less easily hydrolyzed in vivo than those containing only L residues. Moreover, certain D amino acids exert a toxic action, in proportion to the amount absorbed through the intestinal wall.

Severe heat treatments applied to proteins may result in the formation of cyclic derivatives, some of which possess a strong mutagenic action. Thus, above 200°C tryptophan can be transformed, by cyclization, into α, β, or γ carbolines. Derivatives of α-carboline have been isolated from pyrolyzates of tryptophan and proteins ($R_1 = NH_2$; $R_2 =$ H or CH_3), derivatives of β-carbolines from pyrolyzates of proteins (R_3 = H or CH_3; Norharman or Harman derivatives), and derivatives of γ-carbolines from pyrolyzates of tryptophan (R_3 = H or CH_3, $R_5 = NH_2$, and $R_6 = CH_3$).

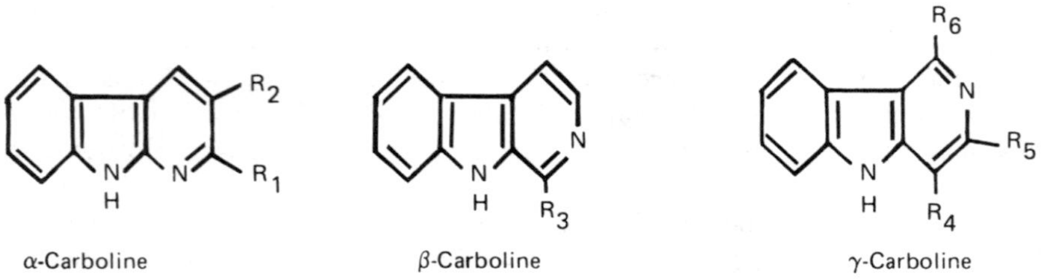

(42)

α-Carboline   β-Carboline   γ-Carboline

When submitted to heat treatments at alkaline pH, certain amino acid residues are destroyed. Arginine is converted into ornithine, urea, citrulline, and ammonia. Cysteine is converted into dehydroalanine. The levels of serine, threonine, and lysine are also reduced during heating at alkaline pH.

Many of the chemical reactions that affect amino acid residues are often accompanied by protein-protein reactions involving the formation of covalent bonds.

4. Protein-Protein Interactions

Some proteins in the native state are but partly digestible because natural covalent bonds are present between polypeptide chains. This is true of collagen, elastin, and keratin. In collagen the type of covalent bond varies. The desmosine residues present in this protein result from the condensation of four molecules of lysine. Covalent isopeptide bonds of the ϵ-N-(γ-glutamyl)lysyl or ϵ-N-(β-aspartyl)lycyl type, as well as disulfide or ester bonds, also have been identified.

Effects of Heat Treatments at Alkaline pH. Thermal treatments carried out at alkaline pH (or severe heat treatments near neutrality) lead to the formation of amino acid residues, such as lysinoalanine, lanthionine, and ornithinoalanine and the formation of intra- or intermolecular covalent cross-links. These cross-links result from the condensation of residues of lysine, cysteine, or orthinine with a residue of dehydroalanine (DHA). The latter is formed through a β-elimination reaction starting with cysteine or phosphoserine residues.

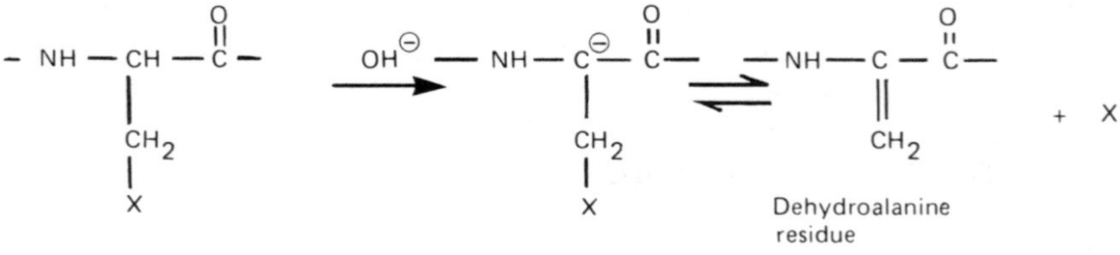

(43)

where X = SH or OPO_3H_2.

The resulting DHA residues are very reactive, combining readily with the ϵ-amino groups of lysine residues, the δ-amino groups of ornithine residues, and the sulfhydryl groups of cysteine residues, thereby producing cross-links involving lysinoalanine, ornithinoalanine, and lanthionine, respectively.

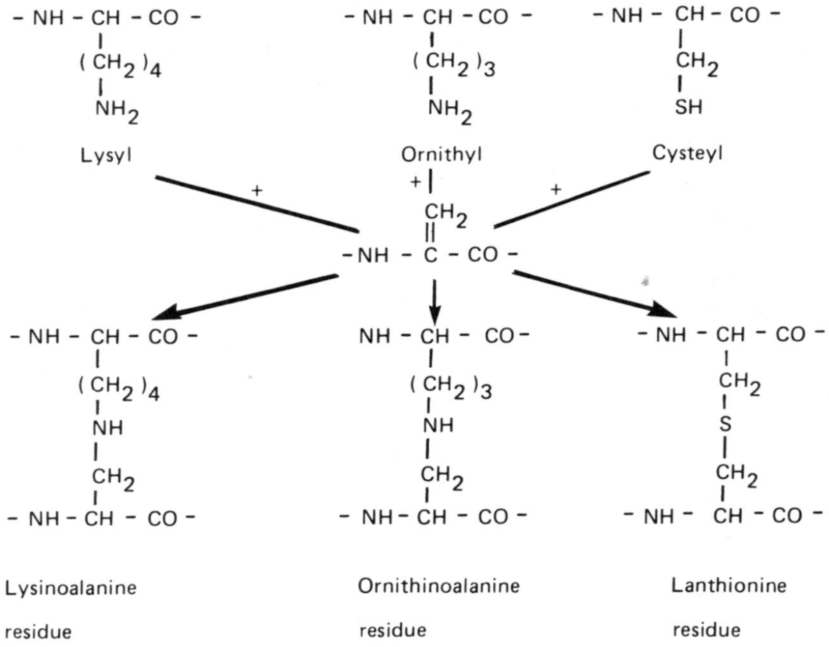

(44)

Lysinoalanine residue

Ornithinoalanine residue

Lanthionine residue

It has been suggested that unusual derivatives also may be formed by condensation between dehydroalanine and other amino acid residues, such as arginine, histidine, threonine, serine, tyrosine, and tryptophan.

The presence of ammonia may prevent the production of these cross-linking compounds, apparently by reacting with DHA to form β-aminoalanine. Cysteine, glucose, and sodium bisulfite or hyposulfite, as well as previous acetylation or succinylation of lysine residues, also decrease the formation of lysinoalanine in proteins during alkaline treatments. The amino acid content and the three-dimensional structure of proteins appear to determine the extent of lysinoalanine formation as much as the severity of the alkaline treatment.

The nutritional value of proteins in which covalent bonds of this kind are formed is often lower than that of the native proteins. Various investigations have shown that the protein efficiency ratio, the net protein utilization, and sometimes the biological value of proteins are lowered in proportion to the severity of the treatment applied (alkalinity, temperature, and time).

In the rat, the ingestion of proteins containing lysinoalanine is often accompanied by diarrhea, pancreatic hyperplasia, and loss of hair. The formation of covalent bonds, the isomerization of essential amino acid residues, and possibly the appearance of toxic substances are responsible for such manifestations. The ingestion of free (100 ppm) or protein-bound (3000 ppm) lysinoalanine by the rat induces nephrocytomegalia, nephrokaryomegalia, and nephrocalcinosis. Protein-bound lysinoalanine is excreted in the feces to the extent of about 50%. Absorbed lysinoalanine is largely eliminated in the urine. It has been shown that $[^{14}C]$ lysinoalanine is partly catabolyzed in the kidneys of rats. Urinary catabolites in the rat are numerous, and some of them are different from those that have been identified in other animal species. The toxic effects of lysinoalanine in the rat may therefore be partly due to the formation of these unusual derivatives, and lysinoalanine formation thus may be of concern only in the rat. Quails, mice, hamsters, and monkeys do not show any renal injury after ingestion of lysinoalanine.

The alkaline treatments used for food processing (solubilization of vegetable proteins, preparation of caseinates, detoxification of oilseed meals containing aflatoxin, and cooking of corn in lime) along with the relatively moderate thermal treatments employed, result in the formation of only small amounts of lysinoalanine and its relatives.

Effects of Severe Heat Treatments. Thermal treatments in excess of those used for sterilizing foods, when applied to proteins in model systems or to protein foods containing only small amounts of carbohydrates (meat and fish), may lead to the formation of covalent isopeptide cross-links of the type ϵ-N-(γ-glutamyl)lysyl or ϵ-N-(β-aspartyl)lysyl between lysine and glutamine or lysine and asparagine residues, respectively. This reaction may involve up to 15% of the lysine residues of a protein. From a nutritional standpoint, formation of these bonds brings about

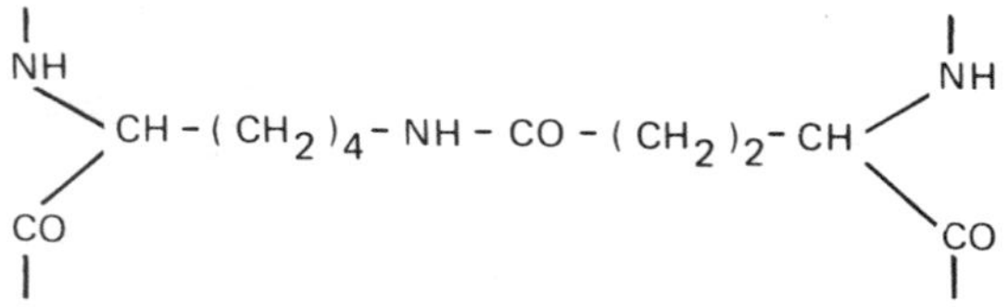

$$\begin{array}{ccc} | & & | \\ NH & & NH \\ \diagdown & & \diagup \\ CH-(CH_2)_4-NH-CO-(CH_2)_2-CH & \\ \diagup & & \diagdown \\ CO & & CO \\ | & & | \end{array} \tag{45}$$

$$\epsilon\text{-}N\text{-}(\gamma-\text{Glutamyl})-\text{L-Lysyl}$$

decreases in nitrogen digestibility, protein efficiency ratio, and biological value of the affected protein. Moreover, the nutritional availability of many amino acids, in addition to lysine, may be severely reduced. These glutamyl-lysyl or aspartyl-lysyl cross-links, by means of steric hindrance, apparently prevent proteases from reaching the sites of hydrolysis, and therefore slow down the in vivo digestion of the proteins. It is known that free ϵ-N-(γ-glutamyl)L-lysine is used as a source of lysine in the rat and the chicken, while ϵ-N-(β-aspartyl)-L-lysine is not (34).

Effects of Other Treatments. Covalent inter- or intramolecular cross-links may arise in proteins submitted to γ-irradiation or stored in the presence of oxidizing lipids. Polymerization starts from free radicals formed mainly from α-carbons of amino acid residues:

$$P_H + LOO\cdot \longrightarrow P\cdot + LOOH \tag{46}$$

| Native protein | Lipid free radical | Lipid peroxide |

The formation of protein free radicals $P\cdot$ is followed by polymerization of the polypeptide chains:

$$P^{\cdot} + P^{\cdot} \longrightarrow P - P \qquad\qquad (47)$$

$$P - P \xrightarrow{\text{LOO}^{\cdot}} P - P^{\cdot} \xrightarrow{P^{\cdot}} P - P - P \qquad\qquad (48)$$

and so on. In low-moisture foods, γ-irradiation may also cause scission of polypeptide chains.

Oxidative cross-linking of tyrosine residues can occur in the presence of H_2O_2 and peroxidase to give dityrosine residues.

5. Interactions Between Protein and Oxidizing Agents (100)

Various technological processes involve oxidizing agents, and these may cause modification of the amino acid residues of proteins. Hydrogen peroxide, for example, is used commercially for both its bactericidal and bleaching properties. In the dairy industry, it is employed as a "cold sterilant" for milk to be used in certain kinds of cheeses, for storage tanks, and for containers of the Tetrapak type. It also improves the color of fish protein concentrate, cereal flours and meals, and protein isolates from oilseeds. Finally, hydrogen peroxide has been suggested for the detoxification of aflatoxin-containing meals and flours and for the husking of seeds, the latter use stemming from its ability to facilitate swelling of seeds in an aqueous medium. Other peroxides, such as benzoyl peroxide, are used for the bleaching of flour and in some instances for the bleaching of whey powder.

Sodium hypochlorite is very widely used because of its bactericidal properties. For example, it has been suggested for use as a bactericidal spray for meat. The chemical properties of this chlorinated derivative also make it useful for detoxifying peanut meals contaminated with aflatoxin. Once the lactone ring of aflatoxin has been opened in an alkaline medium it can be easily oxidized by such agents as sodium hypochlorite, and the derivatives formed appear to be devoid of toxicity.

Lipid peroxides and their degradation products result from lipid oxidation, and they are present in many food systems. These peroxides are often responsible for the degradation of nearby protein constituents. Amino acid residues also can undergo oxidative modifications as a result of photoxidation reactions, irradiation, sulfite-trace metal-oxygen systems, hot air drying, and aeration during fermentation processes.

Finally, it should be mentioned that polyphenols present in many plants are easily oxidized into quinones in the presence of oxygen at neutral or alkaline pH. The peroxides formed from this reaction are strong oxidizing agents.

The amino acid residues most sensitive to oxidative reactions are the sulfur amino acids, tryptophan, and, to a lesser extent, tyrosine and histidine.

Oxidation of Methionine. Strong oxidizing agents, for example, performic acid, are able to oxidize methionine residues into residues of methionine sulfone. Hydrogen peroxide is able to oxidize free or protein-bound methionine into methionine sulfoxide and sometimes into methionine sulfone. Exposure of a 10 mM solution of methionine, pH 5-8, or a 5% casein solution, pH 7, to 0.1 M hydrogen peroxide at 50°C induces after about 30 min a complete conversion of methionine to methionine sulfoxide.

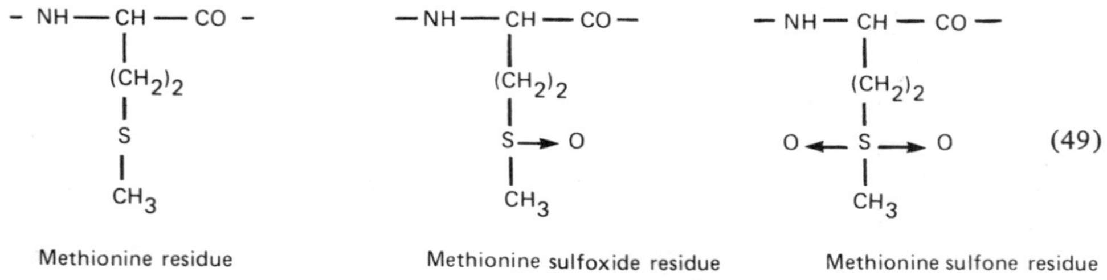

Methionine residue   Methionine sulfoxide residue   Methionine sulfone residue

$$(49)$$

This compound also can be formed by exposing proteins to sodium hypochlorite, to a sulfite-manganese-oxygen system, to oxidizing lipids, or to oxidizing polyphenols. In the presence of light, oxygen, and sensitizing dyes (e.g., riboflavin or methylene blue), methionine residues undergo photoxidation reactions to yield methionine sulfoxide.

The nutritional effects of oxidizing protein-bound methionine have been investigated. Methionine sulfone is biologically unavailable to the rat and may even show a certain degree of toxicity. Free or protein-bound methionine sulfoxide, may, however, replace methionine in the diet of rats or chicks with an efficiency that varies depending on its configuration (L or D). Casein in which all methionine has been oxidized to sulfoxide exhibits, for the rat, a PER or NPU about 10% lower than that of the control casein. It appears that methionine sulfoxide is set free from proteins during digestion, and is absorbed and then reduced to methionine before being used for protein synthesis. Rats receiving oxidized casein show increased levels of free methionine sulfoxide in the blood and muscles, which may indicate that the in vivo reduction of the sulfoxide is slow.

Oxidation of Cysteine and Cystine. The oxidation derivatives of cysteine and cystine, as follows, are numerous. Product identification is difficult because of their instability; however, some have been identified by nuclear magnetic resonance (NMR). From a nutritional standpoint, L-cystine mono- and disulfoxides and cysteine sulfenic acid are able to replace, at least in part, L-cysteine. Cysteic acid and cysteine sulfinic acid cannot replace L-cysteine.

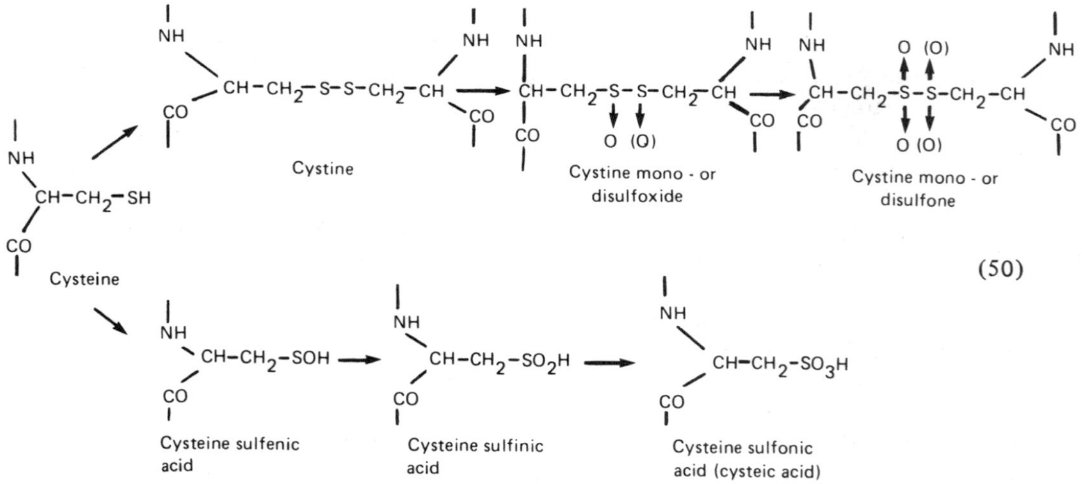

$$(50)$$

The ease with which these derivatives form in foods is not precisely known.

Oxidation of Tryptophan. The reactivity of the indole ring in the presence of strong oxidizing agents generally has been investigated in model systems that bear little relation to the conditions encountered during the commercial processing and handling of food. Thus, the treatment of free tryptophan with peracids leads essentially to the formation of β-oxyindolylalanine and *N*-formylkynurenine.

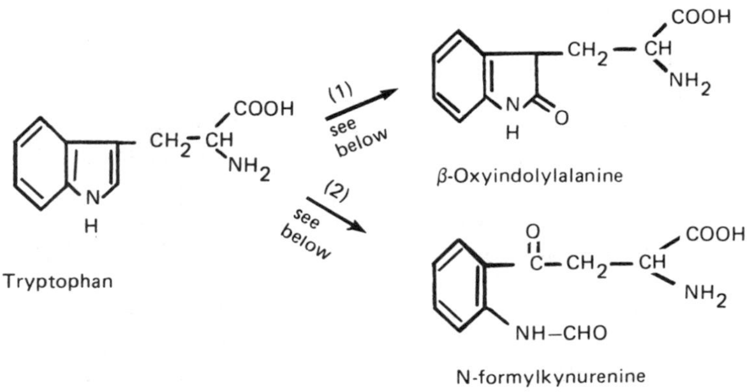

Tryptophan

β-Oxyindolylalanine

N-formylkynurenine

(51)

Notes for reaction (51):

(1) In the presence of RCOOOH, CH_3SOCH_3, and

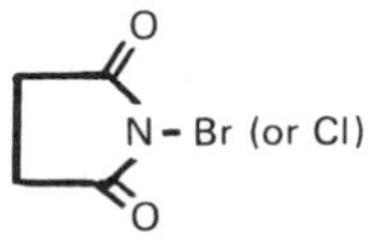

N – Br (or Cl)

(2) In the presence of RCOOOH, $NaIO_4$, O_3, and O_2 + hν

The reactions between tryptophan and dimethylsulfoxide or *N*-bromosuccinimide lead to the formation of β-oxyindolylalanine, as shown in reaction 1. Treatment of a protein with *N*-bromosuccinimide (or *N*-chlorosuccinimide) results first in hydrolysis of the peptide bond in which the carboxyl group of tryptophan is involved and then partial conversion of tryptophan to β-oxyindolylalanine. Free or peptide-bound tryptophan is oxidized to *N*-formylkynurenine in the presence of sodium periodate or ozone or by photoxidation in the presence of riboflavin (reaction 2). In the latter case, such derivatives as β-carboline, hexahydropyrroloindole, and quinazoline also may be formed. Tryptophan is also oxidized in the presence of hydrogen peroxide; by high-performance liquid chromatography it is possible to detect three or four derivatives more polar than tryptophan that absorb at 278 nm. One of the derivatives corresponds to kynurenine. Through ion-exchange chromatography of H_2O_2-oxidized tryptophan, many compounds can be separated that react with ninhydrin. The oxidation of tryptophan by hydrogen peroxide is therefore not a "simple" chemical reaction.

From a nutritional viewpoint, kynurenine, formylated or not, cannot replace tryptophan, at least for the rat. Also, kynurenine is carcinogenic when injected into animal bladders, and tryptophan degradation products, as a group, inhibit growth of cultured mouse embryonic fibroblasts and exhibit mutagenic activities.

The rates of oxidation of free tryptophan and methionine by hydrogen peroxide differ widely. At pHs near neutrality, methionine (10 mM) is oxidized many thousand times faster than tryptophan (10 mM) (experiments at 50°C with 0.01-1 M H_2O_2). It is therefore likely that contacts between peroxides and proteins induce, first, oxidation of methionine and perhaps also cysteine residues, and then tryptophan.

6. Interactions Between Proteins and Carbohydrates or Aldehydes (25,27)

Nonenzymic browning (the Maillard reaction) takes place during the processing or storage of protein foods containing reducing carbohydrates or carbonyl compounds (such as the aldehydes and ketones derived from oxidation of lipids). Since many of the chemical reactions involved in nonenzymic browning possess a high activation energy, they are markedly enhanced during cooking, heat processing, evaporation, and drying. Their rates are greatest in foods with an intermediate moisture content (see Chap. 2). Examples are bakery products, roasted nuts, toasted breakfast cereals, and roller-dried milk powders.

Maillard reactions begin with condensation of a nonionized amino group (ϵ-NH_2 of lysine or terminal α-NH_2) and a reducing sugar. Schematically these reactions may be represented in three steps (see also Chap. 3).

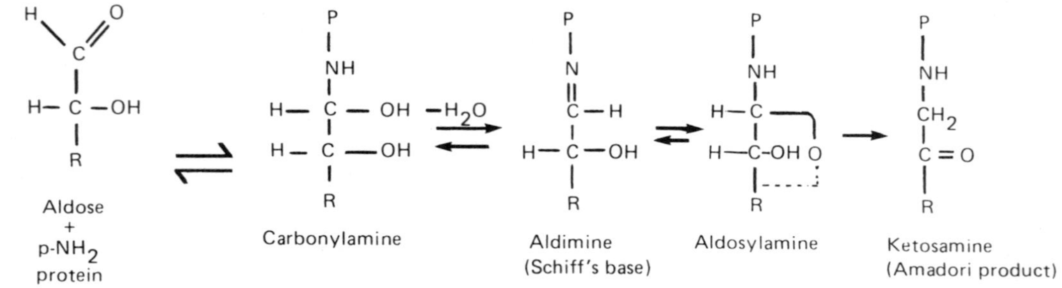

$$(52)$$

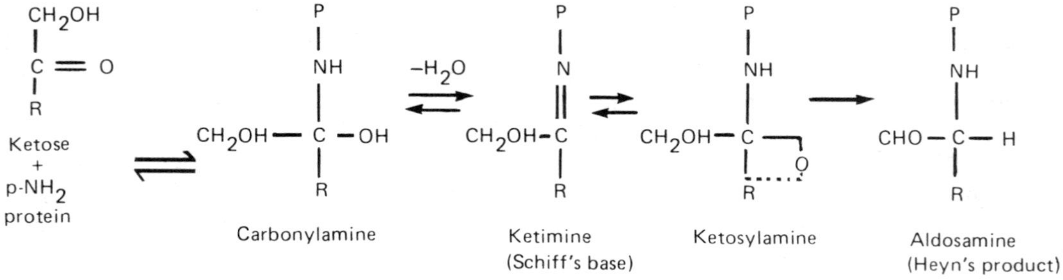

During the first step, Eq. (52), unstable Schiff's bases are formed, which rapidly isomerize into aldosylamines or ketosylamines, depending on the initial reacting carbohydrate. These glycosylamines then transform, by way of an Amadori or Heyn's rearrangement, into stable ketosamines or aldosamines. As an example, ϵ-N-(1-deoxylactulosyl)-L-lysyl, an Amadori product from lactose, represents 70-75% of the Maillard compounds present in overheated milks.

During the second step of this complex set of reactions, ketosamines and aldo-samines evolve into numerous carbonyl and polycarbonyl unsaturated derivatives (such as reductones):

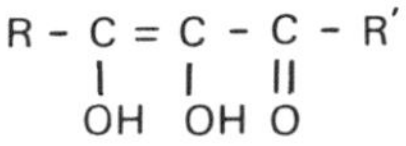

$$R - \underset{\underset{OH}{|}}{C} = \underset{\underset{OH}{|}}{C} - \underset{\underset{O}{\|}}{C} - R' \tag{53}$$

Reductone

Some of these derivatives in Eq. (54) may react with amines and amino acids leading to the formation of ammonia and of new carbonyl compounds (Strecker degradation). Decarboxylation of free amino acids may also take place.

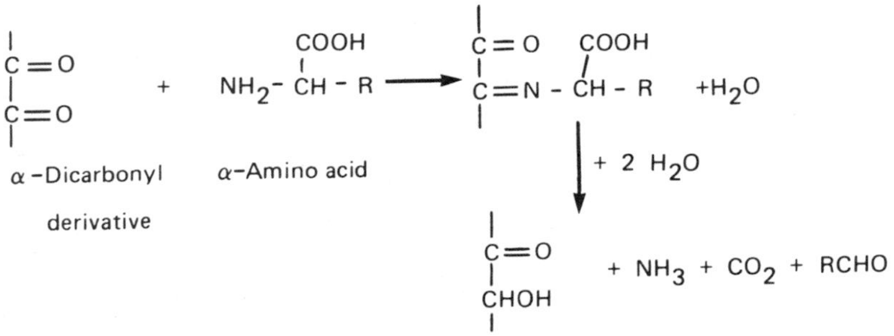

$$\begin{array}{c} \underset{|}{C} = O \\ \underset{|}{C} = O \end{array} \quad + \quad \underset{NH_2-}{\overset{COOH}{|}} \underset{CH-R}{\longrightarrow} \quad \begin{array}{c} \underset{|}{C} = O \quad COOH \\ \underset{|}{C} = N - CH - R \quad +H_2O \end{array} \tag{54}$$

α-Dicarbonyl   α-Amino acid

derivative

$$\downarrow + 2\ H_2O$$

$$\begin{array}{c} \underset{|}{C} = O \\ \underset{|}{CHOH} \end{array} \quad + NH_3 + CO_2 + RCHO$$

During the third step (not shown) the polycarbonyl unsaturated derivatives undergo both scission and polymerization reactions, leading on the one hand to volatile compounds (some with desirable flavors), and on the other hand to brown or black pigments, the melanoidins, with high molecular weights and complex structures. Such pigments are responsible for the color of bread and bakery products.

The nutritional effects of the Maillard reactions are still the object of many investigations. Loss of biologically available lysine is always observed. This may be detected by in vivo experiments, but also by rapid chemical methods. Among the latter, the most frequent approach is to measure reactive lysine residues [protein-bound lysine with non-substituted ε-amino groups by using 1-fluoro-2,4-dinitrobenzene (FDNB)] (13). The reaction of lysine residues with FDNB is followed by acid hydrolysis of the protein, extraction, and determination of the released ε-*N*-dinitrophenyllysine (ε-DNP-lysine) (Fig. 24). If the nondinitrophenylated protein is hydrolyzed with concentrated HCl, Amadori products of lysine, such as ε-*N*-deoxyfructosyllysyl, release free lysine to the extent of about 50% of the initial substituted lysine (Fig. 24). The remainder consists of new derivatives (furosine and pyridosine), which can be determined by ion-exchange chromatography and used as an index of substituted lysine residues and protein damage.

The lysine reacted as Schiff's base (early stage of Maillard sequence) is bioavailable since it can be set free under the acid conditions prevailing in the stomach. Lysine in Amadori or Heyn's products, on the other hand, is no longer available to the rat (but can be regenerated by strong acid to the extent of 50%, as mentioned above). This kind of binding leads, therefore, to a substantial loss of nutritional value (34). The effects of various processing procedures on lysine availability in milk are indicated in Table 18.

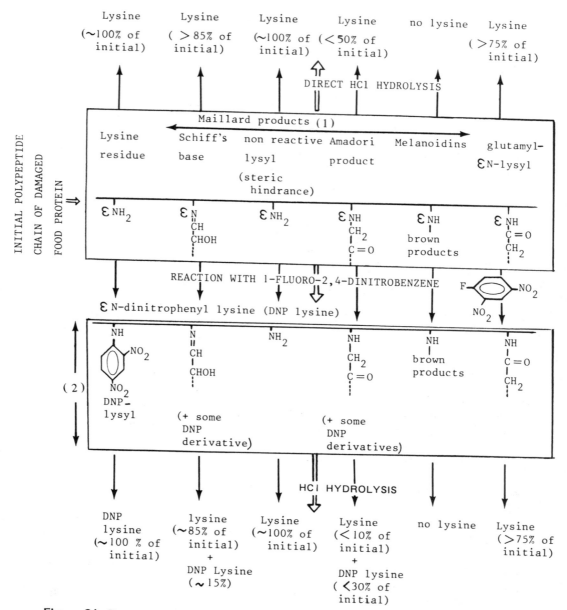

Figure 24 Degree to which lysine can be recovered, by acid hydrolysis, from lysine residues that have participated to various degrees in the Maillard reaction and have or have not been reacted with 1-fluoro-2,4-dinitrobenzene.

Table 18 Lysine Content and Availability in Concentrated or Dried Cow's Milk

Method of preparation	Total lysine (acid hydrolysis) (g/16 g N)	FDNB-reactive (g/16 g N)	Available from in vitro proteolysis (g/16 g N)	Available from rat growth assay (g/16 g N)
Freeze-dried	8.3	8.4	8.3	8.4
Spray-dried	8.0	8.2	8.3	8.1
Evaporated	7.6	6.4	6.2	6.1
Roller-dried (mild)	7.1	4.6	5.4	5.9
Roller-dried (severe)	6.1	1.9	2.3	2.0

Source: From F. Mottu and J. Mauron, *J. Sci. Food Agr. 18*:57-62 (1967).

Moreover, Amadori products exert nutritional effects that are, as yet, incompletely known. They are thought, for example, to inhibit the intestinal absorption of certain essential amino acids.

The formation of inter- or intramolecular covalent bonds in melanoidins (late stage of the Maillard sequence) effectively destroys the digestibility of the protein fraction of these molecules. Melanoidins prepared by heating some protein-carbohydrate model systems also display mutagenic properties, the potency depending on the intensity of the Maillard reaction. Melanoidins are insoluble in aqueous media, and are absorbed only slightly through the intestinal wall. This minimizes the risk of physiological effects. Premelanoidins of lower molecular weights may, however, be absorbed more readily, and their effects are still under investigation.

Various aldehydes, such as gossypol, glutaraldehyde used to protect protein meals against deamination in the rumen of ruminants, malonaldehyde resulting from the oxidation of lipids, and aldehydes in wood smoke, may react with proteins by establishing covalent bonds. These reactions have been the object of many studies in relation to the tanning of leather or the fixation of enzymes on solid supports, but their complex mechanisms have not been completely elucidated. In the presence of formaldehyde, a condensation reaction takes place with the ϵ-amino group of protein-bound lysine:

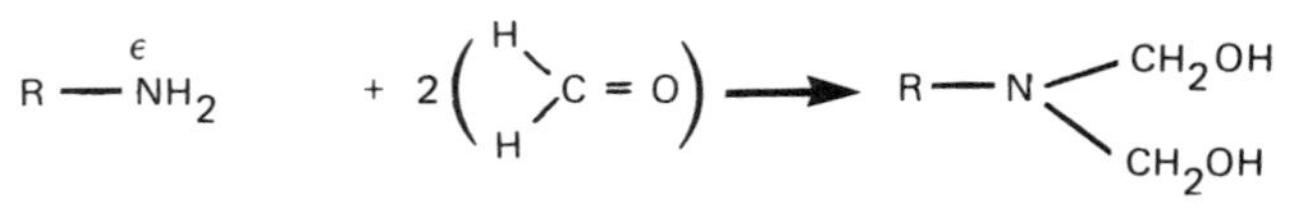

$$R - \overset{\epsilon}{N}H_2 \quad + 2\left(\overset{H}{\underset{H}{}}C = O \right) \longrightarrow R - N\overset{CH_2OH}{\underset{CH_2OH}{}} \tag{55}$$

Dihydroxymethyl derivative

Formaldehyde can be generated during the growth of bacteria on fish, and the subsequent formaldehyde-protein reaction is believed to contribute to the toughening of fish muscle during frozen storage.

Malonaldehyde can react with free amino groups of different polypeptide chains. Covalent bonds of the 1-amino-3-iminopropene type are formed, and this may modify some of the protein's functional properties, such as solubility or water-holding capacity. Casein modified with malonaldehyde is not readily hydrolyzed by proteases.

$$2 \text{ P} - \text{NH}_2 + \overset{O}{\underset{H}{\overset{\|}{\text{C}}}} - \text{CH}_2 - \overset{O}{\underset{H}{\overset{\|}{\text{C}}}} \longrightarrow \text{P} - \text{NH} - \text{CH} = \text{CH} - \text{CH} = \text{N} - \text{P} \qquad (56)$$

$$+ 2 \text{ H}_2\text{O}$$

Malonaldehyde

7. Interactions of Proteins with Other Food Constituents, Contaminants, and Additives

Reactions with Lipids. Lipoproteins, consisting of noncovalent complexes of proteins and lipids, occur widely in living tissues and influence the physical and functional properties of foods. The lipid constituents can, in most instances, be extracted with solvents without affecting the nutritional value of the protein constituents.

There are cases, however, in which covalent bonds are known to occur between oxidation products of lipids and proteins. Lipid oxidation followed by protein-lipid covalent interaction can take place in some foods and feeds, such as frozen or dehydrated fish, fish meals, and oilseeds. Two types of mechanisms appear to be involved in the covalent binding of peroxidizing lipids to proteins and in the lipid-induced polymerization of proteins:

(1) Free radical reaction (94). Lipid free radicals ($LOO^{\cdot}$) can add to protein (P_H), resulting in a lipid-protein free radical:

$$\text{LOO}^{\cdot} + \text{P}_H \longrightarrow {}^{\cdot}\text{LOOP} \qquad (57)$$

This reaction can be followed by measuring polymerization involving cross-linking of protein chains by multifunctional lipid free radicals:

$$^{\cdot}\text{LOOP} + \text{O}_2 \longrightarrow {}^{\cdot}\text{OOLOOP} \qquad (58)$$

$$^{\cdot}\text{OOLOOP} + \text{P}_H \longrightarrow \text{POOLOOP} \qquad (59)$$

and so on.

The reaction of a lipid free radical with a protein also can result in the formation of a protein free radical:

$$\text{LOO}^{\cdot} + \text{P}_H \longrightarrow \text{LOOH} + \text{P}^{\cdot} \qquad (60)$$

Protein free radicals can form at α-carbon atoms and at the sulfur atoms of cysteine residues. Direct polymerization of protein chains can follow:

$$\text{P}^{\cdot} + \text{P} \longrightarrow \text{P} - \text{P}^{\cdot} \quad \text{(dimer)} \qquad (61)$$

$$\text{P} - \text{P}^{\cdot} + \text{P} \longrightarrow \text{P} - \text{P} - \text{P}^{\cdot} \quad \text{(trimer)} \qquad (62)$$

and so on.

(2) Carbonylamine reactions. Aldehyde derivatives resulting from oxidation of unsaturated fatty acids can bind to the amino group of proteins through Schiff's base-type reactions. With malonaldehyde, covalent cross-links are formed (see Sec. II.6).

Lipid-protein reactions can have adverse nutritional effects. When casein is reacted with oxidized ethyl linoleate, this decreases the availability of several amino acids and decreases the values of digestibility, PER, and biological value.

From a practical standpoint, it is likely that proteinaceous foods undergoing lipid oxidation become organoleptically unacceptable before damage to protein nutritive value has occurred to any significant extent. This may not be the case with animal feeds, such as fish meal.

Reactions with Polyphenols. Polyphenols, natural constituents of many plants, can, in the presence of oxygen, be oxidized to the corresponding quinones. This can occur either in an alkaline medium, or through the action of polyphenoloxidase at pH close to neutrality. These quinones can polymerize into brown pigments of large size or can react with the residues of certain amino acids. In the latter case, condensation may take place between quinones and either lysine or cysteine residues, and/or oxidation reactions may occur with methionine residues or possibly with cysteine and tryptophan residues.

During preparation of protein isolates from polyphenol-rich raw materials, such as alfalfa leaves or sunflower seeds, a decrease in the level of available lysine may result from interaction with the oxidizing polyphenols.

Reactions with Halogenated Solvents. Trichloroethylene can combine with the sulfhydryl groups in proteins. The resulting *S*-dichlorovinyl-L-cysteine appears to be the toxic factor in trichlorethylene-extracted soybean meal, and the cause of aplastic anemia in calves.

Lipid solvents, such as dichloromethane, tetrachloroethylene, and fluorocarbons, do not appear to react with proteins.

1,2-Dichloroethane can react with fish proteins under slightly alkaline conditions. As a consequence, the availability of cystine, histidine, and methionine is reduced. Alkylation of the sulfhydryl groups of proteins may take place, and in some instances, cross-linking of protein chains can occur through protease-resistant thioether bonds.

Agene (nitrogen trichloride), once used as a maturing agent for flour, reacts with methionine residues in wheat proteins to form methionine sulfoximine, a toxic substance.

Methyl bromide, used as a fumigant for grain, may react with lysine residues to form *N*-ε-methylated derivatives. Formation of these derivatives probably decreases the digestibility of the protein. Methylation reactions may also occur with residues of histidine and sulfur amino acids.

Reactions with Nitrites. Nitrites can react with secondary or tertiary amines. Some amino acids, such as proline, tryptophan, tyrosine, cysteine, arginine, or histidine (free or protein-bound), may constitute reactive substrates for this reaction, which can occur during the cooking of protein foods or during digestion (at the low pH of the stomach). Some of the resulting nitrosamines or nitrosamides are known to be potent carcinogens.

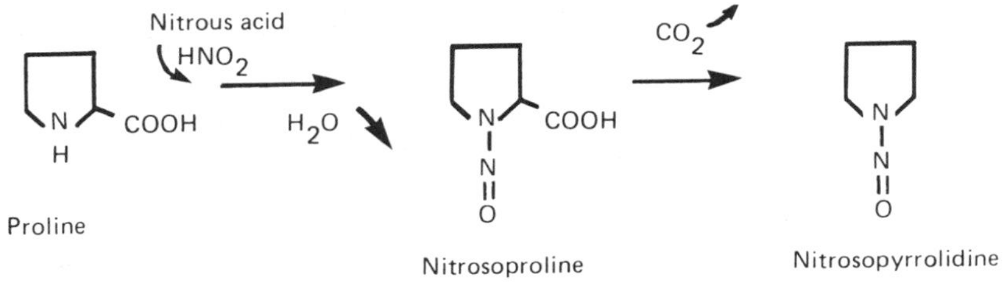

$$\underset{\text{Proline}}{\boxed{}\!\!\overset{}{\underset{\text{N}\;\text{H}}{}}\!\!\text{COOH}} \quad\xrightarrow[\text{H}_2\text{O}]{\overset{\text{Nitrous acid}}{\text{HNO}_2}}\quad \underset{\text{Nitrosoproline}}{\boxed{}\text{COOH}}\quad\xrightarrow{\text{CO}_2}\quad\underset{\text{Nitrosopyrrolidine}}{\boxed{}} \qquad (63)$$

Nitrites are normally present in meats at too low a concentration to bring about a significant decrease in the content or availability of lysine, tryptophan, and cysteine.

Reactions with Sulfites. Sulfite ions react with disulfides to form *S*-sulfonates and thiols. Sulfitolysis is enhanced at pH 7.

$$P - S - S - P + SO_3^{2-} \rightleftharpoons P - S - SO_3^- + P - S^- \tag{64}$$

$$S - Sulfonate$$

Many cystine residues in proteins resist the action of sulfites. *S*-Sulfonates are unstable in strongly acid or alkaline solutions, and usually decompose to disulfides. No detrimental effect of sulfites on the nutritive value of proteins can be inferred from these data. Sulfites may protect proteins from detrimental Maillard reactions since they react with carbonyl compounds.

Reactions with Acetylating Agents and Amino Acid Anhydrides. These reactions can be used to change the overall electric charge of proteins, or to covalently bind amino acids and lipids, for example, to proteins (see Sec. VII.B). The main objectives are to improve either the functional properties or the nutritional value of the protein.

A number of methods are available to covalently bind amino acids to proteins. If the attached amino acid is the limiting amino acid, this can improve the nutritional value of certain proteins. The binding of methionine to soy proteins, for instance, presents the following advantages over the simple addition of free methionine: less methionine flavor, less methionine loss by diffusion and leaching, less methionine degradation in the rumen of ruminants, and greater protection of α- and ϵ-amino groups against Maillard reactions.

Methionine (or lysine or tryptophan) can be bound to the ϵ- and terminal α-amino groups or to the carboxyl groups of glutamic and aspartic acid residues:

$$P - NH_2 + YOOC - \underset{\underset{NHX}{|}}{CH} - R \xrightarrow{pH\ 8\text{-}9} P - NH - CO - \underset{\underset{NHX}{|}}{CH} - R \tag{65}$$

where X is, for instance, benzyl oxycarbonyl and Y is a *p*-nitrophenyl group.

$$P - COO^- + NH_2 - \underset{\underset{COOZ}{|}}{CH} - R \xrightarrow[carbodiimide]{pH\ 4\text{-}6} P - CO - NH - \underset{\underset{COOZ}{|}}{CH} - R \tag{66}$$

where Z is, for instance, a methyl or ethyl group. The elimination of the protecting groups X and Z (by catalytic hydrogenation and by saponification, respectively), yields a protein enriched in the added amino acid. The linkage between the amino acid·and the protein is of the isopeptide type.

Binding the amino acid to the protein may also be obtained directly by heating the protein with *N*-carboxyanhydrides:

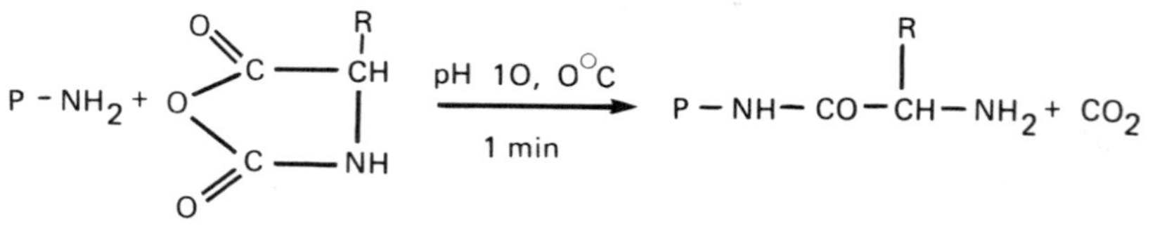

$$P - NH_2 + \underset{\underset{O}{\diagdown}}{\overset{O}{\diagup}} \begin{matrix} C - CH \\ | \\ C - NH \end{matrix} \overset{R}{\underset{}{}} \xrightarrow[1\ min]{pH\ 10,\ 0^\circ C} P - NH - CO - \underset{R}{\overset{R}{\underset{|}{CH}}} - NH_2 + CO_2 \tag{67}$$

By prior polymerization of the *N*-carboxyanhydride of, for example, methionine, it becomes possible to graft polymethionine chains of various lengths onto the ϵ-amino group of the protein. Small amounts of the carrier protein may then be used to complement large amounts of unmodified soy proteins.

The nutritional availability of amino acids grafted in this way is practically identical to that of residues of corresponding amino acids in the native protein.

It has also been suggested that an amino acid can be bound to a polypeptide by using high concentrations of the amino acid ethyl ester in the presence of a protease, such as papain. Under appropriate conditions, the protease catalyzes transpeptidation reactions. This so-called plastein reaction is discussed in the next section.

B. Changes in the Functional Properties of Proteins

Proteins used as food ingredients usually undergo physical or chemical treatments during their preparation or use and these treatments can influence protein functionality. For example, care is taken to minimize damage to a protein's structural and functional properties during extraction and purification. On the other hand, isolated proteins may be intentionally modified to improve existing properties or to develop new ones, and attention must be given to changes in functionality that proteins undergo after incorporation into the final foods.

Mild treatments, either physical or chemical in nature, usually result only in modification of the protein's conformation, whereas severe thermal treatments or the use of highly reactive chemicals usually results not only in conformational changes, but also in modification of the protein's primary structure. The kind and severity of the treatment applied dictates the changes that occur in the protein's functional properties, and these matters are discussed as follows.

1. Changes in Protein Functional Properties Due to Modifications in Secondary, Tertiary, and/or Quaternary Structure

Effects of Variations in the Chemical Environment. The *precipitation of a protein* by adjustment of a solution to the protein's isoelectric pH or by salting out represent simple and efficient methods of separation and purification. These procedures result in reversible aggregation, usually without extensive or irreversible unfolding of the protein, especially if conducted at low temperatures. An exception is casein, in which isoelectric aggregation and precipitation result in a breakdown of the quaternary micellar structure. This occurs because carboxyl groups become protonated, leading to a weakening or rupture of carboxyl-Ca^{2+}-carboxyl linkages, the release of calcium phosphate, and an increased electrostatic attraction among casein molecules. Isoelectric casein is resistant to attack by chymosin and is not affected by calcium, as is native micellar casein.

The *partial removal of water* from protein solutions leads to increased concentrations of all nonaqueous constituents. As a result, increased protein-protein, protein-carbohydrate, and protein-salt interactions may occur and these interactions may markedly alter the protein's functional properties, especially if water removal is carried out at relatively high temperatures. Milk processes for water removal, such as ultrafiltration, give highly soluble protein concentrates. In the case of whey protein concentrates, the extent of ultrafiltration (or diafiltration, i.e., addition of water to the retentate during ultrafiltration) determines the lactose-protein and the salt-protein ratios. These in turn influence the heat-gelling properties of the whey proteins (95).

Treatment of whey with a cation-exchange resin results in *exchange of cations* (Ca^{2+}, K^+, and Na^+) for protons and the production of a low-salt whey. Protein concentrates prepared from low-salt whey display excellent gelling and foaming properties. A product with equally good gelling and foaming properties can be prepared by electrodialysis and concentration.

Acid or alkaline pH enhances the binding of anions or cations, respectively, by proteins. This may affect the functional properties of proteins, especially their solubility. The presence of Ca^{2+} ions decreases the solubility of many proteins at neutral and alkaline pH; NaCl increases protein solubility at isoelectric pH and decreases solubility at alkaline pH. Thus, prior removal of cations through electrodialysis, ion exchange, reverse osmosis, or ultrafiltration facilitates alkaline solubilization of vegetable proteins or caseins.

At *moderately alkaline pH*, as used for the preparation of proteinates (salts of proteins), electrostatic repulsions between ionized carboxyl groups lead to dissociation of oligomeric proteins. Spray-dried sodium and potassium caseinates or soy proteinates therefore display a high degree of solubility and good water-absorption and surface properties.

In the case of soy proteins, alkalinization at pH 10-12 followed by neutralization causes extensive unfolding. After spray-drying, the protein gels when rehydrated at room temperature. The emulsifying capacity and the "precipitability" by Ca^{2+} ions of both 7S and 11S soy globulins appear to be markedly enhanced by such treatments.

At appropriate pH, the *presence of polyvalent ions or of some polyelectrolytes* enhances the formation of ionic cross-links between protein molecules. The Ca^{2+} ions can cross-link proteins through their ionized carboxyl or phosphoseryl groups at neutral or alkaline pH. Calcium proteinates are usually sparingly soluble and often gel when heated. Micellar calcium caseinate is soluble and stable when in the native state but not after chymosin attack. Insoluble calcium proteinates can be solubilized by the addition of Ca^{2+}-complexing agents, such as citrates and polyphosphates, or by alkalinization. The resulting dissociation and unfolding frequently improve water absorption, swelling, and surface properties. In the manufacture of processed cheese, the addition of 3% polyphosphates reduces the size of casein micelles and stabilizes the fat emulsion by enhancing hydrophobic interactions between the submicelles and lipid constituents. These reactions are facilitated at 75-105°C.

It is likely that the improved water retention imparted to cured meat by the addition of polyphosphates results partly from the complexing of calcium and from protein dissociation. Sodium chloride also improves water retention through partial solubilization of myofibrillar proteins. This in turn enhances the effect of polyphosphates.

Various polyelectrolytes also can be used to cross-link and precipitate polypeptide chains. Thus, negatively charged carboxymethylcellulose, alginates, polyacrylic acid, or polyphosphates bind to positively charged proteins at slightly acid pH (68). Proteins from whey and plasma, for example, can be precipitated and recovered by this approach. Although most of these polyelectrolytes can be subsequently separated from the protein constituents, some may remain bound to the proteins in appreciable amounts, and if so, they alter the solubility and functional properties of the protein (15).

During the preparation of protein ingredients, a number of *solvents* with varying degrees of polarity are used to extract and remove: (a) lipids (hexane), (b) chlorophyll and heme pigments (acetone), or (c) phospholipids, water, minerals, and soluble carbohydrates (ethanol and isopropanol). Drying is often necessary before solvent extraction, and treatment with steam may be required to reduce the residual level of solvent. These extraction treatments expose previously buried hydrophobic regions of proteins, often leading to irreversible aggregation and insolubilization (at neutral or isoelectric pH).

The water-absorption capacity of the protein also may be reduced by these treatments. Extraction treatments involving mixtures of water and polar solvents, such as ethanol or isopropanol, are the least detrimental, and protein solubility is often restored after these solvents are removed.

Solvent precipitation of proteins can be used to produce gels. Thus, an 8% solution of soy proteins can be gelled with ethanol present at a 20% concentration. Ethanol concentrations in excess of 40% result in protein precipitation because protein-protein interactions become excessive as compared with protein-water interactions.

Dehydration. Nearly complete removal of water often causes extensive aggregation (protein-protein interactions), especially if high temperatures are used to supply the energy of water removal. This may result in severe losses of protein solubility and surface activity. Since drying often represents the last step in the preparation of a protein ingredient, its effects on functional properties should not be overlooked.

The conditions of drying influence the size and porosity, both internal and surface, of powder particles and, hence, their properties with respect to wettability, water absorption, dispersion, and/or dissolution. High porosities are usually obtained when water is rapidly removed as vapor, resulting in minimal contraction of the particle and migration of salts and/or carbohydrates toward the drying surface. This occurs during freeze-drying and spray-drying. The inclusion of gas bubbles in the protein solution prior to drying, and the controlled agglomeration of dried protein particles, can be used to increase particle porosity.

Mechanical Treatments. Extensive dry milling of protein flours or concentrates results in powders with small particle sizes and large surface areas. This generally results in improved properties of water absorption, protein solubility, fat absorption, and foaming properties as compared with unmilled counterparts. Extensive milling may also permit the preparation, by air classification, of fractions with a high protein content (protein-rich and starch-rich particles separated according to differences in density; see Sec. VI).

Intense shear forces applied to protein suspensions or solutions, as occurs during homogenization of milk, may cause the fragmentation of protein aggregates (micelles) into subunits. The emulsifying capacity of the proteins is generally improved by this treatment. Shear forces applied at an air/water interface often cause denaturation and aggregation of proteins. Partial protein denaturation may stabilize foams, but "over-beating" of some proteins, such as egg white, decreases foam capacity and stability because of protein aggregation.

Mechanical forces also play an important role in protein texturization processes, such as dough or fiber formation and extrusion cooking. Protein alterations enhanced by shear forces are molecular alignment, disulfide bond interchange, and formation of protein networks (see Secs. IV. E and F).

Thermal Treatments. Thermal treatments of proteins may result in structural changes (see Sec. III), hydrolysis of peptide bonds, modifications of amino acid side chains (Sec. VII.A) and condensation with other molecules (Sec. VII.A), depending on the intensity and duration of heat application, water activity, pH, salt content, the kind and concentration of other reactive molecules, and other factors. Side-chain modifications and condensation reactions are detrimental to nutritive value (Sec. VII.A). Structural changes and limited hydrolysis of peptide bonds resulting from mild thermal treatments do not affect nutritional quality (Sec. VII.A) but may influence markedly the functional properties of proteins.

The extent and consequences of thermal denaturation of proteins (conformational changes and aggregation; Secs. III. and IV.D) are highly dependent on the nature of the protein and on the environmental conditions. Mammalian collagen unfolds, dissociates, and dissolves when it is heated above 65°C in the presence of abundant water, similarly treated myofibrillar proteins contract, aggregate, and retain less water than before (Fig. 25). The properties of random coil proteins, such as monomeric caseinates, are little affected even by severe heating.

When starting with a soluble globular protein, heat denaturation generally leads to a loss of solubility. However, the temperature of denaturation and the extent of denaturation are influenced by various factors. The denaturation of 7S and 11S soy proteins, as a function of pH, can be followed by differential scanning calorimetry (Fig. 26). It can be seen that the transition temperature(s) (temperature at which unfolding occurs) and the enthalpy of denaturation (heat absorbed during denaturation; indicative of the extent of strongly endothermal unfolding) both decrease as the pH is adjusted away from the isoelectric pH. At such pH, globular proteins exist in a partly unfolded state and heating causes further unfolding. Electrostatic repulsions existing at these pH levels prevent or reduce aggregation, insolubilization, and gelation upon heating.

Thermal unfolding (50-80°C, 10-15 min) of β-lactoglobulin and whey proteins in the acid (2-4) or slightly alkaline (7-8.5) pH range has been used to improve their functional properties. Even when the pH is restored to 6, these proteins remain largely soluble and display improved thickening, gelling, foaming, and emulsifying properties as compared with the native proteins (24,76). It can be hypothesized that unfolding without aggregation took place and that this increased the amphipolarity of these initially highly hydrophilic proteins.

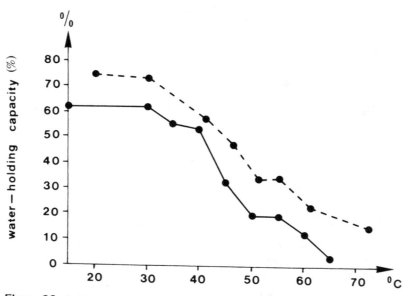

Figure 25 Influence of heating on the water-holding capacity of beef muscle. [R. Hamm, in *The Physiology and Biochemistry of Muscle as a Food* (Briskey, Cassens, Trautman, eds.), University of Wisconsin Press, Madison, 1966, p. 363.]

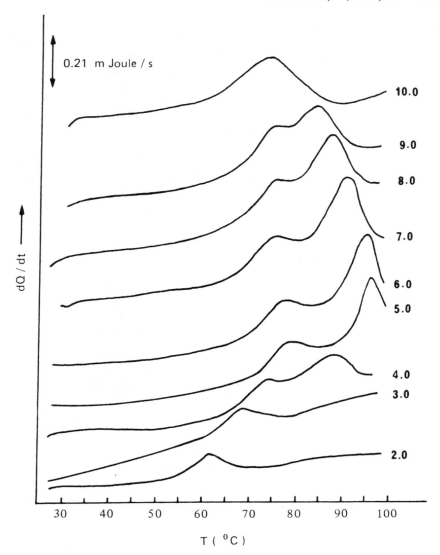

Figure 26 Differential scanning thermograms of 10% soy protein dispersion in distilled water at pH 2-10. Sensitivity: 2.1 mJ/sec. Heating rate: 10°C/min. [A. M. Hermansson, *J. Texture Studies* 9:33 (1978).]

On the contrary, when heat treatments are imposed at the isoelectric pH, they cause extensive protein aggregation. This approach is used to quantitatively precipitate, separate, and purify proteins from liquid whey, blood, or plasma. The insoluble protein isolates thus obtained possess few functional properties except that of high water absorption (Sec. IV.A).

As already mentioned (Sec. IV.D), heating of concentrated protein solutions at pH slightly removed from their isoelectric pHs induces gelation. Upon extended heating, some of these protein gels remain stable; others deteriorate.

In the presence of cations, especially Ca^{2+}, aggregation is favored over unfolding and the temperature of coagulation or gelation may be lowered. These effects are enhanced at pHs above the isoelectric pH, where carboxyl groups are ionized. Calcium ions are also known to harden the texture of heat-set gels, such as soy protein gels.

The water content of a protein solution markedly affects the temperature and enthalpy of denaturation (Sec. III). These relationships are shown for sperm whale myoglobin in Fig. 27. The denaturation temperature (T_m) is minimal (74°C) at a water content of 30-50% and increases sharply in the dry state (122°C at 3% water). The enthalpy of denaturation (ΔH_{app}) decreases markedly when the water content is less than 30%. Such data may have useful applications with respect to controlling functional properties. For example, if a protein solution is to be dehydrated with minimal denaturation and loss of solubility, the curve for denaturation temperature indicates the maximum temperature that can be tolerated at any point (water content) during dehydration. Processes like vacuum concentration or spray drying, in which low product temperatures are maintained when the water content is high, are preferable to plate or roller drying. A positive correlation is usually observed between the solubility of a dried protein ingredient and its enthalpy of denaturation as determined by differential scanning calorimetry after rehydration.

The same curve of denaturation temperature versus water content indicates the lower temperature limit for such processes as texturation by extrusion. Here, extensive protein denaturation and insolubilization is necessary to ensure that the dried, expanded protein matrix rehydrates without breakdown or dissolution.

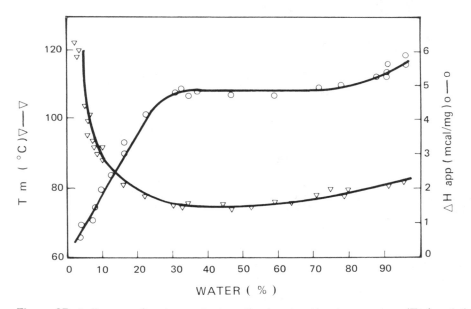

Figure 27 Influence of water content on the denaturation temperature (T_m) and the enthalpy of denaturation (ΔH_{app}) of sperm whale myoglobin. The conformational transition is followed by differential scanning calorimetry. [P. Baardseth, in *Physical, Chemical and Biological Changes in Food Caused by Thermal Processing* (Høyem and Kvåle, eds.), Applied Science Publishers, London, 1977, p. 282.]

Subfreezing temperatures may also cause protein denaturation (Sec. III) and damage to functional properties. The toughening and lessened water retention of fish actomyosin, the precipitation of casein micelles in milk, the thickening and gelation of egg yolk lipoproteins, and the syneresis of certain gels following extended frozen storage and thawing result primarily from a freeze-induced decrease in protein-water hydrogen bonding and an increase in protein-protein interactions.

Limited hydrolysis of peptide bonds, which occurs when vegetable proteins are heated for 10-15 hr in 1-3 M HCl at 100°C, increases threefold the nonprotein nitrogen and may improve markedly the solubility and hence the surface properties of ingredients, such as gluten proteins (50,60). Acid hydrolysis generally causes modifications of protein side chains, for example, by deamidation of asparagine and glutamine residues, dephosphorylation of phosphoserine, and destruction of tryptophan residues. Due to more complex reactions, acid hydrolysis may result in the formation of pigments and derivatives with meaty flavors. Some of these vegetable protein hydrolyzates are neutralized with sodium hydroxide, filtered, and used as flavoring agents.

Limited hydrolysis of peptide bonds also can be obtained by heating proteins in alkaline media, for example, for 20 min to a few hours in NaOH, pH 11-12.5, at 70-95°C. This depolymerization procedure can be used to solubilize and extract sparingly soluble vegetable, microbial, or fish proteins. Partial alkaline hydrolysis of milk proteins is known to greatly improve their foaming properties, and this approach is used to prepare whipping agents. The desirable effect may result from the formation of amphipolar peptides with hydrophobic side chains and a polar sodium carboxylate end group.

It has already been indicated that severe thermal treatments at alkaline pH can cause desulfuration of cysteine or cystine and the formation of nutritionally unavailable lysinoalanine, lanthionine, and D-amino acid residues (Sec. VII.A).

2. Changes Due to Enzyme Action

In Vivo Modification of Proteins. Over 135 types of enzymic modifications of proteins are known to take place in vivo following protein synthesis at the ribosomal level (116). These modifications involve most amino acid side chains, except those of glycine and apolar amino acids. The main reactions are glycosylation, hydroxylation, phosphorylation, acylation, methylation, and cross-linking. In the case of collagen, over 50 types of modifications are known to occur in vivo; some examples are hydroxylation of proline and lysine residues (catalyzed by hydroxylases), binding of oses (glycosyltransferases), and oxidative deamination of some lysine and hydroxylysine side chains to yield aldehyde groups, which then participate in carbonyl-amine condensation. The intra- or intermolecular cross-links thus formed contribute to the increased rigidity of the collagen molecules during cattle aging.

In vivo proteolysis plays a major role in the biological function of many proteins. Turnover of proteins in cells and in the intestine has already been discussed (Sec. V.A.2). Restricted specific proteolysis is necessary for the conversion of enzymically inactive zymogens into enzymes, such as trypsin, chymotrypsin, carboxypeptidases, elastase, and phospholipase. The conversion of trypsinogen into trypsin depends on the hydrolysis by enterokinase of a lysyl-isoleucyl bond close to the N-terminal amino acid. Similar specific proteolysis of protein precursors is required for the formation of several hormonal and other biologically active peptides, and for the triggering of blood clotting. The transfer of secretory proteins from the ribosome across intracellular membranes also involves a proteolytic step. For example, caseins and ovalbumin possess a short hydrophobic section close to their N-terminal amino acids. This section attaches to a membrane receptor and facilitates transfer. Following transfer, this portion is then removed from the protein by proteolysis.

Restricted intracellular proteolysis also takes place in animal muscle after rigor mortis. This affects the proteins involved in the Z lines of myofibrils and contributes to the postrigor tenderization of meat.

Enzymic Treatments of Food Proteins: Proteolysis (92,116). Some enzymic modifications of amino acid side chains have been used for gaining a better understanding of the physicochemical properties of food proteins. Thus, the dephosphorylation of caseins with alkaline phosphatase has helped elucidate the role of phosphate groups in the calcium binding and calcium sensitivity of caseins. Removal of a portion of the carbohydrate unit of κ-casein with a glycosidase has helped to explain the role of the carbohydrate unit in the sensitivity of the molecule to chymosin.

In most cases, improvement of food proteins has been attempted through limited or extensive proteolysis using proteases. Well-known examples are the tenderization of meat with added papain, the precipitation of caseins with chymosin or fungal proteases, the improvement of cheese texture with bacterial or fungal proteases, the prevention of haze formation in beer with papain or microbial proteases, and the preparation of soluble protein hydrolyzates from fish or vegetable proteins.

With protein solutions, the extent of peptide bond hydrolysis can be assessed by measuring the molecular mass of peptides as determined by gel permeation chromatography. It is even simpler to calculate the degree of hydrolysis (D_h) by conducting a potentiometric titration of protons released during proteolysis at a constant pH between 7 and 8 (2). In this pH range, the splitting of each peptide bond leads to the formation of a nonionized α-amino group and an ionized carboxyl group:

$$D_h(\%) = \frac{1}{\alpha} \frac{\text{mol alkali added per g protein}}{\text{number (eq) peptide bonds per g protein}} 100 \qquad (68)$$

where α is the average dissociation coefficient of α-amino groups, and

$$\alpha = \frac{10^{pH-pK}}{1 + 10^{pH-pK}} \qquad (69)$$

The average pK of α-amino groups varies from 7.7 at 25°C to 6.9 at 60°C.

Limited specific proteolysis can cause protein coagulation, as occurs when casein is treated with chymosin or when blood fibrinogen is attacked by thrombin. The glycopeptide released from κ-casein by chymosin is highly hydrophilic, and this may partly explain the subsequent coagulation.

Fish, vegetable, heat-denatured, or solvent-denatured proteins often can be solubilized by partial proteolysis with such enzymes as papain, bromelin, pepsin, or proteases from species of the genera *Aspergillus, Bacillus,* or *Streptomyces.* Partial proteolysis thus may facilitate the extraction and purification of proteins from various sources. The solubility (at pH 3 or 5.5) of denatured soy protein isolates may increase from 10 to 50 or 80% when the degree of hydrolysis is 3 or 8%, respectively (2). Furthermore, the solubility profile of the partially hydrolyzed protein is generally improved over the entire pH range because large aggregates do not form, even at the isoelectric pH. The increased solubility achieved by limited proteolysis is attributable to the formation of smaller, more hydrophilic, and more solvated polypeptide units. These partial hydrolyzates are useful in acid or heat-processed beverages to provide protein enrichment.

Extensive proteolysis often results in the release of bitter-tasting hydrophobic peptides with leucine or phenylalanine terminal residues. Such hydrolyzates may, however, be useful for patients with impaired digestive functions and for those allergic to milk or gluten proteins.

Partial proteolysis also has been used to improve the emulsifying or foaming properties of heat-denatured proteins (Fig. 28). This probably results from the increased solubility of the hydrolyzate, which facilitates diffusion and spreading at oil/water and air/water interfaces. However, when the degree of hydrolysis exceeds 3-5%, the viscosity and thickness of the adsorbed protein films is apparently insufficient to stabilize emulsions and foams.

The reduction of molecular size through proteolysis is clearly detrimental to the gelling and viscoelastic properties of proteins. However, limited proteolysis of gluten proteins may increase dough expansion during bread making and improve the flakiness of crackers.

In so-called plastein formation (42,116), the first step involves proteolysis of a 5% protein suspension with pepsin or papain at neutral pH to obtain peptides of 1-20,000

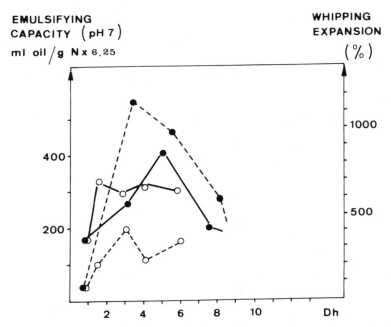

Figure 28 Emulsifying capacity and whipping expansion of a soy protein isolate (Purina 500-E) as influenced by the degree of hydrolysis (D_h): (—) emulsifying capacity at pH 7 (ml oil/g protein); (– – –) whipping expansion (volume %); (●●●) proteolysis carried out with "Alcalase"; (○○○) proteolysis with "Neutrase." [J. Adler-Nissen and H. S. Olsen, in *Functionality and Protein Structure* (Pour El, ed.), Amer. Chem. Soc., Washington, D.C., 1979, p. 125.]

daltons. The second step involves concentrating the hydrolyzate to a 30-40% protein concentration, adding the same or another protease, and adjusting the pH. This allows rearrangement of peptides into "plasteins" through transpeptidation reactions as follows.

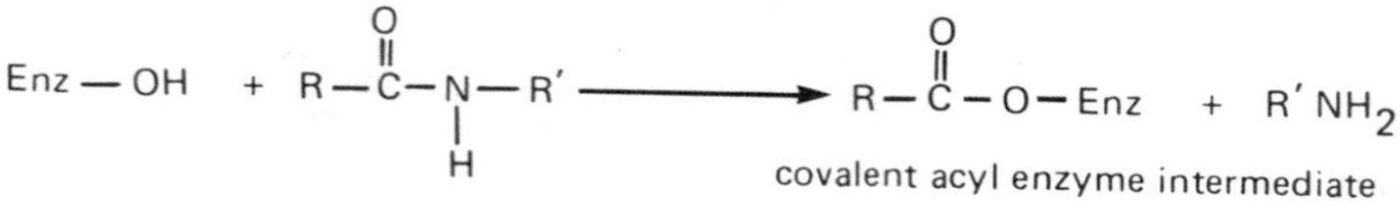

$$
\text{Enz—OH} \;+\; \underset{\underset{\underset{H}{|}}{}}{R-\overset{\overset{O}{\|}}{C}-N-R'} \longrightarrow \underset{\text{covalent acyl enzyme intermediate}}{R-\overset{\overset{O}{\|}}{C}-O-\text{Enz}} \;+\; R'NH_2 \qquad (70)
$$

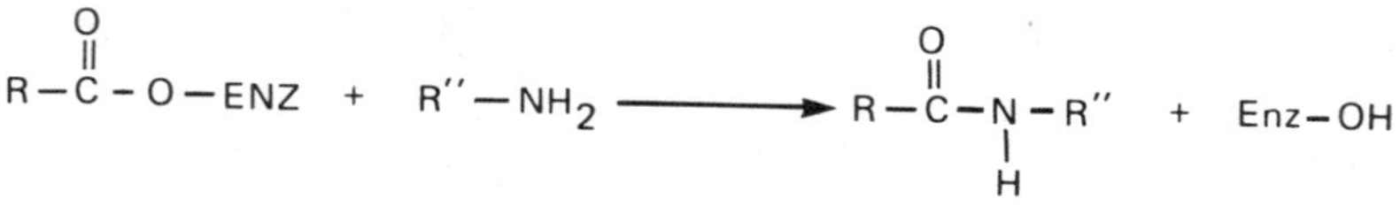

$$
R-\overset{\overset{O}{\|}}{C}-O-\text{ENZ} \;+\; R''-NH_2 \longrightarrow \underset{\underset{H}{|}}{R-\overset{\overset{O}{\|}}{C}-N-R''} \;+\; \text{Enz—OH} \qquad (71)
$$

Fatty or amino acid esters also can react with α-amino groups as does the acyl-enzyme intermediate (42,46).

The first step may be useful to improve the solubility of heat-denatured proteins, as already indicated, and also to help desorb and remove impurities (off-flavors or pigments, for example). The second step has been used, for instance, to covalently attach sulfur-rich peptides to soy protein hydrolyzates (note: direct incorporation of various essential amino acids is also possible using amino acid esters).

By interacting apolar, leucyl, and alkyl esters, succinylated proteins, and papain, it has been possible to obtain amphiphilic molecules with remarkable emulsifying and foaming properties (113). Plastein formation also may be used to modify the bitter peptides resulting from the first proteolysis step.

At this time, the plastein process is probably too expensive for use by the food industry.

3. Specific Chemical Modifications

The primary structures of proteins can be chemically modified to improve their functional properties (15,30,32,43,60,74,93). This approach also has been used with success to study structure-function relationships in proteins (enzymic and other biological functions as well as physicochemical and functional properties).

One should be aware, however, that the intentional chemical modification of food or feed proteins may have serious drawbacks, including damage to nutritional value similar to that observed during severe processing (Sec. VII.A), formation of amino acid derivatives that may be toxic, and the incorporation of reagent residues that also may be toxic.

Modifications of Protein Side Chains. Alteration of amino acid residues can be achieved by heating at acid or alkaline pH. For example, heating of glutenin in an acid environment results in 30% deamidation of the asparagine and glutamine residues, and this improves the solubility and surface properties (especially the emulsifying properties) of the protein. The beneficial effects appear to result from conformational changes stemming from the decrease in hydrogen bonding and the increase in electrostatic repulsion. Alkaline treatments cause conversion of some cysteine (or phosphoserine) residues to dehydroalanine (Sec. VII.A). If the protein-bound dehydroalanine is intentionally reduced to alanine, the hydrophobicity of the protein increases.

The main classes of reactions used for the chemical modification of protein side chains are acylation, alkylation, oxidation, and reduction (Fig. 29). A given side chain may react with reagents of several classes and a given reagent may react with different types of side chains, resulting in the introduction of identical side groups. Indeed, few reagents are specific for one type of side chain, unless very specific conditions of pH and/or temperature are imposed. For example, the alkylation agent iodoacetamide will react with the thioether group of methionine at pHs below 4, whereas thiol, amino, and imidazole groups are unreactive in this pH range because they are protonated. Performic acid oxidation is specific for thioethers and thiol groups at low temperatures ($-10°C$).

Some reagents must be dissolved in small amounts of organic solvents before being added to the aqueous protein solution. Also, mild reagents and mild environmental conditions must be used; otherwise, the changes in protein conformation and functional properties may exceed those that are desired. Furthermore, side chains of a given type do not necessarily display the same reactivity toward a single reagent, since reactivity is a function of the neighboring amino acids in the protein and of protein conformation. Some protein side chains may be buried inside the protein, and in these situations polypeptide chains first must be unfolded if extensive derivatization is desired.

Modification of protein side groups generally results in modification of polarity and, in some cases, of net charge. When these alterations are extensive, the protein may fold, unfold, and/or aggregate with other protein molecules. The protein's behavior toward water and/or other constituents, such as lipids, also may be modified.

The introduction of ionizable carboxyl groups can be achieved by alkylation with iodacetic acid (carboxymethylation of thiol and other groups), by acylation with internal anhydrides of dicarboxylic acids (succinylation, maleylation, and citraconylation of ε-amino groups), or by phosphorylation of serine residues with $POCl_3$ (Fig. 29). The presence of these additional negative charges results in electrostatic repulsion, unfolding, and dissociation, with an improvement in protein solubility and/or dispersibility, even at the isoelectric pH. This may facilitate the extraction of vegetable or microbial proteins and their separation from nucleic acids or other constituents. Introduction of these ionizable groups also generally improves the protein's water-absorption capacity and heat stability and increases its sensitivity to precipitation by calcium ions. Such effects have been observed with fish proteins, soy proteins, gluten, and so on. For example, the gelling properties of fish protein concentrates are markedly improved by succinylation (Fig. 30). Emulsifying and foaming properties are also often improved, and this may result from increased protein solubility, from unfolding with increased amphipolar conformation, and/or from electrostatic repulsions between adsorbed protein films. The properties of succinylated or carboxymethylated proteins are, however, strongly dependent on the pH (Fig. 30), since little ionization takes place at acid pH.

The functional properties also depend on the extent of chemical derivatization. For example, when less than 40% of the ε-amino groups of casein are succinylated, water absorption improves but emulsification ability decreases. The latter improves at higher degrees of succinylation. It is also worthwhile mentioning that the introduction of phosphate or sulfate groups in gluten significantly increases its water absorption, gelation, and film-forming abilities.

A high degree of succinylation reduces the biological availability of lysine residues. It is of interest that maleylation and citraconylation are reversible at pH 2 (such as exists in the stomach) and are, therefore, less detrimental to biological availability than is succinylation.

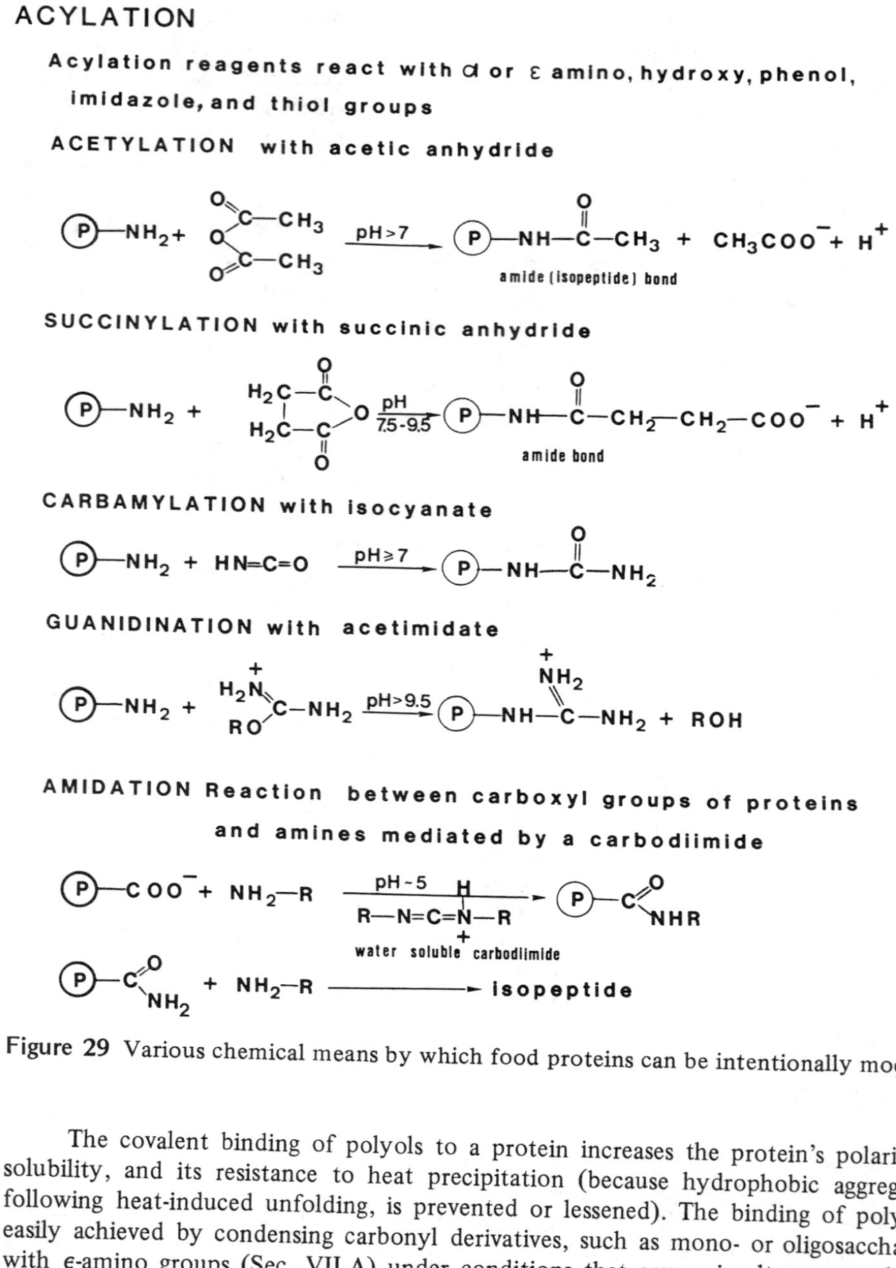

ACYLATION

Acylation reagents react with α or ε amino, hydroxy, phenol, imidazole, and thiol groups

ACETYLATION with acetic anhydride

$$\text{P}-NH_2 + \begin{matrix} O=C-CH_3 \\ | O \\ O=C-CH_3 \end{matrix} \xrightarrow{pH>7} \text{P}-NH-\overset{O}{\underset{\|}{C}}-CH_3 + CH_3COO^- + H^+$$

amide (isopeptide) bond

SUCCINYLATION with succinic anhydride

$$\text{P}-NH_2 + \begin{matrix} H_2C-C \\ | \quad \| \\ H_2C-C \\ O \end{matrix} O \xrightarrow[7.5-9.5]{pH} \text{P}-NH-\overset{O}{\underset{\|}{C}}-CH_2-CH_2-COO^- + H^+$$

amide bond

CARBAMYLATION with isocyanate

$$\text{P}-NH_2 + HN=C=O \xrightarrow{pH \geq 7} \text{P}-NH-\overset{O}{\underset{\|}{C}}-NH_2$$

GUANIDINATION with acetimidate

$$\text{P}-NH_2 + \begin{matrix} H_2N \\ \diagdown \\ RO \end{matrix} C-NH_2 \xrightarrow{pH>9.5} \text{P}-NH-\overset{\overset{+}{NH_2}}{\underset{\|}{C}}-NH_2 + ROH$$

AMIDATION Reaction between carboxyl groups of proteins and amines mediated by a carbodiimide

$$\text{P}-COO^- + NH_2-R \xrightarrow[\underset{+}{R-N=C=N-R}]{pH \sim 5 \quad H} \text{P}-C\overset{O}{\underset{NHR}{\diagup}}$$

water soluble carbodiimide

$$\text{P}-C\overset{O}{\underset{NH_2}{\diagup}} + NH_2-R \longrightarrow \text{isopeptide}$$

Figure 29 Various chemical means by which food proteins can be intentionally modified.

The covalent binding of polyols to a protein increases the protein's polarity, its solubility, and its resistance to heat precipitation (because hydrophobic aggregation, following heat-induced unfolding, is prevented or lessened). The binding of polyols is easily achieved by condensing carbonyl derivatives, such as mono- or oligosaccharides, with ε-amino groups (Sec. VII.A) under conditions that cause simultaneous reduction of the Schiff's base. Such "reductive alkylations" lead to the formation of "neoglyco-proteins."

Proteins can be made more hydrophobic by introducing apolar side groups via acylation or reductive alkylation of ε-amino, thiol, or hydroxyl groups. Thus, it is possible to bind the carboxyl group of a free fatty or amino acid to the ε-amino group of the protein (Fig. 29). The acylating agent can be an intermolecular anhydride of a monocarboxylic acid. The degree of hydrophobicity imparted to the modified protein

ESTERIFICATION

Alcohols react with carboxyl groups

$$\text{(P)}-C\overset{O}{\underset{OH}{}} + R-OH \xrightarrow[0-25°C]{HCl\ <0.1\ M} \text{(P)}-C\overset{O}{\underset{OR}{}} + H_2O$$

ALKYLATION

ALKYLATION of amino, phenol, imidazole, indole, thiol,

and thioether groups with haloacetates or haloalkylamides

Carboxymethylation with iodoacetic acid or iodoacetamide

$$\text{(P)}-S^- + ICH_2COO^- \xrightarrow{pH>7} \text{(P)}-S-CH_2COO^- + I^-$$

$$\text{(P)}-NH_2 + ICH_2C\overset{O}{\underset{NH_2}{}} \xrightarrow{pH>8.5} \text{(P)}-NH-CH_2C\overset{O}{\underset{NH_2}{}} + I^- + H^+$$

REDUCTIVE ALKYLATION of amino, indole thiol, and

thioether groups with an aldehyde or ketone in the presence

of a reducing agent of the hydride donor type

$$\text{(P)}-NH_2 + 2\ HCH=O + \tfrac{1}{2}\ Na\ CN\ BH_3 \xrightarrow[0°C]{pH\ 9}$$

$$\text{(P)}-N(CH_3)_2 + \tfrac{1}{2}\ NaH\ CN\ BO_3 + \tfrac{1}{2}H_2O$$

OXIDATION

Strong oxidation of thiol, disulfide, thioether, and indole

groups with performic acid

$$\text{(P)}-SH + 3\ H-C\overset{O}{\underset{O-OH}{}} \xrightarrow{-10°C} \text{(P)}-SO_3H + 3\ H-C\overset{O}{\underset{OH}{}}$$

$$\text{(P)}-S-S-\text{(P)} + 5\ H-C\overset{O}{\underset{O-OH}{}} + H_2O \longrightarrow 2\ \text{(P)}-SO_3H + 5\ H-C\overset{O}{\underset{OH}{}}$$

REDUCTION

Reductive splitting of disulfide bonds with reducing agents:

2 mercaptoethanol, dithiotreitol, thioglycolic acid

Figure 29 (Continued)

depends on the kind of amino or fatty acid used (long or short hydrocarbon chains; prepolymerized or not) and on the extent of derivatization. Acetylation, for example, improves the emulsifying properties of milk proteins but decreases the water-absorption properties of soy proteins. A high degree of hydrophobicity is likely to enhance intra- and intermolecular hydrophobic interactions and promote protein folding and aggregation. Emulsifying properties of proteins are consistently improved by acetylation, probably because the molecule becomes highly amphiphilic. When casein is reductively alkylated in the presence of aldehydes or ketones so that up to 16 methyl, isopropyl, cyclohexyl, or benzyl groups are introduced per molecule, improved emulsifying properties are observed. Improved emulsifying and foaming properties for soy glycinin are obtained when 5-11 molecules of palmitic acid are introduced per molecule of protein using the *N*-hydroxysuccinimide ester of palmitic acid.

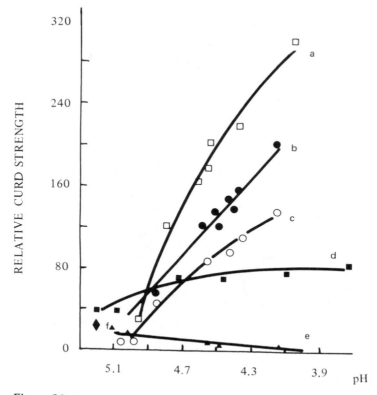

Figure 30 Effect of succinylation on the curd strength of fish protein concentrate (FPC) at various pH levels. Succinylated FPC (b) + corn oil (a) or + Ca^{2+} (c). FPC (high pH extract) (e) + corn oil (d). Raw skim milk curd prepared with rennet (f). Curd strength of gels prepared from 3% protein solutions was measured with a Brookfield viscosimeter. [L. Chen, T. Richardson, and C. H. Amundson, *J. Milk Food Technol. 38*:89 (1975).]

Derivatization of ε-amino groups may have detrimental nutritional effects (see Sec. VII.A). In the first place, the overall rate and extent of protein digestion is often reduced, since proteases such as trypsin, which are specific for lysyl bonds, may be partly inhibited. In the second place, substituted lysine derivatives are often biologically unavailable even when they are released from the polypeptide chain and absorbed. This effect is dependent on the presence or absence of efficient acylases in the kidney or other tissues of different animal species. Thus, formyl-ε-*N*-lysine and acetyl-ε-*N*-lysine can be partly used as a source of lysine by the rat. The PER of methyl casein in rats remains high provided fewer than 50% of the lysine residues are methylated.

When methionine or polymethionine is covalently attached to proteins, as discussed in Sec. VII.A, the isopeptide linkages formed are hydrolyzable by intestinal aminopeptidase. Such binding can, therefore, be used for the amino acid enrichment of proteins and also for the protection of ε-amino groups against Maillard-type reactions.

Formation of Covalent Cross-Links. When protein side chains are converted into very reactive groups, the latter may quickly react to give intra- or intermolecular cross-links. Thus, during alkaline treatment of proteins, stable lysinoalanine and lanthionine cross-links can be produced via the formation of reactive residues of dehydroalanine (Sec. VII.A).

Bifunctional reagents, such as glutaric or malonic aldehydes, or polymerized formaldehyde, can be used to cross-link the ϵ-amino groups of lysine residues. This usually decreases protein solubility and protein digestibility.

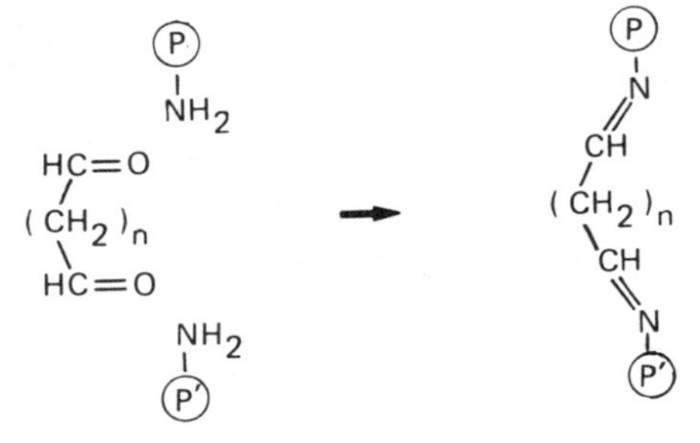

$$
\begin{array}{ccc}
\text{(P)} & & \text{(P)} \\
\mid & & \mid \\
\text{NH}_2 & & \text{N} \\
& & \parallel \\
\text{HC}=\text{O} & & \text{CH} \\
\mid & \longrightarrow & \mid \\
(\text{CH}_2)_n & & (\text{CH}_2)_n \\
\mid & & \mid \\
\text{HC}=\text{O} & & \text{CH} \\
& & \parallel \\
\text{NH}_2 & & \text{N} \\
\mid & & \mid \\
\text{(P')} & & \text{(P')}
\end{array}
\qquad (72)
$$

$n = 1$: malonaldehyde

$n = 3$: glutaraldehyde

Cross-linking of proteins has several nonfood applications, such as conversion of animal hides to leather, formation of attenutated toxins for the production of vaccines, and preparation of surgical implants that are biologically compatible (resistant to proteolysis and devoid of immunogenicity). Proteins in feeds for ruminants also can be cross-linked to reduce their modification in the rumen (Sec. VII.A).

The formation of disulfide cross-links can be promoted by mild oxidation of thiol groups, achieved in the presence of air, chemical oxidizing agents such as bromates, or oxidizing enzymes. This reaction is routinely used in the baking industry to improve the viscoelastic properties of gluten proteins (disulfide interchange predominates over disulfide formation). Disulfide bonds also can be disrupted by adding reducing agents, such as cysteine, or by imposing alkaline conditions. Such reduced proteins are more soluble and therefore often easier to purify. They are also more susceptible to further chemical modification. However, reduction of proteins results in impairment of some functional properties, such as gelation.

Clearly, chemical modification of proteins is a promising approach to improving their functional properties. However, nutritional and toxicological consequences of chemical modifications raise regulatory questions that may prevent rapid implementation of this approach.

REFERENCES

1. Acton, J. C., Ziegler, G. R., and Burge, D. L., Jr. (1983). Functionality of muscle constituents in the processing of comminuted meat products. *CRC Crit. Rev. Food Sci. Nutr. 18*:99-121.

2. Adler-Nissen, J., and H. S. Olsen (1979). The influence of peptide chain length on taste and functional properties of enzymatically modified soy protein, in *Functionality and Protein Structure* (A. Pour-El, ed.), Am. Chem. Soc., Washington, D.C., pp. 125-146.

3. Altschul, A. M. (ed) (1974 and 1976). *New Protein Foods*, Vols. 1A and 2B, *Technology*, Academic Press, New York.

4. Altschul, A. M., and H. L. Wilcke (eds) (1978 and 1981). *New Protein Foods*, vols. 3A and 4B, *Animal Protein Supplies*, Academic Press, New York.

5. Arai, S. (1980). Deterioration of food proteins by binding unwanted compounds such as flavors, lipids and pigments, in *Chemical Deterioration of Proteins* (J. R. Whitaker and M. Fujimaki, eds.), Am. Chem. Soc., Washington, D.C., pp. 195-209.

6. Bender, A. E. (1978). *Food Processing and Nutrition*, Academic Press, London.

7. Bigelow, C. C. (1967). On the average hydrophobicity of proteins and the relation between it and protein structure. *J. Theoret. Biol. 16*:187-211.

8. Blackburn, S. (1978). *Amino Acid Determination*, Marcel Dekker, Inc., New York.

9. Bodwell, C. E., and L. Petit (eds.) (1982). *Plant Proteins for Human Food*, Martinus Nijhoff/Dr. W. Junk Publishers, The Hague, The Netherlands.

10. Bodwell, C. E., J. S. Adkins, and D. T. Hopkins (1982). *Protein Quality in Humans: Assessment and In Vitro Estimation*, AVI Publishing Co., Westport, Conn.

11. Cante, C. J., R. W. Franzen, and F. Z. Saleeb (1979). Proteins as emulsifiers: Methods for assessing the role. *J. Am. Oil Chem. Soc. 56*:71A-77A.

12. Cantor, C. R., and P. R. Schimmel (1980). *Biophysical Chemistry*, Part I. *The Conformation of Biological Macromolecules*, Part II. *The Behavior of Biological Macromolecules*. W. H. Freeman and Co., San Francisco.

13. Carpenter, K. J. (1960). The estimation of available lysine in animal protein foods. *Biochem. J. 77*:604-610.

14. Carpenter, K. J. (1973). Damage to lysine in food processing: Its measurement and its significance. *Nutr. Abstr. 43*:423-451.

15. Cheeseman, J. (1981). Modifications of skimmed milk constituents. *J. Soc. Dairy Technol. 34*:74-77.

16. Cheftel, J. C. (1977). Chemical and nutritional modifications of food proteins due to processing and storage, in *Food Proteins* (J. R. Whitaker and S. R. Tannebaum, eds.), AVI Publishing Co., Westport, Conn., pp. 401-445.

17. Cherry, J. P., and K. H. McWatters (1981). Whippability and aeration, in *Protein Functionality in Foods* (J. P. Cherry, ed.), Am. Chem. Soc., Washington, D.C., pp. 149-176.

18. Chou, D. H., and C. V. Morr (1979). Protein-water interactions and functional properties. *J. Am. Oil Chem. Soc. 56*:53A-62A.

19. Cole, D. J. A., K. N. Boorman, P. J. Buttery, D. Lewis, R. J. Neale, and H. Swan (eds.) (1976). *Protein Metabolism and Nutrition*, Butterworths, London.

20. Cooney, C. L., C. Rha, and S. R. Tannenbaum (1980). Single-cell protein: Engineering, economics and utilization in foods. *Adv. Food Res. 26*:1-52.

21. Damodaran, S., and J. E. Kinsella (1981). Interaction of carbonyls with soy protein: Thermodynamic effects. *J. Agr. Food Chem. 29*:1249-1253.

22. Damodaran, S., and J. E. Kinsella (1981). Interaction of carbonyls with soy protein: Conformational effects. *J. Agr. Food Chem. 29*:1253-1257.

23. Daussant, J., J. Mossé, and J. Vaughan (eds.) (1983). *Seed Proteins*, Academic Press, London.

24. De Wit, J. N. (1981). Structure and functional behavior of whey protein. *Neth. Milk Dairy J. 35*:47-64.

25. Dworschák, E. (1980). Nonenzyme browning and its effect on protein nutrition. *CRC Crit. Rev. Food Sci. Nutr. 13*:1-40.

26. Earl, C. D., and F. M. Ausubel (1983). The genetic engineering of nitrogen fixation. *Nutr. Rev. 41*:1-6.

27. Eriksson, C. (ed.) (1981). *Maillard Reactions in Food: Chemical, Physiological and Technological Aspects*, Pergamon Press, Oxford.

28. FAO (1970). *Amino Acid Content of Foods and Biological Data on Proteins*, FAO, Rome.

29. FAO/WHO (1973). *Energy and Protein Requirements*, Report of a joint FAO/WHO ad hoc expert Committee. World Health Organization Techn. Rep. Ser. 522, WHO, Geneva.

30. Feeney, R. E. (1977). Chemical modification of food proteins, in *Food Proteins* (R. E. Feeney and J. R. Whitaker, eds.), Am. Chem. Soc., Washington, D.C., pp. 3-36.

31. Feeney, R. E. (1980). Overview on the chemical deteriorative changes of proteins and their consequences, in *Chemical Deterioration of Proteins* (J. R. Whitaker and M. Fujimaki, eds.), Am. Chem. Soc., Washington, D.C., pp. 1-49.

32. Feeney, R. E., R. B. Yamasaki, and K. F. Geochegan (1982). Chemical modification of proteins: An overview, in *Modification of Proteins* (R. E. Feeney and J. R. Whitaker, eds.), Am. Chem. Soc., Washington, D.C., pp. 3-55.

33. Fennema, O. R. (ed.) (1979). *Proteins at Low Temperatures*, A. C. S. Symposium series 180, Am. Chem. Soc., Washington, D.C.

34. Finot, P. A. (1982). Nutritional and metabolic aspects of protein modification during food processing, in *Modification of Proteins* (R. E. Feeney and J. R. Whitaker, eds.), Am. Chem. Soc., Washington, D.C., pp. 91-124.

35. Food and Nutrition Board, National Research Council (1980). *Recommended Dietary Allowances*, 9th ed., National Academy of Sciences, Washington, D.C.

36. Fox, P. (ed.) (1982). *Developments in Dairy Chemistry*, vol. 1., *Proteins*, Applied Sci. Pub., London.

37. Franzen, K. L., and J. E. Kinsella (1974). Parameters affecting the binding of volatile flavor compounds in model food system. I. Proteins. *J. Agr. Food Chem. 22:* 675-678.

38. Fraser, T. H. (1982). Microbial factories for the production of animal proteins. *CRC Crit. Rev. Food Sci. Nutr. 16:*217-227.

39. Friberg, S. (ed.) (1976). *Food Emulsions*, Marcel Dekker, Inc., New York.

40. Friedman, M. (ed.) (1975). *Protein: Nutritional Quality of Foods and Feeds, Nutrition and Clinical Nutrition*, vols. 1 and 2, Marcel Dekker, Inc., New York.

41. Friedman, M. (ed.) (1978). *Nutritional Improvement of Food and Feed Proteins*, Plenum Press, New York.

42. Fujimaki, M., S. Arai, and M. Yamashita (1977). Enzymatic protein degradation and resynthesis for protein improvement, in *Food Proteins* (R. E. Feeney and J. R. Whitaker, eds.), Am. Chem. Soc., Washington, D.C., pp. 156-184.

43. Glazer, A. N., R. J. Delange, and D. S. Sigman (1975). *Chemical Modification of Proteins: Selected Methods and Analytical Procedures*, North-Holland Publishing Co., Amsterdam.

44. Graham, H. D. (ed.) (1977). *Food Colloids*, AVI Publishing Co., Westport, Conn.

45. Halling, P. J. (1981). Protein-stabilized foams and emulsions. *CRC Crit. Rev. Food Sci. Nutr. 15:*155-203.

46. Haque, Z., T. Matoba, and M. Kito (1982). Incorporation of fatty acid into food protein: Palmitoyl soybean glycinin. *J. Agr. Food Chem. 30:*481-486.

47. Harper, A. E. (1981). McCollum and directions in the evaluation of protein quality. *J. Agr. Food Chem. 29:*429-435.

48. Hermansson, A. M. (1978). Physico-chemical aspects of soy proteins structure formation. *J. Texture Stud. 9:*33-58.

49. Hermansson, A. M. (1979). Aggregation and denaturation involved in gel formation, in *Functionality and Protein Structure* (A. Pour-El, ed.), Am. Chem. Soc., Washington, D.C., pp. 81-103.

50. Hermansson, A. M., I. Olsson, and B. Hollunberg (1974). Functional properties

of protein for food modification. Studies on rapeseed protein concentrate. *Lebensm. Wiss. u. Technol. 7*:176-181.

51. Horan, F. E. (1977). Protein texturization, in *Food Proteins* (J. R. Whitaker and S. R. Tannebaum, eds.), AVI Publishing Co., Westport, Conn., pp. 484-515.

52. Høyem, T., and O. Kväle (eds.) (1977). *Physical, Chemical and Biological Changes in Food Caused by Thermal Processing*, Applied Science Publishing, London.

53. Hurrell, R. F. (1980). Interaction of food components during processing, in *Food and Health: Science and Technology* (G. G. Birch and K. J. Parker, eds.), Applied Science Publishing, London, pp. 369-388.

54. Hutton, C. W., and A. M. Campbell (1981). Water and fat absorption, in *Protein Functionality in Foods* (J. P. Cherry, ed.), Am. Chem. Soc., Washington, D.C., pp. 177-200.

55. Inglett, G. E., and L. Munck (eds.) (1980). *Cereals for Food and Beverages. Recent Progress in Cereal Science and Technology*, Academic Press, New York.

56. Kato, A., and S. Nakai (1980). Hydrophobicity determined by a fluorescence probe method and its correlation with surface properties of proteins. *Biochim. Biophys. Acta. 624*:13-20.

57. Kazemzadeh, M., J. A. Aguilera, and K. C. Rhee (1982). Use of microscopy in the study of vegetable protein texturization. *Food Technol. 36*:111-118.

58. Khan, K., and W. Bushuk (1979). Structure of wheat gluten in relation to functionality in breadmaking, in *Functionality and Protein Structure* (A. Pour-El, ed.), Am. Chem. Soc., Washington, D.C., pp. 191-206.

59. Kinsella, J. E. (1976). Functional properties of proteins in foods: A survey. *CRC Crit. Rev. Food Sci. Nutr. 7*:219-280.

60. Kinsella, J. E. (1977). Functional properties in novel proteins. Some methods for improvement. *Chem. Ind.* (London) *23*:177-182.

61. Kinsella, J. E. (1978). Texturized proteins: Fabrication, flavoring, and nutrition. *CRC Crit. Rev. Food Sci. Nutr. 10*:147-207.

62. Kinsella, J. E. (1981). Functional properties of proteins: Possible relationships between structure and function in foams. *Food Chem. 7*:273-288.

63. Kinsella, J. E. (1982). Relationships between structure and functional properties of food proteins, in *Food Proteins* (P. F. Fox and J. J. Condon, eds.), Applied Science Publishing, London, pp. 51-103.

64. Kinsella, J. E., and D. Srinivasan (1981). Nutritional, chemical and physical criteria affecting the use and acceptability of proteins in foods, in *Criteria of Food Acceptance* (J. Solms and R. L. Hall, eds.), Forster Publishing, Zurich, pp. 296-332.

65. Lehninger, A. L. (1982). *Principles of Biochemistry*. Worth Publishing, New York.

66. Liener, I. (1979). Significance for humans of biologically-active factors in soybeans and other food legumes. *J. Am. Oil Chem. Soc. 56*:121-129.

67. Liener, I. E. (ed.) (1980). *Toxic Constituents of Plant Foodstuffs*, Academic Press, New York.

68. Lin, C. F. (1977). Interaction of sulfated polysaccharides with proteins, in *Food Colloids* (H. D. Graham, ed.), AVI Publishing Co., Westport, Conn., pp. 320-346.

69. Macritchie, F. (1978). Proteins at interface. *Adv. Prot. Chem. 32*:283-326.

70. Mahler, H. R., and E. H. Cordes (1971). *Biological Chemistry*, 2nd ed., Harper and Row, New York.

71. Matz, M. A. (ed.) (1981). *Protein Food Supplements*, Recent advances food technol. Rev. 54. Noyes Data Corp., Park Ridge, N.J.

72. Mauron, J. (1971). Nutritional evaluation of proteins by enzymatic methods. *Nestlé Res. News*, pages 50-59.

73. Mauron, J. (1972). Influence of industrial and household handling on food protein quality, in *Protein and Amino Acid Function*, vol. 2 (E. J. Bigwood, ed.), Pergamon Press, Oxford, pp. 417-474.

74. Means, G. E., and R. E. Feeney (1971). *Chemical Modification of Proteins*, Holden-Day, Inc., San Francisco.

75. Milner, M., N. S. Scrimshaw, and D. I. C. Wang (eds.) (1978). *Protein Resources and Technology*, AVI Publishing Co., Westport, Conn.

76. Modler, H. W., and V. R. Harwalkar (1981). Whey protein concentrate prepared under acidic conditions. 1. Recovery by ultrafiltration and functional properties. *Milchwissensch. 36*:537-542.

77. Mortimore, G. E. (1982). Mechanisms of cellular protein catabolism. *Nutr. Rev. 40*: 1-12.

78. Munro, H. N. (ed.) (1964, 1964, 1969, 1970). *Mammalian Protein Metabolism*, vols. 1-4, Academic Press, New York.

79. Munro, H. N. (1976). Regulation of body protein metabolism in relation to diet. *Proc. Nutr. Soc. 35*:297-308.

80. Munro, H. N., and M. C. Crim (1980). The proteins and amino acids, in *Modern Nutrition in Health and Disease* (R. S. Goodhart and M. F. Shils, eds.), Lea and Febiger, Philadelphia, pp. 51-98.

81. Neurath, H., and R. L. Hill (eds.) (1975, 1976, 1977, 1979). *The Proteins*, vols. I-IV. Academic Press, New York.

82. New Notes (1982). Proteins. *Food Technol. 36*:54-63.

83. Ney, K. H. (1972). Aminosaüre. Zusammensetzung von Proteinen und die Bitter-keit ihrer Peptide. *Z. Lebensm. Unters. Forsch. 149*:321-330.

84. Norton, G. (ed.) (1978). *Plant Proteins*, Butterworths, London.

85. Pellett, P. L. (1978). Protein quality revisited. *Food Technol. 32*:60-78.

86. Phillips, M. C. (1981). Protein conformation at liquid interfaces and its role in sta-bilizing emulsions and foams. *Food Technol. 35*:50-57.

87. Pirie, N. W. (1979). *Leaf Protein and other Aspects of Fodder Fractionation*, Cambridge University Press, London.

88. Priestley, R. J. (ed.) (1979). *Effects of Heating on Foodstuffs*, Applied Science Publishing, London.

89. Privalov, P. L. (1979). Stability of proteins, in *Advances in Protein Chemistry*, vol. 33 (C. B. Anfinsen, J. T. Edsall, and F. M. Richards, ed.), Academic Press, New York, pp. 167-241.

90. Rhee, K. C., C. K. Kuo, and E. W. Lusas (1981). Texturization, in *Protein Func-tionality in Foods* (J. P. Cherry, ed.), Am. Chem. Soc., Washington, D.C., pp. 51-88.

91. Ribadeau-Dumas, B., G. Brignon, F. Grosclaude, and J. C. Mercier (1972). *Eur. J. Biochem. 26*:505-514.

92. Richardson, T. (1977). Functionality changes in proteins following action of en-zymes, in *Food Proteins* (R. E. Feeney and J. R. Whitaker, eds.), Am. Chem. Soc., Washington, D. C., pp. 185-243.

93. Ryan, D. S. (1977). Determinants of the functional properties of proteins and pro-tein derivatives in foods, in *Food Protein* (R. E. Feeney and J. R. Whitaker, eds.), Am. Chem. Soc., Washington, D.C., pp. 67-91.

94. Schaich, K. M. (1980). Free radical initiation in proteins and amino acids by ion-izing and ultraviolet radiations and lipid oxidation. Part I. Ionizing radiation. Part II. Ultraviolet radiation and photolysis. Part III. Free radical transfer from oxidiz-ing lipids. *CRC Crit. Rev. Food Sci. Nutr. 13*:89-129; 131-159; 189-244.

95. Schmidt, R. H. (1979). Gelation and coagulation, in *Functionality and Protein Structure* (A. Pour-El, ed.), Am. Chem. Soc., Washington, D.C., pp. 131-147.

96. Shen, J. L., and C. V. Morr (1979). Physico-chemical aspects of texturization: Fiber formation from globular proteins. *J. Am. Oil Chem. Soc. 56*:63A-70A.

97. Shimada, K., and S. Matsushita (1980). Thermal coagulation of egg albumin. *J. Agr. Food Chem. 28*:409-412.

98. Shimada, K., and S. Matsushita (1980). Relationship between thermocoagulation of proteins and amino acid compositions. *J. Agr. Food Chem. 28*:413-417.

99. Shimada, K., and S. Matsushita (1981). Effects of salts and denaturants on thermocoagulation of proteins. *J. Agr. Food Chem. 29*:15-20.

100. Simic, M. G., and M. Karel (eds.) (1980). *Autoxidation in Food and Biological Systems*, Plenum Press, New York.

101. Smith, A. K., and S. J. Circle (eds.) (1978). *Soybeans: Chemistry and Technology*, vol. 1, 2nd ed., *Proteins*, AVI Publishing Co., Westport, Conn.

102. Solms, J. F. Osman-Ismail, and M. Beyeleer (1973). The interaction of volatiles with food components. *Can. Inst. Food Sci. Technol. J. 6*:A10-A16.

103. Stanley, D. W., E. D. Murray, and D. H. Lees (eds.) (1980). *Utilization of Protein Resources*, Food Nutrition Press, Westport, Conn.

104. Stryer, L. (1975). *Biochemistry*, W. H. Freeman, San Francisco.

105. Suzuki, T. (1981). *Fish and Krill Protein: Processing Technology*, Applied Science Publishing, London.

106. Swaisgood, H. E., and G. L. Catignani (1982). In vitro measurements of effects of processing on protein nutritional quality. *J. Food Protection 45*:1248-1256.

107. Synge, R. L. M. (1976). Plant foods for human nitrition. Damage to nutritional value of plant proteins by chemical reactions during storage and processing. *Qualitas Plantarum 26*:9-27.

108. Tannenbaum, S. R. (ed.) (1979). *Nutritional and Safety Aspects of Food Processing*, Marcel Dekker, Inc., New York.

109. Tannenbaum, S. R., and D. I. C. Wang (eds.) (1975). *Single-Cell Protein*, vol. II, MIT Press, Cambridge, Mass.

110. Tung, M. A. (1978). Rheology of protein dispersions. *J. Texture Stud. 9*:3-31.

111. United Nations University, World Hunger Programme (1979). Protein-energy requirements under conditions prevailing in developing countries: Current knowledges and research needs. *Food Nutr. Bull.* Suppl. 1. U. N. U., Tokyo.

112. Wall, J. S., and F. R. Huebner (1981). Adhesion and cohesion, in *Protein Functionality in Foods* (J. P. Cherry, ed), Am. Chem. Soc., Washington, D.C., pp. 111-130.

113. Watanabe, M., and S. Arai (1982). Proteinaceous surfactants prepared by covalent attachement of L-leucine n-alkyl esters to food proteins by modification with papain, in *Modification of Protein* (R. E. Feeney and J. R. Whitaker, eds.), Am. Chem. Soc., Wasington, D. C., pp. 199-221.

114. Waterlow, J. C., and P. R. Payne (1975). The protein gap. *Nature 258*:113-117.

115. Waterlow, J. C., and J. M. L. Stephen (eds.) (1982). *Nitrogen Metabolism in Man*, Applied Science Publishers, London.

116. Whitaker, J. R., and A. J. Puigserver (1982). Fundamentals and applications of enzymatic modifications of proteins: An overview, in *Modification of Proteins* (R. E. Feeney and J. R. Whitaker, eds.), Am. Chem. Soc., Washington, D.C., pp. 57-87.

117. Wolf, W. J., and J. C. Cowan (1975). *Soybeans as a Food Source*, CRC Press, Boca Raton, Florida.

118. World conference of soya processing and utilization (1981). *J. Am. Oil Chem. Soc. 58*:121-539.

119. World conference on vegetable food proteins (1979). *J. Am. Oil Chem. Soc. 56*: 99-484.

120. Young, V. R., and N. S. Scrimshaw (1978). Nutritional evaluation of protein and protein requirements with special reference to man, in *Protein Resources and Technology: Status and Research Needs* (M.Milner, N. S. Scrimshaw, and D. I. C. Wang, eds.), AVI Publishing Co., Westport, Conn., pp. 136-173.

BIBLIOGRAPHY

Cherry, J. P. (ed.) (1981). *Protein Functionality in Foods*, ACS Symposium series 147, Am. Chem. Soc., Washington, D.C.

Cherry, J. P. (ed.) (1982). *Food Protein Deterioration.Mechanisms and Functionality*, ACS Symposium series 206, Am. Chem. Soc., Washington, D.C.

Feeney, R. E., and J. R. Whitaker (eds.) (1977). *Food Proteins. Improvement Through Chemical and Enzymatic Modification*, Adv. in Chemistry series 160. Am. Chem. Soc., Washington, D.C.

Feeney, R. E., and J. R. Whitaker (eds.) (1982). *Modification of Proteins. Food, Nutritional and Pharmacological Aspects*, Adv. in Chemistry series 198, Am. Chem. Soc., Washington, D.C.

Fox, P. F., and J. J. Condon (eds.) (1982). *Food Proteins*, Applied Science Publishers, London.

Friedman, M. (ed.) (1977). *Protein Crosslinking. A. Biochemical and Molecular Aspects. B. Nutritional and Medical Consequences*, Plenum Press, New York.

Grant, R. A. (ed.) (1980). *Applied Protein Chemistry*, Applied Science Publishers, London.

Hudson, B. J. F. (ed.) (1982 and 1983). *Developments in Food Proteins*, vols. 1 and 2, Applied Science Publishers, London.

Pour-El, A. (ed.) (1979). *Functionality and Protein Structure*, ACS Symposium series 92, Am. Chem. Soc., Washington, D.C.

Schambye, P. (ed.) (1978). *Biochemical Aspects of New Protein Food*, 11th meeting FEBS, vol. 44, Pergamon Press, Oxford.

Schultz, H. W., and A. F. Anglemier (eds.) (1964). *Symposium on Foods: Proteins and Their Reactions*, AVI Publishing Co., Westport, Conn.

Whitaker, J. R., and M. Fujimaki (eds.) (1980). *Chemical Deterioration of Proteins*, ACS Symposium series 123, Am. Chem. Soc., Washington, D.C.

Whitaker, J. R., and S. R. Tannenbaum (eds.) (1977). *Food Proteins*, AVI Publishing Co., Westport, Conn.

6
ENZYMES

Thomas Richardson* and Douglas B. Hyslop University of Wisconsin-Madison, Madison, Wisconsin

I.	Introduction	372
II.	Enzyme Nomenclature	374
III.	Definitions of Some Terms Used in Enzymology	375
IV.	Specificity, Catalysis, and Regulation	376
	A. Specificity	376
	B. Catalysis	377
	C. Regulation	379
V.	Compartmentalization of Enzymes in Cellular Systems	379
	A. Distribution of Enzymes at the Subcellular Level	380
	B. Distribution of Enzymes in Selected Food Materials	381
VI.	Kinetics of Enzyme-Catalyzed Reactions	381
	A. Steady-State Enzyme Kinetics	381
	B. Enzyme Inhibition Kinetics	384
	C. Immobilized Enzyme Kinetics	390
	D. Immobilized Substrate Kinetics	393
VII.	Enzyme-Related Analyses	395
	A. Enzyme Assays	395
	B. Substrate Assays	399
VIII.	Enzyme Mechanisms	400
IX.	Factors Influencing Enzyme Activity	402
	A. Effect of Temperature on Denaturation and Inactivation of Enzymes	402
	B. pH Effects on Enzymes	416
	C. Water Activity and Enzymic Activity	417
	D. Effects of Electrolytes and Ionic Strength on Enzymes	420

---

*Current affiliation: University of California-Davis, Davis, California

E. Inactivation of Enzymes by Shearing   421
F. Effects of Pressure on Enzymes   421
G. Effects of Ionizing Radiation on Enzymic Activity   422
H. Interfacial Inactivation of Enzymes   423
X. Controlling Enzyme Action   424
A. Factors Useful in Controlling Enzyme Activity   424
B. Enhancing Endogenous Enzyme Activity   425
C. Controlling Exogenous Enzymes   426
XI. Enzymes Added to Foods During Processing   427
XII. Immobilized Enzymes and Microorganisms   434
A. Overview   434
B. Applications of Immobilized Enzymes and Microbial Cells in Food Systems   435
XIII. Modification of Food by Endogenous Enzymes   439
A. Pectic Enzymes   439
B. Amylases   441
C. Cathepsins   441
D. Calcium-Activated Neutral Proteinase (CANP)   442
E. Milk Proteinase(s)   442
F. Lipolytic Enzymes   442
G. Thiaminases I and II   444
H. Phytase   445
I. Myosin ATPase   445
J. Enzymic Browning   445
K. Lipoxygenases   447
L. Peroxidases   451
M. Ascorbic Acid Oxidase   452
N. Antioxidant Enzymes   453
O. Flavor Enzymes   454
P. Pigment-Degrading Enzymes   454
XIV. Enzyme Inhibitors   455
A. Inhibitors in Plants   455
B. Inhibitors in Animal Tissues   457
C. Inhibitors in Microorganisms   458
D. Physiological Significance of Enzyme Inhibitors   458
E. Enzyme Inhibitors as Research Tools   459
XV. Recombinant DNA Technology and Genetic Engineering   459
A. Cloning Strategies   460
B. Host Microorganisms   463
C. Cloning in Plants   464
D. Applications of Recombinant DNA Techniques   464
E. Perspective   465
References   467
Bibliography   476
Food Enzymology   476
Enzyme Chemistry, Kinetics and Mechanism   476
Problem Solving, Enzyme Kinetics   476

I. INTRODUCTION

Enzymes are complex, globular protein catalysts that accelerate chemical reaction rates by factors of 10^{12}-10^{20} over that of uncatalyzed reactions at temperatures around $37°C$

(135). By contrast, nonenzymic catalysis used in industry are orders of magnitude less effective than enzymes under comparable conditions. For example, the disproportionation of hydrogen peroxide catalyzed by catalase occurs 10 million times faster than it does when catalyzed by colloidal platinum at 37°C (185). The molar activity of enzymes is very high, enabling one molecule of enzyme to transform as many as 600,000 molecules of substrate per second (Table 1).

The catalytic activities of analogous enzymes vary with the species (155). In fish, for example, which have no control over body temperature, endogenous enzymes have adapted to function efficiently at the low temperatures of the environment. In general, there is a reciprocal relationship between the molar activities of enzymes and the temperature at which they are adapted to function. For example, thermostable enzymes from rabbit have a lower molar activity than the analogous enzymes in halibut under comparable conditions (155).

It is the catalytic efficiency of enzymes at low temperature that makes them important to the food scientists. This means that foods can be processed or modified by enzymes at moderate temperature, say 25-50°C, where food products would otherwise undergo changes slowly. It also means, however, that endogenous enzymes are active under these conditions as well, and this can be beneficial or deleterious. Continuing enzymic activities in plants after harvest or in animals postmortem have profound effects on food quality (144). Furthermore, the tremendous catalytic power of enzymes is characterized by a reduction in the activation energies for the reactions they catalyze (144). The small activation energies for enzymic reactions indicate that a reduction in temperature has relatively little effect on reaction rates. Consequently, enzymes can be quite active at subfreezing temperatures and, therefore, can be important stimulants of degradative reactions in refrigerated or frozen foods (168).

Of course, one purpose of heat processing is to denature and inactivate enzymes so that the food will not be subjected to continuing enzymic activity. Thus, the food scientist must have an understanding of the denaturation phenomenon to properly process foods.

Another important characteristic of enzymes, in addition to their catalytic power, is their specificity. Industrial catalysts lack this specificity of reaction, which precludes their use for modifying specific components of a food system. The specificity of hydrogen ion catalysis, for example, is very broad, whereas many enzymes perform only a single function, such as hydrolysis of a single bond or bond type. It is this enzymic specificity that allows the food scientist to selectively modify individual food components and not affect others. Specificity also enables food scientists to use enzymes as sensitive, analytical reagents. Analysis for food constituents in many instances can be simplified using the enzymic techniques detailed by Bergmeyer (9,10) and Guilbault (71).

Schwimmer (144) also includes regulation as a characteristic of enzymes in addition to specificity and catalytic efficiency. As we shall see, the control of biochemical processes by enzymes associated with various membranous components of a tissue can have profound effects on regulating life processes as well as in defining the quality of food during postharvest or postmortem processing and storage.

As a generalist, the food scientist should strive to develop a sufficient understanding of food enzymology and enzyme kinetics to bridge the gap between the engineer, using sophisticated mathematical concepts to design enzyme reactors and processes, and the enzymologist, using increasingly arcane methods to study mechanisms of enzymic reactions. The advent of immobilized enzymes and their use in reactors for food process-

Table 1 Molar Activities of Selected Enzymes

Enzyme	Molecules transformed (sec^{-1} per molecule enzyme)	Enzyme	Molecules transformed (sec^{-1} per molecule enzyme)
Carbonic anhydrase	600,000	Lactase	12,500
Catalase	100,000	Lipoxygenase	3,000
β-Amylase	20,000	Phenolase	2,000
Urease	8,000	Invertase	700

Source: From Ref. 144.

ing and the trend toward continuous processing methods require that the food scientist work even more closely with the food engineer. At the same time, the food scientist who has a good, fundamental knowledge of enzyme reaction mechanisms can exploit enzymes to perform desired functions in processing and analysis, and to facilitate conversion of raw materials into higher quality, more desirable foodstuffs.

Enzymes are of obvious importance to the food scientist, and this chapter attempts to introduce some basic concepts relevant to food enzymology. The intent is to provide the reader with sufficient background to pursue a greater understanding of food-related enzymology found in more detailed books by Schwimmer (144), Reed (133), and Whitaker (183). In addition, a number of books derived from symposia on specialized areas of food enzymology are useful (11,124,184). Excellent introductory treatments of enzyme chemistry, kinetics, and mechanisms can be found in a number of recent books and monographs (46,54,125,182). A bio-organic approach to enzyme mechanisms is available in books by Dugas and Penny (40) and by Walsh (180).

As an aid to better understanding enzyme kinetics, the serious beginning student should acquire the relatively simple programmed workbooks of Christensen and Palmer (27) and Tribe et al. (173), and progress to solving problems presented in the more sophisticated treatments of Segel (146) and Montgomery and Swenson (116).

II. ENZYME NOMENCLATURE

Over the years, the number of enzymes isolated and characterized has continued to increase at an enormous rate. Previously it was the custom for the individual who isolated and characterized the enzyme to also name it. However, in many instances the same enzyme was given different names or two different enzymes were given the same name. Consequently, the nomenclature for enzymes became so chaotic that The International Union of Biochemistry instituted a Commission on Nomenclature and Classification of Enzymes to prepare a system of nomenclature that has become standard and should be used in enzyme work (47).

An enzyme is assigned a code number of four numerals, each separated by periods and arranged according to the following principles. The first numeral is the main division to which the enzyme belongs, that is, (a) oxidoreductases, (b) transferases, (c) hydrolases, (d) lyases, (e) isomerases, and (f) ligases; the second is the subclass that identifies the enzyme in more specific terms; the third precisely defines the type of enzymic activity; and the fourth numeral is the serial number of the enzyme in its sub-subclass. Thus, the first three numerals clearly designate the nature of the enzyme. For example, 1.2.3.4

denotes an oxidoreductase with an aldehyde as a donor and O_2 as an acceptor, and it is the fourth numbered enzyme in a particular series. In addition to the code number, each enzyme is assigned a systematic name, which in many instances is too cumbersome to be used in the literature on a routine basis. Consequently, a trivial name has been recommended for common usage. The trivial name is sufficiently short for general use, but is not necessarily very exact or systematic and in a great many instances it is the name already in current use. The International Union of Biochemistry on Nomenclature and Classification of Enzymes (47) catalogued over 2000 enzymes in the 1978 edition.

Aside from enzymes involved in postmortem and postharvest physiology, few of the catalogued enzymes are of direct interest to the food scientist. By far the largest group of enzymes used in food processing is the hydrolases. A few oxidoreductases and isomerases are used, but no transferases, lyases, or ligases. Even in postmortem and postharvest situations the food scientist is concerned largely with controlling hydrolytic and oxidative reactions.

III. DEFINITIONS OF SOME TERMS USED IN ENZYMOLOGY (60,144,180)

1. Enzymes. Proteins that catalyze specific biological reactions with extraordinary catalytic power.
2. Isozymes (isoenzymes). Multiple forms of the same enzyme arising from genetically determined differences in primary amino acid sequences. Other variants should be called multiple forms. Artifacts, such as enzymically active products of hydrolysis (e.g., proteolysis) of an enzyme during handling or storage of biological material, should not be referred to as isozymes.
3. Active site. That portion of an enzyme molecule that consists of both the binding site and a place at which the chemical reaction occurs.
4. Cofactors. (a) Coenzymes that are derivatives of vitamins. (b) Small organic molecules associated with the active site, but not known as vitamins. (c) Organic compounds loosely associated with the enzyme or serving as a cosubstrate to alter one of two enzyme states in metabolic reactions. (d) Inorganic ions—metal cations and anions that participate directly in the reaction or serve as activators.
5. Allosteric effectors. Activators or inhibitors of enzymic activity that alter kinetic properties of oligomeric or monomeric enzymes. Binding of a substrate or other ligand that inhibits or activates the enzyme can result in conformational changes in the enzyme or in enzyme subunits. This, in turn, alters binding of substrate to vacant active sites or the rate of product formation of occupied active sites. Complex kinetics often yield sigmoidal velocity versus substrate curves.
6. Enzyme kinetics. Involves the study of parameters that influence the rate of enzyme-catalyzed chemical reactions and the interpretation of data in terms of possible molecular mechanisms. Steady-state kinetics extending from about 1 sec and beyond are studied in virtually all food-related research.
7. Steady-state kinetics. Kinetics typified by the Michaelis-Menten equation. The enzyme combines with substrate to form a complex, which then decomposes to yield product and free enzyme, the latter able to undergo another catalytic cycle.
8. Standard free energy changes (ΔG°). The value of ΔG (change in free energy) at a concentration of 1 M of each reactant and at a pressure of 1 atm (temperature is not specified). For a given reaction under standard conditions, the ΔG° for conversion of substrate to product is fixed, and thus fixes the point of equilibrium for that reaction. A spontaneous process doing work utilizes free energy ($-\Delta G^\circ$) until it reaches a minimal (equilibrium) value, where $\Delta G^\circ = 0$. The *rate* of a reaction is independent of ΔG° (see Fig. 1).

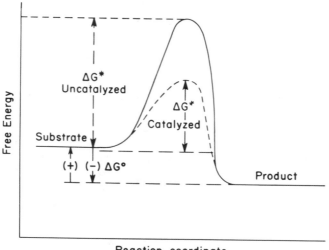

Figure 1 Energy diagram showing the reduction in the free energy of activation (ΔG^*) for a reaction when catalyzed.

9. Transition state. An activated complex formed between molecules (based on probability), which can (a) return to the original state (no reaction), or (b) internally rearrange to form new bonds and product molecules (see Fig. 1).

10. Energy of activation. In transition state theory, this is the amount of free energy, (ΔG^*) required to bring 1 mol of molecules (substrate) in a reaction mixture to the transition state (for conversion to product) and it is expressed as calories (joules) or kilocalories (kilojoules) per mole (kJ/mol). *Enzymes and other catalysts* act to lower this barrier to reaction, but do not alter the point of equilibrium. Thus, it is crucial to distinguish between ΔG° in definition 8, which fixes the equilibrium of a reaction, and ΔG^*, which defines the rate at which equilibrium will be obtained (see Fig. 1 and the discussion in Sec. IV). The energy of activation (E_a) derived from the Arrhenius relationship is only slightly different from ΔG^*.

11. Kilojoule. An SI unit of energy equal to kilocalories/4.184 (1 joule = calories/4.184).

IV. SPECIFICITY, CATALYSIS, AND REGULATION

From the foregoing discussions, we know that the two salient features of enzymic activity are acceleration of reaction rates (without affecting the final equilibrium of the reaction) and the specificity of reaction. The mechanisms behind these two principal characteristics of enzymes have occupied enzymologists for decades and are still incompletely understood. There is, however, a certain logic of sequences of enzymic reactions (180), and as in any series of events, rate-controlling factors exist. Since enzymes can act as regulators and also can be regulated, this aspect of enzymology as it relates to food systems is considered an additional characteristic of enzymes.

A. Specificity (50,144)

The specificities of enzymes distinguish them from nonbiological catalysts. Enzymes show varying degrees of specificity, as follows.

1. Low specificity. The enzyme does not discriminate among substrates, but exhibits specificity only toward the bond being split.
2. Group specificity. The enzyme is specific for a particular chemical bond adjacent to a specific group. For example, trypsin is specific for peptide bonds on the carboxyl side of arginine and lysine.
3. Absolute specificity. The enzyme attacks only one substrate and catalyzes only a single reaction. Most enzymes fall into this category.
4. Stereochemical specificity. As a rule, enzymes demonstrate unerring and complete stereospecificity in catalysis, and can distinguish between optical or geometric isomers. Enzymes almost always use one form of an enantiomeric pair, unless their specific function is to catalyze the isomerization of enantiomers. Enzymes behave as stereospecific catalysts because they are asymmetric or chiral reagents comprised uniquely of L-amino acids. Recall from organic chemistry that a simple molecule containing one asymmetric (chiral) center can exist as two enantiomers (e.g., D- or L-amino acids) with similar physical properties that make separation (resolution) difficult. However, if each of the two enantiomers is derivatized with a common asymmetric compound, two diastereomers result, each with distinctly different physical properties, and this simplifies resolution of the two enantiomers. In general, a similar concept can be used to rationalize the stereospecificity of enzymes. When a chiral enzyme interacts with each enantiomer during reaction, diastereomeric transition states are formed. These have different energies and different reactivities and partition differently between reactants and products. Thus, the differences in physical properties between the diastereomeric transition states favor the reaction of one of the enantiomers with the chiral enzyme.

Obviously, additional factors can be involved in enzymic specificity. For example, chymotrypsin and trypsin possess identical functional groups directly involved in the hydrolysis of polypeptide bonds; however, ancillary binding sites within the active site of each enzyme dictate that the former enzyme hydrolyze peptide bonds on the carboxyl side of hydrophobic amino acid residues, whereas the latter hydrolyze peptide bonds on the carboxyl side of the basic amino acids, lysine and arginine (183). The crucial point to remember is that a relatively limited number of functional groups in the amino acid side chains of enzymes and in associated cofactors can be combined in almost limitless combinations within the three-dimensional matrix of enzymes to catalyze a large number of biochemical reactions in very specific ways (180).

The specificity of enzymes is very important in food processing, where it is often desirable to modify only a single component in the process. In food analysis, the accuracy of enzymic methods is dependent on specificity.

B. Catalysis (60,144,180)

In general terms, catalysis can be characterized as the facilitation of a reaction by a reduction in the activation energy of the reaction in question. It is important to distinguish between the thermodynamic tendency or potential for a reaction to proceed and the actual rate at which product accumulates. The equilibrium concentrations of product and substrate are determined by their difference in free energy content ($\Delta G^\circ = -RT \ln kEq$). At a given temperature R and T are constants requiring ΔG° to be related to $\ln kEq = $ [product]/[substrate]. As depicted in Fig. 1, if the product of a reaction is more stable than the substrate, ΔG° is negative for the reaction. Consider a hypothetical case (180) in which the difference in ΔG° for conversion of a substrate to product is −2.8 kcal/mol = −11.7 kJ/mol. Then, at 30°C,

$$\ln kEq = \frac{-2.8 \text{ kcal/mol}}{-RT}$$

$$kEq = 100$$

$$= \frac{[\text{product}]}{[\text{substrate}]} \tag{1}$$

At chemical equilibrium, there are 100 product molecules for every substrate molecule.

Also, if one compares two possible reactions in which the molar equilibrium ratio is assumed to be 100 in the first reaction and 1000 in the second, one readily understands that $\Delta G°$ for the second reaction is (a) more negative in value (expressed as kcal/mol or kJ/mol), and (b) in this case, more negative by a factor of 3/2 (ln 1000/ln 100). Thus, a slightly more negative value for $\Delta G°$ for one reaction over a different one at the same temperature favors a substantial increase in the chemical equilibrium point for product formation by the first reaction. This type of information is useful when considering coupled enzyme reactions and defines the relative magnitude of each pathway in the coupled reaction.

However, the *rate* at which a conversion occurs, a measure of the *kinetic* lability of substrate molecules, is independent of $\Delta G°$. A molecule can thus be thermodynamically unstable (i.e., $\Delta G°$ is negative for conversion of substrate to product) but kinetically stable if the energy barrier (ΔG^*) for the reaction is sufficiently high (Fig. 1). In most chemical reactions, heat is used to overcome this energy barrier (the activation energy). Heterogenous catalysts (e.g., a nickel catalyst used to hydrogenate vegetable oils) probably lower ΔG^* almost exclusively by selectively stabilizing the transition state (bringing reactants together in favorable juxtaposition on the catalyst surface). Enzymes, however, employ a larger number of interrelated factors to lower the energy of activation (ΔG^*) of a given reaction (144,180).

As a simple chemical reaction proceeds along the reaction coordinate, the highest point on the free energy curve represents the transition state for the reaction (Fig. 1). Chemical catalysts (e.g., a palladium metal catalyst for hydrogenation of an olefin) probably lowers ΔG^* almost exclusively by selective stabilization of the transition state (Fig. 1). In contrast, enzymic catalysts also lower ΔG^* but replace one large activation barrier with multiple lower barriers with small ΔG^* overall (180,183). The transition state is a fleeting entity, lasting perhaps for one molecular vibration ($\sim 10^{-13}$ sec), which decomposes to either reform reactants with which it is assumed to be in equilibrium, or products.

Transition state theory is discussed in greater detail and is contrasted with the Arrhenius relationship in Sec. X. A simplified version of transition state theory for reaction rates relates the rate of a reaction to the magnitude of ΔG^* according to the simple exponential expression (180)

$$k_{obs} = \frac{RT}{nh} e^{-\Delta G^*/RT} \tag{2}$$

where n is Avogadro's number and h is Planck's constant. Those substrate molecules that have a kinetic energy $> \Delta G^*$ can pass over the barrier to products. Reduction of ΔG^* will greatly increase the reaction rate k_{obs} since ΔG^* is an exponential function. A specific example (60) is the decomposition of H_2O_2 at 20°C. The free energy of activation ΔG^* is 75 kJ/mol for the uncatalyzed reaction compared with 54 kJ/mol when the reaction is catalyzed by platinum and 29 kJ/mol when catalyzed by the enzyme catalase. As ΔG^* decreases, the exponential term $e^{-\Delta G^*/RT}$ decreases greatly, leading to a much enhanced rate of reaction. Thus, the term $e^{-\Delta G^*/RT}$ changes greatly and greatly alters the rate k_{obs} because ΔG^* becomes less negative as the free energy of activation is decreased. The increase in rate achieved by platinum as compared with the uncatalyzed reaction is $e^{75/2.43}/e^{54/2.43}$ or 5.7×10^3 fold, and the increase by catalase is $e^{75/2.43}/e^{29/2.43}$ or 1.7×10^8 fold. One can readily see that the rate enhancement is inversely related to ΔG^*.

C. Regulation

It is well understood that closely controlled regulation of enzymic processes is crucial to life (144). In animal tissues postmortem or in plant tissues postharvest, the normal control mechanisms characteristic of living tissues start to change as oxygen and/or energy sources become depleted. The biosynthetic machinery, driven by oxidative metabolism, declines in activity. Such changes may occur rapidly, as in animal tissues, or gradually over months of storage, as in many plant tissues. The equilibrium tends to shift toward a generalized catabolism, which the food scientists seeks to control. Many characteristic food products result from control (contrived or inadvertent) of these processes, such as meats, cheese, and stored fruits and vegetables.

Thus, the activities of endogenous and exogenous (added during processing or from microbial contamination) enzymes can play a role in regulating and controlling the quality attributes of foods during processing and storage. Consequently, the success the food scientist has in regulating enzymic processes has a marked effect on the quality and even quantity (by avoiding spoilage) of foods available to the consumer. There are ample examples in which the enzyme is the regulator or is regulated in controlling the quality of foods (144).

A major determinant in regulation of endogenous enzymic reactions is integrity of cells and tissues, that is, normal compartmentalization of enzymes in subcellular organelles or particles. Once the integrity of the organelles is lost (e.g., death of animal cells, senescence of plant cells postharvest, or disruption of cells during processing) enzyme activity can occur in an uncontrolled fashion.

V. COMPARTMENTALIZATION OF ENZYMES IN CELLULAR SYSTEMS

For the ensuing discussion of enzyme compartmentalization, it is useful for the reader to refer to the cellular structures depicted in Fig. 8 of Chap. 15. Over the years it has become evident that enzymes and enzyme systems are compartmentalized at both the subcellular and the tissue levels. This distribution of enzymes among subcellular particles and tissues of plants and animals is of profound significance to the food scientist (143, 144).

In general, there are several ways in which the maintenance or destruction of the integrity of enzyme localization is important in food systems. The maturation and ripening of plant foods are an integrated and dynamic sequence of biochemical events. Consequently, the integrity of the plant tissue must be maintained to ensure proper ripening. Since most fruits are harvested when mature but still unripe, a wide variety of enzymically induced changes must still occur to obtain acceptable quality (50,140). For example, the loss of some pigments and gains in others, the accumulation of sugars, softening, and biosynthesis of flavor components must occur in a manner that requires the biochemical integrity of the food (80).

In some instances the food scientist has learned to control these integrated biochemical changes to produce better and more abundant food. Controlled atmosphere storage of fruits and vegetables, for example, regulates the rate of metabolism of some fruits (notably, apples and pears) and lengthens the time these products can be stored (50). Another example of enzymic control by the food scientist is accelerated aging of meat (increased hydrolase activity) accomplished by exposing the meat to higher than normal temperatures while surface growth of microorganisms is controlled by ultraviolet light (Tenderay process) (87).

Disruption of cellular and enzymic integrity of foods often occurs during processing operations. Indeed, Schwimmer (143) put forth the thesis that "food processing and technology may be considered as the art and science of the promotion, control and/or the prevention of cellular disruption and its metabolic consequences at the right time and at the right place in the food processing chain." Of course, cellular disruption is accompanied by the release of enzymes that can be beneficial or deleterious. For example, the interaction of released enzymes with substrates not formerly available is important in developing flavor in dehydrated onions (144). On the other hand, disruption of cells can result in release of polyphenoloxidase which then combines with oxygen and phenols to cause the well-known enzymic browning reaction desirable in tea fermentation and undesirable in fruits and vegetables (50). One objective of the food scientist is to control enzymic reactions of this kind. Listed in Sec. X are various methods available to the food scientist for controlling enzymic reactions in foods.

Alterations that have occurred in the normal distribution of enzymes in a cellular system often can be used by the food scientists to determine the history of the food product. For example, freezing and thawing of tissues are known to disrupt biological membranes. A consequence of this is the release of enzymes from formerly intact organelles. Thus, when enzymes characteristic of cellular organelles are dislocated, this may mean that the tissue has been frozen and thawed (65,73). Information of this kind is valuable to regulatory agencies in their efforts to detect fraud.

Furthermore, enzyme activities (132a) and the products of enzymic reactions often can be used as indicators of the freshness of fish (65,156). As autolysis proceeds with storage time and enzymes are released from their binding sites, the increase in products of autolysis often can be used as objective measures of freshness. Thus, enzymic activities can be used to distinguish fresh fish from frozen and to measure deterioration (132a). These few examples give some indication of why knowledge of enzyme localization is important to food scientists.

A. Distribution of Enzymes at the Subcellular Level (19,138,144)

Of the many enzymes in a typical cell, very few are in true solution (19). Most of them exist bound to membranes of the subcellular elements. Furthermore, those enzymes in true solution (in the cytosol) are probably subjected to different conditions than those encountered in a typical in vitro reaction mixture. A thesis of increasing importance has been developed in postharvest and postmortem physiology (81) that implicates a reversible association of enzymes with membranes as a metabolic control mechnism.

Some enzymes are characteristic of a particular subcellular organelle; however, others are ubiquitous in nature. Although these latter enzymes may catalyze the same reaction at different subcellular or tissue locations, they may differ with respect to their primary, secondary, tertiary, or quaternary structures. These *isozymes* thus may yield different kinetics or respond differently to allosteric effectors. The significance of isozymes and isozyme distributions in food tissues is being learned very slowly. It is well known that the enzymes responsible for anaerobic glycolysis (Embden-Meyerhof cycle) in a cell occur in the cytosol, whereas those that catalyze oxidation of substrates via the tricarboxylic acid cycle are compartmentalized in the mitochondria. Detailed discussions on the distribution and latency of enzymes at the subcellular level can be found elsewhere (138).

B. Distribution of Enzymes in Selected Food Materials

In addition to their subcellular and subparticulate distribution, the kind and concentration of enzymes differ among food systems and tissues. Often, the significance of enzyme compartmentalization at the tissue level is obscure; however, at times it may affect the processing procedures or the type of product produced. For example, in the preparation of blue cheese, milk is separated into lipase-rich skim milk and cream. Subsequently the cream is bleached, then recombined with the homogenized skim milk so that the natural lipases can aid in developing the typical blue cheese flavor.

The enzyme distribution in milk (58), eggs, grains, and meats clearly illustrates that enzyme compartmentalization is characteristic of food materials (138).

VI. KINETICS OF ENZYME-CATALYZED REACTIONS (54,146,182)

A. Steady-State Enzyme Kinetics

The simplest scheme for presenting an enzymic reaction kinetically is

$$E + S \underset{k_2}{\overset{k_1}{\rightleftharpoons}} ES \underset{k_4}{\overset{k_3}{\rightleftharpoons}} P + E \tag{3}$$

where E designates enzyme; S, substrate; ES, enzyme-substrate complex; P, product of enzymic reaction; and k_1, k_2, k_3, and k_4, the specific rate constants for the appropriate steps in the total reaction. The cornerstone of the scheme is the formation of the enzyme-substrate complex, often referred to as the Michaelis complex, as it was Michaelis and Menten who first postulated its existence.

The rate of formation of product is given by

$$\frac{d[P]}{dt} = k_3[ES] - k_4[P][E] \tag{4}$$

During the initial part of a typical enzymic reaction the product concentration is essentially zero. Thus we have

$$\frac{d[P]}{dt} = v$$

$$= k_3[ES] \tag{5}$$

where v is the rate of product formation and is usually referred to as the Michaelis-Menten velocity. The concentration of the enzyme-substrate complex initially is zero, but quickly builds to a peak level and then slowly declines if initial concentration of substrate $[S_0]$ is much greater than that of the enzyme $[E_0]$. At any instant of time [ES] is approximately constant, and we can assume steady-state conditions. Mathematically we then have

$$\frac{d[ES]}{dt} = k_1[E][S] - k_2[ES] - k_3[ES]$$

$$\simeq 0 \tag{6}$$

Rearrangement gives

$$\frac{[E][S]}{[ES]} = \frac{k_2 + k_3}{k_1}$$

$$= K_m \tag{7}$$

where K_m is referred to as the Michaelis constant. By definition, the dissociation constant K_s for the enzyme-substrate complex is equal to k_2/k_1. Note that $K_m \geqslant k_s$, and the two are equal when k_3 is zero. The total enzyme condentration $[E_0]$ is given by

$$[E_0] = [E] + [ES] \tag{8}$$

Combining Eqs. (7) and (8) we obtain

$$[ES] = \frac{[E_0][S]}{K_m + [S]} \tag{9}$$

The Michaelis-Menten velocity [from Eq. (4)] then becomes

$$v = k_3[E_0][S]/(K_m + [S]) \tag{10}$$

Experimentally v may be obtained from the slope of a progress curve, that is, from the slope of a plot of [P] versus time (Fig. 2). As the extent of the enzymic reaction increases, the progress curve becomes curvilinear. This means that the rate of formation of product slows down during the course of the reaction because of substrate depletion. Note that in Eq. (10) v becomes smaller when [S] decreases. There are further reasons for a decrease in v as the reaction continues. Since enzymic reactions are reversible, Eq. (3), back reactions become important as the extent of the reaction increases. Moreover, the product may inhibit the enzyme. It is thus customary to take the initial slope of the progress curve as the value of v. If values for v_0 (initial velocity) are obtained at different substrate concentrations and then plotted against those substrate concentrations, a rectangular hyperbola is obtained as predicted by Eq. (10) and as shown in Fig. 3. As [S] increases, v increases significantly at first, but then progressively less so at higher values of [S]. When $[S] \gg K_m$, the Michaelis-Menten velocity approaches a maximum value, referred to as V_{max}. We say that E is saturated with S. As a working rule, a value of [S] 20 times greater than K_m is usually considered sufficient to approach V_{max}. When we have these conditions the denominator in Eq. (10) is approximately equal to [S] since the value of K_m can be ignored. The [S] terms in numerator and denominator then cancel each other, and Eq. (10) becomes

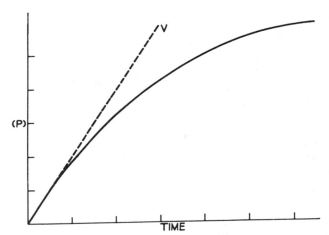

Figure 2 Progress curve of an enzyme-catalyzed reaction. The rate (v) of the reaction (tangent to the curve at the moment of interest) decreases as the reaction proceeds. (Redrawn from Ref. 183.)

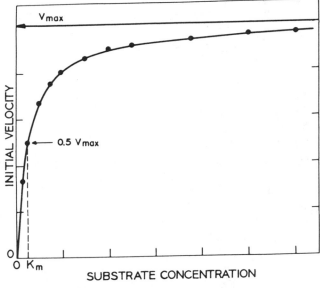

Figure 3 Plot of initial velocity v versus concentration of substrate [S] for a typical enzyme-catalyzed reaction. (Redrawn from Ref. 183.)

$$v = k_3 [E_0]$$

$$= V_{max} \tag{11}$$

From this equation we see that V_{max} is independent of $[S]$; that is, the reaction is zero order with respect to S. On the other hand, V_{max} is first order with respect to E. The term V_{max} is also dependent on k_3, which is usually referred to as the catalytic rate constant or turnover number. For a given enzyme system, an increase in V_{max} may be obtained by an increase in $[E_0]$, but not $[S_0]$.

The Michaelis constant has the units of concentration and specifically refers to a particular substrate concentration. When $[S]$ equals K_m, the term v becomes equal to $V_{max}/2$. The value of the Michaelis constant then tells us the value of substrate concentration required to half-saturate the enzyme. If K_m is large, a large amount of S is required to saturate E; if K_m is small, a small amount of S is required to saturate E. Most enzymic reactions involving single substrates have K_m values between 10^{-5} and 10^{-2} M.

The two most significant parameters in Michaelis-Menten theory are K_m and V_{max}. They are most commonly determined from a Lineweaver-Burk plot. If we take reciprocals of both sides of Eq. (10), we obtain

$$\frac{1}{v} = \frac{K_m}{V_{max}} \frac{1}{[S]} + \frac{1}{V_{max}} \tag{12}$$

This is an equation for a straight line when $1/v$ is plotted against $1/[S]$. Such a plot is shown in Fig. 4. The y intercept is equal to $1/V_{max}$, and the x intercept is equal to $-1/K_m$. The ratio of the slope to the y intercept is equal to K_m.

Lineweaver-Burk plots make K_m and V_{max} very accessible. These parameters are only poorly obtained from a plot of v versus $[S]$ (Fig. 3). However, the plot of v versus $[S]$ graphically illustrates the meaning of K_m and V_{max}.

B. Enzyme Inhibition Kinetics (46,146)

An inhibitor is any compound that reduces the rate of an enzyme-catalyzed reaction. In enzyme inhibition, an inhibitor molecule reacts with the enzyme, the enzyme-substrate complex, or both, resulting in changes in the apparent value of K_m, V_{max}, or both. We say the enzyme is inhibited. If the inhibitor molecule reacts with E, inhibition is referred to as competitive; if the inhibitor reacts with ES, inhibition is referred to as uncompetitive; and if the inhibitor reacts with both E and ES, inhibition is referred to as noncompetitive. Each of these forms of inhibition is reversible; if the inhibitor is removed, enzyme activity is restored. However, there are also irreversible inhibitors. In these cases, when excess inhibitor is removed, the enzyme does not become reactive again. Heavy metals, such as mercury, are good examples of irreversible inhibitors. The following discussion centers around reversible forms of inhibition.

Regardless of whether inhibition is competitive, uncompetitive, or noncompetitive, reaction mechanism (3) is still operative. We merely have additional reactions that occur simultaneously with mechanism (3). Moreover, Eq. (7), which mathematically defines K_m, holds regardless of the type of inhibition, and Eq. (5), which defines the rate of formation of product, likewise holds for all cases.

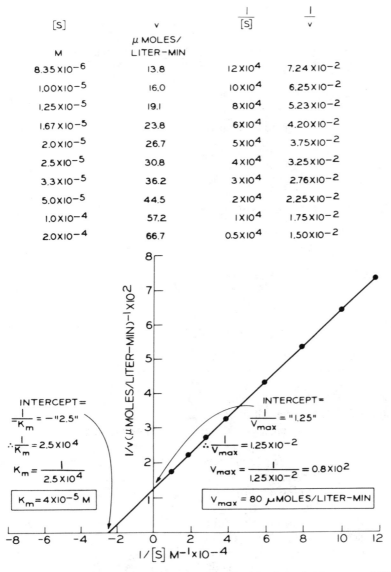

[S] M	v μ MOLES/ LITER-MIN	$\frac{1}{[S]}$	$\frac{1}{v}$
8.35×10^{-6}	13.8	12×10^4	7.24×10^{-2}
1.00×10^{-5}	16.0	10×10^4	6.25×10^{-2}
1.25×10^{-5}	19.1	8×10^4	5.23×10^{-2}
1.67×10^{-5}	23.8	6×10^4	4.20×10^{-2}
2.0×10^{-5}	26.7	5×10^4	3.75×10^{-2}
2.5×10^{-5}	30.8	4×10^4	3.25×10^{-2}
3.3×10^{-5}	36.2	3×10^4	2.76×10^{-2}
5.0×10^{-5}	44.5	2×10^4	2.25×10^{-2}
1.0×10^{-4}	57.2	1×10^4	1.75×10^{-2}
2.0×10^{-4}	66.7	0.5×10^4	1.50×10^{-2}

INTERCEPT =
$$\frac{1}{-K_m} = -\text{"2.5"}$$
$$\therefore \frac{1}{K_m} = 2.5 \times 10^4$$
$$K_m = \frac{1}{2.5 \times 10^4}$$
$$\boxed{K_m = 4 \times 10^{-5} \text{ M}}$$

INTERCEPT =
$$\frac{1}{V_{max}} = \text{"1.25"}$$
$$\therefore \frac{1}{V_{max}} = 1.25 \times 10^{-2}$$
$$V_{max} = \frac{1}{1.25 \times 10^{-2}} = 0.8 \times 10^2$$
$$\boxed{V_{max} = 80 \ \mu\text{MOLES/LITER-MIN}}$$

Figure 4 Double-reciprocal (Lineweaver-Burk) plot of 1/v versus 1/[S] for an enzyme reaction. (Redrawn from Ref. 146, by courtesy of John Wiley and Sons.)

1. Competitive Inhibition

When inhibition is competitive, we have, in addition to reaction mechanism (3),

$$E + I \underset{}{\overset{K_{IE}}{\rightleftharpoons}} EI \tag{13}$$

where I refers to the inhibitor molecule, EI to an enzyme-inhibitor complex, and K_{IE} to the inhibition constant. Strictly speaking, K_{IE} is the dissociation constant for the enzyme-inhibitor complex

$$K_{IE} = \frac{[E][I]}{[EI]} \tag{14}$$

Total enzyme concentration is no longer given by Eq. (8) because EI must be considered. We now have

$$[E_0] = [E] + [ES] + [EI] \tag{15}$$

Equations (14) and (15) may be combined with Eq. (7), which defines K_m. This gives

$$[ES] = \frac{[E_0][S]}{K'_m + [S]} \tag{16}$$

where K'_m, which is referred to as the apparent Michaelis constant, is given by

$$K'_m = K_m \left(1 + \frac{[I]}{K_{IE}} \right) \tag{17}$$

Equation (5), which gives the Michaelis-Menten velocity is now

$$v = \frac{V_{max}[S]}{K'_m + [S]} \tag{18}$$

We see that the Michaelis constant is affected by the competitive inhibitor. The inhibitor molecule competes with S for E. The term K'_m then has a larger numerical value than K_m for a given substrate. The term V_{max}, which equals $k_3[E_0]$, is unaffected by a competitive inhibitor, and is defined by the ability of S to saturate E.

The value of K_{IE} may be obtained with the aid of Lineweaver-Burk plots in the presence and absence of I. The value of K_m may be determined from the plot when I is absent, and the value of K'_m when I is present. Since the value of V_{max} is unaffected by a competitive inhibitor, the different Lineweaver-Burk plots have the same y intercept. Figure 5 shows Lineweaver-Burk plots at different [I] values. Once the values for V_{max}, K_m, and K'_m are determined, the value of K_{IE} may be calculated from Eq. (17), provided the value of [I] is known. It is possible to determine the value of K_{IE} when the value of K'_m is known at a single value of [I]. However, there is less error involved in the determination of K_{IE} when values for K'_m are obtained at a series of [I] values. Then, from Eq. (17) we may plot K'_m versus [I]. The slope of this plot equals K_m/K_{IE}. The x intercept equals $-K_{IE}$.

2. Uncompetitive inhibition

In uncompetitive inhibition the inhibitor molecule reacts with ES but not E. Then, in addition to reaction (3), we have

$$ES + I \xrightleftharpoons{K_{IES}} ESI \tag{19}$$

where ESI is a complex formed by interaction between ES and I, and K_{IES} is the inhibition constant, which is actually the dissociation constant for the ESI complex:

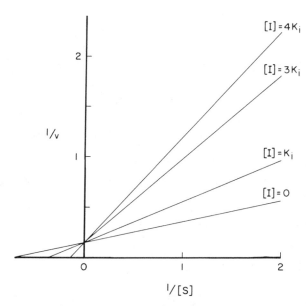

Figure 5 Competitive inhibition as seen in a Lineweaver-Burk plot. Plots are shown for rates measured in the presence of various amounts of competitive inhibitor [I]. (Redrawn from Ref. 46.)

$$K_{IES} = \frac{[ES][I]}{[ESI]} \tag{20}$$

Total enzyme concentration is

$$[E_0] = [E] + [ES] + [ESI] \tag{21}$$

Equations (20) and (21) may be combined with Eq. (7). It can then be shown that [ES] is

$$[ES] = \frac{[E_0][S]/(1 + [I]/K_{IES})}{K_m/(1 + [I]/K_{IES}) + [S]} \tag{22}$$

Equation (5) then becomes

$$v = \frac{V'_{max}[S]}{K''_m + [S]} \tag{23}$$

where

$$V'_{max} = \frac{V_{max}}{1 + [I]/K_{IES}} \tag{24}$$

and

$$K''_m = \frac{K_m}{1 + [I]/K_{IES}} \qquad (25)$$

Both V_{max} and K_m are affected equally by I. Whereas in competitive inhibition K_m is multiplied by an inhibition term in order to obtain the apparent K_m [Eq. (17)], here K_m is divided by the inhibition term [Eq. (25)]. This means that $K''_m < K_m$. Since I combines only with ES, its presence pulls E to S (46). Because I intereferes with ES, $V'_{max} < V_{max}$. The concentration of ES gives the velocity of the enzymic reaction by way of Eq. (5).

Since V_{max} and K_m are affected equally by I during uncompetitive inhibition, Lineweaver-Burk plots in the presence of I should be parallel to one obtained in the absence of I. Lineweaver-Burk plots at different values of [I] are shown in Fig. 6. The y intercept of each plot in the presence of I is equal to $1/V'_{max}$ for the appropriate value of [I]. If we take the reciprocal of both sides of Eq. (24) we get

$$\frac{1}{V'_{max}} = \frac{1}{V_{max}} + \frac{[I]}{K_{IES}V_{max}} \qquad (26)$$

This equation indicates that we may take the y intercept values of the different Lineweaver-Burk plots and plot these values back against the appropriate values of [I]. The slope of this plot is $1/V_{max}K_{IES}$. The y intercept is $1/V_{max}$. The ratio of the y intercept to the slope is then equal to K_{IES}.

3. Noncompetitive Inhibition Kinetics

In noncompetitive inhibition the inhibitor combines with both E and ES. Thus, in addition to reaction (3), reactions (13) and (19) are operative. There is also assumed to be an equilibrium between ES and ESI. However, Eq. (14), which defines K_{IE}, and Eq. (20),

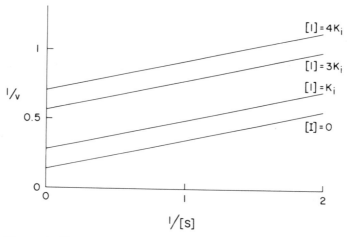

Figure 6 Uncompetitive inhibition as seen in a Lineweaver-Burk plot. (Redrawn from Ref. 46.)

which defines K_{IES}, still apply. Total enzyme concentration is now equal to

$$[E_0] = [E] + [ES] + [EI] + [ESI] \tag{27}$$

If Eqs. (14), (20), (27), and (7) are combined, we can obtain the following for [ES]:

$$[ES] = \frac{[E_0][S]/(1 + [I]/K_{IES})}{K_m(1 + [I]/K_{IE})/(1 + [I]/K_{IES}) + [S]} \tag{28}$$

When $K_{IE} = K_{IES}$, we have "simple" noncompetitive inhibition; when unequal, "mixed" inhibition. If we assume we have simple inhibition and if we refer to the inhibition constant as K_I, Eq. (28) becomes

$$[ES] = \frac{[E_0][S]/(1 + [I]/K_I)}{K_m + [S]} \tag{29}$$

And Eq. (5) is

$$v = \frac{V'_{max}[S]}{K_m + [S]} \tag{30}$$

where V'_{max} is given by Eq. (24), if K_I is substituted for K_{IES}. The value of K_I may be obtained by the same procedure used to obtain K_{IES} in uncompetitive inhibition.

In simple noncompetitive inhibition K_m is unaffected whereas V_{max} is decreased. This means that Lineweaver-Burk plots intersect at a single point on the x-axis and fan out from there. As [I] increases, the slopes of the plots become steeper (Fig. 7).

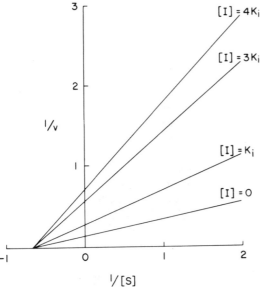

Figure 7 Simple noncompetitive inhibition ($K_{IE} = K_{IES}$) as seen in a Lineweaver-Burk plot. (Redrawn from Ref. 46.)

Table 2 Characteristics of Different Types of Reversible Enzyme Inhibition

Type of inhibition	Inhibitor combines with	Apparent effect on V_{max}	Apparent effect on K_m	Effect on 1/V against 1/[S] plot
Competitive	E	Unchanged	Increased	Convergence on ordinate axis
Uncompetitive	ES	Decreased	Decreased	Parallel lines
Noncompetitive (a) Simple	E and ES			Convergence
$(K_{IE} = K_{IES})$ on abscissa			Decreased	Unchanged
(b) Mixed				
$(K_{IES} > K_{IE})$ (c) Mixed		Decreased	Increased	Above abscissa
$K_{IES} < K_{IE}$		Decreased	Decreased	Below abscissa

Source: From Ref. 46.

Combination of Eqs. (28) and (30) permits derivation of the appropriate relationships that describe mixed inhibition. Here it will only be observed that if $K_{IES} > K_{IE}$, Lineweaver-Burk plots converge above the x-axis. The upshot of this is that the apparent Michaelis constant increases upon addition of I. If $K_{IES} < K_{IE}$, Lineweaver-Burk plots converge below the x-axis, indicating the apparent Michaelis constant decreases with addition of I. With all cases of noncompetitive inhibition V_{max} is decreased upon addition of I.

Table 2 summarizes the effect of I on both K_m and V_{max} for each type of inhibition just discussed.

C. Immobilized Enzyme Kinetics (192)

Enzyme kinetics basically have been developed for reactions in solution where enzyme and substrate are free to move about in a random fashion. The rate of formation of the enzyme-substrate complex might be expected to be diffusion controlled. As a result, the value of k_1 in reaction scheme (3) should be large. Experimental studies have shown that this usually is the case, as values for k_1 generally are found to be in the range 10^6-10^9 liters/mol sec.

However, enzymic reactions often do not occur in solution, but at interfaces, insofar as the cell with its membranes and organelles can be considered a conglomerate of interfaces. Enzymes are often immobilized at these interfaces, and their substrates must come to them in order to be converted to products. Thus, the study of immobilized enzymes is of interest to the food scientist not only for reasons of processing but also for gaining a better understanding of postharvest and postmortem physiology. Developments in enzyme technology have led to the preparation of immobilized enzymes, which in principle can be used repeatedly in batch operations or on a continuous basis.

In a later section we discuss some of the methods for immobilizing enzymes and some of their uses. At present we shall discuss some of the kinetic implications that result as a consequence of immobilization.

First, we shall assume that the support that holds the enzyme is insoluble in the solvent. When this is the case, the enzyme support is surrounded by the Nernst or diffusion layer. Consequently, in the vicinity of an uncharged support the concentration of substrate is less than in the bulk of solution, since there is a concentration gradient across the diffusion layer. With less substrate at the surface of the support, we might expect that immobilized enzymes have less activity and higher apparent substrate K_m values than their soluble counterparts, since concentration will actually be measured in the bulk phase, yielding higher values than those at the interface. This indeed is often so. Thus, when we consider immobilized enzymes, we must take into account that the K_m values we obtain are apparent K_m values. There are various ways of reducing the apparent K_m. One way is to reduce the size of the enzyme support. This effectively reduces the thickness of the Nernst layer. Another way is to increase the stirring rate or flow rate in an immobilized enzyme reactor to reduce the thickness of the Nernst layer. A third way is to solubilize the enzyme support. This eliminates the Nernst layer.

The Nernst layer is not the only factor that affects the activity of immobilized enzymes. There are also electrostatic effects. If the support and substrate have the same charge, we expect mutual repulsion; if they have opposite charges, we expect attraction. When support and substrate have opposite net charges, we might expect to lower the value of the apparent K_m below that obtained when E is not immobilized, since the concentration of substrate at the interface should be higher than in the bulk solution. Experimental studies have shown that the apparent K_m value can be lowered many times below that of the soluble counterpart for some enzymic systems.

Hornby et al. (79) have developed a mathematical expression for the activity of immobilized enzymes that considers both diffusion and electrostatic factors. The Michaelis-Menten velocity is given by

$$v = \frac{V_{max}[S]}{K_m^* + [S]} \tag{31}$$

and

$$K_m^* = \left(K_m + \frac{x V_{max}}{D} \right) \frac{RT}{RT - xzFV} \tag{32}$$

where K_m^* is the apparent Michaelis constant, x is the thickness of the Nernst layer, D is the diffusion coefficient, T is the temperature (Kelvin), z is the valence of substrate, F is Faraday's constant, R is the universal gas constant, and V is the potential gradient around the enzyme support. The term $(K_m + x V_{max}/D)$ may be referred to as a diffusion term. Strictly speaking, K_m and V_{max} are not diffusion parameters, but x and D are. The term $RT/(RT - xzFV)$ may be referred to as an electrostatic term because of the presence of z, F, and V. As for the diffusion term, as the ratio x/D decreases, we expect K_m^* to likewise decrease and approach K_m. The value of x may be reduced by using smaller supports or by increasing the flow or stirring rate, as discussed above. As for the electrostatic term, if z and V have the same sign, that is, if the substrate and the support have the same charge, the term $RT/(RT - xzFV)$ on the right is greater than 1 and K_m^* increases. On the other hand, if z and V have opposite sign, the electrostatic term is less than 1, and K_m^* decreases. If either z or V = O, the electrostatic term equals 1, and K_m^* is affected only by diffusion considerations.

When discussing the mathematical relationships used for determining values for K_m^*, we need to likewise discuss the enzyme reactors used to experimentally determine these values. We shall consider two types of enzyme reactors, the packed-bed reactor and the continuous feed stirred tank reactor. The packed-bed or fixed-bed reactor is a plug-flow or tubular reactor that has no mixing in the direction of flow. The conditions at every point along the reactor are different but constant with time. In a stirred-tank or well-mixed reactor, on the other hand, flow conditions are uniform throughout the reactor.

In a packed-bed reactor the enzyme support is packed in the reactor bed or column, and the enzyme reacts with substrate as it perfuses through the column. Hornby et al. (79) developed and studied kinetic equations to describe the behavior of a packed-bed reactor. The reaction scheme is assumed to be the same as that for soluble enzymes, that is, reaction mechanism (3) applies. Since steady-state conditions are assumed, $dP/dt = -d[S]/dt$. Then, from Eqs. (10) and (11), we have

$$\frac{-d[S]}{dt} = \frac{V_{max}[S]}{K_m^* + [S]} \tag{33}$$

This equation may be rearranged and integrated to obtain

$$[S_0] - [S_t] + K_m^* \ln \frac{[S_0]}{[S_t]} = k[E_0]t \tag{34}$$

where $[S_0]$ is initial substrate concentration and $[S_t]$ is substrate concentration at time t, which in this instance is the time of residence of substrate in the reactor. The term $[E_0]$ is the total number of moles of enzyme per volume of bed. We next call p the fraction of S reacted at any given time

$$p = \frac{[S_0] - [S_t]}{[S_0]} \tag{35}$$

We need to know the void volume V_v, the volume of the pore space in the bed. We also need to know the flow rate Q of the substrate through the column. The ratio of V_v/Q is equal to the residence time t. Introducing this ratio, as well as Eq. (35) into Eq. (34), we obtain

$$[S_0]p = K_m^* \ln(1 - p) + \frac{k_3[E_0]V_v}{Q} \tag{36}$$

The term $k_3[E_0]V_v$ is referred to as the capacity C. Then we have

$$[S_0]p = K_m^* \ln(1 - p) + \frac{C}{Q} \tag{37}$$

an equation for a straight line, if the ratio C/Q is constant. Then, if Q is maintained constant, we may perfuse S through the column and plot $[S_0]p$ versus $\ln(1 - p)$. The slope of this plot equals K_m^*. This relationship has been used successfully to determine values for K_m^*. The value of K_m^* tends to decrease as flow rate increases. One can intuitively feel that a faster flow rate in the column should diminish the thickness of the Nernst layer, thereby resulting in a decrease in K_m^*.

Lilly and Sharp (101) developed a kinetic equation applicable for a continuous feed stirred-tank reactor. We begin by considering substrate balance between input, output, and substrate conversion:

$$Q[S_I] - Q[S_0] = \frac{d[S]}{dt} V_L \tag{38}$$

where $[S_I]$ is concentration of substrate coming into the reactor tank, $[S_0]$ is that coming out, and V_L is the liquid volume in the tank. If we substitute Eq. (33) for $-dS/dt$ and note that the term $k_3[E_0]V_L = C$, and in addition express S in terms of the fraction of substrate converted to product, we have

$$[S_0]p = -\frac{p}{1-p} K_m^* + \frac{C}{Q} \tag{39}$$

This relationship may be used to determine K_m^* in stirred-tank reactors. The slope of a plot of $[S_0]p$ versus $[p/(1-p)]$ equals $-K_m^*$.

D. Immobilized Substrate Kinetics (85,177,178)

With immobilized enzymes we have circumstances in which the enzyme in an enzymic reaction is physically constrained, and this constraint affects enzymic activity. There are also in nature circumstances in which the enzymic substrate is constrained. An example of interest to the food scientist is the group of substrates for lipolytic enzymes. Lipolytic enzymes are known to hydrolyze triacylglycerols. Whereas lipase is soluble in aqueous solution, its substrate is not and generally exists in an emulsoid state. For example, milk triacylglycerols are part of the milk fat globule, which is essentially an oil droplet in an aqueous solution. Because of the physical nature of emulsions, lipase can react only with substrate molecules at the surface of the emulsion droplet. Substrate molecules buried inside the droplet are inaccessible to the enzyme.

Kinetic studies have suggested that lipase activity follows Michaelis-Menten theory, and values for V_{max} and K_m have been calculated. However, a number of years ago Benzonana and Desnuelle (8) observed that, for a given substrate, the value of K_m obtained was dependent on the physical dimensions of the emulsion droplets in suspension. The V_{max}, on the other hand, was independent of emulsion droplet size. In other words, when emulsion particle size was varied, a family of Lineweaver-Burk plots were obtained that had different slopes but a common y intercept. This is the kind of behavior exhibited when we have competitive inhibition (Fig. 5). More specifically, as droplet size decreased (i.e., as total surface area of the emulsion increased), the apparent value of K_m decreased. Clearly, as droplet size decreased, more total S became available to E, less S appeared to be needed to saturate E, and the apparent value of K_m decreased. The value of V_{max} was unaffected because the ES complex was unaffected. Though the behavior of the lipase system resembles the behavior found in competitive inhibition, a competitive inhibitor molecule is not responsible for this behavior. Rather, it is a consequence of the nature of the enzymic system in which the enzyme can react with its substrate only at the surface of the emulsion droplet.

Since E reacts with S only at the droplet surface, it is inappropriate to consider S in terms of its concentration. The important parameter is surface area of the emulsion droplet. When the abcissa for the Lineweaver-Burk plots discussed above is expressed as reciprocal surface area of emulsion droplet rather than reciprocal or substrate concentration, the different curves become superimposable (Fig. 8). This is the appropriate way to handle lipase data.

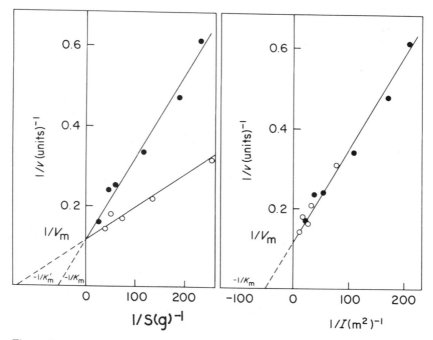

Figure 8 Double-reciprocal plots of lipolytic activities illustrating the effect of interfacial area of triacylglycerol emulsion particles on the kinetics of lipolysis. The large emulsion particles (−●−●−) possessed 0.95 m^2 of interfacial area per gram triacylglycerol, compared with 2.87 m^2 of interfacial area per gram triacylglycerol for the small emulsion particles (−○−○−). When the reciprocal of enzyme velocity (1/v) is plotted against reciprocal of weight(s) of triacylglycerol (1/S), the apparent K_m is larger for the large emulsion particles than it is for the small emulsion particles. On the other hand, when the reciprocal of enzyme velocity (1/v) is plotted against reciprocal of interfacial areas of each emulsion I (m^2), the apparent K_m values are the same. (Redrawn from Ref. 8.)

Although Lineweaver-Burk plots based on surface area of the substrate droplet resolve the problem of the varying values of K_m, there is an additional problem. The Michaelis constants obtained for lipase systems are apparent K_m values. This is true even when K_m is obtained from a Lineweaver-Burk plot in which reciprocal surface area of the emulsion droplets is plotted on the abscissa. Evidence for this comes from the fact that, when lipase is added to solutions that contain its substrate, there is a significant lag time before substrate depletion begins (8). An explanation that has been proposed requires that E must first come to the surface of the emulsion droplet and orient itself there before it interacts with substrate molecules. Various schemes have been proposed to describe the overall enzymic reaction. We shall consider that proposed by Verger et al. (178), as follows:

$$E \xrightleftharpoons[k_d]{k_p} E_s \qquad\qquad (40)$$

and

$$E_s + S \underset{k_2}{\overset{k_1}{\rightleftharpoons}} E_s S \overset{k_3}{\longrightarrow} E_s + P \tag{41}$$

where k_p is referred to as the "penetration" rate constant, k_d the "depenetration" rate constant, and E_s enzyme that has oriented at the surface of the emulsion droplet. The lag phase that exists before substrate depletion begins is explained by the existence of the initial penetration step. This step is believed to consist of more than a simple adsorption process. Rather, it is assumed E_s has a new conformation and may have more catalytic activity than its soluble counterpart. It should be pointed out also that it is assumed that P is soluble in aqueous solution and, once it is formed, rapidly diffuses away from the emulsion surface.

Given the reaction mechanism above, the following may be derived:

$$v = \frac{V^{\#}_{max}[S]}{K^{\#}_m + [S]} \tag{42}$$

where

$$V^{\#}_{max} = \frac{k_3[E_0][S]}{K^{\#}_m + [S]} \tag{43}$$

and

$$K^{\#}_m = \frac{k_d}{k_p} \frac{K_m[S]}{K_m + [S]} \tag{44}$$

The value of $K^{\#}_m$ may be obtained from Lineweaver-Burk plots. However, from Eq. (44) we see that the true value of K_m cannot be determined without information concerning the values of k_d and k_p. At present, more research is needed to obtain more information concerning these two parameters.

VII. ENZYME-RELATED ANALYSES

A. Enzyme Assays (9,10,146)

1. Direct Assays

Enzymic activity generally is determined by measuring the rate at which E acts on S. The most direct approach is to measure the rate of product formation with the aid of a progress curve (plot of [P] versus time). Figure 9 shows progress curves at different values of $[E_0]$. These curves initially are linear, but ultimately become curvilinear. The slope of the linear portion of each curve gives the Michaelis-Menten velocity for the appropriate $[E_0]$ value. We see from Eq. (10) that Michaelis-Menten velocity is dependent on both [S] and $[E_0]$. As the enzymic reaction proceeds, [S] decreases, and v then decreases also. In order to determine enzyme activity we want v to be a function of $[E_0]$ only. Where possible, these conditions can be conveniently obtained by making $[S_0] \gg K_m$. Michaelis-Menten velocity is then equal to V_{max}, which is a function of

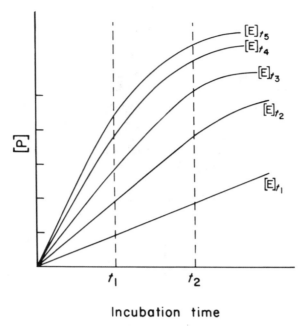

Incubation time

Figure 9 Enzyme assays. Product formation with time at different concentrations of enzyme. (Redrawn from Ref. 146.)

$[E_0]$ but not $[S]$ [see Eq. (11)]. The initial slope of the progress curve then equals V_{max} and is constant for a given value of $[E_0]$.

If progress curves are obtained at different values of $[E_0]$, the initial slopes of each curve may be plotted back against the appropriate $[E_0]$ value to establish a standard curve for the determination of enzyme activity (Fig. 10). Such curves relate rate of product formation to enzyme concentration and are most useful on the linear part of the curve. On the linear part of the curve the slope for each enzyme concentration is constant, so we have

$$\frac{dv_0}{dE_0} = \text{constant} \tag{45}$$

where v_0 is the initial rate of product formation. Equation (45) indicates that v_0 is proportional to $[E_0]$. Since v_0 is a constant for a given value of $[E_0]$, and since it has the units of reciprocal time, Eq. (45) leads to

$$t[E_0] = C \tag{46}$$

where C is a constant. The product of enzyme concentration and time is constant when we have the conditions delineated above for determining enzymic activity.

The following rules should be observed when determining the catalytic activity of an enzyme.

1. It is preferable to measure initial rates with a continuous monitoring technique.
2. Although substrate concentration need not be in excess over the K_m value, enzyme assays usually are carried out with $[S] > 20$ times K_m. These conditions assure maximum velocity of the reaction.

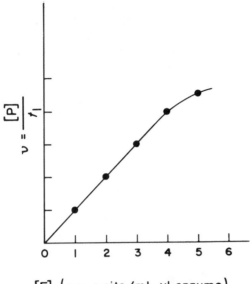

$v = \dfrac{[P]}{t_1}$

[E], (e.g., units/ml, μl enzyme)

Figure 10 Enzyme assays. Initial velocity (calculated as $[P]/t_1$) as a function of enzyme concentration (see Fig. 9). (Redrawn from Ref. 146.)

3. The reaction conditions for the enzyme reaction should be maintained and reported.
4. The rate of the nonenzymic reaction should be determined using a sample heated to inactivate the enzyme.
5. The quantity of enzyme should be measured so that specific activities can be calculated (usually a total protein estimation is made, although the enzyme may constitute only a small part of the total if the preparation is crude).

Once the catalytic activity of the enzyme has been determined, it can be reported as "units of enzyme." The usual unit of enzyme activity is that amount of enzyme that catalyzes the utilization of 1 μmol of substrate per minute under standard conditions. However, with commercial food enzymes the units of activity do not necessarily conform to the above definition. Most enzymes used in food systems are crude, so it is not possible to measure the amount of enzymatically active protein responsible for substrate transformation. Consequently, the concentration of enzyme in a crude preparation is given as units per milliliter solution, or if the protein content of the preparation has been determined, the specific activity of the preparation can be reported as units per mole of enzyme or per milligram total protein. In the latter case, the standard unit of specific activity would be μmoles substrate utilized per minute per milligram protein at stated conditions of temperature, pH, and other variables.

2. Single-Point Assays (9,10,146)

Single-point assays are commonly used by food scientists. They are advantageous in that they save time and money. However, it is important that the person using such an assay understand it, particularly under possible conditions in which it can give erroneous results.

If we are sure the progress curve is linear for a given time period, we can do a single-point assay. If we measure [P] at a chosen time t_1 within this period, then the ratio $[P]/t_1 = v_0$, the initial rate of product formation. This simplifies the assay. It is unnecessary to measure a series of [P] values as a function of time. But, it should be reiterated that the progress curve must be linear up to this chosen time t_1. If it is curvilinear, $[P]/t_1 \neq v_0$.

An interesting example of the use of a single-point assay is found in the milk-clotting test, which has long been used in the dairy industry for the determination of enzymic milk-clotting activity. One of the most significant attributes of milk is that it clots in the presence of small amounts of proteolytic enzymes. Strictly speaking, the total clotting reaction is usually schematized as

$$E + S \underset{k_1}{\overset{k_1}{\rightleftharpoons}} ES \xrightarrow{k_3} P_1 + E + M \tag{47}$$

$$nP_1 \xrightarrow{k_s} P_n \tag{48}$$

where E is the proteolytic enzyme, usually chymosin, though many other clotting enzymes are now used in addition to chymosin. The term S is κ-casein, which exists as a unit of the casein micelle, the clotting particle, and P_1 is a casein micelle particle that contains κ-casein molecules that have been hydrolyzed at the Phe-Met bond by the clotting enzyme. The peptide M is split off κ-casein in the primary phase of the reaction [reaction (47)], and is usually referred to as the glycomacropeptide, or simply the macropeptide. The term n refers to the number of clotting particles that coagulate in the second phase of the reaction [reaction (48)]. The coagulation reaction is kinetically expressed with the aid of k_s, the flocculation rate constant. The term P_n is the final clot.

When we measure clotting times, we do not measure [P] at a chosen time as described for a single-point assay, but we measure the time required for the reaction to progress to a certain point, that is, until we can see clotted flecks of milk. Equation (46) is used to determine enzymic milk-clotting activity. In this case, it is referred to as the Segelke-Storch relationship. Equation (46) indicates that clotting time should be inversely proportional to enzyme concentration. Clotting time is then plotted against reciprocal enzyme concentration, and a straight line is obtained that relates clotting time to enzyme concentration and serves as a standard curve for determination of milk-clotting activity.

Yet, it is well known that the Phe-Met bonds of κ-caseins are about 80% hydrolyzed by the time milk clots. Intuitively, because of the extent of the reaction, one would think that the progress curve for the true activity of the enzyme, the hydrolysis of κ-casein, is not linear up to the clotting time. This means that the measurement of clotting time is not the most precise way of determining enzymic milk-clotting activity. However, as long as a plot of clotting time versus reciprocal enzyme concentration is linear, clotting times should give reasonable estimates of the enzymic activity.

3. Coupled Enzyme Assays (9,10,146)

Sometimes it is inconvenient or even impossible to assay directly for the product of an enzymic reaction. However, the product of the enzymic reaction in question may serve as substrate for a second enzyme, such that a second product is obtained that can conveniently be tested for. When this is the case, it is possible to develop a "coupled" enzyme assay. The enzyme system in question is "coupled" with a second enzyme system, which serves as an indicator for the first.

Schematically we have

$$A \xrightarrow{\ E_1\ } B \xrightarrow{\ E_2\ } C \tag{49}$$

where A is substrate for E_1, the enzyme being assayed, B is product of action of E_1 on A, but also substrate for E_2, the coupled or indicator enzyme, and C is product of the coupled enzyme. The objective is to monitor the rate of formation of C to determine the enzymic activity of E_1. Mathematically we need circumstances where $d[C]/dt \ \alpha [E_1]$.

To obtain these circumstances we need to satisfy a couple of basic conditions. First, the rate of depletion of A should be zero order with respect to A, and the reaction must be irreversible. Second, the rate of formation of C must be first order with respect to B, and this reaction must be irreversible. If $-d[A]/dt$ is zero order with respect to A and is irreversible, then depletion of A is a function of $[E_1]$ only. We can ensure this condition if $[A_0] \gg K_A$ (Michaelis constant for the activity of E_1 on A), or if only a small amount of A is utilized during the assay period. We should be able to assume irreversible conditions because B is removed by the second step of the total reaction. Then we can say

$$- \frac{d[A]}{dt} = \text{constant} \ [E_1] \tag{50}$$

We can be assured that $d[C]/dt$ is first order with respect to B if [B] at steady state $\ll K_B$ (Michaelis constant for the activity of E_2 on B). We can obtain this by assaying with an excess of E_2. We can be assured that the second step of the total reaction is essentially irreversible if only a small extent of reaction proceeds during the assay period.

If the above basic conditions are obtained, then after a short lag time, $d[C]/dt =$ constant and is proportional to (E_1); stated mathematically, since we have steady-state conditions, $d[C]/dt = -d[A]/dt$, and from Eq. (50) we have

$$\frac{d[C]}{dt} = \text{constant} \ [E_1] \tag{51}$$

We may plot [C] versus time for different values of $[E_1]$. The initial slopes of these curves may then be plotted back against the appropriate $[E_1]$ values to obtain a standard curve that relates $d[C]/dt$ to $[E_1]$.

B. Substrate Assays (9,10,146)

Because of their specificity, enzymes often can be used to analyze for specific components in foods without extensive purification procedures. In addition, many enzyme reactions are very sensitive, estimating as little as 10^{-10} g of material depending upon the sensitivity of detection.

Precise control of the reaction conditions is important when enzymes are used for substrate analysis. Since the analyses are for substrates, these must be the limiting component in the system. Thus, activators and cofactors, for example, should be sufficient to saturate the enzyme.

The concentration of substrate can be determined in one of two ways.

1. Total change or equilibrium method: The reaction is allowed to go to completion and the amount of product formed is estimated by some specific chemical or physical method.

2. Kinetic method: When the enzyme concentration is fixed, the reaction rate can be made proportional to substrate, inhibitors, activators, and so on.

The important point to be made here is that the rate of product formation be proportional to [S]. This condition is assured when $[S] \leqslant 0.2K_m$. Then Eq. (10) becomes

$$v = \frac{k_3 [E_0]}{K_m} [S]$$

$$= \text{constant } [S] \tag{52}$$

Thus, when [S] is very small, progress curves (plots of [P] versus time) should be linear. They may be constructed for different values of $[S_0]$. The slopes of these progress curves may then be plotted back against the appropriate $[S_0]$ values to obtain a standard curve that relates rate of product formation to substrate concentration.

VIII. ENZYME MECHANISMS (180,183)

Implicit in the foregoing discussion is the concept of an enzyme-substrate complex formed prior to its conversion to enzyme-product complex, which in turn dissociates into enzyme and product. Indeed, such complexes have been observed experimentally. In many cases, there is formation of an enzyme-substrate covalent intermediate (180, 183). In general, covalent intermediates are the rule for the various hydrolases in which a strong nucleophilic group in the active site attacks, for example, the carbon of a polarized carbonyl function of an ester or amide, for example. This leads to displacement of an alcohol or amine, and formation of a transient acylated enzyme intermediate. Subsequent hydrolysis of this covalent intermediate yields the acid portion of the ester or amide and frees the enzyme for a new hydrolytic cycle. Classic examples of behavior of this kind involve the serine proteases, such as trypsin and chymotrypsin (183), which rely on the strong nucleophilicity of a serine residue in the active site (Fig. 11). After formation of a stereospecific complex between chymotrypsin and substrate, the imidazole group of histidyl-57 acts as a general base to stretch the bond distance between the H and O of the hydroxyl group of the seryl-195 residue (Fig. 11a). Nucleophilic attack of the seryl oxygen at the carbonyl of the substrate leads to expulsion of the R' residue and formation of an acyl enzyme (Fig. 11b,c). Deacylation of the enzyme results when the imidazole group, acting as a general base (Fig. 11d), extracts a proton from water, thereby facilitating the attack of a hydroxide ion at the carbonyl of the acyl enzyme. In the second transition state intermediate (Fig. 11e), imidazole, acting as a general acid, deacylates the enzyme yielding the second product and the regenerated enzyme.

This type of mechanism can be generalized to a variety of hydrolytic enzymes that utilize a limited number of nucleophilic groups on the side chains of amino acids in the active site (180). For example, pancreatic lipase is thought to utilize a system similar to that of the serine proteases to effect hydrolysis of triacylglycerols (85). Papain, ficin, and bromelain utilize the thiol group of cysteine as the nucleophilic agent. In this case, a thiol ester intermediate is formed. In addition to the functional groups of the active site directly involved in the catalytic cycle, ancillary groups, on amino acids in the en-

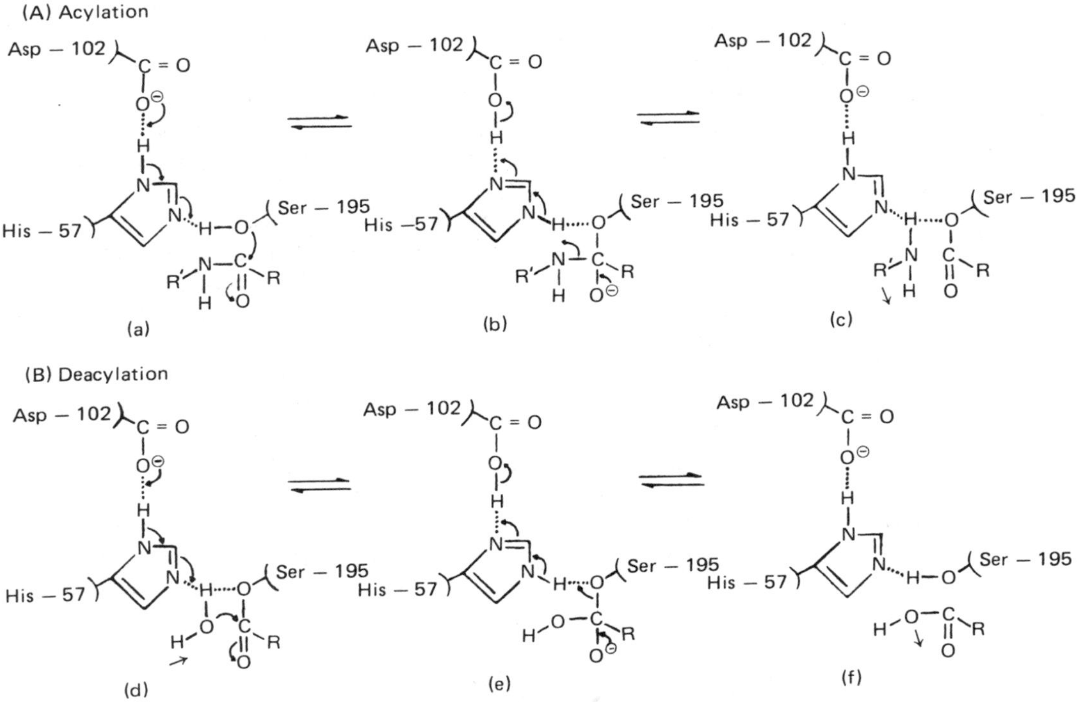

Figure 11 The proposed mechanism of action for chymotrypsin, illustrating the catalytic participation of histidyl, seryl, and aspartyl residues in the active site. (Reprinted from Ref. 183.)

zyme, bind to the substrate to position it for the subsequent reaction. Variations in this first step for forming the Michaelis complex provide a basis for explaining the differing specificities of the proteases (183).

An understanding of enzyme mechanisms at the molecular level may be exploited by the food scientist to develop new processes or new products. For example, if a relatively high concentration of a strong nucleophile can effectively compete with water in Fig. 11 for the acyl group of the intermediate, a novel product might result. Various thiols could yield flavorful thiol esters. Lipolytic enzymes have been used to catalyze interesterification reactions in oils to yield products with different physical properties (86,158,191). The foregoing mechanisms also serve as the basis for the plastein reaction, whereby transamidations result in formation of polypeptides with different characteristics (57). This enables one to incorporate specific amino acids into proteins for nutritional improvement or to enhance surface activities. Although the mechanisms are different, analogous reactions are involved in transglycosidations important in carbohydrate biochemistry and technology.

A novel, commercial exploitation of active-site chemistry of enzymes involves reversible blocking of the cysteine thiol in the active site of papain using thiol-disulfide interchange reactions (88). It is well known that preslaughter, intravenous injection into animals of the sulfhydryl protease papain enhances tenderization of meat via postmortem proteolytic activity. In this case, the enzyme injected before slaughter of the animal is disseminated, by the circulatory system of the animal, throughout its musculature. The uniform distribution of enzyme within the tissue results in postmortem tenderization of the resultant meat. Injection of the free, active enzyme results in stress to the animal accompanied by internal hemorrhaging and rejection of the carcass. However, if the essential thiol group in the active site of papain is reversibly blocked by thiol-disulfide interchange, this inactive enzyme can be administered to the animal without ill effects. The latent papain activity is subsequently regenerated in situ by reducing agents in the meat [perhaps by glutathione while cooking (88)]. If the essential functional groups for enzymic activity are known, it should be possible to introduce this reversible chemical latency into virtually any exogenous enzyme used in food processing.

Thus, an appreciation for enzyme reaction mechanisms and the judicious manipulation of reaction conditions can lead to the formation of novel products. Indeed, it is possible to synthesize specific peptides by reversal of the hydrolytic reactions catalyzed by various proteases (61). Moreover, interchange reactions involving hydrolytic enzymes may be involved naturally in certain food systems. For example, in the aging of cheese, the mixture of exogenous and endogenous enzymes along with the various substrates (at relatively low water concentrations) might react so that new compounds, important to the flavor and texture of the cheese, are obtained. Food scientists should strive to understand and utilize enzyme mechanisms to improve the quality of existing foods and to develop new food products.

IX. FACTORS INFLUENCING ENZYME ACTIVITY

The importance of enzymes to the food scientist is often determined by the conditions prevailing within and outside the food. Control of these conditions is necessary to control enzymic activity during food preservation and processing. In this section, the major factors affecting enzymic activity are discussed. These factors include temperature, pH, moisture, ions and ionic strength, ionizing radiation, shearing, pressure, and interfacial effects.

A. Effect of Temperature on Denaturation and Inactivation of Enzymes

An earlier section was devoted to the kinetics of enzyme-catalyzed reactions. In this section, attention is given to the thermodynamics and kinetics of thermally induced denaturation or inactivation of enzymes.

The effects of temperature on enzymes are very complex; for example, high temperatures may affect dissociation states of functional groups involved in the enzymic reaction, affinity of the enzyme for activators or inhibitors, and ancillary matters, such as solubility of oxygen, which may be a substrate in the reaction. In addition, inactivation of the enzyme can occur, and this is one of the reasons for heating foods.

In general, enzymes function very slowly at subfreezing temperatures and their activities increase as the temperature is increased. Most enzymes show "optimal activity" in the 30-40°C range and begin to denature above 45°C. Enzymes also tend to have a temperature of "maximum resistance" to denaturation, which is usually well below the temperature of maximum activity.

An understanding of these phenomena, as well as that of other behavioral characteristics of enzyme molecules, is useful to the food scientist. To achieve this understanding, the food scientist must know how to obtain and use thermodynamic and kinetic data. For example, the food scientist often needs to know time-temperature profiles for irreversible enzyme inactivation, and this is best studied by kinetic means. However, a full understanding of processes involving enzyme denaturation or inactivation also requires an understanding of thermodynamic principles. For these reasons, the following section is directed to the thermodynamics of enzyme denaturation-inactivation, and the succeeding section deals with the kinetics of inactivating enzymes by heat.

The effects of high temperatures on proteins have received more attention than those of low temperatures, and coverage in the following two sections reflects this situation. However, that enzymes exhibit a temperature of maximum resistance to denaturation indicates that proteins are also subject to denaturation at low temperatures, and this behavior will be discussed in the section on thermodynamics. Aside from this matter, most of the information on the effects of low temperatures on enzymes will be saved for a later section.

1. Thermodynamics of Enzyme Denaturation and Inactivation

Thermal inactivation of enzymes is related to, but not identical with, an important area of protein chemistry—thermal denaturation (see also Chap. 5). Denaturation of a protein has been referred to as "a major change from the original structure, without severance of any of the primary chemical bonds that join one amino acid to another" (183). Enzyme inactivation refers to loss of enzyme activity. Depending on the nature of the active site of the enzyme, loss of activity may require extensive denaturation, or it may require very little.

The following discussion is addressed mainly to protein denaturation. This is because the phenomenon of denaturation provides an ideal context in which to explain the behavior of proteins when subjected to temperature changes. When we take up kinetics of inactivation in the following section, our focus is more specifically on destruction of enzyme activity.

There have been numerous attempts to understand heat denaturation of proteins from a thermodynamic standpoint, and Eq. (53) is one of the most fundamental equations for this purpose:

$$\Delta G = \Delta H - T\,\Delta S \tag{53}$$

where ΔG is the change in Gibbs free energy, ΔH is the change in enthalpy, ΔS is the change in entropy, and T is temperature in degrees Kelvin. Free energy is the work energy available to make a specific process go. If the value of free energy is positive, the process is not energetically favorable (see Fig. 1). If the value is zero, the process is at equilibrium. If the value is negative, the process is energetically favorable, but nothing can be said about the rate.

Enthalpy is the heat absorbed or liberated for a specific chemical process when pressure is constant. If heat is absorbed ($+\Delta H$), the process is endothermic; if liberated ($-\Delta H$), it is exothermic. Values of enthalpy often provide clues as to the kinds and numbers of bonds broken and formed in particular chemical reactions.

Entropy is a measure of the "freedom" or "randomness" that molecules have in a specific situation. An increase in entropy indicates an increase in randomness. This often means that the molecules involved in the reaction have gained freedom of movement they did not have at the outset of the reaction.

Since protein molecules tend to take on a random coil conformation during denaturation, this event is usually associated with large spatial changes by which the amino acid side chains gain freedom of movement they did not have in the native state. Thus, denaturation is accompanied by a significant increase in entropy (+ ΔS). The change in enthalpy during denaturation will also be large and positive because the enzyme must absorb heat to break the numerous electrostatic and hydrogen bonds that stabilize the native structure.

At high temperatures, ΔS for denaturation is large enough that the absolute value of $-T \Delta S$ exceeds the absolute value of $+\Delta H$, making ΔG negative and denaturation favorable. At somewhat lower temperatures the magnitude of $+\Delta H$ exceeds that of $-T \Delta S$, resulting in a positive value for ΔG and unfavorable conditions for denaturation. These observations are based on the behavior of only the protein molecules. We must, however, also consider interactions between protein molecules and the adjacent water. Unfolding of a protein molecule exposes hydrophobic groups to the aqueous environment, resulting in increased structuring of water adjacent to these groups. Increased bonding among water molecules can lead to negative values for both ΔH and ΔS of water, and these values are in opposition to the positive values of ΔH and ΔS for protein unfolding. The water effect is especially pronounced at low temperatures. If it predominates, and the value of $-\Delta H > +T \Delta S$, ΔG is negative and low temperature denaturation is favored. Stated another way, at low temperatures water molecules may form a "shell" around adjacent nonpolar molecules. Different nonpolar entities then lose their ability to interact with each other. Consequently, hydrophobic associations are less stable. In accord with these statements, experimental data demonstrate that some proteins exhibit a temperature of maximum resistance to denaturation. In Fig. 12, a plot of ΔG versus T for α-chymotrypsinogen exhibits a minimum (maximum positive value) at about 10°C, which is the temperature of maximum resistance to denaturation. This important aspect of protein denaturation will be discussed in more detail, but first it is appropriate to describe the means for determining the thermodynamic parameters just mentioned.

Strictly speaking, in thermodynamics, chemical processes are assumed to be reversible, and at equilibrium. In the reversible denaturation of a protein, Eq. (54),

$$P_N \underset{}{\overset{K_D}{\rightleftharpoons}} P_D \tag{54}$$

where P_N is the native protein, P_D is the reversibly denatured protein, and K_D is the ratio $[P_D]/[P_N]$, that is, the denaturation equilibrium constant. The van't Hoff equation gives the thermodynamic relationship between the equilibrium constant and temperature:

$$d \ln K = \frac{\Delta G}{R} \frac{dT}{T^2} \tag{55}$$

Indefinite integration of Eq. (55) results in Eq. (56), which conforms to the generalized equation for a straight line:

$$\ln K = -\frac{\Delta H}{RT} + \text{constant} \tag{56}$$

where the constant is the y intercept of a plot of ln K versus 1/T. The slope of such a plot is equal to $-\Delta H/R$. If the equilibrium concentrations of native and denatured protein are known at different temperatures, ΔH can be readily obtained by use of Eq. (56). Alternatively, if Eq. (55) is integrated between limits, one obtains

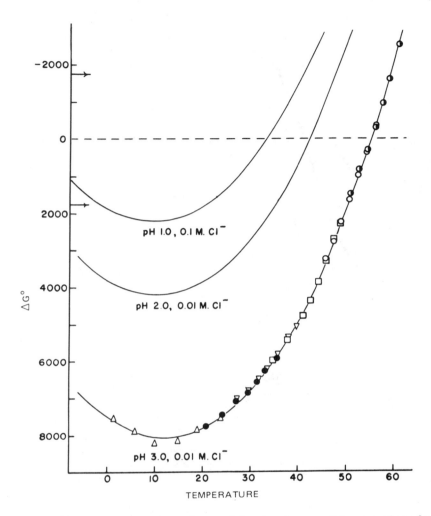

Figure 12 Temperature dependence of the free energy of denaturation of α-chymotrypsinogen at different pH values and ionic strengths. (Redrawn from Ref. 15a.)

$$\ln\frac{K_2}{K_1} = \frac{\Delta H}{R}\frac{T_2 - T_1}{T_2 T_1} \tag{57}$$

where K_1 and K_2 are equilibrium constants at temperatures T_1 and T_2, respectively. Equation (57) permits calculation of ΔH provided values for K are available at two different temperatures.

Values for ΔG may be obtained directly from the well-known equation

$$\Delta G = -RT \ln K \tag{58}$$

Once ΔG and ΔH are obtained, ΔS may be calculated from Eq. (53).

In Eqs. (56) and (57) it is assumed that ΔH is constant within the temperature range studied. However, the enthalpy of denaturation is not always independent of temperature. Figure 13 is a van't Hoff plot for β-lactoglobulin. From the slope of the plot ($-\Delta H/$

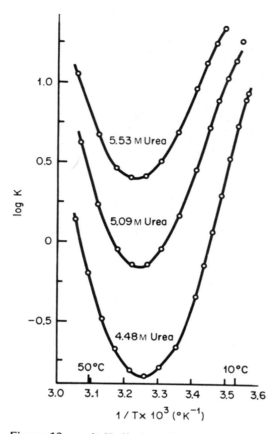

Figure 13 van't Hoff plots (logarithm of the equilibrium constant K = unfolded/native versus reciprocal of absolute temperature) for β-lactoglobulin A in aqueous urea at various concentrations. (Redrawn from Ref. 124a.)

R) it is evident that ΔH approaches zero at about 35°C and increases in absolute value at both higher and lower temperatures in the presence of urea. Thus, protein denaturation in aqueous urea is endothermic at temperatures greater than 35°C and exothermic at temperatures less than 35°C. These data are consistent with the free energy data shown in Fig. 12.

The minimum in the van't Hoff plot represents the temperature of maximum resistance (T_{max}) to denaturation. Listed in Table 3 are T_{max} values and thermodynamic parameters relating to denaturation of several proteins. Included in this table are values for heat capacity at constant pressure (C_P), a thermodynamic parameter important when considering protein denaturation. The term C_P can be determined from the relationship

$$C_P = \frac{d\,\Delta H}{dT}$$

$$= \frac{T\,d\,\Delta S}{dT} \qquad (59)$$

Table 3 Thermodynamic Parameters for Protein Denaturation

Protein (conditions)	ΔG^* (kcal/mol)	ΔH^* (kcal/mol)	ΔS^* (cal/deg)	ΔC_P (cal/mol per deg)	Temperature of maximal stability ($^{\circ}$C)
Ribonuclease (pH 2.5, 30°C)	+0.9	+57	+185	+2,000	−9
Chymotrypsinogen (pH 3, 25°C)	+7.3	+39	+105	+2,600	10
Myoglobin (pH 9, 25°C)	+13.6	+42	+95	+1,400	0
β-Lactoglobulin (5M urea, pH 3, 25°C)	+0.6	−21	−72	+2,150	35

Source: From Ref. 165a.

Heat capacity, then, is a measure of the dependence of ΔH and ΔS on temperature. We have already observed that the unfolding of a protein molecule leads to a large increase in ΔH and ΔS for the protein molecule itself, but to a decrease in ΔS and ΔH for water molecules that constitute the solvent for the protein. Since the values for C_P in Table 3 are likewise large, it follows that ΔH and ΔS are highly dependent on temperature. This is borne out in Fig. 14, which shows that ΔH, ΔS, and C_P increase with increasing temperature.

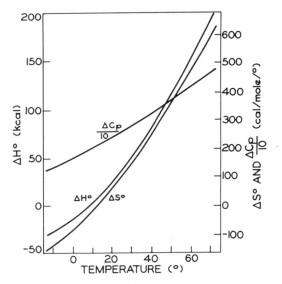

Figure 14 Temperature dependence of enthalpy ΔH°, entropy ΔS°, and heat capacity ΔC_P pertaining to denaturation of chymotrypsinogen. (Redrawn from Ref. 15a.)

The temperature dependence of the thermodynamic parameters is in accord with the existence of a temperature of maximum resistance to denaturation. For the food scientist, who ultimately must be concerned with the quality of proteins in food systems, knowledge of the temperature of maximum resistance to denaturation is of considerable importance.

2. Kinetics of Thermal Inactivation

Thermodynamics deals with reversible processes and equilibrium states. Often the food scientist heats foods with the goal of irreversibly inactivating enzymes. Chemical kinetics provides a powerful tool that can aid in achievement of this goal, especially when it is desired to prepare time-temperature profiles for inactivation treatments.

The objective of this section is to explain thermal inactivation kinetics as applied to enzymes. The first essential piece of information is the rate of a given enzymic process at a specific temperature. Next, rates at different temperatures must be obtained. In this connection both the Arrhenius theory and the transition state theory are discussed. The meaning of Q_{10} values are also discussed since they are widely used. Finally, thermal inactivation curves are considered since they provide the means of obtaining time-temperature profiles for thermal inactivation processes.

Order of the reaction. Protein denaturation often follows first-order kinetics, since by definition a single molecule, the protein, undergoes a conformational change. Likewise, enzyme inactivation is usually assumed to follow first-order kinetics. However, when more than a single kind of enzyme is present, as is frequently the case in food enzyme preparations, the kinetics may be complex.

If first-order kinetics can be assumed, the process is

$$E_N \xrightarrow{k} E_I \tag{60}$$

where E_N is the native or active enzyme, E_I is the inactive enzyme, and k is the specific rate constant for the inactivation process. Mathematically the decrease in active enzyme becomes

$$-\frac{d[E_N]}{dt} = k[E_N] \tag{61}$$

Rearrangement gives

$$\frac{d[E_N]}{[E_N]} = -k \, dt \tag{62}$$

and indefinite integration yields

$$\log [E_N] = -\frac{kt}{2.303} + \text{constant} \tag{63}$$

where the constant is the y intercept of a plot of $\log [E_N]$ versus t, and $-k/2.303$ is the slope. If kinetics are truly first order, such a plot should be linear for at least three half-lives, where the half-life is $0.693/k$.

Temperature dependence. Perhaps the most commonly used mathematical expression for the effect of temperature on rates of chemical processes, including enzyme reactions, is the Arrhenius relationship,

$$k = Ae^{-Ea/RT} \tag{64}$$

Where A is the frequency or Arrhenius factor and E_a is the Arrhenius activation energy. The logarithmic form of Eq. (64) is

$$\ln k = \ln A - \frac{Ea}{RT} \tag{65}$$

which indicates a linear relationship when ln k is plotted against 1/T. The slope of this plot is $-E_a/R$. If inactivation kinetics are first order, k may be obtained from Eq. (63). The magnitude of E_a indicates the temperature dependence of the reaction in question.

In the preceeding section it was mentioned that proteins tend to exhibit denaturation at both high and low temperatures. Similarly, one would expect inactivation to occur at both high and low temperatures. The occurrence of enzyme inactivation at high temperatures is common knowledge, but the behavior of enzymes at low temperatures is less well known. Arrhenius plots for enzyme activity frequently deviate from linearity at some low, nonfreezing temperature characteristic of the enzyme (Fig. 15). This behavior is probably attributable to either unfolding of the protein molecule or to dissociation of subunits if the enzyme is a multiunit protein. This phenomenon will be discussed more fully in the following section.

Transition state theory. The similarity between the Arrhenius and van't Hoff equations should be noted. However, E_a is not identical to ΔH. Yet, the similarity between the two equations suggests that values for E_a may provide some insight into the thermodynamic nature of chemical processes. The transition state, or Eyring theory, is the most successful attempt to relate irreversible kinetic information to thermodynamic information. In brief, this theory suggests that reactant X, in an irreversible reaction, goes

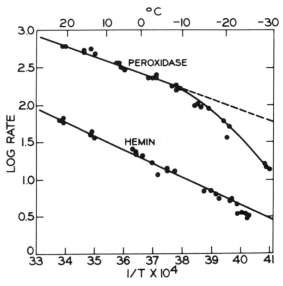

Figure 15 Arrhenius plots of peroxidase- and hemin-catalyzed oxidation of guaiacol. (Reprinted from Ref. 169, by courtesy of Academic Press.)

through a transition state X*, which is in equilibrium with the original reactant. Furthermore, X* has a higher energy level than the original reactant, as well as the final product, P. These ideas are shown graphically in Fig. 1. The height of the energy hill gives the free energy (ΔG^*) of the activated state. With the aid of the Eyring theory it is possible to obtain values for ΔG^*, as well as for enthalpy (ΔH^*) and entropy (ΔS^*) of the activated state. In what follows it is shown how these values can be obtained, and how an explicit relationship between E_a and ΔH^* can be developed. It should be noted that these transition state parameters are obtained from irreversible processes, and, as such, are distinct from true thermodynamic parameters, which are obtained from reversible processes.

The transition state theory can be depicted as

$$X \underset{\longleftarrow}{\overset{K^*}{\rightleftharpoons}} X^* \longrightarrow P \tag{66}$$

The rate of disappearance of X is given by

$$-\frac{d[X]}{dt} = B[X^*] \tag{67}$$

where B is a coefficient. Equation (67) indicates that the loss of reactant molecules is dependent on how fast activated molecules pass over the energy hill in Fig. 1. Since $K^* = [X^*]/[X]$, Eq. (67) becomes

$$-\frac{d[X]}{dt} = BK^*[X] \tag{68}$$

From Eq. (68) it follows that

$$k = BK^* \tag{69}$$

where k is the specific rate constant for the transition process. The value for K^* allows ΔG^* to be calculated by way of Eq. (58). The coefficient (B) is equal to $k_b T/h$, where k_b is Boltzmann's constant and h is Planck's constant. Strictly speaking, B includes a term specifying the fraction of molecules that pass over the energy hill, but this term is usually considered unity. Given the above, Eq. (69) becomes

$$k = \frac{k_b T}{h} e^{-\Delta G^*/RT} \tag{70}$$

which is similar to the Arrhenius relation, Eq. (64). Two distinctions must be made, however. First, the exponential term is now free energy rather than activation energy. Second, the frequency factor ($k_b T/h$) is seen to include a temperature term, which the Arrhenius factor does not include. Equation (70) can be combined with Eq. (53) to relate the specific rate constant with enthalpy and entropy of the transition state. The result is

$$\ln \frac{k}{T} = -\frac{\Delta H^*}{R} \frac{1}{T} + \text{constant} \tag{71}$$

The point to be made here is that a plot of $\ln k/T$ versus $1/T$ yields a straight line with a slope equal to $-\Delta H^*/R$. An Arrhenius plot ($\ln k$ versus $1/T$) has a slope equal to $-E_a/R$. The terms E_a and H^* are related as follows:

$$E_a = \Delta H^* + RT \tag{72}$$

Thus, transition enthalpy may be obtained either from a plot of ln k/T versus 1/T or from an Arrhenius plot with assistance from Eq. (72). The free energy of transition is obtainable from Eq. (70). Finally, the entropy of transition can be calculated from Eq. (53).

Table 4 contains values for the activation energy (E_a) and transition state enthalpy (ΔH^*) for inactivation of a number of enzymes. All these values are quite large, usually in the range 50-150 kcal/mol. This is consistent with the nature of denaturation (and frequently inactivation), in which many hydrogen and electrostatic linkages are broken.

It should be emphasized that the difference between ΔH^* and E_a is equal to RT. This amounts to less than 1 kcal/mol when the temperature is in the range normally used for inactivation studies. Values for ΔH^* and E_a for enzymic catalysis are generally much less than those for inactivation, the former in the range 6-15 kcal/mol. This large difference in the temperature coefficients for catalysis and inactivation explains why enzymes generally have a temperature of maximum catalytic activity. As temperature increases, the rates of both catalysis and inactivation increase, but with different temperature coefficients, the latter are greater. At a critical temperature, the rate of inactivation becomes more rapid than the rate of catalysis. At this temperature the so-called optimum temperature for enzyme activity is encountered.

As was true for the temperature of maximum resistance to denaturation, which was discussed in the preceding section, knowledge of the temperature of optimum activity for an enzyme is important to the food scientist.

The Arrhenius and Eyring models are not the only models used in attempts to describe the effect of temperature on rates of chemical processes. Two other approaches are worthy of mention, the Q_{10} approach and the approach used to construct thermal death time curves for microorganisms.

Table 4 Transition State Denaturation Constants for Various Proteins

Protein	ΔH^* (cal/mol)	ΔS^* (eu)[a]	Bonds broken (No.)[b]	ΔG^* (25°C) (cal/mol)
Lipase (pancreatic)	45,400	68.2	9	25,100
Amylase (malt)	41,600	52.3	8	26,000
Pepsin	55,600	113.3	11	21,800
Egg albumin	132,000	315.7	26	37,900
Hemoglobin	75,600	152.7	15	30,100
Peroxidase (milk)	185,300	466.0	37	46,400
Chymosin	89,300	208.0	18	27,300
Trypsin	40,200	44.7	8	26,900
Invertase (yeast)				
pH 5.7	52,400	84.7	10	27,200
pH 5.2	86,400	185.0	17	31,300
pH 4.0	110,400	262.5	22	32,200
pH 3.0	74,400	152.4	15	29,000

[a]Entropy units in cal/mol per degree.
[b]Number of noncovalent bonds broken on denaturation $+\Delta H^*/5000$, where the average ΔH^* per bond is assumed to be 5000 cal/mol.
Source: Modifed from Ref. 183.

Q_{10} Values. The Q_{10} value is the change in the rate of a reaction that occurs when the temperature is changed 10°C. Thus,

$$Q_{10} = \frac{\text{rate}_T}{\text{rate}_{T-10°C}} \tag{73}$$

For many chemical processes the value for Q_{10} falls in the range 2-3. The Q_{10} value can be related to the Arrhenius equation as

$$Q_{10} = e^{10E_a/RT} \tag{74}$$

Whereas E_a is independent of temperature provided conditions are appropriate, Q_{10} is dependent on temperature. With the aid of Eq. (74) it can be shown that the rate of a typical catalytic reaction, where E_a is about 12,000 cal/mol will approximately double with a 10°C increase in temperature, whereas the rate of a typical inactivation reaction may increase 100-fold during a 10°C rise in a critical temperature range.

Thermal inactivation plots. In the canning industry, microbial inactivation is frequently expressed in terms of thermal death time plots (the logarithm of thermal death time versus temperature). A similar approach can be applied to thermal inactivation of enzymes during blanching. The objective of the following discussion is to describe how thermal inactivation curves are constructed and to define pertinent parameters.

The first of these parameters is the D value, or decimal reduction value, defined as the time required to inactivate 90% of the original enzyme activity at a constant temperature. If an inactivation reaction follows first-order kinetics, the D value equals 2.303/k. The term then may be combined with Eq. (63) to yield

$$\log [E_N] = -\frac{t}{D} + \text{constant} \tag{75}$$

If $\log [E_N]$ is plotted against t, the slope is equal to $-1/D$. The D values can be conveniently obtained from plots on semilogarithmic paper. In this case, $[E_N]$ is plotted on the logarithmic axis (the ordinate) and time on the arithmetic axis (the abscissa). The D value is then the time required for the plot to traverse one log cycle. It is useful to think in terms of D values when dealing with first-order processes. When collecting data, values for enzyme activity should be obtained over a time span at least equal to that of the D value (one log cycle). If the reaction is truly first order, the plot of log activity versus time should be linear over this range.

If D values are available at different temperatures, the logarithm of these D values can be plotted against temperature to obtain a thermal inactivation curve. Such a curve is useful, especially when plotted on semilogarithmic paper, because it readily provides thermal inactivation data at many temperatures. The slope of a thermal inactivation curve, or the temperature dependence of thermal inactivation, is $-1/z$, where z represents a number of degrees (expressed in degrees Fahrenheit) required for the thermal inactivation curve to traverse one log cycle.

If a thermal inactivation plot is linear, it can be expressed mathematically, in terms of two temperatures, as

$$\log \frac{D_2}{D_1} = -\frac{1}{z} (T_2 - T_1) \tag{76}$$

Temperatures T_1 and T_2 are usually expressed in degrees Fahrenheit.

Since D is related mathematically to k by Eq. (74), it is also related to E_a by way of the Arrhenius relation, Eq. (64), and since D and Z are mathematically related, the latter is also related to E_a. By combining Eqs. (64), (74), and (76) it can be shown that

$$z = \frac{RT_1 T_2}{E_a} \frac{9}{5} \tag{77}$$

Thus, a large inactivation energy corresponds to a small z value. Since thermal death time data are usually expressed on the Fahrenheit temperature scale, the term 9/5 must be included to convert to the Kelvin scale, which is used in the Arrhenius equation. It is important to note that, although both E_a and z are independent of temperature, Eq. (77), which relates the two parameters, is dependent on temperature.

The z value can also be related to the Q_{10} value. This can be achieved by combining Eqs. (73) and (77):

$$Q_{10} = e^{(23/z)(9/5)} \tag{78}$$

Again, the term 9/5 transforms the temperature scale from degrees Fahrenheit to degrees Kelvin. The large Q_{10} and E_a values observed for enzyme inactivation are consistent with the fact that, at critical temperatures, a small increase in temperature results in a large increase in the rate of enzyme inactivation.

Inactivation of enzymes serves as a convenient indicator for the effectiveness of heat treatments for some foods. For example, pasteurization of milk at 63°C for 30 min destroys the test organism *Coxiella burnetti* and also just inactivates milk alkaline phosphatase (Fig. 16). Notice that, after adequate pasteurization, some milk enzymes other than alkaline phosphatase remain active. Similarly, fruits and vegetables are sometimes blanched in hot water or steam partly for the purpose of inactivating enzymes, such as lipase, phenolase, lipoxygenase, chlorophyllase, catalase, peroxidase, and ascorbic acid oxidase. Blanching is regarded as adequate when the relatively heat resistant enzyme, peroxidase, is no longer active.

When enzyme activity is undesirable, an attempt is often made to achieve irreversible enzyme inactivation. If the heat treatment given to an enzyme system is marginal and the some of the inactive enzyme exists in a reversibly inactive form (see Chap. 5), partial enzyme reactivation or regeneration occurs upon storage at a lower temperature. This can be an extremely important matter in some heat-processed foods. As shown in Fig. 17, reactivation of peroxidase following various processes of equal thermal severity (for microorganisms) is proportional to the rate of heating. Consequently, in high-temperature-short-time (HTST) processes, the enzyme may not be irreversibly destroyed, but only reversibly inactivated. This creates a dilemma for the processor and regulatory agencies when, for example, milk properly pasteurized by a HTST method exhibits a positive phosphatase test (indicative of improper pasteurization) because of regeneration. Regeneration of enzymes, such as peroxidase and lipoxygenase, may have a detrimental effect on the sensory quality and nutritional value of foods.

3. Some Specific Effects of Low Temperature (53,75,110,168)

The primary reason for exposing foods to low temperatures, especially frozen foods, is to prevent microbial growth while maintaining as many quality attributes of the food as possible.

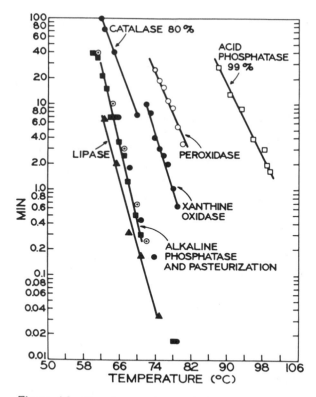

Figure 16 Time-temperature relationships for inactivation of milk enzymes. (Redrawn from Ref. 84, by courtesy of John Wiley and Sons.)

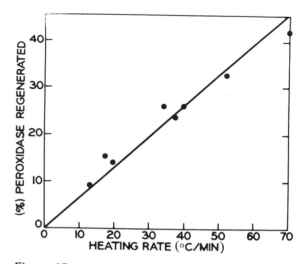

Figure 17 Regeneration of turnip peroxidase as a function of heating rate. (Redrawn from Ref. 143, by courtesy of the Institute of Food Technologists.)

Although some enzymes undergo significant denaturation during freezing and thawing, many are unaffected. In addition, many enzymes exhibit significant activity while present in partially frozen systems. As the temperature of an enzyme solution is decreased from 0 to about 10°C below its initial freezing point, enzymic activity can either increase or decrease depending upon the enzyme and the system. A further decline in temperature almost always results in decreased activity. The following factors appear to be involved in this inconsistent behavior of enzymes in partially frozen systems.

1. Composition of the medium
2. Rate and extent of freezing
3. Concentration effects of freezing
4. Viscosity
5. Complexity of the sample (intact tissue versus simple systems)
6. Changes in phase (e.g., crystallization of water or solidification of triacylglycerols)

Enzymes in simple systems are especially susceptible to permanent damage after freezing and thawing if they are highly purified and in an initially dilute solution. Simple systems thus may behave differently than intact tissues (53,144).

As intimated above, deviations from the Arrhenius relationship do occur with enzyme-catalyzed reactions. Data for oxidation of guaiacol by purified peroxidase and by hemin in 40% methanol from +25°C to −30°C are plotted according to the Arrhenius relationship in Fig. 15. These samples were ice free at all temperatures. A similar non-linear behavior in the Arrhenius relationship has been observed with other enzymes, some of which contained ice and others did not (144). The unexplained decrement in activity between observed rates and the linear extrapolation of the Arrhenius plot may be due to the following factors: change in stability of enzyme conformers, for example, during the change from water to ice the active site may be altered to yield unstable conformers; increased intraenzymic hydrogen bonding yielding inactive enzyme conformers; decreased accessibility of enzyme to substrate at lower temperatures because of phase changes in substrate; increased hydrogen bonding between water and either the substrate or the enzyme active site, resulting in decreased specific activity; formation of enzyme polymers resulting in decreased specific activity; change in mechanism, for example, the rate-determining step in the reaction sequence may be altered; decreased ionization of enzyme, substrate, and buffers as the temperature is decreased; possible shifts in pH of the system and in pH optima of enzymes; and increased viscosity of the system as the temperature is lowered.

During freezing, solutes concentrate in the unfrozen pools of water. The increased concentration of electrolytes and the change in pH that often occurs simultaneously is a major cause of biochemical damage and, no doubt, affects enzymic activities in frozen systems. Whether the effect will be activation, stabilization, or inhibition depends on the enzyme, the nature and concentration of salts, pH, temperature, and the types of other substances present.

The rate of freezing can also influence the activity of some enzymes in partially frozen systems. Rate of freezing influences the size of unfrozen pools and concentration of solutes in unfrozen pools, and this in turn presumably influences enzymic activity.

In frozen systems the high viscosity results in decreased rates of diffusion of enzyme and substrates and thereby tends to limit the rate and extent of enzymic activity.

Slow freezing and slow thawing generally result in greater loss of enzymic activity than fast freezing and thawing.

The freezing and/or thawing of tissues results in damage to the membranes of subcellular organelles, such as mitochondria and lysosomes, and delocalization of enzymes. Enzymes that are released ("activated") by freezing and thawing or other disruptive manipulations, are known as *latent enzymes*. Lysosomes are easily disrupted by changes in osmotic pressure, pH, and salt concentration that accompany freezing and thawing. Since lysosomes contain a wide array of hydrolytic enzymes, tissue alteration can be extensive. To minimize tissue alteration, if this is desired, attention should be paid to minimizing damage to the lysosomes.

Some enzymes can be damaged by exposure to low, nonfreezing (chilling) temperatures, a phenomenon known as cold denaturation. This occurrence, which is the exception rather than the rule, has been observed with purified enzymes, such as catalase, lactate dehydrogenase, glutamate dehydrogenase, and glyceraldehyde phosphate dehydrogenase. Apparently, some enzymes are thermodynamically unstable at low nonfreezing temperatures. Reversible cold inactivation of enzymes can result from reversible dissociation of enzyme subunits at low temperatures or to changes in enzyme conformation.

During low-temperature storage of foods, enzyme-related problems in food quality may result from the following (144).

1. Concentration of solutes attendant upon removal of water by freezing
2. Relative predominance of enzyme action compared with other biological, physical, and chemical reactions because of the relatively low energy of activation of enzyme reactions
3. Fluctuation of temperatures during storage and distribution with a possibility for enhanced activity at higher temperatures
4. Cryptic accumulation of intermediates, side products, and secondary products
5. Cellular disruption and decompartmentalization of enzymes resulting in the formation of nonphysiological food components
6. Shift in physiological function upon lowering of temperature, that is, a response to stress in living tissue
7. Presence of cryoprotective agents in biological tissues that may stabilize enzymes otherwise inactivated at low temperatures
8. Regeneration of enzyme activity (dealt with in the previous section)

B. pH Effects on Enzymes (50,133)

Extremes in pH generally inactivate enzymes. Enzymes usually exhibit maximal activity at a particular pH value, termed the "pH optimum." The relationship between pH and activity is illustrated in Fig. 18. Most enzymes show maximum activities in the pH range 4.5-8.0, and maximum activity is usually, but not always, confined to a rather narrow pH range. There are, however, enzymes with extreme pH optima, such as pepsin, which has a pH optimum of 1.8, and arginase, which has a pH optimum of 10.0. Depending on the enzyme, the activity can correspond to either a sigmoidal (S-shaped) or a bell-shaped curve. At extremes in pH, enzyme activity usually decreases irreversibly because of protein denaturation. However, as shown in Fig. 18 there is a pH range of reversible inactivation. This apparently involves the reversible ionization of functional groups in the active site or in areas that control conformation of the enzyme.

Control of pH as it relates to enzymic activity is important to the food scientist. In an industrial process, the pH can be controlled to maximize, to prevent, or to inhibit an enzymic reaction. For example, unwanted phenolase activity can be avoided by reducing the pH of the system well below 3.0. This is frequently accomplished in fruits by adding natural acidulents, such as citric, malic, or phosphoric acids.

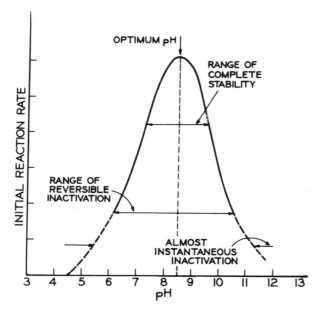

Figure 18 Effect of pH on enzyme activity. (Redrawn from Webb, Ref. 181.)

The pH optimum of an enzyme may occur as a result of (a) a true reversible effect of pH on the maximum velocity of the reaction, (b) an effect of pH on the affinity of substrate for the enzyme, and (c) an effect of pH on the stability of the enzyme.

C. Water Activity and Enzymic Activity (1,144)

Many of the characteristics of enzymic activities in frozen system have parallels in dried foods, in which there is also restricted water activity (a_w). Although it is difficult to believe that enzymes can act at very low water levels, it is well known that dried vegetables soon acquire a haylike aroma if they are not blanched prior to drying. Furthermore, dried oat products become bitter upon storage unless the enzymes are first inactivated by heat. Thus, it is common knowledge that foods protected against microbial spoilage by low water content are still susceptible to enzymic spoilage. For example, as shown in Fig. 19, lipolysis in wheat flour can occur at very low moisture levels. As the moisture level increases so does lipolysis, until at 14% moisture lipolysis becomes very rapid.

Because enzymes are compartmentalized in unaltered tissues they are effectively separated from their substrates. Cereal grains, for example, can be stored for months or years at a water content of 13% without evident deterioration; however, grain that has been mechanically damaged rapidly deteriorates at the same water content. In addition to endogenous enzymes, concern also should be given to adventitious enzymes derived from microorganisms. These enzymes can be important in cereals as well as in egg products, in which bacteria may have proliferated before drying.

The absolute water content of a food is not necessarily a decisive factor, rather, the "boundness" of the water is important (see Chap. 2). Consequently, water activity (a_w) and equilibrium relative humidity are more suitable parameters than water content when considering enzymic activity in dried foods.

Based on data from model systems, as illustrated in Fig. 20, the following generalizations can be made: (a) at a sufficiently high a_w enzymic activity may be measurable in a matter of hours, (b) at a sufficiently low water activity enzymic reactions do not proceed (this is true even though enough water may be present to theoretically cause a hydrolytic reaction to go to completion), and (c) reactions at different water activities tend toward different final values of product accumulation. In this latter instance, one might suspect that enzyme inactivation had occurred; however, this clearly was not so in the example shown in Fig. 20. After storage at a low a_w until product accumulation had ceased, the a_w of the system was increased, with a resultant revival of enzymic activity, producing a higher level of product accumulation characteristic of the new a_w. This apparently occurred because substrate availability (amount dissolved) was a_w dependent. Once the dissolved substrate was used up at a given a_w the reaction stopped. An increase in a_w then increased the amount of free water, which in turn dissolved additional substrate and diluted the reaction products, thereby allowing the reaction to resume.

Enzymic activity, in general, approximately parallels the sorption isotherm (see Fig. 20, Chap. 2) with significant enzymic activity evident only above the region of monomolecular adsorption. Increasing amounts of free water then become available to serve as a vehicle for enzymic processes. However, a few enzymic reactions can occur even below an a_w corresponding to monolayer adsorption of water. This is possible if the substrate is sufficiently mobile and fluid to associate with the enzyme. An example involves the lipolysis of oily, unsaturated triacylglycerols. At very low a_w, oily triacylglycerols, because of their mobilities, are more available and are hydrolyzed much more rapidly than solid triacylglycerols. Therefore, the temperature at which a dried food is stored also has a bearing on the mobility and rate of degradation of triacylglycerols. However, even solid substrates can be attacked at low a_w if they are in intimate contact with the enzyme.

Thus, the rate of enzymic reactions in dried foods is limited by the rate at which the substrate diffuses to the enzyme. This might explain why high-molecular-weight substrates in dried foods are not attacked very readily. For example, wheat gluten in flour containing proteinases does not hydrolyze appreciably even at 65% relative humidity. Diffusion effects involving macromolecular substrates may also result in qualitative changes in the enzymic reaction. For example, in an aqueous medium, amylase attacks soluble starch to form oligosaccharides. However, in the dried state, glucose and maltose appear first, and oligosaccharides appear only at higher water activities. In general, enzyme action at a low a_w tends to prevent accumulation of reaction intermediates or to favor certain reaction pathways, possibly because potential intermediates do not diffuse from the active site; rather, they immediately degrade or react.

The activities of nonhydrolytic enzymes are also influenced by water activity. This behavior has been observed in model systems containing glucose oxidase and polyphenol oxidase (1), and in wheat protein concentrate containing lipoxygenase (179). Water serves as a medium for the reaction and as a vehicle for the substrate. When water is tightly bound, enzymic oxidation is either slow or not possible.

Enzymes are known to be more heat stable in the dry state. As the moisture content increases, susceptibility of the enzyme to heat inactivation increases. This can be of practical significance during dehydration of cereals or malt, for example, and in some baked goods in which not all enzymes are inactivated in baking.

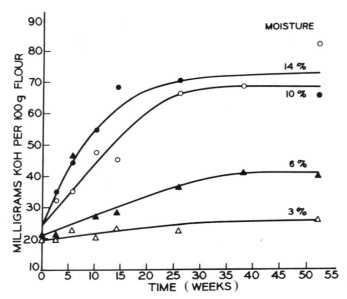

Figure 19 Effect of moisture content and time of storage in sealed containers at 37.8°C on the fat acidity of unbleached patent flour. (Redrawn from Ref. 30, by courtesy of the American Association of Cereal Chemists.)

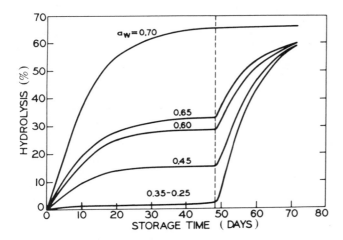

Figure 20 Influence of water activity on rates of enzymic hydrolysis of lecithin in a mixture of ground barley malt and 2% lecithin stored at 30°C. (Redrawn from Ref. 2, by courtesy of J. F. Bergmann-Verlag.)

During dehydration, there may be differential effects on enzymes because of their differing responses to the concentration of various solutes, inhibitors, or activators (144). Some solutes, such as sugars and proteins, often have a protective effect on enzymes. The mutual salting in or salting out of solutes may increase or decrease substrate concentrations, respectively. An increase in substrate concentration can lead to substrate inhibition of some enzymes. On the other hand, an increase in solute concentration coupled with a decrease in water concentration can result in a reversal of hydrolase reactions, the occurrence of transferase reactions, and the production of "abnormal" products, some involving cross-links.

Continued enzymic activity during dried storage is not always detrimental. Rice grains stored for long periods become less sticky after cooking than do unstored rice grains, presumably because amylases act on the amylose chains of the rice starch. Bacterial amylase continuing to act on the starches of baked goods can improve mouth feel. In some dried foods, desirable flavors can develop from the continued action of endogenous enzymes. Also, continued enzymic activities are important in the production of high-quality dates and dehydrated sweet potatoes (144).

D. Effect of Electrolytes and Ionic Strength on Enzymes (36,138)

1. Activation and Inhibition

Some enzymes require certain ions for activity, whereas other enzymes are not particularly affected by ions. Heavy metal ions, such as Hg^{2+}, Pb^{2+}, or Ag^+, usually poison enzymes. Ionic effects vary from one enzyme to another; for example, an ion may be toxic for one enzyme and activate another. Furthermore, a given ion may act as an inhibitor of an enzyme at one concentration, but may activate the same enzyme at another concentration. Although ions have nonspecific effects on the enzyme as a protein, they also may be required as components of the active site.

Anions tend to be fairly nonspecific in their action on enzymes, whereas the cation requirement of any enzyme is sometimes specific. Some enzymes are activated by more than one cation and other enzymes are activated only by one ion. Cations known to activate enzymes include: Na^+, K^+, Rb^+, Cs^+, Mg^{2+}, Ca^{2+}, Zn^{2+}, Cd^{2+}, Cr^{3+}, Cu^{2+}, Mn^{2+}, Fe^{2+}, Co^{2+}, Ni^{2+}, Al^{3+}, and H^+. The ionic radii of these ions fall within a fairly narrow zone in the middle range of observed atomic radii; however, the ions are not always interchangeable.

There are also instances of competitive antagonisms between ions. For example, Ca^{2+}-activated myosin ATPase is inhibited by Mg^{2+}.

Ions may activate enzymes by a variety of mechanisms, including (a) becoming an integral part of the active site, (b) forming a binding link between enzyme and substrate, (c) changing the equilibrium constant of the enzyme reaction, (d) changing the surface charge on the enzyme protein, (e) removing an inhibitor of the reaction, (f) displacing an ineffective metal ion from the active site or from the substrate, and (g) shifting the equilibrium of a less active conformer to a more active conformer.

2. Salting in and Salting out (36)

In general, large quantities of electrolytes have an effect on the solubility of proteins. Salts are often necessary to solubilize or "salt in" a protein in an aqueous system. On the other hand, salts can be used to insolubilize or "salt out" an enzyme, a phenomenon used to isolate enzymes. Ammonium sulfate is often used to fractionate enzymes because it is highly soluble in water and is not harmful to most enzymes.

3. Stability of Enzymes (16)

In food processes, such as brining, enzymes may be inactivated because of the high concentration of electrolytes. In some instances, small concentrations of particular cations and anions can alter the stability of enzymes. The effects are specific and differ from one enzyme to another. These specific effects of electrolytes apparently result from strong preferential binding to the enzyme in either its native or denatured state. However, at high concentrations all electrolytes alter enzyme stability, presumably by nonspecific effects, since all enzymes are affected similarly.

E. Inactivation of Enzymes by Shearing (24)

Figure 21 illustrates the inactivation of rennet by shearing at $4°C$. When the shear value [shear rate (sec^{-1}) times the exposure time (sec)] is greater than 10^4, inactivation becomes detectable. At shear values of the order of 10^7, about 50% inactivation has been observed for solutions of catalase, rennet, and carboxypeptidase (42). Moreover, rennet undergoes reactivation after shear inactivation, reminiscent of reactivation after reversible thermal inactivation.

F. Effects of Pressure on Enzymes (138)

Very high pressures usually inactivate enzymes; however, the necessary pressures are much greater than those normally encountered in food processing. Consequently, pressure effects on enzymes are of little concern to food scientists.

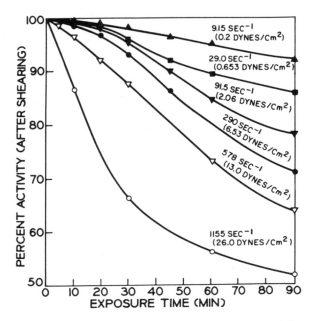

Figure 21 Inactivation of rennet by shearing at $4°C$. (Reprinted from Ref. 24, by courtesy of Academic Press.)

G. Effects of Ionizing Radiation on Enzymic Activity (43,94,133,152,176)

Ionizing radiation, such as γ rays and high-energy electrons, have been proposed as a means for pasteurization or sterilization of foods. Although enzymes can be inactivated by ionizing radiation the dosage required for complete inactivation in situ is about 10-fold greater than that necessary to destroy the microorganisms of concern. Although food can be made microbiologically stable by ionizing radiation, residual enzymic activity in a food can harm its desirable sensory properties and nutritive value.

In general, the destruction of enzymic activity by ionizing radiation is affected by the particular enzyme being tested, water activity, enzyme concentration, enzyme purity, in vitro versus in situ effects, oxygen concentration, pH, and temperature. In the anoxic, dry state, most enzymes possess about the same resistance to inactivation by irradiation. On the other hand, enzymes in dilute aqueous solution show a greater variation in retention of activity after radiation. As shown in Fig. 22 an enzyme, such as trypsin, is most resistant to radiation damage in the dry state, presumably because only the direct effects of irradiation are operative. In the presence of water, inactivation is greater presumably because "indirect effects" become important. This involves the radiolysis of water and the damaging effects of resultant free radicals (see Chap. 5). In such foods as fillet of sole, water contents in the range 20-30% result in optimum enzymic destruction during irradiation. At lower water levels, the inefficient direct-effect mechanism predominates, whereas at higher water levels rapid recombination of resulting free radicals apparently reduces their effectiveness.

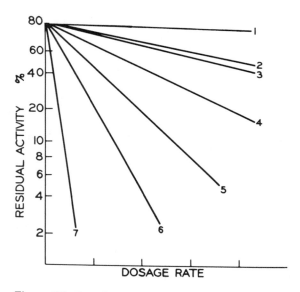

Figure 22 Inactivation of trypsin by irradiation under various conditions. (1) dry trypsin, (2) 10 mg trypsin per milliliter solution at pH 8 and −78°C, (3) 80 mg trypsin per milliliter solution at pH 8 and ambient temperature, (4) 10 mg trypsin per milliliter at pH 2.6 and −78°C, (5) 80 mg trypsin per milliliter at pH 2.7 and ambient temperature, (6) 10 mg trypsin per milliliter at pH 2.6 and ambient temperature (7) 1 mg trypsin per milliliter at pH 2.6 and ambient temperature. (Redrawn from Ref. 94.)

Individual enzymes in a moist environment vary greatly in radioresistance. Catalase is about 60 times more resistant to irradiation inactivation than carboxypeptidase. Other enzymes, which require specific, susceptible functional groups in the active site, such as the thiol group of cysteine in papain, may be particularly susceptible to inactivation by the damaging free radicals that arise from irradiated water.

In general, within certain limits, the more dilute the enzyme solution, the smaller the dose needed to destroy the same percentage of initial enzyme activity. This phenomenon is called the dilution effect. Less pure enzymic preparations are usually more resistant to irradiation damage than are pure preparations. The presence of oxygen in the system often sensitizes the enzyme to radio destruction, perhaps by the formation of unstable intermediate, protein peroxides. The effect of pH on enzyme inactivation by irradiation can be considerable, but is not easily predictable (see Fig. 22). Thus, the effect of radiation on enzymes in meats is influenced by postmortem pH.

In general, inactivation of enzymes by irradiation is more easily accomplished as the temperature is raised. In frozen systems little damage to enzymes is evident, presumably because free radicals are immobilized. Many of the above effects undoubtedly combine to yield differing results in situ (food systems) as compared with in vitro (purified enzymes). In tissues, enzyme compartmentalization is a factor of importance, as are such factors as water activity, the protective effects of other cellular constituents, particularly free radical scavengers, and the effect of pH as influenced by time elapsed postharvest or postmortem.

Combination treatments have been suggested in some instances. For example, a mild heat treatment to inactivate enzymes, coupled with radiation to destroy microorganisms, has been proposed. Alternatively, irradiation following by refrigeration is effective for such foods as fresh fish, strawberries, tomatoes, and mushrooms.

Further details of changes occurring during irradiation of proteins are given in Chap. 5.

H. Interfacial Inactivation of Enzymes (83,85,109)

Proteins in general tend to adsorb at interfaces, and this often results in denaturation. The high interfacial tension at the air/water or oil/water interface is apparently responsible. As a result, the secondary and tertiary structure of the enzyme unfolds rapidly and the molecule spreads over the interface. In an air/water system, hydrophobic amino acid residues orient toward the air and hydrophilic residues toward the water. In an oil/water emulsion, the hydrophobic residues associate with the oil and hydrophilic side chains extend into the water. Simple foaming of enzyme solutions can produce air-water interfaces for surface denaturation. The resulting loss in enzymic activity is not regained readily upon compression or by retrieval of the films in a variety of forms.

Some effects of adsorption at an oil/water interface are similar to adsorption at an air/water interface. However, when enzymes adsorb at an interface already occupied by a close-packed film of another protein or "protective" molecule, unfolding and loss of enzyme activity does not appear to occur. The recoverability of enzymes in active form from oil-in-water emulsions depends on the charge on the oil droplets, the stability of the emulsion, and the concentrations of protein at the interface and in the bulk solution. Even lipase, which is thought to be especially adapted for activity at oil/water interfaces, can be denatured by adsorption at a triacylglycerol/water interface (85). Lipase also can be inactivated at an air/water interface.

X. CONTROLLING ENZYME ACTION

In many situations, food scientists want to either (a) retard or prevent enzyme activity, or (b) enhance or promote the action of enzymes. Schwimmer (144) has discussed the many options available to the food scientist interested in controlling the action of enzymes. Very briefly, the major means for controlling endogenous and, where applicable, exogenous enzymic activities are listed below. Most of these methods have been discussed in more detail in preceding sections. Also, the reader is urged to consult the comprehensive discussions of Schwimmer (144) for additional information.

A. Factors Useful in Controlling Enzyme Activity

1. *Temperature.* The use of low and high temperatures to curb enzyme action in foods is well known and was discussed in detail in the preceding Sec. IX.
2. *Water content and activity.* As we learned in Sec. IX, the amount of water available to an enzyme and its substrate plays a profound role in regulating enzyme activity. As expected, low levels of water can severely restrict enzymic activities and even alter their patterns of activity.
3. *Extremes of pH.* Highly acidic or alkaline conditions can be used to reversibly or irreversibly inactivate enzymes.
4. *Chemicals.* Certain additives may prove useful in inhibiting or preventing enzyme action. For example, chelating agents may bind metal ions essential for enzymic activity.
5. *Alterations of substrates.* At times, leaching of tissues may remove unwanted reactants (or enzymes). Removal of the cosubstrate, oxygen, can be important in minimizing unwanted oxygen reactions catalyzed by various enzymes.
6. *Alterations of products.* Products of a primary enzyme reaction in a sequence of reactions may be selectively channeled into unreactive products or converted back to their original forms. For example, reduction of intermediate quinones back to the original phenols by ascorbic acid can retard enzymic browning in fruits and vegetables.
7. *Preprocessing control.* Schwimmer (144) groups these together as various pretreatments of foods, particularly for raw materials, that can be used to control enzymic activities. Working with the geneticist and breeder, the food scientist may be able to exert some control over levels or types of enzymes that occur in tissues. For example, preprocessing control of enzymes generated during the germination and sprouting of barley is of obvious importance in the preparation of malt in brewing. On the other hand, enzymic activities in sprouting potatoes may be detrimental. For years, dairy cows have been bred to maximize production of milk fat, thereby utilizing biosynthetic enzymes to the fullest.

The food scientist should take advantage, where possible, of naturally existing preprocess control of enzyme distribution (144). For example, the tassel of sugarcane contains invertase, which will hydrolyze sucrose. Thus, great care should be, and is, taken to segregate the tassels from cane tissue to avoid unwanted invertase activity (144). An understanding of such physical separations in raw materials can aid the food scientist in maximizing yields of commodity or in retaining desired quality attributes.

The relative abundance or absence of substrate or enzyme in raw materials may be important to preprocessing control. Sometimes it may not be clear which of these two components limits the rate and extent of enzymic reactions. Browning of avocado depends primarily on levels of phenolase and not on levels of phenols, presence of inhibitors, isozyme patterns, or changes in enzyme properties in different fruit varieties. In contrast, the extent of flavor development in onions is related to the level of flavor precursor and not to the level of flavor forming enzymes.

Very high doses of ionizing radiation may play a role in altering the preprocessing enzymology of certain plant-based foods. However, the level of irradiation needed to effect these changes is 10-20 times greater than that required for asepsis and will adversely affect the quality of most foods (144).

B. Enhancing Endogenous Enzyme Activity (144)

Endogenous enzyme activities in foods may be beneficial and, in fact, may be necessary to impart identity to a food. By judicious use of knowledge of the biology and chemistry of enzymes, and by careful empirical manipulation, the food scientist can exploit the food's enzymes for maximum benefit. Unfortunately, many conditions that favor desirable effects of enzyme activity also promote undesirable effects.

1. Biological Control

Enzyme activities may be potentiated in plants (and animals) by administration of hormones, by horticultural (and husbandry) practices, and by controlling ripening, curing, germination, and sprouting. Although weather- and water-induced stresses may result in deterioration of quality due to enzymic activities, these stresses may also enhance quality in other cases. The effects of stress on plants may lead to deterioration of cellular and organelle membranes, which can result in "unnatural" enzymic reactions.

2. Biochemical Control

Enzyme activity can be evoked or enhanced at the biochemical level by

1. Conversion of inactive zymogens (proenzymes) to active enzymes
2. Activation by specific activators and cofactors or removal of specific inhibitors
3. Disruption or modification of membranes in cells, allowing admixture of enzymes and substrates
4. Modification of substrates, such as native starch, to make them susceptible to enzymic attack

As Schwimmer (144) suggests, events postharvest and portmortem result from a cascade of physical, chemical, and enzymic processes following modification of cellular membrane integrity. As depicted in Fig. 23, physical alterations in cellular membranes allow further enzymic degradation of membrane components. This, in turn, leads to the formation of unstable primary products that subsequently convert to stable secondary products. These secondary products influence the quality attributes of food, either favorably or unfavorably.

In the *Flavorese* approach to flavor potentiation, a crude extract of a given food containing its complex of active enzymes is added to the same food after processing to help restore quality and flavor. The rationale is that processing destroys enzymes, leaving substrates available in excess. The addition of fresh enzymes from the same food or from appropriate microorganisms is intended to generate desirable flavor by suitable enzymic modification of the substrates.

Traditionally, regeneration of enzyme activity after the enzymes have been inactivated by processing has been considered undesirable. However, Schwimmer (144) suggests that, if properly controlled, enzyme regeneration might be exploited by food processors to enhance the quality of foods, especially frozen foods.

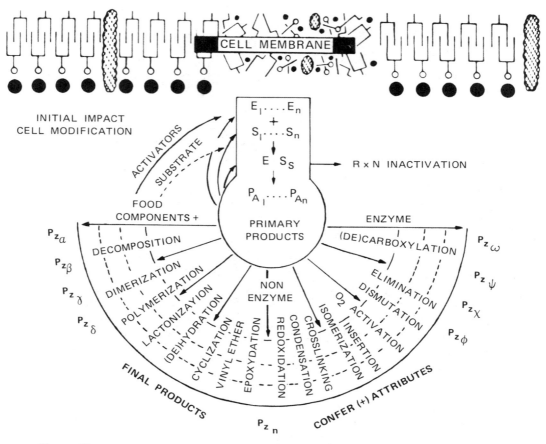

Figure 23 Consequences of enzyme action in foods. Depiction of endogenous enzyme potentiation following modification of cellular membrane integrity. (Reprinted from Ref. 144.)

C. Controlling Exogenous Enzymes

In addition to the endogenous enzymes of foods that affect quality, exogenous enzymes from adventitious microorganisms, or those added by the processor, may be essential for the proper development of desirable quality attributes. For example, in the preparation of certain cheese varieties, endogenous enzymes of milk, plus those added by the processor, as well as those emanating from added and adventitious microorganisms, combine to eventually yield the characteristic flavor and texture of cheese. On the other hand, unwanted exogenous enzymes may adversely affect the quality of the ultimate cheese.

An understanding of the biological mechanisms for controlling enzymic activities, and the biochemical mechanisms of enzyme action, can provide the discerning food scientist with the means to more effectively exploit enzymes in food processing. For example, the physical latency of lysosomal enzymes in cells can be mimicked by microencapsulating enzymes with substrates. Thus, microencapsulated enzymes and substrate(s) can be added to milk during cheese making, and these capsules then become entrapped in the cheese curd and serve as focal points for accelerating flavor development in cheese (103). Systems such as these offer possibilities for controlling the release of enzymes in foods. Furthermore, knowledge that the essential thiol group in the active site of papain can be reversibly blocked by thiol-disulfide interchange led to the development of chemically latent papain derivatives injected into cattle antemortem to effect postmortem tenderization of the meat (88). Activation of the enzyme derivative after slaughter prevents antemortem stress in the animal that would otherwise result from the injection of active enzymes.

Eventually, it may be possible to engineer specific enzymes or entire organisms using recombinant DNA techniques (see Sec. XV) so that the scientist will have better control over desirable food modifications. This will, of course, require close cooperation of food scientists with biotechnologists, bioengineers, producers, and consumers to achieve foods with the most desirable properties.

XI. ENZYMES ADDED TO FOODS DURING PROCESSING (5,11,56,126,130,144,145)

In general, enzymes are purchased and utilized during food processing for recovery of by-products, for fabricating foods, for achieving higher rates and levels of extractions, for improving flavor, and for stabilizing food quality.

Enzymes produced for the food industry are rather crude by biochemical standards. Most enzyme preparations contain a predominating enzyme as well as other enzymes. Enzymes used by food manufacturers are derived from edible and nontoxic plants, animals, and nonpathogenic, nontoxigenic microorganisms. Rigorous testing of cultures to be used as sources of enzymes is essential to establish that they are nonpathogenic and nontoxigenic and do not produce antibiotics (126). In the large-scale manufacture of food-grade microbial enzymes, the microorganisms are grown in an aerated, agitated, liquid medium, or on solid or semisolid media held in large trays or drums. In either system, control of environmental factors, such as pH, temperature, and degree of aeration, is essential. Precautions must be taken to prevent introduction of contaminating organisms and toxicants into the enzyme preparations.

Enzymes secreted into semisolid or solid media, or those retained within the cells, are extracted (before or after drying) with water. The extraction step is obviously not necessary when the enzyme is secreted into a liquid growth medium. The crude extract may be concentrated using ultrafiltration or evaporation, or the enzyme may be precipitated using salts or organic solvents. Food-grade preservatives and stabilizers are added at specified levels before the enzyme preparations are marketed as powders or liquids. Assays for the enzymes listed in Table 5 can be found in the Food Chemicals Codex (56).

There are numerous advantages to obtaining enzymes from microorganisms: (a) microorganisms are very versatile, and it is theoretically possible to find a microorganism to produce any enzyme, (b) microorganisms can be altered by mutations or genetic engineering to produce a greater quantity of enzyme or different enzymes, (c) recovery of enzymes is often easy since many microbial enzymes are extracellular, (d) readily available raw materials are used to produce microbial enzymes, and (e) microorganisms have a very high rate of growth and enzyme production.

The advantages of using enzymes in food processing include (a) they are natural, nontoxic substances, (b) they generally catalyze a given reaction without causing unwanted side reactions, (c) they are active under very mild conditions of temperature and

Table 5 Enzyme Preparations Used in Food Processing (11,56,126,130,144,145)

Enzyme	Sources	Conversions catalyzed	Food use
I. Carbohydrate hydrolases (Carbohydrase)			
A. α-Amylase	1. Barley malt Fungal: 2. *Aspergillus niger*, var. 3. *Aspergillus oryzae*, var. 4. *Rhizopus oryzae*, var. Bacterial: 5. *Bacillus subtilis*, var. 6. *Bacillus licheniformis*, var.	Starch, glycogen + $H_2O \longrightarrow$ dextrins, oligo, monosaccharides (α-1,4-glucan bonds)	Hydrolyzes starch in brewing and distilling industries; provides fermentable sugar for yeast; reduces drying time for baby foods, improves wheat flavor In breadmaking provides fermentable sugar for yeast; improves loaf volume and texture; in brewing, replaces malt in barley brewing; removes starch haze in beer; converts acid-thinned starch into highly fermentable syrups; controls viscosity and stabilizes syrups Liquifies, dextrinizes starch prior to addition of amyloglucosidases for syrup production; accelerates liquefaction of mash in brewing; aids in recovery of candy scrap; favors moisture retention in baked goods
B. β-Amylase	1. Wheat 2. Barley malt 3. Bacterial: *Bacillus polymyxa* *Bacillus cereus*	Starch, glycogen + $H_2O \longrightarrow$ maltose + β-limit dextrins (α-1,4-glucan bonds)	In baking (wheat enzyme) and brewing (barley enzyme), provides fermentable maltose for CO_2 and alcohol production; aids in manufacture of high-maltose syrups (cereal and bacterial enzymes)
C. β-Glucanase	1. *Aspergillus niger*, var. 2. *Bacillus subtilis*, var. 3. Barley malt	β-D-Glucans + $H_2O \longrightarrow$ Oligosaccharides + glucose (β-1,3 and β-1,4 bonds)	Degums mash in brewing; hydrolyzes β-glucan gums in barley used as brewing adjunct to facilitate filtration during brewing; increases yield of extract in coffee substitute manufacture
D. Glucoamylase (amyloglucosidase)	1. *Aspergillus niger*, var. 2. *Aspergillus oryzae*, var. 3. *Rhizopus oryzae*, var.	Starch, glycogen + $H_2O \longrightarrow$ glucose (dextrose) (α-1,4- and α-1,6-glucan bonds)	Converts thinned starches directly to glucose, which can then be converted to fructose using glucose isomerase

Enzyme	Source	Reaction	Application
E. Cellulase(s)	1. *Aspergillus niger*, var. 2. *Trichoderma reesei*, var.	Cellulose + $H_2O \longrightarrow \beta$-dextrins ($\beta$-1,4-glucan bonds)	Complex enzyme system; aids in juice clarification; increases yield in extraction of essential oils and spices; improves beer "body"; improves cookability and rehydratability of dehydrated vegetables; aids in increasing available proteins in seeds; forms fermentable sugar in waste grape and apple pomace; potential for producing glucose from cellulosic waste
F. Hemicellulase	*Aspergillus niger*, var.	Hemicellulose + $H_2O \longrightarrow \beta$-dextrins ($\beta$-1,4-glucan bonds of gums—locust bean, guar, etc).	Aids dehusking seed coats of coffee beans; allows controlled degradation of food gums; removes pentosans from bread; facilitates degermination of corn; improves nutritional availability of plant proteins; facilitates mashing in brewing
G. Invertase (sucrose hydrolase)	*Saccharomyces* sp. (*Kluyveromyces*)	Sucrose + $H_2O \longrightarrow$ glucose + fructose (invert sugar)	Catalyzes formation of invert sugar; prevents crystallization and grittiness in production of confections
H. Lactase (β-galactosidase)	1. *Aspergillus niger*, var. 2. *Aspergillus oryzae*, var.	Lactose + $H_2O \longrightarrow$ galactose + glucose	Hydrolyzes lactose in dairy products; increases sweetness, prevents crystallization of lactose; allows preparation of low-lactose milk for lactose intolerant individuals; improves baking quality of milk-containing bread
I. Pectinase (contains polygalacturonase pectin methylesterase, pectate lyase)	1. *Aspergillus niger*, var. 2. *Rhizopus oryzae*, var.	Pectin methylesterase demethylates pectin; polygalacturonase hydrolyzes β-D-1,4-galacturonide.	Aids in clarification and filtration of fruit juices and wines; prevents gelling in fruit juice concentrates and purees; controls cloud retention in juices; controls pectin levels in jelly; facilitates manufacture of glazed fruit, separation of manadarin orange segments

Table 5 (Continued)

Enzyme	Sources	Conversions catalyzed	Food use
II. *Protein hydrolases* (Proteases)			
A. Bromelain	Pineapples: *Anancis comosus, Ananas bracteatus* (L). Figs: *Ficus* sp.	Generally hydrolyze proteins and polypeptides to yield peptides of lower molecular weight Plant proteases generally hydrolyze polypeptides, amides, and esters (especially at bonds involving basic amino acids or leucine or glycine) yielding peptides of lower molecular weight	Plant proteases tenderize meat; chill-proof beer; improve oil and protein extraction from animals and plants; control and modify protein fuctionalities; allows preparation of protein hydrolysates; improve fish protein processing; improve cookies and waffles and enhance diastase activity of malt; improve hot cereals, marinades
B. Ficin			
C. Papain	Papaya: *Carica papaya* (L).		
D. Fungal protease	1. *Aspergillus niger*, var. 2. *Aspergillus oryzae*, var.	The microbial proteases hydrolyze polypeptides yielding peptides of lower molecular weight	Improves color, texture, and loaf characteristics of bread; controls dough rheology; tenderizes meat; improves dispersibility of dried milk, stability of evaporated milk, and spreadability of cheese spreads
E. Bacterial proteases	1. *Bacillus subtilis*, var. 2. *Bacillus licheiformis*, var.		Improves flavor, texture, and keeping qualities of crackers, waffles, pancakes, and fruit cakes; aids evaporation of fish water; aids filtration during cane sugar manufacture
F. Pepsin	Porcine or other animal stomachs	Hydrolysis of polypeptides, including those with bonds adjacent to aromatic or dicarboxylic acid residues, yielding peptides of lower molecular weight	Bovine pepsin used as a rennet extender and milk coagulant; allows production of protein hydrolysates
G. Trypsin	Animal pancreas	Hydrolysis of polypeptides, amides and esters, at bonds involving the carboxyl groups of L-arginine and L-lysine yielding peptides of lower molecular weight	Suppresses oxidized flavor in milk; allows production of protein hydrolysates

Enzyme	Source	Reaction/Specificity	Application
H. Rennets	1. Fourth stomach of ruminant animals 2. *Endothia parasitica* 3. *Mucor miehei* 4. *Mucor pusillus*	Specificities of the various rennets tend to be similar to that of pepsin; as acid proteases that have optimum pH for activity in acid range and probably possess carboxyl groups of aspartic acids in the active site; very high selectivity for a particular phe-met bond in κ-casein of milk thereby initiating milk coagulation upon cleavage	Coagulates milk in making cheese; aids development of flavor and texture in ripened cheese; chymosin, pepsin, and gastricsin primarily components of conventional animal (calf) rennets; microbial rennets substitute for animal rennets
III. Ester-Triacylglycerol Hydrolases Lipases	1. Edible forestomach tissue of calves, kids, and lambs 2. Animal pancreatic tissues 3. *Aspergillus oryzae*, var. 4. *Aspergillus niger*, var.	Hydrolysis of triacylglycerols or simple fatty acid esters, yielding mono- and diacylglycerols, glycerol, and free fatty acids	Animal lipases develop flavor in manufacture of cheese and in lipolyzed butteroil; microbial lipases catalyze hydrolysis of lipids (e.g., fish oil concentrate)
IV. Oxidoreductases A. Catalase	1. *Aspergillus niger*, var. 2. *Micrococcus lysodeikticus* 3. Bovine liver	$2H_2O_2 \longrightarrow 2H_2O + O_2$	Removes residual H_2O_2 after cold sterilization of milk and egg white; removes H_2O_2 resulting from action of glucose oxidase
B. Glucose oxidase-catalase	*Aspergillus niger*, var.	$Glucose + O_2 \xrightarrow{\text{glucose oxidase}}$ gluconic acid $+ H_2O_2$ $2H_2O_2 \xrightarrow{\text{catalase}} 2H_2O + O_2$	Desugars eggs to prevent browning and off flavor during and after drying; removes O_2 from beverages and salad dressings to prevent off-flavors and improve storage stability; improves color and texture of baked goods and machinability of bread doughs
C. Lipoxygenase	Soybean flour	Linoleate $+ O_2 \longrightarrow$ LOOH (hydroperoxides) (and other 1,4-pentadiene polyunsaturated fatty acids)	Hyperperoxides bleach carotenoids in dough and oxidize thiol groups in gluten to improve dough rheology

Table 5 (Continued)

Enzyme	Sources	Conversions catalyzed	Food use
V. Isomerase A. Glucose isomerase	1. *Actinoplanes missouriensis* 2. *Bacillus coagulans* 3. *Streptococcus olivacens* 4. *Streptococcus olivochromogenes* 5. *Streptococcus rubiginosus*	Glucose ⇌ fructose Xylose ⇌ xylulose	Converts glucose to fructose in preparation of high-fructose corn syrups

pH, (d) they are active at low concentrations, (e) the rate of reaction can be controlled by adjusting temperature, pH, and the amount of enzyme employed, and (f) they can be inactivated after the reaction has proceeded to the desired extent.

Listed in Table 5 are most of the common enzymes currently added to foods in the United States during processing. Inspection of the list indicates that the total number of enzymes of value to the food industry is rather limited considering the thousands of enzymes that have been studied. The overwhelming majority of enzymes employed by the food industry are hydrolases, consisting chiefly of carbohydrases, followed by proteases and lipases (esterases). A few oxidoreductases are of value in food processing, but only one isomerase is used. Also, it is evident that only a few animal and plant enzymes are in use, whereas the majority of enzymes originate from microorganisms.

Additional enzymes useful in food processing include (a) dextranases and α-galactosidases, both used in sugar refining to remove dextran and raffinose, respectively, (b) amino acylases that aid in the resolution of synthetic racemates of amino acids by selectively hydrolyzing the *N*-acylated L-amino acid and thereby enabling its separation from the D isomer using ion exchange, (c) diacetyl reductase for lowering concentrations of unwanted diacetyl in beer, (d) pullulanase to effect high maltose conversion of starch when used with α- and β-amylases, and (e) *O*-methyltransferase to block oxidation of *O*-phenols to inhibit browning (138). In Japan, egg white lysozyme is used as a preservative and to "humanize" cow's milk; also, naringinase and hesperidinase are employed to debitter citrus products (144).

In using enzymes for food modification, one should know the type of substrate one wants to modify and should be aware that extraneous enzymes in the enzyme preparation might result in deleterious effects. For example, lipases in microbial rennets may cause rancidity in cheese. The concentration of enzyme needed depends on the time available, pH, the operating temperature, and the degree of conversion desired. In most instances the substrate is present in excess so that the enzyme is the rate-limiting factor. Because of the enzyme-time relationship, Eq. (47), the time required to complete a process is inversely proportional to the amount of enzyme used. Thus, if twice as much enzyme is used, the processing time should be halved.

Enzymes are marketed by units of activity, rather than by weight or volume, and always contain other substances (salts, preservatives, stabilizers, carriers, nonenzymic organic material, and others) making it difficult to estimate actual quantities of crude enzyme required to achieve a desired result. The concept of total organic solids (TOS) in enzyme preparations allows an estimate to be made of the percent crude enzyme added to a food (126). The TOS for a given enzyme preparation is determined experimentally as

$$TOS(\%) = 100 - A - W - D$$

where A = percentage ash, W = percentage water, D = percentage diluents in the extract or isolated enzyme. The quantity of enzyme (TOS) used, calculated as a percentage of the food, generally varies between 0.052 and 0.00000026%, depending upon the enzyme concentration and the desired effect. Thus, extremely small amounts of enzyme are usually required to effect desired changes.

The two major environmental factors that must be considered when utilizing enzymes are pH and temperature. The activity of the enzyme at the pH of the food must be considered, and the possibility of inactivation of enzymes at extremes in pH must be kept in mind. The best temperature for an enzyme reaction depends on the time available for the reaction, the amount of enzyme that can be afforded, and the temperature sensitivities of the enzyme and the food. Usually, the process is conducted at the highest temperature consistent with good, final quality. The reaction rate approximately doubles for every 10°C rise in temperature until a maximum, termed the "temperature optimum," is reached.

Enzymic reactions are usually carried out at moisture levels ranging from 50 to 95%. More dilute solutions make the reaction inefficient and expensive. However, in some cases, such as chill proofing beer, one must work with a dilute solution of substrate.

It is important to remember that such factors as time, substrate concentration, enzyme concentration, pH, and temperature are interrelated. Thus, the variation of one parameter often affects the enzymic response to the other factors.

XII. IMMOBILIZED ENZYMES AND MICROORGANISMS (25,131)

A. Overview

Ordinarily, after a food process is completed, the added soluble enzymes are inactivated with heat or allowed to continue their activity to develop desirable flavor and texture, but they are never reused. Also, in analytical applications, the enzyme is used once for a single analysis. Recent developments in enzyme technology have led to the preparation of immobilized enzymes, microbial cells, or cellular organelles that in principle can be used repeatedly in batch operations or on a continuous basis.

An immobilized enzyme or cell is chemically or physically restricted in movement so that it can be physically reclaimed from the reaction medium. Major operational advantages of immobilizing enzymes or cells include reusability of the enzyme, the possibility of improved batch or continuous operational modes, rapid termination of reactions, controlled product formation, greater variety of engineering designs for continuous processes, and possible greater efficiency in consecutive multistep reactions. Furthermore, it may be possible to selectively modify the properties of the enzyme or cell by immobilization.

In principle, enzymes, microbial cells or cellular organelles that perform single or sequential reactions can be immobilized by one of five general methods (117):

1. Covalent attachment to an insoluble matrix
2. Inclusion within a gel lattice or (micro)-encapsulation within semipermeable membranes
3. Adsorption onto an insoluble matrix, including hydrophobic and affinity binding
4. Cross-linking to form insoluble aggregates
5. Adsorption followed by cross-linking

The extent to which immobilized enzymes are used in the food processing industry is rather disappointing in view of the extensive effort to exploit this technology (117). There are obvious limitations in using immobilized enzymes or microbial cells for food processing. An obvious constraint is that this technique is restricted to fluid systems. In the case of immobilized microbial cells, there may be additional limitations, such as the necessity for low-molecular-weight substrates and products (25). Furthermore, in potentially important applications involving complex liquid foods, such as the continuous clotting of milk for cheese making (170), a serious problem is encountered with fouling of the immobilized enzyme particles by surface-active constituents in the food, especially proteins. Fouling results in steric prevention of enzymic-substrate interaction. Thus, the

usable lifetime of an immobilized enzyme reactor may be too short for economic operation. At this point, the application of immobilized enzymes to processing complex food
systems, such as milk, is less promising than their use in the simpler systems developed
thus far. Other factors that may result in a decline in the activity of immobilized enzymes
during use are the presence of inhibitory substances in some foods and the detachment
of enzymes from the support.

An additional problem involves maintenance of sanitary conditions in immobilized
enzyme reactors. This is especially troublesome in the presence of nutritious substrates
that foster the growth of undesirable microorganisms at the favorable temperatures of
reactor operations. In addition to conscientious cleaning of reactors, sanitizing agents
should be used. Obviously, the sanitizer should not have a detrimental effect on the
immobilized enzymes in question. Some food-grade sanitizers, such as hydrogen peroxide,
have been employed on immobilized enzymes (139).

Indeed, the few immobilized enzyme systems that have been commercialized depend upon simple, reliable, nontoxic immobilization procedures coupled with the processing of relatively simple substrate solutions. It is unfortunate that many of the numerous methods for immobilizing enzymes are too cumbersome, inefficient, or costly and/or
employ toxic chemicals that render them incompatible with economic and safety standards inherent in food processing (139).

From the difficulties and uncertainties evident in the foregoing discussion, ascer-
taining the costs of immobilized enzyme systems is quite difficult, but some estimates
have been made (29,131,139).

B. Applications of Immobilized Enzymes and Microbial Cells in Food Systems

The Japanese have been quite successful in applying immobilized enzymes and microbial
cells to various processing situations (172). To a lesser extent, American industry has also
been exploiting this technology. Listed in Table 6 are the known commerical applications
of immobilized enzymes.

1. Immobilized, Living Microbial Cells

In the preceeding applications, soluble enzymes as well as nonviable microbial cells have
been immobilized to serve as inexpensive sources of enzymes catalyzing single reactions
(25,131). More recently, living cells have been immobilized to take advantage of the
multistep enzymic reactions characteristic of viable cells (25). A simple, mild procedure
has been developed for entrapping viable cells in a gel of κ-carageenan, which is a natural
stabilizer commonly used in the food industry (25,147). This approach holds the promise
of producing many useful products (Table 7), as well as detoxifying industrial wastes
(Table 8). Hagerdal (72) has recently suggested the coimmobilization of enzymes with
microbial cells. The immobilized enzyme would be used to convert the beginning micro-
bial-resistant substrate into compounds utilizable by the immobilized microorganism.
Thus, immobilization of living microorganisms may offer some new processing possibilities for the food industry.

2. Encapsulation

The retention of soluble enzymes from microbial lysates in ultrafiltration (UF) units and
the microencapsulation of enzymes offer a number of potential applications within the
food industry (147). For example, proteolytic enzymes retained by semipermeable mem-

Table 6 Commercial Applications of Immobilized Enzymes in the Food and Related Industries (20,134,147,172)

Immobilized enzymes or cells	Substrates	Products
Aminoacylase	Synthetic acyl-DL-amino acid + H_2O	L-Amino acid + acyl-D-amino acid
Aspartase, *Escherichia coli*	Fumarate + NH_3	L-Aspartate
Fumarase, *Brevibacterium ammoniagenes*	Fumarate + H_2O	L-Malate
Glucose isomerase *Streptomyces* sp. *Bacillus coagulans* *Actinoplanes missouriensis* *Arthrobacter* sp.	D-Glucose	D-Fructose
α-Galactosidase (used to hydrolyze raffinose in sugar beet juice which hinders crystallization of sucrose; see text)	Raffinose	D-Galactose and sucrose

Table 7 Production of Useful Compounds Using Immobilized Living Microbial Cells

Useful compounds	Microbial cells	Carriers for immobilization
Ethanol	*Saccharomyces carlsbergensis*	Carrageenan
	Saccharomyces cerevisiae	Polyacrylamide
Beer	*Saccharomyces carlsbergensis*	Polyvinyl chloride
	Saccharomyces cerevisiae	Calcium alginate
L-Isoleucine	*Serratia marcescens*	Carrageenan
Acetic acid	*Acetobacter sp.*	Hydrous titanium oxide
Lactic acid	Mixed culture of lactobacilli and yeasts	Gelatin
	Arthrobacter oxidans	Polyacrylamide
α-Ketogluconic acid	*Serratia marcescens*	Collagen
Bacitracin	*Bacillus sp.*	Polyacrylamide
Amylase	*Bacillus subtilis*	Polyacrylamide
Hydrogen gas	*Clostridium butyricum*	Polyacrylamide
	Rhodospirillium rubrum	Agar

Source: From Ref. 25.

Table 8 Decomposition of Poisonous Compounds Using Immobilized Living Microbial Cells

Applications	Poisonous compounds	Microbial cells	Carriers for immobilization
Waste treatment	Phenol	*Candida tropicalis*	Polyacrylamide
	Benzene	*Pseudomonas putida*	Polyacrylamide
	Waste water	Mixed culture of various microorganisms	Bentonite and Mg^{2+} Polyurethane sponge
Denitrification	Nitrate, nitrite	*Micrococcus denitrificans*	Liquid membrane
		Pseudomonas sp.	Active carbon

Source: From Ref. 25.

branes of a UF apparatus can hydrolyze proteins in the retentate to peptides and amino acids, which can then be obtained as products in the UF permeate. Recently, Greco et al. (67) proposed the use of a UF cell as a plurienzymic reactor in which multilayers of various enzymes could be immobilized on the UF membrane to carry out complex reaction sequences.

Olson and coworkers (103) are developing a cheese-ripening system in which cell-free extracts of microorganisms can be encapsulated with substrates and cofactors for generation of such flavors as diacetyl and acetoin. The milk-fat-coated microcapsules containing the appropriate enzyme systems are added to milk and retained in the resulting cheese. The juxtaposition of enzyme, substrates, and cofactors allows for the enhancement and regulation of flavor development in the cheese.

3. Food Analysis

The specificity and sensitivity of enzymes make them ideal reagents for quantifying specific components in complex mixtures of foods (9,71,111,189). One approach to the use of enzymes in food analysis is to immobilize the enzyme onto the surface of a sensing device, such as a specific electrode, which can then be used to measure depletion of substrate or accumulation of a product. Incorporation of the "enzyme electrode" into an appropriate circuit allows recording of the electrode response. A number of enzyme electrodes have been developed, some commercially, that should be of interest to the food industry (153).

Recently developed enzyme immunoassays (EIA), which rely on the specificities of antibodies, utilize enzymes as indicators rather than reagents specific for their appropriate substrates (144). The technique is a conceptual variation of the widely used radioimmunoassay (RIA) and is useful for quantifying trace amounts of food components, particularly toxic proteins such as staphylococcal enterotoxin. Assays of this type can be sensitive to picomolar amounts of the compound under analysis (analyte).

Enzyme immunoassay relies on the development of antibodies specific for the analyte. The analyte itself need not be antigenic, but can usually acquire antigenicity by being covalently coupled to an antigenic protein. Thus, antibodies can be elicited in an appropriate animal against nonantigenic compounds, such as aflatoxin or ochratoxin, when the toxins are injected as protein conjugates. The commerical availability of these haptenic compounds, their associated protein, and the necessary antibodies is currently limited. Furthermore, the preparation of the conjugates and acquisition of the essential antibodies requires a relatively high degree of skill.

As shown in reaction (79), the basis for EIA is competition between analyte and analyte-enzyme for sites on the antibody (144).

$$\text{Analyte} + \text{(analyte-enzyme)} + \text{antibody} \underset{k_2}{\overset{k_1}{\rightleftharpoons}} \text{(analyte-antibody)}$$

$$+ \text{(analyte-enzyme-antibody)}$$

where

$$k_1 \ggg k_2 \tag{79}$$

The analyte coupled to the indicator enzyme (analyte-enzyme) interacts specifically with the antibody and thereby inhibits or prevents the indicator enzyme from reacting with its substrate. A variety of different enzymes can be used as indicators, including peroxidases, various dehydrogenases, lysozyme, lactase, alkaline phosphatase, or glucoamylase. The unknown analyte contained in the food when added to the assay system competes with the analyte-enzyme for the antibody. When the unknown analyte is absent, the analyte-enzyme is maximally bound to the antibody yielding minimum enzymic activity. In the presence of increasing amounts of unknown analyte increasing amounts of active enzyme become available for reacting with its indicator substrate.

In a typical analysis, only the initial concentration of the unknown analyte is variable so that the resultant enzyme activity is proportional to the amount of unknown analyte. From an appropriate calibration curve prepared from known amounts of the analyte, absolute quantities of a given analyte in foods may be obtained.

Enzyme immunoassay is a very flexible analytical tool, and there are several variations of its use (144). The amplification possible from continued enzymic activity is an attractive feature of EIA analyses compared with the fixed amount of radioactive signal available in RIA procedures. Ultrasensitive EIA assays are possible when radioisotopes are incorporated into the assay (144).

4. Control Devices

Enzyme thermistors have been devised in which an enzyme system is immobilized on a thermistor (105). The heat evolved in an enzymic reaction can be used to determine calorimetrically the amounts of substrate reacted. Enzyme thermistors have been designed for routine analysis in clinical chemistry, environmental control, and biotechnology (34). Enzyme thermistors can also be used for continuous monitoring and control during conversion of various substrates in bioreactors (33). For example, flow rates of whey pumped through an immobilized lactase column can be controlled by an enzyme (glucose oxidase or catalase) thermistor sensing for a preset concentration of glucose product in the column effluent. Thus, flow rates can be changed to reflect changes in lactose (in whey) feed concentrations or in lactase enzyme activity.

XIII. MODIFICATION OF FOOD BY ENDOGENOUS ENZYMES

Since foods are complex biological materials, they are subject to a wide variety of modifying agents. Among these are microorganisms that cause undesirable spoilage or beneficial fermentations, endogenous chemicals that undergo such changes as autoxidation, and endogenous enzymes that cause numerous desirable and undesirable changes. Enzymic alterations can encompass entire areas of food science, such as the postharvest physiology of plants and the postmortem physiology of animal tissues. For example, the ripening of fruits or the conversion of muscle to meat are the results of complex biochemical changes mediated by enzymes. The biochemical reactions involved in these processes are discussed in Chaps. 12 and 15 and in specialized texts (50,80,96), so it is not appropriate to deal with them here. Consideration here is limited to those endogenous enzymes of predominant interest to the food scientist in terms of changes in color, texture, flavor, and nutritive value of foods. Thus, the discussion is limited largely to the activities of hydrolases and oxidoreductases, and only the more important ones are considered in detail. Comprehensive discussions on the effects of endogenous enzymes can be found in the treatise by Schwimmer (144).

A. Pectic Enzymes (5,35,128,129,133,141,144)

Pectic substances are a group of heteropolysaccharides in which the major building blocks are units of galacturonic acid linked by α-1,4-glycosidic bonds (Fig. 24). Approximately two-thirds of the carboxylic acid groups are esterified with methanol. A variety of pectic enzymes act on the O-α-(1,4)-polygalacturonopyranoside with activities and specificities depending, in part, upon the degree of methyl esterification (Fig. 24).

Highly esterified polymers (high-methoxyl pectins) are the best substrates for pectin lyases (pectintranseliminases). Pectin lyases split glycosidic bonds adjacent to a methyl ester by a β-elimination reaction, yielding a double bond for each broken glycosidic bond (Fig. 24). Pectin lyases are generally produced in fungi and are not found in bacteria or higher plants.

Pectins in which methoxyl groups have been partially (low-methoxyl pectin) or completely (polygalacturonic acid) hydrolyzed are the best substrates for pectate lyases. In this instance, the β-elimination reaction occurs adjacent to a free carboxyl group. Pectate lyases are typically bacterial enzymes; only a few molds produce them. They are absent in higher plants.

Pectin esterases cleave methanol from esterified carboxyl groups yielding low-methoxyl pectin and polygalacturonic acid. Together with endopolygalacturonase, which hydrolyzes interior glycosidic bonds, and exopolygalacturonase, which hydrolyzes exterior glycosidic bonds, or with pectate lyases, pectinesterase completes an enzyme mixture capable of depolymerizing high-ester pectins.

Since pectic substances are structural elements in the middle lamella and primary cell walls of higher land plants, alterations in their degree of polymerization and esterification can change the texture of vegetables and fruits during ripening, during postharvest storage, or during processing. The pectinesterases and polygalacturonases of the native plant can alter endogenous pectic substances, whereas, exogenous pectic enzymes are produced industrially from fungi and they are used as processing aids, mainly for juice extraction and clarification. The pectic enzymes of microorganisms can be responsible to a large degree for postharvest decay and rotting and textural changes occurring in certain fresh fruits and vegetables. Thus, pectic enzymes are of interest to the food

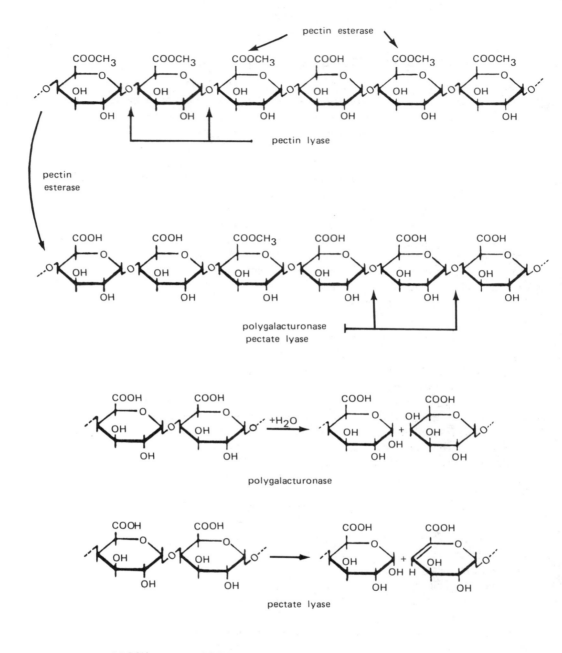

Figure 24 Enzymic degradation of pectin by polygalacturonase, pectate lyase, pectin lyase, and pectinesterase. (Reprinted from Ref. 144.)

scientist for a variety of reasons. The effects of pectin methylesterases on texture are indirect. They demethylate carboxylate groups on pectic substances allowing them to engage in cross-linking reacions via Ca^{2+}. These cross-links increase the firmness of such products as green beans (144).

B. Amylases (35,133,144,164)

Enzymes that hydrolyze starch are termed "amylases." Two amylases have been studied in great detail, namely, α- and β-amylase. α-Amylase, an endo-enzyme, hydrolyzes α-1,4-glucan linkages in an apparent random manner. The random attack of α-amylase on amylose in solution results in a rapid decrease of viscosity, a rapid loss of ability to produce a colored complex with iodine, and an increase in reducing power due to generation of reducing groups. Since relatively few interior glucan linkages are cleaved to effect a large decrease in viscosity, the decrease in viscosity is proportionately greater than the increase in reducing groups.

α-Amylase attacks amylopectin in a similar fashion; however, the α-1,6 branch points are not hydrolyzed, and a small amount of panose, the trisaccharide containing these α-1,6 linkages, is formed. α-Amylase is often referred to as the "liquefying enzyme" because of its rapid action in reducing the viscosity of starch solutions. Treatment of amylose or amylopectin with α-amylase results in a mixture of maltose, glucose and also some panose if the reaction continues for a long period.

β-Amylase is an exoenzyme; that is, it attacks only the end units of starch chains. More specifically, β-amylase removes maltose units from the nonreducing end of the starch chain by hydrolyzing alternate glycosidic linkages. Since maltose increases the sweetness of the starch solution, β-amylase is referred to as a saccharifying enzyme. The β in β-amylase refers to the inversion of the α-1,4 linkage in starch to the β configuration. The occasional 1,3 linkages in amylose and the 1,6 bonds in amylopectin are not attacked by β-amylase, resulting in incomplete degradation of the starch. If a debranching enzyme is present to cleave these bonds, β-amylase activity will continue. With amylopectin, β-amylase activity stops two or three glucose units from the branch point, and the residual molecules are referred to as "limit dextrins." β-Amylases are known to occur only in vegetable tissues.

Amylases are important in fruit ripening, potato processing, production of corn syrup and corn sugar, in brewing, and in bread making. In general, the hydrolysis of starches by these enzymes can (a) provide sugars for subsequent fermentation and utilization by microorganisms, (b) provide reducing sugars to participate in nonenzymic browning, and (c) alter texture, mouth feel, moistness, and sweetness of affected foods. Often acting in conjunction with other endogenous and exogenous carbohydrases, the amylase enzymes endogenous to barley malt and wheat play crucial roles in brewing and baking processes (144).

C. Cathepsins (38,51,119)

Cathepsins are proteinases that exist intracellularly in animal tissue and are active at acidic pH values. These enzymes are located in the lysosomal fraction of the cell and are thus distinguished from proteolytic enzymes such as trypsin and chymotrypsin, which are secreted by cells.

Five cathepsins have been observed and have been designated by the letters A, B, C, D, and E (119). In addition, a catheptic carboxypeptidase has been isolated.

Cathepsins may be involved in changes observed during aging of meats. They are released from lysosomal particles in the muscle cells at the reduced pH of the postmortem tissue. These proteinases presumably permeate the tissue leading to breakdown of myofibrils within the muscle cells, as well as extracellular connective tissues, such as collagen. They are most active in the pH range 2.5-4.5.

D. Calcium-Activated Neutral Proteinase (CANP) (38,51,163)

This enzyme is a metalloproteinase, requiring a cysteinyl thiol group, that occurs in a number of tissues, but has been studied mainly in muscle. The nonlysosomal muscle CANP is present in low concentrations and can function down to a pH of about 6. Possibly, muscle CANP contributes to tenderization of meats by causing breakdown of specific myofibrillar proteins. This leads to disintegration of the Z disk and damage to the M lines, the proteins of which stabilize the three-dimensional structure of the myofibril by locking together the thin and thick filaments, respectively. The CANP is activated by millimolar concentrations of Ca^{2+} and by Mg^{2+}, but there are insufficient concentrations of these ions in tissues postmortem to substantially activate the type of CANP that was originally isolated. However, CANP enzymes have been isolated recently that are activated by micromolar concentrations of Ca^{2+}. These enzymes are more likely to be active in muscle tissues postmortem, and they may act, during conversion of muscle to meat, in conjunction with the lysosomal proteases.

E. Milk Proteinase(s) (39,58)

The major milk proteinase is an alkaline serine proteinase with a specificity similar to that of trypsin. This proteinase is closely related, if not identical, to blood plasmin. It hydrolyzes β-casein to the more hydrophobic γ-caseins and is also capable of hydrolyzing α_s-casein(s), but not κ-casein. Milk proteinase contributes to proteolysis during the ripening of cheese, and since it is relatively stable to heat, if may contribute to gelation in milk that has been subjected to ultra-high temperature processing. Its role in converting β-casein to γ-casein may be important to the physical behavior of milk proteins in various food products.

An acid proteinase with a pH optimum around 4 has been reported in milk; however, it is relatively labile to thermal inactivation.

F. Lipolytic Enzymes (8,39.62,85,177,178)

1. Lipases

Lipases are enzymes that hydrolyze ester linkages of emulsified triacylglycerols at an oil/water interface. These enzymes are widely distributed, occurring in plants, animals, and microorganisms. Lipases are found in many biological fluids, cells, seeds, organs, and various other tissues.

In general, lipases hydrolyze triacylglycerols in a stepwise fashion as follows:

$$\text{Triacylglycerol} \begin{matrix} \nearrow & \text{1,2 Diacylglycerol} & \searrow \\ & & \\ \searrow & \text{2,3 Diacylglycerol} & \nearrow \end{matrix} \text{2-Monoacylglycerol} \qquad (80)$$

However, there are exceptions depending on both positional and fatty acid specificities of the enzyme. This reaction sequence is for lipases with primary ester specificity, which seem to be the predominant type. Some lipases, however, attack all three ester positions.

There are four types of specificities associated with lipases. These are acylglycerol, positional, fatty acid, and stereospecificities. Acylglycerol specificity is characterized by a preferential hydrolysis of low-molecular-weight triacylglycerols over high-molecular-weight substrates.

The best known example of positional specificity is that of pancreatic lipase, which hydrolyzes only the primary esters of triacylglycerols. Lipase preparations that hydrolyze fatty acids from the primary and secondary positions of a triacylglycerol probably contain more than one enzyme, possibly a triacylglycerol lipase and a monoacylglycerol lipase. A lipase that hydrolyzes only the secondary ester from a triacylglycerol molecule has not been reported to date.

Fatty acid specificity occurs when one type of fatty acid is more rapidly hydrolyzed than another type, both attached to the same positions of like triacylglycerol molecules. Although it may be difficult to separate acylglycerol and fatty acid specificities, one principal example of fatty acid specificity exists. This is the specificity of a lipase from the microorganism *Geotrichum candidum* for oleic acid. Oleic acid preferentially is hydrolyzed regardless of the position in the triacylglycerol molecule. However, minor amounts of other fatty acids are also released.

Endogenous lipases are important to the food scientist since fatty acids produced from their hydrolytic action on triacylglycerol can lead to desirable or undesirable flavors. The lipoprotein lipase of milk is extremely important since it releases short-chain fatty acids from milk fat leading to undesirable hydrolytic rancidity (39,58).

2. Phospholipases (138)

A number of phospholipases react with glycerophospholipids. The acyl groups esterified at the sn-1 and sn-2 positions are hydrolyzed in addition to sites on either side of the phosphoryl group [Eq. (81)]. Most work on phospholipases has been done with 3-*sn*-phosphatidylcholine (lecithin) as a substrate, but other glycerophospholipids respond in similar fashion if conditions for hydrolysis are adjusted to the needs of the system.

$$
\begin{array}{c}
\text{(1)} \quad \overset{\displaystyle O}{\overset{\displaystyle \|}{}} \\
\text{C} \quad \text{C} - \text{O} - \text{C} - \text{R}_1 \\
\overset{\displaystyle \|}{\underset{\text{(2)}}{\text{C}}} \quad | \\
\text{R}_2 - \text{C} - \text{O} - \text{C} - \text{H} \quad \overset{\displaystyle O}{} \\
| \quad \text{(3)} \overset{\displaystyle \|}{} \quad \text{(4)} \\
\text{C} - \text{O} - \text{P} - \text{O} - \text{C} - \text{Base} \\
| \\
\text{OH}
\end{array}
\qquad (81)
$$

As shown in Eq. (81), four ester functions can be hydrolyzed. The enzymes responsible for hydrolysis of these ester linkages are referred to as phospholipases, although other terms are encountered in the literature. Phospholipase A, now designated lecithinase A or phospholipase A_2 or phosphatide acylhydrolase (present in snake and bee venoms), hydrolyzes the ester group at the sn-2 position. The resulting compound with a hydroxyl at the sn-2 position is known as lysolecithin, and it is highly surface active.

Other phospholipase A enzymes have been isolated. One catalyzes hydrolysis at position 1, and it is designated phospholipase A_1. Phospholipase B, now designated lysophospholipase or lecithinase B, hydrolyzes acyl ester linkages at positions 1 or 2 of the corresponding lysolecithin, yielding glycerophosphocholine. Pancreatic lipase, capable of catalyzing hydrolysis at the 1,3 positions in a triacylglycerol, also catalyzes hydrolysis at the sn-1 position in 3-*sn*-phosphatidylcholine when suitable conditions exist. Pancreatic lipase preparations frequently contain phospholipase A_2 as a contaminant, and both acyl groups may be split off unless this phospholipase is suppressed.

Phospholipase C, prevalent in molds and bacteria, catalyzes hydrolysis at position 3, giving a diacylglycerol and phosphorylcholine. Phospholipase D, found usually in plants (red cabbage is a good source) is a choline-hydrolyzing enzyme (position 4), yielding phosphatidic acid.

Phospholipase A_2 liberates polyunsaturated fatty acids that predominate at position 2 of glycerophospholipids. These released fatty acids are substrates for lipoxygenase leading to development of aroma in fresh vegetables or in various flavor defects in affected plant foods. The action of phospholipase A_2 in frozen fish may result in undesirable textural changes (144).

G. Thiaminases I and II (15,18,144)

In certain foods, enzymes exist that degrade thiamine. Thiaminase I acts as a transferase to catalyze the fission of thiamine, using amines as cosubstrates.

$$\text{Pyr}-\text{CH}_2-\text{Thiaz}^+ + \text{R}-\text{NH}_2 \longrightarrow \text{Pyr}-\text{CH}_2-\text{NH}-\text{R} + \text{Thiaz} + \text{H}+ \qquad (82)$$

$$\qquad \text{Thiamine} \qquad\qquad \text{Amine}$$

Thiaminase II is a hydrolase that catalyzes the hydrolysis of thiamine.

$$\text{Pyr}-\text{CH}_2-\text{Thiaz} + \text{H}_2\text{O} \longrightarrow \text{Pyr}-\text{CH}_2\text{OH} + \text{Thiaz} + \text{H}^+ \qquad (83)$$

Thiaminase I enzymes are found commonly in most freshwater fish and in many marine fish. The enzymes are either an intrinsic part of the flesh or occur adventitiously from microorganisms. Boiling or cooking the fish in most cases inactivates the enzyme. Some fish possessing thiaminase activity are listed as follows.

Marine species: capelin, herring, menhaden, garfish, whiting
Freshwater species: carp, chub, smelts, minnow

Many shell fish also exhibit thiaminase I activity.

The consumption of raw fish, raw shellfish, salted or cured herring, or unheated fermented fish products, as practiced in Asia, can contribute to thiamine deficiency diseases, such as beri-beri, or at least can raise the dietary requirements for thiamine. Salted herring has been shown to destroy 50-60% of added thiamine within 6 hr.

In addition to fish, certain plant materials exhibit thiaminase activity. Bracken fern and horsetail plants are especially high in thiaminase activity. This enzyme has also been observed in rice polishings, dry beans, and mustard seed. Where people consume an exclusive rice diet, especially of unpolished rice, the thiamine requirements may not be met.

Certain microorganisms that can inhabit the buccal cavity or the intestinal tract produce substantial quantities of antithiamine agents. The persistence of beri-beri- as an endemic condition in certain regions of the world, despite the administration of thiamine, may result from the persistent presence of these organisms in the body.

H. Phytase (144)

Phytic acid, *myo*inositol-1,2,3,4,5,6-hexakis-dihydrogen phosphate, is ubiquitous in plants as a reservoir of phosphorus and occurs in the blood of birds, amphibians, and reptiles. The binding of minerals in protein-phytate-mineral complexes is thought to reduce the nutritional availability of minerals of certain plant foods and some plant protein isolates. Of particular concern has been the effects of phytic acid on the availability of Fe^{3+} and Zn^{2+}. The binding of Ca^{2+} by the monomeric phytate apparently influences the textural properties of certain vegetables, since Ca^{2+} bound to phytate can not participate in the cross-linking of molecules of pectic acid. Thus, the presence of phytate has a tenderizing action undesirable in tender vegetables, such as green beans, and desirable in tough vegetables, such as beans that are dried, then hydrated and cooked.

Phytic acid inhibits certain enzymes, such as pepsin (26), amylase (22,148), and trypsin (150), by combining with the protein and/or by binding cations (e.g., Ca^{2+}) essential for their activity (150). The physiological significance of these observations is unclear.

The enzyme, phytase, hydrolyzes the phosphate residues from phytic acid, thereby essentially destroying its strong affinity for minerals. Thus, phytase may enhance the nutritional availability of minerals and alter the texture of plant foods by releasing Ca^{2+} for participation in cross-linking and other reactions. Phytase, like its substrate phytic acid, is ubiquitous in plants, although occurring sometimes in very low concentrations. Nevertheless, controlling the action of endogenous phytases by manipulating temperature and a_w has been proposed to lower phytic acid concentrations in plant tissues (144). Exogenous phytases may also eventually prove useful in controlling levels of phytic acid in plant based foods.

I. Myosin ATPase (107,120)

This enzyme apparently plays a key role in cold shortening and thaw rigor of muscle, and these occurrences affect the tenderness and water-holding capacity of the resulting meat. Apparently, the postmortem shortening of muscle proceeds by a mechanism similar to normal muscle contraction. Both cold and thaw shortening appear to be attributable to the inability of the sarcoplasmic reticulum (a very fine network of tubules throughout the muscle) and perhaps mitochondria to retain Ca^{2+}. The freed Ca^{2+} diffuses to the contractile proteins where it releases adenosine triphosphate (ATP) from its inert complex with magnesium and also stimulates myosin ATPase. The resultant hydrolysis of ATP furnishes the energy necessary for muscle contraction. For further information on this subject see Chap. 12.

J. Enzymic Browning (50,98,133,144)

Basically, there are four types of browning reactions in foods: Maillard, caramelization, ascorbic acid oxidation, and phenolase browning. The former three are nonenzymic in nature (oxidation of ascorbic acid is sometimes catalyzed by enzymes) and are not considered further (see Chap. 3). Phenolase or enzymic browning (enzyme-catalyzed oxidative browning) is of commercial significance, particularly in fruits and vegetables, in which phenolase enzymes are common (144).

In intact tissue, phenolic substrates are separated from phenolase and browning does not occur. Enzymic browning can be observed on the cut surfaces of light-colored fruits and vegetables, such as apples, bananas, and potatoes. Exposure of the cut surface to air results in rapid browning due to the enzymic oxidation of phenols to orthoquinones, which in turn rapidly polymerize to form brown pigments or melanins. The various enzymes that catalyze oxidations of phenols are commonly known as phenolases, polyphenol oxidases (PPO), tyrosinases, or catecholases. The classification of these enzymes is confusing and controversial. Schwimmer (144) prefers to refer to this enzyme type as "phenolase."

Phenolase enzymes that have been isolated from food sources are oligomers and contain one copper prosthetic group per subunit. Thus, if damage to plant tissue is sufficient, and oxygen and copper are present, browning occurs. Such operations as cutting, peeling, and bruising are sufficient to cause enzymic browning.

Phenolase catalyzes two types of reactions, as illustrated in Fig. 25: hydroxylation A (referred to as phenol hydroxylase or cresolase activity), and oxidation B (referred to as polyphenol oxidase or catecholase activity). The first reaction results in orthohydroxylation of a phenol and the second in oxidation of the diphenol to an orthoquinone. If tyrosine is the substrate (Fig. 25), phenolase catalyzes its hydroxylation to DOPA; and subsequently catalyzes oxidation of DOPA to DOPA quinone.

The remaining portions of the reaction sequence involve nonenzymic oxidations and ultimately polymerization of indole 5,6-quinone to brown melanin pigments. Fur-

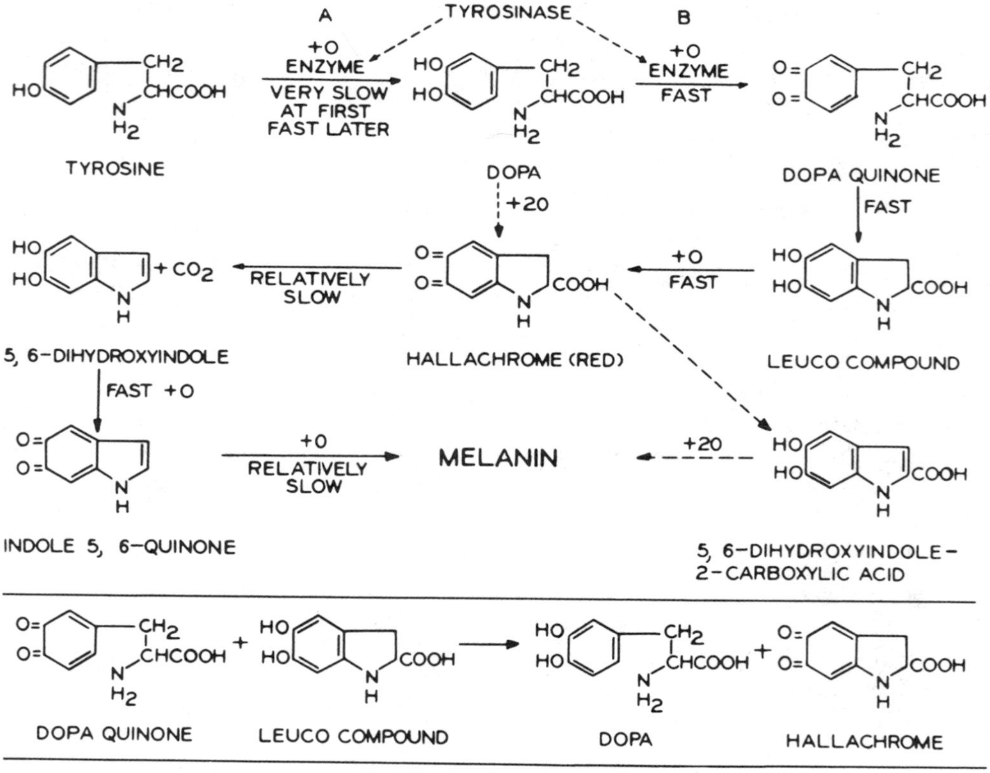

Fig. 25 The formation of melanin pigments resulting from the oxidation of tyrosine by phenolase. (Redrawn from Ref. 98, by courtesy of the American Physiological Society.)

thermore, the melanins can interact with proteins to form complexes. Hydroxylation of monophenols is the slow or rate-determining step. Consequently, monophenols undergo the slow hydroxylation reaction before oxidation to orthoquinone. On the other hand, some phenolase enzymes, such as those in tea, tobacco, and sweet potatoes, do not hydroxylate monophenols. However, those from mushroom and potato perform both functions.

Although tyrosine is a major substrate for certain phenolases, other phenolic compounds of fruit, such as caffeic acid and chlorogenic acid, also serve as substrates. As shown in Fig. 26, these latter compounds are orthodiphenols, readily attacked by the catecholase component of phenolase. However, as is illustrated in Table 9, phenolase does not necessarily attack these substrates at the same rate. Chlorogenic acid is oxidized at 67% the rate of catechol, caffeic acid at only 41% the rate of catechol, and protocatechuic acid is not oxidized at all. The two most prevalent substrates for phenolase in most plant tissues are tyrosine and chlorogenic acid (144).

In general, phenolase is active between pH 5 and 7 and does not have a very sharp pH optimum. At lower pH values of approximately 3 the enzyme is irreversibly inactivated. Furthermore, reagents that complex or remove the copper from the enzyme also inactivate it. Phenolase-induced browning is a major contributor to the desirable color of some foods, such as cider, tea, and cocoa and a minor contributor to the normal color of raisins, prunes, dates, and figs.

For the most part, however, phenolase activities in fruits and vegetables are undesirable because the ensuing brown colors are not pleasing. Consequently, a variety of methods have been developed to inhibit enzymic browning (144). The methods are predicated on eliminating from the reaction one or more of its essential components, that is, oxygen, enzyme, copper, or substrate.

Heat treatments, or the application of sulfur dioxide or sulfites, are commonly used methods of inactivating phenolase. Phenolase activity can be inhibited by the addition of sufficient amounts of acidulents, such as citric, malic, or phosphoric acids to yield a pH of 3 or lower. These agents can also serve as chelators for copper. Oxygen can be excluded from the reaction site by such methods as vacuumization or immersing the plant tissues in a brine or syrup. Phenolic substrates can be protected from oxidation by reaction with borate salts, but these are not approved for food use.

K. Lipoxygenases (4,62,133,167)

At one time, lipoxygenases (LOX) were thought to be restricted to legumes and a few other plants, but it is now evident that LOX occur widely throughout the plant and

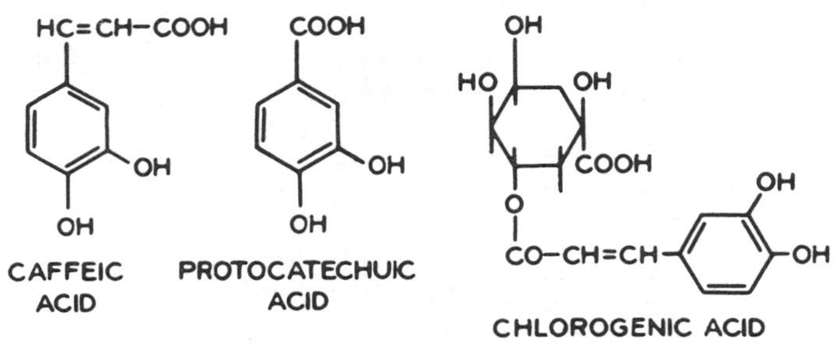

Fig. 26 Some *o*-diphenols that serve as substrates for phenolases.

Table 9 Action of Phenolase from Cling Peaches on Various Phenolic Substrates

Substrate	Conc. at pH 6.2 (M)	Relative activity (%)
Catechol	0.01	100
Caffeic acid	0.01	41
Dopamine	0.01	34
Quinic acid	0.01	5
Shikimic acid	0.01	2
Ferulic acid	0.01	0
Protocatechuic acid	0.01	0
D-catechin	0.003	250
Chlorogenic acid	0.003	67
Isochlorogenic acid	0.003	0

Source: Reprinted from Ref. 102 by courtesy of Institute of Food Technologists.

animal kingdoms (62). Plant LOX are distributed mainly in legumes, such as soybeans, mung beans, navy beans, green beans, peas, and peanuts; in cereals, such as rye, wheat, oats, barley, and corn; in fruits, such as pear, apple, strawberry, and tomato; and in potato (62). In animals the enzymes appear primarily in specialized tissues, including platelets, certain white cells, and testicular tissues.

Lipoxygenases comprise a hetergeneous and diverse group of enzymes. They can exist as isoenzymes in the same plant; for example, soybean contains three principal isozymes designated L-1, L-2, and L-3. This is evident from their substrate specificities. In general, those enzymes with a high pH optimum prefer the anionic fatty acid substrate, whereas those with neutral pH optima attack neutral esters and triacylglycerols (62). In certain foods, lipases may release fatty acids, thereby providing substrates for those LOX enzymes requiring the free acid as a substrate.

More specifically, LOX stereospecifically oxygenate polyunsaturated fatty acids and/or their esters and acylglycerols containing the cis, cis-1,4, pentadiene double-bond system located between carbons 6-10 counting from the methyl terminus. The products then are necessarily optically active hydroperoxides. With linoleic acid, the principal oxygenation products are optically active 9- and 13-hydroperoxide isomers.

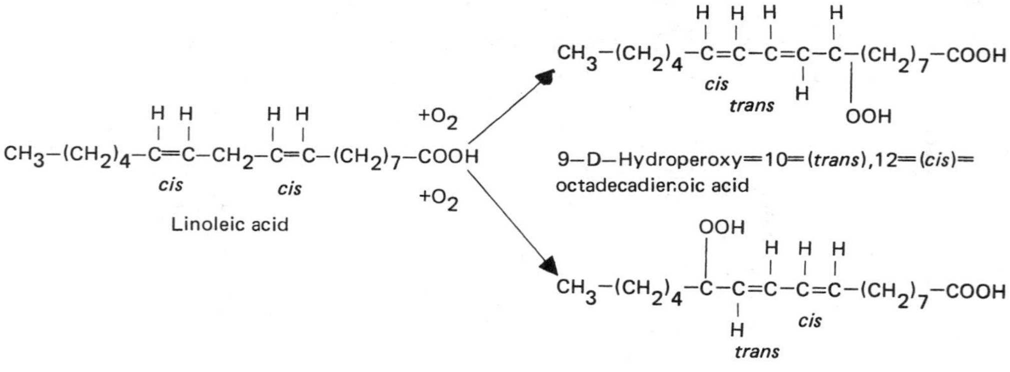

9–D–Hydroperoxy=10=(*trans*),12=(*cis*)= octadecadienoic acid

13–L–Hydroperoxy=9(*cis*),11=(*trans*)= octadecadienoic acid

The enzyme is believed to first abstract a hydrogen atom stereospecifically from position 11 of the linoleic acid molecule and then stereospecifically add oxygen at either end (9 or 13) of the resonating free radical to yield an optically active hydroperoxide. Varying ratios of 9 to 13 hydroperoxide isomers may result as products, depending on the particular isozyme. Also, isoenzymes of LOX within the same plant may attack esters and triacylglycerols in preference to fatty acids yielding different mixtures of hydroperoxide isomers. However, autoxidation of substrate and nonenzymic isomerization of resulting hydroperoxides can lead to artifacts if not properly controlled.

It is now known that soybean LOX, at least, require a nonheme iron in the active site. Thus, LOX catalyze oxidation via a redox cycle driven by oxygen and the polyunsaturated substrate. A possible mechanism for the oxygenation of polyunsaturated fatty acids is shown in Fig. 27. The state of the oxygen involved in LOX activity is unclear, but it may be immobilized superoxide (62).

LOX are of practical importance to the food scientists for a number of reasons, some of which are discussed in more detail below. In Table 10 are listed the various effects that endogenous, as well as exogenous, LOX can have on the color, flavor, texture, and nutritive properties of foods. Most vegetables contain linoleic and linolenic acids subject to peroxidation by LOX. Off-flavor development in soybeans and soybean products is highly dependent on the action of the various endogenous LOX, since subsequent decomposition of the resulting hydroperoxides yields rancid flavors (144). For example, the accumulation of carbonyl compounds in unblanched frozen peas is due to LOX, and failure to blanch plant tissue containing this enzyme may lead to rapid development of off-flavors. In the case of peas, a storage-resistant lipoxygenase isozyme may be the primary source of undesirably flavorful carbonyls, with other isozymes inactivated upon storage. Consequently, to minimize the activity of such enzymes as LOX in stored vegetables, blanching is necessary before they are frozen or dried. However, it is important to remember that the distribution of the enzyme in the food material may affect its heat response. For example, in peas, LOX activity is highest in the center of the pea, but lowest in the skin (49). On the other hand, peroxidase activity is higher in the skin than in the center of the pea (48). Furthermore, the type of flavor developed in affected foods may reflect the availability of a particular unsaturated fatty acid substrate and the specific carbonyls that result from differing decomposition patterns of the individual hydroperoxide isomers (144). Also, products from lipoxygenase activity can undergo secondary enzymic and nonenzymic transformations, yielding a wide variety of products which have marked effects on food quality.

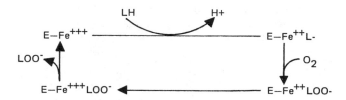

Fig. 27 Proposed mechanism of hydroperoxidation of linoleic acid (LH) by soybean lipoxygenase ($E-Fe^{3+}$). (Reprinted from Ref. 62.)

Table 10 Summary of Changes in Food Quality Mediated Directly or Indirectly by Lipoxygenase

A. Color changes
Bleaching of hard wheat flour via carotene destruction (desirable)
Bleaching of pasta products, also by carotenoid destruction (undesirable)
Participation in loss of desirable green color due to chlorphyll destruction in frozen and perhaps other processed green vegetables
Destruction of xanthophyll and other colored carotenoids important in assessing the quality of alfalfa-derived animal feeds (undesirable)
Destruction of added food colorants (undesirable)
Participation in meat color as prostaglandin synthase as well as lipoxygenase (?)
Destruction of skin pigmentation in some food fishes
Superficial scald in stored apples

B. Flavor changes
Production of components of volatiles responsible for desirable aroma in fruits and fresh vegetables, such as tomatoes, snap beans, bananas, and cucumbers
Production of off-flavors in frozen vegetables and perhaps other processed foods
Production of off-flavors in stored cereals, such as a "cardboard" flavor in barley
Production of off-flavors in high-protein foods, especially legume seeds
Participation in rancidity of meats (as prostaglandin synthase and lipoxygenase)

C. Texture changes
Production of favorable effects on the rheological properties of wheat, flour doughs, and eventually on the texture of baked goods via control of SS-SH balance and hydrophobic bonding of lipids to glutens

D. Nutritional quality changes
Destruction of vitamin A and provitamin A
Destruction of nutritionally essential polyunsaturated fatty acids
Interaction of enzymic product with some essential amino acids of proteins to lower the protein nutritional quality and functionality

E. In vivo functions
Participation in the biogenesis of ethylene by creating active oxygen needed to convert methionine to ethylene
Conversion of carotenoids to plant growth affectors, such as abscissions and growth inhibitors
Conversion of the unsaturated mammalian fatty acid, arachidonic, to hydroxy fatty acids
Destruction, anaerobic, of peroxides arising from wounded tissue

Source: From Ref. 144.

LOX can have an undesirable bleaching effect on pigments during spaghetti processing. LOX cooxidizes such pigments as β-carotene and xanthophylls (108) while peroxidizing its normal substrate(s). LOX also has been found to have a bleaching effect on chlorophyll (144). These destructive effects of LOX are hardly surprising in view of the free radicals generated during peroxidation and peroxide decomposition. Thus, it is very likely that other food constituents, such as vitamins, are also destroyed during these oxidative processes. In fact, the reaction catalyzed by LOX is so destructive that LOX inactivate themselves by a process first order with respect to enzyme (154).

The LOX of wheat are thought to have an important effect on the rheology of flour doughs. In addition, wheat flours are often supplemented with soy flours to gain the bleaching and improving effect of soy LOX. Absorption of atmospheric oxygen during dough mixing improves the rheological properties of the dough The mechanisms of dough improvement are incompletely understood; however, oxidized lipids resulting from lipoxgenase action are thought to assist in dough improvement by changing the rheological properties of gluten. This occurs, in part, through formation of disulfide bonds via complex oxidative and thiol-disulfide interchange reactions involving cysteine residues in the gluten. At the same time lipids are released from the oxidized protein and, presumably, aid dough development by somehow ensuring gas retention during expansion of the dough during baking (144).

L. Peroxidases (133,144)

Peroxidases are a ubiquitous group of enzymes occurring in all higher plants that have been investigated, and in leukocytes. In addition, a peroxidase is present in milk, which is one of the best known sources of this enzyme (58,84). Of the plant peroxidases, horseradish peroxidase has been studied most extensively.

Peroxidases from various sources usually contain a heme (ferriprotoporphyrin IX) prosthetic group. However, other prosthetic groups may also be involved (144).

Peroxidases catalyze the following reaction:

$$ROOH + AH_2 \longrightarrow H_2O + ROH + A \tag{85}$$

The catalytic process of peroxidase appears to result in the transient oxidation of ferric ion (Fe^{3+}) to higher valence (Fe^{5+} or Fe^{4+}) states. The peroxide (ROOH) may be hydrogen peroxide or an organic peroxide, such as methyl or ethyl hydrogen peroxide. In the above reaction, the peroxide is reduced while an electron donor (AH_2) is oxidized. The electron donor may be ascorbate, phenols, amines, or other organic compounds. In many instances, the oxidation product is highly colored, and this serves as a basis for the colorimetric estimation of peroxidase activity. For example, the oxidation of aromatic amines, such as diaminobenzidine and *o*-dianisidine have been used in peroxidase analyses. However, it should be pointed out that aromatic amines of this type can be carcinogenic and should be handled with care. These peroxidatic color reactions are also of value to the food scientists in enzymic analyses of foods. For example, in enzymic reactions in which hydrogen peroxide is a product, this can be coupled to a peroxidase reaction to develop an analytical procedure. For example, in the analysis of glucose using glucose oxidase, hydrogen peroxide is generated, as follows.

$$\text{Glucose} + O_2 \xrightarrow{\text{glucose oxidase}} \text{gluconic acid} + H_2O_2 \tag{86}$$

The H_2O_2 can be estimated with the peroxidatic reaction, and the concentration of glucose then can be estimated from the stoichiometry of the glucose oxidase reaction (9).

Since peroxidase is very resistant to heat inactivation and is widely distributed in plant tissues, and because very sensitive and simple colorimetric tests are available to measure its activity, it has been used as an indicator for the effectiveness of substerilizing heat treatments. The assumption is that destruction of peroxidase is accompanied by destruction of all other enzymes of concern.

Peroxidase also appears to be important from the standpoint of nutrition, color, and flavor. For example, peroxidase activity can lead to the oxidative destruction of vitamin C (13). Also, peroxidase has been shown to catalyze the bleaching of carotenoids in the absence of unsaturated fatty acids (7) and the decoloration of anthocyanins (68). Peroxidase, like most heme pigments, catalyzes the nonenzymic, peroxidative degradation of unsaturated fatty acids, yielding volatile and flavorful carbonyl compounds that contribute to oxidized flavor. Apparently heme pigments, such as those in peroxidase, catalyze the decomposition of hydroperoxides as shown in Fig. 28 (101) with the generation of free radicals. Resultant free radicals could cause the destruction of a wide variety of food components. Lactoperoxidase (LP) may play a particularly important role in the peroxidation of milk lipids and as a component of an antibacterial system in milk.

Xanthine oxidase (XOD) is thought to interact with LP in milk to yield strong oxidizing agents that oxidize lipids and kill microorganisms. These enzymes occur in high concentrations in milk, XOD mostly associated with the membrane surrounding the fat globule and LP present in the aqueous phase (58). Although there are virtually no natural substrates for XOD in milk, the enzyme possesses a rather broad substrate specificity and may utilize aldehydes of differing structures that could arise in milk from various sources. Oxygen, as the electron acceptor, is converted to superoxide and hydrogen peroxide, both of which may act as lipid prooxidants. LP is apparently a strong prooxidant for lipids in milk since it is a ferriheme protein and since it can use H_2O_2 (from XOD) to catalyze various oxidations. Indeed, combined, cyclic reactions of XOD coupled with LP have been proposed as the basis for an antibacterial system in milk and as a source of pro-oxidants for milk lipids (76).

M. Ascorbic Acid Oxidase (13,50)

Ascorbic acid oxidase (AAO) is a copper-containing enzyme that catalyzes oxidation of vitamin C (ascorbic acid), as follows (next page).

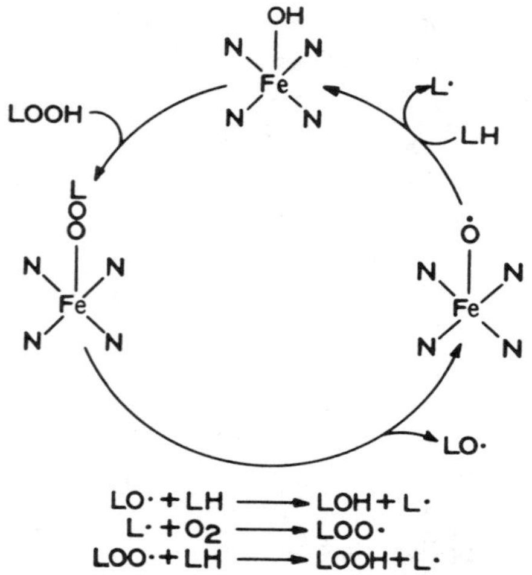

$$LO\cdot + LH \longrightarrow LOH + L\cdot$$
$$L\cdot + O_2 \longrightarrow LOO\cdot$$
$$LOO\cdot + LH \longrightarrow LOOH + L\cdot$$

Fig. 28 Mechanism of hematin-catalyzed peroxidation of unsaturated lipids. (Reprinted from Ref. 167, by courtesy of AVI Publishing Co.)

$$\text{L-Ascorbic acid} + \tfrac{1}{2}O_2 \longrightarrow \text{dehydroascorbic acid} + H_2O \tag{87}$$

Contrary to a number of oxidase enzymes, one product of the reaction is water instead of hydrogen peroxide. However, in the presence of atmospheric oxygen and no enzyme, ascorbic acid is oxidized to dehydroascorbic acid and hydrogen peroxide. This reaction is catalyzed by copper ions, and the resultant hydrogen peroxide leads to further destruction of ascorbic acid.

Ascorbic acid oxidase occurs in all *Cucumis* species (e.g., squash and pumpkin), in seeds, grains, and certain fruits (144). The enzymic oxidation of ascorbic acid is important in the citrus industry. It is present in oranges and other citrus fruits. In the intact fruits the oxidases are presumably balanced by the reductases and the interaction of these two systems determines the final level of ascorbic acid. However, during extraction of juices, the reductases suffer the greatest damage, leaving the oxidases to destroy the ascorbic acid. Thus, it becomes important to limit the action of AAO by holding fresh juices for minimal times and at low temperatures during the blending stage, by deaerating the juice to remove oxygen and finally by pasteurizing the juice to inactivate the oxidizing enzymes.

Enzymic oxidation also has been proposed as a mechanism for the destruction of ascorbic acid in orange peels during the preparation of marmalade. Boiling the grated peel in water substantially reduces the loss of ascorbic acid.

The improving effect of ascorbic acid on bread dough may proceed via coupled oxidation-reduction enzyme reactions leading to cross-linking of gluten by the formation of disulfide bonds (144).

$$2\text{Ascorbate} + O_2 \xrightarrow{\text{oxidase}} 2\text{dehydroascorbate (DHA)} + H_2O \tag{88}$$

$$2\text{DHA} + \text{gluten (SH)}_2 \xrightarrow{\text{reductase}} \text{gluten S–S gluten} + 2\text{ascorbate} \tag{89}$$

Other oxidases, such as cytochrome oxidase, peroxidase, and phenolase, also catalyze destruction of ascorbic acid.

N. Antioxidant Enzymes

As mentioned, certain oxidoreductases can serve as pro-oxidants in foods by direct or indirect action. At the same time, a few oxidoreductases apparently can also function as antioxidants in biological systems. In general, these latter enzymes serve to inactivate oxygen species that have been activated by previous enzymic activity. However, thermal processing may inactivate these enzymes and actually convert them into nonspecific, metalloprotein pro-oxidants, since most of them are metalloenzymes.

1. Catalase (144)

This well-known enzyme destroys H_2O_2 via the disproportionation reaction

$$2H_2O_2 \longrightarrow {}^3O_2 + 2H_2O \tag{90}$$

The resultant oxygen is not an activated form but is in the ground state.

2. Superoxide Dismutase (SOD) (114)

As the name implies, SOD catalyzes the dismutation of superoxide:

$$O_2^{\cdot} + HO_2^{\cdot} \xrightarrow{\text{H}^+} {}^3O_2 + H_2O_2 \tag{91}$$

Although the product oxygen is in the ground state, this enzyme does generate potentially damaging H_2O_2. It has been patented as an antioxidant for foods (112,113), but should be more effective when catalase is present to destroy H_2O_2. Indeed, the cooperative antioxidant effect of these two enzymes can be observed when both are added to a linoleate Fe^{2+} system (175).

The enzyme is ubiquitous and occurs at low levels in most tissues. Depending upon the source, SOD contains Fe, Mn, or Cu and Zn in the active site. Certain low-molecular-weight copper complexes possess substantial SOD activity (99).

3. Glutathione Peroxidase (169)

This selenoenzyme catalyzes the reduction of peroxides by reduced glutathione (GSH):

$$2GSH + ROOH \longrightarrow ROH + H_2O + GSSG \tag{92}$$

Although of apparent importance for destruction of peroxides in vivo, the significance of this enzyme as an antioxidant in foods is unclear. Presumably, depletion of GSH has the effect of inactivating this enzyme.

4. Sulfhydryl Oxidase

This enzyme is bound to membranes in milk and exists as hydrophobic aggregates (161). It catalyzes the oxidation of thiols to yield disulfides and H_2O_2:

$$2R-SH + O_2 \longrightarrow R-S-S-R + H_2O_2 \tag{93}$$

Resultant H_2O_2 can be pro-oxidant; however, it apparently can be utilized by lactoperoxidase in a non-pro-oxidant fashion (162), with this enzymic combination actually exerting a possible antioxidant effect in milk.

The aforementioned antioxidant enzymes, if sufficiently active in foods, may provide protection against oxidative changes and undesirable alterations in color, flavor, and texture.

O. Flavor Enzymes

The development of flavor compounds, particularly in fruits and vegetables, has been linked to enzymes. Enzyme involvement may be direct, by the conversion of flavor precursors to flavors, or indirect, by the generation of oxidizing agents or flavor precursors (138). Flavor-forming mechanisms, including those involving enzymes, are discussed in more detail in Chap. 9.

The Flavorese approach was discussed earlier in Sec. X (see also Chap. 9).

P. Pigment-Degrading Enzymes

Chlorophyllases and anthocyanases both exist in plants and can catalyze destruction of chlorophyll and anthocyanins, respectively, if they are not inactivated in harvested plants (26,149).

XIV. ENZYME INHIBITORS (12,28,52,89,95,100,132,133,157,174)

A variety of naturally occurring inhibitors of hydrolytic enzymes are elaborated by certain animal and plant tissues and by microorganisms. The most studied group of inhibitors exhibits antiproteinase activity. The multitude of proteinase inhibitors have been grouped into families depending upon their biochemical characteristics. Inhibitors for the four mechanistic groups of proteinases (thiol-, serine-, carboxyl-, or metalloproteinases) (180) have been observed; however, those that inhibit serine proteases, such as trypsin, have been examined in most detail (95). For the most part, these inhibitors are proteinaceous in nature and are thought to behave similarly to transition state analog inhibitors (TSAI), in which the trigonal carbonyl carbon of the substrate undergoing attack assumes some tetrahedral character (95,144). In the catalytic cycle, the substrate in its transition state binds much more tightly to the enzyme than does the substrate in its ground state. Therefore, substances with structures that resemble or are analogs of the transition state substrate should be much better inhibitors than analogs of the ground state substrate. Generally, the K_i for a TSAI is much smaller than the K_m of the enzyme for the substrate.

The physiological significance of naturally occurring enzyme inhibitors is unclear, but in general, they probably prevent unwanted enzymic activity (95). Although these endogenous inhibitors may play a role in controlling enzymic activities in foodstuffs, certain of them are thought to present nutritional and safety problems for the consumer.

A. Inhibitors in Plants (12,89)

The principal enzyme inhibitors from plants are those that inhibit mammalian proteolytic enzymes. Of all the inhibitors, soybean trypsin inhibitors have been studied the most. However, inhibitors for pancreatic amylase and lipase have also been isolated from legumes, thus completing the list of polypeptides that inhibit digestion of the three major food components (144).

Proteinase inhibitors are found in virtually all the Leguminosae, in grains, such as wheat, rice, and corn, and in potatoes and beet roots. Because of the importance of legumes, cereals, and potatoes in world nutrition, much research has related to the nutritional significance and health implications of proteinase inhibitors.

The first enzyme inhibitor discovered in plants was a protein in soybeans. Since then, approximately eight proteinaceous proteinase inhibitors have been observed in soybeans. However, the inhibitors that have received the greatest attention are the Kunitz inhibitor and the Bowman-Birk inhibitor. The purified Kunitz inhibitor retains its activity from pH 1 to 12 at temperatures below 30°C. It is reversibly denatured by brief heating at 80°C and is irreversibly denatured by heating at 90°C. Trypsins of various species, including the human, are inhibited by the Kunitz inhibitor.

The Bowman-Birk inhibitor appears to be more stable to denaturation than the Kunitz inhibitor. There is no loss of activity when it is heated in the dry state at 105°C or in 0.02% aqueous solution for 10 min at 100°C. Autoclaving for 20 min at 121°C destroys its activity. The inhibitor is stable to acid (pH 1.5, 2 hr, 37°C) and to peptic and pronase digestion. It inhibits the proteolytic and esterolytic activities of trypsin and chymotrypsin.

Reducing agents such as cysteine facilitate heat inactivation of both purified soybean Kunitz inhibitor and other trypsin inhibitors in soybean extracts (97). Presumably the thiol group of cysteine undergoes thiol-disulfide exchange reactions with the inhibitor(s) causing disruption of the inhibitor structure and inactivation. The fact that only moderate heat treatments are needed to inactivate inhibitors in the presence of cysteine may be useful in designing purification processes that better retain the native functional properties of soy proteins.

1.   Distribution of Inhibitors in Plant Tissues

Most proteinase inhibitors have been found in the seeds of various plants; however, they are not restricted to this part of the plant. For example, in sweet potato, a trypsin inhibitor is found in the tuber as well as in the leaves.

In the white potato, the tuber and the young leaves are rich in a chymotrypsin inhibitor. However, in the field bean, trypsin inhibitors are distributed throughout all parts of the germinating seed and the growing plant, with the levels depending on the stage of growth. Apparently, trypsin inhibitors are absent from leaves, stems, and pods of soybean. This appears to be true as well for the navy bean. Within the cotyledons of soybean, chick pea, and kidney bean, trypsin inhibitors seem to be concentrated in the outer part of the cotyledon mass. In wheat, a trypsin inhibitor is found in the cotyledon as well as in the germ. In addition, a papain inhibitor is present in the germ of soybean, haricot, broad bean, and the garden pea.

2.   Nutritional and Safety Significance of Proteinase Inhibitors

As early as 1917, Osborne and Mendel observed that unless soybeans are cooked for several hours, they do not support the growth of rats, an observation since extended to a variety of experimental animals. Supplementation of unheated soybean meal with methionine (the most limiting amino acid in soybeans) or cystine improves protein utilization to essentially the same extent as proper heating. As expected, heated soybean meal plus methionine have even greater nutritive value than unheated soybean plus methionine. It also has been observed that feeding proteins containing lysine, arginine, isoleucine, or tryptophan at limiting levels in the presence of raw soybean meal results in deficiencies of these amino acids. Thus, it would appear that these inhibitors generally interfere with protein digestion.

On the other hand, there is evidence indicating that the growth-retarding effect of trypsin inhibitor may have little to do with inhibition of protein digestion in the intestines, at least in rodents. For example, raw soybeans or trypsin inhibitor do not depress proteolytic activity in the intestinal tract of rats and mice. Furthermore, active antitryptic preparations have been shown to inhibit the growth of rodents when incorporated into diets containing predigested proteins or free amino acids. How, then, does trypsin inhibitor retard the growth of rats without affecting intestinal proteolysis, and how does it increase the requirement for methionine?

The feeding of raw soybeans to chicks and rats also results in pancreatic hypertrophy. Thus, trypsin inhibitor may cause growth depression by stimulating pancreatic secretions, thereby resulting in excessive endogenous losses of essential amino acids. It has also been postulated that soybean trypsin inhibitors are somehow directly involved in the metabolism of methionine and might, through this means, be responsible for the apparent methionine deficiency.

The detrimental effects of raw soy flour on the pancreas of rats is thought to result from the strong affinity of soybean trypsin inhibitor(s) for rat trypsin, thereby preventing feedback inhibition of trypsin biosynthesis in the pancreas. This stimulates excessive biosynthesis of trypsin resulting in pancreatic hypertrophy. Pancreatic hypertrophy in rats may promote pancreatic cancer. Indeed, rats fed raw soy flour have developed pancreatic cancer in a significant number of cases (55). Because soybean trypsin inhibitors have little affinity for human trypsins, it is thought that they pose little threat to humans (55).

Much of the detrimental effects of feeding raw soybean preparations have been observed in rats. However, it is important to remember that there are marked species differences in response to soybean products in the diet (55,159). For example, the pancreatic hypertrophy observed in rats and chicks exposed to soybean trypsin inhibitor(s) is not evident in dogs, calves, pigs, and monkeys (55,159). Also, raw soybean flour depresses growth rate by 60% in rats and 85% in pigs, but not at all in monkeys. In general, monkeys are least affected by raw soy flour in the diet, suggesting that primates may not be as susceptible to the detrimental effects of raw soybean products as are lower animals.

Other growth inhibitors are also present in raw soybeans, for example, it has been estimated that soybean trypsin inhibitor in rat diets accounts for only 30 and 60% of the decrease in the growth rate and protein efficiency, respectively.

When heated soy flour is fed, it is comparable to feeding casein in rats, pigs, and monkeys, and it produces no deleterious effects (159). Thus, proper heating of soybean products should readily eliminate any potential safety problems in their consumption by humans or animals.

3. Effect of Heat on Trypsin Inhibitor

Most of the plant proteinase inhibitors are destroyed by heat, with a concomitant increase, in some species, in the nutritive value of the protein. Simple cooking is often sufficient to inactivate trypsin inhibitors.

The destruction of trypsin inhibitors is a function of temperature, duration of heating, particle size, and moisture conditions, with high temperatures, long heating, reduced particle size, and increased moisture levels leading to greater destruction of trypsin inhibitors. However, excessive heating has a detrimental effect on the nutritional quality of the major proteins, so a balance must be struck to optimize the heat treatment.

B. Inhibitors in Animal Tissues

Proteolytic enzyme inhibitors have been reported in a variety of body secretions, beef pancreas, blood, and egg whites from various avian species (95).

There are two principal inhibitors of proteinase activity in egg whites: ovomucoid and ovoinhibitor. Purified chicken ovomucoid possesses unusual stability to heat. For example, more than 90% of the activity remains after 30 min at 80°C and pH values between 3 and 7. Ovomucoid complexes porcine, ovine, and bovine trypsins in a 1:1 ratio; however, it has no inhibitory activity on human trypsin. Purified chicken ovoinhibitor reacts with trypsins and chymotrypsins of bovine and avian origin. Ovoinhibitor is quite stable in acid solution, with 93-95% retention of activity after 15 min at 90°C and pH 3 or 5. However, at pH 7 its activity is lost in 15 min at 90°C.

Apparently the proteinase inhibitors of egg white are of limited nutritional significance. Ordinary cooking does not destroy the antitryptic activity of raw egg white. Yet, the feeding of commercial egg white, raw or heated, in the presence of added ovomucoid of known antitrypsin activity has no effect on nitrogen retention by humans. Furthermore, in young rats, the feeding of purified trypsin inhibitor from egg white at a level of 2.5% in a casein diet has no effect on growth or protein efficiency. The dog, however, appears to be sensitive to the antitryptic factors in egg whites.

C. Inhibitors in Microorganisms

The *Streptomyces* spp. are a particularly good source of proteolytic enzyme inhibitors (144,174). Leupeptins inhibit plasmin, trypsin, papain, and cathepsin B; antipain inhibits trypsin, cathepsins A and B, and papain; chymostatins inhibit chymotrypsins; elastatinal inhibits elastase; phosphoramidon inhibits thermolysin; and pepstatin inhibits acid proteinases (174). Of particular interest is pepstatin, an inhibitor of pepsin and, to a lesser extent, of other carboxylproteases, such as chymosin.

Pepstatin is a pentapeptide, isolated from various species of actinomycetes, and it strongly inhibits acid proteinases. The K_i for inhibition of pepsin is 10^{-10} M (136,166). Pepstatin, which contains a unique, central statyl residue involved in interactions with the active sites of acid proteases, has the following structure (106,136):

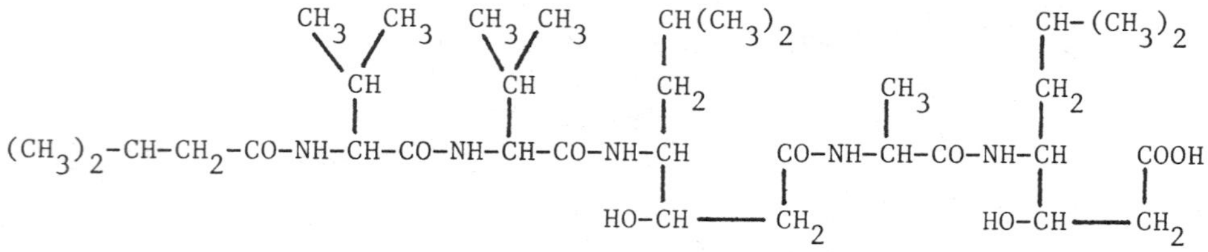

$$
\begin{array}{c}
\text{CH}_3 \ \text{CH}_3 \quad \text{CH}_3 \ \text{CH}_3 \quad \text{CH(CH}_3)_2 \qquad\qquad\qquad \text{CH-(CH}_3)_2 \\
\backslash\diagup \qquad \backslash\diagup \qquad\quad | \qquad\qquad\qquad\qquad\qquad\quad | \\
\text{CH} \qquad\quad \text{CH} \qquad\quad \text{CH}_2 \qquad\quad \text{CH}_3 \qquad \text{CH}_2 \\
| \qquad\qquad | \qquad\qquad | \qquad\qquad | \qquad\quad | \\
(\text{CH}_3)_2-\text{CH}-\text{CH}_2-\text{CO}-\text{NH}-\text{CH}-\text{CO}-\text{NH}-\text{CH}-\text{CO}-\text{NH}-\text{CH} \quad \text{CO}-\text{NH}-\text{CH}-\text{CO}-\text{NH}-\text{CH} \quad \text{COOH} \\
| \qquad\qquad\qquad | \qquad\qquad\qquad\qquad | \qquad\qquad | \\
\text{HO}-\text{CH}---\text{CH}_2 \qquad\qquad \text{HO}-\text{CH}---\text{CH}_2
\end{array}
$$

Several novel features in the structure of pepstatin are relevant to its inhibitory action on pepsin. It is a very hydrophobic peptide that has poor solubility in water. The central statyl residue is thought to combine with the active site of pepsin to mimic the transition state during normal peptic proteolysis. Pepstatin has thus been referred to as a "transition state" inhibitor (166). The central hydroxyl group is essential for assuming a pseudo-transition state and is, therefore, necessary for inhibition of pepsin (106,136, 166). The terminal carboxyl group is not required for inhibitor activity. Under comparable conditions, pepstatin inhibits chymosin 48% of that observed for pepsin, suggesting a certain structural specificity for inhibition.

D. Physiological Significance of Enzyme Inhibitors

1. Ovomucoid and Ovoinhibitor

None of the inhibitors in egg white have been shown to inhibit naturally occurring proteolytic enzymes in egg white. Possibly, they serve some function during embryological development, or they serve an antimicrobial or antiviral function.

2. Plant Enzyme Inhibitors

The inhibition of enzymes by endogenous proteins may represent a regulatory mechanism in plants. For example, potato invertase inhibitor was one of the first inhibitors discovered to act on endogenous enzymes and is one of the most extensively studied. The invertase system is also of economic importance because of the relationship between sugar accumulation in stored tubers and processing quality. The concentration of the invertase inhibitor increases during tuber growth and is in excess over invertase so that invertase activity remains undetectable. Thus, mature tubers are known to contain a low level of invertase relative to a large excess of invertase inhibitor. During curing and storage at a warm temperature, the total amount of invertase increases but excess inhibitor persists. However, when the potatoes are placed in cold storage (below 4°C), the amount of invertase increases dramatically and soon exceeds the amount of inhibitor. The active invertase then converts sucrose to reducing sugars, and they rapidly accumulate until a maximum is reached after several weeks. Subsequently, invertase activity decreases and an excess of invertase inhibitor again develops. Changes in the amount of invertase, invertase inhibitor, and reducing sugars are reversible as the tubers are subjected to alternate cold and warm storage.

3. Ripening of Fruits

Polygalacturonase, which is involved in the degradation of pectin in ripening fruits, is inhibited by a polygalacturonase inhibitor. Thus, degradation of the polygalacturonase inhibitor may trigger polygalacturonase activity in fruits, such as avocados, during ripening.

Selection of plant varieties based on the amount of enzyme inhibitors elaborated may lead to improved storage properties and enable better control to be exercised over ripening.

4. Amylase Inhibitor(s)

It has been proposed that amylase inhibitors may retard the digestion of starch in the diet to provide a basis for weight control in obese individuals (144). For a brief period in 1982, these so-called starch-blockers were marketed in the United States as a dietary aid. However, their efficacy in controlling weight has been questioned, and the U.S. Food and Drug Administration has moved to block their sale over the counter and has ruled that they must be treated as drugs, thereby requiring approval and proof of efficacy.

E. Enzyme Inhibitors as Research Tools

In some cases, it may be desirable to selectively inhibit certain proteolytic (or other enzymic) activity in a complex food product or reaction mixture. Inhibitors might be used, for example, to prevent further proteolysis or to regulate the balance of enzyme activities in a system containing several active enzymes. As more enzyme inhibitors become commercially available, the food scientist should exploit their activities to better understand the effects of enzymes on food systems.

XV. RECOMBINANT DNA TECHNOLOGY AND GENETIC ENGINEERING

In the last decade there has been a veritable explosion in the methods for manipulating the nucleic acids within microbial cells to produce useful foreign proteins, such as animal-derived food enzymes. Unfortunately, most of the information has been developed using *Escherichia coli*, anathema to the food scientist and to the producers of food-grade enzymes since it is considered an indicator organism for fecal contamination or unsanitary practices. Although early research in recombinant DNA technology involved primarily the gram-negative, prokaryote *E. coli*, substantial progress is being made in manipulating the genetic material of gram-positive prokaryotes such as *Bacillus subtilis, Streptococcus lactis*, and *Streptococcus cremoris*. Since the more complex eukaryotic cells (e.g., yeasts and plant and animal cells) process their nucleic acids differently than the simpler prokaryotic microorganisms, it has only been recently that these cells have been engineered to express tailored products through their genetic apparatus. Nonetheless, progress is being made in the use of recombinanat DNA techniques and genetic engineering to provide microbial "strains" and metabolic products of potential use to the food industry. The potential is great for this technology to have a heavy impact on the food industry, both in production of raw materials and in their processing.

Briefly discussed below are some of the principles of recombinant DNA technology. For a more detailed treatment, the reader is referred to the monograph by Old and Primrose (123).

A. Cloning Strategies

A number of strategies are available for cloning foreign DNA into a microorganism. The basic cloning strategy is outlined in Fig. 29 and is described in more detail in the following discussion. In general, the steps in a cloning scheme are

1.   Obtaining desired DNA fragments
2.   Inserting the DNA fragment into a vector (a larger piece of DNA) in which the fragment is stable to biochemical alterations
3.   Introducing the fragment-containing vector into a host cell
4.   Identifying those cells containing the DNA fragment

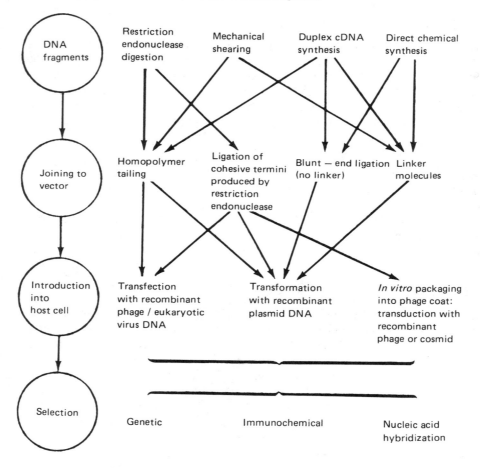

Fig. 29 Schemes for cloning DNA. The arrows indicate the procedures that seem to be of general use currently. (Reprinted from Ref. 123, courtesy of University of California Press.)

1. Obtaining DNA Fragments

The genetic information required for the biosynthesis of specific amino acid sequences is carried in DNA. Therefore, a source of the desired DNA segment or a means of generating a clonable DNA fragment is essential. Fragments can be obtained in a variety of ways (Fig. 29): (a) cleavage of genomic DNA (donor genes) from appropriate cells, (b) enzymic synthesis using isolated messenger RNA (mRNA) as a template or (c) synthesis of DNA fragments using organic chemistry.

Genomic DNA fragments can best be cloned between simple (prokaryotic) microorganisms in which expression of the DNA segments is not complicated by intervening DNA sequences (introns) found in DNA from more complex (eukaryotic) organisms. If the genomic DNA used is from eukaryotes, mRNA transcribed from the gene (Fig. 30) contains intervening sequences that interrupt the structural (coding) region. These sequences must be excised from transcribed mRNA before translation can occur. Since simpler procaryotic genes lack introns, it is unlikely that these organisms contain the necessary splicing enzymes to remove intervening sequences from transcripts of eukaryotic genes. If expression is desired, cloning genetic information from eukaryotic cells into prokaryotes requires that DNA be derived from eukaryotic mRNA, in which the undesired sequences have already been removed by the donor cell.

RNA-directed, DNA polymerase, isolated from oncornoviruses (171), mediates the synthesis of DNA from mRNA. Reverse transcriptase (RTase), the common name for this enzyme, produces a cDNA without introns, obviating potential expression problems in the new host. Reverse transcriptase also directs the synthesis of a second strand of DNA, combined head to tail (3' to 5' ends) with the original cDNA, to yield a double-stranded cDNA (ds-cDNA).

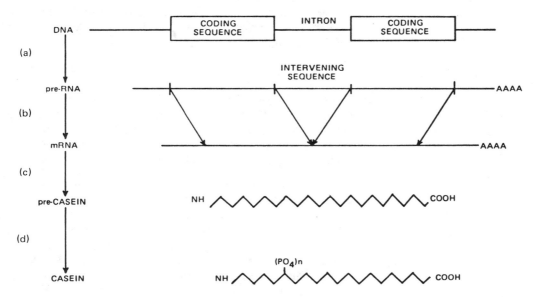

Fig. 30 Biosynthesis of casein as a general example of gene expression in eukaryotic cells. RNA is transcribed from DNA (a) and intervening sequences are removed (b). The resulting polyadenylated mRNA encodes a precasein (c) modified to produce a secreted casein (d).

DNA segments have been synthesized chemically to produce synthetic genes (37, 82). The nucleic acid sequence of these specific genes is based on the amino acid sequence of the desired polypeptides. Using the "triplet code" for each amino acid, the primary amino acid sequence for each polypeptide is converted to the corresponding nucleic acid sequence. Appropriate DNA fragments are then inserted into microorganisms using recombinant DNA techniques and used to direct the production of desired polypeptides by the modified microorganism.

2. Insertion of DNA into a Vector

Before exogenous DNA is introduced into a cell, it must be inserted into a vector or cloning vehicle. A vector is a self-replicating, high-molecular-weight DNA sequence that can be absorbed and retained by the host cell and its progeny. Vectors may contain special nucleotide sequences that promote their integration into the genome of the host cell, or genes that confer unique phenotypes in the host. The function of a cloning vector is to facilitate the introduction of a desired foreign DNA fragment into a host cell and to retain that fragment unchanged when the cell proliferates. Common vectors are microbial plasmids or phage DNA.

3. Introduction of DNA into Host Cells

Different vectors are introduced into cells in different ways. Plasmids are absorbed by "competent" bacterial cells in a process known as transformation. These cells are made competent by pretreatment with Ca^{2+} (121). Viral DNA vectors (phages) are introduced into cells via transfection or transduction depending on whether or not the DNA is packaged into a phage coat. Naked DNA is absorbed into component cells (transfection) in a manner similar to that used in transformation of cells with plasmids (104). On the other hand, packaging of recombinant DNA into a phage protein coat allows its introduction into the host cell through the normal processes of phage infection (transduction), that is, phage adsorption followed by injection of DNA (77,78).

4. Identification of Clones

Immunochemical Techniques. The final step in the general cloning process (Fig. 30) is a selection of clones (i.e., the identification of cells containing the desired DNA fragment). Immunochemical methods are appropriate only when the cloned DNA fragment expresses a gene product. Colonies expressing the gene product are identified using radiolabeled double-antibody techniques and autoradiography (17).

Hybridization to Cloned DNA. Hybridization techniques are used to detect recombinant DNA fragments without requisite expression. This procedure employs the complementary base pairing and duplex formation between homologous strands of DNA or RNA. Radiolabeled DNA or RNA, with base sequences complementary to the cloned DNA being sought, is used as a probe. Bacterial plasmids containing the desired DNA insert bind the probe and are observed as a dark spot on an autoradiograph The advantage of this method is its generality. It does not require gene expression and can be applied to any cloned sequence as long as a suitable radioactive probe is available. The procedure can be used to determine which colony among thousands contains a desired cloned sequence (69).

Hybrid-Arrested Translation (HART). This technique selects for cloned pieces of DNA by correlating their presence with the protein for which they code. The method relies on the principle that mRNA does not translate in a cell-free system when exogenous, complementary DNA is bound to it. In general, actively translating mRNA is treated with cloned DNA fragments under conditions that favor DNA-RNA binding. DNA that inhibits mRNA-directed synthesis of a specific protein is considered to code for that protein and represents a successful clone (127).

5. Restriction Endonucleases

Present recombinant DNA technology is totally dependent on the cleavage of DNA molecules at specific sites. Until 1970, and the discovery of a restriction endonuclease in *Haemophilus influenzae* (91), there were no methods available by which duplex DNA molecules could be cut into discrete fragments. Subsequent discovery of numerous restriction enzymes with varying specificities has allowed the controlled cleavage and joining of the DNA molecules necessary in recombinant DNA technology.

6. Expression

In order for the DNA inserted into the vector to be expressed as desired product by the modified microorganisms, so-called promoter-initiation DNA sequences must also be inserted to precede the DNA coding for product. Likewise, DNA termination sequences code for completion of polypeptide synthesis. A wide variety of expression vectors containing strong DNA promoter sequences have been constructed, particularly for use in *Escherichia coli*.

B. Host Microorganisms

Escherichia coli is the most common bacterial host for recombinant DNA fragments. The use of *E. coli* as a host resulted from the isolation of mutant strains, K12, unable to restrict and degrade foreign DNA (190). Many strains of *E. coli* lacking restriction enzymes have since been described (151). Most *E. coli* strains used as cloning hosts are deficient in the metabolic pathway of DNA recombination. Use of these recA⁻ strains minimizes the probability of inadvertent recombination events that would impare the integrity of cloned DNA. *Escherichia coli* can produce potent endo- and exotoxins. This complicates the use of this bacterium as a final host for the production of food-related products.

Bacillus subtilis is generally recognized as safe (GRAS) in food by the U.S. FDA. This organism is well characterized genetically and biochemically, and it is also capable of excreting extracellular proteins in large amounts. For these reasons *B. subtilis* is widely used in the fermentation industry.

Unfortunately, plasmids found in *B. subtilis* are cryptic (they govern no known host function) and are useless as cloning vectors. The difficulty in using *B. subtilis* is its structural dissimilarily to *E. coli*. As a gram-positive bacterium *B. subtilis* does not retain a plasmid originating from gram-negative bacteria. Therefore, special vectors for the gram-positive bacteria were developed. *Staphylococcus aureus* plasmids, with genes for antibiotic resistance, have been transformed into *B. subtilis*. These plasmids, although they meet most requirements for useful recombinant vectors, are nonamplifiable and cannot be used in *E. coli*. "Bifunctional" plasmids have been constructed to function in both *E. coli* and *B. subtilis* (41,42). However, these plasmids have not completely solved the problem of cloning in *B. subtilis* (64,90). The instability of recombinant plasmids (70) and the lack of expression (42,93) still limit the general use of *B. subtilis* as a host for recombinant DNA technology.

The lack of expression resides in the gene-to-phenotype biochemical pathways. The recognition of different promotor sequences of a gene from *E. coli* by RNA polymerase in *B. subtilis* was demonstrated only when the gene was integrated into the *B. subtilis* chromosome (142). Foreign genes residing in plasmids are not expressed in *B. subtilis* when common *E. coli* promotor sequences are used. However, *B. subtilis* expresses cloned genes linked to *B. subtilis* promotors (74,186,187). Apparently, foreign gene expression in *B. subtilis* requires insertion of *Bacillus* promotor sequences (or *Bacillus*-like promotors from viruses) to initiate transcription of mRNA. In addition, ribosome binding sites specific to *B. subtilis* must precede the foreign structural gene to initiate translation (186,187).

Saccharomyces cerevisiae—baker's yeast—is widely employed in the food industry. It is safe and well characterized. Advantages of using yeast as a final host for cloned DNA are (a) they are eukaryotes and may be able to recognize and delete intervening sequences from eukaryotic mRNA (63), (b) transcription events are similar between yeast and higher eukaryotes, and (c) initiation of translation in yeast is also similar to that of higher eukaryotes. However, some investigators question the ability of yeast to express genes that contain intervening sequences (3). Successful cloning and expression in yeast of human interferon (3) and ovalbumin (63) have been reported. Transcription of the intron-containing, rabbit γ-globin gene in yeast has been achieved. However, the mRNA transcribed containing an intervening sequence (6).

C. Cloning in Plants (123)

Cloning into prokaryotic microorganisms, such as *E. coli*, has become routine in hundreds of laboratories. However, the introduction of foreign DNA into plants is far less advanced primarily because of a lack of suitable vectors. Currently two potential vector systems are under investigation: the DNA from plant viruses and the Ti (tumor-inducing) plasmid from *Agrobacterium tumefaciens*.

D. Applications of Recombinant DNA Techniques

Recombinant DNA technology provides a way to improve microbial and plant processes. No longer are geneticists confined to selection of naturally occurring microorganisms. Induced mutations and genetic rearrangements have resulted in new fermentation and protein-production processes and have increased the efficiency of existing processes.

Recombinant DNA techniques can be used by the food scientists in four principal areas: (a) to study structure-function relationships in proteins and enzymes, (b) to convert inedible biomass to edible biomass, (c) to reprogram metabolic pathways in fermentation organisms, and (d) to produce novel food proteins and peptides.

1. Basic Research

Recombinant DNA techniques can be used in research to isolate genes or specific nuclei acid sequences. This allows study of gene expression. Resulting information might be of value in controlling rates of protein synthesis in the production of specific food proteins.

By altering single nucleic acid bases or segments of bases within the gene (82) one can gain an understanding of how the amino acid sequence of food proteins can affect functionality. More recently, development of synthetic oligonucleotide mutagenesis offers the possibility of systematic alterations in the primary sequences of food proteins for structure-function studies (31). By further understanding structure-function relationships in food proteins, one might ultimately design proteins and enzymes that function better in food systems.

2. Conversion of Biomass

Reprogramming fermentation organisms to more efficiently use existing substrates is of considerable importance. Recombinant DNA techniques were used to improve the metabolic efficiency of *Methylophilus methylotropus* (188). Amylase-expressing genes in yeasts would obviate the need for simple carbohydrates in yeast fermentation media. As fermentation techniques replace chemical synthesis of organic acids and amino acids, recombinant DNA methods can be used to improve production efficiencies.

Reprogramming lactose-fermenting bacteria (*S. lactis* and *S. cremoris*) in milk may yield single strains that can metabolize lactose, hydrolyze casein, produce a cultured milk flavor, and resist phage attack (92). Strains of milk-fermenting bacteria that lyse more rapidly are being designed to accelerate the aging of cheese. Thus, through isolation and recombination of genes coding for many fermentation-related enzymes (from bacterial and other sources) it should be possible to tailor make bacterial strains for any milk-fermenting process.

3. New Proteins from Old Sources

Recombinant DNA technology is best recognized as a means of transplanting genes from one species to another. Successful cloning of genes coding for mammalian interferon, insulin, and growth hormones into bacteria has added a new dimension to research and development in the pharmaceutical industry.

In food-related applications of recombinant DNA technology, genes coding for prochymosin in calves were transferred into *E. coli* with successful expression (44). Genes coding for ovalbumin in chickens have been cloned and are being expressed in *E. coli* (59) and in *S. cerevisiae* (63). The gene coding for phaseolin, the major storage protein in the seed of the French bean, *Phaseolis vulgaris* L., has been cloned and inserted into the sunflower plant (118). This is the first successful insertion and transcription of DNA from one plant into another. The implications of placing foreign DNA into plants are enormous for food scientists. The potential now exists to develop and produce unique functional proteins as commodities. Production of commodity proteins by fermentation is precluded because the yields are too low and the process is too expensive. However, modification or replacement of soy proteins in the soybean, to yield a more functional, more nutritious protein, would greatly benefit the food industry. Functional proteins, like ovalbumin, individual caseins, or casein peptides, each with their own specific functional quality, may eventually be available to food manufacturers.

E. Perspective

Because much of the applied recombinant DNA research occurs in private industry, it is difficult to assess progress in food related applications. However, current areas of research using recombinant DNA techniques include production of amino acids, sweeteners, organic acids, nutritional proteins, flavors, enzymes, and functional proteins. Bethesda Research Labs (BRL) has recently cloned into *E. coli* three genes involved in proline synthesis. Moreover, patents are being sought on genetically engineered organisms that produce increased yields of three amino acids (Aginomoto, Japan). Genes coding for poly-(L-aspartyl-L-phenylalanine), the precursor for aspartame, have been synthesized and cloned (with expression) into bacteria (37). Attempts to clone the protein, thaumatin, an extremely effective sweetener found in plants, are currently being pursued. Addition of polymers of essential amino acids to nutritionally incomplete proteins, through recombinant DNA technology, has been proposed. Genes coding for enzyme systems that produce flavor compounds from lignin substrates are being studied for use in the production of flavors. In addition to the cloned prochymosin gene, attempts are in progress to remove genes for enzymes like pullulanase, cellulase, and glucose isomerase from various undesirable hosts and place them in more desirable organisms.

As recombinant DNA technology matures, it becomes incumbent upon food scientists to conduct both fundamental and applied research on genetic engineering so that food producers, processors, and consumers ultimately benefit. Heretofore, the food industry has been obliged to work with foodstuffs provided by the agricultural industries with little control over the functional qualities, shelf stability, or sensory properties of the raw materials. Food scientists are now in the position to design functional and nutritional proteins and peptides, to use enzymes previously unavilable because of safety considerations, and to engineer enzymes or enzyme systems in microorganisms.

Like any developing technology, genetic engineering is accompanied by many potential problems. The ability to solve or avoid these problems will play an important role in deciding whether these techniques will be used by the food industry.

The U. S. Food and Drug Administration will determine the acceptibility in the United states of foodstuffs resulting from recombinant DNA technology. To date there is no precedent for a decision regarding products from food organisms that have been altered using recombinant DNA techniques. Recently, however, human insulin (Humelin) obtained from microorganisms was approved by the U.S. FDA for use in humans. The rapid approval (within 5 months) of Humelin should encourage continued research into the possibilities of producing modified proteins for the food industry. However, it is reasonable to assume that exhaustive and expensive testing will be required before the U.S. FDA considers any protein resulting from recombinant DNA technology to be suitable for general use in foods. Additional problems surrounding recombinant DNA technology are as follows.

1. Cells can remove cloned genes by the same methods used to insert them into the cell. Selective pressures must be placed on the hosts of recombinant DNA to retain the cloned gene(s).
2. The quality of cloned proteins produced by microorganisms may be affected adversely by the host. Many microorganisms (e.g., *B. subtilis*) secrete proteases and metabolites that contaminate the desired product. This also affects the ease with which the product can be isolated from the culture medium (or plant).
3. Production of foreign proteins may be less efficient than desired because of the metabolic burden placed on the host cells.
4. The product from a cloned gene may be too expensive to produce. Substrate considerations, optimum growth conditions, ease of product recovery, and yields all determine the ultimate production cost of the desired products.
5. The host cell may be unable to make the necessary posttranslational modifications needed to produce an end product with the desired properties. Some examples of important posttranslational modifications of proteins are phosphorylation and glycosylation of caseins and chicken ovalbumin.

These are general problems associated with the use of recombinant DNA technology in the food industry. Each separate case will, no doubt, present unique problems.

REFERENCES

1. Acker, L. W. (1969). Water activity and enzyme activity. *Food Technol 23*:1257-1270.
2. Acker, L., and H. Kaiser (1959). Über den Einflub der Feuchtigkeit auf den Ablauf enzymatischer Reaktionen in wasserarmen Lebensmitteln. *Z. Lebensm. Untersuch. Forsch. 110*:349-356.
3. Ammerer, G., R. Hitzeman, F. Hagie, A. Barta, and B. D. Hall (1981). Functional expression of mammalian genes in yeast, in *Recombinant DNA*, Proceedings of the Third Cleveland Symposium on Macromolecues (A. G. Walton, ed), Elsevier Scientific, Amsterdam, pp. 185-197.
4. Axelrod, B. (1974). Lipoxygenases, in *Food-Related Enzymes*, Adv. Chem. Series 136 (J. R. Whitaker, ed.), Am. Chem. Soc., Washington, D.C., pp. 324-348.
5. Baumann, J. W. (1981). Applications of enzymes in fruit juice technology, in *Enzymes and Food Processing* (G. G. Birch, N. Blakebrough, and K. J. Carpenter, eds), Applied Science Publishers, London, pp. 129-148.
6. Beggs, J. D., J. van den Berg, A. van Oayen, and C. Weissman (1980). Abnormal expression of chromosomal rabbit β-globin gene in *Saccharomyces cerevisiae*. *Nature 283*:835-840.
7. Ben-Aziz, A., S. Grossman, I. Ascarelli, and P. Budouiski (1971). Carotene bleaching activities of lipoxygenase and heme proteins as studied by a direct spectrophotometric method. *Phytochemistry 10*:1445-1452.
8. Benzonana, G., and P. Desnuelle (1965). Etude anetique de l'action de la lipase pancreatique sur des triglycerides en emulsion. Essai d'une enzymologie en milieu heterogene. *Biochim. Biophys. Acta 105*:121-136.
9. Bergmeyer, H. U. (1974). *Methods of Enzymatic Analysis*, 4 vols., 2nd ed., Verlag Chemie, New York.
10. Bergmeyer, H. U. (1978). *Principles of Enzymatic Analysis*, Verlag Chemie, New York.
11. Birch, G. G., N. Blakebrough, and K. J. Parker (eds) (1981). *Enzymes and Food Processing*, Applied Science Publishing, London.
12. Birk, Y. (1976). Proteinase inhibitors from plant sources, in *Methods in Enzymology* (L. Lorand, ed.), Part B, vol. 45, Academic Press, New York, pp. 695-739.
13. Blundstone, H. A. W., J. S. Woodman, and J. B. Adams (1971). A. Changes in vitamin C, in *The Biochemistry of Fruits and Their Products*, vol. 2 (A. C. Hulme, ed.), Academic Press, New York, p. 561.
14. Bolivar, F., R. L. Rodriguez, P. J. Greene, M. V. Betlach, H. L. Heyneker, H. W. Boyer, J. G. Crosa, and S. Falkow (1977). Construction and characterization of new cloning vehicles. II. A multipurpose cloning system. *Gene 2*:95-133.
15. Bergstrom, G. (1968). *Principles of Food Science*, vol. 2, Macmillan, New York, p. 51.
15a. Brandts, J. F. (1964). The thermodynamics of protein denaturation. I. The denaturation of chymotrypsinogen. *J. Amer. Chem. Soc. 86*:4291-4301.
16. Brandts, J. F. (1969). Conformational transitions of proteins in water and aqueous mixtures, in *Structure and Stability of Biological Macromolecules*, vol. 2 (S. N. Timasheff and G. D. Fasman, eds.), Marcel Dekker, Inc., New York, pp. 213-290.
17. Broome, S., and W. Gilbert (1978). Immunological screening method to detect specific translation products. *Proc. Nat. Acad. Sci. U.S. 75*:2746-2749.
18. Broquist, H. P., and T. H. Jukes (1968). Antimetabolites—effect on Nutrition, in *Modern Nutrition in Health and Disease* (M. G. Wohl and R. S. Goodhardt, eds.), Lea and Febiger, Philadelphia, pp. 450-471.

19. Brown, H. D., and S. K. Challopadhyay (1971). Matrix—supported enzymes, in *Chemistry of the Cell Interface* (H. D. Brown, ed.), Academic Press, New York, pp. 185-252.

20. Bucke, C. (1980). Enzymes in fructose manufacture, in *Enzymes and Food Processing* (G. G. Birch, N. Blakebrough, and K. J. Parker, eds.), Applied Science Publishers, London, pp. 51-72.

21. Camus, M.-C., and J. D. Laparte (1976). Wheat as an in vitro inhibitor of peptic proteolysis. Role of phytic acid in by-products. *Biol. Anim. Biochim. Biophys. 16*:719-730.

22. Cawley, R. W., and T. A. Mitchell (1968). Inhibition of wheat amylase by bran phytic acid. *J. Sci. Food Agr. 19*:166-168.

23. Chance, B. (1943). The kinetics of the enzyme-substrate compound of peroxidase. *J. Biol. Chem. 151*:553-577.

24. Charm, S. E., and S. E. Matteo (1971). Scale up of protein isolation, in *Methods in Enzymology*, vol. 21 (W. B. Jakoby, ed.), Academic Press, New York, pp. 476-556.

25. Chibata, I., and T. Tosa (1980). Immobilized microbial cells and their applications. *Trends Biochem. Sci. 5*:88-90.

26. Chichester, C. O., and R. McFeeters (1971). Pigment degneration during processing and storage, in *Biochemistry of Fruits and Their Products*, Part II, vol. 2 (A. C. Hulme, ed.), Academic Press, New York, pp. 707-719.

27. Christensen, H. N., and G. A. Palmer (1967). *Enzyme Kinetics*, W. B. Saunders, Philadelphia.

28. Couch, J. R., and F. G. Hooper (1972). Antitrypsin factors, in *Newer Methods of Nutritional Biochemistry* (A. Salbanese, ed.), Academic Press, New York, pp. 183-195.

29. Coughlin, R. W., and M. Charles (1980). Applications of lactose and immobilized lactase, in *Immobilized Enzymes for Food Processing* (W. H. Pitcher, Jr., ed.), CRC Press, Inc., Boca Raton, Fla., pp. 153-174.

30. Cuendet, L. S., E. Larson, C. G. Norris, and W. F. Geddes (1954). The influence of moisture content and other factors on the stability of wheat flours at 37 8°C. *Cereal Chem. 31*:362-389.

31. Dalbadie-McFarland, G., L. W. Cohen, A. D. Riggs, C. Morin, K. Itakura, and J. H. Richards (1982). Oligonucleotide directed mutagenesis as a general and powerful method for studies of protein functions. *Proc. Nat. Acad. Sci. U.S. 79*:6409-6413.

32. Daniels, F., and Alberty, R. A. (1966). *Physical Chemistry*, 3rd ed., John Wiley and Sons, New York.

33. Danielsson, B., B. Mattiasson, R. Karlsson, and F. Winquist (1979). Use of an enzyme thermistor in continuous measurements and enzyme reactor control. *Biotechnol. Bioeng. 21*:1749-1766.

34. Danielsson, B., B. Mattiasson, and K. Mosbach (1979). Enzyme thermistor analysis in clinical chemistry and biotechnology. *Pure Appl. Chem. 51*:1443-1457.

35. Dilley, D. R. (1971). *Enzymes*, in *The Biochemistry of Fruits and Their Products*, vol. 1 (A. C. Hulme, ed.), Academic Press, New York, pp. 179-207.

36. Dixon, M., and E. Webb (1980). *Enzymes*, 3rd ed., Academic Press, New York.

37. Doel, M. T., M. Eaton, E. A. Cook, H. Lewis, T. Patel, and N. H. Carey (1980). The expression in *E. coli* of synthetic repeating polymeric genes coding for poly (L-aspartyl-L-phenylalanine). *Nucleic Acids Res. 8*:4575-4592.

38. Dransfield, E., and D. Etherington (1981). Enzymes in tenderization of meat, in *Enzymes and Food Processing* (G. G. Birch, N. Blakebrough, and K. J. Parker, eds.), Applied Science Publishers, London, pp. 177-194.

39. Drissen, F. M. (1983). Lipases and proteinases in milk. Occurrence, heat inactivation and their importance for the keeping quality of milk products. Doctoral Thesis, Agricultural University, Wageningen, The Netherlands.

40. Dugas, H., and C. Penney (1981). *Bioorganic Chemistry: A Chemical Approach to Enzyme Action*, Springer-Verlag, New York.

41. Ehrlich, S. D. (1977). Replication and expression of plasmids from *Staphylococcus aureus* in *Bacillus. subtilis. Proc. Nat. Acad. Sci. U.S.* 74:1680-1682.

42. Ehrlich, S. D. (1978). DNA cloning in *Bacillus subtilis. Proc. Nat. Acad. Sci. U.S.* 75:1433-1436.

43. Elias, P. S., and A. J. Cohen (1977). *Radiation Chemistry of Major Food Components*, Elsevier Scientific, New York.

44. Emtage, J. S., S. Angal, M. T. Doel, T. J. R. Harris, B. Jenkins, G. Lelley, and P. A. Lome (1983). Synthesis of calf prochymosin (pro-rennin) in *Escherichia coli. Proc. Nat. Acad. Sci. U.S.* 80:3671-3675.

45. Endo, A. (1964). Studies on pectolytic enzymes of molds. Part X. Purification and properties of endo-polygalacturonase. III. *Agr. Biol. Chem.* (Tokyo) 28;551-558.

46. Engel, P. C. (1981). *Enzyme Kinetics—The Steady State Approach*, 2nd ed., Chapmand and Hall, New York.

47. *Enzyme Nomenclature* (1979). Nomenclature Committee of the International Union of Biochemistry, Academic Press, New York.

48. Ericksson, C. E. (1967). Pea lipoxidase, distribution of enzyme and substrate in green peas. *J. Food Sci.* 32:438-441.

49. Ericksson, C. E., and S. G. Swensson (1970). Lipoxygenase from peas, purification and properties of the enzyme. *Biochim. Biophys. Acta* 198:449-459.

50. Eskin, N. A. M., H. M. Henderson, and R. J. Townsend (1971). *Biochemistry of Foods*, Academic Press, New York.

51. Etherington, D. J. (1981). Enzyme actions in the conditioning of meat, in *Proteinases and Their Inhibitors: Structure, Function and Applied Aspects* (V. Turk and L. J. Vitale, eds.), Pergamon Press, Oxford, pp. 279-290.

52. Feeney, R. E., and R. G. Allison (1969). *Evolutionary Biochemistry of Proteins*, Wiley-Interscience, New York.

53. Fennema, O. (1982). Behavior of proteins at low temperatures, in *Food Protein Deterioration: Mechanisms and Functionality* (J. P. Cherry, ed.), ACS Symposium Series 206, Am. Chem. Soc., Washington, D. C., pp. 109-133.

54. Fersht, A. (1977). *Enzyme Structure and Mechanism*, W. H. Freeman, San Francisco.

55. Flavin, D. F. (1982). The effects of soybean trypsin inhibitors on the pancreas of animals and man: a review. *Vet. Human Toxicol.* 24:25-28.

56. *Food Chemicals Codex*, 3rd ed. (1981). Committee on Codex Specifications, Food and Nutrition Board, National Acad. Sciences, pp. 107-100; 479-499.

57. Fox, P. F. (1981). Exogenous proteinases in dairy technology, in *Proteinases and Their Inhibitors: Structure, Function and Applied Aspects* (V. Turk and L. J. Vitale, eds.), Pergamon Press, Oxford, pp. 245-268.

58. Fox, P. F., and P. A. Morrissey (1981). Indigenous enzymes of bovine milk, in *Enzymes and Food Processing* (G. G. Birch, N. Blakebrough, and K. J. Parker, eds.), Applied Science Publishers, London, pp. 213-238.

59. Fraser, T. H., and B. J. Bruce (1978). Chicken ovalbumin is synthesized and secreted by *Escherichia coli. Proc. Nat. Acad. Sci. U.S.* 75:5936-5940.

60. Freifelder, D. (1982). *Physical Chemistry for Students of Biology and Chemistry*, Van Nostrand Reinhold, New York.

61. Fruton, J. S. (1982). Proteinase-catalyzed synthesis of peptide bonds. *Adv. Enzymology. Relat. Areas Mol. Biol.* 53:329-306.

62. Gardener, H. W. (1980). Lipid enzymes: Lipases, lipoxygenases and hydroperoxidases, in *Autoxidation in Foods and Biological Systems* (M. G. Simic and M. Karel, eds.), Plenum Press, New York, pp. 447-504.

63. Gill, G. S., B. J. Bruce, and T. H. Fraser (1981). Synthesis of higher eukaryotic

proteins in yeast, in *Recombinant DNA*, Proceedings of the Third Cleveland Symposium on Macromolecules, Elsevier Scientific, Amsterdam, pp. 213-227.

64. Goebel, W., J. Kreft, K. Bernhard, H. Schrempf, and G. Weedinger (1978). Replication and gene expression of extrachromosomal replicons in unnatuaral bacterial hosts, in *Genetic Engineering* (H. W. Boyer and S. Nicosea, eds.), Elsevier/North-Holland, Amsterdam, pp. 47-58.

65. Gould, E., and M. J. Medler (1970). Test to determine whether shucked oysters have been frozen and thawed. *J. Assoc. Off. Anal. Chem. 53*:1237-1241.

66. Graveland, A. (1970). Enzymatic oxidation of linoleic acid and glycerol-1-monolinoleate in doughs and flour-water suspensions. *J. Amer. Oil Chem. Soc. 47*:352-361.

67. Greco, Jr., G., D. Albanesi, M. Cantarella, and V. Scardi (1980). Experimental technique for multilayer enzyme immobilization. *Biotechnol. Bioeng. 22*:215-219.

68. Grommeck, R., and P. Markakis (1964). The effect of peroxidase on anthocyanin pigments. *J. Food Sci. 29*:53-57.

69. Grunstein, M., and D. S. Hogness (1975). Colony hybridization: A method for the isolation of cloned DNAs that contain a specific gene. *Proc. Nat. Acad. Sci. U.S. 72*:3961-3965.

70. Gryczan, T. J., and D. Dubman (1978). Construction and properties of chimeric plasmids in *Bacillus subtilis. Proc. Nat. Acad. Sci. U.S. 75*:1428-1432.

71. Guilbault, G. G. (1976). *Handbook of Enzymatic Methods of Analysis*, Marcel Dekker, Inc., New York.

72. Hagerdal, B. (1980). Enzymes co-immobilized with microorganisms for the microbial conversion of non-metabolizable substrates. *Acta Chem. Scand. B34*:611-613.

73. Hamm, R., D. Masic, and L. Telzlaff (1971). Einflub des Gefrierens und auftauens auf die sub cellulāre Vertelung der Aspartat-Aminotransferase in der Sklletmuskulatur verschiedener Geflügelarten. 2. *Lebensm. Unters. Forsch. 147*:71-76.

74. Hardy, K., S. Stahl, H. Küpper (1981). Production in *B. subtilis* of hepatitis β core antigen and of major antigen of foot and mouth disease virus. *Nature 293*:481-483.

75. Heber, U. (1968). Freezing injury in relation to loss of enzyme activities and protection against freezing. *Cryobiology 5*:188-192.

76. Hill, R. D. (1979). Oxidative enzymes and oxidative processes in milk. *CSIRO Food Res. 39*:33-37.

77. Hohn, B. (1975). DNA as substrate for packaging into bacteriophage lambda in vitro. *J. Mol. Biol. 98*:93-106.

78. Hohn, B., and K. Murray (1977). Packaging recombinant DNA molecules into bacteriophage particles in vitro. *Proc. Nat. Acad. Sci. U.S. 74*:3259-3263.

79. Hornby, W. E., M. D. Lilley, and E. M. Crook (1968). Some changes in the reactivity of enzymes resulting from their chemical attachment ot water-insoluble derivatives of cellulose. *Biochem. J. 107*:669-674.

80. Hulme, A. C. (ed.) (1971). *The Biochemistry of Fruits and Their Products*, vol. 2, Academic Press, New York.

81. Hultin, H. O. (1972). Symposium: Biochemical control systems. Enzymic activity and control as related to Subcellular Localization. *J. Food Sci. 37*:524-529.

82. Itakura, K., and A. D. Riggs (1980). Chemical DNA synthesis and recombinant DNA studies. *Science 209*:1401-1405.

83. James, L. K., and L. G. Augenstein (1966). Adsorption of enzymes at interfaces: Film formation and the effect on activity. *Adv. Enzymol. 28*:1-40.

84. Jenness, R., and S. Patton (1959). *Principles of Dairy Chemistry*, John Wiley and Sons, New York, p. 201.

85. Jensen, R. G., and H. Brockerhoff (1974). *Lipolytic Enzymes*, Academic Press, New York.

86. Jensen, R. G., S. A. Gerrior, M. M. Hagerty, and P. McMahon (1978). Preparation of acylglycerols and phospholipids with the aid of lipalytic enzymes. *J. Amer. Oil Chem. Soc. 55*:422-427.

87. Joseph, R. L. (1970). Production of tender beef. *Proc. Biochem. 5*(11):55-58.

88. Kang, C. K., W. D. Warner, and E. E. Rice (1974). Tenderization of meat with proteolytic enzymes. U. S. Patent 3,818,106, June 18.

89. Kassel, B. (1970). Naturally occurring inhibitors of proteolytic enzymes, in *Methods in Enzymology (Proteolytic Enzymes)*, vol. 19 (G. E. Perlman and L. Lorand, eds.), Academic Press, New York, pp. 839-906.

90. Keggins, K. M., P. S. Lovett, and E. J. Duvell (1978). Molecular cloning of genetically active fragments of *Bacillus* DNA in *Bacillus subtilis* and properties of the vector plasmid pUB 110. *Proc. Nat. Acad. Sci. U.S. 75*:1423-1427.

91. Kelly, T. J., and H. O. Smith (1970). A restriction enzyme from *Hemophilus influenzae*. II. Base sequence of the recognition site. *J. Mol. Biol. 51*:393-409.

92. Klaenhammer, T. R. (1982). The impact of biotechnology on food fermentations. *Res. Perspect. 1*(3):19-20.

93. Kreft, J., K. Bernhard, and W. Goebel (1978). Recombinant plasmids capable of replication in *B. subtilis* and *E. coli. Mol. Gen. Genet. 162*:59-67.

94. Kuzin, A. M. (1964). *Radiation Biochemistry*, D. Davey, New York.

95. Laskowski, Jr., M., and I. Kato (1980). Protein inhibitors of proteinases. *Annu. Rev. Biochem. 49*:593-626.

96. Lawrie, R. A. (1979). *Meat Science*, 3rd ed., Pergamon Press, New York.

97. Lei, M.-G., R. Bassette, and G. R. Reeck (1981). Effect of cysteine on heat inactivation of soybean trypsin inhibitors. *J. Agr. Food Chem. 29*:1196-1199.

98. Lerner, A. B., and T. B. Fitzpatrick (1950). Biochemistry of melanin formation. *Physiol. Rev. 30*:91-126.

99. Leuthauser, S. W. C., L. W. Oberly, T. D. Oberly, J. R. J. Sorenson, and G. R. Buettner (1981). Antitumor activities of a copper chelate which has superoxide dismutase activity and an iron chelator, in *Oxygen and Oxy-Radicals in Chemistry and Biology* (M. A. J. Rodgers and E. L. Powers, eds.), Academic Press, New York, pp. 679-682.

100. Liener, I. E., and M. L. Kadade (1969). Protease inhibitors, in *Toxic Constituents of Plant Foodstuffs* (I. E. Liener, ed.), Academic Press, New York, pp. 7-68.

101. Lilly, M. D., and Sharp, A. K. (1968). The kinetics of enzymes attached to water-insoluble polymers. *Chem. Eng.* (London) *215*:CE12-CE18.

102. Luh, B. S., and B. Phithakpol (1972). Characteristics of polyphenoloxidase related to browning in cling peaches. *J. Food Sci. 37*:264-268.

103. Magee, Jr., E. L., N. F. Olson, and R. C. Lindsay (1981). Microencapsulation of cheese ripening systems: Production of diacetyl and acetoin in cheese by encapsulated bacterial cell-free extract. *J. Dairy Sci. 64*:616-621.

104. Mandel, M., and A. Higa (1970). Calcium-dependent bacteriophage DNA infection. *J. Mol. Biol. 53*:159-162.

105. Mandenius, C. F., B. Danielsson, and B. Mattiasson (1980). Enzyme thermistor control of the sucrose concentrations at a fermentation with immobilized yeast. *Acta Chem. Scand. B34*:463-465.

106. Marciniszyn, Jr., J., J. A. Hartsuck, and J. Tang (1977). Pepstatin inhibition mechanism, in *Acid Proteases: Structure, Function and Biology* (J. Tang, ed.), Plenum Press, New York, pp. 199-210.

107. Marsh, B. B. (1972). Post-mortem muscle shortening and meat tenderness, in *Proceedings of the Meat Industry Conference*, Am. Meat Inst. Fdn., Chicago, pp. 109-124.

108. Matsuo, R. R., J. W. Bradley, and G. N. Irvine (1970). Studies on pigment destruction during spaghetti processing. *Cereal Chem. 47*:1-5.

109. McLaren, A. D., and L. Packer (1970). Some aspects of enzyme reactions in heterogeneous systems. *Adv. Enzymol. 33*:245-308.

110. McWeeny, D. J. (1968). Reactions in food systems: Negative temperature coefficients and other abnormal temperature effects. *J. Food Technol. 3*:15-30.

111. Mercier, C. (1982). Enzymes in food analysis, in *Recent Developments in Food Analysis* (W. Baltes, P. B. Czedik-Eysenberg, and W. Pfannhauser, eds.), Basel, pp. 270-285.

112. Michelson, A. M. (1977). Superoxide dismutase and its application as an oxidation inhibitor. U. S. Patent 4,029,819, June 14.

113. Michelson, A. M., and J. Monod (1975). Superoxide dismutases and their applications as oxidation inhibitors. U. S. Patent 3,920,521, Nov. 18.

114. Michelson, A. M., J. M. McCord and I. Fridovich (eds.) (1975). *Superoxide and Superoxide Dismutases*, Academic Press, New York.

115. Miller, W. G., and R. A. Alberty (1958). Kinetics of the reversible Michaelis-Menten mechanism and the applicability of the steady-state approximation. *J. Amer. Chem. Soc. 80*:5146-5151.

116. Montgomery, R., and C. A. Swenson (1976). *Quantitative Problems in Biochemical Sciences*, 2nd ed., W. H. Freeman, San Francisco.

117. Mosbach, K. (1980). Immobilized enzymes. *Trends Biochem. Sci. 5*:1-3.

118. Murai, N. (1983). Personal communication.

119. Mycek, M. J. (1970). Cathepsins, in *Methods in Enzymology (Proteolytic Enzymes)*, vol. 19 (L. Lorand, eds.), Academic Press, New York, pp. 285-315.

120. Newbold, R. P., and P. V. Harris (1972). The effect of pre-rigor changes on meat tenderness. A review. *J. Food Sci. 37*:337-340.

121. Norgard, M. V., K. Keem, and J. J. Monahan (1978). Factors affecting the transformation of *Escherichia coli* strain X1776 by pBR322 plasmid DNA. *Gene 3*: 279-292.

122. Norman, B. E. (1981). New developments in starch syrup technology, in *Enzymes in Food Processing* (G. G. Birch, N. Blakebrough, and K. J. Parker, eds.), Applied Science Publishers, London, pp. 15-50.

123. Old, R. W., and S. B. Primrose (1980). *Principles of Gene Manipulation*, University of California Press, Berkeley.

124. Ory, R. L., and A. J. St. Angelo (eds.) (1977). *Enzymes in Food and Beverage Processing*, ACS Symposium Series 47, American Chemical Society, Washington, D.C.

124a. Pace, N. C., and C. Tanford (1968). Thermodynamics of the unfolding of β-lactoglobulin A in aqueous urea solutions between 5 and 55°. *Biochemistry 7*:198-208.

125. Palmer, T. (1981). *Understanding Enzymes*, Halsted Press, New York.

126. Pariza, M. W., and E. M. Foster (1983). Determining the safety of enzymes used in food processing. *J. Food Protection 46*:453-468.

127. Paterson, B. M., B. E. Roberts, and E. L. Kuff (1977). Structural gene identification and mapping by DNA, mRNA hybrid arrested cell-free translation. *Proc. Nat. Acad. Sci. U.S. 74*:4370-4374.

128. Pilnik, W., and F. M. Rombouts (1981). Pectic enzymes, in *Enzymes and Food Processing* (G. G. Birch, N. Blakebrough, and K. J. Parker, eds.), Applied Science Publishers, London, pp. 105-128.

129. Pilnik, W., and A. G. J. Voragen (1970). Pectic substances and other uronides, in *The Biochemistry of Fruits and Their Products*, vol. 1 (A. C. Hulme, ed.), Academic Press, New York, pp. 53-87.

130. Pintauro, N. D. (1979). *Food Processing Enzymes—Recent Developments*, Noyes Data Corp., Park Ridge, N. J.

131. Pitcher, Jr., W. H. (ed.) (1980). *Immobilized Enzymes for Food Processing,* CRC Press, Boca Raton, Fla.

132. Pressey, R. (1972). Symposium: Biochemical control systems. Natural enzyme inhibitors in plant tissues. *J. Food Sci. 37*:521-523.

132a. Quaranta, H. O., and S. S. Perez (1983). Chemical methods for measuring changes in freeze stored fish: A review. *Food Chem. 11*:79-85.

133. Reed, G. (ed.) (1975). *Enzymes in Food Processing,* Academic Press, New York.

134. Reilly, P. J. (1980). Potential and use of immobilized carbohydreases, in *Immobilized Enzymes for Food Processing* (W. H. Pitcher, Jr., ed.), CRC Press, Boca Raton, Fla., pp. 113-152.

135. Reuben, J. (1971). Substrate anchoring and catalytic power of enzymes. *Proc. Nat. Acad. Sci. U.S. 68*:563-565.

136. Rich, D. H., E. Sun, and J. Singh (1977). Synthesis of dideoxy-pepstatin. Mechanism of inhibition of procine pepsin. *Biochem. Biophys. Res. Commun. 74*: 762-767.

137. Richardson, M. (1980-1981). Protein inhibitors of enzymes. *Food Chem. 6*: 235-253.

138. Richardson, T. (1976). Enzymes, in *Principles of Food Science*, Part I, *Food Chemistry* (O. R. Fennema, ed.) Marcel Dekker, Inc., New York, pp. 285-346.

139. Roland, J. F. (1980). Requirements unique to the food and beverage industry, in *Immobilized Enzymes for Food Processing* (W. H. Pitcher, Jr., ed.), CRC Press, Boca Raton, Fla., pp. 55-80.

140. Romani, R. J. (1972). Symposium: Biochemical control systems. Stress in the post harvest cell: The response of mitochondria and ribosomes. *J. Food Sci. 37*: 513-517.

141. Rombouts, F. M., and W. Pilnik (1972). Research on pectin depolymerases in the sixties—a literature review. *CRC Crit. Rev. Food Technol. 3*:1-26.

142. Rubin, E. N., G. A. Wilson, and F. E. Young (1980). Expression of thymidylate activity in *Bacillus subtilis* upon integration of a cloned gene from *Escherichia coli. Gene 10*:227-235.

143. Schwimmer, S. (1972). Symposium: Biochemical control systems. Cell disruption and its consequences in food processing. *J. Food Sci. 37*:530-535.

144. Schwimmer, S. (1981). *Source Book of Food Enzymology,* AVI Publishing Co., Westport, Conn.

145. Scott, D. (1980). Enzymes, industrial, in *Kirk-Othmer Encyclopedia of Chemical Technology*, 3rd ed., vol. 9, John Wiley and Sons, New York, pp. 173-224.

146. Segel, I. H. (1976). *Biochemical Calculations*, 2nd. ed., John Wiley and Sons, New York.

147. Sharma, B. P., and R. A. Messing (1980). Application and potential of other enzymes in food processing: Amino-acylase, aspartase, fumarase, glucose-oxidase-catalase, in *Immobilized Enzymes for Food Processing* (W. H. Pitcher, Jr., ed.), CRC Press, Boca Raton, Fla., pp. 185-210.

148. Sharma, C. B., M. Goel, and M. Irshad (1978). Myornositol hexaphosphate as a potential inhibitor of α-amylases. *Phytochemistry 17*:201-205.

149. Simpson, K. L., T.-C. Lee, D. B. Rodriguez, and C. O. Chichester (1976). Metabolism in senescent and stored tissues, in *Chemistry and Biochemistry of Plant Pigments*, 2nd ed., vol. 1 (T. W. Goodwin, ed.), Academic Press, New York, pp. 779-842.

150. Singh, M., and A. D. Krikorian (1982). Inhibition of trypsin activity in vitro by phylate. *J. Agr. Food Chem. 30*:799-800.

151. Sinsheimer, R. L. (1977). Recombinant DNA. *Annu. Rev. Biochem. 46*:415-438.

152. Siu, R. G. H. (1957). Action of ionizing radiations on enzymes, in *Radiation Preservation of Food*, U. S. Army Quartermaster Corps., U. S. Gov. Printing

Office, Washington, D.C., pp. 169-176.

153. Skogberg, D., and T. Richardson (1980). Enzyme electrodes for the food industry. *J. Food Protection 43*:808-819.

154. Smith, W. L., and E. M. Lands (1970). The self-catalyzed destruction of lipoxygenase. *Biochem. Biophys. Res. Commun. 41*:846-851.

155. Somero, G. N. (1978). Temperature adaptation of enzymes: Biological optimization through structure-function compromises. *Annu. Rev. Ecol. Syst. 9*:1-29.

156. Spinelli, J. (1971). Biochemical basis of fish freshness. *Proc. Biochem. 6*:36-54.

157. Steiner, R. F., and V. Frattali (1969). Purification and properties of soybean protein inhibitors of proteolytic enzymes. *J. Agr. Food Chem. 17*:513-518.

158. Stevenson, R. W., F. E. Luddy, and H. L. Rothbart (1979). Enzymatic acyl exchange to vary saturation in di- and triglycerides. *J. Amer. Oil Chem. Soc. 56*: 676-680.

159. Struthers, B. J., J. R. MacDonald, R. R. Dahlgren, and D. T. Hopkins (1983). Effects on the monkey, pig and rat pancreas of soy products with varying levels of trypsin inhibitor and comparison with administration of cholecystokinin. *J. Nutr. 113*:86-97.

160. Sutcliffe, J. G. (1979). Complete nucleotide sequence of the *Escherichia coli* plasmid pBR322. *Cold Spring Harbor Symp. Quant. Biol. 43*:77-90.

161. Swaisgood, H. (1980). Sulfhydryl oxidase: Properties and applications. *Enzyme Microb. Technol. 2*:265-272.

162. Swaisgood, H. E., and P. Abraham (1980). Oxygen activation by sulfhydryl oxidase and the enzymes interaction with peroxidase. *J. Dairy Sci. 63*:1205-1210.

163. Szpacenko, A., J. Kay, D. E. Goll, and Y. Otsaka (1981). A different form of th the Ca^{2+}-dependent proteinase activated by micromolar levels of Ca^{2+}, in *Proteinases and Their Inhibitors: Structure, Function and Applied Aspects* (V. Turk and L. J. Vitale, eds.), Pergamon Press, Oxford, pp. 151-162.

164. Talburt, W. F., and O. Smith (1975). *Potato Processing*, 3rd ed., AVI Publishing Co., Westport, Conn.

165. Tanford, C. (1961). *Physical Chemistry of Macromolecules*, John Wiley and Sons, New York, pp. 587-678.

165a. Tanford, C. (1968). Protein denaturation. *Adv. Protein Chem. 23*:121-282.

166. Tang, J. (ed.) (1977). *Acid Porteases: Structure, Function and Biology*, Adv. Exp. Med. Biol., vol. 95, Plenum Press, New York, p. 355.

167. Tappel, A. L. (1962). Hematin compounds and lipoxidase, in *Lipids and Their Oxidation* (H. W. Schultz, E. A. Day, and R. O. Sinnhuber, eds.), AVI Publishing Co., Westport, Conn., pp. 122-138.

168. Tappel, A. L. (1966). Effects of low temperatures and freezing on enzymes and enzyme systems, in *Cryobiology* (H. T. Meryman, ed.), Academic Press, New York, pp. 163-177.

169. Tappel, A. L. (1978). Glutathione peroxidase and hydroperoxides, in *Methods in Enzymology*, vol. 52 (S. Fleischer and Packer, eds.), Academic Press, New York, pp. 506-513.

170. Taylor, M. J., T. Richardson, and N. F. Olson (1976). Coagulation of milk with immobilized proteases: A review. *J. Milk Food Technol. 39*:864-871.

171. Temin, H., and S. Mizutani (1970). RNA-dependent DNA polymerase in virions of Rouse Sarcoma virus. *Nature 226*:1211-1213.

172. Thomas, D. (1978-1979). The future of enzyme technology. *Trends Biochem. Sci. 3-4*:N207-N209.

173. Tribe, M. A., M. R. Eraut, and R. L. Snook (1976). *Enzymes*, Book 7, Cambridge University Press, New York.

174. Umezawa, H. (1976). Structure and activities of protease inhibitors of microbial origin, in *Methods in Enzymology (Proteolytic Enzymes)* Part B, vol. 45 (L. Lorand, ed.), Academic Press, New York, pp. 678-694.

175. Valenzvela, A., H. Adarmes, and R. Guerra (1981). Inhibition of peroxidation of milk fat by the enzymes superoxide dismutase and catalase. Comparison of their effect with that of commercial antioxidants. *Alimentos 6*:5-11; *Dairy Sci. Abstr. 44*:4328 (1982).

176. Vas, K. (1966). Radiation resistance of enzymes in foods irradiated against microbial damage, in *Food Irradiation*, Proc. Int. Symp. Karlsruhe, Germany.

177. Verger, R., M. C. E. Mieras, and G. H. de Haas (1973). Action of phospholipase A at interfaces. *J. Biol. Chem. 248*:4023-4034.

178. Verger, R., J. Rietsch, F. Pattus, F. Ferrato, G. Peroni, G. H. deHaase, and P. Desnuelle (1978). Studies of lipase and phospholipase A_2 acting on lipid monolayers. *Adv. Exp. Med. Biol. 101*:79-94.

179. Wallace, J. M., and E. L. Wheeler (1972). Lipoxygenase inactivation in wheat protein concentrate by heat-moisture treatments. *Cereal Chem. 49*:92-97.

180. Walsh, C. (1979). *Enzymatic Reaction Mechanisms*, W. H. Freeman, San Francisco, p. 978.

181. Webb, F. C. (1964). *Biochemical Engineering*, Van Nostrand, New York, p. 75.

182. Wharton, C. W., and R. Eisenthal (1981). *Molecular Enzymology*, John Wiley and Sons, New York.

183. Whitaker, J. R. (1972). *Principles of Enzymology for the Food Sciences*, Marcel Dekker, Inc., New York.

184. Whitaker, J. R. (ed) (1974). *Food Related Enzymes*, Adv. Chem. Series, 136, American Chemical Society, Washington, D.C.

185. White, A., P. Handler, E. L. Smith, and D. Stetten (1954). *Principles of Biochemistry*, 1st ed., McGraw-Hill, New York, pp. 221-255.

186. Williams, D. M., E. J. Duvall, and P. S. Lovett (1981). Cloning restriction fragments that promote expression of a gene in *Bacillus subtilis. J. Bacteriol. 146*: 1162-1165.

187. Williams, D. M., R. G. Schoner, E. J. Duvall, L. H. Preis, and P. S. Lovett (1981). Expression of *Escherichia coli trp* genes in the mouse dihydrofolate reductase gene clones in *Bacillus subtilis. Gene 16*:199-206.

188. Winass, J. D., M. J. Worsley, E. M. Pioli, P. T. Barth, K. T. Atherton, E. C. Dart, D. Byrom, K. Powell, and P. J. Senior (1980). Improved conversion of methanol to single cell proteins by *Methylophilus methylotrophus. Nature 287*:396-401.

189. Wiseman, A. (1981). Enzymes in analysis of foods, in *Enzymes and Food Processing* (G. G. Birch, N. Blakebrough, and K. J. Parker, eds.), Applied Science Publishers, London, pp. 275-288.

190. Wood, W. B. (1966). Host specificity of DNA produce by *Escherichia coli*: Bacterial mutation affecting restriction and modification of DNA. *J. Mol. Biol. 16*: 118-123.

191. Yokozeki, K., S. Yamanaka, K. Takinami, Y. Hirose, A. Tonaka, K. Sonomoto, and S. Fukui (1982). Application of immobilized lipase to regio-specific intersterification of triglyceride in organic solvent. *Eur. J. Appl. Microbiol. Biotechnol. 14*:1-5.

192. Zaborsky, O. R. (1973). *Immobilized Enzymes*, CRC Press, Boca Raton, Fla., p. 175.

BIBLIOGRAPHY

Food Enzymology

Birch, G. G., N. Blakebrough, and K. J. Parker (eds.) (1981). *Enzymes and Food Processing*, Applied Science Publishers, London.

Ory, R. L., and A. J. St. Angelo (eds.) (1977). *Enzymes in Food and Beverage Processing*, ACS Symposium Series 47, American Chemical Society, Washington, D.C.

Reed, G. (ed.) (1975). *Enzymes in Food Processing*, Academic Press, New York.

Schwimmer, S. (1981). *Source Book of Food Enzymology*, AVI Publishing Co., Westport, Conn.

Whitaker, J. R. (ed.) (1974). *Food Related Enzymes*, Adv. Chem. Series 136, American Chemical Society, Washington, D.C.

Whitaker, J. R. (1972). *Principles of Enzymology for the Food Sciences*, Marcel Dekker, Inc., New York. (Also kinetics and mechanisms.)

Enzyme Chemistry, Kinetics and Mechanism

Dugas, H., and C. Penney (1981). *Bioorganic Chemistry: A Chemical Approach to Enzyme Action*, Springer-Verlag, New York.

Engel, P. C. (1981). *Enzyme Kinetics—The Steady State Approach*, 2nd ed., Chapman and Hall, New York.

Fersht, A. (1977). *Enzyme Structure and Mechanism*, W. H. Freeman, San Francisco.

Palmer, T. (1981). *Understanding Enzymes*, Halsted Press, New York.

Walsh, C. (1979). *Enzymatic Reaction Mechanisms*, W. H. Freeman, San Francisco.

Wharton, C. W., and R. Eisenthal (1981). *Molecular Enzymology*, John Wiley and Sons, New York.

Problem Solving, Enzyme Kinetics

Christensen, H. N., and G. A. Palmer (1967). *Enzyme Kinetics*, W. B. Saunders, Philadelphia.

Montgomery, R., and C. A. Swenson (1976). *Quantitative Problems in Biochemical Sciences*, 2nd ed., W. H. Freeman, San Francisco.

Segel, I. H. (1976). *Biochemical Calculations*, 2nd ed., John Wiley and Sons, New York.

Tribe, M. A., M. R. Eraut, and R. L. Snook (1976). *Enzymes*, Book 7, Cambridge University Press, New York.

7
VITAMINS AND MINERALS

Steven R. Tannenbaum and Vernon R. Young Massachusetts Institute of Technology, Cambridge, Massachusetts

Michael C. Archer University of Toronto, Ontario Cancer Institute, Toronto, Ontario, Canada

I.	Introduction	478
II.	Recommended Dietary Allowances	479
III.	General Causes for Loss of Vitamins and Minerals	482
	A. Genetics and Maturity	482
	B. Handling Postharvest or Immediately Postmortem	482
	C. Trimming	483
	D. Milling	483
	E. Leaching and Blanching	483
	F. Processing Chemicals	485
	G. Deteriorative Reactions	486
IV.	Enrichment, Restoration, and Fortification	486
V.	Water-Soluble Vitamins	488
	A. Ascorbic Acid	488
	B. Thiamine	493
	C. Riboflavin	500
	D. Nicotinic Acid	501
	E. Vitamin B_6	501
	F. Folic Acid	504
	G. Vitamin B_{12}	509
	H. Pantothenic Acid	512
	I. Biotin	512
VI.	Fat-Soluble Vitamins	513
	A. Vitamin A	513
	B. Vitamin K	517
	C. Vitamin D	517
	D. Vitamin E	517

VII. Chemical Properties of Minerals and Their Bioavailability   519
    A. General   519
    B. Chemistry   520
    C. Occurrence   521
    D. Losses and Gains in Processing   523
    E. Availability of Minerals in Foods   523
    F. Safety of the Minerals   527
VIII. Optimization of Nutrient Retention   527
    A. High-Temperature-Short-Time (HTST) Processing 527
    B. Prediction of Vitamin Losses in Storage   529
    References   532
    Bibliography   543

I. INTRODUCTION

One of the major quality aspects of our food supply is its content of vitamins and minerals. From a biological point of view, we eat to survive, and the pattern of our nutrient requirements has developed during a long evolutionary process in which we have adapted to our environment. Although certain food processes, such as cooking, are indeed very old, it is only within the last 50 years that we have begun to consume a significant part of our food in a factory-processed form.

Our modern processed food supply has contributed enormously to the public health status of the population. Certain nutritional diseases, which were common in parts of the United States 50 years ago, such as pellagra, have all but disappeared. A U. S. 10 state nutritional survey (243) demonstrated that although nutritional deficiencies do exist, they are minimal in comparison with what existed prior to the modern era of nutrition and food technology. Consider that it is now possible to eat a diet balanced in all types of foods at any time of year and in any geographic location.

Yet, modern process technology has also introduced its share of problems. Occasionally, this has been the result of inadequate knowledge, but tragic cases of illness and even death have also occurred when essential nutrient value was lost because of ignorance, carelessness, and lack of adherence to "good manufacturing practice." In the past, new food processes have seldom been assessed for their effect on nutrient loss or retention. Multiple processes, such as freezing of reconstituted dehydrated foods, may lead to benefits in process scheduling, but they may also lead to higher than normal losses of vitamins. New forms of food products, such as intermediate moisture foods, may lead to accentuated problems of vitamin stability. A food product that has acceptable nutritional quality should generally be capable of providing those nutrients normally characteristic of its food group.

The purpose of this chapter is to summarize the information available on the chemistry of vitamin and mineral losses in processed and stored foods. This subject has been treated in many review articles, some of which are listed in the Bibliography. Special mention must be made of *Nutritional Evaluation of Food Processing*, which contains a large amount of detailed information on various foods in both the raw and finished state (104). Rather than recapitulate and summarize the many studies carried out on individual foods and processes, this chapter reviews analytically the chemistry of the individual vitamins and the general factors leading to nutrient losses.

Information on the fate of vitamins and minerals in processed foods is reasonably adequate for only a few compounds. Considerable information is available on ascorbic acid, provitamins A, thiamine, and iron; less on riboflavin and vitamin B_6; and very little on folic acid, vitamin B_{12}, and minerals other than iron. The kinds of information available partially reflect the importance past investigators have attached to various nutrients, and also to analytical problems. The amount of space devoted to an individual nutrient in this chapter is influenced by both the complexity of its chemistry and the amount of available information. At some future time, it is hoped that sufficient information will accumulate to fill the more obvious gaps.

II. RECOMMENDED DIETARY ALLOWANCES

In order to understand whether a specific treatment of a specific food leads to acceptable nutrient quality, it is necessary to have an understanding of both human requirements and the amount of a particular nutrient present in the food after normal preparatory procedures.

The quantitative needs of humans for essential nutrients, as determined by scientific approaches, serves as a data base for the U.S. Recommended Dietary Allowances (RDA). These RDA are necessary as guidelines in the planning of diets and food supplies, for nutritional labeling of foods, and for evaluating the nutritional adequacy of foods consumed. The RDA has been defined as "the levels of intake of essential nutrients considered, in the judgment of the Committee on Dietary Allowances of the Food and Nutrition Board on the basis of available scientific knowledge, to be adequate to meet the known nutritional needs of practically all healthy persons" (244).

The minimum requirement for a nutrient varies among apparently similar individuals, and RDA are designed to meet the nutritional needs of practically all healthy persons within a population (15,244,258). Current dietary standards represent judgements based on available studies, and since the data on human nutrient requirements are still limited, it is not surprising that the recommendations of different committees do not always agree. A comparison of some of the dietary allowances proposed by various committees is shown in Table 1 to illustrate this point.

An additional factor that must be considered in setting RDA is the source and availability of nutrients in foods. For some nutrients the dietary requirement may be met by a

Table 1 Comparisons of Recommended Intakes for Selected Nutrients in Young Men, as Proposed by Different Authorities

Nutrient daily requirement	United States, 1974 (68)	Canada, 1974 (71)	West Germany, 1975 (72)	FAO/WHO 1974 (69)
Vitamin C (mg)	45	30	75	30
Iron (mg)	10		12	5-9[a]
Folic acid (μg)	400[b]	200[c]	400	200[c]
Calcium (mg)	800		800	400-500

[a]For diets in which 10-25% energy is derived from animal sources or soybean.
[b]Total food folacin as determined by *L. casei* assay.
[c]Expressed as free folate.
Source: Adapted from Truswell (238).

precursor. For example, in the case of vitamin A an alternative dietary source is provided by the carotenoids. The most common source of vitamin A activity is β-carotene, which in the diets of populations of developing countries may provide 60-100% of vitamin A needs.

The Joint FAO/WHO Expert Group on Requirements of Vitamin A, Thiamine, Riboflavin and Niacin (77) concluded, for guideline purposes, that β-carotene (weight basis) has only one-sixth the activity of retinol when ingested by humans. However, the relative activity varies considerably with different foods and methods of preparation. The RDA for vitamin A is expressed, therefore, as "retinol equivalents," where

$$1 \text{ retinol equivalent} = 1 \ \mu g \text{ retinol}$$

$$= 6 \ \mu g \ \beta\text{-carotene}$$

$$= 12 \ \mu g \text{ other vitamin A active compounds}$$

The availability of nutrients from different foodstuffs depends on the efficiency of absorption in the body. For iron this efficiency is a low 5-25%, depending on the composition of the diet. Iron availability is greatest in meat and lowest in cereals (see Table 2). Therefore, the level of the RDA depends on the nature of the diet.

A further problem in evaluating the nutritional and health significance of dietary intake data is the inadequate status of information on the nutrient composition of foods. Much of the data on food composition are old, and although they are currently under revision, it is frequently unclear how the values were derived and whether the data are sufficiently representative for use in assessing dietary survey data. Furthermore, many important nutrients are not listed in the published tables on food composition, and there is uncertainty regarding the bioavailability of many of the nutrients in foods.

The concentrations of many of the key nutrients for fresh and cooked foods are given in USDA Handbook No. 8 (242). Although data in this compilation are occasionally inaccurate, they afford the only major source of information apart from direct analysis of the product in question. In many instances inaccuracies are caused by analytical procedures of insufficient specificity; consequently, users of these data should be especially cognizant of the methods used and their appropriateness. Special problems in vitamin analysis are briefly noted in this chapter when specific vitamins are discussed.

In Table 3 the most recent U.S. RDA are summarized (244).

Table 2   Absorption of Iron from Different Types of Diet

Type of diet	Assumed upper limit for iron absorption by normal individuals (%)
Less than 10% of energy from foods of animal origin or from soybeans[a]	10
10-25% of energy from foods of animal origin or from soybeans	15
More than 25% of energy from foods of animal origin or from soybeans	20

[a]Soybean is grouped with foods of animal origin owing to the high bioavailability of its iron content.
Source: Beaton and Patwardhan (15).

Table 3 U. S. Recommended Daily Dietary Allowances[a]

	Age (years)	Weight (kg)	Weight (lb)	Height (cm)	Height (in.)	Protein (g)	Fat-soluble vitamins			Water-soluble vitamins							Minerals					
							Vitamin A (μg RE)[b]	Vitamin D (μg)[c]	Vitamin E (mg α-TE)[d]	Vitamin C (mg)	Thiamine (mg)	Riboflavin (mg)	Niacin (mg NE)[e]	Vitamin B6 (mg)	Folacin (μg)[f]	Vitamin B12 (μg)	Calcium (mg)	Phosphorus (mg)	Magnesium (mg)	Iron (mg)	Zinc (mg)	Iodine (μg)
Infants	0.0-0.5	6	13	60	24	kg × 2.2	420	10	3	35	0.3	0.4	6	0.3	30	0.5[g]	360	240	50	10	3	40
	0.5-1.0	9	20	71	28	kg × 2.0	400	10	4	35	0.5	0.6	8	0.6	45	1.5	540	360	70	15	5	50
Children	1-3	13	29	90	35	23	400	10	5	45	0.7	0.8	9	0.9	100	2.0	800	800	150	15	10	70
	4-6	20	44	112	44	30	500	10	6	45	0.9	1.0	11	1.3	200	2.5	800	800	200	10	10	90
	7-10	28	62	132	52	34	700	10	7	45	1.2	1.4	16	1.6	300	3.0	800	800	250	10	10	120
Males	11-14	45	99	157	62	45	1000	10	8	50	1.4	1.6	18	1.8	400	3.0	1200	1200	350	18	15	150
	15-18	66	145	176	69	56	1000	10	10	60	1.4	1.7	18	2.0	400	3.0	1200	1200	400	18	15	150
	19-22	70	154	177	70	56	1000	7.5	10	60	1.5	1.7	19	2.2	400	3.0	800	800	350	10	15	150
	23-50	70	154	178	70	56	1000	5	10	60	1.4	1.6	18	2.2	400	3.0	800	800	350	10	15	150
	51+	70	154	178	70	56	1000	5	10	60	1.2	1.4	16	2.2	400	3.0	800	800	350	10	15	150
Females	11-14	46	101	157	62	46	800	10	8	50	1.1	1.3	15	1.8	400	3.0	1200	1200	300	18	15	150
	15-18	55	120	163	64	46	800	10	8	60	1.1	1.3	14	2.0	400	3.0	1200	1200	300	18	15	150
	19-22	55	120	163	64	44	800	7.5	8	60	1.1	1.3	14	2.0	400	3.0	800	800	300	18	15	150
	23-50	55	120	163	64	44	800	5	8	60	1.0	1.2	13	2.0	400	3.0	800	800	300	18	15	150
	51+	55	120	163	64	44	800	5	8	60	1.0	1.2	13	2.0	400	3.0	800	800	300	10	15	150
Pregnant						+30	+200	+5	+2	+20	+0.4	+0.3	+2	+0.6	+400	+1.0	+400	+400	+150	h	+5	+25
Lactating						+20	+400	+5	+3	+40	+0.5	+0.5	+5	+0.5	+100	+1.0	+400	+400	+150	h	+10	+50

[a]Designed for the maintenance of good nutrition of practically all healthy people in the United States. The allowances are intended to provide for individual variations among most normal persons as they live in the United States under usual environmental stresses. Diets should be based on a variety of common foods to provide other nutrients for which human requirements have been less well defined. (Revised 1980.)

[b]Retinol equivalents. 1 retinol equivalent = 1 μg retinol or 6 μg β-carotene.

[c]As a cholecalciferol. 10 μg cholecalciferol = 400 IU of vitamin D.

[d]α-tocopherol equivalents. 1 mg α-tocopherol = 1 α-TE.

[e]1 NE (niacin equivalent) is equal to 1 mg of niacin or 60 mg of dietary tryptophan.

[f]The folacin allowances refer to dietary sources as determined by $L.$ $casei$ assay after treatment with enzymes (conjugases) to make polyglutamyl forms of the vitamin available to the test organism.

[g]The recommended dietary allowance for vitamin B12 in infants is based on the average concentration of the vitamin in human milk. The allowances after weaning are based on energy intake (as recommended by the American Academy of Pediatrics) and consideration of other factors, such as intestinal absorption.

[h]The increased requirement during pregnancy cannot be met by the iron content of habitual American diets nor by the existing iron stores of many women; therefore, the use of 30-60 mg supplemental iron is recommended. Iron needs during lactation are not substantially different from those of nonpregnant women, but continued supplementation of the mother for 2-3 months after parturition is advisable to replenish stores depleted by pregnancy.

$Source$: From Ref. 244.

III. GENERAL CAUSES FOR LOSS OF VITAMINS AND MINERALS

All foods that undergo processing are subject to some degree of loss in vitamin and mineral content, even though the bioavailability of a nutrient is occasionally increased or some antinutritional factor is inactivated. In general, food processing is accomplished in a manner that attempts to minimize nutrient losses and maximize safety of the product. In addition to losses from processing, preprocessing conditions can also influence nutrient content. These include genetic variation, degree of maturity, soil conditions, fertilizer use, and type, climate, availability of water, light (length of day and intensity), and postharvest or postmortem handling. Some data on geographic variations in ascorbic acid and vitamin A in selected vegetables have been published (127).

A. Genetics and Maturity

Numerous examples of genetic influences on vitamin content of foods are given in Harris and Karmas (104). Data on the effect of maturity are more difficult to find; however, an excellent example is that of tomatoes (Table 4). Not only does the ascorbic acid content vary with period of maturity, but maximum vitamin content occurs when the tomato is in an immature state.

B. Handling Postharvest or Immediately Postmortem

The history of the food from time of harvest or slaughter to time of processing may cause considerable variation in nutritional value. Since many of the vitamins are also cofactors for enzymes or may be subject to degradation by endogenous enzymes, particularly those released after the death of the plant or animal, it is fairly obvious that postharvest or postmortem practices cause substantial fluctuations in nutrient content. In a study on peas, a minor but significant reduction in ascorbic acid took place during the 30 min period it took the truck to reach the processing plant from the field and the additional 30 min required to off-load into a holding hopper and then into the plant (146). A review of the losses of vitamins from fresh vegetables indicates that such losses can be extensive in cases of mishandling, such as when holding times are 24 hrs or more at ambient temperatures (80).

Table 4 Influence of Degree of Maturity on Ascorbic Acid Content of New Yorker Variety Tomatoes

Weeks from anthesis	Average weight (g)	Color	Ascorbic acid (mg%)
2	33.4	Green	10.7
3	57.2	Green	7.6
4	102.5	Green-yellow	10.9
5	145.7	Yellow-red	20.7
6	159.9	Red	14.6
7	167.6	Red	10.1

Source: From Malewski and Markakis (157).

C. Trimming

Plant tissues in particular are subject to trimming and subdividing practices that lead to discarding of some nutrient-rich portions. For example, the skins and peels of fruits and vegetables are usually removed. It has been reported that the concentration of ascorbic acid is greater in the apple peel than in the flesh; furthermore, the waste core of the pineapple contains a greater concentration of vitamin C than the edible portion. Similarly, niacin is reported to be richer in the epidermal layers of the carrot root than in the root that remains after processing (41). It is likely that similar concentration differences can be found in such foods as potatoes, onions, and beets. When peeling is accomplished by a chemically drastic procedure, such as lye treatment, significant losses of nutrients also occur in the outer fleshy layer. Trimming of vegetables, such as spinach, broccoli, green beans, and asparagus, involves discarding bits of stems or tougher portions of the plant that contain significant concentrations of some nutrients.

D. Milling

A special category of trimming involves the milling of cereals. All milled cereals undergo a significant reduction of nutrients, the extent of the loss governed by the efficiency with which the endosperm of the seed is separated from the outer seed coat (bran) and germ. The loss of each nutrient follows its own characteristic pattern, as shown for wheat in Fig. 1.

The loss of certain vitamins and minerals from milled cereals was deemed so relevant to health in the U.S. populace that the concept of replacing nutrients in the final stages of processing was proposed in the 1940s. After a long series of hearings, the Food and Drug Administration issued regulations for a standard of identity for enriched bread. These standards required that four nutrients, namely, thiamine, niacin, riboflavine, and iron, be added to flour, with addition of calcium and vitamin D considered optional. Currently, if bread is to be labeled "enriched," it must meet these standards, but enrichment is mandatory in only some states.

E. Leaching and Blanching

One of the most significant routes for the loss of water-soluble nutrients is via extraction from cut or susceptible surfaces. Food processing operations that lead to losses of this type include washing, flume conveying, blanching, cooling, and cooking. The nature and extent of the loss of course depends on pH, temperature, ratio of water to food, ratio of surface to volume, maturity, and other factors.

Operations of this type may also lead to secondary influences on nutrient content, such as contamination with trace metals and additional exposure to oxygen. In some foods an improvement in mineral content can occur, such as increased calcium from exposure to hard water.

Of the operations listed above, blanching leads to the most important nutrient losses. A discussion of blanching, which includes blanching methods and the influence of blanching time and temperature, was published by Lee (147). Blanching is normally accomplished with steam or hot water, the choice depending upon the type of food and subsequent process. Steam blanching generally results in smaller losses of nutrients, since leaching is minimized in this process. Leaching during blanching or cooking can be almost entirely eliminated by using microwave cookers (193), since the need for a heating medium is eliminated.

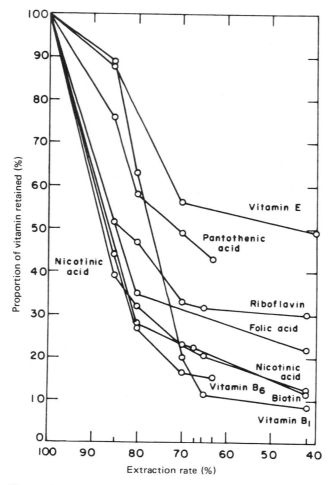

Figure 1 Relation between extraction rate and proportion of total vitamin of the grain retained in flour. (From Ref. 9, courtesy of the FAO of the United Nations.)

An example of typical vitamin losses from broccoli during cooking and the distribution of vitamins between the solid and liquid portions is shown in Table 5. If both liquid and solid portions are considered, there is no measurable loss of B vitamins and a 10-15% loss of vitamin C. If the liquid is discarded, as is the usual practice, substantial losses occur for both boiling and microwave cooking. The indicated losses during microwave cooking are untypically high and could have been minimized by using less cooking water.

Under conditions of good manufacturing practice, leaching, blanching, and cooking losses in the food plant should be no greater, and possibly even smaller, than those found for average practice in the home. A voluminous quantity of data on the vitamin content of canned foods provides verification of this statement (49).

Table 5 Vitamin Content and Distribution in Cooked Broccoli

Cooking method	Solid portion			Liquid portion		
	C	B_1	B_2	C	B_1	B_2
Microwave	64	76	71	23	31	31
Pressure (steam)	72	90	94	6	8	8
Boiling	60	75	69	25	33	33

[a]Expressed as a percentage of broccoli's original vitamin content.
Source: From Thomas et al. (232).

F. Processing Chemicals

A number of different chemicals are added to foods as preservatives or processing aids (Chap. 10), and some of these compounds have a detrimental effect on certain vitamins. For example, oxidizing agents are generally destructive to vitamins A, C, and E; therefore, the use of bleaching or improving agents for flour may lead to a reduction in the content of these vitamins. However, the older process of aging flour through natural oxidative processes undoubtedly leads to similar losses.

Sulfite (SO_2) is used to prevent enzymic and nonenzymic browning of fruits and vegetables. As a reducing agent it protects ascorbic acid, but as a nucleophile it is detrimental to thiamine (see Sec. V.B).

Nitrite is used to preserve meats, and it may be added as such or may be formed by microbial reduction of nitrate. Certain vegetables that naturally contain high concentrations of nitrate, such as spinach and beets, may also contain nitrites because of microbial activity. Nitrite reacts rapidly with ascorbic acid (60) and can also lead to the destruction of carotenoid provitamins, thiamine, and possibly folic acid. Nitrite can act either as an oxidizing agent,

$$NO + H_2O = H^+ + HNO_2 + e^-   E_0 = -0.99 \text{ V}$$

by nucleophilic substitution on N or S, or by addition to double bonds. These reactions are pH sensitive since the reacting species is N_2O_3, formed as

$$H^+ + NO_2^- \rightleftharpoons HNO_2   pK_a = 3.4$$

$$H^+ + NO_2^- + HNO_2 \rightleftharpoons N_2O_3 + H_2O$$

Thus, the reaction with ascorbic acid is pH sensitive, occurring at a negligible rate above pH 6 and proceeding very rapidly at a pH close to or below 3.4.

Ethylene and propylene oxides are used as sterilizing agents, primarily for spices. They are biologically active because of their abilities to alkylate proteins and nucleic acids, and they can react with such vitamins as thiamine via a similar mechanism, resulting in their inactivation.

Alkaline conditions are often employed to extract proteins. Conditions of high pH are also encountered when alkaline baking powders are used. Special instances are also found in which cooked foods, such as eggs, have a pH in the vicinity of 9 because of the loss of CO_2. The destruction of some vitamins (see discussions of individual vitamins), including thiamine, ascorbic acid, and pantothenic acid, is greatly increased under alkaline conditions. Strong acid conditions are only rarely encountered in foods, and few vitamins are sensitive to this condition.

G. Deteriorative Reactions

Many general reactions impair sensory properties of processed and stored foods, and also cause loss of nutrients. Enzymic reactions have already been mentioned, but enzymic contamination by added ingredients should also be considered. For example, the use of plant materials as ingredients can result in the addition of ascorbic acid oxidase to the final food, and the similar use of fishery products as an ingredient can result in the addition of thiaminase.

Lipid oxidation causes the formation of hydroperoxides, peroxides, and epoxides, which will, in turn, oxidize or otherwise react with carotenoids, tocopherols, ascorbic acid, and so on, to cause loss of vitamin activity. The fate of other readily oxidizable vitamins, such as folic acid, B_{12}, biotin, and vitamin D, has not been adequately investigated, but serious losses are not unexpected. The decomposition of hydroperoxides to reactive carbonyl compounds could lead to losses of other vitamins, particularly thiamine, some forms of B_6, and pantothenic acid.

Nonenzymic browning reactions of carbohydrates also lead to highly reactive carbonyl compounds that may be damaging to certain vitamins in the manner just mentioned.

IV. ENRICHMENT, RESTORATION, AND FORTIFICATION

Definitions of the various terms associated with addition of nutrients in the United States are

1. Restoration: Addition to restore the original nutrient content.
2. Fortification: Addition of nutrients in amounts significant enough to render the food a good to superior source of the added nutrients. This may include addition of nutrients not normally associated with the food or addition to levels above that present in the unprocessed food.
3. Enrichment: Addition of specific amounts of selected nutrients in accordance with a standard identity as defined by the U. S. Food and Drug Administration.

A joint policy statement by the Council on Foods and Nutrition of the American Medical Association and the Food and Nutrition Board of the National Academy of Sciences-National Research Council (244) endorsed the continuation of enrichment programs:

> Specifically the following practices in the United States continue to be endorsed: The enrichment of flour, bread, degerminated and white rice (with thiamin, riboflavin, niacin, and iron); the retention or restoration of thiamin, riboflavin, niacin, and iron in processed food cereals; the addition of vitamin D to milk, fluid skimmed milk, and nonfat dry milk, the addition of vitamin A to margarine, fluid skim milk, and nonfat dry milk, and the addition of iodine to table salt. The protective action of fluoride against dental caries is recognized and the standardized addition of fluoride is endorsed in areas in which the water supply has a low fluoride content.

In addition, the Council on Foods and Nutrition and the Food and Nutrition Board, in the same policy statement, continue to endorse the addition of nutrients to foods in keeping with all the following circumstances.

1. The intake of the nutrient is below the desirable level in the diets of a significant number of people.
2. The food used to supply the nutrient is likely to be consumed in quantities that will make a significant contribution to the diet of the population in need.
3. The addition of the nutrient is not likely to create an imbalance of essential nutrients.
4. The nutrient added is stable under proper conditions of storage and use.
5. The nutrient is physiologically available from the food.
6. There is reasonable assurance against excessive intake to a level of toxicity.

Discussion of most of these points are outside the scope of this chapter, but some comment on point 4 is appropriate. The use of cereal grains has been proposed to meet the criterion established under point 2 above (Proposed Policy NAS, 1974), and the stability of vitamins added to cereal grain products (Table 6) and of micronutrients added to breakfast cereals (Table 7) has been established. More recent investigations (75,209) describe the stability of the NAS-NRC proposed micronutrients when added to cereal products, including wheat flour, cornmeal, rice, and bread. A number of prototype products with increased nutrient content have been manufactured by the food industry or are in various stages of development and marketing. There is little doubt that fortified cereal grain products can and should play a positive role in meeting dietary goals and related guidelines (141).

Table 6 Stability of Vitamins Added to Cereal Grain Products

Vitamin	Claim	Found[a]	Months of storage (23°C)		
			2	4	6
White flour					
Vitamin A (IU)	7500	8200	8200	8020	7950
Vitamin E[b] (IU)	15.0	15.9	15.9	15.9	15.9
Pyridoxine (mg)	2.0	2.3	2.2	2.3	2.2
Folacin (mg)	0.30	0.37	0.30	0.35	0.3
Thiamine (mg)	2.9	3.4	—	—	3.4
Yellow corn meal					
Vitamin A (IU)	—	7500	7500	—	6800
Vitamin E[b] (IU)	—	15.8	15.8	—	15.9
Pyridoxine (mg)	—	2.8	2.8	—	2.8
Folacin (mg)	—	0.30	0.30	—	0.29
Thiamine (mg)	—	3.5	—	—	3.6

Vitamin	Claim	After baking[c]	5 Days of bread storage (23°C)
Bread			
Vitamin A (IU)	7500	8280	8300
Vitamin E[b] (IU)	15	16.4	16.7
Pyridoxine (mg)	2	2.4	2.5
Folacin (mg)	0.3	0.34	0.36

[a]Per pound (453.6 g).
[b]As *dl*,α-tocopheryl acetate.
[c]Per loaf (740 g).
Source: From Cort et al. (52).

Table 7 Stability of Micronutrients Added to Breakfast Cereals

| | | Storage periods | |
Micronutrient[a]	Initial value	3 Months, 40°C	6 Months, 23°C
Vitamin A (IU)	5450	4745	5525
Ascorbic acid (mg)	74	69	70
Thiamine (mg)	1.7	1.7	1.8
Riboflavin (mg)	2.0	2.1	1.9
Niacin (mg)	26	24	25
Iron (mg)	20	19	20
Vitamin D (IU)	480	440	470
Vitamin E (IU)	14	14	13
Pyridoxine (mg)	2.4	2.5	2.3
Folacin (mg)	0.5	0.4	0.5
Vitamin B_{12} (μg)	6.1	5.9	5.9
Pantothenic acid (mg)	12	11	11

[a]Micronutrients present in 1 oz. cereal.
Source: From Anderson et al. (3).

V. WATER-SOLUBLE VITAMINS

A. Ascorbic Acid

1. Structure

L-Ascorbic acid is a highly soluble compound that has both acidic and strong reducing properties. These qualities are attributable to its enediol structure, which is conjugated with the carbonyl group in a lactone ring. The natural form of the vitamin is the L-isomer; the D-isomer has about 10% of the activity of the L-isomer and is added to foods for nonvitamin purposes.

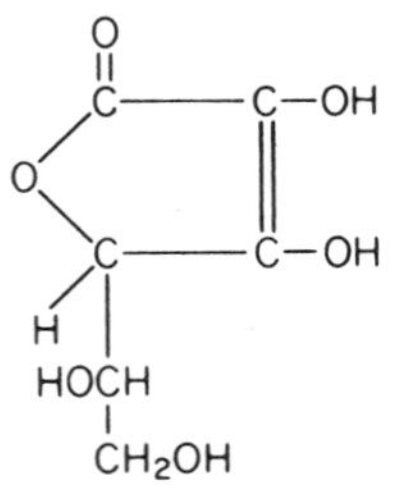

L-ascorbic acid 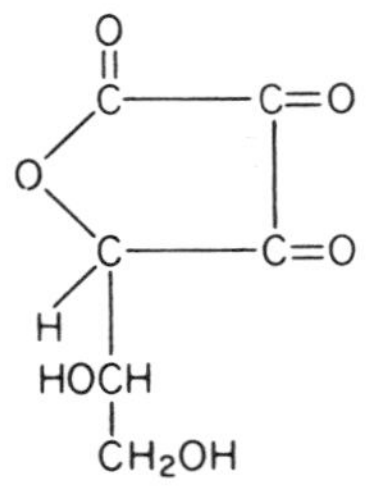   L-dehydroascorbic acid

In solution, the hydroxyl on C3 readily ionizes (pK_1 = 4.04 at 25°C) and a solution of the free acid gives a pH of 2.5. The second hydroxyl is much more resistant to ionization (pK_2 = 11.4).

2. Stability

Ascorbic acid is highly sensitive to various modes of degradation. Factors that can influence the nature of the degradative mechanism include temperature, salt and sugar concentration, pH, oxygen, enzymes, metal catalysts, initial concentration of ascorbic acid, and the ratio of ascorbic acid to dehydroascorbic acid (47,48,81,116,119, 133,134,136,224).

Since so many factors can influence the nature of ascorbic acid degradation, it is not feasible to construct clearly defined precursor-product relationships for any but the earliest products in the reaction pathway. Proposed reaction mechanisms and pathways are based upon both kinetic and physical and chemical measurements, as well as on structure determination of isolated products. Many of these studies have been conducted in model systems at a pH less than 2, or in high concentrations of organic acids, and they therefore may not duplicate the exact degradation pattern that occurs in a particular food product that contains ascorbic acid.

The scheme shown in Fig. 2 demonstrates the influence of oxygen and heavy metals on the route and products of the degradation reaction (116,125,126,133-135). In the presence of oxygen, ascorbic acid is degraded primarily via its monoanion (HA$^-$) to dehydroascorbic acid (A), the exact pathway and overall rate a function of the concentration of metal catalysts (M^{n+}) in the system. The rate of formation of A is approximately first order with respect to [HA$^-$], [O$_2$], and [M^{n+}]. When the metal catalysts are Cu^{2+} and Fe^{3+}, the specific rate constants are several orders of magnitude greater than for spontaneous oxidation, and therefore even a few parts per million of these metals may cause serious losses of vitamin C in food products. The uncatalyzed reaction is not proportional to oxygen concentration at low partial pressures of oxygen, as shown in Table 8. Below a partial pressure of 0.40 atm, the rate seems to level off, indicating a different oxidative path. One possibility is direct oxidation by hydroperoxyl radicals (HO$_2$·) or hydrogen peroxide. In contrast, the rate in the catalyzed pathway is proportional to oxygen partial pressures down to 0.19 atm. The postulated pathway involves formation of a metal anion complex, MHA$^{(n-1)+}$, which combines with oxygen to give a metal-oxygen-ligand complex, MHAO$_2^{(n-1)+}$. This latter complex has a resonance form of a diradical that rapidly decomposes to give the ascorbate radical anion (A·), the original metal ion (M^{n+}), and (HO$_2$·). The radical anion (A·) then rapidly reacts with O$_2$ to give dehydroascorbic acid (A). The oxygen dependence of the catalyzed reactions is a key to establishing this mechanism in which oxidation takes place in the rate-determining step of MHAO$_2^{(n-1)+}$ formation. The oxygen is of considerable importance in explaining the influence of sugars and other solutes on ascorbic acid stability, where at high solute concentrations there is a salting-out effect on dissolved oxygen.

In the uncatalyzed oxidative pathway the ascorbate anion (HA$^-$) is subject to direct attack by molecular oxygen in a rate-limiting step, to give first the radical anions (A·) and (HO$_2$·), followed rapidly by formation of (A) and H$_2$O$_2$. Thus, the catalyzed and uncatalayzed pathways have common intermediates and are indistinguishable by product analysis. Since dehydroascorbic acid is readily reconverted to ascorbate by mild reduction, the loss of vitamin activity comes only after hydrolysis of the lactone to form 2,3-diketogulonic acid (DKG).

The pH-rate profile for uncatalyzed oxidative degradation is an S-shaped curve that increases continuously through the pH corresponding to the pK$_1$ of ascorbic acid and then tends to flatten out above pH 6. This is taken as evidence that it is primarily the monoanion that participates in oxidation. In catalyzed oxidation, the rate is inversely proportional to [H$^+$], indicating that H$_2$A and HA$^-$ compete for O$_2$. However, the specific rate constant for HA$^-$ is 1.5-3 orders of magnitude greater than for H$_2$A (125).

Under anaerobic conditions the rate of ascorbic acid oxidation reaches a maximum at pH 4, declines to a minimum at pH 2, and then increases again with increasing acidity

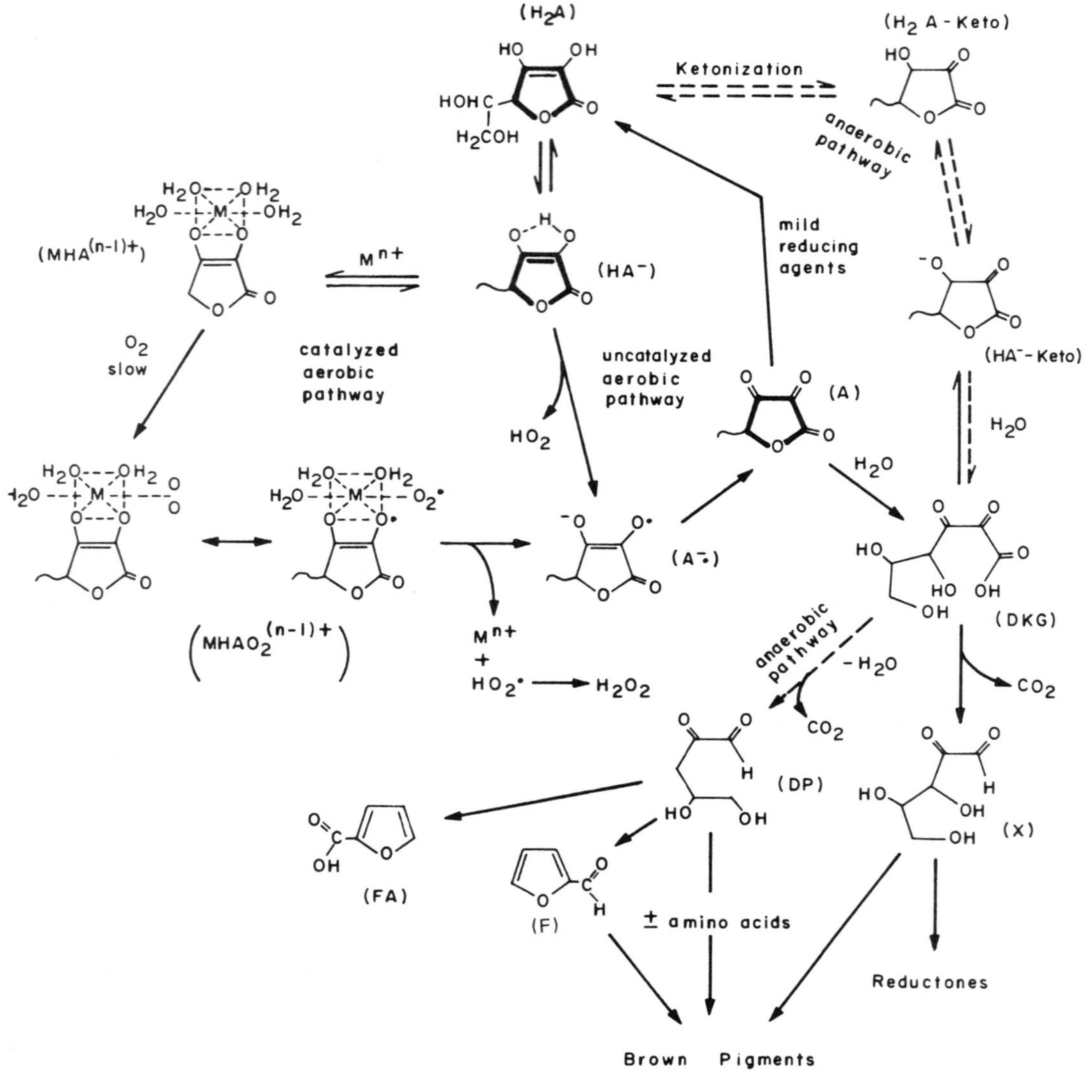

Figure 2 Degradation of ascorbic acid. Bold lines, major forms with vitamin activity. H_2A, reduced ascorbic acid; HA, monoanion of ascorbic acid; A, dehydroascorbic acid; $A^{\bar{\cdot}}$, ascorbate radical anion; DKG, diketogulonic acid; M^{n+}, metal catalyst; $HO_2\cdot$, hydroperoxyl radical; DP, 3-deoxypentosone; X, xylosone; F, furfural; FA, 2-furancarboxylic acid.

(116). Characteristics of the reaction below pH 2 are of little significance in foods, but the maximum at pH 4 is of considerable practical significance and remains the object of much experimentation and speculation. The scheme for anaerobic degradation shown in Fig. 2 is speculative. Following the suggestion of Kurata and Sakurai and their coworkers (133-135), ascorbic acid is shown to react via its keto tautomer (H_2A-keto). The tautomer is an equilibrium with its anion (HA^--keto) which undergoes delactonization to DKG.

Table 8 Variation of Rate Constants (sec^{-1}) of Uncatalyzed Ascorbic Acid Oxidation with Oxygen Partial Pressure

Partial oxygen pressure (atm)	Specific rate constant for ascorbate anion $\times 10^4$
1.00	5.87
0.81	4.68
0.62	3.53
0.40	2.75
0.19	2.01
0.10	1.93
0.05	1.91

Source: From Khan and Martell (126).

Although the anaerobic pathway could also contribute to ascorbic acid degradation in the presence of oxygen, even the uncatalyzed oxidative rate is very much greater than the anaerobic rate at ambient temperatures. Therefore, both pathways may be operative in the presence of oxygen, with the oxidative pathways dominant. In the absence of oxygen there is no added influence of metal catalysts; however, certain chelates of Cu^{2+} and Fe^{3+} are catalytic in a manner independent of oxygen concentration, with catalytic effectiveness a function of metal chelate stability (126).

In Fig. 2 further degradation is shown beyond DKG. Although these reactions are not of nutritional importance (nutritional value is already lost at this point), the decomposition of ascorbic acid is closely tied to nonenzymic browning in some food products. Evidence accumulated thus far tends to indicate a major divergence of products formed, which depends on whether decomposition was oxidative. Since the divergence appears to occur following DKG formation, it is somewhat paradoxical that the reactions themselves do not require molecular oxygen. However, in oxidative degradation, a relatively large proportion of ascorbate is rapidly convered to A, which in turn influences the reaction chemistry via interactions not explicitly shown in the scheme.

Xylosone (X) may be formed by simple decarboxylation of DKG, whereas 3-deoxypentasone (DP) is formed by β-elimination at the C4 of DKG followed by decarboxylation. It may be the accumulation rate of DKG that influences its mode of decomposition, or it may be a more specific interaction with A. In either case, the reaction at this stage begins to assume the characteristics of other carbohydrate, nonenzymic browning reactions (113). Xylosone is further degraded to reductones and ethylglyoxal, whereas DP is degraded to furfural (F) and 2-furancarboxylic acid (FA). Any or all of these compounds may combine with amino acids and contribute to the browning of foods (119).

Spanyar and Kevei (224), who have completed the most critical study of factors influencing ascorbic acid degradation, examined the copper-catalyzed reaction with respect to oxygen concentration; Fe^{3+} catalysis; pH and temperature; ascorbate, dehydroascorbate, and isoascorbate concentrations; cysteine and glutathione; and polyphenols. The most interesting findings of this study involved the interactions of cysteine, Cu^{2+}, and pH. If sufficiently high concentrations of cysteine are present, ascorbate is completely protected, even when it is still in molar excess of cysteine. This protection may be related to the interaction of cysteine with copper, since the pH-rate profile of ascorbate degradation was no longer related to the reciprocal of $[H^+]$, but showed a minimum in the vicinity of pH 4. Although the authors are unaware of any foodstuffs that behave in this manner, an insufficient number of cases have been examined and the cysteine effect may prove to be significant.

In a recent study, volatile degradation products were isolated from a solution of L-dehydroascorbic acid in phosphate buffer solutions at pH 2, 4, 6, or 8 and then heated under reflux for 3 hr or held at 25°C for 200 hr [245]. Of the 15 products identified, the five main degradation products were 3-hydroxy-2-pyrone, 2-furancarboxylic acid, 2-furaldehyde, acetic acid, and 2-acetylfuran. Formation of these compounds depended on pH and temperature; the presence of oxygen had no pronounced effect.

Ascorbic acid is readily oxidized by nitrous acid. Thus, the addition of ascorbic acid to foods has been proposed to prevent nitrosamine formation in products containing sodium nitrite (166). The amount of ascorbic acid required to prevent nitrosamine formation depends markedly on pH and oxygen concentration (7).

3. Assay

A variety of analytical procedures exist for detecting ascorbic acid, but no procedure is entirely satisfactory due to the lack of specificity and problems arising from the numerous interfering substances contained in most foodstuffs. Analysis usually involves oxidation of ascorbic acid by a redox dye, such as 2,6-dichlorophenolindophenol (115). This procedure does not take into account dehydroascorbic acid, which has approximately 80% of the vitamin activity of ascorbic acid. Therefore, redox procedures are often employed in conjunction with treatment of the sample extract with reductants such as H_2S (E_0', pH 7 $= -0.08$ V for $H_2A \rightarrow A$). An alternative approach utilizes the carbonyl properties of A to form bisphenylhydrazone from phenylhydrazine. This procedure obviously is susceptible to errors introduced by the presence of similarly reactive carbonyls that have no vitamin activity.

4. Effect of Processing

Since ascorbic acid is soluble in water, it is readily lost via leaching from cut or bruised surfaces of foods; however, in processed foods the most significant losses result from chemical degradation. In foods that are particularly rich in ascorbic acid, such as fruit products, loss is usually associated with nonenzymic browning. Composition tables may be unreliable for estimating expected concentrations of ascorbic acid, since significant amounts of ascorbic acid are used as processing aids in many types of foods (11).

In such foods as canned juices the loss of ascorbic acid tends to follow consecutive first-order reactions. That is, a rapid initial reaction is oxygen-dependent and proceeds until the available oxygen is completely exhausted, followed by anaerobic degradation. In dehydrated citrus juices degradation of ascorbic acid appears to be only a function of temperature and moisture content (121). The influence of water activity on the stability of ascorbic acid in a variety of foods is shown in Fig. 3 (138). Although ascorbic acid appears to be degraded even at very low water activities, the rate becomes so slow that long storage can be used without excessive ascorbate loss. Data on vitamin C stability in a variety of food and beverage products are summarized in Table 9.

Although the stability of ascorbic acid generally increases as the temperature is lowered, a few investigations have indicated that there might be an accelerated loss upon freezing or in frozen storage. This has been shown to be unlikely for most practical food situations (234); however, storage temperatures above −18°C ultimately lead to

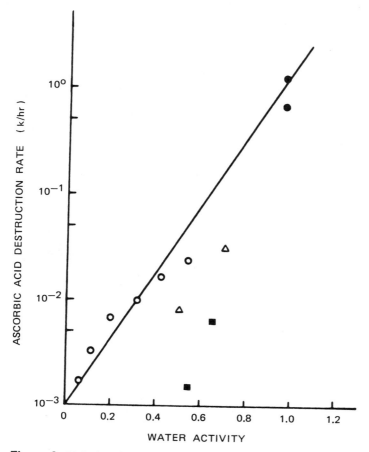

Figure 3 Relation between water activity and rate of destruction of ascorbic acid. ○ orange juice crystals; ● sucrose solution; △ CSM mixture; ■ wheat flour. [From Labuza (138).]

significant losses (108). In general, the largest losses of vitamin C in noncitrus food occur during heating. This is illustrated in Fig. 4 for peas processed by a variety of techniques. It is apparent that the leaching loss during heating far exceeds losses during other process steps. This observation applies to most water-soluble nutrients. Another processing variable that can affect ascorbic acid losses is sulfur dioxide treatment. Fruit products treated with SO_2 have reduced ascorbic acid losses during processing, as well as during storage (17,28).

Studies on ascorbic acid losses in tomato juice (149), green beans (44), and peas (198) have been published recently.

B. Thiamine

1. Chemistry

Thiamine, or vitamin B_1, consists of a substituted pyrimidine linked by a methylene group to a substituted thiazole. It is widely distributed throughout the plant and animal

Table 9 Vitamin C Stability in Fortified Foods and Beverages After Storage at 23°C for 12 Months

Product	No. Samples	Retention (%)	
		Mean	Range
Ready-to-eat cereal	4	71	60-87
Dry fruit drink mix	3	94	91-97
Cocoa powder	3	97	80-100
Dry whole milk, air pack	2	75	65-84
Dry whole milk, gas pack	1	93	—
Dry soy powder	1	81	—
Potato flakes[a]	3	85	73-92
Frozen peaches	1	80	—
Frozen apricots[b]	1	80	—
Apple juice	5	68	58-76
Cranberry juice	2	81	78-83
Grapefruit juice	5	81	73-86
Pineapple juice	2	78	74-82
Tomato juice	4	80	64-93
Vegetable juice	2	68	66-69
Grape drink	3	76	65-94
Orange drink	5	80	75-83
Carbonated beverage	3	60	54-64
Evaporated milk	4	75	70-82

[a]Stored for 6 months at 23°C.
[b]Thawed after storage in a freezer for 5 months.
Source: From Bauernfeind and Pinkert (11), compiled by DeRitter (67).

kingdom, since it plays a key role as a coenzyme in the intermediary metabolism of α-keto acids and carbohydrates. As a result, thiamine can exist in foods in a number of forms, including free thiamine, the pyrophosphoric acid ester (cocarboxylase), and bound to the respective apoenzyme. The details of these various structures are shown in Fig. 5.

Since thiamine contains a quaternary nitrogen, it is a strong base and is completely ionized over the entire range of pH normally encountered in foods. In addition, the amino group on the pyrimidine ring ionizes, the extent depending on pH ($pK_a = 4.8$). The coenzyme role of thiamine is elaborated through position 2 of the thiazole ring, which in its ionized form is a strong nucleophile. Studies on deuterium exchange at this position have shown a half-life of thiamine at room temperature of 2 min at pH 5. At pH 7 the exchange reaction is too fast to follow by ordinary techniques (34).

2. Assay

Thiamine is characterized by strong ultraviolet (UV) absorption bands with a pH-dependent absorption maximum (Table 10). These spectra are useful for analysis under the most limited circumstances only, since many thiamine degradation products also absorb UV radiation. The method of choice for analysis is usually the thiochrome procedure, which involves treatment of thiamine with a strong oxidizing agent (e.g., ferricyanide or hydrogen peroxide) to effect formation of strongly fluorescent thiochrome (see Fig. 5).

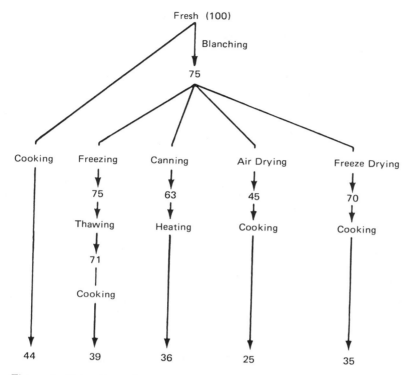

Figure 4   Retention of ascorbic acid in processed peas. [After Bender (18) and Mapson (159).]

Adaptation of this procedure to foods requires enzymic hydrolysis to convert combined forms of the vitamin to free thiamine, and a chromatographic cleanup prior to conversion to thiochrome (115). Analysis by high-performance liquid chromatography (HPLC) is also employed (95).

3. Stability

Thiamine is one of the least stable of all the vitamins. Its stability depends on pH, temperature, ionic strength, buffer type, and other reacting species (72-74,172,173). The typical degradative reaction appears to involve a nucleophilic displacement at the methylene carbon joining the two ring systems. Therefore, strong nucleophiles, such as HSO_3^- (sulfite), readily cause destruction of the vitamin. The similarity of thiamine degradation by sulfite and by alkaline pH is shown in Fig. 5. Both reactions yield 5-(β-hydroxyethyl)-4-methylthiazole and a corresponding substituted pyrimidine. With sulfite, the latter compound is 2-methyl-5-sulfomethylpyrimidine, whereas with alkali it is the corresponding hydroxymethylpyrimidine. The chemistry of thiamine cleavage by sulfite has been studied extensively (150) since this reaction is of particular significance in dehydrated vegetables and fruits in which sulfites are used to inhibit browning.

   Thiamine is also inactivated by nitrite, possibly via reaction with the amino group on the pyrimidine ring. However, it was noticed very early (14) that this reaction is mitigated in meat products as compared with buffer solutions, which implies a protective

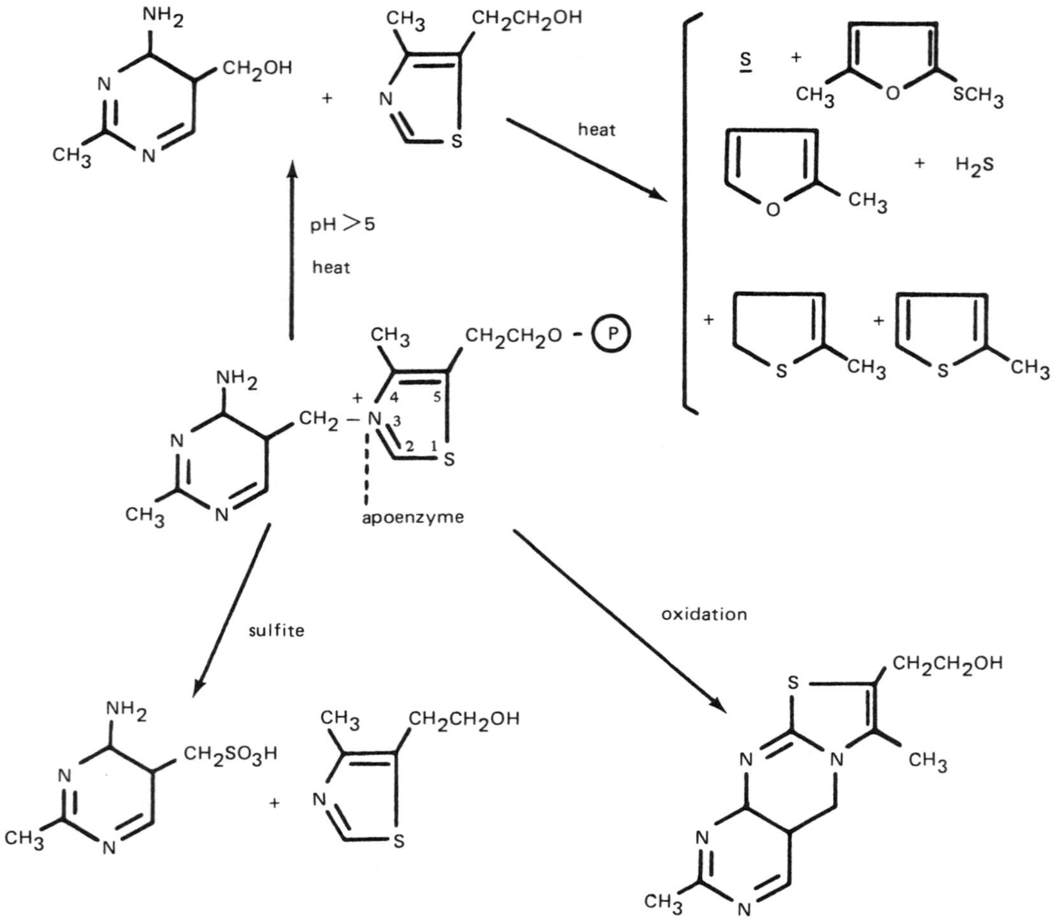

Figure 5 Degradation of thiamine. The circled P is pyrophosphate, which is not essential for vitamin activity. Only the intact vitamin has biological activity.

effect of protein. Destruction of thiamine by sulfite is also lessened when casein and soluble starch are present (150). In the latter case the protective effect was shown to be unrelated to the binding of reactants by the protective substances, to oxidation of reactive sulfhydryl groups, or to competitive oxidation of sulfite by protein. Although the mechanisms of the protective effect are still unclear, it is probable that the protective effect involves other degradative mechanisms, and may be another important source of discrepancies in the literature.

Table 10   Molar Absorptives of Thiamine Solutions

System	λ_{max} (nm)	ϵ
0.005 N HCl	247	14,500
pH 7.0	231	12,000
	267	8,600

Since thiamine can exist in multiple forms its stability depends on the relative concentrations of the various forms. Within a given animal species the ratios of the form depend on the nutritional status prior to death, with variations from one type of muscle to another. The ratios of the forms also depend on physiological stresses that can occur in plants postharvest and in animals immediately postmortem. In the relatively few studies that have been conducted, the enzyme-bound forms (e.g., cocarboxylase) appear to be less stable than the free vitamin. Farrer (78) suggested that the relative concentrations of the various forms of the vitamin may account for some discrepancies in the literature regarding thiamine stability.

The literature through 1953 has been analyzed and summarized by Farrer (78), and only typical examples are used here for didactic purposes. Extensive losses of thiamine occur in cereals as a result of cooking or baking, and in meats, vegetables, and fruits as a result of various processing operations and storage. The stability of thiamine is so strongly influenced by the nature and state of the system that it is difficult to extrapolate between systems, and numerous unexplainable differences exist in the literature (see Ref. 78 for examples).

Temperature is an important factor influencing thiamine stability. Data in Table 11 show differences in thiamine retention among various foods held at two different storage temperatures. Rate constants for thiamine degradation in various foods (including peas, carrots, cabbage, potatoes, and pork) range from 0.0020 to 0.0027 per minute at 100°C. However, these values cannot be extrapolated to other temperatures unless the Arrhenius activation energy is known for the particular system. Activation energies summarized by Farrer (78) vary by a factor of more than 2, depending on the system and the reaction conditions. In phosphate buffer at pH 6.8, Goldblith and Tannenbaum (90) found the activation energy for thiamine degradation to be 22 kcal/mol for both conventional and microwave heating, which is similar to the value in pureed meats and vegetables found by Feliciotti and Esselen (79). Leichter and Joslyn (151) found that the presence of sulfite lowered the activation energy to 13.6 kcal/mol. Since it is also possible that catalytic concentrations (30) of such metals as copper can accelerate the rate of degradation at a given pH, differences in activation energies conceivably can be ascribed to small compositional differences in the systems. However, Feliciotti and Esselen (79) found values similar to that reported above for activation energies in a wide variety of meats and vegetables, irrespective of sample pH. Labuza and Kamman (139) investigated contradictory reports on the stability of different salt forms of thiamine and found the explana-

Table 11 Thiamine Retention in Stored Foods

| System | Retention after 12 months storage | |
	38°C	1.5°C
Apricots	35	72
Green beans	8	76
Lima beans	48	92
Tomato juice	60	100
Peas	68	100
Orange juice	78	100

Source: From Freed et al. (84).

tion to lie with differences in activation energies for destruction of the mononitrate and hydrochloride forms (26.3 versus 22.4 kcal/mol respectively). This leads to a greater stability for the mononitrate form below 95°C and greater stability for the hydrochloride form above 95°C. Stability of both forms was lower at a water activity of 0.86 than at 0.58.

As previously indicated, the rate of thiamine degradation is extremely sensitive to pH. The pH-rate profiles for free thiamine and cocarboxylase at elevated temperature are shown in Fig. 6. It is apparent that either the starch and/or protein components of cereal products exert a protective effect over the pH range examined. Cocarboxylase is more sensitive than thiamine, but the difference in sensitivity is a function of pH, disappearing completely at pHs above 7.5 Since both the amino group on the pyrimidine ring and the 2 position on the thiazole ring are strongly influenced by pH in the region of interest for thiamine stability, either site could be implicated in the degradative reactions. However, based on the nature of the secondary products, it appears the thiazole ring is the most likely site for the degradation reaction.

As is true of other water-soluble vitamins, thiamine can be extensively lost by leaching during cooking operations (see Table 5) and from cereals during milling (see

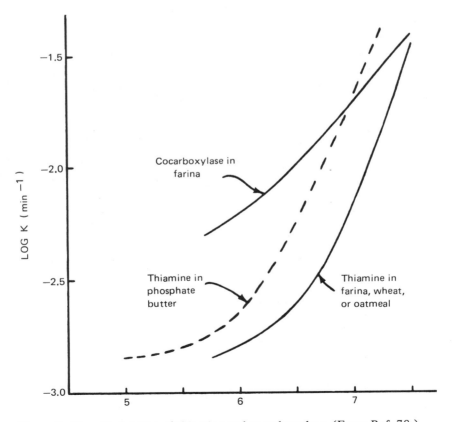

Figure 6 The pH stability of thiamine and cocarboxylase. (From Ref. 78.)

Fig. 1). Little information is available on thiamine stability in dehydrated foods. Studies in dehydrated corn-soymilk (CSM) samples indicate that degradation is influenced strongly by moisture content (29). For example, storage at 38°C for 182 days caused no loss of thiamine when the system was maintained below 10% moisture content, but extensive loss occurred at 13% moisture. Because of the multiple possibilities for physical separation and chemical degradation, substantial losses of thiamine occur in many foods unless great care is taken during all handling, storage, and processing operations.

Extracts from various fish and crustaceans have been found to destroy thiamine (88,94,259), and an enzyme with antithiamine activity was proposed. However, recently the antithiamine factor from carp viscera was shown to be thermostable and probably not an enzyme. The antithiamine factor was identified as hemin or a related compound (132,223). Similarly, various heme proteins from tuna, pork, and beef have been shown to have antithiamine activity (192).

4. Effect of Processing

The results of several studies on thiamine degradation in a variety of foods are summarized in Table 12.

Thermal destruction of thiamine leads to the formation of a characteristic odor, which is involved in the development of "meaty" flavors in cooked foods. Some of the probable reactions (8,71,72,74,169) are summarized in Fig. 5. Following release of the thiazole ring, further degradation is then thought to result in such products as elemental sulfur, hydrogen sulfide, a furan, a thiophene, and a dihydrothiophene. The reactions leading to these products are unclear, but extensive degradation and rearrangement of the thiazole ring must be involved.

Table 12 Studies on Thiamine Degradation in Foods

Product	Treatment	Retention (%)	Reference
Cereals	Extrusion cooking	48-90	Beetner et al. (16)
Potatoes	Soaked in water for 16 hr, then fried	55-60	Oguntona and Bender (183)
	Soaked in sulfite for 16 hr, then fried	19-24	Oguntona and Bender (183)
Soybeans	Soaked in water and then boiled in water or bicarbonate solution	23-52	Perry et al. (189)
Mashed potatoes	Various heating treatments	82-97	Ang et al. (4)
Vegetables	Various heating treatments	80-95	Ang et al. (4)
Meat items	Various heating treatments	83-94	Ang et al. (4)
Frozen fried fish	Various heating treatments	77-100	Ang et al. (4)

C. Riboflavin

1. Structure

Riboflavin (vitamin B$_2$) is an isoalloxazine derivative with a ribitol side chain. In nature it is usually phosphorylated and functions as a coenzyme; one of its forms is flavine mononucleotide (FMN) and the other flavin-adenine dinucleotide (FAD). The enzymes associated with this vitamin are called flavoproteins, and they typically act as hydrogen-transfer agents in the oxidation of such compounds as amino acids and reduced pyridine nucleotides.

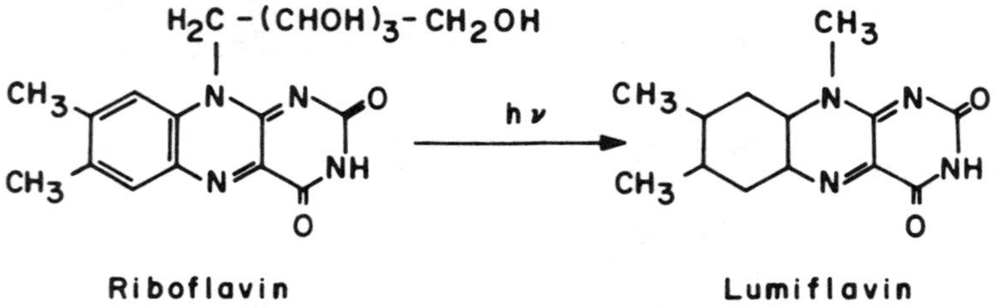

Riboflavin   Lumiflavin

2. Stability

Riboflavin is thermostable and unaffected by atmospheric oxygen. It is stable in strongly acid solution but is unstable in the presence of alkali. It decomposes readily if exposed to light. In alkaline solution, irradiation causes the photochemical cleavage of the ribitol portion, shown above, yielding lumiflavine (131,248). In acid or neutral solution, irradiation leads to production of the blue fluorescent substance lumichrome together with varying amounts of lumiflavin (123).

Lumiflavin, which is apparently a stronger oxidizing agent than riboflavin, can catalyze destruction of a number of other vitamins, particularly ascorbic acid. Some years ago, when milk was sold in transparent bottles, the reaction sequence just described was a significant problem with respect to loss of nutritional value and also with regard to the development of an undesirable flavor known as "sunlight off-flavor." This problem disappeared with the advent of opaque milk containers.

Riboflavin is stable in food under most processing or cooking conditions. In a recent study of the effects of various heating methods on vitamin retention in six fresh or frozen prepared food products, the retention of riboflavin was always greater than 90% (4). Riboflavin retention in peas and lima beans subjected to blanching and processing was in all cases greater than 70% (97). The photochemical destruction of riboflavin has been monitored in macaroni as a function of temperature, water activity, and light (257). The loss of riboflavin occurred in two phases: first, a rapid loss from the irradiated surface, and then a slower phase that followed first-order reaction kinetics. Light intensity was the rate-determining factor in all instances.

3. Assay

Riboflavin is usually assayed fluorometrically by measurement of the characteristic yellowish green fluorescence. It can also be estimated microbiologically using *Lactobacillus casei* (115), or by HPLC (95).

D. Nicotinic Acid

1. Structure

Nicotinic acid is pyridine β-carboxylic acid,

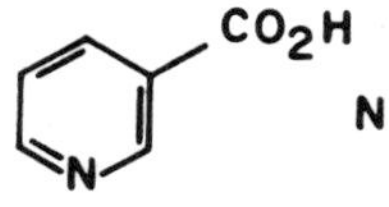

Nicotinic Acid

and nicotinamide is the corresponding amide. The acid and its amide are often given the collective name "niacin." In living systems, nicotinamide is a constituent of the hydrogen-carrying coenzymes nicotinamide adenine dinucleotide (NAD) and nicotinamide adenine dinucleotide phosphate (NADP). Nicotinic acid, which is one of the most stable vitamins, is relatively insensitive to heat, light, air, acid, or alkali.

2. Assay

Niacin is assayed by first hydrolyzing the food with sulfuric acid to liberate nicotinic acid from combined forms (as coenzyme). The pyridine ring of the nicotinic acid is opened with cyanogen bromide, and the fission product is coupled with sulfanilic acid to yield a yellow dye with an absorption maximum at 470 nm (115). Analysis is also possible by HPLC.

3. Stability

Niacin is generally stable in foods, but losses occur in vegetables via nonchemical means, such as trimming and leaching, and these losses parallel those of other water-soluble vitamins (50,208). Considerable losses of niacin occur during the postmortem storage of pork and beef (165,204) due to biochemical reactions. Roasting meat leads to no losses, but drippings contain up to about 26% of the original niacin (55). There appears to be no loss of niacin during milk processing (31).

E. Vitamin B$_6$

1. Structure

There are three closely related compounds that have vitamin B$_6$ activity: pyridoxal (I), pyridoxine or pyridoxol (II), and pyridoxamine (III).

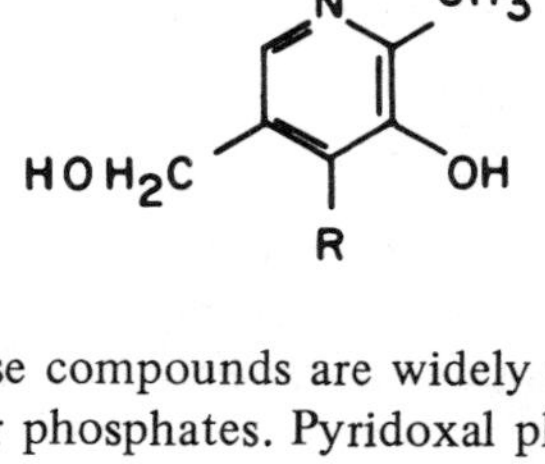

$$\begin{array}{ll}
\underline{R} \\
I & -CHO \\
II & -CH_2OH \\
III & -CH_2NH_2
\end{array}$$

These compounds are widely distributed in the plant and animal kingdoms in the form of their phosphates. Pyridoxal phosphate is a coenzyme for many enzymic transformations of amino acids (e.g., transamination, racemization, and decarboxylation reactions). It exerts its action as a coenzyme via a carbonyl-amine condensation with an amino acid yielding a Schiff base, which is stabilized by chelation to a metal ion (IV).

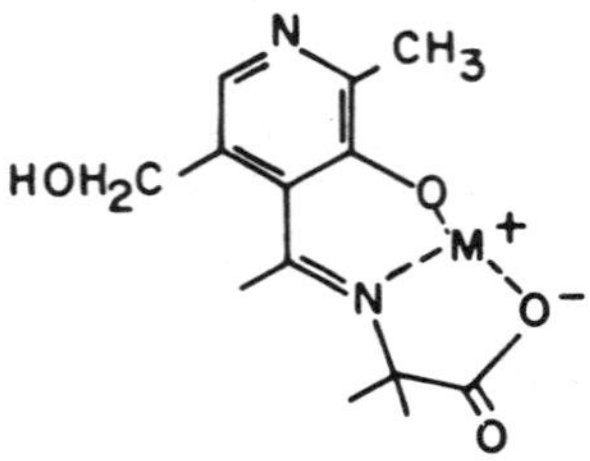

IV

2. Stability

The three forms of vitamin B_6 are stable to heat but are decomposed by alkali. The pyridoxal form of the complex is most stable and is used for fortification of foods. The compounds are converted by UV irradiation in the presence of oxygen to biologically inactive products, such as 4-pyridoxic acid (202). This reaction is probably of little importance in foods, except milk (31).

When a solution of pyridoxal is heated with glutamic acid, a mixture including pyridoxamine and α-ketoglutaric acid results. In fact, heating a mixture of amino acids, pyridoxal, and polyvalent metal ions at 100°C, and at any of a wide range of pH values, leads to the same products that would normally result from the conversion involving the holoenzyme (222).

When cysteine and pyridoxal are allowed to react under conditions similar to those encountered in a sterilization process, the reaction products have no vitamin B_6 activity for rats and approximately 20% activity for *Saccharomyces carlsbergensis*. A product of this reaction appears to be bis-4-pyridoxyldisulfide (251), possibly arising via a thiazolidine (V). A similar reaction sequence is possible by direct reaction of pyridoxal

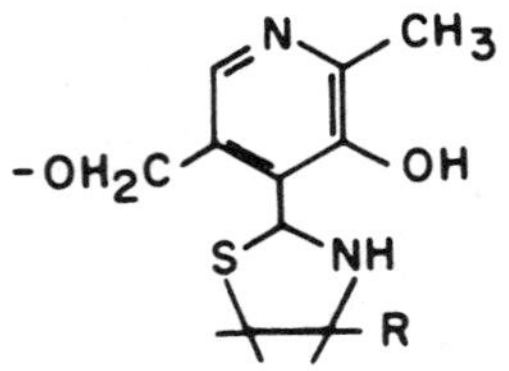

V

with sulfhydryl groups of proteins (225). Since the main result of the interaction of B_6 with amino groups appears to be the interconversion between pyridoxal and pyridoxamine—both with full vitamin B_6 activity—the cysteine reaction may be important to the stability of this vitamin in heat-processed foods.

3. Assay

Vitamin B$_6$ is usually assayed microbiologically using *S. carlsbergensis* as the test organism (115). The individual compounds may be separated by ion-exchange chromatography prior to assay if required, or all forms may be analyzed simultaneously by HPLC (95).

4. Effect of Processing

Information on the distribution of vitamin B$_6$ in its various forms in foods has become available only recently (184,191). Although many foods have not been systematically studied for vitamin B$_6$ destruction during processing, it is nevertheless apparent that both the quantity and form of the vitamin are influenced by the process of heating, concentration, and dehydration.

In view of the chemistry just described, it is not surprising to find an entirely different distribution of vitamin B$_6$ forms in fresh and processed foods. By recalculating some of the data of Polansky and Toepfer (191), it is evident that pyridoxal increases and pyridoxamine decreases during the dehydration of eggs. In fresh milk the main form is pyridoxal; in dried milk pyridoxal is still dominant, but there is more pyridoxamine than in fresh fluid milk. In evaporated milk pyridoxamine is the main form. In raw pork loin the dominant form is pyridoxal, and in fully cooked ham it is pyridoxamine.

A great deal of attention has been paid to the stability of vitamin B$_6$ in heat-processed milk. When pyridoxol is added to milk, it is stable during sterilization. However, the dominant natural form of vitamin B$_6$ in milk is pyridoxal. Studies have shown that sterilized liquid milk and formula milk contain less than half their original vitamin B$_6$ activity, the decrease continuing 7-10 days after processing (107,236). In addition, feeding tests with rats have indicated that the decrease in vitamin B$_6$ activity was even more substantial than that found using the test organism *S. Carlsbergensis*.

In another study (154), both high-temperature-short-time (HTST) pasteurization (2-3 sec at 92°C) and boiling for 2-3 min gave only 3% loss of vitamin B$_6$. Sterilization of milk in bottles (13-15 min at 119-120°C) reduced the vitamin by 84%. Ultra-HTST sterilization of milk by injection of steam at 143°C into preheated milk for 3-4 sec caused negligible loss of vitamin B$_6$. Vitamin B$_6$ is believed to react with substances, possibly cysteine, released from milk proteins upon heating (20,35).

Many processed, edible products have been analyzed for vitamin B$_6$ losses (31,154, 205,213). During canning of vegetables, losses of vitamin B$_6$ range from about 60 to 80%. During the freeze-preservation of vegetables losses range from about 40 to 60%. During canning of seafoods and meats about 45% of the original vitamin B$_6$ is lost. Fruits and fruit juices lose about 15% of their vitamin B$_6$ during freeze-preservation and about 38% during canning. Grains lose about 50-95% of their vitamin B$_6$ during conversion to various cereal products, and meats lose about 50-75% during conversion to comminuted forms.

The susceptibility of vitamin B$_6$ to conventional food processes led in 1952 to a situation that exemplifies the need for concern about the nutrient content of processed foods (177). An infant food, available in liquid and dry form, was identified as the cause of at least 50-60 cases of convulsive seizures. The seizures, which occurred after consumption of the liquid form of the food, were attributed to instability of pyridoxal in the presence of milk proteins. The problem was solved by fortifying the product with the stable form of vitamin B$_6$, pyridoxol.

The kinetics and activation energies for thermal destruction of the different forms of vitamin B_6 have been found to vary between model systems and foods (175). Degradation rates of pyridoxal (I), pyridoxol (II), and pyridoxamine (III) in 0.1 M phosphate buffer (pH 7.2) were determined over the temperature range 110-145°C. Degradation of (III) followed pseudo-first-order kinetics, but (II) was 1.5 order and (I) was second order. The activation energies varied from 40 to 85 kcal/mol depending on the form of the vitamin. Similar experiments in cauliflower puree resulted in a markedly lower activation energy (27 kcal/mol) than was observed in the model system, and nonlinear kinetics that were not first order.

F. Folic Acid

1. Structure

The structures of folic acid and other folates are shown in Fig. 7. Folic acid consists of a 2-amino-4-hydroxypteridine linked to *p*-aminobenzoic acid, which in turn is coupled to glutamic acid (PABG). In biological systems, folic acid can exist in a wide variety of different forms. The pteridine ring can be reduced to yield di- or tetrahydrofolates; five different one-carbon substituents can be present at the N5 and/or N10 position; and the glutamic acid residue can be extended to a poly-γ-glutamyl side chain of varied length. Assuming that the polyglutamyl side chain contains no more than six residues, the theoretical number of possible folic acid compounds exceeds 140, of which about 30 have been isolated and characterized.

Tetrahydrofolic acid is involved in the biological transfer of one-carbon fragments, and its chemistry, function, and biosynthesis have been reviewed in detail (26).

2. Stability

Folic acid is stable to alkali under anaerobic conditions. However, under alkaline aerobic conditions hydrolysis occurs, cleaving the side chain to yield PABG and pterin-6-carboxylic acid (227). Acid hydrolysis under aerobic conditions yield 6-methylpterin. Polyglutamate derivatives of folic acid can be hydrolyzed by alkali in the absence of air to yield folic and glutamic acids (227). Folic acid solutions decompose when exposed to sunlight to yield PABG and pterin-6-carboxaldehyde (153,226). Irradiation of the 6-carboxaldehyde in turn yields the 6-carboxylic acid, which is then decarboxylated to yield pterin (199). These reactions are catalyzed by riboflavine and FMN (207). Only folic acid and its polyglutamate derivatives have vitamin activity.

The interaction of folic acid with two chemicals involved in food processing, sulfite and nitrite, has received some attention. Treatment with sulfurous acid leads to side-chain cleavage with the production of reduced pterin-6-carboxaldehyde and PABG (246). At cold temperatures, nitrous acid reacts with folic acid to yield the N10 nitroso derivative (53). At higher temperatures the 2-amino group also reacts and the 2-hydroxy-10-nitroso derivative is formed (5). It has recently been shown that N10-nitrosofolic acid is a weak carcinogen in mice and hence may represent a hazard to humans (256).

Di- and tetrahydrofolic acid (FH_2 and FH_4, respectively) readily oxidize in air. In neutral solution, FH_4 rapidly oxidizes to yield PABG, pterin, xanthopterin, 6-methylpterin, and other pterin-related compounds (24,200,260), as well as FH_2 and folic acid (185,260,261). Under acid conditions, quantitative cleavage to PABG is observed (261). Air oxidation of FH_4 is substantially reduced in the presence of thiols (25), cysteine (195), or ascorbate (196). Dihydrofolic acid, FH_2, is somewhat more stable than FH_4,

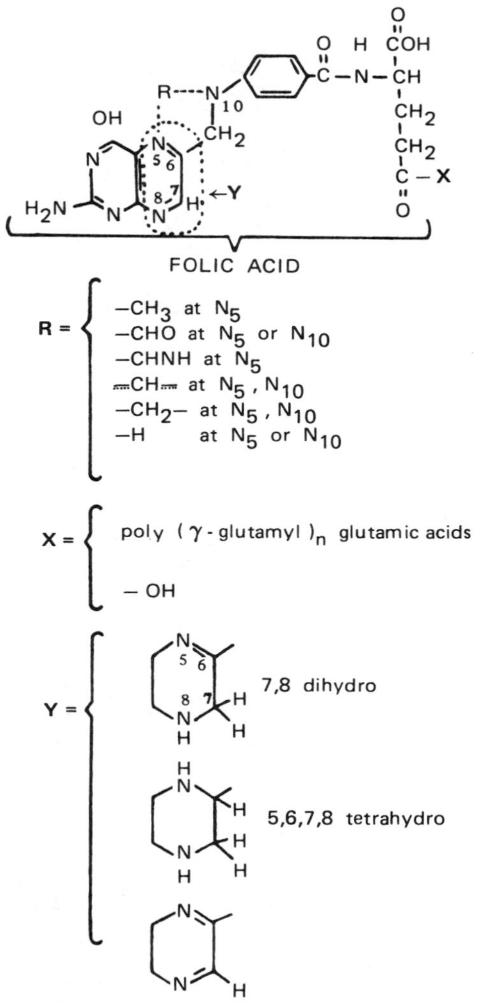

Figure 7 Schematic representations of the theoretical structures of the folates. (From Ref. 12.)

but is also subject to oxidative degradation. FH_2 is oxidized more rapidly in acid than in basic solution, and the products are PABG and 7,8-dihydropterin-6-carboxaldehyde (112,253). Again, reducing agents, such as thiols or ascorbate, retard oxidation.

Many of these studies employed analytical methods that were inadequate to separate the complex products that form from folate degradation. In order to investigate the complex chemistry of FH_4 oxidation, Reed and Archer (201) successfully developed HPLC methods. At pH 4, 7, and 10 the major product resulting from air oxidation of FH_4 in aqueous solution was found to be (*p*-aminobenzoyl)glutamate, indicating that oxidation leads predominantly to cleavage of the tetrahydrofolate. The major products containing the pterin ring were pterin at pH 4 and 6-formylpterin at pH 7 and 10. FH_2 was only detected in the reaction at pH 10. The mechanism of formation of FH_2 and pterins from FH_4 is shown in Fig. 8.

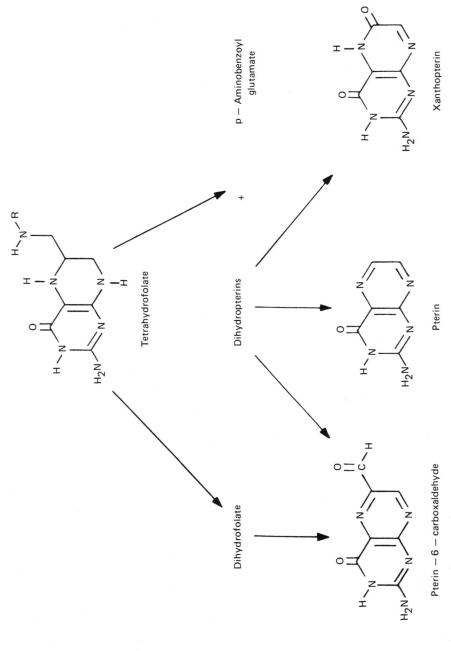

Figure 8 Formation of pterins and 7,8-dihydrofolate from tetrahydrofolate. [Adapted from Reed and Archer (201).]

The compound 5,10-methylene-FH_4 is much more stable to oxidation than FH_4, particularly at high pH (186,196). Stability is decreased by the presence of amines or ammonium salts (196). 10-Formyl-FH_4 is as unstable as FH_4 toward oxygen and other oxidizing agents. On the other hand, 5-formyl-FH_4 is stable to oxygen in neutral or mildly alkaline solution and reacts only slowly with iodine or dichromate in weakly acidic solution. Furthermore, 5-formyl-FH_4 shows considerable stability to alkaline hydrolysis, whereas 10-formyl-FH_4 readily loses its formyl group in alkali (54,160,209). Under acidic conditions, both 5- and 10-formyl-FH_4 lose a molecule of water to form 5,10-methenyl-FH_4. 5-Methyl-FH_4 readily oxidizes at alkaline pH in the presence of air or other oxidizing agents (70,98,124,143). The structure of the oxidation product is probably 5,6-dihydro-5-methylfolate (70). Only reduced forms have vitamin acitivity.

3. Distribution and Assay

Investigation of the distribution of folic acid derivatives in nature has only recently become feasible with the ready synthesis of polyglutamate derivatives (56,130). Stokstad and his group have shown that more than 90% of cabbage folate is in the form of polyglutamates containing more than five glutamic acid residues, mainly as the 5-methyl derivatives (42).

The same workers have shown that soybeans contain mainly monoglutamates (52%), with some diglutamates (16%). Pentaglutamates represent the major portion of the remaining polyglutamates (219). Of the total folate activity in soybeans, 65-70% is in the 5-formyl-FH_4 form. Cow's milk contains about 60% monoglutamates, with the remainder ranging from di to hepta conjugates. In this case, however, 90-95% of the total folate activity is in the 5-methyl-FH_4 form (219).

It has been shown recently that the rate of intestinal absorption of folates is inversely proportional to the length of the γ-glutamyl side chain (13). Thus, before a food can be categorized as a good source of the vitamin, an estimate of chain length is necessary. This type of information is not currently available for most foods.

Folic acid content in foods is usually assayed by microbiological methods after the various glutamate conjugates have been enzymically cleaved by the enzyme conjugase to give free folic acids (115). The free folic acid content of individual foods has been determined using these methods (188,212). In typical American diets the total folate content has been found to be about 180 μg/day (40). These values are significantly lower than the RDA (see Sec. II), and folate deficiency is thought by some to be a significant public health problem (243).

The analysis of folates by HPLC has been described (6).

4. Nutritional Stability

The "nutritional" stabilities of several dietary tetrahydrofolate derivatives in their monoglutamyl forms were recently studied in solution and compared with folic acid itself (182). Nutritional stability was determined by the ability of folate solutions to support growth of *L. casei*. Tetrahydrofolate showed 67% retention of activity when held at 121°C for 15 min in the presence of 1% ascorbate. When the ascorbate concentration was reduced to 0.05%, no activity was left after the same treatment. Complete protection for 5-methyl-FH_4 was afforded by 0.2% ascorbate under these conditions. In the absence of ascorbate, 5-methyl-FH_4 showed its greatest nutritional stability in alkaline conditions, with a half-life of 330 hr in 0.1 M TRIS, pH 9, but only 25 hr in 0.1 M HCl, pH 1.0, at room temperature. The most stable reduced folate proved to be 5-formyl-FH_4, with a half-life of about 30 days at room temperature in the pH range 4-10. With a half-life of about 100 hr at pH 7.0, 10-formyl-FH_4 was more stable than expected; however, this is probably due to conversion to the more stable oxidation product, 10-formylfolic acid, which was nutritionally active in the assay system used. Folic acid itself was extremely stable, with a half-life longer than 700 hr in 0.05 M citrate-phosphate buffer, pH 7.0, at room temperature. However, folic acid was much less stable in the absence of citrate, with a half-life of only 48 hr under the same conditions. No mechanistic interpretation of this result was given, and degradation products were not identified. The stability of folic acid polyglutamate derivatives was not examined.

In another study, pure folic acid in the solid state under normal storage conditions (20°C and 75% relative humidity) had a decomposition rate of 1% per year (237).

5. Effect of Processing

The extent and mechanism of loss of folic acid derivatives in processed foods are not yet clear. Studies of processed and stored milk indicate that the primary inactivation process is oxidative (38,39,83). The destruction of folate parallels that of ascorbate, and added ascorbate can stabilize folate. Both vitamins are greatly stabilized by deoxygenation of the milk, but both inevitably decline during 14 days of storage at 15-19°C.

HTST pasteurization of milk (2-3 sec at 92°C) causes about 12% loss of total folate, and boiling milk for 2-3 min results in a loss of 17% of the total folate (122). Sterilization of milk in bottles (13-15 min at 119-120°C) produces the greatest loss in folate (39%). HTST sterilization of preheated milk by injection of steam at 143°C for 3-4 sec causes only a 7% loss of total folates.

In a study of folic acid in garbanzo beans (chick-peas) (152), washing and soaking resulted in about a 5% loss of folates. In water-blanched beans, free folic acid retention decreased from 75 to 45% as the blanching time increased from 5 to 20 min. Total folates decreased from 77 to 54% under the same conditions. Steam blanching improved folic acid retention to some extent. Folic acid in garbanzo beans was shown to be quite stable toward heat processing. Retorting at 118°C for 30 min resulted in about a 60% retention of free folic acid and about a 70% retention of total folates based on the original amount of folates in dry beans. There was no significant decrease in folate content when the processing time at 118°C was increased from 30 to 53 min.

The fate of folate polyglutamates in meat during storage and processing has been investigated, using chicken liver as a model (200). In the intact tissue, only slight degradation of folate polyglutamates by endogenous conjugase occurred during 48 hr at 4°C, and complete degradation took 120 hr. However, homogenized tissue at 4°C showed complete degradation to folate monoglutamates and only a small amount of diglutamate remained after 48 hr storage. If the liver was heated to over 100°C at any stage prior to or during storage, the polyglutamates were stabilized due to inactivation of the conjugase.

The effects of thermal processing on folacin bioavailability in a liquid model system were examined using a chick bioassay, a microbiological assay, and HPLC analyses (206). These methods yielded essentially the same results. Folacin from cooked beef liver was completely available; that in cooked cabbage was only 60% available.

Date in Table 13 relate to several other studies of folate losses in foods during processing.

Table 13 Folate Losses in Foods Subjected to Various Processes

Food product	Process method	Loss of folate activity (%)	Reference
Eggs	Frying Boiling Scrambling	18-24	Hanning and Mitts (102)
Sauerkraut	Fermentation	None	Cheldin et al (43)
Liver	Cooking	None	Cheldin et al (43)
Halibut	Cooking	46	Cheldin et al (43)
Cauliflower	Boiling	69	Cheldin et al (43)
Carrots	Boiling	79	Cheldin et al (43)
Meats	γ-Irradiation	None	Alexander et al (1)
Grapefruit juice	Canning and storage	Negligible	Krehl and Cowgill (129)
Tomato juice			
Yugoslavia	Canning	70	Suckewer et al (229)
United States	Canning	50	Suckewer et al (229)
	Storage in dark (1 year)	7	Suckewer et al (229)
	Storage in light (1 year)	30	Suckewer et al (229)
Maize	Refining	66	Metz et al (164)
Flour	Milling	20-80	Schroeder (214)
Meat or vegetable stew	Canning and storage (1½ years)	Negligible	Hellendoorn et al (109)
	Canning and storage (3 years)	Negligible	Hellendoorn et al (109)
	Canning and storage (5 years)	Negligible	Hellendoorn et al (109)

Source: From Malin (158).

G. Vitamin B_{12}

1. Structure

Vitamin B_{12} is a red, crystalline substance and is chemically by far the most complex vitamin. Its structure, shown in Fig. 9, has two characteristic components. In the nucleo-tidelike portion, 5,6-dimethylbenzimidazole is bound to D-ribose via an α-glycosidic bond. The ribose contains a phosphate group at the 3′ position. The central ring portion is a "corrin" ring system, which resembles the porphyrins. Coordinated to the four inner nitrogen atoms of the corrin ring is a cobalt atom (vitamin B_{12} is also known as cobalamin). In the form that is usually isolated, the sixth coordination position of the cobalt II

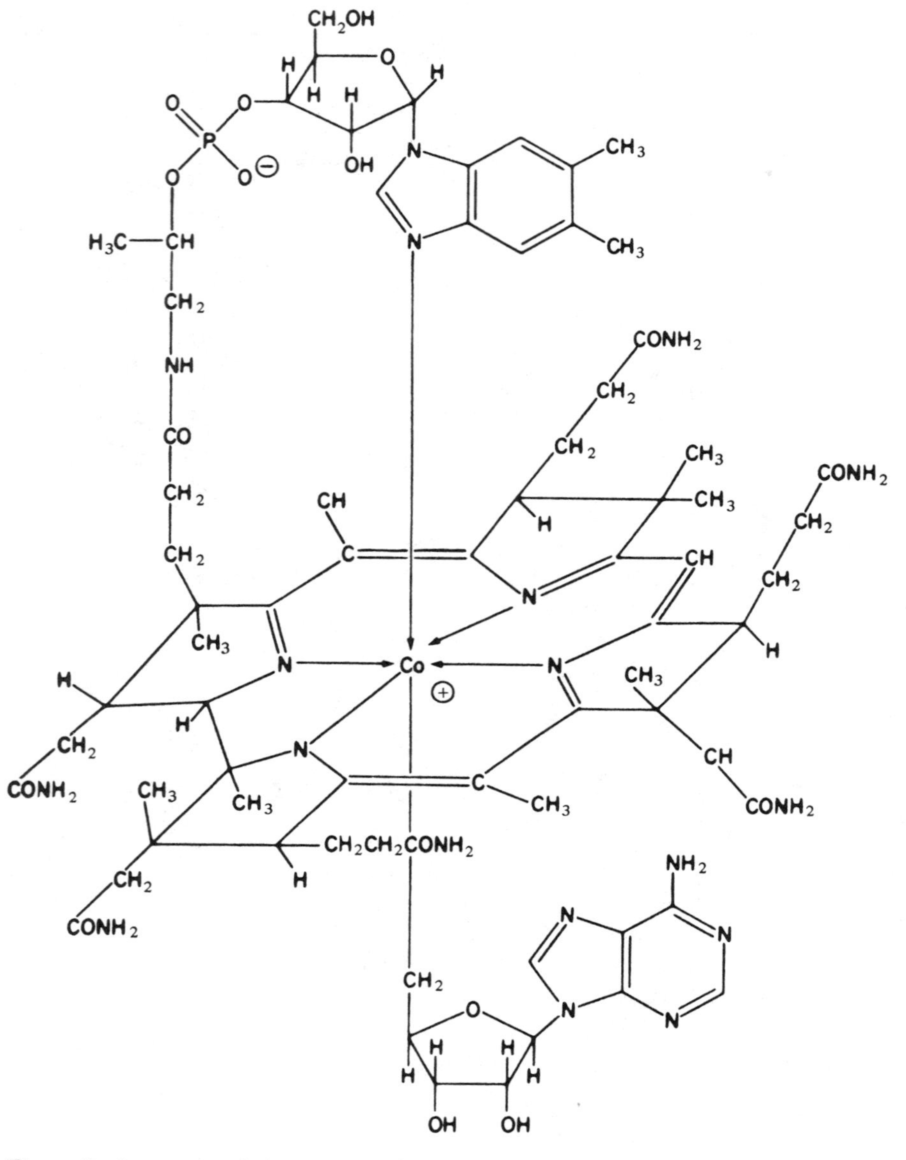

Figure 9   Structure of vitamin B_{12}. [From Whitaker (252).]

atom is occupied by cyanide (cyanocabalamin). In the active coenzyme, the sixth coordination position is filled by 5-deoxyadenosine attached to the cobalt via its methylene group.

Vitamin B_{12} is found in animal tissues but is almost entirely absent from higher plants. It is unique among the vitamins in that it is synthesized exclusively by microorganisms. Vitamin B_{12} coenzyme is required for the action of several enzymes, including methylmalonylmutase and dioldehydratase, and, along with folic acid, for the formation of methionine from homocysteine.

2. Stability

Aqueous solutions of cyanocobalamin are stable at room temperature if they are not exposed to ultraviolet or intense visible light. The pH region of optimal stability is 4-6, and in this region only small losses are found even after autoclaving (31). Heating in alkaline solution can quantitatively destroy vitamin B_{12}. Reducing agents, such as thiol compounds at low concentrations, can protect the vitamin, but in larger amounts they can cause destruction (87,142). Ascorbic acid or sulfite can also destroy vitamin B_{12}. A combination of thiamine and nicotinic acid is slowly destructive to vitamin B_{12} in solution, though neither is harmful by itself (27). Iron protects the vitamin from this pair of chemicals by combining with hydrogen sulfide, the destructive agent that originates from thiamine (217). Ferric salts stabilize vitamin B_{12} in solution, but ferrous salts can cause rapid destruction (218).

3. Assay

Vitamin B_{12} is analyzed microbiologically using *Lactobacillus leichmannii* as a test organism (115).

Prior to absorption, free vitamin B_{12} is bound to a mucoprotein in the gastric juice, called the intrinsic factor. Protein-bound vitamin B_{12} is readily liberated by digestive enzymes.

4. Effect of Processing

Vitamin B_{12} is not destroyed to any appreciable extent by cooking, unless boiled in alkaline solution. In liver, 8% of the vitamin is lost by boiling at 100°C for 5 min; broiling muscle meat at 170°C for 45 min results in a 30% loss (111). In normal oven heating of frozen convenience dinners prior to serving, the retention of vitamin B_{12} ranged from 79 to 100% in products containing fish, fried chicken, turkey, and beef (68). Studies of vitamin B_{12} stability during various heat treatments in the processing of milk are summarized in Table 14.

Table 14   Losses of Vitamin B_{12} During the Heat Treatment of Milk

Treatment	Loss (%)
Pasteurization for 2–3 sec[a]	7
Boiling for 2-5 min[a]	30
Sterilization at 120°C for 13 min[a]	77
Sterilization at 143°C for 3-4 sec (steam injection)[a]	10
Evaporation[b]	70-90
Spray drying[b]	20-35

[a]Karlin et al. (122).
[b]Borenstein (31).

H. Pantothenic Acid

1. Structure, Stability, and Distribution

Pantothenic acid, D(+)–N–(2,4-dihydroxy-3,3-dimethyl-butyryl)-β-alanine (also known as vitamin B_5) is most stable in the pH range 4-7, and it is susceptible to both acid and base hydrolysis. Alkaline hydrolysis yields β-alanine and pantoic acid, whereas acid hydrolysis yields the γ-lactone of pantoic acid (167,250,255).

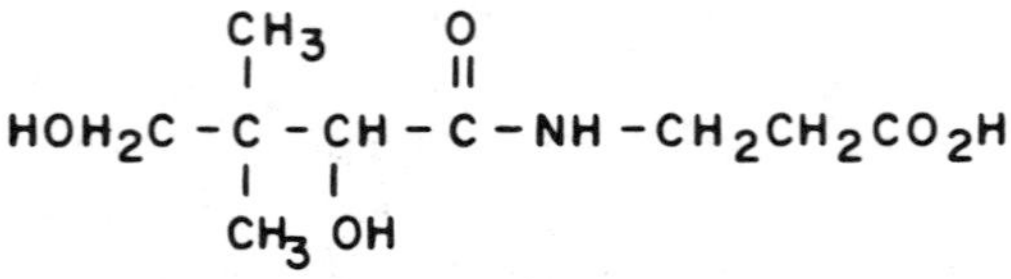

$$HOH_2C - \underset{\underset{CH_3}{|}}{\overset{\overset{CH_3}{|}}{C}} - \underset{\underset{OH}{|}}{CH} - \overset{\overset{O}{\|}}{C} - NH - CH_2CH_2CO_2H$$

Pantothenic Acid

Pantothenic acid is widely distributed in biological material and occurs in foods primarily as a moiety of coenzyme A. The microbial growth factor pantetheine is pantothenylaminoethanethiol.

2. Assay

Because of the low concentration of pantothenic acid in nature, microbiological methods are generally used to assay for this vitamin. *Saccharomyces carlsbergensis* (detects 10 ng) (228) and *Lactobacillus plantarum* (detects 1 ng) are common test organisms (115).

3. Effect of Processing

In a 1971 study, 507 edible products were analyzed for pantothenic acid content (214). In canned foods of animal origin, losses ranged from 20 to 35%, and in vegetable foods, from 46 to 78%. Losses were also large in frozen foods, ranging from 21 to 70% for animal products and from 37 to 57% in vegetable foods. Losses of pantothenic acid from fruits and fruit juices via freezing and canning were 7 and 50%, respectively. Grains lost 37-74% of the original pantothenic acid during conversion to various cereal products, and meats lost 50-75% during conversion to comminuted products.

Losses of pantothenic acid are usually less than 10% during milk pasteurization and sterilization. Cheeses generally have lower levels of pantothenic acid than fresh milk (31).

I. Biotin

1. Structure and Distribution

Biotin, shown below, consists of two fused five-membered rings formed from urea and a thiophene ring. The structure contains three asymmetric centers, and in addition, the rings may be cis or trans fused. Of the eight possible stereoisomers, only one, the cis-fused (+) biotin, is found naturally and has vitamin activity.

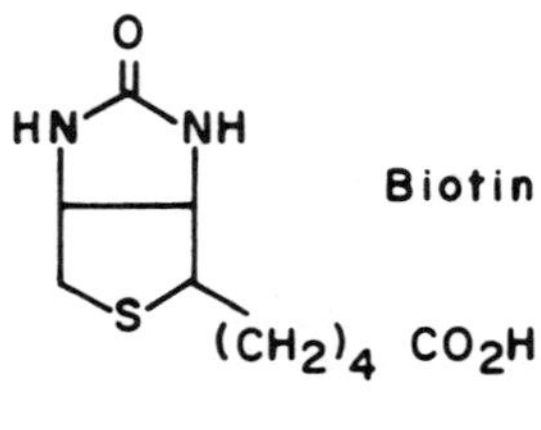

Biotin

Biotin is widely distributed throughout the animal and plant kingdoms. It is a co-enzyme for carboxylation and transcarboxylation reactions. The coenzyme is covalently bound to biotin-dependent enzymes through an amide bond between the ε-amino group of the enzyme's lysine residue and the carboxyl group of biotin.

2. Assay

Microbiological methods using *Allescheria boydii* (detects 0.5 ng) or *Lactobacillus arabinosus* (detects 0.05 ng) are used for the quantitative assay of biotin (228).

3. Stability

Pure biotin is quite stable when exposed to heat, light, air, and moderately acid and neutral conditions (optimum pH 5-8). Alkaline solutions of biotin are reasonably stable up to about pH 9 (162).

Oxidation of biotin with permanganate or hydrogen peroxide in acetic acid yields the sulfone (114,128). Nitrous acid destroys the biological activity of biotin, presumably by forming a nitrosourea derivative. Formaldehyde also inactivates the vitamin.

The limited data available suggest that biotin is very stable during the commercial and domestic preparation of foodstuffs.

Biotin in egg white is strongly bound to the avidin protein complex. Avidin is denatured by heat, however, so that its antagonistic action is destroyed when eggs are cooked. Deficiencies of biotin are unlikely in humans due to the extensive amount of synthesis by bacteria in the intestinal tract.

IV. FAT-SOLUBLE VITAMINS

A. Vitamin A

1. Structure

Vitamin A activity is contained in a series of C20 and C40 unsaturated hydrocarbons that are widely distributed in the plant and animal kingdoms. The structure of the vitamin is shown below, and it may occur as the free alcohol, esterified to fatty acids, or as the aldehyde or acid. In animals, the vitamin is most abundant in the liver, where it is stored generally as the free alcohol or in an esterified form. In plants and fungi, vitamin A activity is contained in a number of carotenoids that are metabolically converted to vitamin A after absorption by the ingesting animal. The structures and provitamin A activities of a number of commonly occurring carotenoids are given in Table 15. The most potent provitamin is β-carotene, which yields two equivalents of vitamin A.

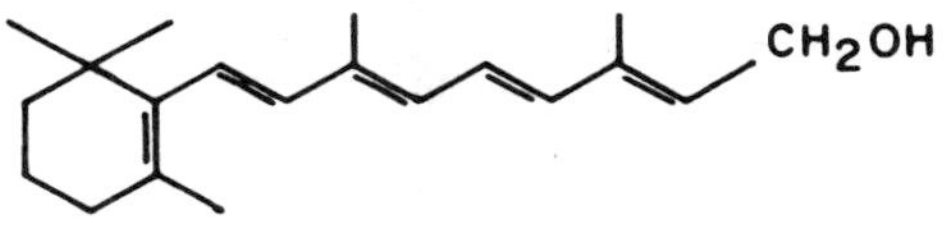

Vitamin A Alcohol

Since the carotenoids are predominatly hydrocarbon in nature, they are fat soluble and water insoluble, and are naturally associated with lipidlike structures. Carotene-pro-

Table 15 Carotenoids: Structure and Provitamin A Activity

Compound	Structure	Relative activity
β-Carotene (widely distributed)	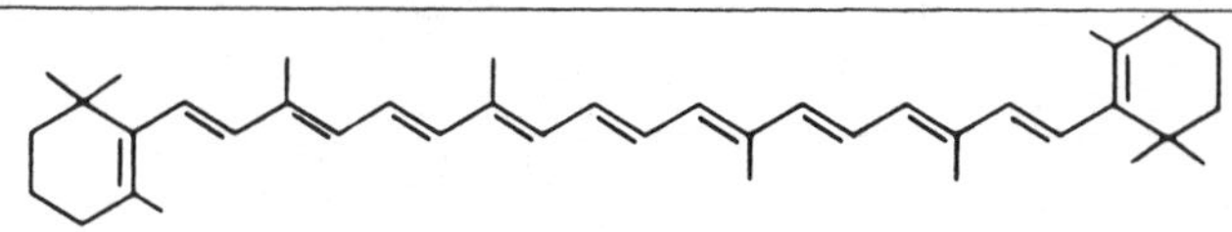	++++
β-Apo-8′-carotenal	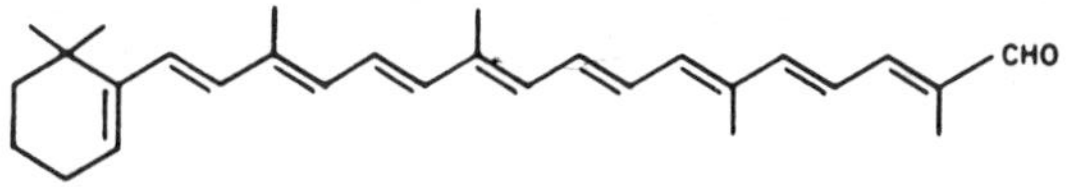	+++
Cryptoxanthin (orange)	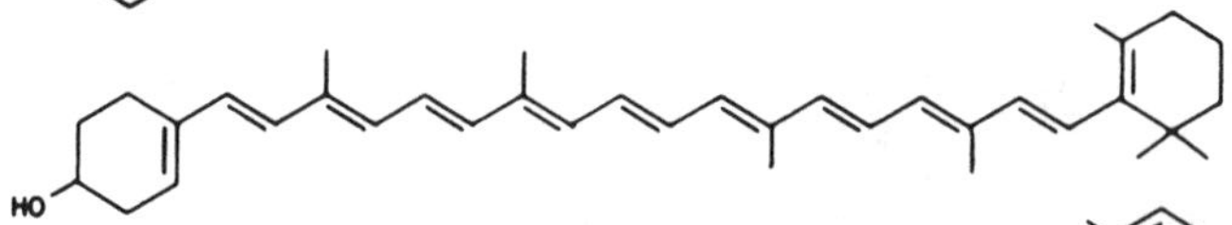	+++
α-Carotene (widely distributed)	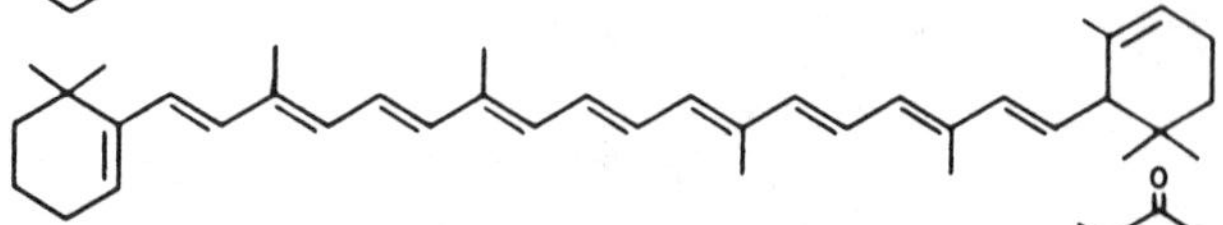	++
Echinenone (sea urchin)	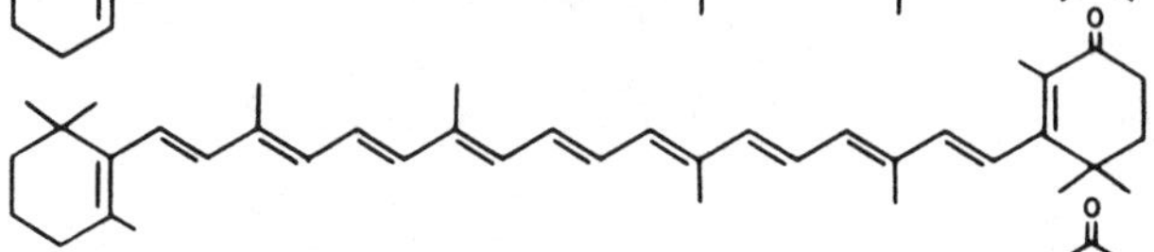	++
Astacene (crustacean)	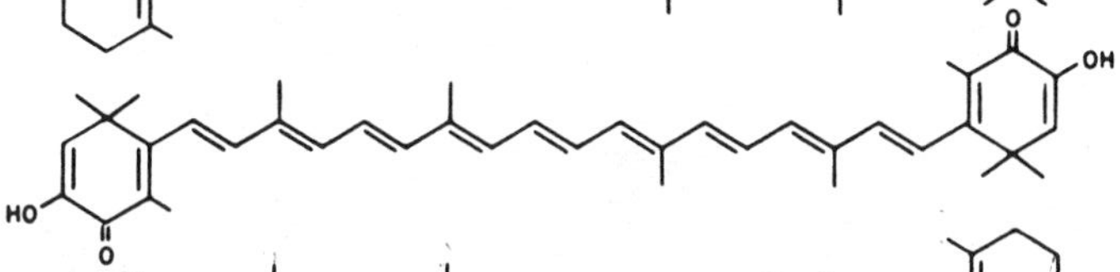	0
Lycopene (tomato)	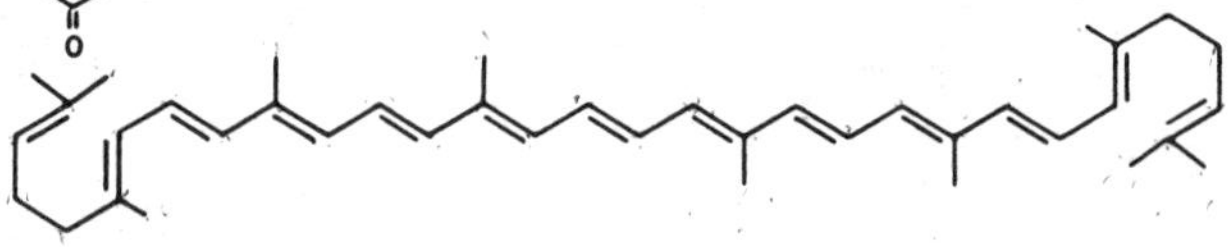	0

tein complexes also occur (178) and may be associated with an aqueous phase. The highly unsaturated carotenoid systems give rise to a series of complex UV and visible spectra (300-500 nm), which account for their strong orange-yellow pigmentation. Extensive listings of carotenoids, their spectra, and provitamin A activities are given by Deuel (69) and Zechmeister (262). A review of the chemistry of carotenoids in foods is given by Borenstein and Bunnell (32) and by Francis in Chap. 8 of this book.

2. Assay

It is now generally accepted that the vitamin A activity of a food is best determined by chromatographic separation of the carotenoids, followed by summation of the activities present in the various geometric and stereoisomers. The early analytical procedures, including the Carr-Price reaction with antimony trichloride, gave results that were too high, and were insensitive to cis-trans isomerization that can occur during processing.

3. Stability

The destruction of provitamins A in processed and stored foods can follow a variety of pathways depending on reaction conditions, and these are summarized in Fig. 10. In the absence of oxygen there are a number of possible thermal transformations, particularly cis-trans isomerization to the neo-β-carotenes. This has been shown to occur in both cooked and canned vegetables (230,249). Overall losses of vitamin A activity during anaerobic sterilization may vary from 5 to 50%, depending on temperature, time, and the nature of the carotenoids. At higher temperatures, β-carotene can fragment to yield a series of aromatic hydrocarbons; the most prominent is ionene (65,156).

If oxygen is present, extensive losses of carotenoids occur, stimulated by light, enzymes, and co-oxidation with lipid hydroperoxides. Chemical oxidation of β-carotene appears to yield primarily the 5,6-epoxide, which may later isomerize to mutachrome, that is, the 5,8-epoxide. Light-catalyzed oxidation yields primarily mutachrome (216). Isomerization of the 5,6-epoxide to mutachrome has been studied in orange juice (59). Further fragmentation of the primary oxidation products yields a complex of compounds (247) similar to those obtained following oxidation of fatty acids. Oxidation of vitamin A results in a complete loss of vitamin activity.

Dehydrated foods are most susceptible to the loss of vitamin A and provitamin A activity during storage (155) because of their propensity to undergo oxidation (140). Overall losses of vitamin A resulting from a number of types of dehydration processes are shown in Table 16.

As is the case in lipid oxidation, the rate of vitamin A loss is a function of enzymes, water activity, storage atmosphere, and temperature (46,91). Therefore, the expected loss depends on the severity of drying conditions (see Table 16) and the degree of protection during storage. The influences of water and oxygen concentrations are shown in Table 17. There is a measurable rate of loss even at very low partial pressures of oxygen. Although the rate of loss in shrimp decreases with increasing moisture, other foods (corn) show more complex behavior (194). In general, the stability of carotenoids is expected to parallel that of unsaturated fatty acids in a given food. The complex interactions of carotenoid stability with water activity and other constituents of food are beyond the scope of this chapter (137).

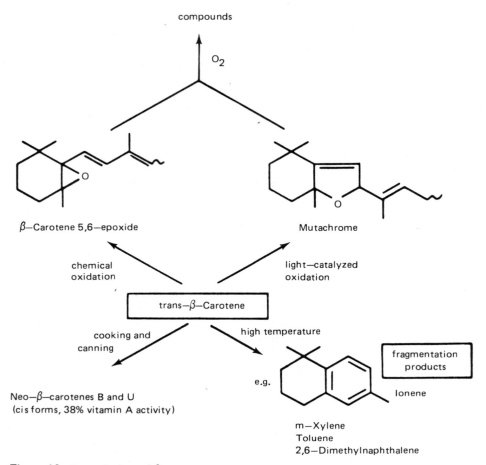

Figure 10　Degradation of β-carotene.

Table 16　Concentration of β-Carotene in Cooked Dehydrated Carrots

Sample	Range of concentration (ppm solids)
Fresh	980-1860
Explosive puff-dried	805-1060
Vacuum freeze-dried	870-1125
Conventional air-dried	636-987

Source: From Dellamonica and McDowell (66).

Table 17　Carotenoid Destruction Rates

Product	H_2O (%)	Gas	k (hr^{-1})
Carrot flakes, 20°C	5	Air	6.1×10^{-4}
		2% O_2	1.0×10^{-3}
Shrimp, 37°C	<0.5	Air	1.1×10^{-3}
	3	Air	0.9×10^{-3}
	5	Air	0.6×10^{-3}
	8	Air	0.1×10^{-3}

$E_a = 19$ kcal/mol

Source: From Labuza (138).

B. Vitamin K

Vitamin K activity is found in a number of fat-soluble naphthoquinone derivatives, and these occur primarily in green plants. Aside from the fact that these compounds are photoreactive, little is known of their chemical behavior in foods. Because this vitamin is present in green leafy vegetables and is synthesized by intestinal bacteria, deficiencies in normal individuals are rare.

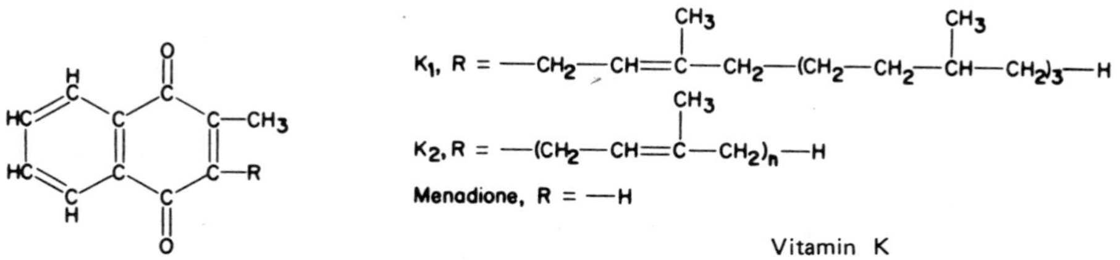

$$K_1, R = -CH_2-CH=C-CH_2-(CH_2-CH_2-CH-CH_2)_3-H$$

$$K_2, R = -(CH_2-CH=C-CH_2)_n-H$$

Menadione, $R = -H$

Vitamin K

C. Vitamin D

Vitamin D activity is found in certain sterol derivatives, and the structure of one member of this group is shown below. These compounds are fat soluble and are sensitive to oxygen and light. A significant fraction of milk sold in the United States is fortified with vitamin D, and stability in foods does not appear to be a significant problem.

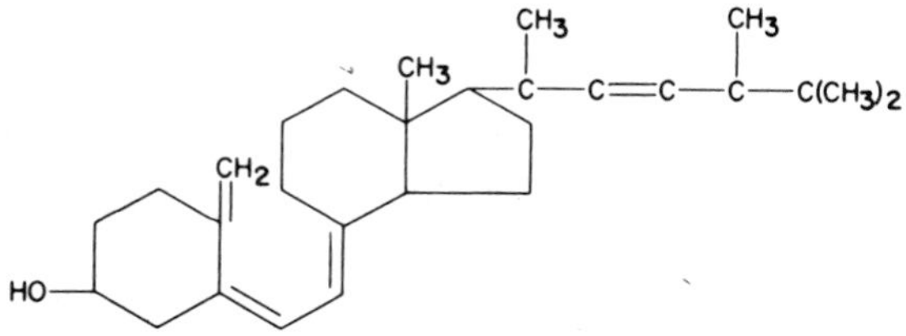

Vitamin D

D. Vitamin E

1. Structure

This fat-soluble vitamin is found in nature in the form of a number of tocopherols and tocotrienols. The most active compound in the group is α-tocopherol. The tocopherols are polyisoprenoid derivatives (chromanol nucleus) and all have a saturated C16 side chain (phytyl); centers of asymmetry at the 2, 4', and 8' positions; and variable methyl substitution at R_1, R_2, and R_3. The naturally occurring stereoisomer, D-α-tocopherol, has a 2D, 4'D, 8'D configuration, and various diastereoisomers are possible via chemical synthesis.

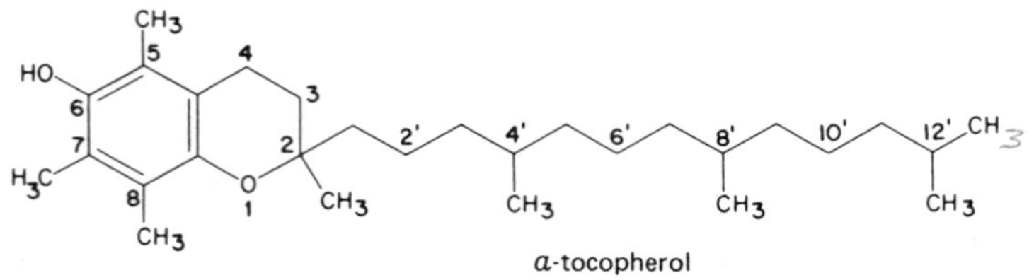

α-tocopherol

	R_1	R_2	R_3
α	CH_3	CH_3	CH_3
β	CH_3	H	CH_3
γ	H	CH_3	CH_3
δ	H	H	CH_3
tocol	H	H	H

2. Assay and Distribution

Vitamin E has multiple functions in human and animal bodies, but its exact biochemical function has still not been defined. Bioassays can be performed on the basis of a variety of biological responses. Ames (2) has categorized the physiological processes involved in bioassay as

1. Bioassays involving a biological function, such as the prevention of fetal resorption in rats or encephalomalacia in chicks
2. Measurement of a physiological parameter, such as prevention of erythrocyte hemolysis
3. Measurement of vitamin E levels in vivo, such as in the liver and in the plasma

Since eight tocopherols are known to exist in nature, a simple colorimetric method has not proven feasible for food analysis. The analytical procedure involves extraction, saponification, and chromatography of the nonsaponifiable fraction by thin-layer (TLC) or gas-liquid chromatography (GLC). The state of the art as of 1971 was reveiwed by Bunnell (36). More recent methodology has involved HPLC (95).

Vitamin E activity is widely distributed among food groups, including seeds and seed oils, vegetable oil, grains, fruits, vegetables, and animal products (63,64,221). Although α-tocopherol is the most important insofar as biological potency is concerned, the other naturally occurring isomers exist in significant concentrations and make an important contribution to both vitamin and antioxidant activity (Table 18). Data similar to those in Table 18 exist for vegetable oils (99). The total tocopherol and α-tocopherol content of typical American diets has been analyzed (37). The estimated range of daily intake is from 2.6 to 15.4 mg, with the most substantial contribution from food products containing vegetable oils or vegetable-oil-derived margarine or shortening. A more recent estimate (21) provided similar values (4.4-12.7 mg/day). An extensive survey of the tocopherol content of foods has been published (10).

Table 18 Tocopherol Content of Cereal Grains

Product	Tocopherols (mg per 100 g)			
	α	β	γ	α-Tocotrienol
Whole yellow corn	1.5	—	5.1	0.5
Yellow corn meal	0.4	—	0.9	—
Whole wheat	0.9	2.1	—	0.1
Wheat flour[a]	0.1	1.2	—	0
Whole oats	1.5	—	0.05	0.3
Oatmeal	1.3	—	0.2	0.5
Whole rice	0.4	—	0.4	—
Milled rice	0.1	—	0.3	—

[a]74% extraction.
Source: From Herting and Drury (110).

3. Stability

The influence of processing and storage on vitamin E has been reviewed by Harris (103). The loss of vitamin E activity can occur either via mechanical loss or by oxidation. An example of mechanical loss by degermination of grain is seen in Fig. 1. Thus, any process causing separation or removal of the lipid fraction, or manufacturing procedures that involve refining or hydrogenation, are likely to cause loss of this vitamin.

Loss by oxidation usually accompanies lipid oxidation, and this can be caused by the use of food processing chemicals, such as benzoyl peroxide or hydrogen peroxide. Dehydrated foods are particularly susceptible for the reasons discussed under vitamin A. The decomposition products of oxidized tocopherols include dimers, trimers, dihydroxy compounds, and quinones (58). A mechanism for forming some of these compounds has been suggested (96), and the effects of pH, water content, and temperature have been investigated for the case of tocopherol oxidation with methyl linoleate hydroperoxides (254). Products arising from oxidation of α-tocopherol with nitric acid resemble those of other oxidation mechanisms, as shown in Fig. 11.

In the absence of oxygen, tocopherol can form addition compounds with linoleate hydroperoxide (89). The initial oxidation product is apparently a semiquinone, which is further oxidized to tocopherolquinone or reacts anaerobically with an alkoxy radical to form the addition compound.

Recently, there has been interest in the use of both ascorbic acid and α-tocopherol as blocking agents to prevent the formation of nitrosamines from nitrite and amine precursors (231). In this context α-tocopherol reacts with nitrite below pH 5 to form products that are identical to those formed by its air oxidation.

VII. CHEMICAL PROPERTIES OF MINERALS AND THEIR BIOAVAILABILITY

A. General

The minerals in foods comprise a large and diverse group of elements and complex ions. Many of these are nutritionally required by humans, and some, particularly the trace elements, are hazardous if consumed in excessive amounts. The recommended dietary allowances are listed in Table 3, and discussions of the biological roles of minerals can be found in several recent surveys (179,203,240,241).

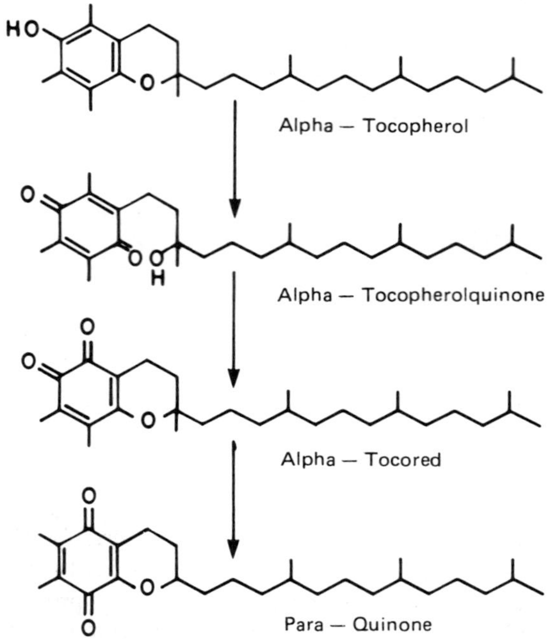

Figure 11 Products of α-tocopherol oxidation by nitric acid. [John and Empte (118).]

The mineral content of foods can vary greatly, depending upon such environmental factors as soil composition for plants or nature of the feed for animals. Losses of minerals from foods occur not so much through destruction by chemical reaction as through physical removal or combination in forms that are not biologically available (213).

The primary mechanism of loss of mineral substances is through leaching of water-soluble materials and trimming of unwanted plant parts, as discussed in Sec. III. Major losses of all minerals occur during the milling of cereals. It is believed by many investigators that the increasing portion of highly refined foods in the diet may lead to the exacerbation of trace mineral deficiencies. Thus, supplementation with trace minerals may become a necessity in the future, but as will be seen, this approach is complicated by the inherent toxicity of the nutrient.

Of equal importance is the interaction of a mineral substance with other constituents of food. Polyvalent anions, such as oxalate and phytate, can form salts with divalent metal cations, and these salts are extremely insoluble, passing through the digestive tract unabsorbed. Thus, the measurement of mineral bioavailability is important and is discussed in a subsequent section of this chapter.

B. Chemistry

A comprehensive discussion of the chemistry of minerals is beyond the scope of this chapter, and the reader is referred to a basic text on inorganic chemistry. Any mineral contains an anionic and a cationic component, but fluoride, iodide, and phosphate are the only anions of particular significance from a nutritional point of view. Fluoride is more often a constituent of water than of foods, and intake is highly dependent on geographic location. Iodine may be present as iodide (I^-) or as iodate (IO_4^-). Phosphate may exist in a variety of forms, including phosphate (PO_4^{3-}), hydrogen phosphate (HPO_4^{2-}), dihydrogen phosphate ($H_2PO_4^-$), or phosphoric acid (H_3PO_4). The respective ionization constants are

$$K_1 = 7.5 \times 10^{-3}$$

$$K_2 = 6.2 \times 10^{-8}$$

$$K_3 = 1.0 \times 10^{-12}$$

Iodide and iodate are relatively strong oxidizing agents compared with the other important inorganic anions in foods (i.e., phosphate, sulfate, and carbonate).

The cations present a far more diverse and complex set of substances than the anions, and their general chemistry can be considered through their elemental subgroups. Some of the metal ions are important from a nutritional viewpoint, and others are significant as toxic contaminants.

Magnesium, calcium, and barium can be considered as a group. They exist only in the +2 oxidation state. Although the halides of this group are soluble, the other important salts, including hydroxides, carbonates, phosphates, sulfates, oxalates, and phytates, are very insoluble. In foods that have undergone some bacterial decomposition, magnesium can form the high insoluble complex $NH_4MgPO_4 \cdot 6H_2O$, also know as struvite.

Copper exists in +1 or +2 oxidation states and forms complex ions. The halides and sulfate are soluble, but the carbonate and phosphate are relatively insoluble. Some other metals with multioxidation states include tin and lead (± 2 and +4), mercury (+1 and +2), iron (+2 and +3), chromium (+3 and +6), and manganese (+2, +3, +4, +6, and +7). Many of these metals form amphoteric ions and can act as oxidizing or reducing substances. Copper and iron are of particular significance in their ability to catalyze oxidation of ascorbic acid and unsaturated lipids (see Chap. 4).

Many of the metal ions also exist as ligands of organic molecules. Examples include iron bound to heme, copper in cytochromes, magnesium in chlorophyll, and cobalt in vitamin B_{12}. The biologically active form of chromium is called the glucose tolerance factor (GTF), which is a complex organic form of trivalent chromium and is about 50 times more effective than inorganic Cr^{3+} in the glucose tolerance bioassay. It contains about 65% chromium in addition to nicotinic acid, cysteine, glycine, and glutamic acid. The exact structure is unknown (163). Cr^{6+} does not have biological activity.

C. Occurrence

Although the mineral content of foods, and thus intake of minerals by humans, is influenced primarily by the nature of the soil and water used for the production of food or animal feed, this is an oversimplification. For example, with regard to copper, the following environmental factors have been suggested as influencing the intake of copper by humans (187): copper content of soil; geographic location; season; water source; use of fertilizers, insecticides, pesticides, and fungicides; and nature of the diet. In addition, it should be realized that minerals can enter foods during processing as direct or indirect additives. This, too, is a highly variable factor. Consequently, the mineral content of foods and water can vary greatly, as is indicated by the data for drinking water (Table 19).

Iodine can serve as a specific example of the type of variability just discussed. First, the iodine content of food is dependent upon geographic location. Thus, indivi-

Table 19   Occurrence of Some Trace Minerals in Drinking Water and Foods

Element	Concentration
Arsenic	0-100 μg/liter, 137-330 μg/day in food
Barium	<1 mg/liter
Beryllium	<1 μg/liter
Cadmium	<10 μg/liter
Chromium	<100 μg/liter
Cobalt	Up to 0.5 mg/kg in green leafy vegetables
Copper	Present in plant and animal foods, 1-280 μg/liter
Lead	20-600 μg/liter; 100-300 μg per capita per day from food
Manganese	0.5-1.5 mg/liter
Magnesium	6-120 mg/liter
Mercury	<1 μg/liter, 10 μg/day intake from food
Molybdenum	<100 μg/liter, 100-1000 μg/kg of diet
Nickel	1-100 μg/liter, 300-600 μg/day in food
Selenium	<10 μg/liter, 100-300 μg/kg in cereals, meats, seafoods
Silver	Traces
Tin	1-2 μg/liter, 1-30 mg/day from food
Vanadium	2-300 μg/liter
Zinc	3-2000 μg/liter

Source: From National Academy of Sciences-National Research Council (211).

duals living near the sea are exposed to relatively high levels of dietary iodine through seafood, dairy products, and vegetable matter. In the latter case, the transfer of iodine to soil occurs largely via atmospheric phenomena, and the consumption of local forage high in iodine leads to high levels in milk.

Second, a major source of iodine in foods is iodized salt, usually in the form of potassium iodide or iodate for increased stability. Third, the loss of various forms of iodine from food has not been extensively investigated, but apparently leaching losses can be extensive. For example, up to 80% of the iodine in fish can be lost in preparation by boiling, with lesser amounts lost by processes that do not require contact with excess water (105).

Compilations of analytical data on the mineral content of foods are not extensive. Agriculture Handbook No. 8 (242) provides information on calcium, phosphorus, sodium, potassium, and iron. An extensive treatment of the elements in a limited number of foods has been compiled in France (117). Information on trace elements is scarce, but is beginning to appear with increasing frequency. Literature sources are: copper (187); zinc (85,100,174); calcium, iron, magnesium, phosphorus, potassium, sodium, and zinc in dairy foods (61); chromium (235); lead and cadmium (45); and mercury (92).

D. Losses and Gains in Processing

The distribution of chromium in wheat products is indicated in Table 20, and the differences among these products appear to be small. In contrast, there are major differences in the mineral contents of egg white and yolk and in various milk products (see Table 21). Thus, there does not appear to be a general rule for predicting patterns of mineral distribution.

Losses of minerals by contact with water, particularly in cooking or blanching, can be considerable. The effect of blanching on the loss of major minerals in spinach is shown in Table 22. The wide range of losses is a function of mineral solubility. In some cases the mineral content may increase during processing, as is the case for calcium in Table 22. The loss of nitrate may be regarded as beneficial from the viewpoint of lessened can corrosion and health problems.

The effect of various food processes on the copper content of potatoes is indicated in Table 23. The unequal distribution of copper in the peel and the interior are also indicated. A somewhat different pattern can be found for the loss of minerals in cooked beans (Table 24). In contrast to spinach (Table 23), calcium is lost from beans to about the same extent as the other major minerals, and the trace elements follow a similar pattern.

Trace elements and minerals can be gained during processing through contact with processing water, equipment, and packaging materials. This is illustrated for tin in Table 25 for vegetables with and without lacquered tinplate. The distribution of metal between solid and liquid portions of the canned contents is also shown.

E. Availability of Minerals in Foods

Measuring the total amount of an element in a particular food source or diet provides only a limited indication of its nutritional value. Of greater practical significance is the amount of the element in a food that is "available" to the body. Various chemical, dietary, and host or physiological factors serve to determine the overall utilization of an ingested element. Thus, Bing (22) has suggested that the availability of iron or iron salt in a food is dependent not only on the form of the mineral assayed but also on the various test conditions that influence its absorption or utilization.

Methods employed to determine bioavailability of minerals include chemical balance studies (233), biological assays in experimental animals (86,168), in vitro tests (106,197), and the use of radioactive tracers. Procedures involving radioisotopes have been widely used for determining the true digestibilities of dietary minerals for farm livestock, and this approach has been reviewed (233). In regard to human nutrition,

Table 20 Distribution of Chromium in Wheat and Wheat Products

Product	Relative biological value
Wheat grain	3
Wheat germ	4
Bread, whole wheat	3.6
Bread, white	3

Source: From Toepfer et al. (235).

Table 21 Distribution of Minerals in Egg and in Dairy Products

Food	mg per 100 g								
	Ca	P	Mg	Na	K	Fe	Cu	Mo	Zn
Whole egg	26	103	5.3	79	54	1.1	29	<20	1
Egg white	1	3	3.1	56	43	0.03	1.6	<10	0.003
Egg yolk	12	43	0.7	3	15	0.36	1.7	<20	0.25
Whole milk	252	197	22	120	348	0.07	12	<10	1.0
Skim milk	259	197	22	134	408	0.07	12	<10	1.1
Cottage cheese	74	159	6	444	89	<0.1	<20	<40	0.4

Source: From Gormian (93).

Table 22 Effect of Blanching on Mineral Loss from Spinach

Mineral	g/100 g		Loss (%)
	Unblanched	Blanched	
Potassium	6.9	3.0	56
Sodium	0.5	0.3	43
Calcium	2.2	2.3	0
Magnesium	0.3	0.2	36
Phosphorus	0.6	0.4	36
Nitrate	2.5	0.8	70

Source: From Bengtsson (19).

Table 23 Copper Content of Processed Potatoes

Type	Copper (mg per 100 g fresh weight)
Raw	0.21 ± 0.10
Boiled	0.10
Baked	0.18
Chips	0.29
Mashed	0.10
French fried	0.27
Instant, uncooked	0.17
Potato peel	0.34

Source: From Pennington and Calloway (187).

Table 24 Mineral Content of Raw and Cooked Navy Beans

Mineral	mg/100 g		Loss (%)
	Raw	Cooked	
Calcium	135	69	49
Copper	0.80	0.33	59
Iron	5.3	2.6	51
Magnesium	163	57	65
Manganese	1.0	0.4	60
Phosphorus	453	156	65
Potassium	821	298	64
Zinc	2.2	1.1	50

Source: From Meiners et al. (161).

considerable investigation has been devoted to the problem of iron availability in food (23,51), and more recently, to zinc (76,176).

Based on radioisotopic methods, two approaches to studying mineral utilization in humans have been taken. Initially, biosynthetically labeled foods were prepared by growing plants in media containing radioactive iron, or by injecting animals with the radioactive tracers (^{55}Fe and ^{59}Fe) prior to slaughter and preparation of the foods. Test meals containing the labeled food item were then ingested, and absorption of the tracer was determined. This is the so-called intrinsic label method. More recently, absorption of iron (and zinc) from food has been studied by the use of the extrinsic label method, in which the radioactive element is added to the food at the time of ingestion.

Table 25 Distribution of Trace Metals in Canned Vegetables

Vegetable	Can[a]	Component[b]	g/kg		
			Lead	Tin	Iron
Green beans	La	L	0.10	5	2.8
		S	0.70	10	4.8
Haricot beans	La	L	0.07	5	9.8
		S	0.15	10	26
Petit pois	La	L	0.04	10	10
		S	0.55	20	12
Celery hearts	La	L	0.13	10	4.0
		S	1.50	20	3.4
Sweet corn	La	L	0.04	10	1.0
		S	0.30	20	6.4
Mushrooms	P	L	0.01	15	5.1
		S	0.04	55	16

[a]La = lacquered, P = plain.
[b]L = liquid, S = solid.
Source: From Crosby (57).

Studies with the extrinsic tag have been validated for the determination of both iron (51) and zinc (76) bioavailability. This latter method has greatly aided quantitative assessment of factors influencing biological utilization of minerals in foods consumed by human subjects.

In the case of iron, chemical form is important, with simple ferrous salts more readily available than ferric salts (239). Furthermore, the size of the elemental iron particle influences bioavailability (171,190). The type of food source also affects the availability of iron, with availability greatest in animal foods and lowest in cereals (see Fig. 12) (144,145). Other dietary factors are also important. For example, vitamin C enhances iron absorption; phosphate, and to a lesser extent calcium, reduce iron absorption; and bran also reduces iron absorption, presumably because of its phytate content. Other dietary constituents, including proteins, amino acids, and carbohydrates, have been suggested as influencing iron availability (33,51), although the quantitative significance of these factors in human nutrition remains uncertain.

Host or physiological factors that serve to modify absorption of dietary iron include the iron nutritional status of the subject, with increased iron absorption occurring in iron-depleted subjects or in patients with iron-deficiency anemia. There is some evidence that iron absorption may be greater in women than in men (33) and that iron absorption decreases in children with increasing age (215). The availability of zinc is similarly affected by various dietary and host factors, and these have been discussed in a number of reviews (101,181).

The problem of availability of calcium, zinc, and iron has been linked to the presence of phytate (inositol hexaphosphate) in certain foods (180). Balance studies on zinc

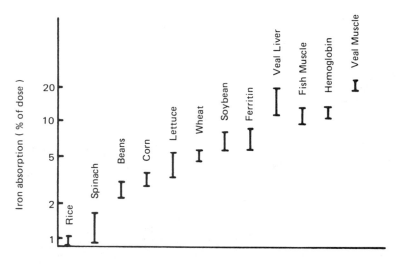

Figure 12 Iron absorption by adults from various foods. Results are expressed as the mean ± standard error. The dose of iron was 2–4 mg, except for lettuce, which was 1–17 mg. (Adapted from Ref. 19.)

in animals have conclusively demonstrated that phytate can impair the absorption of dietary zinc and the reabsorption of endogenously secreted zinc (62). Reduced retention of iron, copper, and manganese was also found in this study. On the other hand, another study has shown that monoferric phytate has the same availability as ferrous ammonium sulfate (170). The situation may be further complicated by the role played by phytases in degrading phytate complexes in the intestine.

The design of rational and effective food fortification programs requires sound information on the availability of minerals in food sources and diets. This information is also important for the evaluation of the nutritional properties of food replacers and food analogs. Additional research is required to determine the bioavailability of the various trace elements essential in human nutrition and to understand the various factors affecting mineral availability in the modern diet. A competent review of the various iron sources available for fortification and their changes due to food processing has been published (148).

F. Safety of the Minerals

Some minerals have no significance from a nutritional perspective but are important because of their potential hazard. Mercury and cadmium are examples of this type. All minerals are toxic at some dose; and the range of safety versus toxicity is highly variable. The relationship between safe and toxic doses of some minerals is shown in Table 26. An explanation of the extreme ranges for a given effect has been given by the U. S. Food and Drug Administration:

> The most important aspect of these tables is the overlap of excess intake ranges that occurs between different degrees of response. This is true for every mineral for which sufficient data exist to make the comparison. The span of values for each response level is frequently very large. For many elements there are examples of small intakes that were very toxic and large intakes that produced no detectable toxic effect. Some of the explanations for these dose-response relationships are identifiable and are discussed under each individual mineral. In other cases the reason is not apparent and reflects the state of uncertainty about the safety of consuming minerals in excess of requirements.

VIII. OPTIMIZATION OF NUTRIENT RETENTION

In the course of processing and storing foods there is an inevitable decline in nutrient value. It is the responsibility of the food manufacturer to ensure that these losses are minimal. The correct approach to optimization of nutrient content of course varies depending on the type of food or process. A few illustrations follow. For a full treatment of this complex subject the reader is referred to a discussion by Karel (120).

A. High-Temperature-Short-Time (HTST) Processing

One of the most significant of recent food processing developments is HTST processing combined with aseptic canning. The aim and purpose of most heat processes (pasteurization or sterilization) is to reduce the microbial population of the product to a point that provides safety and a reasonable shelf life.

Table 26 Relationships Between Excess Oral Intakes[a] of Some Minerals in Humans and Severity of Toxic Responses

Mineral	Response[b]	Toxic effect		
		None	Mild	Severe
Cu	Acute	—	2-16	125-50,000
	Longer	—	—	0.5-4
Co	Longer	—	150	35-600
F	Acute	—	3-19	80-3,000
	Longer	—	3-19	2-160
Fe	Acute	—	—	12-1,500
	Longer	7	—	6-15
I	Longer	8-35,000	1-15,000	1-180,000
Sn	Acute	—	130	23-700
Zn	Acute	13-23	—	8-530
	Longer	2-10	10	—

[a]Excess oral intake is defined as the actual intake required to produce an effect divided by the required intake or an equivalent reference point of average intake.
[b]Acute, within 24 hr; Longer, from 1 day through several generations.
Source: From Food and Drug Administration (82).

It can be demonstrated mathematically that, for a commercial sterilization process of any type, a HTST process is preferable because the Arrhenius activation energy of spore inactivation is large compared with that of vitamin destruction.

Consider two simultaneous reactions:

1. Spore inactivation, $N/N_0 = e^{-k_1 t}$, where N_0 is the initial spore count, N is the count after any heating time, t, and k_1 is the first-order rate constant for spore inactivation.
2. Vitamin destruction, $V/V_0 = e^{-k_2 t}$, where V_0 and V are the respective vitamin concentrations at times 0 and t, and k_2 is the rate constant (assumed first order) for vitamin destruction.

The time for completion of any commercial sterilization process is determined by the level of safety chosen for the product. In low-acid, commercially canned foods the safety level is conventionally taken as a reduction to approximately 10^{-12} the original population of spores of *Clostridium botulinum*. More generally, any indicator organism and any probability level may be chosen. Thus, the time for completion of a process, neglecting heating time and cooling time, will be given by

1. Choosing N/N_0 for the particular organism.
2. Finding k_1 and k_2 as a function of temperature T. The latter condition is usually met by determining the activation energy for the specific reaction

$$k_1 = A_1 \frac{e^{-Ea_1}}{RT}$$

$$k_2 = A_2 \frac{e^{-Ea_2}}{RT}$$

where A_1 and A_2 are collision frequency constants, Ea_1 and Ea_2 are the activation energies in kcal/mole, R is the gas constant, and T is the absolute temperature. Experimentally, these values are usually determined by studying the reaction kinetics over a suitable temperature range and plotting k versus 1/T, the slope equal to $-Ea/R$.

3. Solving the two rate equations simultaneously. Thus, if the time for completion of the process is t_c, then

$$t_c = \left(\ln \frac{N_0}{N} \right) k_1^{-1}$$

$$= \left(\ln \frac{N_0}{N} \right) A_1^{-1} \left(\frac{e^{Ea_1}}{RT} \right)$$

and

$$\ln \frac{V}{V_0} = k_2 k_1^{-1} \left(\ln \frac{N}{N_0} \right), \text{ etc.}$$

Therefore, the vitamin retention V/V_0 can be determined for any process as a function of T if the appropriate constants are known. Since the activation energy for spore inactivation is usually greater than the activation energy for vitamin destruction, the rate constant for spore inactivation increases faster as temperature is raised than does the rate constant for vitamin destruction. As a consequence, a temperature can be found at which the overall rate of spore inactivation is high compared with the rate of vitamin destruction, and the process can be completed with minimal loss of the vitamin.

As a specific example, consider inactivation of thiamine and spores of *Bacillus stearothermophilus*, a common thermophilic contaminant of foods. The appropriate constants are as follows. For thiamine, $k_2 = 0.014$ min^{-1} at 102°C, $E_{a2} = 22$ kcal/mol; for *B. Stearothermophilus*, $k_1 = 0.07$ min^{-1} at 125°C, and $E_{a1} = 69$ kcal/mol. The results of solving the appropriate equations for a variety of processing temperatures and for two levels of commercial sterility are shown in Table 27. It is evident that much less destruction of thiamine occurs when commercial sterilization is conducted under HTST conditions.

B. Prediction of Vitamin Losses in Storage

Another approach to optimization of nutrient content requires knowledge of the nutrient content of a processed food at various times during distribution. With this knowledge, the manufacturer can specify the minimum nutrient content of the product at time of sale, which is essential if nutrient claims are to be made on the label or in advertising associated with the product.

Suppose one wished to predict the nutrient content of a food product at some point in time. The minimal required information would be (a) initial nutrient composition, (b) time-temperature history of the product in all distribution channels (e.g., this would differ for products sold in New Orleans and New England), (c) package performance (permeability to oxygen, water vapor, and light), and (d) influence of other time-dependent environmental factors, such as light, relative humidity, and oxygen.

For products that have a relatively simple composition, it is feasible to guarantee nutrient content under known circumstances. An actual case history for commercial juice

Table 27 Sterility Level $N/N_0 = 10^{-10}$

Temperature (°C)	Time for sterility (min)	Thiamine loss (%)
100	527	99.99
110	47	75
120	4.7	6
130	0.52	2
140	0.067	1
150	0.009	~0
Sterility level $N/N_0 = 10^{-16}$		
100	843	99.99
110	75	89
120	7.6	27
130	0.85	10
140	0.11	3
150	0.015	1

fortified with vitamin C is shown in Fig. 13. The goal of the manufacturer was to maintain at least as much ascorbic acid as specified on the label. If the warehouse temperatures had been above 38°C this would not have been possible with an initial vitamin excess of 33%. Furthermore, in this particular instance ascorbic acid is lost through two different reactions (see Sec. VI.A.1 for details). One reaction is fairly rapid and involves ascorbic acid oxidation by oxygen present in the bottle. The other is a slower reaction in which oxygen is not a reactant.

Figure 13 indicates that the second, slower reaction has a high activation energy; that is, there is little additional loss at ambient temperatures following the initial consumption of residual oxygen, whereas the reaction continues at a significant rate at 38°C. To predict the ascorbic acid content in this product one would require knowledge of the initial oxygen content and of the rates and activation energies of the two reactions. One can then set up equations for the simultaneous reactions in a manner similar to that shown earlier for HTST processing.

$$\text{Ascorbic acid} + O_2 \rightarrow \text{products} \quad \frac{\text{find}}{Ea_1, k_1}$$

$$\text{Ascorbic acid} \rightarrow \text{products} \quad Ea_2, k_2$$

The actual solution of this problem is left as an exercise for the reader. (Hint: the amount of O_2 in the headspace is usually much less than the amount of ascorbic acid in the product.) Numerical solutions for a number of variables clearly require the services of a computer.

For more complex products that have nonuniform structures, and for vitamins that exist in multiple forms, the prediction process is very complex. For this reason it seems illogical to promulgate federal or state regulations that apply across the board to all vitamins and all products. On the basis of the previous discussion it appears more reasonable to specify quality characteristics for individual vitamins in certain classes of foods.

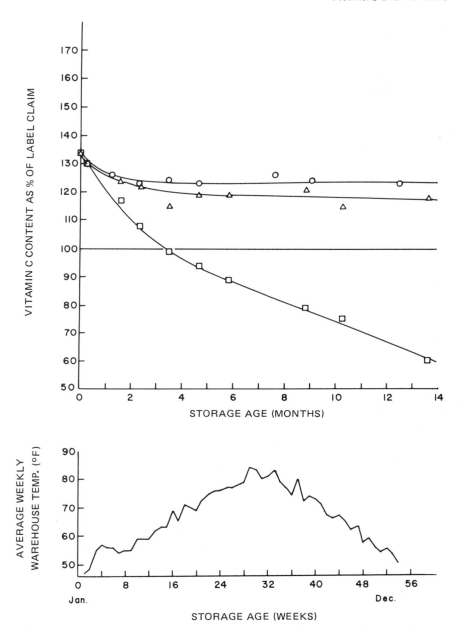

Figure 13 Effect of storage time and storage temperature on the stability of vitamin C in cranberry cocktail. ○ warehouse storage; △ ambient storage; □ 38°C storage. (Courtesy of Ocean Spray, Inc.)

REFERENCES

1. Alexander, H. D., E. J. Day, H. E. Sauberlich, and W. D. Salmon (1956). Radiation effects on water soluble vitamins in raw beef. *Fed. Proc. 15*:921-923.

2. Ames, S. R. (1971). Isomers of alpha-tocopheryl acetate and their biological activity. *Lipids 6*:281-290.

3. Anderson, R. H., D. L. Maxwell, A. E. Mulley, and C. W. Fritsch (1976). Effects of processing and storage on micronutrients in breakfast cereals. *Food Technol. 30*(5):110-114.

4. Ang, C. Y. W., C. M. Chang, A. E. Frey, and G. E. Livingston (1975). Effects of heating methods on vitamin retention in six fresh or frozen prepared food products. *J. Food Sci. 40*:997-1003.

5. Angier, R. B., J. H. Boothe, J. H. Mowat, C. W. Waller, and J. Semb (1952). Pteridine chemistry. II. The action of excess nitrous acid upon pteroylglutamic acid and derivatives. *J. Amer. Chem. Soc. 74*:408-411.

6. Archer, M. C., and L. S. Reed (1980). In *Methods in Enzymology: Vitamins and Coenzymes*, (D. B. McCormick and L. D. Wright, eds.), Vol. 66, Academic Press, New York, pp. 452-459.

7. Archer, M. C., S. R. Tannenbaum, T.-Y. Fan, and M. Weisman (1975). Reaction of nitrite with ascorbate and its relation to nitrosamine formation. *J. Nat. Cancer Inst. 54*(5):1203-1205.

8. Arnold, R. G., L. M. Libbey, and R. C. Lindsay (1969). Volatile flavor compounds produced by heat degradation of thiamine (Vitamin B_1). *J. Agr. Food Chem. 17*: 390-392.

9. Aykroyd, W. R., and J. Doughty (1970). In *Wheat in Human Nutrition*, FAO Nutrition Studies No. 23, Food and Agricultural Organization of the U. N., Rome.

10. Bauernfeind, J. C. (1977). The tocopherol content of food and influencing factors. *Crit. Rev. Food Sci. Nutr. 8*:337-382.

11. Bauernfeind, J. C., and D. M. Pinkert (1970). Food processing with added ascorbic acid. *Adv. Food Res. 18*:220-315.

12. Baugh, C. M., and C. L. Krumdieck (1971). Naturally occurring folates. *Ann. N. Y. Acad. Sci. 186*:7-28.

13. Baugh, C. M., C. L. Krumdieck, H. J. Baker, and C. E. Butterworth (1971). Studies on the absorption and metabolism of folic acid. I. Future absorption in the dog after exposure of isolated intestinal segments to synthetic pteroylpolyglutamates of various chain lengths. *J. Clin. Invest. 50*:2009-2021.

14. Beadle, B. W., D. A. Greenwood, and H. R. Kraybill (1943). Stability of thiamine to heat. I. Effect of pH and buffer salts in aqueous solutions. *J. Biol. Chem. 149*: 339-347.

15. Beaton, G. H., and V. N. Patwardhan (1976). In *Nutrition in Preventive Medicine*, (G. H. Beaton and J. M. Bengoa, eds.), WHO, Geneva, p. 445.

16. Beetner, G., T. Tsao, A. Frey, and J. Harper (1974). A research note. Degradation of thiamine and riboflavin during extrusion processing. *J. Food Sci. 39*:207-208.

17. Bender, A. E. (1958). The stability of vitamin C in a commercial fruit squash. *J. Sci. Food Agr. 9*:754-760.

18. Bender, A. W. (1966). Nutritional effects of food processing. *J. Food Technol. 1*: 261-289.

19. Bengtsson, B. L. (1969). Effect of blanching on mineral and oxalate content of spinach. *J. Food Technol. 4*:141-145.

20. Bergel, F., and D. R. Harrap (1961). Interaction between carbonyl groups and biologically essential substitutents. *J. Chem. Soc.* 4051-4056.

21. Bieri, J. G. and R. P. Evarts (1973). The recommended allowance for vitamin E tocopherols and fatty acids in American diets. *J. Amer. Diet Ass. 62*:147-151.

22. Bing, F. C. (1972). Assaying the availability of iron. *J. Amer. Diet. Ass. 60*:114-122.

23. Bjorn-Rasmussen, E., L. Hallberg, and R. B. Walker (1972). Food iron absorption in man. I. Isotopic exchange between food iron and inorganic iron salt added to food: Studies on maize, wheat and eggs. *Amer. J. Clin. Nutr. 25*:317-323.

24. Blakley, R. L. (1957). The interconversion of serine and glycine: Preparation and properties of catalytic derivatives of pteroylglutamic acid. *Biochem. J. 65*:331-342.

25. Blakley, R. L. (1960). Spectrophotometric studies on the combination of formaldehyde with tetrahydropteroylglutamic acid and other hydropteridines. *Biochem. J. 74*:71-82.

26. Blakley, R. L. (1969). *The Biochemistry of Folic Acid and Related Pteridines*, North Holland, London.

27. Blitz, M., E. Eigen, and E. Gunsberg (1954). Vitamin B_{12} studies. The instability of vitamin B_{12} in the presence of thiamine and niacinamide. *J. Amer. Pharm. Ass. Sci. Ed. 43*:651-653.

28. Bolin, H. R., and A. E. Stafford (1974). Effect of processing and storage on pro-vitamin A and vitamin C in apricots. *J. Food Sci. 39*:1034-1036.

29. Bookwalter, G. N., H. A. Moser, V. F. Pfeifer, and E. L. Griffin, Jr. (1968). Storage stability of blended food products, formula No. 2: A corn-soy-milk food supplement. *Food Technol. 22*:1581-1584.

30. Booth, R. G. (1943). The thermal decomposition of aneurin and cocarboxylase at varying hydrogen ion concentrations. *Biochem. J. 37*:518-522.

31. Borenstein, B. (1968). Vitamins and amino acids, in *Handbook of Food Additives* (T. E. Furia, ed.), CRC Press, Cleveland, pp. 107-137.

32. Borenstein, B., and R. H. Bunnell (1966). Carotenoids: Properties, occurrence and utilization in foods. *Adv. Food Res. 15*:195-276.

33. Bowering, J., A. M. Sanchez, and M. I. Irwin (1976). A conspectus of research on iron requirements of man. *J. Nutr. 106*:985-1074.

34. Breslow, R. (1958). On the mechanism of thiamine action. IV. Evidence from studies on model systems. *J. Amer. Chem. Soc. 80*:3719-3726.

35. Buell, M. V., and R. E. Hansen (1960). Reaction of pyridoxal-5-phosphate with aminothiols. *J. Amer. Chem. Soc. 82*:6042-6049.

36. Bunnell, R. H. (1971). Modern procedures for the analysis of tocopherols. *Lipids 6*:245-253.

37. Bunnell, R. H., J. Keating, A. Quaresimo, and G. K. Parman (1965). Original communications. Alpha-tocopherol content of foods. *Amer. J. Clin. Nutr. 17*: 1-10.

38. Burton, H., J. E. Ford, J. G. Franklin, and J. W. G. Porter (1967). Effects of repeated heat treatments on the levels of some vitamins of the B-complex in milk. *Dairy Res. 34*:193-197.

39. Burton, H., J. E. Ford, A. G. Perkin, J. W. G. Porter, K. J. Scott, S. Y. Thompson, J. Toothill, and J. D. Edwards-Webb (1970). Comparison of milks processed by the direct and indirect methods of ultra-high-temperature sterilization. IV. The vitamin composition of milks sterilized by different processes. *J. Dairy Res. 37*: 529-533.

40. Butterworth, Jr., C. E., R. Santini, Jr., and W. B. Frommeyer, Jr. (1963). The pteroylglutamate components of American diets as determined by chromatographic fractionation. *J. Clin. Invest. 42*:1929-1939.

41. Cain, R. F. (1967). Water-soluble vitamins. Changes during processing and storage of fruit and vegetables. *Food Technol. 21*:998-1007.

42. Chan, C., Y. S. Shin, and E. L. R. Stokstad (1973). Studies of folic acid compounds in nature. III. Folic acid compounds in cabbage. *Can. J. Biochem. 51*: 1617-1623.

43. Cheldin, V. H., A. M. Woods, and R. J. Williams (1943). Losses of B vitamins due to cooking of foods. *J. Nutr. 26*:477.

44. Chen, T. S., and W. L. George (1981). A research note. Ascorbic acid retention in retort pouched green beans. *J. Food Sci. 46*:642-643.

45. Childs, E. A., and J. N. Gaffke (1974). A research note. Lead and cadmium content of selected Oregon groundfish. *J. Food Sci. 39*:853-854.

46. Chou, H.-E., and W. B. Breene (1972). Oxidative decoloration of β-carotene in low-moisture model systems. *J. Food Sci. 37*:66-68.

47. Clegg, K. (1966). Citric acid and the browning of solutions containing ascorbic acid. *J. Sci. Food Agr. 17*:546-549.

48. Clegg, K. M., and A. D. Morton (1965). Carbonyl compounds and the non-enzymic browning of lemon juice. *J. Sci. Food Agr. 16*:191-198.

49. Clifcorn, L. E. (1948). Factors influencing the vitamin content of canned foods. *Adv. Food Res. 1*:39-104.

50. Cook, B. B., B. Gunning, and D. Uchimoto (1961). Variations in nutritive value of frozen green baby lima beans as a result of methods of processing and cooking. *J. Agr. Food Chem. 9*:316-321.

51. Cook, J. D. (1977). Absorption of food iron. *Fed. Proc. 36*:2028-2032.

52. Cort, W. M., B. Borenstein, J. H. Harley, M. Osadca, and J. Scheiner (1976). Nutrient stability of fortified cereal products. *Food Technol. 30*(4):52-62.

53. Cosulich, D. B., and J. M. Smith, Jr. (1949). Communications to the editor. N^{10}-Nitrosopteroylglutamic acid. *J. Amer. Chem. Soc. 71*:3574.

54. Cosulich, D. B., B. Roth, J. M. Smith, Jr., M. E. Hultquist, and R. P Parker (1952). Chemistry of leucovorin. *J. Amer. Chem. Soc. 74*:3252-3263.

55. Cover, S., E. M. Dilsaver, R. M. Hayes, and W. H. Smith (1949). Retention of B-vitamins after large-scale cooking of meat. 2. Roasting by two methods. *J. Amer. Diet. Ass. 25*:949-951.

56. Coward, J. K., P. L. Chello, A. R. Cashmore, K. N. Parameswaran, L. M. DeAngelis, and J. R. Bertino (1975). 5-Methyl-5,6,7,8-Tetrahydropteroyl oligo-γ-L-glutamates: Synthesis and kinetic studies with methionine synthetase from bovine brain. *Biochemistry 14*:1548-1552.

57. Crosby, N. T. (1977). Determination of heavy metals in foods. *Proc. Inst. Food Sci. Technol. 10*:65-70.

58. Csallany, A. S., M. Chiu, and H. H. Draper (1970). Oxidation products of α-tocopherol formed in autoxidizing methyl linoleate. *Lipids 5*:63-70.

59. Curl, A. L., and G. F. Bailey (1956). Part II. Caroetnoids of aged canned valencia orange juice. *J. Agr. Food Chem. 4*:159-162.

60. Dahn, H., L. Loewe, and C. A. Bunton (1960). Über die oxydation von ascorbin-säure durch salpetrige säure. Teil VI. Ürersicht und diskussion der ergebnisse. *Helv. Chim. Acta 43*:320-333.

61. *Dairy Council Digest* (1976). Composition and nutritive value of dairy foods *47*(5).

62. Davies, N. T., and R. Nightingale (1975). Effect of phytate on zinc absorption and faecal zinc excretion and carcass retention of zinc, iron, copper and manganese. *Proc. Nutr. Soc. 34*:8A-9A.

63. Davis, K. C. (1972). Vitamin E: Adequacy of infant diets. *Amer. J. Clin. Nutr. 25*:933-938.

64. Davis, K. C. (1973). Vitamin E content of selected baby foods. *J. Food Sci. 38*:442-446.

65. Day, W. C., and J. G. Erdman (1963). Ionene: A thermal degradation product of β-carotene. *Science 141*:808.

66. Dellamonica, E. S., and P. E. McDowell (1965). Comparison of beta-carotene content of dried carrots prepared by three dehydrated processes. *Food Technol. 19*:1597-1599.

67. DeRitter, E. (1976). Stability characteristics of vitamins in processed foods. *Food Technol. 30*(1):48-54.

68. DeRitter, E., M. Osadca, J. Scheiner, and J. Keating (1974). Vitamins in frozen convenience dinners and pot pies. *J. Amer. Diet. Ass. 64*:391-397.

69. Deuel, H. J. (1951). *The lipids*, Interscience, New York.

70. Donaldson, K. O., and J. C. Keresztesy (1962). Naturally occurring forms of folic acid. III. Characterization and properties of *S*-methyldihydrofolate, an oxidation product of *S*-methyltetrahydrofolate. *J. Biol. Chem. 237*:3815-3819.

71. Dwivedi, B. K., and R. G. Arnold (1971). Hydrogen sulfide from heat degradation of thiamine. *J. Agr. Food Chem. 19*:923-926.

72. Dwivedi, B. K., and R. G. Arnold (1972). Chemistry of thiamine degradation. Mechanisms of thiamine degradation in a model system. *J. Food Sci. 37*:886-888.

73. Dwivedi, B. K., and R. G. Arnold (1973). Chemistry of thiamine degradation: 4-Methyl-5-(β-hydroxyethyl)thiazole from thermally degraded thiamine. *J. Agr. Food Chem. 21*:54-60.

74. Dwivedi, B. K., R. G. Arnold, and L. M. Libbey (1972). Chemistry of thiamine degradation: 4-Methyl-5-(β-hydroxyethyl)thiazole from thermally degraded thiamine. *J. Food Sci. 37*:689-692.

75. Emodi, A. S., and L. Scialpi (1980). Quality of bread fortified with ten micronutrients. *Cereal Chem. 57*(1):1-3.

76. Evans, G. W., and P. E. Johnson (1977). Determination of zinc availability in foods by the extrinsic label technique. *Amer. J. Clin. Nutr. 30*:873-878.

77. FAO/WHO (1967). In Report of a Joint FAO/WHO Expert Group on Requirements of Vitamin A, Thiamine, Riboflavin, and Niacin. WHO Tech. Rep. Ser. No. 362, Geneva.

78. Farrer, K. T. H. (1955). The thermal destruction of vitamin B$_1$ in foods. *Adv. Food Res. 6*:257-311.

79. Feliciotti, E., and W. B. Esselen (1957). Thermal destruction rates of thiamine in pureed meats and vegetables. *Food Technol. 11*:77-84.

80. Fennema, O. (1977). Loss of vitamins in fresh and frozen foods. *Food Technol. 31*(12):32-36.

81. Finholt, P., R. B. Paulssen, and T. Higuchi (1963). Rate of anaerobic degradation of ascorbic acid in aqueous solution. *J. Pharm. Sci. 52*:948-954.

82. Food and Drug Administration (1975). *Toxicity of the Essential Minerals*, DHEW, Washington, D. C.

83. Ford, J. E., J. W. G. Porter, S. Y. Thompson, J. Toothill, and J. Edwards-Webb (1969). Effects of ultra-high-temperature (UHT) processing and of subsequent storage on the vitamin content of milk. *J. Dairy Res. 36*:447-454.

84. Freed, M., S. Brenner, and V. O. Wodicka (1949). Prediction of thiamine and ascorbic acid stability in stored canned foods. *Food Technol. 3*:148-151.

85. Freeland, J. H., and R. J. Cousins (1976). Zinc content of selected foods. *J. Amer. Diet. Ass. 68*:526-529.

86. Fritz, J. C., G. W. Pla, B. N. Harrison, and G. A. Clark (1975). Vitamins and other nutrients. Estimation of the bioavailability of iron. *J. Ass. Offic. Anal. Chem. 58*: 902-905.

87. Frost, D. V., M. Lapidus, K. A. Plaut, E. Scherfling, and H. H. Fricke (1952). Differential stability of various analogs of cobalamin to vitamin C. *Science 116*:119-121.

88. Fujita, A. (1954). Thiaminase. *Adv. Enzymol. 15*:389-421.

89. Gardner, H. W., K. Eskins, G. W. Grams, and G. E. Inglett (1972). Radical addition of linoleic hydroperoxides to α-tocopherol or the analogous hydroxychroman. *Lipids 7*:324-334.

90. Goldblith, S. A., and S. R. Tannenbaum (1966). In *Problems of World Nutrition,*

Proc. Seventh Int. Cong. Nutr., vol. 4, Verlag Friedr. Vieweg & Sohn GmbH Braun-schweig, W. Germany, pp. 3-16.

91. Goldman, M., B. Horev, and I. Saguy (1983). Decolorization of β-carotene in model systems simulating dehydrated foods. Mechanism and kinetic principles. *J. Food Sci. 48*:751-754.

92. Gomez, M. I., and P. Markakis (1974). Mercury content of some foods. *J. Food Sci. 39*:673-675.

93. Gormian, A. (1970). Inorganic elements in foods used in hospital menus. *J. Amer. Diet. Ass. 56*:397-403.

94. Green, R. G., W. E. Carlson, and C. A. Evans (1942). The inactivation of vitamin B_1 in diets containing whole fish. *J. Nutr. 23*:165-174.

95. Gregory, III, J. F. (1983). Methods of vitamin assay for nutritional evaluation of food processing. *Food Technol. 37*(1):75-80.

96. Gruger, Jr., E. H., and A. L. Tappel (1970). Reactions of biological antioxidants. I. Fe(III)-catalyzed reactions of lipid hydroperoxides with α-tocopherol. *Lipids 5*: 326-331.

97. Guerrant, N. B., and M. B. O'Hara (1953). Vitamin retention in peas and lima beans after blanching, freezing, processing in tin and in glass, after storage and after cooking. *Food Technol. 7*:473-477.

98. Gupta, V. A., and F. M. Huennekens (1967). Preparation and properties of crys-talline *S*-methyl tetrahydrofolate and related compounds. *Arch. Biochem. Biophys. 120*:712-718.

99. Gutfinger, T., and A. Letan (1974). Studies of unsaponifiables in several vegetable oils. *Lipids 9*:658-663.

100. Haeflein, K. A., and A. I. Rasmussen (1977). Zinc content of selected foods. *J. Amer. Diet. Ass. 70*:610-616.

101. Halsted, J. A., J. C. Smith, Jr., and M. I. Irwin (1974). A conspectus of research on zinc requirements of man. *J. Nutr. 104*:345-378.

102. Hanning, F., and M. L. Mitts (1949). Effect of cooking on the folic acid content of eggs. *J. Amer. Diet. Ass. 25*:226-228.

103. Harris, R. S. (1962). Influence of storage and processing on the retention of vitamin E in foods. *Vitamins Hormones 20*:603-619.

104. Harris, R. S., and E. Karmas (1975). *Nutritional Evaluation of Food Processing*, AVI Publishing Co., Westport, Conn.

105. Harrison, M. T., S. McFarland, R. McG. Harden, and E. Wayne (1965). Nature and availability of iodine in fish. *Amer. J. Clin. Nutr. 17*:73-77.

106. Hart, H. V. (1971). Comparison of the availability of iron in white bread, fortified with iron powder, with that of iron naturally present in wholemeal bread. *J. Sci. Food Agr. 22*:354-357.

107. Hassinen, J. B., G. T. Durbin, and F. W. Bernhart (1954). The vitamin B_6 content of milk products. *J. Nutr. 53*:249-257.

108. Heid, J. L. (1960). In *Nutritional Evaluation of Food Processing*, (R. S. Harris and H. Von Loesecke, eds.), John Wiley and Sons, New York, pp. 146-148.

109. Hellendoorn, E. W., A. P. de Groot, C. P. van der Mijll Dekker, P. Slump, and J. J. Willems (1971). Nutritive value of canned meals. *J. Amer. Diet. Ass. 58*:434-441.

110. Herting, D. C., and E.-J. E. Drury (1969). Alpha-tocopherol content of cereal grains and processed cereals. *J. Agr. Food Chem. 17*:785-790.

111. Heyssel, R. M., R. C. Bozian, W. J. Darby, and M. C. Bell (1966). Vitamin B_{12} turn-over in man. The assimilation of vitamin B_{12} from natural foodstuff by man and estimates of minimal daily dietary requirements. *Amer. J. Clin. Nutr. 18*:176-184.

112. Hillcoat, B. L., P. F. Nixon, and R. L. Blakley (1967). Effect of substrate decom-position on the spectrophotometric assay of dihydrofolate reductase. *Anal. Bio-chem. 21*:178-189.

113. Hodge, J. E., and E. M. Osman (1976). Carbohydrates, in *Food Chemistry*, (O. R. Fennema, ed.), Marcel Dekker, New York.

114. Hofmann, K., D. B. Melville, and V. deVigneaud (1941). Characterization of the functional groups of biotin. *J. Biol. Chem. 141*:207-214.

115. Horowitz, W. (ed.) (1965). *Official Methods of Analysis of the Association of Official Chemists*, 10th ed., Association of Official Analytical Chemists, Washington, D. C.

116. Huelin, F. E., I. M. Coggiola, G. S. Sidhu, and B. H. Kennett (1971). The anaerobic decomposition of ascorbic acid in the pH range of foods and in more acid solutions. *J. Sci. Food Agr. 22*:540-542.

117. Jaulmes, P., and G. Hamelle (1971). Concentrations of trace elements in food and drink of man. *Ann. Nutr. Alim. 25*:B133-B203.

118. John, W., and W. Emte (1940). Über einige neve oxydations produkte der tokopherole. 8. Mitteilung über antisterilitatsfaktoren (vitamin E). *Hoppe-Seyler's Z. Physiol. Chem. 268*:85-103.

119. Joslyn, M. A. (1957). Role of amino acids in the browning of orange juice. *Food Res. 22*:1-14.

120. Karel, M. (1979). Prediction of nutrient losses and optimization of processing conditions, in *Nutritional and Safety Aspects of Food Processing*, (S. R. Tannebaum, ed.), Marcel Dekker, New York, pp. 233-263.

121. Karel, M., and J. T. R. Nickerson (1964). Effects of relative humidity, air and vacuum on browning of dehydrated orange juice. *Food Technol. 18*:1214-1218.

122. Karlin, R., C. Hours, C. Vallier, R. Bertoye, N. Berry, and H. Morand (1969). Sur la teneur en folates des laits de grand mélange. Effects de divers traitment thermiques sur les taux de folates, B_{12} et B_6 de ces laits. *Int. Z. Vitamin forsch. 39*:359-371.

123. Karrer, P., H. Salomon, K. Schopp, E. Schlittler, and H. Fritzche (1934). Ein neves bestrahlungsprodukt des lactoflavins: Lumichrom. *Helv. Chim. Acta 17*:1010-1013.

124. Keresztesy, J. C., and K. O. Donaldson (1961). Synthetic prefolic A. *Biochem. Biophys. Res. Commun. 5*:286-288.

125. Khan, M. M. T., and A. E. Martell (1967). Metal ion and metal chelate catalyzed oxidation of ascorbic acid by molecular oxygen. I. Cupric and ferric ion catalyzed oxidation. *J. Amer. Chem. Soc. 89*:4176-4185.

126. Khan, M. M. T., and A. E. Martell (1967). Metal ion and metal chelate catalyzed oxidation of ascorbic acid by molecular oxygen. II. Cupric and ferric chelate catalyzed oxidation. *J. Amer. Chem. Soc. 89*:7104-7111.

127. Klein, B. P., and A. K. Perry (1982). Ascorbic acid and vitamin A activity in selected vegetables from different geographical areas of the U.S. *J. Food Sci. 47*:941-948.

128. Kögl, F., and T. J. de Man (1941). Zur chemie des biotins. Nachweis eines schwefelhaltigen ringes. 29. Mitteilung über pelanzliche Wurchsstoffe. *Z. Physiol. Chem. 269*:81-96.

129. Krehl, W. A., and G. R. Cowgill (1950). Vitamin content of citrus products. *Food Res. 15*:179-191.

130. Krumdieck, C. L., and C. M. Baugh (1969). The solid-phase synthesis of polyglutamates of folic acid. *Biochemistry 8*:1568-1572.

131. Kuhn, R., H. Rudy, and T. Wagner-Jauregg (1933). Lactoflavine: Vitamin B_2. *Ber. Bursenges. Phys. Chem. 66*:1950-1956.

132. Kundig, H., and J. C. Somogyi (1967). Isolation of the active moiety of the antithiamine compound from capp viscera. *Int. Z. Vitaminforsch. 37*:476-481.

133. Kurata, T., and Y. Sakurai (1967). Degradation of *L*-ascorbic acid and mechanism of nonenzymic browning reaction. Part II. Nonoxidative degradation of *L*-ascorbic

acid including the formation of 3-deoxy-L-pentosone. *Agr. Biol. Chem. 31*:170-176.

134. Kurata, T., and Y. Sakurai (1967). Degradation of L-ascorbic acid and mechanism of nonenzymic browning reaction. Part III. Oxidative degradation of L-ascorbic acid (degradation of dehydro-L-ascorbic acid). *Agr. Biol. Chem. 31*:177-184.

135. Kurata, T., H. Wakabayashi, and Y. Sakurai (1967). Degradation of L-ascorbic acid and mechanism of nonenzymic browning reaction. Part I. Browning reactivities of L-ascorbic acid and of dehydro-L-ascorbic acid. *Agr. Biol. Chem. 31*:101-105.

136. Kyzlink, V., and D. Curda (1970). Einfluss der saccharose und der zugänglichkeit des saverstoffs auf den oxydationsverlauf der L-ascorbinsäure im flüssigen medium. *Z. Lebensm. Unters. Forsch. 143*:263-273.

137. Labuza, T. P. (1971). Kinetics of lipid oxidation in foods. *Crit. Rev. Food Technol. 2*:355-405.

138. Labuza, T. P. (1972). Nutrient losses during drying and storage of dehydrated foods. *Crit. Rev. Food Technol. 3*:217-240.

139. Labuza, T. P., and J. F. Kamman (1982). A research note. Comparison of stability of thiamin salts at high temperature and water activity. *J. Food Sci. 47*:664-665.

140. Labuza, T. P., S. R. Tannenbaum, and M. Karel (1970). Water content and stability of low-moisture and intermediate-moisture foods. *Food Technol. 24*:543-550.

141. Lachance, P. A. (1981). The role of cereal grain products in the U.S. diet. *Food Technol. 35*(3):49-60.

142. Lang, C. A., and B. F. Chow (1950). Inactivation of microbiological activity of crystalline vitamin B_{12} by reducing agents. *Proc. Soc. Exp. Biol. Med. 75*:39-41.

143. Larrabee, A. R., S. Rosenthal, R. E. Cathou, and J. M. Buchanan (1961). Communications to the editor. A methylated derivative of tetrahydrofolate as an intermediate of methionine biosynthesis. *J. Amer. Chem. Soc. 83*:4094-4095.

144. Layrisse, M., and C. Martinez-Torres (1971). In *Progress in Haematology*, Vol. VII (E. B. Brown and C. U. Moore, eds.), Grune and Stratton, New York, pp. 137-160.

145. Layrisse, M., and C. Martinez-Torres (1972). Model for measuring dietary absorption of heme iron: Test with a complete meal. *Am. J. Clin. Nutr. 25*:401-411.

146. Lee, C. Y., L. M. Massey, Jr., and J. P. Van Buren (1982). Effects of post-harvest handling and processing on vitamin contents of peas. *J. Food Sci. 47*:961-964.

147. Lee, F. A. (1958). The blanching process. *Adv. Food Res. 8*:63-109.

148. Lee, K., and F. M. Clydesdale (1979). Iron sources used in food fortification and their changes due to food processing. *CRC Crit. Rev. Food Sci. Nutr. 11*:117-153.

149. Lee, Y. C., J. R. Kirk, C. L. Bedford, and D. R. Heldman (1977). Kinetics and computer simulation of ascorbic acid stability of tomato juice as functions of temperature, pH and metal catalyst. *J. Food Sci. 42*:640-648.

150. Leichter, J., and M. A. Joslyn (1969). Protective effect of casein on cleavage of thiamine by sulfite. *J. Agr. Food Chem. 17*:1355-1359.

151. Leichter, J., and M. A. Joslyn (1969). Kinetics of thiamin cleavage by sulfite. *Biochem. J. 113*:611-615.

152. Lin, K. C., B. S. Luh, and B. S. Schweigert (1975). Folic acid content of canned garbanzo beans. *J. Food Sci. 40*:562-565.

153. Lowry, O. H., O. A. Bessey, and E. J. Crawford (1949). Photolytic and enzymatic transformations of pteroylglutamic acid. *J. Biol. Chem. 180*:389-398.

154. Lushbough, C. H., J. M. Weichman, and B. S. Schweigert (1959). The retention of vitamin B_6 in meat during cooking. *J. Nutr. 67*:451-459.

155. MacKinney, G., A. Lukton, and L. Greenbaum (1958). Carotenoid stability in stored dehydrated carrots. *Food Technol. 12*:164-166.
156. Mader, I. (1964). Beta-carotene: Thermal degradation. *Science 144*:533-534.
157. Malewski, W., and P. Markakis (1971). A research note. Ascorbic acid content of the developing tomato fruit. *J. Food Sci. 36*:537.
158. Malin, J. D. (1975). Folic acid. *World Rev. Nutr. Diet. 21*:198-223.
159. Mapson, L. W. (1956). Effect of processing on the vitamin content of foods. *Br. Med. Bull. 12*:73-77.
160. May, M., T. J. Bardos, F. L. Barger, M. Lansford, J. M. Ravel, G. L. Sutherland, and W. Shive (1951). Synthetic and degradative investigations of the structure of folinic acid-SF. *J. Amer. Chem. Soc. 73*:3067-3075.
161. Meiners, C. R., N. L. Derise, H. C. Lau, M. G. Crews, S. J. Ritchey, and E. W. Murphy (1976). The content of nine mineral elements in raw and cooked mature dry legumes. *J. Agr. Food Chem. 24*:1126-1130.
162. *Merck Index* (1968). Eighth ed., Merck and Co., Rahway, N. J.
163. Mertz, W. (1975). Effects and metabolism of glucose tolerance factor. *Nutr. Rev. 33*:129-135.
164. Metz, J., A. Lurie, and M. Konidaris (1970). A note on the folate content of uncooked maize. *S. Afr. Med. J. 44*:539-541.
165. Meyer, B., J. Thomas, and R. Buckley (1960). The effect of ripening on the thiamine, riboflavin, and niacin content of beef from grain-finished and grass-finished steers. *Food Technol. 14*:190-192.
166. Mirvish, S. S., L. Wallcave, M. Eagan, and P. Shubik (1972). Ascorbate-nitrite reaction: Possible means of blocking the formation of carcinogenic *N*-nitroso compounds. *Science 177*:65-68.
167. Mitchell, H. K., H. H. Weinstock, E. E. Snell, S. R. Stanberg, and R. J. Williams (1940). Pantothenic acid. V. Evidence for structure of non-β-alanine portion. *J. Amer. Chem. Soc. 62*:1776-1779.
168. Momcilovic, B., B. Belonje, A. Giroux, and B. G. Shah (1975). Total femur zinc as the parameter of choice for a zinc bioassay in rats. *Nutr. Rep. Int. 12*:197-203.
169. Morfee, T. D., and B. J. Liska (1971). Distribution of thiamine degradation products in simulated milk systems. *J. Dairy Sci. 54*:1082-1085.
170. Morris, E. R., and R. Ellis (1976). Isolation of monoferric phytate from wheat bran and its biological value as an iron source to the rat. *J. Nutr. 106*:753-760.
171. Motzok, I., M. D. Pennell, M. I. Davies, and H. U. Ross (1975). Vitamins and other nutrients. Effect of particle size on the biological availability of reduced iron. *J. Ass. Offic. Anal. Chem. 58*:99-103.
172. Mulley, E. A., C. R. Stumbo, and W. M. Hunting (1975). Kinetics of thiamine degradation by heat. A new method for studying reaction rates in model systems and food products at high temperatures. *J. Food Sci. 40*:985-988.
173. Mulley, E. A., C. R. Stumbo, and W. M. Hunting (1975). Kinetics of thiamine degradation by heat. Effects of pH and form of the vitamin on its rate of destruction. *J. Food Sci. 40*:989-992.
174. Murphy, E. W., B. W. Willis, and B. K. Watt (1975). Provisional tables on the zinc content of foods. *J. Am. Diet. Ass. 66*:345-355.
175. Navankasattusas, S., and D. B. Lund (1982). Thermal destruction of vitamin B_6 vitamers in buffer solution and cauliflower puree. *J. Food Sci. 47*:1512-1518.
176. Neathery, M. W., J. W. Lassiter, W. J. Miller, and R. P. Gentry (1975). Absorption, excretion and tissue distribution of natural organic and inorganic zinc-65 in the rat (38731). *Proc. Soc. Exp. Biol. Med. 149*:1-4.
177. Nelson, E. M. (1956). Association of vitamin B_6 deficiency with convulsions in infants. *Public Health Rep. 71*:445-448.

178. Nishimura, M., and K. Takamatsu (1957). A carotene-protein complex isolated from green leaves. *Nature 180*:699-700.

179. The Nutrition Foundation (1976). *Present Knowledge in Nutrition*, Nutrition Foundation, New York.

180. Oberleas, D. (1973). Phytates, in *Toxicants Naturally Occurring in Foods*. NAS/NRC, Washington, D. C., pp. 363-371.

181. Oberleas, D., M. E. Muhrer, and B. L. O'Dell (1966). In *Zinc Metabolism* (A. S. Prasad, ed.), Charles C Thomas, Springfield, Ill., pp. 225-238.

182. O'Broin, J. D., I. J. Temperley, J. P. Brown, and J. M. Scott (1975). Nutritional stability of various naturally occurring monoglutamate derivatives of folic acid. *Amer. J. Clin. Nutr. 28*:438-444.

183. Oguntona, T. E., and A. E. Bender (1976). Loss of thiamine from potatoes. *J. Food Technol. 11*:347-352.

184. Orr, M. L. (1969). Pantothenic acid, vitamin B_6 and vitamin B_{12} in foods, *Home Economics Res. Rept.* No. 36. U. S. Dept. Agric., Washington, D. C.

185. Osborn, M. J., and F. M. Huennekens (1958). Enzymatic reduction of dihydro-folic acid. *J. Biol. Chem. 233*:969-974.

186. Osborn, M. J., P. T. Talbert, and F. M. Huennekens (1960). The structure of "active formaldehyde" (N^5, N^{10}-methylene tetrahydrofolic acid). *J. Amer. Chem. Soc. 82*:4921-4927.

187. Pennington, J. T., and D. H. Calloway (1973). Copper content of foods. *J. Amer. Diet. Ass. 63*:143-153.

188. Perloff, B. P., and R. R. Butrum (1977). Folacin in selected foods. *J. Amer. Diet. Ass. 70*(2):161-172.

189. Perry, A. K., C. R. Peters, and F. O. VanDuyne (1976). Effect of variety and cooking method on cooking times, thiamine content and palatability of soybeans. *J. Food Sci. 41*:1330-1334.

190. Pla, G. W., B. N. Harrison, and J. C. Fritz (1973). Vitamins and other nutrients. Comparison of chicks and rats as test animals for studying bioavailability of iron, with special reference to use of reduced iron in enriched bread. *J. Ass. Offic. Anal. Chem. 56*:1369-1373.

191. Polansky, M. M., and E. W. Toepfer (1969). Vitamin B_6 components in some meats, fish, dairy products, and commercial infant formulas. *J. Agr. Food Chem. 17*:1394-1397.

192. Porzio, M. A., N. Tang, and D. M. Hilker (1973). Thiamine modifying properties of heme proteins from skipjack tuna, port, and beef. *J. Agr. Food Chem. 21*:308-310.

193. Proctor, B. E., and S. A. Goldblith (1948). Radar energy for rapid food cooking and blanching, and its effect on vitamin content. *Food Technol. 2*:95-104.

194. Quackenbush, F. W. (1963). Corn carotenoids: Effects of temperature and moisture on losses during storage. *Cereal Chem. 40*:266-269.

195. Rabinowitz, J. C. (1960). In *The Enzymes* (P. D. Boyer, H. Lardy, and K. Myrback, eds.), Academic Press, New York, p. 185.

196. Ramasastri, B. V., and R. L. Blakley (1964). 5,10-Methylenetetrahydrofolic dehydrogenase from Bakers' yeast. II. Use in assay of tetrahydrofolic acid. *J. Biol. Chem. 239*:106-111.

197. Ranhotra, G. S., F. N. Hepburn, and W. B. Bradley (1971). Availability of iron in enriched bread. *Cereal Chem. 48*:377-384.

198. Rao, M. A., C. Y. Lee, J. Katz, and H. J. Cooley (1981). A research note. A kinetic study of the loss of vitamin C, color, and firmness during thermal processing of canned peas. *J. Food Sci. 46*:636-637.

199. Rauen, H. M., and H. Waldmann (1950). Über die gegenstromverteilung von pterinen. *Z. Physiol. Chem. 286*:180-190.

200. Reed, B., D. Weir, and J. Scott (1976). The fate of folate polyglutamates in meat during storage and processing. *Amer. J. Clin. Nutr. 29*:1393-1396.

201. Reed, L. S., and M. C. Archer (1980). Oxidation of tetrahydrofolic acid by air. *J. Agr. Food Chem. 28*:801-805.

202. Reiber, H. (1972). Photochemical reaction of vitamin B_6 compounds, isolation and properties of products. *Biochim. Biophys. Acta 279*:310-315.

203. Reinhold, J. G. (1975). Trace elements—a selective survey. *Clin. Chem. 21*:476-500.

204. Rice, E. E., E. M. Squires, and J. F. Fried (1948). Effect of storage and microbial action on vitamin content of pork. *Food Res. 13*:195-202.

205. Richardson, L. R., S. Wilkes, and S. J. Ritchey (1961). Comparative vitamin B_6 activity of frozen, irradiated and heat-processed foods. *J. Nutr. 73*:363-368.

206. Ristow, K. A., J. F. Gregory, III, and B. L. Damron (1982). Thermal processing effects on folacin bioavailability in liquid model food systems, liver, and cabbage. *J. Agr. Food Chem. 30*(5):801-806.

207. Roberts, D. (1961). Short communications. Cleavage of folic acid by hydrogen peroxide. *Biochim. Biophys. Acta 54*:572-574.

208. Rockland, L. B., C. F. Miller, and D. M. Hahn (1977). Thiamine, pyridoxine, niacin and folacin in quick-cooking beans. *J. Food Sci. 42*:25-28.

209. Roth, B., M. E. Hultquist, M. J. Fahrenbach, D. B. Cosulich, H. P. Broquist, J. A. Brockman, Jr., J. M. Smith, Jr., R. P. Parker, E. L. R. Stokstad, and T. H. Jukes (1952). Synthesis of leucovorin. *J. Amer. Chem. Soc. 74*:3247-3252.

211. Safe Drinking Water Committee (1977). *Drinking Water and Health*, National Academy of Sciences-National Research Council, Washington, D. C.

212. Santini, R., C. Brewster, and C. Butterworth, Jr. (1964). The distribution of folic acid active compounds in individual foods. *Amer. J. Clin. Nutr. 14*:205-210.

213. Schroeder, H. A. (1971). Losses of vitamins and trace minerals resulting from processing and preservation of foods. *Amer. J. Clin. Nutr. 24*:562-573.

215. Schulz, J., and N. J. Smith (1958). A quantitative study of the absorption of food iron in infants and children. *Amer. J. Dis. Child. 95*:109-119.

216. Seely, G. R., and T. H. Meyer (1971). The photosensitized oxidation of β-carotene. *Photochem. Photobiol. 13*:27-32.

217. Sen, S. P. (1962). Protection of vitamin B_{12} against destruction by aneurin and nicotinamide. *Chem. Ind.* (London) 94-95.

218. Shenoy, K. G., and G. B. Ramasarma (1955). Iron as a stabilizer of vitamin B_{12} activity in liver extracts and the nature of so-called alkali-stable factor. *Arch. Biochem. Biophys. 55*:293-295.

219. Shin, Y. S., E. S. Kim, J. E. Watson, and E. L. R. Stokstad (1975). Studies of folic acid compounds in nature. IV. Folic acid compounds in soybeans and cow milk. *Can. J. Biochem. 53*:338-343.

220. Silverman, M., J. C. Keresztesy, G. J. Koval, and R. C. Gardiner (1957). Citrovorum factor and the synthesis of formyglutamic acid. *J. Biol. Chem. 226*:83-94.

221. Slover, H. T. (1971). Tocopherol in foods and fats. *Lipids 6*:291-296.

222. Snell, E. E. (1958). Chemical structure in relation to biological activities of vitamin B_6. *Vitamins Hormones 16*:77-125.

223. Somogyi, J. C. (1966). Biochemical aspects of antimetabolites of thiamine. *Nutr. Diet. 8*:74-96.

224. Spanyar, P., and E. Kevei (1963). Über die stabilisierung von vitamin C in lebensmitteln. *Z. Lebensm. Unters. Forsch. 120*:1-17.

225. Srncova, V., and J. Davidek (1972). Reaction of pyridoxal and pyridoxal-S-phosphate with proteins reaction of pyridoxal with milk serum products. *J. Food Sci. 37*:310-312.

226. Stokstad, E. L. R., D. Fordham, and A. deGrunigen (1947). Letters to the editors. The inactivation of pteroylglutamic acid (liver lactobacillus casei factor) by light. *J. Biol. Chem. 167*:877-878.

227. Stokstad, E. L. R., B. L. Hutchings, J. H. Mowat, J. H. Boothe, C. W. Waller, R. B. Angier, J. Semb, and Y. SubbaRow (1948). The degradation of the fermentation *Lactobacillus casei* Factor. *J. Amer. Chem. Soc. 70*:5-13.

228. Strohecker, R., and H. M. Henning (1965). *Vitamin Assay*, Verlag-Chemie, Darmstadt, Germany.

229. Suckewer, A., J. Bartinsk, and B. Secomska (1970). Effect of technological processes and storage on folacin in some preserve vegetable products. *Higiene 21*: 619-629.

230. Sweeney, J. P., and A. C. Marsh (1971). Effect of processing on provitamin A in vegetables. *J. Amer. Diet. Ass. 59*:238-243.

231. Tannenbaum, S. R., and W. Mergens (1980). Reaction of nitrite with vitamins C and E. *Ann. N. Y. Acad. Sci. 355*:267-277.

232. Thomas, M. H., S. Brenner, A. Eaton, and V. Craig (1949). Effect of electronic cooking on nutritive value of foods. *J. Amer. Diet. Ass. 25*:39-45.

233. Thompson, A. (1964). Availability of minerals in foods of plant origin. Mineral availability and techniques for its measurement. *Proc. Nutr. Soc. 24*:81-88.

234. Thompson, L. U., and O. Fennema (1971). Inhibition of L-ascorbic acid oxidation by type II gas hydrates. *J. Agr. Food Chem. 19*:232-235.

235. Toepfer, E. W., W. Mertz, E. E. Roginski, and M. M. Polansky (1973). Chromium in foods in relation to biological activity. *J. Agr. Food Chem. 21*:69-73.

236. Tomarelli, R. M., E. R. Spence, and F. W. Bernhart (1955). Biological availability of vitamin B_6 of heated milk. *J. Agr. Food Chem. 3*:338-341.

237. Tripet, F. Y., and U. W. Kesselring (1975). Etude de stabilité de L'acide folique à l'état solide en fontion de la température et de l'humidité. *Pharm. Acta Helv. 50*:318-322.

238. Truswell, A. S. (1976). Symposium on "some aspects of diet and health." *Proc. Nutr. Soc. 35*:1-22.

239. Turnbull, A. (1974). In *Iron in Biochemistry and Medicine* (A. Jacobs and M. Worwood, eds.), Academic Press, New York, pp. 369-403.

240. Underwood, E. J. (1971). *Trace Elements in the Environment*, Academic Press, New York.

241. Underwood, E. J, (1973). In *Toxicants Naturally Occurring in Foods*, NAS/NRC, Washington, D. C., pp. 43-87.

242. USDA (1976 on). *Composition of Foods*, Handbooks 8-1 through 8-9, U. S. Department of Agriculture, Washington, D. C.

243. U.S. Department of Health, Education and Welfare, Health Services and Mental Health Administration (1968-70). *Ten-State Survey*, DHEW Publication No. (HSM) 72-8134.

244. U.S. Food and Nutrition Board (1974). *Recommended Dietary Allowances*, 8th rev. ed., National Academy of Sciences–National Research Council, Washington, D. C.

245. Velisek, J., J. Davidek, D. Jiri, V. Kubelka, Z. Zelinkova, and J. Pokorny (1976). Volatile degradation products of L-dehydroascorbic acid. *Z. Lebensm. Unters. Forsch. 162*:285-290.

246. Waller, C. W., A. A. Goldman, R. B. Angier, J. H. Boothe, B. L. Hutchings, J. H. Mowat, and J. Semb (1950). 2-Amino-4-hydroxy-6-pteridinecarboxaldehyde. *J. Amer. Chem. Soc. 72*:4630-4633.

247. Walter, Jr., W. M., A. E. Purcell, and W. Y. Cobb (1970). Fragmentation of β-carotene in autoxidizing dehydrated sweet potato flakes. *J. Agr. Food Chem. 18*:881-885.

248. Warburg, O., and W. Christian (1933). Über gas gelbe ferment und seine wirkungen. *Biochem. Z. 266*:377-411.
249. Weckel, K. G., B. Santos, E. Hernan, L. Laferriere, and W. H. Gabelman (1962). Carotene components of frozen and processed carrots. *Food Technol. 16*:91-94.
250. Weinstock, H. H., H. K. Mitchell, E. F. Pratt, and R. J. Williams (1939). Pantothenic acid. IV. Formation of β-alanine by cleavage. *J. Amer. Chem. Soc. 61*:1421-1425.
251. Wendt, G., and F. W. Bernhart (1960). The structure of a sulfur-containing compound with vitamin B$_6$ activity. *Arch. Biochem. Biophys. 88*:270-272.
252. Whitaker, J. R, (1972). *Principles of Enzymology for the Food Sciences*, Marcel Dekker, New York.
253. Whiteley, J. M., J. Drais, J. Kirchner, and F. M. Huennekens (1968). Synthesis of 2-amino-4-hydroxy-6-formyl-7,8-dihydropteridine and its identification as a degradation product of dihydrofolate. *Arch Biochem. Biophys. 126*:956-957.
254. Widicus, W. A., and J. R. Kirk (1981). Storage stability of α-tocopherol in a dehydrated model food system containing methyl linoleate. *J. Food Sci. 46*:813-816.
255. Williams, R. J., and R. T. Major (1940). The structure of pantothenic acid. *Science 91*:246.
256. Wogan, G. N., S. Paglialunga, M. C. Archer, and S. R. Tannenbaum (1975). Carcinogenicity of nitrosation products of ephedrine, sarcosine, folic acid, and creatinine. *Cancer Res. 35*:1981-1984.
257. Woodcock, E. A., J. J. Warthesen, and T. P. Labuza (1982). Riboflavin photochemical degradation in pasta measured by high performance liquid chromatography. *J. Food Sci. 47*:545-555.
258. Passmore, R., B. M. Nicol, M. Narayana Rao, G. H. Beaton, and E. M. DeMaeyer (eds.), (1974). *Handbook of Human Nutritional Requirements*, World Health Organization, Geneva. (Also issued as FAO Nutrition Studies, No. 28.)
259. Yudkin, W. H. (1949). Thiaminase, the Chastek-paralysis factor. *Physiol. Rev. 29*:389-402.
260. Zakrezewski, S. F. (1966). On the mechanism of chemical and enzymatic reduction of folate and dihydrofolate. *J. Biol. Chem. 241*:2962-2967.
261. Zakrezewski, S. F. (1966). Evidence for the chemical interaction between 2-mercaptoethanol and tetrahydrofolate. *J. Biol. Chem. 241*:2957-2961.
262. Zechmeister, L. (1962). *Cis-Trans Isomeric Carotenoids, Vitamin A and Arylpolyenes*, Academic Press, New York.

BIBLIOGRAPHY

Archer, M. C., and S. R. Tannenbaum (1979). Vitamins, in *Nutritional and Safety Aspects of Food Processing* (S. R. Tannenbaum, ed.), Marcel Dekker, New York, pp. 47-95.

Bender, A. E. (1978). *Food Processing and Nutrition*, Academic Press, London.

Benterud, A. (1977). Vitamin losses during thermal processing, in *Physical, Chemical and Biological Changes in Food Caused by Thermal Processing* (T. Hoyem and O. Kvale, eds.), Applied Science Publishers, London, pp. 185-217.

Dixon, M., and E. C. Webb (1979). *Enzymes*, 3rd ed., Academic Press, New York.

Eichhorn, G. L. (ed.) (1973). *Inorganic Biochemistry*, 2 vols., Elsevier, Amsterdam.

Food and Nutrition Board (1980). *Recommended Dietary Allowances*, 9th ed., Natl. Research Council, Natl. Academy of Sciences, Washington, D. C.

Goodhart, R. S., and M. E. Shils (1980). *Modern Nutrition in Health and Disease*, 6th ed., Lea & Febiger, Philadelphia.

Harris, R. S., and E. Karmas (eds.) (1975). *Nutritional Evaluation of Food Processing*, AVI Publishing Co., Westport, Conn.

Munson, P. L., J. Glover, E. Diczfalusy, and R. E. Olson. Periodical with annual volumes. *Vitamins and Hormones*. Academic Press, New York.

Prasad, A. S. (ed.) (1982). *Clinical, Biochemical and Nutritional Aspects of Trace Elements*, Alan R. Liss, Inc., New York.

Rechcigl, Jr., M. (ed.) (1982). *Handbook of Nutritive Value of Processed Food*, vols. 1 and 2, CRC Press, Boca Raton, Fla.

Schroeder, H. A. (1971). Losses of vitamins and minerals resulting from processing and preservation of foods. *Amer. J. Clin. Nutr. 24*:562-573.

Underwood, E. J. (1977). *Trace Elements in Human and Animal Nutrition*, 4th ed., Academic Press, New York.

8

PIGMENTS AND OTHER COLORANTS

Frederick J. Francis   University of Massachusetts, Amherst, Massachusetts

I.	Pigments Indigenous to Food	546
	A. Chlorophylls	546
	B. Myoglobin and Hemoglobin	550
	C. Anthocyanins	557
	D. Flavonoids	563
	E. Proanthocyanidins	565
	F. Tannins	567
	G. Betalains	568
	H. Quinones and Xanthones	569
	I. Carotenoids	570
	J. Miscellaneous Natural Pigments	574
II.	Colorants Added to Foods	576
	A. Regulatory Aspects	576
	B. Certified Colorants	577
	C. Colorants Exempt from Certification	580
	D. Possible New Colorants	582
	References	582

The quality of food, aside from the microbiological aspects, is generally based on color, flavor, texture, and nutritive value. Depending on the particular food, these factors may be weighted differently in assessing overall quality. However, one of the most important sensory quality attributes of a food is color. This is because, no matter how nutritious, flavorful, or well textured a food, it is unlikely to be eaten unless it has the right color.

---

*The sections on chlorophylls and meat pigments are condensed versions of those written by Dr. F. M. Clydesdale in the first edition. Appreciation is expressed to Dr. Clydesdale for allowing his materials to be used in modified form.

Acceptibility of color in a given food is influenced by many diverse factors, including cultural, geographic, and sociological aspects of the population. Indeed, color as well as other eating habits may be viewed as a type of culinary anthropology indigenous to a specific region. However, no matter what the biases or habits of a given area, certain food groups are acceptable only if they fall within a certain color gamut. Moreover, acceptability is reinforced by economic worth since in many cases raw food materials are judged as to value by their color.

In this chapter, the word *color* is used to denote the human eye's perception of colored materials, such as red, green, or blue. The term "colorant" is a general term referring to any chemical compound that imparts color. "Pigment," as used here, refers to normal constituents of cells or tissues that impart color. A pigment may have properties beyond that of a colorant, for example, as energy receptors, carriers of oxygen, or protectants against radiation. The term "dye" refers to colorants used in the textile industry and has no place in food usage. A "lake" is a food colorant absorbed on the surface of an inert carrier, such as alumina. A "color additive" refers to any compound or group of compounds that imparts color. It is unfortunate that the term "color" has often been used to mean either color or colorant.

It is obvious that the color of a food is due to the natural pigments present, except in cases in which colorants have been added. Therefore, to achieve the desirable color and acceptability, an understanding of such pigment systems is essential.

I. PIGMENTS INDIGENOUS TO FOOD

A. Chlorophylls

1. Structure

The name "chlorophyll" was originally intended to describe those green pigments involved in the photosynthesis of higher plants. However, it has been extended to all classes of photosynthetic porphyrin pigments. Functionally this appears to be proper except in certain very specific, isolated cases (2).

In any discussion of the chlorophylls, there is a prerequisite of defining structures and chemical terminology. Through the years, terminology has been used, deleted, and revised so that a rather special language has arisen that has become generally acceptable. The following simplified classification of terminology should be an aid in understanding this chemistry (2,15).

1. Pyrrole, one of four cyclic components of the porphyrin nucleus.
2. Porphine, the completely conjugated tetrapyrrole skeleton, consisting of four pyrrole rings connected by four methyne bridges.
3. Porphyrin: in chlorophyll chemistry it is assumed that the term "porphyrins" includes the entire class of closed, completely conjugated tetrapyrroles. The parent compound of this class is porphine, which may be substituted by various groups, such as methyl, ethyl, or vinyl. All other subclasses of porphyrins are referred to the state of oxidation of this compound. Thus, ditetra- or hexahydroporphines may be formed when reduction occurs only on the periphery of the pyrrolic rings. When reduction occurs at the methyne carbons, a class of compounds known as porphyrinogens results.
4. Chlorins, dihydroporphines.
5. Phorbin, a porphyrin with the addition of a C9-C10 ring.
6. Phorbide: all naturally occurring porphyrins have a propionic acid residue at position

7. In the chlorophylls, this position is esterified with a long-chain alcohol (phytol or farnesol). The corresponding structure with a free acid is known as a phorbide if it does not contain magnesium.

7. Phytol, a 20-carbon alcohol with an isoprenoid structure, as shown in Fig. 1.

8. Chlorophyll a: the structure of chlorophyll a is shown in Fig. 1. It is a magnesium-chelated tetrapyrrole structure with methyl substitutions at the 1, 3, 5, and 8 positions; vinyl at the 2; ethyl at the 4; propionate esterified with phytyl alcohol at the 7; keto at the 9; and carbomethoxy at the 10. The empirical formula is $C_{55}H_{72}O_5$-N_4M_g.

9. Chlorophyll b: as may be seen in Fig. 1, chlorophyll b has the same configuration as chlorophyll a except that in the 3 position there is a formyl group rather than a methyl group. The empirical formula is $C_{55}H_{70}O_6N_4M_g$.

10. Pheophytin a, chlorophyll a minus magnesium.

11. Pheophytin b, chlorophyll b minus magnesium.

12. Chlorophyllide a, chlorophyll a minus phytol.

13. Chlorophyllide b, chlorophyll b minus phytol.

14. Pheophorbide a, chlorophyllide a minus magnesium.

15. Pheophorbide b, chlorophyllide b minus magnesium.

This is not intended to be a complete discussion of the chemistry of the chlorophylls but merely an overview of the terminology associated with the chlorophylls normally found in higher plants utilitized for food.

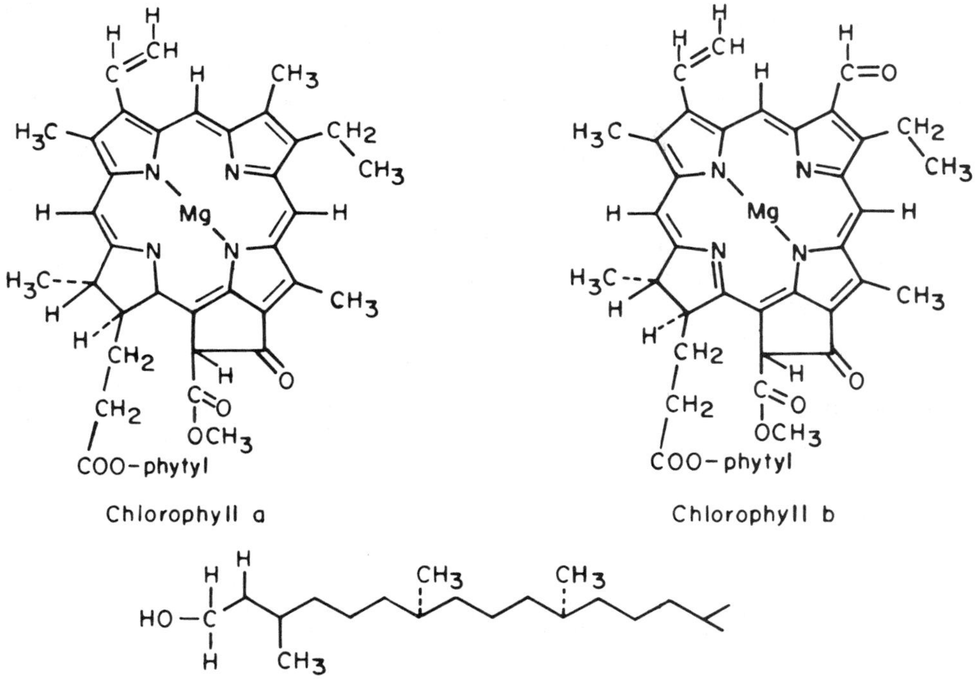

Figure 1 Structural formulas of chlorophylls a and b and the phytol moiety. [From S. Aronoff (1966). In *The Chlorophylls*, L. P. Vernon and G. P. Seely, eds. (15), courtesy of Academic Press.]

A diagrammatic summary of the chlorophylls and some of their conversion products, which may or may not occur during food processing, are shown in Fig. 2.

2. Location in Plants

A variety of chlorophylls has been described, such as chlorophylls a, b, c, and d; bacteriochlorophylls a and b; and chlorobium chlorophylls.

In foods we are mainly concerned with chlorophylls a and b, which occur in the approximate ratio of 3:1 in higher plants. In leaves, these chlorophylls are located within plastid bodies, called chloroplasts. The chloroplasts have an ordered fine structure and appear in the light microscope as green saucer-shaped bodies about 5-10 μm long and 1-2 μm thick. Within the chloroplasts are smaller particles, called grana, which are 0.2-2 μm in diameter and are composed of lamellae that range in size from 0.01 to 0.02 μm. Between the grana are the stroma. Chlorophyll molecules are embedded in the lamellae and are closely associated with lipids, proteins, and lipoproteins. They are held together in a monolayer both by mutual attraction and by the affinity of each molecule's phytol "tail" for lipids and the affinity for each molecule's hydrophobic planar porphyrin ring for proteins. Therefore, within the chloroplast, the chlorophylls may be visualized as embedded between a layer of protein and lipid with a carotenoid positioned alongside the phytol chain of the chlorophyll.

3. Physical Properties

Chlorophyll a and pheophytin a are soluble in alcohols, ether, benzene, and acetone. When they are pure, they are only slightly soluble in petroleum ether. They are insoluble in water. Chlorophyll b and pheophytin b are soluble in alcohols, ether, acetone, and benzene. When pure, they are almost insoluble in petroleum ether and insoluble in water.

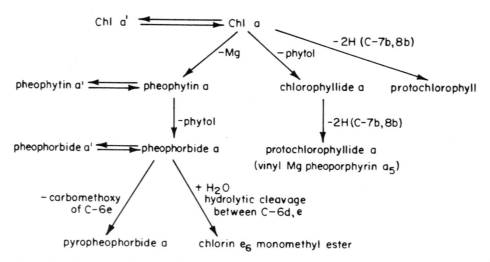

Figure 2 The nomenclature of the chlorophylls and some of their breakdown products. [From S. Aronoff (1966). In *The Chlorophylls*, L. P. Vernon and G. R. Seely, eds. (15), courtesy of Academic Press.]

The chlorophyllides and pheophorbides, which are counterparts to the chlorophylls and pheophytins, respectively, lack only the phytol side chain and are generally oil insoluble and water soluble.

4. Chemical Properties

Chemically, the chlorophylls may be altered in many ways, but in food processing the most common alteration is pheophytinization, which is the replacement of the central magnesium by hydrogen and the consequent formation of the dull olive-brown pheophytins. It is difficult to explain the drastic color shift of green chlorophylls to the dull olive-brown pheophytins by simply visualizing the replacement of magnesium by hydrogen. The accepted structural formula of pheophytin is normally shown this way, but it is likely that some shift in the porphyrin resonance structure is also involved.

Chlorophyllides may be formed by the removal of the phytol chain. These compounds are green and have essentially the same spectral properties as the chlorophylls; however, they are more water soluble than the chlorophylls. If the magnesium in the chlorophyllides is removed, the corresponding pheophorbides are formed, which have the same color and spectral properties as the pheophytins (37).

These relationships may be illustrated diagrammatically as follows.

$$\text{Chlorophyll} \xrightarrow{-\text{Phytol}} \text{Chlorophyllide}$$

$$-\text{Mg} \downarrow \qquad\qquad -\text{Mg} \downarrow$$

$$\text{Pheophytin} \xrightarrow{-\text{Phytol}} \text{Pheophorbide}$$

Many other reactions may occur due to the functional side groups of the chlorophylls, the isocyclic ring, which may be oxidized to form allomerized chlorophyll, and the rupture of the tetrapyrrole ring to form colorless end products. During food processing, it is likely that such reactions do proceed to some extent, but this is limited compared with pheophytinization.

5. Effects of Food Handling, Processing, and Storage

Almost any type of food processing alone or combined with storage causes some deterioration of the chlorophyll pigments. It has been shown that dehydrated products packed in clear containers undergo photooxidation and loss of desirable color. The conversion of chlorophyll to pheophytin occurs in dehydrated foods, and this conversion is directly related to the degree of blanching the foods underwent before dehydration.

With freeze-dried, blanched spinach, the transformation of chlorophyll a to pheophytin a is 2½ times as fast as the corresponding chlorophyll b change and is a function of a_w (30).

Many conditions affect chlorophyll content. Vegetables may show color changes in freezing and subsequent storage. The changes are influenced by the time and temperature of blanching prior to freezing. Walker (36) observed degradation to nonchlorophyll compounds in peas and beans due to the action of lipoxygenase. The lipoxygenases produced free radicals that degraded the chlorophyll. γ-Irradiation and subsequent storage produced degradation of chlorophylls and pheophytins. Fermentation of cucumbers produced pheophytins, chlorophyllides, and pheophorbides.

The color of heat-processed green vegetables turns from bright green to olive-brown due to the conversion of chlorophyll to pheophytin under the influence of acids pro duced during thermal processing. Ten acids are found or formed in spinach, peas, and green beans during thermal processing, but the major acids involved in pigment degradation are acetic and pyrrolidone carboxylic acid (31).

6. Preservation of Green Color

A large amount of research and a number of patents have been directed toward the preservation of green color in thermally processed vegetables. Unfortunately, none of them really function successfully. Perhaps the first serious attempt was the two patents awarded for the "Blair process" (18,19). This process involved the use of alkaline salts, calcium and/or magnesium hydroxide, to maintain the magnesium ion in the chlorophyll molecule. The process produced an attractive product immediately after processing but not after storage. Additional patents based on alkaline treatment were equally unsuccessful.

Borodin in 1882 recognized that chlorophylls could be "fixed" under certain conditions. The conditions favored the action of chlorophyllase, and his fixed chlorophylls were probably chlorophyllides. Thomas, in 1928, and Lesley and Shumate, in 1937, received patents for a process based on blanching for 30 min at 67°C prior to processing. Clydesdale and Francis (22) showed that chlorophyllides in spinach were indeed more stable than chlorophylls but the amount of chlorophyllides obtainable was too low to be of practical significance in color retention.

Processing by high-temperature-short-time (HTST) methods has the general advantage of allowing microbial destruction with less chemical destruction than occurs during conventional thermal processing. Clydesdale et al. (23) attempted to combine alkaline treatment, enzyme conversion to chlorophyllide, and HTST processing to maintain the green color in spinach. Again, an attractive product was obtained immediately after processing but the gains were soon lost on storage.

At the present time, the best way to maintain chlorophyll stability is to start with high-quality materials, process as quickly as possible, and store the product at low temperatures.

B. Myoglobin and Hemoglobin

Meat is an important part of the American diet. In 1978, Americans consumed about 85 kg per capita of beef, pork, veal, and lamb. Beef and pork were the most important, with 55 and 28 kg, respectively. The per capita consumption of 55 kg beef is misleading because this represents carcass weight. It is reduced to 39 kg of raw retail weight and 19 kg of cooked weight. This represents about 15.4 g of protein from beef per day. However, it does represent a large amount of meat and its acceptability is affected by its color and pigment content.

The chemistry of the color of meat is the chemistry of heme pigments. More specifically, it is primarily the chemistry of one pigment, myoglobin. In the live animal, myoglobin accounts for only 10% of the total iron, but during slaughter, the bleeding process removes most of the iron as hemoglobin, and in a well-bled piece of beef skeletal muscle as much as 95% or more of the remaining iron is accounted for as myoglobin. Myoglobin, and its various chemical forms, is not the only pigment in muscle, nor is it the most important biologically, but it is generally the only pigment present in large enough quantities to color meat. Muscle pigments, which are of considerable importance to the living tissue but which contribute little or nothing to the total color, include cytochromes (red heme pigments, which contain iron in a similar porphyrin-protein complex structure); vitamin B_{12}, a much more complex structure than myoglobin and one that contains the same porphyrin ring as the hemes and the cytochromes, but which contains a cobalt atom instead of iron; the flavins, yellow coenzymes involved with the cytochromes in electron transport in the cell; and hemoglobin.

Myoglobin is a complex muscle protein, similar in function to the blood pigment, hemoglobin, in that both serve to complex with the oxygen required for metabolic activity of the animal. The hemoglobin in red corpuscles contains four polypeptide chains and four so-called heme groups, which are planar collections of atoms with an iron atom at the center. The function of the heme group is to combine reversibly with a molecule of oxygen, which is then carried by the blood from the lungs to the tissues. Myoglobin is a junior relative to hemoglobin, being a quarter its size and consisting of a single polypeptide chain of about 150 amino acid units attached to a single hemoglobin group. It is contained within the cells of the tissues and it acts as a temporary storehouse for the oxygen brought by the hemoglobin in blood.

1. Structure

John C. Kendrew, the pioneer who was able to deduce the actual arrangement in space of nearly all the 2600 atoms in myoglobin, compared his first view of the three-dimensional structure of a myoglobin molecule to that of the early explorers of America when they made their first landfall and had the unforgettable experience of glimpsing a new world that no European had seen before. This, indeed, was a pioneering experience for all protein chemistry, and it does underline specifically the complexity of the myoglobin structure. Both hemoglobin and myoglobin are complex proteins, which means that, in addition to the protein portion of the molecule, there is another moiety, nonpeptide in nature, complexed to the peptide chain. The protein moiety is known specifically as globin and the nonpeptide portion is called heme. Heme is composed of two parts, an iron atom and a large planar ring, porphyrin. Porphyrin is made up of four subunits, the heterocyclic compound pyrrole, linked together by methyne bridges.

Hemoglobin may be considered the linking together of four myoglobins; therefore, discussion of the chemistry of these pigments can be limited to myoglobin. Figure 3 shows the isolated heme group, which, when attached to the protein globin, forms myoglobin. Figure 4 shows a vastly simplified diagrammatic structure of myoglobin. In reality a very complex situation exists that only recently has been defined.

In simplified terms, one may view the protein portion folded around the iron of the heme group in eight α-helical segments, forming the total complex molecule. A three-dimensional representation of this myoglobin molecule peptide chain is shown in Fig. 5. The heme iron is represented by the circle in the upper central portion of the figure and the helical peptide segments are labeled A-H, beginning at the amino end and ending at the carboxyl end of the molecule. Obviously, a complete description of this complex structure requires much more space than is available here.

2. Physical Properties

Myoglobin is part of the sarcoplasmic proteins of muscle. It is soluble in water and in dilute salt solutions.

Any attempt to explain the color of myoglobin in a muscle tissue matrix must take into account not only the spectral characteristics of the pigment but also the scatter-

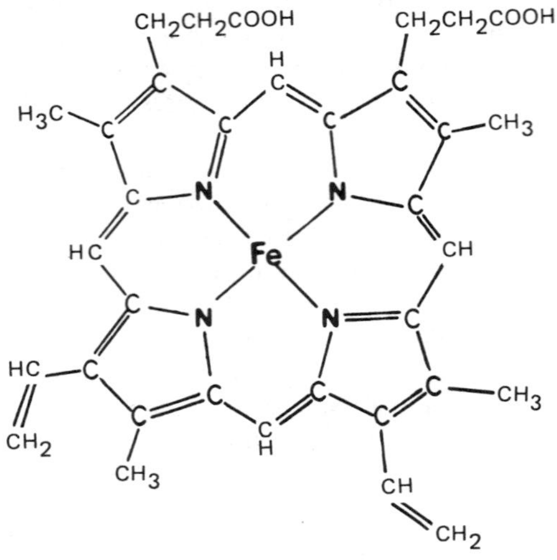

CH₂CH₂COOH and the heme structure...

Figure 3 The isolated heme group.

ing characteristics of the muscle matrix. The total reflectance characteristics of meat are due to two major components. One of these components, absorption by the meat pigments, may be represented by the symbol K; the other, the scattering coefficient of the muscle fiber matrix, may be represented by the symbol S. The ratio K/S then describes the total impact of both absorption and scattering on the eye. In a brightly colored piece of meat, K, the absorption coefficient, would be large in relation to the scattering

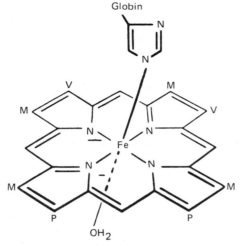

Figure 4 A simplifed diagrammatic structure of myoglobin. [From Price and Schweigert (14), courtesy of W. H. Freeman and Co.]

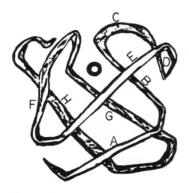

Figure 5 A three-dimensional representation of the myoglobin molecule peptide chain.

coefficient S. As K decreases with respect to S the characteristic peaks on the spectral curve of myoglobin tend to decrease until K becomes relatively small with respect to S and the spectral curve bears little resemblance to a typical myoglobin curve. These are physical parameters that are not normally considered in a discussion of myoglobin pigments but are extremely important when the color of meat is considered as a major quality factor.

3. Chemical Properties

In considering meat pigments in relation to quality one must be concerned mainly with the various complexes of heme, globin, and ligands that surround iron in an oxidized (Fe^{3+}) or a reduced (Fe^{2+}) state. The reaction defining the covalent complex formed between myoglobin and molecular oxygen to form oxymyoglobin is known as oxygenation and is distinct from the oxidation of myoglobin to form metmyoglobin. Furthermore, such complexes may be grouped into ionic and covalent bond types, with the covalent bond types producing the bright red pigments desired in meat. Oxymyoglobin, nitrosomyoglobin, and carboxymyoglobin are examples of the ferrous covalent complexes of myoglobin with molecular oxygen, nitric oxide, and carbon monoxide, respectively. Spectrally, these complexes are characterized by relatively sharp absorption maxima at 535-545 nm and 575-588 nm, such as metmyoglobin and myoglobin, respectively.

Cyanmetmyoglobin and metmyoglobin hydroxide are examples of ferric iron covalent complexes that have the characteristic red color. In the case of such ferric covalent complexes, the negative charge of the ion may be thought of for conceptual purposes as having neutralized the third positive charge of the ferric ion. In general, it may be said that an electron pair forms a stable red covalent complex with myoglobin (ferrous) if it is neutral, or with metmyoglobin (ferric) if it is negatively charged.

In the absence of strong covalent complexers, myoglobin and metmyoglobin form ionic complexes with water. In these complexes, the oxygen of water binds to iron by a dipole-ion interaction since the oxygen atom is not as strong an electron-pair donor as the oxygen molecule.

Myoglobin is characterized by an absorption band with a maximum at 555 nm in the green portion of the spectrum and is purple in color. In metmyoglobin, the major peak is shifted to 505 nm in the blue end of the spectrum, and a smaller peak exists at 627 nm in the red, with the net visual appearance of brown.

The color cycle in fresh meats is reversible and dynamic, with the three pigments, oxymyoglobin, myoglobin, and metmyoglobin, constantly interconverted. Remember that brown metmyoglobin, the oxidized or ferric form of the pigment, cannot bind oxygen even though it is oxidized by the same oxygen that converts myoglobin to the red oxymyoglobin (oxygenation). Therefore, in the presence of oxygen, the purple myoglobin may be oxygenated to the bright red oxygenated pigment oxymyoglobin, producing the familiar "bloom" of fresh meats, or it may be oxidized to metmyoglobin, producing the undesirable brown of less acceptable meats. Figure 6 shows the heme pigment reactions of both fresh and cured meat and meat products. At high oxygen pressures the reaction of myoglobin (Mb) to oxymyoglobin (O_2Mb), shown in Fig. 6, is shifted to the left. The red O_2Mb, once formed, is stabilized by the formation of a highly resonant structure, and as long as the heme remains oxygenated no further color changes take place. However, the oxygen continually associates and dissociates from the heme complex, a process accelerated by a number of conditions, among them low oxygen pressures. When this happens, the reduced pigment (Mb) is oxidized to metmyoglobin (MMb). It is not known whether the oxidation takes place during association or dissociation, as indicated by the dashed arrow in Fig. 6. However, it is known that there is a slow and continuous oxidation to MMb. In fresh meat, the production of indigenous reducing substances constantly reduces MMb to Mb and the cycle continues if oxygen is present.

A summary of the pigments, their mode of formation, and their properties are shown in Table 1.

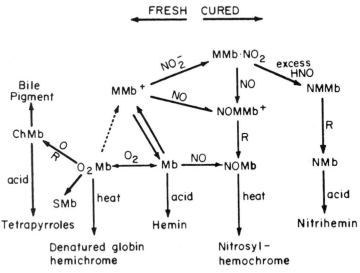

Figure 6 Heme pigment reactions for both fresh and cured meats. ChMb, cholemyoglobin (oxidized porphyrin ring); O_2Mb, oxymyoglobin (Fe^{2+}); MMb^+, metmyoglobin (Fe^{3+}); Mb, myoglobin (Fe^{2+}); MMb · NO_2, metmyoglobin nitrite; NOMMb, nitrosylmetmyoglobin; NOMb, nitrosylmyoglobin; NMMb, nitrimetmyoglobin; NMb, nitrimyoglobin; R, reductants; O, strong oxidizing conditions. [From Fox (25), courtesy of the American Chemical Society.]

4. Effects of Food Handling, Processing, and Storage

Cured Meat Pigments. Certainly, a very important aspect of the chemistry of meat pigments has to do with the curing process. This process is extremely important because of the large volume of meat products that are cured. Figure 6 shows the pathways that form the cured meat pigment nitrosylhemochrome in the presence of nitrate, nitric oxide, and reductants. All these reactions have been observed in vitro, but most of them take place only under fairly strong reducing conditions, since many of the intermediates are unstable in air. If the system in which the cured meat pigment is formed contains nitrite, the heme pigments are in the oxidized state initially, since nitrite is a strong heme pigment oxidant. In general, the formation of the cured meat chromophore or pigment is generally viewed as two processes: (a) biochemical reactions, which reduce nitrite to nitric oxide and reduce the iron in heme to the ferrous state, and (b) thermal denaturation of globin. The latter occurs only when the cured meat product is heated to 66°C or higher and may involve the coprecipitation of the heme pigment with other proteins in meat. Although the mechanisms involved in the complete reaction have not been determined in meat or meat products, it has been established that the end product is either nitrosylmyoglobin if uncooked or the denatured globin nitrosylhemochrome if cooked (25).

Several other factors in addition to those previously mentioned are important in cured meat color. Fox et al. (26), working with frankfurters, found that a lag period in color production, attributed to the presence of oxygen, was reduced by chopping under nitrogen, by vacuum mixing, or by adding ascorbate or cysteine. It was also found that the temperature of cooking was critical to the rate of color formation, to the percentage of cured meat pigment formed, and to the level of color maintained during storage. Traces of copper may produce a black discoloration (17). Excellent reviews of this subject are available (25,27).

Packaging. Packaging is an extremely important consideration when speaking of meat pigments, whether fresh or cured. In the case of fresh meats this may be easily understood because of the reactions the pigments undergo with oxygen to produce either an acceptable oxygenated product or an unacceptable oxidized product. Meat is packaged for five primary reasons: (a) to protect the product from contamination with bacteria and filth, (b) to retard or prevent loss of moisture from the product, (c) to shield the product from oxygen and light, (d) to facilitate handling, and (e) to aid in maintaining attractiveness.

The accelerated rate of oxidation of the heme pigment at low partial pressures of oxygen is of particular importance when considering packaging films for fresh meats. If the oxygen permeability of the film is low but significant, oxygen utilization by the tissue can exceed oxygen penetration of the film and a low partial pressure of oxygen can develop that favors oxidation of O_2Mb to brown MMb. If the packaging material is completely impermeable to oxygen, the heme pigments convert to fully reduced purple Mb and then bloom to red O_2Mb when the package is opened and exposed to oxygen.

There has been an increase in shelf life of fresh meats due in part to constantly improving sanitation techniques, automated methods of meat handling, and pasteurization and sterilization techniques. However, as Fox (25) points out, "To keep pace, information is needed on the factors which govern color stability. The area of greatest concern and interest is that of mechanisms whereby the oxidized pigments are reduced, both by endogenous and exogenous reductants and/or reducing systems."

Table 1 Pigments Found in Fresh, Cured, and Cooked Meat

Pigment	Mode of formation	State of iron	State of hematin nucleus	State of globin	Color
Myoglobin	Reduction of met-myoglobin; deoxy-genation of oxy-myoglobin	Fe^{2+}	Intact	Native	Purplish red
Oxymyoglobin	Oxygenation of myo-globin	Fe^{2+}	Intact	Native	Bright red
Metmyoglobin	Oxidation of myo-globin, oxymyo-globin	Fe^{3+}	Intact	Native	Brown
Nitric oxide myo-globin	Combination of myoglobin with nitric oxide	Fe^{2+}	Intact	Native	Bright red (pink)
Metmyoglobin nitrite	Combination of metmyoglobin with excess nitrite	Fe^{3+}	Intact	Native	Red
Globin hemochro-mogen	Effect of heat, de-naturing agents on myoglobin, oxymyo-globin; irradiation of hemichromogen	Fe^{2+}	Intact	Dena-tured	Dull red
Globin hemochro-mogen	Effect of heat, de-naturing agents on myoglobin, oxy-myoglobin, met-myoglobin, hemo-chromogen	Fe^{3+}	Intact	Dena-tured	Brown
Nitric oxide hemo-chromogen	Effect of heat, salts on nitric oxide myoglobin	Fe^{2+}	Intact	Dena-tured	Bright red (pink)
Sulfmyoglobin	Effect of H_2S and oxygen on myo-globin	Fe^{3+}	Intact but re-duced	Dena-tured	Green
Choleglobin	Effect of hydrogen peroxide on myo-globin or oxymyo-globin; effect of as-corbate or other re-ducing agent on oxymyoglobin	Fe^{2+} or Fe^{3+}	Intact but re-duced	Dena-tured	Green
Verdoheme	Effect of reagents (in excess) as for sulfmyoglobin	Fe^{3+}	Porphy-rin ring opened	Dena-tured	Green

Table 1   (Continued)

Pigment	Mode of formation	State of iron	State of hematin nucleus	State of globin	Color
Bile pigments	Effect of reagents (in large excess) as for sulfmyo-globin	Fe absent	Porphy-rin ring des-troyed; chain of porphy-rins	Absent	Yellow or color-less

Source: From Lawrie (10), courtesy of Pergamon Press.

3. Miscellaneous

Aside from the factors that have been considered so far, a few others should be considered, relative to the stability of meat pigments. For instance, carbon monoxide has been used with air to flush packaged beef prior to sealing. This treatment was found very effective for preserving and stabilizing the color of fresh beef for about 15 days. Several packaging films were also tested, and saran-mylar-polyethylene pouches seemed to be the best. However, when carbon monoxide is used, toxicity must be considered.

The light used in display environments of food is also an important factor influencing fresh meat color. Color changes have been noted in packaged fresh meat exposed to 215 fc of fluorescent light and to heat generated by incandescent light.

Other factors, such as the presence of certain metallic ions, also affect oxidation of O_2Mb and thus the color of fresh meat. It has been shown that copper is extremely active in promoting autoxidation of O_2Mb to MMb; other metals, such as Fe, Zn, and Al, are less active.

Meat pigments and the changes they undergo comprise a very complex and important area of food quality. As of yet there is no way to package fresh meat and obtain long storage life. However, with current basic research, it is hoped that an understanding of the mechanisms of color change will be developed to the point where it is possibe to extend the shelf life of fresh meat.

C. Anthocyanins

Anthocyanins are a group of reddish water-soluble pigments that are very widespread in the plant kingdom. Many fruits, vegetables, and flowers owe their attractive coloration to this group of water-soluble compounds, which exist in the cell sap. The very obvious red color led to an interest in their chemical structure that dates back to the classic work of Paul Karrer. The structures of the anthocyanin group are fairly well known today but the physical chemistry of the pigment complexes and the degradation reactions are less well known.

1. Structure

All anthocyanins are derivatives of the basic flavylium cation structure (I). Twenty anthocyanidins are known, but only six, pelargonidin (II), cyanidin (III), delphinidin (IV), peonidin (V), petunidin (VI), and malvidin (VII), are important in food (Table 2). The

Table 2 Substitution on the Flavylium Cation Structure to Produce the Major Anthocyanidins

Anthocyanidin (structure)	Substituent on carbon number[a]		
	3'	4'	5'
Pelargonidin (II)	H	OH	H
Cyanidin (III)	OH	OH	H
Delphinidin (IV)	OH	OH	OH
Peonidin (V)	OMe	OH	H
Petunidin (VI)	OMe	OH	OH
Malvidin (VII)	OMe	OH	OMe

[a]Compounds (II) to (VII) have OH groups on carbons 3, 5, and 7 and hydrogens on all other carbon atoms.

others are comparatively rare and are found in some flowers and leaves.

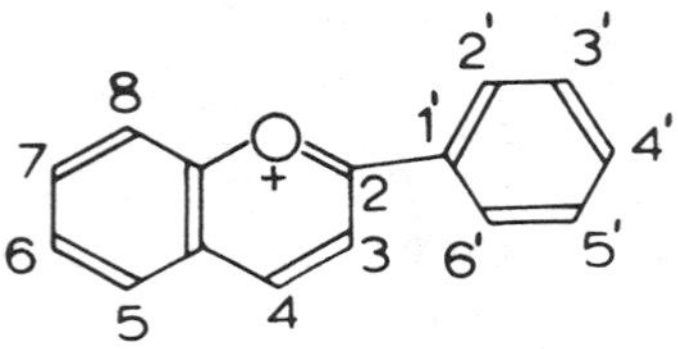

Flavylium cation (I)

An anthocyanin pigment is composed of an aglycone (an anthocyanidin) esterified to one or more sugars. Free aglycones rarely occur in food except possibly as trace components of degradation reactions. Only five sugars have been found as portions of anthocyanin molecules. They are, in order of relative abundance, glucose, rhamnose, galactose, xylose, and arabinose. Anthocyanins may also be "acylated," which adds a third component to the molecule. One or more molecules of *p*-coumaric, ferulic, caffeic, malonic, vanillic, or acetic acids may be esterified to the sugar molecule.

The anthocyanins can be divided into a number of classes depending on the number of sugar molecules. The monosides have only one sugar residue, almost always in the 3 position. The biosides contain two sugars, with either both on the 3 position, or one each on the 3 and 5 positions, or, rarely, in the 3 and 7 positions. The triosides contain three sugars, usually two on the 3 position with one on the 5 position, often three in a branched structure or a linear structure on the 3 position, or, rarely, with two on the 3 position and one on the 7 position. No anthocyanins with four sugar residues have been reported, but there is some evidence that they exist. One anthocyanin with 5 sugar molecules and 4 acyl components has been reported (38). Approximately 140 anthocyanins have been reported in the literature. Methods for the identification and analysis of anthocyanins are well established in the literature (6,7,12,16).

2. Stability in Food

The flavylium nucleus of the anthocyanin pigments is electron deficient and therefore highly reactive. The reactions usually involve decolorization of the pigments and are usually undesirable in fruit and vegetable processing. The rate of anthocyanin destruction is pH dependent and is greater at higher pH values. The rate of reaction is dependent on the amount of pigment in the colorless carbinol base form and is temperature dependent. Meschter calculated the half-life of the pigment in strawberry preserves to be 1300 hr at 20°C and 240 hr at 38°C. The reactivity of the anthocyanin molecule with air and with many components normally present in fruits and vegetables has lead to numerous studies on anthocyanin stability (8).

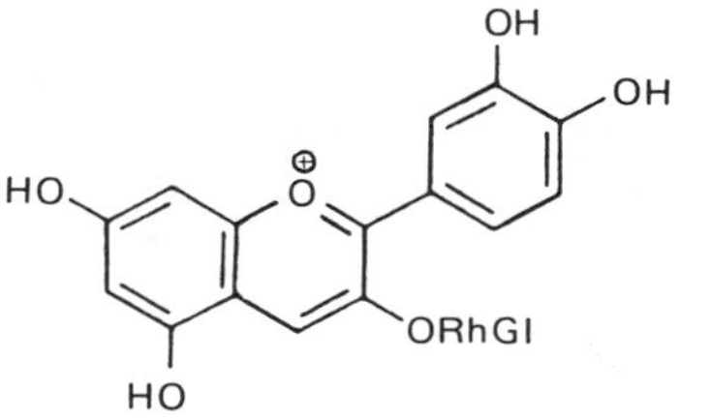

Cyanidin — 3 — rhamnoglucoside (VIII)

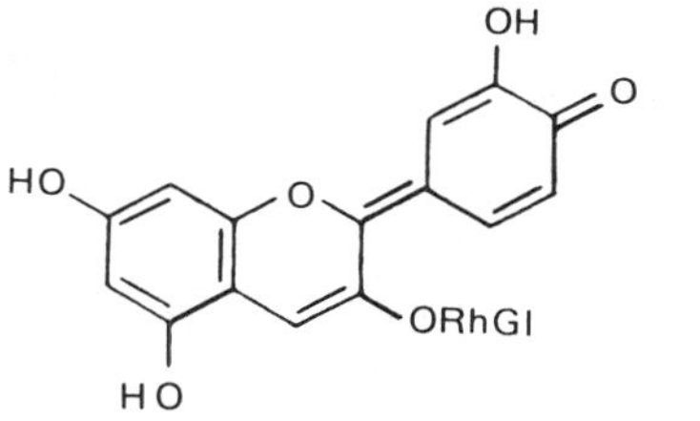

Ouinoidal or anhydrobase (IX)

3. Effect of pH

Anthocyanins show a marked change in color with changes in pH. Figure 7 shows the absorption spectra of cyanidin-3-rhamnoside, Structure (VIII), as influenced by changes in pH. The blue quinoidal base for the same pigment is shown as Structure (IX). Figure 8 shows the structural transformation of malvidin-3-glucoside with pH [Structure (VII), Table 2].

The blue quinoidal base (A in Fig. 8) is protonated to give the red flavylium cation (AH$^+$) which can then hydrate to form a colorless carbinol pseudo-base (B). The carbinol pseudo-base exists in equilibrium with its chalcone (C). The scheme can be summarized as

$$A \underset{\longleftarrow}{\overset{H^+}{\longrightarrow}} AH^+ \underset{\longleftarrow}{\overset{H_2O}{\longrightarrow}} B \underset{\longleftarrow}{\longrightarrow} C$$

The relative proportions of these pigments at different pH values were clarified recently by Brouillard and Delaporte (20) as shown in Fig. 9.

Brouillard (19a) reported on the exception stability of the *Zebrina* anthocyanins, which were reported to occur in the Commelinaceae (33a). He suggested that the stability was due to the inability to convert to the colorless pseudobase and chalcone forms due to

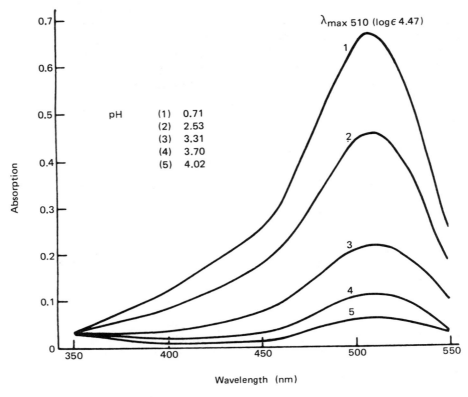

Figure 7 Absorption spectra of cyanidin-3-rhamnoglucoside in buffer solutions at pH values of 0.71–4.02. The concentration of pigment is 1.6×10^2 g/liter. [From Chichester (3), courtesy of Academic Press.]

steric hindrance by the three acyl groups. This group of pigments also has the unusual feature of sugar substitution in the B-ring.

A new type of anthocyanin was recently reported (33a) that shows a different response to pH. These compounds have sugar substitution in the B-ring to the extent of two or more acyl groups. One compound of this type (tricaffeoylcyanidin-3,7,3' triglucoside) is reported to have exceptional stability because the carbinol pseudo-base form (B in Fig. 8) is not formed due to steric hindrance with the acyl groups (20). Five compounds with sugar substitution in the B-ring have been reported to date.

4. Chemical Reactions

Sulfiting of fruits is an important commercial process for bulk storage of fruit prior to making jams, preserves, or maraschino cherries. In recent years, the process has become less important for jams because of the increased use of frozen fruit, which makes a better quality product. The addition of sulfite, or sulfur dioxide, results in rapid bleaching of the anthocyanins, in turn resulting in yellowish colors attributable to other pigments in the fruit. The process is simple sulfite addition reaction at positions 2 or 4,

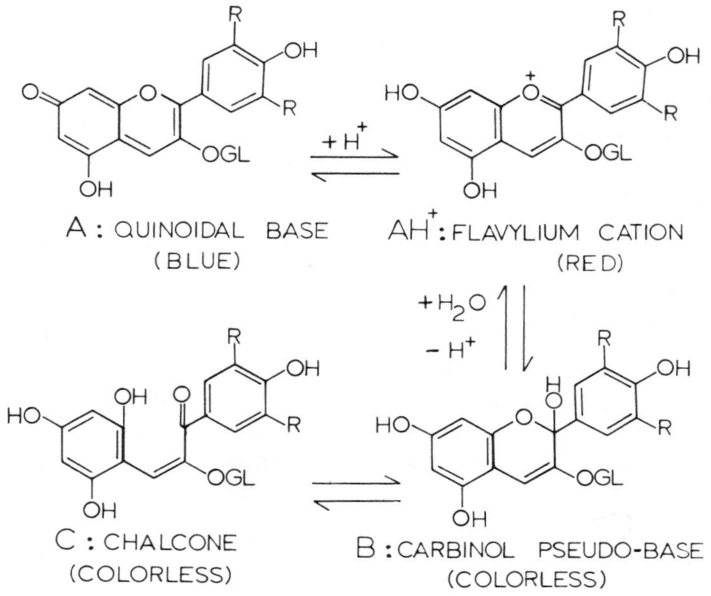

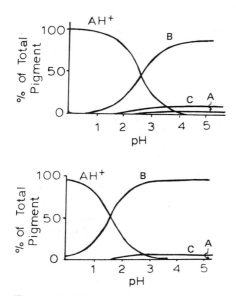

Figure 8 Structural changes in anthocyanins with pH. Malvidin-3-glucoside at 25°C and 0.2 M concentration, Structure (VII).

Figure 9 Effect of pH on distribution of anthocyanin structures. The upper figure refers to malvidin-3-glucoside, Structure (VII). The lower figure refers to malvidin-3,5-diglucoside. A, B, C and AH$^+$ refer to forms in Fig. 8.

producing compounds that are colorless but quite stable (25). Removal of the sulfite by boiling and acidification in jam manufacture results in regeneration of the anthocyanins.

The reaction of anthocyanins with ascorbic acid has been documented by many workers. An interaction occurs that results in degradation of both compounds. For example, cranberry juice cocktail, with approximately 9 mg anthocyanins per 100 g of juice and 18 mg ascorbic acid per 100 g juice, may lose approximately 80% of the anthocyanin in 6 months at room temperature (32). The juice is still brownish red, due in part to the color of degradation products. The actual mechanism of this reaction is not completely clear, but it may involve an intermediate, peroxide, produced by degradation of ascorbic acid. The reaction of peroxide with anthocyanin has been known for a long time and forms the basis of a method to identify the sugar compound on position 3 of the flavylium compound. Peroxide oxidation of flavylium salts, depending on the conditions, produces a series of different compounds (3). Oxidation of ascorbic acid is catalyzed by copper and iron compounds, and this results in a greater rate of destruction of anthocyanins. Even at low pH values (pH = 2.0), at which anthocyanins are more stable, the destruction caused by interaction with ascorbic acid is considerable (32). The presence of metal ions may exert a slight protective effect at low pH values, but this is insignificant compared with the amount of degradation caused by ascorbic acid (33).

Accelerated decoloration of anthocyanins in the presence of ascorbic acid, amino acids, phenols, sugar derivatives, and so on, may be caused by actual condensation reactions with these compounds. The polymers and degradation compounds produced in these reactions are probably quite complex (8). Some of the reaction products form compounds called phlobaphens, which are brownish red in color. For example, strawberry jam after 2 years of storage at room temperature has no detectable anthocyanins left but is still reddish brown in color. Compounds of this type are believed to contribute to the color of aged red wines. Enzyme systems that decolorize anthocyanins have been reported by a number of workers (8). They may be glycosidases, which hydrolyze the protective 3-glycosidic linkage to yield unstable aglycones, or phenolases, which require the interaction of *o*-dihydroxyphenols.

5. Anthocyanin Complexes

The observation that red anthocyanin pigments can be isolated from blue flowers (e.g., cornflowers) has stimulated considerable research on possible structures that can produce the blue color. The colors have been ascribed to "copigmentation" of anthocyanins with yellow flavonoids and other polyphenols and to complex formation with a number of components. A number of such complexes have been isolated and they are reported to contain cations, such as Al, K, Fe, Cu, Ca, and Sn; amino acids; proteins; pectin, carbohydrates; or polyphenols. All these complexes are reported in flowers, but Van Teeling et al. (35) reported a very high molecular weight (77,000,000) complex isolated from canned blueberries.

Some of the anthocyanins possess vicinal hydroxyl groups and this enables them to form complexes with metals. For example, when red, sour cherries, an anthocyanin containing fruit, are placed in plain tinned cans, an anthocyanin-tin complex forms, causing the red color to change to purple. Tinned cans with special organic linings have long been used to avoid this problem.

6. Anthocyanidin Derivatives

Restrictions on the use of artificial red food color have led to interest in the potential of anthocyanin-type compounds as food colorants. Jurd (28) described a whole series of compounds differing in the substituents at the 3 position. These have been patented in the public domain (United States Patents 3,266,903 and 3,314,975), but unfortunately they are subject to reactions with sulfur dioxide, ascorbic acid, pH, and other agents. Timberlake and Bridle (34) recently reported that anthocyanidins with a methyl or a phenyl group at position 4 were very stable in the presence of these compounds, even more so than some of the permitted artificial red colors. However, this class of 4-substituted flavylium compounds would be considered as new compounds and would have to undergo full toxicity clearance procedures.

D. Flavonoids

1. Structure

The yellow flavonoids comprise a diverse, almost ubiquitous class of pigments, with chemical structures similar to the anthocyanins. Approximately 800 flavonoids are known and the number is increasing rapidly (19,20).

One major group is the flavonols, such as kaempferol (X), quercetin (XI), and myricetin (XII). Another group, less common than the flavonols, is the flavones, containing such members as apigenin (XIII), luteolin (XIV), and tricetin (XV). Note the similarity in structure to pelargonidin, cyanidin, and delphinidin. Sixty other aglycones, based on hydroxy and methoxy derivatives of the flavonol and flavone structures, are known.

Five other minor groups are known: the chalcones (XVI), the aurones (XVII), the flavanones (XVIII) and isoflavanones (XIX), and the biflavonyls (XX).

These aglycones usually exist in the glycoside form with glucose, rhamnose, galactose, arabinose, xylose, apiose, or glucuronic acid. The sites of substitution are varied, but the 7, 5, 4', 7, 4', and 3' positions are common. In contrast to the anthocyanins, the 7 position is the most frequent site of substitution because it is usually the most acidic of the hydroxyl groups. The flavonoid compounds also may occur with acyl substituents in a manner similar to the anthocyanins. Compounds have also been isolated in which glucose is attached by a carbon-carbon bond in either the 6 or the 8 position. Such compounds are completely resistant to acid hydrolysis, and although not strictly glycosides, they have closely related properties. The two most common members are vitexin and isovitexin. A number of methods for the isolation and identification of flavonoids have been reported (4,6,8,11).

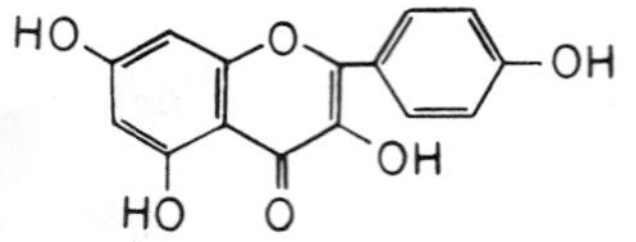

Kaempferol (X)

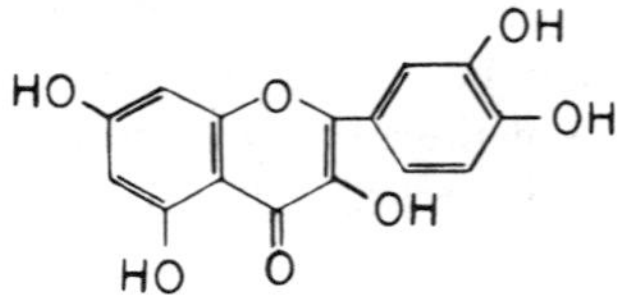

Quercetin (XI)

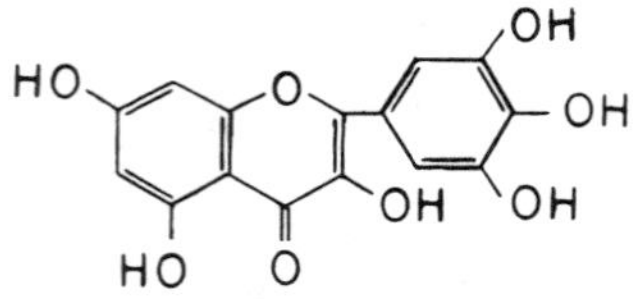

Myricetin (XII)

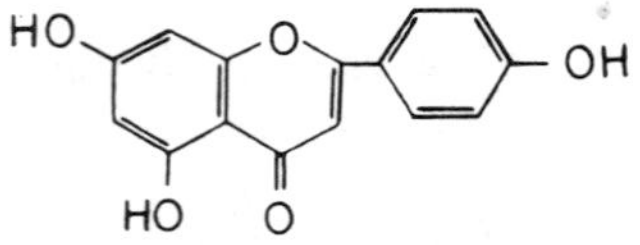

Apigenin (XIII)

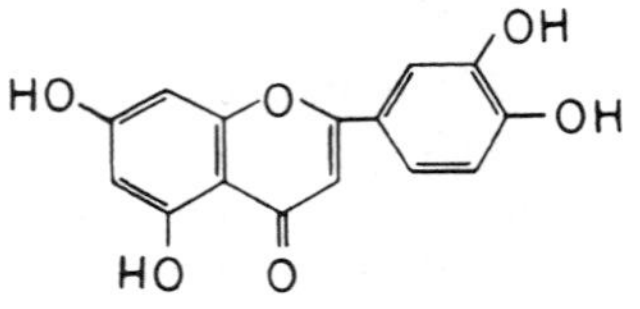

Luteolin (XIV)

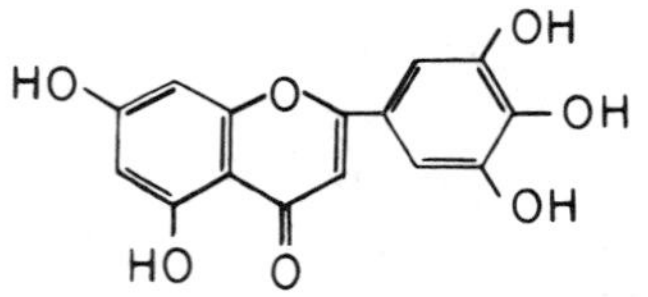

Tricetin (XV)

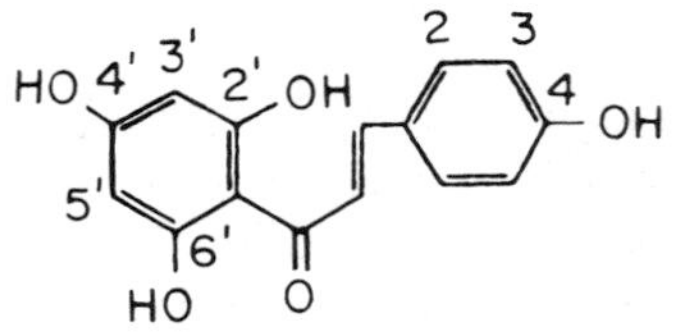

Chalcone (XVI)

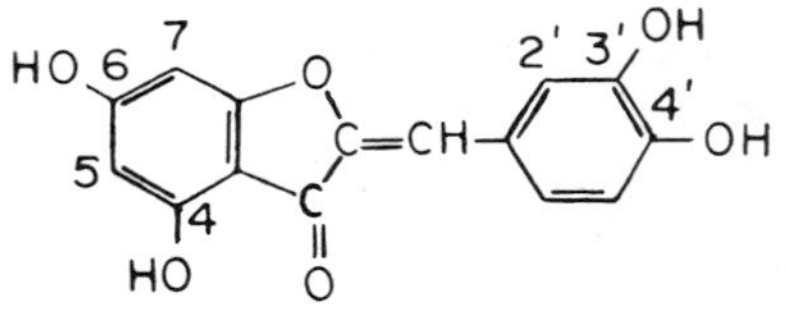

Aurone (XVII)
(Aureusidin)

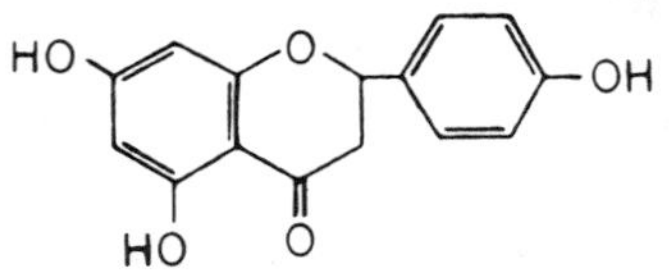

Flavanone (XVIII)
(Naringin)

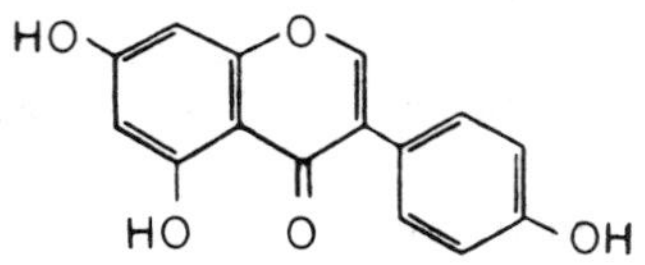

Isoflavanone (XIX)
(Genistein)

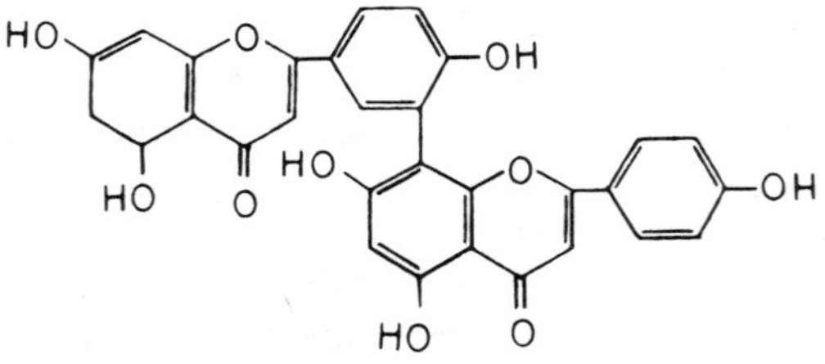

Biflavonyl (XX)
(Amentoflavone)

2. Importance in Foods

The flavonols quercetin and kaempferol glycosides are almost ubiquitous in the plant kingdom, and have been found in over 50% of all plants tested. Glycosides of myricetin have been found in over 10% of all plants tested. Many plants contain flavonoid pigments. There are a large number of research papers on them, primarily because these compounds are useful in plant taxonomy, and this has stimulated a great deal of excellent biochemical research.

With a number of notable exceptions, flavonoids in foods have not been extensively investigated. Possibly, the lack of good quantitative analytical methods has contributed to this, but methods are becoming available. Individual methods based on absorptivity in the 300-400 nm range or fluorescence with aluminum chloride are examples (6).

The three flavonol aglycones, quercetin, kaempferol, and myricetin, are found in considerable amounts in "instant" tea powders, in which they contribute to astringency. In green tea, these three compounds and their glycosides may comprise as much as 30% of the dry weight. One flavonoid, rutin (quercetin-3-rhamnoglucoside), was studied years ago when it became apparent that the compound complexed with iron to form an unsightly dark discoloration in canned asparagus. The tin complex, in contrast, produces a desirable yellow color.

The flavanones, a minor group of flavonoids found mainly in citrus plants, are of interest because of their potential as synthetic sweeteners. Naringin, a flavanone with neohesperidose in the 7 position, is an intensely bitter substance, whereas an isomer of the above compound with rutinose in the 7 position is tasteless. Neohesperidose has rhamnose and glucose in an α-1→2 linkage, whereas rutinose has an α-1→6 linkage. This difference is accentuated when the ring structure is opened to produce chalcone structures with the neohesperidose sugar. One derivative, neohesperidin dihydrochalcone, synthesized from naringen, has a sweetness nearly 2000 times that of sucrose.

The citrus flavonoids were in the news many years ago for their alleged "bioflavonoid" (vitamin P) synergistic activity with ascorbic acid in promoting decreased "capillary fragility." There may well be a small effect of bioflavonoids in maintenance of stronger capillary walls, but the physiological effects have been very largely discounted in normal individuals. The latest promotional effort for citrus flavonoids involves deodorization and disinfection of rooms. It remains to be seen whether this application is successful.

The isoflavones have also received attention lately for their weak estrogenic activity. Three isoflavones have been implicated in the lowered fertility of sheep pastured on clover containing these compounds. There are possible benefits also, because the flush of milk in cows on spring pasture may be due to estrogenic isoflavones. Other isoflavones are also noteworthy, such as the well-known fish toxicant rotenone and the glycosuria-producing phloridzin.

The polyphenolic character of the flavonoids and the ability to sequester metals has aroused interest in their possible value as antioxidants for fats and oils. However, the limited solubility in oil has restricted their use, and several derivatives with increased lipid solubility have been suggested.

The flavonoids are relatively stable toward heat processing in aqueous canned foods, but little research has been done on this aspect.

The bioflavonyls are an interesting group, but their significance in foods is largely unknown. They are dimers, usually of apigenin (XIII), with three parent groups made up of 8→8, 8→3′, and 8→4′ linkages. Two compounds of this class were isolated from the ginkgo tree, the "living fossil" that has been in existence for over 250 million years.

E. Proanthocyanidins

Although proanthocyanidins are colorless, they are considered here because they have structural similarity to anthocyanidins and because under conditions of food handling and processing they can be converted to colored products. An obsolete term for this group of compounds is "leucoanthocyanidins" or "leucoanthocyanins." Other general names, such as anthoxanthins, anthocyanogens, flavolans, flavylans, and flavylogens, have been suggested but proanthocyanidin seems to have the most support. The term "leucoanthocyanidin" should be restricted to the monomeric flavan-3,4-diols.

The basic building block of proanthocyanidins is usually flavan-3,4-diol (XXI) forming a dimer through a 4→8 or a 4→6 linkage, but trimers and higher polymers are common. All produce an anthocyanidin, such as pelargonidin, cyanidin, petunidin, or delphinidin, when heated in the presence of a mineral acid. The main proanthocyanidin in apples is a dimer of two 1-epicatechin units linked through a C4—C8 bond (XXII). This compound gives cyanidin and (-)epicatechin on hydrolysis. It has been found in apples, pears, kola nuts, cocoa beans, hawthorn berries, and other fruits.

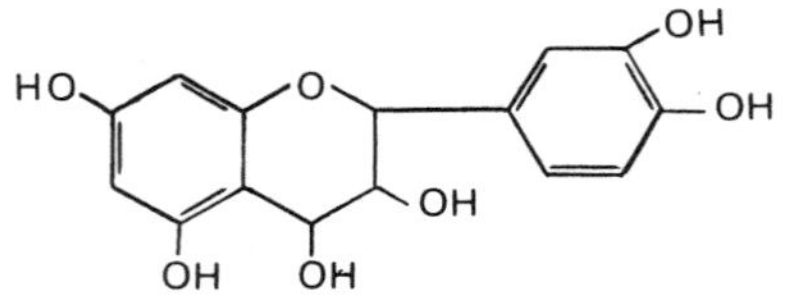

Flavan — 3,4 — diol (XXI)
(catechin and epicatechin
are the trans and cis forms
of carbons 2 and 3)

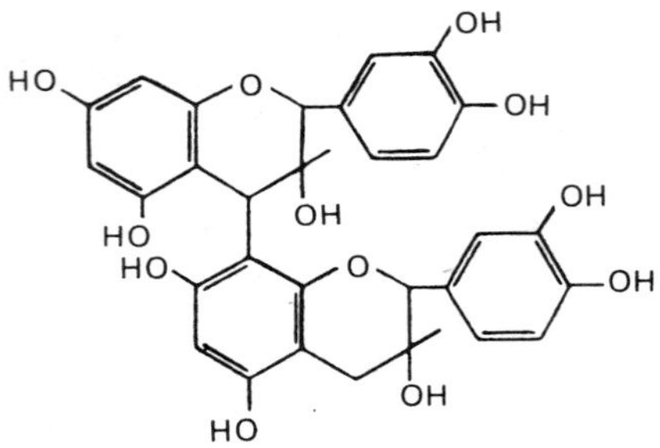

Leucoanthocyanin (XXII)

The proanthocyanidin compounds are interesting in foods because they contribute to the astringency or "puckeriness" of foods. In apple cider they contribute significantly to the characteristic taste, and they are important in the flavor of many other fruits and beverages, such as persimmons, cranberries, olives, bananas, chocolate, tea, and wine. They also contribute to enzymic browning reactions in fruits and vegetables, through their orthohydroxy groups, and to haze formation in beer and wine.

To be astringent, they must have the ability to react with protein, and this restricts the size of the proanthocyanidin polymer. Monomers are apparently too soluble, so the effective molecular size for interaction with protein is usually within the range of two to eight repeating units. However, the actual structure may be more complex than merely a repeating unit of flavan-3,4-diol. The inability of the large polymers to react with protein in a manner analogous to the reaction of tannic acid in the tanning of skin to form leather has led to the term "condensed tannins." This is an unfortunate term but one that is apparently well entrenched.

F. Tannins

The term "tannin" (also tannic acid, gallotannic acid, and incorrectly, digallic acid) is defined in the *Merck Index* as a complex mixture found in the bark of oak, sumac, and myrobalen. Its appearance ranges from colorless to yellow or brown. Turkish and Chinese nutgalls contain 50 and 70%, respectively, of tannin. Tannic acid, as purchased from a chemical supply house has a formula $C_{75}H_{52}O_{46}$, a molecular weight of 1701, and probably consists of nine molecules of gallic acid and one of glucose.

The term "tannin" as used in foods includes two types of compounds. The first type is the "condensed tannins," which may be 4,8 or 2,8 C–C dimers as described previously, or 3,3-ether-linked dimers of catechin and related compounds. Other linkages, such as 2–C, 7–O or 4–C, 7–O compounds have been reported. The second type is the "hydrolyzable tannins" including the gallotannins and the ellagitannins. These compounds are best noted for their ability to tan hides for leather manufacture. The latter are polymers of gallic acid (XXIII) and ellagic acid (XXIV). A typical ellagitannin containing gallic acid, ellagic acid, and one molecule of glucose is chebulagic acid, structure (XXV). Compounds have been reported that are combinations of the two types of tannins, for example, theaflavin gallate, a possible intermediate found in tea fermentation.

Tannins contribute to the astringency of foods and also to enzymic browning reactions, but their mechanisms of action are not well understood.

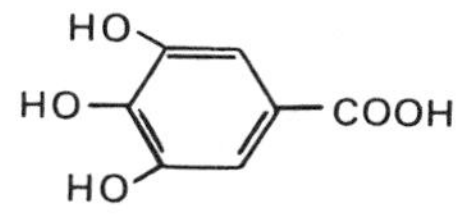

Gallic acid (XXIII)

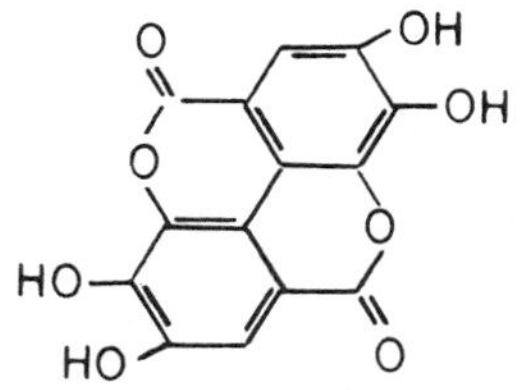

Ellagic acid (XXIV)

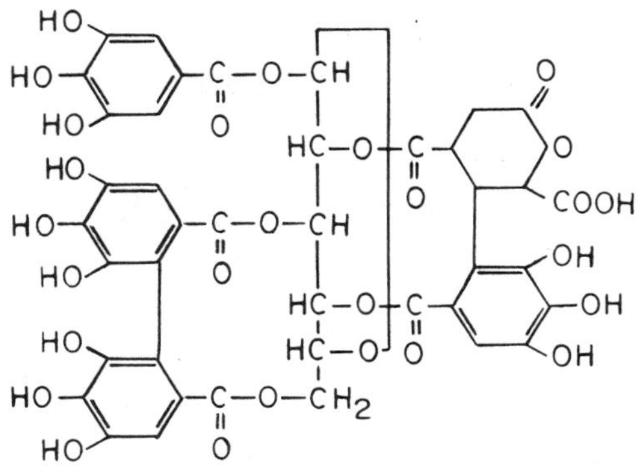

Chebulagic acid (XXV)

G. Betalains

The betalains are a group of compounds resembling the anthocyanins and flavonoids in visual appearance. They were called, incorrectly, in the older literature "nitrogen-containing" anthocyanins. They are found only in 10 families of Centrospermae, of which the best known is the red beet. They are also found as water-soluble compounds in the cell sap of chard, cactus fruits, pokeberries, and a number of flowers, such as bougainvillia and amaranthus.

About 70 betalains are known and they all have the same basic structure (XXVI), in which R and R' may be hydrogen or an aromatic substituent. Their color is attributable to the resonating structures (XXVII) and (XXVIII), as shown in Fig. 10. If R or R' does not extend the resonance, the compound is yellow and is called betaxanthin. If R or R' does extend the resonance, the compound is red and is called a betacyanin. For example, the aglycone of the betacyanin in red beets is called betanidin (XXIX). With glucose, it is called betanin (XXX). The only compounds forming glycosides are glucose and glucuronic acid. The betalains may also occur naturally in the acyl form in a manner similar to the anthocyanins and flavonoids, but the pattern is more complicated. Malonic, ferulic, p-coumaric, sinapic, caffeic, and 3-hydroxy-3-methylglutaric acids have been found attached to the sugar portion of the betalains.

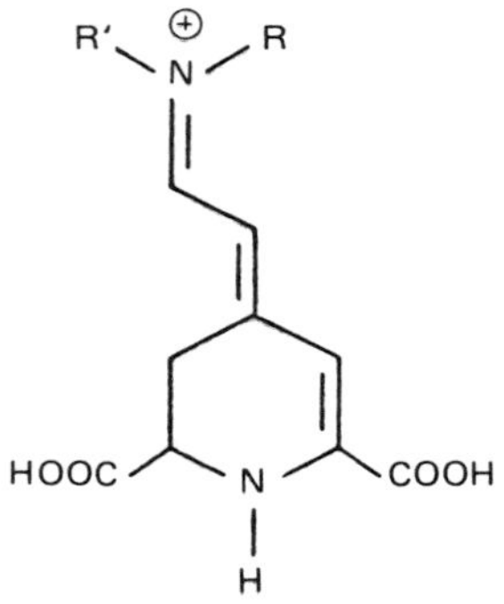

1, 7 – Diazoheptamethin (XXVI)

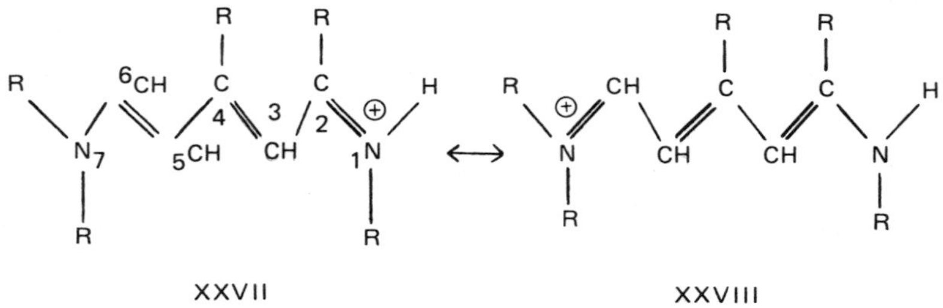

XXVII   XXVIII

Figure 10 Betacyanin resonating structures.

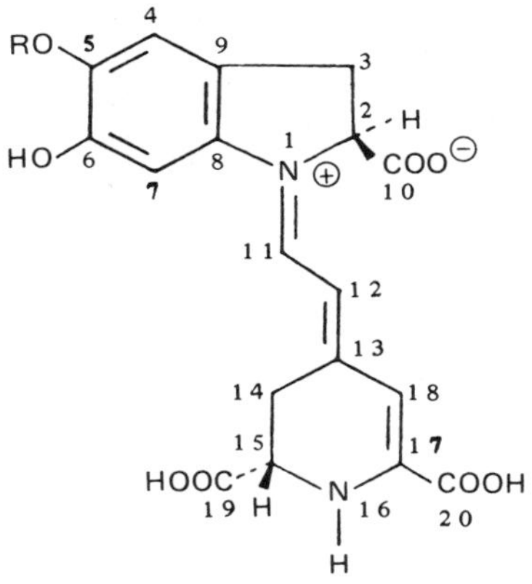

Betanidin, R = H
(XXIX)

Betanin, R = Glucose
(XXX)

The anthocyanins and the betalains have different chemical structures and are easily differentiated. Their spectra are different in that the betalains have no absorption peaks in the 270 nm region. Anthocyanins are extracted easily with methanol and very poorly with water. The reverse is true for betalains. In electrophoresis systems with weakly acidic buffers, the anthocyanins go to the cathode and the betalains to the anode. Electrophoresis is actually the method of choice for separating closely related betalains.

The betalains are subject to degradation by thermal processing, as in canning, but usually sufficient pigment is available so that the food (e.g., beets) has an attractive dark red color. The isolated pigment of beets is stable in the pH range 4-6, and this makes it of interest as a potential colorant in foods.

Members of the Phytolacca family (American pokeberry) had a brief claim to fame as adulterants for wine many years ago, when a pokeberry plant in the corner of the vineyard was insurance against a crop of grapes low in pigment content. The addition of pokeberry juice to wine was outlawed in France in 1892. Apart from being a potent colorant, pokeberry juice also contains a purgatory and emetic substance called phytolaccatoxin. The colorant from pokeberries was formerly called phytolaccanin, but its structure is identical to betanin (24).

H. Quinones and Xanthones

The quinones (5) are large group of yellowish pigments found in the cell sap of flowering plants, fungi, lichens, bacteria, and algae. Over 200 are known, ranging in color from pale yellow to almost black. The largest subgroup is the anthraquinones, of which some members have been used as natural dyestuffs and purgatives for centuries. A typical anthraquinone is emodin (XXXI), which is widely distributed in fungi, lichens, and higher plants. A small subgroup, composed of about 20 pigments, is the napthoquinones (XXXII). Several members of this group are used a dyes, such as henna. Juglone, occurring in walnuts, and plumbagin (XXXII) are typical napthoquinones. The anthraquinones usually occur as glycosides, but the napthoquinones do not. Another larger subgroup is the benzoquinones, which occur mainly in fungi and some flowering plants. An example is the violet-black compound spinulosin (XXXIII). Another subgroup of unique red pigments, the napthacenequinones, confined to the Actinomycetales, is closely related to the tetracycline antibiotics. There are many other subgroups, such as the phenanthraquinones, isoprenoid quinones, and a number with more complex structures.

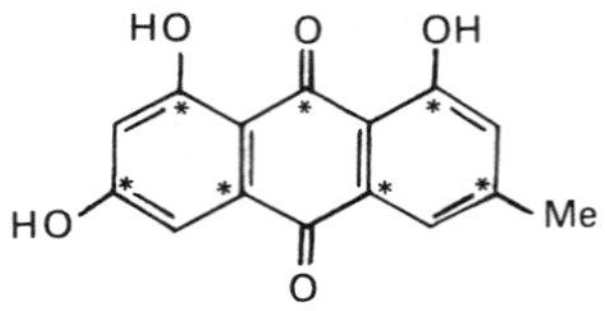

Emodin (XXXI)

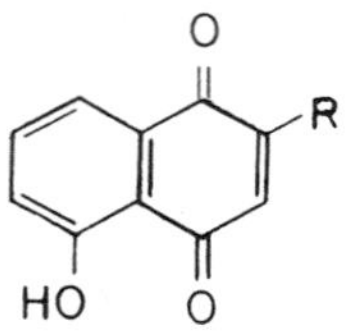

Napthoquinones (XXXII)

Juglone (R = H)
Plumbagin (R = Me)

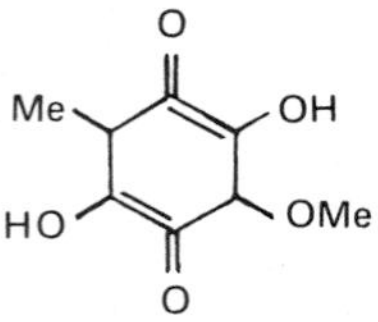

Spinulosin (XXXIII)

The xanthones are a group of 20 yellow pigments that have been confused with the quinones and the flavones. One well-known member is mangiferin (XXXIV), which occurs as a glucoside in mangoes. Xanthones can be easily distinguished from flavones and quinones by spectral data.

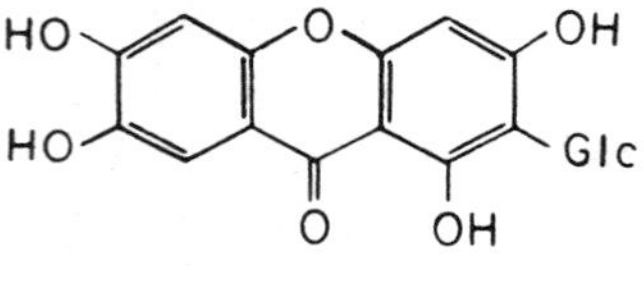

Mangiferin (XXXIV)

I. Carotenoids

The carotenoids are a group of mainly lipid-soluble compounds responsible for many of the yellow and red colors of plant and animal products. They are very widespread and occur naturally in large quantities. Isler (9) has estimated that over 100,000,000 tons of carotenoids are produced annually by nature. Most of this amount is in the form of fuco-xanthin (XXXV) in various algae, and in the three main carotenoids of green leaves: lutein (XXXVI), violaxanthin (XXXVII), and neoxanthin (XXXVIII). Others produced in much smaller amounts but occurring very widely, are β-carotene (Fig. 11) and zeaxanthin (XXXIX). Other pigments, such as lycopene (Fig. 11) in tomatoes, capsanthin (XL) in red peppers, and bixin (XLI) in annatto, predominate in certain plants.

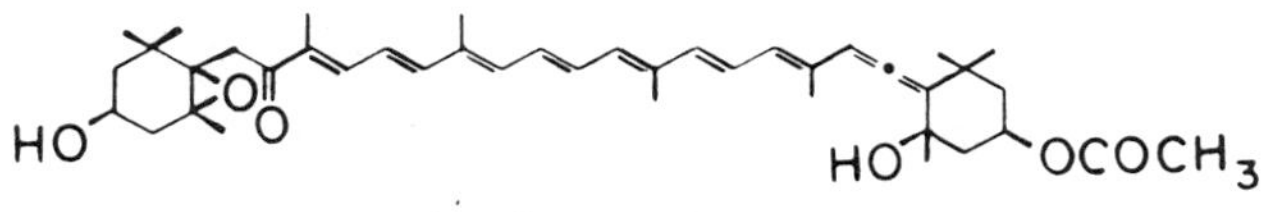

Fucoxanthin (XXXV)

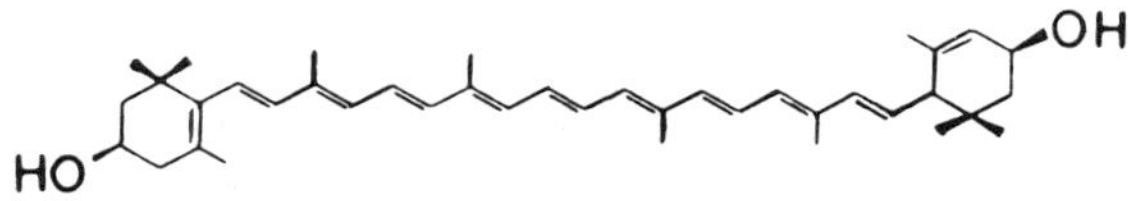

Lutein (XXXVI)

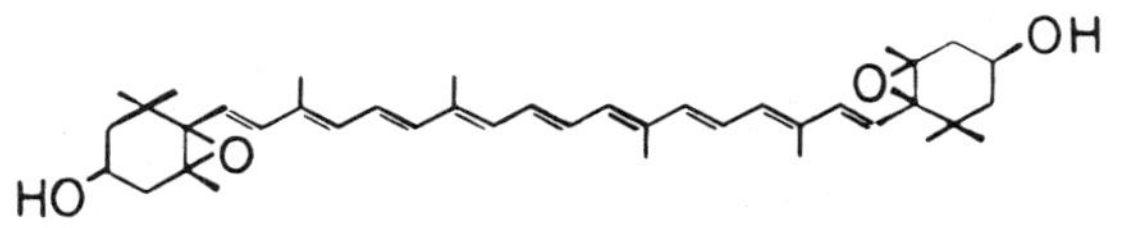

Violaxanthin (XXXVII)

Lycopene

β−Carotene

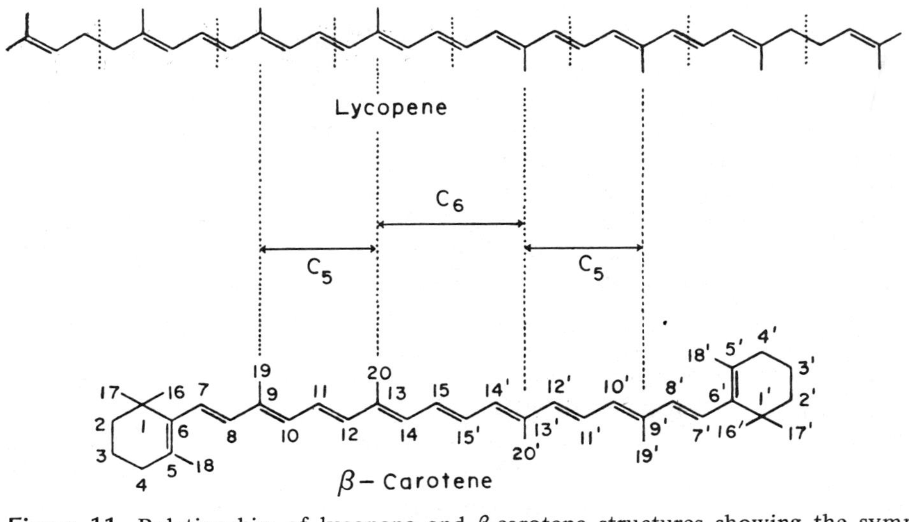

Figure 11 Relationship of lycopene and β-carotene structures showing the symmetry around the 15−15′ carbons and the C_5 (isoprene) repeating units. [From Isler (9), courtesy of Birkhauser Verlag, Basel.]

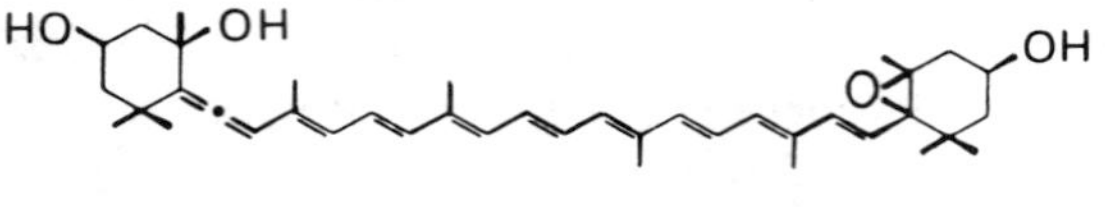

Neoxanthin (XXXVIII)

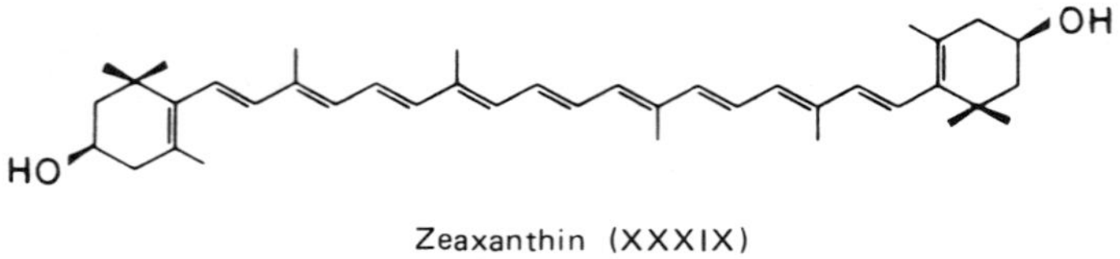

Zeaxanthin (XXXIX)

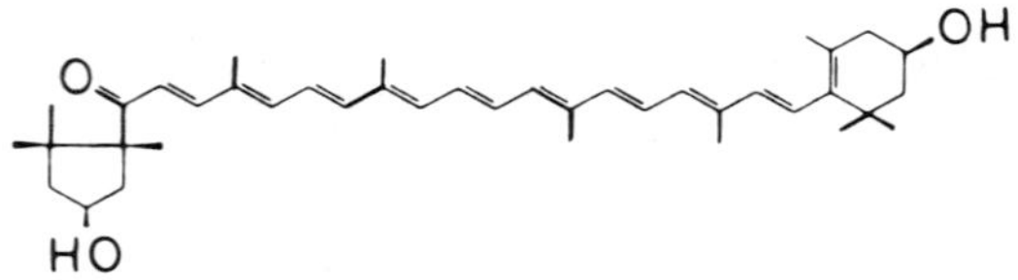

Capsanthin (XL)

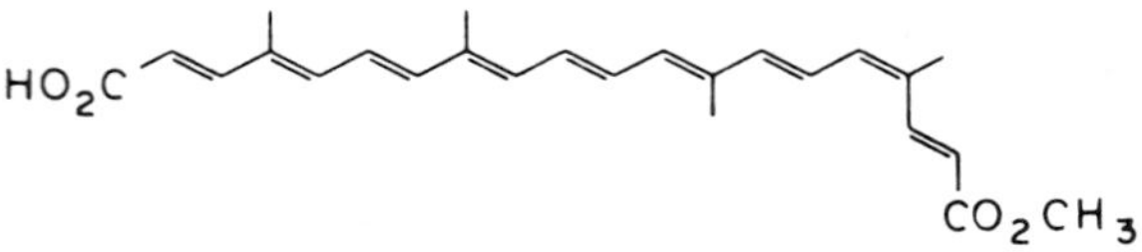

Bixin (XLI)

1. Structure

Carotenoids include a class of hydrocarbons, called carotenes, and their oxygenated derivatives, called xanthophylls. They consist of eight isoprenoid units joined in such a manner that the arrangement of isoprenoid units is reversed in the center of the molecule. The basic carbon structure for lycopene (Fig. 11) shows the symmetrical arrangement around the central pair of carbons. This structure can be cyclyzed to form β-carotene (Fig. 11) and the 15–15′ carbon pair forms the center of the molecule. Other carotenoids have different end groups with the same central structure. About 60 different end groups are known, comprising about 450 known carotenoids, and new ones are continually being reported. Most of the carotenoids reported earlier have a C40 central skeleton, but recently some have been described that contain more than 40 carbons. They are also named as substituted C40 carotenoids. The nomenclature for the carotenoids is too complicated for description here, and the latest modification may be found in the report of the IUPAC Commission.

Carotenoids can exist in the free state in plant tissue (as crystals or amorphous solids), or in solution in lipid media. They also occur as esters or in combination with sugars and proteins. The occurrence of carotenoids as esters of fatty acids has been known for a long time. For example, in autumn leaves, palmitic and linolenic acids are associated with positions 3 and 3′ of lutein. Capsanthin (XL) occurs as the lauric acid ester in paprika. Carotenoid esters have been found in flowers, fruits, and bacteria.

The association of carotenoids with proteins is of more recent investigation, and knowledge has developed primarily from studies of pigments in various invertebrates (21). The association of carotenoids with proteins stabilizes the pigment and also changes the color. For example, the red carotenoid astaxanthin (XLII), when complexed with a protein, forms the blue colorant in lobster shells. Another example is ovoverdin, the green pigment in lobster eggs. Carotenoid-protein complexes also exist in some green leaves, bacteria, fruits, and vegetables.

Carotenoids may also occur in combination with reducing sugars via a glycosidic bond. For many years, crocin, containing two molecules of the sugar gentiobiose united with crocetin (XLIII), was the only known pigment of this type. Crocin is the main pigment in saffron. Lately, a number of carotenoid glycosides have been isolated from bacteria. Detailed procedures for structure determination and analysis of carotenoids are available (1,3,5,6,9,39).

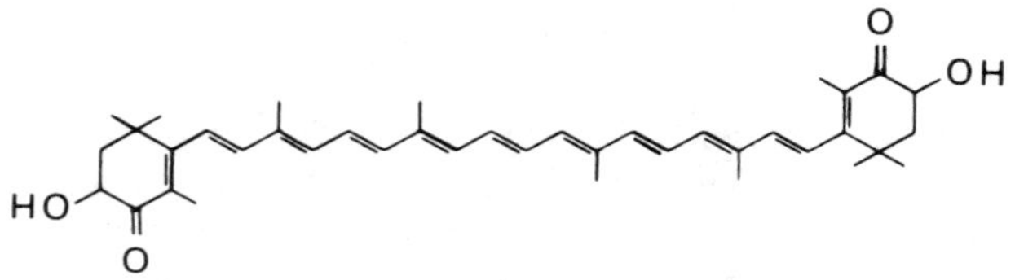

Astaxanthin (XLII)

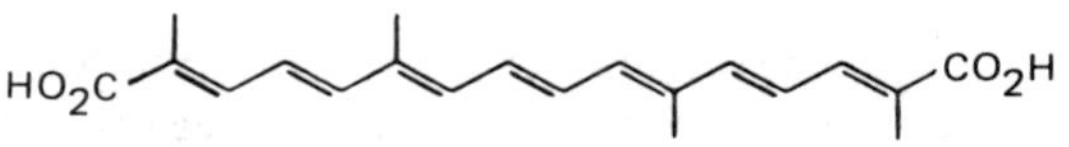

Crocetin (XLIII)

2. Chemical Reactions

Provitamin A. Carotenoids have received a great deal of attention from early researchers because β-carotene is a precursor of vitamin A, a well-known nutrient in the human diet. β-Carotene yields two molecules of vitamin A by cleavage at the center of the molecule. Compounds, such as α-carotene (XLIV), with half their structure identical to β-carotene, function as precursors for one molecule of vitamin A. Lycopene (Fig. 11) has no vitamin A activity. The association of vitamin A with retinene in the visual purple cycle in human vision has been responsible for literally hundreds of papers on carotenoid chemistry and physiology.

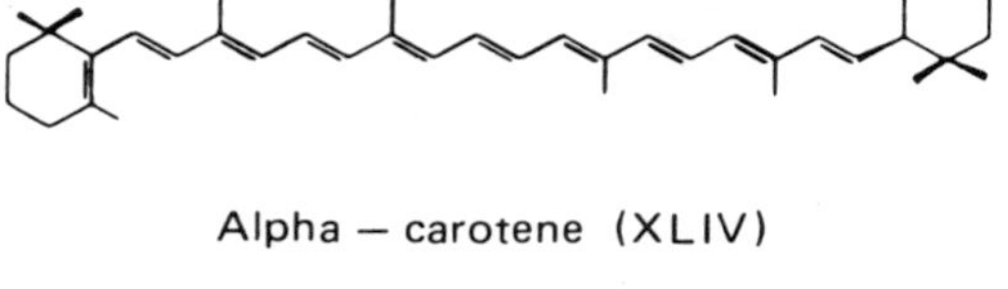

Alpha — carotene (XLIV)

Oxidation Reactions. The main cause of carotenoid degradation in foods is oxidation. The severity of oxidation depends on whether the pigment is in vivo or in vitro and on environmental conditions. In intact living tissue, the stability of the pigments is probably a function of cell permeability and the presence of protective components. For example, lycopene in tomatoes is quite stable but the extracted purified pigment is unstable. Present in many tissues are enzyme systems that degrade carotenoids rapidly. For example, in macerated green leaves, half the carotenoids are lost in 20 min at room temperature. Lipoxygenase interacts to degrade carotenoids in a number of food.

In processed food, the mechanism of oxidation is complex and depends on many factors. The pigments may autoxidize by reaction with atmospheric oxygen at rates dependent on light, heat, and presence of pro- and antioxidants. The reactions are believed to be caused by free-radical formation, in three distinct steps. Carotenoids undergo coupled oxidation in the presence of lipids at rates dependent on the system. Carotenoids are usually more stable in systems with high degrees of unsaturation, possible because the lipid system itself accepts the free radicals more easily than carotene. Conversely, carotenes are less stable in lipid systems with high degrees of unsaturation, but there are several exceptions to this in the literature. Carotenoids can act as antioxidants or pro-oxidants, depending on the system.

With β-carotene in the presence of lipids, the primary site of attack by free radicals is probably the carbon α to the double bond in the ionone ring. The ring is opened and the oxidation proceeds stepwise by the β-oxidation method. In view of the variations in carotenoid structure and the complicated interactions with the surrounding media, it is likely that a number of oxidation mechanisms occur in foods.

J. Miscellaneous Natural Pigments

There are a number of groups of compounds the structures of which differ from the usual carotenoid-flavonoid-quinonoid-porphyrin concept of color. The appearance of a colored molecule may be incidental, resulting from a relatively minor structural variation that shifts the absorption of light into the visible area. An example may be taken from the aromatic ketones, in which the simple compounds are colorless, and complex derivatives, such as gossypol (XLV), are yellow. Gossypol, a toxic substance, has received much attention because it is present in cottonseed meal and the meal is a potential source of protein for humans. Current research has centered around plant breeding efforts to reduce the content of gossypol, and processing efforts—the "liquid cyclone" process—to remove the gossypol. Gossypol also has been in the news for another reason. Clinical trials in China are underway to test its efficiency as a male contraceptive.

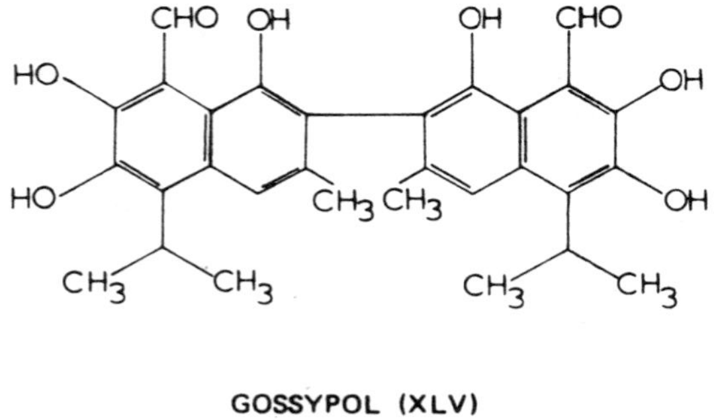

GOSSYPOL (XLV)

Many aromatic dienone compounds are found in molds. One cyclic dienone group resembles the flavonoids, and another (dracorubin) is the pigment of "dragon's blood," a resin exuded from some palm trees. There are many other pigments, of unrelated structure, such as the perinaphthenones, γ-pyrones, sclerotiorins, vulpinic acid pigments, pyrroles, phenazines, phenoxazone, antibiotics, melanins, and riboflavin.

We should not leave the subject of natural pigments without mention of the pale yellow pigment phytochrome, which is so important to overall plant growth. It controls elongation, flowering, germination, induction of dormancy, and the production of anthocyanins and carotenoids. Its structure is unknown at this time.

Table 3 provides a summary of the various natural pigments and their characteristics.

Table 3 Summary of Characteristics of Natural Pigments

Pigment group	Number of compounds	Color	Source	Solubility	Stability
Anthocyanins	150	Orange, red blue	Plants	Water soluble	pH and metal sensitive, heat labile
Flavonoids	800	Colorless, yellow	Most plants	Water soluble	Fairly heat stable
Proanthocyanidins	20	Colorless	Plants	Water soluble	Heat stable
Tannins	20	Colorless, yellow	Plants	Water soluble	Heat stable
Betalains	70	Yellow, red	Plants	Water soluble	Heat sensitive
Quinones	200	Yellow to black	Plants, bacteria, algae	Water soluble	Heat stable
Xanthones	20	Yellow	Plants	Water soluble	Heat stable
Carotenoids	450	Colorless, yellow, red	Plants, animals	Lipid soluble	Heat stable, sensitive to oxidation
Chlorophylls	25	Green, brown	Plants	Organic solvents	Heat sensitive
Heme pigments	6	Red, brown	Animals	Water soluble	Heat sensitive
Riboflavin	1	Greenish yellow	Plants	Water soluble	Stable to heat and pH

II. COLORANTS ADDED TO FOODS

A. Regulatory Aspects

Current legislation pertaining to the regulation and use of colorants added to foods in the United States is the Food, Drug and Cosmetic Act of 1938 as amended by the Color Additive Amendment of 1960. Historically, regulatory interest has centered on the synthetic colorants since they have been far more controversial than the naturally occurring colorants (12a).

The synthesis of mauve in 1856 by Sir William Perkin introduced the era of synthetic colorants. They were much superior to the natural colorants in use at that time, and they rapidly displaced the older dyes. By the year 1900, there were 695 different coal-tar dyes on the market and many of them were used in food. Bernard Hesse of the USDA studied the problem of synthetic colorants in foods and recommended seven, which were legalized under the Pure Food and Drugs Act of 1906. Ten more were added in subsequent years. After many changes and delistings, nine synthetic food colorants are permitted in the United States today (12a).

The number and type of permitted colorants varies widely from country to country, but there is a trend toward using fewer synthetic colorants. It is hoped that the trend to international standardization of food colorants will gain momentum. Actually, there is little scientific justification for a wide variety of synthetic colorants since blends of the approved synthetic colorants allow a manufacturer to provide almost any desired color.

From a regulatory standpoint, colorants permitted for addition to food are classed as either certified or noncertified. With the passage of the 1938 Federal Food, Drug and Cosmetic Act, manufacturers of colorants requiring certification (all are synthetic) had to send a sample of each batch of colorant to the FDA for certification to assure that the preparation was harmless and suitable for food. The 1938 Act referred to "coal-tar colors," which connotes to the consumer an unpleasant black tarry material. These colorants indeed were made from coal derivatives 70 years ago, but the petrochemical industry now provides the materials for synthesis of colorants. The term "coal-tar" has been obsolete for over half a century, and it should be abandoned.

The 1938 Act had other problems, because it allowed colorants to be used in any quantity. FDA also interpreted "harmless" to mean incapable of producing harm in any species of animal in any quantity or under any conditions. Toxicological data under this interpretation led to the delisting of eight FD&C colorants between 1956 and 1960. Obviously, under this arrangement, all added colorants would eventually be delisted. This led to passage of the Color Additives Amendment of 1960 under which FDA was allowed to set a tolerance for the amount of colorant permitted in a food.

The 1938 Act also provided the guide rules for toxicological testing, certification, purity, safety interpretation, and so on. In addition, the law provided for continued use of some colorants through "provisional" listing while additional toxicological data was acquired. The assumption was that compounds listed as provisional would either become "permanently listed" or abandoned. Unfortunately, the provisional listing for some compounds stretched out for many years, but this situation is being rectified.

Noncertified colorants permitted for food use include naturally occurring colorants as well as some synthetic colorants (e.g., β-carotene, either naturally occurring or synthesized, is an approved noncertified food colorant).

Table 4 FD&C Certified Colorants Permitted in the United States

FDA name	Common name	Year listed for food use	Chemical class
FD&C Blue No. 1	Brilliant blue	1929	Triphenylmethane
FD&C Blue No. 2	Indigotine	1907	Indigoid
FD&C Green No. 3	Fast green	1927	Triphenylmethane
FD&C Yellow No. 5	Tartrazine	1916	Pyrazolone
FD&C Yellow No. 6	Sunset yellow	1929	Monoazo
FD&C Red No. 3	Erythrosine	1907	Xanthine
FD&C Red No. 40	Allura red	1971	Monoazo
Orange B[a]	Orange B	1966	Monozao
Citrus Red No. 2[b]	Citrus Red No. 2	1959	Monozao

[a]Allowed only on the surfaces of sausages and frankfurters at concentrations up to 150 ppm by weight.
[b]Allowed only on the skins of oranges, not intended for processing, at concentrations up to 2 ppm by weight.

B. Certified Colorants

The nine currently approved certified colorants are listed in Table 4 along with their common names, the year their use was first approved, and the chemical class to which each belongs. The chemical structures of these colorants appear as structures (XLVI) through (LIV), and their stability characteristics are presented in Table 5. These certified colorants are used in a wide variety of foods in the U.S. market, and a discussion of their reactions under such a broad array of conditions is beyond the scope of this chapter.

Table 5 Stability of FD&C Colors[a]

FD&C color	Light	Acids	Alkalis	Sulfur dixoide	Ascorbic acid	Heat
Blue No. 1	3	5	4	4	4	4
Blue No. 2	1	2	2	2	3	4
Green No. 3	3	5	2	4	4	4
Yellow No. 5	5	5	5	4	5	4
Yellow No. 6	4	5	5	4	5	3
Red No. 33	3	2	5	5	2	5
Red No. 40	5	5	4	5	5	5

[a]1 is very poor stability and 5 is good stability.

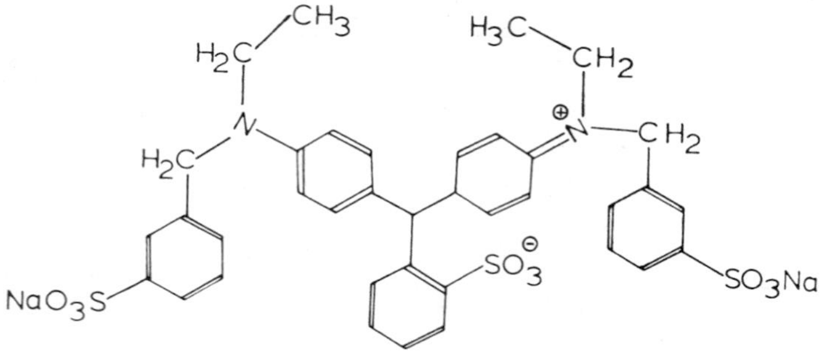

FD&C BLUE NO. 1 (XLVI)

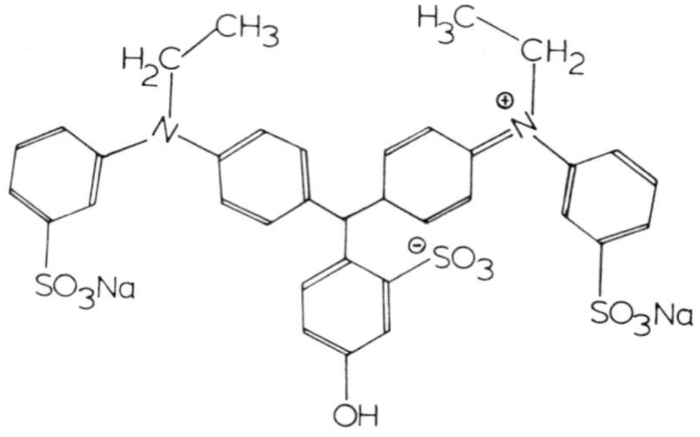

FD&C GREEN NO.3 (XLVII)

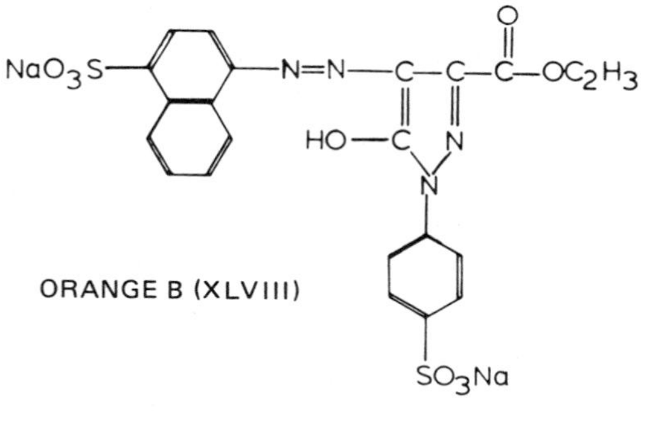

ORANGE B (XLVIII)

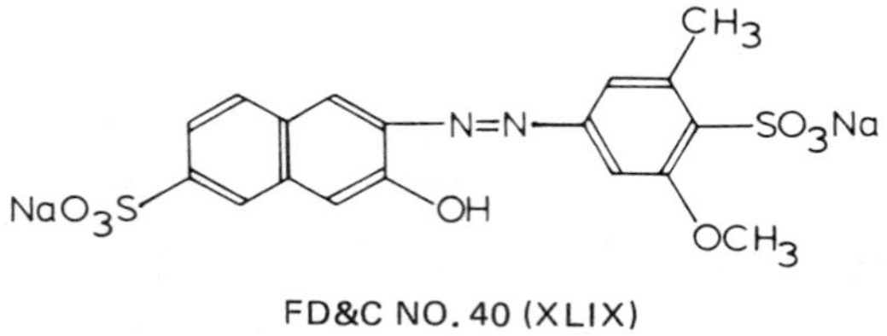

FD&C NO. 40 (XLIX)

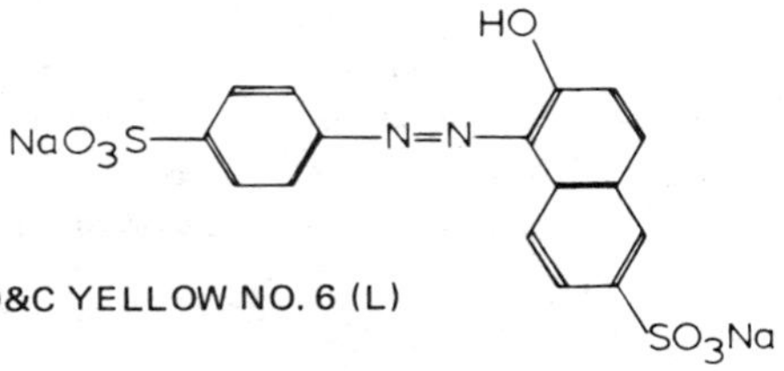

FD&C YELLOW NO. 6 (L)

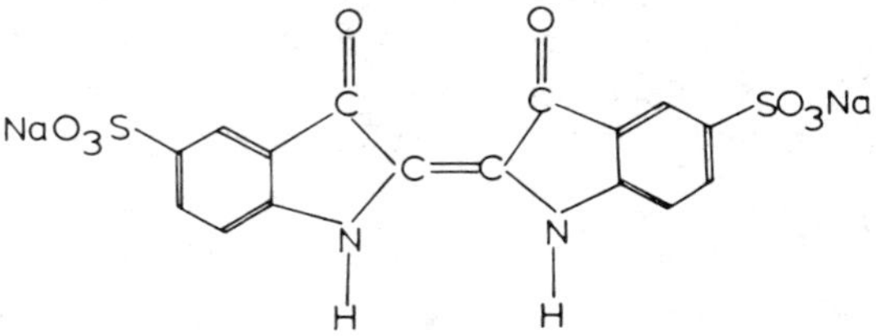

FD&C BLUE NO. 2 (LI)

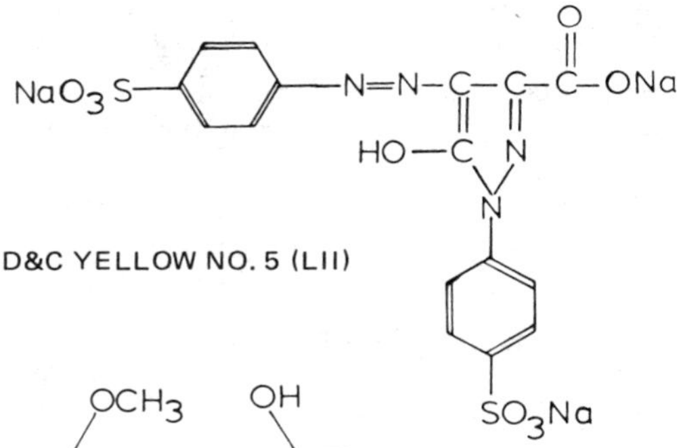

FD&C YELLOW NO. 5 (LII)

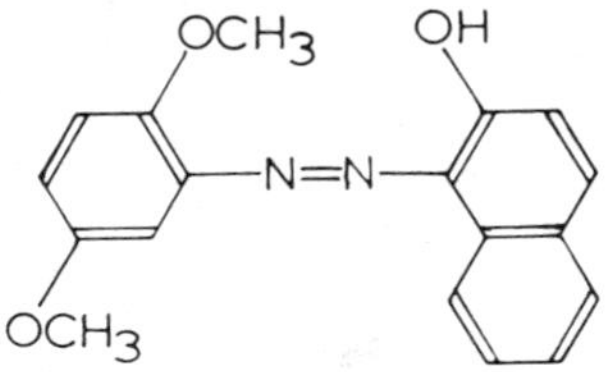

CITRUS RED NO. 2 (LIII)

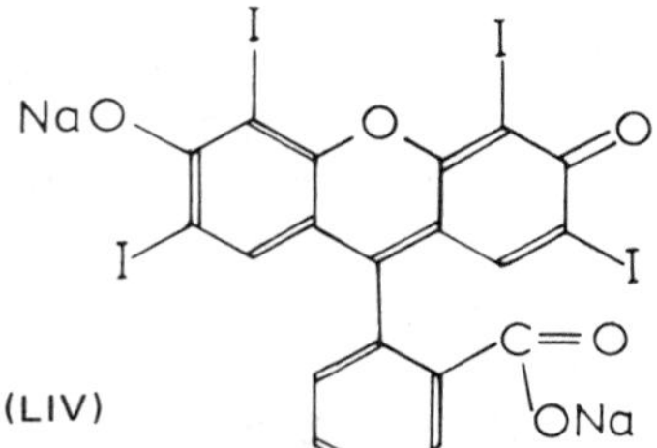

FD&C RED NO. 3 (LIV)

The water soluble FD&C colorants, just discussed, can be transformed to an insoluble powder (lake) by precipitation with aluminum, calcium, or magnesium salts on a substrate of aluminum hydroxide. The precipitates are dried, finely ground, and sold as a separate class of food colorants. The colorant content ranges from 10 to 40%, but the tinctorial power is not proportional to the colorant content. Lakes can be prepared from all the classes of pigments in Table 4.

The insolubility of the lakes in water, oils, and fats, for example, confers several advantages over both water-soluble and oil-soluble colorants. They are more stable to heat and light, and they do not "bleed" or migrate. They are used extensively in confectionery products, bakery products, salad dressings, and chocolate substitutes, in which the presence of water is undesirable. They are also used extensively in the packaging industry, in which films and inks are in contact with food, and in pharmaceutical tablets.

C. Colorants Exempt from Certification

Permissible colorants that can be added to food and are exempt from FDA certification are listed in Table 6, along with restrictions, if any, that exist on their use. Although these colorants are exempt from formal FDA certification requirements, they are, nonetheless, monitored by the FDA to assure that their purity is in accord with specifications and that they are used in accord with regulations. Obviously, use of many of these additives is self-limiting due to flavor, color, expense, and so on. Many preparations have dual uses; for example, fruit and vegetable juices may be classified as food ingredients rather than color additives. There may be a fine line between the two, which may have to be decided on a case-to-case basis. Similarly, some spices (e.g., saffron, turmeric, and paprika) have a role in both flavor and color.

Table 6 Food Color Additives That Do Not Require Certification

Color additive	Restriction
Annatto extract	None
β-Carotene	None
Beet powder	None
β-Apo-8-carotenal	15 mg/lb or pint
Canthaxanthin	30 mg/lb or pint
Caramel	None
Carmine	None
Carrot oil	None
Cochineal extract	None
Cottonseed flour, toasted, partially defatted, cooked	None
Ferrous gluconate	Coloring black olives
Fruit, vegetable juices	None
Grape skin extract (enocianina) bases	Nonalcoholic beverages
Paprika, paprika oleoresin	None
Riboflavin	None
Saffron	None
Titanium dioxide	1% maximum
Turmeric, turmeric oleoresin	None
Ultramarine blue	Coloring salt

Perhaps the most important class of certification-exempt colorants are the carotenoids. Carotenoids are contained in natural extracts from annatto, saffron, paprika, tomatoes, and other sources. Three synthesized carotenoids are currently approved by the FDA for addition to human food. β-Carotene (Fig. 11), which provides a yellow to orange color, is by far the most widely used. β-Apo-8-carotenal (LV) provides an orange to red color, and canthaxanthin (LVI) provides a red color. Canthaxanthin is the most recently approved carotenoid and is rapidly gaining importance, particularly since the use of FD&C Red No. 2 has been banned. The synthesized carotenoids have a real advantage in that their purity and colorant ability can be rigidly controlled.

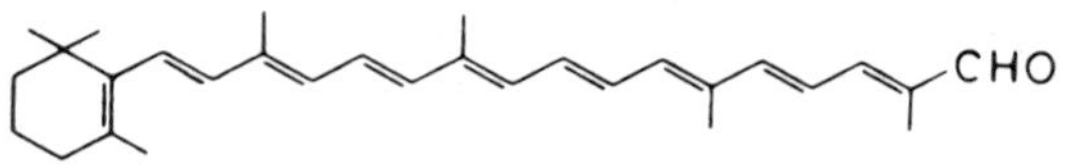

β – apo – 8 – carotenal (LV)

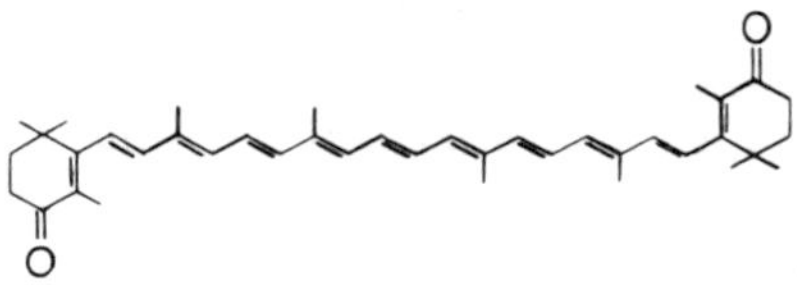

Canthaxanthin   (LVI)

Another carotenoid, bixin (XLI), derived from the seeds of the annatto plant, is widely used in the dairy industry. The exotic and expensive spice, saffron, contains the yellow carotenoid crocin.

Applications of carotenoids as food colorants can be divided into fat-based and water-based categories. The former group includes margarine, butter, and oils and usually requires a 20-30% suspension of a carotenoid in an oil carrier. The stability of carotenoids added in this manner is usually very good. Water-based foods, such as beverages, soups, dairy products, meat products, syrups, or macaroni, require a different carotenoid formulation. It is possible to prepare water-dispersible carotenoids by formation of a colloidal suspension using water-miscible solvents or by emulsification of oily solutions. Synthetic carotenoids are available as oily suspensions, as tiny beadlets coated with gelatin or carbohydrate, or even as microcrystals, depending on the application.

Anthocyanin derivatives have been used for coloring beverages for a long time. The general term "enocianina" refers to a product made from grape skins after production of wine. The blue product is a variable mixture of polymers and usually contains few or no anthocyanins. Recently, red and blue colorants from grape by-products have come on the market. The reds contain low-molecular-weight compounds, usually monomeric anthocyanins, and the blue products contain polymers of greater molecular weight.

Betacyanin preparations from red beets have received considerable attention as food colorants. The colorants are either ground dehydrated beets, single strength juice, or spray-dried juice, and contain a mixture of red betanins (XXX) and yellow betaxanthins. The pigment concentration is about 1%. Purified preparations with higher pigment concentration are desirable, but the necessary toxicity studies have not been done.

A number of miscellaneous compounds are being used as food colorants. Red cochineal is extracted from the insect *Coccus cacti*. The pigment is carminic acid, and its insoluble lake is carmine. Lac is a red pigment from the insect *Laccifer lacca*. The yellow pigment in turmeric is curcumin obtained from the rhizomes of the turmeric plant. The mold *Monascus purpureus* produces a red pigment termed "monascin," which, although illegal in the United States, has been used in the Orient for many years. Chlorophylls and their more stable potassium salts have been used as colorants, particularly for pasta. Replacement of the magnesium ion by copper in chlorophyll, chlorophyllin, pheophytin, or pheophorbide yields colorants that are permissible in Europe. These have been marketed under the general term "Chlorophyll Copper Complex." In the United States chlorophylls can be added to foods only as green vegetables, in which case they are classified as food ingredients. Hemoglobin and some of its more stable chemical derivatives are potential food colorants, but their use in pure form has not been sanctioned.

D. Possible New Colorants

A new concept has emerged in the colorant area. The extensive toxicological data involved in testing food colorants for safety have disadvantages both in cost and in the connotation to the consumer. Obviously, that colorants are absorbed by the body, with possible undesirable consequences, led to the concept of creating colorants that would not be absorbed from the human gut. This concept was pioneered by the Dynapol Company and evolved into the choice of a polymer of high molecular weight with appropriate colorant and solubility components (16).

The linear polymer backbone of 20,000 to 1,000,000 daltons can be made from a variety of materials, such as polyacrylic acid, polyvinylamine, and polyvinyl alcohol. The chromophore groups are usually attached to the polymer by a stable carbon-nitrogen bond. Anthraquinone-type chromophores are preferred, but azo, anthrapyridone, and benzanthrones are suggested in the patents. The solubilizing groups are $-SO_3^-$, $-RSO_3^-$ and $-SO_2-R-O-SO_3^-$, and $-SO_2-R'-SO_3^-$, where R is a 2 to 4 carbon alkyl group and R' is a 2 carbon alkyl group. It is obvious that a wide variety of structures is possible.

The linked polymer concept is not confined to colorants. Actually, the first application permitted by the FDA is likely to be an antioxidant. Three colorants (red, yellow, and blue) are being developed. Other possibilities include sweeteners, flavors, and, possibly, a number of other functional additives.

REFERENCES

1. Bauernfeind, J. C. (1981). *Carotenoids as Colorants and Vitamin A Precursors*, Academic Press, New York.
2. Braverman, J. B. S. (1963). *Introduction to the Biochemistry of Foods*, Elsevier, New York.
3. Chichester, C. O. (1972). *The Chemistry of Plant Pigments*, Academic Press, New York.
4. Geissman, T. A. (1962). *The Chemistry of Flavonoid Compounds*, Macmillan, New York.
5. Goodwin, T. W. (1965). *Chemistry and Biochemistry of Plant Pigments*, Academic Press, New York.
6. Goodwin, T. W. (1976). *Chemistry and Biochemistry of Plant Pigments*, 2nd ed., Vols. 1 and 2, Academic Press, New York.

7. Harborne, J. B. (1967). *Comparative Biochemistry of the Flavonoids*, Academic Press, New York.
8. Harborne, J. B., T. J. Mabry, and H. Mabry (1975). *The Flavonoids*, Chapman and Hall, London.
9. Isler, O. (1971). *Carotenoids*, Birkhauser Verlag, Basel.
10. Laurie, R. A. (1974). *Meat Science*, 2nd ed., Pergamon Press, New York.
11. Mabry, T. J., K. R. Markham, and M. B. Thomas (1970). *The Systematic Identification of Flavonoids*, Springer-Verlag, New York.
12. Markakis, P. (1982). *Anthocyanins as Food Colors*, Academic Press, New York.
12a. Marmion, D. M. (1979). *Handbook of U.S. Colorants for Foods, Drugs and Cosmetics*, John Wiley and Sons, New York.
13. National Academy of Sciences (1971). *Food Colors*, Washington, D.C.
14. Price, J. F., and B. S. Schweigert (1971). *The Science of Meat and Meat Products*, 2nd ed., W. H. Freeman, San Francisco.
15. Vernon, L. P., and G. R. Seely (1966). *The Chlorophylls*, Academic Press, New York.
16. Walford, J. (1980). *Developments in Food Colours*, Vol. 1, Applied Sciences Publishers, London.
17. Baord, B. W., and Ihsan-ul-Haque (1965). Discoloration of canned cured meats caused by copper. *Food Technol. 19*:1721.
18. Blair, J. S. (1940). Color stabilization of green vegetables. U.S. Pat. 2,186,003.
19. Blair, J. S. (1940). Color stabilization of green vegetables. U.S. Pat. 2,189,774.
19a. Brouillard, R. (1981). Origin of the exceptional colour stability of the Zebrina anthocyanins. *Phytochemistry 20*:143.
20. Brouillard, R., and B. Delaporte (1977). Chemistry of anthocyanin pigments. 2. Kinetic and thermodynamic study of proton transfer, hydration and tautomeric reactions of malvidin-3-glucoside. *J. Amer. Chem. Soc. 99*:8461.
21. Cheesman, D. F., W. F. Lee, and P. F. Zagalsky (1967). Carotenoproteins in invertebrates. *Biol. Rev. Cambridge 42*:131.
22. Clydesdale, F. M., and F. J. Francis (1968). A study of chlorophyll changes in thermally processed spinach as influenced by enzyme conversion and pH adjustment. *Food Technol. 22*:135.
23. Clydesdale, F. M., Y. D. Lin, and F. J. Francis (1972). Formation of pyrrolidone-3-carboxylic acid from glutamine during processing and storage of spinach puree. *J. Food Sci. 37*:45.
24. Driver, M. W., and F. J. Francis (1979). Purification of phytolaccanin by removal of phytolaccatoxin from *Phytolacca americana. J. Food Sci. 44*:521.
25. Fox, J. B. (1966). Chemicophysical properties of red meat pigments as related to color. *J. Agr. Food Chem. 14*:207.
26. Fox, J. B., W. E. Townsend, S. A. Ackerman, and C. E. Swift (1967). Cured color development during frankfurter processing. *Food Technol. 21*:386.
27. Fox, Jr., J. B., and S. A. Ackerman (1968). Formation of nitric oxide myoglobin: Mechanisms of the reactions with various reductants. *J. Food Sci. 33*:364.
28. Jurd, L. (1964). Some benzopyrylium compounds potentially useful as color additives for fruit drinks and juices. *Food Technol. 18*:559.
29. Jurd, L. (1964). Reactions involving sulphite bleaching of anthocyanins. *J. Food Sci. 29*:16.
30. LaJollo, F., S. R. Tannenbaum, and T. P. Labuza (1971). Reactions at limited water concentration. 2. Chlorophyll degradation. *J. Food Sci. 36*:850.
31. Lin, Y. D., F. M. Clydesdale, and F. J. Francis (1971). Organic acid profiles in thermally processed spinach puree after storage. *J. Food Sci. 36*:240.
32. Starr, M. S., and F. J. Francis (1968). Oxygen and ascorbic acid effect on relative stability of four anthocyanin pigments in cranberry juice. *Food Technol. 22*:1293.

33. Starr, M. S., and F. J. Francis (1973). Effect of metallic ions on color and pigment content of cranberry juice cocktail. *J. Food Sci. 38*:1043.

33a. Stirton, J. Z., and J. B. Harborne (1980). Two distintive anthocyanin patterns in the Commelinaceae. *Biochem. Sys. Ecol. 8*:285.

34. Timberlake, C. F., and P. Bridle (1968). Flavylium salts resistant to sulphur dioxide. *Chem. Ind.* (London) *43*:1489.

35. Van Teeling, C. G., P. E. Cansfield, and R. A. Gallop (1971). An anthocyanin complex isolated from the syrup of canned blueberries. *J. Food Sci. 36*:1061.

36. Walker, G. C. (1964). Color deterioration in frozen french beans (*Phaseous vulgaris*). *J. Food Sci. 29*:883.

37. White, R. C., I. D. Jones, and E. Gibbs (1963). Determination of chlorophylls, chlorophllides, pheophytins and pheophorbides in plant materials. *J. Food Sci. 28*:431.

38. Yoshimata, K. (1977). An acylated delphinidin 3-rutinoside-5,3^1,5^1-triglucoside from *Lobelia erinus*. *Phytochemistry 16*:1857.

39. Zakaria, M., K. Simpson, P. R. Brown, and A. Krstulovic (1979). Use of reversed phase high performance liquid chromatography analysis for the determination of provitamin A carotenes in tomatoes. *J. Chromatogr. 176*:109.

9

FLAVORS

Robert C. Lindsay   University of Wisconsin-Madison, Madison, Wisconsin

I.	Introduction	586
	A. General Philosophy	586
	B. Methods for Flavor Analysis	586
	C. Sensory Assessment of Flavors	587
II.	Taste and Nonspecific Saporous Sensations	588
	A. Taste Substances: Sweet, Bitter, Sour, and Salty	588
	B. Flavor Enhancers	596
	C. Astringency	598
	D. Pungency	599
	E. Cooling	600
III.	Vegetable, Fruit, and Spice Flavors	601
	A. Sulfur-Containing Volatiles in *Allium* sp.	601
	B. Sulfur-Containing Volatiles in the Cruciferae	602
	C. Unique Sulfur Compound in Shiitake Mushrooms	603
	D. Methoxy Alkyl Pyrazine Volatiles in Vegetables	604
	E. Enzymically Derived Volatiles from Fatty Acids	604
	F. Volatiles from Branched-Chain Amino Acids	606
	G. Volatile Terpenoids in Flavors	606
	H. Flavors Derived from the Shikimic Acid Pathway	609
IV.	Flavors from Lactic Acid-Ethanol Fermentations	609
V.	Flavor Volatiles from Fats and Oils	611
VI.	Flavor Volatiles in Muscle Foods	612
	A. Species-Specific Flavors of Meats from Ruminants	612
	B. Species-Specific Flavors of Meats from Nonruminants	613
	C. Volatiles in Fish and Seafood Flavors	614
VII.	Development of "Process" or "Reaction" Flavor Volatiles	615
	A. Thermally Induced Process Flavors	615
	B. Volatiles Derived from Oxidative Cleavage of Carotenoids	620

VIII. Future Directions of Flavor Chemistry and Technology   620
       References   621
       Bibliography   626

I. INTRODUCTION

A. General Philosophy

Knowledge of the chemistry of flavors is commonly perceived as a relatively recent development in food chemistry that evolved from the use of gas chromatography and fast-scan mass spectrometry. Although the availability of these instrumental tools has provided the means to definitively investigate the entire range of flavor substances, classic chemical techniques were elegantly applied in early studies, especially with regard to essential oils and spice extractives (37). This extensive and somewhat separate focus of attention to perfumery, combined with a rapid, seemingly disorganized development of information about the chemistry of food flavors beginning in the early 1960s, has contributed to the slow evolution of a discipline-oriented identity for the field of flavors. Even though flavor substances represent an extremely wide range of chemical structures derived from every major constituent of foods, their common feature of "stimulating taste or aroma receptors to produce an integrated psychological response known as flavor," remains adequate as a definition of this applied discipline.

Generally, the term "flavor" has evolved to a usage that implies an overall integrated perception of all the contributing senses (smell, taste, sight, feeling, and sound) at the time of food consumption. The ability of specialized cells of the olfactory epithelium of the nasal cavity to detect trace amounts of volatile odorants accounts for the nearly unlimited variations in intensity and quality of odors and flavors (11). Taste buds located on the tongue and back of the oral cavity enable humans to sense sweetness, sourness, saltiness, and bitterness, and these sensations contribute to the taste component of flavor. Nonspecific or trigeminal neural responses also provide important contributions to flavor perception through detection of pungency, cooling, umami or delicious attributes, as well as other chemically induced sensations that are incompletely understood. The nonchemical or indirect senses (sight, sound, and feeling) influence the perception of tastes and smells, and hence food acceptances, but a discussion of these effects is beyond the scope of this chapter. Thus, materials presented in this chapter are confined to discussions about substances that yield taste and/or odor responses, but a clear distinction between the meaning of these terms and that of flavor is not always attempted.

B. Methods for Flavor Analysis

As noted at the beginning of this chapter, flavor chemistry often has been equated with the analysis of volatile compounds using gas chromatography combined with fast-scan mass spectrometry, but this impression is much too restrictive. Here, only limited attention is directed to the analytical approach since extensive discussions can be found in the literature (6,16,50,61,80,81,93).

Several factors make the analysis of flavors somewhat demanding, and these include their presence at low concentrations (ppm, $1:10^6$; ppb, $1:10^9$; and ppt, $1:10^{12}$), complexity of flavor mixtures (e.g., over 450 volatiles have been identified in coffee), extreme volatility (high vapor pressures) and instability of some flavor compounds, and the existence of flavor compounds in dynamic equilibria with other constituents in foods. Identification of flavor compounds requires initial isolation from the bulky constituents of foods combined with substantial concentration, and this should occur with minimal distortion of the native composition if flavor quality is being studied. The adsorption of flavor compounds on porous polymers followed by either thermal desorption or solvent elution has provided a means to minimize destruction of sensitive compounds during isolation. However, higher boiling compounds and some compounds present in very low concentrations still require distillation techniques to assure adequate recovery for identification. Successful identifications of compounds also require separation of flavor isolates into individual components, and recent advances in the technology of preparing fused-silica capillary columns for gas chromatography (see Refs. 31 and 86) have proven highly successful in achieving this goal. Similarly, recent advances in high-performance liquid chromatography have provided powerful means for separating many higher boiling compounds and precursors.

C. Sensory Assessment of Flavors

Sensory assessments of flavor compounds and foods are essential for achieving objectives of flavor investigations regardless of the ultimate goals. Some situations call for sensory characterization of samples by skilled individuals (experienced flavorists or researchers). In other instances, it is necessary to use formal panels for sensory analysis followed by statistical analysis of the data. Excellent reviews and books are available on this subject (1,3,28,29,51,75), and these should be consulted for detailed information on this extremely important aspect of flavor assessment.

Measurement of chemical flavor parameters to provide definitive information about flavor intensity and quality of foods has long been an idealized goal of flavor investigations (78). Much progress has been made in the application of methods for correlation of subjective sensory information with objective flavor chemical data (75), but routine assessment of flavor quality by purely analytical means remains limited.

Attention is usually directed to important flavor compounds that provide the "characterizing" or "character-impact" features of a particular flavor. These substances are reasonably limited in number but are frequently present in extremely low concentrations, and they may be extremely unstable. The seemingly impossible task of defining character-impact compounds led, early on, to the so-called component balance theory for explaining the chemical basis of certain flavors. However, as methods of analysis have become more sensitive, more and more foods with different flavors but similar low-resolution, flavor-compound profiles, have been found to contain one or more character-impact compounds responsible for the principal flavor. Identification of character-impact compounds can be expected for many flavors that remain a mystery for the present, but for some transient, unstable flavor compounds, identification by the methods available currently may not be possible. For example, the key flavor compounds responsible for cheddar cheese aroma, the dihydropyrazines of freshly roasted nuts, and the furfuryl thioaldehyde believed present in freshly brewed coffee may not yield themselves to isolation and identification techniques used now, and their definition probably must await the development of new techniques.

This chapter deals primarily with the chemistry of important character-impact compounds that have been selected to illustrate the chemistry of food systems and the chemical basis for the existence of flavor compounds in foods. Where appropriate, and when information is available, structure-activity relationships for flavor compounds are noted. Limited attention is given to listings of profiles of flavor compounds present in various foods. Comprehensive lists of flavor compounds for foods are available elsewhere

(99), as are tabulations of thereshold concentrations for individual compounds (27). These sources of information should prove useful to the reader in searches for further details about specific flavor compounds. Finally, a choice exists whether information pertaining to the flavor chemistry of major food constituents is dealt with here or in the chapters devoted to those major food constituents. It has been deemed appropriate to conduct these discussions in the major-constituent chapters. For example, flavors deriving from the Maillard reaction are discussed in Chap. 3 and those deriving from free radical oxidation of lipids are discussed in Chap. 4. Information on low-calorie sweeteners and on binding of flavors by macromolecules is, of necessity, partly covered here and partly covered in Chaps. 3 and 5 (binding by macromolecules), and in Chap. 10 (low-calorie sweeteners).

II. TASTE AND NONSPECIFIC SAPOROUS SENSATIONS

Frequently, substances responsible for these components of flavor perception are water soluble and relatively nonvolatile. As a general rule, they are also present at higher concentrations in foods than those responsible for aromas, and have been often treated lightly in coverages of flavors. Because of their important role in the acceptance of food flavors, it is appropriate to examine the chemistry of substances responsible for taste sensations as well as those substances responsible for some of the less well defined flavor-taste sensations.

A. Taste Substances: Sweet, Bitter, Sour, and Salty

Sweet substances have been the focus of much recent attention because of the interest in sugar alternatives and the desire to find suitable replacements for the low-calorie sweeteners saccharin and cyclamate (see Chaps. 3 and 10). The bitterness sensation appears to be closely related to sweetness from a molecular structure-receptor relationship, and as a result much has been learned about bitterness in studies directed primarily toward sweetness. Development of bitterness in protein hydrolysates and aged cheeses is a troublesome problem, and this problem has stimulated research on the causes of bitterness in peptides. With regard to saltiness, national policies that encourage a reduction of sodium in diets have recently stimulated renewed interests in the mechanisms of the salty taste.

1. Structural Basis of the Sweet Modality

Before modern theories on sweetness were advanced, it was popular to deduce that sweetness was associated with hydroxyl (−OH) groups because sugar molecules are dominated by this feature. However, this view was soon subject to criticism because polyhydroxy compounds vary greatly in sweetness, and many amino acids, some metallic salts, and unrelated compounds, such as chloroform ($CHCl_3$) and saccharin (Chap. 10), also are sweet. Still, it was apparent that some common characteristics existed among sweet substances, and over the past 75 years a theory relating molecular structure and sweet taste has evolved that satisfactorily explains why certain compounds exhibit sweetness (45,84).

Shallenberger and Acree (83) first proposed the AH/B theory for the saporous (taste-eliciting) unit common to all compounds that cause a sweet sensation (Fig. 1). The saporous unit was initially viewed as a combination of a covalently bound H-bonding proton and an electronegative orbital positioned at a distance of about 3 Å from the proton. Thus, vicinal electronegative atoms on a molecule are essential for sweetness. Fur-

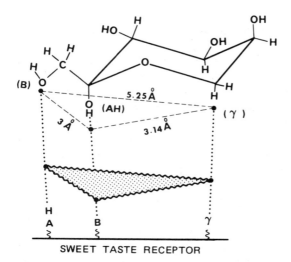

Figure 1 Schematic showing the relationship between AH/B and γ in the saporous sweet unit for β-D-fructopyranose.

ther, one of the atoms must possess a hydrogen-bonding proton. Oxygen, nitrogen, and chlorine atoms frequently fulfill these roles in sweet molecules, and hydroxyl group oxygen atoms can serve either the AH or B function in a molecule. Simple AH/B relationships are shown below for chloroform (I), saccharin (II), and glucose (III).

CHLOROFORM
(I)

SACCHARIN
(II)

D-GLUCOSE
(III)

As indicated in Fig. 1, however, stereochemical requirements are also imposed on the AH/B components of the saporous unit so that they will align suitably with the receptor site (84). The interaction between the active groups of the sweet molecule and the taste receptor is currently envisioned to occur through H bonding of the AH/B components to similar structures in the taste receptor. More recently (25), a third feature has been added to the theory to extend its validity to intensely sweet substances. This addition incorporates appropriate stereochemically arranged lipophilic regions of sweet molecules, usually designated γ, which are attracted to similar lipophilic regions of the taste receptor. Lipophilic portions of sweet molecules are frequently methylene ($-CH_2-$), methyl ($-CH_3$), or phenyl ($-C_6H_5$) groups. The complete sweet saporous structure is geometrically situated so that triangular contact of all active units (AH, B, and γ) with the receptor molecule occurs for intensely sweet substances, and this arrangement forms the rationale for the current tripartite structure theory of sweetness (82).

The γ-site is an extremely important feature of intensely sweet substances, but plays only a limited role in sugar sweetness (7,10). It appears to function through facilitating the accession of certain molecules to the taste receptor site, and, as such, affects the perceived intensity of sweetness. Since sugars are largely hydrophilic, this feature comes into play in a limited sense only for some very sweet sugars, such as fructose, and not at all for less sweet sugars. This component of the saporous sweet unit probably accounts for a substantial portion of the variation in sweetness quality observed between different sweet substances. Not only is it important in the time intensity or temporal aspects of sweetness perception, but it also appears to relate to some of the interactions between sweet and bitter tastes observed for some compounds.

Sweet-bitter sugar structures possess features that apparently allow them to interact with either or both types of receptors, thus producing the combined taste sensation. Bitterness properties in chemical structures depress sweetness (84), even if the concentration in a test solution is below the threshold for the bitter sensation. Bitterness in sugars appears to be imparted by a combination of effects involving the configuration of the anomeric center, the ring oxygen, the primary alcohol group of hexoses, and the nature of any substituents. Often, changes in the structure and stereogeometry of a sweet molecule leads to a loss or suppression of sweetness or the induction of bitterness.

2. The Bitter Taste Modality

Bitterness resembles sweetness because of its dependence on the stereochemistry of stimulus molecules, and the two sensations are triggered by similar features in molecules, causing some molecules to cause both bitter and sweet sensations. Although sweet molecules must contain two polar groups that may be supplemented with a nonpolar group, bitter molecules appear to have a requirement for only one polar group and a hydrophobic group (12). However, some (7,8) believe that most bitter substances possess an AH/B entity identical to that found in sweet molecules, as well as the hydrophobic group. In this concept, the orientation of AH/B units within specific receptor sites, which are located on the flat bottom of receptor cavities, provides the discrimination between sweetness and bitterness for molecules possessing the required molecular features. Molecules that fit into sites that were oriented for bitter compounds give a bitter response; those fitting the orientation for sweetness elicit a sweet response. If the geometry of a molecule were such that it could orient in either direction, it would give bitter-sweet responses. Such a model appears especially valid for amino acids, in which D-isomers are sweet and L-isomers are bitter (48,91). Since the hydrophobic or γ-site of the sweet receptor is nondirectional lipophilicity, it could participate in either sweet or bitter responses. Molecular bulkiness factors serve to provide stereochemical selectivity to the receptor sites located in each receptor cavity. It can be concluded that there is a very broad structural basis for the bitter taste modality, and most empirical observations about bitterness and molecular structure can be explained by current theories.

3. Bitter Compounds That Are Important in Foods

Bitterness may be desirable in food flavors, and because of genetic differences in humans, individuals vary in their ability to perceive certain bitter substances. A compound may be bitter or bittersweet depending on the individual. Saccharin is perceived as purely sweet by some individuals, but others find it to range from only slightly bitter and sweet to quite bitter and sweet. Many other compounds also show marked variations in the manner in which individuals perceive them, and frequently either taste bitter or are not perceived at all. Phenylthiocarbamide (PTC) (IV)

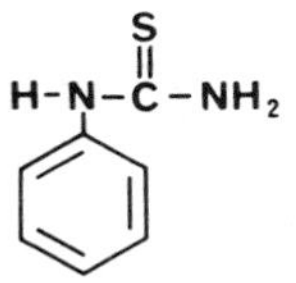

$$H-N-\overset{\overset{S}{\|}}{C}-NH_2$$

**PHENYL
THIOCARBAMIDE
(PTC)**

(IV)

is one of the most notable compounds in this category (1), with about 40% of the Caucasian American population taste blind to the bitterness attribute that is perceived by the other 60% of the Caucasian American population. Although PTC is a novel compound that does not occur in foods, creatine (V)

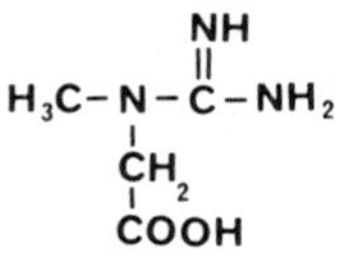

$$H_3C-N-\overset{\overset{NH}{\|}}{C}-NH_2$$
$$\overset{|}{CH_2}$$
$$\overset{|}{COOH}$$

CREATINE

(V)

is a constituent of muscle foods that exhibits similar properties of taste sensitivity in the population. Creatine may occur at levels up to about 5 mg/g in lean meats (1), and this is adequate to make some soups taste bitter to sensitive individuals. As with other bitter substances, these molecules contain AH/B sites suitable for inducing the bitter sensation. Genetic variation in the perception of specific molecules appears to relate to the size of receptor site cavities and to the arrangement and nature of atoms on the cavity walls, these atoms governing which molecules are allowed in the site and which are rejected.

Quinine is an alkaloid that is generally accepted as the standard for the bitter taste sensation. The detection threshold for quinine hydrochloride (VI) is about 10 ppm. In general, bitter substances have lower taste thresholds than other taste substances, and they also tend to be less soluble in water than are other taste-active materials. Quinine

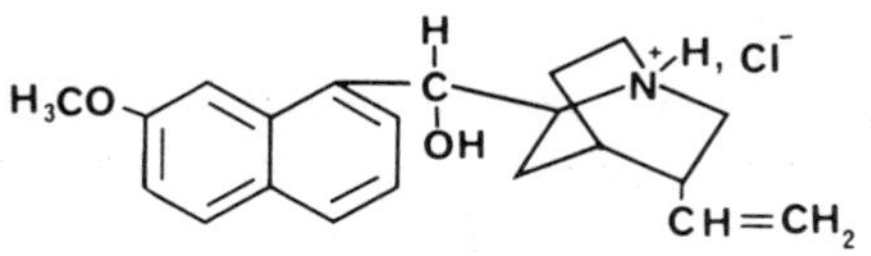

QUININE Cl⁻

(VI)

is permitted as an additive in beverages, such as soft drinks, which have tart-sweet attributes. The bitterness blends well with the other tastes, and provides a refreshing gustatory stimulation in these beverages. The practice of mixing quinine into soft drink beverages apparently stems from past efforts to suppress or mask the bitterness of quinine when it was prescribed as a drug for malaria.

In addition to some soft drinks, bitterness is an important flavor attribute of several other beverages consumed in large quantities, including coffee, cocoa, and tea. Caffeine (VII) is moderately bitter at 150-200 ppm in water, and it occurs in coffee, tea, and kola nuts. Theobromine (VIII)

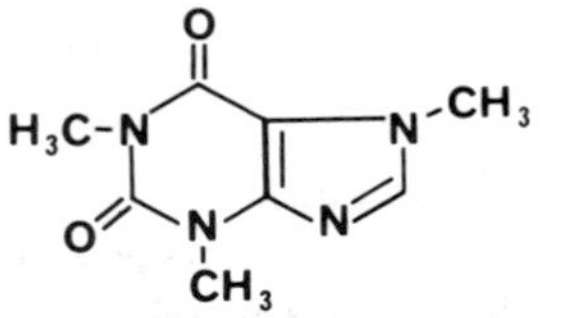

CAFFEINE
(VII)

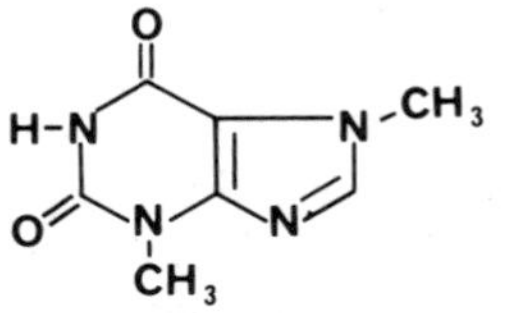

THEOBROMINE
(VIII)

is very similar to caffeine and is present most notably in cocoa, where it contributes to bitterness. Caffeine is added in concentrations up to 200 ppm in soft cola beverages, and much of the caffeine employed for this purpose is obtained by solvent extraction of green coffee beans, which is carried out in the preparation of decaffeinated coffee.

Large amounts of hops are employed in the brewing industry to provide unique flavors to beer. Bitterness, contributed by some unusual isoprenoid-derived compounds, is a very important aspect of hop flavor. These substances can generally be categorized as derivatives of humulone or lupulone, that is, α-acids or β-acids, respectively, as they are known in the brewing industry (46). Humulone is the most abundant substance, and it is converted during wort boiling to isohumulone by an isomerization reaction [Eq. (1)] (19).

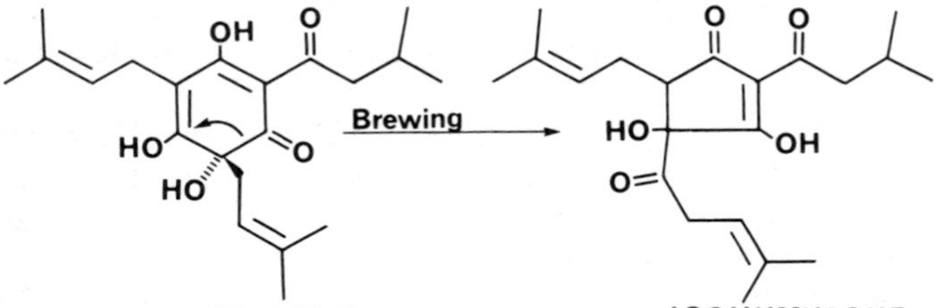

HUMULONE   ISOHUMULONE

(1)

Isohumulone is the precursor for the compound that causes the sunstruck or skunky flavor in beer exposed to light. In the presence of hydrogen sulfide from yeast fermentation, a photocatalyzed reaction occurs at the carbon adjacent to the keto group in the isohexenyl chain. This gives rise to 3-methyl-2-butene-1-thiol (prenylmercaptan), which has a skunky aroma (49). Selective reduction of the ketone in preisomerized hop extracts prevents this reaction, and allows packaging of beer in clear glass without the development of the skunky or sunstruck flavor. Whether volatile hop aroma compounds survive the wort-boiling process was a controversial topic for a number of years. However, it is now well documented that influential compounds do survive the wort-boiling process and others are formed from bitter hop substances; together they contribute to the kettle-hop aroma of beer (74).

The development of excessive bitterness is a major problem in the citrus industry, especially in processed products. In the case of grapefruit, some bitterness is desirable and expected, but frequently the intensity of bitterness in both fresh and processed fruits exceeds that preferred by many consumers. The principal bitter component in navel and Valencia oranges is a triterpenoid dilactone (A- and D-rings) called limonin, and it is also found as a bittering agent in grapefruit. Limonin is not present to any extent in intact fruits, but rather a flavorless limonin derivative produced by enzymic hydrolysis of limonin's D-ring lactone is the predominant form (Fig. 2). After juice extraction, acidic conditions favor the closing of the D-ring to form limonin, and the phenomenon of delayed bitterness occurs, yielding serious economic consequences (25).

Methods for debittering orange juice have been developed using immobilized enzymes from *Arthrobacter* sp. and *Acinetobacter* sp. (40,98). Enzymes that simply open the D-ring lactone provide only temporary solutions to the problem because the ring closes again under acidic conditions. However, the use of limonoate dehydrogenase to convert the open D-ring compound to nonbitter 17-dehydrolimonoate A-ring lactone (Fig. 2) provides an irreversible means to debitter orange juice, although the process has not yet been commercialized.

Citrus fruits also contain a number of flavonone glycosides, and naringin is the predominant flavonone found in grapefruit and bitter orange (*Citrus auranticum*). Juices that contain high levels of naringin are extremely bitter, and are of little economic value except in instances in which they can be extensively diluted with juices containing low bitterness levels. The bitterness of naringen is associated with the configuration of the molecule that develops from the 1→2 linkage between rhamnose and glucose (25). Naringenase is an enzyme that has been isolated from commercial citrus pectin preparations and from *Aspergillus* sp., and this enzyme hydrolyzes the 1→2 linkage (Fig. 3) to yield nonbitter products (25). Immobilized enzyme systems have also been developed to debitter grapefruit juices (35) containing excessive levels of naringen. Naringen also has been

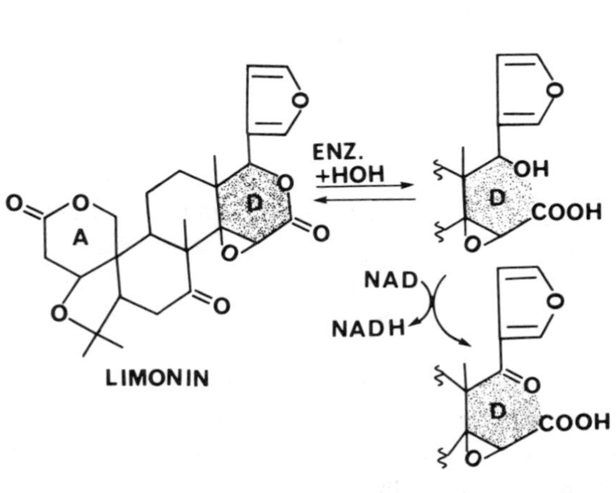

Figure 2 Structure of limonin and reactions leading to enzymically induced nonbitter derivatives (the remainder of the molecule, including the A ring, remains unchanged).

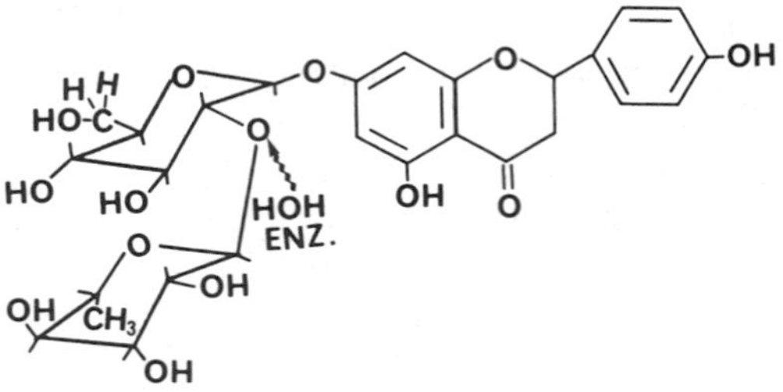

Figure 3 Structure of naringin showing site of enzymic hydrolysis leading to nonbitter derivatives.

commerically recovered from grapefruit peels and is used instead of caffeine for bitterness in some food applications.

Pronounced, undesirable bitterness is frequently encountered in protein hydrolysates and cured cheeses, and this effect is caused by the overall hydrophobicity of amino acid side chains in peptides. All peptides contain suitable numbers of AH-type polar groups that can fit the polar receptor site, but individual peptides vary greatly in the size and nature of their hydrophobic groupings and thus in the ability of these hydrophobic groups to interact with the essential hydrophobic sites of the bitterness receptors. Ney (68) has shown that the bitter taste of peptides can be predicted by a calculation of a mean hydrophobicity value, termed Q. The ability of a protein to engage in hydrophobic associations is related to the sum of the individual hydrophobic contributions of the nonpolar, amino acid side chains, and these interactions contribute mainly to the free energy (ΔG) associated with protein unfolding. Thus, by employing the relationship, $\Delta G = \Sigma \; \Delta g$, where Δg is the contribution of individual amino acid side chains, it is possible to calculate the mean hydrophobicity Q of a protein using the equation

$$Q = \frac{\Sigma \Delta g}{n} \tag{2}$$

where n is the number of amino acid residues. Individual Δg values for amino acids have been determined from solubility data (92), and these are summarized in Table 1. Q values above 1400 indicate that the peptide will be bitter; values below 1300 assure that it will not be bitter. The molecular weight of a peptide also influences its ability to produce bitterness, and only those with molecular weights below 6000 daltons have a potential for bitterness. Peptides larger than this apparently are denied access to receptor sites because of their bulkiness (also see Chap. 5).

The peptide shown in Fig. 4 is derived from cleavage of α_{s1}-casein between residue 144-145 and residue 150-151 (68,88), and has a calculated Q value of 2290. This peptide is very bitter, and is illustrative of the strongly hydrophobic peptides that can be derived readily from α_{s1}-casein. Such peptides are responsible for the bitterness that develops in aged cheeses.

Ney (69) has also used this general approach to predict the bitterness of lipid derivatives and sugars. Hydroxylated fatty acids, particularly some trihydroxy derivatives, frequently are bitter. The ratio of the number of C atoms to the number of OH groups, or the R value (n_C/n_{OH}), of a molecule, gives an indication of the bitterness of these substances. Sweet compounds yield R values of 1.00-1.99, bitter compounds 2.00-6.99, and nonbitter compounds have values above 7.00.

Table 1 Calculated Δg Values for Individual Amino Acids

Amino acids	Δg Value (cal/mol)
Glycine	0
Serine	40
Threonine	440
Histidine	500
Aspartic acid	540
Glutamic acid	550
Arginine	730
Alanine	730
Methionine	1300
Lysine	1500
Valine	1690
Leucine	2420
Proline	2620
Phenylalanine	2650
Tyrosine	2870
Isoleucine	2970
Tryptophan	3000

Source: From Ney (68).

Bitterness in salts appears to involve a different mechanism of reception that seems to be related to the sum of the ionic diameters of the anion and cation components of a salt (6). Salts with ionic diameters below 6.5 Å are purely salty in taste (LiCl = 4.98 Å, NaCl = 5.56 Å, and KCl = 6.28 Å) although some individuals find KCl to be somewhat bitter. As the ionic diameter increases (CsCl = 6.96 Å and CsI = 7.74 Å), salts become increasingly bitter. Magnesium chloride (8.50 Å) is therefore quite bitter.

4. The Salty and Sour Taste Modalities

Classic salty taste is represented by sodium chloride (NaCl), and also by lithium chloride (LiCl). Recent national policies encouraging reduction in sodium consumption have stimulated interest in foods in which sodium salts have been replaced by alternative substances, particularly those containing potassium and ammonium ions. Since these

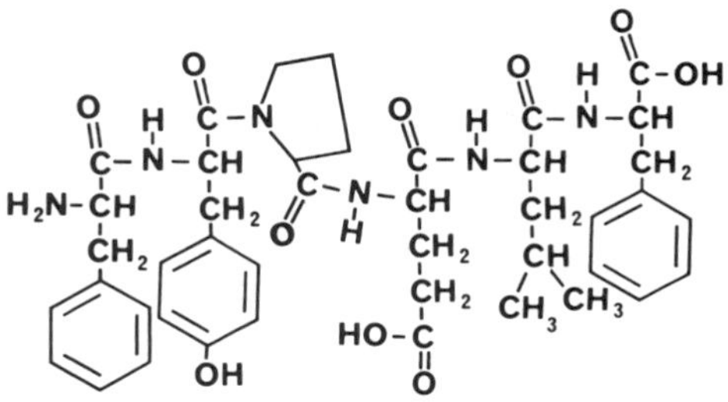

Figure 4 Structure of bitter peptide (phe-tyr-pro-glu-leu-phe) derived from α_{s1}-casein, showing strong nonpolar features.

foods have different, usually less desirable tastes, than those flavored with NaCl, renewed efforts are being expended to better understand the mechanisms of the salty taste, in the hope that low-sodium products with a near normal salty taste can be devised.

Salts have complex tastes, consisting of psychological mixtures of sweet, bitter, sour, and salty perceptual components. Furthermore, it has been shown recently that the tastes of salts often fall outside the traditional taste sensations (79) and are difficult to describe in classic terms. Nonspecific terms, such as chemical and soapy, often seem to more accurately describe the sensations produced by salts then do the classic terms.

Chemically, it appears that cations cause salty tastes and anions inhibit salty tastes (5). Sodium and lithium produce only salty tastes, potassium and other cations produce both salty and bitter tastes. Among the anions, the chloride ion is least inhibitory to the salty taste and it apparently possesses no taste of its own. More complex anions not only inhibit tastes of cations but also contribute tastes of their own. Therefore, soapy tastes caused by sodium salts of long-chain fatty acids (IX) or by long-chain sulfate detergents (X)

$$H_3C-(CH_2)_{10}-C{\overset{O}{\underset{}{\parallel}}}-O^-, Na^+$$

SODIUM LAURATE

(IX)

$$H_3C-(CH_2)_n-{\overset{O}{\underset{O}{\overset{\parallel}{\underset{\parallel}{S}}}}}-O^-, Na^+$$

SODIUM LAURYL SULFATE

(X)

result from specific tastes elicited by the anions, and these tastes can completely mask the taste of the cation (104).

The most acceptable model for describing the mechanism of salty taste perception involves the interaction of hydrated cation-anion complexes (6) with AH/B-type receptor sites (discussed earlier). The individual structures of such complexes vary substantially, so that both water OH groups and salt anions or cations associate with receptor sites.

Similarly, the perception of sour compounds is believed to involve an AH/B-type receptor site (6). However, data are not sufficient to determine whether hydronium ions (H_3O^+), dissociated inorganic or organic anions, or nondissociated molecular species are most influential in the sour response. Contrary to popular belief, the acid strength of a solution does not appear to be the major determinant of the sour sensation (6); rather, other poorly understood molecular features appear to be of primary importance (e.g., weight, size, and overall polarity).

B. Flavor Enhancers

Compounds eliciting this unique effect have been utilized by humans since the inception of food cooking and preparation, but the actual mechanism of flavor enhancement remains largely a mystery. These substances contribute a delicious or umami taste to foods when used at levels in excess of their independent detection threshold, and they simply enhance flavors at levels below their independent detection thresholds. Their effects are prominent and desirable in the flavors of vegetables, dairy products, meats, poultry, fish, and other seafoods. The best known members of this group of substances are monosodium L-glutamate (MSG) (XII) and the 5'-ribonucleotides, of which 5'-inosine monophosphate (5'-IMP) (XI) serves as a suitable example.

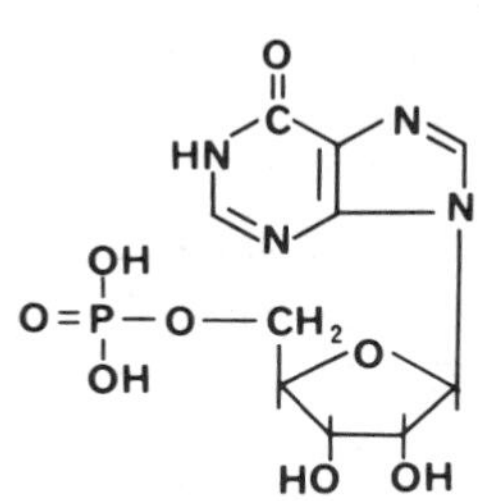

5'-INOSINE
MONOPHOSPHATE
(5'-IMP)

(XI)

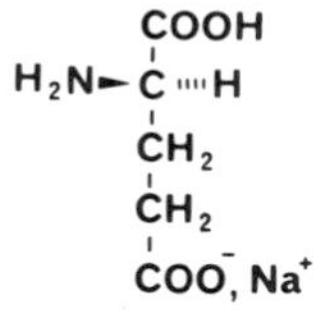

L-GLUTAMATE, Na⁺
(MSG)

(XII)

D-Glutamate and the 2'- or the 3'ribonucleotides do not exhibit flavor-enhancing activity. Although MSG and 5'-IMP and 5'-guanosine monophosphate are the only flavor enhancers used commercially, 5'-xanthine monophosphate and a few natural amino acids, including L-ibotenic acid and L-tricholomic acid, are potential candidates for commercial use (105). Much of the flavor contributed to foods by yeast hydrolysates results from the 5'-ribonucleotides present. Large amounts of purified flavor enhancers employed in the food industry are derived from microbial sources, including phosphorylated (in vitro) nucleotides derived from RNA (48).

Recent investigations have yielded several synthetic derivatives of the 5'-ribonucleotides that have strong flavor-enhancing properties (48). Generally, these derivatives have substitutions on the purine moiety in the 2 position. The strong dependence of flavor-enhancement activity on specific structural characteristics suggests the involvement of a receptor site for these substances, or possibly joint occupancy of receptor sites usually involved in perception of sweet, sour, salty, and bitter sensations. It is well documented that a synergistic interaction occurs between MSG and the 5'-ribonucleotides in providing both the umami taste and in enhancing flavors. This suggests that some common structural features exist among active compounds. Undoubtedly, efforts will remain concentrated on gaining an understanding of the fundamental mechanisms involved in this extremely important component of taste.

Although most attention has been directed toward the 5'-ribonucleotides and MSG, other flavor-enhancing compounds have been claimed to exist. Maltol and ethyl maltol are worthy of mention because they are used commercially as flavor enhancers in sweet goods and fruits. At high concentrations maltol possesses a pleasant, burnt caramel aroma and tastes sweet in dilute solutions. It provides a smooth, velvety sensation to fruit juices when it is employed at concentrations of about 550 ppm. Maltol belongs to a group of unique compounds that exist in a planar enolone form (72). The planar enolone form is completely favored over the cyclic diketone form because the enolone form allows strong intramolecular H bonding to occur [Eq. (3)].

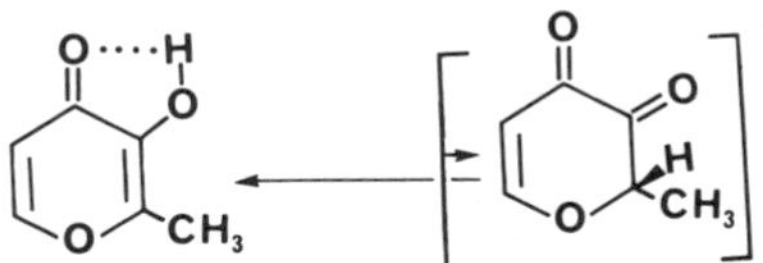

STABLE
ENOLONE FORM   DIKETO FORM

(3)

Maltol and ethyl maltol ($-C_2H_5$ instead of $-CH_3$ on the ring) both could fit the AH/B portion of the sweet receptor (Fig. 1), but ethyl maltol is more effective as a sweetness enhancer than maltol. Maltol lowers the detection threshold concentration for sucrose by a factor of 2 (84). The actual mechanism for the flavor-enhancing effects of these compounds is unknown. Similar compounds of this type are derived naturally from browning reactions and are noted later in this chapter under the section on the development of "reaction" flavors.

Excellent discussions on flavor enhancement can be found in recent reviews (48, 63,105).

C. Astringency

Astringency is a taste-related phenomenon, perceived as a dry feeling in the mouth along with a coarse puckering of the oral tissue. Astringency usually involves the association of tannins or polyphenols with proteins in the saliva to form precipitates or aggregates. Additionally, sparingly soluble porteins such as those found in certain dry milk powders also combine with proteins and mucopolysaccharides of saliva and cause astringency. Astringency is often confused with bitterness because many individuals do not clearly understand its nature, and many polyphenols or tannins cause both astringent and bitter sensations (1).

Tannins (Fig. 5) have broad cross-sectional areas suitable for hydrophobic association with proteins (71). They also contain many phenolic groups that can convert to quinoid structures, and these in turn can cross-link chemically with proteins (67). Such cross-links have been suggested as possible contributors to astringency activity.

Astringency may be a desirable flavor property, such as in tea (77). However, the practice of adding milk or cream to tea removes astringency through the binding of polyphenols with milk proteins. Red wine is a good example of a beverage that exhibits astringency and bitterness, both caused by polyphenols (1). Too much astringency is considered undesirable in wines, and means are often taken to reduce polyphenol tannins, which are related to the anthocyanin pigments. Astringency derived from polyphenols in unripe bananas can also lead to an undesirable taste of products to which the bananas have been added (29).

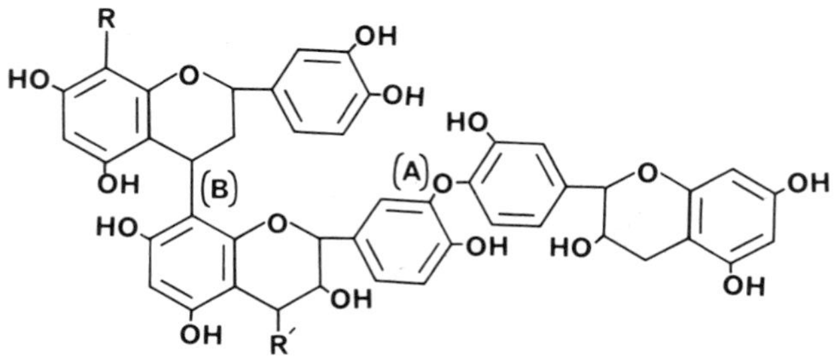

Figure 5 Structure of a procyanidin tannin showing condensed tannin linkage (B) and hydrolyzable tannin linkage (A), and also showing large hydrophobic areas capable of associating with proteins to cause astringency.

D. Pungency

Certain compounds found in several spices and vegetables cause characteristic hot, sharp, and stinging sensations that are known collectively as pungency. Although these sensations are difficult to separate from those of general chemical irritation and lachrymatory effects, they are usually considered separate flavor-related sensations. Some pungent principles, such as those found in chili peppers, black pepper, and ginger, are not volatile and they exert their effects on oral tissues. Other spices and vegetables contain pungent principles that are somewhat volatile, and produce both pungency and characteristic aromas. These include mustard, horseradish, vegetable radishes, onions, galic, watercress, and the aromatic spice, clove, which contains eugenol as the active component. All these spices and vegetables are used in foods to provide characteristic flavors or to generally enhance the palatability. Usage at low concentrations in processed foods frequently provides a liveliness to flavors through subtle contributions that fill out the perceived flavors. Only the three major pungent spices, chili peppers, ginger, and pepper, are discussed in this section, but the chemistry of compounds giving both pungency and aroma is considered later in this chapter in discussions about plant-derived flavors (i.e., eugenol, isothiocyanates, and thiopropanal-S-oxide). Comprehensive reviews on pungent compounds (33,54) should be consulted for in-depth information.

Chili peppers (*Capsicum* sp.) contain a group of substances known as capsaicinoids, which are vanillylamides of monocarboxylic acids with varying chain length (C8-C11) and unsaturation. Capsaicin (XIII)

CAPSAICIN

(XIII)

is representative of these pungent principles. Several capsaicinoids containing saturated straight-chain acid components are synthesized as substitutes for natural chili extractives or oleoresins. The total capsaicinoid content of world *Capsicum* species varies widely (33); for example, red pepper contains 0.06%; cayenne red pepper, 0.2%; Sannam (India), 0.3%; and Uganda (Africa), 0.85%. Sweet paprika contains a very low concentration of pungent compounds and is used mainly for its coloring effect and subtle flavor. Chili peppers also contain volatile aroma compounds that become part of the overall flavor of foods seasoned with them.

Black and white pepper are made from the berries of *Piper nigrum*, and differ only in that black pepper is prepared from immature, green berries, and white pepper is made from more mature berries usually harvested at the time they are changing from green to yellow in color, but before they become red (70). The principal pungent compound in pepper is piperine (XIV), (see p. 600) an amide. The trans geometry of the unsaturated alkyl component is necessary for strong pungency, and loss of pungency during exposure to light and storage is attributed mainly to isomerizations of these double bonds (33). Pepper also contains volatile compounds, including 1-formylpiperdine and piperonal (heliotropin), which contribute to the flavors of foods seasoned with pepper spice or oleoresins. Piperine is also available in a synthesized form for use in flavoring foods (70).

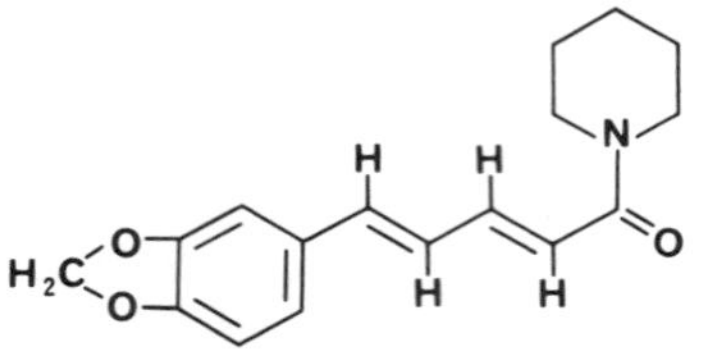

PIPERINE

(XIV)

Ginger is a spice derived from the rhizome of the tuberous perennial, *Zingiber officinale* Roscoe, and it possesses pungent principles as well as some volatile aroma constituents. The pungency of fresh ginger is caused by a group of phenylakyl ketones called gingerols, and [6]-gingerol (XV) is the most active of these compounds (33).

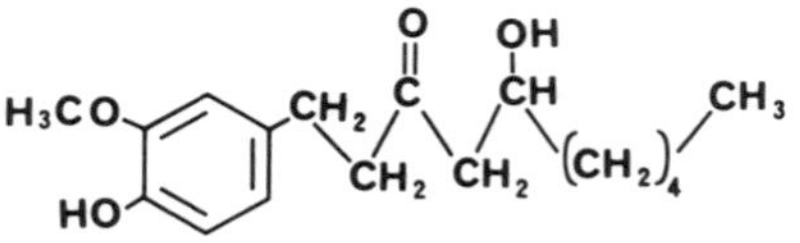

[6]-GINGEROL

(XV)

Gingerols vary in chain length (C5-C9) external to the hydroxyl substituted C atom. During drying and storage gingerols tend to dehydrate to form an external double bond in conjugation with the keto group. This reaction results in a group of compounds known as shogaols, which are even more potent than the gingerols. Exposure of [6]-gingerol to an elevated temperature leads to the cleavage of the alkyl chain external to the keto group, yielding a methyl ketone, zingerone, which exhibits only mild pungency (33).

E. Cooling

Cooling sensations occur when certain chemicals contact the nasal or oral tissues and stimulate a specific saporous receptor (100). These effects are most commonly associated with mintlike flavors, including peppermint, spearmint, and wintergreen. A number of compounds cause the sensation, but (−)-menthol (XVI), in the natural form (*l*-isomer), is most commonly used in flavors. The fundamental mechanism for this response is not known. The aroma component of the total perception is camphoraceous. Camphor (XVII)

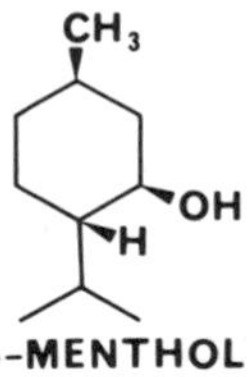

(−)-MENTHOL

(XVI)

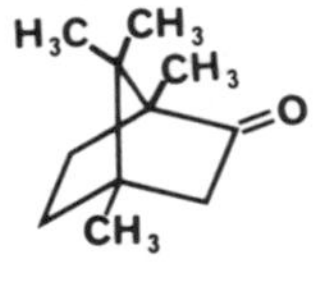

d-CAMPHOR

(XVII)

produces a distinctive camphoraceous odor quality in addition to its cooling sensation (2). The cooling effect produced by the mint-related compounds is mechanistically different from the slight cooling sensation produced when polyol sweeteners (Chaps. 3 and 10), such as xylitol, are tasted as crystalline materials. In the latter case, it is generally believed that endothermic dissolution of the materials gives rise to the effect.

III. VEGETABLE, FRUIT, AND SPICE FLAVORS

Categorization of vegetable and fruit flavors in a reasonably small number of distinctive groups is not easy since logical groupings are not necessarily available for vegetables and fruits. For example, some information on plant-derived flavors was presented in the section on pungency, and some are covered in the section dealing with the development of "reaction" flavors. Emphasis in this section is on the biogenesis and development of flavors in important vegetable and fruits. For information on other fruit and vegetable flavors, the reader is directed to the Bibliography.

A. Sulfur-Containing Volatiles in *Allium* sp.

Plants in the genus *Allium* are characterized by strong, penetrating aromas, and important members are onions, garlic, leek, chives, and shallots. These plants lack the strong characterizing aroma unless the tissue is damaged and enzymes are decompartmentalized so that flavor precursors can be converted to odorous volatiles. In the case of onions (*Allium cepa* L.), the precursor of the compounds responsible for the flavor and aroma is *S*-(1-propenyl)-L-cysteine sulfoxide (30,85). This precursor is also found in leek. Rapid hydrolysis of the precursor by allinase yields a hypothetical sulfenic acid intermediate along with ammonia and pyruvate (Fig. 6). The sulfenic acid undergoes further rearrangements to yield the lachrymator, thiopropanal-*S*-oxide, which is associated with the overall aroma of fresh onions. The pyruvic acid produced during enzymic conversion of the precursor compound is a stable product, and it serves as a good index of the flavor intensity of onion products. Part of the unstable sulfenic acid also rearranges and decom-

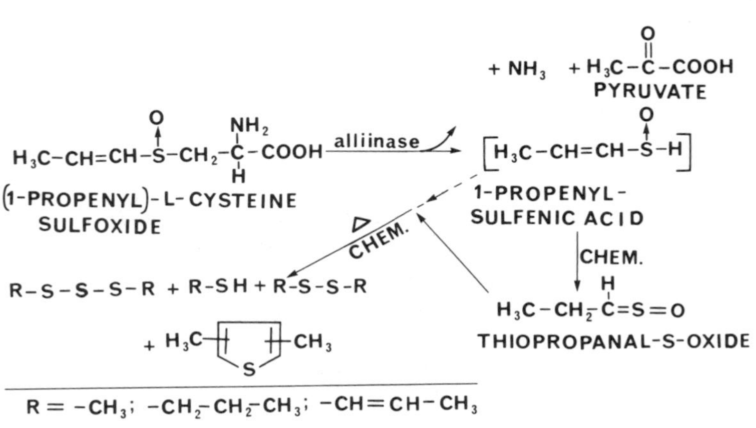

Figure 6 Reactions involved in the formation of onion flavor.

poses to a rather large number of compounds, including mercaptans, disulfides, trisulfides, and thiophenes. These compounds also contribute to the flavor of cooked onions (85).

The flavor of garlic (*Allium sativum* L.) is formed by the same general type of mechanism that functions in onion, except that the precursor is *S*-(2-propenyl)-L-cysteine sulfoxide (85). Diallyl thiosulfinate (allicin) (Fig. 7) contributes to the flavor of fresh garlic, and an *S*-oxide lachrymator similar to that which develops in onion is not formed. The thiosulfinate flavor compound of garlic decomposes and rearranges in much the same manner as indicated for the sulfenic acid of onion (Fig. 6). This results in methyl allyl and diallyl disulfides and other principles of garlic oil and cooked garlic flavors (30).

B. Sulfur-Containing Volatiles in the Cruciferae

The Cruciferae family contains *Brassica* plants such as cabbage (*Brassica oleracea capitata* L.), brussels sprouts (*Brassica oleracea* var. *gemmifera* L.), turnips (*Brassica rapa* var. *rapa* L.), brown mustard (*Brassica juncea* Coss.), as well as watercress (*Nasturtium officinale* R. Br.), radishes (*Raphanus sativus* L.), and horseradish (*Armoracia lapathifolia* Gilib). As noted in the discussion about pungent compounds, the active pungent principles in the Cruciferae are also volatile and therefore contribute to characteristic aromas. Further, the pungency sensation frequently involves irritation sensations, particularly in the nasal cavity, and lachrymatory effects. The flavor compounds in these plants are formed through enzymic processes in disrupted tissues and through cooking. The fresh flavors of the disrupted tissue are caused mainly by isothiocyanates resulting from the action of glucosinolases on thioglycoside precursors. The reaction shown in Fig. 8, yielding allyl isothiocyanate, is illustrative of the flavor-forming mechanism in Cruciferae, and the resulting compound is the main source of pungency and aroma in horseradish and black mustard (33).

A number of other glucosinolates (*S*-glycosides; Chaps. 3 and 11) occur in the Cruciferae (73,85), and each gives rise to characteristic flavors. The mild pungency of radishes is caused by the aroma compound, 4-methylthio-3-*t*-butenylisothiocyanate (XVIII).

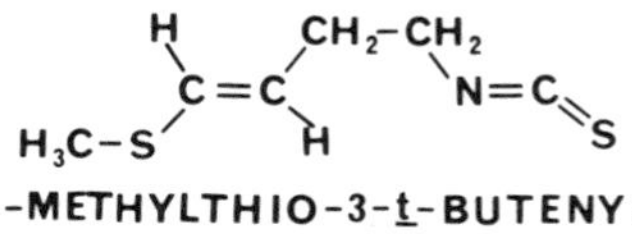

4-METHYLTHIO-3-*t*-BUTENYL
ISOTHIOCYANATE
(XVIII)

In addition to the isothiocyanates, glucosinolates also yield thiocyanates (R—S=C=N) and nitriles (Fig. 8). Cabbage and brussels sprouts contain both allyl isothiocyanate and allyl nitrile, and the concentration of each varies with the stage of growth, location in the edible part, and the severity of processing encountered (85). Processing at temperatures well above ambient (cooking and dehydrating) tends to destroy the isothiocyanates and enhance the amount of nitriles and other sulfur-containing degradation and rearrangement compounds. Several aromatic isothiocyanates occur in Cruciferae; for exam-

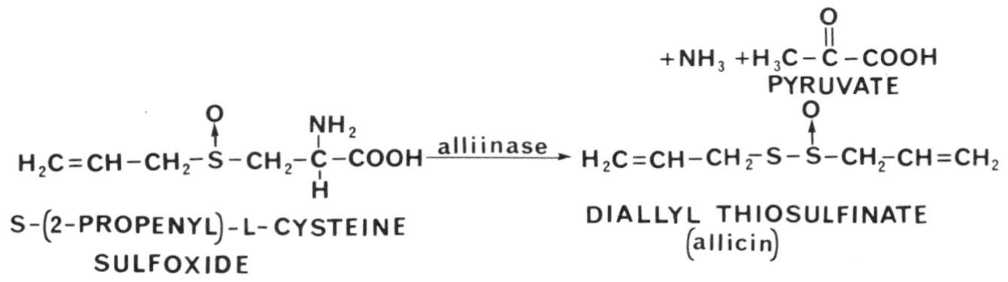

Figure 7 Formation of the principal flavor compound in fresh garlic.

ple, 2-phenylethyl isothiocyanate is one of the main aroma compounds of watercress. This compound also contributes a tingling pungency sensation.

C. Unique Sulfur Compound in Shiitake Mushrooms

A novel C–S lyase enzyme system has been discovered in Shiitake mushrooms (*Letinus edodes*), which are prized in Japan for their delicious flavor. The precursor for the major flavor contributor, lentinic acid, is an S-substituted L-cysteine sulfoxide bound as a γ-glutamyl peptide (25,107). The initial enzyme reaction in flavor development involves γ-glutamyl transpeptidase, which releases a cysteine sulfoxide precursor (lentinic acid). Lentinic acid is then attacked by *S*-alkyl-L-cysteine sulfoxide lyase (Fig. 9) to yield lenthionine, the active flavor compound. These reactions are initiated only after tissue is disrupted, and the flavor develops only after drying and rehydration or after holding freshly macerated tissue for a short period of time. Other polythiepanes in addition to lenthionine are formed (85), but the flavor is ascribed mainly to lenthionine. Flavor roles of other mixed-atom polysulfides with related structural features remain largely unexplored.

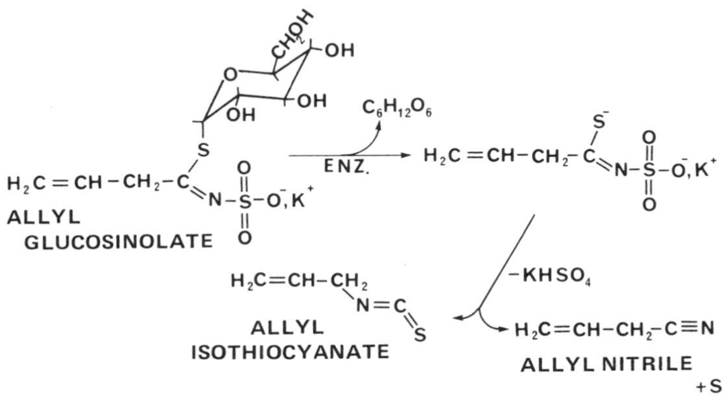

Figure 8 Reactions involved in the formation of Cruciferae flavors.

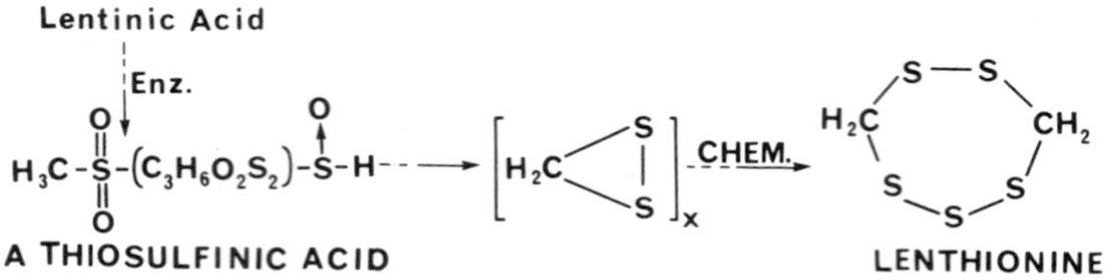

Lentinic Acid

Figure 9 Formation of lenthionine in Shiitake mushrooms.

D. Methoxy Alkyl Pyrazine Volatiles in Vegetables

Many fresh vegetables exhibit green-earthy aromas that contribute strongly to their recognition, and it has been found that the methoxy alkyl pyrazines are frequently responsible for this property (64). These compounds have unusually potent and penetrating odors, and they provide vegetables with strong identifying aromas. 2-Methoxy-3-isobutylpyrazine was the first of this class discovered (14), and it exhibits a powerful bell pepper aroma detectable at a threshold level of 0.002 ppb. Much of the aroma of raw potatoes, green peas, and pea pods is contributed by 2-methoxy-3-isopropyl pyrazine; and 2-methoxy-3-*s*-butylpyrazine contributes to the aroma of raw red beet roots. These compounds arise biosynthetically in plants, and some strains of microorganisms (*Pseudomonas perolens* and *Pseudomonas tetrolens*) also actively produce these unique substances (60). Branched-chain amino acids serve as precursors for methoxy alkyl pyrazine volatiles, and the mechanistic scheme shown in Fig. 10 has been proposed by Murray and coworkers (65).

E. Enzymically Derived Volatiles from Fatty Acids

Enzymically generated compounds derived from long-chain fatty acids play an extremely important role in the characteristic flavors of fruits and vegetables. In addition, these types of reactions can lead to important off-flavors, such as those associated with processed soybean proteins. Further information about these reactions can be found in the chapters dealing with lipids (Chap. 4) and enzymes (Chap. 6).

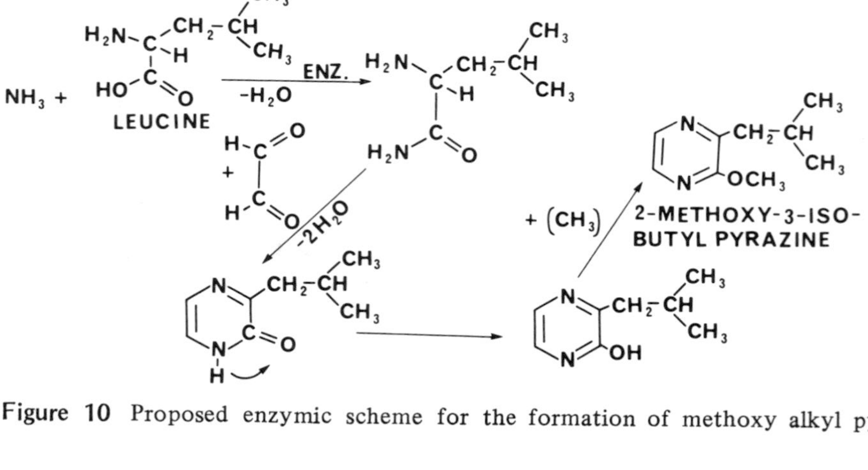

Figure 10 Proposed enzymic scheme for the formation of methoxy alkyl pyrazines.

1. Lipoxygenase-Derived Flavors in Plants

In plant tissues, enzyme-induced oxidative breakdown of unsaturated fatty acids occurs extensively, and this yields characteristic aromas associated with some ripening fruits and disrupted tissues (13,24,32). In contrast to the random production of lipid-derived flavor compounds by purely autoxidizing mechanisms, very distinctive flavors occur when the compounds produced are enzyme determined. The specificity for flavor compounds is illustrated in Fig. 11, where the production of 2-*t*-hexenal and 2-*t*,6-*c*-nonadienal by the site-specific hydroperoxidation of a fatty acid is dictated by a lipoxygenase. Upon cleavage of the fatty acid molecule, oxo acids are also formed, but they do not appear to influence flavors. Decompartmentalization of enzymes is required to initiate this and other reactions, and since successive reactions occur, overall aromas change with time. For example, the lipoxygenase-generated aldehydes and ketones are converted to corresponding alcohols (Fig. 12), which usually have higher detection thresholds and heavier aromas than the parent carbonyl compounds. Although not shown, cis-trans isomerases are also present that convert cis-3 bonds to trans-2-isomers. Generally, C6 compounds yield green, plantlike aromas like those of fresh-cut grass, C9 compounds smell like cucumbers and melons, and C8 compounds smell like mushrooms or violet and geranium leaves. The C6 and C9 compounds are primary alcohols and aldehydes; the C8 compounds are secondary alcohols and ketones (94).

2. Volatiles from β-Oxidation of Long-Chain Fatty Acids

The development of pleasant, fruity aromas is associated with the ripening of pears, peaches, apricots, and other fruits, and these aromas are frequently dominated by medium chain-length (C6-C12) volatiles derived from long-chain fatty acids by β-oxidation (20,97). The formation of ethyl deca-2-*t*,4-*c*-dienoate by this means is illustrated in Fig. 13. This ester is the impact or characterizing aroma compound in the Bartlett pear. Although not included in the figure, hydroxy acids (C8-C12) are also formed by this process, and they cyclize to yield γ- and δ-lactones. Similar reactions occur during the degradation of milk fat, and these reactions are discussed in more detail in Sec. V. The C8-C12 lactones possess distinct coconutlike and peachlike aromas characteristic of these respective fruits.

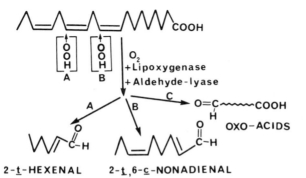

2-*t*-HEXENAL   2-*t*,6-*c*-NONADIENAL

Figure 11 Formation of lipoxygenase-directed aldehydes from linolenic acid: (a) important in fresh tomatoes; (b) important in cucumbers.

2-t-6-c-NONADIENAL   **2-t-6-c-NONADIENOL**

Figure 12 Conversion of aldehyde to alcohol resulting in subtle flavor modifications in cucumbers and melons.

F. Volatiles from Branched-Chain Amino Acids

Branched-chain amino acids serve as important flavor precursors for the biosynthesis of compounds associated with the ripening of some fruits. Bananas and apples are particularly good examples of this process because much of the ripe flavor of each of these fruits is caused by volatiles from amino acids (20,97). The initial reaction involved in this flavor formation (Fig. 14) is sometimes referred to as enzymic Strecker degradation because transamination and decarboxylation occur that parallel those occurring during nonenzymic browning. Several microorganisms, including yeast and malty flavor-producing strains of *Streptococcus lactis*, can also modify most of the amino acids in a fashion similar to that shown in Fig. 14. Plants can also produce similar derivatives from amino acids other than leucine, and the occurrence of 2-phenethanol with a rose or lilaclike aroma in blossoms is attributed to these reactions.

Although the aldehydes, alcohols, and acids from these reactions contribute directly to the flavors of ripening fruits, the esters are the dominant character-impact compounds. It has long been known that isomyl acetate is important in banana flavor, but other compounds are also required to give full banana flavors. Ethyl 2-methylbutyrate is even more applelike than ethyl-3-methylbutyrate (Fig. 14), and the former compound is the dominant note in the aroma of ripe Delicious apples.

G. Volatile Terpenoids in Flavors

Because of the abundance of terpenes in plant materials used in the essential oil and perfumery industries, their importance in other plant-associated flavors is sometimes underestimated. They are largely responsible, however, for the flavors of citrus fruits and many

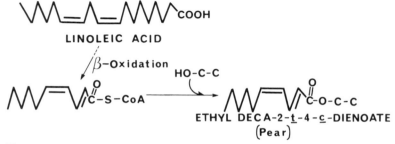

LINOLEIC ACID

β-Oxidation

ETHYL DECA-2-t-4-c-DIENOATE
(Pear)

Figure 13 Formation of a key aroma substance in pears through β-oxidation of linoleic acid followed by esterification.

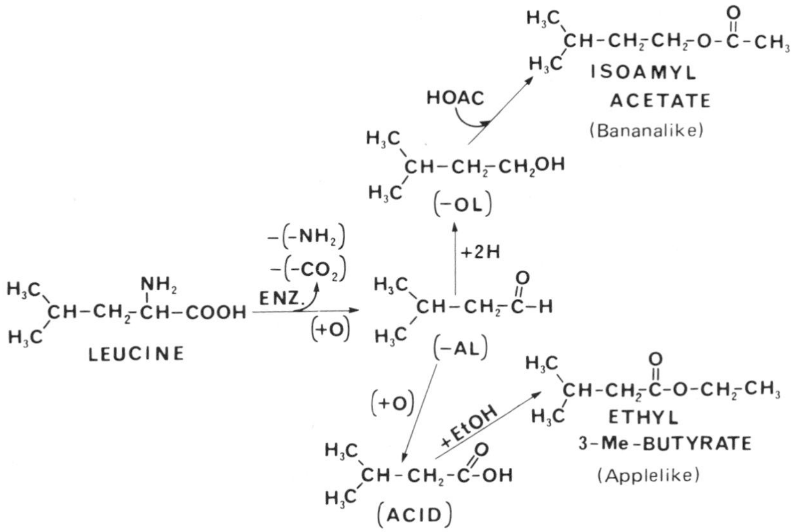

Figure 14 Enzymic conversion of leucine to volatiles illustrating the aroma compounds formed from amino acids in ripening fruits.

seasonings and herbs. Terpenes are present at low concentrations in a number of fruits and are responsible for much of the flavor of raw carrot roots (13).

Terpenes are biosynthesized through the isoprenoid (C5) paths (Fig. 15). Monoterpenes contain 10 C atoms; the sesquiterpenes contain 15 C atoms (25). Terpenes account for some of the branched alkyl compounds found in natural flavor volatiles, but they also may be converted to ring compounds that can become aromaticized. Cumin aldehyde

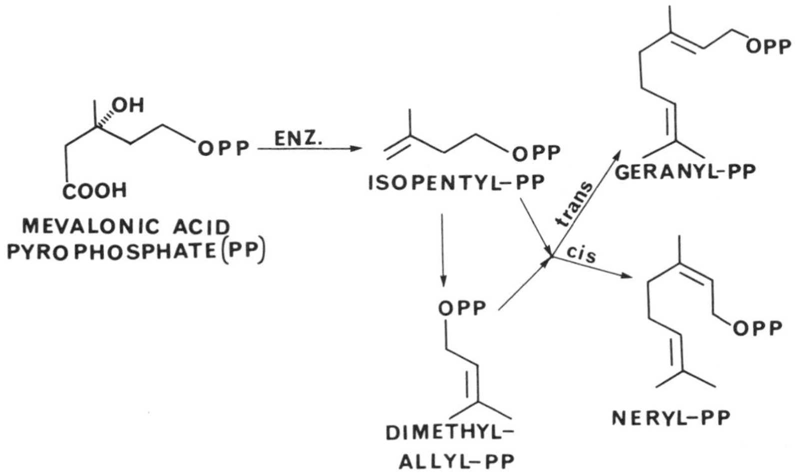

Figure 15 Generalized isoprenoid scheme for the biosynthesis of monoterpenes.

(1-formyl-4-isopropylbenzene), which is a strong characterizing compound in cumin spice, is illustrative of an aromatic derivative of terpenes. Essential oils, or flavor extracts containing terpenes, can be separated into nonoxygenated (hydrocarbon) and oxygenated fractions using silicic acid chromatrography and nonpolar and polar solvent elutions, respectively. Oxygenated terpenes frequently exhibit more desirable flavors than do non-oxygenated terpenes, and the former are therefore preferred for some flavor applications. "Terpeneless" orange oil, for example, contains principally the oxygenated terpene fraction from orange oils.

Terpenes frequently possess extremely strong character-impact properties, and they can be easily identified by one experienced with natural product aromas. For example, the monoterpenes, citral (XIX) with limonene (XX), exhibit distinctive aromas of lemons and limes, respectively.

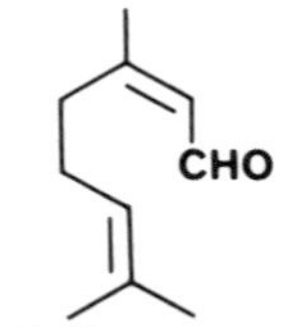

CITRAL (lemon) (XIX) 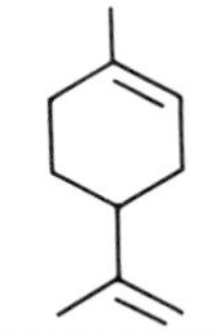   LIMONENE (lime) (XX)

Terpene enantiomers also can exhibit extremely different odor qualities, and the carvones have been studied extensively from this perspective (76). *l*-Carvone [4(R)(−)carvone] (XXII) possesses a strong, characteristic spearmint aroma; *d*-carvone [4(S)(+)carvone] (XXI) has the characteristic aroma of caraway spice. Studies on such pairs of compounds are of interest since they provide information on the fundamental process of olfaction and structure-activity relationships for molecules.

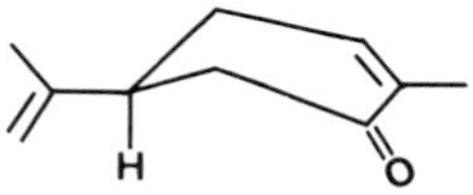

(4S)-(+)-CARVONE (caraway) (XXI) 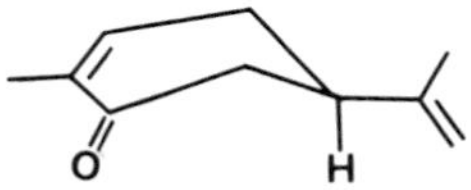   (4R)-(−)-CARVONE (spearmint) (XXII)

Sesquiterpenes are also important characterizing aroma compounds, with β-sinensal (XXIII) and nootkatone (XXIV) providing characterizing flavors to oranges and grape-fruit, respectively. The diterpenes (C20) are too large and nonvolatile to contribute directly to aromas.

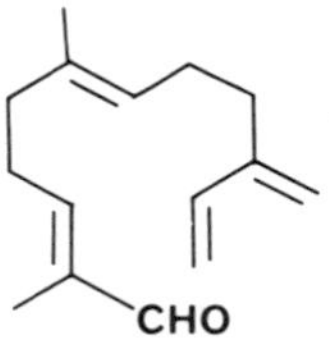

β-SINENSAL (orange) (XXIII) 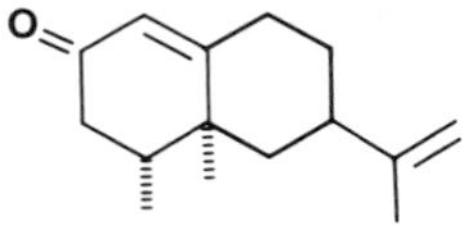   NOOTKATONE (grapefruit) (XXIV)

H. Flavors Derived from the Shikimic Acid Pathway

In biosynthetic systems, the shikimic acid pathway provides the aromatic portion of compounds related to shikimic acid, and the pathway is best known for its role in the production of phenylalanine and other aromatic amino acids. In addition to flavor compounds derived from aromatic amino acids, the shikimic acid pathway provides other volatile compounds (20) frequently associated with essential oils (Fig. 16). It also provides the phenyl propanoid structure to lignin polymers that are the main structural elements of plants. As indicated in Fig. 16, lignin yields many phenols during pyrolysis (9), and the characteristic aroma of smokes used in foods is largely caused by compounds developed from precursors in the shikimic acid pathway. Also apparent from Fig. 16 is that vanillin, the most important characterizing compound in vanilla extracts, can be obtained naturally via the shikimic acid pathway or as a lignin by-product during processing of wood pulp and paper. The methoxylated aromatic rings of the pungent principles in ginger, pepper, and chili peppers, discussed earlier in this chapter, also contain the essential features of those compounds in Fig. 16. Cinnamyl alcohol is an important aroma constituent of cinnamon spice, and eugenol is the principal aroma and pungency element in cloves.

IV. FLAVORS FROM LACTIC ACID-ETHANOL FERMENTATIONS

Involvement of microorganisms in flavor production is extensive, but often their specific or definitive role in the flavor chemistry of fermentations is not well known, or the flavor compounds produced do not have great character impact. Much attention has been given to cheese flavor, but apart from the distinctive flavor properties given by methyl ketones and secondary alcohols to blue-mold cheeses and the moderate flavor properties given

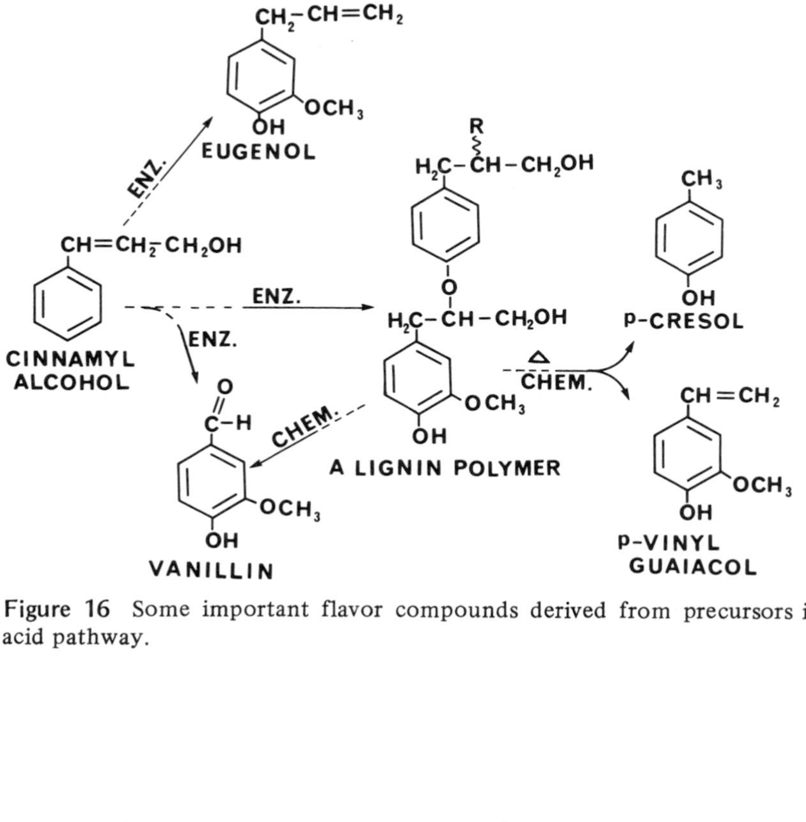

Figure 16 Some important flavor compounds derived from precursors in the shikimic acid pathway.

by certain sulfur compounds to surface-ripened cheeses, microbially derived flavor compounds in cheese generally cannot be classified in the character-impact category (21,56). Similarly, yeast fermentations, carried out extensively for beer, wines, spirits, and yeast-leavened breads, do not appear to yield strong, distinctive character-impact flavor compounds. Ethanol in alcoholic beverages, however, should be considered as having a character impact.

The primary fermentation products of heterofermentative lactic acid bacteria (e.g., *Leuconostoc citrovorum*) are summarized in Fig. 17, and the combination of acetic acid, diacetyl, and acetaldehyde provides much of the characteristic aroma of cultured butter and buttermilk (52). Homofermentative lactic acid bacteria (e.g., *Streptococcus lactis* or *Streptococcus thermophilus*) produce only lactic acid, acetaldehyde, and ethanol. Acetaldehyde is the character-impact compound found in yoghurt, a product prepared by a homofermentative process. Diacetyl is the character-impact aroma compound in most mixed-strain lactic fermentations, and has become universally known as a dairy- or butter-type flavorant. Lactic acid contributes sourness to cultured or fermented dairy products. Acetoin, although essentially odorless, can undergo oxidization to diacetyl.

Viewed in general terms, lactic acid bacteria produce very little ethanol (parts per million levels) and they use pyruvate as the principal final H acceptor in metabolism. On the other hand, yeasts produce ethanol as the major end product of metabolism. Malty strains of *S. lactis* and all brewer's yeasts (*Saccharomyces cerevisiae* and *Saccharomyces carlsbergensis*) also actively convert amino acids to volatile compounds through transaminations and decarboxylations (Fig. 18). These reactions are analogous to those discussed for branched-chain amino acids in Sec. III.F. However, these organisms tend to produce mainly the reduced forms of the derivative (alcohols), although some oxidized compounds (aldehydes and acids) also appear. Wine and beer flavors, which can be as-

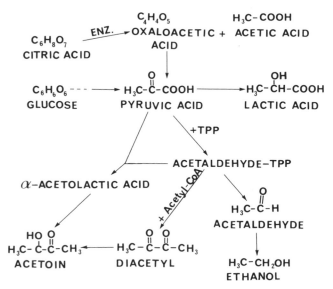

Figure 17 The principal volatile fermentation products from heterofermentative metabolism of lactic acid bacteria.

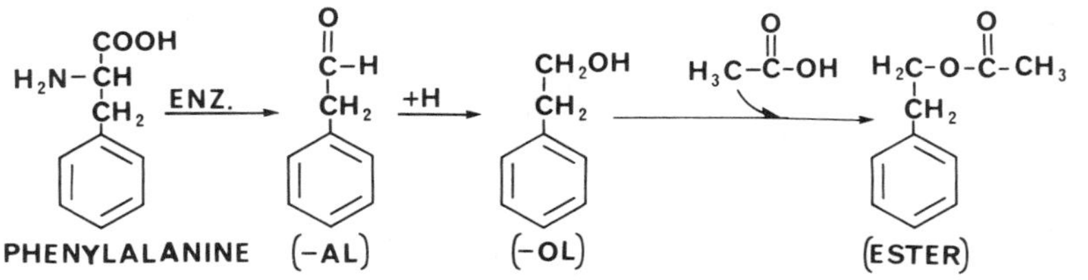

Figure 18 Enzymic formation of volatiles from amino acids by microorganisms using phenylalanine as a model precursor compound.

cribed directly to fermentations, consist of complex mixtures of these volatiles and interaction products of these compounds with ethanol, such as mixed esters and acetals. These mixtures give rise to familiar yeasty and fruity flavors associated with fermented beverages.

V. FLAVOR VOLATILES FROM FATS AND OILS

Fats and oils are notorious for their role in the development of off-flavors through autoxidation, and the chemistry of lipid-derived flavors has been excellently reviewed recently by Grosch (36). Details of lipid autoxidation and other lipid degradations are also presented in Chap. 4. Aldehydes and ketones are the main volatiles from autoxidation, and these compounds can cause painty, fatty, metallic, papery, and candlelike flavors in foods when their concentrations are sufficiently high. However, many of the desirable flavors of cooked and processed foods derive from modest concentrations of these compounds.

Hydrolysis of plant glycerides and animal depot fats leads mainly to the formation of potentially soapy-tasting fatty acids. Milk fat, on the other hand, serves as a rich source of volatile flavor compounds influential in the flavor of dairy products and foods prepared with milk fat or butter. The classes of volatiles obtained by hydrolysis of milk fat are shown in Fig. 19, with specific compounds selected to illustrate each class. The even C-numbered, short-chain fatty acids (C4-C12) are extremely important in the flavor of cheese and other dairy products, with butyric acid the most potent and influential of the

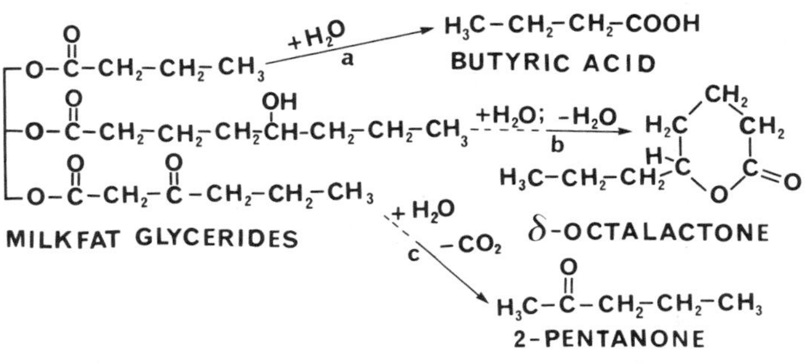

Figure 19 Formation of influential volatile flavor compounds from milk fat by the hydrolytic cleavage of triacylglycerols.

group. Hydrolysis of hydroxy fatty acids leads to the formation of lactones that provide desirable fruitlike flavors in baked goods, but cause staling in stored, sterile concentrated milks. Methyl ketones are produced from β-keto acids by heating after hydrolysis, and they contribute to the flavor of dairy products in much the same manner as do lactones. In blue-mold cheeses, however, methyl ketones are much more abundantly produced by the metabolic activities of *Penicillium roqueforti* on fatty acids than by the conversion of those bound in glycerides (56).

Although hydrolysis of fats other than milk fat does not yield distinct flavors as noted above, animal fats are believed to be inextricably involved in the species-specific flavors of meats. The role of lipids in determining species-specific aspects of meat flavors is discussed in the next section on muscle foods.

VI. FLAVOR VOLATILES IN MUSCLE FOODS

The flavors of meats have attracted much attention, but in spite of considerable research, knowledge about the flavor compounds causing strong character impacts for meats of various species is limited. Nevertheless, the concentrated research efforts on meat flavors have produced a wealth of information about compounds that cause or contribute to cooked meat flavors. The somewhat distinctive flavor qualities of meat flavor compounds that are not species specific are valuable to the food industry, but chemical definitions of lightly cooked and species-specific flavors are still eagerly sought. Some details of the chemistry relating to the development of well-cooked meat flavors are discussed later in Sec. VII.

A. Species-Specific Flavors of Meats from Ruminants

As noted earlier in discussions of flavors of fats and oils, the characterizing flavors of at least some meats are inextricably associated with the lipid fraction (17,53,59). The greatest progress in defining species-related flavors has been made by Wong and coworkers (102,103) in relation to lamb and mutton flavors. These workers have shown that a characteristic sweatlike flavor of mutton is closely associated with some volatile, medium chain-length fatty acids of which several methyl-branched members are highly significant. The formation mechanism of one of the most important branched-chain fatty acids in lamb and mutton, 4-methyloctanoic acid, is shown in Fig. 20. Ruminal fermentations

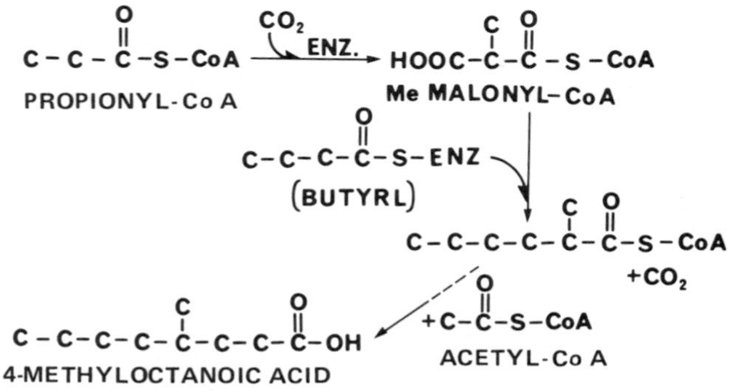

Figure 20 Ruminant biosynthesis of methyl-branched, medium-chain fatty acids.

yield acetate, propionate, and butyrate, but most fatty acids are biosynthesized from acetate, which yields nonbranched chains. Some methyl branching occurs routinely because of the presence of propionate, but when dietary and other factors enhance propionate concentrations in the rumen, greater methyl branching occurs (87). Much remains to be learned about factors governing these flavors among different ruminants, and further efforts should be quite fruitful.

B. Species-Specific Flavors of Meats from Nonruminants

Species-specific aspects of the flavor chemistry of nonruminant meat flavors, particularly those for pork and poultry, are especially vague. Studies have shown that the γ-C5, C9, and C12 lactones are reasonably abundant in pork (22), and these compounds may contribute to some of the sweetlike flavor of pork. However, the distinct porklike or piggy flavor noticeable in lard and cracklings has not been defined. Much interest has centered on the aroma compounds responsible for the swine sex odor, and although these substances are responsible for serious off-flavors in pork, they are not usually regarded as species-specific flavors in pork. The swine sex odor compounds are mainly associated with males, but may occur in castrated males and in females. Two compounds are primarily responsible for the swine sex odor, 5α-androst-16-en-3-one (Fig. 21), which has a urinous aroma, and 5α-androst-16-en-3α-ol, which has a musklike aroma (34). These steroid compounds are particularly offensive to some individuals, especially women, and yet some individuals are genetically odor blind to them (34).

The distinctive flavors of poultry have also been the subject of many studies, and lipid oxidation appears to yield the character impact compounds for chicken (38). The carbonyls *c*4-decenal, *t*2,*c*5-undecadienal, and *t*2,*c*4,*t*7-tridecatriental reportedly may contribute the characteristic flavor of stewed chicken, and they are derived from linoleic and arachidonic acids. Chickens accumulate α-tocopherol (an antioxidant), but turkeys do not, and during cooking, such as roasting, carbonyls are formed to a much greater extent in turkey than in chicken. Additionally, certain chemical environments greatly affect the outcome of some lipid autoxidations. For example, Swoboda and Peers (90) have shown that the presence of copper ions and α-tocopherol results in selective oxidation of milk fat, producing octa-1,*c*-5-dien-3-one, the cause of metallic taints found in butter. Directed lipid oxidations may also occur in poultry, leading to species-related flavors.

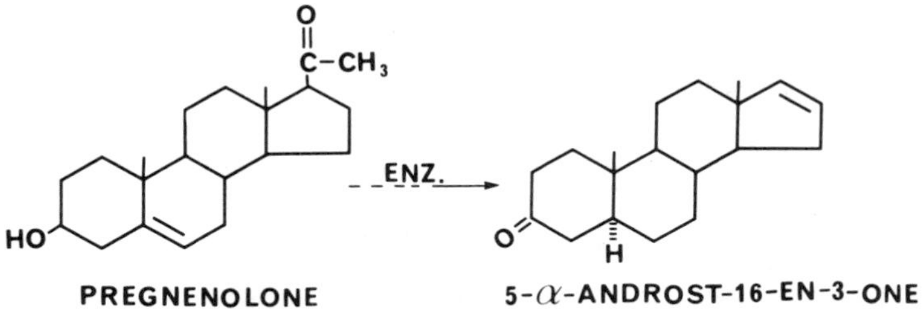

Figure 21 Formation of a steroid compound responsible for the urinous aroma associated with the swine sex odor defect of pork.

C.   Volatiles in Fish and Seafood Flavors

Characterizing flavors in seafoods cover a somewhat broader range of flavor qualities than those occurring in other muscle foods. The broad range of animals involved (finfish, shellfish, and crustaceans) and the variable flavor and aroma qualities related to freshness of each account for the different flavors encountered. Historically, trimethylamine has been associated with fishy and crablike aromas, but alone it usually exhibits only a stale ammoniacal aroma. Trimethylamine and dimethylamine are produced through enzymic degradation of trimethylamine oxide (Fig. 22), which is found in significant quantities only in saltwater species of seafoods. Since very fresh fish contain essentially no trimethylamine, this compound modifies and contributes to the aroma of staling fish, in which it enhances fishhouse-type aromas. Trimethylamine oxide serves as part of the buffer system in marine fish species (42). The formaldehyde produced concurrently with dimethylamine is believed to facilitate protein cross-linking and thereby contribute to the toughening of fish muscle during frozen storage.

Fishy aromas characterized by such terms as "oxidized fish oil" or "cod liver oil-like," arise in part from carbonyls produced from autoxidation of ω3-unsaturated fatty acids. Characteristic aromas appear to result from mixtures of the 2,4-decadienals and 2,4,7-decatrienal (18), but c4-heptenal may contribute to the aroma of frozen cod (57).

Because the fresh flavors and aromas of seafoods frequently have been greatly diminished or lost from fresh, frozen, and processed products available through commercial channels, many consumers associate the fishy flavors described above with all fish and seafoods. However, very fresh seafoods exhibit delicate aromas and flavors quite different from those usually evident in "commercially fresh" seafoods. Recently, a group of enzymically derived aldehydes, ketones, and alcohols have been discovered to provide the characterizing aroma of very fresh fish (44), and these are very similar to the C6, C8, and C9 compounds produced by plant lipoxygenases (see Sec. III.E.1). Collectively, they provide melony, heavy plant like and fresh fish aromas, and these compounds are derived through a lipoxygenaselike system. Animals do not possess lipoxygenase systems like those in plants, but enzymic oxidations involved in prostaglandin-related synthesis appear to accomplish similar results from a flavor standpoint. Hydroperoxidation followed by disproportionation reactions apparently leads first to the alcohol (Fig. 23) and then to the corresponding carbonyl. Some of these compounds contribute to the distinctive flavors of very fresh cooked fish, either directly or as reactants that lead to new flavors during cooking.

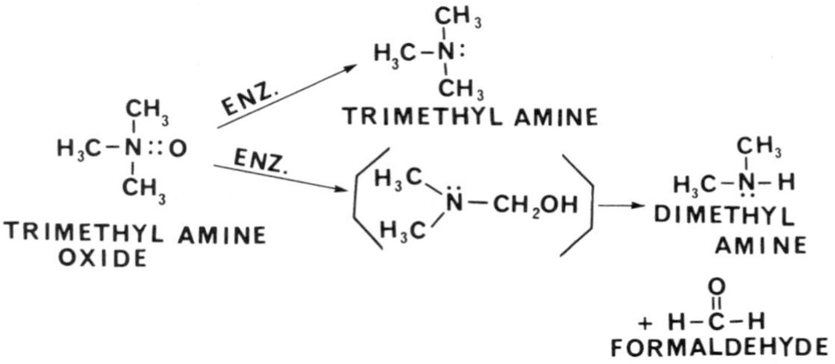

Figure 22   Microbial formation of principal volatile amines in fresh saltwater fish.

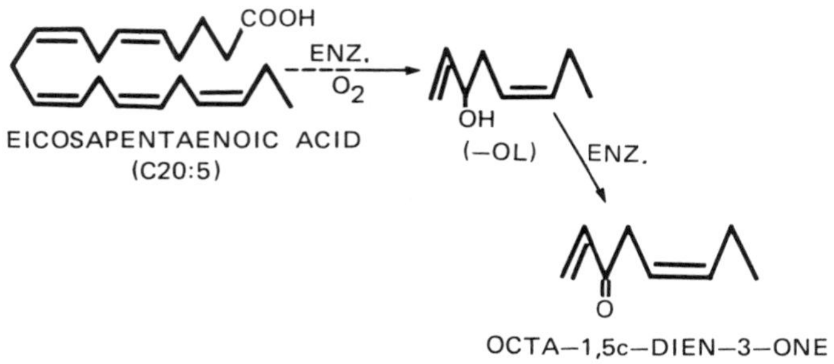

Figure 23 Enzymic formation of influential volatiles in fresh fish aroma from a long-chain ω3-unsaturated fatty acid.

The flavors of crustaceans and mollusks depend heavily on nonvolatile taste substances in addition to contributions from volatiles. For example, the taste of cooked snow crab meat has been largely duplicated by a mixture of 12 amino acids, nucleotides, and salt ions (47). Good imitation crab flavors can be prepared from the taste substances just mentioned, along with some contributions from carbonyls and trimethylamine. Dimethylsulfide provides a characterizing top-note aroma to cooked clams and oysters (58), and this compound arises principally from the thermal degradation of the dimethyl-β-propiothetin present in ingested marine microflora (62).

VII. DEVELOPMENT OF "PROCESS" OR "REACTION" FLAVOR VOLATILES

Many flavor compounds found in cooked or processed foods occur as the result of reactions common to all types of foods regardless of whether they are of animal, plant or microbial origin. These reactions take place when suitable reactants are present and appropriate conditions (heat, pH, and light) exist. These flavors are discussed separately in this section because of their broad importance to all foods, and because they form the basis of some natural flavor concentrates used widely in foods, especially when meat flavors are desired. Related information on this subject can be found in discussions dealing with carbohydrates (Chap. 3), lipids (Chap. 4), and vitamins (Chap. 7).

A. Thermally Induced Process Flavors

Traditionally, these flavors have been broadly viewed as products from browning reactions because of early discoveries showing the role of reducing sugars and amino compounds in the induction of a process that ultimately leads to the formation of brown pigments (see Chap. 3 for details of Maillard browning reactions). Although browning reactions are almost always directly involved in the development of process flavors in foods, the interactions between (a) degradation products of the browning reaction and (b) other food constituents are also important and extensive. By taking a broad approach to discussions of thermally induced flavors, the aforementioned interactions, as well as reactions that occur following the heat treatment, can be appropriately considered.

Although many of the compounds regarded as process flavors possess potent and pleasant aromas, relatively few of these compounds seem to be true character-impact flavor substances. Instead, they often exhibit general nutty, meat, toasted, burnt, floral, plant, or caramel odors. Although some process flavor compounds are acyclic, many are heterocyclic, with nitrogen, sulfur, or oxygen substituents common (Fig. 24). These process flavor compounds occur in many foods and beverages, such as roasted meats, boiled meats, coffee, roasted nuts, beer, bread, crackers, snack foods, cocoa, and most other processed foods. The distribution of individual compounds does, however, depend on such factors as availability of precursors, temperature, time, and water activity (43).

Production of process flavor concentrates is accomplished by selecting reaction mixtures and conditions so that those reactions occurring in normal food processing are duplicated. Selected ingredients (Table 2), usually including a reducing sugar, amino acids, and compounds with sulfur atoms, are processed under elevated temperatures to produce a distinctive profile of flavor compounds (41,101). Thiamine is a popular ingredient because it provides both nitrogen and sulfur atoms already in ring structures (see Chap. 7).

Because of the large number of process flavor compounds produced during normal food processing or process simulation, it is unrealistic to cover the chemistry of their formation in depth. Rather, examples are given to illustrate some of the more important flavor volatiles formed and the mechanisms of their formation. Alkyl pyrazines were among the first compounds to be recognized as important contributors to the flavors of all roasted, toasted, or similarly heated foods (55). The most direct route to their formation results from the interaction of α-dicarbonyl compounds (intermediate products in the Maillard reaction) with amino acids through the Strecker degradation reaction (Fig. 25). Transfer of the amino group to the dicarbonyl provides a means for integrating amino acid nitrogen into small compounds destined for any of the condensation mechanisms envisioned in these reactions. Methionine has been selected as the amino acid involved in the Strecker degradation reaction because it contains a sulfur atom and it leads to formation of methional, which is an important characterizing compound in boiling potatoes and in cheese-cracker flavors. Methional also readily decomposes to yield methanethiol (methyl mercaptan) and dimethyldisulfide, thus providing a source of

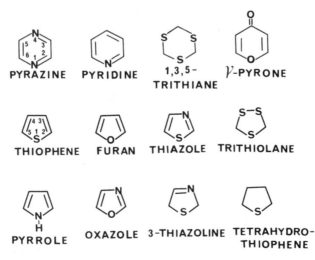

Figure 24 Some heterocyclic structures found commonly in flavor compounds associated with heating or browning of foods.

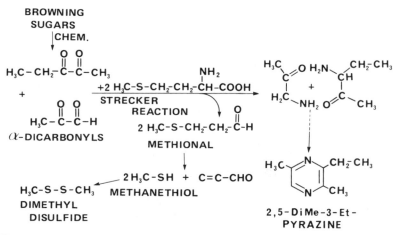

Figure 25 Formation of an alkyl pyrazine and small sulfur compounds in the development of process flavors.

reactive, low-molecular-weight sulfur compounds that contribute to the overall system of flavor development.

Hydrogen sulfide and ammonia are very reactive ingredients in mixtures intended for the development of process flavors, and they are often included in model systems and assist in determining reaction mechanisms. Thermal degradation of cysteine (Fig. 26) yields both ammonia and hydrogen sulfide as well as acetaldehyde. Subsequent reaction of acetaldehyde with a mercapto derivative of acetoin (from the Maillard reaction) gives rise to thiazoline that contributes to the flavor of boiled beef (66,101).

Some heterocyclic flavor compounds are quite reactive and tend to degrade or interact further with components of foods or reaction mixtures. An interesting example of flavor stability and carry-through in foods is provided by the compounds shown in Eq. (4).

$$ (4) $$

2-Me-3-FURAN-
THIOL

(2-Me-3-FURYL)-bis-
DISULFIDE

Table 2 Some Common Ingredients Used in Reaction Systems for the Development of Meat Flavors

Hydrolyzed vegetable protein	Thiamin
Yeast autolysate	Cysteine
Beef extract	Glutathione
Specific animal fats	Glucose
Chicken egg solids	Arabinose
Glycerol	5′-Ribonucleotides
Monosodium glutamate	Methionine

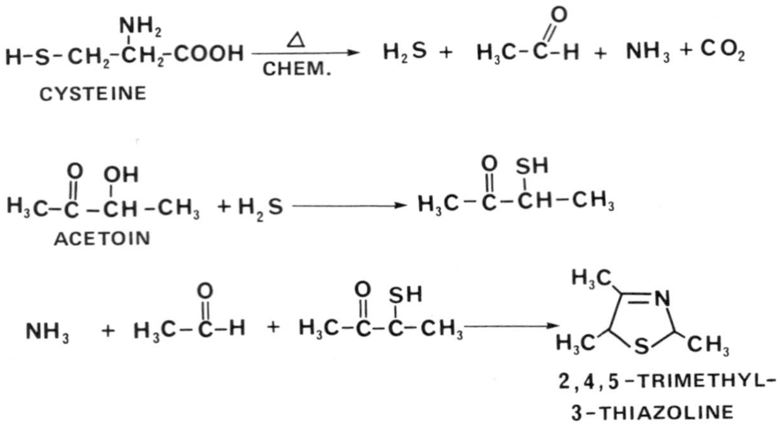

Figure 26 Formation of a thiazoline in cooked beef through the reaction of chemical fragments from cysteine and sugar-amino browning reactions.

both of which provide distinct, but different meatlike aromas (26,101). A roasting meat aroma is exhibited by 2-methyl-3-furanthiol (reduced form), but upon oxidation to the disulfide form, the flavor becomes more characteristic of fully cooked meat that has been held for some time. Chemical reactions, such as the one just mentioned, are responsible for the subtle changes in meat flavor that occur because of the degree of cooking and the time interval after cooking.

During processing of complex systems, sulfur as such, or as thiols or polysulfides, can be incorporated in various compounds, resulting in the generation of new flavors. However, even though dimethylsulfide is often found in processed foods, it usually does not react readily. In plants, dimethylsulfide originates from biologically synthesized molecules, especially methionine via *S*-methylmethionine sulfonium salts (Fig. 27). Methylmethionine is quite labile to heat, and dimethylsulfide is readily released. Dimethylsulfide provides characterizing top-note aromas to fresh and canned sweet corn, tomato juice, and stewing oysters and clams.

Some of the most pleasant aromas derived from process reactions are provided by the compounds shown in Fig. 28. These compounds exhibit caramellike aromas and have been found in many processed foods (43). The planar enol-cyclic-ketone structure [see also Eq. (3)] is derived from the sugar precursors (96), and this structure appears respon-

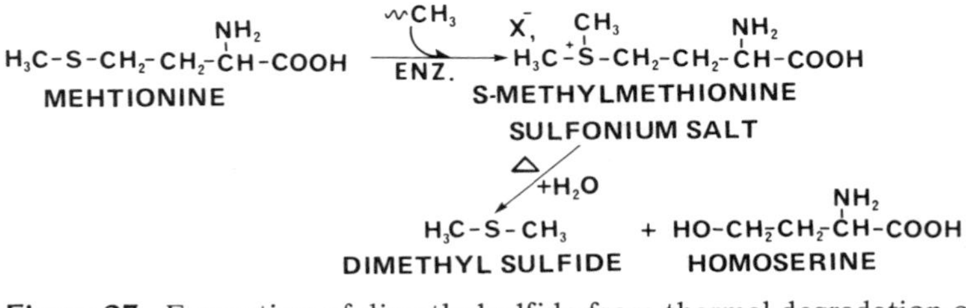

Figure 27 Formation of dimethylsulfide from thermal degradation of *S*-methylmethionine sulfonium salts.

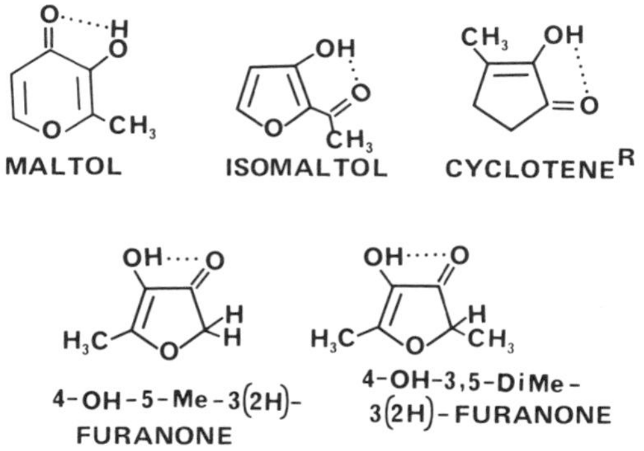

Figure 28 Structures of some important caramellike flavor compounds derived from reactions occurring during processing.

sible for the caramellike odor quality (72). Cyclotene is used widely as a synthesized maple syrup flavor substance (89), and maltol is used widely as a flavor enhancer for sweet foods and beverages (see Sec. II.B). Both furanones have been found in boiled beef where they appear to enhance meatiness. 4-Hydroxy-5-methyl-3(2H)-furanone is sometimes known as the "pineapple compound" because it was first isolated from processed pineapple, where it contributes strongly to the characteristic flavor (96).

The flavor of chocolate and cocoa has received much attention because of the high demand for these flavors. After harvesting, cocoa beans are fermented under somewhat poorly controlled conditions. The beans are then roasted, sometimes with an intervening alkali treatment that darkens the color and yields a less harsh flavor. The fermentation step hydrolyzes sucrose to reducing sugars, frees amino acids, and oxidizes some of the polyphenols (106). During roasting, many pyrazines and other heterocyclics are formed, but the unique flavor of cocoa is derived from an interaction between aldehydes produced from the Strecker degradation reaction. The reactions shown in Fig. 29 between phenylactaldehyde (from phenylalanine) and 3-methylbutanal (from leucine) constitutes an important flavor-forming reaction in cocoa. The product of this aldol condensation, 5-methyl-2-phenyl-2-hexenal, exhibits a characterizing, persistent chocolate aroma. This example also serves to show that reactions in the development of process flavors do not always yield heterocyclic aroma compounds.

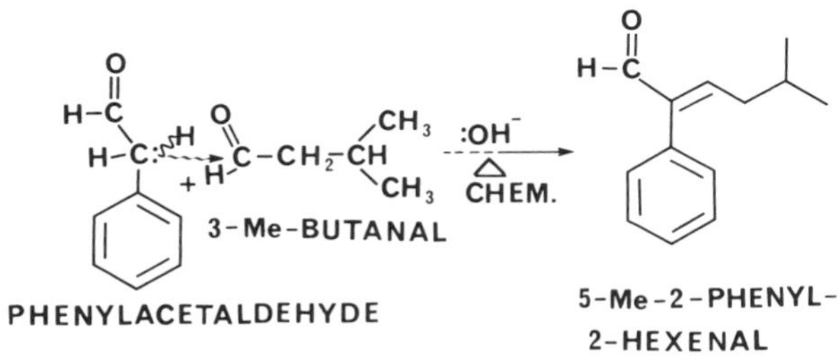

Figure 29 Formation of an important cocoa aroma volatile through an aldol condensation of two aldehydes derived from the Strecker reaction.

B. Volatiles Derived from Oxidative Cleavage of Carotenoids

Oxidations focusing on triacylglycerols and fatty acids are discussed in another section, but some extremely important flavor compounds that are oxidatively derived from carotenoid precursors have not been covered and deserve attention here. Some of these reactions require singlet oxygen through chlorophyll sensitization; others are photooxidation processes (95). A large number of flavor compounds, derived from oxidizing carotenoids (or isoprenoids), have been identified in curing tobacco (23), and many of these compounds are considered important as characterizing tobacco flavors. However, relatively few compounds in this category (three representative compounds are shown in Fig. 30) are currently considered highly important as food flavors. Each of these compounds exhibits unique sweet, floral, and fruit like characteristics that vary greatly with concentration. They also blend nicely with aromas of foods to produce subtle effects that may be either highly desirable or very undesirable. β-Damascenone exerts very positive effects on the flavors of wines, but in beer this compound at only a few parts per billion results in a stale, raisinlike note. β-Ionone also exhibits a pleasant violet, floral aroma compatible with fruit flavors, but it is also the principal off-flavor compound present in oxidized, freeze-dried carrots (4). Furthermore, these compounds have been found in black tea, where they make positive contributions to the flavor. Theaspirane and related derivatives contribute importantly to the sweet, fruity, and earthy notes of tea aroma (104). Although usually present in low concentrations, these compounds and related ones appear to be widely distributed, and it is likely that they contribute to the full, well-blended flavors of many foods.

VIII. FUTURE DIRECTIONS OF FLAVOR CHEMISTRY AND TECHNOLOGY

Knowledge about the chemistry and technology of flavors has expanded greatly in the past 20 years, and information has accumulated to the point where control and manipulation of many flavors in foods is possible. Some areas of research likely to be fruitful are binding of flavors to macromolecules (see Chaps. 3 and 5), flavor development as related to the genetics of plant cultivars and cultured cells (microbial and tissue cultures),

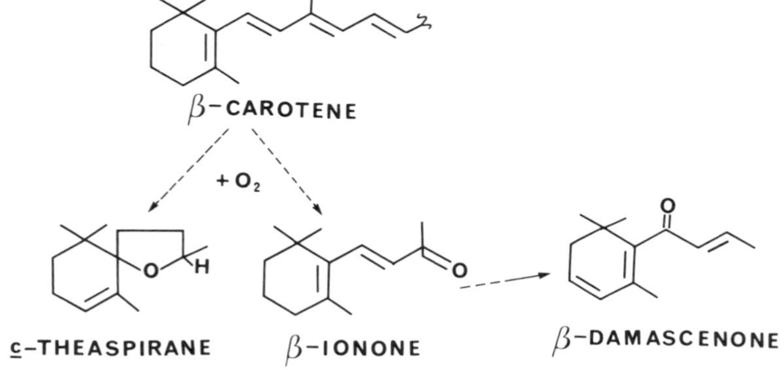

Figure 30 Formation of some important compounds in tea flavor through the oxidative cleavage of carotenoids.

and structure-activity relationships in taste and olfaction employing computer techniques. Enzymic production of flavors within foods and food ingredients will undoubtedly become a significant area of flavor technology in the forseeable future. Early in the modern era of analytical flavor chemistry, that is, 20 years ago, the flavorese concept was offered as a means for regenerating flavors in processed foods, and this type of approach is certain to become more important. However, because of the complex nature of natural flavors as they are now understood, it becomes apparent that the flavorese concept as initially developed would meet with great difficulties when applied to foods. Still, with recent developments in making encapsulated flavor enzyme systems, it is possible to maintain substrate-enzyme proximity and to control the amounts of flavor compounds produced so that unbalanced flavors may be avoided. With the increasing emphasis on high-quality formulated, complex foods, flavor development by enzymes should find a role of increasing importance. Similarly, as process flavors become better understood, more sophisticated practices can be expected to evolve for their production. Finally, efforts to identify flavor compounds continue, particularly in cases in which important characterizing flavor compounds appear to be present.

REFERENCES

1. Amerine, M. A., R. M. Pangborn, and E. B. Roessler (1965). *Principles of Sensory Evaluation of Food*, Academic Press, New York, p. 106.
2. Amoore, J. E. (1967). Stereochemical theory of olfaction, in *Symposium on Foods: Chemistry and Physiology of Flavors* (H. W. Schultz, E. A. Day, and L. M. Libbey, eds.), *AVI Publishing Co., Westport, Conn., pp. 119-147.*
3. *ASTM Manual on Sensory Testing Methods*, STP 434 (1968). American Society for Testing Materials, Philadelphia.
4. Ayers, J. E., M. J. Fishwick, D. G.Land, and T. Swain (1964). Off-flavour of dehydrated carrot stored in oxygen. *Nature 203*:81-82.
5. Bartoshuk, L. M. (1980). Sensory analysis of the taste of NaCl, in *Biological and Behaviorial Aspects of Salt Intake* (R. H. Cagen and M. R. Kare, eds.), Academic Press, New York, pp. 83-96.
6. Beets, M. G. J. (1978). The sweet and bitter modalities, in *Structure-Activity Relationships in Human Chemoreception*, Applied Science Publishers, London, pp. 259-303.
7. Beets, M. G. J. (1978). The sour and salty modalities, in *Structure-Activity Relationships in Human Chemoreception*, Applied Science Publishers, London, pp. 348-362.
8. Belitz, H. D., W. Chen, H. Jugel, H. Stemph, R. Treleano, and H. Wieser (1983). QSAR of bitter tasting compounds. *Chem. Ind.* (London) *1*:23-26.
9. Beltes, W., R. Wittkowski, I. Sochtig, and H. Block (1981). Ingredients of smoke and smoke flavor, in *The Quality of Foods and Beverages* (G. Charalambous and G. Inglett, eds.), Vol. 2, *Chemistry and Technology*, Academic Press, New York, pp. 1-19.
10. Birch, G. G. (1981). Basic tastes of sugar molecules, in *Critreria of Food Acceptance* (J. Solms and R. L. Hall, eds.), Forster Verlag, Zurich, pp. 282-291.
11. Birch, G. G., J. G. Brennan, and K. J. Parker (eds.) (1978). *Sensory Properties of Foods*, Applied Science Publishers, London.
12. Birch, G. G., C. K. Lee, and A. Ray (1978). The chemical basis of bitterness in sugar derivatives, in *Sensory Properties of Foods* (G. G. Birch, J. G. Brennan, and K. J. Parker, eds.), Applied Science Publishers, London, pp. 101-111.

13. Buttery, R. G. (1981). Vegetable and fruit flavors, in *Flavor Research. Recent Advances* (R. Teranishi, R. A. Flath, and H. Sugasawa, eds.), Marcel Dekker, New York, pp. 175-216.

14. Buttery, R. G., R. M. Seifert, R. E. Lundin, D. G. Guadagni, and L. C. Ling (1969). Characterization of an important aroma compound of bell peppers. *Chem. Ind.* (London), 490.

15. Cagan, R. H., and M. R. Kare (eds.) (1980). *Biological and Behaviorial Aspects of Salt Intake*, Academic Press, New York.

16. Charalambous, G. (ed.) (1980). *The Analysis of Control of Less Desirable Flavors in Foods and Beverages*, Academic Press, New York.

17. Cramer, D. A. (1983). Chemical compounds implicated in lamb flavor. *Food Technol. 37*(5):249-257.

18. Crawford, L., and M. J. Kretsch (1976). GC-MS identification of the volatile compounds extracted from roasted turkeys fed a basal diet supplemented with tuna oil: Some comments on fishy flavor. *J. Food Sci. 41*:1470-1478.

19. De Taeye, L., D. De Keukeleire, E. Siaeno, and M. Verzele (1977). Recent developments in hop chemistry, in *European Brewery Convention Proceedings*, European Brewing Congress, Amsterdam, pp. 153-166.

20. Drawert, F. (1975). Biochemical formation of aroma components, in *Proceedings of the International Symposium on Aroma Research*, Zeist (H. Maarse and P. J. Groenen, eds.), Centre for Agricultural Publications and Documents, PUDOC, Wageningen, pp. 13-39.

21. Dumont, J. P., and J. Adda (1979). Flavour formation in dairy products, in *Progress in Flavour Research* (D. G. Land and H. E. Nursten, eds.), Applied Science Publishers, London, pp. 245-262.

22. Dwivedi, B. K. (1975). Meat flavor. *Crit. Rev. Food Technol. 5*:487-535.

23. Enzell, C. R. (1981). Influence of curing on the formation of tobacco flavour, in *Flavour '81* (P. Schreier, ed.), Walter de Gruyter, Berlin, pp. 449-478.

24. Ericksson, C. E. (1979). Review of biosynthesis of volatiles in fruits and vegetables since 1975, in *Progress in Flavour Research* (D. G. Land and H. E. Nursten, eds.), Applied Science Publishers, London, pp. 159-174.

25. Eskin, N. A. M. (1979). *Plant Pigments, Flavors, and Textures: The Chemistry and Biochemistry of Selected Compounds*, Academic Press, New York.

26. Evers, W. J., H. H. Heinsohn, B. J. Mayers, and A. Sanderson (1976). Furans substituted at the three positions with sulfur, in *Phenolic, Sulfur, and Nitrogen Compounds in Food Flavors* (G. Charalambous and I. Katz, eds.), American Chemical Society, Washington, D.C., pp. 184-193.

27. Fazzalari, F. A. (ed.) (1978). *Compilation of Odor and Taste Threshold Values*, American Society for Testing Materials, Philadelphia.

28. Forss, D. A. (1981). Sensory characterization, in *Flavor Research. Recent Advances* (R. Teranishi, R. A. Flath, and H. Sugasawa, eds.), Marcel Dekker, New York, pp. 125-174.

29. Forsyth, W. G. C. (1981). Tannins in solid foods, in *The Quality of Foods and Beverages*, Vol. 1, *Chemistry and Technology* (G. Charalambous and G. Inglett, eds.), Academic Press, New York, pp. 377-388.

30. Freeman, G. G. (1979). Factors affecting flavour during growth, storage and processing of vegetables, in *Progress in Flavour Research* (G. D. Land and H. E. Nurston, eds.), Applied Science Publishers, London, pp. 225-243.

31. Gagliardi, P., and G. R. Verga (1982). Automatic direct head space GC analysis of flavors with capillary column and multidetector systems, in *Chemistry of Foods and Beverages: Recent Developments* (G. Charalambous and G. Inglett, eds.), Academic Press, New York, pp. 49-72.

32. Gardner, H. W. (1975). Decomposition of lineoleic acid hydroperoxides. Enzymic reactions compared with nonenzymic. *J. Agr. Food Chem. 23*:129-136.

33. Govindarajan, V. S. (1979). Pungency: The stimuli and their evaluation, in *Food Taste Chemistry* (J. C. Boudreau, ed.), American Chemical Society, Washington, D.C., pp. 52-91.

34. Gower, D. B., M. R. Hancock, and L. H. Bannister (1981). Biochemical studies on the boar pheromones, 5α-androst-16-en-3-one and 5α-androst-16-en-3α-ol, and their metabolism by olfactory tissue, in *Biochemistry of Taste and Olfaction* (R. H. Cagan and M. R. Kare, eds.), Academic Press, New York, pp. 7-31.

35. Gray, G. M., and A. C. Olson (1981). Hydrolysis of high levels of naringin in grapefruit juice using a hollow fiber naringinase reactor. *J. Agr. Food Sci. 29*:1298-1301.

36. Grosch, W. (1982). Lipid degradation products and flavour, in *Food Flavours*, Part A, *Introduction* (I. D.Morton and A. J. Macleod, eds.), Elsevier Scientific, Amsterdam, pp. 325-398.

37. Guenther, E. (1948). *The Essential Oils*, Vols. 1-6, van Nostrand, New York.

38. Harkes, P. D., and W. J. Begemann (1974). Identification of some previously unknown aldehydes in cooked chicken. *J. Amer. Oil. Chem. Soc. 51*:356-359.

39. Harper, R. (1982). Techniques of analysis of flavours. Sensory methods, in *Food Flavours*, Part A, *Introduction* (I. D. Morton and A. J. Macleod, eds.), Elsevier Scientific, Amsterdam, pp. 79-120.

40. Hasegawa, S., M. N. Patel, and R. C. Snyder (1982). Reduction of limonin bitterness in navel orange juice serum with bacterial cells immobilized in acrylamide gel. *J. Agr. Food Chem. 30*:509-511.

41. Heath, H. B. (1982). *Source Book of Flavors*. AVI Publishing Co., Westport, Conn., p. 110.

42. Hebard, C. E., G. J. Flick, and R. E.Martin (1982). Occurrence and significance of trimethylamine oxide and its derivatives in fish and shellfish, in *Chemistry and Biochemistry of Marine Food Products* (R. E. Martin, G. J. Flick, and D. R. Ward, eds.), AVI Publishing Co., Westport, Conn., pp. 149-304.

43. Hurrell, R. F. (1982). Maillard reaction in flavour, in *Food Flavours*, Part A, *Introduction* (I. D. Morton and A. J. Macleod, eds.), Elsevier Scientific, Amsterdam, pp. 399-437.

44. Josephson, D. B., R. C. Lindsay, and D. A. Stuiber (1983). Identification of compounds characterizing the aroma of fresh whitefish (*Coregonus clupeaformis*). *J. Agr. Food Chem. 31*:326-330.

45. Kier, L. B. (1972). A molecular theory of sweet taste. *J. Pharm. Sci. 61*:1394-1397.

46. Kneen, E. (ed.) (1976). *Methods of Analysis of the American Society of Brewing Chemists* (ASBC), St. Paul, Minn., Appendix II.

47. Konosu, S. (1979). The taste of fish and shellfish, in *Food Taste Chemistry* (J. C. Boudreau, ed.), American Chemical Society, Washington, D.C., p. 203.

48. Kuninaka, A. (1981). Taste and flavor enhancers, in *Flavor Research. Recent Advances* (R. Teranishi, R. A. Flath, and H. Sugisawa, eds.), Marcel Dekker, New York, pp. 305-353.

49. Kuroiwa, Y., and N. Hashimoto (1961). Composition of sunstruck flavor substance and mechanism of its evolution. *ASBC Proc. 19*:28-36.

50. Land, D. G., and H. E. Nursten (1979). *Progress in Flavour Research*, Applied Science Publishers, London.

51. Larmond, E. (1977). *Laboratory Methods for Sensory Evaluation of Food*, Canada Department of Agriculture, Ottawa, Canada.

52. Lindsay, R. C. (1967). Cultured dairy products, in *Symposium on Foods: The Chemistry and Physiology of Flavors* (H. W. Schultz, E. A. Day, and L. M. Libbey, eds.), AVI Publishing Co., Westport, Conn., pp. 315-330.

53. Mabrouk, A. F. (1979). Flavor of browning reaction products, in *Food Taste Chemistry* (J. C. Boudreau, ed.), American Chemical Society, Washington, D.C., pp. 205-245.

54. Maga, J. A. (1978). Simple phenol and phenolic compounds in food flavour. *Crit. Rev. Food Sci. Nutr. 10*(4):323-372.

55. Maga, J. A. (1982). Pyrazines in flavour, in *Food Flavours*, Part A, *Introduction*, (I. D. Morton and A. J. Macleod, eds.), Elsevier Scientific, Amsterdam, pp. 283-323.

56. Margalith, P. Z. (1981). *Flavor Microbiology*, Charles C Thomas, Springfield, Ill., p. 90.

57. McGill, A. S., R. Hardy, and F. D. Gunstone (1977). Further analysis of the volatile components of frozen cold stored cod and the influence of these on flavour. *J. Sci. Food Agr. 28*:200-205.

58. Mendelson, J. M., and R. O. Brooke (1968). Radiation, processing and storage effects on the head gas component of clam meats. *Food Technol. 22*(11):1162-1166.

59. Moody, W. G. (1983). Beef flavor. A review. *Food Technol. 33*(5):227-232.

60. Morgan, M. E., L. M. Libbey, and R. A. Scanlan (1972). Identity of the musty-potato aroma compound in milk cultures of *Pseudomonas taetrolens*. *J. Dairy Sci. 55*:666.

61. Morton, I. D., and A. J. Macleod (1982). *Food Flavours*, Part A, *Introduction*, Elsevier Scientific, Amsterdam.

62. Motohito, T. (1962). Studies on the petroleum odor in canned chum salmon. *Mem. Fac. Fisheries, Hokkaido Univ. 10*:1-5.

63. Motono, M. (1983). Flavor nucleotides' usages in foods, in *Chemistry of Food and Beverages: Recent Developments* (G. Charalambous, ed.), Academic Press, New York, pp. 181-194.

64. Murray, K. E., and F. B. Whitfield (1975). The occurrence of 3-alkyl-2-methoxy-pyrazines in raw vegetables. *J. Sci. Food Agr. 26*:973-986.

65. Murray, K. E., J. Shipton, and F. B. Whitfield (1970). 2-Methoxypyrazines and the flavor of green peas (*Pisum sativum*). *Chem. Ind.* (London) 897-898.

66. Mussinan, C. J., R. A. Wilson, I. Katz, A. Hruza, and M. H. Vock (1976). Identification and flavor properties of some 3-oxazolines and 3-thiazolines isolated from cooked beef, in *Phenolic, Sulfur, and Nitrogen Compounds in Food Flavors* (G. Charalambous and I. Katz, eds.), American Chemical Society, Washington, D.C., pp. 133-145.

67. Neucere, N. J., T. J. Jacks, and G. Sumrell (1978). Interactions of globular protein with simple polyphenols. *J. Agr. Food Chem. 26*:214-216.

68. Ney, K. H. (1979). Bitterness of peptides: Amino acid composition and chain length, in *Food Taste Chemistry* (J. C. Boudreau, ed.), American Chemical Society, Washington, D.C., pp. 149-173.

69. Ney, K. H. (1979). Bitterness of lipids. *Fette, Seifen, Anstrichm. 81*:467-469.

70. Oediger, H. J., and A. Schulze (1980). An economical synthesis of pepper alkaloids, in *Frangrance and Flavor Substances: Proceedings of Second International Haarmann and Reimer Symposium* (R. Croteau, ed.), D&PS Verlag, Pattensen, W. Germany, pp. 83-92.

71. Oh, H. I., J. E. Hoff, G. S. Armstrong, and L. A. Hoff (1980). Hydrophobic interaction in tannin-protein complexes. *J. Agr. Food Chem. 28*:394-398.

72. Ohloff, G. (1981). Bifunctional unit concept in flavour chemistry, in *Flavour '81* (P. Schreier, ed.), Walter de Gruyter, Berlin, pp. 757-770.

73. Olsen, O., and H. Sorensen (1981). Recent advances in the analysis of glucosinolates. *J. Amer. Oil Chem. Soc. 58*:857-865.

74. Peacock, V. E., and M. L. Deinzer (1981). Chemistry of hop aroma in beer. *ASBC J. 39*(4):136-141.

75. Powers, J. J. (1982). Techniques of analysis of flavours, in *Food Flavours*, Part A, *Introduction* (I. D. Morton and A. J. Macleod, eds.), Elsevier Scientific, Amsterdam, pp. 121-168.

76. Russell, G. F., and J. D. Hills (1982). A micro-olfactometer for chemical sensory analyses, in *Chemistry of Foods and Beverages: Recent Developments* (G. Charalambous and G. Inglett, eds.), Academic Press, New York, pp. 101-127.

77. Sanderson, G. W., A. S. Ranadive, L. S. Eisenberg, F. J. Farrell, R. Simons, C. H. Manley, and P. Coggon (1975). Contributions of polyphenolic compounds to the taste of tea, in *Phenolic, Sulfur, and Nitrogen Compounds in Food Flavors* (G. Charalambous and I. Katz, eds.), American Chemical Society, Wasington, D.C., pp. 14-46.

78. Scanlan, R. A. (ed.) (1977). *Flavor Quality: Objective Measurement*, American Chemical Society, Washington, D.C.

79. Schiffman, S. S. (1980). Contribution of the anion to the taste quality of sodium salts, in *Biological and Behaviorial Aspects of Salt Intake* (R. H. Cagan and M. R. Kare, eds.), Academic Press, New York, pp. 99-114.

80. Schreier, P. (ed.) (1981). *Flavour '81*, Walter de Gruyter, Berlin.

81. Schultz, H. W., E. A. Day, and L. M. Libbey (eds.) (1967). *Symposium on Foods: The Chemistry and Physiology of Flavors*, AVI Publishing Co., Westport, Conn.,

82. Shallenberger, R. S. (1977). Chemical clues to the perception of sweetness, in *Sensory Properties of Foods* (G. G. Birch, J. G. Brennan, and K. J. Parker, eds.), Applied Science Publishers, London, pp. 91-100.

83. Shallenberger, R. S., and T. E. Acree (1967). Molecular theory of sweet taste. *Nature* (London) *216*:480-482.

84. Shallenberger, R. S., and G. G. Birch (1975). *Sugar Chemistry*, AVI Publishing Co., Westport, Conn., p. 113.

85. Shankaranarayana, M. L., B. Raghaven, K. O. Abraham, and C. P. Natarajan (1982). Sulphur compounds in flavours, in *Food Flavours*, Part A, *Introduction* (I. D. Morton and A. J. Macleod, eds.), Elsevier Scientific, Amsterdam, pp. 169-281.

86. Shibamoto, T. (1982). Application of fused silica capillary columns for flavor analysis, in *Chemistry of Foods and Beverages: Recent Development* (G. Charalambous and G. Inglett, eds.), Academic Press, New York, pp. 73-99.

87. Smith, A., and W. R. H. Duncan (1979). Characterization of branched-chain fatty acids from fallow deer perinephric triacylglycerols by gas chromatography mass spectrometry. *Lipids 14*:350-355.

88. Sparrer, D., and H. D. Berlitz (1975). Bitter peptide aus casein nach hydrolyse mit α-chymotrypsin and trypsin. *Z. Lebensm. Unters. Forsch. 157*:197-204.

89. Strunz, G. M. (1983). Synthetic routes to 2-hydroxy-3-methylcyclopent-2-3n-1-one and related cyclopentane-1,2-diones: A review. *J. Agr. Food Chem. 31*:185-190.

90. Swoboda, P. A. T., and K. E. Peers (1979). The formation of metallic taint by selective oxidation: The significance of octa-1,cis-5-diene-3-one, in *Progress in Flavour Research* (D. G. Land and H. E. Nursten, eds.), Applied Science Publishers, London, pp. 275-280.

91. T. Tancredi, F. Lelj, and P. Temussi (1979). Three-dimensional mapping of the bitter taste receptor site. *Chem. Senses Flavour 4*:259-265.

92. C. Tanford (1960). Contribution of hydrophobic interactions to the stability of globular conformation of proteins. *J. Amer. Chem. Soc. 84*:4240-4247.

93. Teranishi, R., R. A. Flath, and H. Sugisawa (eds.) (1981). *Flavor Research: Recent Advances*, Marcel Dekker, New York.

94. Tressl, R., D. Bahri, and K. H. Engel (1982). Formation of eight-carbon and ten-

carbon components in mushrooms (*Agaricus campestris*). *J. Agr. Food Chem. 30*: 89-93.

95. Tressl, R., F. Frendesack, and A. Reinke (1982). Changes of aroma components during storage and processing of hops and their contribution to beer flavor, in *Chemistry of Foods and Beverages: Recent Developments* (G. Charalambous and G. Inglett, eds.), Academic Press, New York, pp. 1-24.

96. Tressl, R., K. G. Grunewald, R. Silwar, and D. Bahri (1979). Chemical formation of flavour substances, in *Progress in Flavour Research* (D. G. Land and H. E. Nursten, eds.) Applied Science Publishers, London, pp. 197-213.

97. Tressl, R., M. Holzer, and M. Apetz (1975). Biogenesis of volatiles in fruit and vegetables, in *Aroma Research: Proceedings of the International Symposium on Aroma Research*, Zeist (H. Maarse and P. J. Groenen, eds.), Centre for Agricultural Publishing and Documentation, PUDOC, Wageningen, pp. 41-62.

98. Vaks, B., and A. Lifschitz (1981). Debittering of orange juice by bacteria which degrade limonin. *J. Agr. Food Chem. 29*:1258-1261.

99. Van Straten, S., F. de Vrijer, and J. C. Beauveser (eds.) (1977). *Volatile Compounds in Food*, 4th ed., Central Institute for Nutrition and Food Research, Zeist.

100. Watson, H. R. (1978). Flavor characteristics of synthetic cooling compounds, in *Flavor: Its Chemical, Behavioral, and Commercial Aspects* (C. M. Apt, ed.), Westview Press, Boulder, CO, pp. 31-50.

101. Wilson, R. A. (1975). A review of thermally produced imitation meat flavors. *J. Agr. Food Chem. 23*:1032-1037.

102. Wong, E., C. B. Johnson, and L. N. Nixon (1975). The contribution of 4-methyloctanoic (hircinoic) acid to mutton and goat meat flavor. *N. Z. J. Agr. Res. 18*: 261-266.

103. Wong, E., L. N. Nixon, and B. C. Johnson (1975). Volatile medium chain fatty acids and mutton flavor. *J. Agr. Food Chem. 23*:495-498.

104. Woo, A. H., and R. C. Lindsay (1982). Anionic detergent contamination detected in soapy-flavored butters. *J. Food Protection 45*:1232-1235.

105. Yamaguchi, S. (1979). The umami taste, in *Food Taste Chemistry* (J. C. Boudreau, ed.), American Chemical Society, Washington, D.C., pp. 33-51.

106. Yamanishi, Y. (1981). Tea, coffee, cocoa, and other beverages, in *Flavor Research: Recent Advances* (R. Teranishi, R. A. Flath, and H. Sugisawa, eds.), Marcel Dekker, New York, pp. 231-304.

107. Yashimoto, K., K. Iwami, and H. Mitsuda (1971). A new sulfur-containing peptide from *Lentinus edodes* acting as a precursor for lenthionine. *Agr. Biol. Chem. 35*: 2059-2069.

BIBLIOGRAPHY

Amerine, M. A., R. M. Pangborn, and E. B. Roessler (1965). *Principles of Sensory Evaluation of Food*, Academic Press, New York.

ASTM Manual on Sensory Testing Methods (STP 434) (1969). American Society for Testing Materials, Philadelphia.

Beets, M. G. J. (1978). *Structure-Activity Relationships in Human Chemoreception*, Applied Science Publishers, London.

Fazzalari, F. A. (ed.) (1978). *Compilation of Odor and Taste Threshold Values Data*, American Society for Testing Materials, Philadelphia.

Furia, T. E., and N. Bellanca (eds.) (1975). *Fenaroli's Handbook of Flavor Ingredients*, Chemical Rubber Co., Cleveland.

Heath, H. B. *Source Book of Flavors*, AVI Publishing Co., Westport, Conn.

Land, D. G., and H. E. Nurston (eds.) (1979). *Progress in Flavour Research*, Applied Science Publishers, London.

Morton, I. D., and A. J. Macleod (eds.) (1982). *Food Flavours*, Part A, *Introduction*, Elsevier Scientific, New York.

Schultz, H. W., E. A. Day, and L. M. Libbey (eds.) (1967). *Symposium on Foods: The Chemistry and Physiology of Flavors*, AVI Publishing Co., Westport, Conn.

Teranishi, R., R. A. Flath, and H. Sugisawa (eds.) (1981). *Flavor Research. Recent Advances*, Marcel Dekker, New York.

van Straten, S., F. de Vrijer, and J. C. de Beauveser (eds.) (1977). *Volatile Compounds in Food*, 4th ed., Central Institute for Nutrition and Food Research, Zeist, The Netherlands.

10

FOOD ADDITIVES

Robert C. Lindsay   University of Wisconsin-Madison, Madison, Wisconsin

	I.	Introduction	630
	II.	Acids	630
		A. General Attributes	630
		B. Chemical Leavening Systems	632
	III.	Bases	636
	IV.	Buffer Systems and Salts	638
		A. Buffers and pH Control in Foods	638
		B. Salts in Processed Dairy Foods	640
		C. Phosphates and Water Binding in Animal Tissues	640
	V.	Chelating Agents (Sequestrants)	641
	VI.	Antioxidants	643
	VII.	Antimicrobial Agents	644
		A. Sulfites and Sulfur Dioxide	644
		B. Nitrite and Nitrate Salts	645
		C. Sorbic Acid	646
		D. Natamycin	648
		E. Glyceryl Esters	648
		F. Propionic Acid	649
		G. Acetic Acid	649
		H. Benzoic Acid	649
		I. p-Hydroxybenzoate Alkyl Esters	650
		J. Epoxides	651
		K. Antibiotics	652
		L. Diethyl Pyrocarbonate	652
	VIII.	Nonnutritive and Low-Calorie Sweeteners	653
		A. Cyclamates	653
		B. Saccharin	654

C.   Aspartame   654
D.   Acesulfame K   656
E.   Other Nonnutritive or Low-Calorie Sweeteners   656
IX.   Stabilizers and Thickeners   657
X.   Masticatory Substances   658
XI.   Polyhydric Alcohol Texturizers   658
XII.   Firming Texturizers   660
XIII.   Appearance Control—Clarifying Agents   661
XIV.   Flour Bleaching Agents and Bread Improvers   663
XV.   Anticaking Agents   665
XVI.   Gases and Propellants   666
A.   Protection from Oxygen   666
B.   Carbonation   666
C.   Propellants   667
XVII.   Tracers   681
XVIII.   Summary   681
References   681
Bibliography   687

I. INTRODUCTION

Many substances are incorporated into foods for functional purposes, and in many cases these ingredients also can be found occurring naturally in some food. However, when they are used in processed foods, these chemicals have become known as "food additives." From a regulatory standpoint, each of the food additives must provide some useful and acceptable function or attribute to justify its usage. Generally, improved keeping quality, enhanced nutritional value, functional property provision and improvement, processing facilitation, and enhanced consumer acceptance are considered acceptable functions for food additives. The use of food additives to conceal damage or spoilage to foods or to deceive consumers is expressly forbidden by regulations governing the use of these substances in foods. Additionally, food additive usages are discouraged where similar effects can be obtained by economical, good manufacturing practices.

Since natural counterparts exist for many food additives, and frequently they are derived commercially from natural sources, further discussions about the chemistry of this group of substances can be found in appropriate chapters of this book. For example, natural substances sometimes used as food additives are discussed in Chap. 4 (antioxidants), Chap. 7 (vitamins and minerals), Chap. 8 (colorants), and Chap. 9 (flavors). This chapter focuses on the role of both natural and synthetic substances added to foods and provides an integrating view of their functionalities. Emphasis is given to substances not covered elsewhere in this book.

II. ACIDS

A. General Attributes

Both organic and inorganic acids occur extensively in natural systems where they function in a variety of roles ranging from intermediary metabolites to components of buffer systems. Acids are added for numerous purposes in food and food processing where they provide the benefits of many of their natural actions. One of the most important functions of acids in foods is participation in buffering systems, and this aspect is discussed in a following section. The use of acids and acid salts in chemical leavening systems, the role of specific acidic microbial inhibitors (e.g., sorbic acid or benzoic acid) in food preservation, and the function of acids as chelating agents are also discussed in subsequent sections of this chapter. Acids are important in the setting of pectin gels (Chap. 3), they serve as defoaming agents and emulsifiers, and they induce coagulation of milk proteins (Chaps. 5 and 13) in the production of cheese and cultured dairy products, such as sour cream. In natural culturing processes lactic acid (CH_3–CHOH–COOH) produced by streptococci and lactobacilli causes coagulation by lowering the pH to near the isoelectric point of casein. Cheeses can be produced by adding rennet and acidulants, such as citric acid and hydrochloric acids, to cold milk (4-8°C). Subsequent warming of the milk (to 35°C) produces a uniform gel structure (81,92,105). Addition of acid to warm milk results in a protein precipitate rather than a gel. δ-Gluconolactone also can be used for slow acid production in cultured dairy products (29,84) and chemical leavening systems because it slowly hydrolyzes in aqueous systems to form gluconic acid (Fig. 1). Dehydration of lactic acid yields lactide, a cyclic dilactone (Fig. 2), that also can be used as a slow-release acid in aqueous systems. The dehydration reaction occurs under conditions of low water activity and elevated temperature. Introduction of lactide into foods with high water activity causes a reversal of the process with the production of 2 mol lactic acid.

Such acids as citric are added to some moderately acid fruits and vegetables to lower the pH to a value below 4.5. In canned foods this permits sterilization to be achieved under less severe thermal conditions than is necessary for less acid products and has the added advantage of precluding the growth of hazardous microorganisms (i.e., *Clostridium botulinum*).

Acids, such as potassium acid tartrate, are employed in the manufacture of fondant and fudge to induce limited hydrolysis (inversion) of sucrose (Chap. 3). Inversion of sucrose yields fructose and glucose, which improve texture through inhibition of excessive growth of sucrose crystals. Monosaccharides inhibit crystallization by contributing to the complexity of the syrup and by lowering its equilibrium relative humidity.

One of the most important contributions of acids to foods is their ability to produce a sour or tart taste (48). Acids also have the ability to modify and intensify the taste perception of other flavoring agents. The hydrogen ion or hydronium ion (H_3O^+) is involved in the generation of the sour taste response (Chap. 9). Furthermore, short-chain free fatty acids (C_2-C_{12}) contribute significantly to the aroma of foods. For example, butyric acid at relatively high concentrations contributes strongly to the characteristic

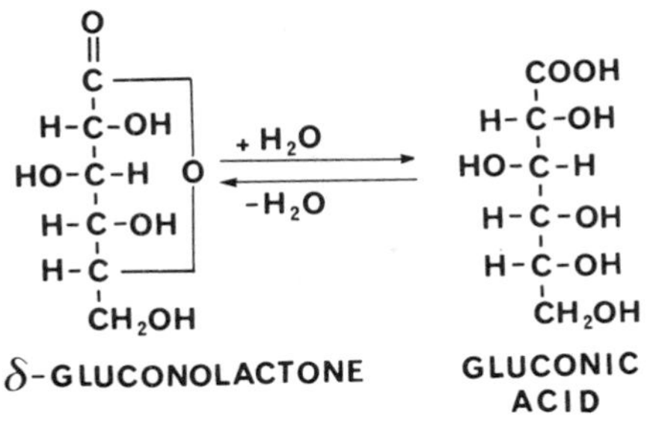

Figure 1   Formation of gluconic acid from the hydrolysis of δ-gluconolactone.

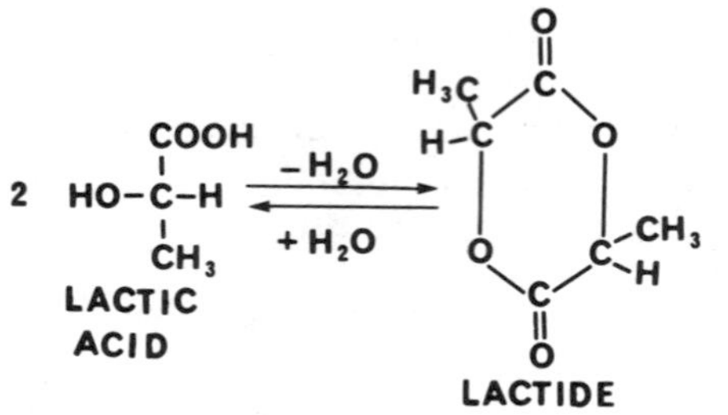

Figure 2 Equilibrium reaction showing formation of lactic acid from hydrolysis of lactide.

flavor of hydrolytic rancidity, but at lower concentrations contributes to the typical flavor of such products as cheese and butter.

Numerous organic acids are available for food applications (49). Some of the more commonly used acids are acetic (CH_3COOH), lactic ($CH_3-CHOH-COOH$), citric ($HOOC-CH_2-COH(COOH)-CH_2-COOH$), malic ($HOOC-CHOH-CH_2-COOH$), fumaric ($HOOC-CH=CH-COOH$), succinic ($HOOC-CH_2-CH_2-COOH$), and tartaric ($HOOC-CHOH-CHOH-COOH$). Phosphoric acid ($H_3PO_4$) is the only inorganic acid extensively employed as a food acidulant. Phosphoric acid is an important acidulant in flavored carbonated beverages, particularly in colas and root beer. The other mineral acids (e.g., HCl and H_2SO_4) are usually too highly dissociated for food applications, and their use may lead to problems with quality attributes of foods.

B. Chemical Leavening Systems

Chemical leavening systems are composed of compounds that react to release gas in a dough or batter under appropriate conditions of moisture and temperature. During baking, this gas release, along with expansion of entrapped air and moisture vapor, imparts a characteristic porous, cellular structure to finished goods. Chemical leavening systems are found in self-rising flours, prepared baking mixes, household and commercial baking powders, and refrigerated dough products (56,57).

Carbon dioxide is the only gas generated from currently used chemical leavening systems, and it is derived from a carbonate or bicarbonate salt. The most common leavening salt is sodium bicarbonate ($NaHCO_3$), although ammonium carbonate [$(NH_4)_2CO_3$] and bicarbonate (NH_4HCO_3) are sometimes used in cookies. Both the ammonium salts decompose at baking temperatures, and thus do not require, as does sodium bicarbonate, an added leavening acid for functionality. Potassium bicarbonate ($KHCO_3$) has been employed as a component of leavening systems in sodium-free diets, but its application is somewhat limited because of its hygroscopic nature and slightly bitter flavor (90).

Sodium bicarbonate is quite soluble in water (619 g per 100 ml) and ionizes completely.

$$NaHCO_3 \rightleftharpoons Na^+ + HCO_3^- \qquad (1)$$

$$HCO_3^- + H_2O \rightleftharpoons H_2CO_3 + OH^- \qquad (2)$$

$$HCO_3^- \rightleftharpoons CO_3^{2-} + H^+ \qquad (3)$$

These reactions, of course, apply only to simple water solutions. In dough systems the ionic distribution becomes much more complex since proteins and other naturally occurring ionic species are available to participate in the reactions. In the presence of hydrogen ions provided mainly by leavening acids, and to some extent by the dough, sodium bicarbonate reacts to release carbon dioxide. The proper balance of acid and sodium bicarbonate

$$R-O^-, H^+ + NaHCO_3 \longrightarrow R-O^-, Na^+ + H_2O + CO_2 \qquad (4)$$

is essential because excess sodium bicarbonate imparts a soapy taste to bakery products; an excess of acid leads to tartness and sometimes bitterness. The neutralizing power of leavening acids is not uniform, and the relative activity of an acid is given by its neutralizing value. The neutralizing value of an acid is determined by calculating the parts by weight of sodium bicarbonate that will neutralize 100 parts by weight of the leavening acid (117). However, in the presence of natural flour ingredients, the amount of leavening acid required to give neutrality or any other desired pH in a baked product may be quite different from the theoretical amount determined for a simple system. Still, neutralizing values are useful in determining initial formulations for leavening systems. Residual salts from a properly balanced leavening process help stabilize the pH of finished products.

Leavening acids are often not easily recognized as acids in the usual sense, yet they must provide hydrogen ions to release carbon dioxide. The phosphates and potassium acid tartrate are metal salts of partially neutralized acids; sodium aluminum sulfate reacts with water to yield sulfuric acid.

$$Na_2SO_4 \cdot Al_2(SO_4)_3 + 6H_2O \longrightarrow Na_2SO_4 + 2Al(OH)_3 + 3H_2SO_4 \qquad (5)$$

As mentioned earlier, δ-gluconolactone is an intramolecular ester (or lactone) that hydrolyzes slowly in aqueous systems to yield gluconic acid.

Leavening acids generally exhibit limited water solubility at room temperature, but some are less soluble than others. This difference in solubility or availability accounts for the initial rate of carbon dioxide release at room temperature and is the basis for classifying leavening acids according to speed. For example, if the compound is moderately soluble, carbon dioxide is rapidly evolved and the acid is referred to as fast acting. Conversely, if the acid dissolves slowly, it is a slow-acting leavening acid. Leavening acids usually release a portion of the carbon dioxide prior to baking and the remainder under the elevated temperatures of the baking process.

General patterns of carbon dioxide release at $27^\circ C$ for fast-acting monocalcium phosphate monohydrate $[Ca(HPO_4)_2 \cdot H_2O]$ and slow-acting 1-3-8 sodium aluminum phosphate $[NaH_{14}Al_3(PO_4)_8 \cdot 4H_2O]$ are shown in Fig. 3. Over 60% of the carbon dioxide is released very quickly from the more soluble monocalcium phosphate monohydrate; only 20% of the potential carbon dioxide is released from the slow-acting 1-3-8 sodium aluminum phosphate during a 10 min reaction period. Because of a hydrated alumina coating, the latter leavening acid reacts to only a small extent until activated by heat. Also shown in Fig. 3 is the low-temperature release pattern of carbon dioxide from coated anhydrous monocalcium phosphate $[Ca(HPO_4)_2]$. The crystals of this leavening acid were coated with compounds of slightly soluble alkali metal phosphates. The gradual release of carbon dioxide over the 10 min reaction period corresponds to the time required for water to penetrate the coating. This behavior is very desirable in some products that encounter a delay prior to baking.

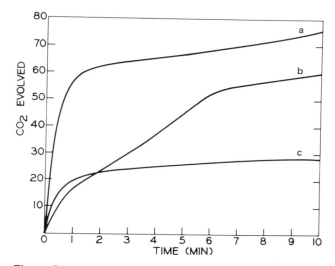

Figure 3 Carbon dioxide production at 27°C from the reaction of $NaHCO_3$ with (a) monocalcium phosphate · H_2O, (b) coated anhydrous monocalcium phosphate, and (c) 1-3-8 sodium aluminum phosphate. [Data from Stahl and Ellinger (117).]

The release of the remainder of the carbon dioxide from leavening systems during baking provides the final modifying action on texture. In most leavening systems the rate at which carbon dioxide is released greatly accelerates as the temperature is elevated. The effect of elevated temperatures on the release rate of carbon dioxide from slow-acting sodium acid pyrophosphate ($Na_2H_2P_2O_7$) is presented in Fig. 4. Even a slight increase in temperature (from 27 to 30°C) noticeably accelerates gas production. Temperatures near 60°C cause a complete release of carbon dioxide within 1 min. Some leavening acids are less sensitive to high temperatures and do not exhibit vigorous activity until temperatures near the maximum baking temperature are obtained. Dicalcium phosphate ($CaHPO_4$) is unreactive at room temperature because it forms a slightly alkaline solution at this temperature. However, upon heating above approximately 60°C, hydrogen ions are released, thereby activating the leavening process. This slow action confines its use to products requiring long baking times, such as some types of cakes. Formulations of leavening acids employing one or more acidic components are common, and systems are often tailored for specific dough or batter applications.

Leavening acids currently employed include potassium acid tartrate, sodium aluminum sulfate, δ-gluconolactone, and ortho- and pyrophosphates. The phosphates include calcium phosphate, sodium aluminum phosphate, and sodium acid pyrophosphate. Some general properties of commonly used leavening acids are given in Table 1. It must be remembered that these are only examples, and that an extensive technology has developed for modification and control of the phosphate leavening acids (117).

Baking powders account for a large part of the chemical leaveners used both in the home and in bakeries. These preparations include sodium bicarbonate, suitable leavening acids, and starch and other extenders. Federal standards for baking powder require that the formula must yield at least 12% by weight of available carbon dioxide, and most contain 26-30% by weight of sodium bicarbonate (90). Traditional baking powders of the

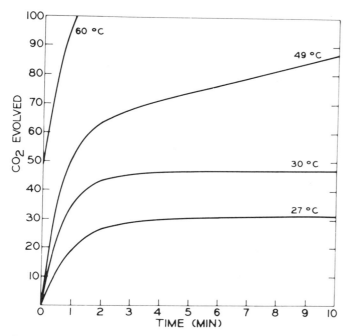

Figure 4 Effect of temperature on the rate of carbon dioxide evolution from the reaction of $NaHCO_3$ and slow-speed acid pyrophosphate. (Reprinted from Ref. 117, p. 201, courtesy of the Avi Publishing Co., Inc.)

Table 1 Some Properties of Common Leavening Acids

Acid	Formula	Neutralizing value[a]	Relative reaction rate at room temperature[b]
Sodium aluminum sulfate	$Na_2SO_4 \cdot Al_2(SO_4)_3$	100	Slow
Dicalcium phosphate dihydrate	$CaHPO_4 \cdot 2H_2O$	33	None
Monocalcium phosphate monohydrate	$Ca(HPO_4)_2 \cdot H_2O$	80	Fast
1-3-8 Sodium aluminum phosphate	$NaH_{14}Al_3(PO_4)_8 \cdot 4H_2O$	100	Slow
Sodium acid pyrophosphate (slow type)	$Na_2H_2P_2O_7$	72	Slow
Potassium acid tartrate	$KHC_4H_4O_6$	50	Medium
δ-Gluconolactone	$C_6H_{10}O_6$	55	Slow

[a]In simple model systems; parts by weight of $NaHCO_3$ that will neutralize 100 parts by weight of the leavening acid.
[b]Rate of CO_2 evolution in the presence of $NaHCO_3$.
Source: From Stahl and Ellinger (117).

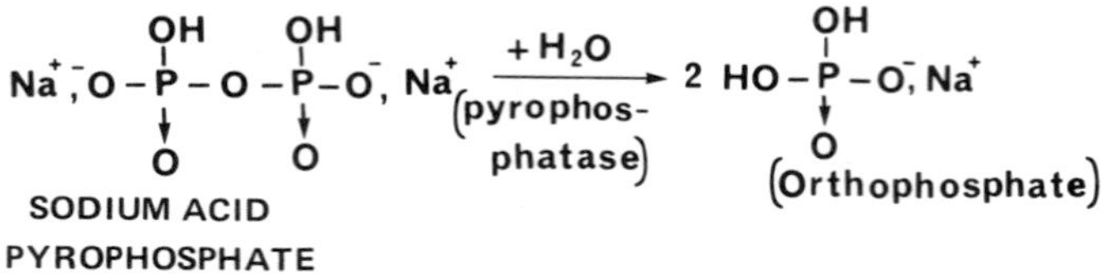

Figure 5   Enzymic hydrolysis of sodium acid pyrophosphate.

potassium acid tartrate type have been largely replaced by double-acting preparations. In addition to $NaHCO_3$ and starch these baking powders usually contain monocalcium phosphate $[Ca(HPO_4)_2 \cdot H_2O]$, which provides rapid action during the mixing stage, and sodium aluminum sulfate $[Na_2SO_4 \cdot Al_2(SO_4)_3]$, which does not react appreciably until the temperature increases during baking.

The increase in convenience foods has stimulated sales of prepared baking mixes and refrigerated dough products. In white and yellow cake mixes, the most widely used blend of leavening acids contains anhydrous monocalcium phosphate $[Ca(HPO_4)_2]$ and sodium aluminum phosphate $[NaH_{14}Al_3(PO_4)_8 \cdot 4H_2O]$; chocolate cake mixes usually contain anhydrous monocalcium phosphate and sodium acid pyrophosphate $(Na_2H_2P_2O_7)$ (117). Typical blends of acids contain 10-20% fast-acting anhydrous monophosphate and 80-90% of the slower acting sodium aluminum phosphate or sodium acid pyrophosphate compounds. The leavening acids in prepared biscuit mixes usually consist of 30-50% anhydrous monocalcium phosphate and 50-70% sodium aluminum phosphate or sodium acid pyrophosphate. The earliest self-rising flours and corn meal mixes contained monoclacium phosphate monohydrate $[Ca(HPO_4)_2 \cdot H_2O]$, but coated anhydrous monocalcium phosphate and sodium aluminum phosphate are in common use today (117).

Refrigerated doughs for biscuit and roll products require limited initial carbon dioxide release during preparation and packaging and considerable gas release during baking. Formulations for biscuits usually contain from 1.0 to 1.5% sodium bicarbonate and 1.4-2.0% slow-acting leavening acids, such as coated monocalcium phosphate and sodium acid pyrophosphate, based on total dough weight. The pyrophosphates are useful in dough because they can be manufactured with a wide range of reactivities. For example, pyrophosphatase in flour is capable of hydrolyzing sodium acid pyrophosphate to orthophosphate (Fig. 5), and the reaction of sodium bicarbonate and pyrophosphate yields some trisodium monohydrogen pyrophosphate, which also can be hydrolyzed to orthophosphates. This enzymic action leads to gas production that assists in sealing packages of refrigerated dough, but it can also lead to formation of large crystals of orthophosphates that may be mistaken for broken glass by the consumer.

III. BASES

Basic or alkaline substances are used in a variety of applications in foods and food processing (1). Although the majority of applications involve buffering and pH adjustments, other functions include carbon dioxide evolution, enhancement of color and flavor, solubilization of proteins, and chemical peeling. The role of carbonate and bicarbonate salts in carbon dioxide production during baking has been discussed previously.

Alkali treatments are imposed on several food products for the purpose of color and flavor improvement. Ripe olives are treated with solutions of sodium hydroxide (0.25-2.0%) to aid in the removal of the bitter principal and to develop a darker color (89). Pretzels are dipped in a solution of 1.25% sodium hydroxide at 87-88°C (186°-190°F) prior to baking to alter proteins and starch so that the surface becomes smooth and develops a deep brown color during baking (90). It is believed that the NaOH treatment used to prepare hominy and tortilla dough destroys disulfide bonds, which are base labile. Soy proteins are solubilized through alkali processing, and concern has been expressed about alkaline-induced racemization of amino acids (Chap. 5) and losses of other nutrients (6). Small amounts of sodium bicarbonate are used in the manufacture of peanut brittle to enhance sugar-amino browning and to provide, through release of carbon dioxide, a somewhat porous structure. Bases, usually sodium bicarbonate, are also used in cocoa processing for the production of dark (Dutch) chocolate. The elevated pH enhances sugar-amino browning reactions and polymerization of flavonoids (Chap. 8) (13), resulting in a smoother, less acid and less bitter chocolate flavor, a darker color, and slightly improved solubility (22).

Food systems sometimes require adjustment to higher pH values to achieve more stable or more desirable characteristics. For example, alkaline salts, such as disodium phosphate, trisodium phosphate, and trisodium citrate, are used in the preparation of processed cheese (1.5-3%) to increase the pH (from 5.7 to 6.3) and to effect protein (casein) dispersion. This salt-protein interaction improves the emulsifying and water-binding capabilities of the cheese proteins (111), apparently because the salts bind the calcium components of the casein micelles, forming either insoluble phosphates or soluble chelates (citrate).

Popular instant milk-gel puddings are prepared by combining dry mixes containing pregelatinized starch with cold milk and allowing them to stand for a short time at refrigerator temperatures. Alkaline salts, such as tetrasodium pyrophosphate ($Na_4P_2O_7$) and disodium phosphate (Na_2HPO_4), in the presence of calcium ions in milk, cause milk proteins to gel in combination with the pregelatinized starch. The optimum pH for acceptable puddings falls between 7.5 and 8.0. Although some of the necessary alkalinity is contributed by alkaline phosphate salts, other alkalizing agents are often added (33).

The addition of phosphates and citrates changes the salt balance in fluid milk by forming complexes with calcium and magnesium ions from casein. The mechanism is incompletely understood, but, depending on the type and concentration of salt added, the milk-protein system can undergo stabilization, gelation, or destabilization (94).

Alkaline agents are used to neutralize excess acid in the production of such foods as cultured butter. Before churning, the cream is fermented by lactic acid bacteria so that it contains about 0.75% titratable acidity expressed as lactic acid (67). Alkalis are then added to achieve a titratable acidity of approximately 0.25%. The reduction in acidity improves churning efficiency and retards the development of oxidative off-flavors. Several materials, including sodium bicarbonate ($NaHCO_3$), sodium carbonate (Na_2CO_3), magnesium carbonate ($MgCO_3$), magnesium oxide (MgO), calcium hydroxide [$Ca(OH)_2$], and sodium hydroxide ($NaOH$) are utilized alone or in combination as neutralizers for foods. Solubility, foaming as a result of carbon dioxide release, and strength of the base influence the selection of the alkaline agent. The use of alkaline agents or bases in excessive amounts leads to soapy or neutralizer flavors, especially when substantial quantities of free fatty acids are present.

Strong bases are employed for peeling various fruits and vegetables. Exposure of the product to hot solutions 60-82°C (140-180°F) of sodium hydroxide (about 3%), with subsequent mild abrasion, effects peel removal with substantial reductions in plant waste-water as compared with other conventional peeling techniques. Differential solubilization of cell and tissue constituents (pectic substances in the middle lamella are particularly soluble) provides the basis for caustic peeling processes (72).

IV. BUFFER SYSTEMS AND SALTS

A. Buffers and pH Control in Foods

Since most foods are complex materials of biological origin, they contain many substances that can participate in pH control and buffering systems. Included are proteins, organic acids, and weak inorganic acid-phosphate salts. Lactic acid and phosphate salts, along with proteins, are important for pH control in animal tissue; polycarboxylic acids, phosphate salts, and proteins are important in plant tissues. The buffering effects of amino acids and proteins and the influence of pH and salts on their functionalities are discussed in Chap. 5. In plants, buffering systems containing citric acid (lemons, tomatoes, and rhubarb), malic acid (apples, tomatoes, and lettuce), oxalic acid (rhubarb and lettuce), and tartaric acid (grapes and pineapples) are common, and they usually function in conjunction with phosphate salts in maintaining pH control. Milk acts as a complex buffer because of its content of carbon dioxide, proteins, phosphate, citrate and several other minor constituents (73).

In situations in which the pH must be altered, it is usually desirable to stabilize the pH at the desired level through a buffer system. This is accomplished naturally when lactic acid is produced in cheese and pickle fermentations. Also, in some instances in which substantial amounts of acids are used in foods and beverages, it is desirable to reduce the sharpness of acid tastes and obtain smoother product flavors without inducing neutralization flavors. This usually can be accomplished by establishing a buffer system in which the salt of a weak organic acid is dominant. The common ion effect is the basis for obtaining pH control in these systems, and the system develops when the added salt contains an ion that is already present in an existing weak acid. The added salt immediately ionizes, resulting in repressed ionization of the acid with reduced acidity and a more stable pH. The effectiveness of a buffer depends on the concentration of the buffering substances. Since there is a pool of undissociated acid and dissociated salt, buffers resist changes in pH. For example, relatively large additions of a strong acid, such as hydrochloric acid, to an acetic acid-sodium acetate system, causes hydrogen ions to react with the acetate-ion pool to increase the concentration of slightly ionized acetic acid, and the pH remains relatively stable. In a similar manner, an addition of sodium hydroxide causes hydroxyl ions to react with hydrogen ions to form undissociated water molecules.

Titration of buffered systems and resulting titration curves (i.e., pH versus volume of base added) reveal their resistance to pH change. If a weak acid buffer is titrated with a base, there is a gradual but steady increase in the pH as the system approaches neutralization; that is, the change in pH per milliliter added base is small. Weak acids are only slightly dissociated at the beginning of the titration. However, the addition of hydroxyl ions shifts the equilibrium to the dissociated species and eventually the buffering capacity is overcome.

In general, for an acid (HA) in equilibrium with ions (H^+) and (A^-), the equilibrium as in Eq. (6).

$$HA \rightleftharpoons H^+ + A^- \tag{6}$$

$$K_a^1 = \frac{[H^+][A^-]}{[HA]} \tag{7}$$

The constant K_a^1 is the apparent dissociation constant and is characteristic of the particular acid. The apparent dissociation constant K_a^1 becomes equal to the hydrogen ion concentration $[H^+]$ when the anion concentration $[A^-]$ becomes equal to the concentration of undissociated acid $[HA]$. This situation gives rise to an inflection point on a titration curve, and the pH corresponding to this point is referred to as the pK_a^1 of the acid. Therefore, for a weak acid, pH is equal to pK_a^1 when the concentrations of the acid and conjugate base are equal:

$$pH = pK_a^1 = -\log[H^+] \tag{8}$$

A convenient method for calculating the approximate pH of a buffer mixture, given the pK_a^1 of the acid, is provided by Eq. (11). This equation is arrived at by first solving Eq. (7) for $[H^+]$ to yield Eq. (9). Since the salt in solution is almost completely dissociated, it is assumed equal to the concentration

$$[H^+] = K_a^1 \frac{[HA]}{[A^-]}$$

$$= K_a^1 \frac{[acid]}{[salt]} \tag{9}$$

of the conjugate base $[A^-]$. The negative logarithms of the terms yield Eq. (10). By substituting pH for $-\log[H^+]$ and pK_a^1 for $-\log K_a^1$, Eq. (11) is obtained. The pH of a buffer system derived from any weak acid that dissociates to H^+ and A^- can be calculated by using Eq. (11).

$$-\log[H^+] = -\log K_a^1 - \log \frac{[acid]}{[salt]} \tag{10}$$

$$pH = pK_a^1 + \log \frac{[salt]}{[acid]}$$

$$= pK_a^1 + \log \frac{[A^-]}{[HA]} \tag{11}$$

In calculating the pH values of buffer solutions, it is important to recognize that the apparent dissociation constant K_a^1 differs from K_a, the true dissociation constant. However, for any buffer, the value of K_a^1 remains constant as the pH is varied, provided the total ionic strength of the solution remains unchanged (134).

The sodium salts of gluconic, acetic, citric, and phosphoric acids are commonly used for pH control and tartness modification in the food industry. The citrates are usually preferred over phosphates for tartness modification since they yield smoother sour flavors (48). When low-sodium products are required, potassium buffer salts may be substituted for sodium salts. In general, calcium salts are not used because of their limited solubilities and incompatibilities with other components in the system. The effective buffering ranges for combinations of common acids and salts are pH 2.1-4.7 for citric acid-sodium citrate, pH 3.6-5.6 for acetic acid-sodium acetate, and pH 2.0-3.0, 5.5-7.5, and 10-12, respectively, for the three ortho- and pyrophosphate anions.

B. Salts in Processed Dairy Foods

Salts are used extensively in processed cheeses and imitation cheeses to promote a uniform, smooth texture. These additives are sometimes referred to as emulsifying salts because of their ability to aid in dispersion of fat. Although the emulsifying mechanism remains somewhat less than fully defined, anions from the salts when added to processed cheese combine with and remove calcium from the *para*-casein complex and this causes rearrangment and exposure of both polar and nonpolar regions of the cheese proteins. It is also believed that the anions of these salts participate in ionic bridges between protein molecules, and thereby provide a stabilized matrix that entraps the fat in processed cheese (111). Salts used for cheese processing include mono-, di-, and trisodium phosphate, dipotassium phosphate, sodium hexametaphosphate, sodium acid pyrophosphate, tetrasodium pyrophosphate, sodium aluminum phosphate, trisodium citrate, tripotassium citrate, sodium tartrate, and sodium potassium tartrate.

The addition of certain phosphates, such as trisodium phosphate, to evaporated milk prevents separation of the butterfat and aqueous phases. The amount required varies with the season of the year and the source of milk. Concentrated milk sterilized by a high-temperature-short-time method frequently gels upon storage. The addition of polyphosphates, such as sodium hexametaphosphate and sodium tripolyphosphate, prevents gel formation through a protein denaturation and solubilization mechanism that involves complexing of calcium and magnesium by phosphates (94).

C. Phosphates and Water Binding in Animal Tissues

The addition of appropriate phosphates increases the water-holding capacity of raw and cooked meats (58), and these phosphates are used in the production of sausages, in the curing of ham and to decrease drip losses in poultry and seafoods. Sodium tripolyphosphate ($Na_5P_3P_{10}$) is the phosphate most commonly added to processed meat, poultry, and seafoods. It is often used in blends with sodium hexametaphosphate [$(NaPO_3)_n$, n = 10-15] to increase tolerance to calcium ions that exist in brines used in meat curing. Ortho- and pyrophosphates often precipitate if used in brines containing substantial amount of calcium.

The mechanism by which alkaline phosphates and polyphosphates enhance meat hydration is not clearly understood despite extensive studies. The action may involve the influence of pH changes (Chap. 5), effects of ionic strength, and specific interactions of phosphate anions with divalent cations and myofibrillar proteins (5). Many believe that calcium complexing and a resulting loosening of the tissue structure is a major function of polyphosphates. It is also believed that binding of polyphosphate anions to proteins and simultaneous cleavage of cross-linkages between actin and myosin results in increased electrostatic repulsion between peptide chains and a swelling of the muscle system. If exterior water is available, it then can be taken up in an immobilized state within the loosened protein network. Further, because the ionic strength has been increased, the interaction between proteins is perhaps reduced to a point at which part of the myofibrillar proteins form a colloidal solution. In comminuted meat products, such as bologna and sausage, the addition of sodium chloride (2.5-4.0%) and polyphosphate (0.35-0.5%) contributes to a more stable emulsion, and after cooking to a cohesive network of coagulated proteins.

If the phosphate-induced solubilization occurs primarily on the surface of tissues, as is the case with polyphosphate-dipped (6-12% solution with 0.35-0.5% retention) fish fillets, shellfish, and poultry, a layer of coagulated protein is formed during cooking and this improves moisture retention (86).

V. CHELATING AGENTS (SEQUESTRANTS)

Chelating agents or sequestrants play a significant role in food stabilization through reactions with metallic and alkaline earth ions to form complexes that alter the properties of the ions and their effects in foods. Many of the chelating agents employed in the food industry are natural substances, such as polycarboxylic acids (citric, malic, tartaric, oxalic, and succinic), polyphosphoric acids (adenosine triphosphate and pyrophosphate), and macromolecules (porphyrins and proteins). Many metals exist in a naturally chelated state. Examples include magnesium in chlorophyll; copper, iron, zinc, and manganese in various enzymes; iron in proteins, such as ferritin; and iron in the porphyrin ring of myoglobin and hemoglobin. When these ions are released by hydrolytic or other degradative reactions, they are free to participate in reactions that lead to discoloration, oxidative rancidity, turbidity, and flavor changes in foods. Chelating agents are sometimes added to form complexes with these metal ions and thereby stabilize the foods.

Any molecule or ion with an unshared electron pair can coordinate or form complexes with metal ions. Therefore, compounds containing two or more functional groups, such as $-OH$, $-SH$, $-COOH$, $-PO_3H_2$, $C=O$, $-NR_2$, $-S-$, and $-O-$, in proper geometric relation to each other, can chelate metals in a favorable physical environment. Citric acid and its derivatives, various phosphates, and salts of ethylenediaminetetraacetic acid (EDTA) are the most popular chelating agents used in foods. Usually, the ability of a chelating agent (ligand) to form a five- or six-membered ring with a metal is necessary for stable chelation. For example, EDTA forms chelates of high stability with calcium because of an initial coordination involving the electron pairs of its nitrogen atoms and the free electron pairs of the anionic oxygen atoms of two of the four carboxyl groups (Fig. 6). The spatial configuration of the calcium-EDTA complex is such that it allows additional coordination of the calcium with the free electron pairs of the anionic oxygen atoms of the remaining two carboxyl groups, and this results in an extremely stable complex utilizing all six electron donor groups.

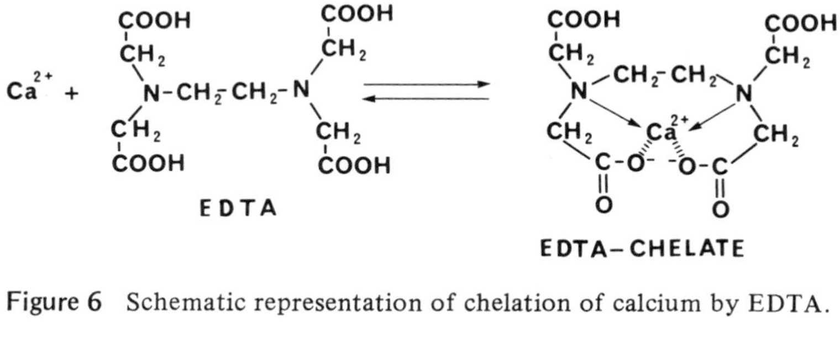

Figure 6 Schematic representation of chelation of calcium by EDTA.

In addition to steric and electronic considerations, such factors as pH influence the formation of strong metal chelates. The nonionized carboxylic acid group is not an efficient donor group, but the carboxylate ion functions effectively. Judicious raising of the pH allows dissociation of the carboxyl group and enhances chelating efficiency. In some instances hydroxyl ions compete for metal ions and reduce the effectiveness of chelating agents. Metal ions exist in solution as hydrated complexes (metal · H_2O^{M+}), and the rate at which these complexes are disrupted influences the rate at which they can be complexed with chelating agents. The relative attraction of chelating agents for different ions can be determined from stability or equilibrium constants (K = [metal · chelating agent]/[metal] [chelating agent]). For example, for calcium the stability constant (expressed as log K) is 10.7 with EDTA, 5.0 with pyrophosphate, and 3.5 with citric acid (46). As the stability constant K increases, more of the metal is complexed, leaving less metal in cation form (i.e., the metal in the complex is more tightly bound).

Chelating agents are not antioxidants in the sense that they arrest oxidation by chain termination or serve as oxygen scavengers. They are, however, valuable antioxidant synergists since they remove metal ions that catalyze oxidation (Chap. 4). When selecting a chelating agent for an antioxidant synergist role, its solubility must be considered. Citric acid and citrate esters (20-200 ppm) in propylene glycol solution are solubilized by fats and oils and thus are effective synergists in all-lipid systems. On the other hand, Na_2EDTA and Na_2Ca-EDTA dissolve to only a limited extent and are not effective in pure fat systems. The EDTA salts (to 500 ppm), however, are very effective antioxidants in emulsion systems, such as salad dressings, mayonnaise, and margarine, because they can function in the aqueous phase.

Polyphosphates and EDTA are used in canned seafoods to prevent the formation of glassy crystals of struvite or magnesium ammonium phosphate ($MgNH_4PO_4 \cdot 6H_2O$). Seafoods contain substantial amounts of magnesium ions, which sometimes react with ammonium phosphate during storage to give crystals that may be mistaken as glass contamination. Chelating agents complex magnesium and minimize struvite formation. Chelating agents also can be used to complex iron, copper, and zinc in seafoods to prevent reactions, particularly with sulfides, that lead to product discoloration.

The addition of chelating agents to vegetables prior to blanching can inhibit metal-induced discolorations and can remove calcium from pectic substances in cell walls and thereby promote tenderness.

Although citric and phosphoric acids are employed as acidulants in soft drink beverages, they also chelate metals that otherwise could promote oxidation of flavor compounds, such as terpenes, and catalyze discoloration reactions. Chelating agents also stabilize fermented malt beverages by complexing copper. Free copper catalyzes oxidation of polyphenolic compounds that subsequently interact with proteins to form permanent hazes or turbidity (59).

The extremely efficient chelating abilities of some agents, notably EDTA, has caused speculation that excessive usage in foods could lead to the depletion of calcium and other minerals in the body. To deal with this concern, levels and applications are regulated, and in some instances calcium is added to food systems through the use of the Na_2Ca salt of EDTA rather than the all-sodium (Na, Na_2, Na_3, or Na_4EDTA) or acid forms. However, there appears to be little concern about using these chelators in the controlled amounts considering the natural concentrations of calcium and other divalent cations present in foods.

VI. ANTIOXIDANTS

Oxidation occurs when electrons are removed from an atom or group of atoms. Simultaneously, there is a corresponding reduction reaction that involves the addition of electrons to a different atom or group of atoms. Oxidation reactions may or may not involve the addition of oxygen atoms or the removal of hydrogen atoms from the substance being oxidized. Oxidation-reduction reactions are common in biological systems and also are common in foods. Although some oxidation reactions are beneficial in foods, others can lead to detrimental effects, including degradation of vitamins (Chap. 7), pigments (Chap. 8), and lipids (Chap. 4) with loss of nutritional value and development of off-flavors. Control of undesirable oxidation reactions in foods is usually achieved by employing processing and packaging techniques that exclude oxygen or involve the addition of appropriate chemical agents.

Before the development of specific chemical technology for the control of free radical-mediated lipid oxidation, the term "antioxidant" was applied to all substances that inhibited oxidation reactions, regardless of the mechanism. For example, ascorbic acid was considered an antioxidant and was employed to prevent enzymic browning of the cut surfaces of fruits and vegetables (Chap. 6). In this application ascorbic acid functions as a reducing agent by transferring hydrogen atoms back to quinones that are formed by enzymic oxidation of phenolic compounds. In closed systems ascorbic acid reacts readily with oxygen and thereby serves as an oxygen scavenger (23). Likewise, sulfurous acid and sulfites are readily oxidized in food systems to sulfonates and sulfate, and thereby function as effective antioxidants in such foods as dried fruits (75). The most commonly employed food antioxidants are phenolic substances. More recently, the term "food antioxidants" often has been applied to those compounds that interrupt the free radical chain reaction involved in lipid oxidation; however, the term should not be used in such a narrow sense.

Antioxidants often exhibit variable degrees of efficiency in protecting food systems, and combinations often provide greater overall protection than can be accounted for through the simple additive effects of each (122). Thus, mixed antioxidants sometimes have a synergistic action, the mechanisms for which are still largely unresolved. It is believed, for example, that ascorbic acid can regenerate phenolic antioxidants by supplying hydrogen atoms to the phenoxy radicals that form when the phenolic antioxidants yield hydrogen atoms to the lipid oxidation chain reaction (7). To achieve this action in lipids ascorbic acid must be made less polar so it will dissolve in fat. This is done by esterification to fatty acids to form compounds, such as ascorbyl palmitate.

The presence of metallic ions, particularly copper and iron, promotes lipid oxidation through a catalytic action (70). These pro-oxidants are frequently inactivated by adding chelating agents, such as citric acid or EDTA (Sec. V). In this role chelating agents are also referred to as synergists since they greatly enhance the action of phenolic antioxidants. However, they are often ineffective as antioxidants when employed alone.

Many naturally occurring substances possess antioxidant capabilities, and the tocopherols are noted examples (Chap. 4). Gossypol, which occurs naturally in cottonseed, is also an antioxidant, but it has toxic properties (Chap. 11). Other naturally occurring antioxidants are coniferyl alcohol (found in plants) and guiaconic and guiacic acid (from gum guaiac) (120). All these are structurally related to butylated hydroxyanisole (BHA), butylated hydroxytoluene (BHT), propylgallate (PG), and di-*t*-butylhydroquinone (TBHQ), which are synthetic phenolic antioxidants currently approved for use in foods.

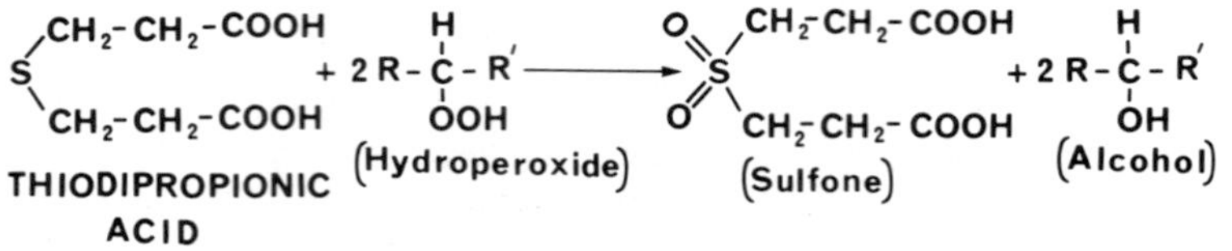

$$S\begin{cases} CH_2\text{-}CH_2\text{-}COOH \\ CH_2\text{-}CH_2\text{-}COOH \end{cases} + 2\,R\text{-}\overset{H}{\underset{OOH}{C}}\text{-}R' \longrightarrow \overset{O}{\underset{O}{S}}\begin{cases} CH_2\text{-}CH_2\text{-}COOH \\ CH_2\text{-}CH_2\text{-}COOH \end{cases} + 2\,R\text{-}\overset{H}{\underset{OH}{C}}\text{-}R'$$

THIODIPROPIONIC (Hydroperoxide)   (Sulfone)   (Alcohol)
ACID

Figure 7   Mechanism of hydroperoxide decomposition by thiodipropionic acid.

Nordihydroguaiaretic acid, a compound related to some of the constituents of gum guaiac, is an effective antioxidant, but its use directly in foods has been suspended because of toxic effects. All these phenolic substances serve as oxidation terminators by participating in the reactions through resonance stabilized free radical forms (122).

Thiodipropionic acid and dilauryl thiodipropionate are approved food antioxidants, but these compounds may be removed from the approved list because they are not being used in foods. Although the presence of a sulfur atom in the thiodipropionates had led to speculation that they could cause off-flavors, this view is unfounded. Thiopropionates have found extensive application in stabilizing polyolefin resins, but the food industry has not readily accepted them.

The thiodipropionates act as secondary antioxidants (35) in that they break down hydroperoxides formed during lipid oxidation to yield relatively stable end products (Fig. 7).

A chemical structure similar to that in the thiodipropionates occurs in methionine (Chap. 3), and accounts, by an analogous mechanism, for some of the antioxidant properties shown by proteins. Reaction of a sulfide with one hydroperoxide yields a sulfoxide; reactions with two hydroperoxides yields a sulfone. Additional aspects of antioxidant chemistry are presented in Chap. 4.

VII. ANTIMICROBIAL AGENTS

Chemical preservatives with antimicrobial properties play an important role in preventing spoilage and assuring safety of many foods (106). Some of these are discussed in the following section.

A. Sulfites and Sulfur Dioxide

Sulfur dioxide long has been used in foods as a general food preservative. Currently, the forms employed include sulfur dioxide gas, and the sodium or potassium salts of sulfide, bisulfite, or metabisulfite. In aqueous solution sulfur dioxide [Eqs. (12) and (13)] and sulfite salts form sulfurous acid, and ions of bisulfite and sulfite (75,77).

$$SO_2 + H_2O \rightleftharpoons H_2SO_3 \tag{12}$$

$$2H_2SO_3 \rightleftharpoons H^+, HSO_3^- + 2H^+, SO_3^{2-} \tag{13}$$

The relative proportion of each form depends on the pH of the solution, and at pH 4.5 or lower the HSO_3^- ion and undissociated sulfurous acid predominate (75). It has been shown that sulfur dioxide is most effective as an antimicrobial agent in acid media, and this effect is believed to result from undissociated sulfurous acid, which is the dominant form below pH 3.0.

The enhanced antimicrobial effect of sulfur dioxide at low pH values may result because undissociated sulfurous acid can more easily penetrate the cell wall. Sulfurous acid inhibits yeasts, molds, and bacteria, but not always to the same degree. This is particularly true at high pH values, where it has been suggested that the HSO_3^- ion is effective against bacteria but not against yeasts (77). Postulated mechanisms by which sulfurous acid inhibits microorganisms include the reaction of bisulfite with acetaldehyde in the cell, the reduction of essential disulfide linkages in enzymes, and the formation of bisulfite addition compounds that interfere with respiratory reactions involving nicotinamide dinucleotide (16).

Of the known inhibitors of nonenzymic browning in foods (Chap. 3), sulfur dioxide is probably the most effective. The chemical mechanism by which sulfur dioxide inhibits nonenzymic browning is not fully understood, but it probably involves bisulfite interactions with active carbonyl groups. Bisulfite combines reversibly with reducing sugars and aldehydic intermediates, and more strongly with α-dicarbonyls and α,β-unsaturated aldehydes (108). These bisulfite addition products appear to retard the browning process, which when coupled with the bleaching action of sulfur dioxide on melanoidin pigments, results in effective inhibition of nonenzymic browning.

Sulfur dioxide also inhibits certain enzyme-catalyzed reactions, notably enzymic browning. The production of brown pigments by enzyme-catalyzed oxidation of phenolic compounds can lead to a serious quality problem during the handling of some fresh fruits and vegetables (Chap. 6). However, the use of sulfite or metabisulfite sprays or dips with or without added citric acid provides effective control of enzymic browning in prepeeled and presliced potatoes, carrots, and apples (26).

Sulfur dioxide also functions as an antioxidant in a variety of food systems, but it is not usually employed for this purpose. When sulfur dioxide is added to beer, the development of oxidized flavors is inhibited significantly during storage. The red color of fresh meat also can be effectively maintained by the presence of sulfur dioxide. However, this practice is not permitted because of the potential for masking deterioration in abused meat products.

When added to flour, sulfur dioxide effects a reversible cleavage of protein disulfide bonds, and this can have desirable effects on the baking properties of bread doughs (108). Prior to drying of fruits, gaseous sulfur dioxide is often applied, and this is sometimes done in the presence of buffering agents (i.e., $NaHCO_3$). This treatment prevents browning and induces oxidative bleaching of anthocyanin pigments. The resulting properties are desired in products, such as those used to make white wines and maraschino cherries (26).

Sulfur dioxide and sulfites are metabolized to sulfate and are excreted in the urine without any obvious pathological results (123). However, the safety-related aspects of sulfur dioxide and its derivatives are undergoing review because of reports of severe reactions in some asthmatics upon consumption of bisulfite, and because of potential mutagenicity. Levels of sulfur dioxide encountered in fruits immediately following drying sometimes approach 2000 ppm; however, concentrations above 500 ppm in most other foods give noticeably disagreeable flavors (16).

B. Nitrite and Nitrate Salts

The potassium and sodium salts of nitrite and nitrate are commonly used in curing mixtures for meats to develop and fix the color, to inhibit microorganisms, and to develop characteristic flavors. Nitrite rather than nitrate is apparently the functional constituent.

Nitrites in meat form nitric oxide, which reacts with heme compounds to form nitro-somyoglobin, the pigment responsible for the pink color of cured meats (Chap. 8). Sensory evaluations also indicate that nitrite contributes to cured meat flavor apparently through an antioxidant role (132), but the details are poorly understood (131). Furthermore, nitrites (150-200 ppm) inhibit *Clostridia* in canned-comminuted and cured meats (17,18,113). In this regard, nitrite is more effective at pH 5.0-5.5 than it is at higher pH values. the antimicrobial mechanism of nitrite is unknown, but it has been suggested that nitrite reacts with sulfhydryl groups to create compounds that are not metabolized by microorganisms under anaerobic conditions (15).

Recently, nitrites have been shown to be involved in the formation of low, but possibly toxic levels, of nitrosamines in certain cured meats. The chemistry and health implications of nitrosamines are discussed in Chap. 11. Nitrate salts also occur naturally in many foods, including such vegetables as spinach (2). The accumulation of large amounts of nitrate in plant tissues grown on heavily fertilized soils is of concern, particularly in infant foods prepared from these tissues. The reduction of nitrate to nitrite in the intestine, with subsequent absorption, could lead to cyanosis due to methemoglobin formation (54). For these reasons, the use of nitrites and nitrates in foods is being questioned. The antimicrobial capability of nitrite provides some justification for its use in cured meats, in which growth of *C. botulinum* is possible. However, in preserved products, in which botulism does not present a hazard, there appears to be little justification for adding nitrates and nitrites.

C. Sorbic Acid

Straight-chain, monocarboxylic, aliphatic fatty acids exhibit antimycotic activity, and α-unsaturated fatty acid analogs are especially effective for this purpose. Sorbic acid (C–C=C–C=C–COOH) and its sodium and potassium salts are widely used to inhibit mold and yeasts in a wide variety of foods, including cheese, baked products, fruit juices, wine, and pickles. Sorbic acid is particularly effective in preventing mold growth, and it contributes little flavor at the concentrations employed (up to 0.3% by weight). The method of application may involve direct incorporation, surface coatings, or incorporation in a wrapping material. The activity of sorbic acid increases as the pH decreases, indicating that the undissociated form is more inhibitory than the dissociated form. In general, sorbic acid is effective up to pH 6.5, which is considerably above the effective pH ranges for propionic and benzoic acids (16).

The antimycotic action of sorbic acid appears to arise because molds are unable to metabolize the α-unsaturated diene system of its aliphatic chain. It has been suggested that the diene structure of sorbic acid interferes with cellular dehydrogenases, which normally dehydrogenate fatty acids as the first step in oxidation (93). This inhibitory effect does not extend to higher animals, and all evidence indicates that animals and humans metabolize sorbic acid in much the same was as they do naturally occurring fatty acids (30). Also, a few molds have been shown to metabolize sorbic acid (39), and it has been suggested that this metabolism proceeds through β-oxidation, similar to that in mammals (93).

Short chain (C2-C12), saturated fatty acids are also moderately inhibitory to many molds, such as *P. roqueforti*. However, some of these molds are capable of mediating β-oxidation of saturated fatty acids to corresponding β-keto acids, especially when the concentration of the acid is only marginally inhibitory (82). Decarboxylation of the resulting β-keto acid yields the corresponding methyl ketone (Fig. 8), which does not exhibit antimicrobial properties. Some believe that antimycotic acids attach to cell surfaces

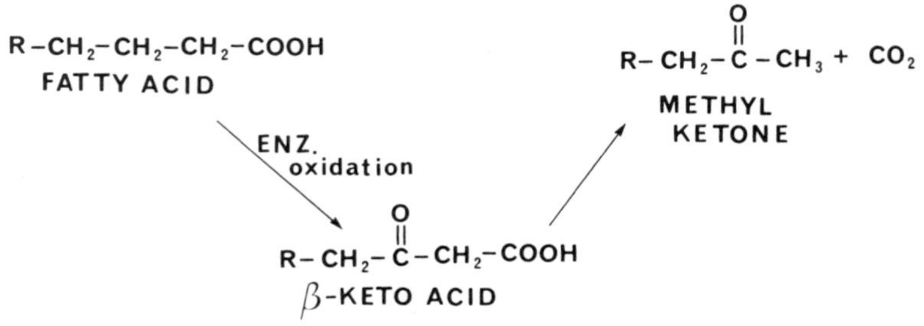

R–CH₂–CH₂–CH₂–COOH
FATTY ACID

$$R-CH_2-CH_2-CH_2-COOH$$
FATTY ACID

$$R-CH_2-\overset{O}{\overset{||}{C}}-CH_3 + CO_2$$
METHYL KETONE

ENZ.
oxidation

$$R-CH_2-\overset{O}{\overset{||}{C}}-CH_2-COOH$$
β**-KETO ACID**

Figure 8 Formation of a methyl ketone via mold-mediated enzymic oxidation of a fatty acid followed by a decarboxylation reaction.

and cause changes in cell permeability. In addition, unsaturated fatty acids may undergo oxidation and the resulting free radicals could exert an inhibitory action by attaching to critical sites of cell membranes. These mechanisms are, however, speculative at this time.

Other mechanisms for deactivating the antimicrobial properties of sorbic acid are shown in Fig. 9. Reaction a in Fig. 9 has been demonstrated in molds, especially *P. roqueforti* (66,87). This involves direct decarboxylation of sorbic acid to yield the hydrocarbon, 1,3-pentadiene. The intense aroma of this compound can cause gasoline or hydrocarbon off-flavors when mold growth occurs in the presence of sorbic acid, especially on the surface of cheese treated with sorbate.

If wine containing sorbic acid undergoes spoilage in the bottle by lactic acid bacteria, an off-flavor described as geraniumlike develops (25). Lactic acid bacteria reduce sorbic acid to sorbyl alcohol, and then, because of the acid conditions they have created, cause a rearrangement to a secondary alcohol (Fig. 9b). The final reaction involves the formation of an ethoxylated hexadiene that has a pronounced, easily recognized aroma of geranium leaves.

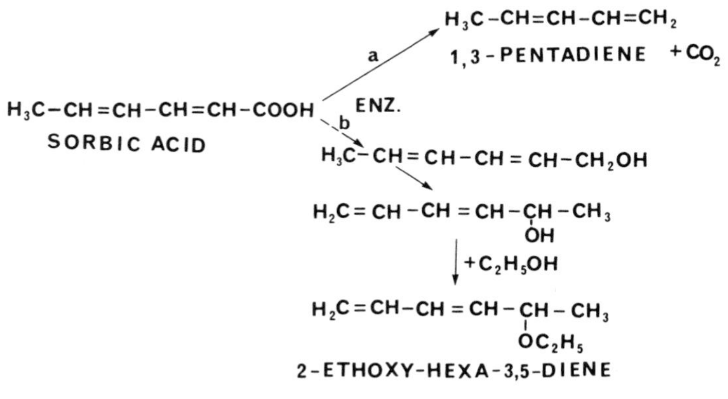

$$H_3C-CH=CH-CH=CH_2$$
1,3-PENTADIENE $+CO_2$

a

$$H_3C-CH=CH-CH=CH-COOH$$
SORBIC ACID

ENZ.

b

$$H_3C-CH=CH-CH=CH-CH_2OH$$

$$H_2C=CH-CH=CH-\underset{OH}{CH}-CH_3$$

$+C_2H_5OH$

$$H_2C=CH-CH=CH-\underset{OC_2H_5}{CH}-CH_3$$
2-ETHOXY-HEXA-3,5-DIENE

Figure 9 Enzymic conversions destroying the antimicrobial properties of sorbic acid: (a) Decarboxylation carried out by *Penicillium* sp.; (b) Formation of ethoxylated diene-hydrocarbon in wine resulting from a reduction of the carboxyl group followed by rearrangement and development of an ether.

Although sorbic acid and potassium sorbate have gained wide recognition as anti-mycotics, more recent research has established that sorbate has broad antimicrobial activity that extends to many bacterial species involved in spoilage of fresh poultry, fish, and meats (107). It is especially effective in retarding toxigenesis of *C. botulinum* in bacon and refrigerated fresh fish packaged in modified atmospheres.

D. Natamycin

Natamycin or pimaricin (CAS Reg. NO. 768-93-8) is a polyene macrolide antimycotic (I) that has recently gained approval in the United States for use against molds on cured cheeses.

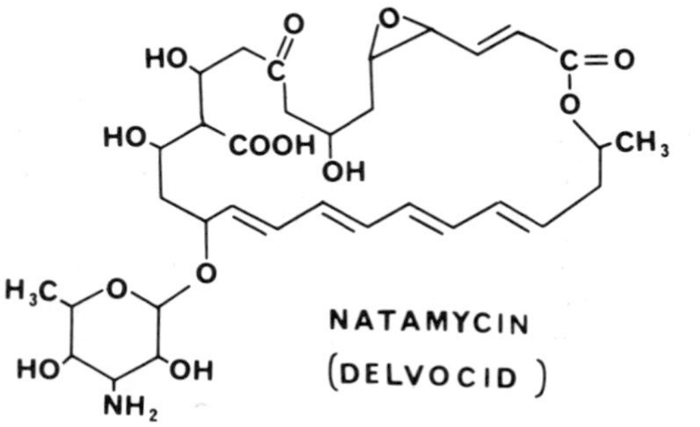

NATAMYCIN
(DELVOCID)

(I)

This mold inhibitor is sold under the tradename Delvocid, and is highly effective when applied to surfaces of foods exposed directly to air, where mold has a tendency to proliferate (50). Natamycin is especially attractive for application on fermented foods, such as cured cheeses, because it selectively inhibits only molds while allowing normal growth and metabolism of ripening bacteria.

E. Glyceryl Esters

Many free fatty acids and monoacylglycerols show pronounced antimicrobial activity against gram-positive bacteria and some yeasts (76). Unsaturated members, especially those with 18-carbon atoms, show strong activity as fatty acids; the medium chain-length members (12-carbon atoms) are most inhibitory when esterified to glycerol. Glyceryl monolaurate (II), also known under the tradename Monolaurin, is inhibitory against several potentially pathogenic staphylococcus and streptococcus when present at concentrations of 15-250 ppm. It is commonly used in cosmetics, and because of its lipid nature can be used in some foods.

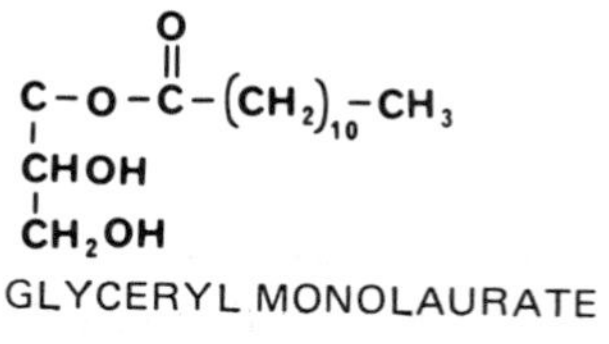

GLYCERYL MONOLAURATE

(II)

Lipophilic agents of this kind also exhibit inhibitory activity against *C. botulinum* (97), and glyceryl monolaurate, serving this function, may find applications in cured meats and in refrigerated, packaged fresh fish. The inhibitory effect of lipophilic glyceride derivatives apparently relates to their ability to facilitate the conduction of protons through the cell membranes, which effectively destroys the proton-motive force needed for substrate transport (43). Cell-killing effects are observed only at high concentrations of these compounds, and death apparently results from the generation of holes in cell membranes.

F. Propionic Acid

Propionic acid (CH_3-CH_2-COOH) and its sodium and calcium salts exert antimicrobial activity against molds and a few bacteria. This compound occurs naturally in Swiss cheese (up to 1% by weight), where it is produced by *Propionibacterium shermanii* (16). Propionic acid has found extensive use in the bakery field, where it not only inhibits molds effectively, but also is active against the ropy bread organism, *Bacillus mesentericus* (99). Levels of use generally range up to 0.3% by weight. As with other carboxylic acid antimicrobial agents, the undissociated form of propionic acid is active, and the range of effectiveness extends up to pH 5.0 in most applications (100). The toxicity of propionic acid to molds and certain bacteria is related to the inability of the affected organisms to metabolize the 3-carbon skeleton. In mammals, propionic acid is metabolized in a manner similar to that of other fatty acids, and it has not been shown to cause any toxic effects at the levels utilized.

G. Acetic Acid

The preservation of foods with acetic acid (CH_3COOH) in the form of vinegar dates to antiquity. In addition to vinegar (4% acetic acid) and acetic acid, also used in food are sodium acetate (CH_3COONa), potassium acetate (CH_3COOK), calcium acetate [$(CH_3-COO)_2Ca$], and sodium diacetate ($CH_3-COONa \cdot CH_3-COOH \cdot \frac{1}{2}H_2O$). The salts are used in bread and other baked goods (0.1-0.4%) to prevent ropiness and the growth of molds without interfering with yeast (16). Vinegar and acetic acid are used in pickled meats and fish products. If fermentable carbohydrates are present, at least 3.6% acid must be present to prevent growth of lactic acid bacilli and yeasts. Acetic acid is also used in such foods as catsup, mayonnaise, and pickles, where it serves a dual function of inhibiting microorganisms and contributing to flavor. The antimicrobial activity of acetic acid increases as the pH is decreased, a property analogous to that found for other aliphatic fatty acids.

H. Benzoic Acid

Benzoic acid (C_6H_5COOH) has been widely employed as an antimicrobial agent in foods, and it occurs naturally in cranberries, prunes, cinnamon, and cloves (16). The undissociated acid is the form with antimicrobial activity, and it exhibits optimum activity in the pH range 2.5-4.0, making it well suited for use in acid foods, such as fruit juices, carbonated beverages, pickles, and sauerkraut. Since the sodium salt of benzoic acid is more soluble in water than the acid form, the former is generally used. Once in the product some of the salt converts to the active acid form. It is most active against yeasts and bacteria and least active against molds. Often benzoic acid is used in combination with sorbic acid or the parabens, and levels of use usually range from 0.05 to 0.1% by weight.

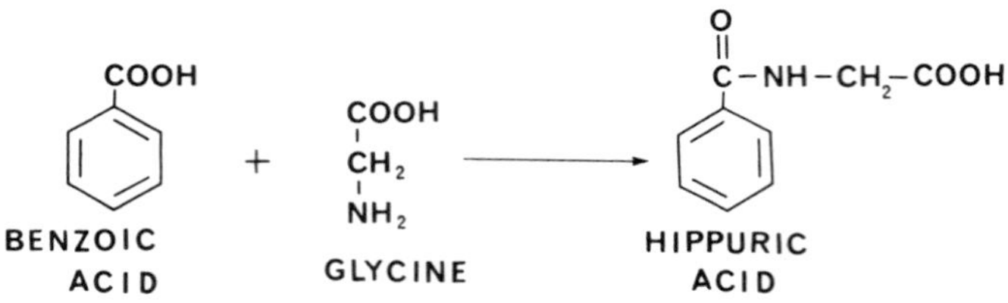

Figure 10 Conjugation of benzoic acid with glycine to facilitate excretion.

Benzoic acid has been found to cause no deleterious effects in humans when used in small amounts (16). It is readily eliminated from the body primarily after conjugation with glycine (Fig. 10) to form hippuric acid (benzoyl glycine). This detoxification step precludes accumulation of benzoic acid in the body.

I. *p*-Hydroxybenzoate Alkyl Esters

The parabens are a group of alkyl esters of *p*-hydroxybenzoic acid that have been used widely as antimicrobial agents in foods, pharmaceutical products, and cosmetics. The methyl (III), propyl, and heptyl (IV) esters are used domestically, and in some other countries the ethyl and butyl esters are used as well (17).

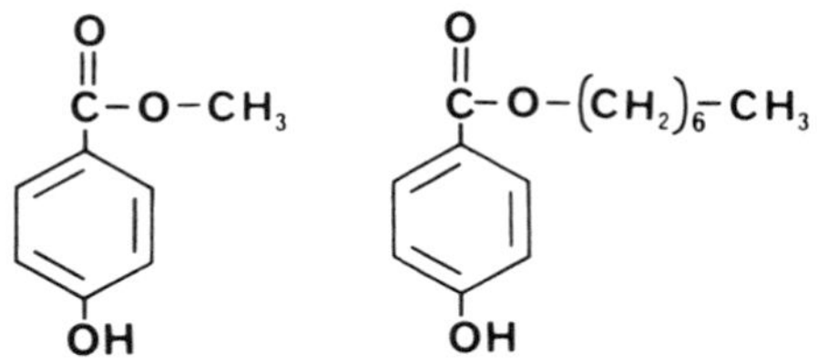

Parabens are used as microbial preservatives in baked goods, soft drinks, beer, olives, pickles, jams and jellies, and syrups. They have little effect on flavor, are effective inhibitors of molds and yeasts (0.05-0.1% by weight), and are relatively ineffective against bacteria, especially gram-negative bacteria (16). The antimicrobial activity of parabens increases and their solubility in water decreases with increases in the length of the alkyl chain. The shorter chain members often are used because of their solubility characteristics. In contrast to other antimycotic agents, the parabens are active at pH 7 and higher, apparently because of their ability to remain undissociated at these pH values (28). The phenolic group provides a weak acid character to the molecule. The ester linkage is stable to hydrolysis even at temperatures used for sterilization. The parabens have many properties in common with benzoic acid and they are often used together. Parabens exhibit a low order of toxicity to humans and are excreted in the urine after hydrolysis of the ester group and subsequent metabolic conjugation.

J. Epoxides

Most antimicrobial agents used in foods exhibit inhibitory rather than lethal effects at the concentrations employed. However, exceptions occur with ethylene (V) and propylene oxides (VI). These chemical sterilants are used to treat certain low-moisture foods and to sterilize

aseptic packaging materials (125). To achieve intimate contact with microorganisms the epoxides are used in a vapor state, and after adequate exposure, most of the residual un-reacted epoxide is removed by flushing and evacuation.

The epoxides are reactive cyclic ethers that destroy all forms of microorganisms, including spores and even viruses, but the mechanism of action of epoxides is poorly understood. In the case of ethylene oxide it has been proposed that alkylation of essential intermediary metabolites with a hydroxyethyl group ($-CH_2-CH_2-OH$) could account for the lethal results (16). The site of attack would be any labile hydrogen in the metabolic system. The epoxides also react with water to form corresponding glycols (Fig. 11). However, the toxicity of the glycols is low, and, therefore, cannot account for the inhibitory effect.

Since the majority of the active epoxide is removed from the treated food and the glycols formed are of low toxicity, it might appear that these gaseous sterilants would be used extensively. Their use, however, is limited to dry items, such as nutmeats and spices, because the reaction of water rapidly depletes the concentration of epoxides in high-moisture foods. Spices often contain high microbial loads and are destined for incorporation into perishable foods. Thermal sterilization of spices is unsuitable because important flavor compounds are volatile and the product is generally unstable to heat. Thus, treatment with epoxides is a suitable method for reducing the microbial load.

The potential formation of relatively toxic chlorohydrins as a result of reactions between epoxides and inorganic chlorides (Fig. 11) is a point of some concern (16). However, there are reports that dietary chlorohydrin in low concentrations causes no

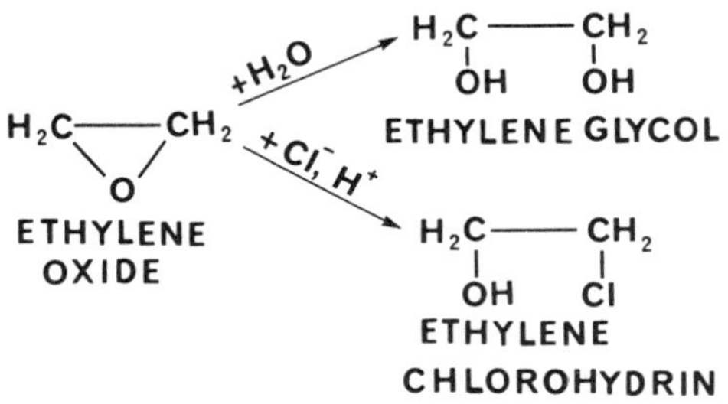

Figure 11 Reactions of ethylene oxide with water and chloride ion, respectively.

ill effect (133). Another consideration in the use of epoxides is their possible adverse effects on vitamins, including riboflavin, niacin, and pyridoxine (5).

Ethylene oxide (boiling point, 13.2°C) is more reactive than propylene oxide and is also more volatile and flammable. For safety purposes, ethylene oxide is often supplied as a mixture consisting of 10% ethylene oxide and 90% carbon dioxide. The product to be sterilized is placed in a closed chamber, the chamber is evacuated, then pressurized to 30 lb with the ethylene oxide-carbon dioxide mixture. This pressure is needed to provide a concentration of epoxide sufficient to kill microorganisms in a reasonable time. When propylene oxide (boiling point, 34.3°C) is used, sufficient heat must be applied to maintain the epoxide in a gaseous state (16).

K. Antibiotics

Antibiotics comprise a large group of antimicrobial agents produced naturally by a variety of microorganisms. They exhibit selective antimicrobial activity, and their applications in medicine have contributed significantly to the field of chemotherapy. The successes of antibiotics in controlling pathogenic microorganisms in living animals have led to extensive investigations into their potential applications in food preservation. However, because of the fear that routine use of antibiotics will cause resistant organisms to evolve, their application to foods is not currently permitted in the United States. The development of resistant strains of organisms would be of particular concern if an antibiotic proposed for use in food is also used in a medical application.

Although antibiotics are not used for food preservation in the United States, some other countries allow limited use of a relatively few antibiotics. These include nisin, chlortetracycline, and oxytetracycline (16). Most actual or proposed applications of antibiotics in foods involve their use as adjuncts to other methods of food preservation. Notably, this includes delaying spoilage of refrigerated, perishable foods and reducing the severity of thermal processes. Fresh meats, fish, and poultry comprise a group of perishable products that could benefit from the action of broad-spectrum antibiotics. In fact, for a few years, the U.S. Food and Drug Administration permitted dipping whole poultry carcasses into solutions of chlortetracycline or oxytetracycline. This increased the shelf life of the poultry, and residual antibiotics were destroyed by usual cooking methods (95).

Nisin has been explored extensively for applications in food preservation. This polypeptide antibiotic is active against gram-positive organisms, especially in preventing the outgrowth of spores (62,112), and it is not used in medical applications. Nisin is produced by lactic streptococci (135), and in some parts of the world it is used to prevent spoilage of dairy products, such as processed cheese and condensed milk. Nisin is not effective against gram-negative spoilage organisms, and some strains of *Clostridia* are resistant. However, nisin is essentially nontoxic to humans, does not lead to cross-resistance with medical antibiotics, and is degraded harmlessly in the intestinal tract (61).

The biochemical modes of actions for antibiotics are just coming into focus, with research efforts emphasizing molecular mechanisms. In addition, there is a continuing search for natural preservatives that, it is hoped, will be suitable for application to foods. However, the necessarily stringent requirements placed on substances for food applications indicate that acceptable substances will be difficult to find.

L. Diethyl Pyrocarbonate

Diethyl pyrocarbonate has been used as an antimicrobial food additive for beverages such as fruit juices, wine, and beer. The advantage of diethyl pyrocarbonate is that it

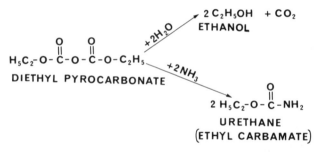

Figure 12 Reactions showing the hydrolysis and amidization of diethyl pyrocarbonate.

can be used in a cold pasteurization process for aqueous solutions, following which it readily hydrolyzes to ethanol and carbon dioxide (Fig. 12). Usage levels between 120 and 300 ppm in acid beverages (below pH 4.0) cause complete destruction of yeasts in about 60 min. Other organisms, such as lactic acid bacteria, are more resistant, and sterilization is achieved only when the microbial load is low (less than 500 ml^{-1}) and the pH is below 4.0. The low pH retards the rate of diethyl pyrocarbonate decomposition and intensifies its effectiveness (116).

Concentrated diethyl pyrocarbonate is an irritant. However, since hydrolysis is essentially complete within 24 hr in acid beverages, there is little concern for direct toxicity. Unfortunately, diethyl pyrocarbonate reacts with a variety of compounds to form carbethoxy derivatives and ethyl esters. Diethyl pyrocarbonate reacts readily with ammonia to yield urethane (ethyl carbamate; Fig. 12). Urethane is a known carcinogen, but until recently there were no reports of urethane formation in foods treated with diethyl pyrocarbonate. Sensitive isotope dilution techniques have revealed that orange juice, beer, and wine previously treated with diethyl pyrocarbonate (250-500 ppm) contained from 0.17 to 2.6 ppm of urethane (85,101). Since ammonia is ubiquitous in plant and animal tissues, it is probable that foods treated with diethyl pyrocarbonate contain some urethane. Because of this, diethyl pyrocarbonate is no longer permitted in foods in the United States.

VIII. NONNUTRITIVE AND LOW-CALORIE SWEETENERS

Nonnutritive and low-calorie sweeteners encompass a broad group of substances that evoke a sweet taste or enhance the perception of sweet tastes (see Chap. 9). The ban on the use of cyclamates in the United States, along with questions raised about the safety of saccharin, have stimulated a search for alternate low-calorie sweeteners to meet the present demand for low-calorie foods and beverages. Although the list of potentially useful low-calorie sweetneers is growing, only a few are currently available for food applications (4).

A. Cyclamates

Both the sodium and calcium salts of cyclamic acid were widely employed before the compounds were prohibited because of evidence suggesting they are carcinogenic. The basic structure of cyclamates (Fig. 13) provides a reason for suspecting the compounds to

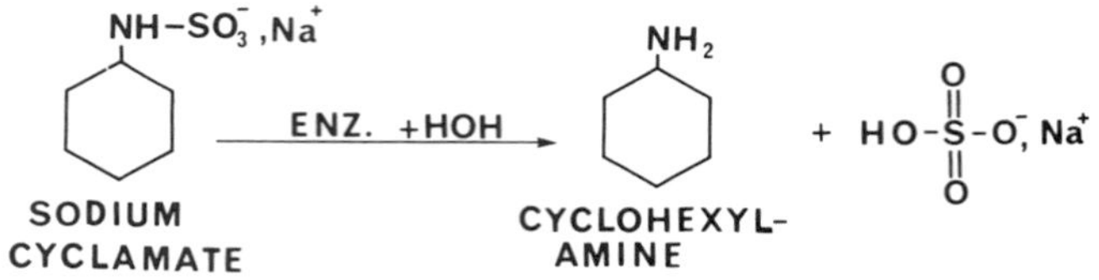

Figure 13   Formation of cyclohexylamine by the hydrolysis of cyclamate.

be potential carcinogens because hydrolysis of the sulfamate ester leads to the formation of cyclohexylamine, a known carcinogen (83,104). This hydrolytic action apparently does not occur as a result of the action of enzymes produced by monogastric digestive systems. However, evidence indicates that some common intestinal microorganisms can readily mediate the reaction (88), and cyclohexylamine can be isolated from the urine (136). For this reason the FDA has refused to reapprove cyclamate sweeteners. The primary route of elimination of cyclamates and absorbed cyclohexylamine is the urine, and this would expose the bladder to an active carcinogen. The initial FDA ban occurred because cyclamates caused bladder cancer in animals (12).

B. Saccharin

Both the calcium and sodium salts of saccharin are used as nonnutritive sweeteners (VII). The level of use depends on the intensity of sweetness desired; however, a slightly bitter flavor often results at relatively high concentrations. The commonly accepted rule of thumb is that saccharin is about 300 times as sweet as sucrose in concentrations up to the equivalent of a 10% sucrose solution (109). Saccharin has been found to cause a low incidence of carcinogenesis in laboratory animals (96).

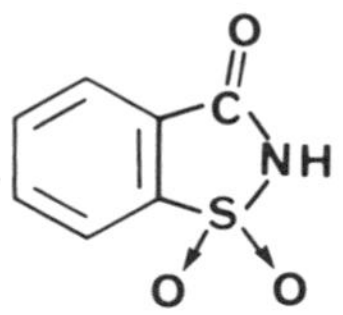

SACCHARIN

(VII)

In humans saccharin is rapidly absorbed, and then is rapidly excreted in the urine (98). Even though current regulations prohibit the use of food additives that cause cancer in any experimental animal, a ban on saccharin in the United States, proposed by the FDA in 1977, has been stayed by congressional legislation pending further research.

C. Aspartame

Very recently, aspartame or L-aspartyl-L-phenylalanine methyl ester (Fig. 14), was approved for use in the United States as a sweetener in dry food mixtures and in soft drinks. Petitions for usage of aspartame in other foods are expected to receive approval soon, and it has been approved for use in a number of other countries.

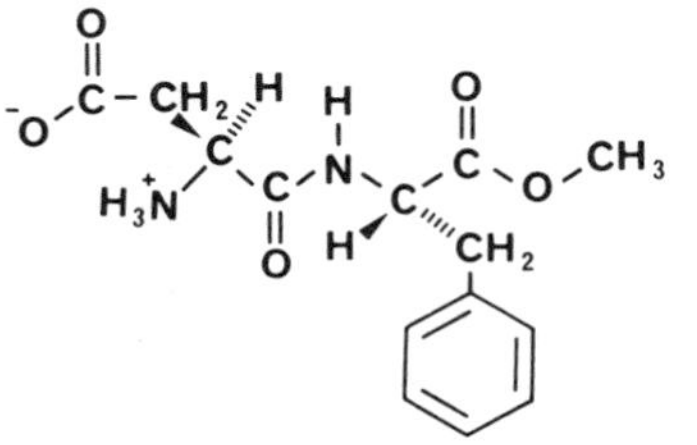

L-ASPARTYL-L-PHENYLALANINE
METHYL ESTER
(ASPARTAME)

Figure 14 Stereochemical configuration of aspartame.

Aspartame is a caloric substance because it is a dipeptide that is completely digested after consumption. However, its intense sweetness (about 200X as sweet as sucrose at a 4% concentration) allows functionality to be achieved at very low levels that provide very few calories. It is noted for a clean, sweet taste similar to that of sucrose (8,9,19,69).

The peptide nature of aspartame makes it susceptible to hydrolysis, other chemical interactions, and microbial degradations, thus limiting its shelf life when used as a general sweetener in aqueous systems. In addition to loss of sweetness resulting from hydrolysis of either the methyl ester on phenylalanine or the peptide bond between the two amino acids, aspartame readily undergoes an intramolecular condensation to yield the diketo-piperazine (5-benzyl-3,6-dioxo-2-piperazine acetic acid) shown in Fig. 15. This reaction is especially favored at neutral and alkaline pH values because nonprotonated amine groups on the molecule are more available for reaction under these conditions. Similarly, alkaline pH values promote carbonyl-amino reactions, and aspartame has been shown to react readily with glucose (118) and vanillin (68) under such conditions. With the glucose reaction, loss of aspartame's sweetness during storage is the principal concern, and loss of vanilla flavor is the main concern in the latter case.

Even though aspartame is composed of naturally occurring amino acids and its daily intake is projected to be very small (0.8 g per person) (37), concern has been expressed about its potential safety as a food additive. However, in approving the use of aspartame in soft drinks, the FDA has cited extensive evidence indicating that it poses no

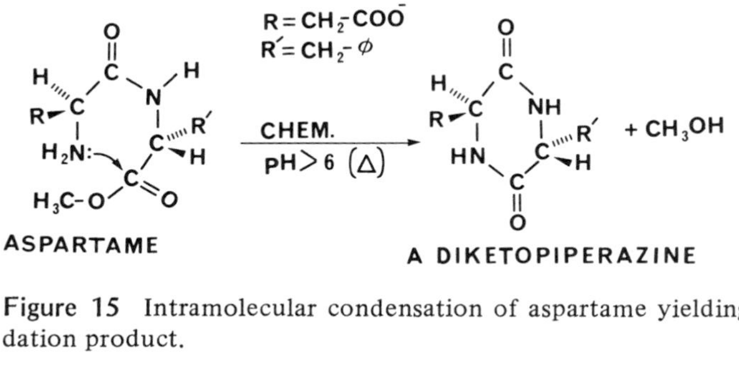

ASPARTAME   A DIKETOPIPERAZINE

Figure 15 Intramolecular condensation of aspartame yielding a diketopiperazine degradation product.

risks when used as approved. Aspartame-sweetened products must be labeled prominently about their phenylalanine content to allow avoidance of consumption by phenylketonuric individuals (37,64). Concerns about the effects of elevated levels of phenylalanine circulating in some other individuals have also been expressed, and research is continuing on this issue. Extensive testing has similarly shown that the diketopiperazine poses no risk to humans at concentrations potentially encountered in foods (38,71,121).

D. Acesulfame K

Very recently, Acesulfame K (6-methyl-1,2,3-oxathiazine-4(3H)-one-2,2-dioxide) was approved for use as a nonnutritive sweetener in the United Kingdom, and its approval is expected soon in a number of other countries, including the United States. The complex chemical name of this substance led to the creation of the trademarked common name, Acesulfame K, which is based on its structural relationships to acetoacetic acid and sulfamic acid, and to its potassium salt nature (Fig. 16).

Acesulfame K is about 200X as sweet as sucrose at a 3% concentration in solution, and it exhibits a sweetness quality between that of cyclamates and saccharin. Extensive testing has shown no toxic effects in animals, and exceptional stability in food applications (24,130). Acesulfame K can be prepared at very high purity by a relatively inexpensive synthesis procedure (Fig. 17). The recent emergence of this compound as a viable nonnutritive sweetener has resulted in limited experiences in food applications, but it should provide an interesting alternative to other nonnutritive sweeteners.

E. Other Nonnutritive or Low-Calorie Sweeteners

Glycyrrhizic acid is a natural, sweet-tasting substance found in licorice root and is approved for use only as a flavor, but not as a sweetener. The root extract contains both the calcium and sodium salts of glycyrrhizic acid. Glycyrrhizic acid is a plant glycoside that on hydrolysis yields 2 mol glucuronic acid and 1 mol glycyrrhetinic acid (a triterpene related to aleanolic acid) (24,69). The sweet taste of glycyrrhizic acid is detectable at 1/50 the threshold taste level of sucrose. Glycyrrhizic acid is used primarily in tobacco products and to some extent in foods and beverages. Its licoricelike flavor influences its suitability for some applications (63,69). Substances structurally related to glycyrrhizic acid have been explored. Stevioside is a naturally occurring glycoside sweetener found in the leaves of *Stevia rebaudiana* Bertoni, and it is 300 times as sweet as sucrose (69).

Neohesperidin dihydrochalcone is a nonnutritive sweetener derived from the bitter flavonones of citrus fruit. This intensely sweet substance, as well as other similar com-

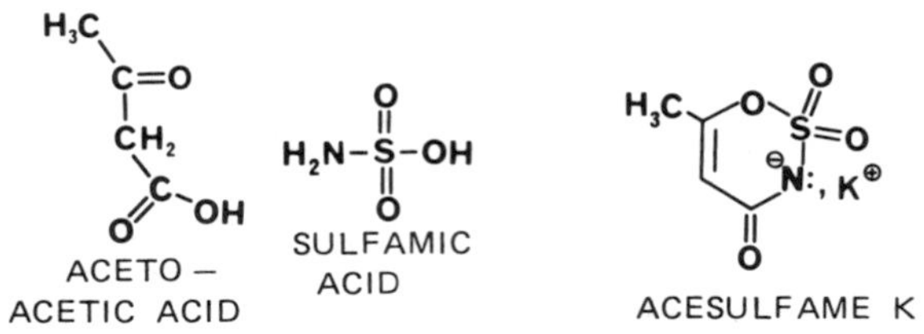

Figure 16 Structurally related compounds that form the basis for the derived name of the nonnutritive sweetener, Acesulfame K.

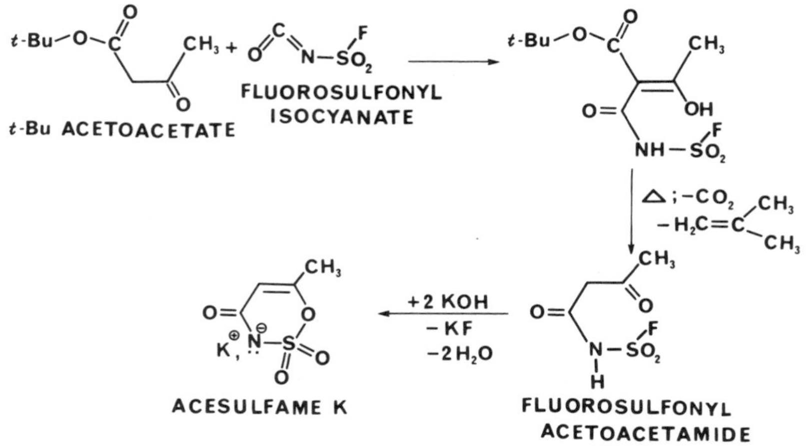

Figure 17 A commercial synthesis procedure for Acesulfame K (130).

pounds, are produced by hydrogenation of: (a) naringin to yield naringin dihydrochalcone, (b) neohesperidin to yield neohesperidin dihydrochalcone, or (c) hesperidin to yield hesperidin dihydrochalcone 4'-*O*-glucoside (63,65). All these sweet substances are related in that they contain (1→2)-linked disaccharides. The dihydrochalcones are undergoing required animal feeding trials, and are not yet allowed in foods (103).

The tropical African fruit katemfe (*Thaumatococcus daniellii*) contains very sweet substances that qualify as low-calorie sweeteners. These are thaumatin I and II, which are alkaline proteins, each with a molecular weight of about 20,000 daltons (129). On a molar basis these proteins are about 10^5 times as sweet as sucrose. An extract of the katemfe fruit is marketed under the tradename Talin in the United Kingdom, and its use as a sweetener and flavor enhancer has been approved in Japan. Talin is up to 5000 times as sweet as sucrose at a 4% concentration in solution, but exhibits a long-lasting sweetness with a slight licoricelike taste. The serendipity berry also contains a sweet protein substance, Monellin, which has a molecular weight of about 11,500 daltons, and it has characteristics similar to those sweeteners found in katemfe. The potential uses of these substances are somewhat limited because the compounds are expensive, unstable to heat, and exhibit a complete loss of sweetness below pH 2 when held in solution at room temperature (69).

Another basic protein, Miraculin, has been isolated from miracle fruit (*Synsepalum dulcificum*). It is tasteless but has the peculiar property of causing sour foods to taste sweet. This material is a glycoprotein with a molecular weight of 42,000 daltons (129). Similar to other protein sweeteners, Miraculin is heat labile and is inactive at low pH values. Further, the taste effects of Miraculin persist for a long time, and this may limit its potential use. It has not been approved for use in the United States.

IX. STABILIZERS AND THICKENERS

Many hydrocolloid materials are widely used for their unique textural, structural, and functional characteristics in foods, where they provide stabilization for emulsions, suspensions, and foams and general thickening properties. Most of these materials, sometimes classed as gums, are derived from natural sources, although some are chemically modified to achieve desired characteristics. Many stabilizers and thickeners are polysaccharides, such as gum arabic, guar gum, carboxymethylcellulose, carrageenan, agar, starch, and pectin. The chemical properties of these and related carbohydrates are discussed in Chap. 3. Gelatin, a protein derived from collagen, is one of the few noncarbohydrate stabilizers used extensively, and it is discussed in Chap. 5. All effective stabilizers and thickeners are hydrophilic and are dispersed in solution as colloids, which leads to the designation hydrocolloid. General properties of useful hydrocolloids include significant solubility in water, a capability to increase viscosity, and in some cases an ability to form gels. Some specific functions of hydrocolloids include improvement and stabilization of texture, inhibition of crystallization (sugar and ice), stabilization of emulsions and foams, improvement (reduced stickiness) of icings on baked goods, and encapsulation of flavors (78). Hydrocolloids are generally used at concentrations of about 2% or less because many exhibit limited dispersibility, and the desired functionality is provided at these levels. The efficacy of hydrocolloids in many applications is directly dependent on their ability to increase viscosity. For example, this is the mechanism by which hydrocolloids stabilize oil-in-water emulsions. They cannot function as true emulsifiers since they lack the necessary combination of strong hydrophilic and lipophilic properties in single molecules.

X. MASTICATORY SUBSTANCES

Masticatory substances are employed to provide the long-lasting, pliable properties of chewing gum. These substances are either natural products or the result of organic synthesis, and both kinds are quite resistant to degradation. Synthetic masticatory substances are prepared by the Fischer-Tropsch process involving carbon monoxide, hydrogen, and a catalyst, and after further processing to remove low-molecular-weight compounds, the product is hydrogenated to yield synthetic paraffin (21). Chemically modified masticatory substances are prepared by partially hydrogenating wood rosin, which is largely composed of diterpenes, and then esterifying the products with pentaerythritol or glycerol (21). Other polymers similar to synthetic rubbers have also been prepared for use as masticatory substances, and these substances are prepared from ethylene, butadiene, or vinyl monomers.

Much of the masticatory base employed in chewing gum is derived directly from plant gums (41). These gums are purified by extensive treatments involving heating, centrifuging, and filtering. Chicle from plants in the *Sapotaceae* (Sapodilla) family, gums from Gutta Katiau from *Palaquium* sp., and latex solids (natural rubber) from *Henea basiliensis* are widely used, naturally derived masticatory substances.

XI. POLYHYDRIC ALCOHOL TEXTURIZERS

Polyhydric alcohols are carbohydrate derivatives that contain only hydroxyl groups as functional groups (Chap. 3), and as a result, they are generally water soluble, hygroscopic materials that exhibit moderate viscosities at high concentrations in water. Although the number of available polyhydric alcohols is substantial, only a few are important in food applications, and these include propylene glycol ($CH_2OH-CHOH-CH_3$), glycerol ($CH_2OH-CHOH-CH_2OH$), sorbitol, and mannitol [$CH_2OH-(CHOH)_4-CH_2OH$].

With a few exceptions, such as with propylene glycol, most polyhydric alcohols occur naturally, although when they do, they usually do not exhibit functional roles in food. For example, free glycerol is known to exist in wine and beer as a result of fermentation, and sorbitol occurs in such fruits as pears, apples, and prunes.

The polyhydroxy structures of these compounds result in water-binding properties that have been exploited in foods. Specific functions of polyhydric alcohols include control of viscosity and texture, addition of bulk, retention of moisture, reduction of water activity, control of crystallization, improvement or retention of softness, improvement of rehydration properties of dehydrated foods, and use as a solvent for flavor compounds (55).

Sugars and polyhydric alcohols are similar, except that sugars contain aldo or keto groups (free or bound) that adversely affect their chemical stability, especially at high temperatures. Many applications of polyhydric alcohols in foods rely on concurrent contributions of functional properties from sugars, proteins, starches, and gums. Polyhydric alcohols generally are sweet but less so than sucrose. Short-chain members are slightly bitter at high concentrations. However, the taste of polyhydric alcohols is generally of limited concern when they are used at low levels (2-10%). When used at high levels, such as in intermediate moisture (IM) foods (glycerol, 25%) and dietary sugar-free candies (sorbitol, 40%), these substances have a substantial influence on product taste.

Recently, attention has been given to the development of polymeric forms of polyhydric alcohols for food applications. Whereas ethylene glycol (CH_2OH-CH_2OH) is toxic, polyethylene glycol 6000 is allowed in some food coating and plasticizing applications. Polyglycerol $[CH_2OH-CHOH-CH_2-(O-CH_2CHOH-CH_2)_n-O-CH_2-CHOH-CH_2OH]$, formed from glycerol through an alkaline-catalyzed polymerization, also exhibits useful properties. It can be further modified by esterification with fatty acids to yield materials with lipidlike characteristics (3). These polyglycerol materials have been approved for food use because the hydrolysis products, glycerol and fatty acids, are metabolized normally.

Intermediate moisture foods deserve some discussion since polyhydric alcohols can make an important contribution to the stability of these products (51). IM foods contain substantial moisture (15-30%), yet are shelf stable to microbiological deterioration without refrigeration. Several familiar foods, including dried fruits, jams, jellies, marshmallows, fruit cake, and jerky, owe their stability to IM characteristics. Some of these items may be rehydrated prior to consumption, but all possess a plastic texture and can be consumed directly. Although moist shelf-stable pet foods have found ready acceptance in recent years, new forms of intermediate moisture foods for human consumption have not as yet become popular. Nevertheless, meat, vegetable, fruit, and combination prepared dishes are under development and may eventually become important forms of preserved foods.

Most IM foods possess water activities of 0.70-0.85, and those containing humectants contain moisture contents of about 20 g water per 100 g of solids (82% H_2O by weight). If IM foods with a water activity of about 0.85 are prepared by desorption, they are still susceptible to attack by molds and yeasts. To overcome this problem, the ingredients can be heated during preparation and an antimycotic agent, such as sorbic acid, can be added. However, in view of recent findings about adsorptive procedures for preparing IM foods, it should be possible to prepare stable intermediate moisture foods without the need for chemical growth inhibitors (79).

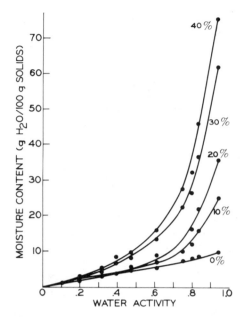

Figure 18 Moisture sorption isotherms of cellulose model systems containing various amounts of glycerol at 37°C. (Reprinted from Ref. 80, p. 86, courtesy of The American Oil Chemists Society.)

To obtain the desired water activity it is usually necessary to add a humectant that binds water and maintains a soft palatable texture. Relatively few substances, mainly glycerol, sucrose, glucose, propylene glycol, and sodium chloride, are sufficiently effective in lowering the water activity while being tolerable organoleptically to be of value in preparing IM foods (10,11). Figure 18 illustrates the effectiveness of the polyhydric alcohol, glycerol, on the water activity of a cellulose model system. Note that, in a 10% glycerol system, a water activity of 0.9 corresponds to a moisture content of only 25 g H_2O per 100 g solids, whereas the same water activity in a 40% glycerol system corresponds to a moisture content of 75 g H_2O per 100 g solids.

The principal flavor criticism of glycerol is its sweet-bitter sensation. Similarly, a problem of excessive sweetness exists for sucrose and glucose when used in IM foods. However, many IM foods show promise when combinations of glycerol, salt, propylene glycol and sucrose are employed.

XII. FIRMING TEXTURIZERS

Thermal processing or freezing of plant tissues usually causes softening because the cellular structure is modified. Stability and integrity of these tissues are dependent on maintenance of intact cells and firm molecular bonding between constituents of cell walls.

The pectic substances (Chaps. 3 and 15) are extensively involved in structure stabilization through cross-linking of their free carboxyl groups via polyvalent cations. Although considerable amounts of polyvalent cations are naturally present, calcium salts (0.1-0.25% as calcium) are frequently added. This increases firmness since the enhanced cross-linking results in increased amounts of relatively insoluble calcium pectinate and pectate. These stabilized structures support the tissue mass, and integrity is maintained even through heat processing. Fruits, including tomatoes, berries, and apple slices, are commonly firmed by adding one or more calcium salts prior to canning or freezing. The most commonly used salts include calcium chloride, calcium citrate, calcium sulfate, calcium lactate, and monocalcium phosphate. Most calcium salts are sparingly soluble, and some contribute a bitter flavor (72).

Acidic alum salts, sodium aluminum sulfate ($NaAl(SO_4)_2 \cdot 12H_2O$), potassium aluminum sulfate, ammonium aluminum sulfate, and aluminum sulfate [$Al_2(SO_4)_3 \cdot 18H_2O$] are added to fermented, salt-brined pickles to make cucumber products that are crisper and firmer than those prepared without these salts (34). The trivalent aluminum ion is believed to be involved in the crisping process through the formation of complexes with pectin substances (89). However, more recent investigations have demonstrated that aluminum sulfate has a softening effect on fresh-pack or pasteurized pickles and should not be included in these products (34). The reasons for the softening are not understood, but the presence of aluminum sulfate counteracts the firming effects normally provided by adjusting the pH to near 3.8 with acetic or lactic acids.

The firmness and texture of some vegetables and fruits can be manipulated during processing without the use of direct additives. For example, an enzyme, pectin methylesterase, is activated during low-temperature blanching (70-82°C for 3-15 min) rather than inactivated, as is the case during usual blanching (88-100°C for 3 min). The degree of firmness produced following low-temperature blanching can be controlled by the holding time prior to retorting (89,127,128). Pectin methylesterase hydrolyzes esterified methanol (sometimes referred to as methoxyl groups) from carboxyl groups on pectin to yield pectinic and pectic acids. Pectin, with relatively few free carboxyl groups, is not strongly bound, and because it is water soluble, it is free to migrate from the cell wall. On the other hand, pectinic acid and pectic acid possess large number of free carboxyl groups and they are relatively insoluble, especially in the presence of endogenous or added calcium ions. As a result they remain in the cell wall during processing and produce firm textures (115). Firming effects through activation of pectin methylesterase have been observed for snap beans, potatoes, cauliflower, and sour cherries. Addition of calcium ions in conjunction with enzyme activation leads to additional firming effects.

XIII. APPEARANCE CONTROL—CLARIFYING AGENTS

In beer, wine, and many fruit juices the formation of hazes or sediments and oxidative deterioration are long-standing problems. Natural phenolic substances are involved in these phenomena. The chemistry of this important groups, including anthocyanins, flavonoids, proanthocyanidins and tannins, is discussed in Chap. 8. Proteins and pectic substances participate with polyphenols in the formation of haze-forming colloids. Specific enzymes have been utilized to partially hydrolyze high-molecular-weight proteins (Chap. 6) and thereby reduce the tendency toward haze formation. However, in some instances excess enzymic activity can adversely affect other desirable properties, such as foam formation in beer.

An important means of manipulating polyphenolic composition to control both its desirable and undesirable effects is to use various clarifying ("fining") agents and adsorbants. Preformed haze can be at least partially removed by filter aids, such as diatomaceous earth. Many of the clarifying agents that have been used are nonselective, and they affect the polyphenolic content more or less incidentally. Adsorption is usually maximal when solubility of the adsorbate is minimal, and suspended or nearly insoluble materials, such as tannin-protein complexes, tend to collect at any interface. As the activity of the adsorbent increases, the less soluble substances still tend to be adsorbed preferentially, but more soluble compounds are also adsorbed.

Bentonite, a montmorillonite clay, is representative of many similar and moderately effective minerals that have been employed as clarifying agents. Montmorillonite is a complex hydrated aluminum silicate with exchangeable cations, frequently sodium ions. In aqueous suspension bentonite behaves as small platelets of insoluble silicate. The bentonite platelets have a negative charge and a very large surface area of about 750 m^2/g. Bentonite is a rather selective adsorbent for protein, and evidently this adsorption results from an attraction between the positive charges of the protein and the negative charges of the silicate. A particle of bentonite covered with adsorbed protein adsorbs some phenolic tannins on or along with the protein (114). Bentonite is used as a clarifying or fining agent for wines to preclude protein precipitation. Doses of the order of a few pounds per thousand gallons usually reduces the protein content of wine from 50-100 mg/liter to a stable level of less than 10 mg/liter. Bentonite rapidly forms a heavy compact sediment and is often employed in conjunction with final filtration to remove precipitated colloids.

The important clarifying agents that have a selective affinity for tannins, proanthocyanidins, and other polyphenols include proteins and certain synthetic resins, such as the polyamides and polyvinyl pyrrolidone (PVP). Gelatin and isinglass (obtained from the swim bladder of fish) are the proteins most commonly used to clarify beverages. It appears that the most important type of linkage between tannins and proteins, although probably not the only type, involves hydrogen bonding between phenolic hydroxyl groups and amide bonds in proteins. The addition of a small amount of gelatin (1.5-6.0 oz. per 100 gal) to apple juice causes aggregation and precipitation of a gelatin-tannin complex, which on settling enmeshes and removes other suspended solids. The exact amount of gelatin for each use must be determined at the time of processing. Juices containing low levels of polyphenolics are supplemented with added tannin or tannic acid (0.005-0.01%) to facilitate flocculation of the gelatin (126).

At low concentrations, gelatin and other soluble clarifying agents can act as protective colloids, at higher concentrations they can cause precipitation and at still higher concentrations they can again fail to cause precipitation. Hydrogen bonding between the colloidal clarifying agents and water accounts for their solubilities. Molecules of the clarifying agent and polyphenol can combine in different proportions to either neutralize or enhance the hydration and solubility of a given colloidal particle. The most nearly complete disruption of H bonding between water and either the protein or the polyphenol gives the most complete precipitation. This would be expected to occur when the amount of dissolved clarifying agent roughly equals the weight of the tannin being removed.

The synthetic resins (polyamides and polyvinyl pyrrolidone)

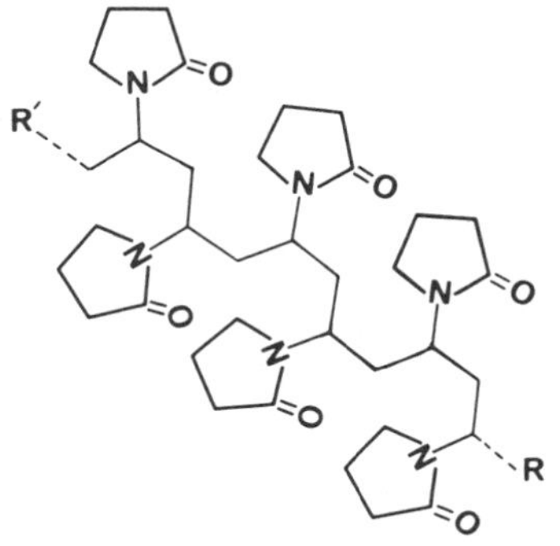

POLYVINYL PYRROLIDONE

(VIII)

have been used to prevent browning in white wines (14) and to remove haze from beers (27). These polymers are available in both soluble and insoluble forms, but requirements for little or no residual polymer in beverages has stimulated use of the high-molecular-weight cross-linked forms that are insoluble. The synthetic resins have been particularly useful in the brewing industry, where reversible refrigeration-induced haze (chill-haze) and permanent haze (that associated with the development of oxidized flavors) are serious problems. These hazes are caused by formation of complexes between native proteins and proanthocyanidins from malted barley. Excessive removal of proteins leads to defective foam character, but the selective removal of polyphenols extends the stability of beer. Initial applications involved polyamides (Nylon 66), but greater efficiency has been achieved with cross-linked polyvinylpyrrolidone (PVP) (VIII). Treatment with 3-5 lb insoluble PVP per 100 barrels of beer provides control of chill haze and improves storage stability (27,91). PVP is added after fermentation and prior to filtration, and it rapidly adsorbs polyphenols. Just as bentonite removes some tannins along with preferentially adsorbed protein, selective tanning adsorbents remove some proteins along with the phenolics.

In addition to the adsorbents already discussed, activated charcoal and some other materials have been employed. Activated charcoal is quite reactive but it adsorbs appreciable amounts of smaller molecules (flavors and pigments) along with the larger compounds that contribute to haze formation. Tannic acid (tannin) is used to precipitate proteins, but its addition can potentially lead to the undesirable effects described previously. Other proteins with low solubility (keratin, casein, and zein) and soluble proteins (sodium caseinate, egg albumen, and serum albumin) also have selective adsorptive capacities for polyphenols, but they have not been extensively employed.

XIV. FLOUR BLEACHING AGENTS AND BREAD IMPROVERS

Freshly milled wheat flour has a pale yellow tint and yields a sticky dough that does not handle or bake well. When the flour is stored, it slowly becomes white and undergoes an aging or maturing process that improves its baking qualities. It is a usual practice to employ chemical treatments to accelerate these natural processes (119), and to use other additives to enhance yeast leavening activity and to retard the onset of staling.

Flour bleaching involves primarily the oxidation of carotenoid pigments. This results in disruption of the conjugated double bond system of carotenoids to a less conjugated colorless system. The dough-improving action of oxidizing agents is believed to involve the oxidation of sulfhydryl groups in gluten proteins. Oxidizing agents employed may participate in bleaching only, in both bleaching and dough improvement, or in dough improvement only. One commonly used flour bleaching agent, benzoyl peroxide $[(C_6H_5CO)_2O_2]$ exhibits a bleaching or decolorizing action but does not influence baking properties. Materials that act both as bleaching and improving agents include chlorine gas (Cl_2), chlorine dioxide (ClO_2), nitrosyl chloride $(NOCl)$, and oxides of nitrogen (nitrogen dioxide, NO_2, and nitrogen tetroxide, N_2O_4) (102). These oxidizing agents are gaseous and exert their action immediately upon contact with flour. Oxidizing agents that serve primarily as dough improvers exert their action during the dough stages rather than in the flour. Included in this group are potassium bromate $(KBrO_3)$, potassium iodate (KIO_3), calcium iodate $[Ca(IO_3)_2]$, and calcium peroxide (CaO_2).

Benzoyl peroxide is usually added to flour (0.015-0.075%) at the mill. It is a powder and is usually added along with diluting or stabilizing agents, such as calcium sulfate, magnesium carbonate, dicalcium phosphate, calcium carbonate, and sodium aluminum phosphate. Benzoyl peroxide is a free radical initiator (see Chap. 4), and it requires several hours after addition to decompose into available free radicals for initiation of carotenoid oxidation.

The gaseous agents for oxidizing flour show variable bleaching efficiencies but effectively improve baking qualities of suitable flours. Treatment with chlorine dioxide improves flour color only slightly but yields flour with improved dough-handling properties. Chlorine gas, often containing a small amount of nitrosyl chloride, is used extensively as a bleach and improver for soft wheat cake flour. Hydrochloric acid is formed from oxidation reactions of chlorine, and the resulting slightly lowered pH values lead to improved cake-baking properties. Nitrogen tetroxide (N_2O_4) and other oxides or nitrogen, produced by passing air through an intense electric arc, are only moderately effective bleaching agents, but they produce good baking qualities in treated flours (90).

Oxidizing agents that function primarily as dough improvers can be added to flour (10-40 ppm) at the mill. They are, however, often incorporated into a dough conditioner mix containing a number of inorganic salts and then added at the bakery. Potassium bromate, an oxidizing agent used extensively as a dough improver, remains unreactive until yeast fermentation lowers the pH of the dough sufficiently to activate it. As a result, it acts rather late in the process and causes increased loaf volume, improved loaf symmetry, and improved crumb and texture characteristics.

Early investigators proposed that the improved baking qualities resulting from treatment with oxidizing agents were attributable to inhibition of the proteolytic enzymes present in flour. However, a more recent belief is that dough improvers, at an appropriate time, oxidize sulfhydryl groups (−SH) in the gluten to yield an increased number of intermolecular disulfide bonds (−S−S−). This cross-linking allows gluten proteins to form thin, tenacious networks of protein films that comprise the vesicles for leavening. The result is a tougher, drier, more extensible dough and finished products with improved characteristics. Excessive oxidation of the flour must be avoided since this leads to inferior products with gray crumb color, irregular grain, and reduced loaf volume (90).

The addition of a small amount of soybean flour to wheat flour intended for yeast-levened doughs has become a common practice. The addition of soybean lipoxygenase (see Chaps. 4 and 6) is an excellent way to initiate the free radical oxidation of caroten-noids (36). Addition of soybean lipoxygenase also greatly improves the rheological properties of the dough by a mechanism not yet elucidated. Although it has been suggested that lipid hydroperoxides become involved in the oxidation of gluten —SH groups, evidence indicates that other protein-lipid interactions are also involved in dough improvement by oxidants (36).

Inorganic salts incorporated into dough conditioners include ammonium chloride (NH_4Cl), ammonium sulfate [$(NH_4)_2SO_4$], calcium sulfate ($CaSO_4$), ammonium phosphate [$(NH_4)_3PO_4$], and calcium phosphate ($CaHPO_4$). They are added to dough to facilitate growth of yeast and to aid in control of pH. The principal contribution of ammonium salts is to provide a ready source of nitrogen for yeast growth. The phosphate salts apparently improve dough by buffering the pH at a slightly lower than normal value. This is especially important when water supplies are alkaline.

Other types of materials are also used as dough improvers in the baking industry. Calcium stearoyl-2-lactylate [$(C_{17}H_{35}COOC(CH_3)HCOOC(CH_3)HCOO)_2Ca$] and similar emulsifying agents are used at low levels (up to 0.5%) to improve mixing qualities of dough and to promote increased loaf volume (124). Hydrocolloid gums have been used in the baking industry to improve the water-holding capacity of doughs and to modify other properties of doughs and baked products (52). Carrageenan, carboxymethylcellulose, locust bean gum, and methylcellulose are among the more useful hydrocolloids in baking applications. Methylcellulose and carboxymethylcellulose have been found to retard retrogradation and staling in bread, and they also retard migration of moisture to the product surface during subsequent storage. Carrageenan (0.1%) softens the crumb texture of sweet dough products. Several hydrocolloids (e.g., carboxymethylcellulose at 0.25%) may be incorporated into doughnut mixes to significantly decrease the amount of fat absorbed during frying. This benefit apparently arises because of improvements in the dough and because a more effective hydrated barrier is established on the surface of the doughnuts (78).

XV. ANTICAKING AGENTS

Several conditioning agents are used to maintain free-flowing characteristics of granular and powdered forms of foods that are hygroscopic in nature. In general, these materials function by readily absorbing excess moisture, by coating particles to impart a degree of water repellency, and/or by providing an insoluble particular diluent (53). Calcium silicate ($CaSiO_3 \cdot X H_2O$) is used to prevent caking in baking powder (up to 5%), table salts (up to 2%), and other foods and food ingredients. Finely divided calcium silicate absorbs liquids in amounts up to 2½ times its weight and still remains free flowing. In addition to absorbing water, calcium silicate also effectively absorbs oils and other nonpolar organic compounds. This characteristic makes it useful in complex powdered mixes and in certain spices that contain free essential oils (120).

Food-grade calcium and magnesium salts of long-chain fatty acids, derived from tallow, are used as conditioning agents in dehydrated vegetable products, salt, onion and garlic salt, and in a variety of other food ingredients and mixes that exist in powder form. Calcium stearate is often added to powdered foods to prevent agglomeration, to promote free flow during processing, and to ensure freedom from caking during the shelf life of the finished product. Calcium stearate is essentially insoluble in water but adheres well to particles and provides a partial water-repellent coating for the particles. Commercial stearate powders have a high bulk density (about 20 lb/ft³) and possess large surface areas that make their use as conditioners (0.5-2.5%) reasonably economical. Calcium stearate is also used as a release lubricant (1%) in the manufacture of pressed tablet-form candy (110).

Other anticaking agents employed in the food industry include sodium silicoaluminate, tricalcium phosphate, magnesium silicate, and magnesium carbonate (42). These materials are essentially insoluble in water and exhibit variable abilities to absorb moisture. Their use levels are similar to those for other anticaking agents (e.g., about 1% sodium silicoaluminate is used in powdered sugar). Microcrystalline cellulose powders are used to prevent grated or shredded cheese from clumping together (40). Anticaking agents are either metabolized (starch and stearates) or exhibit no toxic actions at levels employed in food applications (45).

XVI. GASES AND PROPELLANTS

Gases, both reactive and inert, play important roles in the food industry. For example, hydrogen is used to hydrogenate unsaturated fats (Chap. 4), chlorine is used to bleach flour (see bleaching agents and dough improvers in this chapter) and sanitize equipment, sulfur dioxide is used to inhibit enzymic browning in dried fruits (see sulfites and sulfur dioxide in this chapter), ethylene gas is used to promote ripening of fruits (Chap. 15), ethylene oxide is used as a sterilant for spices (see epoxides in this chapter), and air is used to oxidize ripe olives for color development. However, the functions and properties of essentially inert gases used in food are the topics of primary concern in the following sections.

A. Protection from Oxygen

Some processes for oxygen removal involve the use of inert gases, such as nitrogen or carbon dioxide to flush a headspace, to strip or sparge a liquid, or to blanket a product during or after processing. Carbon dioxide is not totally without chemical influence because it is soluble in water and can lead to a tangy, carbonated taste in some foods. The ability of carbon dioxide to provide a dense, heavier-than-air, gaseous blanket over a product makes it attractive in many processing applications. Nitrogen blanketing requires thorough flushing followed by a slight positive pressure to prevent rapid diffusion of air into the system. A product that is thoroughly evacuated, flushed with nitrogen, and hermetically sealed exhibits increased stability against oxidative deterioration (74).

B. Carbonation

The addition of carbon dioxide (carbonation) to liquid products, such as carbonated soft drinks, beer, some wines, and certain fruit juices, causes them to become effervescent, tangy, slightly tart, and somewhat tactual. The quantity of carbon dioxide used and the method of introduction varies widely with the type of product (74). For example, beer becomes partially carbonated during the fermentation process, but is further carbonated prior to bottling. Beer usually contains 3-4 volumes of carbon dioxide (1 volume of beer at 16°C and 1 atm pressure contains 3-4 volumes of carbon dioxide gas at the same temperature and pressure). Carbonation is often carried out at lowered temperatures (4°C) and elevated pressures to increase carbon dioxide solubility. Other carbonated beverages contain from 1.0 to 318 volumes of carbon dioxide depending upon the effect desired. The retention of large amounts of carbon dioxide in solutions at atmospheric pressure have been ascribed to surface adsorption by colloids and to chemical binding. It is well established that carbamino compounds are formed in some products by rapid, reversible reactions between carbon dioxide and free amino groups of amino acids and proteins (44). In addition, formation of carbonic acid (H_2CO_3) and bicarbonate ions (HCO_3^-) also aid in stabilizing the carbon dioxide system. Spontaneous release of carbon dioxide from beer, that is, gushing, has been associated with trace metallic impurities and with the presence of oxalate crystals that provide sites for nucleation of gas bubbles.

C. Propellants

Some fluid food products are dispensed as liquids, foams, or sprays from pressurized aerosol containers. Since the propellant usually comes into intimate contact with the food, it becomes an incidental food component or ingredient. The principal propellants for pressure dispensing of foods are nitrous oxide, nitrogen, and carbon dioxide (31,74). Foam and spray products are usually dispensed by nitrous oxide and carbon dioxide because these propellants are quite soluble in water and their expansion during dispensing assists in the formation of the spray or foam. Carbon dioxide is also employed for such products as cheese spreads, in which tanginess and tartness are acceptable characteristics. Nitrogen, because of its low solubility in water and fats, is used to dispense liquid streams in which foaming should be avoided (catsup, edible oils, and syrups). The use of all these gases in foods is regulated, and the pressure must not exceed 100 psig at 21°C or 135 psig at 54°C. At these conditions, none of the gases liquefy and a large portion of the container is occupied by the propellant. Thus, as the product is dispensed, the pressure drops, and this can lead to difficultes with product uniformity and completeness of dispensing. The gaseous propellants are nontoxic, nonflammable and economical and usually do not cause objectionable color or flavors. However, carbon dioxide, when used alone, imparts an undesirable taste to some foods.

Liquefied propellants also have been developed and approved for food use. Although these have been used extensively in nonedible products, the common liquefied propellants are not acceptable in foods because of problems with off-flavors, corrosiveness, or toxicity. Those approved for foods are octafluorocyclobutane, or Freon C-318 ($\overline{CF_2-CF_2-CF_2-CF_2}$), and chloropentafluoroethane, or Freon 115 ($CClF_2-CF_3$) (32). These propellants exist in the container as a liquid layer situated on top of the food product. An appropriate headspace containing vaporized propellant is also present. Use of a liquefied propellant enables dispensing to occur at a constant pressure, but the contents first must be shaken to provide an emulsion that will foam or spray upon discharge from the container. Constant pressure dispensing is essential for good performance of spray aerosols. These propellants are nontoxic at levels encountered and they do not impart off-flavors to foods. They are, however, expensive compared with compressed gas propellants. In addition to applications in spray products, liquefied propellants are used along with nitrous oxide to dispense whipped cream and foamed toppings (45). They yield particularly good foams because they are highly soluble in any fat that may be present, and they can be effectively emulsified.

Please note that the text continues after Table 2, on p. 681.

Table 2 Selected Food Additives[a]

Class and general function	Chemical name	Additional or more specific function	Source of discussion
I. Processing additives			
Aerating and foaming agents	Carbon dioxide	Carbonation, foaming	10
	Nitrogen	Foaming	10
	Sodium bicarbonate	Foaming	10
Antifoam agents	Aluminum stearate	Yeast processing	—
	Ammonium stearate	Beet sugar processing	—
	Butyl stearate	Beet sugar, yeast	—
	Decanoic acid	Beet sugar, yeast	—
	Dimethylpolysiloxane	General use	—
	Dimethylpolysilicone	General use	—
	Lauric acid	Beet sugar, yeast	—
	Mineral oil	Beet sugar, yeast	—
	Oleic acid	General use	—
	Oxystearin	Beet sugar, yeast	—
	Palmitic acid	Beet sugar, yeast	—
	Petroleum waxes	Beet sugar, yeast	—
	Silicon dioxide	General use	—
	Stearic acid	Beet sugar, yeast	—
Catalysts (including enzymes)	Nickel	Lipid-reducing reactions	4
	Amylase	Starch conversion	6
	Glucose oxidase	Oxygen scavenger	6
	Lipase	Dairy flavor developer	6
	Papain	Chill-proofing beer, meat tenderizer	6

	Substance	Function	
	Pepsin	Meat tenderizer	6
	Rennin	Cheese production	6
Clarifying and flocculating agents	Bentonite	Adsorbs proteins	10
	Gelatin	Complexes polyphenols	10
	Polyvinylpyrrolidone	Complexes polyphenols	10
	Tannic acid	Complexes proteins	10
Color control agents	Ferrous gluconates	Dark olives	—
	Magnesium chloride	Canned peas	—
	Nitrate, nitrite (potassium, sodium)	Cured meat	8,10
	Sodium erythrobate	Cured meat color enhancer	8,10
Freezing and cooling agents	Carbon dioxide	—	10
	Liquid nitrogen	—	—
	Freezant-12, Cl_2CF_2	—	—
Malting and fermenting aids	Ammonium chloride	Yeast nutrients	10
	Ammonium phosphate (dibasic)		
	Ammonium sulfate		
	Calcium carbonate		
	Calcium phosphate		
	Calcium phosphate (dibasic)		
	Calcium sulfate		
	Potassium chloride		
	Potassium phosphate (dibasic)		
Material handling aids	Aluminum phosphate	Anticaking, free flow	10
	Calcium silicate	Anticaking, free flow	10
	Calcium stearate	Anticaking, free flow	10
	Dicalcium phosphate	Anticaking, free flow	10

670 Lindsay

Table 2 (Continued)

Class and general function	Chemical name	Additional or more specific function	Source of discussion
	Dimagnesium phosphate	Anticaking, free flow	10
	Kaolin	Anticaking, free flow	10
	Magnesium silicate	Anticaking, free flow	—
	Magnesium stearate	Anticaking, free flow	—
	Sodium carboxymethylcellulose	Bodying, bulking	10
	Sodium silicoaluminate	Anticaking, free flow	10
	Starches	Anticaking, free flow	3
	Tricalcium phosphate	Anticaking, free flow	10
	Tricalcium silicate	Anticaking, free flow	10
	Xanthan (and other gums)	Bodying, bulking	3,10
Oxidizing-reducing agents	Acetone peroxides	Free radical initiator	10
	Benzoyl peroxide	Free radical initiator	
	Calcium peroxide	Free radical initiator	
	Hydrogen peroxide	Free radical initiator	
	Sulfur dioxide	Dried fruit bleach	
pH Control and modification agents Acidulants (acids)	Acetic acid	Antimicrobial agent	10
	Citric acid	Chelating agent	4,10
	Fumaric acid	Chelating agent	10
	δ-Gluconolactone	Leavening agent	10
	Hydrochloric acid	—	10
	Lactic acid	—	10
	Malic acid	Chelating agent	10
	Phosphoric acid	—	10
	Potassium acid tartrate	Leavening agent	10
	Succinic acid	Chelating agent	10
	Tartaric acid	Chelating agent	10

Category	Compound	Function	
Alkalies (bases)	Ammonium bicarbonate	CO_2 source	10
	Ammonium hydroxide	—	
	Calcium carbonate	—	
	Magnesium carbonate	—	
	Potassium bicarbonate	CO_2 source	
	Potassium hydroxide	—	
	Sodium bicarbonate	CO_2 source	
	Sodium carbonate	—	
	Sodium citrate	Emulsifier salt	
	Trisodium phosphate	Emulsifier salt	
Buffering salts	Ammonium phosphate (mono, dibasic)	—	10
	Calcium citrate	—	
	Calcium gluconate	—	
	Calcium phosphate (mono, dibasic)	—	
	Potassium acid tartrate	—	
	Potassium citrate	—	
	Potassium phosphate (mono, dibasic)	—	
	Sodium acetate	—	
	Sodium acid pyrophosphate	—	
	Sodium citrate	—	
	Sodium phosphate (mono, di, tribasic)	—	
	Sodium potassium tartrate	—	
Release and antistick agents	Acylated monoacylglycerols	—	4,10
	Beeswax	—	4,10
	Calcium stearate	—	4,10
	Magnesium silicate	—	4,10
	Mineral oil, white	—	—
	Mono- and diacylglycerols	Emulsifiers	4

Table 2 (Continued)

Class and general function	Chemical name	Additional or more specific function	Source of discussion
Release and antistick agents	Starches	—	3
	Stearic acid	—	4
	Talc	—	—
Sanitizing and fumigating agents	Chlorine	Oxidant	—
	Methyl bromide	Insect fumigant	—
	Sodium hypochlorite	Oxidant	—
Separation and filtration aids	Diatomaceous earth	—	—
	Ion-exchange resins	—	—
	Magnesium silicate	—	—
Solvents, carriers, and encapsulating agents	Acetone	Solvent	—
	Agar-agar	Encapsulation	3
	Arabinogalactan	Encapsulation	3
	Cellulose	Carrier	3
	Glycerine	Solvent	4
	Guar gum	Encapsulation	3
	Methylene chloride	Solvent	—
	Propylene glycol	Solvent	—
	Triethyl citrate	Solvent	—
Washing and surface removal agents	Sodium dodecylbenzene-sulfonate	Detergent	—
	Sodium hydroxide	Lye peeling	—
II. Final product additives			
Antimicrobial agents	Acetic acid (and salts)	Bacteria, yeast	10
	Benzoic acid (and salts)	Bacteria, yeast	—
	Ethylene oxide	General	—

Category	Substance	Function / Target	
Antioxidants	p-Hydroxybenzoate alkyl esters	Mold, yeast	4,7,10
	Nitrates, nitrites (K, Na salts)	*C. botulinum*	—
	Propionic acid (and salts)	Mold	
	Propylene oxide	General	
	Sorbic acid (and salts)	Mold, yeast, bacteria	
	Sulfur dioxide and sulfites	General	
	Ascorbic acid (and salts)	Reducing agent	
	Ascorbyl palmitate	Reducing agent	
	BHA	Free radical terminator	4,10
	BHT	Free radical terminator	4,10
	Gum guaiac	Free radical terminator	4,10
	Propylgallate	Free radical terminator	4,10
	Sulfite and metabisulfite salts	Reducing agents	4,10
	Thiodipropionic acid (and esters)	Hydroperoxide decomposer	4,10
Appearance control agents			
Colors and color modifiers	Annatto	Cheese, butter, baked goods	8
	Beet powder	Frosting, soft drinks	8
	Caramel	Confectionary	8
	Carotene	Margarine	8
	Cochineal extract	Beverages	—
	FD&C Green No. 3	Mint Jelly	8
	FD&C Red No. 3 (erythrosine)	Canned fruit cocktail	8
	Titanium dioxide	White candy, Italian cheeses	8
	Turmeric	Pickles, sauces	8

Table 2 (Continued)

Class and general function	Chemical name	Additional or more specific function	Source of discussion
Other appearance agents	Beeswax	Gloss, polish	4
	Glycerine	Gloss, polish	4
	Oleic acid	Gloss, polish	4
	Sucrose	Crystalline glaze	3
	Wax, caranuba	Gloss, polish	—
Flavors and flavor modifiers			
Flavoring agents[b]	Essential oils	General	9
	Herbs and spices		
	Plant extractives		
	Synthetic flavor compounds		
Flavor potentiators	Disodium guanylate	Meats and vegatables	9
	Disodium inosinate	Meats and vegetables	9
	Maltol	Bakery goods, sweets	9
	Monosodium glutamate	Meats and vegetables	9
	Sodium chloride	General	—
Moisture control agents	Glycerin	Plasticizer, humectant	3,10
	Gum acacia		3,10
	Invert sugar		3
	Propylene glycol		10
	Mannitol		3,10
	Sorbitol		3,10
Nutrient, dietary supplements			
Amino acids	Alanine	—	5
	Arginine	Essential	
	Aspartic acid	—	
	Cysteine	—	

7

Cystine	—
Glutamic acid	—
Histidine	—
Isoleucine	Essential
Leucine	Essential
Lysine	Essential
Methionine	Essential
Phenylalanine	Essential
Proline	—
Serine	—
Threonine	Essential
Valine	Essential

Minerals

Boric acid	Boron source
Calcium carbonate	Breakfast cereals
Calcium citrate	Cornmeal
Calcium phosphates	Enriched flour
Calcium pyrophosphate	Enriched flour
Calcium sulfate	Bread
Cobalt carbonate	Cobalt source
Cobalt chloride	Cobalt source
Cobalt sulfate	Cobalt source
Cupric chloride	Copper source
Cupric gluconate	Copper source
Cupric oxide	Copper source
Cupric sulfate	Copper source
Calcium fluoride	Water fluoridations
Ferric phosphate	Iron source
Ferric pyrophosphate	Iron source
Ferrous gluconate	Iron source
Ferrous sulfate	Iron source
Iodine	Iodine source
Iodide, cuprous	Table salt

Table 2 (Continued)

Class and general function	Chemical name	Additional or more specific function	Source of discussion
Minerals	Iodate, potassium	Iodine source	
	Magnesium oxide	Magnesium source	
	Magnesium phosphates	Magnesium source	
	Magnesium sulfate	Magnesium source	
	Magnesium chloride	Magnesium source	
	Manganese citrate	Manganese source	
	Manganese oxide	Manganese source	
	Molybdate, ammonium	Molybdenum source	
	Nickel sulfate	Nickel source	
	Phosphates, calcium	Phosphorus source	
	Phosphates, sodium	Phosphorus source	
	Potassium chloride	NaCl substitute	
	Zinc chloride	Zinc source	
	Zinc stearate	Zinc source	
Vitamins	p-Aminobenzoic acid	B complex factor	7
	Biotin	—	
	Carotene	Provitamin A	
	Folic acid	—	
	Niacin	—	
	Niacinamine	Enriched flour	
	Pantothenate, calcium	B complex vitamin	
	Pyridoxine hydrochloride	B complex vitamin	
	Riboflavin	B complex vitamin	
	Thiamine hydrochloride	Vitamin B_1	
	Tocopherol acetate	Vitamin E_1	
	Vitamin A acetate	—	

Miscellaneous nutrients	Vitamin B$_{12}$	—	—
	Vitamin D	—	—
	Betaine hydrochloride	—	Dietary supplement
	Choline chloride	—	Dietary supplement
	Inositol	—	Dietary supplement
	Linoleic acid	4	Essential fatty acid
	Rutin	—	Dietary supplement
Sequestrants (chelating agents)	Calcium citrate	10	—
	Calcium disodium EDTA	—	—
	Calcium gluconate	—	—
	Calcium phosphate (monobasic)	—	—
	Citric acid	—	—
	Disodium EDTA	—	—
	Phosphoric acid	—	—
	Potassium citrate	—	—
	Potassium phosphate (mono, dibasic)	—	—
	Sodium acid pyrophosphate	—	—
	Sodium citrate	—	—
	Sodium gluconate	—	—
	Sodium hexametaphosphate	—	—
	Sodium phosphate (mono, di, tri)	—	—
	Sodium potassium tartrate	—	—
	Sodium tartrate	—	—
	Sodium tripolyphosphate	—	—
	Tartaric acid	—	—
Surface tension control agents	Dioctyl sodium sulfosuccinate	—	—
	Ox bile extract	—	—
	Sodium phosphate (dibasic)	—	—

Table 2 (Continued)

Class and general function	Chemical name	Additional or more specific function	Source of discussion
Sweeteners			
Nonnutritive	Ammonium saccharin	—	9,10
	Calcium saccharin	—	
	Saccharin	—	
	Sodium saccharin	—	
Nutritive	Aspartame	—	9,10
	Glucose	—	3
	Sorbitol	—	3
Texture and consistency control agents			
Emulsifiers and emulsifier salts	Calcium stearoyl-2-lactylate	Dried egg white, bakery	4,10
	Cholic acid	Dried egg white	4
	Desoxycholic acid	Dried egg white	4
	Dioctyl sodium sulfosuccinate	General	—
	Fatty acids (C_{10}–C_{18})	General	4
	Lactylic esters of fatty acids	Shortening	4
	Lecithin	General	4
	Mono- and diacylglycerols	General	4
	Ox bile extract	General	4
	Polyglycerol esters	General	—
	Polyoxyethylene sorbitan esters	General	—
	Propyleneglycol, mono, diesters	General	4
	Potassium phosphate, tribasic	Processed cheese	10,13
	Potassium polymetaphosphate	Processed cheese	10,13
	Potassium pyrophosphate	Processed cheese	10,13
	Sodium aluminum phosphate, basic	Processed cheese	10,13
	Sodium citrate	Processed cheese	10,13
	Sodium metaphosphate	Processed cheese	10,13

Sodium phosphate, dibasic	Processed cheese	10,13
Sodium phosphate, monobasic	Processed cheese	10,13
Sodium phosphate, tribasic	Processed cheese	10,13
Sodium pyrophosphate	Processed cheese	10,13
Sorbitan monooleate	Dietary products	4
Sorbitan monopalmitate	Flavor dispersion	4
Sorbitan monostearate	General	4
Sorbitan tristearate	Confection coatings	4
Stearoyl-2-lactylate	Bakery shortening	4
Stearyl monoglyceridylcitrate	Shortenings	4
Taurocholic acid (salts)	Egg whites	4
Firming agents		
Aluminum sulfate(s)	Pickles	10
Calcium carbonate	General	
Calcium chloride	Canned tomatoes	
Calcium citrate	Canned tomatoes	
Calcium gluconate	Apple slices	
Calcium hydroxide	Fruit products	
Calcium lactate	Apple slices	
Calcium phosphate, monobasic	Tomatoes, canned	
Calcium sulfate	Canned potatoes, tomatoes	
Magnesium chloride	Canned peas	
Leavening agents		
Ammonium bicarbonate	CO_2 source	10
Ammonium phosphate (dibasic)	—	
Calcium phosphate	—	
Glucono-δ-lactone	—	
Sodium acid pyrophosphate	—	
Sodium aluminum phosphate	—	
Sodium aluminum sulfate	—	
Sodium bicarbonate	CO_2 source	
Masticatory substances		
Paraffin (synthetic)	Chewing gum base	10
Pentaerythritol ester of rosin	Chewing gum base	

Table 2 (Continued)

Class and general function	Chemical name	Additional or more specific function	Source of discussion
Propellants	Carbon dioxide	—	10
	Freon-115	—	
	Nitrous oxide	—	
Stabilizers and thickeners	Acacia gum	Foam stabilizer	3,10
	Agar	Ice cream	
	Alginic acid	Ice cream	
	Carrageenan	Chocolate drinks	
	Guar gum	Cheese foods	
	Hydroxypropylmethylcellulose	General	
	Locust bean gum	Salad dressing	
	Methylcellulose	General	
	Pectin	Jellies	
	Sodium carboxymethylcellulose	Ice cream	
	Tragacanth gum	Salad dressing	
Texturizers	Carrageenan	—	3
	Mannitol	—	3
	Pectin	—	3
	Sodium caseinate	—	5,13
	Sodium citrate	—	10
Tracers	Titanium dioxide	Vegetable protein extenders	10

[a]For additional information see Refs. 20, 42, 47, and 60.
[b]Individual members comprising flavoring agents are too numerous to mention. See Refs. 47 and 60 for comprehensive listings.

XVII. TRACERS

Substances in this category are added to food constituents that are difficult to detect in finished foods when routine monitoring of the constituent's concentration in a food is desired and when normal analytical procedures are inadequate. Although the concept seems attractive from a regulatory standpoint, this practice has become unnecessary because analytical procedures for adequate specificity and sensitivity for individual substances in finished foods have been satisfactorily developed in response to needs. Currently in the United States, titanium dioxide (0.1%) must be added to vegetable proteins used as meat extenders. However, removal of this requirement has been requested because gel electrophoresis techniques can quantitatively distinguish between soy and meat protein fractions in foods (41). Adequate regulation of soy protein additions to meat should be possible by monitoring inventory and processing records and by analyzing meat products with electrophoretic techniques.

XVIII. SUMMARY

Summarized in Table 2 are various kinds of food additives and their functions in food.

REFERENCES

1. Allied Chemical Co. (1982). *Sodium Bicarbonate*, Morristown, N.J.
2. Aston, M. R. (1970). The occurrence of nitrates and nitrites in foods; Literature survey No. 7 (April) British Food Manufacturing Industries Research Association, Leatherhead, Surrey, England.
3. Babayan, V. K. (1968). Versatile intermediates and new foods: The polyfunctional polyglycerols. *Food Prod. Dev. 2*(2):58-64.
4. Bakal, A. I. (1983). Functionality of combined sweeteners in several food applications. *Chem. Ind.* (London) *18*:700-708.
5. Bakerman, H., M. Romine, J. A. Schricker, S. M. Takahashi, and O. Mickelson (1956). Stability of certain B-vitamines exposed to ethylene oxide in the presence of choline chloride. *J. Agr. Food Chem. 4*:956-959.
6. Bankhead, R. R., K. E. Weingartner, D. A. Kuntz, and J. W. Erdman (1978). Effects of sodium bicarbonate blanch on the retention of micronutrients in soy beverage. *J. Food. Sci. 43*:345-348.
7. Bauerfeind, J. C., and D. M. Pinkert (1970). Food processing with added ascorbic acid, in *Advances in Food Research* (C. O. Chichester, E. M. Mrak, and G. F. Stewart, eds.), Academic Press, New York, pp. 219-315.
8. Beck, C. I. (1974). Sweetners, character, and applications of aspartic acid-based sweeteners, in *Symposium: Sweeteners* (G. E. Inglett, ed.), AVI Publishing Co., Westport, Conn., pp. 164-181.
9. Beck, C. I. (1978). Application potential for aspartame in low calorie and dietetic foods, in *Low Calorie and Special Dietary Foods* (B. K. Dwividi, ed.), CRC Press, West Palm Beach, Fla., pp. 61-114.
10. Bone, D. P. (1973). Water activity in intermediate moisture foods. *Food Technol.* (Chicago) *27*(4):71-76.
11. Brockman, M. C. (1970). Development of intermediate moisture foods for military use. *Food Technol.* (Chicago) *24*:896-899.
12. Bryan, G. T., and E. Erturk (1970). Production of mouse urinary bladder carcinomas by sodium cyclamate. *Science 167*:996-998.
13. Cacaofabriek De Zaan, B. V. (1977). *Color*, 1540 AA KOOG AAN DEZAAN, The Netherlands.

14. Caputi, A., and R. G. Peterson (1965). The browning problem in wines. *Amer. J. Enol. Viticult.* *16*(1):9-13.

15. Castellani, A. G., and C. F. Niven, Jr. (1955). Factors affecting the bacteriostatic action of sodium nitrite. *Appl. Microbiol.* *3*:154-159.

16. Chichester, D. F., and F. W. Tanner (1972). Antimicrobial food additives, in *Handbook of Food Additives* (T. E. Furia, ed.), CRC Press, Cleveland, Ohio, pp. 115-184.

17. Christiansen, L. N., R. W. Johnston, D. A. Kautter, J. W. Howard, and W. J. Aunan (1973). Effect of nitrite and nitrate on toxin production by *Clostridium botulinum* and on nitrasamine formation in perishable canned comminuted cured meat. *App. Microbiol.* *25*:357-366.

18. Christiansen, L. N., R. B. Tompkin, A. B. Shaparis, T. U. Kueper, R. W. Johnston, D. A. Kautter, and O. J. Kolari (1974). Effect of sodium nitrite on toxin production by *Clostridium botulinum* in bacon. *Appl. Microbiol.* *27*:733-737.

19. Cloninger, M. R., and R. E. Baldwin (1974). L-Aspartyl-L-phenylalanine methyl ester (aspartame) as a sweetener. *J. Food Sci.* *39*:347-349.

20. Committee on Food Protection, Food and Nutrition Board (1965). *Chemicals Used in Food Processing*, Publ. 1274, National Academy of Sceinces Press, Washington, D.C.

21. Considine, D. M. (ed.) (1982). Masticatory substances, in *Foods and Food Production Encyclopedia*, Van Nostrand Reinhold, New York, pp. 1154-1155.

22. Cook, L. R. (1972). *Chocolate Production and Use*, Books for Industry, New York.

23. Cort, W. M., J. W. Scott, and J. H. Harley (1975). Proposed antioxidant exhibits useful properties. *Food Technol.* *29*(11):46-50.

24. Crasky, G. A., and R. E. Winegard (1979). A survey of less common sweeteners, in *Developments in Sweeteners—1* (C. A. M. Hough, K. J. Parker, and A. J. Vlitos, eds.), Applied Science Publishers, London, pp. 135-164.

25. Crowell, E. A., and J. F. Guyman (1975). Wine constituents arising from sorbic acid addition, and identification of 2-ethoxyhexa-3,5-diene as source of geranium-like off-odor. *Amer. J. Enol. Viticult.* *26*(2):97-102.

26. Cruess, W. V. (1948). *Commerical Fruit and Vegetable Products*, 4th ed., McGraw-Hill, New York, p. 248.

27. Dahlstrom, R. W., and M. R. Sfat (1972). The use of polyvinylpyrrolidone in brewing. *Brewer's Dig.* *47*(5):75-80.

28. Davidson, P. M., and A. L. Branen (1981). Antimicrobial activity of non-halognated Phenolic Compounds. *J. Food Protection* *44*:623-632.

29. Deane, D. D., and E. G. Hammond (1960). Coagulation of milk for cheese making by ester hydrolysis. *J. Dairy Sci.* *43*:1421-1429.

30. Duel, H. J., C. E. Calbert, L. Anisfeld, H. MeKeehan, and H. D. Blunden (1954). Sorbic acid as a fungistatic agent for foods. II. Metabolism of α, β-unsaturated fatty acids with emphasis on sorbic acid. *Food Res.* *19*:13-26.

31. E. I. du Pont de Nemours and Co. (1960). *General Background on Food Aerosols*, Wilmington, Del.

32. E. I. du Pont de Nemours and Co. (1965). *Food Propellent "Freon" 115*, Wilmington, Del.

33. Ellinger, R. H. (1972). Phosphates in food processing, in *Handbook of Food Additives* (T. E. Furia, ed.), CRC Press, Cleveland, Ohio, pp. 617-780.

34. Etchells, J. L., T. A. Bell, and L. J. Turney (1972). Influence of alum on the firmness of fresh-pack dill pickles. *J. Food Sci.* *37*:442-445.

35. Evans Chemetics, Inc. (1966). *Thiodipropionate Antioxidants*, Darien, Conn.

36. Faubian, J. M., and R. C. Hoseney (1981). Lipoxygenase: Its biochemistry and role in breadmaking. *Cereal Chem.* *48*:175-180.

37. Federal Register (1983). *Food Additives Permitted for Direct Addition to Food for Human Consumption; Aspartame*, Vol. 48, 132, July 8, pp. 31376-31382.

38. Filer, L. J., G. L. Baker, and D. Stegink (1983). Effect of aspartame on plasma and erythrocyte free amino acid concentrations in one-year-old infants. *J. Nutr. 113*: 1591-1599.

39. Finol, M. L., E. H. Marth, and R. C. Lindsay (1982). Depletion of sorbate from different media during growth of *Pencicillium* species. *J. Food Protection 45*:398-404.

40. FMC Corp. (1982). *Avicel Micro-crystalline Cellulose: An Anticaking Agent for Grated and Shredded Cheese*, Philadelphia.

41. *Food Chemical News* (1983). USDA petitioned to remove titanium dioxide marker requirements for soy protein. March 21, pp. 37-38.

42. *Food Chemicals Codex* (1981). Food and Nutrition Board, National Research Council, National Academy Press, Washington, D.C., 3rd ed.

43. Freese, E., and B. C. Levin (1978). Action mechanisms of preservatives and antiseptics, in *Developments in Industrial Microbiology* (L. A. Underkofler, ed.), Society of Industrial Microbiology, Washington, D.C., pp. 207-227.

44. Fruton, J. S., and S. Simmonds (1958). *General Biochemistry*, 2nd ed., Wiley, New York, p. 919.

45. Furia, T. E. (1972). Regulatory status of direct food additives, in *Handbook of Food Additives* (T. E. Furia, ed.), CRC Press, Cleveland, Ohio, pp. 903-966.

46. Furia, T. E. (1972). Sequestrants in food, in *Handbook of Food Additives* (T. E. Furia, ed.), CRC Press, Cleveland, Ohio, pp. 271-294.

47. Furia, T. E., and N. Bellanca (eds.) (1976). *Fenaroli's Handbook of Flavor Ingredients*, Chemical Rubber Co., Cleveland, Ohio.

48. Gardner, W. H. (1966). *Food Acidulants*, Allied Chemical Corp., New York.

49. Gardner, W. H. (1972). Acidulants in food processing, in *Handbook of Food Additives* (T. E. Furia, ed.), CRC Press, Cleveland, Ohio, pp. 225-270.

50. GB Fermentation Industries, Inc. (1982). *Delvocid Mold Prevention for Food Products*, Charlotte, N.C.

51. Gee, M., D. Farkas, and A. R. Rahman (1977). Some concepts for the development of intermediate moisture foods. *Food Technol.* (Chicago) *31*(4):58-64.

52. Glicksman, M. (1962). Utilization of natural polysaccharide gums in the food industry, in *Advances in Food Research* (C. O. Chichester, E. M. Mrak, and G. F. Stewart, eds.), Academic Press, New York, pp. 109-200.

53. Glidden Pigments Division SCM Corp. (1980). *Silicron Fine Particle Silica for the Food Industry*, Baltimore, Maryland.

54. Govindarajan, S. (1972). Nitrites and nitrates in foods. *Food Prod. Dev. 6*(6):33-34.

55. Griffin, W. C., and M. J. Lynch (1972). Polyhydric alcohols, in *Handbook of Food Additives*, 2nd ed. (T. E. Furia, ed.), CRC Press, Cleveland, Ohio, pp. 431-455.

56. Griswold, R. M. (1962). Leavening agents, in *The Experimental Study of Foods* (R. M. Griswold, ed.), Houghton Mifflin Co., Boston, pp. 330-352.

57. Halliday, E. G., and I. T. Noble (1943). The chemistry of baking powders and their use in baking, in *Food Chemistry and cookery* (E. G. Halliday and I. T. Noble, eds.), University of Chicago Press, Chicago, pp. 206-222.

58. Ham, R. (1971). Interactions between phosphates and meat proteins, in *Symposium: Phosphates in Food Processing* (J. M. DeMan and P. Melnychyn, eds.), AVI Publishing Co., Westport, Conn., pp. 65-84.

59. Harborne, J. B. (1967). *Comparative Biochemistry of Flavonoids*, Academic Press, New York.

60. Heath, H. B. (1981). *Source Book of Flavors*, AVI Publishing Co., Westport, Conn.

61. Heinemann, B., and R. Williams (1966). Inactivation of nisin by pancreatin. *J. Dairy Sci. 49*:312-314.

62. Heinemann, B., L. Vorio, and C. R. Stumbo (1965). Use of nisin in processing food products. *Food Technol.* (Chicago) *19*:160-166.

63. Hodge, J. E., and G. E. Inglett (1974). Structural aspects of glycosidic sweeteners containing (1'→2)-linked disaccharides, in *Symposium: Sweeteners* (G. E. Inglett, ed.), AVI Publishing Co., Westport, Conn., pp. 216-234.

64. Horowitz, D. L., and J. K. Bauer-Nehrling (1983). Can aspartame meet our expectations. *J. Amer. Diet. Ass. 83*(2):142-146.

65. Horowitz, R. M., and B. Gentili (1974). Dihydrochalcone sweeteners, in *Symposium: Sweeteners* (G. E. Inglett, ed.), AVI Publishing Co., Westport, Conn., pp. 182-193.

66. Horwood, J. F., G. T. Lloyd, E. H. Ramshaw, and W. Stark (1981). An off-flavor associated with the use of sorbic acid during Feta cheese maturation. *Aust. J. Dairy Technol. March*: 38-40.

67. Hunziker, O. F. (1940). *The Butter Industry*, 3rd ed., Printing Products Corp., Chicago, p. 238.

68. Hussein, M. M., R. P. D'Amelia, A. L. Manz, H. Jacin, and W. T. C. Chen (1984). Determination of reactivity of aspartame with flavor aldehydes by gas chromatography, HPLC, and GPC. *J. Food Sci.* 49:520-524.

69. Inglett, G. E. (1981). *Sweeteners: A Review. Food Technol.* (Chicago) *35*(3): 37-41.

70. Ingold, K. U. (1962). Metal catalysis, in *Lipids and Their Oxidation* (H. W. Schultz, E. A. Day, and R. O. Sinnhuber, eds.), AVI Publishing Co., Westport, Conn., pp. 93-121.

71. Ishii, H., and T. Koshimizu (1981). Toxicity of aspartame and its diketopiperazine for Wistar rats by dietary administration for 104 weeks. *Toxicology 21*:91-94.

72. Jacobs, M. B. (1951). *The Chemistry and Technology of Food and Food Products*, 2nd ed., Interscience, New York.

73. Jenness, R., and S. Patton (1959). *Principles of Dairy Chemistry*, Wiley, New York, p. 218.

74. Joslyn, M. A. (1964). Gassing and deaeration in food processing, in *Food Processing Operations: Their Management, Machines, Materials, and Methods*, (M. A. Joslyn and J. L. Heid, eds.), AVI Publishing Co., Westport, Conn., pp. 335-368.

75. Joslyn, M. A., and J. B. S. Braverman (1954). The chemistry and technology of the pretreatment and preservation of fruit and vegetable products with sulfur dioxide and sulfites, in *Advances in Food Research* (E. M. Mrak and G. F. Stewart, eds.), Academic Press, New York, pp. 97-160.

76. Kabara, J. J., R. Varable, and M. S. Jie (1977). Antimicrobial lipids: Natural and synthetic fatty acids and monoglycerides, *Lipids 12*:753-759.

77. King, A. D., J. D. Ponting, D. W. Sanshuck, R. Jackson, and K. Mihara (1981). Factors affecting death of yeast by sulfur dioxide. *J. Food Protection 44*:92-97.

78. Klose, R. E., and M. Glicksman (1972). Gums, in *Handbook of Food Additives*, (T. E. Furia, ed.), CRC Press, Cleveland, Ohio, pp. 295-359.

79. Labuza, T. P., S. Cassil, and A. J. Sinskey (1972). Stability of intermediate moisture foods. *J. Food Sci. 37*:2.

80. Labuza, T. P., N. D. Heidelbaugh, M. Silver, and M. Karel (1971). Oxidation at intermediate moisture contents. *J. Amer. Oil Chem. Soc. 48*:86-90.

81. Larson, W. A., N. F. Olson, C. A. Ernstrom, and W. M. Breene (1967). Curd forming techniques for making pizza cheese by direct acidification procedures. *J. Dairy Sci. 50*:1711-1716.

82. Lawrence, R. C. (1966). The oxidation of fatty acids by spores of *Penicillium roqueforti. J. Gen. Microbiol. 44*:393-405.

83. Legator, M. S., K. A. Palmer, S. Green, and K. W. Peterson (1969). Cytogenetic

studies in rats of cyclohexylamine, a metabolite of cyclamate. *Science 165*:1139-1140.

84. Lindsay, R. C., E. A. Day, and L. A. Sather (1967). Preparation and evaluation of butter culture flavor concentrates. *J. Dairy Sci. 50*:25-31.

85. Loforth, G., and T. Gejvall (1971). Diethyl pyrocarbonate: Formation of urethan in treated beverages. *Science 174*:1248-1250.

86. Mahon, J. H., K. Schlamb, and E. Brotsky (1971). General concepts applicable to the use of polyphosphates in red meat, poultry, seafood processing, in *Symposium: Phosphates in Food Processing* (J. M. DeMan and P.Malnychyn, eds.), AVI Publishing Co., Westport, Conn., pp. 158-181.

87. Marth, E. H., C. M. Capp, L. Hasenzahl, H. W. Jackson, and R. V. Hussang (1966). Degradation of potassium sorbate by *Penicillium* species. *J. Dairy Sci. 49*:1197-1205.

88. Matsui, M., A. Tanimura, H. Kurata, A. Ozaki, Y. Benno, and T. Mitsuoka (1982). Studies on metabolism of food additives by microorganisms inhabiting gastrointestinal tract. VII. Formation of cyclohexylamine from sodium cyclamate in germ-free and conventional mice with cyclamate converting bacteria. *Serokuhin Eiseigaker Zasshi 23*(3):270-277 (*CA 97*:108672n, 1982).

89. Matz, S. A. (1962). *Food Texture*, AVI Publishing Co., Westport, Conn., p. 241.

90. Matz, S. A. (1972). *Bakery Technology and Engineering*, 2nd ed., AVI Publishing Co., Westport, Conn., p. 40.

91. McFarlane, W. D. (1968). Biochemistry of beer oxidation. *MBAA Tech. Q. 5*(1): 87-93.

92. McNurlin, T. F., and C. A. Ernstrom (1962). Formation of curd by direct addition of acid to skimmilk. *J. Dairy Sci. 45*:647-656.

93. Melnick, D., F. H. Luckmann, and C. M. Gooding (1954). Sorbic acid as a fungistatic agent for foods. VI. Metabolic degradation of sorbic acid in cheese by molds and the mechanism of mold inhibition. *Food Prep. 19*:44-58.

94. Melnychyn, P., and J. M. Wolcott (1971). Interactions of milk proteins and phosphates, in *Symposium: Phosphates in Food Processing* (J. D. deMan and P. Melnychyn, eds.), AVI Publishing Co., Westport, Conn., pp. 49-64.

95. Meredith, W. E., H. H. Weiser, and A. R. Winter (1965). Chlortetracycline and oxytetracycline residues in poultry tissues and eggs. *Appl. Microbiol. 13*:86-88.

96. Nakanishi, K. (1980). Effects of sodium saccharin on urinary bladder tumorigenesis. *Nogoya-shiritsu Dargaku Igakkai Zasshi 31*(3/4):310-326 (*CA 94*:75108M, 1981).

97. Notermans, S., and J. DuFreene (1981). Effect of glyceryl monolaurate on toxin production by *Clostridium botulinum* in meat slurry. *J. Food Safety 3*(2):83-88.

98. O'Brien, L., and R. C. Gelardi (1981). *Alternative Sweeteners*, Chemtech, May, pp. 275-278.

99. O'Leary, D. K. and R. D. Karlovec (1941). Development of *B. mesentericus* in bread and control by calcium and phosphate or calcium propionate. *Cereal Chem. 18*:730-741.

100. Olson, Jr., J. C., and H. Macy (1945). Propionic acid, sodium propionate, and calcium propionate as inhibitors of molds. *J. Dairy Sci. 28*:701-710.

101. Ough, C. S. (1976). Ethylcarbamate in fermented beverages and foods. II. Possible formation of ethylcarbamate from diethyl dicarbonate addition to wine. *J. Agr. Food Chem. 24*:328-331.

102. Pomeranz, Y., and J. A. Shellenberger (1971). *Bread Science and Technology*, AVI Publishing Co., Westport, Conn.

103. Pratter, P. J. (1980). Neohesperidin dihydrochalcone: An updated review on a naturally derived sweetener and flavor potentiator. *Perfumer Flavorist 5*(6):12-18.

104. Price, J. M., C. G. Biana, B. L. Oser, E. E. Vogin, J. Steinfeld, and H. L. Ley (1970). Bladder tumors in rats fed cyclohexylamine or high doses of a mixture of cyclamate and saccharin. *Science 167*:1131-1132.

105. Quarne, E. L., W. A. Larson, and N. F. Olson (1968). Effect of acidulants and milk clotting enzymes on yield, sensory quality and proteolysis of pizza cheese made by direct acidification. *J. Dairy Sci. 51*:848-856.

106. Robach, M. C. (1980). Use of preservatives to control microorganisms in foods. *Food Technol. 34*(10):81-84.

107. Robach, M. C., and J. N. Sofos (1982). Use of sorbates in meat products, fresh poultry and poultry products: A review. *J. Food Protection 45*:374-383.

108. Roberts, A. C., and D. J. McWeeny (1972). The uses of sulphur dioxide in the food industry. *J. Food Technol. 7*(3):221-238.

109. Salant, A. (1972). Nonnutritive sweeteners, in *Handbook of Food Additives*, 2nd ed. (T. E. Furia, ed.), CRC Press, Cleveland, Ohio, pp. 523-586.

110. S. B. Penick & Co. (1983). *Food Plus-a-tives for Better Food Processing*, New York.

111. Scharpf, L. G. (1971). The use of phosphates in cheese processing, in *Symposium: Phosphates in Food Processing* (J. M. deMan and P. Melnychyn, eds.), AVI Publishing Co., Westport, Conn., pp. 120-157.

112. Scott, U. N., and S. L. Taylor (1981). Effect of nisin on the outgrowth of *Clostridium botulinum* spores. *J. Food Sci. 46*:117-126.

113. Sebranek, J. G., and R. G. Cassens (1973). Nitrosamines: A review. *J. Milk Food Technol. 36*:76-88.

114. Singleton, V. L. (1967). Adsorption of natural phenols from beer and wine. *MBAA Tech. Q. 4*(4):245-253.

115. Sistrunk, W. A., and R. F. Cain (1960). Chemical and physical changes in green beans during preparation and canning. *Food Technol.* (Chicago) *14*:357-362.

116. Splittstoesser, D. F., and M. Wilkison (1973). Some factors affecting the activity of diethylpyrocarbonate as a sterilant. *Appl. Microbiol. 25*:853-857.

117. Stahl, J. E., and R. H. Ellinger (1971). The use of phosphates in the baking industry, in *Symposium: Phosphates in Food Processing* (J. M. Deman and P. Melnychyn, eds.), AVI Publishing Co., Westport, Conn., pp. 194-212.

118. Stamp, J. A., and T. P. Labuza (1983). Kinetics of the Maillard reaction between aspartame and glucose in solution at high temperatures. *J. Food Sci. 48*:543-544.

119. Stauffer, C. E. (1983). Dough conditioners. *Cereal Foods World 28*:729-730.

120. Stecher, P. G. (1968). *The Merck Index*, 8th ed., Merck & Co., Rahway, N. J.

121. Stegink, L. D., R. Koch, M. E. Blaskouics, L. J. Filer, G. L. Baher, and J. E. McDonnell (1981). Plasma phenylalanine levels in heterozygous phenylketonurics and normal adults administered aspartame at 34 mg/kg baby weight. *Toxicology 20*:81-90.

122. Stukey, B. N. (1972). Antioxidants as food stabilizers, in *Handbook of Food Additives* (T. E. Furia, ed.), CRC Press, Cleveland, Ohio, pp. 185-224.

123. *Sulfites as Food Additives* (1975). Scientific Status Summary, Institute of Food Technologists, Chicago.

124. Thompson, J. E., and B. D. Buddemeyer (1954). Improvement in flour mixing characteristics by a steryl lactylic acid salt. *Cereal Chem. 31*:296-302.

125. Toledo, R. T. (1975). Chemical sterilants for aseptic packaging. *Food Technol.* (Chicago) *29*(5):102-112.

126. Tressler, D. K., and M. A. Joslyn (1954). *The Chemistry and Technology of Fruit and Vegetable Juice*, AVI Publishing Co., Westport, Conn., p. 536.

127. van Buren, J. P. (1973). Improves firmness without additives. *Food Eng. 45*(5):127.

128. van Buren, J. P., J. C. Moyer, D. E. Wilson, W. B. Robinson, and D. B. Hand (1960).

Influence of blanching conditions on sloughing, splitting, and firmness of canned snap beans. *Food Technol.* (Chicago) *14*:233-236.

129. van der Wel, H. (1974). Miracle fruit, katemfe, and serendipity berry, in *Symposium: Sweeteners* (G. Inglett, ed.), AVI Publishing Co., Westport Conn., pp. 194-215.

130. von Rymon Lipinski, G. W., and B. E. Huddart (1983). Acesulfame K. *Chem. Ind.* (London) *11*:427-432.

131. Wasserman, A. E., and F. Talley (1972). The effect of sodium nitrite on the flavor of frankfurters. *J. Food Sci. 37*:536-538.

132. Wasserman, A. E., W. Kimoto, and J. G. Phillips (1977). Consumer acceptance of nitrite-free bacon. *J. Food Protection 40*:683-685.

133. Wesley, F., F. Rourke, and O. Darbishire (1965). The formation of persistent toxic chlorohydrins in foodstuffs by fumigation with ethylene oxide and with propylene oxide. *J. Food Sci. 30*:1037-1042.

134. White, E. H. (1970). *Chemical Background for the Biological Sciences*, 2nd ed., Prentice-Hall, Englewood Cliffs, N.J., p. 45.

135. Whitehead, H. R. (1933). A substance inhibiting bacterial growth produced by certain strains of lactic streptococci. *Biochem. J. 27*:1792-1795.

136. Wills, J. H., D. M. Serrone, and F. Coulston (1981). A 7-month study of ingestion of sodium cyclamate by human volunteers. *Reg. Toxicol. Pharmacol. 1*:163-176.

BIBLIOGRAPHY

1. Branen, A. L., and P. M. Davidson (eds.) (1983). *Antimicrobials in Foods*, Marcel Dekker, New York.

2. Committee on Food Protection, Food and Nutrition Board (1965). *Chemicals Used in Food Processing*, Publ. 1274, National Academy of Sciences Press, Washington, D.C.

3. *Food Chemicals Codex* (1981). Food and Nutrition Board, National Research Council, National Academy Press, Washington, D.C., 3rd ed.

4. Furia, T. E. (ed.) (1972). *Handbook of Food Additives*, CRC Press, Cleveland, Ohio.

5. Furia, T. E., and N. Bellanca (eds.) (1976). *Fenaroli's Handbook of Flavor Ingredients*, Chemical Rubber Co., Cleveland, Ohio.

6. Heath, H. B. (1981). *Source Book of Flavors*, AVI Publishing Co., Westport, Conn.

7. Packard, V. S. (1976). *Processed Foods and the Consumer; Additives, Labeling, Standards, and Nutrition*, University of Minnesota Press, Minneapolis.

8. *Source Listing of Natural and Synthetic Raw Materials* (1980). Chemical Sources Association, Suite 600, 900 Seventeenth St., N.W., Washington, D.C.

9. Taylor, R. J. (1980). *Food Additives*, John Wiley and Sons, New York.

11

UNDESIRABLE OR POTENTIALLY UNDESIRABLE CONSTITUENTS OF FOODS

Gerald N. Wogan and Michael A. Marletta Massachusetts Institute of Technology, Cambridge, Massachusetts

I.	Introduction	689
II.	Food Safety	690
III.	Natural Toxic Constituents of Foods	692
	A. Plant Foodstuffs	692
	B. Animal Foodstuffs	700
IV.	Intentional Food Additives	702
	A. Nitrites and N-Nitroso Compounds	702
	B. Carrageenan	703
	C. Antioxidants	704
	D. Safrole	704
	E. Azo Dye Color Additive (Butter Yellow)	704
	F. Sulfites	705
V.	Products of Microbial Growth	706
	A. Mycotoxins	706
	B. Bacterial Toxins	710
VI.	Unintentional Additives	712
	A. Chemicals from Processing	712
	B. Accidental Contaminants	714
VII.	Summary and Conclusions	716
	References	716
	Bibliography	723

I. INTRODUCTION

A very diverse group of nonnutrient chemicals found in foods or food raw materials are of interest with respect to food safety by virtue of their undesirable or potentially hazardous properties. Chemically, these substances represent many classes of inorganic and organic compounds, ranging from elemental metals and simple inorganic salts to complex macromolecules. These characteristics give the area of food toxicology a multifaceted character that might logically be discussed by organizing information around any of several general themes. Such themes can focus on the chemistry of the toxicants or on the biological nature of the toxicological responses elicited by them. Alternatively, substances can also be grouped by sources and routes of exposure, and this organization is used in this chapter, in keeping with that of the remainder of the book. Within this framework, the discussion includes a general summary of presently known problem areas, with detailed presentations only of especially important topics. Available information is summarized concerning the chemical nature of the compounds; their sources and routes of entry into the food supply; their effects in relevant biological systems; their occurrence and consequences in human diets, if such information is known; and the nature of control measures to reduce or eliminate hazards from these compounds, if such hazards exist.

II. FOOD SAFETY

Knowledge about undesirable constituents of foods has stimulated the development of extensive control measures designed to minimize the likelihood of significant health risks resulting from consumption of foods or substances used in food processing. It is instructive to review the general nature of these control measures as they are currently carried out in the United States. Programs of the U.S. federal government are specifically used as examples, because food protection programs in various states (and in some other countries as well) are often organized along similar lines.

The legal framework for food protection in the United States is provided by a series of laws dealing with various aspects of the field, with specific responsibilities divided among several agencies of the federal government whose collective purpose is to assure the availability of a safe and nutritious food supply. Federal regulations pertaining to food safety began in 1908 with enactment of the Pure Food Act, and were expanded in the Food, Drug and Cosmetic Act enacted in 1938. This Act, together with a series of amendments enacted in 1954 (Pesticides), 1959 (Food Additives), 1960 (Color Additives), and 1968 (Animal Drugs), embody the majority of regulations applying to foods. The remaining federal regulations containing provisions dealing specifically with food protection include the Federal Insecticide, Fungicide and Rodenticide Act (FIFRA) of 1948 (with Amendments in 1972 and 1975), the Wholesome Meat Act of 1967, and the Wholesome Poultry Products Act of 1968. Agencies charged with the responsibility of administering and enforcing the various provisions of these laws are the U.S. Food and Drug Administration (FDA), the Environmental Protection Agency (EPA), and the U.S. Department of Agriculture (USDA).

The primary regulatory mechanism involved in administering these laws is the establishment of "tolerances," that is, formal rules specifying the levels of contaminants or additives that will render a food "adulterated" and unfit for consumption. Tolerance levels for food additives, colorants, residues of animal drugs, and contaminants are set by the FDA, whereas those for pesticide residues are determined by the EPA. Enforcement to ensure compliance on the part of food processors and manufacturers is carried out by the FDA and the USDA through various inspection and surveillance programs in which samples of foods are subjected to analysis to determine whether specific constituents are present at levels exceeding established tolerance values.

From a regulatory viewpoint, food chemicals fall into three broad categories: (a) direct additives, that is, substances deliberately added for a specific purpose; (b) indirect additives, such as residues from pesticides or animal drugs used in food production and chemicals that migrate from packaging materials, and (c) unavoidable contaminants, for example, mycotoxins on various crops, insect fragments and undesirable naturally occurring food constituents. For each of these classes of substances, regulation involves establishing tolerence levels (or so-called action levels in the case of unavoidable contaminants) above which the food is removed from commerce.

Tolerance levels are the estimated upper limits of intake at which a food additive or contaminant can be safely ingested. Numerically, tolerances are derived from two other parameters: the no observed effect level (NOEL) determined experimentally in appropriate animal and other test systems; and the acceptable daily intake (ADI) calculated by dividing the NOEL by a safety factor (conventionally 100 or greater) to account for uncertainty in extrapolating animal data to humans, together with assumed variations in human susceptibility.

Tolerance is then defined as the ratio of maximum permissible intake [ADI × average body weight of consumer (assumed to be 70 kg by the FDA, 60 kg by the EPA)] to maximum potential exposure [food factor (estimated % of diet consisting of food in question) × average food consumption (assumed to be 1.5 kg per capita per day)].

More than 2000 chemicals are intentionally added to processed foods, and most of these are regulated through tolerances set for individual compounds or small groups of related substances. Several hundred of the intentional additives have been designated as generally recognized as safe (GRAS) based on extensive safety evaluations and may be added to foods at levels consistent with good manufacturing practices to achieve the desired effect.

The pivotal importance of toxicological data in tolerance setting is evident from the foregoing summary, for it provides the basis for establishment of the NOEL. The requisite information is derived from quantitative dose-response experiments designed to provide statistically valid estimates of doses at which adverse effects are not observed. Safety evaluation for tolerance setting typically includes a broad spectrum of bioassays whose toxicological end points include evidence of acute, subchronic, or chronic toxicity in animals, that is, adverse reactions detectable after exposure times measured in periods of approximately 7 days, 3 months, or 2 years, respectively.

This approach for establishing NOEL values is used for all toxicological end points except carcinogenicity, which is specifically addressed by the so-called Delaney clause [Sections 409(c)(3)] of the Food Drug and Cosmetic Act. This clause excludes the intentional use of any food additive found to induce cancer in humans or animals. In recent years, uses of a major food additive (cyclamate) and a pesticide leaving excessive residues in food (DDT) have been banned on the basis of evidence of carcinogenicity in animals. Bans of saccharin and sodium nitrite based on violation of the Delaney clause have been proposed but not yet put into effect. It is a matter of great current controversy whether the regulatory action required by the Delaney clause should be modified to permit restricted use of carcinogenic food additives under conditions deemed safe, as determined by quantitative estimation of risk.

The spectrum of potential risks associated with toxic substances in food includes the danger of acute (sometimes fatal) poisoning, such as with botulism or paralytic shellfish poisoning, in which the hazards are clear and well understood. At the other extreme lies the undefined risk of lifetime exposure to substances (some of which are naturally occurring constituents of foods) that have powerful activity as mutagens in experimental test systems. The extent to which such substances might contribute to the disease burden of humans cannot yet be assessed. The succeeding sections of this chapter include examples from each of the major problem areas encountered in food protection, and they have been selected to provide an indication of the nature and quality of the available information base upon which control measures must be constructed.

III. NATURAL TOXIC CONSTITUENTS OF FOODS

A. Plant Foodstuffs

Nature has provided plants with the capability of synthesizing a multitude of chemicals that cause toxic reactions when eaten by humans or animals. In the course of their evolution, humans must have learned by trial and error to avoid those plants that cause acute, easily recognizable poisoning (63) or to develop processing methods that reduce or eliminate the toxicity. Nonetheless, many foodstuffs that are still regularly consumed, including some of the major sources of plant protein of nutritional value, contain substances that are deleterious if consumed in sufficient quantities. For the most part, their existence has been recognized for a long time, and ordinary levels of intake are low enough to avoid apparent signs of toxicity. However, grossly increased intakes sometimes occur, for example, when acute food shortages arise in developing countries, and mass poisonings have been observed under these circumstances. The effects that may result from long-term, low-dose exposure to naturally occurring toxicants in foods are unknown. Extrapolation of laboratory in vivo carcinogenicity studies to real-life risks is difficult to do accurately, and this is a timely example of a problem facing regulatory officials. Nonetheless, cognizance must be taken of these toxicants when plant proteins are used in the formulation of novel products.

Listed in Table 1 are most of the major toxicants found in plant foodstuffs, along with descriptions of their essential features. A few of these toxicants are dealt with in greater detail in the following sections.

1. Protease Inhibitors, Hemagglutinins, and Saponins

It is useful in this discussion to consider these three groups of substances together. Although they are not all related chemically or toxicologically, they are often present simultaneously in the same groups of pulses, legumes, and cereals. Indeed, much of what we know about them has resulted from research stimulated by early observations that the nutritive value of soybean meal can be improved by heating, or that raw kidney beans cause weight loss and death when fed to rats.

Protease inhibitors (64,85,88,126) are relatively small proteins that have the property in vitro of binding to and inhibiting proteolytic enzymes. In general, association occurs very rapidly and the complex formed is very stable. The Kunitz inhibitor, originally isolated from soybeans, binds to trypsin with 1:1 stoichiometry (59), exhibiting a second-order rate constant of 2×10^7 liters/mole per sec (43). The dissociation constant at pH 6.5 is 10^{-11} M (60). These kinetic parameters are typical for protease inhibitors. The Kunitz inhibitor is a so-called single-headed inhibitor, binding, as mentioned above, in a 1:1 fashion with trypsin. Another extensively studied protein, the Bowman-Birk inhibitor, also from soy, is a double-headed inhibitor capable of binding one molecule of trypsin and one molecule of chymotrypsin at two independent sites (79,95-98).

Most enzyme-inhibitor studies have been carried out with bovine trypsin and chymotrypsin; however, the assumption that these enzymes adequately represent mammalian proteases in general has recently been shown to be faulty. Studies with homogeneous cationic trypsin purified from humans have shown that the enzyme does rapidly bind the Bowman-Birk inhibitor but that the dissociation constant is much greater than that of the bovine complex, leading to less enzyme inhibition (61,126). Furthermore, the Kunitz inhibitor is only weakly active to inactive with human trypsin.

Although a great deal of sophisticated physical biochemistry has been done on the structure and mode of action of these inhibitors, their role in animal nutrition and toxicology remains largely undefined. Their ability to impair protein hydrolysis is possibly related to the impaired nutritive value of raw products that contain them, but this has not been conclusively proved. When fed to animals in purified form, the chief toxicological response is pancreatic hypertrophy, the significance of which is not clear. Since these inhibitors are inactivated by heating, their destruction may be related to the improvement in nutritive value that moist soybean meal undergoes during moderate heating.

Hemagglutinins (53,59) are also proteins that have in common the ability to cause agglutination of red blood cells in vitro. This effect, which is highly specific for each protein, results from binding to the erythrocyte plasma membrane, and the hemagglutinins have been referred to as "lectins" because of their specificity of binding. Lectins also have the ability to stimulate mitosis in cell cultures and have become useful tools in the study of membrane structure and function.

Although many hemagglutinins are known to exist, only a few have been isolated in pure form. Most are glycoproteins with about 4-10% carbohydrate content; however, the most thoroughly characterized lectin, concanavalin A, does not contain a carbohydrate moiety (53). Some purified proteins in this class are lethal when fed or injected into animals; the most toxic is ricin isolated from the castor bean, which has an LD_{50} of 5 μg/kg in rats. By contrast, soybean and kidney bean lectins are less toxic by a factor of 1000, and those from lentils and peas are nontoxic. The toxicity of all hemagglutinins is destroyed by moist (but not dry) heat.

There have been cases of human intoxication in which processing did not inactivate the lectins. A massive poisoning occurred in Berlin in 1948 after partially cooked bean flakes were consumed (44). As food becomes more scarce in underdeveloped countries and alternative sources of protein are sought, there is a fear that the ground bean and cereal mixtures being considered, if improperly cooked, will lead to adverse effects from lectins (57).

Saponins (11) are glycosides that occur in a wide variety of plants and are characterized by three properties: bitter taste, foaming in aqueous solutions, and hemolysis of red blood cells. They are highly toxic to fish and other aquatic cold-blooded animals, but their effects in higher animals are variable. Chemically, they occur in two groups according to the nature of the sapogenin moiety conjugated with hexoses, pentoses, or uronic acids. The sapogenins are steroids (C27) or triterpenoids (C30). Interest in this group of substances has been initiated mainly by their hemolytic activity, but this property seems to be unimportant with respect to in vivo toxicity.

2. Glucosinolates

Glucosinolates (119) are thioglucoside antithyroid agents that occur in plants of the family Cruciferae, most representatives of importance as foods being in the genus *Brassica.*

Table 1 Toxic Constituents of Plant Foodstuffs

Toxins	Chemical nature	Main food sources	Major toxicity symptoms
Protease inhibitors	Proteins (mol. wt. 4000–24,000)	Beans (soy, mung, kidney, navy, lima); chick-pea; peas, potato (sweet, white); cereals	Impaired growth and food utilization; pancreatic hypertrophy
Hemagglu-tinins	Proteins (mol. wt. 10,000–124,000)	Beans (castor, soy, kidney, black, yellow, jack); lentils; peas	Impaired growth and food utilization; agglutination of erythrocytes in vitro; mitogenic activity to cell cultures in vitro
Saponins	Glycosides	Soybeans, sugarbeets peanuts, spinach, asparagus	Hemolysis of erythrocytes in vitro
Glucosino-lates	Thioglycosides	Cabbage and related species; turnips; ruta-baga; radish; rapeseed; mustard	Hypothyroidism and thyroid enlargement
Cyanogens	Cyanogenetic glucosides	Peas and beans; pulses; linseed; flax; fruit kernels; cassava	HCN poisoning
Gossypol pigments	Gossypol	Cottonseed	Liver damage; hemorrhage; edema
Lathyrogens	β-Aminopropi-onitrile and derivatives	Chick-pea; vetch	Osteolathyrism (skeletal deformities)
	β-N-Oxalyl-L-α,β-diamino-propionic acid	Chick-pea	Neurolathyrism (CNS damage)
Allergens	Proteins?	Practically all foods (particularly grains, legumes, and nuts)	Allergic responses in sensitive individuals
Cycasin	Methylazoxy-methanol	Nuts of *Cycas* genus	Cancer of liver and other organs
Favism	Vicine and con-vicine (pyrimi-dine-β-glucosides)	Fava beans	Acute hemolytic anemia
Phytoalex-ins	Simple furans (ipomeamarone)	Sweet potatoes	Pulmonary edema; liver and kidney damage
	Benzofurans (psoralins)	Celery; parsnips	Skin photosensitivity
	Acetylenic furans (wyerone)	Broad beans	

Table 1 (Continued)

Toxins	Chemical nature	Main food sources	Major toxicity symptoms
	Isoflavonoids (pisatin and phaseollin)	Peas, french beans	Cell lysis in vitro
Pyrrolizi-dine alka-loids	Dihydropyrroles	Families Compositae and Boraginaccae; herbal teas	Liver and lung damage carcinogens
Safrole	Allyl-substituted benzene	Sassafras; black pepper	Carcinogens
α-Amanitin	Bicyclic octapeptides	*Amanita phalloides* mushrooms	Salivation; vomiting; convulsions; death
Atractylo-side	Steroidal glyco-side	Thistle (*Atractylis gummifera*)	Depletion of glycogen

Representative examples of foods that contain glucosinolates are cabbage, broccoli, turnips, rutabaga, and mustard greens, and the glucosinolates, in addition to their antithyroid properties, are responsible, upon hydrolysis, for the pungent nature of these plants.

All natural thioglucosides, of which about 70 have been identified, occur in association with an enzyme(s) that can hydrolyze them to an organic aglycone, glucose, and bisulfate. However, this enzyme is inactive in intact tissue, and activation requires tissue disruption, such as crushing of the wet, unheated tissue. Cooked or boiled food, such as cabbage, contains intact glucosinolates. Intramolecular rearrangements can also take place in the aglycone to yield isothiocyanate, nitrile, or thocyanate (Fig. 1).

Although the role of these antithyroid substances in the etiology of human endemic goiter is apparently minimal, their presence in commodities used as animal feeds can have negative effects on animal growth.

All glucosinolates contain β-D-thioglucose as the sugar moiety (31). The aglycone can undergo further metabolism in the crushed plant to yield episulfide derivatives. Recently, a protein has been purified and named the epithiospecifier protein (ESP) (118). It has a molecular weight of 30,000-40,000, and when present it directs the reaction of sulfur into a terminal unsaturated position of the glucosinolate intermediate, yielding an episulfide (Fig. 2). Iron is also required for this reaction.

Again, not much is known regarding long-term, low-dose exposure to glucosinolates and their breakdown products. In vivo testing has recently shown allylisothiocyanate, a hydrolysis product of the mustard glucosinolate sinigrin, to be a carcinogen in rats. Isocyanates and isothiocyanates are alkylating agents and so are episulfides, which have a reactivity akin to epoxides, especially under slightly acidic conditions. Thiocyanate inhibits iodine uptake, hence, the antithyroid activity, and the nitrile breakdown products also have been shown to be toxic (119).

3. Cyanogens

Cyanide in trace amounts is widely distributed in plants and occurs mainly in the form of cyanogenetic glucosides. All of which occur in the β-configuration (76). Three gluco-

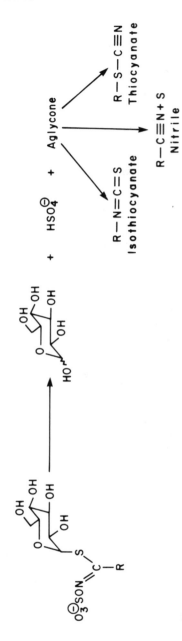

Figure 1 Hydrolysis products of thioglucosides.

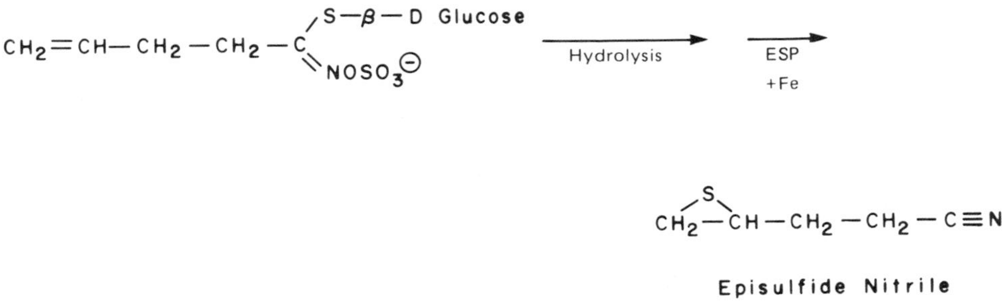

$$CH_2=CH-CH_2-CH_2-C \overset{\displaystyle /S-\beta-D \text{ Glucose}}{\underset{\displaystyle \diagdown NOSO_3^{\ominus}}{}}$$ →Hydrolysis→ →ESP +Fe→

$$CH_2 \overset{\displaystyle S}{\overbrace{}} CH-CH_2-CH_2-C\equiv N$$

Episulfide Nitrile

Figure 2 Episulfide nitrile formation.

sides have been identified in edible plants: amygdalin (benzaldehyde cyanohydrin gluco-side), dhurrin (*p*-hydroxybenzaldehyde cyanohydrin glucoside), and linamarin (acetone cyanohydrin glucoside). Amygdalin is present in bitter almonds and other fruit kernels; dhurrin, in sorghum and related grasses; and linamarin, in pulses, linseed, and cassava. HCN (degradation product of a cyanogenetic glucoside) yields as high as 245 mg per 100 g from immature bamboo shoots have been reported. The lethal dose of HCN for humans is of the order of 0.5-3.5 mg/kg body weight, and occasionally sufficient quan-tities of cyanogenetic foods are consumed to cause fatal poisoning in humans. The possi-bility of chronic toxicity resulting from regular consumption of low levels has been sug-gested but not proved.

Hydrolysis of linamarin is shown in Fig. 3. The enzymic process can be initiated by cell disruption, which occurs when the food is bruised. The hydrolase is extracellular, but once the cell wall barrier is broken it can bind to the intracellular cyanogen, yielding HCN. It is well known that bruised cassava root can be quite toxic (75,132).

It had previously been thought that the primary function of the enzyme rhodanese was to protect against cyanide poisoning; however, it has since been discovered that it can only protect against low doses of HCN. Other observations, such as the distribution of rhodanese in tissues and its subcellular location, have led to a reexamination of the

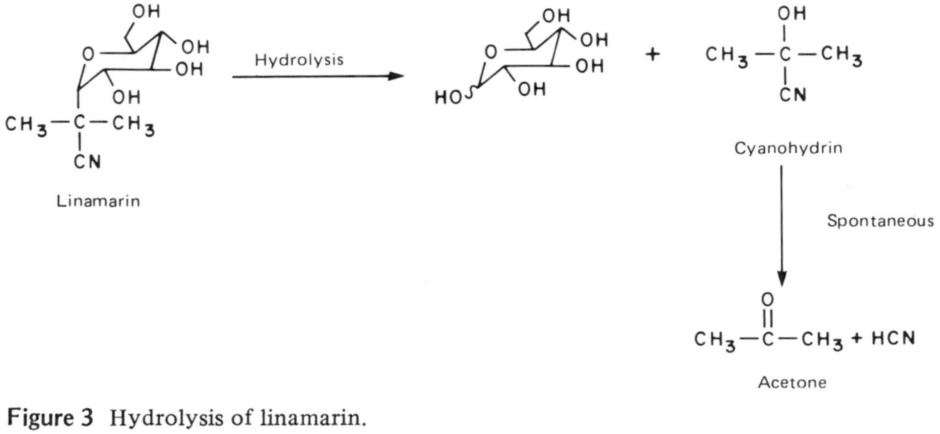

Figure 3 Hydrolysis of linamarin.

role of this enzyme. It is now thought to be part of a multienzymatic system involved in the regulation of the sulfane pool (125).

4. Gossypol

Gossypol (Fig. 4) and several closely related pigments occur at levels of 0.4-1.7% in pigment glands of cottonseeds (8). It is a highly reactive substance and causes a variety of toxic symptoms in domestic and experimental animals. It also causes a reduction in the nutritive value of cottonseed flour, a protein source of increasing importance for human feeding. Glandless, gossypol-free cottonseed is now being developed by selective plant breeding.

5. Phytoalexins

Phytoalexins, often referred to as "stress metabolites," are secondary metabolites in plants and they are produced in response to various stresses such as fungal infection, ultraviolet (UV) light, cold, treatment with heavy metal salts, and physical injury (4). A number of them have been isolated and structurally characterized. Almost all the studies to date have involved certain species of the Legummosae and Solanaceae. Initial studies were carried out in peas (*Pisum sativum*) and beans (*Phaseoulus vulgaris*) infected with fungi, and the compounds pisatin and phaseollin, respectively, were isolated from these vegetables (Fig. 5) (26,83). Since then many phytoalexins have been isolated including ipomeamarone (sweet potatoes), wyerone (fava beans), and the psoralins (celery). Psoralins induce skin photosensitivity, and the feeding of blighted (psoralin-containing) sweet potatoes can cause pulmonary edema and death in livestock. Pisatin and phaseollin are known to be effective in lysing erythrocytes in vitro; however, not much is known regarding in vivo toxicity. The effect of long-term, low-dose exposure to phytoalexins is unknown.

6. Pyrrolizidine Alkaloids

These alkaloids are widely distributed in the plant kingdom and are found in a large number of genera, including Boragenaceae, Compositae, and Legummosae, to name a few (18). Over 150 have been isolated and structurally characterized. The basic ring structure is shown in Fig. 6.

Plants containing these substances grow easily in the southeastern and western United States and are responsible for substantial loss of livestock. In 1972, the value of horses and cattle lost in Oregon due to consumption of *Senecio jacobea* (ragwort) was estimated to be 20 million dollars (109). The responsible alkaloids cause venoocclusive

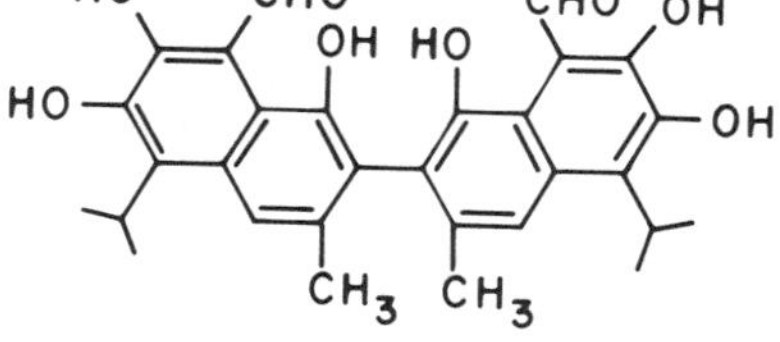

Figure 4   Gossypol.

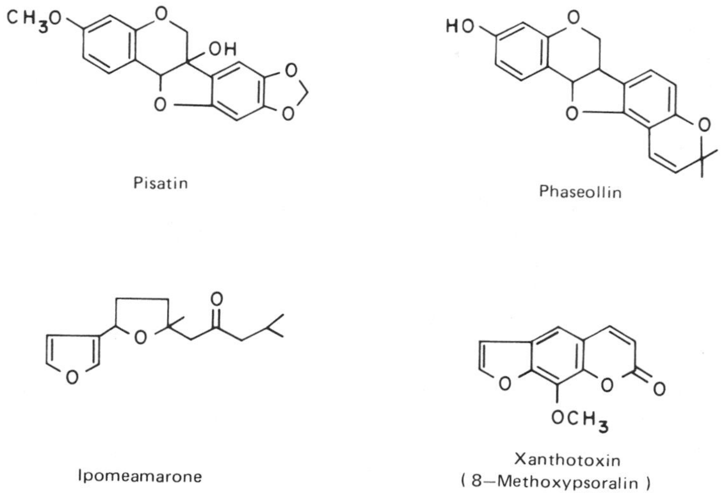

Figure 5 Examples of phytoalexins.

disease in the liver, sometimes exhibit toxicity to the lungs, and recently some have been shown to be carcinogenic. They find their way to humans via herbal teas and as contaminants in wheat fields and recently some have been found in honey (27). Large-scale poisonings from pyrrolizidine alkaloids have occurred in Africa and Afganistan, where in the latter case 1700 people were adversely affected (74). The disease seems to mimic Reyes syndrome. It is thought that the parent compound must be metabolized to a pyrrole intermediate to exert its toxicity.

7. Other Plant Toxicants

The other plant toxicants listed in Table 1 tend to be of concern to more restricted populations, based on unusual patterns of intake or extraordinary sensitivity to the toxic agents. Human neurolathyrism (80), a crippling disease that results from degenerative lesions of the spinal cord, is known to occur only in India. Although it is associated with ingestion of certain varieties of *Lathyrus sativus*, the causative agent is unknown. The toxic amino acids listed in Table 1 occur in the offending plant and they have been shown to induce lesions in animals reminiscent in some respects of the symptoms of lathyrism in humans. They are therefore suspected, but not proved, to be involved in causing the disease.

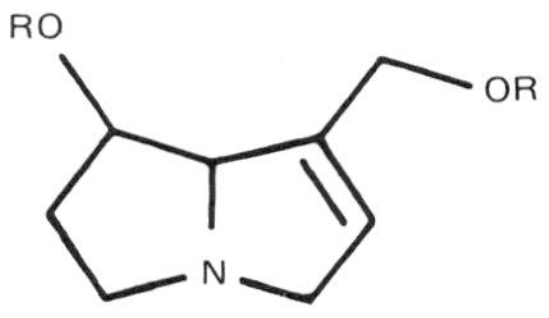

Figure 6 General structure of pyrrolizidine alkaloids.

Unlike true toxins that have undesirable effects on anyone who consumes them, food allergens exert their undesirable effects not through innate toxicity, but rather through their ability to provoke undesirable reactions (i.e., allergy) in susceptible individuals (82). The range of food constituents known to cause allergic responses in sensitive individuals is very wide, and virtually all kinds of foods are involved.

Cycasin (131) represents still another kind of problem. This compound, the glucoside of methylazoxymethanol, is a normal component of a number of plants that serve as an emergency source of starch for some populations in the Pacific and Japan. Although it has very potent carcinogenic activity in animals, it appears that traditional methods of processing the starch effectively remove the toxic substance. Its importance to humans is therefore uncertain.

Favism (67) is a clinical syndrome in humans consisting of acute hemolytic anemia and related symptoms resulting from the ingestion of fava beans (*Vicia faba*) or inhalation of pollen from the plant. The disease shows a remarkable localization in the insular and littoral regions of the Mediterranean area and has been attributed to an inborn error of metabolism with an ethnic distribution. Susceptible individuals have a deficiency of glucose-6-phosphate dehydrogenase in erythrocytes, sensitizing them to the active agents in the bean and resulting in acute hemolytic damage. Pyrimidine glucosides have been isolated from the fava bean that after hydrolysis to the aglycone cause the depletion of erythrocyte glutathione. This ultimately leads to a cell that cannot respond to a further oxidative challenge. The structures of the β-glucosides are shown in Fig. 7.

B. Animal Foodstuffs

Poisonous animals, that is, those whose tissues are toxic and cause adverse responses when eaten, are restricted almost entirely to marine forms. Their presence among the edible species of marine animals creates a problem that seems certain to become increasingly important as people turn to the ocean for additional sources of animal protein. The problem is especially difficult because present knowledge about the nature of the toxic agents and the factors determining their occurrence is so limited that it is impossible to predict with any certainty when and where the toxins occur.

More than 1000 species of marine organisms are known to be poisonous or venomous, and many of these are edible forms or otherwise enter the food chain (47-49). Toxins causing them to be poisonous vary considerably in their chemistry and toxicology. Some appear to be proteins of high molecular weight; others are guanidinium-type compounds of small size. Most have not been isolated or purified.

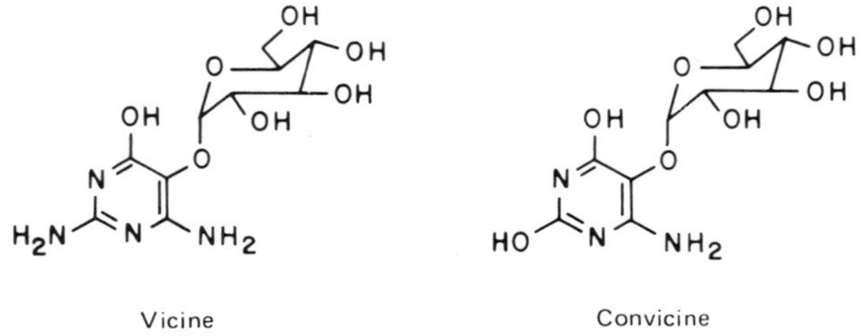

Figure 7 Pyrimidine glucosides isolated from fava beans.

Two main types of poisoning by marine animals are recognized: fish poisoning (resulting from eating fish containing poisonous tissues), and shellfish poisoning (resulting from ingestion of shellfish that have concentrated toxins from the plankton they have consumed). These are known, respectively, as icthyotoxism and paralytic shellfish poisoning (29).

1. Paralytic Shellfish Poisoning

This syndrome is caused by eating clams or mussels that have ingested toxic dinoflagellates and have effectively concentrated the toxic agents contained in them. Shellfish become toxic when local conditions favor growth (blooms) of the dinoflagellates beyond their normal numbers, such circumstances being referred to as "red tides." The organisms involved along the American coastlines are usually species of *Gonyaulax*, although other genera and species are also toxic (91).

The toxic agent has been isolated and purified from cultures of the dinoflagellate and from toxic shellfish. The purified toxin, with an empirical formual of $C_{10}H_{17}N_7O_4 \cdot$ 2HCl, is known as saxitoxin (15,92). Other toxins from various dinoflagellates, many of which are very closely related to saxitoxin, have been isolated and structurally characterized (108). Saxitoxin also has been isolated from fresh water blue-green algae (77). Saxitoxin is stable to heat and is not destroyed by cooking.

The purified toxin has an LD_{50} of 9 $\mu g/kg$ body weight in mice (intraperitoneal injection), and the estimated total lethal dose for humans is thought to be between 1 and 4 mg. The toxin depresses respiratory and cardiovascular regulatory centers in the brain, and death usually results from respiratory failure. The fatality rate of affected individuals is 1-10% in most outbreaks.

2. Icthyotoxism

About 500 species of marine fishes are known to be poisonous when eaten, and many of these species are among the edible varieties. Poisoning syndromes resulting from their ingestion are variable in character and are usually named after the kind of fish involved, such as ciguatera, tetraodon, scombroid, clupeoid, cyclostome, or elasmobranch. The general character of the problem can be illustrated by selected examples.

Ciguatera poisoning is the most common form of fish poisoning. It can occur following ingestion of a wide variety of commonly used food fishes, such as groupers, sea basses, and snappers. This form of toxicity is associated with the food chain relationship of the fish. The toxic agent apparently originates in a blue-green algae and is then passed directly to herbivorous fish and indirectly to carnivorous species (39). The toxic agent has been isolated in pure form and has an empirical formula of $C_{35}H_{65}NO_8$. Its LD_{50} in mice is 80 $\mu g/kg$ body weight, but its precise mode of action is unknown. Death of poisoned individuals appears to be due to cardiovascular failure.

Clupeoid poisoning sometimes occurs following eating of certain herring, anchovies, tarpons, or bonefishes and is particularly prevalent in the Caribbean. The mode of development may be similar to that of ciguatera poisoning, but the source and character of the toxin are unknown. The clinical syndrome, however, is well characterized and death is common.

Tetraodon (puffer fish) poisoning is probably the most widely known and studied type of fish poisoning (87). Puffer fish are not commonly used for food but are consumed under special circumstances in Japan, where fatal poisonings are occasionally reported. Tetrodotoxin is probably the most lethal of all fish poisons.

These few examples serve to emphasize the need for additional research on the occurrence and nature of toxins in marine animals. Such information is imperative as an aid to determining which marine animals can be safely utilized as food and which cannot.

IV. INTENTIONAL FOOD ADDITIVES

Intentional food additives are chemicals added to foods to accomplish one or more objectives: improvement of nutritive value, maintenance of freshness, creation of some desirable sensory property, or to aid in processing. At present, more than 2000 chemicals are added to foods for these purposes (38). Since they are numerous and some are consumed in relatively large quantities over a lifetime, it is important that conditions of safe use be accurately established.

In most technologically developed countries, the use of food additives is regulated through legal mechanisms that not only specify the precise conditions under which the additive can be used but also require evidence from appropriate studies with experimental animals indicating that the compound does not have a significant adverse effect on human health. During the last two decades, greater amounts and kinds of animal data have been required by regulatory agencies and other groups charged with the responsibility of safety evaluation, and the testing procedures have become more sophisticated. Evidence indicating a lack of hazard must be based on studies of acute and chronic toxicity in at least two species of animals, results from a variety of biochemical tests for toxicity, results from studies of reproduction, results from tests for carcinogenicity, mutagenicity, and teratogenicity, and results from studies of the metabolic fate of the material.

Every newly proposed food additive must satisfactorily pass this complete toxicological evaluation before a regulation permitting its use can be issued. Once approved under these rules, an additive can be considered safe within the limits of the ability of animal testing to detect toxicological hazard. The great majority of food additives fall into this category, although some additives of long-standing use have not undergone the intensive testing described above.

New evidence has become available, however, that indicates the existence of toxicological problems with certain intentional food additives. Since it is not feasible to attempt a detailed discussion of all classes of food additives, some examples of current interest are discussed as illustrative of the kinds of problems that can arise.

A. Nitrites and N-Nitroso Compounds

Sodium nitrite has a long history of use as a preservative and color stabilizer, particularly in meat and fish products. Regulations governing maximum use levels were based on no-effect levels with respect to acute toxicity in animals. Recently, however, it has been discovered that the use of nitrite may present another kind of hazard through its ability to interact with amines or amides, thus forming N-nitroso derivatives of considerable toxicological concern.

Nitrosamines can form by the reaction of secondary or tertiary amines with N_2O_3—the active nitrosating reagent in most food products—through the following type of reaction:

$$R_2NH + N_2O_3 \rightarrow R_2N \cdot NO + HNO_2 \tag{1}$$

$$R_3N + N_2O_3 \rightarrow R_2N \cdot NO + R \tag{2}$$

Nitrosation of amines may take place in foods during storage or processing, and nitrosamines may then be ingested as such. Alternatively, it has been shown that nitrosation can take place in the strongly acid conditions of the human stomach, in which case ingestion of the precursors can lead to the in vivo formation of nitrosamines or nitrosamides.

The kinetics of nitrosamine formation from secondary amines is third order with a pH optimum of 3.4, the pK_a of nitrous acid; thus,

$$v = k_n [HNO_2]^2 [amine] \tag{3}$$

Detailed investigations of the prevalence of this type of food contamination have been conducted utilizing very sensitive and accurate methods for determining the presence of nitrosamines in food. The thermal energy analyzer is capable of detecting 0.1 mg of a nitrosamine and is used as a detector with liquid and gas chromatography.

Meats, fish, and cheese are the principal types of products likely to contain significant amounts of nitrosamines, and representative data so far indicate that the levels of nitrosamines generally found are in the range 10-40 ppb. Dimethylnitrosamine is the predominant compound in most instances, but other kinds of nitrosamines have also been detected (32,33,38,42,99,100,115,121).

N-Nitroso compounds are of toxicological concern because many of the known representatives of the class are potent carcinogens in animals. Approximately 80% of more than 100 N-nitroso compounds so far tested are carcinogenic for one or more tissues of experimental animals (66). Furthermore, formation of carcinogenic levels of nitrosamines in vivo has been achieved by the simultaneous feeding of high levels of nitrate and amines to animals (72,101). Recent studies have shown that reducing agents, such as sodium ascorbate and sodium erythrobate, inhibit the formation of dimethylnitrosamine in frankfurters (34). Other studies indicate that the extent to which these compounds inhibit the reaction depends on the particular reducing agent and the experimental conditions (73).

In general, the levels of nitrosamines so far detected in foods have been far below the critical dose in animals. However, these observations have stimulated reconsideration of the use of nitrite as a food additive and also have promoted additional research to determine the conditions under which nitrosation can take place in vivo and in foods. The possible role of nitrosamines in the etiology of human cancer incidence at widely different geographic locations is also under investigation.

B. Carrageenan

"Carrageenan" is a family of naturally occurring hydrocolloids consisting of high-molecular-weight, linear, sulfated polysaccharides. They are commercially extracted from several related species of red seaweeds and are widely used by food and other industries as stabilizers, thickeners, and gelling agents.

Chemically, carrageenans share the common feature of a regularly repeating linear structure of alternating $\alpha(1,3)$-linked and $\beta(1,4)$-linked galactose units. The 1,3-linked units occur as monosulfates and the 1,4-linked units as mono- and disulfates, the anhydride and the anhydride sulfate. Food-grade carrageenans have average molecular weights of 100,000 daltons or higher, and those less than 10,000 are not used as food additives.

These substances have a long history of use, and applications have been increasing, particularly during the last decade. Estimated average daily intake per capita in the

United States currently is about 0.5 mg/kg body weight. Infants on formula diets ingest as much as 50-80 mg/kg · day.

The toxicology of carrageenan in animals is influenced by the route of administration. It is a strongly inflammatory substance when injected subcutaneously or intraperitoneally. Early feeding studies in rats indicated that it was toxicologically inert when orally ingested, apparently owing to lack of absorption. However, somewhat more recent studies have linked oral ingestion of partially hydrolyzed carrageenan to ulcerative colitis and other tissue changes in the gastorintestinal tract of some species of laboratory animals (62,122-124). Feeding studies with natural carrageenan used in food applications show that it is not ulcerogenic in monkeys and rats but has some undesirable effects in guinea pigs (1,7,12,68,107). It thus appears that native, undegraded carrageenan is not ulcerogenic to most laboratory species, apparently due to its lack of absorption through the intestinal mucosa. The ulcerogenic action in guinea pigs is thought to indicate that herbivorous animals have the ability to absorb the high-molecular-weight material.

Further investigation of this problem is now in progress, involving studies on the metabolic fate of carrageenan, as well as further evaluation of the effects of chronic feeding of the substance as it is used in foods.

C. Antioxidants

Butylated hydroxy toluene (BHT) and butylated hydroxy anisole (BHA) are two of the most widely used chemicals to retard oxidation. At present, some controversy surrounds the use of these antioxidants; however, there is a growing literature on the anticarcinogenic activity of these compounds. Many reducing agents, e.g., ascorbate, α-tocopherol, BHT, and BHA, inhibit nitrosamine formation, and BHA inhibits bracken fern carcinogenesis (81).

D. Safrole

Safrole (4-allyl-1,2-methylenedioxybenzene) is a component of many essential oils, such as star anise oil, camphor oil, and sassafras oil. It also occurs in oil of mace, ginger, California bay laurel, black pepper, and cinnamon leaf oil (52). It was widely used as a flavoring agent in soft drinks but was discontinued when it was shown to induce liver tumors in rats (46,65). High levels (0.5%) in the rat diet are required to induce a high incidence of malignant tumors. Related compounds, isosafrole and dihydrosafrole, are also carcinogens in rats, inducing liver and esophageal tumors, respectively. Mice do not show this organotropic effect.

Safrole must be metabolized to exert its carcinogenic effect. Many mammals oxidize safrole to 1′-hydroxysafrole, a demonstrated carcinogen in animals (13,14). Conjugation and other routes of metabolism lead to yet more potent carcinogens (128, 129).

E. Azo Dye Color Additive (Butter Yellow)

Butter yellow (4-dimethylammoazobenzene), so called because of its use in some countries as a coloring agent in butter and margarine, induces liver and bladder tumors when fed at suitably high levels to rats. Structure-activity relationships developed from the many compounds studied have shown that relatively minor alterations in structure (e.g., ethyl substiution for methyl) lead to toxicologically inactive compounds. In general, compounds that are highly polar and contain only one azo linkage are not carcinogenic (e.g., amaranth, FD&C Red No. 2); those that are polar but contain two azo bonds that can be converted to benzidine by in vivo reductive cleavage are carcinogenic (e.g., direct blue 6) (94).

F. Sulfites

Sulfur dioxide and SO_2-generating compounds are added to foods to act as antimicrobial preservatives, to inhibit browning, and as antioxidants. Several chemical forms are used for specific purposes, including SO_2 gas and salts of sulfite, bisulfite, and metabisulfite. These compounds have a long history of use as effective food additives. Based on toxicological evaluations that indicated an apparently low order of toxicity, the U.S. Food and Drug Administration proposed to affirm GRAS status for these agents as direct food additives for specific applications [*Federal Register 47*(142):29956, July 9, 1982].

The safety of sulfites continues to be a matter of some debate because of several observations suggesting potential health hazards. The most significant of these is the apparent sensitivity of certain individuals, especially asthmatics, who develop severe bronchospasms after exposure to sulfites. It has been estimated that as many as 500,000 to 1 million persons in the U.S. population (5-8% of asthma sufferers) may exhibit some degree of sensitivity to sulfite-induced bronchospasm.

A number of case reports have been published attributing this response to eating sulfite-containing foods, particularly salad bar meals (110), which can contain between 100 and 200 mg sulfites. Wines and beers may also represent an important source, since they contain sulfites at levels of about 200 and 20 mg/l, respectively, and have also produced symptoms in sensitive individuals (5,110). Clinical symptoms, which include faintness, weakness, tightness of the chest, and dyspnea, develop with rapid onset following exposure.

The mechanisms underlying these exposures have not been fully elucidated, but experimental studies have produced clear evidence that SO_2 exposure can produce mild bronchoconstriction even in normal individuals, and that asthmatics develop symptoms of greater magnitude and at lower levels of exposure than healthy subjects.

A second concern relates to conflicting evidence relating to the possible carcinogenicity of sulfites. Whereas these compounds were found not to be carcinogenic in several animal studies, evidence does exist that SO_2 may act as a promoting agent in producing lung tumors in rats treated with the carcinogen benzo[a]pyrene (45). Furthermore, sulfites are mutagenic and genotoxic to a variety of bacteria and fungi and therefore represent hypothetical hazards on this basis, although the magnitude of risk resulting from mutagenicity per se cannot now be assessed.

Sulfites are also capable of destruction of thiamine by cleavage, and when premixed into a diet can result in significant thiamine loss upon prolonged storage. However, from the standpoint of human nutrition, the impact of thiamine loss in persons consuming a typical diet would be minimal. Attention has also been directed to the possibility that sulfites may represent a hazard to individuals deficient in the enzyme sulfite oxidase, which is responsible for conversion of sulfites to inorganic sulfates, the form excreted in urine. A genetically transmitted deficiency in this enzyme has been identified, but is thought to be a very rare disease. In such individuals, exogenous sulfites might represent a toxicological hazard, but in view of the fact that endogenous sulfite generation exceeds estimated exogenous intake by a large factor, the magnitude of the added risk would appear to be very small.

V. PRODUCTS OF MICROBIAL GROWTH

A. Mycotoxins

Mold spores are ubiquitously distributed in nature, and they easily germinate and grow on foods and feeds, especially under moist conditions. That moldiness generally results in unpleasant flavors and other undesirable changes in products has been known for a long time. However, another possible consequence of mold growth, first noted long ago, has received appropriate attention only within the last two decades. This consequence results from the ability of some molds to manufacture, during their growth period, poisonous substances that produce toxic symptoms of various kinds when foods or feeds containing them are eaten by humans or animals. These poisonous substances are referred to generically as mycotoxins, and the toxicity syndromes produced by them are called mycotoxicoses.

Contamination of the food supply by mycotoxins gives rise to problems of several kinds. A direct hazard to human health can result when mycotoxin-contaminated foods are eaten by humans. It is important to note that mycotoxins remain in the food after the mold that has produced them has died and can therefore be present in foods that are not visibly moldy. Furthermore, many kinds of mycotoxins, but not all, are relatively stable substances that survive the usual conditions of cooking or processing.

Problems of a somewhat different character can be created if livestock feed becomes contaminated with mycotoxins. In addition to economic losses through the death or unhealthfulness of livestock, mycotoxins or metabolic products of them can remain as residues in meat or be passed into milk and eggs and thus eventually be consumed by people.

Historically, mass poisoning of human populations by mycotoxins has occurred under two circumstances. Ergotism, a toxicosis resulting from eating grains contaminated with *Claviceps purpurea*, occurred in epidemic proportions during the Middle Ages, and more restricted outbreaks have occurred as recently as 1951 in France. Alimentary toxic aleukia (ATA) is a mycotoxicosis caused by eating grain that becomes moldy as a result of overwintering in the field. Both toxicoses are of acute onset and are clearly associated with consumption of large doses of mycotoxins.

In contrast to the small number of documented cases of mycotoxicoses in humans, literally hundreds of reports have appeared in the literature regarding toxicity syndromes in livestock caused by consumption of moldy feeds. In a few such instances, the causative fungi and toxic agents have been identified, but in most cases they remain unknown.

It is important to consider mycotoxins from the perspective of their real or possible significance in human health. In this context, the toxins and fungi producing them fall into one or more of the following general categories:

1. Mycotoxins known, by direct evidence of exposure and response, to have caused some form of toxicity in humans
2. Mycotoxins known to be toxic to animals, to occur on occasion in human foods or foodstuffs, and to constitute possible but unproven hazards to humans
3. Fungi, isolated from human foods or foodstuffs, that produce mycotoxins under laboratory conditions and constitute possible but unproven hazards to humans
4. Mycotoxins that have been observed in feeds or forage of domestic animals, that are known to cause toxicity syndromes in these animals, and present a possible hazard to humans through residues in edible tissues or in products made from them

5. Fungi isolated from animal feeds, litter, or various other sources that produce metabolites with toxic properties in experimental bioassays, but have not been observed in foods and probably do not pose a hazard to animals or humans

1. Mycotoxins That Sometimes Occur in Human Foods

Information in Table 2 relates the first four categories, just discussed, to those situations in which there is direct evidence or reasonable expectation of a human hazard. For more detailed information on mycotoxins, the reader is referred to several comprehensive reviews (19,22,41,55,56,120).

It is clear from the evidence summarized in Table 2 that mycotoxins present public health hazards of a wide range of types and severity. Aside from ergotism, as mentioned earlier, alimentary toxic aleukia is the only mycotoxicosis for which there is extensive direct evidence of mass human poisoning. That syndrome, for which the exact chemical agents have never been identified, occurred in various areas of the USSR during the latter years of World War II. Epidemics of the syndrome involving many thousands of persons were associated with the use of millet and other grains that could not be harvested at the usual time and were allowed to overwinter in the field. The grains became heavily molded during the spring thaws, were used because of food shortages, and resulted in widespread toxicosis.

The significance to humans of the remaining mycotoxins listed in Table 2, except for the aflatoxins, must be assessed from inferences drawn from limited data on the occurrence and biological activity of these substances. Based on evidence available at present, it is not possible to make a meaningful generalization concerning the hazard these substances represent to human health.

2. Aflatoxins

Because research on aflatoxins has stimulated additional studies in the mycotoxin field, and because much is known about them, including their significance to humans, this group of mycotoxins warrants further brief elaboration. Various aspects of this subject have been comprehensively surveyed (28,41).

Aflatoxins are produced by a few strains of *Aspergillus flavus* or *Aspergillus parasiticus*, the spores of which are widely disseminated, especially in soil. Although toxin-producing fungi usually produce only two or three aflatoxins under a given set of conditions, 14 chemically related toxins or derivatives have been identified. One of these, aflatoxin B_1 (Fig. 8), is most frequently found in foods and is also the most potent toxin of the group.

With respect to substrate, requirements for toxin production are relatively non-specific, and appropriate molds can produce toxins on virtually any food (or synthetic media) that supports growth. Thus, any food material must be considered liable to aflatoxin contamination if it becomes moldy. However, experience has shown that the frequency and levels of aflatoxin found in foods depend greatly on the type of food and on the region where the food is grown.

As regards their toxic and other biological effects, the aflatoxins are very interesting compounds. Acute or subacute poisoning can be produced in animals by feeding aflatoxin-contaminated diets or by dosing them with purified preparations of the toxins. Although there are species differences in susceptibility to acute toxicity, a completely refractory species of animal is not known to exist. Symptoms of poisoning are produced

Table 2 Mycotoxins and Toxin-Producing Fungi That Sometimes Occur in Human Foods

Toxin or syndrome	Fungal sources	Foods mainly affected	Chief pharmacological effects after ingestion	References
Aspergillus toxins				
Aflatoxins	*A. flavus*, *A. parasiticus*	Peanuts, oil seeds, grains, pulses, and others; residues in food of animal origin	Toxic to liver; carcinogenic to liver of several animals and possibly humans	3, 28, 41, 130
Sterigmato-cystin	*A. nidulans*, *A. versicolor*	Cereal grains	Toxic and carcinogenic to liver of rats	3
Ochratoxins	*A. ochraceous* *P. viridicatum*	Cereal grains; green coffee; residues in animals	Toxic to kidney of rats	20, 84, 111
Penicillium toxins				
Luteoskyrin	*P. islandicum*	Rice and other grains	Toxic, possibly cancinogenic to liver of rats	58, 59
Patulin	*P. articae*; *P. claviformi*; others	Apple products, cereals, wheat	Edema; toxic to kidney of rats	21, 30
Fusarium toxins				
Zearalenone	*Gibberella-zeae*	Corn, wheat barley, oats	Hyperestrogenism in swine and laboratory animals	70, 93
Alimentary toxic aleukla (ATA)	*F. poae, F. sporotrichioides*	Millet and other cereal grains	Panleukocyto-penia due to bone marrow damage; mortality up to 60% in human epidemics	71
12,13-Epoxy-tricothecanes	*Fusarium* spp., *irichoderma* spp., *Glio-iricothecium* spp.	Corn; other cereal grains	Cardiovascular collapse; increased clotting time, leukopenia; may have been involved in ATA in humans	6, 54

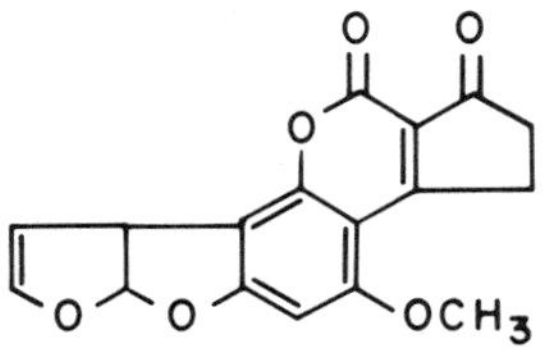

Figure 8 Aflatoxin B_1.

in most domestic animals by feed containing as little as 10-100 ppm or less of aflatoxin. Cattle can tolerate relatively high levels of the toxin, and contaminated cattle secrete milk containing aflatoxin M_1.

Aflatoxin B_1 is among the most potent chemical carcinogens known, and it is this property that has provided an important stimulus for research on mycotoxins in general. Carcinogenic activity of various aflatoxins has been demonstrated experimentally in ducks, rainbow trout, ferrets, rats, mice, and rhesus monkeys. Rainbow trout are an extremely sensitive species, developing cancer when fed diets containing less than 1 ppb alfatoxin B_1. A high incidence of liver cancer is induced in rats by feeding diets containing more than 15 ppb aflatoxin B_1.

A very important aspect in the history of the aflatoxin problem is the early and continuing emphasis that has been placed on the development of assay methods for detection and quantifying levels of these toxins in foods. Availability of such methods, coupled with the application of appropriate handling procedures for feeds and foods, has served to minimize the risk of human exposure to aflatoxins in those countries possessing the necessary technological capabilities. It follows that the hazard of aflatoxins to humans is greater in developing countries where advanced agricultural and scientific practices are less available. It is indeed from those regions that the currently available information on the implications of aflatoxins to human health has come.

This information derives from field studies designed on the basis of the following rationale. With respect to both toxic and carcinogenic consequences in animals, aflatoxins affect mainly, although not exclusively, the liver. It is reasonable to assume that humans respond similarly and that liver disease, particularly liver cancer, is the principal illness associated with aflatoxin exposure. It has been known for a long time that liver cancer, a relatively uncommon form of cancer in the United States and Europe, occurs at a much greater frequency in other populations, particularly those located in central and southern Africa and Asia. Several field studies have been conducted to determine whether the elevated incidences of liver cancer in these areas is associated with aflatoxin exposure.

In one such study, conducted in Uganda, foods eaten by several tribal groups were analyzed for aflatoxin (86). Tribal groups with the highest incidence of the disease were found to have consumed food that was most frequently contamined with aflatoxin.

A study of similar general design and purpose was conducted in Thailand, but in this case cooked samples of foods "as eaten" were systematically collected, and actual intake levels of aflatoxin were calculated (2,102-106). Again, an elevated incidence of liver cancer was associated with high levels of aflatoxin intake. Results of a similar character recently have been obtained from a third study conducted in Kenya.

These epidemiological data do not necessarily establish a cause-effect relationship between induction of liver cancer and ingestion of aflatoxin. Indeed, it is very likely that many factors are involved in the induction of liver cancer. The above data do, however, strongly suggest that it is prudent to minimize the exposure of humans to aflatoxins.

Practically nothing is known about the acute effects of aflatoxins in humans. However, certain lines of evidence suggest that aflatoxins and/or other mycotoxins may be involved in some types of acute poisoning in children. During the study in Thailand, mentioned above, it was possible to study children dying of a form of acute encephalitis, known as Reye's syndrome, involving degeneration of the liver and other visceral organs. When tissues from children fatally affected by this disease were analyzed, it was found that aflatoxin B_1 was identifiable in liver, brain, and other tissues in far greater amounts and frequency than in tissues of children from the same region who died of other causes. This circumstantial evidence suggests that aflatoxins may have been involved in the disease. Of considerable interest is that a sample of rice eaten by one child, immediately before he became ill, contained some aflatoxin B_1 and also contained viable fungal spores that proved to be capable of producing not only aflatoxins, but four other toxic materials of unknown character.

One facet of the study in Thailand also included the isolation of fungi from food samples and the screening of them for the ability to produce toxic materials. Fungi were isolated from the samples, cultured on sterile medium, and extracts of the culture were administered to rats. Using this screening procedure, more than 50 fungi were isolated that proved capable of producing toxic materials other than aflatoxins or other known mycotoxins. This kind of result has been reported by several other investigators studying both foods and feeds.

The presence of toxigenic molds on human foods presents an obvious potential risk to public health, but it is not known to what extent this potential is expressed. However, its existence provides strong motivation to utilize all reasonable approaches to minimize contamination of foods by mycotoxins.

B. Bacterial Toxins

Just as the growth of spoilage molds can introduce toxic chemicals into foods, the growth of certain bacteria can also result in the production of pharmacologically active agents. Botulism, arising from the growth of *Clostridium botulinum*, is probably the best known form of food poisoning because of the extraordinary potency of the toxin involved and the high fatality rate. Poisoning by the enterotoxins produced by *Staphylococcus aureus*, although a much less serious illness, occurs far more frequently. These two forms of bacterially induced food poisoning are the most clearly recognizable syndromes in this category, but a variety of less well defined food-borne bacterial toxins are known to exist (23).

1. Botulism

Botulism is the disease resulting from the ingestion of foods contaminated with the toxin produced by *C. botulinum*, an anaerobic, spore-forming bacillus. Various aspects of the conditions for growth and toxin production, as well as isolation and characterization of the toxin, have been comprehensively reviewed (16,50).

Clostridia with the capacity to produce toxin are ubiquitously distributed, with a natural habitat in the soil, where they are present as spores. Contamination of food with these organisms can readily occur. However, in spite of the wide distribution of various types of botulinum spores in nature, the prevalence of one kind of botulinum poisoning in a given locality is apparently attributable mainly to the dietary habits of the local inhabitants.

Six serological types of toxin-producing *C. botulinum* are known: A, B, (Cα, Cβ), D, E, and F. Of these types, only A, B, and E have been frequently associated with botulism in humans. The toxins produced by these organisms are proteins. The precise molecular weight of the molecule in its physiologically active form is unknown. However, crystalline toxin A as isolated from a culture of the bacterium has a molecular weight of 900,000. This molecule can be chromatographically separated into two fractions of molecular weight 128,000 (α), and 500,000 (β), both of which retain toxicity. Therefore, the actual size of the toxic entity in food poisoning is not clear.

The site of action of botulinum toxin has been established as those synapses in the peripheral nervous system that depend on acetylcholine for the transmission of nerve impulses. Death is a consequence of suffocation resulting from paralysis of the diaphragm and other muscles involved in respiration.

The potency of botulinum toxin is well known. In its purified form, 1 μg of the toxin contains about 200,000 minimum lethal doses for a mouse, and it is suspected that not much more than 1 μg of the toxin may be fatal for humans. *Botulinum* toxins are thermolabile, losing their biological activity upon heating for 30 min at 80°C. This characteristic is of great practical significance, since the toxins can be inactivated by commercial heat processing or most ordinary cooking conditions.

Today, the most frequent cause of botulism in humans is inadequately heated or cured food prepared at home. Commercially prepared foods have been remarkably safe, with the exception of small-scale outbreaks over the least decade involving such products as smoked whitefish, canned tuna, canned liver paste, and canned vichysoisse. Much work is now in progress to prevent similar recurrences in the future.

2. Enterotoxins Produced by *Staphylococcus aureus*

Staphylococcal food poisoning is perhaps the most commonly experienced form of foodborne toxicity. Characteristics of this type of toxicity have been comprehensively reviewed (9,10,117), and only the main points relevant to this discussion are presented here.

Symptoms of the poisoning generally appear 2-3 hr after eating and consist of profusive salivation followed by nausea, vomiting, abdominal cramps, and diarrhea. Most patients return to normal in 24-48 hr, and death is rare. Because so few affected individuals seek medical treatment, the actual incidence of poisoning is unknown but is thought to be very large. Only outbreaks affecting large numbers of individuals usually come to the attention of public health officials.

The cause of the disease is growth and toxin production by *Staphylococcus auerus*, a common organism on the skin and external epithelial tissues of humans and animals. Only a few of the various subtypes of *S. auerus* produce enterotoxin, and this capacity has been difficult to determine except by direct isolation of toxin from contaminated foods. However, easier methods of detection have been developed more recently involving analysis for thermostable nuclease (114).

At least four immunologically distinct enterotoxins are known to exist, and they are designated types A, B, C, and D. Physicochemical studies of enterotoxins A, B, and C have shown them to be proteins of similar composition, with molecular weights in the range 30,000-35,000.

The pharmacological mode of action of the toxins is not clearly understood, but they are very potent in producing their effects. The emetic dose of enterotoxin B in monkeys has been shown to be 0.9 mg/kg body weight, and it has been estimated that humans respond to as little as 1 μg enterotoxin A.

Based on serological procedures, it has been found that enterotoxin A is most frequently encountered, followed by D, B, and C in that order. The emetic activity for monkeys is retained in crude culture extracts even after 1 hr of boiling. Purified toxins are somewhat more sensitive to heat inactivation but still must be regarded as comparatively stable from the viewpoint of their toxicological activity.

Three conditions must be met for staphylococcal food poisoning to occur: (1) enterotoxin-producing organisms must be present; (2) the food must support growth of the pathogen and toxin production (baked beans, roast fowl, potato salad, chicken salad, custards, and cream-filled bakery products are common vehicles); and (3) the food must remain at a suitable temperature for sufficient time for toxin production (4 hr or more at ambient temperature).

VI. UNINTENTIONAL ADDITIVES

Unintentional food additives comprise chemicals that become part of the food supply inadvertently, through any of several different routes. For the most part, these unintentional food additives are residues that enter foods from normal production, processing, and handling procedures, or are contaminants that enter foods through purely accidental circumstances (133). Major categories of the more important examples of unintentional food additives are summarized in the following sections.

A. Chemicals from Processing

1. Fumigants

Ethylene oxide is commonly used as a fumigant to sterilize foods when heat is impractical. This epoxide and others can produce adverse effects by destroying essential nutrients or by reacting chemically with food components to produce toxic products. For example, ethylene oxide can combine with inorganic chlorides to form the corresponding chlorohydrin. Ethylene chlorohydrin, at concentrations up to 1000 ppm, has been found in whole and ground spices that have been fumigated commercially. The chlorohydrins are relatively toxic to animals but effects of chronic exposure to low levels have not been evaluated and no tolerance limits have been set.

Methyl bromide, a widely used fumigant for wheat and cereals, can react with nitrogen or side chains of sulfur-bearing amino acids to yield methylated products (127).

Many studies have been carried out with a variety of fumigants, and in general, some residues almost always remain in the treated food. These residues, although usually not acutely toxic, need further testing to ascertain their chronic effects (90).

2. Solvent Extraction and the Production of Toxic Factors

Extraction of various oil seeds with trichloroethylene was once practiced in several countries, until it was found that the extracted residue was toxic when fed to animals. Soybean oil meal, for example, when extracted in this fashion, invariably caused aplastic anemia when fed to cattle. Ultimately it was shown that the toxic factor was *S*-(dichlorovinyl)-L-cysteine, which was produced by the reaction of the solvent with cysteine in the soybean meal. When trichloroethylene is used to extract caffeine from coffee beans the same residue forms (17). This example clearly demonstrates that a chemical, such as trichloroethylene, even though it is regarded as nontoxic at the levels encountered in foods, can, upon chemical interaction with food constituents, generate highly toxic products.

3. Products of Lipid Oxidation

A large number of changes can be induced in food lipids during commercial processing. Important among these from a toxicological point of view are some of the oxidative and polymerization reactions that take place, particularly after prolonged heating. For example, several investigators have shown that fatty acid monomers and dimers, differing from their natural counterparts, accumulate during abusive deep-fat frying and similar heating conditions. Such heated oils when fed to rats result in depressed growth and food efficiency and also in liver enlargement. The mechanisms responsible for these changes are unknown (133).

Oxidation of lipids can occur quite readily at low temperatures (autoxidation). The primary early oxidation products of unsaturated fatty acids are hydroperoxides resulting from a free radical reaction involving at some point molecular oxygen. For details the reader should refer to Chap. 4. Numerous secondary products of lipid oxidation have been characterized. The lipid hydroperoxide may decompose to aldehydes, ketones, and alcohols or in many instances enter into reactions with proteins. There is a growing body of evidence suggesting that it is the secondary products that are of the greatest toxicological significance (133).

4. Carcinogens in Smoked Foods

The smoking of food for preservation and flavoring is one of the oldest forms of food processing. Despite its long use, surprisingly little is known about the toxicological connotations of the practice. For example, it is well known that products exposed directly to wood smoke become contaminated with polycyclic aromatic hydrocarbons, many of which are known to be carcinogenic in animals (116). It has, in fact, been suggested (but not proved) that the high incidence of stomach cancer in Iceland may be associated with the habitual consumption of heavily smoked meats and meat products. In addition to polycyclic hydrocarbons, wood smoke condensate contains many other classes of compounds (phenolics, acids, carbonyls, and alcohols), most of which have not been adequately investigated for toxicological activity.

5. Pyrolysis Products

Reports that charred meat contained polycyclic aromatic hydrocarbons led to studies of the mutagenicity of various cooked foods, pyrolyzed amino acids, and beef extract. Analysis of the pyrolysis mixture has led to the isolation of potent mutagens as assessed by the *Salmonella* testing system (113). Some of these pyrolysis products have also been shown to be carcinogenic (112), but again the importance to the human population has yet to be evaluated.

6. Pesticides and Herbicides

Fatalities have resulted from the exposure of humans to large doses of very toxic pesticides and herbicides. These instances have been, by and large, occupationally related,

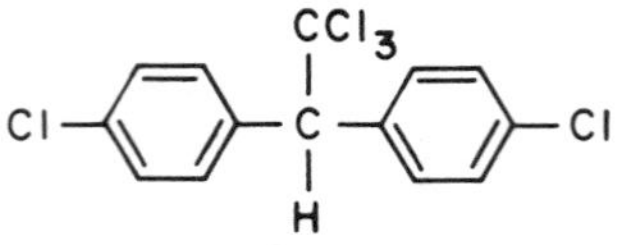

Figure 9   DDT.

but accidental contamination of food sources has also occurred. The organophosphorus insecticides, which act as acetylcholinesterase inhibitors, are used widely and have been involved in human poisoning (78). Chlorophenoxy herbicides, such as 2,4-dichlorophenoxyacetic acid (2,4-D) and 2,4,5-trichlorophenoxyacetic acid (2,4,5-T), are themselves quite toxic and have been found at times to contain the very toxic TCDD (2,3,7,8-tetrachlorodibenzo-*p*-dioxin) as a contaminant.

Toxic agents that resist degradation in the environment have created a different and quite difficult problem. DDT, an organochlorine insecticide, is a well-known example of this type of compound (Fig. 9), and chlordane, aldrin, dieldrin, and lindane are additional examples. The primary problem with such compounds is that they are resistant to biological and chemical breakdown and consequently can build up to high concentrations in the environment. It was not until the third decade of use that environmental problems with DDT surfaced. The stability of the compound and its very great lipid solubility resulted in its appearance in organisms positioned high in the food chain. During the 1950s and 1960s, a period of intensive use of DDT, average DDT concentrations in the lipids of humans were 5 ppm (50). During that period, it has been estimated that an adult ingested 0.2 mg DDT per day and that in 1970, after use of DDT was restricted, ingestion levels dropped to 0.04 mg/day (51).

It is for these reasons that great care must be exercised in selecting the types of pesticides allowed for use with foods and in controlling the levels of pesticides that carry through to the final food product and to the environment.

B. Accidental Contaminants

1. Heavy Metals

Heavy metals have been receiving increased attention as widespread environmental contaminants and as accidental food contaminants. They enter the environment mainly as a result of industrial pollution and find their way into the food chain through a number of routes. The two metals of principal concern in this connection are mercury and cadmium (35,36).

The toxicological importance of mercury depends heavily on its chemical form. Exposure to organic mercurials, especially methyl mercury, is more dangerous than exposure to inorganic salts of mercury. Both forms attack the central nervous system, but exposure to methyl mercury is much more ominous, since lesions induced by it are irreversible. Human response to mercury was well documented after the mass-poisoning episode in the Minimata Bay area of Japan. In this disaster, clinical signs of poisoning were evident when intakes of methyl mercury exceeded 4 μg/kg body weight. Taking into account the total dietary sources of mercury, a "tolerable weekly intake" of 300 μg mercury per person, of which not more than 200 μg should be methyl mercury, has been established.

Practically all the methyl mercury in the diet comes from contaminated fish; other foods generally contain less than 100 ppb of total (inorganic and organic) mercury. In recent years, market-basket surveys reveal a daily intake of 1-20 μg total mercury per individual in the United States and Western Europe.

Cadmium is widely distributed in the enviroment and is readily absorbed when eaten. All available information deals with inorganic salts, even though organic compounds containing cadmium do exist.

A small proportion of ingested cadmium is stored in the kidneys in the form of a metal-protein complex. Long-term exposure to excessive amounts results in damage to the renal tubules in animals and humans. Other long-term effects include anemia, liver dysfunction, and testicular damage.

Human exposure to cadmium takes place mainly through foods, most of which contain less than 50 ppb of the metal. Representative intakes in various parts of the world are in the range 40-80 μg per person day, and a tolerance of 400-500 μg per week has been established.

2. Polychlorinated and Polybrominated Biphenyls (PCB and PBB)

PCB and PBB are stable inert molecules that serve effectively as insulators (e.g., in transformers) and fire retardants and consequently they have had wide industrial use. Their resistance to biological as well as chemical breakdown and their lipophilic properties parallel the troubling characteristics of DDT, discussed earlier.

The discovery in 1966 of PCB in fish attracted attention among the scientific community, since it was thought that these chemicals were used only in closed controlled systems. Subsequent research revealed that they are in fact widespread environmental contaminants representing a potential hazard to human health (62).

With respect to PCB in foods, current information suggests that (a) they rarely appear in fresh fruits and vegetables; (b) they are frequently found in fish, poultry, milk, and eggs; and (c) they can enter foods through migration from packaging materials. Levels generally encountered are in the range 1-40 ppm, with an average of less than 2 ppm.

The toxicological implications of these residues is not entirely clear. Although these chemical do not have a high order of acute or chronic toxicity to animals, they accumulate in adipose tissues, and evidence of human poisoning at very high levels of intake has been reported.

The best example of large-scale poisoning was the accidental contamination of animal feed in Michigan in 1973 (40). About 1 year elapsed before it was determined that a PBB mixture sold as a flame retardant, Fire Master FF-1, was accidentally added to animal feed in place of Nutrimaster, a magnesium oxide feed additive. The Farm Bureau Service distributed this feed within the state of Michigan. Several dairy herds suffered illness (decreased milk production and anorexia), and people ingested contaminated foodstuffs on the farm and via commercial preparations. By the time it was over the Michigan Department of Agriculture determined that livestock on more than 500 farms had been contaminated with PBB. Intensive investigations followed this disaster. To illustrate the persistence of these compounds, two cows that were in the highly contaminated group had 174 and 200 ppm PBB in their body fat when first examined, and 8-12 months later the levels in this herd still ranged from 108 to 2480 ppm (37). These cattle were determined to be clinically normal when they were killed 9-19 months after exposure. It is still unknown what effects the human population has suffered or will suffer as a result of this accident.

Humans were exposed to high concentrations of PCB in Japan when a heat exchanger leaked into rice oil. The consumed oil led to clinical symptoms of chloracne, increased skin pigmentation, numbness, and headaches. PCB and PBB are ubiquitous in the environment, as evidenced by the fact that 31.1% of 637 samples of human adipose tissue collected from the general population in 1971 were positive for PCB. Of the total sample, 5.2% had levels of PCB and PBB in excess of 5.2 ppm (25,134).

3. Chlorinated Naphthalenes

These substances are of interest because they were shown to be the causative agents in a previously unknown disease of cattle known as bovine hyperkeratosis. This toxic syndrome is of great economic importance since it kills large numbers of cattle in affected herds. These compounds are commonly used in wood preservatives and lubricating oils, and the problem arises when feedstuffs become contaminated with machine oil during processing. Other domestic animal species are also susceptible to poisoning, and the compounds are toxic to humans. No evidence exists as to the possible presence of these contaminants in other portions of the food chain.

4. Radionuclides

As concern grows over exposure to low-level radiation, food must be considered a contributor to the total body burden. The National Academy of Science examined this question in 1973 and concluded that radiation from food sources amounted to 20-25 mrad/year, which is about 25% of the average natural background and about 10% the average background when medical and other sources in the United States are included (24).

VII. SUMMARY AND CONCLUSIONS

As illustrated by the examples summarized in this chapter, protecting the safety of the food supply requires control measures to deal with problems that are highly variable in respect to their nature and the amount and quality of available toxicological information concerning them. In some instances, as in the case of alfatoxin contamination, for example, the problem has a long history, has been thoroughly studied, and control measures that minimize exposure of consumers have been devised even though the problem is recurrent and seemingly impossible to eliminate completely. In contrast, discovery of the generation of highly mutagenic substances through pyrolysis in the course of ordinary cooking procedures indicates the existence of hypothetical health hazards that are not yet defined and that cannot be easily eliminated.

In all such instances, progress toward problem recognition and definition, as well as the development of adequate control measures, can be most effectively made by multidisciplinary research efforts, involving the joint participation of chemists, biologists, microbiologists, toxicologists, and epidemiologists. In all such problems, food chemistry plays a role of pivotal importance.

REFERENCES

1.   Abraham, R., L. Goldberg, and F. Coulston (1972). Uptake and storage of degraded carrageenan in lysosomes of reticulendothelial cells of the Rhesus monkey (*Macaca mulatta*). *Exp. Mol. Pathol. 17*:177-184.

2. Alpert, M. E., M. S. R. Hutt, G. N. Wogan, and C. S. Davidson (1971). Association between aflatoxin content of food and hepatoma frequency in Uganda. *Cancer* *28*:253-260.

3. Areos, J. C. (1978). Criteria for selecting chemical compounds for carcinogenicity testing: an essay. *J. Environ. Pathol. Toxicol. 1*:433-450.

4. Bailey, J. A., and J. W. Mansfield (1982). *Phytoalexins*, John Wiley and Sons, New York.

5. Baker, G. J., P. Collett, and D. H. Allen (1981). Bronchospasm induced by metabisulfite-containing foods and drugs. *Med. J. Aust. 2*:614-617.

6. Bamburg, J. R., and F. M. Strong (1971). 12,13-Epoxytrichothecenes, in *Microbial Toxins*, Vol. 7 (S. Kadis, A. Ciegler, and S. J. Ajl, eds.), Academic Press, New York, pp. 207-292.

7. Benitz, K. F., L. Goldberg, and F. Coulston (1973). Intestinal effects of carrageenan in the Rhesus monkey (*Macaca mulatta*). *Food Cosmet. Toxicol. 11*:565-574.

8. Berardi, L. C., and L. A. Goldblatt (1980). Gossypol, in *Toxic Constituents of Plant Foodstuffs*, 2nd ed. (I. E. Liener, ed.), Academic Press, New York, pp. 184-238.

9. Bergdoll, M. S. (1970). Enterotoxins, in *Microbial Toxins*, Vol. 3, (T. C. Montie, S. Kadis, and S. J. Ajl, eds.), Academic Press, New York, pp. 265-326.

10. Bergdoll, M. S. (1980). Staphlococcal food poisoning, in *Safety of Foods*, 2nd ed. (H. D. Graham, ed.), AVI Publishing Co., Westport, Conn., pp. 108-175.

11. Birk, Y., and I. Peri (1980). Saponins, in *Toxic Constituents of Plant Foodstuffs*, 2nd ed. (I. E. Liener, ed.), Academic Press, New York, pp. 161-183.

12. Bonfils, S. (1970). Carrageenan and the human gut. *Lancet 2*:414.

13. Borchert, P., J. A. Miller, E. C. Miller, and T. Shires (1973). 1'-Hydroxysafrole, a proximate carcinogenic metabolite of safrole in rat and mouse. *Cancer Res. 33*: 590-600.

14. Borchert, P., P. G. Wislocki, J. A. Miller, and E. C. Miller (1973). The metabolism of the naturally occurring hepatocarcinogen safrole to 1'-hydroxysafrole and the electrophilic reactivity of 1'-acetoxysafrole. *Cancer Res. 33*:575-589.

15. Bordner, J., W. E. Thiessen, H. A. Bates, and H. Rapoport (1975). The structure of a crystalline derivative of saxitoxin. *J. Amer. Chem. Soc. 97*:6008-6012.

16. Boroff, D. A., and B. R. DasGupta (1971). Botulinum toxin, in *Microbial Toxins*, Vol. 2A (S. Kadis, C. Montie, and S. J. Ajl, eds.), Academic Press, New York, pp. 1-68.

17. Bradenberger, H., and H. Bader (1967). Uber den Einbau von Trichlorathylen in Inhaltsstoffe des Kaffees bei dessen Decoffeinierung. *Helv. Chem. Acta 50*:463-466.

18. Bull, L. B., C. C. J. Culvenor, and A. T. Dick (1968). *The Pyrrolizidine Alkaloids*, North Holland, Amsterdam.

19. Busby, Jr., W. F., and G. N. Wogan (1979). Food-borne mycotoxins and alimentary mycotoxicoses, in *Food-Borne Infections and Intoxications*, 2nd ed. (H. Riemann and F. L. Bryan, eds.), Academic Press, New York, pp. 519-612.

20. Chu, F. S. (1974). Studies on ochratoxins. *CRC Crit. Rev. Toxicol. 2*:499-524.

21. Ciegler, A., R. W. Detroy, and E. B. Lillehoj (1971). Patulin, penicillic acid, and other carcinogenic lactones, in *Microbial Toxins*, Vol. 6, (A. Ciegler, S. Kadis, and S. J. Ajl, eds.), Academic Press, New York, pp. 409-434.

22. Ciegler, A., S. Kadis, and S. J. Ajl (eds.) (1971). *Microbial Toxins*, Vol. 6, Academic Press, New York.

23. Collins-Thompson, D. L. (1980). Food spoilage and food-borne infection hazards, in *Safety of Foods*, 2nd ed. (H. D. Graham, ed.), AVI Publishing Co., Westport, Conn., pp. 15-53.

24. Comar, C. L., and J. H. Rust (1973). Natural radioactivity in the biosphere and food stuffs, in *Toxicants Occurring Naturally in Foods* (A. C. Leopold and R. Ardrey, eds.), National Academy of Sciences, Washington, D.C., pp. 88-105.

25. Cordle, F., P. Corneliussen, C. Jelinek, B. Hackley, R. Lehman, J. McLaughlin, R. Rhoden, and R. Shapiro (1978). Human exposure to polychlorinated biphenyls and polybrominated biphenyls. *Environ. Health Perspect. 24*:157-172.

26. Cruickshank, I. A. M., and D. R. Perrin (1960). Isolation of a phytoalexin from *Pisum sativum* L. *Nature 187*:799-800.

27. Deinzer, M. L., P. A. Thompson, D. M. Burgett, and D. L. Isaacson (1977). Pyrrolizidine alkaloids: Their occurrence in honey from tansy ragwort (*Senecio jacobea* L.). *Science 195*:497-499.

28. Detroy, R. W., E. B. Lilehoj, and A. Ciegler (1971). Aflatoxin and related compounds, in *Microbial Toxins*, Vol. 6 (A. Ciegler, S. Kadis, and S. J. Ajl, eds.), Academic Press, New York, pp. 4-178.

29. Dymsza, H., Y. Shimizu, F. E. Russell, and H. D. Graham (1980). Poisonous marine animals, in *The Safety of Foods*, 2nd ed. (H. D. Graham, ed.), AVI Publishing Co., Westport, Conn., pp. 625-651.

30. Enomoto, M., and Y. Ueno (1974). *Penicillium islandicum* (toxic yellowed rice)-lueteoskyrin-islanditoxin-cyclochlorotine, in *Mycotoxins* (I. F. H. Purchase, ed.), Elsevier, New York, pp. 303-326.

31. Ettlinger, M. G., and A. Kjaer (1968). Sulfur compounds in plants. *Rec. Adv. Phytochem. 1*:59-144.

32. Fazio, T., J. N. Damico, J. W. Howard, R. H. White, and J. O. Watts (1971). Gas chromatographic determination and mass spectrometric confirmation of n-nitrosodimethylamine in smoke-processed marine fish. *J. Agr. Food Chem. 19*:250-253.

33. Fiddler, W., R. C. Doerr, J. R. Ertel, and A. E. Wasserman (1971). Gas-liquid chromatographic determination of *n*-nitrosodimethylamine in ham. *J. Ass. Offic. Anal. Chem. 54*:1160-1163.

34. Fiddler, W., J. W. Pensabene, E. G. Piotrowski, R. C. Doerr, and W. E. Wasserman (1973). Use of sodium ascorbate or erythrobate to inhibit formation of *N*-nitrosodimethylamine in frankfurters. *J. Food Sci. 38*:1084-1085.

35. Friberg, L., and J. Vostal (1972). *Mercury in the Environment*, Chemical Rubber Company Press, Cleveland, Ohio.

36. Friberg, L., M. Piscator, and G. Nordberg (1971). *Cadmium in the Environment*, Chemical Rubber Company Press, Cleveland, Ohio.

37. Fries, G. F., and G. S. Marrow (1975). Excretion of polybrominated biphenyls into the milk of cows. *J. Dairy Sci. 58*:947-951.

38. Furia, T. E. (1972). *Handbook of Food Additives*, Chemical Rubber Company Press, Cleveland, Ohio.

39. Gentile, J. H. (1971). Blue-green and green algal toxins, in *Microbial Toxins*, Vol. 7 (S. Kadis, A. Ciegler, and S. J. Ajl, eds.), Academic Press, New York, pp. 27-66.

40. Getty, S. M., D. E. Rickert, and A. L. Trapp (1977). Polybrominated biphenyl (PBB) toxicosis: An environmental accident. *CRC Crit. Rev. Environ. Control 7*: 309-323.

41. Goldblatt, L. A. (1969). *Alfatoxin—Scientific Background, Control, and Implications*, Academic Press, New York.

42. Gough, T. A., K. S. Webb, and R. F. Coleman (1978). Estimate of the volatile nitrosamine content of UK food. *Nature 272*:161-163.

43. Green, N. M. (1957). Kinetics of the reaction between trypsin and the pancreatic trypsin inhibitor. *Biochem. J. 66*:407-415.

44. Griebel, C. (1950). Erkrankungen durch Bohnenflocken (*Phaseolus vulgaris* L.) und Platterbsen (*Lathyrus tingitanus* L.). *Z. Lebensm Unter. Forsch 90*:191-196.

45. Gunnison, A. F. (1981). Sulfite toxicity: A critical review of in vitro and in vivo data. *Food Cosmet. Toxicol.* 19:667-

46. Hagan, E. C., P. M. Jenner, W. I. Jones, O. Fitzhugh, E. L. Long, J. G. Brouwer, and W. K. Webb (1965). Toxic properties of compounds related to safrole. *Toxicol. Appl. Pharmacol.* 7:18-24.

47. Halstead, B. W. (1965). *Poisonous and Venomous Marine Animals*, Vol. I, United States Government Printing Office, Washington, D.C.

48. Halstead, B. W. (1967). *Poisonous and Venomous Marine Animals*, Vol. II, United States Government Printing Office, Washington, D.C.

49. Halstead, B. W. (1970). *Poisonous and Venomous Marine Animals*, Vol. III, United States Government Printing Office, Washington, D.C.

50. Hauschild, A. 'H. W. (1980). Microbial problems in food safety with particular reference to *Clostridium botulinum*, in *The Safety of Foods* (H. D. Graham, ed.), AVI Publishing Co., Westport, Conn., pp. 68-107.

51. Hayes, Jr., W. J., W. E. Dale, and C. I. Pirkel (1971). Evidence of safety of long-term, high, oral doses of DDT for man. *Arch. Environ. Health* 22:119-135.

52. Homberger, F., and E. Boger (1968). The carcinogenicity of essential oils, flavors, and spices. A review. *Cancer Res.* 28:2572-2374.

53. Jaffe, W. G. (1980). Hemagglutinins (lectins), in *Toxic Constituents of Plant Foodstuffs*, 2nd. ed. (I. E. Liener, ed.), Academic Press, New York, pp. 73-102.

54. Joffe, A. Z. (1971). Alimentary toxic aleukia, in *Microbial Toxins*, Vol. 7, (S. Kadis, A. Ciegler, and S. J. Ajl, eds.), Academic Press, New York, pp. 139-189.

55. Kadis, S., A. Ciegler, and S. J. Ajl (eds.) (1971). *Microbial Toxins*, Vol. 7, Academic Press, New York.

56. Kadis, S., A. Ciegler, and S. J. Ajl (eds.) (1972). *Microbial Toxins*, Vol. 8, Academic Press, New York.

57. Korte, R. (1972). Heat resistance of phytohemagglutinins in weaning food mixtures containing beans (*Phaseolus vulgaris*). *Ecol. Food Nutr.* 1:303-307.

58. Krogh, P. (1976). Mycotoxic nephropathy, in *Advances in Veterinary Science and Comparative Medicine*, Vol. 20 (C. A. Brandly, C. E. Cornelius, and W. I. B. Beveridge, eds.), Academic Press, New York, pp. 147-170.

59. Kunitz, M. (1947). Isolation of a crystalline protein compound of trypsin and of soybean trypsin-inhibitor. *J. Gen. Physiol.* 30:311-320.

60. Laskowski, Jr., M., and R. W. Sealock (1971). Protein proteinase inhibitors: Molecular aspects, in *The Enzymes*, 3rd ed., Vol. III (P. D. Boyer, ed.), Academic Press, New York, pp. 376-473.

61. Laskowski, Jr., M., I. Kato, T. R. Leary, J. Schrode, and R. W. Sealock (1974). Evolution of specificity of protein proteinase inhibitors, in *Protease Inhibitors* (H. Fritz, H. Tschesche, L. J. Greene, and E. Truscheit, eds.), Springer-Verlag, New York, pp.

62. Lee, D. H. K., and H. L. Falk (eds.) (April 1972). *Environ. Health Perspect.*, experimental issue No. 1.

63. Leopold, A. C., and R. Ardrey (1972). Toxic substances in plants and the food habits of early man. *Science* 176:512-514.

64. Liener, I. E., and M. L. Kakade (1980). Protease inhibitors, in *Toxic Constitutents of Plant Foodstuffs*, 2nd ed. (I. E. Liener, ed.), Academic Press, New York, pp. 7-71.

65. Long, E. L., A. A. Nleson, O. G. Fitzhugh, and W. H. Hansen (1963). Liver tumors produced in rats by feeding safrole. *Arch. Pathol.* 75:594-604.

66. Magee, P. N., and J. M. Barnes (1967). Carcinogenic nitroso compounds. *Adv. Cancer Res.* 10:163-246.

67. Mager, J., M. Chevion, and G. Glaser (1980). In *Toxic Constituents of Plant Foodstuffs*, 2nd ed. (I. E. Liener, ed.), Academic Press, New York, pp. 266-294.

68. Maillet, M., S. Bonfils, and R. E. Lister (1970). Carrageenan: Effects in animals. *Lancet 2*:414-415.

69. Marcus, R., and J. Watt (1969). Seaweeds and ulcertative colitis in laboratory animals. *Lancet 2*:489-490.

70. Mirocha, C. J., C. M. Christensen, and G. H. Nelson (1971). F-Z (Zearalenone) estrogenic mycotoxin from *Fusarium*, in *Microbial Toxins*, Vol. 7 (S. Kadis, A. Ciegler, and S. J. Ajl, eds.), Academic Press, New York, pp. 107-138.

71. Mirocha, C. J., S. V. Pathre, and C. M. Christensen (1977). Zearalenone, in *Micotoxins in Human and Animal Health* (J. V. Rodricks, C. W. Hesseltine, and M. A. Mehlman, eds.), Pathotox Publishers, Park Forest South, Ill., pp. 345-364.

72. Mirvish, S. S., M. Greenblatt, and V. R. C. Kommineni (1972). Nitrosamide formation in vivo: Induction of lung adenomas in Swiss mice by concurrent feeding of nitrite and methyluren or ethyluren. *J. Nat. Cancer Inst. 48*:1311-1315.

73. Mirvish, S. S., L. Wallcave, M. Eagen, and P. Shubik (1972). Ascorbate-nitrite reaction: Possible means of blocking the formation of N-nitroso compounds. *Science 177*:65-68.

74. Mohabbat, O., M. Shafig Younos, A. A. Merrar, R. N. Srivastave, Sedig Ghaos Gholam, and G. N. Aram (1967). An outbreak of hepatic veno-occlusive disease in northwestern Afganistan. *Lancet* 269-271.

75. Montgomery, R. D. (1965). The medical significance of cyanogen in plant foodstuffs. *Am. J. Clin. Nutr. 17*:103-

76. Montgomery, R. D. (1980). Cyanogens, in *Toxic Constituents of Plant Foodstuffs*, 2nd ed. (I. E. Liener, ed.), Academic Press, New York, pp. 143-160.

77. Moore, R. E. (1981). Constituents of blue-green algae, in *Marine Natural Products*, Vol. IV (P. J. Scheuer, ed), Academic Press, New York, pp. 1-52.

78. Murphy, S. D. (1980). Pesticides, in *Casarett and Doull's Toxicology, The Basic Science of Poisons*, 2nd ed. (J. Doull, C. D. Klaassen, and M. O. Amdur, eds.), Macmillan, New York, pp. 357-408.

79. Odani, S., and T. Ikenaka (1972). Studies on soybean trypsin inhibitors. IV. Complete amino acid sequence and the anti-proteinase sites of Bowman-Birk soybean inhibitor. *J. Biochem.* (Tokyo) *71*:839-848.

80. Padmanaban, G. (1980). Lathrogens, in *Toxic Constituents of Plant Foodstuffs*, 2nd ed. (I. E. Liener, ed.), Academic Press, New York, pp. 239-265.

81. Pamuku, A. M., S. Yalciner, and G. T. Bryan (1977). Inhibition of carcinogenic effect of bracken fern (*Pteridium aquilinum*) by various chemicals. *Cancer 40*: 2450-2454.

82. Perlman, F. (1980). Allergens, in *Toxic Constituents of Plant Foodstuffs*, 2nd ed. (I. E. Liener, ed.), Academic Press, New York, pp. 295-328.

83. Perrin, D. R., and W. Bottomley (1962). Studies on phytoalexins. V. The structure of pisatin from *Pisum sativum* L. *J. Amer. Chem. Soc. 84*:1919-1922.

84. Purchase, I. F. H. (1972). Aflatoxin residues in food of animal origin. *Food Cosmet. Toxicol. 10*:531-544.

85. Rackis, J. J., and M. R. Gumbmann (1981). Protease inhibitors: Physiological properties and nutritional significance, in *Antinutrients and Natural Toxicants in Foods* (R. L. Ory, ed.), Food and Nutrition Press, Westport, Conn., pp. 203-237.

86. Rodricks, J. V., C. W. Hesseltine, and M. A. Mehlman (eds.) (1977). *Mycotoxins in Human and Animal Health*, Pathotox Publishers, Park Forest South, Ill.

87. Russell, F. E. (1965). Marine toxins and venomous and poisonous marine animals, in *Advances in Marine Biology*, Vol. 3 (F. S. Russell, ed.), Academic Press, London, pp. 255-384.

88. Ryan, C. A., and G. M. Haas (1981). Structural, evolutionary and nutritional properties of proteinase inhibitors from potatoes, in *Antinutrients and Natural Toxi-*

cants in Foods (R. L. Ory, ed.), Food and Nutrition Press, Westport, Conn., pp. 169-185.

89. Saito, M., M. Enomoto, and T. Tatsuno (1971). Yellowed rice toxins: Luteoskyrin and related compounds, chlorine-containing compounds and citrinin, in *Microbial Toxins*, Vol. 6 (A. Ciegler, S. Kadis, and S. J. Ajl, eds.), Academic Press, New York, pp. 299-380.

90. Salunkhe, D. K., and Z. M. T. Wu (1977). Toxicants in plants and plant products. *CRC Crit. Rev. Food Sci. Nutr. 9*:265-323.

91. Schantz, E. J. (1971). The dinoflagellate poisons, in *Microbial Toxins*, Vol. 7 (S. Kadis, A. Ciegler, and S. J. Ajl, eds.), Academic Press, New York, pp. 3-26.

92. Schantz, E. J., V. E. Ghazarossian, H. K. Schnoes, F. M. Strong, J. P. Springer, J. O. Pezzanite, and J. Clardy (1957). The structure of saxitoxin. *J. Amer. Chem. Soc. 97*:1238-1239.

93. Scott, P. M. (1974). Patulin, in *Mycotoxins* (I. F. H. Purchase, ed.), Elsevier, New York, pp. 383-403.

94. Searle, C. E. (ed.) (1976). *Chemical Carcinogens*, ACS monograph 173, American Chemical Society, Washington, D.C.

95. Seidel, D. S., and I. E. Liener (1971). Identification of the trypsin-reactive site of the Bowman-Birk soybean inhibitor. *Biochim. Biophys. Acta 251*:83-93.

96. Seidel, D. S., and I. E. Liener (1971). Guanidination of the Bowman-Birk soybean inhibitor: Evidence for the tryptic hydrolysis of peptide bonds involving homo-arginine. *Biochem. Biophys. Res. Commun. 42*:1101-1107.

97. Seidel, D. S., and I. E. Liener (1972). Isolation and properties of complexes of the Bowman-Birk soybean inhibitor with trypsin and chymotrypsin. *J. Biol. Chem. 247*:3533-3528.

98. Seidel, D. S., and I. E. Liener (1972). Identification of the chyrotrypsin-reactive site of the Bowman-Birk soybean inhibitor. *Biochim. Biophys. Acta 258*:303-309.

99. Sen, N. P., B. Donaldson, J. R. Iyengar, and T. Panalaks (1973). Nitrosopyrrolidine and dimethylnitrosamine in bacon. *Nature* (London) *241*:473-474.

100. Sen, N. P., L. A. Schwinghamer, B. A. Donaldson, and W. F. Miles (1972). *J. Agr. Food Chem. 20*:1280-1281.

101. Shank, R. C., and P. M. Newberne (1972). Nitrite-morpholine-induced hepatomas. *Food. Cosmet. Toxicol. 10*:887-888.

102. Shank, R. C., N. Bhamarapravati, J. E. Gordon, and G. N. Wogan (1972). Dietary aflatoxins and human liver cancer. 4. Incidence of primary liver cancer in two municipal populations in Thailand. *Food Cosmet. Toxicol. 10*(2):171-179.

103. Shank, R. C., J. B. Gibson, A. Nondasuta, and G. N. Wogan (1972). Dietary afla-toxins and human liver cancer. 2. Aflatoxin in market foods and foodstuffs of Thailand and Hong Kong. *Food Cosmet. Toxicol 10*(1):61-69.

104. Shank, R. C., J. B. Gibson, and G. N. Wogan (1972). Dietary aflatoxins and human liver cancer. I. Toxigenic molds in foods and foodstuffs in tropical Southeast Asia. *Food Cosmet. Toxicol. 10*(1):51-60.

105. Shank, R. C., J. W. Gordon, A. Nondasuta, B. Subhamani, and G. N. Wogan (1972). Dietary aflatoxins and human liver cancer. III. Field survey of rural Thai families for ingested aflatoxin. *Food Cosmet. Toxicol. 10*(1):71-84.

106. Shank, R. C., P. Siddhichai, B. Subhamani, N. Bhamarapravati, J. E. Gordon, and G. N. Wogan (1972). Dietary aflatoxins and human liver cancer. V. Duration of primary liver cancer and prevalence of hepatomegaly in Thailand. *Food Cosmet. Toxicol 10*(2):181-191.

107. Sharratt, M., P. Grasso, F. Carpanini, and S. Gangolli (1971). Carrageenan and ul-cerative colitis. *Gastroenterology 61*:410-411.

108. Shimizu, Y. (1978). Dinoflagellate toxins, in *Marine Natural Products*, Vol. I (P. J. Scheuer, ed.), Academic Press, New York, pp. 1-42.

109. Snyder, S. P. (1972). Livestock losses due to tansy ragwort poisoning. *Oregon Agrirec. 255*:3-4.

110. Stevenson, D. D., and R. A. Simon (1981). Sensitivity to ingested metabisulfites in asthmatic subjects. *J. Allergy Clin. Immunol. 68*:26-32.

111. Steyn, P. S. (1971). Ochratoxin and other dihydroisocoumarins, in *Microbial Toxins*, Vol. 6 (A. Ciegler, S. Kadis, and S. J. Ajl, eds.), Academic Press, New York, pp. 179-205.

112. Sugimura, T. (1982). Mutagens, carcinogens, and tumor promoters in our daily food. *Cancer 49*:1970-1984.

113. Sugimura, T., and M. Nagao (1979). Mutagenic factors in cooked foods. *CRC Clin. Rev. Toxicol. 6*:189-210.

114. Tatini, S. R. (1981). Thermonuclease as an indicator of staphylococcal entero-toxins in food, in *Antinutrients and Natural Toxicants in Foods* (R. L. Ory, ed.), Food and Nutrition Press, Westport, Conn., pp. 53-76.

115. Telling, G. M., T. A. Bryce, and J. Althorpe (1971). Use of vacuum distillation and gas chromatography-mass spectrometry for determination of low levels of volatile nitrosamines in meat products. *J. Agr. Food Chem. 19*:937-940.

116. Tilgner, D. J., and H. Daun (1969). Polycyclic aromatic hydrocarbons (polynuclears) in smoked foods. *Residue Rev. 27*:19-41.

117. Tood, E. C. D. (1978). Foodborne disease in six countries—a comparison. *J. Food Protection 41*:559-565.

118. Tookey, H. L. (1973). Crambe thioglucoside glucohydrolase (EC 3.2.3.2): Separation of a protein required for epithibutane formation. *Can. J. Biochem. 51*:1654-1660.

119. Tookey, H. L., C. H. Van Etten, and M. E. Daxenbichler (1980). Glucosinolates, in *Toxic Constituents of Plant Foodstuffs*, 2nd ed. (I. E. Liener, ed.), Academic Press, New York, pp. 103-142.

120. Ueno, Y. (1980). Trichothecene mycotoxins: Mycology, chemistry, and toxicology. *Adv. Nutr. Res. 3*:301-353.

121. Wasserman, A. E., W. Fiddler, R. C. Doerr, S. F. Osman, and C. J. Dooley (1972). Dimethylnitrosamine in frankfurters. *Food Cosmet. Toxicol. 10*:681-684.

122. Watt, J., and R. Marcus (1969). Ulcerative-colitis in guinea-pigs caused by seaweed extract. *J. Pharm. Pharmacol. 21*:187S-188S.

123. Watt, J., and R. Marcus (1970). Ulcertative-colitis in rabbits fed degraded carrageenan. *J. Pharm. Pharmacol. 22*:130-131.

124. Watt, J., M. C. Path, and R. Marcus (1970). Hyperplastic mucosal changes in the rabbit colon produced by degraded carrageenin. *Gastroenterology 59*:760-768.

125. Westley, J. (1980). Rhodonese and the sulfane pool, in *Enzymatic Basis of Detoxication*, Vol. II (W. B. Jakoby, ed.), Academic Press, New York, pp. 245-262.

126. Wilson, K. A. (1981). The structure, function and evolution of legume proteinase inhibitors, in *Antinutrients and Natural Toxicants in Foods* (R. L. Ory, ed.), Food and Nutrition Press, Westport, Conn., pp. 187-202.

127. Winteringham, F. P. W., A. Harrison, R. G. Bridges, and P. M. Bridges (1955). The fate of labelled insecticide residues in food products. II. The nature of methyl bromide residues in fumigated wheat. *J. Food Sci. Agr. 6*:251-261.

128. Wislocki, P. G., P. Borchert, J. A. Miller, and E. C. Miller (1976). Metabolic activation of the carcinogen 1'-hydroxysafrole in vivo and in vitro and electrophilic reactivities of the possible ultimate carcinogens. *Cancer Res. 36*:1686-1695.

129. Wislocki, P. G., E. C. Miller, J. A. Miller, E. C. McCoy, and H. S. Rosenkrautz (1977). Carcinogenic and mutagenic activities of safrole, 1'-hydroxysafrole, and some known or possible metabolies. *Cancer Res. 37*:1883-1891.

130. Wogan, G. N. (1973). Aflatoxin carcinogenesis, in *Methods in Cancer Research* (H. Busch, ed.), Academic Press, New York, pp. 309-344.
131. Wogan, G. N., and W. F. Busby, Jr. (1980). Naturally occurring carcinogens, in *Toxic Constituents of Plant Foodstuffs*, 2nd ed. (I. E. Liener, ed.), Academic Press, New York, pp. 329-370.
132. Wood, T. (1966). The isolation, properties, and enzymic breakdown of linamarin from cassava. *J. Sci. Food Agr. 17*:85-90.
133. Yannai, S. (1980). Toxic factors induced by processing, in *Toxic Constituents of Plant Foodstuffs*, 2nd ed. (I. E. Liener, ed.), Academic Press, New York, pp. 371-427.
134. Yobs, A. R. (1972). Levels of polychlorinated biphenyls in adipose tissue of the general population of the nation. *Environ. Health Perspect. 1*:79-81.

BIBLIOGRAPHY

Coulston, F. (ed.) (1969). *Regulatory Aspects of Carcinogenesis and Food Additives: The Delaney Clause*, Academic Press, New York.
Galli, C. L., R. Paoletti, and G. Vettorazzi (1978). *Chemical Toxicology of Food*, Elsevier/North Holland, Amsterdam.
Graham, H. D. (ed.) (1980). *Safety of Foods*, 2nd ed., AVI Publishing Co., Westport, Conn.
Leonard, B. J. (ed.) (1978). *Toxicological Aspects of Food Safety*, Springer-Verlag, New York.
Liener, I. E. (ed.) (1974). *Toxic Constituents of Animal Foodstuffs*, Academic Press, New York.
Liener, I. E. (ed.) (1980). *Toxic Constituents of Plant Foodstuffs*, 2nd ed., Academic Press, New York.
National Research Council, Committee on Food Protection (1973). *Toxicants Occurring Naturally in Foods*, 2nd ed., National Academy of Sciences, Washington, D.C.
Riemann, H., and F. C. Bryan (eds.) (1979). *Food-Borne Infections and Intoxications*, 2nd ed., Academic Press, New York.
Witechi, H.-P. (ed.) (1980). *The Scientific Basis of Toxicity Assessment*, Elsevier/North Holland, Amsterdam.

12

CHARACTERISTICS OF MUSCLE TISSUE

Herbert O. Hultin   University of Massachusetts Marine Station, Gloucester, Massachusetts

I.	Introduction	726
II.	Nutritive Value	726
III.	Similarities and Differences Among Muscles of Various Animal Species	729
IV.	Muscle Structure	730
	A. Skeletal Muscle	730
	B. Smooth or Involuntary Muscle	734
	C. Cardiac (Heart) or Striated Involuntary Muscle	734
V.	Proteins of the Muscle Cell	735
	A. Contractile Proteins	735
	B. Contraction	740
	C. Roles of Thick and Thin Filaments in Contraction	740
	D. Relaxation	742
	E. Soluble Components of the Muscle Cell	742
	F. Insoluble Components of the Muscle Cell	742
VI.	Muscle Types	743
VII.	Connective Tissue	745
	A. Collagen	746
	B. Conversion of Collagen to Gelatin	748
VIII.	Biochemical Changes in Muscle Postmortem	750
	A. Biochemical Changes Related to Energy Metabolism	750
	B. Consequences of ATP Depletion	753
	C. Increase in Tenderness on Aging Postrigor	754
	D. ATP Breakdown to Hypoxanthine	756
	E. Loss of Calcium-Sequestering Ability	757
	F. Changes in Lipid Postmortem	758
	G. Breakdown of Other Components Critical to Quality	759

IX.   Effect of Postmortem Changes on Quality Attributes of Meats   760
      A.   Texture and Water-Holding Capacity   760
      B.   Color   764
      C.   Flavor   765
      D.   Nutritional Quality   765
 X.   Antemortem Factors Affecting Postmortem Biochemical Changes   765
XI.   Effects of Processing on Meat Components   766
      A.   Hot Boning   767
      B.   Electrical Stimulation   767
      C.   Refrigeration   768
      D.   Freezing   769
      E.   Heating   772
      F.   Dehydration   773
      G.   Curing   776
      H.   Other Chemical Additives   777
      I.   Modified (Controlled) Atmospheric Storage   777
      J.   Preparation of Gel Meat Products   779
      K.   Preservation by Ionizing Radiation   781
      L.   Packaging   781
      References   782
      Bibliography   788

I. INTRODUCTION

Since earliest times, humans have had a desire to satisfy hunger with animal food. There is evidence that numerous people in many cultures have lived chiefly on this type of food. It is difficult to explain our desire for animal products. It may stem from evolutionary reasons related to the high performance and good health of people who have been able to obtain sufficient amounts of these products, or it simply may be due to the sensory appeal of these tissues. Whatever the cause, these foods have been very important in cultural traditions; for example, meat was often served to the most prominent individuals in a society. In our day, meals are still often designed around meat or animal products.

Many types of animal tissues have been, and are, used as foods. Of these, muscle tissues, milk and milk products, and eggs are the most important from an economic and quantitative standpoint. The term "meat" refers to muscle, especially that from mammals, which has undergone certain chemical and biochemical changes following death. However, because of the long-standing importance of meat in the diet, the term often has been used synonymously with food. Many muscle tissues are not generally referred to as meat, such as fish muscle. In this chapter, the terms "muscle" and "meat" are used interchageably in recognition that muscle is generally considered a more appropriate term for the functional tissue and meat a more appropriate term for the tissue after it has passed through certain changes following the death of the animal. Furthermore, meat often implies a product that includes some adipose (fat) tissue and bone.

II. NUTRITIVE VALUE

In addition to its aesthetic appeal, meat is important because of its high nutritive content. The composition of lean muscle tissue is given in Table 1, and it is relatively constant for a wide variety of animals. Meat composition varies mostly in lipid content, which may be evident as different degrees of marbling in some mammalian meats. The muscle tissues of

Table 1 Composition of Lean Muscle Tissue

Species	Composition (%)			
	Water	Protein	Lipid	Ash
Beef	70-73	20-22	4-8	1
Pork	68-70	19-20	9-11	1.4
Chicken	73.7	20-23	4.7	1
Lamb	73	20	5-6	1.6
Cod	81.2	17.6	0.3	1.2
Salmon	64	20-22	13-15	1.3

Source: Compiled from Ref. 130.

fatty fish show large seasonal variations in lipid, based in large part on the reproductive cycle of the fish. For example, the lipid content of mackerel is known to vary between 5.1 and 22.6% during the course of a year (73). The adjustment in total composition of the fish is made by increasing or decreasing water content; that is, the percentage of lipid plus water remains essentially constant in fatty fish (80). The composition of meat varies considerably depending on the amount of fat, bone, and skin included in the sample.

As shown in Table 1, about 18-23% of the lean portion of meat and fish muscle is protein. A significant exception to this occurs in starving fish or in fish in certain stages of the reproductive cycle. The latter is a form of starvation since sufficient dietary protein cannot be provided to compensate for protein synthesis in gonadal tissue. Fish utilize muscle tissue during starvation, and it is replaced by water; that is, in lean fish, the percentage of protein plus water remains essentially constant. The muscle from a living but severely starved cod has been observed to contain 95% water (80). Not only is the protein content of muscle tissue very large, but the quality of this protein is also very high, containing kinds and ratios of amino acids very similar to those required for maintenance and growth of human tissue. Of the total nitrogen content of muscle, approximately 95% is protein and 5% is smaller peptides, amino acids, and other compounds.

The lipid components of muscle tissue vary more widely than do the amino acids. In fish muscle, the differences have been "institutionalized" in the concept of lean or white fish and fatty fish. In lean fish, storage fat is carried in the liver. Muscle of lean fish contains less than 1% lipid, mostly phospholipid, located in the membranes. In fatty fish, depot fat apparently occurs as extracellular droplets in the muscle tissue (61). In white muscle, fat is apparently diffusely located among the muscle cells; in red muscle, distinct fat droplets exist within the cells (17). In addition to species variations, the lipid components of muscle can be markedly influenced by diet.

Basically, the lipid composition of meat (mammalian and avian muscle) can be categorized into lipids from muscle tissue and lipids from adipose tissue. The lipid composition of these two tissues can be quite different. Lipids in the lean portion contain greater portions of phospholipids than lipids in adipose tissue. Lean muscle contains about 0.5-1% phospholipids, and the fatty acids of phospholipids are more unsaturated than those of triacylglycerols. Consequently, lipids in the lean portion of meat have a higher degree of unsaturation than those in adipose tissue. Oxidation of the highly unsaturated fatty acids found in the membrane fractions of muscle may be very important in some of the deteriorative reactions of meat.

Within a species, red muscle contains more lipid than white muscle. The type of fatty acids present in muscle tissue is also dependent on species. The degree of unsaturation of fatty acids in cold-blooded fish is much greater than that of fatty acids in avian and mammalian muscles (Table 2). Presumably, the high concentration of polyunsaturated fatty acids in fish is necessary to keep lipids fluid at the low muscle temperatures sometimes encountered. The much greater percentage of polyenoic fatty acids in lean fish (cod) compared with fatty fish (mackerel) reflects differences in phospholipid-triacylglycerol ratios, as mentioned above. Poultry fat is more unsaturated than pork fat, which in turn is more unsaturated than beef or mutton fat.

In recent years, the relatively large amount of saturated fatty acids in mammalian muscle tissue has been a source of controversy as to its role in producing certain forms of atherosclerosis. As yet, however, there has been no firm scientific evidence demonstrating that fats from muscle tissue are deleterious to health when eaten in reasonable quantities. The cholesterol content of meat and fish, about 75 mg per 100 g (127), is well below that considered undesirable for humans.

Muscle tissue is an excellent source of some of the B-complex vitamins, especially thiamine, riboflavin, niacin, B_6, and B_{12}. However, the B-vitamin content of muscle varies considerably depending on the species and on the type of muscle within a species. In addition, the levels of the B vitamins are also influenced by breed, age, sex, and general health of the animal. Less work has been performed on the fat-soluble vitamins than on the water-soluble vitamins in meat, but the levels of vitamins D, E, and K in meats are generally rather low, although the amount of vitamin E (tocopherols) can be influenced by the diet of the animal. The content of vitamin A in muscle foods is somewhat greater than that of the other fat-soluble vitamins, and these foods contribute about 23% of the average intake of this vitamin for the U.S. consumer (103). Ascorbic acid is present only at very low levels in meat.

Meat is a good source of iron and phosphorus and a rather poor source of calcium (approximately 10 mg per 100 g of meat), except in certain deboned or mechanically separated meats in which the calcium content may be greatly increased due to the presence of small bone fragments in the edible product. Meat generally contains from 60 to 90 mg sodium per 100 g and about 300 mg potassium per 100 g lean tissue. Minerals and the water-soluble B-complex vitamins are found in the lean portion of the meat. The concentration of these substances therefore varies, depending on the amount of fat tissue and bone in a particular cut of meat, as well as on the cooking process.

Table 2 Degree of Saturation of Fatty Acid Components of Lipid from Muscle Tissue of Various Species

Species	% Saturated	% Monenoic	% Polyenoic
Beef	40–71	41–53	0–6
Pork	39–49	43–70	3–18
Mutton	46–64	36–47	3–5
Poultry	28–33	39–51	14–23
Cod (lean fish)	30	22	48
Mackerel (fatty fish)	30	44	26

III. SIMILARITIES AND DIFFERENCES AMONG MUSCLES OF VARIOUS ANIMAL SPECIES

It would be educationally desirable to compare in detail the properties of a wide variety of muscle tissues from all animal species that are important sources of food. Economic and practical considerations have, however, resulted in highly selective research so that the amount and type of information available vary greatly with species and with muscles within a species. This is especially apparent when one attempts to compare in detail the properties of muscles from cold-blooded animals from an aquatic environment with the properties of muscles from warm-blooded mammals and birds. Nonetheless, it is useful to consider available information and compare the nature of muscles from the groups just mentioned.

Throughout the animal kingdom, muscle function is similar; that is, it provides locomotion. There are, however, important differences among species and, indeed, among muscles of the same animal, especially when muscle function is different. Nowhere is the species difference more obvious than between fish as compared with land animals and birds. The differences observed are due to three basic factors. First, the fish body is supported by water; thus, fish do not require extensive, strong connective tissues to maintain and support the muscles. Second, since most commercially important fish are poikilothermic animals and they live in a cold environment, the proteins of fish muscles have properties different from those of warm-blooded species. Finally, the structural arrangement of fish muscle is markedly different from that of land animals and birds. This arrangement is related to the peculiar movement of fish.

Generally speaking, food scientists have studied those properties of muscle tissues that have important influences on their value as food. For example, in land animals and birds, texture is the major property that can be controlled during slaughtering and post-slaughtering procedures. Since textural properties of these muscles are due in part to the state of contraction of the myofibrillar proteins, considerable attention has been directed toward understanding the cause of contraction in these muscles postmortem. The muscle proteins of land animals and birds are also greatly influenced by the rate of change of pH postmortem, and this aspect has, therefore, also received considerable attention.

On the other hand, textural properties of fresh fish are less of a problem than they are with warm-blooded animals; thus, comparatively little information is available on the postmortem biochemistry of fish as related to texture.

When considering frozen animal tissues, a quite different picture emerges. Important textural changes occur in fish during frozen storage, and these changes result from the intolerance of fish proteins to conditions induced by freezing, such as increased salt concentrations and decreased pH values. Consequently, textural changes in frozen fish have received considerable study. On the contrary, the proteins of land animals and birds are reasonably stable during frozen storage; consequently, research in this area has been somewhat limited.

The importance of collagen also differs among species. In mammals and birds, both the amount and type of collagen have important influences on textural properties of the muscle. In fish, however, collagen is readily softened by normal cooking procedures, is not an important factor in the final product, and has, therefore, received little attention.

Furthermore, flavor changes are very important in fish, and extensive studies have been conducted on the degradation of lipids and amino compounds postmortem since these degradation compounds contribute greatly to flavor. This area is less important for most land animals and birds, and has, therefore, received comparatively little attention.

In the remainder of this chapter, the structural and chemical features of muscle tissues are considered, with emphasis on the changes that occur following death and during storage and processing. These changes are related to their effects on texture, color, flavor, and nutritional attributes of muscle tissues.

IV. MUSCLE STRUCTURE

A. Skeletal Muscle

Skeletal muscle is composed of long, narrow, multinucleated cells (fibers) that range from a few to several centimeters in length and from 10 to 100 μm in diameter. Although there are differences in muscle fibers with regard to the amount of sarcoplasm and the amount and location of cellular membrane components, there is a close resemblance, at the cellular level, of muscles from a wide variety of organisms. There are, however, different arrangements of the muscle cells and connective tissue depending on the means of locomotion of the animal and its environment.

Figure 1 represents a typical arrangement of muscle components in mammals and birds. The fibers are arranged in a parallel fashion to form bundles, and groups of bundles form a muscle. Surrounding the whole muscle is a heavy sheath of connective tissues, called the "epimysium." From the inner surface of the epimysium, other connective tissues penetrate the interior of the muscle, separating the groups of fibers into bundles. This connective tissue layer is termed the "perimysium," and extending from this are finer sheaths of connective tissue that surround each muscle fiber. These last sheaths are termed "endomysia." The connective tissue sheaths merge with large masses of connec-

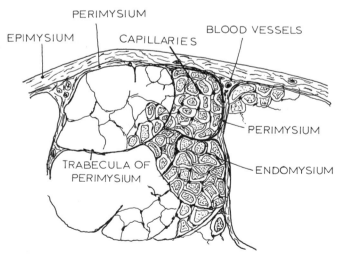

Figure 1 Muscle cross section, illustrating the arrangement of connective tissue into epimysium, perimysium, and endomysium and the relationship to muscle fibers and fiber bundles. The typical position of blood vessels is also shown. (Reprinted from Ref. 42, p. 563, by courtesy of J. B. Lippincott Co.)

tive tissue tendons at the termini of the muscle, these merger points serving to anchor the muscle to the skeleton. The long components of the circulatory system are located in the perimysium, whereas the smallest units (capillaries) are within the endomysium.

The arrangement of muscle fibers in fish is quite different from that in birds and mammals and is based on their necessity to flex their bodies for propulsion through water. The arrangement of the muscle tissue in a typical bony fish is shown in Fig. 2. The W-shaped segments standing on end are called myotomes and they have one forward and two backward flexures. These myotomes are not perpendicular to the vertical midplane of the fish but typically intercept this plane at sharp angles (67). Thus, if a cut is made at right angles to the central skeleton, multiple myotomes are exposed. The myotomes are one cell deep, and the muscle cells are roughly perpendicular to the surface of the myotome. The myotomes are connected one to another by thin layers of collagenous connective tissue called myosepta (myocommata).

In squid, a cephalopod, the mantle (external body wall) is made up of circular muscle fibers. At more or less regular intervals, these fibers are interrupted by sections of radially oriented fibers that function to "thin" the mantle and cause the mantle cavity to expand following the power stroke of the circular fibers (14). In some species of squid, superficial fibers also run longitudinally along the mantle. The myofibrils of squid mantle muscle are helically wound and are referred to as obliquely striated (50).

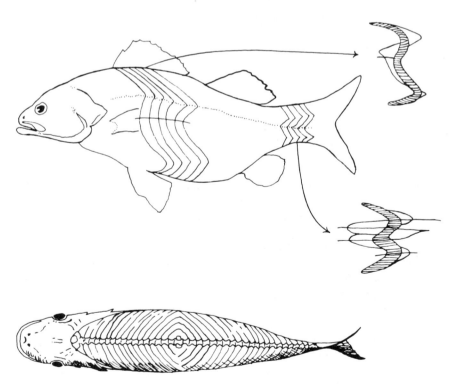

Figure 2 Myotome pattern of musculature of bony fish, with detailed lateral views of a single myotome. (Reprinted from Ref. 67, p. 74, by courtesy of John Wiley & Sons.)

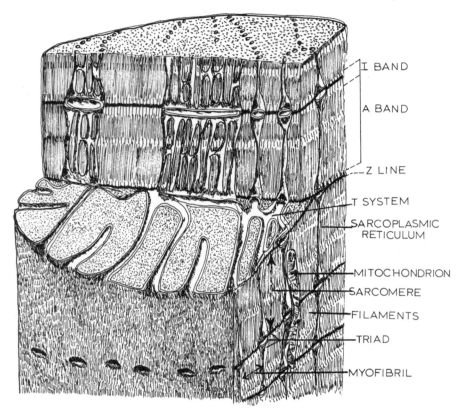

Figure 3 Cutaway of a muscle fiber showing the outer membrane and its invaginations (T system), which run horizontally and meet with two terminal sacs of the longitudinal sarcoplasmic reticulum in the triad. Repetitive cross sections are also indicated. (Adapted from Ref. 107.)

Figure 3 is a diagrammatic view of a typical muscle fiber, with the major components being the sarcolemma, the contractile fibrils, the cell fluid (sarcoplasm), and the organelles. The surface of the muscle fiber is termed the "sarcolemma." The sarcolemma is believed to consist of three layers: an outer network of collagen fibrils, a middle or amorphous layer, and an inner plasma membrane. Invaginations of the plasma membrane form the transverse (T) system. The ends of this T system meet in the interior of the cell close to two terminal sacs of the sarcoplasmic reticulum. The sarcoplasmic reticulum is a membranous system located in the cell (fiber) and generally arranged parallel to the main axis of the cell. The meeting of the T system and the sarcoplasmic reticulum (triadic joint) occurs at different intrafiber locations in different muscles. The triadic joint appears to exist most frequently at or near the Z disks in fish and frog muscle and at the junction of the A and I bands in reptiles, birds, and mammals (129). The terminal cisternae of the longitudinal elements (sarcoplasmic reticulum) surround each sarcomere (see below) like a hollow collar. The collar is perforated with holes and is termed a "fenestrated collar."

The T system extends the plasma membrane into the interior of the muscle cell, and it is this phenomenon that allows the muscle cell to respond as a unit (essentially no lag period in the interior of the cell). Depolarization of the plasma membrane and its intracellular extension (T system) triggers the liberation of calcium from terminal sacs of the sarcoplasmic reticulum. This liberation of calcium activates ATPase of the contractile proteins and allows contraction to occur. The calcium functions by overcoming magnesium inhibition of muscle ATPase. Relaxation is achieved in part by a reversal of the process and by the sequestering of calcium by the sarcoplasmic reticulum (51).

Mitochondria serve as prime energy transducers for the muscle cell, and these organelles are located throughout the cell. In some cases there is a concentration of mitochondria near the Z line or near the plasma membrane. Nuclei are distributed near the surface of the muscle cell and have an important role in protein synthesis. The Golgi apparatus is believed to have a secretory role.

Lysosomes are subcellular fractions that contain large quantities of hydrolytic enzymes and serve a digestive role in the cell. Lysosomes exist as part of the muscle cell, per se, and also originate from phagocytic cells present in the circulatory system. The proteolytic enzymes of these bodies are termed "cathepsins," and cathepsins with differing activities have been isolated.

Glycogen particles and lipid droplets also occur in some muscle cells, depending on the state of the muscle.

All these cellular components are bathed in the sarcoplasm, a semifluid material that contains soluble components, such as myoglobin, some enzymes, and some metabolic intermediates of the cell.

The major inner components of the muscle fiber are the fibrils (myofibrils), which constitute the contractile apparatus. The myofibrils are surrounded by the sarcoplasm and some of the elements discussed above, such as mitochondria, the T system, and the sarcoplasmic reticulum.

The characteristic striated appearance of skeletal muscle is due to a specific repetitive arrangement of proteins in the myofibrils. This arrangement is shown in Fig. 4. The dark bands of the fibrils are anisotropic or birefringent when viewed in polarized light. The bands that appear lighter are isotropic, and are therefore termed "I bands." The darker bands, being anisotropic, are termed "A bands." In the center of each of the I bands is a dark line, called the Z line or Z disk. In the center part of each A band a light zone exists, and this is called the H zone. Frequently, at the center of the H zone there exists a darker M line. The contractile unit of the fibril is termed the "sarcomere," defined as the material located between and including two adjacent Z disks.

A sarcomere is comprised of thick and thin longitudinal filaments. The A band is comprised of thick filaments (mostly myosin), whereas the I band is composed of thin filaments (mostly actin). The thin filaments extend outward from the Z disks in both directions, and in parts of the A band the thin filaments overlap the thick filaments. The lighter zone of the A band, the H zone, is that area where the thin filaments do not overlap the thick filaments. The contractile state of the muscle has an important bearing on the size of these various bands and zones, since during contraction the thin and thick filaments slide past each other. During contraction, the length of the A band remains constant, but the I band and the H zone both shorten.

Thin filaments are apparently imbedded in or connected to the Z disk material, and because of this, the Z disk presumably serves as an anchor during the contractile process. According to a recent model, each thin filament in the I band is linked to the four closest

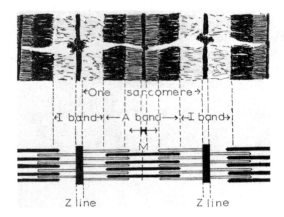

Figure 4 Striated muscle fibril in longitudinal section. The I band consists only of thin filaments. The A band is darker where it consists of overlapping thick and thin filaments and lighter in the H zone, where it consists solely of thick filaments. The M line is caused by a bulge in the center of each thick filament, and the pseudo-H zone is a bare region on either side of the M line. (Adapted from Ref. 55.)

thin filaments in the adjacent sarcomere. This linkage is believed to occur by means of bridging structures that consist in part of α-actinin and tropomyosin.

The M line is located in the area of the myosin thick filaments where projections on the myosin headpieces are not present (Sec. V.A.1). The M line probably serves to keep the filaments in the correct geometric position (105).

Striated muscle from other vertebrates as well as invertebrates has essentially the same structure as that discussed above for mammalian muscle. However, there may be some differences in the arrangement of the myofibrils, the amount of sarcoplasm, the relationship of the nuclei and mitochondria to the other components in the muscle cell, the arrangement of the sarcoplasmic reticulum, and the location of the triadic joint. In fish muscle cells, peripheral myofibrils are perpendicular to the sarcolemma (surface) but the inner myofibrils have the normal arrangement parallel to the long axis of the cell (13). Furthermore, the solubility characteristics of the contractile proteins of striated scallop muscle (86) differ from those of vertebrate muscle. Some molluskan muscles, such as scallop adductor and squid muscles, are obliquely striated instead of cross-striated (96).

B. Smooth or Involuntary Muscle

Smooth muscle fibers do not show the characteristic striations of voluntary or skeletal muscle. Certain organs containing smooth muscle, such as the gizzards of birds and intestinal tissue, are used for food.

C. Cardiac (Heart) or Striated Involuntary Muscle

Heart tissue is used as food directly or may be incorporated into sausage products. The myofibrillar structure of heart muscle is similar to that of striated skeletal muscles, but cardiac fibers contain larger numbers of mitochondria than do skeletal fibers. The fiber arrangement of cardiac muscle is also somewhat less regular than that observed in skeletal muscle.

V. PROTEINS OF THE MUSCLE CELL

The proteins of muscle can be roughly categorized into contractile, soluble, and insoluble fractions. The soluble fraction can be extracted from muscle with water or dilute salt solution and is made up of enzymes, such as those involved in glycolysis, and the muscle pigment, myoglobin. Although there is some question of whether all these proteins are soluble in situ (51), they are nevertheless easily extractable. The contractile proteins are soluble in salt solutions of high ionic strength but not in water or dilute salt solutions. The insoluble fraction remains after treatment with salt solutions of high ionic strength, and it consists of connective tissue proteins, membrane proteins, and usually some unextracted contractile proteins. These fractions are discussed separately.

A. Contractile Proteins

1. Myosin

The major protein of the thick filaments is myosin, which comprises 50-60% of the myofibrillar contractile proteins. It is an elongated protein molecule with a molecular weight of approximately 470,000 daltons. Myosin contains two identical polypeptide chains, each with a high degree of α-helical structure. In addition, the two chains are supercoiled, that is, wound around each other as illustrated in Fig. 5. The molecule has two globular heads, which are responsible for its enzymic (ATPase) activity and its ability to interact with actin. The globular heads contain the amino residue termini of the two polypeptide chains making up this molecule.

Associated with each globular head section are two "light chains," so that four light chains are associated with each myosin molecule. The light chains exist in two distinct chemical classes. One class is designated the DTNB light chain, so-called because treatment of myosin with the thiol reagent DTNB [(5,5'-dithiobis)-2-(nitrobenzoic acid)] causes its removal. About 50% of the DTNB light chain can be removed without a significant change in the Ca^{2+}-dependent ATPase activity of the molecule. The second class of light chain is called the "alkali light chain" because it is released under alkaline conditions. When it is removed there is loss of myosin ATPase activity. The two types of light chains are different chemically. The DTNB light chain has a molecular weight of about 18,000 daltons; the alkali light chains are heterogeneous, one of the chains with a molecular weight of 25,000 daltons, and the other 16,000 daltons (45). Alkali light chains are essential for ATPase activity, and it has been suggested that the DTNB chains may be involved in calcium regulation of muscle activation. Each globular head of myosin contains one alkali light chain and one DTNB light chain.

Myosin can be cleaved near the head region by proteolytic enzymes, such as trypsin, producing two fractions of the protein. One of these is called light meromyosin and the other, which contains the globular head structures of the myosin molecule, is called heavy meromyosin (Fig. 5). Separated heavy meromyosin retains its ability to interact with actin and its ATPase activity. The globular section of heavy meromyosin is designated "subfragment 1," is normally attached to the rod section with rather flexible linkages, and can assume a wide variety of positions in relation to the rod section.

When separated from subfragment 1, the rodlike tail section of heavy meromyosin, called "subfragment 2," shows no tendency to associate with itself or with light meromyosin. This indicates that light meromyosin (which associates readily) is responsible for forming the backbone structure of the thick filament, and that subfragment 2 (rod

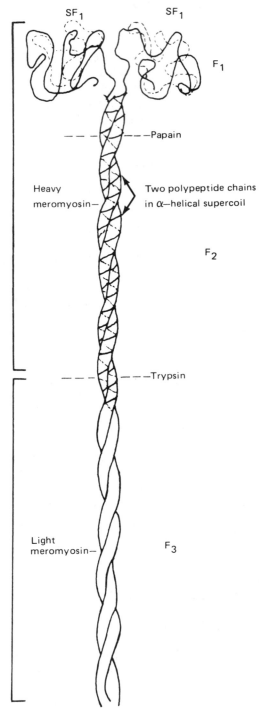

Figure 5 Schematic representation of the myosin molecule showing its globular head and long tail and the points of enzymic fragmentation. F_1 and F_2 represent subfragments 1 and 2, respectively, and F_3 is light meromyosin. SF_1 represents the globular head, and the dotted lines represent the light chains. (Adapted from Ref. 72.)

section of heavy meromyosin) is free to swing out from the surface to allow the globular heads to make contact with the thin filaments. When the myosin molecule is cleaved by papain this occurs near the head, releasing the two globular sections from the rest of the rodlike tail of the molecule (Fig. 5). Further action by papain causes a splitting of this rod into subfragment 2 and light meromyosin.

Each thick filament contains some 400 molecules of myosin (53). These molecules polarize when they interact, joining in head-to-tail fashion in two directions, as illustrated in Fig. 6. It is undoubtedly this polarity that allows contraction to occur. Thick filaments can be re-formed from isolated myosin molecules.

Myosin can be extracted from muscle with a salt solution (ionic strength about 0.6) of slightly alkaline pH. It can then be purified by repeated cycles of precipitation, followed by resolubilization in salt solution of high concentration. Solutions commonly used for extraction are 0.3 M KCl and 0.15 M phosphate at pH 6.5, or 0.47 M KCl, 0.1 M phosphate, and 0.01 M pyrophosphate at pH 6.5 (51). If a short period of time is used for the extraction, a crude myosin, called myosoin A, is produced. If longer periods of extraction (e.g., overnight) are used, then a crude actomyosin preparation, called myosin B, is obtained.

2. Actin

A major protein of the thin filaments is actin, which comprises 15-30% of myofibrillar protein of muscle (53). Actin is bound to the structure of the muscle much more firmly than is myosin. Extraction of actin can be accomplished by prolonged exposure of a muscle powder (obtained by acetone extraction) with an aqueous solution of ATP (8).

Actin probably exists in muscle as a double-helical structure called fibrous actin, or F-actin (see Fig. 7). Globular actin, or G-actin, is the monomeric form of the protein with a molecular weight of 43,000-48,000 daltons. It is stable in water, where it can also exist as a dimer. Globular actin binds ATP very firmly and, in the presence of magnesium, spontaneously polymerizes to form F-actin with the concurrent hydrolysis of bound ATP to give bound ADP and inorganic phosphate. Globular actin also polymerizes in the presence of neutral salts at a concentration of approximately 0.15 M. Filaments of F-actin interact with myosin, heavy meromyosin, and subfragment 1 in a similar and characteristic way. Each actin subunit in the double helix binds to one molecule of subfragment 1, that is, the globular head of the myosin molecule. When negatively stained and examined in the electron microscope, myosin or its subfragments appear as long strings of arrowheads pointing in the same direction. This indicates that the actin mono-

Figure 6 Possible arrangement of myosin molecules in a filament. Globular regions, ≳; tail, −. The polarities of the myosin molecules are reversed on either side of the center, but all molecules on the same side have the same polarity. (Reprinted from Ref. 54, p. 289, courtesy of Academic Press, Inc.)

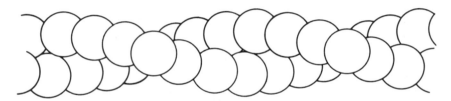

Figure 7 Arrangement of actin in double-helical structure.

mers in the F-actin filament are oriented in opposite directions on either side of the Z disk, which allow the thick filaments in adjacent sarcomeres to move toward the Z disk.

3. Actomyosin

When purified actin and myosin are mixed in vitro, a complex, called actomyosin, is formed. Actomyosin has a high viscosity and also exhibits a large flow birefringence. Although actin by itself has no enzymic activity, it significantly modifies the ATPase activity of myosin in the actomyosin complex. Pure myosin requires Ca^{2+} for its activity and is inhibited by Mg^{2+}. The ATPase activity of actomyosin, however is stimulated by Mg^{2+}. Sulfhydryl groups are involved in the interaction between myosin and actin (9). The complex of actomyosin can be dissociated in the presence of ATP and/or ADP and magnesium ions (65). It is most likely that a similar type of combination between actin and myosin occurs in muscle cells and is probably intimately involved with contraction. The interaction of actin and myosin in the absence of ATP and ADP, and the plasticizing effect of these nucleotides when they are present, is a very important factor influencing the quality of meat.

4. Other Contractile Proteins

Tropomyosin is a two-stranded coiled-coil of an α-helix with a molecular weight of 65,000-70,000 daltons and a length of some 400 Å. It has a very high α-helical content (over 90%), and there are indications that the structure is stabilized by hydrophobic interactions between the two strands. In rabbit muscle it has been shown that the two subunits have different molecular weights, 37,000 and 33,000 daltons, and that their chemical composition is different (45). This heterogeneity might be due in part to the presence of both red and white fibers in the muscle source. Tropomyosin shows a strong tendency to aggregate by end-to-end bonding of the individual molecules. It is believed that these polymerized tropomyosin molecules lie along each groove of the actin double helix such that each molecule interacts with seven G-actin monomers (Fig. 8).

Troponin is often isolated with tropomyosin. This protein consists of three subunits designated troponin C (17,000-18,000 daltons), troponin I (20,000-24,000 daltons), and troponin T (37,000-40,000 daltons) (45). Each subunit of troponin has distinct functions. Troponin C contains a very high concentration of acidic amino acid residues and is the subunit responsible for binding calcium and conferring calcium sensitivity to the contractile process. Troponin I strongly inhibits the ATPase activity of actomyosin. Troponin T functions to provide a strong association site for binding troponin to tropomyosin. The complex of tropomyosin and troponin is designated the "relaxing factor" (112).

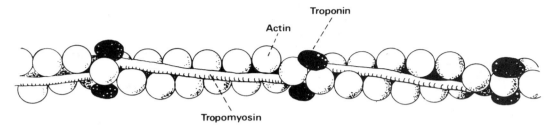

Figure 8 Thin filament complex with tropomyosin lying in grooves of the actin double helix. Troponin complex is situated with a periodicity of 400 Å. (Reprinted from Ref. 30, p. 364, by courtesy of Cambridge University Press.)

C-protein makes up approximately 3% of the thick filament mass. It has been suggested that the C-protein molecule, a single polypeptide chain of 140,000 daltons, may be wrapped around the thick filament surface with its major axis in a plane normal to that of the thick filament. These protein rings may protect the filament against destruction by tensile forces or changes in the ionic environment.

α-Actinin is located exclusively in the Z disk. It has a molecular weight of 180,000 daltons and consists of two polypeptide subunits of similar mass (45). α-Actinin forms a strong complex with F-actin, but not with G-actin, the tropomyosin-troponin complex, or tropomyosin. The ability of α-actinin to cause gelation of F-actin can be reversed by tropomyosin.

β-Actinin is a dimeric protein made up of polypeptides of 37,000 and 34,000 daltons. β-Actinin binds to F-actin and inhibits the recombination of fragmented F-actin. It is located at the free end of the thin filaments and may be involved in regulating the length of thin filaments.

Several other proteins have recently been isolated from Z disks (37,39,118). One of these proteins, Z-nin, with a molecular weight of some 300,000-400,000 daltons, is a necessary component of the Z disk along with α-actinin and tropomyosin (118).

One of the structural proteins of the M line has turned out, surprisingly, to be creatine kinase, an enzyme involved in ATP regeneration. That such a catalytic agent is also an important structural component of the myofibril has important implications with respect to operating mechanisms of the contractile protein system (45). An additional protein of 165,000 daltons, called myomesin, also has been shown to be a component of the M line.

Connectin is an elastic protein that may be responsible for the elasticity and mechanical continuity of the myofibrils of striated muscle (84). Connectin is a highly insoluble protein that forms three-dimensional nets of very thin filaments in a myofibril. The connectin nets may be identical to the "gap filaments" observed between thin and thick filaments when muscle is stretched beyond the point of filament overlap (77). Connectin is probably identical to the protein identified as titin.

Another protein that forms filaments in the muscle cell is desmin or skeletin. Desmin filaments link adjacent myofibrils at their Z-disk level and connect the myofibrils into the cellular cytoskeleton (109). Other as yet unidentified proteins are probably also present as part of the contractile apparatus.

Paramyosin is found in muscles of mollusks and other invertebrates, accounting for as much as 50% of the total structural protein of these muscle fibers. It has a molecular weight of 220,000 daltons and is essentially 100% α-helical. It forms the central core of the myosin-containing thick filaments in these systems. The major contractile proteins of muscle are summarized in Table 3.

B. Contraction

Although it is not the purpose of this chapter to delve into the physiology of muscle in the living animal, it is necessary to have some knowledge of the normal functioning and interactions of contractile proteins in the living cell to understand postmortem changes.

Muscle is stimulated by an electrical nervous impulse responsible for depolarization of the muscle cell membrane. The transverse tubular system of the muscle is an extension of the cell membrane (129), and its role therefore is to transmit the stimulus (depolarization) to the interior of the muscle cell so that the whole cell can react as a unit. As described in Sec. III.A, the transverse system joins two other bulbous projections from the sarcoplasmic reticulum near the myofibrils. In some manner, depolarization of the transverse tubular system causes a release of calcium from the terminal sacs of the sarcoplasmic reticulum. It seems probable that the released calcium (effective concentration is about 0.5-1 μM) modifies the conformation of the troponin complex, which then in some way switches on the contractile apparatus. Although the exact molecular mechanism of this process is not understood, one possibility is the following (45). Ordinarily, troponin I is bound strongly to both actin and troponin C. When the released calcium interacts with troponin C, the troponin complex undergoes a structural change, binding troponin I more tightly to troponin C and weakening its interaction with actin. This weakening of the interaction of troponin I causes it to move, which in turn allows the tropomyosin to move into the groove of the actin superhelix. This movement of tropomyosin has the ability to interfere with the binding of Mg-ATP at the relaxing site of the myosin fibril. This relaxing site is, in fact, an inhibitory site responsible for preventing myofibrillar ATPase activity. When the inhibition is overcome, ATP is hydrolyzed by the actomyosin complex. This hydrolysis provides energy for the systematic changes that take place in the contractile proteins and leads to the phenomenon of contraction. Thus, the contractile process is regulated by the amount of free calcium ions released into the sarcoplasm.

C. Roles of Thick and Thin Filaments in Contraction

It is generally agreed that muscle contracts by a process that allows the thin and thick filaments to slide past each other, as was originally proposed by Huxley and Hanson (see Fig. 9) (51). However, the exact mechanism by which the energy released during the hydrolysis of ATP is converted into the mechanical energy of muscle contraction is not fully understood at the molecular level. In some way, the globular head portion of the myosin molecule interacts with actin and then pulls the actin filaments parallel to the fiber axis. Since interaction of the filaments occurs through the cross-bridges, tension must be developed and carried by means of the cross-bridges. It seems likely that an important factor that allows this to occur is the flexible nature of the rodlike portion (subfragment 2) of the heavy meromyosin section of myosin, which would allow both for the attachment of the globular head to the thin filament and the ability to change the

Table 3   Proteins of the Myofibril

Thick filaments	Z disk
Myosin	α-Actinin
C-Protein	Desmin
I-Protein	Eu-actinin
Creatine kinase	Filamin
Myomesin	55,000 dalton protein
	Vimentin
Thin filaments	Synemin
Actin	
Tropomyosin	Gap filaments
Troponins	Connectin (titin)
β-Actinin	
α-Actinin	

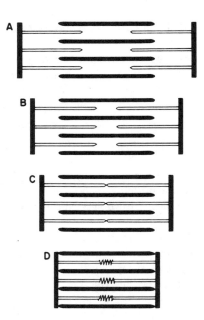

Figure 9 Contraction states of a sarcomere showing thick (dark) filaments and thin (light) filaments attached to the Z disks. If B is taken as the schematic representation of a sarcomere from resting muscles, A represents that from a stretched muscle, and C and D those from muscle contracted to different degrees. The jagged lines in the thin filaments in D represent those portions of the thin filaments that overlap each other.

position of the globular portion of the head in relation to the thick and thin filaments. Maximal reversible contraction varies from 20 to 50% of the rest length of the sarcomere.

D. Relaxation

Relaxation occurs when calcium ions are removed from the sarcoplasm. This takes place by active, energy supported transport or by sequestering of calcium by the sarcoplasmic reticulum.

When the calcium ion is reduced to a concentration below about 0.5 μM, the myofibrils lose their capacity to hydrolyze ATP. ATP then functions as a plasticizing agent, causing separation of actin and myosin, and the sarcomeres relax to their rest lengths.

E. Soluble Components of the Muscle Cell

Soluble proteins of the muscle cell are known by various terms, for example, myogen, which is a water extract of muscle, or simply sarcoplasmic extract. In any case, the soluble proteins of the muscle cell constitute a very significant portion of the proteins of the cell, usually from 25 to 30% of the total proteins.

Most of these soluble proteins are enzymes, principally glycolytic enzymes, but other enzymes are present as well, including those of the pentose shunt and such auxiliary enzymes as creatine kinase and AMP deaminase. The oxygen storage component of the muscle cell, myoglobin, is also a water-soluble protein. The high viscosity of the sarcoplasm is probably attributable to the presence of soluble proteins in concentrations as high as 20-30%. In some species, glyceraldehyde-3-phosphate dehydrogenase comprises approximately 10% of this soluble protein (24). As noted, these proteins may not be soluble in vivo but they are easily solubilized.

Other soluble nonprotein constituents of the sarcoplasm include various nitrogen-containing compounds, such as amino acids and nucleotides; some soluble carbohydrates (intermediates in glycolysis); lactate, which is the major end product of glycolysis; a number of enzyme cofactors; and inorganic ions, including inorganic phosphate, potassium, sodium, magnesium, and calcium.

F. Insoluble Components of the Muscle Cell

Some components of the muscle cell are not soluble in water or in dilute or concentrated salt solutions. This category includes unextractable contractile proteins, such as connectin and desmin, portions of membrane systems, glycogen granules, fat droplets, and the largest fraction, connective tissue proteins.

Most of the lipid material in the cell is associated with the membranes, and these lipids are typically rich in phospholipids. The phospholipid content of a membrane varies from one membrane to another. There is approximately 90% phospholipid in the lipid fraction of mitochondria (32) and about 50% phospholipid in the lipid fraction of plasma membrane (62). The total lipid content of the muscle cell is small, only about 3-4% of the total weight of mammalian cells (36) and less than 1% of the muscle cell of lean fish (1). It is extremely important, however, since it is involved in the structure and function of the membranes, and as we shall see later, it is very important in some deteriorative reactions.

The major phospholipids of muscle are lecithin (phosphatidylcholine), phosphatidylethanolamine, phosphatidylserine, phosphatidylinositol, and some of the acidic glycerol phosphatides, such as cardiolipin. The amounts of each of these substances vary with the particular subcellular fraction examined. This is also somewhat true of the kinds of fatty acids that make up the phospholipids. Furthermore, compositional differences of the sort just mentioned also exist among muscles of a given species and among comparable muscle of different species.

The principal neutral lipids in muscle are triacylglycerols and cholesterol.

Proteins of the membrane fractions are usually very insoluble unless they are treated with detergents (surface-active agents). These fractions are therefore usually not extractable with most aqueous solvents.

VI. MUSCLE TYPES

Differences in pigmentation among muscles are obvious, and muscles are accordingly classified as either white or red. There may be differences even among muscles of the same organism, for example, the white breast muscle of the chicken or turkey and the dark muscles of the leg. Moreover, there is often a gradation of color in an individual muscle, with redness increasing from periphery to axis. In general, white muscles are capable of fast contractions, whereas many red muscles contract much more slowly.

It is also known that a simple classification based on overall appearance of the muscle is not very accurate or useful. This is because most muscles are made up of both red and white fibers, with red muscles containing a preponderance of so-called red fibers and white muscles containing a preponderance of so-called white fibers. Therefore, muscles may differ in the degree of "redness" or "whiteness."

In addition to the difference in color, many morphological and biochemical differences exist between red and white fibers (34). Red fibers tend to be smaller, contain more mitochondria, possess greater concentrations of myoglobin and lipid, and are more generously supplied with blood than are white fibers. However, glycogen and many of the enzymes related to glycolysis tend to be less abundant in red fibers than in white. In addition, red fibers have thicker sarcolemma, much less extensive and more poorly developed sarcoplasmic reticulum, less sarcoplasm, and a less granular appearance than do white fibers. The Z disks of red muscle fibers also tend to be rougher and thicker than those observed in white muscle fibers. In some cases, it has been shown that there are differences in the isoenzyme contents of red and white muscle fibers.

The classification of fibers as either red or white is arbitrary and incomplete. Histochemically, there is another type of fiber, called the intermediate fiber, which has properties (e.g., size and level of enzymes) intermediate between those of white and red fibers.

It is generally considered that red and white muscle fibers have different functions in the living animal. Red fibers, because of their well-developed vascular system, copious supply of oxygen, and high content of myoglobin, are ideally suited for oxidative metabolism. This role is further attested to by their low content of glycolytic and related enzymes and by their relatively high content of mitochondria, which function in respiratory utilization of oxygen. It has been suggested, therefore, that red muscles function best for sustained activity (34). In contrast, white muscles are presumably best suited for vigorous activity for a short period of time.

Our understanding of muscle fiber types is gradually evolving. More recent work has led to the classification of muscle fibers into three categories: fast-twitch red, fast-twitch white, and slow-twitch intermediate. These categories are based on speed of contraction and histochemical assessment of oxidative and glycolytic capacities (11). Another classification scheme (106) characterizes muscle types as fast-twitch oxidative glycolytic, fast-twitch glycolytic, and slow-twitch oxidative. These categories are based on a more careful examination of some of the histochemical, morphological, and biochemical properties of muscle.

The red and white muscles of fish have certain characteristics that are worthy of note. First, in fish muscle, the red, white, and intermediate fibers are more distinctly segregated than in mammalian or avian muscles. The arrangement of these muscles in codfish is shown in Fig. 10. The dark muscle is concentrated superficially, particularly around the lateral line—a group of skin sensory organs located on the surface of the fish body, along each side, in a single row from head to tail. In a few fish, such as tuna, which are very active and fast swimmers, deep-seated dark muscle also is present. A thin layer of intermediate muscle fibers separates the dark from the light fibers. In some species of fish, particularly salmonids, some isolated dark fibers are scattered throughout the main bulk of the white muscle. Modifications of the pattern in Fig. 10 are seen in most species of fish.

Generally, the proportion of dark muscle in fish increases toward the tail region. The fraction of red muscle fibers in the total muscle tissue varies greatly among species. Migratory (fatty) fish have abundant red tissue. At a point one-third of the way forward from the tail, the amount of red fibers is 15–30% of the total. For bottom-feeding lean fish, the percentage of red fibers is low, ranging from 2 to 12% (41), and some fish species have no detectable red fibers at this point in their bodies.

White muscle from fish is very uniform in composition no matter where it is located. Dark muscle, however, varies in composition as a function of its location, containing more lipid in the anterior part of the fish and more water and protein in the posterior

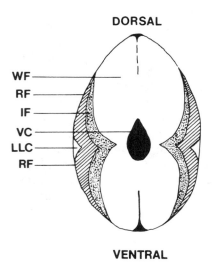

Figure 10 Transverse section of cod at point of maximal flexure: WF, white fibers; RF, red fibers; IF, intermediate fibers; VC, vertebral column, LLC, lateral line canal. (Reprinted from Ref. 40, p. 229, by courtesy of Conseil International pour l'Exploration de la Mer.)

part (80). Also, when fish fatten, lipids accumulate mostly in the anterior dark muscle. A larger fraction of dark muscle tissue is present in fish that live in a low-temperature environment than in fish accustomed to higher temperatures (80). Like other animals, the white muscle in fish appears to be used mostly for short, intense expenditures of energy, whereas the dark muscle is used for longer term exertion (swimming), which is why one finds more dark muscle tissue in migratory fish.

It has been suggested that red muscle tissue in fish serves many of the same functions as the liver in warm-blooded vertebrates. Although this hypothesis is not widely accepted, the red muscle tissue of fish does have characteristics that give this hypothesis some validity. For example, unlike warm-blooded vertebrates, the content of glycogen is greater in red muscle tissue than in white. Another characteristic worth noting is that the content of trimethylamine oxide in migratory fish is greater in dark muscle than in white, but the reverse is true in the muscle of nonmigratory lean fish (80).

An understanding of the differences among muscle fibers is important in food science since many of the characteristics of muscle postmortem are a function of the type of muscle fiber. For example, muscle color and susceptibility to cold shortening (Sec. IX.A) are influenced greatly by fiber type. Muscles with a preponderance of white fibers generally are much less susceptible to cold shortening than the muscles that have a majority of red fibers.

VII. CONNECTIVE TISSUE

Connective tissue consists of various fibers, several different cell types, and amorphous ground substances. Connective tissues hold and support the muscles through the component tendons, epimysium, perimysium, and endomysium. The relationship between connective tissue and muscle cells is so intimate that most likely all substances passing in or out of the muscle cell diffuse through some type of connective tissue.

The amorphous ground substance is a nonstructured mixture of carbohydrates, proteins, and lipids. Some of this mixture surrounds each cell as part of the sarcolemma.

Several types of cells are found in the connective tissue. These include fibroblasts, mesenchyme cells, macrophages, lymphoid cells, mast cells, and eosinophilic cells.

The fat cell deserves special attention. Fat cells appear to arise from undifferentiated mesenchyme cells, usually developing around the blood vessels in muscle. As a fat cell develops, it begins to accumulate droplets of lipid. These droplets grow by coalescence, and eventually, in the mature fat cell, only a single large drop of fat exists. The cytoplasm surrounds the fat droplet, and the subcellular organelles are located in this thin cytoplasmic layer. Fat cells are located outside the primary muscle bundles, that is, in the perimysial spaces or subcutaneously, but not in the endomysial area. Lipid in the latter area is usually associated with membranes or it exists as tiny fat droplets distributed in the muscle fiber itself.

The amount of fat that accumulates in an animal depends on age, the level of nutrition, exercise, and other physiological factors that function in the specific muscle in question. Initially, fat generally tends to accumulate abdominally or under the skin and only later does it accumulate around the muscle. The latter accumulation leads to the so-called marbling of meat. Since this is the last fat to be deposited, a large quantity of feed is necessary to develop marbling. The amount of fat surrounding and within muscles is not constant but can increase or decrease depending on the food intake of the animal and other factors. The amount of fat surrounding the major muscle influences the appearance or "finish" of the meat.

The membranes surrounding fat cells contain small amounts of phospholipids and some cholesterol, but the most abundant lipids in adipose tissue are triacylglycerols and free fatty acids. This fat can be extracted from the carcass by rendering or heating to melt the fat and rupture fat cells. This fat can be used as such, as lard, for example, or it can be chemically modified for use in the manufacture of shortenings and margarines. The fat component of meat is very important since some is needed to produce a satisfactory flavor and mouth feel; however, excessive amounts decrease the usable lean portion without improving these quality attributes.

A. Collagen

The major fraction of connective tissue is collagen. This component is important because it contributes significantly to toughness. Also, the partially denatured product of collagen, gelatin, is a useful ingredient in many food products since it serves as the functional ingredient in temperature-dependent gel-type deserts. Collagen is abundant in tendons, skin, bone, the vascular system of animals, and the connective tissue sheaths surrounding muscle. Collagen comprises one-third or more of the total protein of mammals (72). A portion of collagen is soluble in neutral salt solution, part is soluble in acid, and another fraction is insoluble.

The collagen monomer is a long cylindrical protein about 2800 Å long and 14-15 Å in diameter. It consists of three polypeptide chains wound around each other in a suprahelical fashion. Collagen occurs in several polymorphic forms. The most common type is known as type I collagen and it consists of three polypeptide chains. Two of the polypeptide chains, designated $\alpha 1(I)$, are identical and are hydrogen bonded to one another and to a third chain, $\alpha 2$, which has a different amino acid sequence (117). Each chain has a molecular weight of about 100,000 daltons, yielding a total molecular weight of about 300,000 daltons for collagen. Another type of collagen found in muscle is referred to as type III, which is made up of three identical α chains designated $\alpha 1(III)$. This type of collagen is unusual in that intramolecular disulfide bonds exist in the nonhelical, carboxyterminal peptide. Another class of collagen molecules, type IV, is more complex and appears to consist of polypeptide chains of dissimilar sizes. The composition of type V collagen is also not very clear and may in fact represent more than a single type.

"Collagen" is a family of related molecules, and the complete elucidation of all of the members has not yet been achieved. The presence of type I collagen in the epimysium of muscle, types I and III in the perimysium, and types III, IV, and V in the endomysium have been reported (117). Type III collagen appears to have an especially important role in imparting toughness to muscles. In some instances, the single strands of peptides that constitute collagen are cross-linked with covalent bonds. When two of the peptides are joined in this fashion it is called a β component, and when all three are so joined, the product is known as a γ component. The solubility of collagen decreases as intermolecular cross-linking increases.

The peptides of collagen are mostly helical with the exception of the few residues at each end. However, the helices differ from the typical α-helix due to the abundance of hydroxyproline and proline, which interfere with α-helical structure. Collagen molecules are linked end to end and adjacently to form the collagen fibril, as shown in Fig. 11. There is a periodicity in the cross-striations of collagen at about 640-700 Å intervals. Col-

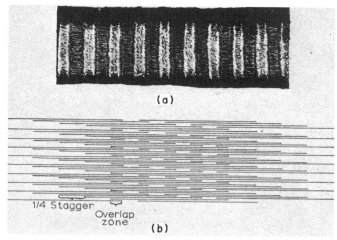

(a)

1/4 Stagger

Overlap
zone

(b)

Figure 11 Diagrammatic (a) and schematic (b) representations of the alignment of collagen molecules in the native collagen fibril. (Reprinted from Ref. 8, p. 1001, by courtesy of Dr. A. J. Bailey.)

lagen fibrils are sometimes arranged in parallel fashion to give great strength, as in tendons, or they may be highly branched and disordered as in skin.

The amino acid composition of collagen is unusual in that glycine represents nearly one-third of the total residues, and this amino acid is distributed uniformly at every third position throughout most of the molecule. This repetitive occurrence of glycine does not include the first 14 or so amino acid residues from the N terminus and the first 10 or so from the C terminus of the molecule (51), which are termed the telopeptides. Collagen is also unique as the only protein that has a large content of hydroxyproline (up to 10%) and contains the amino acid hydroxylysine. The content of hydroxyproline is less in fish collagens than in mammalian collagens. Hydroxyproline stabilizes the collagen molecule, and the collagens with low hydroxyproline contents denature at low temperatures. Since hydroxyproline occurs in so few other proteins, it is often taken as a measure of the amount of collagen in a food sample. Collagen also contains a large amount of proline. Collagen is nutritionally unbalanced with respect to amino acid composition; for example, tryptophan is almost totally absent.

Two important oxidations take place in the polypeptides that comprise collagen. These are the conversion of proline to hydroxyproline and the conversion of lysine to hydroxylysine. This latter compound, and lysine, can be oxidized, in some instances to α-amino adipic Δ-semialdehyde. Oxidation of proline and lysine to their hydroxylated residues is catalyzed by proline hydroxylase and lysine hydroxylase, respectively. These enzyme systems utilize molecular oxygen, α-ketoglutarate, ferrous ion, and a reducing substance, such as ascorbate. Ascorbate aids in wound healing through its role in the synthesis of collagen. Production of the α-amino adipic Δ-semialdehyde is catalyzed by a copper-containing enzyme called lysyl oxidase. These compounds are involved in collagen cross-linking.

The covalent cross-links involved in the β and γ components of collagen and the intermolecular cross-links between collagen molecules form spontaneously by the condensation of aldehyde groups. This may involve an aldol condensation-type reaction, or formation of a Schiff base when the aldehyde reacts with an amino group. When hydroxylysine reacts with hydroxylysine aldehyde, the reaction product undergoes an Amadori-type rearrangement to form a "keto" structure, hydroxylysino-5-keto-*nor*-leucine. Examples of these reactions are presented in Fig. 12.

With increasing age of the animal, the collagen cross-links are converted from a reducible form to a more stable nonreducible form. The nature of this "mature" nonreducible cross-link is not known, although several hypotheses have been proposed (117). It is likely that the reactions leading to nonreducible cross-links result in cross-links involving several functional groups and an extensive polymeric network. Intermolecular cross-links are confined to the end-overlap region involving lysine aldehyde in the telopeptide of one chain and a hydroxylysine in the helical region of an adjacent chain (28 nm from the N or C terminus). Noncollagenase proteinases, since they can attack the telopeptide region of collagen, can effectively reverse the cross-linking effect.

Cross-links in collagen increase with increasing age, and this may partially explain why meat from older animals is tougher than that from younger animals, even though muscles from younger animals generally contain more collagen. The situation in fish is very different. The collagen of the myocommata of older fish is weaker and has fewer cross-links than that of younger fish. On the other hand, older fish have more collagen (evident as thicker myocommata) than younger fish.

As collagen develops cross-links it becomes less soluble in a variety of solvents, such as salt and acid solutions. Whereas the amount of insoluble mammalian collagen often increases manyfold with age, the amount of insoluble collagen of cod increases only slightly and the amount of soluble collagen actually increases (80). Starving fish produce more collagen and greater collagen cross-linking than fish that are well fed.

Most proteolytic enzymes have little activity against native collagen, although they will readily degrade denatured collagen. Collagenase has been identified in several animal tissues. Its presence is difficult to detect since it exhibits low activity, apparently because of control mechanisms operative in the tissue. Little work has been done on collagenases that occur naturally in mammalian muscle tissue (66).

For the most part, animal collagenases cleave a single bond in each of the three chains of the collagen molecule. Several microorganisms, particularly *Clostridia* species, produce collagenases. These enzymes differ from the animal collagenases in that they degrade collagen extensively. Noncollagenase proteinases can cleave the collagen molecule in the telopeptide region. As noted earlier, they can contribute to destabilization of the collagen molecule by disrupting the region in which intermolecular cross-links are formed. Indeed, it has been suggested that noncollagenase proteases are required for the initial denaturation of collagen as well as for the effective and complete degradation of collagen fibers (15). The importance of collagenase enzymes in the breakdown of collagen in situ is not known, but they may have some importance in postmortem tenderization.

B. Conversion of Collagen to Gelatin

Collagen fibrils shrink to less than one-third their original length at a temperature, known as the shrinkage temperature, characteristic of the species from which the collagen is derived (51). This shrinkage involves a disassembly of molecules in the fibrils and a collapse of the triple-helical collagen molecule. Essentially the same type of molecular changes occurs when collagen is heated in solution. The midpoint of the collagen-to-gelatin transition is defined as the *melting temperature*. In addition to the noncovalent bonds af-

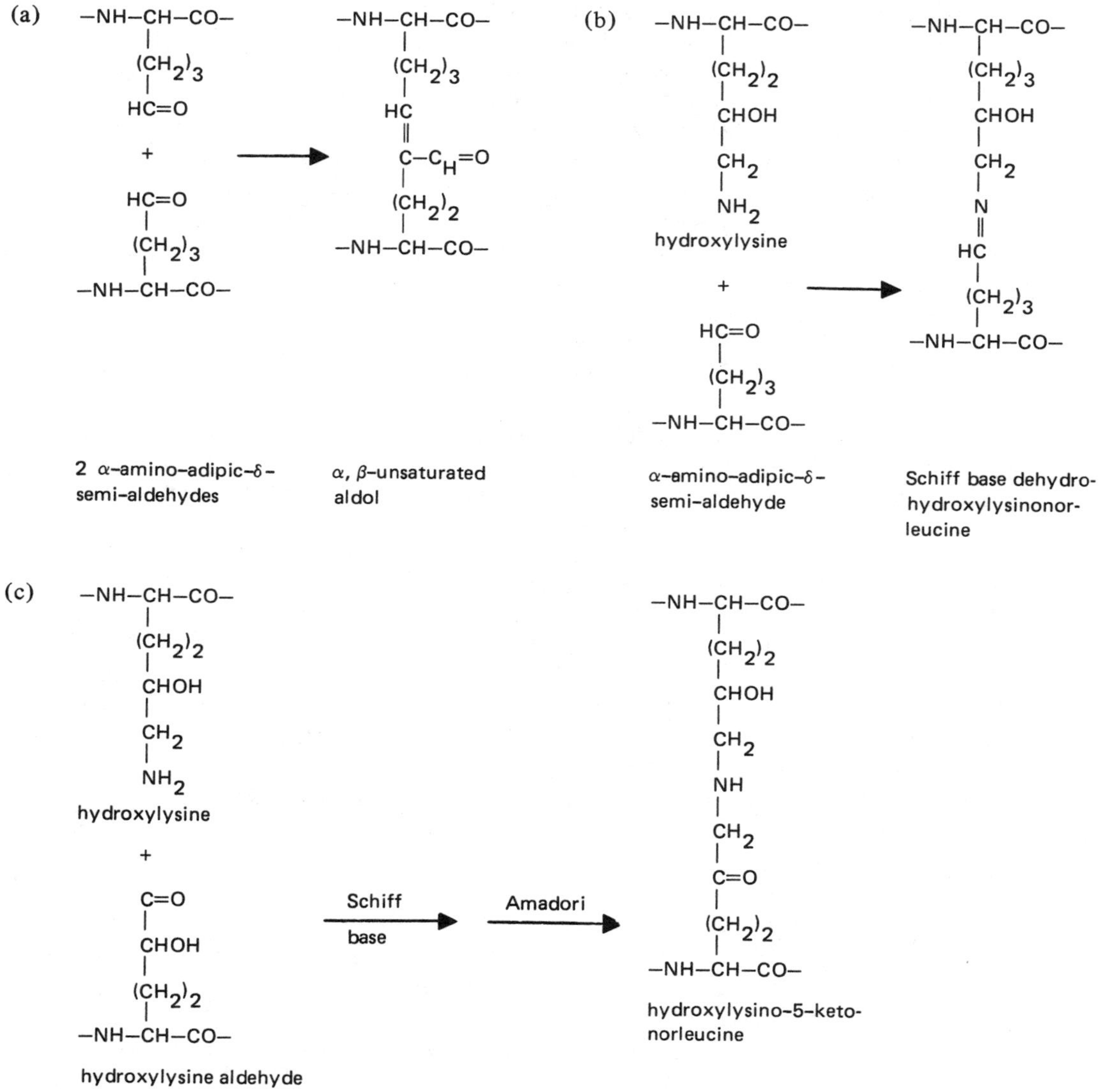

(a)

```
-NH-CH-CO-          -NH-CH-CO-
     |                   |
  (CH2)3              (CH2)3
     |                   |
  HC=O                 HC
                        ‖
     +        ──────▶  C-CH=O
                        |
  HC=O                (CH2)2
     |                   |
  (CH2)3              -NH-CH-CO-
     |
  -NH-CH-CO-
```

2 α-amino-adipic-δ-   α, β-unsaturated
semi-aldehydes   aldol

(b)

```
-NH-CH-CO-          -NH-CH-CO-
     |                   |
  (CH2)2              (CH2)3
     |                   |
  CHOH                CHOH
     |                   |
  CH2                 CH2
     |                   |
  NH2                 N
hydroxylysine           ‖
                      HC
     +                  |
                     (CH2)3
  HC=O                  |
     |        ──────▶  -NH-CH-CO-
  (CH2)3
     |
  -NH-CH-CO-
```

α-amino-adipic-δ-   Schiff base dehydro-
semi-aldehyde   hydroxylysinonor-
                          leucine

(c)

```
-NH-CH-CO-                                    -NH-CH-CO-
     |                                             |
  (CH2)2                                        (CH2)2
     |                                             |
  CHOH                                          CHOH
     |                                             |
  CH2                                           CH2
     |                                             |
  NH2                                           NH
hydroxylysine                                     |
                                                CH2
     +              Schiff      Amadori           |
                   ──────▶     ──────▶           C=O
  C=O               base                          |
     |                                         (CH2)2
  CHOH                                            |
     |                                         -NH-CH-CO-
  (CH2)2
     |                                   hydroxylysino-5-keto-
  -NH-CH-CO-                             norleucine
```

hydroxylysine aldehyde

Figure 12 Cross-link formation in collagen by side chain groups. (a) Aldol condensation followed by loss of water. (b) Schiff base formation. Lysine reacts in a manner analogous to hydroxylysine. (c) Schiff base formation followed by Amadori rearrangement.

fected, some intermolecular bonds, some intramolecular bonds (Schiff base and aldol condensation bonds), and a few main-chain peptide bonds are hydrolyzed. This results in conversion of the tri stranded collagen structure to a more amorphous form, known as gelatin (128). These changes constitute denaturation of the collagen molecule but not to the point of a completely unstructured product. If the latter event happens, glue instead of gelatin is produced.

After gelatin is produced and the temperature is lowered to below the critical value, there is a partial renaturation of the collagen molecule, involving what is called the "collagen fold" (128). Apparently, those parts of collagen that are rich in the amino acid residues proline and hydroxyproline regain some of their structure, following which they can apparently interact. When many molecules are involved, a three-dimensional structure is produced which is responsible for the gel observed at low temperatures. This collagen fold is absolutely dependent on temperature. Above the melting temperature, the structure degrades and so does the gel. The strength of the gel formed is proportional to the square of the concentration of gelatin and directly proportional to molecular weight (128). Ionic strength and pH seem to have only small influences on gel structure over the range of values encountered in food products.

Processing of collagen into gelatin involves three major steps. There is first the removal of noncollagenous components from the stock (skin and bones), then the conversion of collagen to gelatin by heating in the presence of water, and finally recovery of gelatin in the final form. The molecular properties of gelatin depend on the number of bonds that are broken in the parent collagen molecules. Removal of noncollagenous material may be done with acid or with alkali, and the procedures used influence the properties of the gelatin.

Conversion of collagen to gelatin occurs during normal cooking of meat, and this accounts for the gelatinous material that is sometimes evident in meat after heating and cooling. As aspic type of product, for example, results from the conversion of collagen to gelatin. Also, conversion of the collagen molecule to the much less structured gelatin can play an important role in the tenderness of meat, especially in meat of poor quality. This kind of meat (e.g., pot roast or stew meat) has a high content of collagen and is usually cooked for a long time in liquid to achieve the desired tenderness, that is, the desired conversion of collagen to gelatin.

Since collagenase enzymes are apparently present in muscle and some collagen cross-linkages are easily broken, one may expect that significant changes in collagen occur during postmortem aging and that these changes have a desirable effect on tenderness. Information on this point, however, is inconclusive. Several workers have reported that there is little change in collagen solubility (which is based in part on the number of cross-linkages in the collagen) on aging postmortem. Furthermore, it has been reported that collagen content is not affected significantly by postmortem conditions (49). Other workers (64,87), however, have reported either a significant change in the amount of intramuscular perimysial and endomysial collagen, or a change in the molecular structure of intramuscular collagen during postmortem holding. Furthermore, collagen undergoes a slight increase in shrinkage temperature with increasing age of beef cattle, and a postmortem decline in shrinkage temperature over the period of 45 min to 7 days postmortem (56). The exact meaning of these results remains uncertain.

VIII. BIOCHEMICAL CHANGES IN MUSCLE POSTMORTEM

A. Biochemical Changes Related to Energy Metabolism

Muscle is a highly specialized tissue, and its functioning represents a classic example of conversion of chemical to mechanical energy in living systems. Muscle requires a large outlay of energy to rapidly operate the contractile apparatus (Sec. V.B). This energy is derived immediately from the high-energy compound ATP. Creatine phosphate then rapidly transfers its high-energy phosphate to ADP to prevent excessive decreases in ATP levels during periods of vigorous muscle activity. Resynthesis of ATP from ADP and creatine phosphate is catalyzed by the enzyme creatine kinase. The enzyme adenylate kinase converts two molecules of ADP to one molecule of ATP and one molecule of AMP; this reaction also serves as a source of ATP for the muscle cell.

For long-term activity, the muscle must rely on the oxidation of substrates, usually either carbohydrate or lipid in nature, to maintain an appropriate level of ATP. Lipid metabolism seems to be an especially important source of utilizable energy in muscles that exhibit sustained activity, for example, those that support sustained running of animals or sustained flights of migratory birds. Carbohydrates are also an important energy source for muscle. Glycogen is the most important source of carbohydrate energy, but free glucose is also utilized. In red muscles and in muscles that are not working extremely hard, it is likely that most of the energy is supplied via the Krebs cycle and the mitochondrial electron transport system (aerobic respiration). This system provides a large quantity of ATP molecules per molecule of substrate utilized and allows for the complete conversion of substrate to carbon dioxide and water.

The mitochondrial system, however, requires oxygen and in some instances, when the muscle is under heavy stress, the oxygen available is not sufficient to maintain mitochondrial function. The anaerobic glycolytic system may then become predominant. This is especially likely in white muscles, which are generally involved in sporadic bursts of activity requiring very large amounts of energy. During anaerobic glycolysis, glycogen is converted by the Emden-Meyerhoff pathway through a series of phosphorylated six-carbon and three-carbon intermediates to pyruvate, which is then reduced to lactate. The system requires the cofactor NAD^+, and it continually regenerates the NAD^+ required. The terminal enzyme of the sequence, lactate dehydrogenase, is principally responsible for regeneration of NAD^+. In anaerobic glycolysis, ATP production is much less efficient than it is in aerobic respiration. For example, anaerobic glycolysis yields only 2 or 3 mol ATP per mole of glucose, compared with 36 or 37 when aerobic respiration is operative. Also, anaerobic glycolysis results in incomplete oxidation of substrates and accumulation of lactate. Lactate can penetrate the cell membrane, and much of it is removed to the blood, where it goes to the liver and is used in the resynthesis of glucose. The glucose is then carried back to the muscle, where glycogen is eventually reproduced. Lactate transport out of fish muscle is very slow, and a very long period is required to deplete it (80). Hydrolysis of ATP during contraction may lead to a temporary reduction of pH in the living muscle cell (90). Anaerobic glycolysis is favored only under conditions in which the mitochondrial system cannot function, that is, under conditions of limited oxygen. The amount of anaerobic glycolysis that occurs in a muscle is, therefore, dependent on many factors, including the vascular system, the type of muscle, the amount of myoglobin, and the number of mitochondria present.

We are interested in these reactions mainly because many of them proceed after the death of the animal and have an important bearing on the food quality of the muscle. Muscle tissue is often referred to as "dead," as opposed to postharvest fruits and vegetables, which are said to be "alive." The distinction is mostly due to differences in the physiology and morphology of muscle tissue as compared with plant tissue. Animal tissues are more highly organized than plant tissues, and life processes of animals rely heavily on a highly developed circulatory system. After death, all circulation ceases, and this rapidly brings about important changes in the muscle tissue. The principal changes are attributable to a lack of oxygen (anaerobic conditions) and the accumulation of certain waste products, especially lactate and H^+. Plant tissue is less dependent on a highly developed circulatory system than animal tissue, and although certain substances are no longer available to the fruit or vegetable after harvest (see Chap. 15), oxygen can diffuse in, CO_2 can diffuse out, and waste products can be removed from the cytoplasm by accumulation in the large vacuoles present in mature parenchymatous tissue. In contrast, although the enzymes of muscle postmortem are still subject to the same control mechanisms as in the live tissue, they are faced with an entirely different environment.

It is believed that cells in general try to maintain a high energy level. The energy level of a cell can be expressed as "energy charge," which is defined as (7)

$$\text{Energy charge} = \frac{\text{ATP} + 1/2\text{ADP}}{\text{ATP} + \text{ADP} + \text{AMP}} \qquad (1)$$

The muscle cell postmortem tries to maintain a high energy charge, but it has restrictions imposed on it because the circulatory system has been disrupted. In a short time, the mitochondrial system ceases to function in all but surface cells, since internal oxygen is rapidly depleted. Oxidative metabolism of some substrates, such as lipids, ceases at this time. ATP is gradually depleted through the action of various ATPases, some contributed by the contractile proteins but most coming from the membrane systems (12). Some ATP is temporarily regenerated by the conversion of creatine phosphate to creatine and the transfer of its phosphate to ADP. The adenylate kinase reaction discussed above may also function here. After the creatine phosphate has been used up, which occurs fairly rapidly, anaerobic glycolysis continues to regenerate some ATP with the end product, lactate, accumulating. Glycolytic activity may cease because of exhaustion of substrate or, more likely, because of the decrease in pH caused by ATP hydrolysis. As glycolytic activity slows down, ATP concentration decreases, with most of the ATP depleted in 24 hr or less, depending on the species and the circumstances. Although the hydrogen ions generated in muscle come from hydrolysis of ATP and not from production of lactate, there is a close correlation between the amount of lactate produced and the pH decline. This relationship occurs because there is an almost linear relationship between ATP produced by the glycolytic system (and hence the ATP that may be hydrolyzed later) and the amount of lactate produced.

The decrease in pH that accompanies postmortem glycolysis has an important bearing on meat quality. One point of concern is the rate of glycolysis and the concomitant pH decrease. Since there are some 20 enzymes involved, glycolysis conceivably can be controlled at many points, depending on the particular subcellular conditions. However, there seems to be certain points that are especially important in controlling the rate of glycolysis. In some muscles the phosphorylase step, that is, the conversion of glycogen to glucose-1-phosphate, is the controlling step (93). This is true, for example, in "slow-glycolyzing" muscles of the pig. However, in "fast-glycolyzing muscles," control is also exerted by phosphofructokinase. In muscle minces, in which inorganic phosphate is included, the eventual rate-limiting step is the one controlled by glyceraldehyde-3-phosphate dehydrogenase. This is attributed to hydrolysis of NAD catalyzed by endogenous nucleotide pyrophosphatase (94). The decrease in concentration of this cofactor limits the activity of the dehydrogenase.

The rate of pH decline also may be affected by the buffering capacity of muscle. Generally, white muscles have relatively large contents of the amino acids, carnosine and anserine, and these probably function as buffering agents in the muscle cell.

A second aspect important in postmortem muscle is the extent of glycolysis, that is, how much glycolysis occurs. Under ordinary circumstances, the factor that limits the extent of postmortem glycolysis is pH. When the pH is low enough, certain critical enzymes, especially phosphofructokinase, are inhibited and glycolysis ceases. The pH at which this occurs is generally around 5.1-5.5.

The final pH attained is called the "ultimate pH," and this value has an important influence on the textural quality of meat, its water-holding capacity, its resistance to growth of microorganisms, and its color. If an animal, prior to slaughter, is stressed or exercised vigorously its glycogen content decreases substantially and a higher ultimate pH is likely to result postmortem since there is insufficient substrate for glycolysis.

The ultimate pH of fish is not related to stress or the exercise to which a fish is subjected prior to death. In living fish, lactate produced during exercise, such as struggling during harvest, is only very slowly removed from the muscle; thus, when a fish struggles extensively before death, much lactate can be produced and this lactate, for the most part, is present in the muscle postmortem. On the other hand, a fish that struggled only slightly prior to harvest would contain only a small amount of lactate at the point of death, but normal postmortem glycolysis would cause muscle lactate to increase to essentially the same level as that existing in fish that struggled vigorously prior to death. Since the amount of lactate formed is roughly proportional to the amount of ATP produced by anaerobic glycolysis, and since all the ATP in postrigor fish is broken down to produce hydrogen ions, the net effect is that struggling or stress of fish prior to death does not affect its ultimate pH. The ultimate pH of red meat fish, like tuna and mackerel, is around 5.5, that of lean white fish is 6.2-6.6.

Depletion of glycogen can be achieved experimentally in mammalian and avian muscles by the antemortem injection of certain hormones, such as epinephrine. It is possible that other phenomena also have a bearing on ultimate pH, for example, hydrolysis of the cofactor NAD or inactivation of a key enzyme by some handling stress.

B. Consequences of ATP Depletion

The consequences of ATP depletion in a muscle cell are in part similar to those that presumably occur in any cell in which energy sources become depleted. Biosynthetic reactions come to a halt, and there is a loss of the cell's ability to maintain its integrity, especially with respect to membrane systems. There is an additional unique phenomenon that occurs with muscle cells. ATP and ADP serve as plasticizers for actin and myosin (65). When these nucleotides are present, they prevent interaction of these two proteins. When ATP and ADP are depleted postmortem, actin and myosin interact and the muscle passes into a state known as rigor mortis. This means literally the "stiffness of death." The stiffness is due to the tension developed by antagonistic muscles. This tension cannot be relieved because ATP and/or ADP is absent. Individual muscles may or may not become stiff, but all become inextensible when ATP is depleted. When there is sufficient ATP and/or ADP, muscle can be stretched a reasonable length before it breaks. In the state of rigor, however, the muscle ATP has been depleted and it consequently cannot be stretched significantly without breaking.

The rate of loss of ATP postmortem is directly and causatively related to the decline in pH and is affected by such factors as species, animal to animal variation within a species, type of muscle, and temperature of the carcass. The ultimate pH of many species postmortem is close to 5.5, although it may be somewhat higher for some; for example, 5.9 is common in chicken muscle and 6.2-6.6 in lean fish. At the moment of slaughter or harvest, average ATP content of many muscles is in the range 3-5 mg/g of fresh tissue, and the time at which most of the ATP is degraded can be used as a rough indicator of the time of rigor onset. The time postmortem for rigor onset (incipient loss of extensibility) may be as little as 2-4 hr in chicken muscle, 10-24 hr in beef muscle, and up to 50 hr in whale muscle (51). The time for rigor onset is greatly dependent on temperature. Beef muscle held at 37°C (higher than normal for postmortem handling) can go into rigor at 4 hr postmortem.

The relationships between several chemical and physical properties of muscle postmortem are illustrated in Fig. 13. The decrease in creatine phosphate slightly precedes that of acid-labile phosphates (principally ATP) since creatine phosphate is used up in an attempt to maintain the physiological concentration of ATP. The decline in pH closely follows the net loss in acid-labile phosphate in the latter stages of the curve. pH declines somewhat faster initially because ATP from several regenerating sources is being hydrolyzed. The reciprocity of the acid-labile phosphate and the extensibility curves illustrates the role of ATP is maintaining extensibility of the muscle.

C. Increase in Tenderness on Aging Postrigor

It has long been known that muscle tissue from freshly slaughtered animals is tender prior to the development of rigor mortis, but is less tender immediately thereafter. When muscle is allowed to stand postrigor at refrigerated temperatures it becomes more tender than muscle in rigor. This tenderization process is often called "resolution of rigor." This process is extremely important, and much research has been done to define the mechanisms of this tenderization. Some workers have reported changes in actin-myosin interactions with time postrigor (5,123); other workers have not been able to observe such changes (133). The amount of troponin and tropomyosin also have been reported to decrease with time of postrigor storage, with the former decreasing faster than the latter (136). This is taken as evidence that structural alterations occur in the myofibril as well as in the Z disk.

An event of prime importance in the process of postrigor tenderization is the disintegration of the Z disk (26,122), It has been demonstrated that this disruption of the Z disk is a primary process in the development of tenderness in poultry (134) and in beef (102). Disintegration of the Z disk renders myofibrils more susceptible to fragmentation during homogenization (122), and results of this technique, expressed as the myofibril fragmentation index, are used as a measure of meat tenderness (100).

There is good evidence that destruction of the Z disk comes about by the action of proteolytic enzymes in the muscle. In particular, a calcium-activated neutral protease, termed the calcium-activated factor (CAF), has been isolated form several species of animals, and this protease is capable of causing disintegration of the Z disk in vitro (19). Treatment of muscle or isolated myofibrils with CAF leads to the release, but not degradation, of α-actinin. CAF has also been shown to be active against troponin, tropomyosin, and C protein but does not hydrolyze actin, myosin, α-actinin, or connectin (84). However, recent work suggests the existence of a species of "Z-disk actin" that is susceptible to hydrolysis by CAF (91). When CAF hydrolyses troponin T, a 30,000 dalton species is produced, and the amount of this component correlates positively with the tenderness of beef (100). CAF appears to exist in two forms in muscle, one form requiring a millimolar concentration of calcium for activity, and the other activated by a micromolar concentration of calcium, that is, a concentration similar to that present in living muscle. The physiological reasons that two forms of this enzyme exist are not known. An endogenous

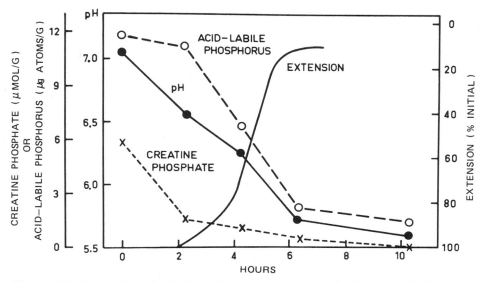

Figure 13 Chemical and physical changes during rigor development in beef sterno-mandibularis muscle held at 37°C. (Reprinted from Ref. 92, p. 215, by courtesy of the University of Wisconsin Press.)

protein inhibitor of CAF exists in skeletal muscle, presumably to aid in regulating the activity of this enzyme (125). It has been suggested that CAF is involved in the turnover of myofibrillar proteins in the living animal by untying the myofibril. This may occur through its degradative actions on the Z disk, on the C protein (which may be responsible for tying the thick filaments together), and on desmin, which presumably serves as an intermyofibrillar link.

It has also been demonstrated that the presence of Ca^{2+} at concentrations known to occur physiologically leads to a weakening of the Z disk in postrigor muscle (46,47). The effectiveness of Ca^{2+} in accomplishing this is dependent on the tension developed during postmortem contraction of the muscles. Fragmentation of the myofibrils results from weakening of the Z disk, but α-actinin is not released, nor is troponin T degraded. Furthermore, inhibition of CAF does not lessen this Ca^{2+}-induced weakening of the Z disks, indicating that this is a different process from the CAF-catalyzed reaction. This may be the major factor responsible for the weakening of Z disks normally seen in aged muscle (i.e., muscle held postrigor for several days at refrigerator temperatures). In addition to the weakening of the Z disks, Ca^{2+} also causes the release of a protein from the myofibrils, possibly a Z-disk component, which weakens the interaction between actin and myosin (124). This weakening of the actin-myosin interaction may be involved in resolution of rigor and the increase in sarcomere length during postrigor holding (aging) of muscle.

Cathepsins are another group of proteases known to occur in muscle tissue. Their role in postmortem changes in muscle has been the subject of much dispute because, like most lysosomal enzymes, they have pH optima in the acid range, compatible with intra-lysosomal pH but not with the pH of muscle sarcoplasm. The cathepsins of muscle tissue may be located in the lysosomes of the muscle cell itself (cathepsins B, D, H, and L), or they may be associated with lysosomes from components of the connective tissue or from those of the circulatory system (35). In postmortem muscle tissue, breakdown of some of the cellular membranes may occur, allowing cathepsins from nonmuscle cells to act on muscle cell components. Muscle cathepsins have acid pH optima and may not function in muscle tissue postmortem. This is true, for example, of cathepsins D and L. However, some cathepsins appear to exhibit activity at the ultimate pH found in some postmortem muscles. At pH 6, cathepsin B hydrolyzes several myofibrillar proteins (95). Myosin is most readily attacked by cathepsin B followed by troponin, tropomyosin, and actin, with the last proteins only slightly susceptible. α-Actinin is not affected by cathepsin B. This enzyme degrades troponin T rapidly and troponin I more slowly and does not affect troponin C. When myofibrils are incubated with cathepsin B, disappearance of the Z disk is first observed, followed by loss of the M line, and then a decrease in the density of the A band. In rabbit muscle, a catheptic protease apparently exists that is maximally active in the ultimate pH range of muscle, about 5.5, and cleaves myosin into heavy and light meromyosin (5). This protease also has the ability to degrade α-actinin.

It is possible to induce the onset of rigor at elevated temperatures. When beef or lamb muscles are held under these conditions and restrained to prevent shortening, increased tenderness is observed. There is also a release of lysosomal enzymes. It has been suggested that the release of these enzymes combined with the low pH at a high temperature causes hydrolysis of myofibrillar proteins and increases tenderness. Both myosin and troponin T are affected by this process (29). Even without restraint, the small extent of contraction that occurs in these muscle tissues compared with those that undergo cold shortening, would result in better tenderness. Proteolysis by lysosomal enzymes would serve to improve this further. It has been reported that the content of muscle connectin, the protein that forms a network of longitudinal filaments between Z disks, decreases with time postmortem. This loss of connectin is dependent on calcium concentration and is closely related to the loss of elasticity in the muscle tissue. Thus, the loss of connectin may be involved in the postmortem tenderization of meat (121).

D. ATP Breakdown to Hypoxanthine

Another important consequence of the loss of ATP in muscle postmortem is its conversion to hypoxanthine, as shown in Fig. 14. Certain 5'-mononucleotides, intermediates in the production of hypoxanthine, are important flavor enhancers in muscle foods. For these compounds to have flavor-enhancing properties, the ribose component must be phosphorylated at position 5' and the purine component must be hydroxylated at position 6. Compounds that fall into this category are IMP (inosinic acid) and GMP (guanylic acid). ATP is first converted to ADP and then to AMP by a disproportionation reaction. AMP is then deaminated to IMP, a flavor enhancer. IMP, however, can degrade to inosine and eventually to hypoxanthine. Hypoxanthine has a bitter flavor, and it has been suggested as a cause of off-flavors in stored fish.

Hypoxanthine also may cause off-flavors in unheated irradiated beef since some enzyme activity remains (enzymes are more difficult to destroy by ionizing radiations than are microorganisms). Irradiated beef therefore should be exposed to a heat treatment sufficient to inactivate enzymes so that hypoxanthine production and other detrimental enzyme-catalyzed changes do not occur.

The kinetics of ATP breakdown to IMP, inosine, and hypoxanthine in cod muscle at 0°C are shown in Fig. 15. In this figure, it is also possible to compare the aforemen-

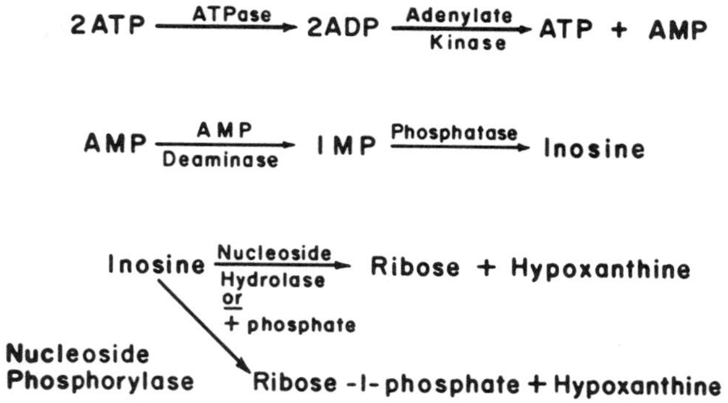

$$2\,ATP \xrightarrow{\text{ATPase}} 2ADP \xrightarrow[\text{Kinase}]{\text{Adenylate}} ATP + AMP$$

$$AMP \xrightarrow[\text{Deaminase}]{\text{AMP}} IMP \xrightarrow{\text{Phosphatase}} Inosine$$

$$Inosine \xrightarrow[\substack{\text{or} \\ \text{+ phosphate}}]{\text{Nucleoside Hydrolase}} Ribose + Hypoxanthine$$

Nucleoside Phosphorylase → Ribose -I- phosphate + Hypoxanthine

Figure 14   Conversion of ATP to hypoxanthine in postmortem muscle.

tioned changes with those of glycogen, lactate, and pH at any given time. The production of IMP from ATP (through AMP) is rapid, that of inosine is less so, and production of hypoxanthine from inosine is slow. Thus, measurement of the levels of different breakdown products at a given time can give an indication of the degree of spoilage. The breakdown products may be measured individually; however, in some cases, it is advantageous to know the concentrations of more than one component. For example, the K value, which determines the relationship between inosine, hypoxanthine, and the total amount of ATP-related compounds, is often used to determine the freshness of fish (110). These products also may have significance in canned muscle tissue, such as herring, mackerel, or salmon. Since the enzymic reactions are stopped during processing, analysis of the canned product for these components should indicate the quality of the raw fish that was used. The use of levels of ATP degradation products as a measure of freshness has proven less successful for muscle tissues from warm-blooded animals than for muscle tissues from cold-blooded animals.

E.   Loss of Calcium-Sequestering Ability

A possible reason for increased ATPase activity postmortem and the tendency for muscle to contract is a loss in the ability of the sarcoplasmic reticulum (38) and the mitochondria (18) to sequester calcium. The sarcoplasmic reticulum and mitochondria release calcium during aging postmortem. The inability of these organelles to control the calcium content of the sarcoplasm postmortem is probably due both to a deterioration in the energy-generating systems of the muscle and to an increase in permeability that the sarcoplasmic reticulum and mitochondria exhibit with postmortem age (this permits greater leakage of the calcium from the membrane). Mitochondria lose their ability to maintain their integrity postmortem because of anoxic conditions and the attendant disruption in energy production. Increased levels of calcium in the sarcoplasm may cause contractile proteins to more fully display their ATPase potential and may initiate activity of ATPases in membrane systems. This, in turn, causes increased activity of the glycolytic system and a more rapid decline in pH. The sarcoplasmic reticulum of postmortem chicken, however, does not lose its ability to sequester calcium (48).

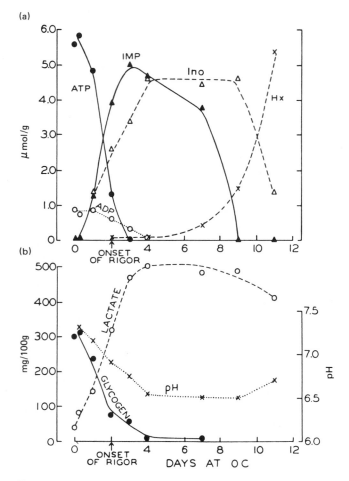

Figure 15 (a) Nucleotide degradation in cod muscle stored at 0°C: ATP, ●; ADP, ○; IMP, ▲; inosine, Δ; hypoxanthine, x. (b) Glycolytic changes. (Reprinted from Ref. 33, p. 1839, by courtesy of the Fisheries Research Board of Canada.)

F. Changes in Lipid Postmortem

Changes occur in the lipid components of muscle tissue postmortem, and these changes have been more extensively studied in fish than in warm-blooded animals. This is probably true for the following reasons: (a) fatty acids of fish lipids are much more highly unsaturated than those of mammals and birds and thus would be expected to undergo more rapid oxidation with associated development of off-odors and flavors; (b) free fatty acids in fish tissues are expected to have a greater effect on the contractile proteins since these proteins are less stable in fish muscle than they are in warm-blooded animals; and (c) a larger proportion of fish tissue is marketed in the frozen state. Freezing encourages lipid oxidation, partly because the competing reactions of microbiological spoilage are avoided. Thus, lipid oxidation is relatively more important in frozen muscle tissue than in fresh tissue.

Lipases and phospholipases have been isolated from muscle tissue of many species of fish, both lean and fatty. In general, lipase activity is greater in red muscle than in white muscle of the same fish species. In lean fish, the enzyme of principal concern is phospholipase, since most of the lipid in this type of tissue is phospholipid. Extremely rapid hydrolysis of lipids occurs in sterile cod stored at 20°C (99). It is also possible that lipases and phospholipases produce free fatty acids during the early stages of cooking Hydrolysis of phospholipids appears to occur more rapidly than hydrolysis of triacylglycerols.

Oxidation of lipids also occurs during postmortem storage of fish muscle tissue. This is not surprising considering the large concentration of highly polyunsaturated fatty acids present in fish lipids. The exact mechanism of lipid oxidation in fish tissue is not known. A "microsomal" fraction has been isolated from lean and fatty fish tissue that catalyzes oxidation of membrane lipids in vitro (88). The role of this system in the oxidation of fish lipids postmortem and in situ has not yet been determined. The relationship between lipolytic activity and oxidation is not clear. Some authors have reported that lipid oxidation occurs more rapidly in tissues containing free fatty acids (116); other workers have shown that free fatty acids produced from phospholipids inhibit lipid oxidation in muscle tissue (114).

G. Breakdown of Other Components Critical to Quality

1. Reaction Products from Amino Acids

Microorganisms can utilize amino acids in muscle postmortem leading to a large number of breakdown products, such as hydrocarbons, aldehydes, ketones, sulfides, disulfides, mecaptans, and amines (see Chap. 5). Microbial conversion of amino acids to a variety of di- and polyamines is a particular problem in certain fatty fish, since these products have unpleasant sensory characteristics and have been implicated in an allergenic type of food poisoning called scombroid poisoning, the symptoms of which often involve nausea and headaches. Although these symptoms are usually attributed to histamine, there is some evidence that other di- and polyamines produced also may be involved (108).

2. Breakdown of Trimethylamine Oxide

Marine fish, and indeed most animals from the marine environment, contain a significant concentration of trimethylamine oxide (TMAO) in the body tissues. In certain species of fish the content of trimethylamine oxide in the aqueous phase of the muscle tissue can exceed 0.1 M. In postmortem fish muscle, microorganisms may reduce the TMAO to trimethylamine, a compound responsible in part for the "fishy" odor that develops in fish on storage.

Trimethylamine oxide also decomposes in fish flesh to dimethylamine and formaldehyde, as indicated in Eq. (2).

$$(CH_3)_3-N{\rightarrow}O \longrightarrow (CH_3)_2-\overset{H}{N} + HCHO \qquad (2)$$

This reaction is catalyzed by an enzyme and proceeds particularly fast in muscle tissue from gadoid species, such as cod, haddock, and hakes (2).

The rate of this reaction is increased by freezing the muscle tissue or by disruption of the tissue, such as occurs in mincing. It has been hypothesized that the formaldehyde produced in this reaction cross-links the muscle proteins and this contributes to the tough texture that developes in gadoid fish during frozen storage.

IX. EFFECT OF POSTMORTEM CHANGES ON QUALITY ATTRIBUTES OF MEATS

A. Texture and Water-Holding Capacity

1. Rate and Extent of pH Change

One of the important biochemical events in muscle tissue postmortem is the reduction in pH brought about principally by glycolysis. Both the extent and the rate of pH change are important. The rate of change of pH is important because if the pH drops low enough while the temperature of the carcass is still high, for example, pH 6.0 and 35°C (76), considerable denaturation of contractile and/or sarcoplasmic proteins may occur. If the sarcoplasmic proteins are denatured they may adsorb to contractile proteins, thus modifying the physical properties of the contractile proteins. This phenomenon, along with the denaturation of the contractile proteins, decreases the ability of the contractile proteins to bind water. This too rapid drop in pH postmortem has been associated with the pale-soft-exudative (PSE) condition that is especially common in pork. A similar condition may also occur in red-meat fish, like tuna. The Japanese term this "yakeniku." Characteristic attributes are a soft texture, poor water-holding capacity, and a pale color.

Also, lysosomal enzymes are released at high temperatures, and an accompanying low pH may allow the lysosomal enzymes to act on contractile or connective tissue proteins producing greater tenderness. The contraction that may occur at elevated temperatures (see below) and the accompanying toughness developed from this may obscure the tenderizing effect of low pH and high temperature. However, when contraction is controlled, tenderization does occur under conditions of low pH and high temperature (29). Under normal conditions, the toughening effect of shortening, which occurs if muscle is held at a high temperature, overrides the tenderizing process.

The "ultimate" or final pH the muscle tissue reaches is also important. A low ultimate pH in beef provides resistance to the growth of microorganisms and a normal color. Sometimes a high ultimate pH occurs. This may be produced by some antemortem stress, such as vigorous exercise or fright. This causes glycogen levels to be low, and there is insufficient substrate and therefore insufficient ATP hydrolysis to produce a low pH. This is fairly common in beef, and the final meat may be dark, firm, and dry. These "dark-cutting" muscles exhibit excellent water-holding properties but have poor resistance to the growth of microorganisms. It is possible that this high-pH meat may be dark due to the continuance of mitochondrial activity for a longer than normal period of time (6). Active mitochondria might compete with myoglobin for oxygen at the surface of the meat, thus reducing the amount of oxymyoglobin (bright red) formed.

The ultimate pH of fish meat is generally higher than that of land animals and birds. It is affected by season, primarily because feeding patterns change with the season. Low ultimate pH values in fish muscle are occasionally observed, and this generally occurs in fish that recently resumed feeding following a period of starvation. This usually occurs in early summer and may lead to a defect known as "gaping."

Gaping occurs when the myotomes of fish muscle separate. Normally, the myotomes (see Fig. 2) are connected to the myocommata by thin tubular sections that are part of the myocommata. When these tubules break, gaping occurs. Gaping is particularly troublesome when fish are frozen, thawed, and then filleted, since freezing produces ice crystals that are physically disruptive to the connective tissue, which is weak at the existing pH (80).

The texture of cooked fish is closely related to the postmortem pH of the flesh: that is, the lower the ultimate pH, the tougher the texture. pH is the most important factor governing texture in cooked fish, and the effect of pH is accentuated if the fish is preserved by freezing (80). pH apparently exerts its effect on the texture of fish muscle by influencing the contractile elements since fish collagen is disrupted by normal cooking.

Although the ultimate pH of a mammalian muscle has an important influence on toughness, other factors are also significant (16).

2. Actin-Myosin Interaction and Contraction

A major phenomenon that affects the texture of meat is the interaction of actin and myosin after the depletion of ATP and ADP. If meat is cooked and eaten before it passes into rigor, that is, before actin and myosin interact, it is very tender. However, once actin and myosin interact, the meat exhibits increased toughness and a loss of extensibility. Meat is sometimes aged to partially overcome this effect.

The development of toughness in meat is particularly influenced by the state of the sarcomere when actomyosin is formed, that is, the extent to which thin and thick filaments overlap. This is controlled by several factors in the early postmortem period. Since critical changes in sarcomere length require the presence of ATP, either as a plasticizing agent so thin and thick filaments can slip past each other or as an energy source for contraction, it is necessary to consider principally the prerigor period when ATP and ADP are still abundant.

If muscle is excised in a prerigor state, it is susceptible to rapid contraction. The extent of shortening depends on many factors, including the nature of the specific muscle, elapsed time between death and excision, and the physiological state of the muscle at the time of death (Sec. XI). At high (physiological) temperatures, postmortem contraction is great. As the temperature is reduced, contraction of excised, prerigor muscle decreases, and for some muscles, minimal contraction occurs at about 10-20°C. If these muscles are exposed to still lower temperatures in the range 10-0°C, increased contraction again occurs (Fig. 16) (77). This behavior is particularly noticeable and of commercial importance with muscles that consist primarily of red fibers, such as beef and lamb. Shortening that occurs at low, nonfreezing temperatures is termed "cold-shortening." Muscles that are highly susceptible to cold-shortening (beef and lamb) should be maintained at temperatures above 10°C until they pass into rigor. Otherwise, the shortening results in undesirable changes in tenderness.

The occurrence of a minimum in the contraction-temperature curve indicates that at least two processes are involved in the phenomenon of cold-shortening. One of these is the normal decrease in contraction that is expected as the temperature is lowered, similar in nature to the behavior of most chemical reactions. At some point, however, it appears an additional factor comes into play that overcomes the natural tendency for contraction to decrease with decreasing temperature. This phenomenon appears to be the release of calcium into the sarcoplasm at concentration sufficient to induce contraction. It has been suggested that calcium ions are released from the mitochondria as a response to the loss of oxygen postmortem (18). In addition, low, nonfreezing tempera-

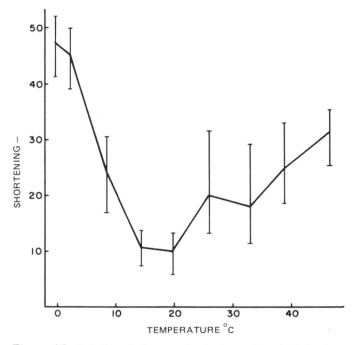

Figure 16 Relation between shortening of excised beef muscle and storage temperature postmortem. Vertical lines represent standard deviations. (Reprinted from Ref. 76, p. 788, by courtesy of Dr. R. H. Locker.)

tures may impair the ability of the sarcoplasmic reticulum to sequester calcium, thus causing calcium levels in the sarcoplasm to rise to activating levels.

Marsh and Leet (82) have shown (Fig. 17) that toughness increases as shortening increases, up to a point at which the muscle is approximately 60% of its rest length (40% contraction). This length represents the point at which the myosin filaments just reach the Z disk; that is, the I band just disappears. At this state of contraction, 94-100% of the myosin heads are bound to actin in rabbit skeletal muscle (23). With greater shortening, the muscle loses toughness rapidly. The reason for this tenderization is not completely clear, but it may result because thin and thick filaments slide past each other to an extent exceeding that in vivo, thus leading to severe, irreversible disruption of the sarcomere.

Muscles on the carcass also can undergo some cold-shortening. The extent to which this occurs depends on whether the muscle has been severed during slaughter (e.g., sternomandibularis or neck muscle), whether the temperature is conducive to cold shortening, whether a temperature gradient exists along the muscle during prerigor cooling causing some parts to contract while other parts stretch, whether the muscle is naturally anchored at both ends or at one end only (e.g., longissimus dorsi), the species being handled, and how the carcass was positioned during the prerigor phase. With regard to the last point, improvements in tenderness can be accomplished by holding the carcass so that muscles are stretched during the postmortem prerigor period. The success of this approach is in accord with the known existence of a strong positive correlation between sarcomere

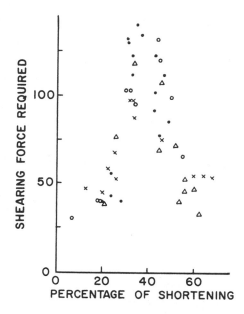

Figure 17 The relationship between shortening of prerigor beef muscle and tenderness following cooking. Shortening was produced by treatments of varying temperature and time and by freezing-thawing. (Reprinted from Ref. 82, p. 635, by courtesy of Macmillan Journals, Ltd.)

length and degree of tenderness. Excision of a muscle prerigor, of course, removes the physical restraint to contraction, and the muscle is then much more susceptible to substantial shortening, sometimes leading to final lengths only 30-40% of the original rest length (60-70% contraction) (82). This results in muscle that is highly disrupted (tender), misshapen and highly exudative (poor water-holding capacity).

Data relating muscle shortening to water-holding capacity are shown in Fig. 18. Drip losses do not become excessive (poor water-holding capacity) until shortening exceeds about 40% of the original length. This behavior is in accord with the previously stated belief that severe disruption of the muscle fiber occurs when contraction exceeds about 40%.

3. Thaw Rigor

A phenomenon related to cold-shortening is freeze-thaw contraction, or "thaw rigor." If a muscle is frozen prerigor it often undergoes considerable shortening during rapid thawing. This can be accompanied by toughening of the muscle (if contraction is less than 40%) and poor water-holding capacity (if contraction exceeds 40%). For this reason meat that is to be frozen generally should be in rigor or postrigor if optimal quality is desired. However, thaw rigor can be avoided in fish if frozen prerigor samples are held at normal temperatures of frozen storage for about 2 months. Thaw rigor in frozen prerigor lamb, beef, or poultry can be avoided by holding the frozen muscle at $-3°C$ for a time equal to that normally used during chilling. Thaw rigor presumably occurs because ice crystals, by direct or indirect means, disable the sarcoplasmic reticulum and/or mitochondria in pre-

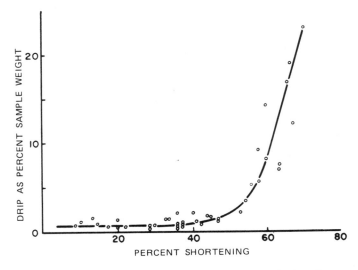

Figure 18 The relation between "drip" exudation (as percentage of frozen sample weight) and extent of thaw shortening (as percentage of initial excised length). (Reprinted from Ref. 83, p. 456, by courtesy of Institute of Food Technologists.)

rigor muscle, thus destroying their ability to maintain low levels of calcium in the sarcoplasm and triggering thaw contraction.

4. Structural Changes

Dissolution and disintegration of the Z disk is related to increased tenderness of meat during postmortem aging. However, it is probably not the only factor of importance since the Z disk is intact in tender prerigor muscle. Furthermore, disintegration of the Z disk appears to occur well in advance of substantial tenderization during aging. It appears very probable that, in addition to disintegration of the Z disk, other changes, such as a change in the interaction of the thin and thick filaments or changes in the networks of desmin and connectin filaments, are also involved in postmortem tenderization (75).

5. Fatty Acid Production

If fatty acids are produced during meat storage (e.g., by the actions of lipases or phospholipases), they may interact with contractile proteins and denature them. This may be an important cause of toughening of fish muscle during frozen storage. Oxidation products of fatty acids, including free radicals, can interact with proteins and may cause insolubilization of proteins by intermolecular cross-linking. It is not clear, however, whether free fatty acids accelerate or inhibit the oxidation process (see Sec. VII.F).

B. Color

Color of muscle is affected principally by pH. The pH controls the physical state of the myofibrils (and hence the reflection of light from muscle), the functioning of mitochondria and the ability to compete with myoglobin for oxygen, and the metmyoglobin-reducing systems in muscle. If these reducing systems are inactivated by an unfavorable pH, one can expect more rapid oxidation of bright red oxymyoglobin to brown met-myoglobin.

During most slaughtering procedures, some hemoglobin from the blood remains in the muscle. Estimates of the amount of hemoglobin range from 5 to 40% of the total pigment. With respect to meat color, the principal difference between myoglobin and hemoglobin is that binding of O_2 to hemoglobin is sensitive to pH, whereas with myoglobin it is not. The colors formed from each type of pigment and the reactions required to produce these colors are identical. Meat pigments are discussed further in Chap. 8.

C. Flavor

Little is known concerning the effect of postmortem handling on the development of a normal cooked flavor in meat. However, proteolysis and nucleotide metabolism may produce compounds that contribute significantly to flavor or that react on cooking to produce flavor (104). Five oligopeptide hydrolases active at neutrality have been shown to occur in rabbit muscle. These enzymes cause an increase in amino acids from oligopeptides and could contribute substantially to the development of flavor in cooked meat (98). The major flavor components of haddock muscle are of relatively low molecular weight, eluting predominantly with the amino acid fraction (126). Flavor is highly dependent on the particular cut of meat and its chemical composition. Meat flavors are also discussed in Chap. 9.

D. Nutritional Quality

The changes that occur in meat postmortem have little effect on nutritional quality. The only possible exception is when muscle loses large amounts of fluid (e.g., during thaw rigor or with PSE pork). Soluble nutrients may then be lost if the exuded fluid is not collected and consumed with the final product. Amino acids and vitamins, expecially those of the B complex, are most affected.

X. ANTEMORTEM FACTORS AFFECTING POSTMORTEM BIOCHEMICAL CHANGES

For any given animal, several antemortem factors influence the course of postmortem events. The factors of greatest importance are those that influence the level of muscle glycogen, although rough handling can lead to significant mechanical damage (bruising).

Glycogen is depleted by several stress conditions, and in general, it is desirable to minimize these stress conditions as much as possible. Stress conditions include exercise, fasting, hot and cold temperatures, and fear (74). Increased glycogenolysis and lipolysis generally accompany stress, but not all animals react in the same way or to the same degree to any one of these stresses. It is very difficult, for example, to deplete cattle of glycogen by exercising, whereas significant lowering of glycogen in pig muscles occurs with relatively mild exercise. Cattle must be subjected to both psychological and physical stress to achieve important modification of meat properties. Within a given species, the response to stress may vary among breeds or between sexes. It is possible literally to frighten some stress-susceptible animals sufficiently to cause death.

Heat stress (38°C) affects the quality of chicken breast muscle. Heat-stressed birds exhibit lower ultimate pH values and are tougher than control birds (71).

The method of slaughtering also has an effect on meat quality. It is known that death achieved by means of carbon dioxide brings about more rapid postmortem changes and poorer meat quality than other techniques of slaughter. Captive bolt stunning causes strong muscle contractions throughout the animal, presumably due to brain damage.

Generally, it is desirable to reach a moderately low ultimate pH, since this gives improved resistance to microbial spoilage and, in the case of red meats, provides a more desirable color. As mentioned earlier (Sec. VIII.A), the ultimate pH should be achieved slowly; that is, glycolysis should be slow and complete. Many stress-susceptible animals show a tendency to exhibit a rapid decrease in pH, even though the ultimate pH is no lower, or may even be considerably higher than that of non-stress-susceptible animals. This is undesirable since low pH values are attained before the muscle is cooled adequately, resulting in protein denaturation.

The food quality of fish is more dependent on antemortem factors than is the food quality of warm-blooded animals. This dependence is partly attributable to the fact that most fish of commerce are wild animals, hence genetically diverse, and partly attributable to the fact that fish respond to the environment more than do domesticated mammals and birds. The water temperature, the nature of the area in which they are caught, the amount of exercise they have had, and the diet they have consumed are important and uncontrollable factors that influence the ultimate quality of fish. For example, the color of dark muscle intensifies when fish become more active (80). Species also differ in quality, as is true of mammals and birds.

One important difference between fish and mammals is that the ultimate pH of fish is not affected by exercise prior to death. The reasons for this were discussed earlier in Sec. VIII.A. Thus, fish captured with a great deal of struggle have the same ultimate pH as those that are captured and killed quickly, provided other factors remain the same. A low ultimate pH leads to toughening of the cooked fish muscle and may also cause the phenomenon of gaping (see Sec. IX.A).

The development of oxidized "cold-store" flavor in fish is dependent on the nutritional state of the fish at the time of capture. This odor is attributable mainly to the compound cis-4-heptenal and is produced by oxidation of the phospholipid fraction of fish muscle in both lean and fatty fish. When a lean fish, such as cod, is starved, some of the phospholipid is utilized, presumably for energy. This results in a reduction of the C22:6 fatty acid, the source of the cis-4-heptenal. Therefore, off-flavor development is less severe in starved cod than in well-fed cod (89).

XI. EFFECTS OF PROCESSING ON MEAT COMPONENTS

The purpose of many food-processing techniques is to slow down or prevent deleterious changes in food materials. These deleterious changes are often caused by contaminating microorganisms, by chemical reactions among natural components of the food tissues, or by simple physical occurrences (dehydration). The reactions may be chemically catalyzed by enzymes indigenous to the tissue.

The procedures used to prevent changes in foods caused by microorganisms or chemical and physical reactions are (a) removal of water; (b) removal of other active components, such as oxygen or glucose; (c) use of chemical additives; (d) lowering of temperature; (e) input of energy; and (f) packaging. Most of the unit operations in food processing are based on combinations of two or more of these fundamental procedures. Very often, however, when one desirable objective is achieved other undesirable consequences occur. This is especially true when the principal objective of the processing procedure is to inactivate the microbial population. Conditions designed to do this very often lead to extensive chemical and physical changes in the foods. In this section, the effects of various food processing techniques on chemical components and quality attributes of meat are considered. This discussion includes the effects of such treatments as hot boning and electrical stimulation that are used in the early postmortem period to conserve energy or improve quality, and the effects of processes that intentionally disrupt the native tissue so that new products with significantly different characteristics can be obtained, such as gel-type products.

A. Hot Boning

The meat industry is a large consumer of energy. Any decrease in the amount of energy required for meat processing not only benefits the industry economically, but also helps conserve national resources. The standard industrial procedure after slaughtering and dressing is to place the carcass or half-carcass in a cold room and allow it to attain a low, nonfreezing, temperature. The major purpose of this step is to reduce the growth of microorganisms. However, this procedure involves the unnecessary cooling of bone and fat along with muscle. The technique of "hot boning," consisting of removal of warm meat from the skeleton, can be accomplished after a short period of storage postmortem (minimum of about 3 hr), which allows for partial cooling of the carcass (to 14-20°C, at which shortening is minimal) and the accomplishment of early postmortem biochemical events. This technique thus results in energy savings when the excised meat is later cooled to refrigerator temperatures.

Fundamental studies of the effects of hot boning have mostly involved beef and lamb, although technological developments in this area have also included pork. Beef of satisfactory tenderness can be attained by this procedure (131). The holding time at 14-20°C has a beneficial effect on tenderness (possibly due to proteolysis) that tends to overcome the small degree of toughening that usually accompanies hot boning. This small amount of toughening could be due to "rigor shortening," the phenomenon whereby a small degree of shortening occurs just before depletion of ATP in muscle held at a temperature at which cold-shortening is minimal. This rigor shortening is thought to be caused by the inability of the sarcoplasmic reticulum to maintain low calcium levels as the ATP concentration begins to rapidly decrease. Enough ATP remains, however, for some contraction to occur.

B. Electrical Stimulation

Benjamin Franklin observed in 1749 that stimulation of turkeys by electricity improved their tenderness (79). However, major developments in electrical stimulation came in response to the discovery of cold-shortening. To prevent cold-shortening, it is important that rigor occur before the carcass is cooled to temperatures that would stimulate contraction of sensitive, prerigor tissue. Electrical stimulation involves subjecting the carcass to an electrical current before the muscle goes into rigor. Typical conditions are 500-600 V, 2-6 A, applied in 16-20 pulses of 1.5-2.0 sec duration within 1 hr of slaughter. The electrical current functions in the same way as an electrical nerve impulse in the living animal and causes strong contraction of the muscles. This process leads to a faster development of rigor mortis, a more rapid postmortem decline in pH, and improved color and tenderness of some muscles. High-energy phosphate intermediates, such as creatine phosphate, ATP, and ADP, decrease more rapidly during electrical stimulation (20). Electrically st.mulated muscles display contracture bands in certain areas and stretched sarcomeres in other areas.

Three major theories have been advanced to explain the increased tenderness of muscle tissue that has been electrically stimulated: (a) the stimulation occurs before the temperature of the carcass is lowered; thus, cold-shortening is markedly reduced; (b) the more rapid glycolysis causes the pH to fall rapidly, causing release of lysosomal hydrolases that can then modify structural components of the muscle; and (c) there is a physical disruption of the myofibrils caused by supercontractions (25), and this disruption prevents the toughening that would ordinarily occur with contraction. These hypotheses are not mutually exclusive, and more than one phenomenon may occur simultaneously.

It has also been reported that electrical stimulation lowers the shrinkage temperature of collagen, which is an indication of a reduction in its thermal stability (57). Electrical stimulation also improves the color of beef muscle. White fibers are more susceptible to electrical stimulation than red fibers, and intermediate fibers respond in an intermediate fashion (120).

Electrical stimulation is often used prior to hot boning since the former greatly lessens any problems with toughening that might be encountered during the hot-boning process. It is used principally for those species that undergo cold-shortening, that is, beef and lamb.

C. Refrigeration

Refrigeration is the most widely used technique to preserve meat. This technique is not only used for meat to be sold "fresh," but is also commonly employed, at least temporarily, for meat to be marketed in forms other than fresh.

Refrigeration preserves muscle tissue by retarding the growth of microorganisms and by slowing many chemical and enzymic reactions. However, the greater rate of glycolysis in prerigor beef muscle at $0°C$ than at $5°C$ and the occurrence of cold-shortening in some muscles should be adequate warning that biological systems do not always obey simple rules.

Generally speaking, a temperature as low as possible without freezing is desirable for fresh meat. During postmortem cooling and subsequent refrigerated storage, control of relative humidity is very important. If considerable moisture is lost from the surface, the weight reduction becomes of economic importance and the meat pigment, myoglobin, oxidizes to brown metmyoglobin. However, a small amount of moisture loss from the surface is desirable since this tends to retard growth of microorganisms.

Postrigor beef muscle is sometimes aged at refrigerated temperatures for 1-2 weeks to improve its tenderness. After relatively long-term storage at refrigerator temperatures, growth of microorganisms on the meat surface may necessitate some trimming. Mold growth may impart a musty flavor considered desirable by some.

The susceptibility of meat to microbial spoilage depends on the ratio of cut surface to weight, with small cuts spoiling more rapidly than large cuts. Ground meat is especially susceptible to spoilage by microorganisms.

It should not be assumed that lowering the temperature to a value close to $0°C$ necessarily reduces the activity of all enzymes in muscle tissue. Although a reduction in muscle temperature from about $40°C$ (close to that of living mammals or birds) to $0°C$ lowers the rate of most enzyme-catalyzed reactions, it is not reasonable to expect that enzyme activity in an organism accustomed to living at refrigerated temperatures, such as fish, decreases when the organism, postmortem, is stored at the same temperature. It is well established that cold-blooded organisms that live in a cold environment have enzymes that function well at those temperatures (52). This adaptation is not obvious, however, unless one measures the activity of the enzyme at low substrate concentrations, similar to those that exist in tissue. Thus, the Michaelis constant K_m not the maximum initial velocity V_{max} is the appropriate value to monitor.

It has also been demonstrated that "adaptation" of enzymes to low temperature sometimes occurs in warm-blooded animals. For example, it has been demonstrated that the Michaelis constant of lactate dehydrogenase from chicken breast muscle is much smaller at refrigerated temperatures than it is at the body temperature of the bird, which is around 40°C (51). Since a small Michaelis constant means that the enzyme retains a high proportion of its maximal activity at low substrate concentrations, lowering the temperature from 40 to 4°C combined with a decline in K_m may have little effect on the activity of the enzyme in situ. This behavior may have a bearing on the temperature dependence of cold-shortening in beef muscle [shortening is minimal between 14 and 20°C (76)] and on the rate of ATP loss in chicken muscle [rate is minimal at 20°C (27)]. The major point is that decreases in rates of enzyme-catalyzed reactions in situ may not always correlate well with decreases in temperature.

D. Freezing

Freezing is an excellent process for preserving the quality of meat and fish for long periods. Its effectiveness stems from internal dehydration (ice crystal formation) and lowering of temperature. Although some microorganisms survive storage at very low temperatures, there generally is no opportunity for growth of microorganisms if recommended storage temperatures are maintained. Quality retention (or loss) during freeze-preservation of animal tissues can relate to how the freezing process is conducted and/or to innate characteristics of the tissue.

1. Quality Retention Influenced by Methods of Freezing

A major problem arising from improper freezing (slow) and storage procedures (too long or too high a temperature) is excessive thaw exudate. When muscle tissue is frozen rapidly, small ice crystals form both intra- and extracellularly. When muscle is frozen slowly ice crystals form first in the extracellular space, presumably because the freezing point is higher there (concentration of solutes is lower) than it is inside the fibers. Since the vapor pressure of ice crystals is lower than that of supercooled water, there is a tendency during further cooling for the water inside fibers to migrate to ice crystals that already exist in extracellular areas, resulting eventually in very large extracellular ice crystals and compacted muscle fibers. The increase in salt concentration and the concomitant change in pH may cause extensive denaturation of muscle proteins. If exposure to the high salt concentrations and unfavorable pH values occurs at a relatively high subfreezing temperature, the process of denaturation can be accelerated, with a resulting decrease in water-holding capacity of the tissue. This loss in water-holding capacity of the protein, along with mechanical damage to cells by ice crystals, is responsible in large part for the thaw exudate.

Although freezing rate has some influence on the quality of thawed cooked meat from warm-blooded animals, it is generally a factor of secondary importance since changes during retail frozen storage frequently obscure any benefits of rapid freezing (minimal drip) that exist initially. Exceptions to this statement can occur when the time of frozen storage is short, or the temperature of frozen storage is below −18°C, where changes usually occur at a very slow rate (44). Since fish muscle is more sensitive to protein denaturation and lipid oxidation than are muscles from mammals or birds, it is recommended that fish be stored at −30°C (22). Furthermore, the rate at which fish muscle is frozen influences the degree of protein denaturation (85). Although rapid freezing generally results in less denaturation than slow freezing, intermediate freezing rates can be more detrimental than slow freezing as judged by textural changes and solubility of actomyosin. Cod fillets frozen at intermediate rates show intracellular ice crystals large enough to damage cellular membranes.

Freezing rate also can influence the color of animal tissues in the frozen state. Rapid freezing causes tissues to become very pale. This is undesirable for red meats but desirable for poultry. This effect is apparently related to the number and size of ice crystals produced and the thickness and nature of the tissue matrix. Once thawed, the appearance of rapidly frozen and slowly frozen tissues cannot be distinguished.

The tenderness of beef is influenced by freezing and subsequent handling procedures. When postrigor muscle is frozen and stored at −3°C for 28 days, there is a greater decrease in shear force than when the muscle is aged at 15°C for 3 days. Also, when postrigor muscles are frozen, thawed, and then aged at 15°C for 24 hr, the tenderness of these samples is greater than that of samples that have been stored for 48 hr at 15°C, then frozen, thawed, and evaluated (same treatments but different sequence). It is possible that enzymic reactions are accelerated during freezing because of the concentration of cofactors or because disruption of membranes allows enzymes to interact with substrates more easily than they would in the intact muscle (132).

Dehydration of the surface of animal tissue during frozen storage (freezer burn) can occur if improper packaging techniques are employed. Freezer burn occurs when the equilibrium vapor pressure above the surface of the meat is greater than that in the air, causing ice crystals to sublime. Oxidation of the heme pigment (darkening) usually occurs during freezer burn.

2. Quality Retention Related to Innate Characteristics of the Tissue

As discussed in Sec. IX.A.3, the meat of warm-blooded animals should not be frozen prior to the development of rigor. If muscle is frozen prerigor it contracts greatly on thawing and is tough and exudative. This thaw exudate (drip), in addition to being unsightly, represents economic and nutritional losses if it is not utilized.

Enzymes generally are not inactivated by the freezing process, and some may continue to function in frozen animal tissue. Enzymes with low activation energies retain considerable activity in the frozen state, often more than expected by extrapolation of Arrhenius plots from above-freezing temperatures. Freezing damage to the tissue (enzyme delocalization), changes in environmental conditions (pH, ionic strength, and substrate concentration), and changes in the Michaelis constant may account for the retention of high enzymic activity, which sometimes exceeds that in the unfrozen tissue.

As just mentioned, freezing can result in a delocalization of certain enzymes in tissue. The location of mitochondrial isoenzymes of glutamate-oxaloacetate transaminase has been used to differentiate frozen and thawed beef, pork and chicken from the unfrozen products. However, this procedure may not be suitable for pork since procedures other than freezing and thawing can damage pig muscle mitochondria and cause redistribution of this enzyme (51). Freezing and thawing also cause release of cytochrome oxidase from muscle mitochondria. The soluble activity of the enzyme increased about

2½-fold in frozen and thawed trout muscle and 4-fold in frozen and thawed beef muscle as compared with unfrozen samples. This might be useful in developing an improved method for distinguishing unfrozen from frozen and thawed meat and fish (10).

The major enzymes involved in the spoilage of animal tissues during long-term frozen storage are those that act on lipids. Lipases and phospholipases release free fatty acids from the lipids of animal tissues, and these free fatty acids may then undergo oxidation. When free fatty acids are produced from triacylglycerols, the rate of oxidation is faster than that of the original triacylglycerols (36); however, this is not the case when the free fatty acids are produced from phospholipids. In the latter instance, generation of free fatty acids in either fish or beef inhibits the rate of oxidation (114). The reason for this anomalous behavior is not clear, but it might be that enzyme-catalyzed oxidation of membrane lipids is slowed either by direct inhibition of the enzyme by the fatty acids or, indirectly, by disruption of the membrane structure (115).

In the isolated microsomal membrane fraction from fish skeletal muscle, enzymic oxidation of lipid components occurs at temperatures as low as $-20°C$ (3). Lipid oxidation is especially important in products that contain highly unsaturated fatty acids, such as pork and seafoods.

3. Quality Retention in Frozen Fish

Fish may undergo extensive textural changes during frozen storage because of the very high sensitivity of the fish proteins to the denaturing factors active during frozen storage. Tropomyosin is the most stable protein, actin is the next most stable, and myosin is the least stable. Since this instability is a problem of commercial importance, much attention has been given to the denaturation of fish muscle proteins at subfreezing temperatures. In addition to the denaturing effects of salt concentration and changes in pH that occur during freezing, it is also possible that fish muscle proteins may be denatured by interaction with fatty acids produced during frozen storage. The latter mechanism may be a major cause of the undesirable toughening that codfish muscle undergoes during frozen storage. There is evidence that the major source of free fatty acids in lean fish is phospholipids of the membrane system, not triacyglycerols (51). It is also possible that fish muscle proteins may interact with products of lipid oxidation.

A problem unique to marine species is the breakdown of trimethylamine oxide, by an endogenous enzyme, forming dimethylamine and formaldehyde. It has been suggested that formaldehyde causes cross-linking of muscle proteins, rendering them insoluble and causing the flesh to be tough. It also has been suggested that the formation of new disulfide bonds in fish muscle proteins may contribute to denaturation (85). An integrated scheme for some of the reactions that can cause denaturation of fish proteins and toughening of fish muscle is shown in Fig. 19.

Minced fish is often stabilized for frozen storage by removal of water-soluble components and the addition of stabilizing agents. In addition to antioxidants, polyhydroxy compounds, such as sorbitol and sugar, are used as protein stabilizers. These compounds promote hydration of proteins which results in a positive free energy change. The positive free energy change increases with an increase in surface area of the protein. Therefore, the denatured form, which has a larger surface area, is thermodynamically less stable than the undenatured forms and so the latter is the preferred structure (4). The product obtained by washing and adding stabilizers is called surimi (see Sec. X.J) and is used in the manufacture of fish sausage.

Polyhydroxy compounds are only one class of substances that impart stability to proteins under adverse conditions, such as the low water and high salt concentrations

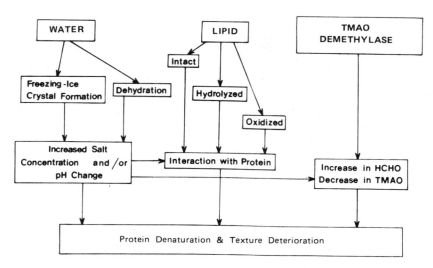

Figure 19 Some factors influencing the denaturation of fish proteins during frozen storage. (Adapted from Ref. 113.)

encountered during freezing. Others include some free amino acids and amino acid derivatives, such as taurine and β-alanine, and certain nitrogen compounds, such as methylamines, including trimethylamine oxide, betaine, and sarcosine (135). Presumably these compounds exert stabilizing effects whether present naturally or added during processing.

That frozen fish sometimes undergo gaping was mentioned in Sec. IX.A. Gaping is apparently caused in part by the formation of small ice crystals in the myocommata. This problem may become severe if freezing is done more than 1 day after the death of the fish.

E. Heating

Heating of muscle tissue brings about extensive changes in its appearance and physical properties and these changes are dependent on the time-temperature conditions imposed. α-Actinin is the most heat labile muscle protein, becoming insoluble at 50°C. The heavy and light chains of myosin become insoluble at about 55°C, and actin at 70-80°C. Tropomyosin and troponin are the most heat resistant proteins of muscle, becoming insoluble at about 80°C (21). The 30,000 dalton component also increases during heating, and this may occur because of heat activation of the calcium-activated protease (CAF). Apparently, the connectin filaments, which are situated parallel to the myofibrils, are the only links in the I band that survive heating, and it is largely the tensile strength of these elements that is responsible for the residual toughness of myofibrillar proteins in cooked meat (78).

As muscle tissue is cooked, it first becomes firmer as the proteins are denatured. This process continues as the temperature is raised. However, this toughening of contractile proteins is counteracted by concomitant alterations in collagen. The temperature at which the change in collagen begins to occur is about 50°C for young animals and about 60°C for older animals (16). In the case of cod muscle, any temperature sufficiently high to cook the product results in sufficient disruption of collagen so that it is not a factor in the toughness of the cooked material (81).

The amount of change observed during heating of animal tissue depends on the physical property being measured. Peak shear force, initial shear force, and adhesion values all give different responses as a function of temperature (16).

Heating also deactivates enzymes. Some of the enzymes of muscle survive short times at temperatures of 60-70°C. The enzymic activities of myosin and actomyosin are destroyed during conventional cooking and even by less severe heat treatments.

Denaturation of muscle proteins decreases their water-holding capacity. This decrease in water-holding capacity may produce desirable juiciness (free moisture), provided the free water is not expelled from the tissue. Heat denaturation and surface dehydration that take place on cooking cause a surface layer to form that tends to retard loss of fluid from the meat until it is cut and eaten. The emulsifying capacities of muscle proteins are also greatly decreased by heating.

During heating, fat is melted, adipose tissue cells are ruptured, and there is a significant redistribution of fat. When the meat is eaten warm, the melted fat serves to increase palatability of the product by giving a desirable mouth feel, especially at the end of the chewing period when most of the aqueous juices have been lost.

Myoglobin also undergoes denaturation during heating. The susceptibility of the heme pigment to oxidation in the denatured protein is much greater than it is in undenatured myoglobin. On heating, therefore, red meat generally turns brown due to the formation of the oxidized pigment, hemin (see Chap. 8).

With severe heating of meat, there are further changes in proteins and free amino acids with production of some volatile breakdown products. Sulfur-containing compounds are produced, including hydrogen sulfide, mercaptans, sulfides, and disulfides, as well as aldehydes, ketones, alcohols, volatile amines, and others (see Chap. 5). Lipid components may also break down into volatile products, such as aldehydes, ketones, alcohols, acids, or hydrocarbons (see Chap. 4). Some of these volatile compounds, in both the fat and the lean portions of the meat, contribute to the flavor and odor of cooked meat.

Commercial heating generally has moderately detrimental effects on the vitamin content of meat. Thiamine is sensitive to heat, and it is partially destroyed during cooking or thermal processing. Certain amino acids may interact with glucose and/or ribose of the meat (Maillard reaction), and the nutritional values of these amino acids can be impaired when this occurs (see Chaps. 3 and 5). Lysine, arginine, histidine, and methionine seem to be particularly susceptible to degradation via this route. However, the conditions used during commercial sterilization generally have little detrimental effect on the nutritive value of proteins and lipids.

Some muscles cannot withstand the extreme conditions of temperature needed for sterilization without undergoing extensive changes in quality. Pork is an example of a meat that does not tolerate high temperatures well. Thus, a moderate heat treatment is usually used for pork preservation, often along with a curing process and subsequent refrigeration of the product.

F. Dehydration

Preservation of muscle tissue by dehydration is an ancient practice, originating at least 3000 years ago. The quality resulting from the early techniques was, however, extremely poor by our standards, and this technique was used only because of dire necessity.

Drying can be successfully employed for both raw and cooked meat; however, the quality of the final reconstituted product is superior when the meat is cooked prior to dehydration. The reason for this is not clear, but it may be that protein denaturation, which occurs during heating in the presence of the original moisture, is less severe than that during the combined process of heating and moisture removal.

The type of dehydration also has a very significant effect on the quality of the final product. Conventional air drying, which is done at a relatively high temperature, is more detrimental to muscle tissue than a process, such as freeze dehydration, that involves a low temperature and sublimation of ice. In addition to the difference in temperature between these two kinds of dehydration, there is also a difference in salt distribution. During drying an increase in salt concentration occurs. As moisture evaporates, more moisture is drawn to the surface, bringing salts with it (70). Thus, a high concentration of salt develops at the surface. The resulting increase in ionic strength and change in pH affect protein stability. In freeze dehydration, in which water is converted directly from the solid state to the gaseous state, the increase in solute concentration is more evenly distributed, and the product is at a lower temperature than during dehydration by conventional means. The lower the temperature at which the muscle proteins are exposed to high salt concentrations or unfavorable pH, the less the extent of denaturation.

Even freeze-dried meat shows a lower water-holding capacity than fresh tissue. If the flesh of freshly slaughtered animals is freeze-dried before the onset of rigor, the breakdown of ATP and glycogen can be inhibited. During rehydration of this tissue rapid hydrolysis of ATP occurs along with rapid glycolysis. This can result in strong rigor contraction and a considerable loss of water-holding capacity. However, if the muscle tissue is ground and salted prerigor and then freeze-dried, rigor is prevented during rehydration and the high water-holding capacity of the tissue is retained (43). A possible explanation for this behavior relates to the binding of salt ions by proteins and the resulting increase in electrostatic repulsion between proteins. This apparently prevents the proteins from interacting after ATP has been depleted. If rigor occurs initially then the proteins are irreversibly bound and cannot be dissociated by treatment with salt.

The effects of heat on denaturation of proteins in raw muscle occur in stages. From 0 to 20°C, there is a loss in native structure of protein as measured by water-holding capacity. This is presumably caused by denaturation of sarcoplasmic proteins. The next major loss in water-holding capacity begins in the range 40-50°C and is most likely caused by denaturation of contractile proteins. This denaturation continues up to about 80°C. Collagen changes substantially only at relatively high temperatures and is rapidly converted to gelatin at around 100°C. The heat-induced changes in collagen tend to increase water-holding capacity, but the effect is slight and not enough to offset the decrease caused by denaturation of sarcoplasmic and contractile proteins. Maximal denaturation of myofibrillar proteins occurs when the temperature is high and the moisture content of the tissue reaches a value below about 20-30% (113).

During high-temperature drying, tissue fat melts. Fat in excess of about 35-40% (fresh weight basis) usually cannot be retained in dehydrated tissue.

Of all the quality attributes, texture is the one most severely altered by dehydration. The change in texture is determined principally by the extent of denaturation of the muscle proteins, and this in turn is highly dependent on the pH of the meat at the start of dehydration. Even a very good freeze-drying procedure results in some lowering of the water-holding capacity of muscle if the process is carried out near the isoelectric points of the proteins (about 5.5). If one starts at a high ultimate pH (about 6.7), a relatively small decrease in the water-holding capacity of muscle occurs during dehydration. Measurements of rehydration, extractability of proteins, and ATPase activity all have been used to determine in a rough way the amount of denatured protein in dehydrated muscle.

Some changes in dehydrated animal tissue also occur during storage. The extent of these changes depend, in large part, on the conditions of storage (temperature, oxygen level, and moisture content), but also on the processing procedure used and the initial quality of the raw material.

Lipid oxidation is a major factor limiting the shelf life of dehydrated muscle tissue. The highly unsaturated nature of pork lipids and lipids from fatty fishes causes these products to be unstable in dehydrated form unless some kind of special protective treatment is applied. The addition of antioxidants and the use of appropriate packaging techniques are the most common approaches to retarding oxidation.

Lipid oxidation results principally in undesirable flavors that can make the animal tissue unacceptable. There also may be some destruction of oxidizable nutrients, such as essential fatty acids, some amino acids, and some vitamins; oxidation of heme pigments (discoloration); and protein cross-linking.

Another major type of deteriorative reaction in dehydrated muscle tissues is non-enzymic browning. Carbonyl sources, such as glucose, phosphorylated sugar derivatives, and other aldehydes, and ketones; and a large number of amino groups, from either free amino acids or the ϵ-amino groups of lysine residues, are present in meat. These can interact by means of the Maillard reaction (discussed in Chaps. 3 and 5), resulting in dark pigments and decreased nutritive value of amino acids and proteins. Moreover, when sugars react with proteins, the physical properties of the proteins can change, resulting in a toughening or hardening of the texture. Nonenzymic browning can also lead to desirable changes in flavors, some of which are typical of cooked meats.

One approach to overcoming the tough texture of dehydrated meat is to retain more water in the product, that is, to prepare products with intermediate levels of water. Intermediate moisture meat has a water content low enough to prevent the growth of bacteria and other properties that minimize the growth of yeast and molds. Food-borne pathogens cannot grow at water activities of 0.85 or below, and even higher water activities may be used if the pH of the product is adjusted to less than 5 (70). To produce an intermediate moisture meat with a water activity of about 0.85, one can either remove water from the meat or add substances to reduce the water activity of the native water. The latter course of action is most desirable from the standpoint of avoiding problems with muscle toughness. Many compounds can be used for this purpose, although glycerol has the advantages of being soluble, relatively stable, and nonvolatile, of contributing little to odor and color, and of having a less undesirable impact on flavor than either salt or sugars. A mixture of sodium chloride, glycine, lactic acid, and glycerol is a particularly promising mixture for lowering a_w and retaining desirable sensory properties (70). Sorbic acid is also often added to inhibit the growth of certain molds and yeasts that can grow at water activities of less than 0.85.

During storage of intermediate moisture meats, some loss of protein solubility occurs. This may be due in part to sugar-amino reactions (Maillard) of the meat, although the substances mentioned above (humectants) shift the water activity at which the maximum rate of browning occurs to values well below that of the product (70). Cross-linking of proteins induced by lipid oxidation probably also occurs to some extent.

The humectant glycerol and other polyols can also promote protein cross-linking (70), and this lessens protein solubility. Glycerol probably degrades to aldehydes and peroxides in the presence of oxygen, and the peroxides and aldehydes then have the ability to interact with protein molecules.

Collagen seems to be the most susceptible protein to degradation during storage of intermediate moisture meats, and it becomes more soluble with storage time. However, heating of the product must occur before changes in collagen are evident. The hemoproteins of muscle tissue also degrade during storage of intermediate moisture foods in the presence of oxygen.

G. Curing

The production of cured meat is an ancient process. Curing involves the addition to meat of sodium chloride, sodium nitrite, and/or sodium nitrate. Sugar may or may not be added along with other ingredients to improve flavor. Sodium chloride significantly inhibits microorganisms, including *Clostridium botulinum*, and nitrite has a small inhibitory effect on *C. botulinum*. As discussed in Chap. 8, nitrite produces the cured meat pigment, nitric oxide myoglobin. It is believed that microorganisms of meat convert some sodium nitrate to nitrite.

Endogenous low-molecular-weight components in the sarcoplasm promote formation of nitric oxide myoglobin and nitrite decomposition (97). Reduced glutathione is active in this respect, and its effectiveness is enhanced by ATP, IMP, and ribose.

During the curing process—whether it be dry curing, in which salt is rubbed on the surface of the meat, or curing with a curing pickle, in which the meat is submerged in a strong salt solution—the high osmotic pressure of the external fluid initially draws water and soluble proteins out of the meat. Later, salt diffuses into the meat, causing some of the expelled protein to diffuse back in and the meat to swell. The salt-protein complex that forms binds water well; thus, the water-holding capacity of proteins generally increases during curing. The final meat contains increased ash due to the absorbed salts. Curing appears to have little effect on either the quality of proteins or the stability of the B vitamins.

If meat with a high ultimate pH is used for curing, salt penetration is difficult since the fibers are enlarged. In pork that is pale, soft, and exudative (a low pH is reached too quickly postmortem), the water-holding capacity of the cured product may be low. Even though salt penetration can occur easily, this advantage is apparently offset by the large amount of denatured protein, which has poor water-holding capacity. Freezing of meat prior to curing also increases the penetration rate of salt because freezing partially disrupts the meat structure.

It has been claimed that nitrite has a significant beneficial effect on the flavor of cured meats by preventing oxidation through the antioxidative activity of nitric oxide myoglobin (60) and *S*-nitrosocysteine (59), a compound formed during the curing process [Eq. (3)].

$$NO_2^- + HS-CH_2-\underset{\underset{NH_2}{|}}{\overset{\overset{H}{|}}{C}}-COOH \xrightarrow{H^+} ONS-CH_2\underset{\underset{NH_2}{|}}{\overset{\overset{H}{|}}{C}}-COOH \tag{3}$$

S—Nitrosocysteine

A major detrimental change that can occur in cured meat during storage is the oxidation of nitric oxide hemochromagen (pink) or nitric oxide myoglobin (pink) to brown metmyoglobin. The rate of oxidation increases with increasing oxygen content; therefore, cured meat preferably should be packaged in a container from which oxygen is excluded.

The presence of salt in cured meat enhances oxidation of lipid components, thus considerably reducing the shelf life. Ascorbic acid in the presence of a metal chelating agent, such as polyphosphate, can effectively forestall this oxidative reaction.

Recently, concern has arisen over the potential hazard of using nitrite in cured meats since this compound reacts with amines, especially secondary amines, to form *N*-nitrosamines, which may be carcinogenic [Eq. (4)].

$$R_2NH + NO_2^- \xrightarrow{\text{H}^+} R_2NNO + H_2O \tag{4}$$
$$\text{Nitrosamine}$$

Although *N*-nitrosamines do not form in meat if the concentration of nitrite does not exceed U.S. Federal regulations, commercial samples of frankfurters have occasionally been found that contain this compound at low levels. In three instances, it is suspected that improper mixing may have resulted in high local concentrations of nitrite, thereby favoring the formation of nitrosamines. High temperatures may also induce nitrosamine formation, and so a product such as bacon, which is subjected to very high temperatures, is especially vulnerable. Careful control of nitrite concentration at levels just high enough to inhibit *C. botulinum* is the current practice. Even lower levels of nitrite might be satisfactory if compounds in addition to NaCl were present to help inhibit *Clostridia*. Sorbate has been suggested as one such compound. Reducing agents, such as ascorbate, and antioxidants, such as tocopherol, are effective in reducing the conversion of nitrite to nitrosamines and are added to cured meat products.

H. Other Chemical Additives

Many chemical additives in addition to those used for curing are added to meat products. Most of these chemicals are antimicrobial agents, flavoring agents, or antioxidants, and they may be added intentionally (e.g., vinegar for pickling meat or antioxidants in sausage), or they may be produced by fermentation (e.g., lactic acid in certain types of dry sausage).

Smoking is often used in conjunction with salting and curing, and it is considered desirable because of the flavor it imparts to the meat. Also, many of the chemical components of smoke are effective antimicrobial agents if present in sufficient concentration. Furthermore, wood smoke may have other effects that depend on its direct interaction with meat components. For example, many of the components of smoke are effective antioxidants. In using the same smoking procedure, different flavors are produced in different meat products. Therefore, the interaction of smoke with meat components does take place.

I. Modified (Controlled) Atmospheric Storage

Fresh meat and fish held at a refrigerated temperature has a limited shelf life primarily because of microbial growth. It is possible to slow down this growth by storing muscle foods in modified or controlled atmospheres. The term "modified atmosphere" refers to an initial adjustment in the composition of the atmosphere surrounding the product, with no further control other than the package. In contrast, the term "controlled atmosphere" indicates that the atmosphere is continuously controlled. Modified atmospheric storage is more practical than controlled atmospheric storage in commercial applications involving animal tissues.

Some studies have been conducted on animal tissue stored under hypobaric conditions, and this approach is effective but expensive. Most studies have involved modification of atmospheric composition at normal pressures.

The preservative effect of carbon dioxide on stored muscle tissue has been known for a century. The effect of CO_2 is not due to a simple displacement of oxygen since nitrogen atmospheres do not offer the same extension of storage life that carbon dioxide does. At high concentrations of carbon dioxide, surface browning of red meat is often observed. This presumably occurs because the myoglobin and hemoglobin pigments undergo oxidation to their ferric states. The most desirable concentration of CO_2 to use in a modified atmosphere is a compromise between bacterial inhibition (the more CO_2 the greater the inhibition) and product discoloration. It is possible, therefore, in nonred meats and many fish species, to use very high concentrations of carbon dioxide and achieve the benefit of long shelf life. A small amount of carbon monoxide gas stabilizes the color of red meats, since the carboxymyoglobin pigment is more resistant than oxymyoglobin to oxidation in the presence of CO_2. Carbon monoxide also has been shown to be effective in preventing the discoloration of rockfish fillets. Carbon monoxide cannot, however, be legally used.

Oxygen has an inhibitory effect on the breakdown of trimethylamine oxide to dimethylamine and formaldehyde in refrigerated red hake muscle. Thus, in species in which this is a potential problem, such as gadoids, conditions that exclude oxygen, whether hypobaric, modified atmosphere, or vacuum storage, could be potentially harmful to the quality of the product (see Sec. VIII.G.2).

It is known that carbon dioxide lowers the internal pH of tissues, and it may be that carbon dioxide causes intracellular acidification of spoilage bacteria (101). This could upset the major H^+/K^+ transport systems in the cell, which would have marked effects on gross metabolism. One of these effects could be a shift in equilibrium of the decarboxylating enzymes, such as isocitrate dehydrogenase, making them rate limiting. A decrease in surface pH has been observed with rockcod fillets stored in an atmosphere containing 80% CO_2 (101), but a decrease in pH of red hake and salmon flesh was not observed in an atmosphere containing 60% CO_2 (31).

CO_2 and its hydrated bicarbonate ion can directly affect membranes, causing hydration or dehydration of the surface. This can result in changes in intermolecular distances of membrane components, making the membranes either more or less permeable (101). These effects could have major effects on attacking microorganisms.

Since modified atmospheres containing carbon dioxide slow the growth of normal spoilage microorganisms in both meat and fish, there is some danger that organisms of public health significance may benefit. *Clostridium botulinum* is of particular concern in this situation, particularly in cases in which the oxygen content is low. Maintenance of low temperatures (close to 0°C), the presence of some oxygen in the atmosphere, and the use of additional agents such as potassium sorbate, help guard against this potential problem. It remains, however, a matter of considerable concern and requires close attention in commercial situations in which modified atmospheres are used.

J. Preparation of Gel Meat Products

Gel-type products made from comminuted muscle tissue are popular items in many parts of the world and include such items as bologna-type sausages, chicken hot dogs, and kamaboko, the Japanese fish sausage. The important characteristics of these products are provided by the muscle proteins and by the particular treatments imposed to improve their emulsification properties, water-holding capacity, and textural properties.

The first step typically involves comminution of the muscle tissue in the presence of salt. This partially solubilizes the myofibrillar proteins and yields a thick sol, or paste. This is followed by a heating process that establishes a gelatinous structural network of proteins. This protein network is most likely stabilized by hydrogen and hydrophobic interactions and induces extensive structural order in the entrapped water (137). Emulsification of lipid most likely results from interaction of lipids and protein through hydrophobic associations, the number of protein hydrophobic groups increasing as the protein denatures during heating. During mixing, the proteins of the muscle coat the surface of each fat droplet as it forms, and the emulsion is further stabilized when the proteins gel during heating. Myosin is the most important protein in stabilizing the emulsion (63).

The specific properties of any product formed depend on the species from which the muscle tissue is derived and on the particular muscles used. In addition, processing variables such as rate and severity of heating, postmortem biochemical state of the muscle, salt concentration, and pH, can markedly influence the final properties of the protein gel. Nevertheless, the basic mechanism of gel formation is probably similar for muscle from many animal species. Myosin is the most important component of the muscle tissue with respect to gel-forming ability. Actin assists in this process by forming F-actomyosin, which in turn interacts with free myosin (138). F-actomyosin acts as a cross-link between the tail portions of actin-bound and free myosin molecules, adding strength to the gel. With rabbit muscle, optimal gelation is observed when the ratio of myosin to F-actomyosin by weight is approximately 4. Native tropomyosin (the tropomyosin-troponin complex) has no effect on gel structure (111).

The tail and head portions of the myosin molecule play different roles in the heat-induced gelation of the muscle minces. The head portion of the myosin molecule undergoes irreversible aggregation involving the oxidation of −SH groups. This aggregation contributes to formation of the three-dimensional protein network of the gel. The tail portion of the myosin molecule undergoes a partially irreversible helix-to-coil transition during the heating process and then participates in the formation of a three-dimensional network. The formation of the protein gel occurs by means of protein denaturation under strictly defined conditions. However, denaturation of the muscle proteins, especially myosin, under other conditions, such as in the original tissue, can interfere with gel formation. Thus, any condition that encourages incidental denaturation, such as low pH, freeze denaturation, or dehydration, produces a weak gel structure.

The ability of muscle tissue to bind water and function well in comminuted animal tissue is greater when actomyosin is in the dissociated state. This is the normal condition of prerigor muscle, and prerigor muscle has a very high water-holding capacity. If, however, the muscle tissue goes into rigor either in the intact or minced form, it loses much of its ability to bind water, and this loss cannot be completely restored by the addition of salt. However, if salt is added to minced muscle tissue before rigor occurs, then the water-holding capacity of the proteins is maintained at a high level. Binding salt to muscle proteins results in electrostatic repulsion among molecules, thus loosening the protein network and allowing it to bind more water. The protein interactions that occur during onset of rigor, that is, after the loss of ATP, hinder this swelling effect of salt (43). The extent to which muscle sarcomeres shorten is irrelevant. Thus, very satisfactory sausage products can be made from prerigor muscle, but it must be ground and salted before the ATP level declines sufficiently to allow rigor mortis.

Satisfactory muscle protein gels can be formed from the entire minced tissues of land animals and birds. However, production of the Japanese fish sausage, kamaboko, requires an extremly elastic gel that can be formed only after the water-soluble proteins have been removed from fish muscle. This modified muscle mince is called surimi, and its use had led to a large increase in the sale of fish-gel products in Japan. When water-soluble proteins are not removed they apparently associate with the denatured contractile proteins during and after heating, thereby interfering with the latter's ability to form a strong gel (119).

The likely mechanism of gel formation in kamaboko is shown in Fig. 20. The fish muscle is ground with salt to produce the actomyosin paste, which, upon heating to 50°C, then converts to a gel with a relatively open structure. Raising the temperature to approximately 60°C partially degrades this gel. Alkaline proteases in the fish muscle or in contaminating organ tissues may contribute to this occurrence, but this is a matter of uncertainty (68,119). At temperatures greater than 60°C a gel more rigid than the original is formed, and this comprises the final kamaboko.

Due to the sensitivity of fish myofibrillar proteins to freeze denaturation, stabilizing agents, such as polyhydroxy compounds (sugars and sugar alcohols), polyphosphates, or salt, must be added to fish muscle if it is to be stored frozen before being used to manufacture kamaboko.

Two mechanical processes that aid in the formation of gels from animal tissues are massaging and tumbling. In the former, pieces of salted animal tissue are continually struck by moving metal arms. The same type of mechanical stress is obtained by tumbling salted animal tissue. These processes tend to rupture some muscle cells and assist penetration of salt into the cells. The salt then solubilizes part of the myofibrillar proteins, and some of these proteins migrate from the cells and form a viscous, sticky paste that engulfs the cells.

With relatively small pieces of meat, the tissue can be simply salted and used as such. The protein sol that forms on the surface can then serve as a binding agent for the

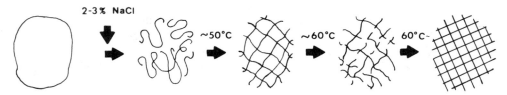

Figure 20 Mechanism of gel formation in the process of making kamaboko. The NaCl extracts the myofibrillar protein forming a sol (paste). With time at 50°C, a slightly transparent gel (suwari) forms (loose net in figure). Between 50 and 60°C, part of the gel structure is destroyed (modori; broken net in figure). Above 60°C, the nontransparent elastic kamaboko gel is formed (ordered net). (Reprinted from Ref. 119, p. 106, by courtesy of Applied Science Publishers Ltd.)

small pieces of meat. This technique is commonly used to prepare restructured meats. The same principle is used to improve the texture of large pieces of meat, such as hams. In this case, the salt and other curing agents are injected into the tissue and the meat is then massaged or tumbled to disrupt some of the muscle cells. This allows for better distribution of the curing ingredients by rupturing some cells and results in internal extraction of some myofibrillar proteins. This often produces improved slicing characteristics of the meat because of better adhesion among cells.

K. Preservation by Ionizing Radiation

Although ionizing radiation has not been approved for processing of most food products, it does constitute a potentially useful form of preservation. Aside from its desirable ability to inactivate microorganisms, it also has the undesirable effect of altering meat pigments. If raw or cured meat is exposed to large doses of ionizing radiation, it turns brown. The color of cooked meat is not affected by ionizing radiation if air is present, but a pink color develops in the absence of oxygen. However, on exposure to air the pink color reverts to the original brown color. The pink color of irradiated cooked meat is attributable to a denatured globin-hemochrome pigment. In a model system, irradiation of NO-myoglobin causes denitrosylation and formation of metmyoglobin (58). If the radiation dose is sufficiently high, the metmyoglobin is, in turn, degraded to other products, including myoglobin, and ferriperoxide derivative, and small amounts of choleglobin-type pigments.

Sterilizing doses of ionizing radiation also result in the breakdown of various lipids and proteins to compounds that have distinct and often undesirable odors. Tenderization of muscle may also occur during this treatment. The odors vary with the particular muscle tissue treated and range all the way from extremely unpleasant odors in beef to rather mild odors in pork and chicken. Furthermore, irradiated beef steak is sometimes bitter, and this may result from the conversion of ATP to hypoxanthine. Many of these reactions can be prevented by lowering the temperature during irradiation. Temperatures of $-80°C$ or below greatly lessen undesirable side effects without seriously decreasing destruction of microorganisms. This occurs because microorganisms are primarily destroyed by direct interaction with ionizing radiation, whereas most of the undesirable changes in chemical components result from the indirect effects of free radicals produced during irradiation.

Generally, enzymes are not completely inactivated by irradiation treatments sufficient to sterilize the product. Therefore, for long-term storage it is necessary to heat the meat to approximately $70°C$, that is, cook it, prior to irradiation and storage.

L. Packaging

The basic reasons for packaging processed muscle tissue were discussed earlier. In summary, frozen meat should be packaged to prevent loss of moisture and freezer burn. Dehydrated products must be packaged to prevent the entrance of water and oxygen, so that nonenzymic browning and oxidative reactions do not occur. Cured meat should be packaged to exclude oxygen so that nitric oxide hemochromagen does not oxidize to brown denatured metmyoglobin. Packages for fresh meat must be reasonably impermeable to moisture to prevent weight loss but are usually permeable to oxygen so that the fresh red bloom of the oxygenated myoglobin pigment can be maintained. The shelf-life of fresh meat is increased considerably if the meat is packaged under vacuum to exclude oxygen since this greatly decreases microbial growth. However, this results in development of a bluish purple color (nonoxygenated myoglobin) regarded by consumers as undesirable, even though the red color of oxymyoglobin returns when the package is opened.

REFERENCES

1. Ackman, R. G. (1980). Fish lipids, Part 1, in *Advances in Fish Science and Technology* (J. J. Connell, ed.), Fishing News Books, Surrey, England, pp. 86-103.

2. Amano, K., and K. Yamada (1964). A biological formation of formaldehyde in the muscle tissue of gadoid fish. *Bull. Jpn. Soc. Sci. Fish. 30*:430-435.

3. Apgar, M. E., and H. O. Hultin (1982). Lipid peroxidation in fish muscle microsomes in the frozen state. *Cryobiology 19*:154-162.

4. Arakawa, T., and S. N. Timasheff (1982). Stabilization of protein structure by sugars. *Biochemistry 21*:6536-6544.

5. Arakawa, N., S. Fujiki, C. Inagaki, and M. Fujimaki (1976). A catheptic protease active in ultimate pH of muscle. *Agr. Biol. Chem. 40*:1265-1267.

6. Ashmore, C. R., W. Parker, and L. Doerr (1972). Respiration of mitochondria isolated from dark-cutting beef: Postmortem changes. *J. Anim. Sci. 34*:46-48.

7. Atkinson, D. E., and G. M. Walton (1967). Adenosine triphosphate conservation in metabolic regulation. *J. Biol. Chem. 242*:3239-3241.

8. Bailey, A. J. (1972). The basis of meat texture. *J. Sci. Food Agr. 23*:995-1007.

9. Bailin, G., and M. Barany (1967). Studies on actin-actin and actin-myosin interaction. *Biochim. Biophys. Acta 140*:208-221.

10. Barbagli, C., and G. S. Crescenzi (1981). Influence of freezing and thawing on the release of cytochrome oxidase from chicken's liver and from beef and trout muscle. *J. Food Sci. 46*:491-493.

11. Barnard, R. J., V. R. Edgerton, T. Furukawa, and J. B. Peter (1971). Histochemical, biochemical, and contractile properties of red, white, and intermediate fibers. *Amer. J. Physiol. 220*:410-414.

12. Bendall, J. R. (1951). The shortening of rabbit muscle during rigor mortis: Relation to the breakdown of the adenosine triphosphate and creatine phosphate and to muscular contraction. *J. Physiol. (London) 114*:71-88.

13. Bishop, C. M., and P. H. Odense (1967). The ultrastructure of the white striated myotomal muscle of the cod, *Gadus morhua. J. Fish. Res. Bd. Can. 24*:2549-2553.

14. Bone, Q., A. Pulsford, and A. D. Chubb (1981). Squid mantle muscle. *J. Mar. Biol. Ass. U.K. 61*:327-342.

15. Bornstein, P., and W. Traub (1979). The chemistry and biology of collagen, in *The Proteins*, Vol. IV, 3rd ed. (H. Neurath and R. L. Hill, eds.), Academic Press, New York, pp. 411-632.

16. Bouton, P. E., P. V. Harris, and D. Ratcliff (1981). Effect of cooking temperature and time on the shear properties of meat. *J. Food Sci. 46*:1082-1087.

17. Braekkan, O. R. (1959). A comparative study of vitamins in the trunk muscles of fishes, in *Reports on Technological Research Concerning Norwegian Fish Industry*, Vol. III, No. 8, A. S. John Griegs Boktrykkeri, Bergen, Norway, pp. 23-30.

18. Buege, D. R., and B. B. Marsh (1975). Mitochondrial calcium and postmortem muscle shortening. *Biochem. Biophys. Res. Commun. 65*:478-482.

19. Busch, W. A., M. H. Stromer, D. E. Goll, and A. Suzuki (1972). Ca^{2+}-specific removal of Z-lines from rabbit skeletal muscle. *J. Cell Biol. 52*:367-381.

20. Calkins, C. R., T. R. Dutson, G. C. Smith, and Z. L. Carpenter (1982). Concentration of creatine phosphate, adenine nucleotides and their derivatives in electrically stimulated and nonstimulated beef muscle. *J. Food Sci. 47*:1350-1353.

21. Cheng, C.-S., and F. C. Parrish (1979). Heat-induced changes in myofibrillar proteins of bovine longissimus muscle. *J. Food Sci. 44*:22-24.

22. Clucas, I. J., and P. S. Sutcliff (1981). An introduction to fish handling and processing, *Rep. Tropical Products Institute*, G143, London, p. 8.

23. Cooke, R., and K. Franks (1980). All myosin heads form bonds with actin in rigor rabbit skeletal muscle. *Biochemistry 19*:2265-2269.

24. Cori, G. T., M. W. Selin, and C. F. Cori (1948). Crystalline D-glyceraldehyde-3-phosphate dehydrogenase from rabbit muscle. *J. Biol. Chem. 173*:605-618.

25. Cross, H. R. (1979). Effects of electrical stimulation on meat tissue and muscle properties—a review. *J. Food Sci. 44*:509-514.

26. Davey, C. L., and K. V. Gilbert (1969). Studies in meat tenderness. 7. Changes in the fine structure of meat during aging. *J. Food Sci. 34*:69-74.

27. deFremery, D., and M. F. Pool (1963). Biochemistry of chicken muscle as related to rigor mortis and tenderization. *Food Res. 25*:73-87.

28. Dugan, L. R. (1978). Fats, in *The Science of Meat and Meat Products*, 2nd ed. (2nd printing) (J. F. Price and B. S. Schweigert, eds.), Food and Nutrition Press, Westport, Conn., pp. 133-145.

29. Dutson, R. T. (1983). Relationship of pH and temperature to disruption of specific muscle proteins and activity of lysosomal proteases. *J. Food Biochem. 7*:223-245.

30. Ebashi, S., M. Endo, and I. Ohtsuki (1969). Control of muscle contraction. *Q. Rev. Biophys. 2*:351-384.

31. Fey, M. S., and J. M. Regenstein (1982). Extending shelf-life of fresh wet red hake and salmon using CO_2-O_2 modified atmosphere and potassium sorbate ice at $1°C$. *J. Food Sci. 47*:1048-1070.

32. Fleischer, S., and G. Rouser (1965). Lipids and subcellular particles. *J. Amer. Oil Chem. Soc. 42*:588-607.

33. Fraser, D. I., J. R. Dingle, J. A. Hines, S. C. Nowlan, and W. J. Dyer (1967). Nucleotide degradation, monitored by thin-layer chromatography, and associated postmortem changes in relaxed cod muscle. *J. Fish Res. Bd. Can. 23*:1837-1841.

34. George, J. C., and A. J. Berger (1966). In *Avian Myology*, Academic Press, New York, pp. 25-114.

35. Goll, D. E., Y. Otsuka, P. A. Nagainis, J. D. Shannon, S. K. Sathe, and M. Muguruma (1983). Role of muscle proteinases in maintenance of muscle integrity and mass. *J. Food Biochem. 7*:137-177.

36. Govindarajan, S., H. O. Hultin, and A. W. Kotula (1977). Myoglobin oxidation in ground beef: Mechanistic studies. *J. Food Sci. 42*:571-582.

37. Granger, B. L., and E. Lazarides (1980). Synemin: A new high molecular weight protein associated with desmin and vimentin filaments in muscle. *Cell 22*:727-738.

38. Greaser, M. L., R. G. Cassens, W. G. Hoekstra, and E. J. Briskey (1969). The effect of pH-temperature treatments on the calcium-accumulating ability of purified sarcoplasmic reticulum. *J. Food Sci. 34*:633-637.

39. Greaser, M. L., S.-M. Wang, and L. F. Lemanski (1981). New myofibrillar proteins. In Proc. 34th Ann. Reciprocal Meat Conf., National Live Stock and Meat Board, Chicago, Ill., pp. 12-16.

40. Greer-Walker, M. (1970). Growth and development of the skeletal muscle fibers of the cod (*Gadus morhua* L.). *J. Cons. Cons. Int. Explor. Mer 33*:228-244.

41. Greer-Walker, M., and G. A. Pull (1975). A survey of red and white muscle in marine fish. *J. Fish Biol. 7*:295-300.

42. Ham, A. W. (1969). *Histology*, 6th ed., Lippincott, Philadelphia, p. 563.

43. Hamm, R. (1981). Post-mortem changes in muscle affecting the quality of comminuted meat products, in *Developments in Meat Science*, (R. Lawrie, ed.), 2. Applied Science Publishers, London, pp. 93-124.

44. Hamm, R. (1982). Postmortem changes in muscle with regard to processing of hot-boned beef. *Food Technol. 36*(No. 11):105-115.

45. Harrington, W. F. (1979). Contractile proteins of muscle, in *The Proteins*, Vol. IV, 3rd ed. (H. Neurath and R. L. Hill, ed.), Academic Press, New York, pp. 245-409.

46. Hattori, A., and K. Takahashi (1979). Studies on the post-mortem fragmentation of myofibrils. *J. Biochem. 85*:47-56.

47. Hattori, A., and K. Takahashi (1982). Calcium-induced weakening of skeletal muscle Z-disks. *J. Biochem. 92*:381-390.

48. Hay, J. D., R. W. Currie, F. H. Wolfe, and E. J. Sanders (1973). Effect of postmortem aging on chicken breast muscle sarcoplasmic reticulum. *J. Food Sci. 38*:700-704.

49. Herring, H. K., R. G. Cassens, and E. J. Briskey (1967). Factors affecting collagen solubility in bovine muscles. *J. Food Sci. 32*:534-538.

50. Howgate, P. (1979). Fish, in *Food Microscopy* (J. G. Vaughan, ed.), Academic Press London, pp. 343-392.

51. Hultin, H. O. (1976). Characteristics of muscle tissue, in *Principles of Food Science*, Part 1, *Food Chemistry* (O. R. Fennema, ed.), Marcel Dekker, New York, pp. 577-617.

52. Hultin, H. O. (1980). Enzymes from organisms acclimated to low temperature, in *Enzymes: The Interface Between Technology and Economics* (J. P. Danehy and B. Wolnak, eds.), Marcel Dekker, New York, pp. 161-178.

53. Huxley, H. E. (1960). Muscle cells, in *The Cell*, Vol. IV (J. Brachet and A. E. Mirskey, eds.), Academic Press, New York, pp. 365-481.

54. Huxley, H. E. (1963). Electron microscope studies on the structure of natural and synthetic protein filaments from striated muscle. *J. Mol. Biol. 7*:281-308.

55. Huxley, H. E. (1965). The mechanism of muscular contraction. *Sci. Amer. 213*(No. 6):18-27.

56. Judge, M. D., and E. D. Aberle (1982). Effects of chronological age and postmortem aging on thermal shrinkage temperature of bovine intramuscular collagen. *J. Anim. Sci. 54*:68-71.

57. Judge, M. D., E. S. Reeves, and E. D. Aberle (1981). Effect of electrical stimulation on thermal shrinkage temperature of bovine muscle collagen. *J. Anim. Sci. 52*:530-534.

58. Kamarei, A. R. (1982). Effects of ionizing radiation on nitric oxide myoglobin, Ph.D. Thesis, Massachusetts Institute of Technology.

59. Kanner, J. (1979). S-nitrosocysteine (RSNO), an effective antioxidant in cured meat. *J. Amer. Oil Chem. Soc. 56*:74-76.

60. Kanner, J., I. Ben-Gera, and S. Berman (1980). Nitric-oxide myoglobin as an inhibitor of lipid oxidation. *Lipids 15*:944-948.

61. Ke, P., R. G. Ackman, B. A. Linke, and D. M. Nash (1977). Differential lipid oxidation in various parts of frozen mackerel. *J. Food Technol. 12*:37-47.

62. Kidwai, A. M., M. A. Radcliffe, E. Y. Lee, and E. E. Danual (1973). Isolation and properties of skeletal muscle plasma membrane. *Biochim. Biophys. Acta 298*:593-607.

63. Kinsella, J. E. (1982). Relationships between structure and functional properties of food proteins, in *Food Proteins* (P. F. Fox and J. J. Condon, eds.), Applied Science Publishers, London, pp. 51-102.

64. Kruggel, W. G., and R. A. Field (1971). Soluble intramuscular collagen characteristics from stretched and aged muscle. *J. Food Sci. 36*:1114-1117.

65. Kushmerick, M. J., and R. E. Davies (1968). The role of phosphate compounds in thaw contraction and the mechanism of thaw rigor. *Biochim. Biophys. Acta 153*: 279-287.

66. Laakkonen, E., J. W. Sherbon, and G. H. Wellington (1970). Low-temperature, long-time heating of bovine muscle. 3. Collagenolytic activity. *J. Food Sci. 35*: 181-183.

67. Lagler, K. F., J. E. Bardach, R. R. Miller, and D. R. M. Passino (1977). *Ichthyology*, 2nd ed., John Wiley and Sons, New York.

68. Lanier, T. C., T.-S. Lin, D. D. Hamann, and F. B. Thomas (1981). Effects of alkaline protease in minced fish on texture of heat-processed gels. *J. Food Sci. 46*: 1643-1645.

69. Lawrie, R. A. (1974). *Meat Science*, 2nd ed., Pergamon Press, Oxford.

70. Ledward, D. A. (1981). Intermediate moisture meats, in *Developments in Meat Science*, 2 (R. Lawrie, ed.), Applied Science Publishers, London, pp. 159-194.

71. Lee, Y. B., G. L. Hargus, E. C. Hagberg, and R. H. Forsythe (1976). Effect of antemortem environmental temperatures on postmortem glycolysis and tenderness in excised broiler breast muscle. *J. Food Sci. 41*:1466-1469.

72. Lehninger, A. L. (1970). *Biochemistry*, Worth, New York.

73. Leu, S.-S., S. N. Jhaveri, P. A. Karakoltsidis, and S. M. Constantinides (1981). Atlantic mackerel (*Scomber scombrus*, L): Seasonal variation in proximate composition and distribution of chemical nutrients. *J. Food Sci. 46*:1635-1638.

74. Lister, D., N. G. Gregory, and P. D. Warriss (1981). Stress in meat animals, in *Developments in Meat Science*, 2 (R. Lawrie, ed.), Applied Science Publishers, pp. 61-92.

75. Locker, R. H. (1982). A new theory of tenderness in meat, based on gap filaments, in Proc. 35th Ann. Reciprocal Meat Conf., National Live Stock and Meat Board, Chicago, pp. 92-100.

76. Locker, R. H., and C. J. Hagyard (1963). A cold-shortening effect in beef muscle. *J. Sci. Food Agr. 14*:787-793.

77. Locker, R. H., and N. G. Leet (1976). Histology of highly stretched beef muscle. IV. Evidence for movement of gap filaments through the Z-line, using the N_2-line and M-line as markers. *J. Ultrastruct. Res. 56*:31-38.

78. Locker, R. H., G. J. Daines, W. A. Carse, and N. G. Leet (1977). Meat tenderness and the gap filaments. *Meat Sci. 1*:87-104.

79. Lopez, C. A., and E. W. Herbert (1975). *The Private Franklin, the Man and His Family*, W. W. Norton and Company, New York, p. 44.

80. Love, R. M. (1980). *The Chemical Biology of Fishes*, Vol. 2, *Advances 1968-1977*, Academic Press, London.

81. Love, R. M., J. Lavety, and F. Vallas (1982). Unusual properties of the connective tissues of cod (*Gadus morhua* L), in *Chemistry and Biochemistry of Marine Food Products* (R. E. Martin, G. J. Flick, C. E. Hebard, and D. R. Ward, eds.), AVI Publishing Co., Westport, Conn., pp. 67-73.

82. Marsh, B. B., and N. G. Leet (1966). Resistance to shearing of heat-denatured muscle in relation to shortening. *Nature 221*:635-636.

83. Marsh, B. B., and N. G. Leet (1966). Studies in meat tenderness. III. The effects of cold-shortening on tenderness. *J. Food Sci. 31*:450-459.

84. Maruyama, K., M. Kimura, S. Kimura, K. Ohashi, K. Suzuji, and N. Katunuma (1981). Connectin, an elastic protein of muscle. Effects of proteolytic enzymes in situ. *J. Biochem. 89*:711-715.

85. Matsumoto, J. J. (1979). Denaturation of fish muscle proteins during frozen storage, in *Proteins at Low Temperatures* (O. Fennema, ed.), Adv. in Chem. Ser. 180, Am. Chem. Soc.., Washington, D.C., pp. 205-224.

86. Matsumoto, J. J., W. J. Dyer, J. R. Dingle, and D. G. Ellis (1967). Proteins in extracts of prerigor and postrigor scallop striated muscle. *J. Fish Res. Bd. Can. 24*: 873-882.

87. McClain, P. E., G. J. Creed, E. R. Wiley, and I. Hornstein (1970). Effect of postmortem aging on isolation of intramuscular connective tissue. *J. Food Sci. 35*:258-259.

88. McDonald, R. E., S. D. Kelleher, and H. O. Hultin (1979). Membrane lipid oxidation in a microsomal fraction of red hake muscle. *J. Food Biochem. 3*:125-134.

89. McGill, A. S., R. Hardy, and F. D. Gunstone (1977). Further analysis of the volatile components of frozen cold-stored cod and the influence of these on flavour. *J. Sci. Food Ag. 28*:200-205.

90. McLoughlin, J. V. (1970). Muscle contraction and postmortem pH changes in pig skeletal muscle. *J. Food Sci. 34*:717-719.

91. Nagainis, P., and F. H. Wolfe (1982). Calcium activated neutral protease hydrolyzes Z-disc actin. *J. Food Sci. 47*:1358-1364.

92. Newbold, R. P. (1966). Changes associated with rigor mortis, in *The Physiology and Biochemistry of Muscle as Food* (E. J. Briskey, R. G. Cassens, and J. C. Trautman, eds.), University of Wisconsin Press, Madison, pp. 213-224.

93. Newbold, R. P., and C. A. Lee (1965). Post-mortem glycolysis in skeletal muscle. The extent of glycolysis in diluted preparation of mammalian muscle. *Biochem. J. 97*:1-6.

94. Newbold, R. P., and R. K. Scopes (1971). Post-mortem glycolysis in ox skeletal muscle: Effects of mincing and of dilution with or without addition of orthophosphate. *J. Food Sci. 36*:209-214.

95. Noda, T., K. Isogai, H. Hayashi, and N. Katunuma (1981). Susceptibilities of various myofibrillar proteins to cathepsin B and morphological alteration of isolated myofibrils by this enzyme. *J. Biochem. 90*:371-379.

96. Nunzi, M. G., and C. Franzini-Armstrong (1981). The structure of smooth and striated portions of the adductor muscle of the valves in a scallop. *J. Ultrastruct. Res. 76*:134-148.

97. Okayama, T., N. Ando, and Y. Nagata (1982). Low-molecular weight components in sarcoplasm promoting the color formation of processed meat products. *J. Food Sci. 47*:2061-2063.

98. Okitani, A., Y. Otsuka, R. Katakai, Y. Kondo, and H. Kato (1981). Survey of rabbit skeletal muscle peptidases active at neutral pH regions. *J. Food Sci. 46*: 47-51.

99. Olley, J., and J. A. Lovern (1960). Phosphilipid hydrolysis in cod flesh stored at various temperatures. *J. Sci. Food Ag. 11*:644-652.

100. Olson, D. G., and F. C. Parrish (1977). Relationship of myofibril fragmentation index to measures of beefsteak tenderness. *J. Food Sci. 42*:506-509.

101. Parkin, K. L., and W. D. Brown (1982). Preservation of seafood with modified atmospheres, in *Chemistry and Biochemistry of Marine Food Products* (R. E. Martin, G. J. Flick, C. E. Hebard, and D. R. Ward, eds.), AVI Publishing Co., Westport, Conn., pp. 453-465.

102. Parrish, F. C., R. B. Young, B. E. Miner, and L. D. Anderson (1973). Effect of postmortem conditions on certain chemical, morphological and organoleptic properties of bovine muscle. *J. Food Sci. 38*:690-695.

103. Pearson, A. M. (1981). Meat and health, in *Developments in Meat Science*, 2 (R. Lawrie, ed.), Applied Science Publishers, London, pp. 241-292.

104. Pearson, A. M., A. M. Wolzak, and J. I. Gray (1982). Possible role of muscle proteins in flavor and tenderness. *J. Food Biochem. 7*:189-210.

105. Pepe, F. A. (1971). Structure of the myosin filament of striated muscle, in *Prog-*

ress in Biophysics and Molecular Biology, Vol. 221 (J. A. V. Butler and D. Noble, eds.), Pergamon Press, New York, pp. 77-96.

106. Peter, J. B., R. J. Barnard, V. R. Edgerton, C. A. Gillespie, and K. E. Stempel (1972). Metabolic profiles of three fiber types of skeletal muscle in guinea pigs and rabbits. *Biochemistry 11*:2627-2633.

107. Porter, K. R., and C. Franzini-Armstrong (1965). The sarcoplasmic reticulum. *Sci. Amer. 212*(3):72-78, 80.

108. Ritchie, A. H., and I. M. Mackie (1980). The formation of diamines and polyamines during storage of mackerel (*Scomber scombrus*), in *Advances in Fish Science and Technology*, (J. J. Connell, ed.), Fishing News Books, Surrey, England, pp. 489-494.

109. Robson, R. M., M. Yamaguchi, T. W. Huiatt, F. L. Richardson, J. M. O'Shea, M. K. Hartzer, W. E. Rathbun, P. J. Schreiner, L. E. Kasang, M. H. Stromer, Y.-Y. S. Pang, R. R. Evans, and J. F. Ridpath (1981). Biochemistry and molecular architecture of muscle cell 10-nm filaments and Z-line: roles of desmin and α-actin, in Proc. 34th Ann. Reciprocal Meat Conf., National Live Stock and Meat Board, Chicago, Ill., pp. 5-11.

110. Saito, T., K. Arai, and M. Matsuyoshi (1959). A new method for estimating the freshness of fish. *Bull. Jpn. Soc. Sci. Fish 24*:749-750.

111. Samejima, K., M. Ishioroshi, and T. Yasui (1982). Heat induced gelling properties of actomyosin: effect of tropomyosin and troponin. *Ag. Biol. Chem. 46*:535-540.

112. Schaub, M. C., S. V. Perry, and W. Hacker (1972). The regulatory proteins of the myofibril. Characterization and biological activity of the calcium-sensitizing factor (troponin A). *Biochem. J. 126*:237-249.

113. Shenouda, S. Y. K. (1980). Theories of protein denaturation during frozen storage of fish flesh, in *Advances in Food Research*, Vol. 26 (C. O. Chichester, ed.), Academic Press, New York, pp. 275-311.

114. Shewfelt, R. L. (1981). Fish muscle lipolysis—a review. *J. Food Biochem. 5*:79-100.

115. Shewfelt, R. L., and H. O. Hultin (1983). Inhibition of enzymic and non-enzymic lipid peroxidation of flounder muscle sarcoplasmic reticulum by pretreatment with phospholipase A_2. *Biochim. Biophys. Acta 751*:432-438.

116. Shono, T., and M. Toyomizu (1973). Lipid alteration in fish muscle during cold storage. II. Lipid alteration pattern in jack mackerel muscle. *Bull. Jpn. Soc. Sci. Fish. 39*:417-421.

117. Sims, T. J., and A. J. Bailey (1981). Connective tissue, in *Developments in Meat Science, 2* (R. Lawrie, ed.), Applied Science Publishers, London, pp. 29-59.

118. Suzuki, A., M. Saito, A. Okitani, and Y. Nonami (1981). Z-nin, a new high molecular weight protein required for reconstitution of the Z-disk. *Ag. Biol. Chem. 45*: 2535-2542.

119. Suzuki, T. (1981). *Fish and Krill Protein: Processing Technology*, Applied Science Publishers, London.

120. Swatland, H. J. (1981). Cellular heterogeneity in the response of beef to electrical stimulation. *Meat Sci. 5*:451-455.

121. Takahashi, K., and H. Saito (1979). Post-mortem changes in skeletal muscle connectin. *J. Biochem. 85*:1539-1542.

122. Takahashi, K., T. Fukazawa, and T. Yasui (1967). Formation of myofibrillar fragments and reversible contraction of sarcomeres in chicken pectoral muscle. *J. Food Sci. 32*:409-413.

123. Takahashi, K., F. Nakamura, and A. Inoue (1981). Postmortem changes in the actin-myosin interaction of rabbit skeletal muscle. *J. Biochem. 89*:321-324.

124. Takahashi, K., F. Nakamura, and M. Okamoto (1982). A myofibrillar component that modifies the actin-myosin interaction in postrigor skeletal muscle. *J. Biochem. 92*:809-815.

125. Takahashi-Nakamura, M., S. Tsuji, K. Suzuki, and K. Imahori (1981). Purification and characterization of an inhibitor of calcium-activated neutral protease from rabbit skeletal muscle. *J. Biochem. 90*:1583-1589.

126. Thomson, A. B., A. S. McGill, J. Murray, R. Hardy, and P. F. Howgate (1980). The analysis of a range of non-volatile constituents of cooked haddock (*Gadus aeglefinus*) and the influence of these on flavour, in *Advances in Fish Science and Technology* (J. J. Connell, ed.), Fishing News Books, Surrey, England, pp. 484-488.

127. Tu, C., W. D. Powrie, and O. Fennema (1967). Free and esterified cholesterol content of animal muscles and meat products. *J. Food Sci. 32*:30-34.

128. Veis, A. (1965). *The Macromolecular Chemistry of Gelatin*, Academic Press, New York.

129. Walker, S. M., and G. R. Schrodt (1966). T system connexions with the sarcolemma and sarcoplasmic reticulum. *Nature 211*:935-938.

130. Watt, B. K., and A. L. Merrill (eds.) (1963). *Composition of Foods*, Agriculture Handbook No. 8, U.S. Department of Agriculture, Washington, D.C.

131. Will, P. A., R. L. Henrickson, and R. D. Morrison (1976). The influence of delay of chilling and hot boning on tenderness of bovine muscle. *J. Food Sci. 41*:1102-1106.

132. Winger, R. J., and O. Fennema (1976). Tenderness and water holding properties of beef muscle as influenced by freezing and subsequent storage at −3 or 15°C. *J. Food Sci. 41*:1433-1438.

133. Wolfe, F. H., and K. Samejima (1976). Further studies of postmortem aging effects on chicken actomyosin. *J. Food Sci. 41*:244-249.

134. Yamamoto, K., K. Samejima, and T. Yasui (1977). A comparative study of the changes in hen pectoral muscle during storage at 4°C and −20°C. *J. Food Sci. 42*:1642-1645.

135. Yancey, P. H., M. E. Clark, S. C. Hand, R. D. Bowlus, and G. N. Somero (1982). Living with water stress: Evolution of osmolyte systems. *Science 217*:1214-1222.

136. Yang, R., A. Okitani, and M. Fujimaki (1978). Postmortem changes in regulatory proteins of rabbit muscle. *Ag. Biol. Chem. 42*:555-563.

137. Yasui, T., M. Ishioroshi, H. Nakano, and K. Samejima (1979). Changes in shear modulus, ultrastructure and spin-spin relaxation times of water associated with heat-induced gelation of myosin. *J. Food Sci. 44*:1201-1204.

138. Yasui, T., M. Ishioroshi, and K. Samejima (1982). Effect of actomyosin on heat-induced gelation of myosin. *Ag. Biol. Chem. 46*:1049-1059.

BIBLIOGRAPHY

Allen, C. E., and E. A. Foegeding (1981). Some lipid characteristics and interactions in muscle foods—a review. *Food Technol. 35*(5):253-257.

Connell, J. J. (ed.) (1980). *Advances in Fish Science and Technology*, Fishing News Books, Surrey, England.

Forrest, J. C., E. D. Aberle, H. B. Hedirck, M. D. Judge, and R. A. Merkel (1975). *Principles of Meat Science*, W. H. Freeman, San Francisco.

Goll, D. E., Y. Otsuka, P. A. Nagainis, J. D. Shannon, S. K. Sathe, and M. Muguruma (1983). Role of muscle proteinases in maintenance of muscle integrity and mass. *J. Food Biochem. 7*:137-177.

Lawrie, R. A. (1974). *Meat Science*, Pergamon, New York.

Lawrie, R. A. (ed.) (1980). *Developments in Meat Science*, 1, Applied Science Publishers, London.

Lawrie, R. A. (ed.) (1981). *Developments in Meat Science*, 2, Applied Science Publishers, London.

Love, R. M. (1970). *The Chemical Biology of Fishes*, Academic Press, London.

Love, R. M. (1980). *The Chemical Biology of Fishes,* Vol. 2, *Advances 1968-1977,* Academic Press, London.

Martin, R. E., G. J. Flick, C. E. Hebard, and D. R. Ward (eds.) (1982). *Chemistry and Biochemistry of Marine Food Products*, AVI Publishing Co., Westport, Conn.

Neurath, H., and R. L. Hill (1979). *The Proteins*, 3rd ed., Vol. IV, Academic Press, New York.

Price, J. F., and B. S. Schweigert (1978). *The Science of Meat and Meat Products*, 2nd ed, Food and Nutrition Press, Westport, Conn.

Romans, J. R., and P. T. Ziegler (1977). *The Meat We Eat*, 11th ed., Interstate Publishers, Danville, Ill.

Rust, R. E. (1976). *Sausage and Processed Meats Manufacturing*, AMI Center for Continuing Education, American Meat Institute, Washington, D.C.

Suzuki, T. (1981). *Fish and Krill Protein: Processing Technology*, Applied Science Publishers, London.

13

CHARACTERISTICS OF EDIBLE FLUIDS OF ANIMAL ORIGIN: MILK

Harold E. Swaisgood North Carolina State University, Raleigh, North Carolina

I.	Introduction	791
II.	Milk Biosynthesis	792
III.	Chemical Composition	795
	A. Milk Proteins	796
	B. Milk Lipids	797
	C. Milk Salts and Sugars	799
	D. Enzymes	801
IV.	Structural Organization of Milk Components	802
	A. Structure of Milk Proteins	802
	B. Casein Micelle and Milk Salts	808
	C. The Fat Globule	812
V.	Use of Milk Components as Food Ingredients	813
	A. Effects of Processing on Milk Components	813
	B. Use in Formulated Foods	817
VI.	Nutritive Value of Milk	820
	A. Nutrient Composition of Milk and Milk Products	820
	B. Lactose Tolerance	821
	C. Effect of Processing on Nutritive Value	822
	References	825
	Bibliography	827

I. INTRODUCTION

For human infants, as well as other young mammals, milk is the first, and for most, the only food ingested for a considerable period of time. With the domestication of animals, it became possible to include milk in the diet of adult humans as well. For much of the world, particularly the West, milk from cattle (*Bos taurus*) accounts for nearly all the milk produced for human consumption. Consequently, in the United States the dairy industry is based primarily on cow's milk. Therefore, unless otherwise noted, the discussion of milk in this chapter refers to bovine milk.

The total production of milk in the United States for selected years between 1970 and 1980 is given in Table 1. Of the 120 billion pounds produced in 1978, 97% was processed and distributed by the industry, 1% was sold directly to consumers, and 2% was consumed on the farm. The share of product sales attained by various milk products is also listed in this table. Nearly 50% of the total milk production is sold as fluid milk or related products. Cheese accounts for roughly one-quarter of the production, so that three-fourths of the total milk is consumed as fluid milk or cheese. The latter may actually rise in consumption in conjunction with the growing popularity of wine. Of these products, the most significant increase in consumption was observed for yoghurt, whereas the percentage of sales for fluid milk has actually decreased slightly.

Recently a new fluid milk product, ultra-high temperature processed milk (UHT milk) has been introduced in the United States. Although it has been sold in Europe for some time (it represents 50% of the fluid milk market in Germany), it has not previously been sold in the United States. This product, which is commercially sterile and can be distributed and merchandised without refrigeration, is expected to achieve a place in the U.S. market, particularly in the form of flavored milk drinks.

With this brief introduction to the commercial aspects of milk production and processing, let us now examine the biological origin and chemical nature of milk.

II. MILK BIOSYNTHESIS

In the early years of dairy chemistry, the product of the mammary gland was more thoroughly investigated and characterized than the gland itself because of the economic importance of milk products. More recently, however, with advances in molecular and cellular biology, the mammary gland has been the subject of intensive research. A number of excellent reviews have appeared in the past decade to which the reader is referred for a complete discussion of our understanding of the biosynthesis of milk (17-20,24,32). Some people have argued that in the future, because of expanding human population, competition for available grain will eliminate animal agriculture. The fallacy of this argument, however, lies in the presence of a rumen in the cow's digestive system that allows this animal to synthesize nutrients from crude cellulosic and fibrous plant materials and from simple forms of nitrogen, such as urea. Consequently, these animals need not compete with, but actually add to the human food supply by providing high-quality nutrients from otherwise unusable feedstocks.

The bovine mammary gland, resulting from years of genetic selection, is an amazingly productive organ for biosynthesis. An average cow in the United States produces 5400 kg (12,000 lb) of milk in a 305 day lactation. The cellular machinery of this gland is shown at various levels of magnification in Fig. 1. Milk originating in the secretory tissue collects in ducts that increase in size as the teat region is approached. The smallest complete milk factory, which includes a storage room, is the alveolus. It is a roughly spherical micro-organ consisting of a central storage volume (the lumen) surrounded by a single layer of secretory epithelial cells, which is connected to the duct system. These cells are directionally oriented such that the apical end with its unique membrane is positioned next to the lumen, and the basal end is separated from blood and lymph by a basement membrane. Consequently, a directional flow of metabolites occurs through the cell, with

Table 1 U.S. Milk Production and Distribution of Product Sales for Selected Years Between 1970 and 1980

Year	Milk production (kg in billions)	Product sales (% total processed)					
		Fluid milk and related products	Condensed and evaporated milk	Cheese	Ice cream and frozen dessert	Yoghurt	Butter
1970	53.1	52.4	11.9	18.9	10.8	0.3	5.7
1975	52.3	48.3	12.5	24.5	10.1	0.7	3.9
1976	54.6	47.7	13.0	26.4	9.4	0.9	3.9
1978	55.2	45.0	12.6	26.8	9.6	1.0	4.9
1980	58.2	44.3	12.1	27.8	9.4	1.0	5.3

Source: From Ref. 2.

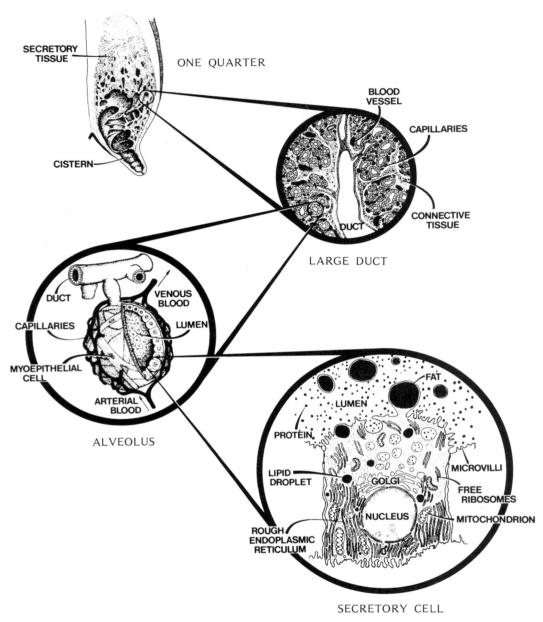

Figure 1 Bovine mammary gland at various magnifications.

the building blocks of milk entering from the blood through the basement membrane. These basic components are manufactured (synthesized) into milk components on the production lines of the endoplasmic reticulum, which is supplied with energy from the "boilers" of the mitochondria. The components are then packaged in secretory vesicles by the Golgi apparatus or as lipid droplets in the cytoplasm. Finally, the vesicles and lipid droplets pass through the apical plasma membrane and are stored in the lumen. The layer of secretory epithelial cells surrounding the lumen of the alveolus is in turn surrounded by a layer of myoepithelial cells and blood capillaries. When oxytocin, a pituitary hormone circulating in the blood, binds to the myoepithelial cells, the alveolus contracts, expelling the milk stored in the lumen into the duct system.

It is currently thought that the intracellular membranes (endomembranes) also exhibit directional flow to the apical plasma membrane, with concurrent transformation of membranes from endoplasmic reticulum to Golgi, to secretory vesicles, to apical plasma membrane within the epithelial secretory cell. The endoplasmic reticulum, the endomembrane production line where active synthesis occurs, appears like a cellular plumbing system, with the inside of these tubes, known as the cisternae, emptying into the Golgi apparatus. The Golgi apparatus transforms into Golgi vesicles, the packages that carry the aqueous phase milk components to the apical plasma membrane.

Ribosomes, the machinery for protein synthesis, exist both free in the cytoplasm and bound to the endoplasmic reticulum. Near the basement membrane, the endoplasmic reticulum is covered with ribosomes making this membrane appear "rough." However, moving in the direction of the apical end, amino acids are depleted and synthesis slows, allowing ribosomes to dissociate. Hence, this membrane becomes smooth as it is transformed into Golgi membrane. Synthesis is completed in the lumen of the Golgi; for example, proteins are glycosylated and phosphorylated and lactose is synthesized. It is here and in the Golgi vesicles that casein micelles first appear. Furthermore, this membrane is impermeable to lactose, so that the major secretory products of the cell are now segregated from all other cellular constituents. Secretion of the products of synthesis is completed when the Golgi vesicles merge with the apical plasma membrane, fuse to become part of that membrane, and empty the contents of the package into the alveolar lumen for storage. Note that, in the membrane transformation, the inside or lumen side of the vesicle membrane becomes the outside of the cell plasma membrane.

On the other hand, fat, which is also synthesized by the endoplasmic reticulum, is directed to the cytoplasmic side of the membrane where it collects as lipid droplets. These droplets move to the apical plasma membrane where they are expelled into the alveolar lumen by pinocytosis, thus acquiring on their surface a coat of plasma membrane. The presence of this membrane has important consequences for the manufacturing characteristics of milk products. Apical plasma membrane, lost in this process, is continually replenished, at least in part, by the fusion of vesicles as previously described.

In closing this discussion, it should be appreciated that many, if not all, of the characteristics of milk and its constituents are a consequence of the mechanism of synthesis and secretion.

III. CHEMICAL COMPOSITION

The composition of milk reflects the fact that it is the sole source of food for the very young mammal. Hence, it is composed of a complex mixture of lipids, proteins, carbohydrates, vitamins, and minerals. The average composition of milk and the range of average

Table 2 Composition of Bovine Milk from Western Cattle

Component	Average percentage	Range for breed[a] (average percentages)
Water	86.6	85.4-87.7
Fat	4.1	3.4-5.1
Protein	3.6	3.3-3.9
Lactose	5.0	4.9-5.0
Ash	0.7	0.68-0.74

[a]Western breeds including Guernsey, Jersey, Ayshire, Brown Swiss, Shorthorn, and Holstein.
Source: Adapted from Ref. 7.

compositions for milks of Western cattle are given in Table 2. The greatest variability in composition is exhibited by the lipid fraction. Largely due to the economic value of milk fat, breeders have selected animals producing higher percentages of this constituent. Consequently, a wide range of fat percentages is observed when comparing individual milks. Due to the contribution of lactose and milk salts to osmolality, and the required matching of milk's osmotic pressure with that of blood, very little variability is observed for these constituents. The pH of freshly drawn milk is 6.6-6.7, which is slightly lower than that of blood.

A. Milk Proteins

Milk contains 30-35 g/liter of total protein that rates very high in nutritive quality (Table 3). There are six major gene products of the mammary gland: α_{s1}-caseins, α_{s2}-caseins, β-caseins, κ-caseins, β-lactoglobulins, and α-lactalbumins, all of which exhibit genetic

Table 3 Concentrations of the Major Milk Proteins

Protein	Concentration (g/liter)	Approximate percentage of total protein	
Caseins:	24-28		80
α_s-caseins	15-19		42
α_{s1}	12-15	34	
α_{s2}	3-4	8	
β-caseins	9-11		25
κ-caseins	3-4		9
γ-caseins	1-2		4
Whey proteins:	5-7		20
β-lactoglobulins	2-4		9
α-lactalbumin	1-1.5		4
proteose-peptones	0.6-1.8		4
blood proteins:			
serum albumin	0.1-0.4		1
immunoglobulins	0.6-1.0		2
		100	100

polymorphism since they are products of codominant, allelic, autosomal genes. Milk proteins are classified as either caseins or whey proteins. All the caseins exist with calcium phosphate in a unique, highly hydrated spherical complex known as the *micelle*. These complexes vary in size from 30 to 300 nm in diameter. Due to the unusual characteristics of the caseins and the micellar complex, milk proteins can be readily separated into casein and whey protein fractions. Historically, this separation by means of acid precipitation or rennet coagulation has formed the basis for many milk products, such as cheese, whey products, and food ingredients. Casein comprises the largest fraction of bovine milk proteins; consequently, the curd formed by agglomeration of casein micelles during cheese manufacture retains most of the total milk protein. Until recently, whey was often discarded; however, it is rapidly becoming economically necessary and advantageous to recover the whey proteins for food use.

In addition to the primary gene products several proteins of milk result from post-translational proteolysis, perhaps mostly due to the presence of small amounts of plasmin, a blood enzyme. Thus, the γ-caseins and some of the proteose-peptones are derived from β-casein. Milk also contains small amounts of protein derived directly from blood, such as serum albumin and some of the immunoglobulins.

The amino acid compositions and the covalent molecular weights of milk proteins are very accurately known (Table 4). A key distinction of the caseins is their large content of phosphoseryl residues and their moderately large proline content. Although the caseins are rather low in half-cystine, the whey proteins contain considerably more of this residue.

For more complete discussion of the characteristics of individual milk proteins, the reader is referred to several excellent reviews (5,6,14,22,23,25-27,39,45,46,48,50,51).

B. Milk Lipids

Bovine milk contains the most complex lipid known. For detailed characteristics of the various fractions and a discussion of their biosynthesis, the reader should consult one of the reviews (3,4,15,16,32). Triacylglycerols (triglycerides) comprise by far the greatest proportion of lipids, making up 97-98% of the total lipid (Table 5). In milk, the triacylglycerols are present in globules 2-3 μm in diameter surrounded by a membrane derived from the cellular apical plasma membrane. The concentrations given in Table 5 represent that for freshly drawn milk. During storage some lipolysis occurs giving higher concentrations of free fatty acids and mono- and diacylglycerols.

Over 400 different fatty acids have been identified in bovine lipids. Consequently, more than 64×10^6 different triacylglycerols are possible; however, only 20 individual fatty acids account for most of the residues. Twelve of the major fatty acids and their weight percentages in the total lipid are listed in Table 6. Based on all of the fatty acids identified, saturated acids account for 62.83% of the total, monoenoic acids, 30.75%, dienoic acids, 2.97%, polyenoic acids, 0.85%, monobranched acids, 1.36%, multibranched acids, 0.83%, and miscellaneous acids, 0.40%. In addition to being subject to the greatest variability in total amount, the lipids are more subject than are other milk constituents to compositional changes influenced by environmental factors, such as diet. For example, the concentration of unsaturated fatty acids can be greatly increased by feeding these acids in a form protected by a denatured protein coat so they pass through the rumen without modification.

Triacylglycerols containing different fatty acids have a chiral carbon at the 2 position (Fig. 2). The glycerol carbons are assigned stereospecific numbers (sn), as shown in

Table 4 Amino Acid Composition of the Major Proteins in the Milk of Western Cattle

Acid	α_2-Casein B	α_{s2}-Casein A[a]	κ-Casein B	β-Casein A²	γ_1-Casein A²	γ_2-Casein A²	γ_3-Casein A	β-Lactoglobulin A	α-Lactalbumin B
						Amino Acids in			
Asp	7	4	4	4	4	2	2	11	9
Asn	8	14	7	5	3	1	1	5	12
Thr	5	15	14	9	8	4	4	8	7
Ser	8	6	12	11	10	7	7	7	7
SerP	8	11	1	5	1	0	0	0	0
Glu	24	25	12	18	11	4	4	16	8
Gln	15	15	14	21	21	11	11	9	5
Pro	17	10	20	35	34	21	21	8	2
Gly	9	2	2	5	4	2	2	3	6
Ala	9	8	15	5	5	2	2	14	3
½ Cys	0	2	2	0	0	0	0	5	8
Val	11	14	11	19	17	10	10	10	6
Met	5	4	2	6	6	4	4	4	1
Ile	11	11	13	10	7	3	3	10	8
Leu	17	13	8	22	19	14	14	22	13
Tyr	10	12	9	4	4	3	3	4	4
Phe	8	6	4	9	9	5	5	4	4
Trp	2	2	1	1	1	1	1	2	4
Lys	14	24	9	11	10	4	3	15	12
His	5	3	3	5	5	4	3	2	3
Arg	6	6	5	4	2	2	2	3	1
Pyr Glu	0	0	1	0	0	0	0	0	0
Total residues	199	207	169	209	181	104	102	162	123
Molecular weight	23,612	25,228	19,005	23,980	20,520	11,822	11,557	18,263	14,174

[a]This composition, including 11 phosphoserine residues, corresponds to α_{s4}-casein in the α_{s2}-casein family.

Source: From Ref. 46.

Table 5 Lipid Composition of Bovine Milk

Lipid	Weight percent	g/liter[a]
Triacylglycerols (triglycerides)	97-98	31.2
Diacylglycerols (diglycerides)	0.3-0.6	0.14
Monoacylglycerols (monoglycerides)	0.02-0.04	0.01
Free fatty acids	0.1-0.4	0.08
Free sterols	0.2-0.4	0.10
Phospholipids	0.2-1.0	0.19
Hydrocarbons	Trace	Trace
Sterol esters	Trace	Trace

[a]Based on the usual butterfat percentage of commercial pasteurized whole milk, 3.2%.
Source: From Patton and Jensen (32).

the figure. Asymmetry is a distinguishing feature of milk triglycerides with the short-chain saturated fatty acids, 4:0, 6:0, and 8:0, specifically occupying the sn-3 position (Table 7).

The phospholipids and sterols occur in the cellular membrane fractions that carry over into milk. Phosphatidylcholine (34.5 mol%), phosphatidylethanolamine (31.8 mol%), and sphingomyelin (25.5 mol%) account for most of the phospholipid fraction. Due to the publicity surrounding a proposed correlation between cholesterol and atherosclerosis, the content of this lipid in foods has received much attention. Actually, milk contains relatively little cholesterol; for example, a 227 g glass of milk contains 27 mg, but a large egg has 275 mg, 10 small shrimp have 125 mg, and a 3.5 oz. fresh water fish contains 70 mg. Since cholesterol occurs in the fat globule membrane, its concentration is related to fat content.

C. Milk Salts and Sugars

The salts in milk consist principally of chlorides, phosphates, citrates, and bicarbonates of sodium, potassium, calcium, and magnesium (35). Salts, inorganic components of milk, should not be confused with ash, which is the oxides of the minerals resulting from combustion. The milk salts are distributed between a soluble phase, such as might be obtained in a dialysate or a membrane permeate (ultrafiltrate), and a colloidal phase (Table 8). The

Table 6 Major Fatty Acid Constituents of Bovine Milk Fat

Fatty acid	Weight percent	Fatty acid	Weight percent
4:0	3.6	15:0	1.5
6:0	2.2	16:0	25.2
8:0	1.1	18:0	11.9
10:0	1.9	16:1	1.8
12:0	3.0	18:1	25.5
14:0	11.2	18:2	2.1

Source: Jensen et al. (15).

sn **POSITION**

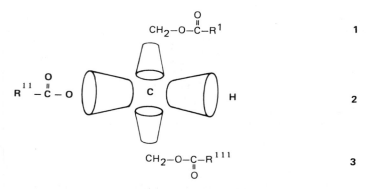

Figure 2 Stereochemical characteristics of a triacylglycerol molecule.

distribution of calcium, magnesium, phosphate, and citrate between soluble and colloidal phases and their interactions with milk proteins are important factors in the stability of dairy products.

From consideration of the requirement of constant osmolality (a result of biosynthesis conditions), we would expect a reciprocal relationship between milk salts and lactose. In fact, such an inverse relationship has been documented between sodium and lactose contents and between sodium and potassium contents (35). Consequently, milk has an essentially constant freezing point (-0.53 to $-0.57°C$), and this property is used to detect adulteration (addition of water).

Table 7 Distribution of the Individual Fatty Acids in Bovine Milk Triacylglycerols

Fatty acid	sn-Position (mol %)		
	1	2	3
4:0	10	5	85
6:0	16	26	58
8:0	17	42	41
10:0	21	50	29
12:0	25	47	28
14:0	27	54	19
16:0	45	42	13
16:1	48	36	16
18:0	58	26	16
18:1	42	28	30
18:2	22	78	0

Source: From Pitas et al. (34), Patton and Jensen (32), and Jenness (13).

Table 8 Concentration of the Principal Salts and Lactose in Milk and Distribution in the Soluble and Colloidal Phases

Component	Total in milk (mg/100 g)		Percentage in soluble phase	Percentage in colloidal phase
	Mean	Range of values		
Total calcium	117.7	110.9–120.3	33	67
Ionized calcium	11.4	10.5–12.8	100	0
Magnesium	12.1	11.4–13.0	67	33
Sodium	58	47–77	94	6
Potassium	140	113–171	93	7
Citrate	176	166–192	94	6
Total phosphorus	95.1	79.8–101.7	45	55
Inorganic phosphorus	62.9	51.9–70.0	54	46
Chloride	104.5	89.8–127.0	100	0
Lactose	4800	–	100	0

Source: From White and Davies (49) and Davies and White (9).

Lactose (4-O-β-D-galactopyranosyl-D-glucopyranose) is the predominant carbohydrate in bovine milk, accounting for 50% of the solids in skim milk. Its synthesis is associated with that of a major whey protein, α-lactalbumin, which acts as a modifier protein for UDP-galactosyltransferase. Thus, α-lactalbumin changes the specificity of this enzyme such that the galactosyl group is transferred to glucose rather than to glycoproteins. Lactose occurs in both α and β forms, with an equilibrium ratio of $\beta/\alpha = 1.68$ at 20°C (30). The β form is far more soluble than the α form, and the rate of mutarotation is rapid at room temperature but very slow at 0°C. The α-hydrate crystal form, which crystallizes under ordinary conditions, occurs in a number of shapes, but the most familiar is the "tomahawk" shape, which imparts a "sandy" mouth feel to dairy products, such as sandy ice cream. Lactose, with a sweetness about one-fifth that of sucrose, contributes to the characteristic flavor of milk.

D. Enzymes

Although present in small amounts, enzymes can have important influences on the stability of dairy products. Their effects may become even more important with the move toward high-temperature-short-time processing treatments and long product storage periods, since these conditions provide the potential for survival of enzymes or their reactivation. Enzymes may be introduced into milk as a result of microbial growth; however, we shall limit our consideration to those enzymes indigenous to milk (Table 9). Of these, protease(s) and lipase(s) are particularly important due to their effects on flavor and protein stability. The oxidoreductases can also have effects on flavor stability due to their influence on the oxidative state, particularly in the lipid fraction.

Milk also contains a complement of vitamins, which will be discussed in a later section.

Table 9 Enzymes in Bovine Milk, Partial List

Oxidoreductases

Xanthine oxidase
  (xanthine:O_2 oxidoreductase)
Sulfhydryl oxidase
  (protein, peptide $-SH:O_2$ oxidoreductase)
Lactoperoxidase
  (donor:H_2O_2 oxidoreductase)
Superoxide dismutase
  ($O_2^-:O_2^-$ oxidoreductase)

Catalase
  ($H_2O_2:H_2O_2$ oxidoreductase)
Diaphorase
  (NADH:lipoamide oxidoreductase)
Cytochrome c reductase
  (NADH:cytochrome c oxidoreductase)

Transferases

UDP-Galactosyltransferase
  (UDP galactose:D-glucose-1-galactosyltransferase)
Ribonuclease
  (polyribonucleotide 2-oligonucleotide transferase)
γ-Glutamyltransferase

Hydrolases

Protease
  (peptidyl peptide hydrolase)
Lipase
  (glycerol-ester hydrolase)
Lysozyme
  (mucopeptide N-acetylneuraminyl hydrolase)
Alkaline phosphatase
  (orthophosphoric monoester phosphohydrolase)
Phosphoprotein phosphatase
  (phosphoprotein phosphohydrolase)

B-Esterase
  (carboxylic ester hydrolase)
Cholinesterase
  (acylcholine acylhydrolase)
α-Amylase
  (α-1,4-glucan-4-glucanohydrolase)
β-Amylase
  (β-1,4-glucan maltohydrolase)
5'-Nucleotidase
  (5'-ribonucleotide phosphohydrolase)

Lyases

Aldolase
  (fructose-1,6-diphosphate
  D-glyceraldehyde-3-phosphate lyase)

Carbonic anhydrase
  (carbonate hydrolyase)

Source: Compiled in part from Shahani et al. (41).

IV. STRUCTURAL ORGANIZATION OF MILK COMPONENTS

A. Structure of Milk Proteins

Scientific knowledge of milk holds a unique position among foods in that the primary structures of the major milk proteins have been determined (5,45,46,50,51). Although sufficient understanding has not yet been achieved to completely correlate structural information with food functionality, progress in this direction is occurring. Consequently, the food scientist should bear these considerations in mind while examining the structure of food proteins.

1. The Caseins

The primary structure of the principal genetic variant of the major protein (α_{s1}-casein B, >40%) in bovine milk is given in Fig. 3. A number of unique features result from this structure. First, there are eight sites of posttranslational phosphorylation; consequently, this protein exhibits interactions with calcium typical of the caseins. Another important characteristic discernible from the primary structure is the clustering of polar and nonpolar residues. For example, residue numbers 41-80 contain the cluster of phosphoseryl residues, and this peptide carries a net negative charge of about −21 at the normal pH of milk. The remaining C-terminal portion is rather hydrophobic. These characteristics suggest a unique dipolar structure composed of a highly solvated, charged domain and a hydrophobic globular domain. Most likely the polar domain approaches random coil type behavior and the hydrophobic domain apparently possesses a mixture of α-helix, β-structure, β-turns, and unordered structure. The flexible nature of the polar domain causes the molecular dimensions to be very sensitive to ionic strength and to the binding of ions, particularly protons (H^+) and Ca^{2+}. In addition, intermolecular interactions between hydrophobic domains leads to molecular self-association, or association with other caseins. These hydrophobic interactions become more important as the polar domain is discharged by binding of Ca^{2+} to the orthophosphate groups, since this binding probably greatly reduces the dimensions of the polar domain. The intermolecular interactions then result in precipitation of isolated α_{s1}-casein or formation of micelles by interaction with κ-casein.

The general characteristics of α_{s1}-casein are shared by other calcium-sensitive caseins (those caseins precipitated by Ca^{2+}). Their similarities are predicted by comparison of their primary structures (Figs. 4 and 5) with that of α_{s1}-casein. In fact, structural homologies, particularly in the region of containing clusters of phosphoseryl residues, suggest that all these proteins, α_{s1}-, α_{s2}-, and β-caseins, may have evolved from a common ancestral gene. Structures of both α_{s2}- and β-casein are characterized by charged polar domains and hydrophobic domains. Like α_{s1}-casein, sequences in the polar domains, which may approach random coil secondary and tertiary structure, are such that clusters of seryl residues are phosphorylated. α_{s2}-Casein contains several phosphoseryl clusters and thus is the most hydrophilic, whereas β-casein contains only a single phosphoseryl cluster in the N-terminal sequence and the remaining large C-terminal sequence is very hydrophobic. Thus, β-casein is the most hydrophobic of all the milk proteins and its characteristics are consequently the most temperature dependent. Predictably, the properties of α_{s2}-casein are more sensitive to ionic strength than to temperature.

The flexible, highly solvated polar domains of the calcium-sensitive caseins should be the most susceptible locale to proteolysis. Thus, limited proteolysis of β-casein by the indigenous milk proteinase yields: (a) the γ-caseins, derived from the C-terminal portion, and (b) components 5,8-fast, and 8-slow of the proteose-peptone fractions, derived from the N-terminal part. Likewise, peptides in the λ-casein fraction are derived from limited proteolysis of α_{s1}- and β-casein.

The amphiphilic nature of the structure of κ-casein (Fig. 6) is a key factor in the unique ability of this protein to stabilize the milk casein micelle. κ-Casein does not contain clusters of phosphoseryl residues in its polar domain as do the calcium-sensitive caseins; hence, it does not bind as much as Ca^{2+} (1-2 versus 8-9 mol/mol for α_{s1}- and 4-5 mol/mol for β-casein), so the polar domain is not discharged or dehydrated by addition of this ion. Consequently, κ-casein is not precipitated by Ca^{2+}. Furthermore, the charge and hydration of the polar domain of some κ-casein molecules is increased by

1

H.Arg-Pro-Lys-His-Pro-Ile-Lys-His-Gln-Gly-Leu-Pro-Gln-[Glu-Val-Leu-Asn-Glu-Asn-Leu-

21
Leu-Arg-Phe-Phe-Val-Ala]-Pro-Phe-Pro-Gln-Val-Phe-Gly-Lys-Glu-Lys-Val-Asn-Glu-Leu-

41
Ser-Lys-Asp-Ile-Gly-*SerP*-Glu-*SerP*-Thr-Glu-Asp-Gln-[Ala]-Met-Glu-Asp-Ile-Lys-Gln-Met-

61
Glu-Ala-Glu-*SerP*-Ile-*SerP*-*SerP*-*SerP*-Glu-Glu-Ile-Val-Pro-Asn-*SerP*-Val-Glu-Gln-Lys-His-

81
Ile-Gln-Lys-Glu-Asp-Val-Pro-Ser-Glu-Arg-Tyr-Leu-Gly-Tyr-Leu-Glu-Gln-Leu-Leu-Arg-

101
Leu-Lys-Lys-Tyr-Lys-Val-Pro-Gln-Leu-Glu-Ile-Val-Pro-Asn-*SerP*-Ala-Glu-Glu-Arg-Leu-

121
His-Ser-Met-Lys-Glu-Gly-Ile-His-Ala-Gln-Gln-Lys-Glu-Pro-Met-Ile-Gly-Val-Asn-Gln-

141
Glu-Leu-Ala-Tyr-Phe-Tyr-Pro-Glu-Leu-Phe-Arg-Gln-Phe-Tyr-Gln-Leu-Asp-Ala-Tyr-Pro-

161
Ser-Gly-Ala-Trp-Tyr-Tyr-Val-Pro-Leu-Gly-Thr-Gln-Tyr-Thr-Asp-Ala-Pro-Ser-Phe-Ser-

181   199
Asp-Ile-Pro-Asn-Pro-Ile-Gly-Ser-Glu-Asn-Ser-[Glu]-Lys-Thr-Thr-Met-Pro-Leu-Trp.OH

Figure 3 Primary structure of α_{s1}-casein B. Residues in brackets are substituted or deleted in other genetic variants of α_{s1}-casein. Italicized residues denote sites of posttranslational modification; for example, Ser 41 is phosphorylated in α_{s0}-casein.

1
H.Lys-Asn-Thr-Met-Glu-His-Val-*SerP*-*SerP*-*SerP*-Glu-Glu-Ser-Ile-Ile-*SerP*-Gln-Glu-Thr-Tyr-

21
Lys-Gln-Glu-Lys-Asn-Met-Ala-Ile-Asn-Pro-Ser-Lys-Glu-Asn-Leu-Cys-Ser-Thr-Phe-Cys-

41
Lys-Glu-Val-Val-Arg-Asn-Ala-Asn-Glu-Glu-Glu-Tyr-Ser-Ile-Gly-*SerP*-*SerP*-*SerP*-Glu-Glu-

61
SerP-Ala-Glu-Val-Ala-Thr-Glu-Glu-Val-Lys-Ile-Thr-Val-Asp-Asp-Lys-His-Tyr-Gln-Lys-

81
Ala-Leu-Asn-Glu-Ile-Asn-Glu-Phe-Tyr-Gln-Lys-Phe-Pro-Gln-Tyr-Leu-Gln-Tyr-Leu-Tyr-

101
Gln-Gly-Pro-Ile-Val-Leu-Asn-Pro-Trp-Asp-Gln-Val-Lys-Arg-Asn-Ala-Val-Pro-Ile-Thr-

121
Pro-Thr-Leu-Asn-Arg-Glu-Gln-Leu-*SerP*-Thr-*SerP*-Glu-Glu-Asn-Ser-Lys-Lys-Thr-Val-Asp-

141
Met-Glu-*SerP*-Thr-Glu-Val-Phe-Thr-Lys-Lys-Thr-Lys-Leu-Thr-Glu-Glu-Glu-Lys-Asn-Arg-

161
Leu-Asn-Phe-Leu-Lys-Lys-Ile-Ser-Gln-Arg-Tyr-Gln-Lys-Phe-Ala-Leu-Pro-Gln-Tyr-Leu-

181
Lys-Thr-Val-Tyr-Gln-His-Gln-Lys-Ala-Met-Lys-Pro-Trp-Ile-Gln-Pro-Lys-Thr-Lys-Val-

201
Ile-Pro-Tyr-Val-Arg-Tyr-Leu.OH   207

Figure 4 Primary structure of α_{s2}-casein A. Italicized residues denote sites of potential postranslational modification.

1
H.Arg-Glu-Leu-Glu-Glu-Leu-Asn-Val-Pro-Gly-Glu-Ile-Val-Glu-*SerP*-Leu-*SerP*-*SerP*-*SerP*-Glu-

21
Glu-Ser-Ile-Thr-Arg-Ile-Asn-Lys-Lys-Ile-Glu-Lys-Phe-Gln-[*SerP*]-Glu-[Glu]-Gln-Gln-Gln-
                           ➤ γ₁-caseins

41
Thr-Glu-Asp-Glu-Leu-Gln-Asp-Lys-Ile-His-Pro-Phe-Ala-Gln-Thr-Gln-Ser-Leu-Val-Tyr-

61
Pro-Phe-Pro-Gly-Pro-Ile-[Pro]-Asn-Ser-Leu-Pro-Gln-Asn-Ile-Pro-Pro-Leu-Thr-Gln-Thr-

81
Pro-Val-Val-Val-Pro-Pro-Phe-Leu-Gln-Pro-Glu-Val-Met-Gly-Val-Ser-Lys-Val-Lys-Glu-

101
Ala-Met-Ala-Pro-Lys-[His]-Lys-Glu-Met-Pro-Phe-Pro-Lys-Tyr-Pro-Val-Gln-Pro-Phe-Thr-
                        ➤ γ₃-caseins
                        ➤ γ₂-caseins

121
Glu-[Ser]-Gln-Ser-Leu-Thr-Leu-Thr-Asp-Val-Glu-Asn-Leu-His-Leu-Pro-Pro-Leu-Leu-Leu-

141
Gln-Ser-Trp-Met-His-Gln-Pro-His-Gln-Pro-Leu-Pro-Pro-Thr-Val-Met-Phe-Pro-Pro-Gln-

161
Ser-Val-Leu-Ser-Leu-Ser-Gln-Ser-Lys-Val-Leu-Pro-Val-Pro-Glu-Lys-Ala-Val-Pro-Tyr-

181
Pro-Gln-Arg-Asp-Met-Pro-Ile-Gln-Ala-Phe-Leu-Leu-Tyr-Gln-Gln-Pro-Val-Leu-Gly-Pro-

201   209
Val-Arg-Gly-Pro-Phe-Pro-Ile-Ile-Val.OH

Figure 5 Primary structure of β-casein A². Residues in brackets are substituted in other genetic variants. The arrows indicate bonds hydrolyzed by plasmin, yielding the γ-caseins and proteose-peptones. Italicized residues denote sites of posttranslational modification.

posttranslational glycosylation with charged oligosaccharide residues. The N-terminal hydrophobic domain, however, is very hydrophobic and probably contains a considerable amount of β-sheet and α-helical structure. Specific cleavage of the Phe_{105}-Met_{106} peptide bond by chymosin (rennet) or in the same region of the sequence by other proteinases separates the two domains. The liberated hydrophobic domain (*para-κ*-casein) is not soluble; the polar domain (macropeptide) is extremely soluble. The whole molecule is thus somewhat analogous to a detergent. In physiological buffers, isolated κ-casein exists in the form of large spherical aggregates resembling soap micelles, each held together by lateral interactions among the hydrophobic domains. The hydrophobic domain of κ-casein interacts with similar domains of other caseins when they are present.

1
PyroGlu-Glu-Asn-Gln-Glu-Gln-Pro-Ile-Arg-Cys-Glu-Lys-Asp-Glu-Arg-Phe-Phe-Ser-Asp-

21
Lys-Ile-Ala-Lys-Tyr-Ile-Pro-Ile-Gln-Tyr-Val-Leu-Ser-Arg-Tyr-Pro-Ser-Tyr-Gly-Leu-

41
Asn-Tyr-Tyr-Gln-Gln-Lys-Pro-Val-Ala-Leu-Ile-Asn-Asn-Gln-Phe-Leu-Pro-Tyr-Pro-Tyr-

61
Tyr-Ala-Lys-Pro-Ala-Ala-Val-Arg-Ser-Pro-Ala-Gln-Ile-Leu-Gln-Trp-Gln-Val-Leu-Ser-

81
Asp-Thr-**Val**-Pro-Ala-Lys-Ser-Cys-Gln-Ala-Gln-Pro-Thr-Thr-Met-Ala-Arg-His-Pro-His-

101   105↓106
Pro-His-Leu-Ser-Phe-Met-Ala-Ile-Pro-Pro-Lys-Lys-Asn-Gln-Asp-Lys-Thr-Glu-Ile-Pro-

121
Thr-Ile-Asn-Thr-Ile-Ala-Ser-Gly-Glu-Pro-*Thr*-Ser-*Thr*-Pro-*Thr*-[*Thr*]-Glu-Ala-Val-Glu-

141
Ser-Thr-Val-Ala-Thr-Leu-Glu-[Asp]-*SerP*-Pro-Glu-Val-Ile-Glu-Ser-Pro-Pro-Glu-Ile-Asn-

169
Thr-Val-Gln-Val-Thr-Ser-Thr-Ala-Val.OH

Figure 6 Primary structure of κ-casein A. Residues in brackets are substituted in genetic variant B. Italicized residues indicate potential sites of posttranslational modification. The bond between residues 105 and 106 is specifically hydrolyzed by chymosin (rennet).

2. The Whey Proteins

Primary structures for the principal genetic variants of the two major whey proteins are shown in Figs. 7 and 8. These structures are typical of compact globular proteins, with a rather uniform sequence distribution of nonpolar, polar, and charged residues. Hence, these proteins fold intramolecularly, burying most of their hydrophobic residues so that extensive self-association or interaction with other proteins does not occur. β-Lactoglobulin does undergo limited self-association; at the pH of milk a dimer is formed with a geometry resembling two impinging spheres. The structure of β-lactoglobulin is dependent on the pH; thus, below pH 3.5 the dimer associates to a slightly expanded monomer, between pH 3.5 and 5.2 the dimer tetramerizes to give an octomer, and above pH 7.5 the dimer dissociates and undergoes a conformational change giving an expanded monomer. The functionality of β-lactoglobulin in foods is influenced greatly by the presence of both a sulfhydryl group and disulfide bonds. The relative importance of the sulfhydryl group is influenced by conformational changes since this determines the availability of the sulfhydryl group for reaction. Thus, under appropriate conditions β-lactoglobulin readily participates in sulfhydryl-disulfide interchange reactions and this affects many of its characteristics, such as solubility.

The three-dimensional structure of α-lactalbumin is probably very similar to that of lysozyme, a protein with homologous primary structure. It has a very compact, nearly spherical overall shape, although in comparison to lysozyme, the structure is slightly less stable and more expanded. It has been suggested that interaction of α-lactalbumin with galactosyltransferase brings its monosaccharide binding site (specificity for glucose) in proximity to the monosaccharide binding site of the enzyme.

B. Casein Micelle and Milk Salts

As a result of their phosphorylation and amphiphilic nature, the caseins interact with each other and with calcium phosphate to form large spherical complexes (micelles) ranging in diameter from about 30-50 nm to 300 nm (Fig. 9). Micelles contain 92% protein, composed of α_{s1}-, α_{s2}-, β-, and κ-casein at an average ratio of 3:1:3:1, and 8% inorganic constitutents composed primarily of calcium phosphate. Since the characteristics of micelles determine the behavior of milk and milk products during industrial processes, such as pasteurization, sterilization, concentration, freezing, and cheesemaking, the properties of both natural micelles and model micelle systems have received considerable attention (8,33,40,42,43). All evidence indicates that micelles are formed from nearly spherical submicelles, each with a diameter of 15-20 nm. Hence, the micelle has a raspberrylike appearance (Fig. 9).

Submicelles result from interactions among the hydrophobic domains of the individual caseins, and these submicelles probably have a variable composition with respect to the component proteins. The structure of calcium phosphates in the micelle is a matter of controversy, although many investigators believe it exists as amorphous tertiary calcium phosphate interspersed throughout the micelle. The presence in milk of other ions, particularly Mg^{2+}, and of the caseins, is believed to prevent transformation of amorphous calcium phosphate to more stable forms, such as hydroxyapatite.

All components of the micelle are apparently in equilibria with the serum phase. Lowering the temperature to near 0°C, for example, causes some β-casein, κ-casein, and colloidal calcium phosphate to reversibly dissociate from the micelle. Also, addition of κ-casein to milk causes the average micelle size to decrease, whereas addition of β-casein

1

H.Leu-Ile-Val-Thr-Gln-Thr-Met-Lys-Gly-Leu-Asp-Ile-Gln-Lys-Val-Ala-Gly-Thr-Trp-Tyr-

21

Ser-Leu-Ala-Met-Ala-Ala-Ser-Asp-Ile-Ser-Leu-Asp-Ala-Gln-Ser-Ala-Pro-Leu-Arg-

41

Val-Tyr-Val-Glu-[Glu]-Leu-Lys-Pro-Thr-Pro-Glu-Gly-Asp-Leu-Glu-Ile-Leu-Leu-[Gln]-Lys-

61

Trp-Glu-Asn-[Gly]-Glu-Cys-Ala-Gln-Lys-Lys-Ile-Ile-Ala-Glu-Lys-Thr-Lys-Ile-Pro-Ala-

81

Val-Phe-Lys-Ile-Asp-Ala-Leu-Asn-Glu-Asn-Lys-Val-Leu-Val-Leu-Asp-Thr-Asp-Tyr-Lys-

101

Lys-Tyr-Leu-Leu-Phe-Cys-Met-Glu-Asn-Ser-Ala-Glu-Pro-Glu-Gln-Ser-Leu-[Ala]-Cys-Gln-

121 [SH]

Cys-Leu-Val-Arg-Thr-Pro-Glu-Val-Asp-Asp-Glu-Ala-Leu-Glu-Lys-Phe-Asp-Lys-Ala-Leu-

141

Lys-Ala-Leu-Pro-Met-His-Ile-Arg-Leu-Ser-Phe-Asn-Pro-Thr-Gln-Leu-Glu-Glu-Gln-Cys-

161 162

His-Ile.OH

Figure 7 Primary structure of β-lactoglobulin B. Residues in brackets are substituted in other genetic variants. Location of the free thiol appears to vary between residues 119 and 121.

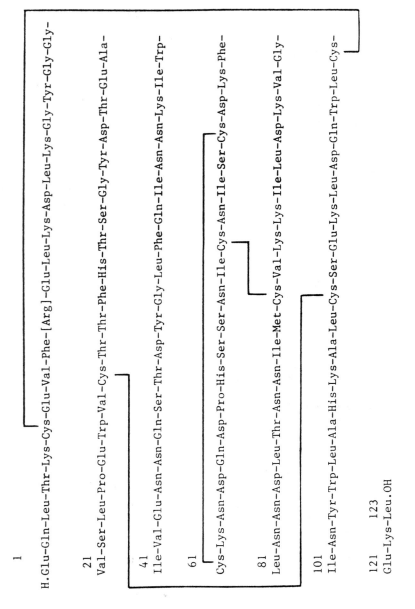

1

H.Glu-Gln-Leu-Thr-Lys-Cys-Glu-Glu-Val-Phe-[Arg]-Glu-Leu-Lys-Asp-Leu-Lys-Gly-Tyr-Gly-Gly-

21

Val-Ser-Leu-Pro-Glu-Trp-Val-Cys-Thr-Thr-Phe-His-Thr-Ser-Gly-Tyr-Asp-Thr-Glu-Ala-

41

Ile-Val-Glu-Asn-Asn-Gln-Ser-Thr-Asp-Tyr-Gly-Leu-Phe-Gln-Ile-Asn-Asn-Lys-Ile-Trp-

61

Cys-Lys-Asn-Asp-Gln-Asp-Pro-His-Ser-Ser-Asn-Ile-Cys-Asn-Ile-Ser-Cys-Asp-Lys-Phe-

81

Leu-Asn-Asn-Asp-Leu-Thr-Asn-Asn-Ile-Met-Cys-Val-Lys-Lys-Ile-Leu-Asp-Lys-Val-Gly-

101

Ile-Asn-Tyr-Trp-Leu-Ala-His-Lys-Ala-Leu-Cys-Ser-Glu-Lys-Leu-Asp-Gln-Trp-Leu-Cys-

121   123
Glu-Lys-Leu.OH

Figure 8 Primary structure of α-lactalbumin B. The residue in brackets is substituted by Gln in the A variant.

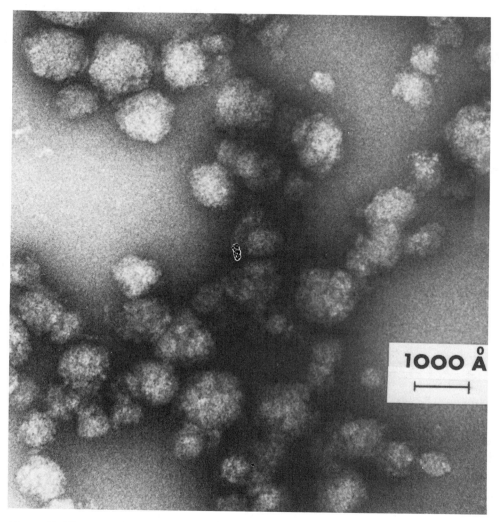

Figure 9 Electron photomicrographs of native casein micelles. Samples was fixed in 1% glutaraldehyde and prepared by negative staining with uranyl acetate. (From Ref. 20.)

or α_{s1}-casein causes an increase in micelle size. However, under conditions existing in milk, free submicelles have not been found in the serum phase.

The major milk salt, colloidal calcium phosphate, exists in equilibria between the micellar and serum phases. Substantial amounts of magnesium and some citrate are also in the micelle. In addition, Ca^{2+} is bound to the orthophosphate groups of the caseins and small amounts of other cations may be bound to these proteins. Consequently, the properties of micelles and associated salts cannot be considered independently.

Both κ-casein and Ca^{2+} are absolute requirements for micelle formation. Studies of model micelles also indicate that phosphate is necessary for formation of micelles that are temperature stable. Micelles are characterized by a porous, "spongy" structure, which allows equilibration of all constituents with the serum phase and contributes to an exceptional protein hydration of 3.7 g H_2O per g dry protein. This hydration, determined from hydrodynamic measurements, is an order of magnitude greater than that of typical globular proteins. A fundamental characteristic of micelles is their stability to close approach; for example, pellets formed by sedimentation redisperse spontaneously upon standing. Thus, the micellar surface must be highly solvated and unreactive, a key to its stability under many harsh processing conditions. However, removal of the polar domain of κ-casein molecules completely changes the surface characteristics of the micelle. Micellar surfaces thus modified are very reactive, leading to association of the micelles and formation of gels, such as those formed in cheesemaking. In fact, only a portion of the κ-casein need be modified for coagulation to occur.

Use of micellar models is helpful in rationalizing existing data and in predicting the behavior of milk. Numerous models for the micelle have been proposed (40,42); however, two models appear to satisfy most of the requirements. The major difference between Slattery's model (42,44) and Schmidt's model (40) is the relative importance of hydrophobic interactions and tertiary calcium phosphate in binding the submicelles together. Both models view the submicelle as a strictly protein complex of variable composition, held together predominantly by interactions between the hydrophobic domains of the individual caseins. These models also suggest a nonuniform distribution of κ-casein in the submicelle. This nonuniform distribution could result from specific intermolecular interaction between β-sheet structures in the hydrophobic domains of κ-casein. The question of the relative contribution of hydrophobic interactions and salt bridge formation in cementing the submicelles together has been carefully considered by those who develop micelle models. In Slattery's view the submicelle is itself amphiphilic so that subsequent aggregation produces micelles, resembling a soap micelle, with a predominance of κ-casein's polar domain on the micelle surface. Presumably, the primary function of the salts is to neutralize the repulsive negative charges of the orthophosphate groups on α_{s1}-, α_{s2}-, and β-caseins, thus allowing this portion of the submicelle to interact by hydrophobic association. The numerous hydrophobic interactions, singly weak but collectively strong, constitute a force of great importance in stabilizing micelles. Moreover, this repression of long-range, electrostatic repulsive forces, enabling numerous hydrophobic groups to interact, is relevant not only to milk proteins, but also to many other macromolecular interactions important in food chemistry.

Alternatively, in Schmidt's model the entire surface of the submicelle is hydrophilic, allegedly composed of the polar domains of all the caseins. However, individual proteins may not be represented at the surface of the complete micelle in accord with their proportion in the total caseinate system. Polar domains of α_{s1}-, α_{s2}-, and β-casein contain orthophosphate groups that can form salt linkages via tertiary calcium phosphate, thereby binding the submicelles together. As a consequence, submicelles rich in these caseins tend to be on the inside of the complete micelle. Conversely, polar domains of κ-casein do not contain orthophosphate groups and cannot form salt bridges. Consequently, submicelles rich in κ-casein tend to be on the outer surface of the micelle.

Since micelles are stable at low temperatures, hydrophobic interactions (weakened as the temperature is lowered) alone cannot account for submicellar bonding.

C. The Fat Globule

To lessen the surface free energy for lipid in an aqueous medium, the lipid molecules associate to form large spherical globules. Most of the globules vary in diameter from

2 to 10 μm, so the largest fat globules are 30-fold larger than the largest casein micelles. During secretion through the apical plasma membrane, the globule acquires a coat of this membrane. Roughly 60% of the phospholipid and 85% of the cholesterol in milk are associated with fat globules, undoubtedly located within the membrane. Many of the enzymes with activities of importance to milk properties, such as alkaline phosphatase, xanthine oxidase, 5′-nucleotidase, sulfhydryl oxidase, and phosphodiesterase, are also associated with this membrane. The lipid composition of this membrane (Table 10) is similar to that of plasma membrane. However, the milk fat globule membrane apparently exists in a state of flux since its appearance and its composition change during the aging of milk.

The lipid core of fat globules consists almost exclusively of triacylglycerols. Prior to secretion, some components of the cellular cytoplasm become adsorbed to the droplet surface. Consequently, these components lie beneath the adsorbed plasma membrane and, perhaps, on patches of the fat globule surface not covered with plasma membrane. Thus, the surface of native fat globules probably does not consist entirely of plasma membrane.

Since fat globules have a lower density than the aqueous phase, they should rise at a rate proportional to the square of their radius. However, it is observed that creaming occurs much more rapidly than would be predicted from the size of individual fat globules. This occurs because protein components of whey, possibly immunoglobulins, associate with the fat globule, providing surface characteristics that favor globule aggregation. The large clusters, of course, rise much more rapidly than single globules.

V. USE OF MILK COMPONENTS AS FOOD INGREDIENTS

A. Effects of Processing on Milk Components

1. The Fat Phase

Milk fat is utilized in three major forms: (a) homogenized, as in normal whole milk, (b) concentrated, as in creams obtained by centrifugal separation, and (c) isolated, as in but-

Table 10 Composition of the Lipid Portion of the Milk Fat Globule Membrane

Component	Percentage of membrane lipids
Carotenoids	0.45
Squalene	0.61
Cholesterol esters	0.79
Triacylglycerols[a]	53.41
Free fatty acids	6.30
Cholesterol	5.20
Diacylglycerols	8.14
Monoacylglycerols	4.70
Phospholipids	20.40

[a]Contains a large portion of high-melting glycerides (melting point, 52-53°C).
Source: From Thompson et al. (47).

ter fat obtained by churning. An excellent review of the characteristics of the fat phase and products derived from it is available (4). Homogenization stabilizes the fat phase against gravity separation. Fat globules are reduced in size from 3-10 μm to less than 2 μm in diameter by forcing the liquefied fat through restricted passages under high pressure (~2500 psi) and at high velocity. The resulting new exposed fat surface area, representing a four- to sixfold increase, is stabilized by adsorption of milk proteins, including casein micelles and subunits. Gravity separation of homogenized milk is prevented not only by the reduced size of the globules but also by stabilization against cluster formation due to denaturation of immunoglobulins and interaction of the globule surface with casein. The size reduction and concomitant alteration in the lipid/water interface imparts a number of other characteristics to homoginized milk, some desirable and others undesirable. For example, homogenized milk is more susceptible to lipolysis and light-induced lipid oxidation. Thus, milk must be heated sufficiently to inactivate casein-associated lipases prior to homogenization. Also, homogenized milk exhibits reduced heat stability, has a greater foaming capacity, and yields protein gels with lower curd tension than does unhomogenized milk. However, homogenization increases the whiteness of milk, as would be expected, and gives milk a more bland flavor.

Cream can be obtained by centrifugal separation of cold (5-10°C) or warm milk. Cold separation is less disruptive to the fat globule membrane, and the resulting cream contains more immunoglobulins and is more viscous than cream separated at a higher temperature.

Clumping of fat globules, disruption of the fat globule membrane, and release of free fat occur from violent agitation during churning. Consequently, about 50% of the membrane is released into the buttermilk phase. Clumping and coalescence of fat globules during churning does not properly occur when the fat is either all solid or all liquid; thus, the temperature for churning must be carefully controlled to give an appropriate ratio of solid to liquid fat. The churning process involves destabilization of the oil-in-water emulsion to form a continuous fat phase, including some intact fat globules (representing 2-46% of the total fat), and 16% water as finely dispersed droplets.

2. The Protein Micelle and Milk Salt System

Conditions imposed by processing, such as temperature, pH, and concentration, alter equilibria within the salt system, within the protein system, and between salts and proteins. Often these changes are not completely reversible, so the final characteristics of the product depend on the processing conditions.

The milk serum phase is saturated with both calcium phosphate and calcium citrate; consequently, small changes in environmental conditions result in significant changes (35). Pasteurization (71.7°C for 15 sec) or sterilization (142-150°C for 3-6 sec) irreversibly increases the amount of colloidal calcium phosphate at the expense of both soluble and ionized calcium and soluble phosphate. As a result, the pH also decreases due to the release of protons from primary and secondary phosphate. However, the calcium transferred to tertiary calcium phosphate does not come entirely from the serum phase since heating also shifts the equilibrium for calcium binding to casein. Pateurization and especially sterilization also affect the size distribution of micelles, with an increase in abundance of both small and large micelles. These changes occur during even moderate increases in temperature and may not be completely reversible. Cooling to 0-4°C has the opposite effect, causing an increase in soluble calcium and phosphate and an increase in pH. As noted previously, β- and κ-casein also dissociate from the micelle upon cooling. However, these kinds of dissociations are largely reversible upon warming.

Pasteurization or UHT sterilization also cause denaturation of substantial fractions of the whey proteins, particularly β-lactoglobulin, which interacts with κ-casein. This interaction, which probably involves a thiol-disulfide interchange, alters the micellar surface properties. These changes affect the interaction of micelles with calcium phosphate, and their stability. For example, preheating unconcentrated milk at 90°C for 10 min reduces its subsequent stability to treatments at still higher temperatures. However, the same heat treatment, when applied to concentrated milk, nearly doubles the ability of the micelles to resist destabilization during subsequent processing at higher temperatures. In concentrated milk the destabilizing effect of micelle interaction with denatured serum proteins presumably is more than compensated for by heat-induced reduction of calcium ion concentration, which is much greater in concentrated milk than in unconcentrated milk. However, if unconcentrated milk is preheated at higher temperatures than just stated, the micelles become more stable to subsequent more severe heat processing, presumably due to formation of hydroxyapitite. Such phenomena illustrate the importance, complex nature, and interdependence of changes in milk salts and milk proteins during exposure of milk to various conditions.

Heat treatments of increasing severity are accompanied by increased production of dehydroalanyl residues, due to β-elimination of disulfide bonds and phosphoseryl residues, and increased Maillard browning. Reaction of dehydroalanyl residues with ε-amino groups to form lysinoalanine, or with thiol groups to form lanthionine, results in cross-linking of the proteins. Continued heating (e.g., 20-40 min at 140°C) destabilizes micelles, thus leading to gel formation. Micelle destabilization results from a combination of factors, the most important being a decrease in pH. The decrease in pH results from degradation of lactose, a shift in the equilibria of primary and secondary phosphates to hydroxyapitite, and hydrolysis of phosphoserine (10). The shift in phosphate equilibria with increasing temperature (as just stated) occurs rapidly, lowering the pH from about 6.7 to 5.5. Lactose degradation and phosphoserine hydrolysis continue more slowly, yielding an estimated pH of 4.9 at the point of milk coagulation. The colloidal calcium phosphate that forms upon heating is not the tertiary calcium phosphate-citrate that occurs in the native micelle; consequently, this change in salt equilibria is mostly irreversible.

It is interesting to note that, at pH 4.9 and ambient temperature, the original colloidal calcium phosphate is completely dissolved. Furthermore, the high temperature (140°C) and declining pH result in a progressive dissociation of the micelles. Thus, the coagulum that eventually occurs most likely consists of a matrix of interacting individual protein chains or submicelles and not whole micelles, as occurs in a rennet curd. Thiol-disulfide interchange and cross-linking with dehydroalanyl residues may contribute to the stability of such protein-chain interactions.

Addition of salts, acids, or alkalis to milk affects the calcium phosphate equilibrium and the binding of Ca^{2+} and tertiary calcium phosphate-citrate to the micelle, thus influencing micelle stability. For example, addition of secondary sodium phosphate stabilizes the micelle system by (a) increasing the pH and thus the micellar net charge, (b) reducing the soluble calcium concentration by shifting the equilibrium toward tertiary calcium phosphate, and (c) competitive displacement of micellar calcium by sodium. Consequently, this salt is frequently used as a milk stabilizer. Likewise, polyvalent anions, such as citrate or polyphosphates, which bind and increase micellar net charge and lower soluble calcium, are excellent milk stabilizers. Conversely, addition of calcium salts increases the ionized calcium concentration and the colloidal calcium phosphate, resulting in a lower pH and less stable micelles. Concentration of milk has a similar effect since the equilibrium shifts toward colloidal calcium phosphate and the pH falls; for example, 11-fold concentration of milk increases the soluble calcium and phosphate only 8-fold, and the Ca^{2+} only 6-fold, whereas the pH is lowered from 6.7 to 5.9.

These principles must be recognized in the manufacture of milk products and in the incorporation of milk into formulated foods. Because of the instability of micelles in concentrated milk, stabilizers are usually added prior to concentration. Also, milk is preheated prior to concentration or drying to shift the calcium phosphate equilibrium and thereby lessen the concentration of soluble calcium. On the other hand, the curd-setting properties of milk can be improved by adding calcium salts.

3. The Whey Proteins

Since denaturation of whey proteins occurs rapidly at temperatures above 70°C, normal commercial heat treatments denature at least a portion of these proteins. Immunoglobulins are the least heat stable of the whey proteins, with serum albumin, β-lactoglobulin, and α-lactalbumin showing increasingly greater stability. Denatured whey proteins, particularly β-lactoglobulin, are considerably less soluble and more sensitive to calcium ions than are their undenatured counterparts. Furthermore, denatured β-lactoglobulin complexes with κ-casein in the micelle, thus altering the stability of the micelle. If whey proteins have not been heat denatured they are quite soluble at acid pH, a characteristic that facilitates their incorporation into carbonated beverages.

4. Lactose

Reaction of the aldehyde group of lactose with the ε-amino group of lysine (onset of Maillard reactions) occurs even under very mild heat treatment, and the reaction continues slowly during storage. The degree of Maillard reaction and attendant browning is very sensitive to the severity of heat treatment, and the effects are desirable in some products and undesirable in others. Very severe heat treatments, such as those used for in-can sterilization of concentrated milks, causes lactose to partially degrade via Maillard reactions to yield organic acids, principally formic acid. This source of acidity is one of the major causes of protein destabilization in such products.

The very low solubility of the α-anomer of lactose results in its crystallization in frozen products. Crystallization of α-lactose in frozen milk is typically accompanied by prompt destabilization and precipitation of casein, the reasons for which are not fully understood. Soluble lactose may have a direct stabilizing effect on casein, and if so, this is lost with lactose crystallization. Lactose crystallization may also have an indirect effect on casein stability, since its removal from the unfrozen phase decreases solute concentration, resulting in additional ice formation. The effect is similar to that observed during concentration of milk. Thus, calcium ion concentration in the unfrozen phase increases and more tertiary calcium phosphate precipitates, yielding a decline in pH and casein instability.

A common defect in ice cream, known as sandiness, is caused by the formation of large α-lactose crystals. This defect has been largely overcome by the use of stabilizers of plant origin that impede formation of lactose nuclei.

B. Use in Formulated Foods

1. Functionality of Proteins

Proteins from traditional sources are being increasingly utilized as ingredients in a growing number of formulated foods. The benefits of milk proteins as ingredients in other foods stem from their excellent nutritional characteristics and their ability to contribute unique and essential functional properties to the final foods (28,29). Some essential characteristics of a high-quality protein ingredient are enumerated in Fig. 10. That milk proteins have evolved as the primary source of mammalian nutrition, at least for the very young, ensures excellent nutritional and flavor properties and freedom from antinutritional factors. The functional properties of a protein ingredient depend on its structural characteristics, which for the caseins are unique.

The functionality of some proteins can be improved by limited proteolysis. However, it is extremely important to carefully control the extent of the reaction to optimize desirable properties and prevent off-flavors. Such control can most readily be accomplished through the use of immobilized enzyme reactors.

2. The Caseins

Caseins exhibit excellent solubility and heat stability above pH 6. Due to their unique amphiphilic structure, these proteins also have very good emulsifying characteristics. The compositions of various casein preparations are given in Table 11. Since the properties of these products differ due to the different methods used for isolation, the particular product used in a food must be chosen according to the function required. Roughly 120,000 metric tons of casein products are produced annually world wide.

Acid (hydrochloric, lactic, and sulfuric) caseins are simply isoelectric precipitates (pH 4.6), and they are not very soluble. Likewise, rennet casein, which is not soluble in the presence of Ca^{2+}, does not contain amphiphilic κ-casein since the hydrophilic C-terminal macropeptide has been released. Rennet casein has a high ash content since colloidal calcium phosphate-citrate is included in the clotted micelles, whereas acid casein contains very little ash because the calcium phosphate has been released in the supernatant whey. Coprecipitates of casein and denatured whey proteins are more soluble than acid or rennet casein but are not as soluble as the caseinates. Solubilization of coprecipitates can be achieved by adjustment to alkaline pH and addition of polyphosphates. Because of their low solubility, these products are best suited for such products as breakfast cereals, snack and pasta products, and baked foods, to which they contribute texture and dough-forming characteristics.

The caseinates (sodium, potassium, and calcium) are prepared by neutralizing acid casein with the appropriate alkali prior to drying. These isolates, especially sodium and potassium, are very soluble and extremely heat stable over a wide range of conditions. Because of the detergentlike structure, these proteins, when used in products above pH 6, exhibit excellent emulsification, water binding, thickening, whipping-foaming, and gelling characteristics. Thus, these isolates have found wide acceptance as emulsifiers and water binders in formulated meat products, texturized vegetable protein products, margarine, toppings, cream substitutes, and coffee whiteners, and as foam stabilizers.

3. The Whey Proteins

Whey protein concentrates or isolates are very desirable as nutritional ingredients due to their high concentration of sulfur-containing amino acids. However, the conventional

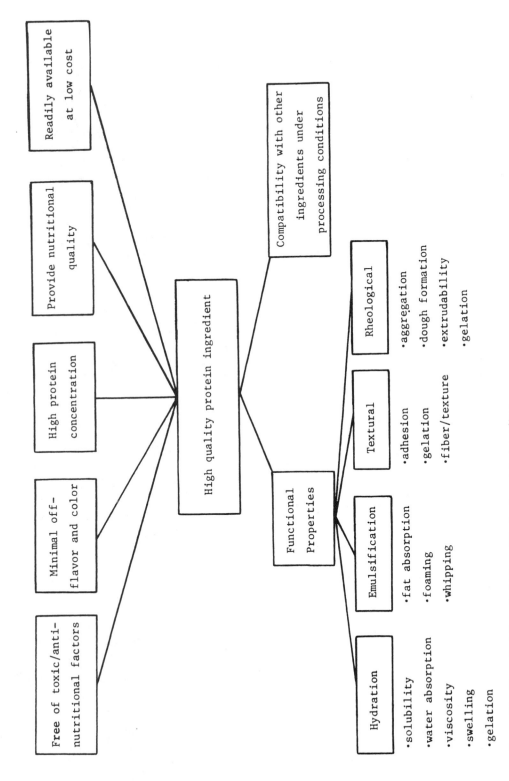

Figure 10 Characteristics of a high-quality food protein ingredient.

Table 11 Composition of Casein and Whey Protein Isolates Commercially Available as Protein Ingredients for Formulated Foods

Protein product	Component (%)			
	Protein	Ash	Lactose	Fat
Casein products				
Sodium caseinate	94	4.0	0.2	1.5
Calcium caseinate	93.5	4.5	0.2	1.5
Acid casein	95	2.2	0.2	1.5
Rennet casein	89	7.5	—	1.5
Coprecipitate[a]	89-94	4.5	1.5	1.5
Whey protein concentrates				
Ultrafiltrate	50-62	0.5-6	15-40	1.5-15
Gel filtrate	50-54	11-13	25-37	0.8-2.0
Metaphosphate complex	54-58	10-15	13	3.3-7.3
CMC complex[b]	50	8	20	1.2
Electrodialysate	27-37	1.4-2.0	40-80	2.4-4.3
Lactalbumin[c]	80-86	2.5	7	5

[a]Prepared by heating sufficiently to denature and complex whey proteins with casein followed by precipitation with acid or calcium ion.
[b]Carboxymethylcellulose, whey protein complex.
[c]Precipitate formed by heat denaturation of whey proteins at pH 4.5-5.2; this product should not be confused with α-lactalbumin.
Source: Adapted from Morr (29).

methods of isolation cause protein denaturation, and the resulting products (lactalbumin, protein-metaphosphate complex and protein-CMC complex) have functional properties that are not impressive. Until recently, less harsh methods for isolation, such as ultrafiltration, gel filtration, electrodialysis, and ion exchange with an inorganic matrix, were not economically feasible. However, some of these techniques have been developed to a point where they soon may be competitive. The compositions of some commercially available whey protein products are listed in Table 11. These products vary greatly in composition, extent of denaturation, and, consequently, in functional properties. Most applications require the protein ingredient to be highly soluble with low levels of lactose, milk salts, and other whey components.

Whey proteins, like soy proteins, lack an amphiphilic structure and are, therefore, generally inferior to casein with respect to emulsification properties. Lactalbumin, because of its degree of denaturation and low solubility, functions best in products requiring textural and rheological properties, such as meat products, ground meat extenders, processed cheese, pasta products, and snack foods. Whey protein concentrates, and especially isolates (which contain low lactose and high protein levels) that are not substantially denatured, exhibit much better hydration and emulsification properties than the more severely denatured whey protein products. For example, some isolates are effective foam stabilizers and form gels that compare favorably to those from egg white proteins. Because undenatured forms of whey proteins are soluble under acid conditions, they should have applications in acid-type food formulations, such as carbonated, fortified beverages.

4. Lactose

Uses of lactose are based on its low relative sweetness, protein stabilizing properties, crystallization habit, ability to accentuate flavor, nutritional attributes (inclusion of lactose in the diet improves utilization of calcium and other minerals), and ability to engage in Maillard browning (31). The relative sweetness of lactose is commonly reported as one-sixth that of sucrose; however, more recent studies indicate that its relative sweetness varies with concentration, ranging from one-half to one-fourth that of sucrose. Since sugars are often used to increase viscosity and/or mouth feel, or to improve texture, lactose can function in these capacities without being too sweet. Thus, lactose can replace a portion of the sucrose in toppings and icings to improve appearance and stability and to reduce sweetness. Unlike many other food sugars, addition of lactose does not reduce the solubility of sucrose. Furthermore, as the concentration of lactose is increased the crystal habits of both sucrose and lactose change—crystals of both become smaller and thereby producing a softer, smoother product. Hence, the quality of certain candy and confectionery products can be substantially improved by the addition of lactose.

One of the principal functions of lactose in baked goods is to improve crust color and toasting qualities via the Maillard browning reactions. When added to biscuits, it also increases volume and tenderness and improves texture. Moreover, in baked goods containing yeast, lactose is not utilized by the yeast and thus remains available for browning.

In common with other polyols, lactose stabilizes protein structure in solution. Thus, it has been used, for example, to reduce the insolubilization of lipovitellin during freeze-drying and to preserve the activity of such enzymes as rennin during spray-drying.

VI. NUTRITIVE VALUE OF MILK

A. Nutrient Composition of Milk and Milk Products

Dairy foods make a significant contribution to the total nutrient diet of the U.S. population. For example, Americans obtain one-fourth or more of their protein, calcium, phosphorus, and riboflavin from dairy products (Fig. 11). The nutrient composition of milk (Table 12) indicates that it is nearly a complete food. The composition listed, which applies to fresh raw or pasteurized whole milk, does not include vitamin D, since this nutrient is usually added to fluid milk products at a level of 41 IU per 100 g, which is equivalent to about 35% of the RDA (Recommended Dietary Allowance) per 250 ml. Thus, only iron, vitamin C, and folacin are present in somewhat deficient amounts.

Calculation of nutrient density, that is, the quantity of a nutrient relative to the food's caloric contribution, provides a meaningful method of comparing the nutritive quality of various foods (12,21). Choosing 2000 kcal as the basis for comparison, which is roughly the average requirement for individuals in the U.S. population, the percentages of the RDA contained in 2000 kcal portions of various milk products can be calculated, and these values are shown in Table 13. Milk is obviously an excellent source of protein, riboflavin, vitamin B_{12}, calcium, and phosphorus and an adequate source of vitamin A, thiamine, niacin equivalents, and magnesium. Removal of fat from milk removes most of the fat-soluble vitamins (A, D, and E), but the nutrient density of other nutrients is significantly increased. In the preparation of cheese, the water-soluble vitamins are significantly lowered due to their partial elimination in the whey. Calcium, however, is not reduced in renneted cheeses, as opposed to acid cheeses, since it remains in the clotted casein micelles. Although milk is usually not considered a good source of vitamin C, proper

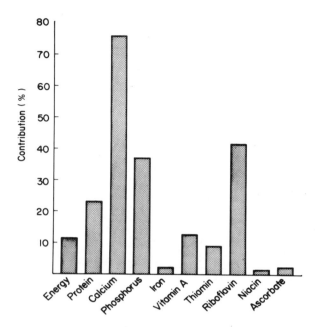

Figure 11 Contribution of dairy foods to total nutrient consumption. (From Ref. 45.)

processing can result in vitamin C being retained sufficiently so that milk contains a nearly adequate amount of this vitamin. In fact the vitamin C values given in Tables 12 and
13 may be too conservative, since some reports indicate that fresh raw milk contains 2
mg vitamin C per 100 g milk (37,38).

The nutritional quality of a protein is usually a more important consideration than
its quantity. Milk proteins correspond very well to human requirements and therefore are
regarded as of high quality. The biological value of protein in raw milk is 0.9, based on
a value of 1.0 for whole egg protein (38). Milk proteins are slightly deficient in the sulfur
amino acids, methionine and cysteine, causing the biological value to be slightly less than
ideal. Since sulfur amino acids are present in greater amounts in whey proteins than in
caseins, the former have a higher biological value (1.0) than the latter (0.8). It is also
noteworthy that caseins are readily digestible as a result of their structure.

B. Lactose Tolerance

A rather high percentage of the African and Asian population produce less intestinal
β-galactosidase (lactase) than do Europeans or North Americans. Consequently, lactose
malabsorption is encountered more frequently in the African and Asian population than
in other populations. The symptoms of lactose malabsorption are diarrhea, bloating, and
abdominal cramps (12). Lactase deficiency, however, is usually one of degree rather than
totality (11,12,21). Since the common lactose intolerance test is conducted by administering a very large dose of 1-2 g lactose per kg body weight, the reported frequency of
lactose intolerance does not adequately portray milk tolerance. Thus, intolerance to a
240 ml glass of milk is infrequently observed among individuals classified as lactose intol-

Table 12 Nutrient Composition of Whole Milk (3.3 % fat)[a]

Nutrient[b]	Amount in 100 g	% RDA[c] in 250 ml
Protein	3.29 g	17.2
Vitamin A	31 RE[d]	8.9
Vitamin C	0.94 mg	4.2
Thimaine	0.038 mg	8.2
Riboflavin	0.162 mg	30.0
Niacin	0.85 NE[e]	13.9
Vitamin B_6	0.042 mg	5.4
Folacin	5 μg	3.2
Vitamin B_{12}	0.357 μg	30.7
Calcium	119 mg	32.0
Phosphorus	93 mg	25.0
Magnesium	13 mg	10.2
Iron	0.05 mg	0.9
Zinc	0.38 mg	6.5

[a]Calculated from the recommended daily allowances (RDA) given by the National Academy of Sciences (36) and the composition given in USDA Handbook No. 8-1 (1).
[b]Includes all nutrients for which RDA are available, with the exception of vitamin D, vitamin E, and iodine, for which compositional data were not given in USDA Handbook No. 8-1.
[c]Average RDA for all males and females above age 11. A 250 ml volume is slightly more than 1 cup.
[d]Retinol equivalents: 1 μg retinol or 6 μg β-carotene.
[e]Niacin equivalents: 1 mg niacin or 60 mg dietary tryptophan. Only 10% of the NE in milk corresponds to niacin.

erant (11). Furthermore, in certain cases milk intolerance is apparently not caused by lactose intolerance. Nevertheless, a significant percentage of the population benefits from preconsumption hydrolysis of lactose in milk. The means for doing this industrially, using soluble or immobilized microbial lactase, have been developed. Since relatively few persons are intolerant to lactose in low doses, and since those who are can be provided with lactose-hydrolyzed milk, milk and milk products should be regarded as excellent foods for remedying dietary deficiencies on a worldwide basis.

C. Effect of Processing on Nutritive Value

1. Effect on Proteins

Heating is involved in most types of processing, including pasteurization, sterilization, concentration, and drying. Effects of processing can be divided into two categories: (a) those that alter secondary, tertiary, and quaternary structure of proteins, and (b) those that alter the primary structure of proteins. The former effects, which lead to protein unfolding, may actually improve the biological value of a protein since peptide bonds become more accessible to digestive enzymes. However, modification of the primary structure may lower digestibility and produce residues that are not biologically available. Heat treatment of milk can cause β-elimination of cystinyl and phosphoseryl

Table 13 Relative Nutritive Quality of Various Milk Products

Product	Protein	Vitamin A (RE)[b]	Vitamin C	Thiamine	Riboflavin	Niacin (NE)[c]	Vitamin B$_6$	Folacin	Vitamin B$_{12}$	Ca	P	Mg	Fe	Zn
Milk														
Whole (3.3% fat)	219	113	53	104	379	177	69	41	390	406	318	129	11	83
Skim	395	–[d]	97	171	571	321	114	71	720	732	601	190	15	152
Cheese														
Cheddar	251	167	0	11	133	170	18	22	137	373	265	42	23	103
Cottage (2% fat)	619	49	0	44	294	379	84	72	527	175	347	40	24	62
Yoghurt														
Plain, whole milk	231	109	30	79	333	85	52	57	407	413	324	119	11	129
Fruit variety[e]	174	24	22	60	249	63	39	44	305	310	243	89	9	97
Ice cream	73	110	9	32	175	60	23	5	155	136	104	42	6	70

[a]See Table 12 for the basis of calculation. Compositions of various products were taken from USDA Handbook No. 8-1 (1).
[b]Retinol equivalents: 1 μg retinol or 6 μg β-carotene.
[c]Niacin equivalents: 1 mg niacin or 60 mg dietary tryptophan.
[d]Vitamin A is often added to milk.
[e]A low-fat fruit variety containing 10 g protein per 8 oz. serving.

residues forming dehydroalanine. This substance readily reacts with lysyl residues to give lysinoalanine cross-links in the protein chain. Lysinoalanine is not biologically available, and the cross-linking lowers digestibility of the protein. Furthermore, since the nutritive value of milk proteins is limited by the low content of sulfur amino acids, such changes are particularly significant. Fortunately, pasteurization or UHT processing does not result in significant formation of lysinoalanyl residues; however, in-can sterilization or boiling does.

Even mild heat treatment initiates Maillard reactions, forming lactuloselysine and other compounds and thus reduces the amount of available lysine. Loss of available lysine is not significant for pasteurization (1-2%) or UHT sterilization (2-4%); however, more severe treatments, such as high-temperature evaporative concentration or in-can sterilization, can cause losses of more than 20% (37). Storage of UHT products for long periods at temperatures above 35°C can also significantly reduce available lysine. Since milk proteins contain an abundance of lysine, small losses are not nutritionally significant except in cases in which milk products are used to compensate for lysine-deficient diets.

2. Effect on Vitamins

The fat-soluble vitamins, A (also carotene), D, and E, and the water-soluble vitamins, riboflavin, pantothenic acid, biotin, and nicotinic acid, are quite stable (37). Accordingly, these vitamins do not sustain any detectable loss during pasteurization or UHT sterilization (Table 14). However, thiamine, B_6, B_{12}, folic acid, and ascorbate (vitamin C) are more susceptible to heat and/or oxidative degradation. Vitamins C, B_{12}, and folic acid are particularly susceptible to oxidative degradation during processing and subsequent storage. The oxidized form of vitamin C (dehydroascorbate) is very heat sensitive,

Table 14 Effects of Processing on Nutrients in Milk[a]

Nutrient	Pasteurization[b]	Ultra-high[c] temperature sterilization	Spray-drying[d]	Evaporated[e] milk
Vitamin A	0	0	0	0
Thiamine	10	10	10	40
Riboflavine	0	0	0	0
Nicotinic acid	0	0	0	5
Vitamin B_6	0	10	0	40
Vitamin B_{12}	10	10	30	80
Vitamin C	10-25	25	15	60
Folic acid	10	10	10	25
Pantothenic acid	0	0	0	0
Biotin	0	0	10	10

[a]The value listed in the table are the percentage loss resulting from processing.
[b]Heated at 71-73°C for 15 sec.
[c]Heated at 130-150°C for 1-4 sec.
[d]Pretreated by heating at 80-90°C for 10-15 sec, homogenized, and evaporated at reduced pressure. Spray-dried by exposing a fine mist of milk concentrate to air at 90°C for 4-6 sec.
[e]Pretreated by heating at 95°C for 10 min and concentrated by evaporation at 50°C under reduced pressure. In-can sterilization by autoclaving at 115°C for 15 min.
Source: Taken from *Handbook of Nutritive Value of Processed Food* (38).

whereas ascorbate is quite heat stable. Hence, methods that exclude or remove oxygen during processing, and packages that exclude oxygen during storage, serve to protect these vitamins.

In general, pasteurization and UHT sterilization of milk under proper conditions cause much less vitamin loss than those that occur during normal household food preparation. Furthermore, if the product is stored for long periods, the type of packaging and storage conditions are very important. In addition to exclusion of oxygen, the exclusion of light is also important, not only to protect against development of off-flavor, but also to prevent loss of riboflavin.

Severe heat treatments, such as in-can sterilization or drying, can cause significant losses of many of the vitamins. This is particularly true for vitamin B_{12}, of which milk is an important dietary source. Knowledge of the causes of milk degradation and the use of methods to minimize degradation, such as the use of high-temperature-short-time processing and exclusion of oxygen and light during processing and storage, should, in the future, lead to milk products that are shelf stable, microbiologically safe, and almost unchanged with respect to original nutrients.

REFERENCES

1. *Agriculture Handbook* No. 8-1 (1976). *Composition of Foods, Dairy and Egg Products*, United States Department of Agriculture, Agricultural Research Service.
2. Anonymous (1982). Dairy market trends. *Dairy Field 165*:39-57.
3. Bauman, D. E., and C. L. Davis (1974). Biosynthesis of milk fat, in *Lactation—A Comprehensive Treatise*, Vol. II (B. L. Larson and V. R. Smith, eds.), Academic Press, New York, pp. 31-75.
4. Brunner, J. R. (1965). Physical equilibria in milk: The lipid phase, in *Fundamentals of Dairy Chemistry* (B. H. Webb and A. H. Johnson, eds), AVI Publishing Co., Westport, Conn., pp. 403-505.
5. Brunner, J. R. (1981). Cow milk proteins: Twenty-five years of progress. *J. Dairy Sci. 64*:1038-1054.
6. Brunner, J. R., C. A. Ernstrom, R. A. Hollis, B. L. Larson, R. McL. Whitney, and C. A. Zittle (1960). Nomenclature of the proteins of bovine milk—first revision. *J. Dairy Sci. 43*:901-911.
7. Corbin, E. A., and E. O. Whittier (1965). The composition of milk, in *Fundamentals of Dairy Chemistry* (B. H. Webb and A. H. Johnson, eds.), AVI Publishing Co., Westport, Conn., pp. 1-36.
8. Creamer, L. K., G. P. Berry, and O. E. Mills (1977). A study of the dissociation of β-casein from the bovine casein micelle at low temperature. *N. Z. J. Dairy Sci. Technol. 12*:58-66.
9. Davies, D. T., and J. C. D. White (1960). The use of ultrafiltration and dialysis in isolating and in determining the partition of milk constituents between the aqueous and disperse phases. *J. Dairy Res. 27*:171-190.
10. Fox, P. F. (1981). Heat-induced changes in milk preceding coagulation. *J. Dairy Sci. 64*:2127-2137.
11. Garza, C. (1979). Appropriateness of milk use in international supplementary feeding programs. *J. Dairy Sci. 62*:1673-1684.
12. Hansen, R. G. (1974). Milk in human nutrition, in *Lactation—A Comprehensive Treatise*, Vol. III (B. L. Larson and V. R. Smith, eds.), Academic Press, New York, pp. 281-308.
13. Jenness, R. (1974). The composition of milk, in *Lactation—A Comprehensive Treatise*, Vol. III (B. L. Larson and V. R. Smith, eds.), Academic Press, New York, pp. 3-107.

14. Jenness, R., B. L. Larson, T. L. McMeekin, A. M. Swanson, C. H. Whitnah, and R. McL. Whitney (1956). Nomenclature of the proteins of bovine milk. *J. Dairy Sci. 39*:536-541.

15. Jensen, R. G., J. G. Quinn, D. L. Carpenter, and J. Sampugna (1967). Gas-liquid chromatographic analysis of milk fatty acids: A review. *J. Dairy Sci. 50*:119-125.

16. Kinsella, J. E., and J. P. Infante (1978). Phospholipid synthesis in the mammary gland, in *Lactation—A Comprehensive Treatise*, Vol. IV (B. L. Larson and V. R. Smith, eds.), Academic Press, New York, pp. 475-502.

17. Larson, B. L. (ed.) (1978). *Lactation—A Comprehensive Treatise*, Vol. IV, Academic Press, New York.

18. Larson, B. L., and V. R. Smith (eds.) (1974). *Lactation—A Comprehensive Treatise*, Vol. I, Academic Press, New York.

19. Larson, B. L., and V. R. Smith (eds.) (1974). *Lactation—A Comprehensive Treatise*, Vol. II, Academic Press, New York.

20. Larson, B. L., and V. R. Smith (eds.) (1974). *Lactation—A Comprehensive Treatise*, Vol. III, Academic Press, New York.

21. Leveille, G. A. (1979). United States food and nutrition issues—roles of dairy foods. *J. Dairy Sci. 62*:1665-1672.

22. Lindquist, B. (1963). Casein and the action of rennin, Part 1. *Dairy Sci. Abstr. 25*: 257-265.

23. Lindquist, B. (1963). Casein and the action of rennin, Part 2. *Dairy Sci. Abstr. 25*: 299-308.

24. Linzell, J. L., and M. Peaker (1971). Mechanism of milk secretion. *Physiol. Rev. 51*: 564-597.

25. McKenzie, H. A. (1967). Milk proteins. *Adv. Protein Chem. 22*:55-234.

26. McKenzie, H. A. (ed.) (1970). *Milk Proteins*. Vol. I, Academic Press, New York.

27. McKenzie, H. A. (ed.) (1970). *Milk Proteins*, Vol. II, Academic Press, New York.

28. Morr, C. V. (1979). Utilization of milk proteins as starting materials for other foodstuffs. *J. Dairy Res. 46*:369-376.

29. Morr, C. V. (1982). Functional properties of milk proteins and their use as food ingredients, in *Developments in Dairy Chemistry. I. Proteins* (P. F. Fox, ed.), Applied Science Publishers, London, pp. 375-399.

30. Nickerson, T. A. (1965). Lactose, in *Fundamentals of Dairy Chemistry*, (B. H. Webb and A. H. Johnson, eds.), AVI Publishing Co., Westport, Conn., pp. 224-260.

31. Nickerson, T. A. (1976). Use of milk derivative, lactose, in other foods. *J. Dairy Sci. 59*:581-587.

32. Patton, S., and R. G. Jensen (1976). *Biomedical Aspects of Lactation*, Pergamon Press, New York.

33. Payens, T. A. J. (1979). Casein micelles: The colloid-chemical approach. *J. Dairy Res. 46*:291-306.

34. Pitas, R. E., J. Sampugna, and R. G. Jensen (1967). Triglyceride structure of cow's milk fat. I. Preliminary observations on the fatty acid composition of positions 1, 2, and 3. *J. Dairy Sci. 50*:1332-1336.

35. Pyne, G. T. (1962). Some aspects of the physical chemistry of the salts of milk. *J. Dairy Res. 29*:101-130.

36. *Recommended Dietary Allowances* (1980). Ninth Revised Edition, National Academy of Sciences, Washington, D.C.

37. Renner, E. (1980). Nutritional and biochemical characteristics of UHT milk, in *Proceedings, International Conference on UHT Processing and Aseptic Packaging of Milk and Milk Products*, Department of Food Science, North Carolina State University, Raleigh, pp. 21-52.

38. Rolls, B. A. (1982). Effect of processing on nutritive value of food: *Milk and Milk*

Products, in *Handbook of Nutritive Value of Processed Food,* Vol. I (M. Rechcigl, ed.), CRC Press, Boca Raton, pp. 383-399.

39. Rose, D., J. R. Brunner, E. B. Kalan, B.L. Larson, P. Melnychyn, H. E. Swaisgood, and D. F. Waugh (1970). Nomenclature of the proteins of cow's milk: Third revision. *J. Dairy Sci. 53*:1-17.

40. Schmidt, D. G. (1980). Colloidal aspects of casein. *Neth. Milk Dairy J. 34*:42-64.

41. Shahani, K. M., W. J. Harper, R. G. Jensen, R. M. Parry, Jr., and C. A. Zittle (1973). Enzymes in bovine milk: A review. *J. Dairy Sci. 56*:531-543.

42. Slattery, C. W. (1976). Review: Casein micelle structure; an examination of models. *J. Dairy Sci. 59*:1547-1556.

43. Slattery, C. W. (1979). A phosphate-induced sub-micelle equilibrium in reconstituted casein micelle systems. *J. Dairy Res. 46*:253-258.

44. Slattery, C. W., and R. Evard (1973). A model for the formation and structure of casein micelles from subunits of variable composition. *Biochim. Biophys. Acta 317*:529-538.

45. Swaisgood, H. E. (1973). The caseins. *CRC Crit. Rev. Food Technol. 3*:375-414.

46. Swaisgood, H. E. (1982). Chemistry of milk protein, in *Developments in Dairy Chemistry. I. Proteins* (P. F. Fox, ed.), Applied Science Publishers, London, pp. 1-59.

47. Thompson, M. P., J. R. Brunner, C. M. Stine, and K. Lindquist (1961). Lipid components of the fat-globule membrane. *J. Dairy Sci. 44*:1589-1596.

48. Thompson, M. P., N. P. Tarassuk, R. Jenness, H. A. Lillevik, U. S. Ashworth, and D. Rose (1965). Nomenclature of the proteins of cow's milk—second revision. *J. Dairy Sci. 48*:159-169.

49. White, J. C. D., and D. T. Davies (1958). The relation between the chemical composition of milk and the stability of the caseinate complex. *J. Dairy Res. 25*:236-255.

50. Whitney, R. McL. (1977). Milk proteins, in *Food Colloids* (H. D. Graham, ed.), AVI Publishing Co., Westport, Conn., pp. 66-151.

51. Whitney, R. McL., J. R. Brunner, K. E. Ebner, H. M. Farrell, R. V. Josephson, C. V. Morr, and H. E. Swaisgood (1976). Nomenclature of the proteins of cow's milk: Fourth revision. *J. Dairy Sci. 59*:785-815.

BIBLIOGRAPHY

Brunner, J. R. (1965). Physical equilibria in milk: The lipid phase, in *Fundamentals of Dairy Chemistry* (B. H. Webb and A. H. Johnson, eds.), AVI Publishing Co., Westport, Conn., pp. 403-505.

Brunner, J. R. (1981). Cow milk proteins: Twenty-five year of progress. *J. Dairy Sci. 64*:1038-1054.

Larson, B. L. (ed.) (1978). *Lactation—A Comprehensive Treatise,* Vol. IV, Academic Press, New York.

McKenzie, H. A. (ed.) (1970). *Milk Proteins,* Vol. I, Academic Press, New York.

McKenzie, H. A. (ed.) (1970). *Milk Proteins,* Vol. II, Academic Press, New York.

Morr, C. V. (1982). Functional properties of milk proteins and their use as food ingredients, in *Developments in Dairy Chemistry. I. Proteins* (P. F. Fox, ed.), Applied Science Publishers, London, pp. 375-399.

Patton, S., and R. G. Jensen (1976). *Biomedical Aspects of Lactation,* Pergamon Press, New York.

Swaisgood, H. E. (1973). The caseins. *CRC Crit. Rev. Food Technol. 3*:375-414.

Swaisgood, H. E. (1982). Chemistry of milk protein, in *Developments in Dairy Chemistry. I. Proteins* (P. F. Fox, ed.), Applied Science Publishers, London, pp. 1-59.

Whitney, R. McL. (1977). Milk proteins, in *Food Colloids* (H. D. Graham, ed.), AVI Publishing Co., Westport, Conn., pp. 66-151.

14

CHARACTERISTICS OF EDIBLE FLUIDS OF ANIMAL ORIGIN: EGGS

William D. Powrie and Shuryo Nakai University of British Columbia, Vancouver, British Columbia, Canada

I.	Introduction	830
II.	Structure and Composition of the Shell and Shell Membranes	830
	A. Egg Shell	830
	B. Shell Membranes	831
III.	Composition of Albumen and Yolk	832
	A. Albumen	832
	B. Yolk	832
IV.	Proteins in Albumen	833
	A. Ovalbumin	834
	B. Conalbumin (Ovotransferrin)	834
	C. Ovomucoid	836
	D. Lysozyme	836
	E. Ovomucin	836
	F. Other Albumen Proteins	837
V.	Microstructure of Yolk Particles	837
VI.	Proteins and Lipoproteins in Granules and Plasma	838
	A. Proteins and Lipoproteins in Granules	841
	B. Proteins and Lipoproteins in Plasma	843
VII.	Chemistry of Deterioration of Shell Eggs During Storage	844
VIII.	Chemical Changes During Processing of Liquid Egg Products	845
	A. Pasteurization of Egg Products	845
	B. Freezing of Yolk	846
	C. Spray-Dried Egg Products	847
IX.	Functional Properties of Whole Egg Magma, Albumen, and Yolk	847
	A. Emulsifying Power of Egg Yolk	848
	B. Foaming Power of Albumen	848
	C. Heat Coagulability of Albumen, Yolk, and Egg Magma	850

References   851
Bibliography   854

I. INTRODUCTION

Infertile eggs from hens classified as *Gallus domesticus* are used almost exclusively for human consumption. Liquid whole egg (egg magma), yolk, and albumen (egg white) are excellent sources of some nutrients, and they possess valuable functional properties. A review on the nutritional value of eggs has been presented by Cook and Briggs (19), and a summary of the functional properties has been prepared by Baldwin (4). The average weight of shell eggs produced by common selectively-bred strains of hens is about 58 g. Shell eggs consist of 8-11% shell, 56-61% albumen, and 27-32% yolk. According to Cotterill and Geiger (22), the solids content of liquid whole egg decreased from 24.7 to 24.1% over the 1966-1975 period. Commercially prepared liquid yolk contains about 15-20% albumen since albumen adheres to the vitelline membrane of separated intact yolk.

Egg products are commonly marketed in the following forms: (a) refrigerated, (b) frozen, or (c) spray-dried. In compliance with U.S. FDA regulations, all egg products must be pasteurized to destroy viable *Salmonella* organisms. For refrigerated and frozen egg products, pasteurization is carried out in a heat exchanger prior to cooling or freezing. In the case of spray-dried albumen, the liquid product may be pasteurized prior to drying or the dried product may be stored at a high temperature for *Salmonella* inactivation. The addition of 10% sucrose or 10% NaCl to commercial frozen yolk is common practice to prevent gelation (viscosity increase). Frozen salted yolk is used for the manufacture of mayonnaise and salad dressing, whereas frozen sugared yolk is employed for bakery products and ice cream. Frozen plain yolk without added sugar or salt is used in noodles and baby foods. Prior to spray drying, albumen is usually desugared by a glucose oxidase treatment to obviate nonenzymic browning in the dry albumen during storage.

Reviews on the chemistry of egg components have been written by Baker (3), Brooks and Taylor (9), Cook (20), Feeney (27), Parkinson (55), Powrie (57), and Vadehra and Nath (74).

II. STRUCTURE AND COMPOSITION OF THE SHELL AND SHELL MEMBRANES

A. Egg Shell

The egg shell is composed of (a) a matrix of interwoven protein fibers; and (b) interstitial calcite (calcium carbonate) crystals, in a proportion of about 1:50. In addition, the surface of the shell is covered by a cuticle, a foamy protective layer of protein. Figure 1 is a diagram of a radial section of the shell (67). The matrix consists of two regions, the mammillary knob and the spongy matrix. Microscopic studies have shown that the matrix fibers pass through calcite instead of simply surrounding them. Undoubtedly the matrix has a significant influence on shell strength. The matrix is made up of protein-mucopolysaccharide complexes. The polysaccharides consist of chondroitin sulfates A and B, galactosamine, glucosamine, galactose, mannose, fucose, and sialic acid.

The elemental composition of the egg shell is 98.2% calcium, 0.9% magnesium, and 0.9% phosphorus (present in the shell as phosphate) (63). Shell hardness increases with an increase in magnesium content (7).

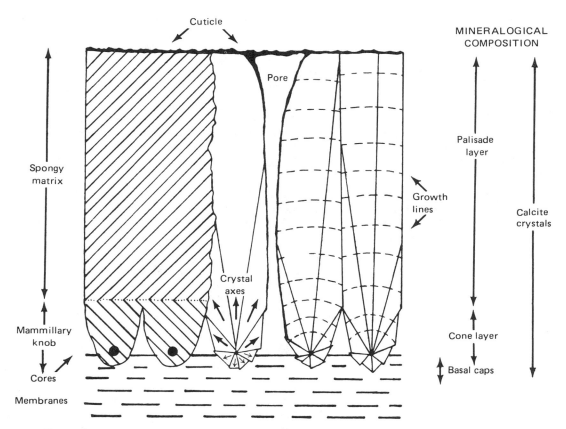

Figure 1 Schematic cross section of an egg shell.

Numerous funnel-shaped pore canals (7000-17,000 per egg) are distributed at right angles to the shell surface and form passages between the shell membrane and the cuticle. The pore canals are filled with protein fibers that retard microbial entry into the egg.

The water-insoluble protein cuticle forms a protective coating (about 10-30 μm thick) on the entire surface of the egg. The cuticle is composed primarily of mucoproteins with polysaccharides representing about 5% of the total organic matter. The polysaccharide constituents are mainly glucose, galactose, mannose, and fucose.

B. Shell Membranes

Outer and inner membranes with thicknesses of 48 and 22 μm, respectively, reside between the inner surface of the shell and the albumen. Mammillary cores are discrete bodies embedded in the outer shell membrane. The membranes are composed of protein-polysaccharide fibers. Galactosamine, glucosamine, sialic acid, glucose, mannose, and fucose are present in both membranes (21). Mammillary cores, the initiation centers of calcite crystals, are rich in hexosamine, sialic acid, and hexose. The outer shell membrane is attached firmly to the shell by numerous cones on the inner shell surface and by fiber associations. The outer membrane has six layers of fibers oriented alternately in different directions, whereas the inner membrane has three layers of fibers parallel to the shell and at right angles to each other.

III. COMPOSITION OF ALBUMEN AND YOLK

A. Albumen

Albumen is made up of four distinct layers: outer fluid albumen (thin), viscous albumen (thick), inner fluid albumen (thin), and a chalaziferous layer. The proportions and moisture contents of the layers are presented in Table 1. The wide range in the proportion of layers has been attributed to breed, environmental conditions, size of egg, and rate of egg production (63).

The major constituent of the albumen layers is water, which decreases somewhat from the outer to the inner albumen layers (Table 1). The water content of mixed albumen ranges from 87 to 89% and is dependent on the strain and age of the hens (57).

The proximate analysis of albumen is listed in Table 2. Protein is the major constituent of albumen solids. The variability of protein content (9.7-10.6%) in albumen can be attributed primarily to age of the bird. The amount of lipid in albumen (about 0.3%) is negligible compared with that of yolk. The carbohydrates of albumen exist in combined (with protein) and free forms and can reach a level of about 1%. Approximately 98% of the free carbohydrates (about 0.5%) of the albumen is glucose. The elemental composition of the ash of albumen is presented in Table 3, and the total amount does not vary appreciably. The predominant cations are potassium and sodium.

B. Yolk

The solids content of yolk is about 50%. Yolk from fresh eggs has a solids content of 52-53%, but yolk solids decrease about 2% when eggs are stored at refrigerated temperatures for 1-2 weeks because water migrates into the yolk from the albumen. Proteins and lipids are the major constituents of yolk, with minor amounts of carbohydrates and minerals also present (Table 2). The protein content of yolk is about 16%, with limited variability. The amount of lipid varies between about 32 and 35% and this variation can be attributed to the strain of bird rather than to diet. The yolk lipid fraction contains approximately 66% triacylglycerol, 28% phospholipid, 5% cholesterol, and minor amounts of other lipids. Rhodes and Lea (61) estimated the composition of yolk phospholipids as 73.0% phosphatidylcholine, 15.5% phosphatidylethanolamine, 5.8% lysophosphatidylcholine,

Table 1 Proportion and Moisture Content of Albumen Layers

Layer	Total albumen layers (%)		Moisture (%)
	Mean	Range	
Outer thin white	23.2	10-60	88.8
Thick white	57.3	30-80	87.6
Inner thin white	16.8	1-40	86.4
Chalaziferous (including chalazae)	2.7		84.3

Table 2   Composition of Albumen, Yolk, and Whole Egg (Wet Basis)

Egg component	Solids content (%)	Protein (%)	Lipid (%)	Carbohydrate (%)	Ash (%)
Albumen	11.1	9.7-10.6	0.03	0.4-0.9	0.5-0.6
Yolk	52.3-53.5	15.7-16.6	31.8-35.5	0.2-1.0	1.1
Whole egg	25-26.5	12.8-13.4	10.5-11.8	0.3-1.0	0.8-1.0

2.5% sphingomyelin, 2.1% lysophosphatidylethanolamine, 0.9% plasmalogen, and 0.6% inositol phospholipid.

The fatty acid composition of yolk lipid is influenced by the types of fatty acids in the diet of the hen. The total amount of saturated fatty acids, primarily palmitic and stearic, does not change with an alteration of dietary fatty acid composition, but linoleic acid increases and oleic acid decreases when the level of dietary polyunsaturated fatty acids is elevated.

The fatty acid content of yolk lipid fractions is listed in Table 4 (58). Palmitic and stearic acids amount to about 30% of the total fatty acids in triacylglycerols, whereas they amount to about 49% in phosphatidylcholine (lecithin) and 54% in phosphatidylethanolamine (cephalin).

The total amount of free and combined carbohydrates in yolk (0.2-1.0%) is similar to that found in albumen (Table 2). The content of free carbohydrates has been estimated to be 0.2%. Protein-bound carbohydrates are mannose-glucosamine polysaccharides.

As shown in Table 3, the major elements in yolk ash (total ash about 1.1%) are calcium, potassium, and phosphorus.

IV. PROTEINS IN ALBUMEN

Albumen can be considered a protein system consisting of ovomucin fibers in an aqueous solution of numerous globular proteins. The protein composition of the thin and thick layers of albumen differs only in the ovomucin content, with the ovomucin content of the thick layer four times greater than that of the thin layer. Protein fractions can be separated by stepwise addition of ammonium sulfate and by ion-exchange chromatography. Several types of electrophoretic techniques also have been used to separate albumen pro-

Table 3   Elemental Composition of Albumen and Yolk (Wet Basis)

Element	Percentage in albumen	Percentage in yolk
Sulfur	0.195	0.016
Potassium	0.145-0.167	0.112-0.360
Sodium	0.161-0.169	0.070-0.093
Phosphorus	0.018	0.543-0.980
Calcium	0.008-0.02	0.121-0.262
Magnesium	0.009	0.032-0.128
Iron	0.0009	0.0053-0.011

Table 4   Fatty Acid Composition of Lipid Fractions of Yolk and Dietary Lipid

Fatty acid	Total fatty acids (%)				
	Crude lipid	Triacylglycerol	Lecithin	Cephalin	Dietary lipid[a]
16:0	23.5	22.5	37.0	21.6	14.0
16:1	3.8	7.3	0.6	trace	2.7
18:0	14.0	7.5	12.4	32.5	2.4
18:1	38.4	44.7	31.4	17.3	29.1
18:2	16.4	15.4	12.0	7.0	44.4
18:3	1.4	1.3	1.0	2.0	3.2
20:4	1.3	0.5	2.7	10.2	0.8
20:5	0.4	0.2	0.8	3.0	0.8
22:5					
22:6	0.8	0.6	2.1	6.4	1.3

[a]Of hen.

teins. Seven major fractions, including ovalbumins A_1 and A_2, globulins G_1, G_2, and G_3, ovomucoid, and conalbumin, have been obtained with moving-boundary electrophoresis (28). With polyacrylamide disk-gel electrophoresis, Chang et al. (13) were able to separate albumen into 12 protein bands, whereas 19 protein bands were obtained by several investigators using starch-gel electrophoresis. Variation in the number of fractions of proteins obtained from albumen of various strains and inbred lines has been attributed to genetic polymorphism. The relative amounts of various proteins in albumen and some of their properties are presented in Table 5 (3).

A. Ovalbumin

Ovalbumin, the major protein fraction in albumen, is classed as a phosphoglycoprotein since phosphate and carbohydrate moieties are attached to the polypeptide chain. Purified ovalbumin is made up of three components, A_1, A_2, and A_3, which differ only in phosphorus content. The A_1, A_2, and A_3 components have two, one, and no phosphate, respectively, existing in a relative proportion of 85:12:3. The polypeptide chain of ovalbumin has 385 amino acid residues, 4 sulfhydryl groups, 1 disulfide group, and a molecular weight of 42,699 (53). The carbohydrate moiety is located at Asn-292, and the two phosphorylated serines are residues 68 and 344. The single carbohydrate moiety consists of a core of two N-acetylglucosamine and four mannose units with variable numbers of additional residues of the same sugars. The carbohydrate moiety of ovalbumin has a molecular weight between 1560 and 1580.

During storage of eggs, ovalbumin is converted to S-ovalbumin, a more heat-stable protein, conceivably through sulfhydryl-disulfide interchange (68,69).

Ovalbumin in solution is readily denatured and coagulated by exposure to new surfaces but is resistant to thermal denaturation. For example, when albumen at pH 9 is heated for 3.5 min at 62°C, only 3-5% of the ovalbumin is altered significantly (41).

B. Conalbumin (Ovotransferrin)

Conalbumin and ovotransferrin are synonymous terms. Conalbumin, prepared by fractional precipitation of albumen with ammonium sulfate, contains neither phosphorus nor sulfhydryl groups. About 0.8% hexose and 1.4% hexosamine are present in this protein.

Table 5 Proteins in Egg Albumen

Protein	Albumen (%, dry basis)	pI[a]	Molecular weight	Amino acid residues[b]	HΦ[c] (cal/res)	T_d[d] (°C)	Characteristics
Ovalbumin	54	4.5	44,500	385	1110	84.0	Phosphoglycoprotein
Ovotransferrin	12	6.1	76,000	–	1080	61.0	Binds metalic ions
Ocomucoid	11	4.1	28,000	185	920	70.0	Inhibits trypsin
Ovomucin	3.5	4.5–5.0	$5.5–8.3 \times 10^6$	–	–	–	Sialoprotein, viscous
Lysozyme	3.4	10.7	14,300	129	970	75.0	Lyses some bacteria
G_2 Globulin	4.0?	5.5	$3.0–4.5 \times 10^4$	–	–	92.5	–
G_3 Globulin	4.0?	4.8	–	–	–	–	–
Ovoinhibitor	1.5	5.1	49,000	–	–	–	Inhibits serine proteases
Ficin inhibitor	0.05	5.1	12,700	–	–	–	Inhibits thioproteases
Ovoglycoprotein	1.0	3.9	24,400	–	–	–	Sialoprotein
Ovoflavoprotein	0.8	4.0	32,000	–	–	–	Binds riboflavin
Ovomacroglobulin	0.5	4.5	$7.6–9.0 \times 10^5$	–	–	–	Strongly antigenic
Avidin	0.05	10	68,300	512	1060	–	Binds biotin

[a]Isoelectric point.
[b]Number of residues in the published sequence.
[c]Average hydrophobicity.
[d]Denaturation temperature.

Conalbumin consists of a single polypeptide chain and can exist in equilibrium as three forms with different Fe^{3+} contents (zero, one, or two Fe^{3+} atoms per molecule). It possesses a single oligosaccharide unit composed of four mannose residues and eight *N*-acetylglucosamine residues. Two atoms of Fe^{3+}, Al^{3+}, Cu^{2+}, or Zn^{2+} per molecule of conalbumin form a heat-stable complex at pH 6. These complexes are red, colorless, yellow, and colorless, respectively.

Considerable effort has been made to identify the ligands involved in the binding of metal ions and anions. Tyrosylphenolate groups have been implicated in metal binding, and imidazole side chains of histidine residues have been implicated in the binding of both metal ions and anions.

Conalbumin is more heat sensitive but less susceptible to surface denaturation than ovalbumin. Cunningham and Lineweaver (24) showed that 1% conalbumin in phosphate-bicarbonate buffer had a minimum heat stability near pH 6, at which value 40% of the conalbumin was altered during heating for 10 min at 57°C. Conalbumin at pH 9 was not altered significantly when the solution was heated under the same conditions. The heat stability of conalbumin in albumen is similar to that of conalbumin in buffer.

C. Ovomucoid

Ovomucoid is a glycoprotein consisting of three separate domains, each cross-linked by three disulfide bonds. The polypeptide is composed of 26% α-helix, 46% β-structure, 10% β-turn, and 18% random coil structure (78). Ovomucoid contains about 20-25% carbohydrate, which consists of D-mannose, D-galactose, glucosamine, and sialic acid. The carbohydrate is present as three polysaccharide chains, each attached to the polypeptide through an asparaginyl residue.

Ovomucoid in acidic and in moderately alkaline media is very resistant to heat coagulation, but in the presence of lysozyme in alkaline solution, the protein coagulates at temperatures above 60°C (48).

D. Lysozyme

Lysozyme, an enzyme able to lyse bacterial cell walls, consists of three components that can be separated by cation-exchange chromatography. Its isoelectric point of 10.7 is much higher than that of other albumen proteins, and its molecular weight (14,600) is the lowest. Lysozyme consists of 129 amino acid residues, and its primary structure has been successfully elucidated. Four disulfide linkages but no free-SH groups are present on each polypeptide molecule.

Thermal inactivation of lysozyme is dependent on pH and temperature. Lysozyme is much more heat sensitive in egg albumen than when present alone in a phosphate buffer between pH 7 and 9. When egg albumen is heated for 10 min at 63.5°C, lysozyme is inactivated to a greater degree as the pH is increased above 7.

E. Ovomucin

Ovomucin is a glycoprotein that contributes to the gelatinous structure of thick albumen. This protein is not soluble in water but can be solubilized in a dilute salt solution at pH 7 or above. Ovomucin consists of three components that can be separated by moving boundary electrophoresis. The carbohydrate content of the purified protein is around

30%, with 10-12% as hexosamine, 15% as hexose, and 3-8% as sialic acid. Ovomucin has been separated into a carbohydrate-rich fraction (50%) and a carbohydrate-poor fraction (15%). These fractions have been named β- and α-ovomucin, respectively (62). Two types of carbohydrate polymers have been reported to exist in association with ovomucin (70).

Purified ovomucin in solution is resistant to heat denaturation. Ovomucin solutions with pH values between 7.1 and 9.4 do not change in absorbance during heating for 2 hr at about 99°C (24).

Ovomucin and lysozyme can form a water-soluble complex through electrostatic interaction over the pH range 7.2-10.4. Interaction decreases as the pH is elevated within this range. Presumably, the ovomucin-lysozyme complex contributes to the gelatinous structure of thick albumen, but contradictory hypotheses have been presented. Cotterill and Winter (23) have postulated that the thinning of thick albumen during egg storage is caused by a reduction in the amount of naturally occurring ovomucin-lysozyme complex as the pH rises to 9.5.

F. Other Albumen Proteins

Avidin, a biotin-binding protein, is a basic glycoprotein (pI, 10) consisting of 128 amino acid residues. Ovoglobulins G_2 and G_3 are excellent foaming agents. Ovoinhibitor is capable of inhibiting trypsin and chymotroypsin. Flavoprotein is made up of an apoprotein that strongly binds riboflavin, but negative nutritional consequences have not been observed.

V. MICROSTRUCTURE OF YOLK PARTICLES

Yolk can be classed as a dispersion containing a variety of particles distributed uniformly in a protein (livetin) solution. The types of particles are (a) yolk spheres, (b) granules, (c) low-density lipoprotein, and (d) myelin figures. The electron micrograph (Fig. 2) provides a view of a variety of particles in yolk.

Chang et al. (16) found numerous spheres with diameters of about 20 μm in white yolk, but only a few spheres (about 40 μm in diameter) were observed in yellow yolk (about 99% of the total yolk). Bellairs (5) noted three types of sphere surfaces: (a) lamellated capsules (several layers of membrane), (b) unit membranelike structures, and (c) naked surfaces. The majority of spheres have naked surfaces.

Granules are smaller and more numerous than yolk spheres. Although granule diameters can be as large as 1.7 μm, the majority of granules have diameters between about 1.0 and 1.3 μm (16). Electron micrographs (Fig. 3) indicate that granules are made up of electron-dense subunits that can be dispersed by diluting yolk with 0.34 M NaCl (Fig. 4). At higher salt concentrations, the native structure of the subunits is disrupted with the appearance of small and large globules (30-100 nm), myelin figures (56-130 nm), and electron-dense fluffy masses made up of entangled strands to which electron-dense microparticles are attached (Fig. 5). The microparticles, which are roughly circular, have average diameters of about 10 nm and are considered high-density lipoproteins.

Round profiles (diameters of 25 nm) were noted in electron micrographs of yolk. These profiles are considered low-density lipoproteins that exist as individual particles uniformly distributed throughout the yolk.

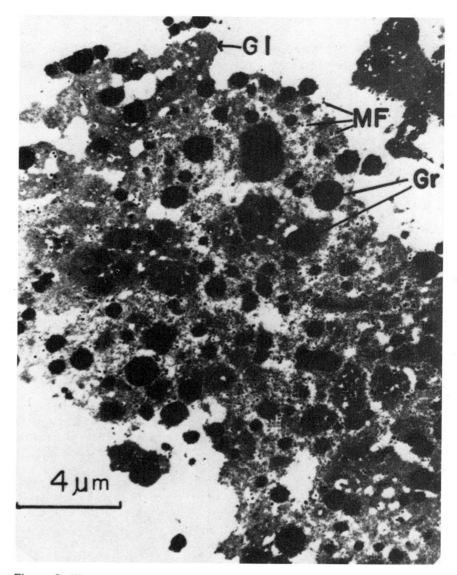

Figure 2 Electron micrograph of yolk composed of globules (GI), granules (Gr), and myelin figures (MF).

VI. PROTEINS AND LIPOPROTEINS IN GRANULES AND PLASMA

By high-speed centrifugation, yolk can be separated into sedimented granules and a clear supernatant, called plasma. The granules make up about 19-23% of the yolk solids. Granules on a moisture-free basis contain about 35% lipid, 60% protein, and 6% ash. About 37% of the total lipid consists of phospholipids, mainly phosphatidylcholine (82%) and phosphatidylethanolamine (15%).

After granules are disrupted in a 1.71 M NaCl solution, two floating and three sedimenting fractions form during ultracentrifugation. The floating fractions, referred to as pellicle FI and subpellicle FII, contain most of the low-density components. FI consists

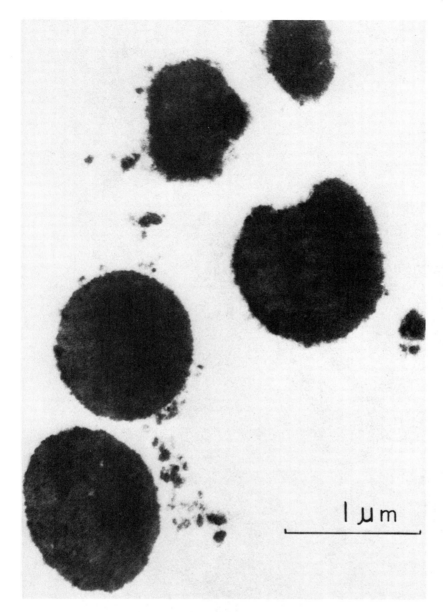

Figure 3 Electron micrograph of yolk granules.

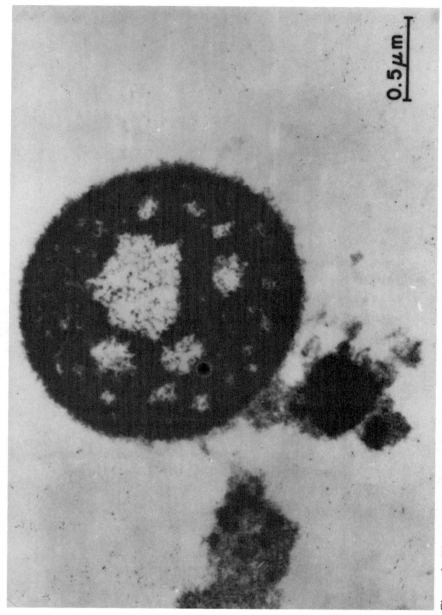

Figure 4 Electron micrograph of yolk diluted with 0.34 M NaCl.

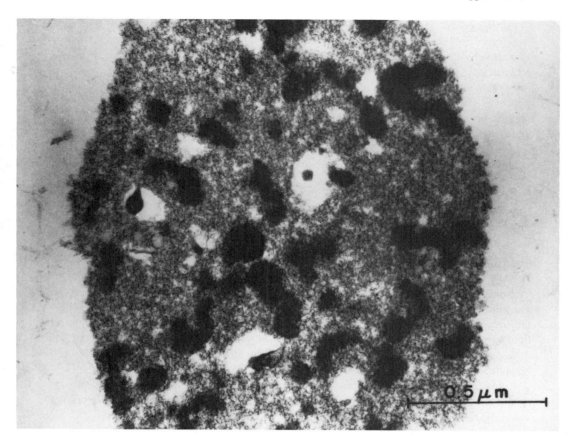

Figure 5   Electron micrograph of yolk diluted with 1.71 M NaCl.

of small and large globules (30-100 nm) and myelin figures (MF), whereas FII is particularly rich in MF and low-density lipoproteins (LDL) with diameters around 27 nm.

Plasma represents about 78% of liquid yolk and has a moisture content of about 49%. On a dry weight basis, plasma consists of 77-81% lipid, 2% ash, and about 18% protein.

A. Proteins and Lipoproteins in Granules

Granules are composed of 70% α- and β-lipovitellins, 16% phosvitin, and 12% low-density lipoprotein (11). Lipovitellins and phosvitin have an affinity for each other, and a phosvitin-lipovitellin complex is probably the basic unit of the granule (59,60). The electron-dense subunits in the granule are composed of low-density lipoproteins surrounded by phosvitin strands with attached lipovitellin micelles (16).

1. Phosvitin

Phosvitin contains about 10% phosphorus and represents about 80% of the protein phosphorus in yolk. This phosphoprotein can be separated by moving-boundary electrophoresis into three peaks but is homogeneous when analyzed by ultracentrifugation. Abe et al.

(1), using gel filtration with Sephadex G-200, separated phosvitin into two components, designated α-phosvitin and β-phosvitin, with molecular weights of 160,000 and 190,000, respectively. In the presence of sodium dodecylsulfate (SDS), α- and β-phosvitins dissociate into polypeptides. The β-phosvitin consists mainly of one polypeptide with a molecular weight of 45,000, whereas the molecular weights of the three different polypeptides of α-phosvitin are 37,500, 42,500, and 45,000. Phosvitin polypeptides react readily in aqueous solution to form aggregates. The phosphorus contents of α- and β-phosvitins are 9.2 and 3.0%, respectively. Little or no cysteine and cystine have been found in phosvitin, but an exceptionally high concentration of serine (31% of the total amino acid residues) is present. Phosphate in the protein is presumably esterified with serine.

Ferric ions bind tightly to phosvitin to form a soluble complex, and phosvitin is thus the iron carrier of yolk (31,71).

2. Lipovitellins

The high-density lipoprotein fraction, lipovitellin, can be separated into α- and β-lipovitellins by moving-boundary electrophoresis or by chromatography using a column of hydroxyapatite, Dowex-1, or triethylaminoethylcellulose. The apparent lipid and protein phosphorus contents of these proteins depend on the type of column used in the chromatographic separation (Table 6). The lipids in α- and β-lipoproteins include 40% neutral lipids and 60% phospholipids (47). The phospholipid fraction of each protein contains 75% phosphatidylcholine, 18% phosphatidylethanolamine, and 7% sphinogomyelin and lysophospholipids.

At pH values below 7, lipoproteins exist as dimers. With an increase in pH, however, monomers gradually form. The molecular weights of α- and β-lipoprotein dimers are about 400,000.

3. Low-Density Lipoproteins (LDL) and Myelin Figures (MF)

LDL and MF in the granule floating fraction FII have been separated by gel filtration with Sepharose 2B (30). LDL from yolk of eggs from 39-week-old hens contains about 84% lipid made up to about 31% phospholipids, 3.7% cholesterol, and 65% triacylglycerols.

Table 6   Lipid and Protein Phosphorus Contents of α- and
β-Lipovitellins (LV) Isolated by Column Chromatography

Chromatographic column	Lipid content (%)		Protein P (%)	
	α-LV	β-LV	α-LV	β-LV
Hydroxyapatite (HA)	20	20	1.20	0.45
Dowex 1, then HA	22	22	0.50	0.27
TEAE cellulose	14.6	16.5	0.54	0.28

Yolk MF are made up of one or more electron-dense lamellae and a less electron dense core (16). The lamellae are equidistant from one another with a repeat period of around 4.5 nm (Fig. 6). The yolk MF structure is similar to phospholipid myelin figures. The lipid content of yolk MF is approximately 86%. About 35% phospholipids and 11.5% cholesterol are present in the lipid fraction of MF (30). Concanavalin A affinity chromatography of delipidated MF produces retained and unretained fractions with 1 and 4.8% total carbohydrates, respectively (39).

B. Proteins and Lipoproteins in Plasma

Plasma is composed of globular protein fraction, called livetin, and a low-density lipoprotein fraction. The livetin and lipoprotein fractions represent about 11 and 66%, respectively, of the total yolk solids.

1. Livetin

Three components, (α-, β-, and γ-livetins) can be separated from the livetin fraction by moving-boundary and paper electrophoresis, but with disk-gel electrophoresis 15 protein

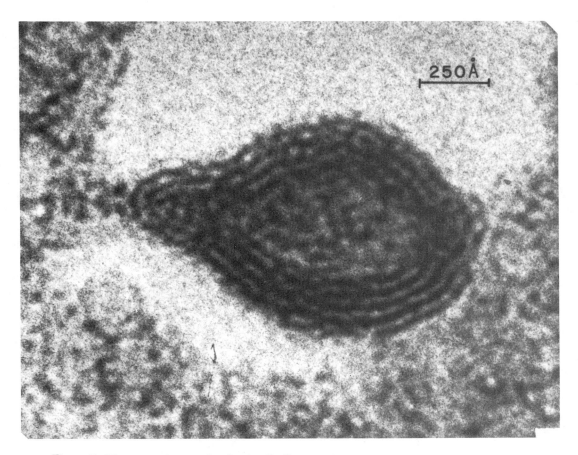

Figure 6 Electron micrograph of a myelin figure.

Table 7 Composition of Livetins

Component	α-Livetin	β-Livetin	γ-Livetin
Nitrogen (%)	14.3	14.3	15.6
Hexose (%)	—	7	2.6
Hexosamine (%)	—	—	1.8
Molecular weight	80,000	45,000	150,000

bands can be obtained from the livetin fraction (57). The chemical composition and molecular weights of α-, β-, and γ-livetins are presented in Table 7. Livetins are presumably derived from the blood of hen. Williams (79) identified α-livetin as serum albumin, β-livetin as α_s-glycoprotein, and γ-livetin as a γ-globulin.

2. Low-Density Lipoprotein

The plasma low-density lipoprotein (LDL) fraction of egg yolk has a density of 0.98 and can be isolated by flotation procedures. In some publications, plasma LDL is called very low density lipoproteins. The purified LDL fraction contains between 84 and 89% lipids, which are made up of 75% neutral lipids and 26% phospholipids, the latter consisting of 71-76% phosphatidylcholine, 16-20% phosphatidylethanolamine, and 8-9% sphingomyelin and lysophospholipids (47).

With differential flotation techniques, two components, LDL_1 and LDL_2, have been separated from the plasma LDL fraction. The composition of these fractions is presented in Table 8. The average molecular weights of LDL_1 and LDL_2 have been estimated to be 10 and 3 million, respectively (46).

A LDL micelle is considered to be a sphere composed of a triacylglycerol core upon which phospholipids and proteins are layered (36). The diameters of these LDL micelles range between 12 and 50 nm. Presumably the micelles are stabilized by ultraparticles adsorbed in the surfaces of the triacylglycerol cores (16). These ultraparticles can be dislodged from the surfaces by proteolysis with pepsin. Proteins in the LDL micelles exist in the antiparallel β-conformation.

VII. CHEMISTRY OF DETERIORATION OF SHELL EGGS DURING STORAGE

During the storage of shell eggs, the pH of albumen increases at a temperature-dependent rate from about 7.6 to a maximum value of about 9.7 (38). In one study, the pH of albumen in shell eggs rose to 9.2 during a 3 day storage period at $3°C$. The rise in pH is caused by a loss of carbon dioxide through the pores in the shell. The pH is dependent on the equilibria among dissolved CO_2, HCO_3^-, CO_3^{2-}, and proteins. The concentrations of HCO_3^- and CO_3^{2-} ions are governed by the partial pressure of CO_2 in the external environment. The rise in pH of albumen causes a breakdown in the gel structure of the thick white.

Fresh yolk has a pH of about 6, and this changes very little, even over prolonged storage of shell eggs. In one study involving egg storage temperatures of 2 and $37°C$, the yolk pH rose to 6.4 in about 50 and 18 days, respectively (65).

One of the most important quality attributes of eggs is the consistency of the albumen; consumers desire eggs with a firm, gelatinous albumen. It is generally agreed that the gelatinous character of the albumen (thick white) is dependent on the presence of

Table 8 Composition of Low-Density Lipoproteins

Component	LDL$_1$	LDL$_2$
Total lipid (%)	89	86
Lipid phosphorus	1.0	1.14
Lipid molar N/P ratio	1.04	1.14
Cholesterol, % lipid	3.3	3.4
Free fatty acid, % lipid	5	4.5
Phospholipid composition (%)		
Phosphatidylcholine	84	83
Phosphatidylethanolamine	13	14
Nonlipid residue (%)	11	14
Nitrogen, % nonlipid residue	14.8	13.7
Phosphorus, % nonlipid residue	0.15	0.16

ovomucin, but the chemical and physical changes in ovomucin that bring about the decline in albumen quality during storage of shell eggs are only partially understood.

Chemical scission of O-glycosides from the protein moiety of ovomucin seems to be the most plausible explanation for albumen thinning. Studies have shown that, when eggs are stored at 30°C for 20 days, the hexosamine and hexose contents of ovomucin decrease by 50% and the sialic acid content decreases to 12% of the original value. β-Elimination of O-glycosidically linked carbohydrates in glycoproteins, such as ovomucin, can occur under alkaline conditions (37).

A change in ovalbumin to a more stable form (*S*-ovalbumin) during storage of eggs has been reported (69). The denaturation temperature for *S*-ovalbumin is 92.5°C, compared with 84.5° for ovalbumin. During the ovalbumin → *S*-ovalbumin transition, an intermediate protein with a denaturation temperature of 88.5°C is formed (25). The increased heat stability of *S*-ovalbumin apparently cannot be explained by an increase in net negative charges (40). Reported decreases in the intrinsic viscosity and Stokes radius of ovalbumin in stored eggs suggest increases in compactness and hydrophobicity of this protein.

VIII. CHEMICAL CHANGES DURING PROCESSING OF LIQUID EGG PRODUCTS

A. Pasteurization of Egg Products

Liquid egg products are pasteurized by heat to inactivate *Salmonella*, a pathogenic organism sometimes present as a result of contamination. Pasteurization of albumen in a commercial heat exchanger for 2-3 min at 53°C causes an increase in whipping time, little destruction of *Salmonella*, and an insignificant influence on angel food cake volume (38). As the temperature is increased above 53°C, damage to the foaming capability of albumen increases. When albumen is heated for 2 min at 58°C or flash heated to 60°C, albumen turbidity and viscosity increase and angel food cake volume decreases. Whole egg and yolk, in contrast, can be pasteurized at temperatures between 60 and 63°C without significant changes in physical and functional properties.

Conalbumin is a heat-sensitive protein insolubilized rapidly at temperatures above 58°C when the albumen pH is in the region 6-7 (24). Ovalbumin, ovomucoid, lysozyme, and ovomucin, however, are heat stable in the vicinity of pH 7. The iron-conalbumin complex is much more heat stable than metal-free conalbumin (2). Cunningham and Lineweaver (24) carried out detailed studies on the influence of polyvalent cations on the heat stability of conalbumin in albumen at pH 7 and 60°C. The Fe^{3+}, Cu^{2+}, and Al^{3+} ions are very effective stabilizers at pH 7 but not at pH 9. When albumen (pH 7) containing 10^{-3} mol aluminum ions is heated for 13 min at 60°C, the volume and texture of angel food cakes prepared from the heated albumen is comparable to those of cakes prepared from unheated albumen, but the whipping time increases. This increase in whipping time is attributable to heat denaturation of the ovomucin-lysozyme complex (29). Apparently, weak lamellae are formed from heat-damaged ovomucin-lysozyme, and these are easily broken by the mechanical action of the whipping beaters.

Sodium hexametaphosphate (SHMP), at a 2% level, is effective in inhibiting thermal coagulation of conalbumin in albumen at pH values between 7 and 9, with the greatest stabilization at about pH 7. The foam performance of SHMP-treated albumen heated for 3 min at 61.7°C is damaged only slightly.

B. Freezing of Yolk

When yolk is frozen and stored below −6°C, the viscosity of the thawed product is much greater than that of native yolk (56). This irreversible change in fluidity has been termed "gelation." The functional properties of yolk are altered during the gelation process. For example, the volume of sponge cakes made with gelled yolk is considerably smaller than that of cakes prepared with unfrozen yolks (34).

The rate and extent of yolk gelation is influenced by the rate of freezing, temperature and duration of frozen storage, and the rate of thawing. Rapid freezing of yolk in liquid nitrogen effectively inhibits gelation provided the frozen product is thawed rapidly (42). As the temperature of frozen storage is lowered from −6 to −50°C, the rate of gelation increases.

Gelation of yolk can be minimized by prefreezing treatments, such as addition of cryoprotection agents or proteolytic enzymes, or by colloid milling. Sugars, such as sucrose, glucose, and galactose, at a 10% level, are effective cryoprotective (antigelling) agents that do not significantly alter the viscosity of the unfrozen yolk. NaCl at levels of 1-10% also prevents gelation during freezing, although NaCl increases the viscosity of unfrozen yolk. Addition of NaCl to LDL solutions apparently increases the unfrozen water content through the formation of a LDL-water-NaCl complex (77).

Treatment of native yolk with proteolytic enzymes, such as trypsin and papain, is an additional approach to preventing gelation during the freezing process. Unfortunately, enzyme-treated yolk has a low emulsifying action compared with untreated yolk, thus preventing commercial use of this method. Colloid milling lessens but does not prevent yolk gelation.

The mechanism of yolk gelation has been explored by several investigators, but complete details are still lacking (17,35,45,74,76). The basic requisites for initiating yolk gelation are the formation of ice crystals and the lowering of the storage temperature below −6°C (the freezing point of yolk is about −0.58°C). At −6°C, about 81% of the initial water in yolk is frozen. The fivefold increase in the concentration of soluble salts that occurs in yolk during freezing to −6°C may be responsible in part for gelation. Freezing also reduces the average distance between reactants, such as lipoproteins, in the unfrozen phase, and this enhances their aggregation. Constituents in both plasma and granules are involved in gelation (17).

Electron micrographs of frozen and thawed yolk have shown that granules are disrupted and low-density lipoproteins are released during the freezing process. The large lipid masses, which are apparent in micrographs of frozen yolk, probably form by coalescence of low-density lipoproteins.

C. Spray-Dried Egg Products

During spray-drying and subsequent storage of dry whole egg, yolk, and albumen, the following types of chemical and physical changes can occur: a decrease in solubility or dispersibility, undesirable changes in color, a decrease in whipping power, and development of off-odors.

Unstored, spray-dried whole egg is quite soluble (95-98% dispersibility in 10% KCl) but has about half the foaming power of fresh whole egg in the making of sponge cake (9). The loss of foaming power is apparently caused by the liberation of free fat (8). Microscopic studies have shown that large fat droplets are present in whole egg powder, whereas none have been noted in native whole egg. After spray-dried whole egg is stored for only 3 hr, the foaming capability of the reconstituted product is almost completely lost (38).

During storage of spray-dried whole egg for 3-4 months at about 27°C, common changes are the development of a brown color, development of an off-flavor, and a decrease in dispersibility. The Maillard reaction, involving glucose and the amino group of phosphatidylethanolamine, contributes to both color and flavor changes. Removal of naturally occurring glucose in whole egg by yeast fermentation prior to drying retards both discoloration and loss of dispersibility. Off-flavor development in stored whole egg powder also can occur because of oxidation of fatty acids in phospholipids.

The addition of sucrose and corn syrup solids to whole yolk before spray-drying retards the Maillard reaction in the dried product during storage and minimizes the loss of foaming power brought about by drying. Flavor stability also improves gradually as the level of sugar solids is increased to 5% for sucrose and 10% for corn syrup solids. At higher levels of sugars (10-15%), the rate of autoxidation of egg lipids increases and an off-flavor develops gradually.

Similarly, when egg yolk is spray-dried, the foaming ability of the rehydrated product is lower than that of the native yolk (64). Further, yolk stored at 13°C develops an off-flavor in about 4 months. Spray-drying of yolk may cause irreversible changes in the structure of low-density lipoproteins, which result in the release of foam-inhibiting lipids (64). As the concentration of added sucrose is raised to the 15% level, the foaming ability of reconstituted spray-dried yolk increases. Low-density lipoproteins are apparently protected by sucrose during drying (64).

Functional properties of albumen are not altered appreciably during the spray-drying process (10). However, during storage of the dried albumen, brown discoloration develops, solubility decreases, and its ability to produce angel food cakes of acceptable volume declines (12). A decrease of 26% in the volume of angel food cake was observed when dried albumen was stored for 2 weeks at 40°C. When glucose is removed prior to drying, none of the above-mentioned changes occur during normal storage.

IX. FUNCTIONAL PROPERTIES OF WHOLE EGG MAGMA, ALBUMEN, AND YOLK

Proteins and lipoproteins are functional constituents of whole egg magma, albumen, and yolk. These constituents, being surface-active, can lower the interfacial tension between water and a hydrophobic phase, such as vegetable oil or air, and can aggregate at the interface to form large particles. The surface activity of proteins and lipoproteins are important in the creation of food emulsions and foams. Heat treatment of egg products can lead to the creation of gels with elastic properties. Gel properties are important in bakery products, souffles, and omelets.

A. Emulsifying Power of Egg Yolk

Egg yolk is an important emulsifying ingredient in the manufacture of mayonnaise, salad dressing, and cakes. The emulsifying components in yolk are phospholipids, lipoproteins, and proteins. Although phospholipids of egg yolk are surface-active agents, they are inefficient in stabilizing an emulsion (81).

Yolk particles, such as low-density lipoproteins, myelin figures, and high-density lipoproteins, are important emulsion stabilizers because of their ability to interact at the surfaces of oil droplets to form protective layers (15).

Mayonnaise is a stable oil-in-water emulsion containing between 65 and 80% oil. The key to the preparation of a stable mayonnaise is the formation of small oil droplets in a continuous water phase that has a viscosity sufficiently great to impede coalescence of the oil droplets. The initial step in the manufacture of mayonnaise involves the preparation of a yolk-salt mix, the salt serving to increase viscosity by the breakdown of granules to smaller particles with greater water-binding capacity. During the slow addition of oil to the agitated, viscous yolk mix, small oil droplets are formed and they become coated with protective layers consisting of various kinds of yolk particles. The adherence of the particle-containing layers to the oil droplets is exceptionally strong. Water-washed oil droplets from commercial mayonnaise contain about 45% of the total mayonnaise N and 43% of the total P (75), indicating that yolk proteins and phospholipids are major constituents of the protective layers. When studied by means of electron mciroscopy, the protective layers appear as numerous electron-dense particles distributed in a less dense matrix (Fig. 7) (15). Coalesced low-density lipoproteins and microparticles (lipovitellins) from salt-disrupted granules in yolk presumably are the major constituents of the protective layers.

B. Foaming Power of Albumen

Egg white foam contributes to the lightness of meringues, souffles, angel food cakes, fluffy omelets, and frappe candies. With these foods, the key property of an albumen foam is the prolonged retention of a large volume of gas in the form of small bubbles surrounded by stable, semirigid, elastic, protein-containing thin walls or lamellae (4,6). Destabilization of an egg white foam involves fluid drainage from the lamellae and gas loss upon lamella breakage.

During the beating of egg white, large air bubbles are incorporated initially and these are subsequently subdivided by the shearing action of beater blades. The volume of egg white foam increases with beating time to a maximum and then declines. The surface appearance of an albumen foam changes from moist and glossy to dry and dull as air is incorporated into the system (6).

Native egg white contains water-soluble proteins as surface-active compounds, and these can migrate to air/water interfaces (32,49). Proteins at these interfaces orient themselves with hydrophobic groups directed toward the air phase and hydrophilic

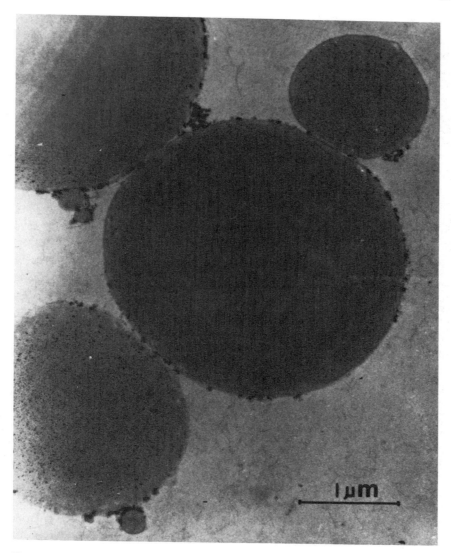

Figure 7 Electron micrograph of oil droplets of mayonnaise.

groups directed toward the aqueous phase. Such an orientation promotes the unfolding of the macromolecules to random coil conformations (see Chap. 5).

The unfolding of proteins at interfaces is called surface denaturation. The rate of surface denaturation is faster for flexible, highly hydrophobic secondary-structured proteins than it is for proteins with tertiary and quaternary structures (18,32). Foaming ability of egg white is related to surface denaturation of the globulin fraction of albumen (44,50,73).

Denatured proteins interact readily through a variety of physical and chemical bonds to produce aggregated protein films that enhance the entrapment of air bubbles in beaten egg white. Hydrophobic associations are apparently important in protein aggregation during foam formation (75). As beating of egg white proceeds, denaturation of proteins and aggregation of denatured protein molecules increase (50). Aggregated protein particles play an important role in the stability of egg white foam by holding water in the lamellae and providing structural rigidity and elasticity (32,50,51). Aggregated ovomucin is an especially important contributor to foam stability (44,50).

C. Heat Coagulability of Albumen, Yolk, and Egg Magma

The thermal setting of cake batters, custards, and puddings has been attributed in part to denaturation and coagulation of egg proteins (4,6). When egg yolk, albumen, and egg magma (mixed whole egg) are heated to critical temperatures, their rheological properties change considerably. For example, when albumen is heated to 62°C, the viscosity and opacity of this protein sol increases. When the temperature is elevated to about 65°C, a weak gel is formed and the strength of the gel increases with subsequent increases in temperature. As the storage time of shell eggs is increased from 0 to 10 days, the elasticity and rigidity of heat-induced gels of albumen from these eggs also increase (33). In the case of yolk, the viscosity begins to increase appreciably at about 65°C, and at 70°C, fluidity is lost completely with the formation of a semisolid crumbly mass (4,6).

The term "heat coagulation" has been used to designate the process of thermal denaturation and aggregation of proteins in albumen and in yolk. Denaturation of proteins involves the breakage of hydrogen bonds, the uncoiling of polypeptide chains and the exposure of reactive groups. Denatured proteins interact readily through chemical and physical bonds to form protein aggregates that may participate in the formation of a three-dimensional gel network.

Heat denaturation of proteins in albumen has been studied by differential scanning calorimetry (26). Three major endotherms at 65, 74, and 84°C are produced by denaturation of conalbumin lysozyme, and ovalbumin, respectively, when present in albumen at pH 7. Electrophoretic studies of proteins in heated albumen (61.7°C for 3 min) indicated that the heat stability of conalbumin increases with a rise in albumen pH to 9, whereas ovalbumin in structurally stable regardless of albumen pH (13).

Aggregation of denatured albumen proteins occurs by the formation of hydrophobic, electrostatic, and disulfide linkages (43). Raman spectral shifts of heated ovalbumin in solution indicate the development of intermolecular β-sheets (54). Thermocoagulation of albumen proteins can be greatly lessened by increasing the net charges on the macromolecules (66).

Proteins and lipoproteins in yolk are altered significantly at temperatures of about 60°C or higher. These changes can be ascertained by rheological measurements and by disk-gel electrophoresis. For example, yolk heated for 3 min at 60, 62.8, and 65.6°C has, respectively, viscosities of about 30, 38, and 200 P (25°C) at a shear rate of 1.9 sec^{-1} (14). Thermally induced viscosity increases of yolk can be explained by the formation of protein and lipoprotein complexes by the aggregation process (13). Low-density lipoproteins, the major solid component in yolk, play an important role in the development of heat-induced solidification of yolk (52).

When whole egg magma (about 60-65% albumen and 35-40% yolk) is heated, a slight opacity becomes apparent at 60°C and increases as the temperature rises to 72°C (80). At 75°C, whole egg magma possesses a soft curd consistency, and the firmness increases with heat input to 87°C. Electrophoretic studies on heated whole egg magma indicate that livetin and some globulins are altered at 60-63°C, whereas ovalbumin and conalbumin are not changed appreciably by this heat treatment (13,72,80). The increased heat stability of conalbumin in the magma can be attributed to the formation of an iron-conalbumin complex with the iron ions coming from the yolk. When magma is heated to 87°C, almost all the proteins lose their electrophoretic mobility. The addition of 10% sucrose or 10% sodium chloride to magma increases the heat stability of all the proteins.

REFERENCES

1. Abe, Y., T. Itoh, and S. Adachi (1982). Fractionation and characterization of hen's egg yolk phosvitin. *J. Food Sci. 47*:1903-1907.
2. Azari, P. R., and R. E. Feeney (1958). Resistance of metal complexes of conalbumin and transferrin to proteolysis and to thermal denaturation. *J. Biol. Chem. 232*:293-302.
3. Baker, C. M. A. (1968). The proteins of egg white, in *Egg Quality, A Study of the Hen's Egg* (T. C. Carter, ed.), Oliver and Boyd, Edinburgh, Scotland, pp. 67-108.
4. Baldwin, R. E. (1977). Functional properties in foods, in *Egg Science and Technology* (W. J. Stadelman and O. J. Cotterill, eds.), AVI Publishing, Westport, Conn., pp. 246-277.
5. Bellairs, R. (1961). The structure of the yolk of the hen's egg as studied by electron microscopy. 1. The yolk of unincubated egg. *J. Biophys. Biochem. Cytol. 11*: 207-225.
6. Bennion, M. (1980). *The Science of Food*, Harper and Row, San Francisco.
7. Brooks, J., and H. P. Hale (1955). Strength of the shell of the hen's egg. *Nature 175*:848-849.
8. Brooks, J., and J. R. Hawthorne (1944). Eggs. *J. Soc. Chem. Ind. 29*:434-435.
9. Brooks, J., and D. J. Taylor (1955). *Eggs and Egg Products*. Special Rept. Food Invest. Bd. DSIR No. 60.
10. Brown, S. L., and M. E. Zabik (1967). Effect of heat treatments on the physical and functional properties of liquid and spray-dried egg albumen. *Food Technol. 21*:87-92.
11. Burley, K. W., and W. H. Cook (1961). Isolation and composition of avian egg yolk granules and their constituents α- and β-lipovitellins. *Can. J. Biochem. Physiol. 39*: 1295-1307.
12. Carlin, A. F., and J. C. Ayres (1951). Storage studies on yeast-fermented dried egg white. *Food Technol. 5*:172-175.
13. Chang, P. K., W. D. Powrie, and O. Fennema (1970). Disc-gel electrophoresis of proteins in native and heat-treated albumen, yolk and centrifuged whole egg. *J. Food Sci. 35*:774-778.
14. Chang, P. K., W. D. Powrie, and O. Fennema (1970). Effect of heat treatment on viscosity of yolk. *J. Food Sci. 35*:864-867.
15. Chang, C. M., W. D. Powrie, and O. Fennema (1972). Electron microscopy of mayonnaise. *Can. Inst. Food Sci. Technol. J. 5*:134-137.
16. Chang, C. M., W. D. Powrie, and O. Fennema (1977). Microstructure of egg yolk. *J. Food Sci. 42*:1193-1200.
17. Chang, C. M., W. D. Powrie, and O. Fennema (1977). Studies on the gelation of egg yolk and plasma upon freezing and thawing. *J. Food Sci. 42*:1658-1665.
18. Cherry, J. P., and K. H. McWatters (1981). Whippability and aeration, in *Protein Functionality in Foods* (J. P. Cherry, ed.), ACS Symposium Series, 147, American Chemical Society, Washington, D.C., pp. 149-196.
19. Cook, F., and G. M. Briggs (1977). Nutritive value of eggs, in *Egg Science and Technology* (W. J. Stadelman and O. J. Cotterill, eds.), AVI Publishing, Westport, Conn., pp. 92-108.

20. Cook, W. H. (1968). Macromolecular components of egg yolk, in *Egg Quality, A Study of the Hen's Egg* (T. C. Carter, ed.), Oliver and Boyd, Edinburgh, Scotland, pp. 109-132.

21. Cooke, A. S., and D. A. Balch (1970). Studies of membrane, mammillary cores and cuticle of the hen egg shell. *Br. Poultry Sci. 11*:345-352.

22. Cotterill, O. J., and G. S. Geiger (1977). Egg product yield trends from shell eggs. *Poultry Sci. 56*:1027-1031.

23. Cotterill, O. J., and A. R. Winter (1955). Egg white lysozyme. 3. The effect of pH on the lysozyme-ovomucin interaction. *Poultry Sci. 34*:679-686.

24. Cunningham, F. E., and H. Lineweaver (1965). Stabilization of egg-white proteins to pasteurizing temperatures above 60°C. *Food Technol. 19*:1442-1447.

25. Donovan, J. W., and C. J. Mapes (1976). A differential scanning calorimetric study of conversion of ovalbumin to S-ovalbumin in eggs. *J. Sci. Food Agr. 27*:197-204.

26. Donovan, J. W., C. J. Mapes, J. G. Davis, and J. A. Garibaldi (1975). A differential scanning calorimetric study of the stability of egg white to heat denaturation. *J. Sci. Food Agr. 26*:73-78.

27. Feeney, R. E. (1964). Egg proteins, in *Symposium on Foods: Proteins and Their Reactions* (H. W. Schultz and H. F. Anglemier, eds.), AVI Publishing, Westport, Conn., pp. 209-224.

28. Forsythe, R. H., and J. F. Foster (1949). Note on the electrophoretic composition of egg white. *Arch. Biochem. 20*:161-163.

29. Garibaldi, J. A., J. W. Donovan, J. G. Davis, and S. L. Comino (1968). Heat denaturation of the ovomucin-lysozyme electrostatic complex—a source of damage to the whipping properties of pasteurized egg white. *J. Food Sci. 33*:514-524.

30. Garland, T., and W. D. Powrie (1978). Chemical characterization of egg yolk myelin figures and low-density lipoproteins isolated from egg yolk granules. *J. Food Sci. 43*:1210-1214.

31. Greengard, O., A. Sentenac, and H. Mendelsohn (1964). Phosvitin, the iron carrier of egg yolk. *Biochim. Biophys Acta 90*:406-407.

32. Halling, P. J. (1981). Protein—stabilized foams and emulsions. *Crit. Rev. Food Sci. Nutr. 15*:155-203.

33. Hickson, D. W., E. S. Alford, F. A. Gardner, K. Diehl, J. O. Sanders, and C. W. Dill (1982). Changes in heat-induced rheological properties during cold storage of egg albumen. *J. Food Sci. 47*:1908-1911.

34. Jordan, R., B. N. Luginbill, L. E. Dawson, and C. J. Echterling (1952). The effect of selected pretreatments upon the culinary qualities of eggs frozen and stored in a home-type freezer. *Food Res. 17*:1-7.

35. Kamat, V., G. Graham, M. Barrat, and M. Stubbs (1976). Freeze-thaw gelation of hen's egg yolk low density lipoprotein. *J. Sci. Food Agr. 27*:913-927.

36. Kamat, V. B., G. A. Lawrence, M. D. Barratt, A. Darke, R. B. Leslie, G. G. Skipley, and J. M. Stubbs (1972). Physical studies of egg yolk low density lipoproteins. *Chem. Phys. Lipids 9*:1-25.

37. Kato, A., K. Ogino, Y. Kuramoto, and K. Kobayashi (1979). Degradation of the O-glycosidically linked carbohydrate units of ovomucin during egg white thinning. *J. Food Sci. 44*:1341-1344.

38. Kline, L., T. F. Sugihara, M. L. Bean, and K. Ijichi (1965). Heat pasteurization of raw liquid egg white. *Food Technol. 19*:105-114.

39. Kocal, J. T., S. Nakai, and W. D. Powrie (1980). Preparation of apolipoprotein of very low density lipoprotein from egg yolk granules. *J. Food Sci. 45*:1761-1267.

40. Kurisaki, J., Y. Murata, S. Kaminogawa, and K. Yamauchi (1982). Heterogeneity and properties of heat-stable ovalbumin from stored egg. *J. Agr. Food Chem. 30*:349-353.

41. Lineweaver, H., F. E. Cunningham, J. A. Garibaldi, and K. Ijichi (1961). *Heat Stability of Egg White Proteins Under Minimal Conditions That Kill Salmonellae*, ARS 74-39, U.S. Department of Agriculture, Washington, D.C.

42. Lopez, H., C. R. Fellers, and W. D. Powrie (1954). Some factors affecting gelation of frozen egg yolk. *J. Milk Food Technol. 17*:334-339.

43. Ma, C.-Y., and J. Holme (1982). Effect of chemical modifications on some physico-chemical properties and heat coagulation of egg albumen. *J. Food Sci. 47*:1454-1459.

44. MacDonnell, L. R., R. E. Feeney, H. L. Hanson, A. Campbell, and T. F. Sugihara (1955). The functional properties of the egg white proteins. *Food Technol. 9*: 49-53.

45. Mahadevan, S., T. Satyanarayana, and S. A. Kumar (1969). Physico-chemical studies on the gelation of hen's egg yolk. Separation of gelling protein components from yolk plasma. *J. Agr. Food Chem. 17*:767-771.

46. Martin, W. G., J. Augustnyniak, and W. H. Cook (1964). Fractionation and characterization of the low-density lipoproteins of hen's egg yolk. *Biochim. Biophys. Acta 84*:714-720.

47. Martin, W. G., W. G. Tattrie, and W. H. Cook (1963). Lipid extraction and distribution studies of egg yolk lipoproteins. *Can. J. Biochem. Physiol. 41*:657-666.

48. Matsuda, T., K. Watanabe, and Y. Sato (1982). Interaction between ovomucoid and lysozyme. *J. Food Sci. 47*:637-641.

49. Nakamura, R. (1963). Studies on the foaming property of the chicken egg white. VI. Spread monolayer of the protein fraction of the chicken egg white. *Agr. Biol. Chem. 27*:427-430.

50. Nakamura, R., and Y. Sato (1964). Studies on the foaming property of the chicken egg white. Part IX. On the coagulated proteins under various whipping conditions (The mechanism of foaminess). *Agr. Biol. Chem. 28*:524-529.

51. Nakamura, R., and Y. Sato (1964). Studies on the foaming property of the chicken egg white, Part X. On the role of ovomucin (B) in the egg white foaminess (The mechanism of the foaminess). *Agr. Biol. Chem. 28*:530-534.

52. Nakamura, R., T. Fukano, and M. Taniguchi (1982). Heat-induced glation of hen's egg yolk low-density lipoprotein (LDL) dispersion. *J. Food Sci. 47*:1449-1463.

53. Nisbet, A. D., R. H. Saundry, A. J. G. Moir, L. A. Fothergill, and J. E. Fothergill (1981). The complete amino acid sequence of hen ovalbumin. *Eur. J. Biochem. 115*:335-345.

54. Painter, P. C., and J. L. Koenig (1976). Raman spectroscopic study of the proteins of egg white. *Biopolymers 15*:2155-2166.

55. Parkinson, T. L. (1966). The chemical composition of egg. *J. Sci. Food Agr. 17*: 101-111.

56. Powrie, W. D. (1968). Gelation of egg yolk upon freezing and thawing, in *Low Temperature Biology of Foodstuffs* (J. Hawthorn and E. J. Rolfe, eds.), Pergamon Press, Oxford, England, pp. 319-331.

57. Powrie, W. D. (1977). Chemistry of eggs and egg products, in *Egg Science and Technology* (W. J. Stadelman and O. J. Cotterill, eds.), AVI Publishing, Westport, Conn., pp. 65-91.

58. Privett, O. S., M. L. Bland, and J. A. Schmidt (1962). Studies on the composition of egg lipid. *J. Food Sci. 27*:463-468.

59. Radomski, M. W., and W. H. Cook (1964). Fractionation and dissociation of the avian lipovitellins and their interaction with phosvitin. *Can. J. Biochem. 42*:349-406.

60. Radomski, M. W., and W. H. Cook (1964). Chromatographic separation of phosvitin, α- and β-lipovitellin of egg yolk granules on TEAE-cellulose. *Can. J. Biochem. 42*:1203-1312.

61. Rhodes. D. N., and C. H. Lea (1956). Phospholipdes. 4. On the composition of hen's egg phospholipides. *Biochem. J. 65*:526-533.

62. Robinson, D. S., and J. B. Monsey (1971). Studies of the composition of egg white ovomucin. *Biochem. J. 121*:537-547.

63. Romanoff, A. L., and H. Romanoff (1949). *The Avian Egg*, John Wiley and Sons, New York.

64. Schultz, J. R., H. E. Snyder, and R. H. Forsythe (1968). Co-dried carbohydrates effect on the performance of egg yolk solids. *J. Food Sci. 33*:507-513.

65. Sharp, P. F., and C. K. Powell (1931). Increase in the pH of the white and yolk of hen's eggs. *Ind. Eng. Chem. 23*:196-199.

66. Shimada, K., and S. Matsushita (1980). Thermal coagulation of egg albumen. *J. Agr. Food Chem. 28*:409-412.

67. Simkiss, K. (1968). The structure and formation of the shell and shell membranes, in *Egg Quality, A Study of the Hen's Egg* (T. C. Carter, ed.), Oliver and Boyd, Ediburgh, Scotland, pp. 3-25.

68. Smith, M. B. (1964). Studies on ovalbumin. I. Denaturation by heat and the heterogeneity of ovalbumin. *Aust. J. Biol. Sci. 17*:261-270.

69. Smith, M. B., and J. F. Back (1965). Studies on ovalbumin. II. The formation and properties of S-ovalbumin, a more stable form of ovalbumin. *Aust. J. Biol. Sci. 18*:365-377.

70. Smith, M. B., T. M. Reynolds, C. P. Buckingham, and J. F. Back (1974). Studies on the carbohydrate of egg white ovomucin. *Aust. J. Biol. Sci. 27*:349-360.

71. Taborsky, G. (1963). Interaction between phosvitin and iron and its effect on a rearrangement of phosvitin structure. *Biochemistry 2*:266-271.

72. Torten, J., and J. Eisenberg (1983). Studies on colloidal properties of whole egg magma. *J. Food Sci. 47*:1423-1428.

73. Townsend, A., and S. Nakai (1983). Relationship between hydrophobicity and foaming characteristics of proteins. *J. Food Sci. 48*:588-594.

74. Vadehra, D. V., and K. R. Nath (1973). Eggs as a source of protein. *Crit. Rev. Food Technol. 4*:193-309.

75. Vincent, R. F. (1964). Some physico-chemical properties of egg yolk. Master's Thesis, University of Wisconsin, Madison.

76. Wakamatu, T., Y. Sato, and Y. Saito (1982). Identification of the components responsible for gelation of egg yolk during freezing. *Agr. Biol. Chem. 46*:1495-1503.

77. Wakamatu, T., Y. Sato, and Y. Saito (1983). On sodium chloride action in the gelation process of low density lipoprotein (LDL) from hen egg yolk. *J. Food Sci. 48*:507-512.

78. Watanabe, K., M. Tsukasa, and Y. Sato (1981). The secondary structure of ovomucoid and its domains as studied by circular dichroism. *Biochim. Biophys. Acta 667*:242-250.

79. Williams, J. (1962). Serum proteins and the livetins of hen's egg yolk. *Biochem. J. 83*:346-355.

80. Woodward, S. A., and O. J. Cotterill (1983). Electrophoresis and chromatography of heat-treated plain, sugared and salted whole egg. *J. Food Sci. 48*:501-506.

81. Yeadon, D. A., L. A. Goldblatt, and A. M. Altschul (1958). Lecithin in oil-in-water emulsions. *J. Amer. Oil. Chem. Soc. 36*:435-438.

BIBLIOGRAPHY

Carter, T. C. (ed.) (1968). *Egg Quality, A Study of Hen's Eggs*, Oliver and Boyd, Edinburgh, Scotland.

Halling, P. J. (1981). Protein-stabilized foams and emulsions, *Crit. Rev. Food Sci. Nutr.* *15*:155-203.

Stadelman, W. J., and O. J. Cotterill (eds.) (1977). *Egg Science and Technology*, AVI Publishing, Westport, Conn.

Vadehra, D. V., and K. R. Nath (1973). Eggs as a source of protein, *Crit. Rev. Food Technol. 4*:193-309.

15

CHARACTERISTICS OF EDIBLE PLANT TISSUES

Norman F. Haard   Memorial University of Newfoundland, St. John's, Newfoundland, Canada

I.	Introduction	858
II.	Chemical Composition	858
	A. Water	858
	B. Carbohydrates	859
	C. Proteins and Other N Compounds	861
	D. Lipids and Related Compounds	863
	E. Organic Acids	863
	F. Pigments	867
	G. Mineral Elements	867
	H. Vitamins	868
III.	Structure	869
	A. Organ Structure	869
	B. Tissue and Cell Structure	872
	C. Subcellular Structure	877
IV.	Physiology and Metabolism	878
	A. Respiration	879
	B. Gene Expression and Protein Synthesis	885
	C. Ancillary Metabolic Processes	885
	D. Control Mechanisms	892
V.	Handling and Storage of Fresh Fruits and Vegetables	895
	A. Mechanical Injury	895
	B. Temperature	898
	C. Controlled Atmospheres	900
	D. Humidity	903
	E. Ionizing Radiation	903
	F. Other Factors	904

VI. Effects of Processing on Fruits and Vegetables   904
    A. Thermal Processing   904
    B. Freezing   906
    C. Dehydration   908
VII. Conclusion   910
    References   910
    Bibliography   911

I. INTRODUCTION

Plant tissues, though the agency of photosynthesis, are ultimately the supplier of all human's food. Although thousands of crops fit into this dietary picture, some 100-200 species are of major importance in world trade and 14 species provide the bulk of world food crops: rice, wheat, sorghum, barley, sugarcane and sugar beets, potato, sweet potato, cassava, bean, soybean, peanut, coconut, and banana (12). Many of these crops are associated with legends and song that illustrate their fascinating roles in the history of different cultures. It has been said that crop production, the management of useful plants, is the basis of civilization.

The economic plant groupings that enter our diet include the Gramineae (cereals), Leguminosae (pulses), root crops, stem and leaf crops, fruit and seed vegetables, and a variety of plants used for extractives and derivatives (e.g., spices, beverages, and oils). Many volumes have been written to describe the characteristics of edible plant tissues, and it is beyond the scope of this chapter to provide complete details on individual commodities, much less all these crops. The intent of this chapter is to outline the compositional and physiological diversity of crop plants and to relate these properties to postharvest quality, storage, and processing. Postharvest physiology remains one of the most important and challenging sectors of science and technology.

II. CHEMICAL COMPOSITION

Detached plant parts, like all living organisms, contain a wide range of different chemical compounds and show considerable variation in composition. Apart from the obvious interspecies differences, an individual plant organ, largely composed of living tissues that are metabolically active, is more or less constantly changing in composition. The rate and extent of chemical change can depend on the growing conditions prior to harvest, the physiological role of the plant part, the genetic pool of the cell, and the postharvest environment.

A. Water

Given an unlimited supply of water, the moisture content of viable plant tissue assumes a characteristic maximum value associated with a state of complete turgor of the component cells. At turgor, the internal pressure (up to 9 atm or more) developed by osmotic forces is balanced by the inward pressure of the extended cell wall. Accordingly, the maximum water content of a given tissue is influenced by structural and chemical characteristics as well as extrinsic factors. The susceptibility of a commodity to water loss is dependent on the water vapor deficit of the surrounding atmosphere and the extent to which the external surface of the tissue is structurally and chemically modified to reduce transpiration of water loss. Water generally represents about 70-90% of the fresh weight of "wet" crops and less than 20% of the weight of "dry" crops such as cereals and pulses. Apart from the influence of water loss on wilting and weight loss, one must recognize that moisture balance can indirectly evoke undesirable or desirable physiological changes in some crops. The interaction of water with other molecular constituents and its role in cellular metabolism are discussed in Chaps. 2 and 16.

B. Carbohydrates

In general, approximately 75% of the solid matter of plants is carbohydrate. The total carbohydrate content can be as low as 2% of the fresh weight in some fruits or nuts, more than 30% in starchy vegetables, and over 60% in some pulses and cereals. Total carbohydrates are generally regarded as consisting primarily of simple sugars and polysaccharides, but they also properly include pectic substances and lignin. In plants, carbohydrates are localized in the cell wall and intracellularly in plastids, vacuoles, or the cytoplasm.

1. Cell Wall Constituents

The principal cell wall constituents are cellulose, hemicelluloses, pectins, and lignin (Table 1). Cellulose is one of the most abundant substances in the biosphere, is largely insoluble, and is indigestible by human beings. The hemicelluloses are a heterogeneous group of polysaccharides that contain numerous kinds of hexose and pentose sugars and in some cases residues of uronic acids. These polymers are classified according to the types of sugar residues predominating and are individually referred to as xylans, arabino-galactans, glucomannans, and so on. Pectin is generally regarded as $\alpha 1,4$-linked galac-turonic acid residues esterified to varying degrees with methanol. However, purified pectin invariably contains significant quantities of covalently linked nonuronide sugars, such as rhamnose, arabinose, and xylose. The albedo (white spongy layer) of citrus peel is an especially rich source of pectin, containing up to 50% of this constituent on a dry weight basis. Some other waste vegetable tissues are also excellent sources of pectin, which is used as a gelation agent. Lignin, always associated with cell wall carbohydrates, is a three-dimensional polymer comprised of phenyl propane units, such as syringaldehyde and vanillin, and these are linked through aliphatic three-carbon side chains. Lignification of cell walls, notably those of xylem and sclerenchymatous tissues, confers considerable rigidity and toughness to the wall.

The relative proportion and contents of cell wall constituents vary considerably among species, with maturity at harvest, and with elapsed time after harvest. Cell wall constituents are the primary components of "dietary fiber." The biological availability of protein and other nutrients can be reduced by these constituents, but there is considerable evidence for the beneficial role played by fiber in health and disease (11). The structure and chemistry of these substances is discussed in more detail in Chap. 3.

2. Starch and Other Polysaccharides

The principal carbohydrate of plant tissues, which is not associated with cell walls, is starch, a linear ($\alpha 1,4$) or branched ($\alpha 1,4;1,6$) polymer of D-glucose (see Chap. 3). Starch is localized in intracellular plastids that have species-specific shape, size, and optical properties. Industrially, starch is obtained from such crops as potato, sweet potato, or cereals that may contain up to 40% starch on a fresh weight basis (18). Although starch contributes more calories to the normal human diet than any other single substance, it is absent or negligible in most ripe fruits and many vegetables.

Table 1 Principal Cell Wall Constituents

Common name	Chemical nature	Linkages and cross-linkages with carbohydrates
Cellulose	$\beta(1\rightarrow4)$-linked glucose	Polymer of up to 12,000 glucose units; shows crystalline regions in which the molecules are arranged in fibrils, parallel groups of which form bundles; interactions with other wall constituents are likely
Hemicellulose xylans	$\beta(1\rightarrow4)$-linked xylose	Polymer of 150–200 units, mostly xylose with uronic acid residues, one of which is generally acetylated; may contain side chain $\beta(1\rightarrow3)$ links with arabinose
Glucomannans	$\beta(1\rightarrow4)$ randomly linked glucose and mannose	Strongly absorbed to cellulose microfibrils; may contain $\beta(1\rightarrow6)$ linkages with galactose
Arabinogalactans	$\beta(1\rightarrow3)$ linked galactose	Arabinose units linked $\beta(1\rightarrow3)$ or $\beta(1\rightarrow6)$ to galactose; also occurs as a constituent of pectic substances
Pectin substances	$\alpha(1\rightarrow4)$ linked galacturonic acid	Contain $\alpha(1\rightarrow6)$-linked rhammose and lesser quantities of fucose, xylose, and galactose linked $\beta(1\rightarrow4)$ to main and side chains; variable numbers of uronide carboxylate groups are esterified as methyl esters; regarded as filler substances within cell wall matrix and influence water distribution
Lignin	Insoluble, high-molecular-weight polymer of coumaryl, coniferyl, and sinapyl alcohols	Contains 100 or more aromatic units high in methoxyl groups; penetrates wall, producing secondary thickening, and acts as a hydrophobic filler; forms covalent links with wall carbohydrates
Cell wall protein, extensin	Hydroxyproline-rich protein	Various *O*-glycosidic, *N*-glycosidic, ester, and Schiff base links with polysaccharides; hydroxyquinone chelate bridges with lignin

Examples of other polysaccharides that occur as intracellular constituents of edible plants include α-D-1,4-linked glycopyranose of sweet corn, β-glucans of mango; and fructans, which contain D-fructose as the main unit linked at the 2,1 and 2,6 positions. Gums and mucilages are also found in some plants. These carbohydrates are chemically related to the hemicelluloses, and some, such as carageenan, gum arabic, locust bean gum, and gum tragacanth, are isolated and used as food additives (see Chap. 3).

3. Simple Sugars

The sugar content of fruits and vegetables varies from negligible (e.g., avocado) to over 20% (e.g., ripe banana) on a fresh weight basis. Sucrose, glucose, and fructose are the principal sugars of most commodities. In general, fruits and vegetables contain more reducing sugars than sucrose, although in many cases the reverse is true. Other sugars, such as xylose, mannose, arabinose, galactose, maltose, sorbose, octulose, and cellobiose, may also be present, and in some instances they may constitute a major portion of the total sugars. Plant tissues may also contain sugar alcohols, such as sorbitol, and sugar acids, such as ascorbic acid.

C. Proteins and Other N Compounds

The protein content of edible plants varies considerably, although it generally represents only a small percentage of the fresh weight. Certain crops, such as cereals, pulses, tubers, and bulbs, may contain appreciable quantities of protein. When appreciable quantities of proteins accumulate, they are normally referred to as "storage proteins," although it is not clear that they are synthesized by the plant for the purpose of storage. These substances may function to lower the osmotic pressure of the amino acid pool, bind ammonia without changing in situ acidity, or they may serve as macromolecular shields to protect other compounds against enzymic action (21). Regardless of the role of storage proteins in the physiology or metabolism of the plant, they are very important to the food scientist because of their contribution to the nutritive value and functional properties of various crops. Plant storage proteins are generally categorized based on their solubility, but this type of classification is somewhat arbitrary. It also should be recognized that the isolation of plant proteins may be hindered by the presence of tannins and phenolic oxidizing enzymes. A general feature of storage proteins from seeds and tubers is the occurrence of amide linkages. For example, rapeseed protein has an isoelectric point of pH 11, and almost all the aspartic and glutamic acid residues are amidated. Generally, plant storage proteins decrease in molecular weight during senescence. This decrease in molecular weight is partly due to dissociation of larger complexes. Finally, it should be noted that the storage proteins from different cultivars of a given species may exhibit different electrophoretic patterns. Accordingly, electrophoresis or isoelectric focusing can be used to "fingerprint" or identify cultivars of unknown origin (16).

1. Cereal Proteins

The protein content of wheat is about 12%, and extensive research has shown that the baking performance of wheat flour is related to the properties of certain storage proteins. Early workers divided the proteins of wheat into four solubility classes: albumins, which are soluble in water; globulins, which are soluble in salt solutions but insoluble in water; gliadins, which are soluble in 70-90% alcohol; and glutenins, which are insoluble in neutral aqueous solutions, saline solutions, or alcohol. In wheat, the principal storage proteins are the gliadin and glutenin fractions. These proteins represent about 80-85% of the endosperm protein.

The gliadins are a heterogeneous mixture containing as many as 40-60 components. The molecular weights of these proteins are approximately 36 kilodaltons, the amino acid sequences are homologous, and they appear to exist as single polypeptide chains rather than as associated units.

Glutenin proteins consist of at least 15 distinct molecular weight fractions ranging from 12 to 133 kilodaltons for unassociated components. Associated forms of glutenin have average molecular weights of 150-3000 kilodaltons. Subunits associate by intra- and intermolecular disulfide bonds.

The physical characteristics of wheat flour doughs can be changed by adding small amounts of agents that oxidize or reduce disulfide linkages. Dough viscosity and elasticity can also be modified by the addition of proteolytic enzymes. The use of such additives to control mixing times and product uniformity is standard practice in continuous bread-making operations.

The albumin and globulin protein fractions have relatively low molecular weights of approximately 12 kilodaltons and remain unchanged in size following reduction of disulfide bonds.

Gluten is a complex formed from gliadins and glutenins following hydration and mixing. Gluten may be prepared from wheat flour by gentle washing with an excess of water. The rubbery residue obtained by this procedure contains about 75-85% protein and 5-10% lipid on a dry weight basis, and may contain varying amounts of starch or albumin and globulin proteins. The hydrated gluten complex forms a three-dimensional viscoelastic network that gives wheat flour its valued dough and bread-making characteristics. The rheological properties of the gluten complex apparently relate to the high content of amide groups, which are responsible for hydrogen bonding; the high content of proline, which disrupts secondary structures; an appropriate number and location of hydrophobic amino acids, which contribute to glutenin-gliadin interaction; and the presence of half-cystine groups, which enable the formation of intra- and intermolecular disulfide linkages.

2. Tuber Proteins

Approximately 70-80% of the extractable proteins in potato tuber (about 2-4% protein and amino acids on a fresh weight basis) are classified as storage protein. Up to 40 individual proteins have been separated from potatoes by electrophoresis. In mature potato tubers, the main group of proteins has subunit molecular weights of 17-20 kilodaltons. The heterogeneity of these storage proteins appears to be due to charge differences, and it is likely that this is a reflection of different amounts of amide in the polypeptide chain.

3. Pulse Storage Proteins

The storage proteins in pulses, such as soybean, are also characterized by heterogeneity in molecular weight and charge. Soybean proteins do not contain gliadin and glutenin, and they have a relatively high solubility in water or dilute salt solutions at pH values above or below the isoelectric points. The water-soluble proteins have a molecular weight range of approximately 20-600 kilodaltons. The 11S globulin fraction (350 kilodaltons) contains 12 subunits that show a complex pattern of association-dissociation. Interaction of these proteins is partly stabilized by hydrophobic bonds. Dissociation of subunits may be affected by appropriate conditions of ionic strengh, pH, or solvent (23). The solubility and food-related functional properties of soybean proteins can be improved by limited hydrolysis with proteolytic enzymes, such as trypsin. Excessive hydrolysis of soybean proteins is normally undesirable since it gives rise to bitter-tasting peptides (see Chaps. 5 and 9).

4. Nonprotein Nitrogen

Plant tissues may contain appreciable quantities of nonprotein nitrogen. For example, more than two-thirds of the nitrogen in potato tuber or apple fruit may be in the form of free amino acids and other constituents. Accordingly, one must use caution in extrapolating nitrogen content to protein content. Besides amino acids, plants may contain appreciable quantities of nonprotein nitrogen in amines, purines, pyrimidines, nucleosides, betaines, alkaloids, porphyrins, and nonproteinogenic amino acids (Table 2).

D. Lipids and Related Compounds

Lipids may be largely confined to cellular membranes or they may be present as reserve material in certain crops (Table 3). Storage lipids consist of triacylglycerols with characteristic fatty acid components, notably palmitic, oleic, and linoleic acids (Table 4). In plant tissues that contain no storage lipid, the majority of lipid is membrane-associated phospholipid or glycolipid. Phospholipids may also accumulate in certain storage organs; for example, soybeans contain up to 8% phospholipd. Lipid substances are also prominent in the protective epidermal cells of certain plant organs. Some examples of these protective substances are illustrated in Fig. 1. Other lipid or lipophilic substances, although present in trace quantities, may contribute to the characteristic flavor of edible crops. Odoriferous substances of fruits are mainly oxygenated compounds (esters, alcohols, acids, aldehydes, and ketones), many of which are derivatives of terpenoid hydrocarbons or lower aliphatic acids and alcohols. The range of such volatile constituents present in vegetables is generally more limited than that found in fruits. In some instances, these substances are dissolved in terpenoid hydrocarbons and localized in special oil sacs, as are the essential oils of peppermint leaf. In other cases, the components are formed as a result of cellular decompartmentation during wounding, cooking, or chewing the tissue.

E. Organic Acids

Small quantities of organic acids occur in plants as metabolic intermediates (e.g., tricarboxylic acid cycle, glyoxylate cycle, or shikimic acid pathway) and they may also accumulate in vacuoles. The accumulation of organic acids gives rise to an acidic or sour taste (8). Acid levels range from very low (e.g., sweet corn) to high in such crops as currant, cranberry, or spinach. An acidic crop may contain over 50 mEq acid per 100 g of tissue, and a pH less than 2.

1. Aliphatic Plant Acids

The most widely occurring and most abundant acids in plants are citric and malic, each of which can constitute up to 3% of the tissue on a fresh weight basis. Instances in which

Table 2 Nonprotein Nitrogenous Compounds Found in Plant Tissues

Substance	Structure	Occurrence
γ-Aminobutyric acid	CH$_2$–CH$_2$–CH–COOH 　　　　　| 　　　　NH$_2$	Grapefruit, lemon, grape, raspberry, avocado, potato
Choline	CH$_3$ | CH$_3$–N–CH$_2$CH$_2$OH | CH$_3$	Grapefruit, grape, potato, eggplant; universally distributed as part of lecithin
Glutathione	SH | CH$_2$ | HOOC–C–CH$_2$–CH$_2$–C–NH–C–C–N–CH$_2$–COOH 　　|　　　　　　‖　　|　‖　| 　　NH$_2$　　　　O　　H　O　H	Citrus fruit, grape, potato
Pipecolic acid		Apple, cherry, peach, grape, raspberry, avacado

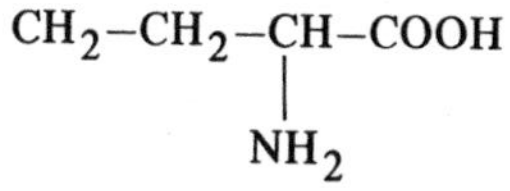

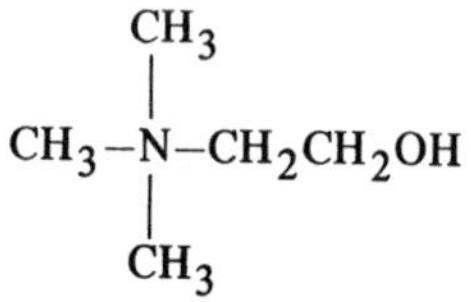

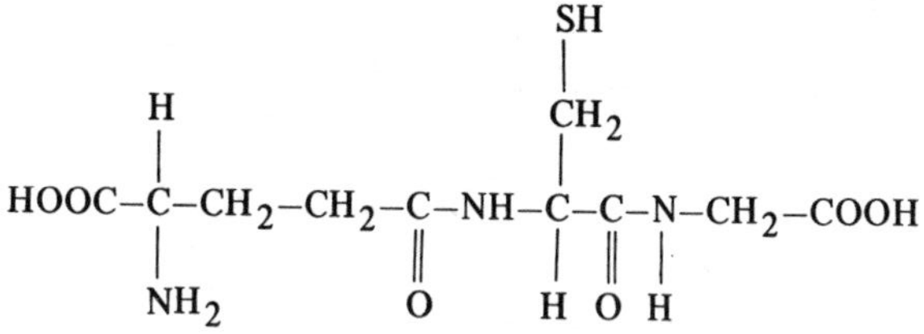

Name	Structure	Distribution
Asparagine	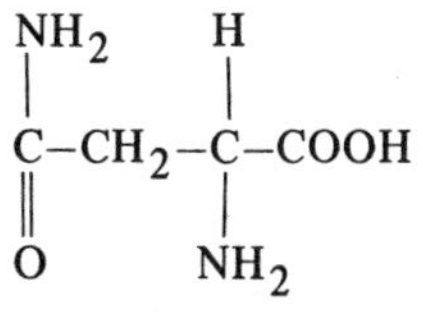 $NH_2-C-CH_2-\overset{\displaystyle H}{\underset{\displaystyle NH_2}{C}}-COOH$, $\overset{\displaystyle O}{}$	Widely distributed
Solaninidine		Potato, toxic alkaloid
Dopamine		Banana fruit
Ornithine	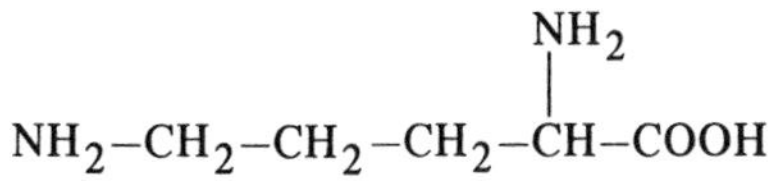 $NH_2-CH_2-CH_2-CH_2-\overset{\displaystyle NH_2}{CH}-COOH$	Cherry, plum, lime, grape
S-Methyl-L-cysteine sulfoxide	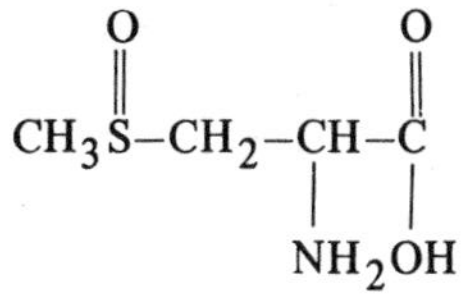 $CH_3\overset{\displaystyle O}{S}-CH_2-\overset{\displaystyle O}{\underset{\displaystyle NH_2OH}{CH-C}}$	Onion, cabbage, turnip
Taurine	$NH_2-CH_2-CH_2SO_3H$	Peach, papaya, fig, potato

Table 3 Lipid Contents of Some Fruits

Source	Lipid content (% DWB)[a]
Oil-palm trees	74–81
Avocado	35–70
Olive	30–70
Laurel	24–55
Grape	0.2
Banana	0.1
Apple	0.06

[a]DWB = dry weight basis.

neither malic or citric acids predominate include grapes and avocado (tartaric), spinach (oxalic), and blackberry (isocitric). The total acidity of many fruits declines during ripening, although specific acids may actually increase.

2. Carbocyclic Plant Acids

Various aromatic acids are also found in plant tissues (Fig. 2). The alicyclic acids, quinic and shikimic, are widely distributed in the plant kingdom as key intermediary metabolites. Certain aliphatic acids, such as chlorogenic, are important food constituents because of their contribution to enzymic and nonenzymic browning reactions. Others, such as benzoic, play an important role in such crops as cranberry, by virtue of their antifungal acitivity.

Table 4 Principal Fatty Acids of Reserve Triglycerides[a]

Source	Fatty acids (kind and % total fatty acid content of triacylglycerols in each source)							
	Lauric	Myristic	Palmitic	Stearic	Oleic	Linoleic	Linolenic	Arachidonic
Coconut	45	20	5	3	6	–	–	–
Palm kernel	55	12	6	4	10	–	–	–
Olive	–	–	15	–	75	10	–	–
Peanut	–	–	9	6	51	26	–	–
Cotton-seed	–	–	23	–	32	45	–	–
Corn	–	–	6	2	44	48	–	–
Soybean	–	–	11	2	20	64	3	–
Sunflower	–	–	6	4	31	57	–	–
Wheat	–	3	18	7	16	51	4	2
Rye	–	6	11	4	18	35	7	2

[a]The spectrum of fatty acids in a vegetable oil influences its functional properties (e.g., solidification temperature, tendency to oxidize, smoke point, flash point, flavor, and nutritive value).

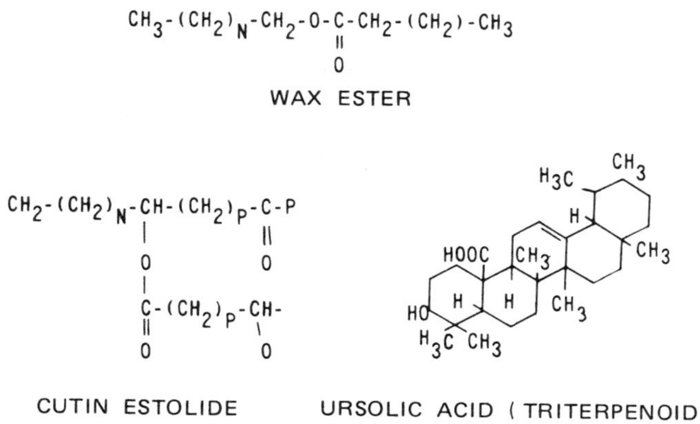

$$CH_3-(CH_2)_N-CH_2-O-C-CH_2-(CH_2)-CH_3$$
$$\underset{O}{\overset{\|}{}}$$

WAX ESTER

$$CH_2-(CH_2)_N-CH-(CH_2)_P-C-P$$

CUTIN ESTOLIDE   URSOLIC ACID (TRITERPENOID)

Figure 1 Examples of lipid substances that accumulate in the outer epidermal membranes of certain vegetative tissues. These compounds have properties of water repellance, melting points between 40 and 100°C, and paracrystallinity at ambient temperature.

F. Pigments

The principal pigments of plant tissues—the chlorophylls, carotenoids, and flavonoids—are discussed in Chap. 8. The extent of pigment synthesis and degradation in detached fruits and vegetables can be influenced by storage conditions, such as light, temperature, and relative humidity, and by volatile substances such as ethylene. Ethylene is formed in plant tissue during ripening, following wounding, or by exposure of the plant to air pollutants. By virtue of its hormone action, ethylene initiates the degradation of chlorophyll. The exact mechanism by which ethylene action leads to chlorphyll degradation is not known, although it appears that oxidative reactions are contributory.

G. Mineral Elements

The total mineral content of plant tissues is sometimes expressed as ash content (residue remaining after incineration). The ash content of plant tissue varies from less than 0.1% to as much as 5% of the fresh weight. Mineral content is influenced by species characteristics as well as by agronomic practices. Natural variations in the contents of several minerals in vegetables are shown in Table 5. The distribution of particular minerals in a given tissue is known to be nonuniform. In peas, phosphorus is many times richer in cotyledons than in testa, whereas the difference is reversed for calcium. The most abundant mineral elements in plants are potassium, calcium, magnesium, iron, phosphorus, sulfur, and nitrogen. Potassium, the single most abundant element, occurs in many plants; for example, parsley contains over 1% on a fresh weight basis. Mineral elements occur mainly as salts of organic acids.

It has become increasingly evident that the mineral content of a given species can have a profound influence on physiological disorders that arise pre- and postharvest. In recent years there has been considerable study of the role of calcium as related to the postharvest storage life of fruits and vegetables. If calcium is increased by agronomic practices or by postharvest treatments, this can result in improved storage life and product quality.

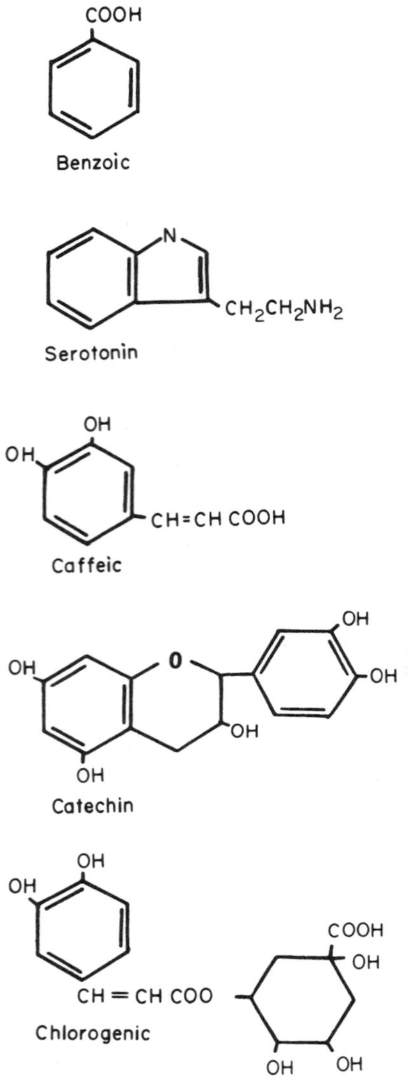

COOH

Benzoic

Serotonin
CH₂CH₂NH₂

OH
OH
CH=CH COOH
Caffeic

O
OH
OH
OH
OH
OH
Catechin

OH
OH
CH = CH COO
Chlorogenic
COOH
OH
OH   OH

Figure 2 Examples of carbocyclic acids that occur in plant tissues.

H. Vitamins

Plant tissues are an important source of several vitamins. Some examples of edible plant tissues that provide significant quantities of vitamins are summarized in Table 6. The vitamin content of a given species can vary considerably with variety, growing conditions, maturity, postharvest storage conditions, and processing. Vitamins are usually distributed nonuniformly in plant tissues. The old adage "it's a sin to eat the potato and throw away the skin," makes reference to the greater concentrations of thiamine and ascorbic acid in the cortex of the tuber as compared with the interior. The influence of postharvest storage on the vitamin content of different crops is incompletely understood, although it is known that the levels of ascorbic acid and β-carotene can fluctuate considerably.

Table 5 Normal Range of Some Macronutrients in Vegetative Tissues

Element	Normal range (mg per 100 g FWB)[a]	Especially rich sources
Ca	3–300	Spinach may contain up to 600 mg per 100 g FWB
Mg	2–90	Sweet corn
P	7–230	Seeds and young growing parts
Na	0–124	Celery
Cl	1–180	Celery
S	2–170	High-protein tissues
Fe	0.1–4	Parsley contains up to 8 mg per 100 g FWB

[a]FWB = fresh weight basis.
Source: Adapted from Duckworth (5).

III. STRUCTURE

Knowledge of the structure of edible plant tissues is important to the food chemist for several reasons. It is important to recognize that an excised mass of tissue is not necessarily homogeneous with respect to cell type, cellular organization and distribution of chemicals. In addition, the nature and extent of chemical changes that occur in the postharvest tissue are partly, if not wholly dependent on cellular organization. Each tissue is structurally adapted to carry out a particular function. Much of the metabolic activity of the plant is carried out in relatively unspecialized tissue called parenchyma, which generally makes up the bulk of edible plant tissues. The outer layer of a plant is called the epidermis and is structurally adapted to provide protection against biological or physical stress. Specialized tissues, called collenchyma and sclerenchyma, provide structural support. Water, minerals, and solute molecules are transported through the vascular tissues, xylem and phloem.

A. Organ Structure

Consideration is given to four categories of edible plant tissues, with categorization based on their appearance: roots, stems, leaves, and fruits. Foods are also classified on the basis of their economic use, for example, fruits, vegetables, nuts, grains, berries, bulbs, or tubers. However, the latter type of classification is based on custom rather than on botanical systematics, and accordingly varies in different cultures. Indeed, there are legal precedents that the tomato is a vegetable and not a fruit, much to the dismay of the botanist.

1. Roots

The economically important root crops include sugar beets, sweet potatoes, yams, and cassava. The basic anatomical structure of root tissue is illustrated in Fig. 3. Fleshy roots are formed by secondary growth of the cambia (Fig. 3b, c). The secondary vascular tissues of such crops consist of small groups of conducting elements scattered throughout a matrix of parenchyma tissue. Multiple cambia are formed in certain tissues, such as beet root, (Fig. 3c) and sweet potato.

Table 6 Contribution of Edible Plants to Vitamins in the U.S. Diet

Group	Vitamin A (%)	Thiamine (%)	Riboflavin (%)	Niacin (%)	Ascorbic acid (%)
Fruits	7.3	4.3	2.0	2.5	35.0
Potatoes, sweet potatoes	5.7	6.7	1.9	7.6	20.9
Vegetables	36.4	8.0	5.6	6.8	38.3
Dry beans and peas	TR	5.5	1.8	7.0	TR
Flour (cereal)	0.4	33.6	14.2	22.7	0.0
Total (plants)	49.8	58.1	25.5	46.6	94.2

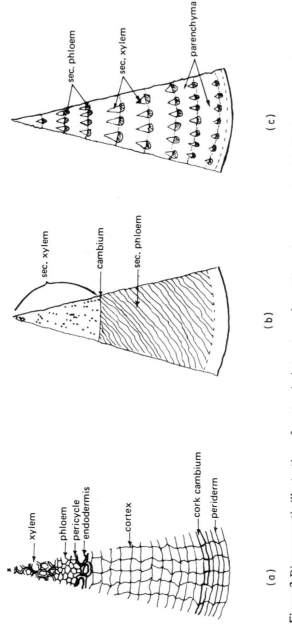

Figure 3 Diagrammatic illustration of anatomical structure of root tissue in cross-section (a). Fleshy roots are formed by secondary growth of meristematic tissue called cambia. A typical example is carrot (b), in which the cambia forms phloem on the outside and xylem on the inside. The conducting elements are scattered throughout a mass of parenchyma tissue. Beet root shows a different type of secondary growth, in which a series of concentric cambia are formed (c).

2. Stems

Examples of stem crops are potato, sugarcane, asparagus, bamboo shoot, rhubarb, and kohlrabi. Underground stems (rhizomes and bulbs) are modified structures and are exemplified by potatoes and onions. Topographically, a stem consists of four distinct regions: the cortex, located between the epidermis and the vascular system; and the pith, which forms the central core of cells (Fig. 4a). The secondary tissues exhibit different forms as illustrated in Fig. 4b and c. In bulbs, such as onion and garlic, the organism consists of a short stem that bears a series of fleshy leaves above and around the stem.

3. Leaf Crops

Lettuce, mustard greens, tea, chard, spinach, watercress, parsley, and cabbage are some examples of leaf crops. A leaf is typically a flat and expanded organ, primarily involved with photosynthesis, but deviations in morphology and function may occur. The leaf is composed of tissues that are fundamentally similar to those found in stem and root, but their organization is usually different. The epidermis contains a well-developed cuticle in most instances and stomata are characteristically present. Beneath the epidermal layer lies a series of elongated, closely packed palisade cells rich in chloroplasts. The vascular system consists of netlike veins in dicots and parallel veins in monocots.

4. Fruits

Botanically, a fruit is a ripened ovary or the ovary with adjoining parts; that is, it is the seed-bearing organ. The flesh of the fruit may be developed from the floral receptacle, for carpellary tissue, or from extrafloral structures, such as bracts. Whatever its origin, it is generally largely composed of parenchymatous tissue. The anatomical features of some fruit types are illustrated in Fig. 5. Legumes, capsules, caryopses, and nuts are not thought of as fruit by the layperson. Fruits are usually eaten raw as a dessert item, and they possess characteristic aromas due to the presence of various organic esters.

B. Tissue and Cell Structure

Some of the food-related properties of plant organs are intimately related to their substructure. Accordingly, some of the salient features of substructure are briefly discussed.

1. Parenchyma Tissue

The Greek word *parenchyma*, meaning "that which is poured in beside," is descriptive of the soft or succulent tissue that surrounds the hardened or woody tissues. This tissue is the most abundant type present in edible plants. The mature parenchyma cell possesses a thin cell wall, is occupied by large vacuoles, and has a size and shape that vary with different species. The extent of cellular contact or lack of same also varies with different commodities and has an important influence on texture. The spaces between adjacent cells can contain air (e.g., 25 vol% in some apples), or pectic substances. Pectins serve as intercellular cement, and the degree of polymerization, methylation and interaction with wall components can account for the smooth (e.g., peach) or coarse characteristics of different parenchyma tissues. Some examples of parenchymatous tissues are diagramatically illustrated in Fig. 6.

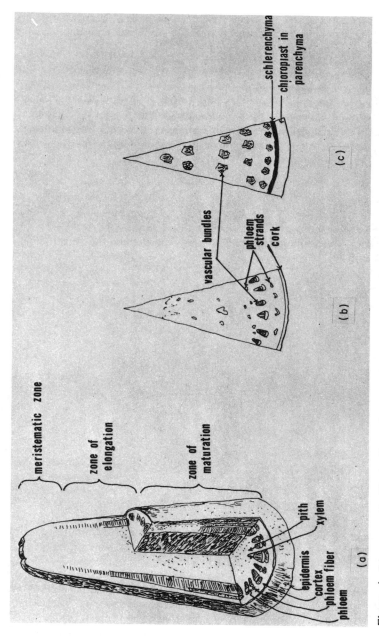

Figure 4 Diagrammatic illustration of anatomical structure of stem tissue (a). In dicotyledonous stem structures, such as potato (b), the vascular bundles are arranged in a single ring, as seen in transverse section. Monocotyledons, represented by asparagus (c), show a characteristic scattering of vascular bundles throughout the parenchymatous ground tissue.

2. Protective Tissues

Protective tissues develop at the surface of plant organs, forming the "peel" or "skin" of the commodity. The outermost multilayer region of cells is called the epidermis. These cells usually fit together compactly with no air spaces, and their outer walls are usually rich in wax or cutin (Fig. 1). An extracellular layer, called the cuticle, is usually present on the outer surface of the epidermis (Fig. 7). More elaborate structural modification of epidermal tissue is found in seeds, such as cereals, and in pericarps of some fruits, such as citrus and banana. The practical importance of protective tissues of fruits and vegetables is evident when postharvest crop losses are measured as a function of surface abrasions.

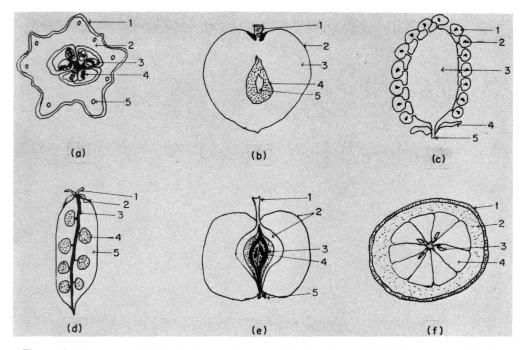

Figure 5 Diagrammatic illustrations of anatomical structures of different types of fruit. (a) Pepo (cucumber, squash, and pumpkin) in cross-section: (1) rind (receptacular), (2) flesh (ovary wall), (3) placenta, (4) seed, and (5) vascular bundle. (b) Drupe (cherry, peach, and plum) in longitudinal section: (1) pedicel, (2) skin (ovary wall), (3) flesh (ovary wall), (4) pit (stony ovary wall), and (5) seed. (c) Aggregate (raspberry, strawberry, and blackberry) in longitudinal section: (1) fleshy ovary wall, (2) seed (stony ovary wall plus seed), (3) fleshy receptacle, (4) sepal, and (5) pedicel. (d) Legume (pea, soybean, and lima bean) in longitudinal section: (1) pecidel, (2) sepal, (3) vascular bundles, (4) seed, and (5) pod (ovary wall). (e) Pome (apple and pear) in longitudinal section: (1) pedicel, (2) skin and flesh (receptacle), (3) leathery carpel (ovary wall), (4) seed, and (5) calyx (sepals and stamens). (f) Hespiridium (citrus) in cross-section: (1) collenchymatous exocarp (the flavedo), (2) parenchymatous mesocarp (the albedo), (3) seed, and (4) endocarp of juice sacs formed by breakdown of groups of parenchyma like cells. (g) Kernel of wheat, longitudinal and cross-sectional views. (Courtesy of the Wheat Flour Institute.)

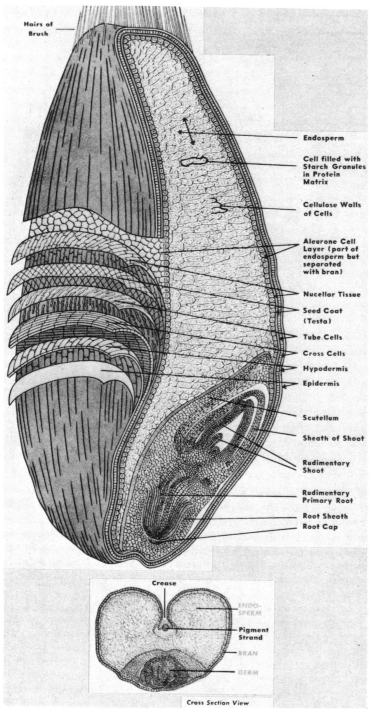

Hairs of Brush

Endosperm

Cell filled with Starch Granules in Protein Matrix

Cellulose Walls of Cells

Aleurone Cell Layer (part of endosperm but separated with bran)

Nucellar Tissue

Seed Coat (Testa)

Tube Cells

Cross Cells

Hypodermis

Epidermis

Scutellum

Sheath of Shoot

Rudimentary Shoot

Rudimentary Primary Root

Root Sheath
Root Cap

Crease

ENDO-SPERM

Pigment Strand

BRAN

GERM

Cross Section View

Figure 5 (continued)

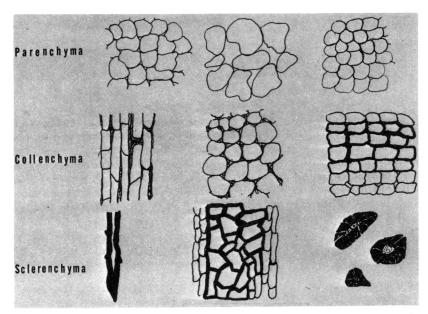

Figure 6 Diagrammatic illustrations of parenchymatous and supporting tissues.

Accordingly, it is important to "wound heal" certain commodities (e.g., potatoes) under suitable conditions prior to storage.

3. Supporting Tissues

Collenchyma tissue is characterized by elongated cells that have unevenly thickened walls (Fig. 6). This type of tissue is found particularly in petioles, stems, and leaves, occupying the ridges that are often situated on the surface of the organ (e.g., strands on the outer edge of the celery petiole). Thickened primary walls are especially rich in pectin and hemicelluloses, giving the walls unusual plasticity. Collenchyma tissue, although relatively soft, resists degradation during cooking and mastication.

Sclerenchyma cells are uniformly thick-walled cells that are normally lignified. Mature sclerenchyma tissue is nonliving and consists mainly of cell walls (Fig. 6). The two main types of sclerenchyma cells are "fibers" and sclereids. These cells are not perturbed by cooking and they give rise to stringy textures in asparagus and green beans or gritty textures in fruit, such as quince or pear. In certain commodities, rapid lignification is stimulated by the physical stresses associated with harvest, handling, or storage: hence, the folklore that, "water in the pot should be boiling before going to the garden for certain vegetables."

4. Vascular Tissue

The vascular tissues, xylem and phloem, are complex tissues composed of simple and specialized cells. Although it is not clear whether vascular tissue has significant direct impact on food quality, it is possible that the vascular system has an influence on physiological events (e.g., storage disorders) in postharvest crops.

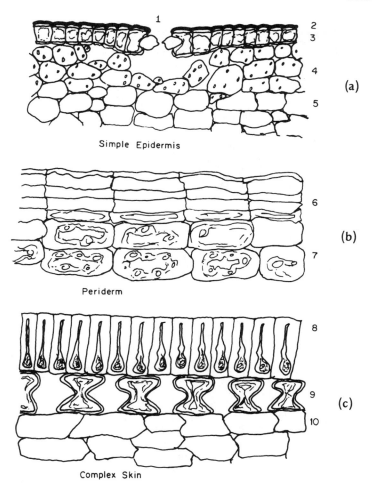

Figure 7 Diagrammatic illustration of protective tissues. (a) Simple epidermis, such as asparagus stem. (b) Periderm, such as potato tuber. (c) Complex "skin," such as testa of a pear. (1) Stoma, (2) cuticle, (3) epidermis, (4) chloroplast containing parenchyma, (5) unspecialized parenchyma, (6) corky tissue, (7) parenchyma cells containing developed starch granules, (8) compact layer of elongated epidermal cells with heavily thickened walls, (9) hypodermal layer of hourglass-shaped cells with large intercellular spaces, and (10) parenchymatous inner layer.

C. Subcellular Structure

The main components of typical plant cells are illustrated in Fig. 8. The reader should refer to books on botany and plant physiology for detailed information on the structural characteristics of organelles. Organelles such as nuclei, mitochondria, microsomes, plastids, vacuoles, and lysosomes have been studied to varying degrees in the context of postharvest events, and important findings from these studies are discussed in the next section.

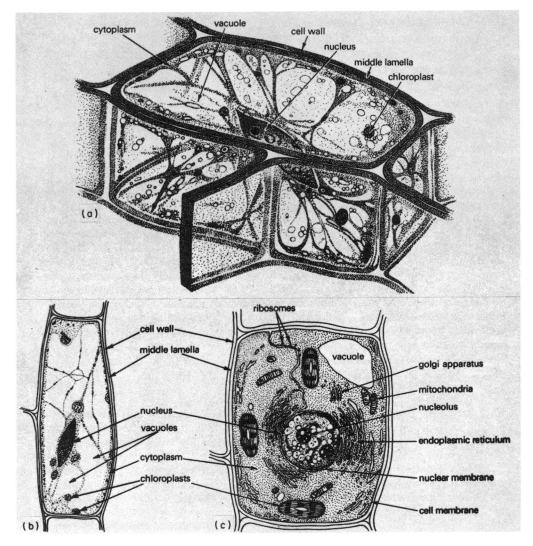

Figure 8 Diagrammatic representation of a mature plant cell. (a) Light microscope, (b) light microscope section, (c) electron microscope image. (Reprinted from Ref. 12, courtesy of W. A. Freeman.)

IV. PHYSIOLOGY AND METABOLISM

If we imagine that a circle encloses the set of all biochemical reactions occurring in plant tissues, we can, on the basis of current knowledge, define a subset consisting of reactions that are basically identical in all plant tissues. This common subset of reactions includes basal events, such as activation of amino acids preparatory to protein synthesis, mitochondria-linked respiration, and anabolism of certain cell wall polysaccharides. In addition to reactions common to all detached plant parts, we can imagine additional subsets of reactions common to large groups of organs, such as photosynthesis in green plants or organic acid metabolism. Finally, there may be aspects of metabolism that are highly specific to individual families, genera, species, or cultivars. An example of the latter type of subset is the biogenesis of a family of esters that contribute to the distinctive aroma of a given plant. Unfortunately, our knowledge of metabolism in plant tissues is shallow in comparison to well-studied systems, such as those of the rat and *Escherichia coli*.

A. Respiration

Respiration is an aspect of basal metabolism of primary concern to postharvest physiologists. The mechanism of respiration in detached plant parts is basically the same as that in organs of animal origin. The importance of respiration in postharvest crops is best illustrated by comparing the relationship between respiratory rate and storage life of fleshy plant parts (Table 7; Fig. 9). Commodities that exhibit rapid carbon dioxide evolution or oxygen consumption rates are generally quite perishable, whereas those with slow respiratory rates may be stored satisfactorily for relatively long periods of time (19). Moreover, the shelf life of a given commodity can be greatly extended by placing it in an environment that appropriately retards respiration (e.g., refrigeration, or controlled atmosphere).

The relative importance of different oxidation cycles (Krebs cycle, glyoxylate shunt, Embden-Meyerhof pathway, and pentose phosphate cycle) appears to vary among species, among organs, and with the ontogeny of the plant. Metabolic pathways involving glycolysis and Krebs cycle oxidation are most common, although pentose phosphate oxidation can account for approximately one-third of glucose metabolism in some instances. The relative importance of the pentose cycle appears to increase in mature tissues, and it is known to be operative in such fruit as pepper, tomato, cucumber, lime, orange, and banana.

The respiratory quotient (RQ) is a useful qualitative guide to changes in metabolic pathways or the emergence of nonrespiratory decarboxylase or oxygenase systems (Table 8). The RQ is defined as the ratio of volumes of carbon dioxide evolved and oxygen consumed during a given period of respiration. The RQ for fruit respiration increases at the climacteric stage of ripening.

The rise in RQ during climacteric respiration is indicative of increased decarboxylation or decreased carboxylation. Decarboxylation has been attributed to the increased activity of malic enzyme in ripening pear and apple fruit.

Table 7 Respiration Rates and Perishability of Fruits and Vegetables[a]

Commodity	Respiration (mg CO_2 per kg hr)		Storage life (weeks) at 5°C
	5°C	25°C	
Peas	50	475	1
Asparagus	45	260	2-3
Avocado	10	400	2-4
Turnips	6	17	16-20
Apples	3	30	12-32

[a]Data are representative and vary with cultivar and cultivation practice of a particular commodity. Note that the Q_{10} for respiration may differ with different commodities.

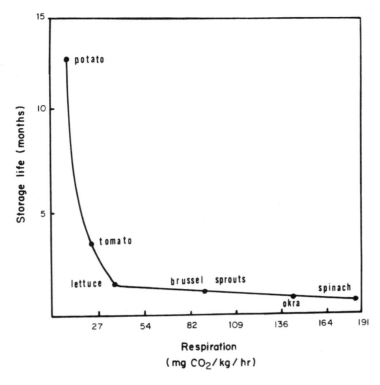

Figure 9 Relationship between initial respiration rate and storage life to unsalable condition.

Table 8 Respiratory Quotients of Some Detached Plant Parts[a]

System	Respiratory quotient
Leaves rich in carbohydrates	1.0
Germinating starch seeds	1.0
Germinating linseed	0.64
Germinating buckwheat seed	0.48
Maturing linseed	1.22
Mature apples in air	1.0
Overripe apples in air	1.3
Apples browning in air containing chloroform vapor	0.25
Apples treated with HCN	2.0
Wheat seedlings in 5-20% O_2	0.95
Wheat seedlings in 3% O_2	3.34

[a]Suggestion: Using a set of metabolic charts, rationalize the differences in respiratory quotients on the basis of storage materials being oxidized and metabolic pathways.

Although respiratory rates of different commodities differ widely, it is not always clear why they do so. Cellular respiration may be coupled to various metabolic reactions related to wounding, harvesting, normal cellular maintenance, or the normal ontogeny of the organ. Catabolism of sugars is clearly important in such commodities as sweet corn and peas which undergo a dramatic loss in sweet taste within hours after harvest. Similarly, anabolic reactions (such as the synthesis of lignin), which are dependent on energy made available by respiration, are equally important in postharvest systems.

Although much of the activity in postharvest plants is related to the energy-conserving reactions of oxidative phosphorylation, some of this activity results in generation of heat (thermogenesis). For example, the respiratory burst in senescing flowers and in wounded storage organs, such as potato tuber, is not completely coupled to ATP synthesis. Thermogenesis appears to operate via an alternate electron transfer chain, which is insensitive to cyanide and other inhibitors of cytochrome oxidase (Fig. 10). The plant hormone ethylene appears to evoke cyanide-insensitive respiration in a variety of postharvest crops (20). Although the relationship between ethylene action and the stimulation of respiration may be indirect, it is nonetheless an important one from the viewpoint of postharvest metabolism. Evocation of the alternate electron transfer chain may function to decontrol respiration causing: (a) rapid catabolism of storage polysaccharides, (b) electron transfer without expenditure of reducing power, or (c) generation of heat, thereby differentially affecting chemical reactions or causing volatilization of biologically active molecules or accelerating autolytic reactions.

1. Respiratory Patterns

Most fleshy fruits show a characteristic rise in respiratory rate more or less coincident with the obvious changes in color, flavor, and texture that typify ripening. Fruit that exhibit this respiratory rise are classified as "climacteric fruit" (Table 9). This respiratory climacteric has been the subject of considerable study because it marks the onset of senescence [deterioration or old age (3)]. The relative magnitude of the climacteric burst varies considerably in different fruits, and many fruits (and all vegetables fail to exhibit an autonomous rise after harvest (Fig. 11; Table 9). This latter group has traditionally been categorized as nonclimacteric, although it may be that some of these fruits are postclimacteric when harvested.

No clear metabolic difference has been demonstrated between climacteric and nonclimacteric fruits, although ripening in fruits categorized as nonclimacteric often proceeds more slowly. The relation between climacteric rise and ripening indices is not always perfect. Although the climacteric rise and ripening occur simultaneously in many fruits under a wide variety of conditions, evidence exists that the events can be separated.

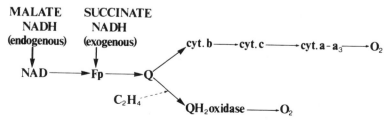

Figure 10 Pathways for oxidation of various substrates in mitochondria.

Table 9 Classification of Edible Fruits According to Respiratory Patterns

Climacteric		Nonclimacteric	
Common name	Scientific name	Common name	Scientific name
Apple	*Pyrus Malus*	Blueberry	*Prunus avium*
Apricot	*Prunus armeniaca*	Cacao	*Theobroma cacao*
Avocado	*Persea gratissima*	Cherry	*Prunus avium*
Banana	*Musa sapientum*	Cucumber	*Cucumis sativus*
Cherimoya	*Ammona cherimola*	Fig	*Ficus carica*
Feijoa	*Feifoa sellowiana*	Grape	*Vitus vinifera*
Mango	*Mangifera indica*	Grapefruit	*Citrus paradisi*
Papaya	*Carica papaya*	Lemon	*Citrus limon*
Passion fruit	*Passiflora edulis*	Litchi	*Litchi chinensis*
Papaw	*Asimina triloba*	Melon	*Cucumis melo*
Peach	*Prunus persica*	Olive	*Oleo europei*
Pear	*Pyrus communis*	Orange	*Citrus sinesis*
Plum	*Prunus americana*	Pineapple	*Ananas comosus*
Sapote	*Casimiroa adulis*	Rin-tomato	*Lycopersicum esculentum*
Tomato	*Lycopersicum esculentum*	Strawberry	*Frugaria vesc americana*
Watermelon	*Citrullus lanatus*	Tamarillo	*Chphomandra betacea*

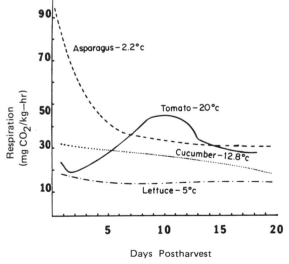

Figure 11 Respiratory patterns of a stem (asparagus), a leafy vegetable (lettuce), a non-climacteric fruit (cucumber), and a climacteric fruit (tomato). [Adapted from Ryall and Lipton (19).]

Detached stem, root, and leaf tissues normally respire at a steady rate or show a gradual decline in respiration rate with the onset of senescence under constant environmental conditions (Fig. 10). Although stress—imposed by mehcanical injury, temperature extremes, chemicals, or biological agents—can result in a burst of respiratory activity, an autonomous climacteric is not apparent.

2. Respiratory Control

The respiration process in plant tissues is markedly influenced by temperature. Within the "physiological" temperature range for a species, the rate of respiration usually increases as the temperature is raised, and the extent of the change in respiration rate can be expressed in terms of a Q_{10} value. The Q_{10} is defined as the rate of a reaction at T°C divided by the rate at $T - 10$°C. For most fruit and vegetable species Q_{10} values range from 7 to less than 1, although values between 1 and 2 are most common (Fig. 12). Lowering the temperature of a climacteric fruit delays the time of climacteric onset, as well as its magnitude. In certain commodities, such as potato tuber, transfer of the tissue from a

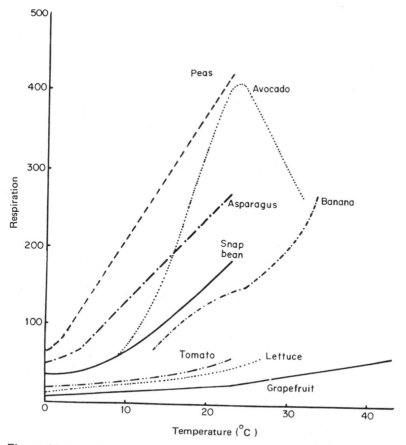

Figure 12 Respiration rate as a function of storage temperature for various commodities. [Adapted from Duckworth (5).]

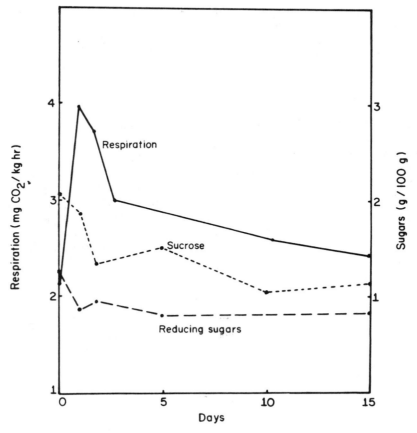

Figure 13 Changes in the amounts of reducing and nonreducing sugars and the rate of respiration when potatoes previously stored at 10°C are moved to 21°C. (Adapted from Ref. 17.)

cold to a warm environment results in a temporary burst in respiratory activity followed by equilibration to an intermediate level of activity (Fig. 13).

In general, either a reduction in oxygen tension (below 21%) or an increase in carbon dioxide tension (above 0.03%) slows respiration and associated deteriorative reactions. Certain commodities can be successfully held at very low oxygen tension (e.g., less than 1%) or at a very high partial pressure of carbon dioxide (e.g., greater than 50%); others are adversely affected by such extremes (see Sec. V.C). Furthermore, it should be recognized that the concentration of a given gas inside the tissue may differ considerably from that of the surrounding environment because of solubility, diffusion, and metabolic characteristics of the tissue. Cytochrome oxidase, the final electron acceptor in "normal" respiratory metabolism, has a relatively high affinity for oxygen and is saturated at relatively low concentrations of oxygen. On the other hand, the affinity of the alternate chain oxidase for oxygen is less than that of cytochrome oxidase, and accordingly respiration via the alternative oxidase is influenced more profoundly by low concentrations of oxygen in situ.

Certain chemical treatments have also been successfully employed to restrict the respiration of postharvest plant organs, and they act to lessen deteriorative reactions. Ventilation or scrubbing of storage atmospheres to minimize atmospheric ethylene accumulation can also be an effective means of maintaining low respiratory rates.

B. Gene Expression and Protein Synthesis

It is now clear that the expression of genetic information is an event of general and fundamental importance in postharvest crops. Increased levels of specific enzymes have been noted to occur in numerous plant organs following harvest. For example, aldolase, carboxylase, chlorophyllase, phosphorylase, peroxidase, phenolase, transaminase, invertase, pectic enzymes, phosphatase, o-methyltransferase, catalase, and indoleacetic acid oxidase are among the enzymes known to emerge in ripening fruit. It is becoming increasingly clear that many of these enzymes arise as a result of de novo synthesis and have a relatively short half-life in situ. It is generally observed that plant senescence is accompanied by increased RNA synthesis and an increased rate of transcription. Indeed, there appears to be an absolute requirement for continued protein synthesis in ripening fruit. For example, cycloheximide, an inhibitor of protein synthesis, also inhibits degreening, softening, and ripening of ethylene-treated banana fruit (Fig. 14). Increased protein synthesis and nucleic acid synthesis are most pronounced at the early stages of fruit ripening (3).

Recent studies have also demonstrated that slow-to-ripen or ripening-impaired mutant varieties of fruit can be propagated. These findings indicate that postharvest deterioration of plant tissues can be controlled by manipulation of gene expression. With recent advances in mutant selection techniques, gene cloning, and transfer of genetic material, it may now be possible to genetically engineer crops for improved storage life.

C. Ancillary Metabolic Processes

Many anabolic and catabolic reactions and reaction sequences exist that are more or less confined to certain groups or species. Unfortunately, other than for a few reactions, such as starch metabolism and pectin catabolism, relatively little information is available concerning metabolic events that specifically influence the quality of postharvest crops. Those postharvest metabolic events known to influence food quality are discussed in the next section.

1. Changes in Cell Wall Constituents

Quantitatively the most important biochemical changes taking place in harvested vegetative tissues involve carbohydrates. Fruit ripening and the softening of vegetative tissues is usually accompanied by catabolism of cell wall polysaccharides. Pectin-degrading enzymes have been studied extensively in ripening fruit. The two major groups of pectic enzymes found in mature plant tissues are esterases and hydrolases. Although hydrolase activity (polygalacturonase) has been demonstrated in relatively few fruits (tomato, pear, pineapple, and avocado), pectinesterase is commonly encountered, and its activity is often observed to increase during ripening. The amounts and types of pectic enzymes in fruits have an important bearing on the clarity or cloud stability of fruit juices and other derived beverages. In orange and tomato juices, in which a stable colloidal suspension is desired, the presence of pectinesterases and polygalacturonases is undesirable. In con-

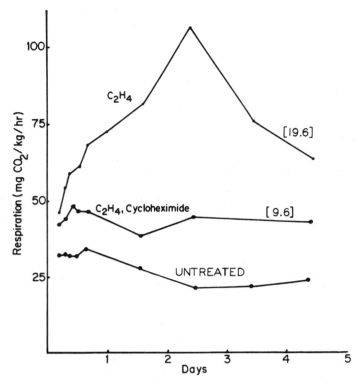

Figure 14 Affect of cycloheximide (2 μg/ml) on the respiration rate and sugar content of banana slices vented continuously with 10–15 ppm ethylene. Numbers in brackets are the percentage of soluble solids at the completion of the experiment. (Reprinted from Ref. 4, courtesy of Pergamon Press.)

trast, the presence of pectic enzymes is beneficial in such products as lime, apple, and grape juices, in which a clear beverage is considered desirable. The methylation of pectin via *S*-adenosylmethionine also appears to contribute to the softening of ripening fruit (14).

Other cell wall-degrading enzymes, such as cellulases and hemicellulases, are presumably active in vegetative tissues, although their contribution to changes in texture are not always clear. Although cellulase activity increases in ripening tomato fruit, it does not appear to have a detectable effect on softening. However, enzymes apparently do act on cell wall polysaccharides sufficiently to produce significant quantities of sugars in postharvest products, such as citrus and in some fruits that contain negligible starch. The toughening of vegetative tissues, such as asparagus and the pods of many leguminous species, which occurs shortly after harvest, is related to the biosynthesis of cell wall constituents, notably lignin.

2. Starch-Sugar Transformations

Synthesis of starch or its degradation to simple sugars are also important metabolic events in postharvest commodities. In potatoes, sugars are undesirable since they can cause poor texture after cooking, an undesirable sweet taste, and/or excessive browning during frying. The conversion of starch to sucrose and reducing sugars is promoted in most potato cultivars when they are stored at nonfreezing temperatures below 5°C. It is not known why cold exposure of potatoes stimulates starch catabolism, nor is it entirely clear which enzymes are involved. Suggested loci of control of this event are summarized in Fig. 15. Accumulation of sucrose may serve to rapidly divert hexose-6-phosphates, which would otherwise restrict the initial reactions of glycolysis. Amylases, which are important in starch catabolism in germinating tubers, do not appear to be very active in the dormant tuber. Some investigators have found that phosphorylase exhibits increased activity in tubers during storage at low temperatures, whereas others have reported that this enzyme exhibits no significant change in quantity or proportion of isoenzymes during cold storage. This is a moot point since there are a multitude of other modes by which the enzyme may be activated (in situ) in response to cold stress. For example, potato tubers contain a protein inhibitor of the enzyme invertase that becomes active at low, nonfreezing temperatures (cold stress). According to one scheme, sucrose, or perhaps sucrose phosphate, functions to limit starch hydrolysis via feedback control on phosphorylase or by promotion of starch synthesis via ADP glucose-starch glucosyltransferase. Although phosphorylase catalyzes synthesis of α1,4-glycosidic linkages, in vivo, it is involved primarily in starch breakdown. It has also been recently suggested that the major reason

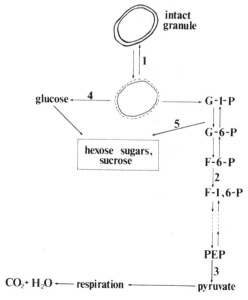

Figure 15 Some possible control loci of starch catabolism in plant tissues: (1) disruption of the plastid membrane by low temperature or other factors, (2) feedback control on phosphorylase, (3) feedback control on pyruvate kinase and entry of pyruvate into the tricarboxylic acid cycle, (4) amylase action to form dextrins and glucose, (5) activation of hexose phosphatases, and interconversion of hexose sugars to sucrose with the influence of invertase inhibitor.

for sugar accumulation in potatoes stored at low, nonfreezing temperatures is the cold lability of phosphofructokinase and possibly pyruvate kinase (1).

Regardless of why sugars accumulate during cold storage of plant commodities, this does represent a problem during commercial storage of potatoes. In practical situations, this problem is overcome by "reconditioning" (holding for several days at warm temperatures to reduce the sugar content through catabolism and conversion to starch) or by leaching out unwanted sugar by exposure of cut potatoes to water.

In some commodities, notably seeds (peas, corn, and beans) and underground storage organs (sweet potato, potato, and carrot), synthesis of starch, rather than degradation, may predominate after harvest. Starch synthesis is generally optimal at temperatures above ambient. Diminution of sugars accompanies starch synthesis in such crops and is normally detrimental to quality. The principal enzymes in starch synthesis appear to be those in photosynthetic tissues, namely, fructose transglycosidase, adenosine diphosphate glucose:starch glucosyltransferase, UDPG pyrophosphorylase, and ADPG pyrophosphorylase. However, the details of starch synthesis in postharvest commodities is not completely understood. Decreases in sugars may also result from their oxidation via mitochondria-linked reactions.

3.   Metabolism of Organic Acids

Organic acids are in a constant state of flux in postharvest plant tissues and tend to diminish during senescence. Much of the loss is undoubtedly attributable to their oxidation in respiratory metabolism as suggested by an increase in respiratory quotient. The respiratory quotient is approximately 1.0 when sugars are substrates, increases to 1.3 when malate or citrate are substrates, and further increases to 1.6 when tartrate is the substrate. Some suggested pathways associating respiratory metabolism with organic acid metabolism are shown in Fig. 16. Certain enzymes, (e.g., malic enzyme and pyruvate decarboxylase) and the respiratory quotient increase concomitantly with climacteric respiration of certain fruits.

The metabolism of aromatic organic acids via the shikimic acid pathway (Fig. 17) is important because of its relationship to protein metabolism (aromatic amino acids), accumulation of precursors of enzymic browning (e.g., chlorogenic acid), and lignin deposition (phenyl propane residues). It also should be mentioned that acetyl-CoA participates in the synthesis of phenolic compounds, lipids, and volatile flavor substances. Apart from their general importance to flavor, texture, and color, certain organic acids appear to modify the action of plant hormones and are thereby implicated in control of ontogenic changes. For example, the finding that chlorogenic acid modifies the activity of indole-3-acetic acid oxidase may well relate to the increased concentration of this acid in the green blotchy areas of defective tomatoes.

4.   Lipid Metabolism

Although the lipid components of most fruits and vegetables are present at relatively low levels, the metabolism is important during postharvest storage of plant tissues, especially when handling and processing conditions are not ideal. Changes in membrane lipids appear to be important in ontogenic events, such as ripening and senescence. The aging of various plant tissues is associated with a decline in polyunsaturated fatty acids, and such changes appear to relate to autolysis of membranes and associated loss in cell integrity. Oxidative enzymes, such as lipoxygenase, and hydrolytic enzymes, such as lipase

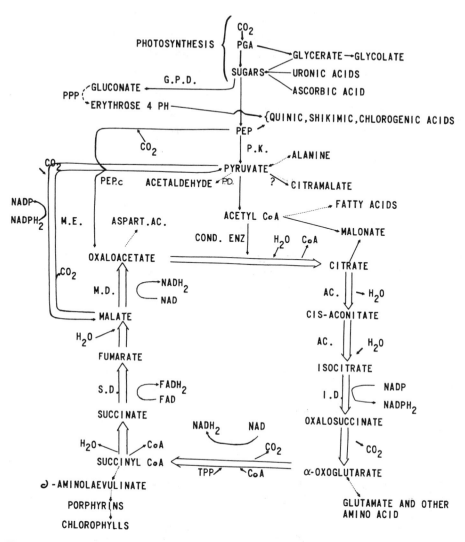

Figure 16 Krebs cycle (⇉) with some other modes of organic acid synthesis (→) or synthesis of other important constituents (⇢). Abbreviations: AC, aconitase; Cond. enz, condensing enzyme; GPD, glucose-6-phosphate dehydrogenase; ID, isocitric dehydrogenase; MD, malate dehydrogenase; ME, malic enzyme; PEPc, phosphoenolpyruvate carboxykinase; PGA, phosphoglyceric acid; PD, pyruvate decarboxylase; PK, pyruvate kinase; PPP, pentose phosphate pathway; SD, succinic dehydrogenase; TPP, thiamine pyrophospate. (Reprinted from Ref. 3, p. 89, courtesy of Academic Press.)

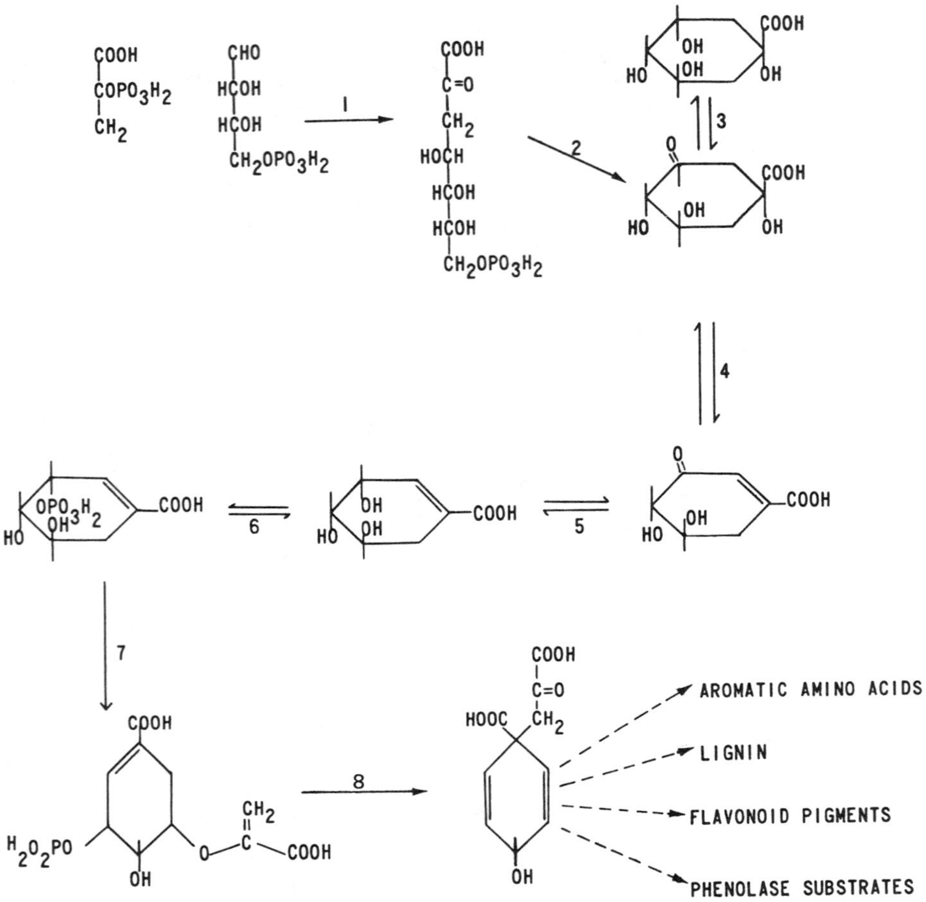

Figure 17 Biosynthesis of some important cell constituents via the shikimic acid pathway. (1) Phosphoenolpyruvate + D-erythrose-4-phosphate to 3-deoxy-D-arabinoheptulosonic acid 7-phosphate, (2) to 5-dehydroquinic acid, (3) to quinic acid, (4) to 5-dehydroshikimic acid, (5) to shikimic acid, (6) to 5-phosphoshikimic acid, (7) to 3-(enolpyruvate ether) of phosphoshikimic, (8) to prephenic acid. Although this scheme is based on studies with microorganisms, there is ample evidence that this pathway operates in higher plants.

and phospholipase, have been shown to be important in postharvest metabolism. It is also possible that enzymic systems that yield hydrogen peroxide and species of activated oxygen, such as the superoxide anion, also contribute to oxidative reactions in plants. For example, hydrogen peroxide and other organic peroxides appear during the ripening of certain fruits, and the ripening process can be arrested by treatment with antioxidants (7). Such processes as freezing and dehydration can activate lipoxygenase in unheated plant tissues and thus lead to off-flavor development.

The reserve lipids (e.g., in palm, olive, and avocado) have compositions markedly different from those that are part of the functional cell framework. Little is known concerning the metabolic control of triacylglycerol accumulation in developing plant tissues, nor is it clear whether the enzyme systems that participate in synthesis of reserve lipids are distinct from those involved with the synthesis of membrane lipids. The enzymic hydrolysis of triacylglycerols in postharvest oilseeds has an important bearing on their quality.

5. Pigment Metabolism

Carotenoids. The biosynthesis of carotenoids and related terpenoids is an important event in many edible plant tissues. Biosynthesis of these compounds occurs through the universal system for isoprene compounds, and it involves the formation of mevalonic acid from acetyl-CoA. Carotenoid biosynthesis and catabolism occur in detached plant parts, and these reactions may be dramatically influenced by storage conditions. In certain instances, these reactions are stimulated by oxygen, inhibited by light and high temperatures, and affected by hormones, such as ethylene and abscisic acid. The most common catalytic agents for carotenoid destruction appear to be the lipoxygenases (indirectly via lipid oxidation) and the peroxidases (apparently promote β-carotene degradation directly).

Flavonoids and Related Pigments. The pathway for synthesis of the basic C6:C3 structure that comprises the A phenyl ring of the flavonoids is shown in Fig. 17. Relatively little is known about reactions involved with the formation of specific flavonoid pigments, although there have been recent advances in this field (22). Synthesis of anthocyanins—colorants related to flavonoids—occurs in postharvest plant organs and is stimulated by light and influenced by temperature. The purple anthocyanin pigments of red cabbage are synthesized and accumulate when cabbage is placed in storage below 10°C. Preharvest treatments, such as application of *N*-dimethylaminosuccinamic acid, can induce early formation of anthocyanins in certain crops.

The catabolism of anthocyanin pigments is not very well understood. Anthocyanins can be oxidized by degradation products resulting from the oxidation of phenolic compounds. Decolorizaton of anthocyanins can also be initiated by endogenous or fungal glycosidases.

Chlorophylls. One of the more obvious changes that occurs in senescing plant tissues containing chlorophyll is the loss of a characteristic green color. Peel and sometimes pulp degreening is associated with the ripening of most fruit, and "yellowing" is a characteristic of most stem and leaf tissues consumed as vegetables. Chlorophyll degradation is accompanied by the synthesis of other pigments in a number of detached plant tissues. Chlorophyll catabolism is markedly influenced by environmental parameters, such as light, temperature and humidity, and the effects of these factors are specific for the tissue. For example, light accelerates the degradation of chlorophyll in ripening tomatoes and promotes the formation of the pigment in potato tubers.

Chlorophyll degradation in plant tissues often can be promoted by application of parts per million levels of the hormone ethylene. It is a common commercial practice to utilize this principle for degreening citrus fruits and other commodities such as banana and celery. Chlorophyllase emergence in plant tissues is associated with the buildup of endogenous ethylene. However, the relationship between chlorophyllase action and degreening is not clear at this time. Chlorophyllase is a hydrolytic enzyme that converts chlorophylls a and b to their respective chlorophyllides (see Chap. 8). Present evidence suggests that chlorophyllase does not catalyze the initial step in chlorophyll degradation in senescing plant tissues. The initial steps in this process appear to require molecular oxygen, and removal of the phytol side chain occurs in later steps. Lipoxygenase, which emerges in many plants during senescence, is known to contribute to the loss of chlorophyll in frozen vegetables, probably through the action of lipid hydroperoxides.

6. Aroma Compounds—Biogenesis and Degradation

The characteristic flavor of fruits and vegetables is determined by a complex spectrum of organic compounds. Aroma of fruits and vegetables is generally caused by esters, aldehydes, ketones, and so on (see Chap. 9). Our understanding of the metabolism of such compounds is relatively shallow at the present time. In Bartlett pears, the loss in characteristic flavor of overripe fruit is associated with the rapid diminution of 2,4-decadienoic acid esters and a concomitant increase in esterase activity.

In certain tissues, flavor precursors are enzymically converted to flavor compounds when the cells are disrupted by chewing or other means of mechanical injury. A patented process exists whereby flavor-potentiating enzymes are added to processed foods just prior to consumption, to regenerate the fresh flavor that was lost during heating or dehydration. This method takes advantage of the fact that precursor molecules may be stable to processing, whereas the flavor compounds and enzymes are labile. The concept of flavor precursors and their conversion to flavor compounds also has an important bearing on the characteristic flavors that may develop as a consequence of different cooking and food preparation techniques (see also Chap. 8).

D. Control Mechanisms

The dynamic chemical changes in postharvest fruits and vegetables, as illustrated above, are subject to a wide array of biochemical control mechanisms. Hence, our ability to control specific desirable or undesirable quality attributes of these crops is dependent on our understanding of how such changes are orchestrated at the cellular level. The mere presence of an enzyme does not alone indicate its functionality in association with a physiological event. For example, invertase is present in potato tubers, but its activity is arrested by a proteinaceous invertase inhibitor when potatoes are stored at low, nonfreezing temperatures. Flavor precursors are not necessarily converted to flavor compounds in situ, despite the presence of appropriate flavorases, since the reactants may be physically separated within the cell microstructure. Similarly, subtle differences in microconstituents, such as inorganic ions, can profoundly influence the direction of metabolism by modifying the activity of a particular enzyme. For example, calcium ions appear to limit peroxidase-catalyzed reactions associated with lignification of cell walls. Factors involved with the control of enzyme-catalyzed reactions are discussed in more detail in Chap. 6.

Specific mention should be made of the plant hormones, which are important regulators of physiological events in detached plant organs. Of the five known categories of plant hormones (four are shown in Fig. 18) ethylene has received the most attention by postharvest physiologists. The ability of this agent to "trigger" ripening of fruit and generally promote plant senescence in plant tissues has led to the descriptive name "ripening hormone."

The mechanism of ethylene biogenesis in plant tissues has been the subject of extensive research. There is now convincing evidence that L-methionine is the precursor of ethylene and that S-adenosylmethionine (SAM) and 1-aminocyclopropane-1-carboxylic acid (ACC) are key intermediates in the conversion of methionine to ethylene in situ

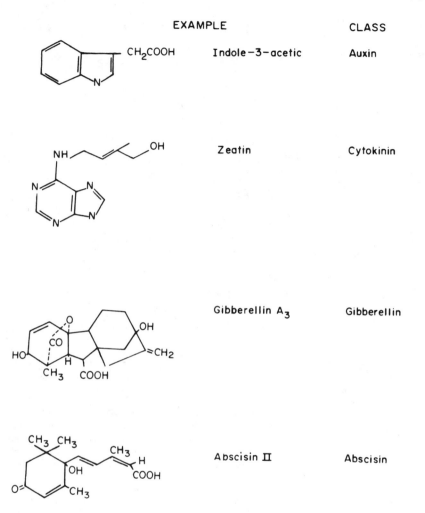

EXAMPLE   CLASS

Indole-3-acetic   Auxin

Zeatin   Cytokinin

Gibberellin A$_3$ Gibberellin

Abscisin II   Abscisin

Figure 18 Some plant hormones and their structures.

(24). Since ethylene biogenesis is autocatalytic, it is sometimes necessary, to avoid excessive rates of fruit ripening, to remove even trace quantities of this gas from the storage environment. It has long been recognized that ethylene acts to stimulate respiration, and this is of both practical and theoretical interest (Fig. 19). The stimulation of respiration increases heat output by produce in storage and hence increases refrigeration requirements.

Recent evidence indicates that ethylene gas acts to stimulate cyanide-resistance respiration in plant tissue. Cyanide-resistance respiration is mediated by an alternate mitochondrial electron transfer chain. Although a direct relationship between ethylene action and alternate-chain respiration has not been established, it appears that the hormone indirectly, if not directly, influences alternate-chain respiration.

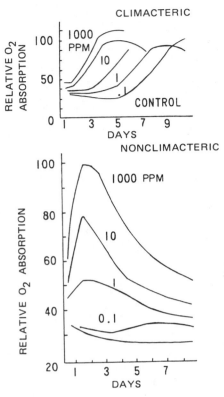

Figure 19 Influence of ethylene on the respiratory rates of climacteric and nonclimacteric fruits. Note that increased ethylene concentration results in the early onset of climacteric respiration in climacteric fruits and in an increased respiratory burst in the nonclimacteric fruit.

A number of substances believed to act as membrane perturbers, such as halogens, hydrogen cyanide, and carbon monoxide, exert an ethylene-like effect on plant tissues. Hence, the membranes of specific organelles, as well as the plasma membrane, may be sites of ethylene action. However, the specific mechanism or mechanisms by which ethylene gas exerts its physiological effects remains as a challenging problem to postharvest physiologists.

It is known that ethylene gas can be converted in situ to ethylene glycol, with ethylene oxide as an intermediate (2). Since both ethylene oxide and ethylene glycol appear to be inactive in evoking a physiological response, this conversion may be a mechanism for metabolic inactivation of the hormone. Evidence also indicates that hormones, in addition to ethylene, are involved as control agents for ripening and other physiological events in postharvest fruits and vegetables.

Fruit ripening is not always accompanied by a clearly defined increase in ethylene to threshold levels. The efficacy of exogenous ethylene to stimulate such events as ripening can depend on the receptivity state of the tissue. For example, the concentration of ethylene required to initiate fruit ripening declines as the fruit approaches full development of growth. Similarly, fruit attached to the plant (e.g., banana and avocado) display less sensitivity to ethylene than detached fruit. Also, certain varieties of pear are sensitized to ethylene only after the harvested fruit are stored at low, nonfreezing temperatures. These observations indicate that "juvenility factors" desensitize fruit to ethylene and perhaps otherwise control physiological events in fruits and vegetables. Application of exogenous sources of hormones and plant growth regulators, including auxins, cytokinins, or gibberellins, can simulate the action of juvenility factors in desensitizing plants to exogenous or endogenous ethylene.

There is also evidence that the abscisins, which play a role in leaf or organ abscission from the parent plant, also act to promote other senscence phenomena such as fruit ripening. As with other physiological events, such as seed germination, it appears that plant hormones act in concert to control physiological events in postharvest commodities. Although the mechanisms by which plant hormones exert their physiological effects are not well understood, some suggested loci of hormone action are outlined in Fig. 20. Although it is clear that hormones frequently exert an influence on gene expression it has not yet been possible to establish a definite corollary between a control site of DNA metabolism or protein synthesis and hormone action. For example, if a given plant hormone appears to trigger DNA transcription in situ, one must ask whether this is an indirect consequence of promoting feedback inhibition by stimulating accumulation of a particular end product.

V. HANDLING AND STORAGE OF FRESH FRUITS AND VEGETABLES

Throughout its life cycle, the intact plant may respond to various physical and chemical challenges from the environment. A plant's reactions to stress or environmental change are normally in the direction of protection, for example, prevention of water loss, adaptation to temperature extremes, or the development of physical and chemical barriers to pathogens. When the challenge is such that the tissue cannot adapt to the situation, the response may be one of hypersensitivity manifested as gene-directed death of particular groups of cells. It is important to recognize that species of plants may differ markedly in response to physical, chemical, and biological challenges. This is important because it means that no single condition or treatment is ideal for extending the storage lives of all fruits and vegetables. The following sections deal with the influence of physical and chemical factors on the physiological behavior and storage life of fresh fruits and vegetables.

A. Mechanical Injury

Wounding of plant organs generally results in a temporary, localized burst of respiration and cell division, formation of ethylene gas, and rapid turnover of certain cellular constituents, and sometimes results in the accumulation of certain secondary metabolies or products that appear to have a protective function. In flesh organs the wounded cells commence synthesis of messenger RNA and proteins (Fig. 21). Although the impact of wound physiology on plant food commodities is not fully understood, there are sufficient examples to illustrate the general and practical significance of these events.

1. Protective Substances

The response of plant organs to stress may be accompanied by "wound healing," that is, the formation of a physical barrier of protective substances. The formation of a waxy or

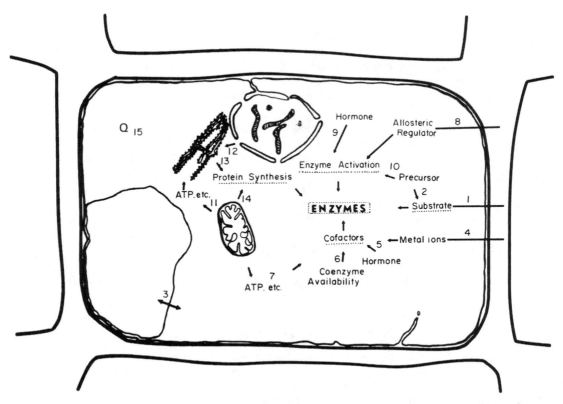

Figure 20 Conceivable loci of hormone action at the cell level: (1) transport or release of substrate, (2) conversion of substrate precursor, (3) transport of solute from intracellular compartments, (4) transport or release of metal ions, (5) hormone functioning directly as cofactor, (6) transport or release of cofactor, (7) biosynthesis of cofactor, (8) availability or release of allosteric effector, (9) hormone acting directly as enzyme effector, (10) conversion of precursor to active enzyme, (11) provision of utilizable energy for protein synthesis, (12) transcription, (13) translation, (14) cytoplasmic coded protein synthesis, and (15) release of autolytic enzymes from lysosome.

suberized barrier or of a lignified layer is important in certain commodities since these barriers can have a profound influence on preventing the invasion by saprophytic microorganisms and subsequent spoilage. Accordingly, following harvest of certain crops, such as potatoes, it is desirable to cure or store the crop under environmental conditions conducive to rapid wound healing. These conditions may differ from those yielding maximum storage life after wound healing is complete. Wound healing has become of increasing concern with the advent of mechanical harvesting techniques, since these procedures can result in extensive surface abrasions. In certain crops, such as asparagus, the cut injury associated with harvesting may be especially undesirable since the resulting stimulation of lignin deposition is directly associated with the loss in tenderness and succulence.

2. Stress Metabolites

It is now clear that wound injury can also lead to the synthesis and accumulation of a diverse family of substances called "stress metabolites" (9). In certain cases, accumula-

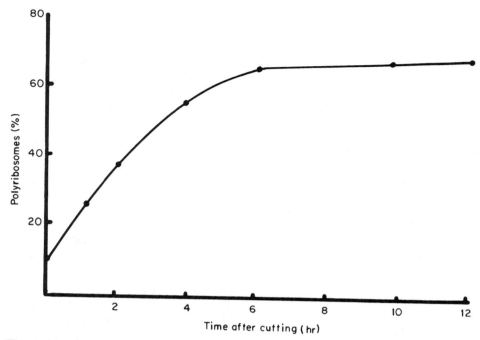

Figure 21 The polyribosome content of carrot disks after cutting. Accompanying the increase in polyribosomes there are increases in both the rate of amino acid incorporation into proteins and the protein content of the tissue.

tion of a stress metabolite occurs only when mechanical injury is accompanied by other conditions such as the presence of ethylene or exposure to chilling temperatures. Stress metabolites encompass a broad family of chemical compounds, including isoflavonoids, diterpenes, glycoalkaloids, and polyacetylenes (Fig. 22), and they appear to serve a protective function for the plant by virtue of their antibiotic properties. However, they may also detract from food quality because of their bitter taste (e.g., coumarins in carrot root, furanoterpenes in sweet potato root, or glycoalkaloids in white potato tuber), and because some exhibit toxic properties. Certain stress metabolites are also precursors of enzymic browning reactions.

3. Browning

Enzyme-catalyzed browning may also occur in certain plant tissues as a result of mechanical injury. These reactions, catalyzed by the enzyme polyphenoloxidase, are discussed in detail elsewhere (Chap. 6). The propensity of a given tissue to brown may vary considerably from one cultivar of a crop to another. These differences appear to relate to variations in enzyme content and sometimes to the kinds and amounts of phenolic substrates present. These reactions may be considered desirable in certain instances, e.g., tea fermentation, but are generally undesirable in postharvest fruits and vegetables.

4. Wound Ethylene

The importance of ethylene synthesis as a result of wounding of postharvest crops should not be overlooked. Ethylene, as mentioned, accelerates wound healing, stimulates respira-

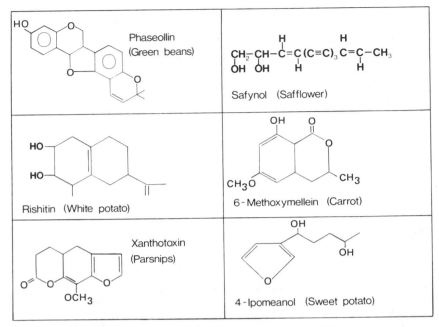

Figure 22   Examples of stress metabolites identified in edible plants.

tion and associated heat output, and hastens ripening of fruits. The mechanism of ethylene biogenesis following tissue wounding and the full consequences of wound ethylene on quality are not yet fully understood.

B. Temperature

Some of the reactions that take place in postharvest crops are essential for maintenance of tissue integrity, such as respiration, and some are more or less incidental to this primary purpose. The latter category of reactions may, however, influence food quality in desirable (e.g., flavor biogenesis) or undesirable (e.g., loss of free sugars) ways. All reactions may have slightly different temperature coefficients, thus making the influence of temperature change on overall quality quite complex. Moreover, other tissue attributes that indirectly influence biochemical transformation, such as membrane integrity and solubility and diffusivity of oxygen, are influenced by temperature. Within certain limits of temperature change, vegetative tissues accept the resulting imbalances, and though the metabolism alters, there is no obvious adverse reaction to this stress. Within these limits, changes associated with ripening, senescence, or breaking of dormancy proceed normally, albeit at different rates depending on the temperature. Hence, lowering of temperature can retard undesirable physiological reactions as well as spoilage caused by growth of microorganisms. Generally, commodities that have relatively fast respiratory rates respond best to lowered temperature (see Fig. 12). In such crops, precooling prior to shipping or storage can contribute greatly to an extended storage life.

1. Low-Temperature Injury

Exposure of a particular commodity to a temperature outside the recommended range for more than short periods causes injury and a decrease in both quality and shelf-life. The disorders known as "low-temperature injuries" or "chilling injuries" occur in some plant tissues at temperatures above the freezing point of the tissues but below the minimum recommended storage temperature. Some plant tissues, such as pear fruit or lettuce, do not normally undergo chilling injury at any temperatures above freezing, but for many tisssues there is a critical, nonfreezing temperature below which chilling injury occurs (Table 10).

The extent of low-temperature injury is dependent on both temperature and the duration of exposure. Recovery is known to take place from short exposures to potentially harmful temperatures. Visible effects of chilling injury include localized tissue necrosis (e.g., apples), failure to ripen (e.g., banana), development of areas of tissue that do not soften on cooking (e.g., "hardcore" in sweet potato), surface pitting (e.g., grapefruit), and wooliness in texture (e.g., peach).

Chilling injury can be generally attributed to an imbalance in metabolic reactions resulting in the underproduction of certain essentials and the overproduction of metabolites that become toxic to the tissue. For example, accumulations of ehtanol, acetaldehyde and oxaloacetic acid are associated with low-temperature breakdown of certain fruits.

Low temperatures may also influence the balance of metabolic processes by altering the integrity and permeability properties of biological membranes. There appears to be a relationship between chilling sensitivity and the fluidity of membranes as temperature is reduced. These changes in cellular membranes have been observed to occur coincidently with temperatures at which chilling injury occurs in the intact plant organ. In some instances these phase changes may cause an alteration in the flow of metabolites through the membrane. In other instances, there is a direct relationship between a membrane phase change and the rate of a specific enzymic reaction (Fig. 23). The oxidation of succinic acid via mitochondrial enzymes is coupled to the electron transfer chain and

Table 10 Chilling Injury of Commodities Stored Below Critical Nonfreezing Temperatures[a]

Commodity	Minimum storage temperature ($^\circ$C)	Character of injury
Apples	3	Internal browning, soft rot, scald, etc.
Bananas	13	Dull color, poor flavor, failure to ripen
Cranberries	2	Rubbery texture, red flesh
Olives	7	Internal browning
Oranges	3	Pitting, browning stain
Potatoes	4	Accumulation of reducing sugars
Tomatoes	9	Water soaking and soft rot

[a]The chilling temperature often varies with the variety, the cultivation practices, and the gaseous atmospheres present during exposure to cold.

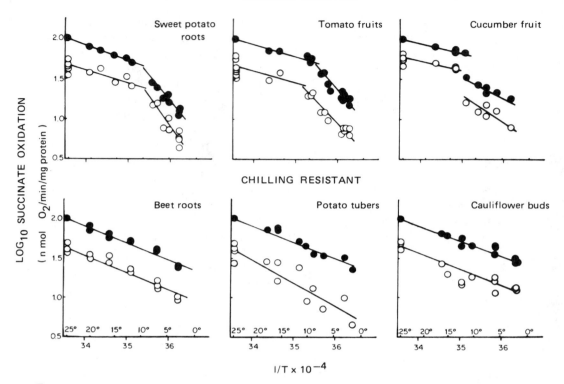

Figure 23 Rate of succinate oxidation in mitochondria (catalyzed by succinate dehydrogenase, which is bound to the mitochondrial membrane) as influenced by temperature. Discontinuities in the Arrhenius plots for the chill-sensitive species correspond to phase changes in the membrane lipids. Light and dark symbols represent different experimental conditions. (Courtesy of the American Association of Plant Physiologists.)

respiratory metabolism. It is therefore reasonable to suggest that the influence of temperature on the fluidity of biological membranes is one factor that can cause an inbalance in cellular metabolism. In view of the diverse ways in which chilling injury can occur in postharvest crops it is likely that the specific mechanisms underlying the disorder may vary in different tissues.

2. High-Temperature Injury

Exposure of plant tissues to abnormally high temperatures may also produce characteristic injuries in specific commodities. For example, storage of many fruits above 30°C results in failure to ripen normally and exposure of apples to 55°C for a few minutes gives the fruit an appearance similar to that resulting from chilling injury.

C. Controlled Atmospheres

Controlled atmosphere storage usually refers to storage in which the composition of the atmosphere has been altered with respect to the proportions of oxygen and/or carbon dioxide, and in which the proportions of these gases are carefully controlled, usually within ±1% of the desired value. Modified atmosphere storage does not differ in principle from controlled atmosphere, except that control of gas concentration is less accurate. For example, carbon dioxide may be allowed to simply accumulate in railroad cars, either by normal respiratory activities of plant tissue or by sublimation of dry ice (solid CO_2) used for cooling.

Hypobaric storage is another form of controlled atmosphere storage in the sense that the partial pressure of the gaseous environment is precisely controlled by reducing the pressure, and by continuously administering small amounts of gas (usually air). Typical pressure conditions prevailing in the chamber are 0.1-0.3 atm air. In addition, hypobaric storage creates additional benefits by providing a saturated relative humidity, and by increasing the diffusivity of gases within the internal spaces of the tissue.

The use of controlled atmospheres (CA) for the maintenance of high-quality produce during storage and transportation grew out of research by Kidd and West (13), who discovered over 60 years ago that certain varieties of apples remained in better condition in atmospheres that contained less oxygen and more carbon dioxide than air.

1. Benefits

The efficacy of these controlled atmosphere techniques for extending storage life has been demonstrated for many crops, but their commercial use has been somewhat limited, either for economic reasons or because rapid turnover of a commodity in the marketplace lessens the need for improved storage procedures. Conventional controlled atmosphere storage is used routinely for storage of apples and pears, and there has been increased use of controlled atmospheres in consumer packs. The latter approach is achieved by product-generated gases and the use of packaging materials with the desired gas permeability.

Hypobaric storage has been employed successfully by commercial growers of cut flowers, but at the time of this writing it has not been employed commercially for the storage or transportation of fruits or vegetables.

Collectively, these techniques appear to retard respiration, reduce the autocatalytic production and accumulation of ethylene and reduce decay caused by microorganisms. In effect, they generally retard ripening and senescence-related processes, such as degreening of vegetables.

2. Disorders

Closely related species, and even different cultivars of the same species, have different, very specific, and often unpredictable tolerances to low concentrations of oxygen and high concentrations of carbon dioxide (Table 11). The maturity of the tissue at harvest and the temperature of the tissue during storage can also influence the response to controlled atmosphere conditions. With respect to maturity, McIntosh apples, when stored in an atmosphere containing 1% oxygen, exhibit less internal browning if immature than if mature.

If the composition of the storage atmosphere is not properly controlled it can have disastrous results since the produce may develop physiological injuries. The degree of susceptibility to injury and the specific symptoms vary, not only among cultivars, but even among growing areas for the same cultivars and among years for a given location. Some crops, such as sweet cherries, plums, and peaches, are not very susceptible to car-

Table 11 Susceptibility of Various Crops to Extremes in CO_2 and O_2 Concentrations[a]

Commodity	Minimum O_2 tolerance (%)	Maximum CO_2 tolerance (%)	Symptoms
Lettuce	0.5	2–4	Brownish red discoloration of midrib
Broccoli	0.25	20	Tolerant
Cauliflower	–	5	Excessive softening, discolors on cooking.
Potatoes	5	10	Prevents wound healing

[a]Values vary with cultivar, cultivation practice, and storage temperature.

bon dioxide injury and can be stored at carbon dioxide concentrations up to at least 20%, whereas other crops, such as lettuce, can show discoloration ("brown stain") when exposed to 2% or greater concentrations of this gas. Similarly, atmospheres containing large amounts of carbon dioxide and small amounts of oxygen interfere with wound healing of potato tubers. Thus, wounded potatoes, when exposed to controlled atmospheres immediately after harvest, decay very rapidly.

The mechanism by which physiological disorders are induced by extremes in atmospheric composition is not entirely clear. However, available evidence indicates that the same principles discussed in reference to low-temperature disorders also apply to carbon dioxide injury or damage caused by oxygen deprivation. Studies have shown that toxic levels of metabolites, such as succinic acid, ethanol, and acetaldehyde, accumulate in certain commodities prior to injury symptoms.

A further similarity to low temperature injury is that CO_2 injury in pears has been attributed to alterations in the structure and function of organelle membranes. It is, of course very difficult to distinguish primary sites of injury from secondary consequences of primary action. However, it is reasonable to believe that alterations in respiratory metabolism and the inability of the tissue to cope with the resulting imbalances are a primary event in such injuries.

Although the final electron acceptor for respiratory metabolism (cytochrome oxidase) has a relatively high affinity for oxygen and hence should not be influenced even by substantial reductions in oxygen concentration, it is important to recognize that the gaseous concentrations in situ are not the same as those existing in the storage atmosphere. Thus, the diffusivity and solubility of a critical gas in the tissue may be determining factors. In view of recent findings that injury caused by low oxygen is minimal when carbon dioxide levels are retained at low levels, it appears that damage previously attributed to oxygen deprivation may, in fact, be caused by elevated levels of carbon dioxide. For example, apples can be successfully stored in an atmosphere consisting of only 1% oxygen and 99% nitrogen. Hypobaric storage also results in a large decrease in oxygen with an accompanying reduction in the partial pressure of carbon dioxide. Reports that hypobaric storage extends the storage life of many commodities beyond that possible with conventional CA storage (low oxygen and high carbon dioxide) may relate to the markedly different concentrations of carbon dioxide used in the two techniques. However, other differences also exist; namely, plant tissues stored under hypobaric conditions contain very low concentrations of ethylene, exhibit improved diffusion properties for gases, and are exposed to very high relative humidities.

D. Humidity

It is generally recommended that fruits and vegetables be stored at relative humidities sufficiently high to minimize water loss and to maintain cell turgor, but not so high as to cause condensation and accompanying microbial growth. The tendency of a commodity to lose moisture at a given relative humidity is related to the nature and amount of specialized cells at the tissue surface. Techniques that use a saturated relative humidity without condensation have been developed and employed successfully for vegetable storage in Canada and North European countries. Although such techniques are relatively expensive, they can be very effective in minimizing both physiological deterioration and microbial decay. Successful use of storage atmospheres that are saturated with water vapor requires a minimal temperature gradient throughout the storage chamber so that all atmospheric water remains in the vapor state.

Low relative humidity can influence physiological processes of postharvest crops. For example, banana fruit fails to ripen and undergo climacteric respiration when held at 25% relative humidity, whereas green tomatoes and pears, when exposed to an atmosphere with 25% relative humidity, exhibit a bimodal climacteric and intensification of ripening indices. Atmospheric moisture levels can also influence the severity of chilling injury in certain fruits. Core breakdown and scald of apples is minimized by storage at low relative humidity, whereas low-temperature breakdown of citrus fruits is lessened when fruits are wrapped in wax paper to minimize moisture loss.

E. Ionizing Radiation

Low doses of ionizing radiation have been successfully applied experimentally and commercially to prolong the storage life of fresh fruits and vegetables. The inhibition of sprouting in potatoes, onions, and carrots, as well as the destruction of insect pests in stored grains, may be accomplished with 10-25 K rad doses. Control of mold growth requires doses in excess of 100 K rad, and destruction of pathogenic bacteria necessitates doses of 500-1000 K rad. Megarad doses of ionizing radiation can temporarily or permanently prevent the onset of fruit ripening. Studies indicate that delay of ripening is associated with the uncoupling and eventual de novo biosynthesis by mitochondria. High doses of radiation, which irreversibly prevent normal fruit ripening, appear to interfere with the ability of the tissue to repair radiation damage.

Unfortunately, the quality of fruits and vegetables is usually adversely affected by doses of ionizing radiation in excess of 200 K rad. Moreover, low doses employed for sprout inhibition can interfere with wound healing, and this has created problems with commercial attempts to use this technique to prevent sprouting of potatoes,

The physiological changes induced by irradiation, especially high doses, can bring about the death of cells. Irradiated commodities often show characteristic symptoms of damage. For example, apples exposed to 1000 K rad exhibit a ring of necrotic tissue, not unlike core-flush, adjacent to the core. Citrus and banana exhibit peel lesions at doses as low as 25 K rad. Many commodities also exhibit extensive pulp softening, and this has been related to the activation of pectic enzymes. Despite these difficulties, there appears to be renewed interest in the use of ionizing radiation for the preservation of fresh fruits and vegetables. As with other techniques, the above limitations can probably be overcome when the influence of variables, such as maturity, temperature, and gaseous environment, and more fully controlled.

F. Other Factors

Postharvest treatments with fungicides, bactericides, and senescence inhibitors can be applied to fruits and vegetables to indirectly or directly minimize decay. Various experimental studies indicate that senescence inhibitors, such as hormone analogs (e.g., cytokinins and auxins) and respiratory inhibitors (e.g., benzimidazole and vitamin K analogs) are effective in prolonging the storage life of certain crops, although they are not approved for use commercially. Calcium treatment inhibits normal ripening of several fruits and prevents the development of various physiological disorders. Calcium in plants is involved in a number of fundamental physiological processes that influence the structure of cell walls, membranes, and chromosomes and the activity of enzymes. In addition, some studies indicate that endogenous calcium may also serve an important regulatory function in fruit ripening. Little is known about how calcium exerts a beneficial effect when applied to detached plant tissues.

Exposure of fruits and vegetables to various other chemicals can influence physiological processes in an undesirable way. For example, postharvest commodities show different susceptibilities to air pollutants, and this can be of practical concern in urban areas. Automobile emissions, especially ethylene, can accelerate tissue necrosis at parts per million levels. The volatile components of photochemical smog [O_3, PANS (peroxyacyl nitrates), and NO_2] are known to create undesirable injury symptoms in a wide array of fruits and vegetables. Other stresses to plants include SO_2 (e.g., from coal-fired plants for generating electricity) and ammonia from leaks in refrigeration systems.

VI. EFFECTS OF PROCESSING ON FRUITS AND VEGETABLES

The basic purpose of food processing is to curtail the activity of microorganisms and retard chemical and physical changes that would otherwise adversely affect the edible quality of food, and to do so with minimal damage to native qualities. Food processing techniques differ in principle from the preservation techniques described in the previous section in that viability of the living tissue cells is lost during processing. An important advantage of processing is it enables foods to be stored in edible condition for longer periods than would generally be possible by other means. However, because of the severity of these processes, the product usually undergoes rather extensive changes, and as a result the quality characteristics of a given commodity are often detectably altered. Such alterations may be desirable (e.g., inactivation of antinutritional factors by heat) or undesirable (e.g., loss of vitamins by heat) and may be acceptable to consumers depending on local customs and eating preferences. The reader is referred to other chapters in this book for detailed accounts of how processing influences the various chemical components of food.

A. Thermal Processing

Blanching and thermal sterilization are the most common types of heat processes used for fruits and vegetables. Blaching is a brief heat treatment involving exposure of the tissues to water or steam for a few minutes at about 100°C. Prior to freezing, dehydration, or irradiation, blanching is done primarily to inactivate enzymes, whereas prior to heat sterilization blanching is done for several reasons, the most important to remove air from tissue. Under certain conditions, blanching may serve to activate enzymes in such a way as to promote desirable changes. For example, blanching of some green vegetables can stabilize their bright green color and this is believed to result from activation of chlorophyllase, which in turn catalyzes the conversion of chlorophylls to the respective chlorophyllides. Similarly, the texture-firming effect of blanching on certain vegetables has been attributed to activation of pectin methylesterase, which in turn catalyzes the conversion of pectins to pectinic acids. The ionic nature of the product allows for interaction of pectinic acids with divalent ions, such as calcium, and this results in a more rigid structure.

More severe thermal treatments, such as sterilization or cooking, have the general effect of causing (a) leakage of solutes into the surrounding medium, (b) mechanical loss of particles, (c) loss of some volatile compounds, (d) oxidation, and (e) hydrolysis and other degradative reactions. The extent to which changes take place in fruits and vegetables during heat treatment depends on the type and quality of the tissue and the conditions of processing. Processing variables that may influence chemical changes include time, temperature, mode of heat transfer, type of container, and the characteristics of the cooking medium (e.g., water quality and content of oxygen, acid, sugar, salt, and other minerals). Proper preparation and pretreatment of the food prior to thermal processing can also influence product quality. For example, soaking can improve flavor, as in olive processing, in which a lye treatment removes the bitter principle. Also, following peeling and cutting light-colored plant tissues, it is often desirable to add inhibitors of enzymic browning to avoid excessive browning in the interval prior to further processing.

1. Texture

The texture of fruits and vegetables is likely to soften as a result of heat treatment. This is due to various factors including the loss in cell turgor pressure, loss of extracellular and vascular air, and the denaturation or degradation of cell wall constituents and other polysaccharides. Failure to inactivate pectic-hydrolyzing enzymes early in the process can result in undesirable texture changes in certain products, such as canned tomatoes. Slightly alkaline conditions (unusual) cause extensive degradation of pectins by a trans-elimination reaction. Starch granules gelatinize during heating in an aqueous environment, and this reaction undoubtedly influences the texture of certain commodities. Schlerenchyma tissues retain their rigid structure even after prolonged heating.

The deterioration of texture can be minimized by choosing less mature produce, lessening the severity of thermal processes, choosing cultivars that are less susceptible to softening, and by adding acids, pectic enzyme inhibitors, or calcium salts to the canning medium. Fruits heated in sucrose syrups tend to be firmer than those heated in the absence of this solute. Sucrose firms the texture of canned fruit by osmotic dehydration of the tissue, which in turn appears to facilitate cohesion of cell wall polysaccharides.

2. Flavor

It is clear that heating fruits and vegetables can cause dramatic changes in flavor, but our knowledge of the chemical basis for these changes in incomplete. It is generally true that flavor changes can be attributed to the loss of certain compounds by volatilization or by formation of new substances via degradation or addition reactions. For example, the major carbonyl component of the fresh banana, 2-hexenal, is not detected in the heat-processed puree. Volatile losses are also important to flavor change in other canned foods, including pear, cabbage, and onion. Volatile losses may also continue through chemical mechanisms during postprocessing storage. For example, in plain tin cans this may occur at the tin/product interface by reduction of aldehydes and ketones to primary and secondary alcohols.

Conversely, new volatile substances, which influence flavor, may arise in canned foods. Formation of hydrogen sulfide and other volatile sulfur compounds in heated products, such as cabbage and corn, are examples of this kind. Methyl sulfide is generated during heat processing of tomatoes, and an off-flavor in canned apple sauce has been attributed to the formation of caproic acid.

Nonvolatile substances may also contribute to flavor changes in heated foods. Off-flavors in some canned fruits and vegetables are attributed to the formation of ammonium pyrolidonecarboxylate from the cyclization of glutamine (Fig. 24). Moreover, failure to completely inactivate enzymes may also result in flavor changes during processing and subsequent storage. In some instances, enzymes are naturally heat resistant (e.g., peroxidase), and in others the enzyme acquires heat resistance because of the manner in which it is physically situated in the tissue. Examples of the latter are β-glucosidase in plum kernels and lipoxygenase in the cob of sweet corn. Peroxidase, which can catalyze production of off-flavors in vegetables, can partially "regenerate" or "reactivate" following high-temperature-short-time thermal processing. This behavior is also true of numerous other enzymes. Enzyme regeneration is discussed further in Chap. 6.

3. Color

The influence of thermal processing on plant pigments is discussed in Chap. 8. In addition to the changes in the major natural pigments, other changes in coloration can occur during heating. For example, the interaction of proanthocyanidins (colorless) with metal ions can result in pink discoloration of fruits, such as pear. Furthermore, if precautions are not taken to prevent polyphenoloxidase activity following tissue disruption and prior to and during the initial stages of heating, the product may develop a brown color. Finally, nonenzymic or Maillard browning may contribute to discoloration of plant commodities during heat processing and subsequent storage when conditions are favorable for these reactions (see Chap. 3).

4. Nutrient Losses

The influence of heat on nutrients is fully discussed in Chap. 7. Leaching losses of water-soluble vitamins and minerals can be substantial if plant products are exposed to hot water. Chemical degradation can also occur; for example, ascorbic acid and thiamine become increasingly heat labile as the pH is raised to near neutral, or above. However, in some instances the nutritional value of certain plant tissues can be improved by heat treatment. For example, the digestability coefficient of legume crops improves substantially after cooking.

B. Freezing

Lowering the temperature well below the initial freezing point of a product provides an environment in which microbial growth is not possible and reaction rates are generally retarded. On the one hand, freezing kills plant tissue and causes damage, primarily textural, which varies greatly with the commodity. On the other hand, freezing stops the growth of microorganisms and preserves color, flavor, and nutritive value quite well provided the temperature of subsequent storage is low (−18°C or lower). The cause of damage to frozen plant tissues is not entirely clear, although it is generally agreed that ice formation directly causes physical disruption of the relatively rigid plant cell structure and indirectly causes damage due to the concentration of solute molecules in the un-

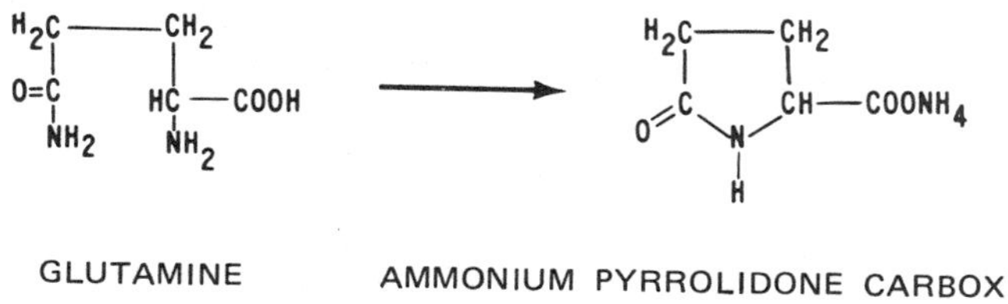

$$\begin{array}{ccc}
\text{H}_2\text{C}-\!\!-\!\!-\text{CH}_2 & & \text{H}_2\text{C}-\!\!-\!\!-\text{CH}_2 \\
| \quad\quad | & \longrightarrow & | \quad\quad\quad | \\
\text{O}=\text{C}\quad\text{HC}-\text{COOH} & & \text{O}=\text{C}\quad\;\text{CH}-\text{COONH}_4 \\
| \quad\quad | & & \backslash\;\;\text{N}\;/ \\
\text{NH}_2\quad\text{NH}_2 & & |\\
& & \text{H}
\end{array}$$

GLUTAMINE AMMONIUM PYRROLIDONE CARBOXYLATE

Figure 24 Formation of ammonium pyrrolidone carboxylate from glutamine. The reaction proceeds slowly during processing and more slowly during storage.

frozen phase. It is generally necessary to blanch or otherwise pretreat tissue to prevent enzymes from catalyzing detrimental reactions during frozen storage.

1. Texture

The most prominent change occurring in properly frozen, stored, and thawed fruits and vegetables involves texture. Damage to texture is most evident in such tissues as lettuce or strawberry, which lose cell turgor and crispness. Damage to texture can be primarily attributed to physical damage caused by ice formation. This leads to loss of cell turgor, dislocation of endogenous water, and a separation of cells because of strain on the middle lamella. Ultrarapid freezing can result in a thawed product with superior textural properties, provided frozen storage is avoided or occurs only briefly and at very low temperatures. Although chemical reactions do not normally have a significant role in textural damage to freeze-preserved plant tissues, as occurs for example with frozen fish, this has not been carefully studied. Some firming of texture can be accomplished by treating plant tissues with calcium salts to increase the ionic interaction of pectic substances. It has also been suggested that ultraslow freezing can minimize freeze damage to texture (6).

2. Color

Chemical degradation of pigments in frozen plant tissues is not a problem unless degradative enzymes are not sufficiently inactivated by blanching or other means. Enzyme-catalyzed degradation of pigments appears to be most common in low-acid commodities. Enzymes that have been implicated in the fading of color include the oxidoreductases, peroxidase, lipoxygenase, and polyphenoloxidase. These enzymes probably promote oxidative deterioration of chlorophylls, anthocyanins, and carotenoids by both direct and indirect (e.g., intermediary free radical or oxidant species) mechanisms. Anthocyanase may also be present in certain tissues and contribute to hydrolysis of glycosidic bonds of anthocyanin pigments. Physical damage by ice formation can also lead to excessive leaching of water-soluble pigments from the thawed product.

The appearance of certain fruit juices, notably those from citrus fruits, can also be adversely influenced if the juice is not pasteurized. Endogenous pectin methylesterase can demethylate pectin, allowing interlinking of pectin molecules via divalent ions, followed by precipitation and loss of "cloud." Other commodities, such as some varieties of strawberry, undergo gelation during frozen storage due to the same reactions.

Commodities susceptible to enzymic browning, such as potato or apple, can undergo extensive darkening prior to freezing or after thawing unless polyphenoloxidases are inactivated. Postthawing discoloration of tissues containing anthocyanins can also occur in some plant tissues because of interaction of metal ions with phenolic substances and flavonoids. This can be avoided by treating the susceptible tissue with chelating agents, such as citric acid, prior to freezing.

3. Flavor

Significant changes in the flavor of properly frozen plant tissues are usually evident only after prolonged storage. Enzymic or nonenzymic oxidative reactions, which proceed slowly at normal temperatures of frozen storage, appear to be largely responsible for changes in flavor. Deterioration in flavor may result from the loss of the characteristic fresh aroma and the concomitant development of off-flavors. Certain crops, such as limes, are more sensitive to flavor loss than others. Other crops, such as other citrus fruit juices or squash, develop characteristic unpleasant off-flavors when improprerly pretreated, when oxygen is not properly excluded, or when they are stored above $-18°C$. Off-flavor development sometimes occurs together with color loss, and both can be largely avoided by inactivation of oxidative enzymes, such as peroxidase and lipoxygenase. The specific chemical mechanism of such flavor changes are not clear at this time.

4. Nutrients

Blanching of vegetables prior to freezing, especially if done in water, can result in significant losses of water-soluble vitamins and minerals. Little change in the nutrient content of plant tissues occurs as a result of freezing alone. On thawing, significant amounts of water-soluble vitamins and minerals can be lost in the thaw exudate and in the cooking medium for vegetables. Vitamins, notably ascorbic acid and folacin, may undergo substantial decreases during long-term storage of plant tissues in the frozen state. These vitamin degradation reactions can be catalyzed by enzymes, if they are allowed to remain active, or by metals. The presence of oxygen accelerates some of these reactions.

C. Dehydration

Prior to harvest, the water content of an edible plant tissue is maintained by the parent plant until the onset of senescence. During senescence, water may be lost at a faster rate than it is supplied by the plant, causing fruit and other plant organs to dehydrate. If the relative humidity of the environment is sufficiently low (e.g., 60%), a product water activity eventually is established that precludes spoilage by microorganisms, physiological deterioration, or significant chemical change. Such crops as cereals, pulses, grapes, figs, and dates can be naturally preserved by allowing them to dry in the field. Once removed from the parent plant, these crops and others can be further dried, if necessary, by direct action of the sun or by a variety of "artificial" techniques.

   The extents of physical, chemical, and biochemical changes that plant tissues undergo during dehydration depend greatly on the method of dehydration employed and the nature of the commodity. A given commodity dried by sun drying, vacuum drying, osmotic drying, air drying, or freeze-drying may have quite different characteristics due to differences in the stresses imposed and the drying conditions. Slow processes, like sun drying can be desirable in specific instances when it allows for various physical and chemical changes that are considered desirable. In such instances, the sun-drying process may be viewed as a combination of dehydration and fermentation. However, in most instances it is desirable to maintain the native quality of the commodity, and for this reason artificial drying techniques are generally superior to sun drying. As with other methods of preservation, a successful result depends on proper choice of raw material (e.g., maturity and cultivar) and appropriate procedures prior to, during, and following dehydration.

1. Texture

The texture of dried tissue is greatly dependent on the method of moisture removal, but in most instances native texture is severely damaged. Freeze-dried vegetables may possess textural characteristics similar to those achieved by simple freezing (reasonably good), whereas a comparable sample that has been air dried often bears little resemblance to the fresh tissue. If dehydration involves high temperatures, an impervious layer may develop on the outer surface of the commodity (case hardening), and this may cause a rubbery and otherwise intractable texture. The mechanism of texture change in dehydrated fruits and vegetables is undoubtedly complex, involving both chemical and physical processes (15). The native state of macromolecules, such as polysaccharides and proteins, is dependent on the aqueous environment and is expected to change substantially and to a large extent irreversibly during the less delicate methods of dehydration.

2. Color

The principal color change of dried plant tissues, especially fruits, is browning, and this may be of enzymic (if improperly blanched) or nonenzymic origin. Sulfur dioxide is applied principally to prevent nonenzymic browning, although it is also effective against enzymic browning. Chlorophyll may be converted to pheophytin if hot air is used during dehydration, and this results in a change in appearance from bright green to dull olive-brown. Carotenoids appear to be stable in many dehydrated products provided blanching has effectively inactivated oxidative enzymes. Anthocyanins are generally unstable during conventional dehydration, and they assume a grayish color if sulfuring is not employed. The chemical nature of these pigment changes are not well understood, although both enzymic and nonenzymic reactions can be expected to be involved. Further details on this subject are presented in Chap. 8.

3. Flavor

As would be expected, there is considerable loss of native volatile flavor compounds during conventional dehydration, and new flavor substances may be generated as a result of the Maillard and other chemical reactions. For example, during dehydration of prunes, crotonaldehyde is formed from acetaldehyde and this contributes to the dehydrated flavor. Nonvolatile and heat-stable flavor precursors are retained in dehydrated cabbage and other dried vegetables so that the fresh flavor can be partially regenerated by addition of an appropriate enzyme extract during rehydration. The preservation of fresh flavor is best with freeze drying, in which chemical reactions and volatile loss are minimized.

4. Nutrients

Dehydrated vegetative tissues may undergo considerable loss in nutritive value if precautions are not taken to minimize oxidation. Treatment with sulfur dioxide to retard browning lessens destruction of vitamin C but accelerates destruction of thiamine. Nonenzymic browning reactions, if allowed to occur, cause a decrease in the biological value of proteins (see Chap. 3).

VII. CONCLUSION

As utilizable agricultural lands diminish, the problems of food availability are certain to increase and botanical sources of food will tend to supplant animal sources. There are many gaps in our knowledge of the postharvest physiological attributes of edible plant tissues, and continued research is needed to improve our ability to better utilize postharvest crops. Totally new approaches to plant food production, quality improvement, and preservation will also evolve from basic research on plant physiology. For example, recent advances in recombinant DNA technology have the potential of improving the quality of plants used as food and for solving various problems related to postharvest storage.

REFERENCES

1. Ap Rees, T., W. L. Dixon, C. J. Pollack, and F. Franks (1981). Low temperature sweetening of higher plants, in *Recent Advances in the Biochemistry of Fruits and Vegetables* (J. Friend and M. J. C. Rhodes, eds.), Academic Press, New York, pp. 41-60.
2. Beyer, E. M. (1981). Ethylene action and metabolism, in *Recent Advances in the Biochemistry of Fruits and Vegetables* (J. Friend and M. J. C. Rhodes, eds.), Academic Press, New York, pp. 107-120.
3. Biale, J., and R. E. Young (1981). Respiration and ripening of fruits—restrospect and prospect, in *Recent Advances in the Biochemistry of Fruits and Vegetables* (J. Friend and M. J. C. Rhodes, eds.), Academic Press, New York, pp.
4. Brady, C. J., P. B. H. O'Connel, J. K. Palmer, and R. M. Smille (1970). An increase in protein synthesis during ripening of the banana fruit. *Phytochemistry 9*:1037-1047.
5. Duckworth, R. B. (1966). *Fruits and Vegetables*, Pergamon Press, New York.
6. Finkle, B. J., E. Sa B. Pereira, and M. S. Brown (1971). Prevention of freeze damage to vegetables and fruit tissues by ultra slow cooling. *Paper 5.19 Proceedings (1971) 13th Int. Congr. of Refrig.*, pp. 497-499.
7. Frenkel, C. (1978). The role of hydroperoxides in the onset of senescence processes in higher plant tissues, in *Postharvest Biology and Biotechnology* (H. O. Hultin and M. Milner, eds.), Food and Nutrition Press, Westport, Conn., pp. 433-448.
8. Gardener, W. H. (1966). *Food Acidulants*, Allied Chemical Corporation, New York, p. 546.
9. Haard, N. F. (1983). Stress metabolites, in *Post-harvest Physiology and Crop Protection* (M. Lieberman, ed.), Plenum, New York, pp. 299-314.
10. Hughes, O., and M. Bennion (1970). *Introductory Foods*, Macmillan, London, pp. 1-15.
11. Expert Panel on Food Safety (1979). *Dietary Fiber: A Scientific Status Summary*. Institute of Food Technologists, Chicago, Ill.
12. Janick, J., R. W. Schery, F. W. Woods, and V. M. Ruttan (1969). Food and human needs, in *Plant Science* (J. Janick, R. W. Schery, F. W. Woods, and V. M. Ruttan, eds.), W. H. Freeman, San Francisco, pp. 37-52.
13. Kidd, F., and C. West (1921). *Food Inv. Bd. Rep. 1921* 14-16.
14. Knee, M., and I. M. Bartley (1981). Composition of cell wall polysaccharides in ripening fruits, in *Recent Advances in the Biochemistry of Fruits and Vegetables* (J. Friend and M. J. C. Rhodes, eds), Academic Press, New York, pp. 131-145.
15. Kon, S. (1968). Pectic substances of dry beans and their possible correlation with cooking time. *J. Food Sci. 35*:46-51.

16. Macko, V., and H. Stegmann (1969). Mapping of potato proteins by combined electrofocusing and electrophoresis. Identification of varieties. *Hoppe-Seyler's Z. Physiol. Chem. 350*:917-919.

17. Paez, L., and H. O. Hultin (1970). Respiration of potato mitochondria and whole tubers and relationship to sugar accumulation. *J. Food Sci. 35*:46-51.

18. Radley, J. A. (1968). *Starch and Its Derivatives*, Chapman and Hall, London.

19. Ryall, L., and W. Lipton (1971). *Handling, Transportation, and Storage of Fruits and Vegetables*, Vol. 1, *Vegetables and Melons*, AVI Publishing Co., Westport, Conn.

20. Solomos, T., and G. Laties (1976). Induction by ethylene of cyanide-resistant respiration. *Biochem. Biophys. Res. Commun. 70*:663-671.

21. Stegemann, H. (1975). Properties and physiological changes in storage proteins, in *The Chemistry and Biochemistry of Plant Proteins* (J. B. Harborne and C. F. Van Sumere, eds.), Academic Press, New York, Chap. 3.

22. Timberlake, C. F. (1981). Anthocyanins in fruits and vegetables. *Recent Advances in the Biochemistry of Fruits and Vegetables* (J. Friend and M. J. C. Rhodes, eds.), Academic Press, New York, pp. 221-247.

23. Wolf, W. J. (1972). Purification and properties of soybeans, in *Soybeans: Chemistry and Technology* (A. K. Smith and S. J. Circle, eds.), AVI Publishing Co., Westport, Conn.

24. Yang, S. F. (1981). Biosynthesis of ethylene and its regulation, in *Recent Advances in the Biochemistry of Fruits and Vegetables* (J. Friend and M. J. C. Rhodes, eds.), Academic Press, New York, pp. 89-104.

BIBLIOGRAPHY

Friend, J., and M. J. C. Rhodes (eds) (1981). *Recent Advances in the Biochemistry of Fruits and Vegetables*, Academic Press, New York.

Haard, N. F. (1984). Postharvest physiology and biochemistry of fruits and vegetables. *J. Chem. Educ. 61*:277-290.

Haard, N. F., and D. K. Salunkhe (eds.) (1975). *Postharvest Biology and Handling of Fruits and Vegetables*, AVI Publishing Co., Westport, Conn.

Hulme, A. C. (ed.) (1971). *Biochemistry of Fruits and Their Products*, Academic Press, New York.

Hultin, H. O., and M. Milner (eds.) (1978). *Postharvest Biology and Biotechnology*, Food and Nutrition Press, Westport, Conn.

Lieberman, M. (ed.) (1983). *Postharvest Physiology and Crop Protection*, Plenum, New York.

Ryall, A. L., and W. J. Lipton (1971). *Handling, Transportation and Storage of Fruits and Vegetables*, Vol. 1, *Vegetables and Melons*, AVI Publishing Co., Westport, Conn.

16

AN INTEGRATED APPROACH TO FOOD CHEMISTRY: ILLUSTRATIVE CASES

Theodore P. Labuza   University of Minnesota, St. Paul, Minnesota

I.	Introduction	913
II.	Shelf Life Test Design Using Kinetics	914
	A. Introduction	914
	B. Order of Reaction	914
	C. Arrhenius Approach	915
	D. The Simple Shelf Life Plot Approach	919
	E. Integration of Humidity Effect—Dehydrated Foods	922
	F. Steps in Shelf Life Testing	923
III.	Case Studies of Shelf Life	924
	A. Frozen Pizza	924
	B. Dehydrated Mashed Potatoes	929
	References	936
	Bibliography	938

I. INTRODUCTION

An attempt is made in this final chapter to show how the subject matter presented in this book and other information relevant to the field of food science can be integrated and thereby aid in product development, in determination of product shelf life, and in solving many other problems related to foods. The subject matter areas that must be included in this integration are food microbiology, food chemistry, analytical chemistry, physical chemistry, and food regulations, and it becomes evident during the remainder of this chapter how knowledge from these various subject matter areas can be utilized in total.

Before discussing the case studies of two illustrative products, it is necessary to provide some background information on the design of shelf-life tests.

II. SHELF LIFE TEST DESIGN USING KINETICS

A. Introduction

Only a brief review of the critical principles relating to shelf life test design using kinetics is presented here, since an in-depth review is inappropriate for present purposes, and since detailed information is available elsewhere (1,19,21). What should be made clear is that there are two general ways to predict product shelf life. The most common method is to select some single abuse condition, expose the food to it, test it two or three times during some specified period, generally by sensory methods, and then extrapolate the results (often educated speculation) to normal storage conditions. Another approach is to assume that certain principles of chemical kinetics apply with respect to temperature dependency, such as the Arrhenius relationship, and to utilize a more elaborate design, which is more costly but is likely to provide better results. The key to the kinetic basis is that certain principles are generally followed. These are discussed in this chapter.

B. Order of Reaction

Quality loss for most foods, as found by Labuza (15), conforms to the following general equations.

$$\text{Decrease of desirable attribute} \qquad \frac{-dA}{d\theta} = k(A)^n \qquad (1)$$

or

$$\text{Increase of undesirable attribute} \qquad \frac{+dB}{d\theta} = k(B)^n \qquad (2)$$

where $dA/d\theta$ or $dB/d\theta$ is the change in quantity of A or B with time, (A) or (B) is the measured amount of the attribute at any time, k is the rate constant in appropriate units, and n is the order of the reaction, generally 0, 1, or 2.

Most shelf life data for change in a quality attribute, based on some chemical reaction or microbial growth, follows a zero-order ($n = 0$) or first-order ($n = 1$) pattern. For zero-order data, a linear plot is obtained using linear coordinates (Fig. 1), whereas for first-order data, semilogarithmic coordinates (log A or log B) are needed to produce a linear plot (Fig. 2). For second-order data, a plot of 1/A or 1/B versus time produces a linear relationship. Thus, if the order is known, one can extrapolate, on the basis of a few measurements and on appropriate plots, to the value of A_s or B_s, the values the attributes reach at the specified end of the shelf life, θ_s. Knowing A_0, θ or A can be calculated as follows for a certain amount of time θ or conversely for a decrease in amount of desirable attribute A (for increase of an undesirable attribute B, substitute B for A and + for −):

$$\text{Zero order} \qquad A = A_0 - k_z\theta \qquad (3)$$

$$\text{First order} \qquad \ln A = \ln A_0 - k_1\theta \qquad (4)$$

where A is the attribute value at time θ, k_z is the zero-order rate constant [=slope of Eq. (3) in amount/time], and k_1 is the first-order rate constant [= slope of Eq. (4) in units of time^{-1}].

Various statistical methods can be used to determine the value of k (the slope) from the appropriate plot. The better the precision of the method for measuring the attribute value, the smaller the extent of change needed in the attribute value to arrive at an accu-

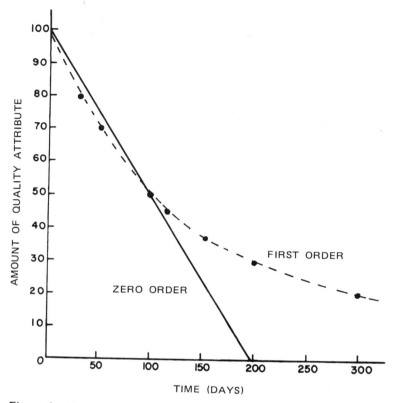

Figure 1 Change in quality versus time showing effect of order of reaction on extent of change.

rate value of k. Generally, precision is defined as

$$\text{Precision} = \pm \frac{\sigma}{\overline{X}} 100\% \qquad (5)$$

where σ is the standard deviation and $\overline{X}$ is the mean value of replicate analyses. Based on the calculations of Benson (1) in Table 1, it is apparent that one must allow a large change in the reactant species monitored or have a small percentage of precision to obtain a reliable estimate of the rate constant. Unfortunately, because of normal variabilities, many food reactions, such as nonenzymic browning, have a precision error of more than ±10%. Thus, this requires one to hold the product for a long time or at a high temperature to obtain accurate estimates of the rate constant. Since many food products become unacceptable after only a 20-30% change from the initial condition (19), this puts further limits on the accuracy of the measured rate constant.

C. Arrhenius Approach

Most textbooks on physical chemistry provide excellent explanations of the Arrhenius approach to interrelating temperature and the reaction rate constant. Labuza (17) has applied this approach to rates of chemical reactions in foods. The well-known Arrhenius relation is

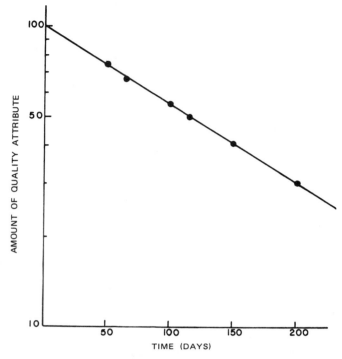

Figure 2 Semilogarithmic plot for a first-order reaction of quality loss versus time.

Table 1 Effect of Analytical Precision of the Method on the Estimation of the Rate Constant for Chemical Reactions

Analytical precision (%)	% Error in reaction rate constant k at the following % change in reactant species monitored						
	1%	5%	10%	20%	30%	40%	50%
± 0.1	14	2.8	1.4	0.7	0.5	0.4	0.3
± 0.5	70	14	7	3.5	2.5	2	1.5
± 1.0	>100	28	14	7	5	4	3
± 2.0	>100	56	28	14	10	8	6
± 5.0	>100	>100	70	35	25	20	15
±10.0	>100	>100	>100	70	50	40	30

Source: From Benson (1).

$$k = k_0 e^{-E_A/RT} \tag{6}$$

where k is the rate constant, k_0 is the preexponential constant, E_A is the activation energy in kilojoules per mole, R is the gas constant, and T is the absolute temperature (degrees Kelvin). A semilog plot of k (on the log scale) versus 1/T, as shown in Fig. 3, gives a straight line. Under some instances, if two critical reactions occur in a food and they have different rates and activation energies, it is possible that one predominates above some critical temperature T_c and the other below this temperature (Fig. 3). The main value of the Arrhenius plot is that one can collect data at high temperatures (low 1/T) and then extrapolate for the shelf life at some lower temperature, as shown in Fig. 4. Rate con-

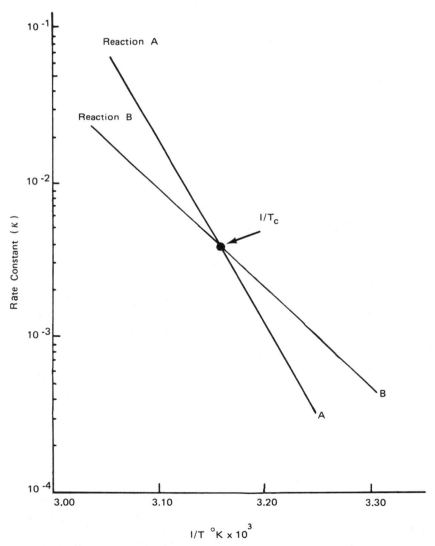

Figure 3 Arrhenius plot of reaction rate k versus inverse absolute temperature for two reactions shown crossover at T_c.

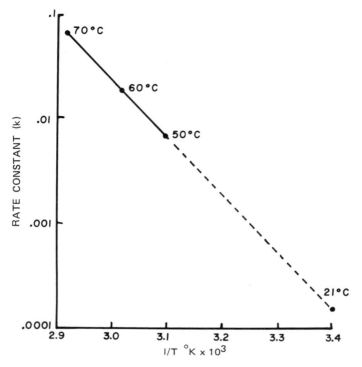

Figure 4 Prediction of shelf life by extrapolation from high temperature to low on an Arrhenius plot.

stants at three or more temperatures are needed to construct a plot of reasonable accuracy. From a statistical standpoint at least six temperatures should be used, but because of the cost of analysis and the difficulty of obtaining incubators at six constant temperatures, fewer temperatures are often used. Labuza and Kamman (21) have described several statistical methods to obtain accuracy when less than six temperatures are used, one of which is to consider each data point as an independent estimation of k. This approach has worked well in several studies (2,23,29). It should be noted that it has been quite common in the food science literature to derive an E_A value from three temperatures, which generally result in an r^2 of ~1. However, the statistical 95% confidence levels of E_A obtained from such a plot are generally as large as E_A because of the large $t_{95\%}$ value. But, this does not mean the data are wrong. In these cases the *average* E_A is probably an adequate measure of the temperature dependence of the reaction.

One should also note that both k_0 and E_A are dependent on water activity (10,16) and, thus, a_w must be held constant.

Recently, Labuza and Riboh (22) enumerated potential causes for nonlinear Arrhenius plots and these include (a) a change in a_w or moisture; (b) a change of physical state; (c) a change in the critical reaction with a change in temperature, as noted in Fig. 3; (d) a change in pH with a change in temperature; (e) a decrease in dissolved oxygen as the temperature is raised (slowing oxidation); (f) a partitioning of reactants between two phases; and (g) a concentration of reactants upon freezing. These should all be kept in mind when a Arrhenius plot is used.

It should be noted that nutrient losses in products held in both sine wave and square wave temperature fluctuations (2,10,23,26,29,30) can be predicted from steady-state (constant temperature data for each trial) generated Arrhenius plots with an accuracy of ±20%. In the case of lipid oxidation in potato chips, the prediction is good to ±3% (2). Thus, using this approach, that is, doing studies of shelf life at several high-temperature abuse conditions, allows prediction of accumulated shelf life loss under almost any condition. For this purpose, Eqs. (7) and (8) are set up in a series expansion, such as

$$B = B_0 \pm \sum_{i=0}^{i=n} (k_i \Delta\theta_i)_{T_i} \tag{7}$$

$$\ln A = \ln A_0 \pm \sum_{i=0}^{i=n} (k_i \Delta\theta_i)_{T_i} \tag{8}$$

where k_i is the rate at temperature T_i for the $\Delta\theta_i$ interval during any period along the plot in Fig. 5. This is very similar to the TTT approach (8).

D. The Simple Shelf Life Plot Approach

For a given extent of deterioration and reaction order, the rate constant is inversely proportional to the time to reach some degree of quality loss (12). Thus, a plot of log shelf life (log θ_s) versus $1/T$ is a straight line, as shown in Fig. 6a. Furthermore, if only a small temperature range is considered, then one finds that most food data yield a linear plot when log θ_s is plotted versus temperature T (Fig. 6b). The equation of the line in Fig. 6b is

$$\theta_s = \theta_0 e^{-bT} \tag{9}$$

where θ_0 is the shelf life at T = 0, θ_s is the shelf-life at T, and

$$b = \frac{\ln Q_{10}}{10}$$

$$= \frac{0.503 E_A}{T(T + 10)}$$

$$= \text{slope of Eq. (9)}$$

By taking the ratio of shelf-life between any two temperatures 10°C apart, the Q_{10} of the reaction can be found, where

$$Q_{10} = \frac{\theta_s \text{ at } T}{\theta_s \text{ at } T + 10°C} \tag{10}$$

In determing Q_{10}, it is assumed that Eq. (9) is valid. Whereas this is true for small temperature ranges, it is usually not true for large temperature ranges (15,22). Thus, Q_{10} values based on a large temperature range will often be somewhat inaccurate. In addition, the factors that cause deviations in the Arrhenius plot also influence the shelf life plot (Fig. 6b) and the calculated value of Q_{10}.

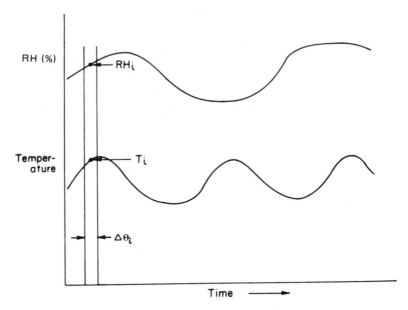

Figure 5 Possible temperature and humidity distribution cycle.

The data in Table 2 indicate the importance of accurate estimates of the Q_{10} value when making shelf life predictions. For example, if a product has a shelf life of 2 weeks at 50°C and a Q_{10} of 2.5, it has a shelf life of 31 weeks at 20°C. If, however, the Q_{10} value were 3.0 instead of 2.5, the shelf life at 20°C would be 54 weeks instead of 31. Thus, a small difference in the Q_{10} value can make a large difference in the answer when one is predicting shelf-life at low temperatures from accelerated tests at higher temperatures. A 0.5 difference in Q_{10} for the 30-40°C temperature range is a difference in E_A of only 4 kcal/mol. At best, in the many studies in which a good estimate of the statistical variation of E_A has been made, it has a 95% confidence limit of ±5 kcal/mol. Thus, although this approach to shelf-life prediction is sound, the quality of the data are such that predictions usually cannot be made with great accuracy. Unfortunately, it is generally impossible to improve the quality of the data because of cost and time limitations.

Another useful property of the shelf life plot is shown in Fig. 7. Suppose we want to be sure that a product has at least 18 months shelf life at 23°C (73°F) and we want to determine whether this is so by using an accelerated test at 40°C (104°F). How long must the product keep at 40°C to be equivalent to 18 months at 23°C? The answer is determined by placing a point on the plot at 18 months and 23°C and drawing a straight line to the 40°C vertical line, using a slope dictated by the Q_{10} value. Lines based on Q_{10} values of 2-5 are shown in Fig. 7, indicating a shelf life of 1 month if the Q_{10} is 5 and 5.4 months if the Q_{10} is 2. Provided the true Q_{10} is known, and the Arrhenius relation is valid, the answer is correct. The major problem stems from not accurately knowing the Q_{10}. According to Labuza (19), canned products have Q_{10} values ranging from 1.1. to 4, dehydrated foods from 1.5 to 10, and frozen foods from about 3 to as much as 40. Thus, the only way to achieve accurate results is to determine the Q_{10} by conducting shelf life studies at two or more temperatures. One must also not forget to hold a_w constant for dehydrated foods since a_w affects Q_{10} values.

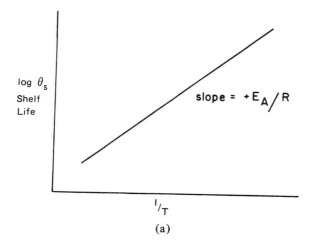

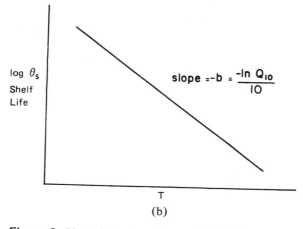

Figure 6 Plot of the logarithm of shelf life versus inverse absolute temperature (a) and versus temperature (shelf life plot) (b).

Table 2 Effect of Q_{10} on Shelf Life

Temperature (°C)	Shelf life (weeks)			
	$Q_{10} = 2$	$Q_{10} = 2.5$	$Q_{10} = 3$	$Q_{10} = 5$
50	2[a]	2[a]	2[a]	2[a]
40	4	5	6	10
30	8	12.5	18	50
20	16	31.3	54	4.8 years

[a]Arbitrarily set at 2 weeks at 50°C. Shelf lives at lower temperatures are calculated on this arbitrary assumption.

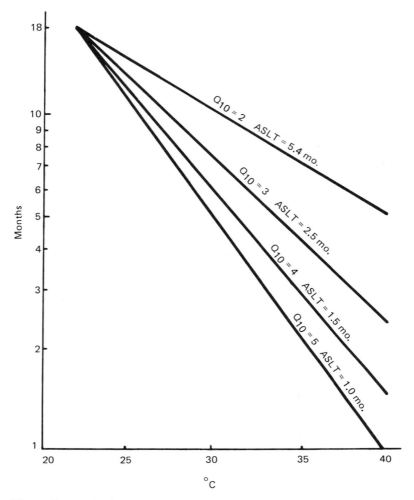

Figure 7 Shelf life plot for a dehydrated food with desired shelf life of 18 months at 23°C and equivalent accelerated shelf life test (ASLT) times at 40°C.

E. Integration of Humidity Effect—Dehydrated Foods

The classic studies of Oswin (27), Heiss (9), Karel (11), and other work at MIT (24,25, 28,32) led to mathematical models for predicting moisture gain with time in a packaged food. This has been reviewed recently (18,20). Basically, as the a_w of the food changes, the rates of chemical reactions either rise or fall. If the rate of loss of quality is known as a function of a_w and the water vapor transmission rate for the film is known, the quality change with time can be easily predicted. The problem is that in many cases this effect of a_w is not considered. Food is placed into environmental cabinets with different temperatures and, thus, different humidities. Thus, under one condition drying occurs, but in the other the product gains moisture leading to a confounding of the estimate of θ_s or the rate constant. The dehydrated product should therefore always be stored at the same constant humidity at all temperatures to eliminate this discrepancy. Today, with the modern programmable calculators and minicomputers, rapid solutions using the simplified equations in the above references can be obtained. Nakabayashi et al. (26) for drugs and Cardoso and Labuza (5) for pasta have, in fact, further expanded the solution of predicting quality loss to include both variable temperature and humidity (Fig. 5), the effect of percent relative humidity (%RH) and temperature on the film permeability, and the effect of temperature on the sorption isotherm of the product. Thus, if a time-temperature-humidity distribution were known, the shelf life could be predicted to within 10-20%. The problem is that since the temperature and humidity distribution in the United States can be so different depending on the region of the country and the time of the year, the true shelf life of any individual package is unknown. All one can do then is to decide whether to protect for the worst condition or for some average condition.

F. Steps in Shelf Life Testing

The following steps should be followed in designing a shelf life test for quality loss in food.

   1. Determine the microbiological safety and quality parameters for the proposed formulation and process.

   2. Determine from an analysis of the ingredients and the process which chemical reactions are likely to be the major causes of quality loss. A quick literature search is mandatory. If major potential problems exist at this stage, the formula should be changed, if possible.

   3. Select the package to be used for the shelf life test. Frozen and canned products can be packaged in the final container. Dry products should be stored open in chambers at selected %RH or in sealed jars at the desired moisture and a_w.

   4. Select the storage temperatures (at least two). Commonly, the following are good choices.

Product	Test temperatures (°C)	Control (°C)
Canned	25, 30, 35, 40	4
Dehydrated	25, 30, 35, 40, 45	−18
Frozen	−5, −10, −15	< −40

   5. Using the shelf life plot (Fig. 6b) and knowing the desired shelf life at the average distribution temperature, determine how long the product must be held at each test temperature. If no information is available on the probable Q_{10} value, then more than two temperatures are needed.

   6. Decide which tests will be used and how often they will be conducted at each temperature. A good rule of thumb is that the time interval between tests, at any temperature below the highest temperature, should be no longer than

$$f_2 = f_1 Q_{10}^{\Delta/10} \qquad (11)$$

where f_1 is the time between tests (e.g., days, weeks) at highest test temperature T; f_2 is the time between tests at any lower temperature T_2; and Δ is the difference in degrees Celsius between T_1 and T_2. Thus, if a dry product is held at 45°C and tested once a month, then at 40°C with a Q_{10} of 3, the product should be tested at least every

$$f_2 = 1 \times 3^{(5/10)}$$

$$= 1.73 \text{ months}$$

Of course, more frequent testing is desirable, especially if the Q_{10} is not accurately known. Use of needlessly long intervals may result in an inaccurate determination of shelf life and, thus, a danger of rendering the experiment totally useless. At each storage condition, at least six data points are needed to minimize statistical errors; otherwise, the confidence in θ_s is significantly diminished.

7. Plot the data as it is collected to determine the order and to decide whether test frequency should be increased or decreased. All too often the data are not analyzed until the experiment is over and then the scientist finds that nothing can be concluded.

8. From each test storage condition, estimate k or θ_s, make the appropriate shelf life plot, and estimate the potential shelf life at the desired (final) storage condition. Of course, one can also store product at the final condition and determine its shelf life to test the validity of the prediction. In academia this is commonly done; however, in industry this is uncommon because of the pressure of time. If the shelf life plot indicates that the product meets stability expectations, then the product has a change of performing satisfactorily in the marketplace.

III. CASE STUDIES OF SHELF LIFE

In this section two types of actual shelf life problems are examined, and attention is given to the factors influencing stability, to the design of the shelf life test, and, in one instance, to modeling of product behavior in a distribution system.

A. Frozen Pizza

1. General Aspects of Quality Loss

A frozen pizza consists of a crust on top of which is placed a mixture of tomato sauce, cheese, spices, vegetables, and sausage. The product is then partially baked, cooled, shrink-wrapped, and blast-frozen. The major problems that occur during storage are development of package ice, development of off-flavors, loss of the deep-red tomato color, loss of spice impact, and development of crust sogginess. Each of these changes influences shelf life and must be considered in any test design.

2. Specific Changes That Limit Aspects of Shelf Life

Microbiological Changes. Both the cheese and sausage, if naturally fermented, have high total counts of bacteria. Thus, standard plate counts are not good predictors of whether off-flavors will arise from microbial causes. Since the product is prebaked and then frozen, the numbers of vegetative microorganisms declines until thawing occurs. Unfortunately, pathogens such as *Staphylococcus aureus* are not totally inactivated by these treatments. If the product is then abused during distribution (repeated freeze and thaw), the pathogens may grow; however, the product would need to exceed 7°C near the surface for this to occur. Obviously, if even moderately good control of the distribution system is achieved, this will not be a problem.

In a shelf life study, one can determine the potential magnitude of the problem with pathogens. This can be achieved by preinoculating the meat and cheese with organisms of public health significance and abusing the product by cycling it through a freeze and thaw sequence. A typical test sequence used by some food companies is as follows:

Day	Abuse temperature cycle	No. packages remaining
1	24 hr at −18°C	5
2	20 hr at −18°C 4 hr at 38°C	4
3	20 hr at −18°C 4 hr at 38°C	3
4	20 hr at −18°C 4 hr at 38°C	2
5	20 hr at −18°C 4 hr at 38°C	1

The study begins with five packages placed at −18°C for 1 day. At the end of the day a single package is removed and is tested for microbiological indicators. All the other packages are returned to −18°C for the next 20 hr and then abused at 38°C. Another package is then removed for testing, and the cycle is repeated for the other package. If there is no significant increase in spoilage or pathogenic organisms after day 5, the food is deemed safe microbiologically. Of course, since it takes some time before the analysis can be done, these products should not be tested for taste. This abuse sequence is rather severe and some product development people feel that the tests can be terminated after the third day.

Chemical Changes. When pizza is frozen, temperature lowering usually causes reaction rates to decrease dramatically. The Q_{10} for frozen foods have been reported to vary from 3 to 40 (19). However, as noted in Chap. 2, reactants may concentrate in the unfrozen aqueous phase, and this increase in concentration may cause some reactions to increase in rate or to decrease in rate less than expected. This type of behavior is most likely to occur just below the freeze-thaw point of the pizza (−5 to −3°C). This means that data collected under abuse conditions, for example, at −5°C, may not abide by a straight-line relationship assumed for shelf life and Arrhenius plots, and that straight-line extrapolation of −5°C data to lower temperatures may result in substantial errors. In addition, if the product is freeze-thaw abused, an inordinately great time is spent at or near −3 to −5°C, where uncommonly large reaction rates are often observed. This propensity for the product to tarry at −3 to −5°C occurs because such products as pizza freeze more rapidly than they thaw (equal but opposite temperature differentials). That ice has a much greater thermal diffusivity than immobilized water is the mechanistic explanation for this behavior.

Another accelerating factor can arise during freezing. Because of membrane damage to cells, enzymes that are normally compartmentalized are released into the cellular fluids (see Chap. 6). These enzymes can then contact substrate and accelerate loss of quality.

In pizza, enzymes of the greatest potential concern are lipases and lipoxygenases. Lipases exist in natural cheese, and these enzymes might produce sufficient amounts of fatty acids to cause a hydrolytic rancid flavor, especially if butyric acid is released (from milk-based products). One way to avoid this problem is to use an imitation cheese made with soybean oil. Lipases are also present in vegetables (green peppers) and in the crust. Prebaking of the crust generally is not sufficient to destroy lipase, so the crust could develop a soapy flavor or an off-odor from the free fatty acids produced. The vegetables and tomato sauce should be no problem if they are blanched or cooked; however, spices contain many enzymes, including lipases. It should be noted that the standard catalase or peroxidase test used to determine the adequacy of vegetable blanching does not detect lipase. One approach to monitoring lipase activity is to measure total free fatty acids during storage of pizza. In addition, volatile fatty acids can be isolated and characterized by gas-liquid chromatography (GLC) or high-performance liquid chromatography techniques. The big problem is sampling. The food scientist must decide whether to sample each ingredient, a slice, and/or an entire pizza.

Lipoxygenase, present in spices, sausage, wheat flour, and vegetables, catalyzes oxidation of unsaturated fats, producing peroxides as well as volatile breakdown products. As with lipase, it is heat stable and may survive baking or precooking. Several tests can be used to determine the consequences of lipoxygenase activity, including determination of peroxide value and measurement of volatile aldehydes or ketones by GLC (e.g., hexanal or pentanal). If peroxides are measured, this should be done at least twice weekly; otherwise, substantial changes in peroxides, which are indicators of the onset of oxidative rancidity, may be missed. This is true since peroxides increase to a maximum and then fall to near zero soon thereafter. Thus, a low peroxide value does not necessarily mean a nonrancid food. The peroxide value at which a food is considered rancid varies from food to food. Because sausage is cured, it may become rancid first. Thus, the addition of antioxidants may be desirable, and should be done if allowed. It is also possible that addition of EDTA would be helpful.

In addition to causing off-odors, oxidative rancidity can lead to a bleaching of the deep-red tomato color, giving rise to an orangish color. Thus, one should test for color changes with either a color meter or by extracting the pigments and measuring them in a spectrophotometer (see Chap. 8). Obviously, to retard color loss and rancidity, lipoxygenase should be destroyed during processing. Unfortunately, heat treatments needed to destroy lipoxygenase may also damage desirable flavors and textures, so a balance must be sought. Vacuum packaging or nitrogen flushing is also helpful (32). However, because of the irregular surface of a pizza this is difficult to adequately accomplish at high speeds.

Another cause of deterioration in pizza quality is loss of spice impact. This can result from volatilization and/or chemical reactions. Since many flavor compounds are aldehydes and ketones, they can react with proteins or amino acids, thereby decreasing flavor. Others may be oxidized directly by oxygen or indirectly by the free radicals produced during oxidative rancidity. Although sensory testing is the best approach to monitoring loss of spice impact, this approach is expensive. If available, GLC analysis of the headspace coupled with mass spectroscopy to identify key volatile spice flavor compounds is a rapid, useful, but not inexpensive, alternative.

The last major chemical reaction leading to off-flavors or color changes in pizza is nonenzymic browning, and the cheese component is especially susceptible. Since the a_w drops as the freezer temperature is decreased, this could accelerate browning. At $-15°C$ the a_w is 0.86, which is near the level for the maximum rate of browning. Browning can be measured easily by extracting the pigments and measuring the optical density at 400-450 nm, or by measuring the amount of available lysine by the fluorodinitrobenzene procedure (30). Browning also could be significant in imitation cheese if high-fructose corn syrup is used as the sugar source.

In summary, tests for the following constituents or properties should be considered for monitoring chemical changes in pizza during frozen storage/:

Total free fatty acids
Specific volatile free fatty acids by GLC
Peroxides
Oxidative volatiles (e.g., hexanal) by GLC
Spice volatiles by GLC
Lysine
Color (decrease in red color or increase in brown)
Sensory properties: taste and flavor

Nutrient Loss. Chemical reactions can also cause a loss of vitamins. Of special concern, if nutritional labeling is done, are vitamins C and A, both of which are subject to oxidation. Since vitamin C is quite unstable at pH values above 5.0, its loss is sometimes used as an index of loss of overall quality. For example, when 15-20% of the vitamin C is lost from frozen vegetables, they become unacceptable from a sensory standpoint (19).

Physical Changes. One of the major complaints consumers have about frozen pizza is that the crust is not as crisp as that of fresh pizza. The reason for this is not entirely clear. If the crust has a high enough moisture content it certainly freezes when held at $-18°C$. Since all the other ingredients also exhibit frozen water, then at any temperature below about $-18°C$ all pizza components have the same a_w (see Chap. 2) and there should be no moisture migration from the sauce to the crust. At higher temperatures (-7 to $-1°C$), however, the crust thaws and achieves a lower a_w than the other ingredients, and might absorb the water from the sauce by capillary suction, leading to a higher moisture and a loss of crispness.

For dry foods, crispness is lost at a_w's above about 0.4-0.5. Unfortunately, a pizza crust with that a_w would not have a desirable chewiness. One way to impart chewiness while retaining crispness is to fry the crust in fat. Research is needed to determine if the starch in flour can be chemically modified to yield a more desirable texture without losing other desirable functional properties. If successful, this would overcome the need to fry the crust in fat.

Another undesirable physical change in pizza is a loss of meltability and functionality of the cheese. As the water in the cheese freezes, concentration of the lactose and salts can cause irreversible chemical and structural changes in the protein. In addition, nonenzymic browning can lead to loss of protein functionality. These reactions reduce the ability of the protein to bind water after thawing. In addition, irreversible protein-protein interactions, occurring at the high solute concentrations in frozen cheese, tend to reduce the flow (meltability) of the cheese when heated. The concentrated solutes also affect the pectin and starches in the tomato sauce, resulting in loss of water binding and, thus, weeping (exudation of water) upon thawing.

Finally, package ice is a problem in stored frozen pizza. As it goes through inadvertent freeze-thaw cycles, the air space in the package warms up faster than the product, and the air then can hold more water. Water acquired by the air comes from the frozen ingredients. As the air temperature declines again, the water vapor crystallizes out as ice particles on the product surface. Repeated cycling causes the crystals to grow, become visible, and eventually convey an undesirable appearance to the product. A freeze-thaw cycling test, such as for the microbiological test, is generally adequate to determine if package ice will be a problem. The abuse temperature should be at 20°C instead of 38°C.

To prevent package ice, the wrapper should be as tight as possible to minimize air spaces, something not easily done with pizza. The wrapper should also contain a release agent to precent the topping from sticking to it.

3. Shelf Life Test

Temperature Range. Given that a minimum shelf life of 12 months at −18°C is desired, a shelf life plot can be constructed as in Fig. 8, showing equivalent times at higher temperatures. From Fig. 8 it is evident that the product must, depending on the Q_{10} value, retain satisfactory quality for the following minimum times at −4°C.

Q_{10}	Time at −4°C that equates to 12 months at −18°C
2	4.5 months
5	1.2 months
10	14 days
20	6 days

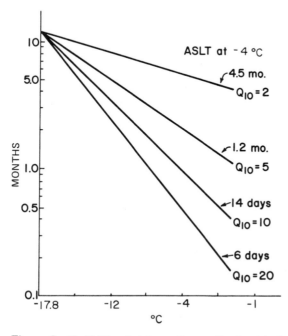

Figure 8 Shelf life plot for a frozen food with desired shelf life of 12 months at −18°C and equivalent ASLT at −4°C.

Most published results suggest that Q_{10}'s for vitamin C loss and quality loss in frozen vegetables range from 2 to 20 and that the shelf life of vegetables is only 6-8 months at $-18°C$. Considering these Q_{10} values, a product that does not retain good quality for 4.5 months at $-4°$ may not retain good quality for 12 months at $-18°C$. Given the above calculations, this suggests use of the sampling frequency shown in Table 3. The various temperatures are necessary to ascertain whether the Arrhenius relationship is conformed to. All simple tests should be conducted at each sampling time; sensory testing should be concentrated mainly toward the end of the test sequence, with a few near the beginning. In addition to this simple, constant-temperature study, a variable temperature (cycling) study, as described before, should be done to determine freeze-thaw effects. Data should be analyzed as previously discussed.

Controls. Whenever sensory analysis is to be used or rates of change of some quality factors are to be determined, both initial values and stored controls are needed. It is desirable, from a statistical standpoint, to perform numerous measurements (replicates) on the initial product so the precision of the method can be accurately established. Recent work (2,21,23,29,30) suggests that 6-10 initial samples should be tested, but this is not reasonable for sensory testing because of the time and expense. For sensory tests involving a comparison triangle method, a stored control is also needed. Whereas one can use suitably processed, refrigerated controls for shelf life studies on dehydrated and canned foods, this is not appropriate for frozen foods. A fresh product is also not appropriate for this purpose since it has not been frozen and thawed. Thus, pizza held at -40 to $-70°C$ (to minimize change during storage) is a suitable control in shelf life studies of frozen pizza. Temperature fluctuations in the freezer used for control samples must also be kept to a minimum.

B. Dehydrated Mashed Potatoes

1. General Aspects of Quality Loss

Dehydrated potatoes are made by drum drying a slurry of previously cooked, mashed potatoes that may include added sulfite and antioxidants. Dehydrated potatoes are susceptible primarily to browning at temperatures above 30-32°C and primarily to lipid oxidation below 30-32°C. In addition, the rate of vitamin C loss generally exceeds the rates of either of the other aforementioned reactions if the a_w is above the monolayer value (see Chap. 2).

2. Specific Aspects

Microbiological Changes. Since this is a dehydrated product, the a_w should be in the range 0.2-0.4, and growth of microorganisms should not occur (see Chap. 2). Growth

Table 3 Sampling Frequency for Frozen Pizza Shelf Life Study

$-3.9°C$ (25°F)	1, 2*, 3, 4, 5, 8, 12, 14, 16*, 20* weeks
$-6.7°C$ (20°F)	2, 4*, 10, 15*, 20* weeks
$-9.4°C$ (15°F)	4*, 10, 15*, 20* weeks

* = Sensory test times.

of typical spoilage organisms does not occur unless the product reaches an a_w of at least 0.7. Most likely if the product a_w did increase to this level during storage, chemical reactions would render the product unsatisfactory before microbial growth became a factor of importance.

Chemical Changes. Potatoes may contain up to about 4% free reducing sugars, and almost 50% of the protein nitrogen is in the form of free amino acids. Thus, unless the product is dried and kept near or below the monolayer value, nonenzymic browning occurs. Because of the presence of a large amount of free amino acids, both the typical Shiff's base-Amadori rearrangement sequence and the Strecker degradation pathway occur. The latter reaction produces CO_2 (6) and other volatile compounds, such as 2-methylpropanal, 2- and 3-methylbutanal, and isovaleraldehyde, which are detectable in the parts per billion range as extremely charred off-odors (3,4,31). The browning reaction can utilize vitamin C as one of the substrates and it can also reduce the subsequent hydratability of the potato.

At or below about 32°C, and especially if the product is dried below the monolayer value, oxidation of unsaturated lipids occurs. Dried potatoes contain only 1% lipid, but only 2-3 ppm hexanal, an oxidative degradative product, renders them unacceptable (7,31). Since about 60-70% of the fatty acids are polyunsaturated, production of volatile off-flavors occurs readily unless antioxidants and vacuum packaging are used. Gas chromatography-mass spectrometry (GC-MS) is an easy, effective method to detect and correlate the changes in headspace compounds with sensory acceptability. Generally, only about 5-10% of the unsaturated fat present needs to be oxidized to produce an unacceptable product.

Kreisman (13) has published a shelf life plot for dehydrated potatoes (Fig. 9) that correlates available shelf life data with lipid oxidation and nonenzymic browning. The a_w values for these studies varied between 0.2 and 0.4, and this accounts, in part, for the breadth of the confidence limits.

Physical Changes. In general, there should be no physical changes in stored dehydrated potatoes except loss of rehydratability. However, the extent of browning or oxidation needed to produce poor rehydration probably would make the dry product unacceptable from an odor standpoint.

3. Shelf Life Design

It is obvious from Fig. 9 that the Arrhenius or shelf life approach cannot be used over a wide temperature range and, although chemical reactions can be easily measured, sensory acceptability is the key factor in determining the end of shelf life. The following test conditions are reasonable when dehydrated potatoes are stored in moisture-impermeable pouches (* = sensory test).

Temperature (°C)	Test times (months)
25	6, 9, 12,* 18*
30	6, 8, 10,* 12,* 14
35	1, 2, 3,* 4*
40	1, 2,* 3*

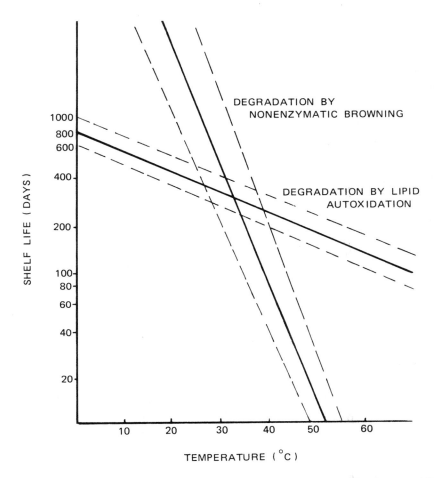

Figure 9 Shelf life plot for dehydrated potatoes with quality loss resulting from non-enzymic browning or lipid oxidation.

As was true in the frozen food case study, the frequency factor equation cannot be used with dehydrated potatoes because of the change in degradative mechanism with a change in temperature. Obviously, 25°C is also too low to achieve significant acceleration of the reactions compared with normal storage conditions, but since the primary mode of degradation changes at about 35°C, tests at 25°C are advisable. Control samples in this instance should be nitrogen flushed, sealed in cans, and held at 0°C. Other than sensory testing, GC-MS measurements of specific volatiles, and perhaps measurement of brown pigment development, are all that is needed.

4. Distribution Model

To determine the effects of product distribution on quality loss in dehydrated potatoes, Kresiman (13) developed a simulated distribution model. This pertained to a plant in the

Midwest that shipped product to a central distribution center, which in turn then distributed product to several locations, including Boston, New York, Minneapolis, St. Louis, Los Angeles, and Phoenix. Average, minimum, and maximum times and temperatures at each location along the route were estimated (Table 4). The shelf life consumed was then calculated for each of the following conditions:

1. Maximum time, maximum temperature (θ_{max}, T_{max})
2. Maximum time, average temperature (θ_{max}, T_{avg})
3. Maximum time, minimum temperature (θ_{max}, T_{min})
4. Average time, maximum temperature (θ_{avg}, T_{max})
5. Average time, average temperature (θ_{avg}, T_{avg})
6. Average time, minimum temperature (θ_{avg}, T_{min})
7. Minimum time, maximum temperature (θ_{min}, T_{max})
8. Minimum time, average temperature (θ_{min}, T_{avg})
9. Minimum time, minimum temperature (θ_{min}, T_{min})

The zero-order model using Eq. (3) was used. Thus, the percentage shelf life consumed can be calculated from

$$\% \text{ shelf life consumed} = \left(\frac{\theta_i}{\theta_s}\right)_{T_i} \times 100 \qquad (12)$$

where θ_i is the time at condition T_i and θ_s is the shelf life at T_i (from Fig. 9).

    Two methods can be used to determine the percentage shelf life consumed. One can consider that only lipid oxidation or only browning is limiting, make calculations for each, and then select the degradation model that exhausts the shelf life most rapidly. Alternatively, one can assume, as is actually true, that both degradative mechanisms function simultaneously and that data from the mechanism yielding the shortest shelf life θ_s at any T_i should be used in Eq. (12). In the present simulation, the latter alternative was used and the average values (solid line in Fig. 9) were chosen to represent θ_s. Table 5 shows the results for distribution to Boston (generally cool and humid), and Table 6 is for distribution to Phoenix. It is apparent that the quality of the product at the point of purchase can become unacceptable (e.g., 172% of the shelf life is consumed if the product is held in Boston under "maximum" conditions). On the other hand, it can be very acceptable with only 10-20% of the shelf life lost if the product moves through the distribution scheme in the minimum time.

    An exercise like this can help the food scientist decide what changes in ingredients and/or packaging are needed or it can help the manufacturer to identify troublesome distribution conditions that should be eliminated. The data can also be used to determine the effect of seasonal changes. Table 7 shows calculations for the Phoenix distribution system, assuming average times from Table 4 for each leg of the trip, and 90 day seasons with maximum temperatures in summer, minimum in winter, and average in fall and spring. It is evident that only product made in the summer or spring is acceptable at point of purchase. It must be remembered, however, that these data are only simulations; food scientists must collect actual time-temperature data for their own products to obtain realistic results.

    It should now be evident that food scientists must wisely integrate many skills if they are to succeed in developing and distributing food products that are wholesome, nutritious, appealing, stable, and economical.

Table 4 Maximum, Average, and Minimum Times and Temperatures in a Theoretical Distribution System

	New York max – avg – min			Boston max – avg – min			Mpls. max – avg – min			St. Louis max – avg – min			Los Angeles max – avg – min			Phoenix max – avg – min		
Plant: θ[a]	10	3	0	10	3	0	10	3	0	10	3	0	10	3	0	10	3	0
°C	33	2	−16	33	2	−16	33	2	−16	33	2	−16	33	2	−16	33	2	−16
Rail: θ	6	4	1	6	4	1	6	4	1	6	4	1	6	4	1	6	4	1
°C	34	3	−14	34	3	−14	34	3	−14	34	3	−14	34	3	−14	34	3	−14
Chgo: θ	250	158	38	240	148	30	230	140	25	245	145	30	290	156	42	290	166	50
°C	26	10	4	16	10	4	16	10	4	16	10	4	16	10	4	16	10	4
Rail: θ	8	6	3	8	6	3	5	3	1	7	3	1	10	8	4	10	8	4
°C	35	11	−13	35	1	−14	34	3	−14	34	3	−14	34	4	−14	34	4	−14
D. Ctr: θ	357	139	28	232	116	25	175	83	18	187	88	23	330	170	46	260	151	39
°C	27	14	−2	27	14	−2	32	16	−6	32	16	−6	35	23	10	35	23	10
Rail: θ	7	3	1	8	6	3	6	4	2	7	5	2	7	5	2	6	4	2
°C	43	16	−16	37	13	−13	34	3	−14	34	4	−13	43	18	−15	49	24	−12
Whsle: θ	55	25	3	43	20	4	41	16	4	51	23	4	43	11	3	42	8	4
°C	31	20	7	30	14	4	27	12	−1	38	24	13	38	19	7	43	24	13
Retail: θ	81	24	4	68	17	3	68	17	3	76	23	4	70	20	4	68	17	3
°C	32	24	18	32	24	18	32	24	18	32	24	18	32	24	18	32	24	18
Home: θ	180	90	30	280	90	30	280	90	30	190	90	30	280	90	30	280	90	30
°C	35	10	4	35	10	4	32	10	0	33	12	2	38	14	7	40	16	8
Total θ	954	452	108	895	410	99	821	360	84	869	384	95	1046	467	132	972	451	133

[a]Days.
Source: From Kreisman (13).

Table 5 Percentage of Shelf Life (θ_s) Lost in Each Link of the Simulated Distribution Chain to Boston Using Nine Combinations of Time and Temperature[a]

Link of chain	$\theta_{mx}T_{mx}$	$\theta_{mx}T_{av}$	$\theta_{mx}T_{mn}$	$\theta_{av}T_{mx}$	$\theta_{av}T_{av}$	$\theta_{av}T_{mn}$	$\theta_{mn}T_{mx}$	$\theta_{mn}T_{av}$	$\theta_{mn}T_{mn}$
Plant	4.5	1.3	0.7	1.4	0.4	0.2	—	—	—
Rail	3.2 (7.7)	0.8 (2.1)	0.5 (1.2)	2.1 (3.5)	0.5 (0.9)	0.3 (0.5)	0.5	0.1	0.1
Chicago	48.0 (55.7)	40.0 (42.1)	32.9 (34.1)	29.6 (33.1)	24.7 (25.6)	20.3 (20.8)	6.0 (6.5)	5.0 (5.1)	4.1 (4.2)
Rail	5.0 (60.7)	1.4 (43.5)	0.6 (34.7)	3.8 (35.8)	1.0 (26.6)	0.5 (21.3)	1.9 (8.4)	0.5 (5.7)	0.2 (4.4)
Distribution center	66.2 (126.9)	43.0 (86.4)	26.4 (61.0)	33.1 (70.0)	21.5 (48.1)	13.2 (24.5)	7.1 (15.6)	4.6 (10.3)	2.8 (7.3)
Rail	7.0 (133.9)	1.5 (87.9)	0.6 (61.7)	5.2 (75.2)	1.1 (49.2)	0.5 (34.9)	2.2 (19.7)	0.6 (10.8)	0.2 (7.5)
Wholesale warehouse	13.2 (147.1)	8.1 (95.8)	5.9 (67.6)	6.2 (81.3)	3.7 (52.9)	2.7 (37.7)	1.2 (18.9)	0.7 (11.6)	0.6 (8.1)
Retail	25.2 (172.3)	17.4 (113.3)	14.5 (82.1)	6.3 (87.6)	4.4 (57.3)	3.6 (41.3)	1.1 (20.1)	0.8 (12.3)	0.6 (8.7)
Home	175.0	46.7	38.4	56.3	15.0	12.3	18.8	5.0	4.1
Total %θ_s consumed	347.3	159.9	120.4	143.9	72.3	53.6	38.8	17.3	12.8

[a]Values in parentheses are accumulated percent θ_s lost.
Source: From Kreisman (13).

Table 6 Percentage of Shelf Life (θ_s) Lost in Each Link of the Simulated Distribution Chain to Phoenix Using Nine Combinations of Time and Temperature[a]

Link of chain	$\theta_{mx}T_{mx}$	$\theta_{mx}T_{av}$	$\theta_{mx}T_{mn}$	$\theta_{av}T_{mx}$	$\theta_{av}T_{av}$	$\theta_{av}T_{mn}$	$\theta_{mn}T_{mx}$	$\theta_{mn}T_{av}$	$\theta_{mn}T_{mn}$
Plant	4.5	1.3	0.7	1.4	0.4	0.2	—	—	—
Rail	3.2 (7.7)	0.8 (2.1)	0.5 (1.2)	2.1 (3.5)	0.5 (0.9)	0.3 (0.5)	0.5	0.1	0.1
Chicago	58.0 (65.7)	48.3 (50.4)	39.7 (40.9)	33.2 (36.7)	27.7 (28.6)	22.7 (23.3)	10.0 (10.5)	8.3 (8.5)	6.9 (6.9)
Rail	5.3 (70.9)	1.4 (51.8)	0.8 (41.7)	4.2 (40.9)	1.1 (29.7)	0.6 (23.9)	2.1 (12.6)	0.6 (9.1)	0.3 (7.2)
Distribution center	162.5 (233.4)	65.0 (116.8)	43.3 (85.0)	94.4 (135.3)	37.8 (67.5)	25.2 (29.1)	24.4 (37.0)	9.8 (18.8)	6.5 (13.7)
Rail	37.5 (270.9)	1.5 (118.3)	0.5 (85.5)	25.0 (160.3)	1.1 (68.5)	20.3 (49.4)	12.5 (49.5)	0.5 (19.3)	0.2 (13.9)
Wholesale warehouse	97.7 (369.6)	10.8 (129.1)	7.6 (93.2)	18.6 (178.9)	2.1 (70.6)	1.5 (50.9)	9.3 (58.8)	1.0 (20.3)	0.7 (14.6)
Retail	25.2 (393.8)	17.4 (146.5)	14.5 (107.6)	6.3 (185.2)	4.4 (74.9)	3.6 (54.5)	1.1 (59.9)	0.8 (21.1)	0.6 (15.3)
Home	400.0	56.0	43.1	128.6	18.0	13.9	42.8	6.0	4.6
Total %θ_s consumed	793.8	202.5	150.7	313.7	92.9	68.3	102.8	27.1	19.9

[a]Values in parentheses are accumulated percent θ_s lost.
Source: From Kreisman (13).

Table 7 Percentage of Shelf Life Lost ($\%\theta_s$) at Each Point in the Simulated Distribution Chain to Phoenix for a Product Undergoing Seasonal Temperature Changes and Held for the Average Time at Each Point (See Table 6)

Link of chain	Summer $\%\theta_s$ lost	Summer Days per season	Fall $\%\theta_s$ lost	Fall Days per season	Winter $\%\theta_s$ lost	Winter Days per season	Spring $\%\theta_s$ lost	Spring Days per season
Plant	1.36	3/sum	0.40	3/fall	0.22	3/wtr	0.40	3/spr
Rail	2.10	4/sum	0.53	4/fall	0.31	4/wtr	0.53	4/spr
Chicago	16.60	83/sum	13.83	83/fall	11.37	83/wtr	13.83	83/spr
	13.83	83/fall	11.37	83/wtr	13.83	83/spr	16.60	83/sum
Rail	0.96	7/fall	0.54	7/wtr	0.96	7/spr	3.68	7/sum
	0.08	1/wtr	0.14	1/spr	0.53	1/sum	0.26	1/fall
D.C.	14.83	89/wrt	22.25	89/spr	55.63	89/sum	22.25	89/fall
	15.50	62/spr	38.75	62/sum	15.50	62/fall	10.36	62/wtr
Rail	1.06	4/spr	25.00	4/sum	1.04	4/fall	0.33	4/wtr
Whsle.	2.05	8/spr	18.60	8/sum	2.05	8/fall	1.45	8/wtr
Retail	4.10	16/spr	5.93	16/sum	4.10	16/fall	3.40	16/wtr
	0.37	1/sum	0.26	1/fall	0.21	1/wtr	0.26	1/spr
$\%\theta_s$ lost commercially	72.84%		137.60%		105.77%		73.23%	
Home	127.14	89/sum	17.80	89/fall	13.69	89/wtr	17.80	89/spr
	0.20	1/fall	0.15	1/wtr	0.20	1/spr	1.48	1/sum
Total $\%\theta_s$ lost	200.18%		155.55%		119.66%		92.51%	

Source: From Kreisman (13).

REFERENCES

1. Benson, S. W. (1960). *Foundations of Chemical Kinetics*, McGraw-Hill, New York.
2. Bergquist, S., and T. P. Labuza (1983). Kinetics of peroxide formation in potato chips undergoing a sine wave temperature fluctuation. *J. Food Sci. 43*:712.
3. Boggs, M. M., R. G. Buttery, D. W. Venstrom, and M. L. Belote (1964). Relation of hexanal in vapor above stored potato granules to subjective flavor estimates. *J. Food Sci. 29*:487.
4. Buttery, R. G., and R. Teranaski (1963). Measurement of fat oxidation and browning aldehydes by direct vapor injection GLC. *J. Agr. Food Chem. 11*:504.
5. Cardoso, G., and T. P. Labuza (1983). Effect of temperature and humidity on moisture transport for pasta packaging material. *Br. J. Food Technol. 18*:587.
6. Cole, S. J. (1967). The Maillard reaction in food products: Carbon dioxide production. *J. Food Sci. 32*:245.
7. Fritsche, C. W. and J. A. Gale (1977). Hexanal as a measure of oxidative deterioration in low fat foods. *J. Amer. Oil Chem. Soc. 54*:225.
8. Guadagni, D. G. (1968). Cold storage life of frozen fruits and vegetables as a function of temperature and time, in *Low Temperature Biology of Foodstuffs* (J. Haw-

thorne and E. J. Rolfe, eds.), Pergamon Press, New York.

9. Heiss, R. (1958). Shelf life determinations. *Mod. Pkg. 31*(8):119.

10. Kamman, J., and T. P. Labuza (1981). Kinetics of thiamine and riboflavin loss in pasta as a function of constant and variable storage conditions. *J. Food Sci. 46*: 1457.

11. Karel, M. (1967). Use-tests. Only real way to determine effect of package on food quality. *Food Can. 27*:43.

12. Kramer, A. (1974). Storage retention of nutrients. *Food Technol. 28*:50.

13. Kreisman, L. (1980). Application of the principles of reaction kinetics to the prediction of shelf life of dehydrated foods. M. S. thesis, University of Minnesota, Minneapolis.

14. Labuza, T. P. (1971). Kinetics of lipid oxidation of foods. *C. R. C. Rev. Food Technol. 2*:355.

15. Labuza, T. P. (1979). A theoretical comparison of losses in foods under fluctuating temperature sequences. *J. Food Sci. 44*:1162.

16. Labuza, T. P. (1980). The effect of water activity on reaction kinetics of food deterioration. *Food Technol. 34*:36.

17. Labuza, T. P. (1980). Temperature/enthalpy/entropy compansation in food reactions. *Food Technol. 34*(2):67.

18. Labuza, T. P. (1982). Moisture gain and loss from packaged foods. *Food Technol. 36*(4):92.

19. Labuza, T. P. (1982). Scientific investigation of shelf life, in *Shelf Life Dating of Foods*. Food and Nutrition Press, Westport, Conn., Chapter 3.

20. Labuza, T. P., and R. Contreras-Medellin (1981). *Cereal Foods World 26*(7):335.

21. Labuza, T. P., and J. Kamman (1983). Reaction kinetics and accelerated tests simulation as a function of temperature, in *Applications of Computers in Food Research* (I. Saguy, ed.), Marcel Dekker, New York, Chapter 4.

22. Labuza, T. P., and D. Riboh (1982). Theory and application of Arrhenius kinetics to the prediction of nutrient losses in foods. *Food Technol 36*(10):66.

23. Labuza, T. P., K. Bohnsack, and M. N. Kim (1982). Kinetics of protein quality loss stored under constant and square wave temperature distributions. *Cereal Chem. 59*:142.

24. Labuza, T. P., S. Mizrahi, and M. Karel (1972). Mathematical models for optimization of flexible film packaging of foods for storage. *Trans. Amer. Soc. Agr. Eng. 15*:150.

25. Mizrahi, S., T. P. Labuza, and M. Karel (1970). Computer aided predictions of food storage stability. *J. Food Sci. 35*:799.

26. Nakabayashi, K., T. Shimamoto, and H. Mina (1981). Stability of package solid dosage forms. *Chem. Pharm. 28*(4):1090; *29*(7):2027; 2051; 2057.

27. Oswin, C. R. (1946). The kinetics of package life. *J. Soc. Chem. Ind. 64*:67; 224; *65*:419.

28. Quast, D. G., and M. Karel (1972). Computer simulation of storage life of foods undergoing spoilage by two interactive mechanisms. *J. Food Sci. 37*:679.

29. Riboh, D. K., and T. P. Labuza (1982). Kinetics of thiamine loss in pasta stored in a sine wave temperature condition. *J. Food Proc. Preserv. 6*(4):253.

30. Saltmarch, M., and T. P. Labuza (1982). Kinetics of browning and protein quality loss in sweet whey powders under steady and non-steady state storage conditions. *J. Food Sci. 47*:92.

31. Sapers, G. M. (1970). Flavor quality in explosion puffed dehydrated potato. *J. Food Sci. 35*:731.

32. Simon, I., T. P. Labuza, and M. Karel (1971). Computer aided prediction of food storage stability: Oxidation of a shrimp food product. *J. Food Sci. 36*:280.

BIBLIOGRAPHY

Benson, S. W. (1960). *Foundations of Chemical Kinetics*, McGraw-Hill, New York.

Kramer, A. (1974). Storage retention of nutrients. *Food Technol. 28*:50.

Labuza, T. P. (1980). Temperature/enthalpy/entropy compensation in food reactions. *Food Technol. 34*(2):67.

Labuza, T. P. (1982). Moisture gain and loss from packaged foods. *Food Technol. 36*(4): 92.

Labuza, T. P., and D. Riboh (1982). Theory and application of Arrhenius kinetics to the prediction of nutrient losses in foods. *Food Technol. 36*(10):66.

SUBJECT AND CHEMICAL INDEX

A

A-band, 732–734, 756
Abscisic acid, 891
Abscisins, 893, 895
Acacia gum, 680
Acceptable daily intake (ADI), 691
Acesulfame K, 656, 657
Acetaldehyde, 186, 610, 617, 645
Acetates as antimicrobial agents, 649
Acetic acid, 436, 492, 550, 558, 610, 632, 649, 661, 670, 672
   antimicrobial agent, 649, 670, 672
   chlorophyll, effect on, 550
   component of anthocyanins, 558
   firming agent, 661
   food additive, 632
Acetic acid–sodium acetate buffer, 638
Aceto-acetic acid, 656
Acetoin, 610, 617, 618
Acetone, 273, 672
Acetone cyanohydrin glucoside, 697
Acetone peroxides in dough, 670
Acetylated tartaric acid monoacylglycerol, 173
Acetylating agents, reaction with proteins, 348
Acetylcholine, 711
Acetylcholinesterase inhibitors, 714
Acetyl-CoA, 612
2-Acetylfuran, 492
N-Acetylglucosamine, 834, 836
Acidic alum salts for firming plant tissue, 661

Acids
   as chelating agents, 641
   as food additives, 630–636
   formed during thermal processing, 550
   from thermal decomposition of lipids, 205–209
Acidulants, 670
Acrolein, 205, 207
Actin, 246, 247, 258–261, 733, 737–739, 740–742, 753–756, 761, 762, 771, 772, 779
   characteristics, 258–261
   contraction, role in, 740, 741
   in muscle relaxation, 742
   quarternary structure, 258–260
α-Actinin, 258, 734, 739, 754, 755, 756
β-Actinin, 739, 741
Actinomycin D, 254
Activated carbon, 218
Activation energy
   enzymes, effect of, 376, 378
   for food deterioration, 918
   for spore inactivation, 528, 529
   for thiamine degradation, 497, 498
   for vitamin B_6 degradation, 504
   for vitamin destruction, 528–530
   water activity, dependence on, 918
Active oxygen method (AOM), 196
Actomyosin, 258–261, 738, 761, 770, 773, 779, 780
   characteristics, 258–261
   quarternary structure, 258–260

Acylases, 362
Acylated monoacylglycerols, 671
Acylation of proteins, 360
Acylglycerols (*see also* Triacylglycerols)
    definition, 143, 146
    dimerization, 189
    mesomorphic phases (liquid crystals),
        163-165, 169
    nomenclature, 143-145
    oxidative crosslinks in, 189, 192
    polymerization, 189, 192
    stereospecific numbering, 144, 145
    subgroups, 146, 147
    trimerization of, 189
Acyl radicals, exchange during interesterifica-
        tion, 223
Adenosine diphosphate glucose: starch glu-
        cosyltransferase, 888
S-Adenosylmethionine
    in ethylene synthesis, 892
    in ripening fruit, 886
Adenylate kinase, 751, 752, 757
ADP (adenosine diphosphate) in muscle, 737,
        738, 751-753, 756-758, 761, 768
ADP glucose-starch glucosyltransferase, 887
ADPG pyrophosphorylase, 888
Adulteration of food
    Accum's comments, 5
    in 1800s, 4-6
    historical examples, 5, 6
Aerating agents, 668
Aerosols, 667
Aflatoxin B$_1$ structure, 709
Aflatoxins, 218, 328, 338, 339, 707-710
    alkaline pH, effect of, 338
    hydrogen peroxide, effect of, 339
    removal during refining of fat, 218
    sodium hypochlorite, effect of, 339
Agar, 129, 134, 175, 658, 672, 680
    as emulsion stabilizer, 175
    encapsulation agent, 672
    stabilizer, 680
Agaran, 129
Agaropectin, 129
Agene, 347
Aging of meat (*see* Muscle aging)
Aglycones, 558, 562
Agrobacterium tumefaciens, in recombinant
        DNA technology, 464
AH/B theory of sweetness, 588, 589
Air spaces in plant tissue, 872
Alanine, 250, 252, 253
β-Alanine, 512, 772
Albedo, 859
Albumins, 273, 298, 411, 861, 862 (*see also*
        Serum albumin)
Alcohol dehydrogenase, 246
Alcoholic beverages, 610, 611
Aldehyde dehydrogenase, 312
Aldehyde-lyase, 605
Aldehydes
    decomposition of, 185, 186
    interaction with proteins, 331
    from ionizing irradiation of lipids, 214

[Aldehydes]
    from lipid oxidation, 183-187, 195
    oxidation of, 185, 186
    from thermal decomposition of lipids,
        206-208
Aldehydes, α,β-unsaturated, 645
Aldo condensation, 749
Aldolase
    in milk, 802
    in ripening fruit, 885
Aldosylamine, 342
Aldrin, 714
Algae as a source of protein, 331
Alginates, 130-132, 134, 350
Alginic acid, 680
Alimentary toxic aleukia (ATA), 706-708
Alkalies added to food, 671
Alkaline phosphatase, 413, 802, 813
Alkaline substances as food additives, 636-638
Alkaloids, 328
1-Alkene, 187
Alkoxy radicals, 179, 183, 187-189, 194, 519
    epoxides from, 188
    from hydroperoxides, 179, 183, 187
    from lipid oxidation, 179
    reactions of, 187-189
    from thermal decomposition of lipids, 208
Alkylation of proteins, 361
S-Alkyl-L-cysteine sulfoxide lyase, 603
Alkyl gallates, 203, 204
Alkyl radical, 187, 189
Allergens, 694, 700
Allescheria boydii for biotin assay, 513
Allium sp. aromas, 601
Allyl glucosinolate, 602, 603
Allylisothiocyanate, 602, 603, 695
Allyl nitrile, 602, 603
Aluminum phosphate, 669
Aluminum silicate, 661
Aluminum stearate, 668
Aluminum sulfate, 661, 679
Alveolus, 792, 794, 795
Amadori products, 342-345
Amadori rearrangement
    in collagen, 749
    in Maillard browning, 99, 101, 102
α-Amanitin, 695
Amaranth, 705 ✓
Amentoflavone, 564
Ames test, 232
Amidation of proteins, 360
Amino acid anhydrides, 348
Amino acid antagonism, 320
Amino acid availability, 320, 321
Amino acid nutritive value
    alkaline hydrolysis, effect of, 355
    alkaline pH, effect of, 336
    after grafting to proteins, 348, 349
    halogenated solvents, effect of, 347
    heat, effect of, 335, 336
    Maillard browning, effect of, 343-345
    oxidation, effect of, 339-342
    processing, effect of, 332
Amino acid reactions

[Amino acid reactions]
  acylation, 269
  with aldehydes, 269
  amidation, 268
  with 1,2-benzene dicarbonal, 267
  with *p*-chloromercuribenzoate, 269
  with dansyl chloride, 269, 271
  deamination, 269
  decarboxylation, 268
  with Ellman's reagent, 269
  esterification, 268
  with fluorescamine, 267, 269
  with 1-fluoro-2,4-dinitrobenzene, 270
  with hypochlorite, 268
  with iodoacetic acid, 269, 270
  with *o*-methyl isourea, 270
  with ninhydrin, 266, 267, 269
  with Orange G, 270
  oxidation, 270
  with performic acid, 269, 270
  with phenylisothiocyanate, 269, 271
  with phosgene, 268
  reduction, 268
  with vitamin B_6, 502
Amino acid requirements in diet, 323
Amino acid residues, isomerization, 335
Amino acids (*see also* specific amino acids)
  absorption spectra, 254, 313
  acidic and basic properties, 249
  bitterness of L-isomers, 590
  bitterness versus hydrophobicity, 594, 595
  from cell cultures, 331
  chemical names, 250, 251
  chemical synthesis, 331
  chirality, 253
  classification, 248
  D and L configurations, 246, 248, 253, 254
  diet, requirements in, 317–319
  in dehydrated potatoes, 930
  as dietary supplements, 674, 675
  essential, 317–319
  flavors from, 107, 606, 607, 609–611, 615
  fluorescence, 254
  general description, 247, 248
  general structure, 248, 249
  genetic engineering, 331
  heat, destruction by, 336
  hydrophobicity, 248, 249, 252, 253, 594, 595
  imbalance, 319, 320
  ionization, 248, 249
  iron bioavailability, effect on, 526
  isoelectric points, 252
  limiting, 320, 322, 328
  Maillard browning, involvement in, 99
  and meat flavor, 765
  in milk, 797, 798, 821
  molecular weights, 250, 251
  movements in body, 314–316
  in muscle, 742, 752
  muscle postmortem, degradation in, 759
  with negatively charged side chains, 248
  with nonpolar side chains, 248
  number of kinds, 247–248

[Amino acids (*see also* specific amino acids]
  oxidation, 339–342
  oxidation, sensitivity to, 339
  pK_a values, 249, 252
  plasma, levels in, 315
  with polar side chains, 248
  with positively charged side chains, 248
  properties, 248
  in proteins, 323
  pyrolysis products, safety of, 713
  racemization, 637
  separation processes, loss by, 334
  stereochemistry, 253
  structures, 250, 251
  to supplement proteins, 320
  sweetness of D-isomers, 590
  symbols, 250, 251
  tissue, levels in, 315
  toxic, 694
  vitamin B_6, reaction with, 502
  zwitterions, 248, 249
D-Amino acids, 335, 355
Amino acylases, 433, 436
α-Amino adipic-Δ-semialdehyde, 747
α-Amino-adipic-δ-semialdehyde, 749
β-Aminoalanine, 337
2-Amino aldose, 99, 101
p-Aminobenzoic acid, 504
p-Aminobenzoic acid-glutamic acid, 504
(*p*-Aminobenzoyl) glutamate, 505, 506
γ-Aminobutyric acid, 864
1-Aminocyclopropane-1-carboxylic acid, 892
e-Amino group, 270
2-Amino-4-hydroxypteridine, 504
1-Amino-2-keto sugar, 99, 101
Aminopeptidases, 313, 362
β-Aminopropionitrile, 694
Amino sugars, 99
Ammonia, 315, 316, 325, 330, 335–337, 343, 601, 617, 618, 653
Ammonium aluminum sulfate, 661
Ammonium bicarbonate, 632, 671, 679
Ammonium carbonate, 632
Ammonium chloride, 665, 669
Ammonium hydroxide, 671
Ammonium phosphate
  buffer salt, 671
  dough conditioner, 665
  leavening agent, 679
  malting-fermenting aid, 669
Ammonium pyrrolidonecarboxylate, 906, 907
Ammonium stearate, 668
Ammonium sulfate, 420, 665, 669
AMP (adenosine monophosphate), 751, 756, 757
AMP deaminase, 742, 757
Amygdalin, 697
α-Amylase
  in carbohydrate digestion, 72
  for hydrolyzing corn starch, 90, 91
  in milk, 802
  pancreatic, 72
  in saliva, 72
β-Amylase, 72, 91, 374, 802

[β-Amylase]
  in carbohydrate digestion, 72
  for hydrolyzing corn starch, 91
  in milk, 802
Amylase inhibitors, 455, 459
Amylases, 72, 87, 90, 91, 116, 374, 411, 418,
      420, 428, 436, 441, 668, 802, 887
  action of, 441
  food, function in, 668
  importance, 441
  in potatoes, 887
  Schardinger dextrin, formation of, 87
  starch granules, activity against, 116
  types, 441
Amylopectin
  polymerization, degree of, 114
  properties, 115
  repeating units, 89
  staling, role in, 118
  starch granules, distribution in, 112, 113
  structure, 113
Amylose
  acid hydrolysis, susceptibility to, 92
  films and fibers from, 114
  gelatinization, behavior during, 115
  properties, 113–115
  repeating units, 89
  role in staling, 118
  starch granules, distribution in, 112, 113
Anaerobic glycolysis in muscle postmortem
      750–755
Analysis of food using immobilized enzymes,
      437, 438
5-α-Androst-16-en-3-one, 613
Anhydrobase structure, 559
Animal fat characteristics, 147
p-Anisidine, 195
Anisidine value, 195
Annatto, 673
Annatto extract, 580–581
Anserine, 752
Antibiotics, 575, 652
Anticaking agents, 665, 666, 669
Antifoam agents, 668
Antimicrobial agents, 644–653, 672, 673
  acetic acid, 649
  antibiotics, 652
  benzoic acid, 649, 650
  diethyl pyrocarbonate, 652, 653
  epoxides, 651, 652
  glyceryl esters, 648
  p-hydroxybenzoate alkyl esters, 650
  natamycin, 648
  nitrites, 645, 646
  parabens, 649, 650
  propionic acid, 649
  sorbic acid, 646–648
  sulfites, 644, 645
Antimony trichloride, 515
Antinutritional factors, 321, 328, 329
Antioxidants, 198–205, 643, 644, 673, 704,
      777, 929 (*see also* specific chemicals)
  application, modes of, 205
  definition of, 198

[Antioxidants]
  in dehydrated potatoes, 929
  factors influencing effectiveness, 200–203
  food examples, 204, 205
  foods, levels permitted in, 204, 205
  mechanism of action, 198, 200–202
  nonabsorbable type, 205
  regulation of use, 204, 205
  safety, 204, 205, 704
  selection of, 202, 203
  in smoke, 777
  synergism of, 202
  types, 199, 200, 203, 204
Antistick agents, 671, 672
Antithyroid substances, 694, 695
Anthocyanases, 454, 907
Anthocyanidins
  ascorbic acid, effect of, 563
  derivatives, 563
  pH, effect of, 563
  sulfites, effect of, 563
Anthocyanins, 452, 557–563, 891
  ascorbic acid, effect of, 562
  carbinol, pseudo-base form of, 559–561
  chalcone, form of, 559, 561
  copigmentation of, 562
  description, 557
  enzymes, effect of, 562
  flavonoids, interaction with, 562
  flavylium, cation form of, 561, 562
  a food colorant, 581
  major types, 557, 558
  metal ions, effect of, 562
  pH, effect of, 559, 560
  phlobaphens from, 562
  in plants postharvest, 891
  polyphenols, interaction with, 562
  properties, 575
  quinoidal base form, 559, 561
  stability, 558–560
  structures, 557, 558
  sulfites, effect of, 560, 562
  various substances, effect of, 562
Anthocyanogens (*see* Proanthocyanidins)
Anthoxanthins (*see* Proanthocyanidins)
Anthraquinones, 569
Apigenin, 563–565
Apiose, 563
Aplastic anemia, 712
Apolipoproteins, 235
Appearance control agents (other than color),
      674
Aqueous clathrates, 44, 45
Arabinogalactans
  encapsulation agent, 672
  in plant tissue, 859, 860
Arabinose
  component of anthocyanins, 558
  component of flavonoids, 563
  in plant pectins, 859
  structure, 80
Arachidonic acid
  cyclization of, 189, 192
  essential in diet, 229, 232

[Arachidonic acid]
  nomenclature, 142
  oxidation, rate of, 196
  precursor of prostaglandins, 229
Arginase, 416
Arginine, 250, 252, 253
Aromas (*see also* Flavor compounds and sensations)
  from *Allium* sp. 601, 602
  from amino acid degradation products, 606, 607
  caramel-like, 619
  of cocoa and chocolate, 619
  from cruciferae family, 602, 603
  enzyme-derived, 601–606
  from fats and oils, 611, 612
  from fatty acid degradation products, 604–606
  from fruit, 872
  from lipid oxidation, 613–615
  from methyl alkyl pyrazine volatiles, 604
  from oxidation of carotenoids, 620
  in plant tissues postharvest, 863, 892
  pungent types, 599, 600
  from roasted meat, 618
  from shiitake mushrooms, 603, 604
  from shikimic acid pathway, 609
  from smoke, 609
  from sulfur volatiles, 601–604
  from terpenoids, 606–608
  from thermal processing, 615–619
Arrhenius activation energy, 409, 412, 413
  decimal reduction value, relation to, 413
  temperature independence, 412, 413
Arrhenius approach to shelf life testing, 915, 917–919
Arrhenius equation, 412, 917
Arrhenius plot, 415, 529, 770, 917, 918
  nonlinearity, causes of, 415, 918
  for shelf life testing, 917, 918
Arrhenius theory, 408, 410
Arsenic, 522
Ascorbate, 555, 704
Ascorbic acid (*see* Vitamin C)
Ascorbic acid oxidase, 413, 452, 453, 486
Ascorbyl palmitate, 200, 643, 673
Aseptic canning, 527, 529
Ash
  of eggs, 832, 833
  of milk, 796
  of plant tissue, 867
Asparagine, 250, 252, 253, 865
Aspartame, 332, 655–657, 678
Aspartase, 436
Aspartic acid, 250, 252, 253
L-Aspartyl-L-phenylalanine methyl ester, 655–657
Aspergillus flavus, 707, 708
Aspergillus nidulans, 708
Aspergillus ochraceous, 708
Aspergillus parasiticus, 707, 708
Aspergillus versicolor, 708
Astacene, 514
Astaxanthin, 573

Astringency, 598
Atherosclerosis, 728
ATP (adenosine triphosphate) in muscle, 737–739, 742, 750–753, 756–758, 760, 761, 767–769, 774, 776, 780, 781
ATPase, 733, 735, 738, 740, 752, 757, 775
ATP chelating properties, 641
Atractyloside, 695
Aureusidin structure, 564
Aurones, 563, 564
Autoxidation (*see* Lipid autoxidation)
Auxins, 893, 895
Avicel, 121
Avidin
  biotin, effect on, 513
  in egg albumen, 835, 837
Azo dyes, safety, 704, 705

B

Bacillus mesentericus, 649
Bacillus stearothermophilus, 529
Bacillus subtilis in recombinant DNA technology, 459, 463, 464, 466
Bacitracin, 436
Bacteria
  flavors from, 610
  protein, source of, 330
Bacterial proteases, 430
Bacterial toxins, 710–712
Bacteriochlorophylls, 548
Baking powder, 634, 665
Barium, 521, 522
Bases as food additives, 636–638
Beaumon, William, 4
Beef composition, 727
Beef fat fatty acids, 150, 151
Beer, 436, 592
Beeswax, 671
Beet powder, 580, 673
Bentonite, 662, 663, 669
Benzaldehyde cyanohydrin glucoside, 697
Benzene, 437
1,2-Benzene dicarbonal, 267
Benzidine, 705
Benzoic acid, 631, 649–651, 672, 866, 868
  antimicrobial agent, 649–651, 672
  in plant tissue, 866, 868
  structure, 868
Benzoquinones, 569
Benzoyl glycine, 650
5-Benzyl-3, 6-dioxo-2-piperazine acetic acid, 655
Benzyl oxycarbonyl, 348
Benzoyl peroxide, 339, 519, 664, 670
Beryllium, 522
Berzelius, Jons Jacob, 4
Betacyanin
  betaine, 772
  betaine hydrochloride, 677
  food colorant, 581
  resonating structure of, 568
Betalains, 568, 569, 575
Betanidin, 568, 569

Betanins, 568, 569, 581
Beta pleated sheet structure of proteins, 256–258, 263
Betaxanthin, 568, 581
BET plot, 59
Beverages, 632, 661–663
Biflavonyl structure, 564
Bile pigments, 557
Bioassays for protein nutritive value, 321, 322
Bioavailability, inadequate information, 480
Bioflavonoid (vitamin P), 565
Bioflavonyls, 563–565
Biological value of proteins, 322, 323, 337, 338, 346
    alkali and heat, effect of, 337
    oxidation, effect of, 346
Biotin, 512, 513, 837
Birefringence of starch granules, 112
Bisphenylhydrozone, 492
Bitterness
    foods, importance in, 590–595
    hydrophobicity of bitter substances, 590
    of L-isomers of amino acids, 590
    of peptides, 863
    removal with immobilized enzymes, 593
    of stress metabolites in plants, 897
    theory, 590
    of various compounds, 590–595
Bitter principal in olives, 637
Bitter taste, 590–595
Bixin, 571, 572, 580, 581
Blair process, 550
Blanching
    for enzyme inactivation, 413
    minerals, loss of, 523
    nutrient content of food, effect on, 483
Bleaching agents for flour, 663–665
Bleaching of fat, 218
Bloom of fresh meat, 554
Bolland and ten Have theory of antioxidant action, 200
Borate salts, 447
Botulinum toxin, 333
Botulism, 710, 711
Bound water, 36–38, 285 (*see also* Protein hydration, Protein water holding capacity, Protein-water interactions, Water holding capacity)
Bovine hyperkeratosis, 716
Bowman-Birk inhibitor, 333, 455, 692, 693
Bread enrichment, 483
Bread improvers, 663–665
Bread staling, 663, 665
Breakfast cereals, stability of nutrients in, 488
Bromates in dough, 297
Bromelain, 356, 400, 430
Browning, ascorbic acid oxidation, 445
Browning, caramelization, 445
Browning, enzymic (oxidative), 96, 380, 424, 445–447, 888, 897, 907, 909
    ascorbic acid, effect of, 424
    after cell disruption, 380
    inhibition, 447

[Browning, enzymic (oxidative)]
    in injured plants, 897
    in plant tissues during freeze-preservation, 907
    in plant tissues postharvest, 888
    plant tissue, processing of, 909
Browning, Maillard (sugar amine), 92, 96–107, 342–346, 348, 445, 486, 491, 492, 615–618, 637, 663, 773, 775, 815, 816, 824, 847, 887, 888, 909, 926, 929–931
    Amadori rearrangement, 99, 101, 102
    control, 100, 103, 104, 348
    in dehydrated potatoes, 929–931
    detection, methods of, 98, 99
    in dried egg products, 847
    flavors from, 106, 107, 615–618
    in frozen pizza, 926
    heat, effect of, 342
    Heyns rearrangement, 99, 101
    melanoidins from, 99, 102, 104
    in milk, 815, 816, 824
    in muscle, 773, 775
    nutritional consequences, 99, 102, 104
    in peanut brittle, 637
    plant tissue, processing of, 909
    in processed potatoes, 887, 888
    products from, 99–105
    and protein nutritive value, 92
    requirements for, 98
    sugar structure, effect of, 103, 104
    sulfites, effect of, 99, 104
    toxicity from, 103–105
    vitamin C degradation, relation to, 491, 492
    vitamins, effect on, 486
    water activity, effect of, 342
    in wine, 663
Browning, nonenzymic (*see* Browning, carmelization; Browning, Maillard; Strecker degradation)
Bubble pressure, 305
Buffering salts added to food, 671
Buffer systems for foods, 638–640
2-Butenal, 187
2-Butene-1,4-dial, 187
Butter, 637
Butter yellow, 704, 705
t-Butyl acetoacetate, 657
Butylated hydroxyanisole (BHA)
    antioxidant properties, 200, 203, 643, 673
    quencher of singlet oxygen, 182
    safety, 704
    structure, 199
Butylated hydroxytoluene
    antioxidant properties, 200, 201, 203, 643, 673
    linoleate, reaction with, 201
    quencher of singlet oxygen, 182
    safety, 704
    structure of, 199
Di-t-Butylhydroquinone, 643
Butyl stearate, 668
Butyric acid
    flavor, 611, 631
    nomenclature, 142

C

Cadmium, 552, 527, 714, 715
Caffeic acid, 447, 448, 558, 568, 868
  anthocyanins, component of, 558
  in betalains, 568
  in plant tissue, 868
  structure, 868
Caffeine
  bitterness, 592
  extraction, 713
  structure, 592
Cahn-Ingold-Prelog procedure, 143
Calcite (*see* Calcium carbonate)
Calcium, 125, 479, 481, 483, 521-524, 526,
    527, 733, 735, 738, 740, 742, 754-756,
    761, 762, 764, 767, 801, 803, 811, 815,
    820, 830, 867, 904, 905, 907
  in casein micelle, 811
  in egg shell, 830
  firming of plant tissues and gels, 125, 905,
    907
  fruits and vegetables postharvest, effect
    on, 904
  in milk, 801, 803, 815, 820
  in muscle, 733, 735, 738, 740, 754-756,
    761, 762, 764, 767
  recommended intake, 479
  RDA, 481
  storage life of plant tissue, effect on, 867
Calcium-activated factor (CAF), 754
Calcium-activated neutral proteinase, 442
Calcium-activated protease, 772
Calcium carbonate
  alkaline additive, 671
  in egg shell, 830
  firming agent, 679
  for flour, 664
  malting-fermenting aid, 669
Calcium chloride, 679
Calcium citrate
  buffer salt, 671
  chelating agent, 677
  firming agent, 679
  in milk, 814
Calcium disodium EDTA, 677
Calcium gluconate
  buffer salt, 671
  chelating agent, 677
  firming agent, 679
Calcium hydroxide, 637, 679
Calcium iodate, 664
Calcium lactate, 679
Calcium peroxide, 664, 670
Calcium phosphate
  buffer salt, 671
  dough conditioner, 665
  firming agent, 679
  leavening agent, 634, 679
  malting-fermenting aid, 669
  in milk, 797, 808, 811, 812, 814, 815, 817
Calcium-sequestering ability of muscle post-
    mortem, 757
Calcium stearate, 665, 666, 669, 671

Calcium stearoyl-2-lactylate, 665, 678
Calcium sulfate
  dough conditioner, 665
  firming agent, 679
  for flour, 664
  malting-fermenting aid, 669
Caloric intake and protein requirements, 317
Camphor structure, 600
Cancer and consumption of polyunsaturated
    fatty acids, 236
Canthaxanthin, 580, 581
Capillaries, 730, 731
Capric acid, 142
Caproic acid, 142
Caprylic acid, 142
Capsaicin, 599
Capsaicinoids, 599
Capsanthin, 571, 572
Caramel, 580, 673
Caramelization
  flavors from, 106
  furans from, 98
  humic substances formed, 98
  levoglucosan from, 98
  reaction rate, 98
  taste and fragrance from, 98
Carbamino compounds, 667
Carbamylation of proteins, 360
Carbohydrases, 433
Carbohydrate degradation products (*see* Poly-
    saccharide degradation products, Sugar
    degradation products)
Carbohydrate hydrolysis, 90-92
Carbohydrates (*see also* Polysaccharides, Sugars)
  abundance on earth, 74
  amount in various vegetables and cereals,
    328
  caloric importance, 70
  caloric value, 72, 73
  classes, 70, 72
  and dental caries, 73, 74
  diet, function in, 72, 73
  digestion, 72, 73
  in egg albumen, 833
  in egg yolk, 833
  elemental composition, 70
  and fat utilization, 73
  foods, function in, 133
  iron bioavailability, effect on, 526
  as sweeteners, 73
  in whole egg, 833
Carbolines, 335, 336, 341
Carbonates, 521
Carbonation, 666, 667
Carbondiimide, 348
Carbon dioxide
  for animal slaughtering, 766
  for carbonation, 666, 667
  for controlled atmosphere storage of plants,
    901, 902
  in eggs, 844
  food, function in, 668
  freezant, 669
  leavening agent, 632-635, 637

[Carbon dioxide]
    for muscle storage, 778
    plant respiration, effect on, 884
    propellant, 667, 680
Carbonic acid, 667
Carbonic anhydrase, 246, 259, 374, 802
Carbon monoxide and meat, 557, 778
Carbonyl addition mechanism of fat inter-
        esterification, 224
Carbonyls
    from lipid oxidation, 195
    lipid oxidation, test for, 195
    from thermal decomposition of lipids, 210
    vitamins, effect on, 486
N-Carboxyanhydride, 268, 348
Carboxylase, 885
Carboxylproteases, 458
Carboxymethylation of proteins, 359
Carboxymethylcellulose, 121, 122, 175, 350,
        658, 665
    in baked goods, 665
    as emulsion stabilizer, 175
Carboxymyoglobin, 553, 778
Carboxypeptidases, 313, 355, 421
Carcinogenicity studies on heated fats, 232
Cardiolipin, 743
Carmine, 580, 582
Carminic acid, 581
Carnosine, 752
β-Apo-8-Carotenal, 514, 580, 581
α-Carotene, 514, 573
β-Carotene, 182, 480, 513–516, 571, 572, 573,
        576, 580, 581, 620
    degradation products, 515, 516
    in dehydrated carrots, 516
    exempt from certification, 576
    and flavor development, 620
    oxidation, susceptibility to, 515
    quencher of singlet oxygen, 182
    structure, 514, 571
    vitamin A activity, 480, 513, 573
Neo-β-carotene, 515, 516
β-Carotene 5,6-epoxide, 516
Carotenes
    colorant, 673
    in protein from leaves and grass, 330
Carotenoids (*see also* Vitamin A)
    as antioxidants, 574
    assays, 515
    association with various substances, 572, 573
    characteristics, 570–574
    chemical reactions, 573, 574
    chlorophyll, association with, 548
    cis-trans isomerization, 515
    degradation, 515, 516
    destruction, rate of, 517
    exempt from certification, 580, 581
    and flavor development, 620
    in flour, 664, 665
    forms used in foods, 581
    lipid oxidation, effect of, 486
    in milk fat globule membrane, 813
    number, 572

[Carotenoids (*see also* Vitamin A)]
    oxidation, 573, 574
    in plants postharvest, 891
    as pro-oxidants, 574
    properties, 575
    solubility, 515
    stability, 515
    vitamin A activity, 513, 514
Carrageenan, 129, 130, 134, 175, 658, 680,
        703, 704
    in baked goods, 665
    as emulsion stabilizer, 175
    safety, 703, 704
    stabilizer, 680
    texturizer, 680
Carriers for food ingredients, 672
Carrot oil, 580
Carr-Price reaction, 515
(4R)-(-)-Carvone, 608
(4S)-(+)-Carvone, 608
Casein, 255, 261, 265, 266, 292, 294, 295,
        300–303, 306, 309, 310, 340, 349, 442,
        461, 465, 594, 595, 631, 637, 640, 663,
        796–798, 803–808, 811, 812, 814–817,
        819
    as adsorbent, 663
    biosynthesis, 461
    bitter peptides from, 594, 595
    carrageenan, interaction with, 292
    in cheese, 637
    emulsifying properties, 300–303
    foaming power, 306, 309, 310
    instability in frozen milk, 816
    functional properties, 817, 819
    gel characteristics, 294
    heat, effect of, 814
    hydrophobicity, 309
    isoelectric precipitation, 631
    oxidized, 340
    structure alteration during precipitation,,
        349
    structures, 803–808, 811, 812
para-κ-Casein, 640, 806
α_{S_2}-Casein A, 805
κ-Casein A, 807
β-Casein A², 806
α_{S_1}-Casein B, 804
α_S-Caseins, 796, 798, 803, 808, 811, 812
β-Caseins, 796–798, 803, 808, 812, 815
γ-Caseins, 796–798, 806
κ-Caseins, 796, 798, 803, 808, 811, 812, 815
Caseinates, 817
Casein micelle, 263, 309, 637, 795, 797, 803,
        808, 811, 812, 814–816
    in cheese, 637
    concentration, effect of, 815, 816
    electron photomicrograph, 811
    foaming agents, 309
    and gel formation, 812
    heat, effect of, 814–816
    during homogenization, 814
    hydration, 812
    hydrophobic interactions, 808, 812

[Casein micelle]
models, 812
synthesis, 795
Catalase, 246, 374, 413, 414, 421, 423, 431, 438, 453, 802, 885
in milk, 802
in ripening fruit, 885
Catalysts
added to food, 668, 669
hydrogenation, effect on selectivity, 221
for hydrogenation of fat, 219, 222
for interesterification of fat, 223, 224
poisons for, 222
Catechin, 448, 567, 868
Catechol, 447, 448
Catecholase (*see* Phenolase, Polyphenol oxidase)
Cathepsins, 315, 441, 442, 733, 755, 756
Cell cultures from amino acids and proteins, 331
Cellobiose structure, 86
Cellulase, 332, 429, 466
Cellulose
abundance, 70, 71
acid hydrolysis, susceptibility to, 92
carrier, 672
chemical modification, 121, 122
crystallinity, 109, 110
food texture, role in, 121
moisture sorption isotherm, 660
in plant tissue, 859, 860
properties, 121
repeating units, 89
water, interaction with, 109
Cereal proteins, 861, 862
antinutritional factors, 328
as food source, 325
limiting amino acids, 328
protein nutritive value, 328
water content, 328
Cerebroside composition, 146
Certified colorants, 576–580
Chalcones, 563, 564
Charcoal, 663
Chebulagic acid structure, 567
Cheese, 637, 640, 927
Chelating agents, 198, 202, 641, 642, 677
Chemical equilibrium, 378
Chemical leavening systems (*see* Leavening systems, chemical)
Chemical methods for protein nutritive value, 324
Chemical modification of proteins (*see* Proteins, intentional chemical modification)
Chemical reactions
Arrhenius relationship, 12
and food deterioration, 8–11
among major food constituents, 11
Chemical scores of proteins, 323, 324
Chevreul, Michel Eugene, 4
Chicken composition, 727
Chicle, 658
Chili peppers, 599

Chilling injury of fruits and vegetables, 899, 900
Chirality of amino acids, 253
Chlordane, 714
Chlorhydrin, 712
Chlorides in milk, 801
Chlorinated naphthalenes, 716
Chlorine, 664, 672
Chlorine dioxide, 664
Chlorins, 546
Chlorobium chlorophylls, 548
2-Chloroethanol, 263, 279
Chloroform, AH/B theory of taste, 589, 590, 596, 598
Chlorogenic acid, 447, 448, 866, 868, 888
Chlorohydrins, 651
p-Chloromercuribenzoate, 269
Chloropentafluoroethane, 667
Chlorophyll
acids, effect of, 549, 550
blanching, effect of, 905
copper derivatives of, 582
defined, 546
degradation, 548, 549
ethylene, degradation by, 867
food colorant, 582
food processing, effect of, 549
iron in, 521
lipid oxidation, catalysis of, 177
lipoxygenases, effect of, 549
plants, location in, 548
in plant tissue postharvest, 891, 892
preservation of, 550
properties, 548, 575
in proteins from leaves and grass, 330
ratio of kinds, 548
stability of, 549
structure of, 547
Chlorophyll a
and formation of singlet oxygen, 182
structure of, 547
Chlorophyll a and b, properties, 548
Chlorophyll b, structure, 547
Chlorophyllase, 413, 414, 454, 550, 885, 891, 905
Chlorophyllides, 548, 549, 550
Chlorophyllides a and b, 547
Chlorophyllin, 582
Chloroplasts, 330, 548
Chlortetracycline, 652
Chocolate bloom, 158
Cholemyoglobin, 554, 556, 781
Cholesterol
and coronary heart disease, 233–235
in lipoprotein of egg yolk, 842
in low-density lipoproteins of egg yolk, 845
in milk, 799, 813
in milk fat globule membrane, 813
in muscle, 728
in myelin figures of egg yolk, 843
source of in body pool, 234
Cholic acid, 678
Choline, 864

Choline chloride, 677
Cholinesterase in milk, 802
Chondroitin sulfates in egg shell, 830
Chromium, 222, 521, 522
Churning of cream, 814
Chylomicrons, 234, 235
Chymosin, 349, 350, 356, 398, 411, 458,
    806, 807
Chymotrypsin, 313, 355, 400, 401, 692
Chymotrypsin inhibitor, 321, 333, 456
Chymotrypsinogen, 280, 281, 407
α-Chymotrypsinogen
    free energy of denaturation, 404, 405
    temperature dependence of thermodynamic
        parameters during denaturation, 407
Ciguatera poisoning, 701
Cinnamyl alcohol, 609
Citraconylation of proteins, 359
Citral, 608
Citrates
    Ca^{2+} complexing, 350
    in cheese, 640
    in milk, 801, 811, 815
Citric acid, 200, 202, 218, 416, 447, 610,
    631, 632, 638, 642, 643, 670, 863
    buffer effect in plants, 638
    in cheese, 631
    chelating properties, 642, 643, 670
    food additive, 632
    metal chelation by, 202, 218
    in plant tissues, 863
Citrulline, 336
Clarifying and flocculating agents, 661–663,
    669
Clathrate hydrates, 44, 45
Claviceps purpurea, 706
Climacteric fruits, 881, 882, 894
Clostridium botulinum, 528, 646, 649, 710,
    711, 776–778
    indicator organism in thermal processing,
        528
    inhibition, 646, 649
Clupeoid poisoning, 701
Cobalamin (*see* Vitamin B_{12})
Cobalt, 197, 521, 522, 528
    lipid oxidation, catalyst for, 197
    safety, 528
    in vitamin B_{12}, 510, 521
Cocarboxylase, 494, 497, 498
Cochineal extract, 580, 581, 673
Cocoa, 637
Cocoa butter
    fatty acid distribution, 149, 150
    polymorphic forms, 158
Coconut oil fatty acid distribution, 149, 150
Coefficient of protein digestibility, 322
Coenzyme A, 512
Coenzymes, 375
Cogelation, 292
Cold denaturation of enzymes, 416
Cold shortening, 745, 761, 767–769
Collagen, 246, 248, 261, 334, 336, 352, 729,
    746–750, 761, 768, 772–774, 776

[Collagen]
    comparison of muscle species, 729
    conversion to gelatin, 748–750
Collagenase, 748, 750
Collagen fibrils, 732
Collagen fold, 750
Collenchyma tissue, 869, 876
Color (*see also* Colorants, pigments)
    control agents, 669
    definition, 546
    frozen pizza, alteration in, 926
    importance, 545, 546
    from Maillard browning, 343
    of plant tissue after dehydration, 909
    of plant tissue after freezing, 907, 908
    of plant tissues after heating, 906
Color additive amendment, 576
Color additive defined, 546
Color and color modifiers, 673
Colorant
    certified, 567–580
    defined, 546
    exempt from certification, 580
    history of in U.S., 576
    nonabsorbable types, 582
    U.S. regulations, 576
Comminuted muscle products, 779–781
Conalbumin, 294, 302, 309, 834–836, 845,
    846, 850, 851
Concanavalin A, 693
Condensed tannins, 566, 567
Conglycinin, 261
Coniferyl alcohol, 643
Conjugation from lipid oxidation, 178
Connectin, 739, 741, 742, 754, 756, 764, 772
Connective tissue, 730, 731, 745–750
Connective tissue proteins, 742
Consistency control agents, 678, 679
Consumer enlightenment, 14, 15
Contaminants, 691, 713–716
Contractile proteins, 247
Controlled atmosphere storage
    of fruits and vegetables, 380, 900–902
    of muscle tissue, 778, 779
Convicine, 328, 694, 700
Cooling sensation, 600, 601
Copper, 197, 222, 489–491, 521–524, 527,
    528, 555
    fat hydrogenation, catalyst for, 222
    lipid oxidation, catalyst for, 197
    meat, discoloration of, 555
    in potatoes, 524
    safety, 528
    vitamin C degradation, role in, 489–491
Corn oil, 150, 211
Corn syrup production, 91
Coronary heart disease
    causative factors, 233, 235
    and dietary lipids, 233, 237
Corrin ring system, 509
Cottonseed flour, 580
p-Coumaric acid, 588, 568
Coumarins, 328

Covalent bonding in proteins, 264, 265
Coxiella burnetti, 413
C-protein, 739, 741, 754
Cream, 814
Creaming of milk, 813
Creatine, 591, 752
Creatine kinase, 739, 741, 742, 751
Creatine phosphate, 750–752, 754, 755, 767
Creatinine, 316
Cresolase, 446
Crocetin, 573
Crocin, 573, 580, 581
Crotonaldehyde formation in plants during
    dehydration, 909
Cruciferae aromas, 602
Cryoprotective agents, 416
Crystalline hydrates of monoacylglycerols,
    172
Crystallization of lipids (*see* Lipid crystalliza-
    tion, Triacylglycerol crystallization)
Cumin aldehyde, 607
Curcumin, 582
Cured meat pigments (*see* Pigments, cured
    meat)
Curing of meat, 776, 777
Cuticle
    of egg, 831
    of plant tissue, 874
Cutin estolide, 867
Cyanide in vitamin B_{12}, 510
Cyanidin, 557, 558, 566
Cyanin-3-rhamnoglucoside, 559, 560
Cyanmetmyoglobin, 553
Cyanogen bromide, 501
Cyanogenetic factors, 328
Cyanogenetic glucosides, 85, 694, 696
Cyanogens, 694, 695, 697
Cyanohydrin structure, 697
Cycasin, 694, 700
Cyclamates, 653, 654, 691 (*see also* Sodium
    cyclamate)
Cyclic compounds from thermal decomposi-
    tion of lipids, 206, 208–210
Cyclic dimers from Diels-Alder reaction, 189
Cyclic esters from heated fat, 230
Cyclic monomers from heated or oxidized fats,
    189, 192, 230, 231
Cyclohexylamine structure, 654
Cyclosome poisoning, 701
Cysteic acid, 335, 340
Cysteine, 250, 252, 253, 269, 340, 342, 347,
    491, 502, 503, 555, 618, 821
    cured meat color, effect on, 555
    determination of, 269
    a flavor, 618
    in milk, 821
    oxidation, 340, 342, 347
    pyridoxal, reaction with, 502, 503
    vitamin C, protective effect on, 491
Cysteine sulfenic acid, 340
Cysteine sulfinic acid, 340
Cystine, 254, 340, 342, 456
    oxidation, 340, 342
    supplement for soybean meal, 456

Cystine disulfoxide, 340
Cystine sulfoxide, 340
Cystinyl residues in milk, 822
Cytochrome C, 246
Cytochrome C reductase in milk, 802
Cytochrome oxidase, 453, 770, 884
Cytochromes, 521, 550
Cytokinins in plant tissues, 893, 895

D

Dairy product production, 793
Dairy products, mineral content, 524
β-Damascenone, 620
Dansyl chloride, 271
Dark-cutting beef, 760
Davy, Sir Humphry, 3
DDT structure, 714
Deamidation of proteins, 335, 355
2,4-Decadienal, 184, 186, 211, 614
Decanoic acid, 668
2,4,7-Decatrienal, 614
2-Decenal, 183
*c*4-Decenal, 613
Decimal reduction value, 412, 413
Decompartmentalization of enzymes, 416
Deep fat frying
    behavior of food during, 211
    calories from, 210
    changes in fat properties during, 210–212
    compounds from, 210, 211
    consequences, 210–212
    control of fat quality, 213
    and fat absorbed by food, 210, 211
    tests for fat quality, 212
    volatiles from, 210, 211
Degumming of fat, 218
Dehydration of muscle, 773–776
Dehydroalanine, 336, 337, 358, 363, 824
Dehydroalanyl residues in milk, 815, 824
Dehydroascorbic acid, 489, 490, 492
Dehydrogenases, 646
17-Dehydrolimonoate, 593
Delaney clause, 691
Delphinidin, 557, 558, 566
Delvocid, 648
Denaturation of proteins (*see* Protein denatura-
    tion)
Dental caries, 73, 74
Deodorization of fat, 218
5-Deoxyadenosine, 510
ϵ-N-Deoxyfructosyllysyl, 343
ϵ-N-(1-Deoxylactulosyl)-L-lysyl group, 342
3-Deoxypentosone, 490, 491
Dephosphorylation of proteins, 355
Desmin, 739, 741, 742, 764
Desmosine, 248, 336
Desoxycholic acid, 678
Desulfuration of proteins, 335, 355
Dextran, 132, 134
Dextranases, 433
Dhurrin, 697
Diacetyl, 610
Diacetyl reductase, 433

Diacylglycerols (*see also* Acylglycerols, Triacylglycerols)
    emulsifiers, 671
    in milk fat globule membrane, 813
Diafiltration and protein functionality, 349
Dialdehydrase, 510
Diallyl thiosulfinate, 602, 603
Diaminobenzidine, 451
o-Dianisidine, 451
Diaphorase, 802
Diatomaceous earth, 661, 662, 672
1,7-Diazoheptamethin structure, 568
Dicalcium phosphate
    anticaking agent, 669
    for flour, 664
    leavening agent, 634, 635
α-Dicarbonyl compounds, 99, 645
Dichlorodifluoromethane (Freezant-12), 669
1,2-Dichloroethane, 347
2,6-Dichlorophenolindophenol, 492
2,4-Dichlorophenoxyacetic acid (2,4-D)
S-Dichlorovinyl-L-cysteine, 347, 712
Dieldrin, 714
Dielectric constant, a test for fat quality, 212
Diels-Alder reaction in lipid oxidation, 188
Dietary fiber
    from plant cell walls, 859
    polysaccharides, 71, 73, 74
Dietary supplements, 674, 675
Diethyl pyrocarbonate, 652, 653
Digalactosyl diglycerides, 298
Diglycerides (*see* Diacylglycerols)
Diglycerol monostearate, 171
7,8-Dihydrofolate, 506
Dihydrogen phosphate, 521
5,6-Dihydro-5-methylfolate, 507
Dihydroporphines, 546
7,8-Dihydropterin-6-carboxaldehyde, 505
Dihydropyrazines, 587
Dihydropyrroles, 695
Dihydrosafrole, 704
Dihydrothiophene, 499
D(+)-N-(2,4-Dihydroxy-3,3-dimethylbutyryl)-β-alanine, 512
2,2-Dihydroxy-1,3-indanedione, 266
o-Dihydroxyphenols, 562
2,3-Diketogulonic acid, 489, 490
Dilatometric curves, 161–163
Dilauryl thiodipropionate, 644
Dimagnesium phosphate, 670
Dimer esters, a test for fat quality, 212
Dimerization from lipid oxidation, 188
Dimers
    from heated or oxidized lipids, 231
    from ionizing irradiation of lipids, 214, 216
    from thermal decomposition of lipids, 206, 209, 210
Dimethyl-allyl pyrophosphate, 607
Dimethylamine, 614, 759, 771, 772, 778
N-Dimethylaminosuccinamic acid, 891
4-Dimethylammoazobenzene, 704, 705

5,6-Dimethylbenzimidazole, 509
Dimethyl disulfide, 616, 617
Dimethylformamide, 226
2,6-Dimethylnaphthalene, 516
Dimethylnitrosamine, 703
Dimethylpolysilicone, 668
Dimethylpolysiloxane, 668
Dimethyl-β-propiothetin, 615
Dimethylsulfide, 335, 615, 618
2,4-Dinitrophenylhydrazine, 195
ε-N-Dinitrophenyllysine, 343, 344
Dinoflagellates, 701
Dioctyl sodium sulfosuccinate, 677, 678
Dipeptidases, 313
Disaccharides, 71, 85, 86
Disodium EDTA, 677
Disodium phosphate, 637
Dissociation constant for acids, 639, 640
Disulfide bonds, 263, 265, 292, 297, 645, 664
    in dough, 297, 664
    in protein gelation, 292
    in protein structure, 263, 265
Disulfides, 602, 773
5,5′-Dithio-bis-(2-nitrobenzoic acid), 269, 735
Dithiothreitol, 270, 279
Dityrosine residues from protein oxidation, 339
DOPA, 446
Dopamine, 448, 865
DOPA quinone, 446
Dough conditioners, 664, 665, 862
Dough formation, 296–298
Dracorubin, 575
Dumas, Jean Baptiste, 7
D value (*see* Decimal reduction value)
Dyes, 312, 546
Dynapol colorants, 582

E

Echinenone, 514
Edman's reaction, 271
Egg albumen
    as adsorbent, 663
    composition, 832–837
    consistency, 844, 845
    foaming properties, 848–850
    heat denaturation, 850, 851
Egg albumin
    emulsifying ability, 303
    gel characteristics, 294
Egg pH, 844
Egg processing, 845–847
Egg products, functional properties, 847–851
Egg proteins, 832–837, 849–851
    during foaming, 849, 850
    heat denaturation, 850, 851
Eggs
    general properties, 830
    mineral content, 524
Egg shell, 830, 831

Egg shell membranes, 831, 832
Egg white, 300, 306 (*see also* Egg albumen)
Egg white proteins as foaming agents, 307,
    308, 310
Egg yolk
    composition, 832–834
    emulsifying properties, 848
    functional properties, 846–848, 850
    gelation, 846
    granules, 838, 839, 841–843
    microstructure of particles, 837–841, 843
    plasma, 838, 841, 843, 844
Eicosanoic acid, 142
Eicosapentaenoic acid, 615
Elaidic acid, 232
Elasmobranch poisoning, 701
Elastase, 313, 355
Elastatinal, 458
Elastin, 246, 248, 336
Electrical charge and emulsion stability, 167
Electrical stimulation of muscle, 767, 768
Electrostatic interactions, 262–265, 292
    in protein gelation, 292
    in proteins, 262, 264, 265
Ellagic acid, 567
Ellagitannins, 567
Ellman's reagent, 269
Emden-Meyerhoff pathway, 751, 879
Emodin, 569, 570
Emulsification versus protein hydration,
    285
Emulsifiers, 116, 167–175, 678, 679
    classes, 169, 170
    definition, 167
    esters of monoacylglycerols with hydroxy-
        carboxylic acids, 172, 173
    fatty acid monoesters of ethylene or propy-
        lene glycol, 174
    glycerol esters, 172
    HLB system, 170, 171, 175
    importance, 167
    ionic surfactants, 168
    lactylated monoacylglycerols, 172
    monoacylglycerols, 172
    phase inversion temperature (PIT), 171
    phospholipids, 175
    polyglycerol esters, 172
    safety, 171
    sorbitan fatty acid esters, 174, 175
    starch gelatinization, effect on, 116
    types, 171–175
    water-soluble gums, 175
Emulsifiers, proteinaceous
    deamidation, effect of, 358
    effectiveness, factors influencing, 299–303
    effectiveness, methods of assessing, 298,
        299
    efficacy of various kinds, 301–303
    efficacy versus concentration, 300
    efficacy versus heat, 300
    efficacy versus hydrophobicity, 300, 301
    efficacy versus pH, 300, 301
    efficacy versus solubility, 299, 300
    efficacy versus unfolding, 301

[Emulsifiers, proteinaceous]
    egg yolk, 848
    functions, 299
    heat, effect of, 773
    intentional chemical modification, effect of,
        359, 361
    mechanisms of action, 298
    from milk, 298, 817, 819
    proteinaceous foaming agents, comparison
        with, 310
    surfactants, efficacy in presence of, 299
    types, 298, 300–302
    unwanted, 302, 303
    w/o emulsions, ineffective for, 298
Emulsifying activity index for proteins, 302,
    303
Emulsion capacity, 299
Emulsions (*see also* Emulsifiers)
    in butter, 814
    coalescence, 166, 167
    continuous phase, 166
    creaming or sedimentation, 166, 299
    definition, 166
    dispersed phase, 166
    factors influencing characteristics, 299
    flocculation, 166
    HLB system, 170, 171
    importance, 167
    importance of liquid crystals, 166
    interfacial area, 166
    in milk, 814
    phase inversion, 299
    phase inversion temperature (PIT), 171
    proteins, effect on, 275–278
    Stokes law, 166
    volume percent of phases, 166
    work to produce, 166
Emulsion stability, 166–169, 299, 641
    contact angle of solid particles, 168
    DLVO theory, 167, 168
    electrical charge, 167, 168
    electric double layer, 168
    ionic surfactants, 168
    interfacial tension, 167
    liquid crystals, 169
    macromolecules, 169
    phase density, 169
    salts, effect of, 641
    tests for, 299
    viscosity of the continuous phase, 169
Enantiomorphs of amino acids, 253, 254
Encapsulating agents, 672
Encapsulation of enzymes, 435
Endomysium, 730, 731, 745
Endoperoxides, 191, 194
Endoplasmic reticulum, 795
Endothelial cell injury, role in coronary heart
    disease, 234
Ene reaction, 182
Energy of activation (*see* Activation energy)
Energy charge in muscle, 752
Enocianina, 581
Enolate ion formation, mechanism of inter-
    esterification of fat, 224

Enolization of sugar, 92–94
Enrichment, 483, 486
Enterokinase, 355
Enterotoxins from *Staphylococcus aureus*, 711
    712
Enthalpy
    definition, 403
    of enzyme inactivation, 407
    of protein denaturation, 407
    protein denaturation, change during,
        403–406
Entropy
    definition, 403
    of enzyme inactivation, 407
    of protein denaturation, 407
    protein denaturation, change during,
        403–405
Enzyme activity
    altered products, effect of, 424
    altered substrates, effect of, 424
    chemicals, effect of, 424
    control, 424–427
    cryoprotective agents, 416
    during dehydration, 420
    electrolyte dependence, 420
    foaming, effect of, 423
    freezing, effect of, 415, 416
    in frozen muscle, 770
    inhibition, 420
    interfaces, effect of, 423
    ionic activators, 420
    ionic strength dependence, 420
    ionizing radiation, effect of, 421, 422
    at low temperatures, 415
    pH, effect of, 416, 417, 424
    preprocessing treatments, effect of, 424
    pressure, effect of, 421
    shearing, effect of, 421
    substrate mobility, 418
    temperature, effect of, 402–406, 408, 424,
        768, 769
    units, 397
    water activity, effect of, 417–419, 424
Enzyme assays
    coupled, 398, 399
    direct, 395, 396
    initial velocity as a function of enzyme con-
        centration, 397
    progress curves, 395, 396
    single point, 397, 398
Enzyme compartmentalization
    in cereal grains, 417
    foods, significance in, 379–381
    ionizing radiation, effect of, 423
    metabolic control mechanism, 380
    of phenolase, 445
    sugar cane, processing of, 424
Enzyme decompartmentalization
    in fish, 759
    flavor generation in plants, 601–603, 605
    during freezing, 770, 925
    and hydrogen cyanide, 697
    and linamarin, 697
    and thioglucosides, 695

Enzyme denaturation, 403, 406 (*see also*
    Protein denaturation)
Enzyme distribution (*see* Enzyme compart-
    mentalization)
Enzyme generated flavor, 601–616
Enzyme inactivation
    catalytic activity, rules for determining,
        396, 397
    by cold denaturation, 416
    and denaturation, 402, 403
    by foaming, 423
    by heating muscle, 773
    heat treatment, indicator of, 413
    at interfaces, 423
    by ionizing irradiation, 422, 423, 425
    lysozyme, 836
    molar activity, 373, 374
    order of reaction, 408
    pH, effect of, 416
    by pressure, 421
    salting in and out, 420
    by shearing, 421
    specificity, 373, 376, 377
    for substrate assays, 399, 400
    temperature dependence, 402–405, 407–416
    thermodynamics, 403–405, 407–409
    transition state theory, 376, 378
    units, 397
Enzyme inhibition
    competitive, 384–387, 390, 393
    by electrolytes, 420
    by heavy metals, 420
    K_m and V_{max}, effect on, 384
    Lineweaver-Burk plot, 386–389
    noncompetitive, 384, 388–390
    reversibility, 384
    uncompetitive, 384, 386–388, 390
Enzyme inhibitors
    in animals, 457
    and food safety, 456, 457
    heat inactivation of, 455, 456, 458
    importance, 455
    mechanism of action, 455, 456
    mechanistic groups, 455
    in microorganisms, 458
    nutritional significance, 458, 459
    physiological significance, 458, 459
    in plants, 455–459
    as research tools, 459
    sources, 455, 456
    specific response to, 456
Enzyme kinetics
    Arrhenius theory, 408
    catalytic activity, rules for determining,
        396, 397
    definition, 375
    enzyme-substrate complex, 381
    half-life, 408
    initial velocity, 382, 383, 396
    Lineweaver-Burk plot, 384–386, 393–395
    Michaelis complex, 381
    Michaelis constant, 382, 384, 386–395
    Michaelis-Menten theory, 384
    Michaelis-Menten velocity, 381, 382, 386

[Enzyme kinetics]
progress curve, 383
steady state, 375, 381
transition state theory, 408
units, 397
V_{max}, 382, 384, 386-390, 392-396
zero order, 384
Enzyme mechanisms, 400-402
Enzyme nomenclature, 374, 375
Enzyme-product complex, 400
Enzyme reactivation, 413, 414, 416, 425, 801
of lipoxygenase, 414
at low temperatures, 416
in milk, 801
of peroxidase, 413, 414
Enzyme regeneration (*see* Enzyme reactivation)
Enzymes (*see also* Proteins)
active site, 375
activity at subfreezing temperatures, 277
activity versus species, 373
allosteric effectors, 375
as analytical reagents, 373
catalysis, 377-379
catalytic effectiveness, 373
cofactors, 375
definition, 375
delocalization during freezing, 416
energy of activation, 376
for flavor development in plant tissues, 892
food, effects on, 333, 373
function, 372, 373
heat, inactivation by, 333, 373
isoenzymes, 375
in milk, 801, 802
reaction rates, acceleration of, 376
stability, 418, 421
Enzymes added to food
advantages, 427, 433
carbohydrate hydrolases, 428, 429
conditions of use, 433, 434
ester-triacylglycerol hydrolases, 431
functions, 427-434
isomerase, 432
listing of, 428-433
oxidoreductases, 431
preparation, 427
preservation, 427
protein hydrolases, 430, 431
purity, 427, 433
total organic solids in, 433
units of activity, 433
Enzymes, endogenous
amylases, 441
antioxidant enzymes, 453, 454
ascorbic acid oxidase, 452, 453
calcium-activated neutral proteinase, 442
cathepsins, 441, 442
control, 425, 426
encapsulated, 426
flavor, effect on, 425
flavor enzymes, 454
food, effect on, 425, 426
lipolytic enzymes, 442

[Enzymes, endogenous]
lipoxygenases, 447-451
in meat tenderization, 427
milk proteinases, 442
myosin ATPase, 445
pectic enzymes, 439-441
peroxidases, 451, 452
phospholipases, 443
phytase, 445
pigment-degrading enzymes, 454
by shearing, 421
temperature dependence, 402-405, 407-416
thermodynamics, 403-405, 407-409
thiaminases, 444
Enzymes, immobilized
advantages and disadvantages, 434
for cheese, 437
costs of, 435
for debittering orange juice, 593
to decompose poisonous compounds, 437
definition, 434
diffusion factors, 391
electrostatic effects, 391
by encapsulation, 435, 437
for food analysis, 437, 438
foods, use in, 434-436
in living microbial cells, 435
methods of making, 434
Michaelis constant, 392
Michaelis-Menten velocity, 391
Nernst layer, 391, 392
for process control, 438
reaction kinetics, 390-393
reactor bed, capacity of, 392
on ultrafiltration membranes, 435, 437
Enzyme-substrate complex, 381, 400
Enzymically modified proteins (*see* Proteins, enzymically modified)
Enzymic methods for protein nutritive value, 324, 325
Enzymic reactions
and cell integrity, 379
control, 379
food quality, effect on, 379
Eosinophilic cells, 745
Epicatechin, 566
Epidermis of plant tissue, 874, 877
Epimysium, 730, 745, 746
Epinephrine, 753
Episulfide nitrile, 695, 697
2,3-Epoxide, 186
5,6-Epoxide, 515
Epoxides
from alkoxy and peroxy radicals, 188
antimicrobial agents, 651, 652
from oxidation of aldehydes, 186
from thermal decomposition of lipids, 210
vitamins, effect on, 486, 652
12,13-Epoxytrichothecanes, 708
Equilibrium constants for metal chelates, 642
Equilibrium relative humidity, 46, 47
Ergotism, 706, 707
Erucic acid, 142, 149

Erythrocytes, 693, 694, 698, 700
Erythrosine
    colorant, 673
    formation of singlet oxygen, 182
Escherichia coli in recombinant DNA technol-
        ogy, 459, 463, 464
Essential amino acids, 317–319
Essential fatty acids, 229
Essential oils, 609
β-Esterase in milk, 802
Esterification of proteins, 361
Esters
    from ionizing irradiation of lipids, 214
    of monoglycerides, 171
    from thermal decomposition of lipids,
        206, 207, 209
Estrogens, 328
Ethanol, 273, 436, 610
Ethyl carbamate, 653
Ethyl deca-2-*t*-4-*c*-dienoate, 604, 605
Ethylene
    and carotenoid metabolism, 891
    chlorophylls, degradation of, 867
    chlorophylls, effect on, 891
    for fruit ripening, 666
    in plant respiration, 881
    in plant ripening, 892–895
    in plants stored under controlled condi-
        synthesis in wounded plant tissue, 895,
        897, 898
Ethylene chlorohydrin, 651, 712
Ethylenediaminetetraacetic acid (EDTA),
        641–643
Ethylene glycol, 894
Ethylene oxide
    antimicrobial agent, 651, 672
    from ethylene in plants, 894
    proteins and nucleic acids, effect on,
        485
    safety, 712
    spice sterilant, 666
    structure, 651
    vitamins, effect on, 485
Ethylglyoxal, 491
Ethyl linoleate, 176, 198
Ethyl maltol, 106, 597, 598
Ethyl 2-methylbutyrate, 606
Ethyl-3-methylbutyrate, 606, 607
Eu-actinin, 741
Eugenol, 609
Eutectic point of lipids, 160, 161
Excess water methods for determining protein
        hydration, 285
Extensin in plant tissue, 860
Eyring theory (*see* Transition state theory)

F

FAD (Flavin-adeninedinucleotide), 500
Farnesol, 547
Fat and oil processing, 218–222
Fat cells, 745

Fat interesterification (*see* Interesterification of
        fat)
Fats (*see* Lipids, Triacylglycerols)
Fatty acid autoxidation (*see* Lipid autoxida-
        tion)
Fatty acid dimers, 713
Fatty acid monomers, 713
Fatty acids
    anticaking agents, 665
    aroma of food, 631
    carotenoids, association with, 572
    cis, trans, 142
    and coronary heart disease, 233
    crystallographic structure of, 151–154
    in egg yolk, 833, 834
    enzymic, 176
    essential, 229
    flavors from, 611–613
    food additives, 678
    in frozen pizza, 926
    generation versus water activity, 419
    hexagonal (α) packing, 154
    interaction with proteins in frozen fish, 771
    from ionizing irradiation of lipids, 214
    from lipolysis, 175, 176
    in milk, 797, 799, 800
    in milk fat globule membrane, 813
    nomenclature of saturated, 141, 142
    nomenclature of unsaturated, 141–143
    off-flavor, 637
    orthorhombic (β') packing, 153, 154
    oxidation products, 448, 449, 771
    peroxidase, effect of, 452
    in plant lipids, 866
    polymorphism of, 152–158
    rate of oxidation versus that of acylgly-
        cerols, 197
    selectivity of hydrogenation, 219, 220
    solid solution, 152
    subcell lattice, 153, 154
    triclinic (β) packing, 153, 154
    unit cell, 152, 154
Fatty acid thermal decomposition
    conditions for, 205, 207, 209
    general scheme, 206
    mechanisms, 207, 208
    nonoxidative, 205, 206, 209
    oxidative, 207–210
    products from, 206, 207, 209
Favism, 694, 700
FD&C colorants
    general information, 577–580
    structures, 578, 579
FD&C Green No. 3, 673
FD&C Red No. 2, 705
FD&C Red No. 3, 673
Fermentations and flavor development,
        609–611
Ferricyanide, 494
Ferrous ammonium salts, 527
Ferrous gluconate, 580, 669
Ferulic acid, 448, 558, 568
    anthocyanins, component of, 558

[Ferulic acid]
    in betalains, 568
Fiber, dietary (*see* Dietary fiber)
Fibers (*see* Muscle fibers)
Fibrinogen, 246, 247, 356
Fibroblasts, 745
Fibroin, 246
Fibrous proteins, 246
Ficin, 400, 430
Ficin inhibitor, 835
Filaments (*see* Muscle filaments)
Filamin, 741
Filter aids, 672
Firming agents, 660, 661, 679
Firmness of plant tissues, 660, 661
Fischer-Tropsch process, 658
Fish (*see also* Muscle, seafood)
    cholesterol, 728
    composition, 727
    fatty acids, saturation of, 728
    lipids, changes postmortem, 758, 759
    nutritive value, 727, 728
    off-flavors, 766
    protein concentrate, 289
    proteins, 359, 362
    texture, 760, 761
Flatulence, 71
Flavan-3,4-diol, 566
Flavanones, 563, 564
Flavins in muscle, 551
Flavolans (*see* Proanthocyanidins)
Flavones, 563–565
Flavonoids
    acyl substituents, 563
    as antioxidants, 565
    interaction with anthocyanins, 562
    isolation and identification, 563, 565
    numbers, 563
    in plants postharvest, 891
    properties, 575
    stability, 565
    types, 563–565
Flavonols, 563–565
Flavoproteins, 500, 837
Flavor (*see also* Flavor compounds and sensations)
    alterations in frozen pizza, 925, 926
    from caramelization, 98, 106, 107
    character-impact type, 587
    comparative changes in various species, 729, 730
    definition, 586
    of dried vegetables, 417
    dry foods, development in, 420
    enzymes in cheese, effect of, 426
    enzymes in onions, role of, 380
    from Maillard browning, 106, 107, 343
    onions, development in, 425
    of plant tissue after dehydration, 909
    of plant tissue after freezing, 908
    of plant tissue after heating, 905–907
    potentiation, 425
    from protein hydrolysis, 355
    from Strecker degradation, 100
    from sugar decomposition products, 107

Flavor analysis, 586–588
Flavor binding
    by gum arabic, 106
    by proteins, 310–312
    by Schardinger dextrins, 106
    by sugars and carbohydrates, 106
Flavor chemistry, future directions, 620
Flavor compounds and sensations (*see also*
        Aromas, Flavor, Taste sensations)
    from alcoholic beverages, 610, 611, 616, 620
    from amino acids, 606, 607, 612, 615–618
    from apples, 606
    from apricots, 605
    from bakery products, 612, 616
    from bananas, 606
    from beer, 592
    from bread, 610
    from brussels sprouts, 602
    from butter, 610, 613
    from cabbage, 602
    caramel-like, 619
    from carrots, 607, 620
    change during heating of plant tissue, 905–907
    character-impact type, 587, 605, 606, 608, 609
    from cheese, 609, 610
    from chives, 601
    from chocolate, 619
    from citrus fruits, 606, 608
    from cocoa, 616, 619
    from coffee, 616
    cooling, 600, 601
    from corn, 618
    from cucumbers, 605
    from dairy products, 611–613, 616
    of dried eggs, 847
    enzyme generated, 601–606, 614, 615, 621
    from fats and oils, 611, 612
    from fatty acids, 604–606, 612, 613
    from fish, 766
    from garlic, 599, 601, 602
    from ginger, 599, 608, 609
    from green peas, 604
    from horseradish, 599, 602
    from lactic acid-ethanol fermentations, 609–611
    from leek, 601
    from lipid oxidation, 613–615
    from melons, 605
    from methyl alkyl pyrazine volatiles, 604
    from muscle foods, 612–615, 616, 617, 618
    from mustard, 599, 602
    neutralizer induced, 637, 638
    from nuts, 616
    from onions, 599, 601, 602
    from oxidation of carotenoids, 620
    from paprika, 599
    from peaches, 605
    from pears, 605
    from pepper, 599, 609

[Flavor compounds and sensations]
from pineapple, 619
in plant tissues postharvest, 892
from potatoes, 604, 616
from processing, 615–620
pungency, 599, 600
from radishes, 599, 602
from red beets, 604
from ruminal fermentations, 612, 613
from seafoods, 618
from shallots, 601
from shiitake mushrooms, 603, 604
from Shikimic acid pathway, 609
from smoke, 609
from spearmint, 608
from spices, 599, 608, 609
from sulfur volatiles, 601–604
from tea, 620
from terpenoids, 606–608
from tobacco, 620
from tomato juice, 618
from turnip, 602
from watercress, 599, 601, 602
from yogurt, 610
Flavor enhancement, 596, 597
Flavor enzymes, 454
Flavorese approach, 425, 621
Flavor potentiators, 674
Flavor precursors in plant tissues, 892
Flavor reversion of oil
hydroperoxides in, 228
linolenate in, 221, 227–229
singlet oxygen in, 228
Flavors and flavor modifiers, 674
Flavor volatiles (*see* Aromas)
Flavylans (*see* Proanthocyanidins)
Flavylium cation structure, 557, 558, 561
Flavylogens (*see* Proanthocyanidins)
Flour bleaching, 663–665
Fluorescamine, 267
Fluorescence, for measuring lipid oxidation, 195
Fluoride, 520
Fluorine safety, 528
Fluorocarbons, 347
1-Fluoro-2,4-dinitrobenzene, 270, 343, 344
Fluorodinitrobenzene procedure for non-enzymic browning, 926
Fluorosulfonyl acetoacetamide, 657
Fluorosulfonyl isocyanate, 657
FMN (Flavinmononucleotide), 500, 504
Foam formation
from dried egg products, 847
factors influencing, 307, 308
methods, 304
protein properties favoring, 308–310
Foaming (*see also* Foams)
in beer, 663
during deep fat frying, 211
enzymes, effect on, 423
and protein hydration, 285
Foaming agents, 668
Foaming agents, proteinaceous

[Foaming agents, proteinaceous]
alkaline hydrolysis, effect of, 355
egg albumen, 848–850
factors influencing, 307
intentional chemical modification, effect of, 359, 361
from milk, 819
proteinaceous emulsifiers, comparison with, 310
proteolysis, effect of, 357
types, 306–310
Foaming capacity, 306
Foaming power, 305, 306
Foams (*see also* Foaming)
from aerosols, 667
characteristics, 304
definition, 303
density, 306
destabilizing mechanisms, 304, 305
drainage measurements, 305
examples, 303
gas diffusion in, 305
methods of forming, 304
overrun, 305, 306
phase volume, 305, 306
pressure in, 305
protein-lipid interaction, effect of, 274
proteins, effect on, 275–278
rupture of lamellae in, 305
schematic representation of formation, 305
stability, assessment of, 306
stabilizing factors, 305
strength or stiffness, 306
volume, 305, 306
Foam stability
assessing, 306
factors influencing, 307, 308
protein properties favoring, 308–310
Foam volume defined, 305, 306
Folacin, 481, 487, 488, 820, 908 (*see also* Folates, Folic acid)
loss in frozen plant tissue, 908
in milk, 820
RDA, 481
Folates (*see also* Folacin, Folates)
active forms, 504
degradation products, 504–507
flavin mononucleotides, effect of, 504
functions, 504
processing, effect of, 508, 509
riboflavin, effect of, 504
stability, 504, 505, 507–509
structures, 504–506
Folic acid (*see also* Folacin, Folates)
nitrite, effect of, 485
recommended intake, 479
structure, 505
Folic acid, 10-formyl, 508
Food additives (*see also* Chap. 10 and specific chemicals)
direct or intentional, 691, 702–705
function, 630
indirect, 691, 712–716

[Food additives]
  safety, 702
Food analysis (*see* Analysis of food using immobilized enzymes)
Food and Agriculture Organization
  requirements for amino acid intake, 317–319, 323
  requirements for protein intake, 317–319
Food chemistry
  definition, 2
  history, 2–7
  study of, 7–13
Food chemists, societal role, 13–15
Food laws, 690
Food, processed
  antinutritional factors, effect on, 482
  health, effect on, 482
  nutrients, effect on, 482
  nutrients in, 478
Food safety, 18, 690–692
Food supply, safety of, 15–17
Formaldehyde, 345, 513, 614, 771, 772, 778
  biotin, effect on, 513
  reaction with proteins, 345
Formaldehyde, polymerized, protein crosslinking reagent, 363
Formic acid in milk, 816
1-Formyl-4-isopropylbenzene, 608
1-Formylpiperdine, 599
6-Formylpterin, 505
Fortification of food, 486
Fragmentation index, 754
Free energy
  and chemical equilibrium, 377, 378
  of enzyme inactivation, 407
  equation for, 377
  of protein denaturation, 403–405, 407
Free radicals
  addition to a double bond, 189
  cyclic compounds, source of, 189
  from ionizing irradiation of lipids, 216
  in lipid oxidation, 198, 202
  noncyclic dimers, source of, 189
Freeze-preservation of muscle, 769–772
Freezer burn, 770, 781
Freezing
  concentration of solutes, 61–64
  of egg yolk, 846, 847
  and enzymic reactions, 62–64
  and nonenzymic reactions, 62, 63
  plant tissues, effect on, 906–908
  properties of the unfrozen phase, effect on, 62
  reaction rates, effect on, 61–64
  relative effects of temperature and concentration of solutes, 63
Freezing and cooling agents, 669
Freezing of muscle, distinguishing between frozen and unfrozen, 770, 771
Freon-115, 667, 680
Freon C-318, 667
Fructans in plant tissue, 681
β-D-Fructopyranose, 589

Fructose, 74, 75, 93, 94, 631, 861
  amount in foods, 74, 75
  by enolization of D-glucose, 93, 94
  in plant tissue, 861
Fructose transglycosidase, 888
Fruit (*see also* Plant tissues postharvest)
  firmness, 660, 661
  lipid content, 866
  sulfites, use of, 645
Fruit structure, 872, 874
Frying fats
  consumption of, 232
  safety of, 231, 232
Fucose in eggs, 830, 831
Fucoxanthin, 570, 571
Fuller's earth, 218
Fumarase, 436
Fumaric acid
  chelating agent, 670
  food additive, 632
Fumigants, 672, 712
Fungal protease, 430
2-Furaldehyde (*see* Furfural)
Furan, 499, 616
2-Furancarboxaldehyde (*see* Furfural)
2-Furancarboxylic acid, 490–492
Furanone, 4-OH-3,5-dimethyl-3(2H), 619
Furanone, 4-OH-5-methyl-3(2H), 619
Furfural, 94, 490–492
Furfuryl thioaldehyde, 587
Furosine, 343·
Fusarium poae, 708
Fusarium sporotrichioides, 708

G

Galactomannan, 76
Galactosamine in egg shell, 830, 831
Galactose, 78, 558, 563, 830, 831, 836, 846
  anthocyanins, component of, 558
  in egg cuticle, 831
  eggs, additive for, 846
  in egg shell, 830
  flavonoids, component of, 563
  in ovomucoid, 836
  structure, 78
α-Galactosidase, 433, 436
β-Galactosidase, 429, 821, 822
α-Galactosides, 328, 329
Galactosyltransferase in milk, 808
Galacturonic acid, 439
Gallic acid
  antioxidant properties, 204
  structure, 567
Gallotannic acid, 567
Gallus domesticus, 830
Gangliosides, 146
Gap filaments, 739, 741
Gaping, 760, 761
Gaping of fish, 766, 772
Gases added to food, 666, 667
Gas hydrates, 44, 45
Gay-Lussac, Joseph Louis, 3

Gelatin, 169, 292, 300, 306, 309, 658, 662, 669, 746, 748–750, 774
  clarifying agent, 662
  as an emulsion stabilizer, 169
  foaming agent, 309
  formation from collagen, 748–750
  function in food, 669
Gelatinization of starch, 114
Gelation (*see also* Protein gelation)
  of egg yolk during freezing, 846, 847
  of frozen eggs, 830
  gel characteristics, 294
  of proteins, 283, 285
Gels, pectin, 123–126
Gels, polysaccharide
  cross-links in, 112
  junction zones in, 110, 111
  strength of, 110, 111
  sugars, effect of, 112
Gel-type muscle products, 779–781
Gene expression in fruit postharvest, 885
Generally recognized as safe (GRAS), 691
Genetic engineering (*see* Recombinant DNA technology)
Genistein, 564
Gentiobiose, 86, 573
Geranyl pyrophosphate, 607
Gibberella zeae, 708
Gibberellins in plant tissues, 893, 895
[6]-Gingerol, 600
Gliadins, 247, 261, 274, 296, 297, 861, 862
Globin, 307, 551
  (denatured) hemichrome, 554
  hemichromogen, 556
  hemochrome, 781
  hemochromogen, 556
Globular proteins, 246, 301
γ-Globulin, 246
Globulins, 274, 298, 834, 835, 849, 850, 861, 862
  in bread, 298
  in egg albumen, 834, 835, 849, 850
Glucagon, 254
Glucanase, 428
β-Glucans in plant tissue, 861
Glucoamylase, 90, 91, 428, 438
Glucomannans in plant tissue, 859, 860
Gluconic acid, 631, 633
δ-Gluconolactone
  leavening agent, 633–635, 670, 679
  structure, 631
Glucosamine
  in egg shell, 830
  in egg shell membranes, 831
  in ovomucoid, 836
Glucose, 70–72, 74, 75, 78, 79, 84, 92–94, 97, 100, 418, 558, 563, 565, 568, 589, 590, 596, 598, 660, 678, 751, 773, 801, 831, 832, 846, 847, 861
  abundance of, 71
  AH/B theory of taste, 589, 590, 596, 598
  amount in foods, 74, 75
  anomeric structures, 79
  in anthocyanins, 558

[Glucose]
  in betalains, 568
  caramelization of, 98
  digestion of, 72
  in egg albumen, 832
  in egg cuticle, 831
  eggs, additive for, 846
  in egg shell membranes, 831
  eggs, removal from, 847
  enolization of, 93
  equilibrium structures, 92, 93
  in flavonoids, 563
  isomers, 92, 93
  Lobry de Bruyn-Alberda van Edenstein reaction, 94
  in Maillard browning, 100
  in milk, 801
  mutarotation of, 92, 93
  in naringin, 565
  in plant tissue, 861
  pyrolysis of, 84
  structure, 70, 78
  sweetener, 678
  thermal degradation products of, 94, 97
  water activity, effect on, 660
Glucose isomerase, 90, 92, 432, 436, 466
Glucose oxidase, 418, 438, 668, 830
  activity at low water activity, 418
  for desugaring egg albumen, 830
  function in food, 668
Glucose oxidase-catalase, 431
Glucose-1-phosphate, 752
Glucose-6-phosphate dehydrogenase, 700
Glucosinolases, 602
Glucosinolates, 83, 84, 328, 693, 694, 697
  enzymic decomposition of, 84
  sinigrin, 83
Glucuronic acid, 563, 568, 656
Glutamate dehydrogenase, 416
Glutamate-oxaloacetate transaminase, 770
Glutamic acid, 250, 252, 253, 502, 504
Glutamic oxaloacetic transaminase, 231
Glutamic pyruvic transaminase, 231
Glutamine, 250, 252, 253, 906, 907
γ-Glutamyl-ϵ-N-lysine, 277
ϵ-N-(γ-Glutamyl)-lysyl bond, 296, 338
γ-Glutamyltransferase in milk, 802
γ-Glutamyl transpeptidase, 603
Glutaraldehyde
  in protein texturization, 296
  reaction with proteins, 345
Glutaric aldehyde, 363
Glutathione, 402, 617, 700, 776, 864
Glutathione peroxidase, 454
Glutelins, 274
Gluten, 296, 307, 357, 359, 664, 665, 862
  foaming agent, 307
  intentional chemical modification, 359
  oxidation, 664, 665
  proteolysis, effect of, 357
Glutenins, 261, 296, 297, 310, 861, 862
Glycans, 71 (*see also* Polysaccharides)
Glyceraldehyde-3-phosphate dehydrogenase, 416, 742, 752

Glycerides (*see* Acylglycerols)
Glycerine (*see* Glycerol)
Glycerol, 205, 658, 659, 660, 672, 775, 776
   antioxidant carrier, 205
   in intermediate moisture foods, 659
   texturizer, 658
   water activity, effect on, 660
Glycerol monostearate, 171
sn-Glycerol-1-stearate-2-oleate-3-myristate, 145
Glyceryl monolaurate, 648, 649
Glycine, 250, 252, 253, 747
   in collagen, 747
   hydrophobicity, 252, 253
Glycinin, 261, 334
Glycogen, 76, 120, 121, 751, 753, 758, 760, 765, 774
   decrease in muscle postmortem, 751, 753, 758
   in muscle, 76
Glycogen particles, 733, 742
Glycolytic enzymes of muscle, 742
Glycomacropeptide, 398
Glycoproteins, 246, 693
Glycopyranose, 1,6-Anhydro-β-D (*see* Levoglucosan)
Glycosidase, 356, 562
Glycosides, 80–83, 85, 91, 93, 328, 602, 693, 694, 845
   aglycon, 80
   alkali labile, 93
   alkyl D-glucopyranoside, 80, 82
   aminoglycosides, 81
   amount in diet, 81
   characteristics, 83
   cyanogenetic, 85
   in egg albumen, 845
   ethyl β-D-mannopyranoside, 81
   hydrolysis, 91
   methyl β-D-fructopyranoside, 81
   naturally occurring, 81
   nomenclature, 80, 81, 83
   physiological potency, 81
   sensory properties of, 81
   stability of, 83
   thioglycosides, 81
   toxic, 693, 694
Glycosylamine in Maillard browning, 99–101
Glycosyltransferases, 355
Glycyrrhetinic acid, 656
Glycyrrhizic acid, 656
Glyoxal, 186
Glyoxylate shunt in plants postharvest, 879
GMP in muscle, 756–758
Golgi apparatus, 733, 795
Gossypol, 218, 328, 329, 345, 574, 643, 694, 698
   antioxidant, 643
   removal during refining of fat, 218
   structure, 574, 698
Gramicidin, 254
Grana, 548
Grape skin extract bases, 580
Guanidation of proteins, 360

Guanidine salts, 279
Guanosine 5'-monophosphate
   flavor enhancer, 597
   structure, 83
Guanylic acid in muscle, 756
Guaran (*see* Guar gum)
Guar gum, 89, 111, 126, 127, 134, 672, 680
   encapsulation agent, 672
   repeating units of, 89
   solution viscosity, 111
   stabilizer, 680
Guiacic acid, 643
Guiaconic acid, 643
Gum arabic, 111, 127, 128, 134, 175, 658
   as emulsion stabilizer, 175
   solution viscosity, 111
Gum guaiac, 200, 203, 643
Gum, microbial (*see* Dextran, Xanthan gum)
Gums, 169, 657, 658, 861
   as emulsion stabilizers, 169
   in plant tissue, 861
Gums, plant (*see also* specific gums)
   description and properties of, 125–134
   properties, 134
Gum tragacanth, 128, 129, 134

H

Hanneberg, W., 6
Harman derivatives, 336
Hassall, Arthur Hill, 6
Hatch, William H., 7
Haze in beverages, 661–663
Heat
   antinutritional factors, effect on, 333, 334
   and Maillard browning, 342
   and protein crosslinking, 338
   proteins, effect on, 275, 277, 333–336
   proteins at alkaline pHs, effect on, 336–338
Heat capacity
   of enzyme inactivation, 406, 407
   of protein denaturation, 406, 407
Heat preservation of muscle, 772, 773
Heavy meromyosin, 735–737, 740, 756
Heavy metals, 420, 714, 715
Helical structures of proteins, 256
Heliotropin, 599
α-Helix
   in proteins, 255, 258, 260, 263
   structure, 256
Hemagglutinins, 328, 329, 693, 694
Hematoporphyrin and formation of singlet oxygen, 182
Heme, 521, 551, 552
Heme compounds, 646
Heme pigments
   in muscle, 550, 551
   properties, 575
Heme proteins, effect on thiamine, 499
Hemicellulase, 429, 886
Hemicellulose, 123, 859–861
Hemin, 409, 773
Hemoglobin, 246, 247, 303, 309, 411, 551, 582, 765, 778

[Hemoglobin]
    blood, function in, 551
    emulsifying ability, 303
    foaming agent, 309
    legal status as food colorant, 582
    in muscle, 551
    structure, 551
Hemoproteins, 246
Heptane, 184
2-Heptenal, 184
cis-4-Heptenal, 614, 766
Heptenal from heating fat, 211
Heptyl paraben, 650
Herbicides, 713, 714
Hesperidin, 657
Hesperidinase, 433
Hesperidin dihydrochalcone 4'-0-glucoside, 657
Heteroproteins, 246
Hexahydropyrroloindole, 341
Hexanal, 184, 187, 196, 211, 312, 313
    from 4,5-epoxide, 187
    from heating fat, 211
    from lipid oxidation, 196
Hexane, 187, 327
2-Hexenal, 605, 905
3-Hexenal, role in flavor reversion of oil, 228
Hexosamine
    in egg shell membranes, 831
    in ovomucin, 836, 845
Hexoses, 71, 79, 80, 94 (*see also* Sugars; specific sugars)
    chiral center, 79, 80
    degradation products of, 94
Heyn's product, 342, 343
High-fructose corn syrup, 90–92
High-methoxyl pectins, 439
High-temperature–short-time processing for optimizing nutrient retention, 527–529
Hippuric acid, 650
Histamine, 759
Histidine, 250, 252, 253
HLB system, 170, 171
Hofmeister series, 273
Homogenization
    of milk, 814
    protein-lipid interactions, effect on, 274
Homoproteins, 246
Hops, 592, 593
Hot boning of meat, 767
Humectancy (*see* Hydrophilicity)
Humectants in intermediate moisture foods, 659, 660
Humidity
    plants postharvest, effect on, 903
    shelf life of dehydrated food, effect on, 922
Humulone, 592
Hydrocarbons
    from heated or oxidized fat, 230
    from ionizing irradiation of lipids, 214
    from thermal decomposition of lipids, 205–209

Hydrochloric acid
    acidulant, 670
    in cheese, 631
Hydrocolloids, 657, 658 (*see also* Carbohydrates; Gums, plant; Pectic substances, Polysaccharides; Proteins; Stabilizers; Starches)
Hydrogen, 219, 221, 436
    concentration, effect on fat isomerization, 221
    for hydrogenation of fat, 219
Hydrogenation of fat, 218–222, 232, 233, 237
    catalysts for, 219, 222
    destruction of phytosterols during, 237
    formation of isoacids during, 219
    isomerization during, 219, 221, 222, 232
    mechanisms, 221, 222
    objectives, 219
    safety of, 232, 233
    schematic course of, 219
    selectivity of, 219–222
    and serum cholesterol, 237
    and *trans* fatty acids, 232
Hydrogenation selectivity
    conditions influencing, 221, 222
    definition, 219
    for fatty acids, 219, 220
    influence on isomerization, 222
    selectivity ratio, 219
Hydrogen bonding
    in protein gelation, 292
    in proteins, 263–265
    in protein-water interactions, 271, 272
    temperature, effect of, 286
Hydrogen cyanide, 695–697
Hydrogen peroxide, 339, 489, 494, 519, 670, 890
    free radical initiator, 670
    in plant tissue postharvest, 890
    and protein crosslinking, 339
    vitamin C, effect on, 489
Hydrogen phosphate, 521
Hydrogen sulfide, 277, 335, 492, 499, 511, 571, 617, 773
Hydrolases, 375, 433, 697, 802
Hydrolytic enzymes, 400, 402
11-Hydroperoxide, 183
10-Hydroperoxide, role in flavor reversion of oil, 228
Hydroperoxides, 176–183, 185, 187, 191, 193, 195, 197, 202, 207–210, 217, 486, 665, 713
    decomposition of, 182, 185, 197
    in dough, 665
    formation by singlet oxygen, 182
    interference with carbonyl measurement, 195
    from ionizing irradiation of lipids, 217
    from lipid oxidation, 176–182, 187, 191
    in lipid oxidation, 202
    and protein oxidation, 191, 193
    from thermal oxidation of fatty acids, 207–210

[Hydroperoxides]
vitamins, effect on, 486
α-Hydroperoxyaldehyde from oxidation of oleate, 185
Hydroperoxyl radicals, effect on vitamin C, 489, 490
Hydrophilicity of sugars, 105
Hydrophobic interactions, 44–46, 263–265, 272, 274, 292, 297, 310, 311, 350, 738, 803, 808, 812, 849, 862
in casein, 803
in casein micelles, 808, 812
in dough formation, 297
in egg foams, 849
in flavor binding by proteins, 310, 311
in muscle, 738
in processed cheese, 350
in protein gelation, 292
in proteins, 263–265, 272
between proteins and lipids, 274
in soybeans, 862
Hydrophobicity (*see also* Protein hydrophobicity)
of bitter substances, 590
of proteins, 265, 266
of protein side chains, 263
Hydrophobicity value and bitterness, 594, 595
Hydroquinone, 198
Hydrostatic pressure, effect on proteins, 277
Hydroxides, 521
3-Hydroxy-2-acetylfuran, 98
Hydroxyapatite, 808, 815
p-Hydroxybenzaldehyde cyanohydrin glucoside, 697
p-Hydroxybenzoate alkyl esters, 650, 673
5-(β-Hydroxyethyl)-4-methylthiazole, 495
Hydroxy fatty acids from heated or oxidized fats, 230
Hydroxylases, 355
Hydroxylysine, 747, 748, 749
Hydroxylysine aldehyde, 749
Hydroxylysino-5-keto-*nor*-leucine, 748, 749
4-Hydroxymethyl-2,6-ditertiarybutyl-phenol, 199, 200, 204
5-Hydroxymethyl-2-furaldehyde, 94, 97, 99, 102
5-Hydroxymethyl-2-furan carboxaldehyde (*see* 5-Hydroxymethyl-2-furaldehyde)
2-H-4-Hydroxy-5-methyl-furan-3-one, 98
5-Hydroxymethyl-2-furfural (*see* 5-Hydroxymethyl-2-furaldehyde)
3-Hydroxy-3-methylglutaric acid, 568
3-Hydroxy-2-methylpyran-4-one, 98
Hydroxymethylpyrimidine, 495
Hydroxyproline, 746, 747, 750
Hydroxypropylmethyl cellulose, 680
3-Hydroxy-2-pyrone, 492
1'-Hydroxysafrole, 704
Hygroscopicity (*see* Hydrophilicity of sugars)
Hypobaric storage for fruits and vegetables, 901, 902
Hypochlorite, 268

Hypoxanthine, 756–758, 781
H-Zone, 733, 734

I

I-Band, 732–734, 762, 772
L-Ibotenic acid, 597
Ice
amorphous, 33
basal plane, 31
consequences in food, 61
coordination number, 33
crystal symmetry, 33
defects in, 33, 34
dynamic nature of, 32–34
hexagonal symmetry, 31
hydrogen atoms, location of, 32, 33
in package, 927, 928
physical constants, 26
properties, 25, 26
reaction rates at subfreezing temperatures, role in, 61–64
solutes, effects on structure, 34, 35
structure, 30–35
unit cell, 30
vapor pressure, 26, 49
Icthyotoxism, 701
Immobilized enzymes (*see* Enzymes, immobilized)
Immobilized substrate (*see* Substrate, immobilized)
Immunoglobulins, 247, 796, 799
Immunoglobulins in milk, 813, 814, 816
IMP in muscle, 756–758, 776
Inclusion structures
fatty acid-amylose, 116
Schardinger dextrins, 87
Indole-3-acetic acid oxidase
in plant tissue postharvest, 888
in ripening fruit, 885
Inorganic phosphate in muscle, 742
Inosine, 758
Inosine 5'-monophosphate, 83, 596, 5
Inosinic acid in muscle, 856
Inositol, 175
Inositol hexaphosphate, 527
Insulin, 466
Intentional additives (*see* Food
Interesterification of fat, 223–
Interfaces, effect on proteins
Interfacial absorption, effec 423
Interfacial tension
and emulsion stabilit
of protein solutions
Intermediate fibers, 7
Intermediate moistu
Intermediate moist
Intrinsic factor, 5
Introns, 561
Invertase, 374, 892
inhibitor,
in plant

[Invertase]
  in potatoes, 887
  in ripening fruit, 885
Iodate, 521
Iodide, 520, 521
Iodine
  in food, 522
  forms, 521
  in iodized salt, 522
  RDA, 481
  safety, 528
  value, 195
Iodoacetamide, 359
Iodoacetic acid, 269, 270, 359
Ionene, 515
Ion exchange, effect on protein functionality, 349, 350
Ion-exchange resins, 672
Ionic interactions of proteins, 272
Ionic strength
  emulsions, effect on, 299, 300
  protein solubility, effect on, 273
  protein solution viscosity, effect on, 290
Ionizing irradiation of lipids
  degradation mechanisms, 215–217
  doses used, 213
  nutritive effects, 217
  objectives of, 213
  in presence of oxygen, 217
  products from, 214–217
  safety of, 213, 217
Ionizing irradiation of muscle, 781
Ionizing radiation
  enzymes, effect on, 422, 423, 425
  plants postharvest, effect on, 903
  proteins, effect on, 277, 338, 339
β-Ionone, 620
Ions
  binding of flavors by proteins, effect on, 312
  protein hydration, effect on, 287
meamarone, 694, 698, 699
omeanol, 898
tein, 741
197, 479–481, 483, 488–491, 511, 521–523, 525–528, 820, 842, 846, 351
vailability, 523, 525, 526
st for lipid oxidation, 197
ncy of absorption, 480
olk, 842
chment of bread, 483
on with conalbumin, 846, 851
on with phosvitin, 842
20
ded intake, 479, 481
, effect on, 511
C degradation, 489–491
06, 607
, 448
enase, 778

Isodesmosine, 248
Isoelectric point, 249, 252, 262, 272, 274, 286, 292, 294, 327, 335, 349, 774, 775
  of amino acids, 252
  definition, 249
  and gelation of proteins, 292, 294
  for isolation of soy proteins, 327
  for protein precipitation, 349
  and protein-protein interactions, 286
  of proteins, 262, 272
  of proteins versus heating, 335
  versus strength of protein-lipid interaction, 274
  and water holding capacity of muscle, 286
Isoenzymes, 375
Isoflavanones, 563–565
Isoflavonoids, 695
Isohumulone, 592
Isoionic point of proteins, 262
Isoleucine, 250, 252, 253, 436
Isomaltol, 86, 98, 107, 619
Isomerase, 433, 605
Isomerization of amino acid residues, 335
Isopentyl pyrophosphate, 607
Isopeptide bonds, 336, 338, 362
Isoprenoid quinones, 569
Isopropyl citrate, 200
Isosafrole, 704
Isotherms (*see* Moisture sorption isotherms)
Isothiocyanates, 602, 695, 696
Isovaleraldehyde, 930
Isovitexin, 563
Isozymes, 375, 381

J

Juglone, 569, 570

K

Kaempferol, 563–565
Kamaboko, 779, 780
Katemfe, 657
Keratin, 246, 336, 663
α-Ketogluconic acid, 436
α-Ketoglutarate, 747
α-Ketoglutaric acid, 502
Ketones
  and flavor reversion of oil, 229
  from ionizing irradiation of lipids, 214, 216
  from lipid oxidation, 195
  from thermal decomposition of lipids, 205–208
K_m (*see* Michaelis constant)
Kolin, 670
Kraft temperature, 163, 164
Krebs cycle in plants postharvest, 878, 889
Kreis test for lipid oxidation, 195
Kunitz inhibitor 33, 455, 692, 693
Kwashiorkor, 319, 330
Kynurenine, 341
Kynurenine, N-formyl, 341

L

α-Lactalbumin, 274, 796, 798, 801, 816, 819
α-Lactalbumin B, 261, 810
Lactase, 73, 374, 429, 438, 821, 822
Lactate
    in muscle, 742, 751, 752
    in muscle postmortem, 751–753, 758
Lactate dehydrogenase, 416, 751, 769
Lactic acid, 436, 631, 632, 637, 638, 661, 670
    acidulant, 670
    in butter, 637
    firming agent, 661
    food additive, 632
    and pH of animal tissue, 638
    in milk, 631
    structure, 632
Lactide, 631, 632
Lactobacillus arabinosus for biotin assay, 513
Lactobacillus casei
    folates, analysis of, 507
    riboflavin, analysis of, 500
Lactobacillus leichmannii for assay of vitamin B₁₂, 511
Lactobacillus plantarum for assay of pantothenic acid, 512
β-Lactoglobulin, 261, 274, 281, 292, 294, 302, 303, 309, 352, 406, 407, 796, 798, 808, 809
    gel characteristics, 294
    van't Hoff plot for denaturation, 406
β-Lactoglobulin A, 261
β-Lactoglobulin B, 809
Lactones
    and flavor reversion of oil, 229
    from ionizing irradiation of lipids, 214
    from thermal decomposition of lipids, 206–209
δ-Lactones, 605
γ-Lactones, 605
Lactoperoxidase, 452, 802
Lactose, 73, 76, 438, 816, 820
    behavior during freezing, 816
    crystallization, 816
    digestion of, 73
    functional properties, 820
    intolerance, 73
    and Maillard browning, 816, 820
    milk, amount in, 76, 796
    sweetness, 820
    synthesis, 795
    tolerance, 821, 822
Lactoselysine in milk, 824
Lactylated monoacylglycerols, 172
Lakes
    carmine, 582
    defined, 546
    of FD&C colorants, 580
Lamb
    composition, 727
    flavor, 612
Land-Grant College Act, 7

Lanthionine, 336, 337, 355, 363, 815
Lanthionyl bond, 296
Laplace's capillary pressure equation, 305
Lard interesterification, 225
Larrea divaricata, 204
Latent enzymes, 416
Lathyrogens, 694
Lauric acid, 142, 668
Lauric acid fat, 146
Lavoisier, Antoint Laurent, 3
Leaching
    iodine loss by, 522
    of minerals from food, 520, 523
    nutrient content of food, effect on, 483
    of thiamine during processing, 498
    of vitamins during processing, 485, 492
Lead, 521, 522, 525
Leaf structure, 872
Leavening acids, 633–635
Leavening agents, 679
Leavening systems, chemical, 631–636
Lecithin, 145, 175, 200, 419, 742
Lecithinase, 419
Lectins, 333, 693
Lenthionine, 603, 604
Lentinic acid, 603, 604
Leucine, 251–253, 604, 607
Leucoanthocyanidins, 566
Leucoanthocyanins (*see* Proanthocyanidins)
Leuconostoc citrovorum, 610
Leupeptins, 458
Levoglucosan, 84, 85
Liebig, Justus von, 4
Light meromyosin, 735, 736, 756
Lignin
    flavors from, 609
    in plant tissue, 859, 860
    structure, 609
Limit dextrins, 441
Limonene, 608
Limonin, 593
Limonoate dehydrogenase, 593
Linamarin, 697
Lindane, 714
Lineweaver-Burk plots, 384, 385, 387–390, 393–395
    and competitive inhibition, 387
    for immobilized substrates, 393–395
    and noncompetitive inhibition, 389, 390
    and uncompetitive inhibition, 388
Linoleate
    autoxidation of, 178, 180
    in Diels-Alder reaction, 189
    dimerization of, 191
    oxidation of, 185
    products from thermal decomposition, 209
    reaction with BHT, 201
Linoleate hydroperoxides, 519
Linoleic acid, 142, 196, 219, 220, 229, 232, 448, 449, 606, 677
    in diet, 229, 232
    nomenclature, 142

[Linoleic acid]
    nutrient added to food, 677
    rate of hydrogenation, 220
    rate of oxidation, 196
    in pear flavor, 606
    selectivity of hydrogenation, 219, 220
Linoleic acid fats, 147
Linolenic acid, 142, 196, 219, 220, 226, 449, 605
    nomenclature, 142
    in plant flavors, 605
    rate of hydrogenation, 220
    rate of oxidation, 196
    reduction of concentration in soybean oil, 226
    selectivity of hydrogenation, 219, 220
Lipase inhibitors, 455
Lipases, 333, 393, 394, 411, 413, 414, 423, 433, 442-444, 668, 759, 764, 771, 801, 802, 814, 888, 925, 926
    in frozen pizza, 925, 926
    function in food, 668
    in milk, 801, 802, 814
    in plant tissues postharvest, 888
    and tracyglycerol emulsions, 393, 394
Lipid autoxidation (*see also* Oxidation)
    aldehydes from 183-187
    in biological systems, 189-191, 193
    browning, role in, 191
    catalysis by chlorophyll, 177
    catalysis by myoglobin, 177
    in complex systems, 189-191, 193
    consequences of, 176
    in cured meat, 777
    cyclic monomers, source of, 189, 192
    in dehydrated potatoes, 929-931
    Diels-Alder reaction during, 189
    dimers from, 188, 189
    in dried egg products, 847
    in fish, 758, 759
    formation of conjugated dienes, 178
    in frozen products, 771
    flavors from, 611-615
    free radical mechanism of, 176-178
    generalized scheme, 179
    initiation, 177
    malonaldehyde from, 186, 193, 194
    measurement of, 193-196
    in milk, 814
    photosensitized, 181, 182
    polymers from, 188, 189
    propagation, 177
    and protein crosslinking, 338
    and protein denaturation, 771, 772
    rate of, 196-198
    rates for fatty acids versus acylglycerols, 197
    safety of products, 713
    singlet oxygen, involvement of, 177, 181, 182
    and storage of dehydrated muscle, 775
    termination, 177
    vitamins, effect on, 486
Lipid crystallization, 151-164
Lipid degradation by heat, 205-213

Lipid derivatives determining bitterness, 594
Lipid droplets, 733
Lipid metabolism in plant tissue postharvest, 888, 890, 891
Lipid oxidation (*see* Lipid autoxidation)
Lipids (*see also* Triacylglycerols)
    ability to associate with proteins, 303
    autoxidation of, 176-193
    binding of flavors by proteins, effect on, 312
    Cahn-Ingold Prelog procedure, 143, 144
    changes in muscle postmortem, 758, 759
    chirality, 143
    classification of, 145-147
    and coronary heart disease, 233-237
    crystallographic structure of, 151, 152
    definition, 140
    in dough formation, 296, 297
    in egg albumen, 832, 833
    in egg yolk, 832-834
    emulsifiers for, 169-175
    emulsions from, 166-175
    flavors from 611, 612
    foams, effect on, 307
    hydrolysis of, 175, 176, 925, 926
    importance, 140
    ionizing irradiation of (*see* Ionizing irradiation of lipids)
    in lipoprotein of egg yolk, 842
    in low-density lipoprotein of egg yolk, 844, 845
    mesomorphic phases (liquid crystals) of, 163-165, 169
    in milk, 796, 797, 799, 800
    in muscle, 727, 742, 743, 745, 746
    in myelin figures of egg yolk, 843
    nomenclature, 140-145
    phase diagram for, 160, 161
    in plant tissues, 328, 863, 866, 867
    polymorphism, 152-158
    protein denaturation, effect on, 274
    proteins, interaction with, 274
    proteins, reactions with, 346
    R/S system, 143, 144
    solid solution, 152
    sources, 140
    space lattice, 152
    starch gelatinization, effect on, 115
    stereospecific numbering, 144, 145
    triacyglycerol distribution patterns, 147-149
Lipolysis
    desirability of, 175, 176
    during deep fat frying, 175
    at low water activity, 418
    of milk, 814
    oxidation, effect on, 175
    smoke point, effect on, 175
    in wheat flour, 417
Lipoproteins, 234, 235, 246, 346
Lipovitellins in egg yolk, 841, 842, 848
Lipoxygenases, 191, 204, 333, 374, 413, 418, 431, 447-451, 549, 574, 605, 664, 665, 888, 890-892, 907, 908, 925, 926

[Lipoxygenases]
  activity at low water activity, 418
  carotenoids, effect on, 574, 891
  chlorophylls, effect on, 549, 892
  damage to frozen plant tissue, 908
  and flavor generation, 605
  flour, added to, 664, 665
  in frozen pizza, 925, 926
  inhibition by propyl gallate, 204
  and lipid oxidation, 191
  plant color, effect on, 907
  in plant tissues postharvest, 888, 890
  reactivation, 413
  regiospecificity, of, 191
  stereospecificity of, 191
Liquefying enzyme, 441
Liquid crystals, 163–165, 169, 172, 173,
  175
Liquid nitrogen, 669
Livetin, 837, 843, 844, 850
Locust bean gum, 127, 134, 665, 680
  in baked goods, 665
  stabilizer, 680
Low-density lipoproteins in egg yolk,
  842–845, 850
Low-methoxyl pectin, 439
Low-temperature injury (*see* Chilling injury
  of fruits and vegetables)
Lumiflavin, 500
Lutein, 571, 572
Luteolin, 563–565
Luteoskyrin, 708
Lyases, 603, 802
Lycopene, 514, 571, 573, 574
  structure, 514, 571
  susceptibility to oxidation, 574
  vitamin A activity, 514
Lymphoid cells, 745
Lysine, 251–253, 343–345, 348, 747, 824
  binding to proteins, 348
  in collagen, 747
  loss from Maillard browning,
    343–345
  in milk, 824
Lysine aldehyde, 748
Lysine hydroxylase, 747
Lysinoalanine, 336, 337, 355, 363, 815, 824
Lysinoalanyl bond, 296
Lysosomes, 416, 733, 755, 756
Lysozyme, 266, 302, 303, 309, 433, 438,
  802, 808, 835, 836, 845, 846, 850
  in egg albumen, 835, 836, 845, 846, 850
  in milk, 802, 808
Lysyl oxidase, 747
Lysyl residues in milk, 824

M

Macrophages, 745
Magnesium, 481, 521, 522, 524, 733, 737,
  738, 742, 801, 808, 811, 820, 830
  in casein micelles, 808
  in egg shell, 830
  in milk, 801, 811, 820

[Magnesium]
  in muscle, 733, 737, 738, 742
  RDA, 481
Magnesium ammonium phosphate, 642
Magnesium-ATP, 740
Magnesium carbonate
  alkaline additive, 671
  anticaking agent, 666
  for flour, 664
  neutralizer in food, 637
Magnesium chloride
  firming agent, 679
  function in food, 669
Magnesium oxide, 637, 715
Magnesium silicate
  anticaking agent, 666, 670
  antistick agent, 671
  separation aid, 672
Magnesium stearate, anticaking agent, 670
Maillard reaction (*see* Browning, Maillard)
Maleylation
  of proteins, 359
  protein solubility, effect on, 272
Malic acid, 416, 447, 632, 638, 670, 863
  buffer effect in plants, 638
  chelating agent, 670
  food additive, 632
  in plant tissue, 863
Malic enzyme, 888
Malonaldehyde
  from lipid autoxidation, 186, 193, 194
  mechanism of formation, 194
  protein crosslinking reagent, 363
  reaction with proteins, 345, 346
  structure, 346
Malonic acid
  anthocyanins, component of, 558
  in betalains, 568
Malting and fermenting aids, 669
Maltol, 98, 106, 597, 598, 619
  caramel-like flavor, 619
  flavor enhancer, 597, 598
  from pyrolysis, 98
  structure, 619
Maltose, 86, 418
Malvidin, 557, 558
Malvidin-3,5-diglucoside, 561
Malvidin-3-glucoside, 559, 651
Mammary gland, 794
Manganese, 197, 521, 522, 527
Mangiferin, structure of, 570
Mannitol, texturizer, 658, 680
Mannose
  in eggs, 830, 831, 836
  by enolization of D-glucose, 93, 94
  structure, 78
Maragoni effect, 309
Marasmus, 319
Marbling of meat, 745
Marine oils
  characteristics of, 147
  fatty acids in, 151
Mast cells, 745
Masticatory substances, 658, 679

Mauve, 576
Mayonnaise, 848
Meat (*see also* Muscle)
 aging, 380
 cholesterol, 728
 color (*see* Myoglobin)
 composition, 727
 consumption, per capita, 550
 fatty acids, saturation of, 728
 hydrolases, 380
 nutritive value, 726–728
 pH (*see* Muscle pH)
 pigments (*see* specific pigments; Pigments,
  cured meat)
 processing (*see* Muscle processing)
 tenderness (*see* Muscle tenderness)
 water holding capacity (*see* Muscle water
  holding capacity, Water holding
  capacity)
Mechanical injury to fruits and vegetables,
 895–898
Mechanical treatments, effect of proteins,
 277, 351
Melanins, 446, 575
Melanoidins, 99, 102, 104, 343–345, 645
Menadione, 517
Menthol, 600
Mercaptans, 602, 773
β-Mercaptoethanol, 270, 279
Mercury, 521, 522, 527, 714, 715
Mesenchyme cells, 745
Mesomorphic structure (*see* Liquid crystals)
Metabolism (*see* Muscle biochemistry post-
  mortem, Plant biochemistry postharvest,
  Plant respiration postharvest)
Metal catalysis of lipid oxidation, 197, 198
Metal chelates, stability of, 642
Metal ions, chelation, 641, 642
Metalloproteins, 246
Methanethiol, 616, 617
Methemoglobin, 646
Methional, 616, 617
Methionine, 251–253, 270, 339, 340, 342,
  347, 348, 456, 618, 821
 binding to proteins, 348
 determination, 270
 in flavor development, 618
 in milk, 821
 oxidation, 339, 340, 342, 347
 supplement for soybean meal, 456
Methionine sulfone, 339, 340
Methionine sulfoxide, 339, 340
Methionine sulfoximine, 347
2-Methoxy-3-s-butylpyrazine, 604
2-Methoxy-3-isobutyl pyrazine, 604
2-Methoxy-3-isopropyl pyrazine, 604
6-Methoxymellein, 898
8-Methoxypsoralin, 699
Methylazoxymethanol, 694, 700
Methyl bromide, 347, 672, 712
Methylbutanal, 619, 930
3-Methyl-2-butene-1-thio, 592
Methylcellulose, 122, 665, 680

S-Methyl-L-cysteine sulfoxide in plant tissue,
 865
Methylene chloride, 672
4-Allyl-1,2-methylenedioxybenzene, 704
2-Methyl-3-furanthiol, 617, 618
(2-Methyl-3-furyl)-bis-disulfide, 617
Methylheptanoate, 183
Methylhistidine, 248
Methyl-9-hydroperoxy-10,12-octadecadienoate,
 184
o-Methyl isourea, 270
Methyl ketones, 609, 612, 646, 647
ε-N-Methyllysine, 248
Methyl-malonyl-CoA, 612
Methylmalonylmutase, 510
Methyl mercaptan, 616
Methyl mercury, 714, 715
s-Methylmethionine sulfonium salt, 618
Methyl octanoate, 183, 184
4-Methyloctanoic acid, 612
Methylophilus methylotropus, 465
6-Methyl-1,2,3-oxathiazine-4(3H)-one-2,2-
  dioxide (Acesulfame K), 656
Methyl-10-oxodecanoate, 184
Methyl-10-oxo-8-decenoate, 183, 184
Methyl-12-oxo-9-dodecenoate, 184
Methyl-9-oxononanoate, 183, 184
Methyl-8-oxooctanoate, 183
Methyl-11-oxo-9-undecenoate, 184
Methyl paraben, 650
5-Methyl-2-phenyl-2-hexanal, 619
2-Methylpropanal, 930
6-Methylpterin, 504
2-Methyl-5-sulfomethylpyrimidine, 495
2-Methylthio-3-t-butenylisothiocyanate, 602
o-Methyltransferase, 433, 885
Methyl-9-undecenoate, 184
Metmyoglobin, 553, 554, 556, 764, 765, 768,
  777, 781
Metmyoglobin hydroxide, 553
Metmyoglobin nitrite, 554, 556
Mevalonic acid pyrophosphate, 607
Michaelis complex, 381, 401
Michael constant, K_m, 382, 384, 387–396,
  399, 769, 770
Michaelis-Menten theory, 393, 394
Michaelis-Menten velocity, 386, 395
Microbial cells, immobilized, 435, 436
Microbial gums (*see* Dextran, Xanthan gum)
Microbial spoilage of muscle, 768
Microorganisms
 for determining protein nutritive value,
  324, 325
 in flavor development, 609
 source of protein, 330, 331
Microwave blanching, 483
Microwave cooking, 484, 485
Milk
 biosynthesis, 792, 794, 795
 buffering properties, 638
 clotting, 398 (*see also* Milk, gelation)
 coagulation, 398 (*see also* Milk, gelation)
 enzymes, 414, 801, 802

[Milk]
    gelation, 812, 815, 816 (*see also* Milk, clotting)
    lipids, 797, 799, 800
    nutritive value, 820, 822, 825
    pH, 796, 814–816
    production, 792, 793
    products, nutrient composition, 823
    proteinases, 442
    salts and stability, 637, 640
    vitamins, effects of processing, 824, 825
Milk composition, 795–802
    enzymes, 801, 802
    lipids, 797, 799, 800
    proteins, 796–798
    salts, 799–801
Milk fat
    characteristics, 146
    fatty acid distribution, 150
    flavors from, 611
    globule membrane, 795, 813, 814
    globules, 812, 813
    uses, 813, 814
Milk processing
    versus functionality, 813–816
    versus nutritive value, 822, 824, 825
Milk proteins
    concentration in milk, 822
    functionality in foods, 817–819
    heat, effect of on nutritional value, 822, 824
    kinds, 796–798
    nutritional quality, 821
    structures, 802–808
    water binding capacity, 637
Milk salts, 799–801, 808, 811, 812, 814–816
    in casein micelle, 808, 811, 812
    concentration, effect of, 816
    freezing, effect of, 816
    heat, effect of, 814, 815
Milling of cereals versus nutritive value, 483, 484, 520
Mineral bioavailability, 523, 525–527
Mineral content of food, 482–486, 520–525
    blanching, effect of, 483
    cereal products, 523
    eggs, 523, 524
    factors influencing, 521
    gains by processing, 523, 525
    importance, 482
    leaching, effect of, 483
    literature sources, 522
    loss, causes of, 482–486
    losses by processing, 523
    milk, 523, 524
    milling, effect of, 483
    navy beans, 525
    potatoes, 524
    trimming, effect of, 483
    variability, 520
Mineral oil, 668, 671
Minerals (*see also* specific minerals)
    added to food, 675, 676

[Minerals]
    amounts in food and water, 521, 522
    in conalbumin, 836, 846
    in dairy foods, 821
    dietary role, 519
    in egg albumen, 833
    in egg yolk, 841, 842
    important anions, 520
    important cations, 521
    intake of, 521
    interaction with other constituents, 520
    loss by leaching, 520
    loss during milling of cereals, 520
    loss, mode of, 520
    in milk and milk products, 822, 823
    in muscle, 728, 742
    in plant tissue, 867, 869
    RDA, 481
    safety, 527, 528
Miracle fruit, 675
Miraculin, sweetener, 657
Mitochondria, 733, 734, 742, 743, 751, 757, 760, 761, 763, 764, 770, 795
M-line, 733, 734, 739, 756
Mobility of solutes versus stability of foods, 60, 61
Modified atmospheric storage for muscle, 778, 779
Modified starch, 118, 119
Moisture control agents, 674
Moisture sorption isotherms, 39–41, 50–56, 660
    of cellulose and glycerol, 660
    versus Clausius-Clapeyron equation, 54
    definition, 50
    desoprtion type, 51
    examples, 51–56
    versus food stability, 53
    hysteresis, 54, 55
    hysteresis versus food stability, 55
    monolayer value, 51
    resorption type, 51
    temperature dependence, 54
    usefulness, 50
    zones, 39–41, 51, 53
Molds as a source of protein, 330
Molybdenum, 522
Monascin, 582
Monellin, 332, 657
Monoacylglycerols
    antimicrobial agent, 648
    antioxidant carrier, 205
    crystalline hydrates from, 172
    emulsifiers, 671
    liquid crystals from, 169, 172
    in milk fat globule membrane, 813
Mono- and diacylglycerols, 678
Monocalcium phosphate, leavening agent, 633–636
Monogalactosyl diglycerides, 298
Monoglycerides, (*see* Monoacylglycerols)
5′-Mononucleotides, 756
Monosaccharides
    anomers of, 78

[Monosaccharides]
  conformation of, 78
  digestion of, 71–73
  epimers of, 78
  Haworth ring structure, 78
  nomenclature, 78, 80
  origin of, 76–78
  structure of, 78, 79
Monoserine, 618
Monosodium L-glutamate
  flavor enhancer, 596
  structure, 597
Muscle (*see also* Meat; specific kinds of muscle)
  added phosphates on hydration, effect of, 640, 641
  comparison of species, 729, 730
  flavors from, 612–615
Muscle aging, 754–756, 768
Muscle biochemistry postmortem
  amino acid degradation, 759
  anaerobic glycolysis, 750–755
  ATP to hypoxanthine, 756–758
  compared to plant biochemistry postharvest, 751, 752
  degradation of trimethylamine oxide, 759, 760
  energy charge, 752
  lipids, effect on, 758, 759
  loss of calcium sequestering ability, 757
  and pH, 751–755, 757, 758, 760, 761
  versus quality, 753
Muscle color
  after curing, 777
  after freezing, 770
  and electrical stimulation, 767, 768
  and ionizing radiation, 781
  and pH, 764, 765
Muscle contraction, 733, 761–764, 767, 774
  and meat quality, 761–764
  versus temperature, 762
Muscle emulsions, 779
Muscle extensibility, 753–755, 761
Muscle fibers, 730, 732–734
Muscle fibrils, 733, 734, 740, 742, 764
Muscle filaments, 733, 737, 740
Muscle flavor, 765, 773, 775–777, 781
  as affected by ionizing irradiation, 781
  after curing, 776
  changes during storage of dry muscle, 775
  heat, effect of, 773
  smoking, effect of, 777
Muscle lipids, 742, 743, 758, 759
Muscle nutritive value, 726–728, 765, 773, 775
Muscle pH (*see also* pH)
  changes postmortem, 751–755, 757, 758, 760, 761
  and color, 764, 765
  during curing, 776
  during dehydration, 774, 775
  and electrical stimulation, 767, 768
  during freezing, 769
  and quality, 760, 761, 765, 766
  ultimate, 753, 755, 760

Muscle processing
  chemical, 777
  controlled atmosphere, 777, 778
  curing, 776, 777
  dehydration, 773–776
  electrical stimulation, 767, 768
  freezing, 769–772
  gel-producing, 779–781
  heating, 772, 773
  hot boning, 767
  ionizing irradiation, 781
  major types, 766
  packaging, 781, 782
  refrigeration, 768, 769
Muscle proteins, 258–261, 729, 730, 735–743, 746–750 (*see also* specific proteins)
  collagen, 746–750
  comparison of species, 729, 730
  contractile, 735–741
  insoluble, 742, 743
  in membranes, 743
  soluble, 742
Muscle quality and antemortem events, 765, 766
Muscle quality and postmortem events
  actin-myosin interaction, 761–764
  fatty acid increase, 764
  nucleotide metabolism, 765
  proteolysis, 765
  thaw rigor, 763, 764
  Z-disk alterations, 764
Muscle relaxation, 733, 742
Muscle shortening (*see* Muscle contraction)
Muscle structure, 730–735
  cardiac (heart) or striated involuntary muscle, 734
  skeletal, 730–734
  smooth or involuntary muscle, 734
Muscle tenderness
  and actin-myosin interaction, 761–763
  aging, effect of, 754–756
  versus contraction, 761–763
  and electrical stimulation, 767
  and fatty acids, 764
  after freezing, 770
  of frozen fish, 771, 772
  heat effect of, 772, 773
  and hot boning, 767, 768
  and ionizing irradiation, 781
  and pH, 760, 761
  and thaw rigor, 763, 764
  and Z-disk degradation, 764
Muscle texture, effect of drying, 774, 775
Muscle types, 743–745
Muscle water holding capacity (*see also* Water holding capacity)
  and contraction, 763, 764
  in cured products, 776
  dehydration, effect of, 774
  after freezing, 769, 770
  heat, effect of, 773
  and pH, 760, 761
  and thaw rigor, 763

Mutachrome, 515, 516
Mutarotation of sugar, 92
Mycotoxins, 706–710
Myelin figures in egg yolk, 842, 843
Myocommata, 731, 748, 760, 772
Myofibrillar proteins, 300
Myofibrils (*see* Muscle fibrils)
Myogen, 742
Myoglobin, 177, 246, 247, 280, 354, 550–557,
    407, 733, 735, 743, 751, 760, 764, 765,
    768, 773, 778, 781
  catalysis of lipid oxidation, 177
  changes in fresh and cured meat, 554
  chemical alterations of, 553, 554
  CO, effect of, 557
  color changes, 554–556
  color of, 551–553
  denaturation versus water activity, 354
  in muscle, 551
  light absorption by, 553, 554
  light, effect of, 557
  in live animals, 550
  in meat, 550
  metal ions, effects of, 557
  oxidation of, 554, 556
  oxygenation of, 554, 556
  packaging, effect of, 555
  properties of, 556
  in sarcoplasm, 551
  solubility of, 551
  structure of, 551–553
  thermodynamic parameters of denaturation,
    280
NO-Myoglobin, 781
Myomesin, 739, 741
Myosepta, 731
Myosin, 247, 258–261, 733–738, 740–742,
    753–756, 761, 762, 771, 773, 779
  characteristics, 258–261
  contraction, role in, 740, 741
  in muscle relaxation, 742
  quarternary structure, 258–260
Myosin ATPase, 445
Myotomes, 731, 760
Myricetin, 563–565
Myristic acid
  coronary heart disease, role in, 236
  nomenclature, 142
Myrosinase, 333

N

NAD in muscle, 752, 753
Napthacenequinones, 569
Napthoquinones, 517, 569, 570
Naringin, 564, 565, 593, 594, 657
Naringinase, 433
Naringin dihydrochalcone, 657
Natamycin, 648
National Academy of Sciences
  amino acid intake requirements, 317–319,
    323
  protein intake requirements, 317–319

Navy beans, cooking versus mineral content,
    525
Neoglycoproteins, 360
Neohesperidin, 657
Neohesperidin dihydrochalcone, 565, 656,
    657
Neohesperidose, 565
Neoxanthin, 571, 572
Nernst layer for immobilizing enzymes, 391,
    392
Neryl pyrophosphate, 607
Net protein ratio, 321
Net protein utilization, 322, 323, 337
Neurolathyrism, 699
Neutralization of fat, 218
Neutralizers for food, 637
Neutralizing value of leavening acids, 632
Newtonian fluids, 289
Niacin, 481, 483, 488, 501, 511, 820 (*see also*
    Nicotinic acid)
  assay, 501
  forms, 501
  losses by trimming, 483
  in milk, 820
  RDA, 481
  stability, 501
  effect on vitamin B_{12}, 511
Nickel, 197, 219, 221, 222, 522, 668
  catalyst for hydrogenation of fat, 219, 221,
    222
  catalyst for lipid oxidation, 197
  function in food, 668
Nicotinamide, 501
Nicotinamide adenine dinucleotide, 501
Nicotinamide adenine dinucleotide phosphate,
    501
Nicotinic acid, 501, (*see also* Niacin)
Ninhydrin, 266, 267, 341
Nisin, 652
Nitrate, 437, 523, 524, 669
Nitric acid, 519
Nitric oxide hemochromogen, 556, 777, 781
Nitric oxide myoglobin, 556, 776, 777
Nitrimetmyoglobin, 554
Nitrimyoglobin, 554
Nitrite, 347, 437, 485, 492, 504, 555, 645,
    646, 669, 673, 691, 702, 703
  antimicrobial agents, 673
  in cured meat pigments, 555
  folates, effect on, 504
  formation in meat, 645, 646
  function in food, 669
  oxidizing action, 485
  proteins, reaction with, 347
  safety, 646, 702, 703
  and vitamin C, 492
  vitamins, effect on, 485
Nitrogen
  balance, 316, 317
  function in food, 668
  propellant, 667
  requirement in diet, 316
  solubility index, 288

Nitrogenase, 331
Nitrogen dioxide, 664
Nitrogen-fixing genes, 332
Nitrogen tetroxide, 664
Nitrogen trichloride, 347
p-Nitrophenyl group, 348
Nitrosamides from proteins, 347
Nitrosamines, 347, 492, 519, 646, 702, 703, 777
    inhibition by tocopherols, 519
    inhibition by vitamin C, 492, 519
    from meat curing, 777
    from proteins, 347
    safety, 702, 703
N-Nitroso compounds, 702, 703
S-Nitrosocysteine, 776
N10-Nitrosofolic acid, 504
Nitrosomyoglobin, 553, 554, 646
Nitrosoproline, 347
Nitrosopyrrolidine, 347
Nitrosyl chloride, 664
Nitrosylhemochrome, 554, 555
Nitrosylmetmyoglobin, 554
Nitrosylmyoglobin, 554, 555
Nitrous acid
    biotin, effect on, 513
    vitamin C, effect on, 492
Nitrous oxide, 667, 680
2-t-6-c-Nonadienal, 605, 606
2-t-6-c-Nonadienol, 606
Nonaethylene glycol monoester, 174
Nonanal, 183, 185
3-Nonenal, 184
6-Nonenal, 229
Nonprotein nitrogen in plant tissues, 863–865
No observed effect level (NOEL), 691
Nootkatone, 608
Nordihydroguaiaretic acid, 643
Norharman derivatives, 336
Nuclease, 711
Nuclei in muscle, 733, 734
Nucleic acids in single-cell proteins, 330
Nucleoproteins, 246
Nucleoside hydrolase, 757
Nucleoside phosphorylase, 757
5'-Nucleotidase in milk, 802, 813
Nucleotide pyrophosphatase, 752
Nucleotides
    changes in muscle postmortem, 753–758
    in muscle, 742
Nutrient composition of foods, inadequate information, 480
Nutrient loss
    and loss in sensory quality, 927
    from plant tissues during heating, 906
Nutrient retention, optimization, 527–531
    based on content during distribution, 529–531
    by HTST processing, 527–529
Nutrients added to food, 674–677
Nutrient stability in breakfast cereals, 487
Nutrient supplements, 675
Nutritive value of meat, 726–728
Nutritive value of plant tissue
    after dehydration, 909
    after freeze-preservation, 908

O

Ochratoxins, 708
2,6,Octadecadienal, 229
3,5,-Octadiene-2-one, 184
Octa-1,c-5-dien-3-one, 613, 615
Octafluorocyclobutane, 667
δ-Octalactone, 611
Octanal, 184
Octane, 183, 211
2-Octenal, 186, 211
2-Octene, 184, 186
Odors (see Aromas)
Oil extraction from plants, 327
Oils (see Lipids, Triacylglycerols)
Oleate
    autoxidation of, 178
    dimerization of, 189, 190
    products from thermal decomposition, 209
Oleic acid
    crystal structure of, 155
    function in food, 668
    hydrogenation rate, 220
    hydrogenation selectivity, 219, 220
    nomenclature, 142
    oxidation of, 185
    oxidation rate, 196
    unit cell, 154
Oleic-linoleic acid fats, 147
Oligosaccharides
    definition, 70
    as dietary fiber, 71
    laxative effect, 71
    nomenclature, 85
    reducing types, 85
    Schardinger dextrins, 87, 88
    synthesis, 85
    types, 85, 86
Olive oil, 150
Olives, 637
Optimization of mineral retention (see Nutrient retention, optimization)
Optimization of vitamin retention (see Nutrient retention, optimization)
Orange G, 270
Order of reaction
    enzyme inactivation, 408
    protein denaturation, 408
    and shelf life testing, 914–916
Organelles of plant tissue, 877, 878
Organic acids
    in plants postharvest, 888–890
    in plant tissue, 863, 866, 868
    sensory role in plant tissue, 888
Organoleptic evaluation of lipid oxidation, 196
Ornithine, 336, 337, 865
Ornithinoalanine, 336, 337
Orosomucoid, 246
Orthophosphate
    in casein, 803
    leavening agent, 634, 636
    in milk, 811, 812
    structure, 636

Ovalbumin, 247, 261, 274, 281, 294, 302, 307, 309, 332, 334, 465, 834, 835, 845, 850
  denaturation, 845
  foam stabilizer, 307
Overrun, 305, 306
Ovoflavoprotein, 835
Ovoglobulins, 837
Ovoglycoprotein, 835
Ovoinhibitor, 457, 458, 835, 837
Ovomacroglobulin, 835
Ovomucin, 833, 835–837, 845, 846, 850
Ovomucoid, 307, 834–836, 845
Ovomucoid inhibitor, 457, 458
Ovotransferrin (*see* Conalbumin)
Ovoverdin, 573
Oxalate, 520, 521, 667
Oxalic acid
  buffer effect in plants, 638
  in plant tissue, 866
Oxaloacetic acid, 610
β-N-Oxalyl-L-α,β-diaminopropionic acid, 694
Oxazole, 616
Ox bile extract, 677, 678
Oxidation (*see also* Lipid autoxidation, Protein oxidation)
  of amino acids, 270, 339–342
  of carotenoids, 573, 574
  fatty acid flavors from, 605
  of fatty acids, 605
  of proteins, 270, 335, 361
  value, 195
Oxidative deamination, 315
Oxidative rancidity in frozen pizza, 926
Oxidized lipids, safety, 713
Oxidizing agents, effect on vitamins, 485
Oxidizing-reducing agents, 670
Oxidoreductases, 433, 801, 802, 907
  in milk, 801, 802
  plant color, effect on, 907
Oxirane test, 195
Oxoacid, 183, 605
Oxoester, 183
Oxygen
  absorption for measuring lipid oxidation, 196
  concentration versus lipid oxidation, 197
  for controlled atmosphere storage of plants, 901, 902
  Pauli exclusion in, 181
  plant respiration, effect on, 884
  singlet state, 177, 181
  triplet state, 181
  vitamin C stability, effect on, 489
Oxygen bomb method, 196
β-Oxyindolylalanine, 341
Oxymyoglobin, 553, 554, 556, 760, 765, 778, 782
Oxystearin, 668
  tetracycline, 652
  in, 795

ducts, 355, 781, 782
760, 776

Palladium catalyst for fat hydrogenation, 222
Palmitic acid
  and coronary heart disease, 236
  function in food, 668
  nomenclature, 142
Palmitoleic acid, 142
1-Palmitoyl-2-linoleyl-sn-glycero-3-phospho-choline, 145
Palm oil, effects of refining on, 218
Pancreatic hypertrophy, 693, 694
Panose, 441
Pantetheine, 512
Pantoic acid, 512
Pantothenic acid, 485, 486, 488, 512
  alkaline conditions, effect of, 485
  assay for, 512
  distribution, 512
  lipid oxidation, effect of, 486
  processing, effect of, 512
  stability, 512
  structure, 512
Pantothenylaminoethanethiol, 512
Papain, 349, 356, 357, 400–402, 423, 427, 430, 668, 736, 737, 846
  function in food, 668
  to prevent gelation of egg yolk, 846
Paprika and paprika oleoresin, 580
Parabens, 649, 650
Paraffin, 679
Paralytic shellfish poisoning, 701
Paramyosin, 740
Parenchyma tissue, 752, 869, 871, 872, 876, 877
Pasteurization
  of egg products, 845, 846
  of milk, 814, 815
Peanut oil, fatty acids in, 150
Peanut proteins, 300, 301
Pectate lyases, 439, 440
Pectic acids, 124, 661
Pectic enzymes, 124, 439–441, 885
Pectic substances, 123–126, 439–441, 638, 642, 660, 661, 859, 872, 960
  during alkali peeling, 638
  chemical constituents, 123
  in parenchyma tissue, 872
  and plant texture, 660, 661
  in plant tissue, 859, 960
  properties, 123, 125, 126
  removal of calcium from, 642
  structure, 124
Pectin, 89, 124, 175, 658, 680, 859 (*see also* Pectic substances)
  in plant tissue, 859
  repeating units of, 89
  stabilizer, 175, 680
  texturizer, 680
Pectinase, 429
Pectin esterases, 439, 440, 441
Pectin gels (*see* Gels, pectin)
Pectinic acids, 124, 661
Pectin lyases, 439
Pectin methylesterase, 124, 661, 885, 886, 905, 907

[Pectin methylesterase]
activity after blanching, 905
and firmness of plant tissues, 661
and loss of "cloud", 907
in ripening fruit, 885, 886
Pectin transeliminases, 439
Peeling, alkali, 638
Pelargonidin, 557, 558, 566
Penicillium islandicum, 708
Penicillium roqueforti, 612, 646, 647
Penicillium viridicatum, 708
Pentadecane, 211
1,3-Pentadiene, 647
Pentaerythritol, 658
Pentaerythritol ester of rosin, 679
Pentane, 196
2-Pentanone, 611
Pentosans, 296
Pentose phosphate pathway, 879, 889
Pentoses, 71 (*see also* Sugars; specific sugars)
chiral center, 79, 80
degradation products of, 94
Pentose shunt in muscle, 742
Pentylfuran, 211
2-*n*-Pentylfuran, 227, 228
Peonidin, 557, 558
Pepper flavor, 599
Pepsin, 294, 313, 356, 357, 411, 416, 430,
458, 669
function in food, 669
pH optimum, 416
Pepstatin, 458
Peptide bond, 246, 254, 255
characteristics, 254
structure, 255
Peptides, absorption, 313
Performic acid, 269, 270, 359
Periderm of plant tissue, 877
Perimysium, 730, 731, 745, 746
Perinaphthenones, 575
Perishability of food
versus solute mobility, 60, 61
versus water activity, 55–60
Peroxidase, 281, 339, 409, 411, 413, 414, 438,
449, 451–453, 885, 891, 907, 908
carotenoids, effect on, 891
damage to frozen plant tissue, 908
plant color, effect on, 907
and protein crosslinking, 339
reactivation, 413, 414
in ripening fruit, 885
Peroxides, 194, 198, 228, 230, 486, 519
decomposers, 198
in flavor reversion of oil, 228
from heated or oxidized fat, 230
in lipid oxidation, 198
vitamins, effect on, 486
Peroxide value
of frozen pizza, 926
and rancid flavor, 193
test for lipid oxidation, 193
and totox value or oxidation value, 195
Peroxy radical
epoxides from, 188

[Peroxy radical]
in lipid oxidation, 200–202
malonaldehyde, source of, 194
reactions of, 189
Pesticide safety, 713, 714
Petroleum ether insolubles, a test for fat
quality, 212
Petroleum waxes, 668
Petunidin, 557, 558, 566
pH (*see also* Muscle pH)
amino acids, effect on, 321, 336
and Arrhenius plots, 918
control agents, 670
of eggs, 844
emulsions, effect on, 299, 300
flavor-binding by proteins, effect on, 312
foaming ability of proteins, effect on, 307
and food deterioration, 11
gelation, effect on, 354
protein crosslinking at alkaline pH, 336
protein denaturation, effect on, 351
protein dissociation, effect on, 350
protein hydration, effect on, 286, 287
protein gelation, effect on, 293, 294
protein racemization, effect on, 335
protein solubility, effect on, 272, 274, 289
thiamine, effect on, 494, 498
Phages, 462
Phase diagram for lipids, 160, 161
Phaseollin, 465, 695, 698, 699, 898
a stress metabolite, 898
structure, 699
Phenanthraquinones, 569
Phenazines, 575
2-Phenethanol, 606
Phenolase, 333, 347, 374, 380, 413, 416, 418,
424, 445, 446, 448, 453, 562, 885 (*see
also* Polyphenoloxidase)
activation during cell disruption, 380
activity at low water activity, 418
anthocyanins, effect on, 562
control of activity, 416
in ripening fruit, 885
Phenol hydroxylase, 446
Phenols, 424, 437, 446
Phenoxazone, 575
Phenylacetaldehyde, 619
Phenylalanine, 251–254, 611, 655, 656
in aspartame, 655, 656
flavors from, 611
hydrophobicity, 252
structure, 252
2-Phenylethyl isothiocyanate, 603
Phenylhydrazine, 492
Phenylisothiocyanate, 271
4-Phenylspiro[furan-2(3H),1'-phthalan] 3,3'-
dione, 267
Phenyl thiocarbamide, 591
Pheophorbides, 547–549, 582
copper derivative of, 582
description of, 549
Pheophytins, 182, 547–549, 582
from chlorophyll, 549
copper derivative of, 582

[Pheophytins]
and formation of singlet oxygen
properties of, 548
Phlobaphens, 562
Phloem, 869, 871, 876
Phloridzin, 565
Phorbide, 546
Phorbin, 546
Phosgene, 268
Phosphatase, acid, 414
Phosphatase, alkaline, 414, 438
Phosphatases, 356, 414, 438, 885
Phosphates, 520, 521, 526, 640–642, 677–679
chelating properties, 642, 677
emulsion stability, effect on, 641
as food additives, 640, 641
texture control agents, 678, 679
water holding capacity of muscle, effect on, 641
Phosphatidylcholine, 145, 175, 742, 799
Phosphatidylethanolamine, 175, 742, 799
Phosphatidylinositol, 742
Phosphatidylserine, 742
Phosphodiesterase, 813
Phosphofructokinase, 752, 888
Phospholipases, 74, 355, 443, 444, 759, 771, 890
Phospholipids, 145, 146, 235, 727, 742, 743, 759, 766, 771, 799, 813, 832, 833, 842–845
and coronary heart disease, 235
definition and nomenclature, 145
in egg yolk, 832, 833
glycerophospholipid type, 145, 146
lecithin, 145
in lipoprotein of egg yolk, 842
in lipovitellins, 842
in low-density lipoproteins of egg yolk, 844, 845
in milk, 799, 813
in milk fat globule membrane, 813
in muscle, 727, 742, 743
in myelin figures of egg yolk, 843
phosphoacylglycerol, 146
sphingomyelins, 146
Phosphoprotein phosphatase, 802
Phosphoproteins, 246, 841
Phosphoric acid, 200, 202, 416, 447, 521, 632, 670, 677
acidulant, 670
chelating agent, 202, 677
food additive, 632
Phosphorus, 481, 522, 524, 801, 820, 830, 841, 842
in eggs, 830, 841, 842
in milk, 801, 820
RDA, 481
Phosphorylase, 752, 885, 887
in potatoes, 887
in ripening fruit, 885
Phosphoserine, 815
Phosphoseryl residues in milk, 797, 803, 815, 822
Phosvitin, 841, 842
Phytase, 445, 527
Phytates, 520, 527, 328, 521

Phytoalexins, 694, 698, 699
Phytochrome, 575
Phytoglycogen, 121
Phytohemagglutinins, 333
Phytol
from chlorophyll, 549
hydrophobic nature of, 548
structure of, 547
Phytolaccanin (*see* Betanin)
Phytosterols
and coronary heart disease, 237
hydrogenation, effect of, 237
serum cholesterol, effect on, 237
Pickles, 661
Pig fat fatty acids, 150, 151
Pigment metabolism, 891, 892
Pigments
binding to proteins, 312
defined, 546
in plant tissues, 867
in production of singlet oxygen, 181, 182
Pigments, cured meat
formation of, 555
processing conditions, effect of, 555
Pimaricin, 648
Pipecolic acid, 864
Piperine, 599
Piperonal, 599
Pisatin, 695, 698, 699
Pizza, frozen
loss of crust crispness, 927
shelf life testing, 924–929
tests for monitoring quality, 927
pK_a values of amino acids, 249, 252
Plant aromas, 863
Plant biochemistry, postharvest (*see also* Plant tissue, postharvest)
aroma compounds, 892
cell wall changes, 885, 886
compared to muscle biochemistry postmortem, 751, 752
control mechanisms, 892–896
ethylene, 881
gene expression, 885
metabolic enzymes, 889
metabolic oxidation, 881
metabolic pathways, 879
metabolism of lipids, 888, 891
metabolism of organic acids, 888–890
pigment metabolism, 891, 892
protein synthesis, 885
starch–sugar transformations, 886–888
Plant color
dehydration, effect of, 909
freezing, effect of, 907, 908
heat, effect of, 906
Plant flavor, effect of heat, 905–907
Plant hormones, 892–896
Plant lipids, 863, 866, 867 (*see also* Lipids, Triacylglycerols)
Plant nutritive value
dehydration, effect of, 909
freezing, effect of, 908
heat, effect of, 906

Plant proteins, 860–863
Plant respiration, postharvest
    chilling injury, effect of, 899, 900
    climacteric, 881, 882
    controlled atmosphere storage, effect of,
        901, 902
    cycloheximide, effect of, 886
    and ethylene, 881, 884, 866
    heat-generated, 881
    mechanical injury, effect of, 895
    metabolic pathways, 879
    O_2 and CO_2, effect of, 884
    patterns, 881–883
    Q_{10} value, 883
    rate versus storage life, 879, 880
    respiratory quotient, 879, 880
    temperature, effect of, 883, 884
Plant structure
    cell, 878
    fruits, 872, 874
    leaf crops, 872
    organelles, 887, 878
    parenchyma tissue, 872, 876
    protective tissues, 874, 876, 877
    roots, 869, 871
    stems, 872, 873
    subcellular, 877, 878
    supporting tissue, 876
    vascular tissue, 876
Plant taste, 863
Plant texture
    dehydration, effect of, 909
    freezing, effect of, 907
    heat, effect of, 905
Plant tissue, postharvest (*see also* Plant bio-
        chemistry, postharvest)
    carbohydrates, 858, 861
    chemicals, effects of, 904
    controlled atmosphere storage, 900–902
    dehydration, effects of, 908, 909
    freeze-processing, effect of, 906–908
    humidity, effect of, 903
    hypobaric storage, 901, 902
    ionizing irradiation, effect of, 903
    mechanical injury, 895–898
    modified atmosphere storage, 901
    storage temperature, effect of, 898–900
    thermal processing, effect of, 904–906
    tolerance to O_2 and CO_2, 901, 902
    water content, 858, 859
Plant vitamins, 868, 870
Plasma membrane, 733
Plasma proteins, 316
Plasmids, 462, 463
Plasmin, 797
Plastein reaction, 357, 401
Plastids, 859
Platinum for hydrogenation, 222
Plumbagin, 569, 570
Polar compounds in fat, 212
Polyacrylic acid, 350
Polyamide clarifying agents, 662, 663
Polybrominated biphenyls, 715, 716
Polychlorinated biphenyls, 715, 716

Polycyclic aromatic hydrocarbons, 713
Polyethylene glycol, 659
Polygalacturonase, 124, 439, 440, 459, 885,
        886
Polygalacturonase inhibitor, 459
Polygalacturonic acid, 439
Polygalacturonopyranoside, 439
Polyglycerol, 659
Polyglycerol esters, 678
Polyhydric alcohols, 658–660
Polymerization of oxidizing lipids, 188
Polymers from heated or oxidized fats, 210,
        231
Polymorphism of lipids, 152–158
    chocolate bloom, 158
    enantiotropic forms, 153
    functional properties, effect on, 157, 158
    monotropic form, 153
    transition point, 153
Polyols, 360
Polyoxyethylene sorbitan esters, 678
Polyoxyethylene sorbitan monooleate (Tween
        80), 171
Polyoxyethylene sorbitan monostearate (Tween
        60), 171
—Polyphenoloxidase (*see also* Phenolase)
    frozen plant tissue, effect on, 907
    plant color, effect on, 907
    in plant tissues, 897
Polyphenols, 328, 329, 339, 347, 562, 598,
        661–663
    anthocyanins, reaction with, 562
    astringency, 598
    in beer, 663
    and haze, 662
    proteins, reaction with, 347
Polyphosphates, 350, 780, 815, 817
    Ca^{2+} complexing, 350
    in milk, 815, 817
    proteins, effect on, 350
Polyproline structure of proteins, 257
Polysaccharide degradation products
    anhydro sugars, 96, 97
    by transglycosylation, 96
Polysaccharide gums (*see* Gums, plant)
Polysaccharides (*see also* Carbohydrates)
    definition, 70
    as dietary fiber, 71, 73, 74
    digestibility, 108
    function in foods, 108, 109, 111
    gel formation, 110–112
    laxative effect, 71
    linear and branched, 111
    nomenclature, 71
    properties, 88
    retrogradation of, 110
    size, 71
    solubility of, 109, 112
    solution viscosity, 110
    structure, 71, 72
    syneresis of, 110
    terminology, 87, 88
    types of, 89
Polysorbate liquid crystals, 169

Polyunsaturated fat
  and cancer, 237
  and ceroid production, 237
  and coronary heart disease, 235–237
  and free radical damage, 237
  and serum uric acid, 237
  and skin lesions, 237
Polyvinyl pyrrolidone
  clarifying agent, 662, 663
  function in food, 669
  structure, 663
Pork composition, 727
Pork flavor, 613
Porphine, 546
Porphyrin, 546, 551, 641
  chelating properties, 641
  component of myoglobin, 551
Porphyrinogens, 546
Potassium, 522, 524, 801
Potassium acid tartrate
  buffer salt, 671
  for hydrolysis of sucrose, 631
  leavening agent, 632, 634–636, 670
Potassium aluminum sulfate, 661
Potassium bicarbonate for leavening, 632, 671
Potassium bromate, 664
Potassium chloride, 669
Potassium citrate
  buffer salt, 671
  chelating agent, 677
Potassium hydroxide, 671
Potassium iodate, 522, 664
Potassium phosphate
  buffer salt, 671
  malting-fermenting aid, 669
Potassium sorbate, 778
Potatoes
  processing versus copper content, 524
  reconditioning, 888
  removal of sugars, 888
  starch–sugar transformations, 886–888
Potatoes, dehydrated, and shelf life testing,
  929–936
Poultry (*see* Muscle)
Poultry flavor, 613
Pregelatinized starch, 118
Pregnenolone, 613
Prenylmercaptan, 592
Proanthocyanidins
  astringency, 566
  in browning, 566
  components of, 566
  and haze in beer and wine, 566, 662, 663
  properties, 575
  terminology, 565
Process flavors, 615–620
Processing versus nutrients (*see* Specific nutrients)
Prochymosin, 465
Product accumulation versus water activity, 418
Prolamins, 274
Proline, 251–253, 255, 746, 747, 750, 797, 862
  in collagen, 746, 747, 750
  in gluten, 862
  in milk, 797

Proline hydroxylase, 747
Propane, 194
Propellants for food, 667, 680
S-(1-Propenyl)-L-cysteine sulfoxide, 601, 603
S-(2-Propenyl)-L-cysteine sulfoxide, 602, 603
1-Propenyl-sulfenic acid, 601
Propionibacterium shermanii, 649
Propionic acid, 649, 673
Propionyl-CoA, 612
Propylene glycol
  antioxidant carrier, 205
  solvent, 672
  texturizer, 658, 659
  water activity, effect on, 660
Propylene glycol esters, 678
Propylene glycol monoester, 174
Propylene glycol monostearate, 171
Propylene oxide
  antimicrobial agent, 651, 652, 673
  proteins and nucleic acids, effect on, 485
  structure, 651
  vitamins, effect on, 485
Propyl gallate
  antioxidant properties, 200, 203, 204, 643,
    673
  structure, 199
Prostaglandins, 229
Prosthetic group, 246
Protease inhibitors, 455, 692–694
Proteases, 258, 333, 356, 358, 362, 418, 430,
    431, 664, 693, 748, 801, 802
  in flour, 418, 664
  in milk, 801, 802
Protective tissue in plants, 874
Proteinaceous emulsifiers (*see* Emulsifiers, proteinaceous)
Proteinaceous foaming agents (*see* Foaming
  agents, proteinaceous)
Protein aggregation
  in comminuted meat products, 779
  definition, 290, 291
  by dehydration, 351
  after denaturation, 275
  during foaming of egg albumen, 849, 850
  heat, effect of, 273, 286, 352, 353
  during heating of eggs, 850
  ions, effect of, 354
  by mechanical treatments, 351
  versus solubility, 288
Protein anabolism, 315
Protein antinutritional factors
  heat inactivation, 333, 334
  in leguminosae, 333, 334
  loss during separation processes, 334
  in soybeans, 333, 334
Proteinase, acid, 442
Protein association, definition, 290
Protein binding, effect on amino acid availability, 320
Protein-calorie malnutrition, 319
Protein catabolism, 315
Protein coagulation
  definition, 291
  mechanism, 293

Protein concentrate, 327, 329
Protein conformation
    amino acid availability, effect on, 320
    hydrophobic interactions, 45, 46
    ions, effect of, 43
    and water structure, 45, 46
Protein crosslinking
    at alkaline pH, 336, 337
    breakage and release of H_2S, 277
    in collagen, 749
    disulfide type, 363
    in fish, 759, 760
    by formaldehyde in fish, 771
    formation of dityrosine residues, 339
    by gamma irradiation, 338
    by halogenated solvents, 347
    and heating, 338
    by hydrogen peroxide, 339
    intentional, 362, 363
    in vivo, 355
    with malonaldehyde, 346
    in milk, 815, 824
    in muscle by oxidized fatty acids, 764
    by oxidizing lipids, 338, 346
    by peroxidase, 339
    by polyvalent ions or polyelectrolytes, 350
    reagents for, 363
    and storage of dry muscle, 775, 776
Protein degradation products, 335, 336, 338–346, 355
    from acid hydrolysis, 355
    from alkaline pH, 336, 355
    from gamma irradiation, 338, 339
    from heat, 338
    from hydrogen peroxide, 339
    from Maillard browning, 342–346
    from oxidation, 339–342
    from oxidizing lipids, 338, 346
    from peroxidase, 339
Protein denaturation
    by acids and bases, 278
    activation energy, 281
    in baked goods, 117
    in bread, 298
    causes, 274–279
    chemical alterations, 277
    cold, effect of, 277, 355
    collagen, 748
    in comminuted meat products, 779, 780
    of conalbumin, 836
    contractile proteins, 760
    definition, 274, 403
    differential scanning thermograms, 353
    drying of muscle, effect of, 774, 775
    effects, 274, 275
    egg proteins, 846, 849–851
    energetics, 279–282
    enthalpy, 354
    flavor-binding by proteins, effect on, 312
    foaming, effect of, 351
    freezing, effect of, 769–772
    in gelation, 292, 293
    in heated milk, 815
    in heated muscle, 772, 773

[Protein denaturation]
    homogenization of milk, effect of, 814
    hydrostatic pressure, effect of, 277, 421
    interfaces, effect of, 275–278, 351
    ionizing irradiation, effect of, 277, 422
    measurement, 275
    mechanical treatments, 277
    metal ions' protection against, 278, 279
    organic compounds, effect of, 279
    organic solvents, effect of, 279, 356
    ovalbumin, 834, 845
    ovomucin, 837
    ovomucoid, 836
    pH alteration, effect of, 351, 416
    protein-lipid interaction, effect of, 274
    rate, 281
    reducing agents, effect of, 279
    reversibility, 275, 279, 280
    shearing, effect of, 421
    temperature alteration, effect of, 273, 275, 277, 286, 352, 355, 402–416
    thermodynamics, 280, 281, 403–407
    versus solubility, 288, 351
    water activity, effect of, 275, 354
    in whey, 816, 819
Protein digestibility, 324
Protein digestion, 313, 314
Protein efficiency, 323
Protein efficiency ratio, 321, 324, 325, 328, 334, 337, 338, 346, 362
    alkali and heat, effect of, 337
    of methyl casein, 362
    protein oxidation, effect of, 346
    of soybean meal versus heat, 334
    of vegetables and cereals, 328
Protein, 55,000 dalton component, 741
Protein film, 294
Protein flocculation
    definition, 291
    mechanism, 293
Protein flours, 329
Protein functionality (*see also* specific properties)
    chemical modifications, effect of, 358–362
    classification, 282
    deamidation, effect of, 358
    definition, 282
    dehydration, effect of, 351, 354
    desirable properties, 818
    in dough formation, 296–298
    evaluation methods, 282, 283
    heat, effect of, 351–355
    ion exchange, effect of, 349
    low temperature, effect of, 355
    mechanical treatments, effect of, 351
    of milk proteins, 817–819
    pH, effect of, 350, 359
    plastein formation, effect of, 358
    polyvalent ions, effect of, 350
    precipitation, effect of, 349
    proteolysis, effect of, 357, 358
    requirements in various foods, 283
    solvents, effect of, 350, 351
    water removal, effect of, 349

Protein gelation (*see also* Gelation)
  amount of water entrapped, 292
  bonding forces, 292–294
  calcium ions, effect of, 354
  classification, 294
  cold, effect of, 355
  conditions for, 291, 292
  definition, 290, 291
  of egg proteins during heating, 850, 851
  emulsion stability, effect of, 300
  heat-set gelation, 293
  hydrogen bonds, 293
  hydrophobic associations, 293
  importance, 291
  mechanisms, 293
  in milk, 812, 815, 816
  from milk proteins, 819
  pH. effect of, 293, 294, 354
  polysaccharides, effect of, 292
  protein concentration, effect of, 292
  protein solubility, effect of, 288
  proteolysis, effect of, 357
  reduction, effect of, 363
  solubility, relation to, 292
  solvents, effect of, 351
  temperature, effect of, 273, 286
Protein gels in meat, 779, 781
Protein hydration (*see also* Water holding
    capacity)
  in casein micelle, 812
  dehydration, effect of, 286, 351
  in emulsification, 285
  in foaming, 285
  freezing, effect of, 355
  in frozen fish, 771
  gelation, impediment to, 292
  hydrolysis, effect of, 357
  ions, effect of, 287
  malonaldehyde, effect of, 345
  mechanical treatments, effect of, 351
  methods for determining, 285
  pH, effect of, 286, 287
  polyphosphates, effect of, 350
  protein solution viscosity, effect on, 290
  relation to other functional properties, 287
  sequence of, 285
  solvents, effect of, 351
  temperature, effect of, 286, 287, 351, 352
  terms relating to, 283, 285
  versus time, 285
Protein hydrophobicity, 261, 309
  versus bitterness, 265, 594
  calculation, 594, 595
  of casein, 265, 266
  emulsifying ability, effect on, 399–302
  equation, 265
  gelation, effect on, 294
  intentional chemical modification, 358,
    360, 361
  protein-lipid interaction, effect on, 274
  and side chains, 263
  versus structure, 265
  versus volume of hydrophobic groups, 266
Protein isolates, 327, 329

Protein-lipid interactions
  homogenization, effect of, 274
  versus isoelectric point, 274
  protein hydrophobicity, effect of, 274
Protein metabolism, 313–316
Protein nutritive value, 323
  acid hydrolysis, effect of, 355
  acylating agents, effect of, 348
  alkaline hydrolysis, effect of, 355
  alkaline pH, effect of, 337
  amino acid anhydrides, effect of, 348, 349
  crosslinking, effect of, 338
  determination, 321–325
  factors influencing, 319, 321
  halogenated solvents, effect of, 347
  heat, effect of, 333, 337
  intentional chemical modification, effect of,
    358, 359, 362, 363
  lipids, effect of, 346
  Maillard browning, effect of, 343–345
  of milk, 821
  of muscle, 726–728
  nitrites, effect of, 347
  oxidation, effect of, 340–342
  plastein formation, effect of, 358
  processing, effect of, 332
  proteolysis, effect of, 357
  racemization, effect of, 335
  sulfites, effect of, 348
  of vegetables and cereals, 328
Protein oxidation
  amino acid residues, effects on, 339–342
  and dityrosine residues, 339
  by heating, 335
  by hydrogen peroxide, 339
  intentional, 359, 361
  by irradiation, 339
  by oxidizing lipids, 191, 193
  by peroxidase and H_2O_2, 339
  by peroxides, 339
  by photoxidation, 339
  by quinones, 347
  by sulfites, 339
Protein precipitation
  definition, 290, 291
  at low temperature, 355
  by solvents, 351
  temperature, effect of, 273
Protein polymerization, definition, 290
Protein quality, 319
Protein reactions (*see also* Proteins, intentional
    chemical modification)
  acetylation, 348, 360, 361
  acylation, 269, 359–361
  with aldehydes, 269, 342–346
  at alkaline pHs, 336, 338
  alkylation, 359
  amidation, 268, 360
  with amino acid anhydrides, 348
  with amino acids, 348
  binding of polyols, 360
  carbamylation, 360
  with carbonyls, 342–346
  with carboxyanhydrides, 348

[Protein reactions]
carboxymethylation, 359, 361
with *p*-chloromercuribenzoate, 269
citraconylation, 359
crosslinking, 362, 363
with dansyl chloride, 269, 271
deamidation, 358
deamination, 269
decarboxylation, 268
with Ellman's reagent, 269
enzyme catalyzed, 355–358
esterification, 268, 361
with fluorescamine, 267, 269
with 1-fluoro-2,4-dinitrobenzene, 270
with formaldehyde, 345
glycosylation, 795
glycosylation of casein, 806
guanidination, 360
with halogenated solvents, 347
by heating, 335, 336, 338
hydrolysis, 355
in vivo, 355, 356
with iodoacetic acid, 269, 270
with ionizing radiation, 338, 339
with lipid free radicals, 346
with lipids, 346
maleylation, 359
with malonaldehyde, 346
with metal ions, 278, 279
with *o*-methyl isourea, 270
with nitrites, 347
with Orange G, 270
oxidation, 270, 339–342, 359, 361
with oxidizing lipids, 338, 339
with performic acid, 270
with phenylisothiocyanate, 269, 271
phosphorylation, 359, 795, 803
photoxidation, 339
plastein formation, 357, 358
with polyphenols, 347
proteolysis, 356–358
with quinones, 347
reduction, 268, 359, 361, 363
reductive alkylation, 360, 361
succinylation, 359, 360, 362
with sulfites, 348
with various agents, 339
Protein requirements in diet, 316–319
according to various organizations, 317–319
consequences of exceeding or not meeting, 317
factors influencing, 317
Proteins, 246, 247 (*see also* Cereal proteins, Egg
proteins, Enzymes, Milk proteins, Muscle
proteins)
and aldehydes, 331
amino acids in, 254 (*see also* Amino acids)
bakery products, added to, 298
bentonite, absorption to, 662
beta bends, 256
biological activity, 247
bitterness, 863
carotenoids, association with, 572, 573
from cell cultures, 331

[Proteins]
chelating properties, 641
chemical synthesis, 331
classification, 246
crosslinking (*see* Protein crosslinking)
in dairy foods, 821 (*see also* Milk proteins;
specific proteins)
dyes, binding of, 312
in eggs, 830–837 (*see also* Egg proteins;
specific proteins)
as emulsion stabilizers, 169 (*see also* Emulsifiers, proteinaceous)
flavor binding, 310–312
foam formation, interactions during, 309
as foaming agents, 304–310 (*see also*
Foaming agents, proteinaceous)
foaming properties, assessment of, 306
genetic engineering, 331
hydrolysis, 356, 357
ion exchange, effect of, 349, 350
ionizing irradiation, effect of, 338
iron bioavailability, effect on, 526
hydrophobicity, 261, 263 (*see also* Protein
hydrophobicity)
interfacial tension of w/o system, 309
ionizable groups, prevalence of, 262
isoelectric point, 774, 775 (*see also* Isoelectric point)
isoionic point, 262
lipid-binding ability, 303
metals, binding of, 263
in milk, 796, 822
in milk products, 823 (*see also* Milk
proteins)
molecular weights, 261
movements in body, 314–316
needs in world, 325
oxidation (*see* Protein oxidation)
pigments, binding of, 312
pK_a values of ionizable groups, 262
in plant tissue, 860–863
production, ways to increase, 325, 327,
329–332
in pulses, 862, 863
qualities desired in food, 247
racemization, 335
reactions (*see* Protein reactions)
salting in and out, 420
single-cell, 330, 331
solubility in water, 271–274
stabilization by disulfide crosslinks, 263
stabilization by ions, 263
structural properties, 254–265
structure (*see* Protein structure)
surface tension of aqueous solutions, 309
tannins, interaction with, 331
TBA test, interference with, 194
thiamine, protective effect on, 496, 498
thiol-disulfide interchange, 263, 401, 402,
808, 815, 834
titration curves, 262
toxicants, binding of, 312
toxic types, 692–695, 700, 711
in tubers, 862

[Proteins]
  ultrafiltration, effect of, 349
  in vegetables and cereals, 327, 328
  volume, 266
  volumic mass, 266
  world production, 325, 326
  yield per hectare, 325, 326
Proteins, enzymically modified
  in vivo, 355, 356
  plastein formation, 357, 358
  by proteases, 356
Proteins, intentional chemical modification
  acetylation, 360, 361
  acylation, 359–361
  alkylation, 359
  amidation, 360
  carbamylation, 360
  carboxymethylation, 359
  citraconylation, 359
  crosslinking, 362, 363
  deamidation, 358
  esterification, 361
  guanidination, 360
  maleylation, 359
  oxidation, 359, 361
  phosphorylation, 359
  polyols, binding of, 360
  reduction of, 359, 361, 363
  reductive alkylation, 360, 361
  succinylation, 359, 360, 362
Protein solubility
  at alkaline pH, 350
  classification based on, 273, 274
  alkaline pH, effect of, 338
  deamidation, effect of, 358
  dehydration, effect of, 351, 354
  denaturation, effect of, 274, 275
  dry muscle, loss in, 775, 776
  electrical charge, effect of, 272
  emulsifying ability, effect on, 299, 300
  flavor binding, effect of, 311
  functionality, effect on, 288
  heat, effect on, 288, 333
  and Hofmeister series, 273
  hydrolysis, effect of, 355–357
  intentional chemical modification, effect of,
    359, 360, 363
  ionic strength, effect of, 273
  ions, effect of, 350
  at the isoelectric point, 272, 350
  maleylation, effect of, 272
  malonaldehyde, effect of, 345
  measure of, 288
  mechanical treatments, effects of, 351
  nonaqueous solvents, effect of, 273
  pH, effect of, 272, 274, 288, 350
  and protein extractability, 329
  salting in and out, 273, 420
  sodium chloride, effect of, 350
  sodium dodecylsulfate, effect of, 272
  solvents, effect of, 351
  succinylation, effect of, 272
  temperature, effect of, 273
  texturization, loss during, 296
  thermodynamics, 271

Protein solution viscosity
  equation for, 288, 289
  factors influencing, 289, 290
  importance in foods, 290
  non-Newtonian, 289
  shear thinning, 290
  solubility, relation to, 290
  thixotropic behavior, 290
  yield stress, 290
Proteins, plant (*see* Plant proteins)
Protein stability at interfaces, 423
Protein stabilization during freezing, 771, 772
Protein structure, 261
  beta bends, 256, 258
  beta-pleated sheets, 263
  beta structure, 256
  bonding types, 258
  caseins, 803–808, 811, 812
  chemical modification, 358
  chloro-2-ethanol, effect of, 279
  cold, effect of, 277
  covalent bonding, 264, 265
  disulfide crosslinks, 263, 265
  electrostatic interaction, 262–265
  and functional properties, 358
  and gelation, 292, 293
  heat, modification by, 333
  α-helix, 255, 258, 263
  γ-helix, 256
  π-helix, 256
  3_{10} helix, 256
  hydrogen bonding, 263–265
  hydrophobic groups, location of, 258
  hydrophobic interactions, 263–265
  hydrostatic pressure, effect of, 277
  intentional chemical modification, 359
  interactions, 265
  interfaces, effect of, 275–278
  ion exchange, effect of, 349, 350
  ionizing irradiation, effect of, 277
  mechanical treatments, effect of, 277, 351
  in muscle, 258–261
  native, 255
  organic compounds, effect of, 279
  pH, effect of, 278, 350
  β-pleated sheets, 256
  polar groups, location of, 258
  polyproline, 257
  precipitation, effect of, 349
  primary, 246, 254–257
  quaternary, 246, 258–260, 263
  random coil, 256, 258
  reducing agents, effect of, 279
  secondary, 246, 254–257, 262, 263
  side chains, effect of, 256
  solvents, effect of, 351
  stabilization by lactose, 820
  temperature, effect of, 273, 351–353
  tertiary, 246, 258
  van der Waals interactions, 262, 264, 265
  water removal, effect of, 349
  weakness, 275
Protein texturization
  by fiber formation, 294, 295
  and glutaraldehyde, 296

[Protein texturization]
    importance in foods, 294
    by mechanical treatments, 351
    nutritive consequences, 296
    and protein insolubilization, 296
    and temperature, 354
    by thermal coagulation, 294
    by thermoplastic extrusion, 295, 296
Protein toxins, heat inactivation, 333
Protein turnover, 315, 316
Protein water holding capacity (*see* Protein hydration, Water holding capacity)
Protein-water interactions
    bonding forces, 271, 272
    denaturation, effect of, 275
    extent, 271
    factors influencing, 285–287
    freezing, effect of, 355
    gelation, impediment to, 292
    and Hofmeister series, 273
    ionic strength, effect of, 273
    measurement, 285
    nonaqueous solvents, effect of, 273
    pH, effect of, 272, 274, 286
    protein denaturation, effect of, 274
    protein solution viscosity, effect of, 290
    schematic representation, 272
    sequence of, 284
    and solubility, 271–274, 288
    temperature, effect of, 273
    terms relating to, 283, 285
Protein wholesomeness
    halogenated solvents, effect of, 347
    heat, effect of, 335, 336
    intentional chemical modification, effect of,
        358, 363
    and lysinoalanine, 338
    Maillard browning, effect of, 345
    nitrites, effect of, 347
    after oxidation, 341
    processing, effect of, 332
Proteolysis
    in animal muscle postrigor, 356
    bitterness, effect on, 357
    and coagulation, 356
    and flavor binding by proteins, 312
    measuring extent, 356
    solubility, effect on, 355
Proteose-peptones, 796, 797, 799, 803, 806
Protocatechuic acid, 447, 448
Protochlorophyll, 548
Protochlorophyllides, 548
Protopectin, 124
Protopectinase, 124
Pseudomonas perolens, 604
Pseudomonas tetrolens, 604
D-Psicose, 93
Psoralins, 694, 698
Pteridine, 504
Pterin, 504, 505
Pterin-6-carboxaldehyde, 504, 506
Pterin-6-carboxylic acid, 504
Publication credibility, 15
Puddings, 637

Puffer fish poisoning, 701
Pullulanase, 433, 466
Pulse proteins, 862, 863
Pungency, 599, 600
Pyrazine, 616
Pyrazine, 2,5-dimethyl-3-ethyl, 617
Pyridine, 616
Pyridine, β-carboxylic acid, 501
Pyridine ring, 501
Pyridosine, 343
Pyridoxal, 501–504
Pyridoxal phosphate, 501
Pyridoxamine, 501–504
4-Pyridoxic acid, 502
Pyridoxine, 487, 488, 501
Pyridoxol, 501, 503, 504
bis-4-Pyridoxyldisulfide, 502
Pyrimidine ring, 494, 496, 498
Pyrolysis products, safety, 713
$\dot{\gamma}$-Pyrones, 575, 616
Pyropheophorbides, 548
Pyrophosphatase, 636
Pyrophosphates, 634
Pyrroles, 546, 575, 616
Pyrrolidine, 248
Pyrrolidone carboxylic acid, 550
Pyrrolizidine alkaloids, 695, 698, 699
    structure, 699
Pyruvate
    in garlic flavor, 603
    in onion flavor, 601
    structure, 603
Pyruvate decarboxylase, 888
Pyruvate kinase, 887, 888
Pyruvic acid, 610

Q

Quality attributes of food, 7–13
Quercetin, 563–565
Quercetin-3-rhamnoglucoside, 565
Quinazoline, 341
Quinic acid, 448, 866
Quinine, 591
Quinoidal base structure, 559, 561
Quinones, 201, 339, 347, 519, 520, 569, 575,
    643
    colors of, 569
    description of, 569
    oxidation, 643
    properties of, 575
    structure, 520
Q_{10} value, 412, 413, 883, 919–922, 925, 929
    defined, 919
    for plant respiration, 883
    for quality loss in frozen foods, 925
    for shelf life prediction, 920-922
    for vitamin C loss in frozen vegetables, 929
    z value, relation to, 413

R

Racemization
    of amino acids, 637
    of proteins, 335

Radiant energy, influence on lipid oxidation, 198
γ-Radiation, effect on chlorophylls, 549
Radionuclides, 716
Rancidity of fat, 227
Random coil structure of proteins, 256, 258
Randomization of fat (*see* Interesterification of fat)
Rapeseed oil, 149
Reaction flavors, 615–620
Reaction kinetics, independence of $\Delta G°$, 378
Reaction rate versus solute mobility, 418
Reactions, energy barrier to 378, 379
Reactions of amino acids (*see* Amino acid reactions)
Reactions of proteins (*see* Protein reactions)
Recombinant DNA, 427
Recombinant DNA technology
  applications, 464, 465
  clones, identification of, 462, 463
  cloning in plants, 464
  cloning strategies, 460–463
  and conversion of biomass, 465
  DNA fragments, obtaining, 461
  DNA inserted into vector, 462
  DNA introduced into host cells, 462
  microorganisms used, 459, 460, 463
  perspective, 465
  for protein production, 331, 332, 465
  and restriction endonucleases, 463
  for synthesis of food ingredients, 332
Recommended dietary allowances, 479, 481
Red fibers, 738, 743–745, 768
Red muscle, 751, 759
Red tide, 701
Reduction of proteins, 361
Reductive alkylation of proteins, 361
Reductone, 343
Refining of fat, 217, 218
Refractive index, change during hydrogenation of fat, 219
Regulatory agencies, 690
Relative humidity (*see* Humidity)
Relative humidity method for determining protein hydration, 285
Relative net protein ratio, 322
Relative protein values, 322
Relaxing factor, 738
Release and antistick agents, 671, 672
Rennet, 421, 431, 631 (*see also* Chymosin)
Rennin, 669
Resolution of rigor, 754, 755
Respiration (*see* Plant biochemistry, postharvest; Plant respiration, postharvest)
Respiration, postmortem versus postharvest 751, 752
Respiratory quotient of plants postharvest, 879, 882, 888
Restoration of nutrients, 486
Restriction endonuclease, 460, 463
Retinene, 573
Retinol equivalent, 480
Retrogradation
  of polysaccharides, 110
  of starch in bread, 665
  of starch gel, 115, 118

Reverse transcriptase, 462
Reyes syndrome, 699, 710
Rhamnose
  anthocyanins, component of, 558
  flavonoids, component of, 563
  in naringin, 565
  in plant pectins, 859
Rhodanese, 85, 697
Ribitol, 500
Riboflavin, 341, 481, 483, 485, 488, 500, 504, 575, 580, 820
  assay, 500
  folates, effect on, 504
  forms, 500
  losses during cooking, 485
  in milk, 820
  properties of, 575
  RDA, 481
  stability, 500
  structure, 500
Ribonuclease, 280, 281, 407, 802
5'-Ribonucleotides, 596, 617
Ribose, 509, 756, 757, 773, 776
D-Ribose oxidase, 104
Ribose-1-phosphate, 757
Ribosomes, 795
Ricin, 247, 693
Rigor mortis, 753, 767
Rigor onset, 754
Rishitin, 898
RNA, 461
Root structure, 869, 871
Rotenone, 565
R/S system, 143, 144
Ruhemann's purple, 267
Rutin, 565, 677
Rutinose, 565

S

Saccharifying enzyme, 441
Saccharin, 589, 590, 596, 598, 654, 677, 678, 691
  AH/B theory of taste, 589, 590, 596, 598
  bitterness, 590
  structure, 654
Saccharomyces carlsbergensis
  flavors from, 610
  pantothenic acid, analysis of, 512
  vitamin B_6, analysis of, 502, 503
Saccharomyces cerevisiae
  flavors from, 610
  in recombinant DNA technology, 464, 465
Safety attributes of food, 7–13
Safety of foods, 15–20, 690–692
  disagreement about, 15–17
  perspective, 17–20
Saffron, 580
Safrole, 695, 704
Safynol, 898
Salmonella, 845
Saltiness
  cations and anions, role of, 596
  theory, 596
  of various substances, 595–596

Salting-in of proteins, 273
Salting-out of proteins, 273
Salts
    bitterness, 595
    for food, 639–641
    polysaccharide solubility, effect on, 112
Salty taste, 595, 596
Saussure, (Nicolas) Theodore de, 3
Sandiness of ice cream, 816
Sanitizing agents, 672
Saponins, 328, 693, 694
Saporous sensations (*see* Flavor compounds and
        sensations, Taste)
Sarcolemma, 732, 734, 743, 745
Sarcomere, 732–734, 741, 742, 755, 761, 762,
        768, 780
Sarcoplasm, 730, 732–734, 742, 743, 757, 761,
        762, 764, 776
Sarcoplasmic proteins, 760
Sarcoplasmic reticulum, 732–734, 740, 742,
        743, 757, 762, 763, 767
Sarcosine, 772
Saturated fats and coronary heart disease, 233,
        235–237
Sausage (*see* Gel-type muscle products)
Saxitoxin, 701
Scatchard equation as related to flavor binding
        by proteins, 311
Schaal oven test, 196
Schardinger dextrins
    a flavor binder, 107
    in fragrance stabilization, 88
    formation of, 88
    structure, 107
Scheele, Carl Wilhelm, 3
Schiff's base, 193, 269, 311, 342–344, 346,
        360, 748, 749
Sclerenchyma tissue, 869, 876, 905
Sclerotiorins, 575
Scombroid poisoning, 701, 759
Seafood flavors, 614, 615
Seaweed gums (*see* Agar, Algin, Carrageenan)
Secretin, 254
Segelke-Storch relationship, 398
Selenium, 522
Semiquinones, 201
Senescence of fruit, 881, 882, 885, 888
Sensory assessment of flavors, 587, 588
Sensory evaluation for lipid oxidation, 196
Separation aids, 672
Sequestrants (*see* Chelating agents)
Serendipity berry, 657
Serine, 251, 252, 253
Serine proteinase, alkaline, 442
Serotonin, 315, 868
Serum albumin, 261, 273, 300, 303, 308, 309,
        663, 796, 799, 816
    as adsorbent, 663
    as a foaming agent, 308, 309
    hydrophobicity, 309
    in milk, 816
Settling of fat, 218
Shelf life case studies
    dehydrated mashed potatoes, 929–936
    frozen pizza, 924–929

Shelf life test design
    Arrhenius approach, 915, 917–919
    general approaches, 914
    humidity, effect of, 922, 923
    order of reaction, 914–916
    simple plots, 919–922
    steps and considerations, 923, 924
Shikimic acid, 448, 866
Shikimic acid pathway, 609, 863, 888, 890
    flavors from, 609
    in plant tissue postharvest, 888, 890
Shogaols, 600
Sialic acid
    in egg, 830, 831
    in ovomucin, 836, 845
    in ovomucoid, 836
Silicon dioxide, 668
Silver, 522
Sinapic acid, 568
Sinapine, 328
β-Sinensal
    in orange flavor, 608
    structure, 608
Single-cell proteins, 330, 331
Singlet oxygen
    in flavor reversion of oil, 228
    inhibitors, 198
    in lipid oxidation, 181, 182, 184
    quenchers of, 182
Sinigrin, 83, 695
Skeletal muscle, 730–734
Skeletin, 739
Skin of plant tissue, 877
Smoked foods, carcinogens in, 713
Smoking of muscle, 777
Societal role of food chemists, 13–15
Sodium, 522, 524, 801
Sodium acetate, 671
Sodium acid pyrophosphate
    buffer salt, 671
    leavening agent, 634, 635, 679
    structure, 636
Sodium aluminum phosphate
    for flour, 664
    leavening agent, 633–636, 679
Sodium aluminum sulfate
    firming agent, 661
    leavening agent, 632, 634–636, 679
Sodium ascorbate, 703
Sodium bicarbonate
    function in food, 668
    leavening agent, 632, 633–636, 671,
        679
    in peanut brittle, 637
Sodium carbonate
    alkaline additive, 671
    neutralizer in food, 637
Sodium carboxymethylcellulose
    bulking agent, 670
    stabilizer, 680
Sodium caseinate, 680
Sodium chloride
    eggs, added to, 830, 846, 851
    emulsion stability, effect on, 641
    water activity, effect on, 660

Sodium citrate
buffer salt, 671
chelating agent, 677
emulsifier, 671
texture control agent, 678, 680
Sodium cyclamate, 654
Sodium dodecylbenzenesulfonate, 672
Sodium dodecylsulfate, 272, 279
Sodium erythrobate, 669, 703
Sodium gluconate, 677
Sodium hexametaphosphate, 846
Sodium hydroxide
neutralizer in food, 637
for olives, 637
peeling agent, 672
Sodium hypochlorite, 339, 672
Sodium laurate, 596
Sodium lauryl sulfate, 596
Sodium methoxide, 223
Sodium nitrate, 776
Sodium nitite, 776
Sodium phosphate
buffer salts, 671
in milk, 815
Sodium potassium tartrate, 671
Sodium silicoaluminate, 666, 670
Solanine, 328
Solaninidine, 865
Solid fat index, 159, 160, 163–165
and consistency of fat, 164, 165
by NMR, 163
Solids, finely divided, 168, 169
Solid solutions of lipids, 152
Solubility (*see* Protein solubility)
Solvents for extraction, 712, 713
Solvents for food, 672
Solvents, halogenated, 347
Sorbic acid, 631, 646–648, 659, 673, 674, 775
antimicrobial agent, 659, 673
flavors of degradation products, 647
structure, 674
Sorbitan emulsifiers, 679
Sorbitan monooleate (Span 80), 171
Sorbitan monostearate (Span 60), 171
Sorbitan tristearate (Span 15), 171
Sorbitol, 174, 658, 659, 678, 771
in sugar-free candies, 659
sweetener, 678
texturizer, 658
Sorbyl alcohol, 647
Sourness, theory, 596
Sour taste, 595, 596
Soybean flour, 664
Soybean meal, 334, 692, 712, 713
Soybean oil
fatty acid distribution in, 150
flavor reversion in, 221
linoleate content, reduction of, 226
selectivity ratio during hydrogenation, 219–221
Soybean proteins, 287, 289, 300, 303, 310–313, 353, 356, 357, 359, 637, 862, 863
binding of 1-hexanal, 312, 313

[Soybean proteins]
denaturation, 353
emulsifying properties, 300, 303
flavor binding, 311
foaming agent, 306, 310
hydration as influenced by pH and temperature, 287
intentional chemical modification, 359
lipid binding ability, 303
pH, effect on solubility, 289
proteolysis, effect on functionality, 357
solubility versus hydrolysis, 356
Soybeans, 327, 328, 333
Soy product texturization, 294, 295
Spans, 174
Sphingomyelin, 146, 799
Spice impact, 926
Spinach, blanching and mineral content, 524
Spinulosin, 569, 570
Squalene, in milk fat globule membrane, 813
Stability constants for metal chelates, 642
Stability of food
versus solute mobility, 60, 61
versus water activity, 55–60
Stabilizers (*see also* Carbohydrates; Gums, plant; Pectic substances; Polysaccharides; Proteins; Starches)
Stabilizers for food, 657, 658, 680
Stachyose, flatulence caused by, 71
Staling of baked goods, 118
Standard free energy change, definition, 375
Staphylococcal food poisoning, 771
Staphylococcus aureus, 333, 710, 711, 924
Starch, 70, 76, 112, 113, 115–117, 658, 670, 672
amylopectin content, 113
amylose content, 113
anticaking agents, 670
antistick agent, 672
caloric importance, 70
gelatinization of, 113, 115–117
granules, 112
in plant tissue, 859
storage in plants, 76
source, 112
types, 112
Starch blockers, 459
Starch catabolism in plant tissues, 886–888
Starch, corn, 120
Starch granules, 112, 114, 116, 296, 877
amylases, resistance to, 116
characteristics, 112, 114
membrane of, 114
in plant cells, 877
Starch, modified, 118, 119
Starch, pregelatinized, 118
Starch-protein interactions in baked goods, 117
Starch-sugar transformations in plants postharvest, 74, 76, 886–888
Starch synthesis in plant postharvest, 888
Stearic acid
antistick agent, 672
β form of, 154
in coronary heart disease, 236

[Stearic acid]
   function in food, 668
   nomenclature, 142
   unit cell of, 152, 154
Stearoyl-2-lactylate, 169, 171, 173, 174, 679
Stearoyl monoglyceridylcitrate, 679
1-Stearoyl-2-oleoyl-3-myristoyl-sn-glycerol, 145
Stem structure, 872, 873
Stereoisomers of amino acids, 253, 254
Stereospecific numbering of lipids, 144, 145
Steric strains in proteins, 262
Sterigmatocystin, 708
Sterilization of milk, 792, 814, 815
Stevioside, 656
Stohmann, F., 6
Stokes law, 166
Storage proteins, 247
Strecker degradation, 99, 100, 103, 104, 343,
      606, 616, 617, 619, 930
   in dehydrated potatoes, 930
   and α-dicarbonyl compounds, 99
   flavors from, 100, 606, 616, 617, 619
   nutritional consequences, 99, 100, 103, 104
   sulfites, effect of, 103
   toxicity from, 103
   of L-valine, 103
Streptococcus cremoris, 459, 465
Streptococcus lactis
   flavors from, 606, 610
   in recombinant DNA technology, 459, 465
Streptococcus thermophilius, 610
Stress metabolites in plant tissues, 896–898
Structural proteins, 246
Struvite, 521, 642
Subcellular structure of plant tissue, 877
Substrate assays, 399, 400
Substrate, immobilized
   in emulsions, 393–395
   lipase kinetics, 393
   reaction kinetics, 393
   triacylglycerols, 393
Succinic acid
   chelating agent, 670
   food additive, 632
Succinylation
   of proteins, 359, 360
   protein solubility, effect on, 272
Sucrose, 73–76, 92, 98, 124, 194, 631, 660,
      830, 846, 847, 851, 861, 905
   amount in foods, 74, 75
   canned fruit, effect on, 905
   caramelization of, 98
   and dental caries, 74
   digestion of, 73
   eggs, additive for, 830, 846, 847, 851
   in fruits during ripening, 76
   hydrolysis, ease of, 92
   in plant tissue, 861
   role in pectin gels, 124
   source in diet, 74
   in sweet corn, 74, 76
   TBA test, interference with, 194
   water activity, effect on, 660
Sucrose esters, 298

Sugar (*see also* specific sugars)
   bitterness, 594
   carotenoids, association with, 572, 573
   egg yolk, additives for, 830, 846, 847, 851
   foams, effect on, 307
   gels, effect on, 112
   hydrophilicity of, 105
   in ovomucin, 836
   in plant tissue, 861
   respiring plants, change in, 884
   starch gelatinization, effect on, 115
   sweetness of, 108
Sugar alcohols, sweetness of, 109
Sugar degradation products, 94–97 (*see also*
      Browning, Maillard; Caramelization)
   by aldose-ketose isomerization, 95
   by anomerization, 95
   3-deoxy-D-glucosone, 95
   by inter- and intramolecular dehydration, 95
   toxicity of, 94
Sugar reactions
   acyclic form required by, 92–94
   dehydration, 94–96
   enolization, 92–94
   mutarotation, 92
   thermal degradation, 95
Sugar-starch conversions in sweet corn, 74, 76
Sulfamic acid, 656
Sulfanilic acid, 501
Sulfates, 521
Sulfhydryl-disulfide interchange (*see* Thiol-di-
      sulfide interchange)
Sulfhydryl groups (*see* Thiols)
Sulfhydryl oxidase, 454, 802, 813
Sulfides, 773
Sulfite oxidase, 705
Sulfites, 104, 337, 348, 447, 485, 493, 495,
      504, 511, 560, 562, 644, 645, 666, 673,
      705, 909, 929
   anthocyanins, effect on, 560, 562
   antimicrobial properties, 644, 645, 673
   antioxidant, 645, 673
   browning inhibitor, 104, 348, 485, 645,
      666, 909
   in dehydrated potatoes, 929
   enzyme inhibitor, 645
   folates, effect on, 504
   functions in food, 705
   lysinoalanine formation in proteins, effect
      on, 337
   proteins, reaction with, 348
   safety, 705
   thiamine, effect on, 485, 495
   metabolism, 645
   vitamin B_{12}, effect on, 511
   vitamin C, effect on, 485, 493
Sulfmyoglobin, 556
S-Sulfonates, 348
Sulfone, 644
Sulfoxide, 644
Sulfur dioxide, 670 (*see also* Sulfites)
Sunflower flour, 329
Sunflower protein concentrate, 329
Superoxide dismutase, 454, 802

Surface area and rate of lipid oxidation, 197, 202
Supporting tissues in plants, 876
Surface removal agents, 672
Surface tension
    control agents, 677
    of protein solutions, 309
Sweet corn, starch in, 74, 76
Sweeteners, nonnutritive and low-calorie,
    565, 653–657, 677, 678
    acesulfame K, 656, 657
    aspartame, 654–656
    cyclamates, 653, 654
    saccharin, 654
    Naringin, 565
    neohesperidin dihydrochalcone, 565
Sweetness
    AH/B theory, 588, 589
    of D-amino acids, 590
    of corn syrups, 92
    of high fructose corn syrups, 92
    of lactose, 820
    of seafoods, 615
    of sugar alcohols, 108, 109
    of sugars, 108
Sweet taste, 588–590
Swelling method for determining protein hy-
    dration, 285
Swelling of proteins, 283
Swine sex odor, 613
Synemin, 741
Syneresis, 39, 110
Synergism of antioxidants, 202
Synsepalum dulcificum, 657
Syringaldehyde in lignin, 859

T

Talc, 672
Talin, 657
Tannic acid, 567, 663, 669
Tannins, 328, 331, 567, 575, 598, 662, 663
    astringency of, 567, 598
    browning, role in, 567
    and haze, 662, 663
    properties of, 575
    proteins, interaction with, 331
    structure, 567, 598
    types of, 567
Tartaric acid, 200, 632, 638, 670, 863
    buffer effect in plants, 638
    chelating agent, 670
    food additive, 632
    in plant tissue, 863
Tartrates
    in cheese, 640
    chelating agents, 677
Taste
    astringency, 598
    basic sensations, 586
    bitter, 590–595
    cooling, 600, 601
    of plant tissue, 863
    pungency, 599, 600
    salty, 595, 596

    sour, 595, 596, 631
    sweet, 588–590
Taurine, 772, 865
Taurocholic acid, 679
TBA test (*see* Thiobarbituric acid test)
Telopeptides, 747, 748
Temperature
    lipid oxidation, effect on, 197
    plant respiration, effect on, 883, 884
    plants postharvest, effect on, 898–900
    protein hydration, effect on, 286, 287
    protein solubility, effect on, 273
    protein solution viscosity, effect on, 290
Tenderness (*see* Muscle tenderness)
Tendons, 745–747
Terpenoid flavors, 606–608
Tertiary butylhydroquinone (TBHQ), 199,
    200, 203, 204
    antioxidant properties of, 203, 204
    structure of, 199
2,3,7,8-Tetrachlorodibenzo-*p*-dioxin, 714
Tetrachloroethylene, 347
Tetraglycerol monostearate, 171
Tetrahydrothiophene, 616
Tetraodon poisoning, 701
Tetrasodium pyrophosphate, 637
Tetrodotoxin, 701
Texture control agents, 678, 679
Texture of plant tissue
    after dehydration, 909
    after freezing, 907
    after heating, 905
Texturization (*see* Protein texturization, Soy
    texturization)
Texturizers for food, 680
Texturizing agents, 658–661
Thaumatin, 332, 466
Thaumatococcus daniellii, 657
Thaw exudate, 769, 770
Thaw rigor, 763
Theaflavin gallate, 567
c-Theaspirane, 620
Thenard, Louis-Jacques, 3
Theobromine, 592
Thermal death time plots, 412
Thermal processing
    flavors from, 615–619
    plant tissue, effects on, 904
Thermal treatments, effect on protein function-
    ality, 351
Thermodynamics
    of enzyme denaturation, 404, 405
    of protein denaturation, 279–281, 404, 405
    of water structure during protein denatura-
    tion, 404
Thiaminase, 444, 486
Thiamine, 481, 483, 485, 487, 488, 493–499,
    511, 529, 530, 616, 617, 705, 773
    alkaline conditions, effect of, 485
    analysis, 494, 495
    antithiamine factors in seafood, 499
    chemical nature, 493, 494
    degradation product, 495–496, 499

[Thiamine]
    and flavor development, 616, 617
    forms in food, 494
    form versus stability, 497
    HTST, retention by, 529, 530
    loss during cooking, 485
    in milk, 820
    molar absorptivity, 496
    in muscle, 773
    nitrite, effect of, 485, 495
    prevalence in foods, 493
    processing, effect of, 497
    RDA, 481
    role in intermediary metabolism, 494
    stability, 495–499
    stability factors, 495, 497
    stability in various foods, 499
    stability versus, pH, 494
    structure, 496
    sulfite, effect of, 495, 496, 497, 705
    thermal processing, retention during, 530
    vitamin B_{12}, effect on, 511
Thiazole, 616
Thiazole ring, 494, 496, 498, 499
Thiazolidine, 502
3-Thiazoline, 616
Thickeners, 657, 658, 680
Thick filaments, 733-735, 737-742, 761, 764
Thin filaments, 733, 734, 737, 740-742, 761, 764
Thiobarbituric acid (TBA) test
    correlation with flavor scores, 193
    interfering substances, 194
    for lipid oxidation, 186, 193, 194
Thiochrome, 494
Thiocyanates, 602, 695, 696
Thiodipropionic acid, 200, 644, 673
β-D-Thioglucose, 695
Thioglucosides, 696
Thioglycosides, 693, 694
Thiol-disulfide interchange, 263, 401, 402, 808,
        815, 834
    in milk, 808, 815
    in ovalbumin, 834
    in proteins, 263
Thiols
    in flour, 664
    folates, effect on, 504
    in muscle, 738
    pyridoxal, reaction with, 502
    vitamin B_{12}, effect on, 511
Thiophene, 499, 512, 602, 616
Thiopropanal-S-oxide, 601
Thiosulfinic acid, 603, 604
Thixotropic solutions, 290
Thomson, Thomas, 4
Threonine, 251-254
L-Threonine deaminase, 277
Thrombin, 247, 356
Tin, 521, 522, 525, 528
Titanium dioxide, 580, 673, 680, 681
    colorant, 673
    tracer, 680, 681
Titin, 739, 741
α-Tocopherol, 518, 520, 613, 704

Tocopherolquinone, 519, 520
Tocopherols (*see also* Vitamin E)
    antioxidant, 200, 203, 643
    formation of nitrosamines, effect on , 519
    intake of, 518
    lipid oxidation, effect of, 486
    structures of, 203
    types of, 203
α-Tocored, 520
Tocotrienols, 518
Tofu, 291
Tolerance levels for food additives and contam-
        inants, 691
Toluene, 516
Topomyosin, 738, 739, 741
Tortillas, 637
Totox value, 195
Toxicants, binding to proteins, 312
Toxic constituents of food, naturally occurring
    in animal tissues, 700-702
    in plant tissues, 692-700
Toxic proteins, 692-695, 700, 711 (*see also*
        Protein degradation products, Protein
        wholesomeness)
Toxins (*see* specific kinds)
Tracers, 667, 681
Tracers in food, 680
Tragacanth gum, 175, 680
Tranfection, 462
Transaminase, 885
Transamination, 315
Transduction, 462
Trans fatty acids
    in adipose tissue, 233
    in butter, 232
    in hydrogenated fats, 219, 221-223
    nutritional effects of, 232
Transferase reactions during dehydration, 420
Transferases in milk, 802
Transfer proteins, 247
Transferrin, 247
Transglycosidation, 401
Transition state analog inhibitors, 455
Transition state inhibitor, 458
Transition state theory, 408-411
Transpeptidation, 349
Transverse system, 732, 733, 740
α, α-Trehalose, 86
Triacylglycerols
    amount in diet, 229
    autoxidation (*see* Lipid autoxidation)
    chocolate bloom, 158
    commercial processing (*see* Fat and oil proc-
        essing)
    consistency, factors influencing, 164, 165
    consistency of, 160-165
    consumption trends, 229
    and coronary heart disease, 233-237
    crystallization, 158
    crystal structure of, 157
    dilatometric curves for, 161-163
    dimerization of, 189
    enantiomers of, 145
    fatty acid distribution in, 149-151

[Triacylglycerols]
   fatty acid distribution patterns, theories
     of, 147–149
   flavor, effect on, 226, 229
   flavors from, 611, 612
   α form of, 155, 156
   β form of, 155–158
   β′ form of, 155–158
   hydrogenation (*see* Hydrogenation of fat)
   hydrolysis of, 175, 176
   hydrolysis at low water activity, 418
   interesterification (*see* Interesterification of
     fat)
   ionizing irradiation of (*see* Ionizing irradia-
     tion of lipids
   K points of, 161
   melting, 158–160
   in milk, 797, 799, 800
   in milk fat globule membrane, 813
   monoacid type, 145
   mouth feel of, 226
   nomenclature, 143–145
   nutritional functions of, 229
   oxidation rate versus free fatty acid rate,
     197
   oxidative crosslinks in, 189, 192
   polymerization of, 189, 192
   polymorphic forms of, 154–158
   polymorphic form versus functionality, 157,
     158
   β-position of, 145
   racemic mixture of, 145
   R/S system of, 143, 144
   safety after heating and oxidation, 229–232
   safety after hydrogenation, 232, 233
   solid fat index for, 159, 160, 163
   solidification point of, 161
   structure, 143, 145
   subgroups, 146, 147
   tests for quality during frying, 212, 213
   triclinic (β) packing of, 153, 154
   trimerization of, 189
Triacylglycerol thermal degradation, 206
   conditions for, 205, 207
   during deep fat frying, 210–212
   mechanism of, 207, 208
   oxidative, 207–209
   products from, 205–209
   tests for, 212, 213
Triadic joint, 732, 734
Tributyrin, anaerobic thermal decomposition
   of, 205, 206
Tricaffeoylcyanidin-3,7,3′-triglucoside, 560
Tricalcium phosphate, 666, 670
Tricalcium silicate, 670
Tricaproin, anaerobic thermal decomposition
   of, 205, 206
Tricaprylin, anaerobic thermal decomposition
   of, 205, 206
Tricetin, 563–565
Trichloroethylene, 347, 712, 713
2,4,5-Trichlorophenoxyacetic acid (2,4,5-T),
   714

L-Tricholomic acid, 597
*t*2,*c*4,*t*7-Tridecatriental, 613
Triethyl citrate, 672
Triglycerides (*see* Triacylglycerols)
2,4,5-Trihydroxybutyrophenone (THBP), 199,
   200, 204
Trilaurin, 156
Trimers, from ionizing irradiation of lipids, 214
Trimethylamine, 614, 615, 759
Trimethylamine oxide, 614, 745, 759, 760,
   771, 772, 778
ε-N-Trimethyllysine, 248
2,4,5-Trimethyl-3-thiazoline, 618
Trimming, effect on nutrient content of food,
   483
2,4,6-Trinitrobenzene sulfonic acid, 270
Tripentyltrioxane, 185
Trisaccharides, 86
   maltotriose, 86, 88
   manninotriose, 86, 88
   raffinose, 86, 88
Trisodium citrate, 637
Trisodium monohydrogen pyrophosphate, 636
Trisodium phosphate
   alkali in cheese, 637
   emulsifier, 671
Trisulfides, 602
Trithiane, 616
Trithiolane, 616
Triticale, 325
Tropomyosin, 258, 260, 734, 740, 754, 756,
   771, 772, 779
Troponin, 258, 260, 738–741, 754–756, 772
Trypsin, 281, 302, 309, 313, 355, 362, 400,
   411, 422, 430, 692, 693, 735, 736, 846,
   863
   for modifying soy proteins, 863
   to prevent gelation of egg yolk, 846
Trypsin inhibitor, 247, 321, 328, 329, 333,
   334, 455–457
Trypsinogen, 355
Tryptophan, 251–254, 335–336, 341, 342,
   348, 747
   binding to proteins, 348
   carbolines, formation of, 335, 336
   in collagen, 747
   oxidation of, 335, 341, 342
T system (*see* Transverse system)
Tuber proteins, 862
Tubulin, 247
Turmeric and turmeric oleoresin, 580, 673
Tyrocidin A, 254
Tyrosinase, 446 (*see also* Phenolase, Polyphe-
   noloxidase)
Tyrosine, 251–254, 446, 447

U

UDP-galactosyltransferase, 801, 802
UDPG pyrophosphorylase, 888
Ultrafiltration, effect on protein functionality,
   349, 350
Ultramarine blue, 580

Ultraviolet spectrophotometry for measuring lipid oxidation, 195
*t*2,*c*5-Undecadienal, 613
Undecane, 211
2-Undecenal, 183
Unintentional additives (*see* Food additives, indirect)
Urea, 279, 315, 316, 322, 325, 331, 336, 512
Urease, 374
Urethane, 653
Uric acid, 316
Urinary nitrogen, 316
Ursolic acid, 867

V

Valine, 251–253
Vanadium, 522
van der Waals interactions
    in flavor binding by proteins, 310
    in proteins, 262, 264, 265
Vanillic acid, 558
Vanillin
    in lignin, 859
    structure, 609
    in vanilla, 609
Vanillylamides, 599
van't Hoff equation and protein denaturation, 404
van't Hoff plot for denaturation of β-lactoglobulin, 405, 406
Vascular tissue of plants, 876
Vegetable butters, 146
Vegetables (*see also* Plant tissues, postharvest)
    and amino acids, 328
    antinutritional factors, 328
    firmness, 660, 661
    protein nutritive value, 328
    water content, 328
Verbascose, 71
Verdoheme, 556
Vicine, 328, 694, 700
Vimentin, 741
p-Vinyl guaiacol, 609
5-Vinyl-2-thiooxazolidone, 333
Violaxanthin, 571
Viscosity
    equation for, 288, 289
    of polysaccharide solutions, 111, 117
    of protein solutions, 288–290
Vitamin A, 229, 480, 481, 485–488, 513, 515, 516, 820 (*see also* β-Carotene, Carotenoids)
    assay, 515
    degradation, 515, 516
    distribution, 513
    forms, 513
    lipid oxidation, effect of, 486
    lipid, transport in, 229
    in milk, 820
    nitrite, effect of, 485
    oxidizing agents, effect of, 485
    RDA, 481

[Vitamin A]
    solubility, 515
    source, 480
    stability, 515
    structure, 513
Vitamin B$_1$ (*see* Thiamine)
Vitamin B$_2$ (*see* Riboflavin)
Vitamin B$_5$ (*see* Pantothenic acid)
Vitamin B$_6$ (*see also* Pyridoxal, Pyridoxamine, Pyridoxine, Pyridoxol)
    activation energies for degradation, 504
    analysis, 503
    deficiency symptoms, 503
    degradation products, 502
    forms, 501
    function, 501, 502
    lipid oxidation, effect of, 486
    processing, effect of, 503, 504
    RDA, 481
    reactions, 502, 503
    stability, 502–504
    structure, 501
Vitamin B$_{12}$, 481, 486, 488, 509–511, 521, 551, 820
    assay, 511
    binding to intrinsic factor, 511
    cobalt in, 521
    functions, 510
    lipid oxidation, effect of, 486
    in milk, 820
    in muscle, 551
    processing, effect of, 511
    RDA, 481
    sources, 510
    stability, 511
    structure, 509, 510
Vitamin C
    ability to retard formation of nitrosamines, 492
    alkaline conditions, effect of, 485
    analyses, 492
    anthocyanins, effect on, 562
    antioxidant properties, 200, 202, 643, 673
    and ascorbic acid oxidase, 452, 453
    bread dough, effect on, 453
    chemicals, effect on stability, 491
    concentration versus plant maturity, 482
    degradation products, 489–492
    degradation scheme, 490
    folates, effect on, 504, 507
    iron bioavailability, effect on, 526
    lipid oxidation, effect of, 486
    loss during cooking, 484
    loss during food storage, 530, 531
    loss in frozen plant tissue, 908
    loss, mechanism of, 530
    loss postharvest, 482
    loss by trimming, 483
    Maillard browning, relation to, 491, 492
    metal-catalyzed oxidation, 489, 521
    metal chelation by, 202
    in milk, 820
    nitrosamines, effect on formation, 519
    oxidation by lumiflavin, 500

[Vitamin C]
    oxidizing agents, effect of, 485
    pH, effect on stability, 489, 491
    processing, effect of, 492, 493, 495
    properties, 488
    protein structure, effect on, 279
    RDA, 481
    recommended intake, 479
    stability, 488–495
    stability, factors influencing, 488, 489, 495
    stability in various products, 494
    structure, 488
    sulfites, effect of, 493
    temperature, effect of, 492, 493
    vitamin B_{12}, effect of, 511
    water activity, effect of, 492, 493
Vitamin content of food
    blanching, effect of, 483
    deteriorative reactions, effect of, 486
    enzymes, effect of, 486
    genetics, effect of, 482
    leaching, effect of, 483
    lipid oxidation, effect of, 486
    loss, causes of, 482–486
    milling, effect of, 483
    plant maturity, effect of, 482
    postharvest handling, effect of, 482
    postmortem handling, effect of, 482
    processing chemicals, effect of, 485
    storage, effect of, 529, 530, 531
    trimming, effect of, 483
Vitamin D, 229, 481, 483, 488, 517, 820
    in milk, 820
    properties, 517
    RDA, 481
    structure, 517
    transported in lipid, 229
Vitamin E, 229, 481, 485, 487, 488, 518–520,
    820 (*see also* Tocopherols)
    assay, 518
    in cereal grains, 519
    degradation products, 519
    distribution, 518
    intake of, 518
    loss by processing, 519
    in milk, 820
    nitrosamines, effect of formation, 519
    oxidation agents, effect of, 485
    oxidation products, 520
    oxidation susceptibility, 519
    properties, 518
    RDA, 481
    stability, 519
    structure, 518
    transported in lipid, 229
Vitamin K
    properties, 517
    structure, 517
    transported in lipid, 229
Vitamins (*see also* specific kinds)
    added to food, 676
    in dairy foods, 820–823
    epoxides, effect of, 652
    ionizing radiation, effect of, 217

[Vitamins]
    lipid oxidation, effect of, 486
    losses during freeze preservation of plant
      tissue, 908
    in milk, 820–822
    milk processing, effect of, 824, 825
    in muscle, 728
    in plant tissue, 868, 870
    RDA, 481
    stability in cereal products, 487, 488
Vitexin, 563
V_{max}, 382, 384, 386–390, 393, 395, 396, 769
    in enzyme assays, 396
    for lipases, 393
    substrate concentration needed, 382
Vulpinic acid pigments, 575

W

Washing agents, 672
Water
    abundance and forms, 24
    amount in foods, 24, 25
    amount in milk, 796
    amount in various vegetables and cereals,
      328
    bonding forces, 28, 29
    bound, 36–38, 417
    bulk phase, 39, 41, 42
    cellulose, interaction with, 109, 110
    classification, 39–41
    constitutional, 38, 39
    coordination number of, 35, 36
    density of, 26, 36
    dipole moment, 28
    distance to nearest neighbors, 36
    in egg albumen, 832, 833
    in egg yolk, 832, 833
    entrapment in protein gels, 292, 293
    entrapped, 38, 39, 41
    in foods, 24, 63, 64, 115
    free, 41
    hydrogen bonding of, 28–30
    hydrogen bonding to proteins, 43
    hydrogen bonding, temperature depend-
      ence, 35
    ionic forms, 27, 28
    ions, interaction with, 39, 42
    irradiation, effect of, 277
    isotopic variants, 26, 28
    in life, 24
    minerals in, 522
    molecules, association of, 28–30
    multilayer, 38–40, 43
    neutral solutes with H-bonding capabilities,
      interaction with, 43
    nonpolar substances, interaction with, 43–
      46
    physical constants, 26
    in plant tissue, 858, 859
    polysaccharides, interaction with, 109, 110
    properties, 25, 26, 30, 36, 39–41
    properties of a single molecule, 25–27
    prose, 24

[Water]
proteins, absorption by, 283
proteins, interaction with, 271–274
structure of pure bulk water, 35, 36
uptake by proteins, 283
vapor pressure, 26, 49
vicinal, 38–40, 43
viscosity of, 26, 36
Water activity
activation energy, effect of, 918
BET plot, 59
versus bond dissociation energy, 37
versus chemical reaction rates, 55, 56, 58, 59
versus chlorophyll degradation, 56
and Clausius-Claperyron equation, 47, 48
definitions, 46
in dehydrated potatoes, 929
enzyme activity, effect on, 417, 418
versus enzymic hydrolysis, 56
equilibrium relative humidity, relation to, 46, 47
versus fatty acid generation rate, 419
flavor binding by proteins, effect of, 312
and flavor development, 616
in food deterioration, 11, 46
versus food stability, 55–60
freezing point, relation to, 46, 47
of frozen pizza, 927
and good manufacturing practices, 46
of intermediate moisture foods, 659, 660
in intermediate moisture meat, 775
intrinsic property of food, 47
versus lecithin enzymic hydrolysis rate, 419
lipid oxidation rate, effect of, 197
versus Maillard reaction, 56, 58, 103, 342
measurement of, 47
versus microbial growth, 55–57
versus minimum reaction rates, 58, 59
mole fraction of solvent, relation to, 46, 47
monolayer value, 58, 59
versus oxidation, 56, 58
and product accumulation, 418
Q_{10} value, effect on, 920
versus Raoult's law, 37
shelf life of dehydrated food, effect on, 922
starch gelatinization, effect on, 115
at subfreezing temperatures, 48, 50
sugars, effect of, 105
temperature coefficients, 48
temperature dependence, 47–50
versus texture, 59, 60
versus thiamine loss, 56, 58
thiamine stability, effect on, 499
vitamin C stability, effect on, 492
Water binding, 36 (*see also* Water holding capacity, Protein hydration)
Water binding by proteins, 283 (*see also* Protein hydration, Protein-water interactions)
Water binding capacity (*see* Water holding capacity)
Water bridges, 43, 44

Water holding capacity, 38, 283, 286, 637, 640, 665, 753, 760, 761, 763, 764, 817, 819 (*see also* Protein hydration)
of cheese and cheese proteins, 637
of dough, 665
of meat, 640, 753
of milk proteins, 817, 819
of muscle, 753, 760, 761, 763, 764, 779, 780
of muscle as influenced by pH, 286
of proteins, 283
Water saturation method for determining protein hydration, 285
Water structure
bridges, 43, 44
change during protein denaturation, 404
at hydrophobic surface, 45
ions, effect of, 39, 42, 43
molecular details, 28–30
and neutral solutes with H-bonding capabilities, 43
pictoral effect of ions, 42
of pure bulk water, 35, 36
structure-breaking ions, 42
structure-forming ions, 42
temperature dependence, 36
thermodynamics during protein denaturation, 404
Waxes, composition, 146
Wax ester, 867
Wettability of proteins, 283
Wheat kernel structure, 875
Wheat proteins, 861, 862
Whey proteins, 289, 300, 303, 307, 309, 352, 796, 808, 817–819 (*see also* specific milk proteins)
denaturation, 819
foaming agent, 307, 309
functional properties, 817–819
in milk, 796
pH, effect on solubility, 289
White fibers, 738, 743–745, 768
White muscles, 751, 752, 759
Wiley, Harvey Washington, 7
Wood rosin, 658
Wound healing in plant tissue, 895, 896
Wyerone, 694, 698

X

Xanthan gum, 132–134, 175, 670
bulking agent, 670
as emulsion stabilizer, 175
5'-Xanthine monophosphate, 597
Xanthine oxidase, 414, 452, 802, 813
Xanthones, 569, 570, 575
Xanthophylls, 330, 572
Xanthopterin, 504, 506
Xanthosine-5'-monophosphate, 83
Xanthotoxin, 699, 898
Xylans, 71, 123, 859, 860
Xylem, 869, 871, 876
m-Xylene, 516

Xylitol
  cooling sensation, 601
  and dental caries, 74
Xylose
  in anthocyanins, 558
  in flavonoids, 563
  in plant pectins, 859
Xylosone, 490, 491

Y

Yeast
  in dough, 665
  flavor from, 610
  as protein source, 330
Yeast protein emulsifying ability, 303
Yuba, 294

Z

Z-Disk, 732, 733, 738, 739, 741, 754–756, 762, 764
Z-disk actin, 754
Zearalenone, 708
Zeaxanthin, 571, 572
Zein, 274, 663
Zinc, 481, 522, 525, 528
  bioavailability, 525
  RDA, 481
  safety, 528
Zingerone, 600
Z-line (*see* Z-disk)
Z-nin, 739, 741
z value, 412, 413
Zwitterions, 248